Autodesk® Inventor® 2012 Essentials Plus

Autodesk® Inventor® 2012 Essentials Plus

DANIEL T. BANACH
TRAVIS JONES

DELMAR
CENGAGE Learning™

Australia • Brazil • Japan • Korea • Mexico • Singapore • Spain • United Kingdom • United States

Autodesk® Inventor® 2012 Essentials Plus

Daniel T. Banach, Travis Jones

Vice President, Editorial: Dave Garza

Director of Learning Solutions: Sandy Clark

Acquisitions Editor: Stacy Masucci

Managing Editor: Larry Main

Senior Product Manager: John Fisher

Editorial Assistant: Andrea Timpano

Vice President, Marketing: Jennifer Baker

Marketing Director: Deborah Yarnell

Associate Marketing Manager: Jillian Borden

Production Director: Wendy Troeger

Senior Content Project Manager: Angela Sheehan

Senior Art Director: David Arsenault

Technology Project Manager: Joe Pliss

Compositor: PreMediaGlobal

Cover Image: stock-photo-metal-background. Image copyright mehmetcan 2012. Used under license from Shutterstock.com

Library of Congress Control Number: 2011924973

ISBN-13: 978-1-111-64665-3

ISBN-10: 1-111-64665-1

Delmar
5 Maxwell Drive
Clifton Park, NY 12065-2919
USA

Cengage Learning is a leading provider of customized learning solutions with office locations around the globe, including Singapore, the United Kingdom, Australia, Mexico, Brazil and Japan. Locate your local office at: **international.cengage.com/region**

Cengage Learning products are represented in Canada by Nelson Education, Ltd.

To learn more about Delmar, visit **www.cengage.com/delmar**

Purchase any of our products at your local college store or at our preferred online store **www.cengagebrain.com**

Printed in the United States of America
1 2 3 4 5 6 7 15 14 13 12 11

CONTENTS

CHAPTER 4 CREATING PLACED FEATURES 143

CHAPTER 5 CREATING AND EDITING DRAWING VIEWS 206

CHAPTER 6 CREATING AND DOCUMENTING ASSEMBLIES 308

CHAPTER 7 ADVANCED PART MODELING TECHNIQUES 417

CHAPTER 8 iCOMPONENTS AND PARAMETERS 509

CHAPTER 9 ADVANCED ASSEMBLY MODELING TECHNIQUES 598

CHAPTER 10 SHEET METAL DESIGN 670

INTRODUCTION

INTRODUCTION

Welcome to the *Autodesk Inventor 2012 Essentials Plus* manual. This manual provides a thorough coverage of the features and functionalities offered in Autodesk Inventor.

Each chapter in this manual is organized with the following elements:

Objectives. Describes the content and learning objectives.

Topic Coverage. Presents a concise, thorough review of the topic.

Exercises. Presents the workflow for a specific command or process through illustrated, step-by-step instructions.

Checking Your Skills. Tests your understanding of the material using True/False and multiple-choice questions.

NOTE TO THE LEARNER

Autodesk Inventor is designed for easy learning. Autodesk Inventor's help system provides you with ongoing support as well as access to online documentation.

As described above, each chapter in this manual has the same instructional design, making it easy to follow and understand. Each exercise is task-oriented and based on real-world mechanical engineering examples.

WHO SHOULD USE THIS MANUAL?

The manual is designed to be used in instructor-led courses, although you may also find it helpful as a self-paced learning tool.

Recommended Course Duration

Four days (32 hours) to seven days (56 hours) are recommended, although you may use the manual for specific Autodesk Inventor topics that may last only a few hours.

User Prerequisites

It is recommended that you have a working knowledge of Microsoft® Windows XP Professional®, Windows Vista™ or Windows 7™ as well as a working knowledge of mechanical design principles.

Manual Objectives

The primary objective of this manual is to provide instruction on how to create part and assembly models, document those designs with drawing views, and automate the design process.

Upon completion of all chapters in this manual, you will be proficient in the following tasks:

- Basic and advanced part modeling techniques
- Drawing view creation techniques
- Assembly modeling techniques
- Sheet metal design

While working through these materials, we encourage you to make use of the Autodesk Inventor help system, where you may find solutions to additional design problems that are not addressed specifically in this manual.

Manual Description

This manual provides the foundation for a hands-on course that covers basic and advanced Autodesk Inventor features used to create, edit, document, and print parts and assemblies. You learn about the part and assembly modeling tools through online and print documentation and through the real-world exercises in this manual.

ESSENTIALS EXERCISE FILES

The exercise files for each chapter can be downloaded from http://www.cengagebrain.com.

Accessing a Student Companion site from CengageBrain:

1. Go to: http://www.cengagebrain.com.
2. TYPE Banach, Autodesk® Inventor® 2012 Essentials Plus or 1111646651 in the **Search** window.
3. LOCATE the desired product and click on the title.
4. When you arrive at the Product Page, CLICK on the **Free Stuff** tab.
5. Use the "**Click Here**" link to be brought to the Companion site.
   - Note: you will only see the Click Here link if there is a companion product available.
6. Click on the **Student Resources** link in the left navigation pane to access the resources.

Projects

Most engineers work on several projects at a time, with each project consisting of a number of files. To accommodate this, Autodesk Inventor uses projects to help organize related files and maintain links between files.

Each project has a *project file* that stores the paths to all files related to the project. When you attempt to open a file, Autodesk Inventor uses the paths in the current project file to locate other necessary files.

For convenience, a project file is provided with the *exercises.*

Using the Project File

Before starting the exercise, you must complete the following steps:

1. Start Autodesk Inventor.
2. On the Get Started tab > Launch panel, click Projects.
3. In the Projects window, select Browse. Navigate to the folder where you placed the Essentials Exercises, and double-click the file "Inv 2012 Ess Plus.ipj".
4. The "Inv 2012 Ess Plus" project will become the current project.
5. You can now start doing the exercises.

NOTE

Projects are reviewed in more detail in Chapter 1.

AUTODESK CERTIFICATION

Autodesk certifications are industry-recognized credentials that can help you succeed in your design career by providing to both you and your employer benefits such as accelerated professional development, improved productivity, and enhanced credibility. Through certification you demonstrate your knowledge and skills to current and prospective employers.

Autodesk Inventor 2012 Associate Exam Objectives

The Autodesk Inventor 2012 Certified Associate exam is the first level of certification, consisting of questions that assess your knowledge of the commands, features, and common tasks. Following are the possible exam objectives and the section that each exam objective is covered in the book.

- Control a project file
  Chapter 1—Projects in Autodesk Inventor
- Identify how to use visual style to control the appearance of a model
  Chapter 1—Visual Style
- Use sketch constraints
  Chapter 2—Constraining the Sketch
- Create dynamic input dimensions
  Chapter 2—Dynamic Input and Adding Dimensions Manually
- Create extrude features
  Chapter 3—Extruding a Sketch
- Create revolve features
  Chapter 3—Create Revolve Features
- Use the Project Geometry and Project Cut Edges commands
  Chapter 3—Projecting Part Edges
- Create fillet features
  Chapter 4—Fillets
- Create hole features
  Chapter 4—Holes
- Create a shell feature
  Chapter 4—Shelling
- Create work features
  Chapter 4—Creating Work Planes, UCS—User Coordinate System

- Create a pattern of features
  Chapter 4—Patterns
- Edit a section view
  Chapter 5—Creating Section Views
- Edit base and projected views
  Chapter 5—Editing Drawing Views
- Create and edit dimensions in a drawing
  Chapter 5—Adding Dimensions to a View
- Create and edit a hole table
  Chapter 5—Creating Hole Tables
- Use assembly constraints
  Chapter 6—Assembly Constraints
- Find minimum distance between parts and components
  Chapter 6—The Minimum Distance Command
- Modify a bill of materials
  Chapter 6—Modify a bill of materials
- Animate a presentation file
  Chapter 6—Creating Presentation Files
- Emboss text and a profile
  Chapter 7—Embossed Text and Closed Profiles
- Describe and use Shrinkwrap
  Chapter 7—Shrinkwrap
- Create an iPart
  Chapter 8—Create an iPart
- Use iLogic
  Consult the Help System
- Create and constrain sketch blocks
  Consult the Help System
- Create a level of detail
  Chapter 9—Levels of Detail Representations
- Use the frame generator commands
  Chapter 9—The Frame Generator
- Create sheet metal features
  Chapter 10—Contour Flange, Contour Roll, Flange, Lofted Flange
- Create and edit a sheet metal flat pattern
  Chapter 10—Flat Pattern
- Annotate a sheet metal part in a drawing
  Chapter 10—Detailing Sheet Metal Designs

AUTODESK INVENTOR 2012 PROFESSIONAL EXAM OBJECTIVES

The Autodesk Inventor 2012 Professional Certification exam is a performance-based test. Performance-based testing is defined as testing by performing tasks. To earn the credential of Certified Professional, you must pass the Certified Associate and Professional exam. It is recommended that you pass the Associate exam first. Following are the possible exam objectives and the section that each exam objective is covered in the book.

- Control a project file
  Chapter 1—Projects in Autodesk Inventor
- Create extrude features
  Chapter 3—Extruding a Sketch
- Create hole features
  Chapter 4—Holes
- Edit a section view
  Chapter 5—Creating Section Views
- Modify a style in a drawing
  Chapter 5—Drawing Standards and Styles
- Create and edit a hole table
  Chapter 5 –Creating Hole Tables
- Create a part in the context of an assembly
  Chapter 6—Creating Parts in Place, Exercise 6-2: Designing Parts in the Assembly Context
- Apply and use assembly constraints
  Chapter 6—Assembly Constraints
- Create a sweep feature
  Chapter 7—Sweep Features
- Create a 3D path using the Intersection Curve and the Project to Surface commands
  Chapter 7—3D Sketch from Intersection Geometry, Project to Surface, and Project to 3D Sketch
- Create a loft feature
  Chapter 7 –Loft Features
- Create a multi-body part
  Chapter 7—Multi-Body Parts
- Create a part using surfaces
  Consult the Help System
- Create an iPart
  Chapter 8—Creating iParts
- Create a level of detail
  Chapter 9—Levels of Detail Representations
- Create a positional representation
  Chapter 9—Positional Representations
- Use the Frame Generator commands
  Chapter 9—The Frame Generator
- Create components using the Design Accelerator commands
  Chapter 9—Design Accelerator
- Create sheet metal features
  Chapter 10—Contour Flange, Contour Roll, Flange, Lofted Flange

ACKNOWLEDGEMENTS

Copyedit

The authors would like to thank PreMediaGlobal for the comprehensive and attentive copyediting. Their expertise, knowledge, and attention to detail have added a great deal to this book.

CHAPTER 1

Getting Started

INTRODUCTION

This chapter provides a look at the user interface, application options, starting commands, creating projects, and how to view models in Autodesk Inventor.

OBJECTIVES

In this chapter, you will gain an understanding of the following:

- The user interface
- How to open files
- How to create new files
- Different file types used in Autodesk Inventor
- Save options
- Application options
- How to issue commands
- The Help system
- Reasons for which a project file is used
- How to create a project file for a single user
- Autodesk Vault
- Different viewing and appearance commands

GETTING STARTED WITH AUTODESK INVENTOR

The default Autodesk Inventor screen looks similar to the following image. From here you can open existing files or create new files. The default screen can be changed to display the Open dialog box or the New dialog box, or to start a specified file, which can be set via the Application Options in the General tab under Start-up action.

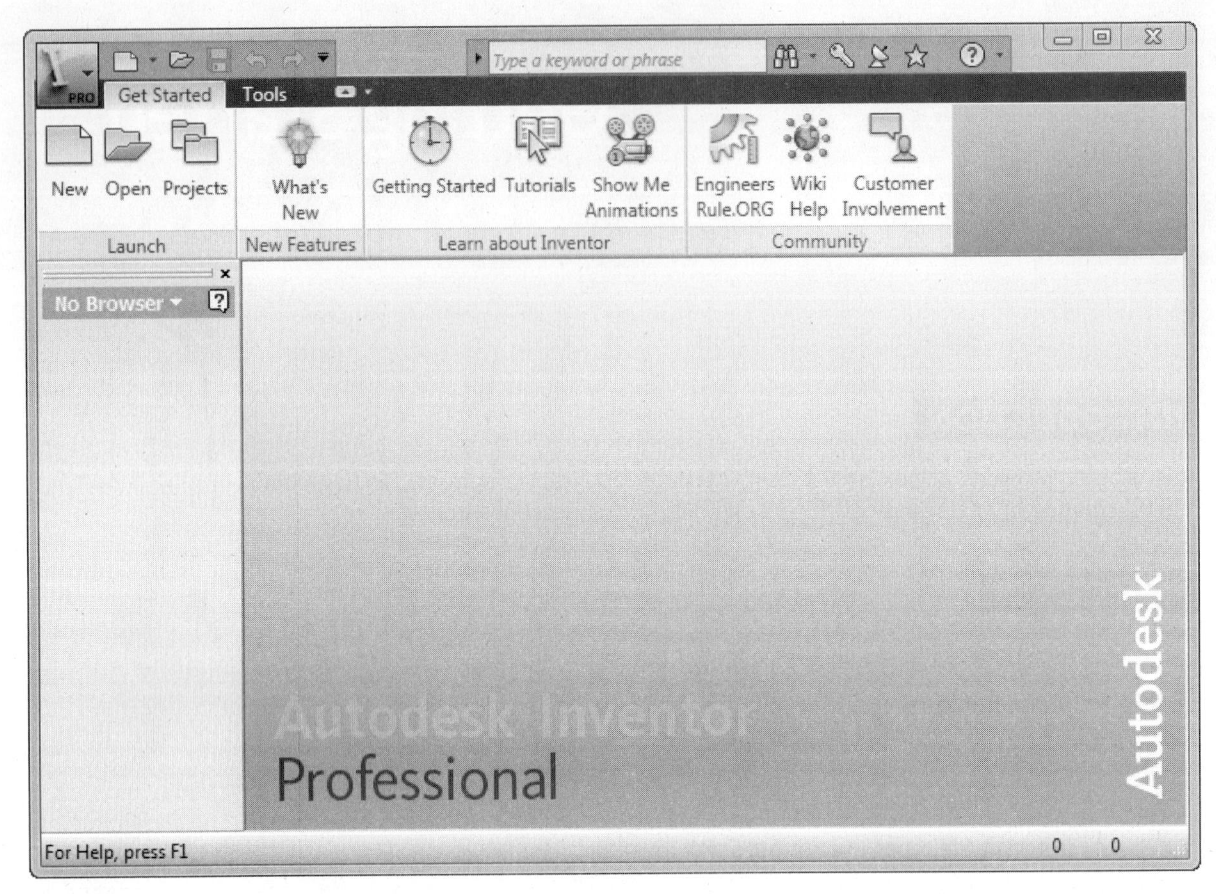

FIGURE 1.1

USER INTERFACE

The default sketch environment of a part (*.ipt*) in the Autodesk Inventor application window is shown in the following image.

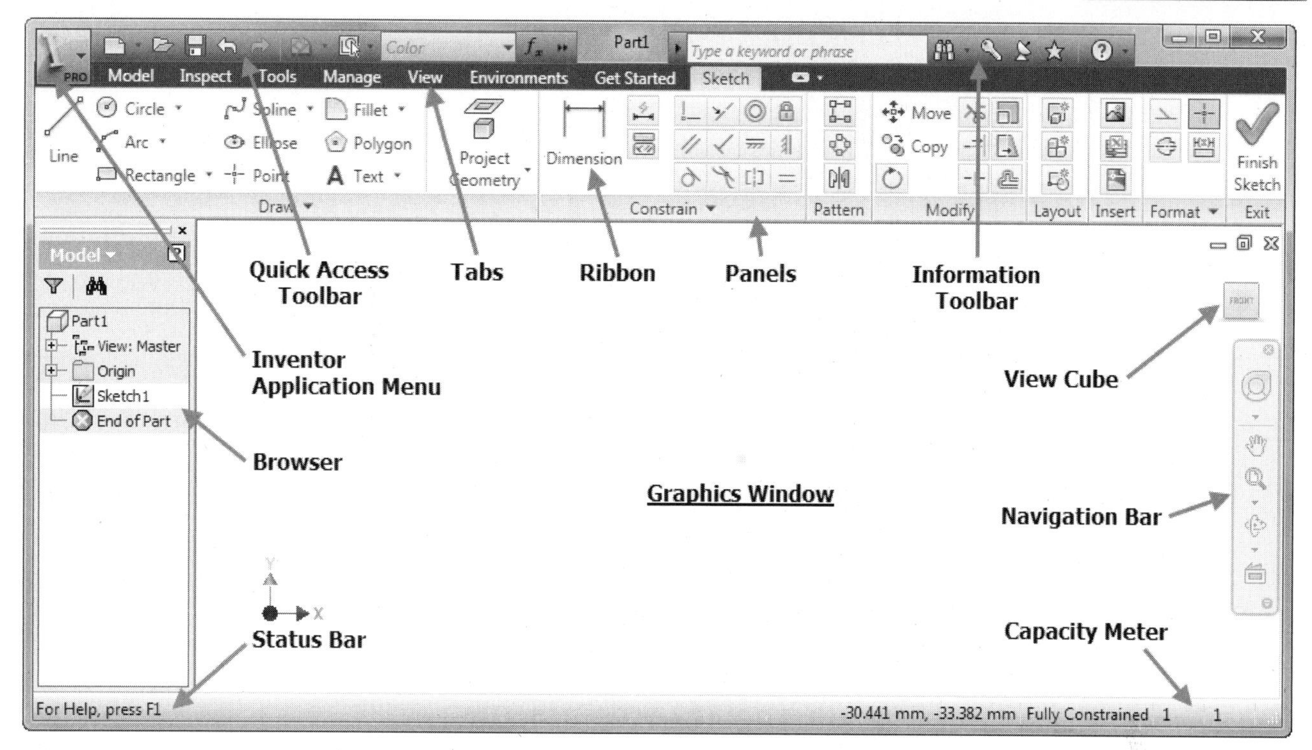

FIGURE 1.2

The screen is divided into the following areas:

Inventor Application Menu

Contains common commands for working with files.

Quick Access Toolbar

Accesses common commands as well as commands that can be added or removed.

Tabs

Changes available commands by clicking on a tab.

Ribbon

Accesses basic Windows and Autodesk Inventor commands. The set of commands in the ribbon changes to reflect the environment in which you are working.

Panels

Changes to show available commands for the current tab, click on a tab to display a set of new panels and commands.

Information Toolbar

Displays common help commands as well as Subscription services.

ViewCube

Displays current viewpoint and allows you to change the orientation of the view.

Navigation Bar

Displays common viewing commands. Viewing commands can be added by clicking the bottom drop arrow.

Capacity Meter

Displays how many occurrences (parts) are in the active document, the number of open documents in the current session, and how much memory is being used. Note: The capacity meter that shows memory usage is available only on 32-bit computers.

Status Bar

Views text messages about the current process.

Browser

Shows the history of how the contents in the file were created. The browser can also be used to edit features and components.

Graphics Window

Displays the graphics of the current file.

INVENTOR APPLICATION MENU

Besides selecting commands for working with files you can control how the recent or open documents are listed in the menu. The following image shows the functionality available from the Inventor Application Menu.

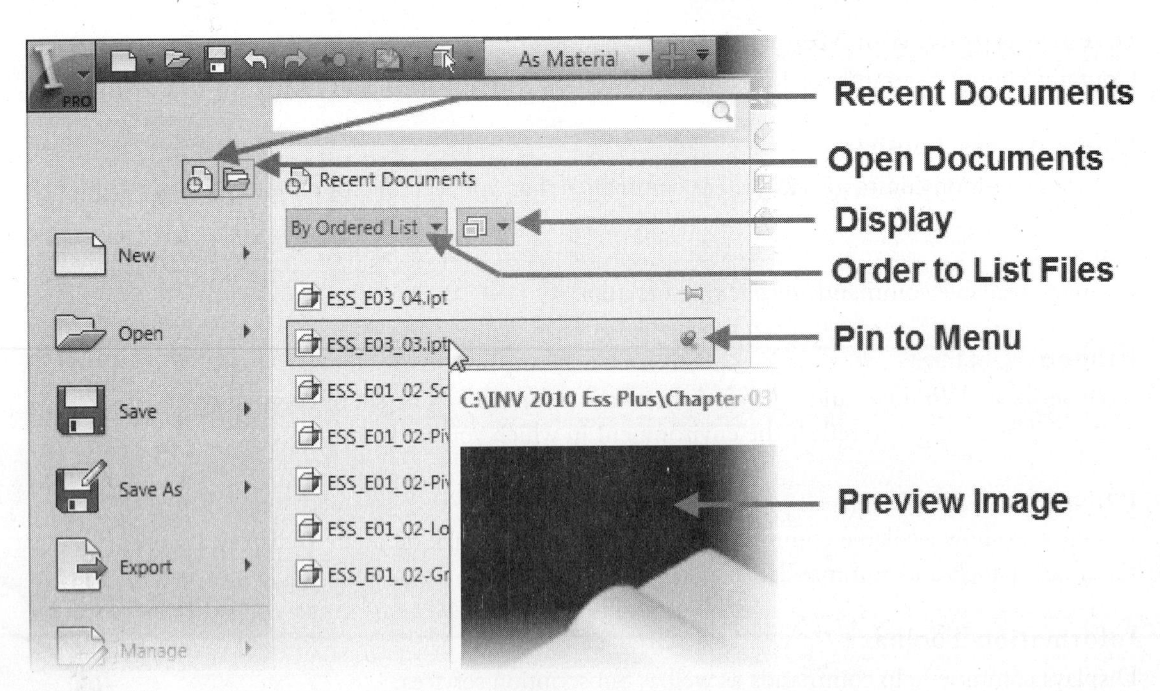

FIGURE 1.3

Recent Documents

Displays documents that were previously opened.

Open Documents
Displays documents that are opened.

Display
Controls what is displayed in the list: icons or images and what size.

Order to List Files
Controls the order that the files are listed: by Ordered List, by Access Date, by Size, by Type opened.

Pin to Menu
Click the push pin to keep the file in the list no matter when it was last opened.

Preview Image
Hovers the cursor over a file to see a larger image of the file when it was last saved.

NOTE

Double-click on the Inventor Application button, and a dialog box will appear asking you to save each unsaved document and then Inventor will close.

RIBBON
The ribbon displays commands that are relevant to the selected tab. The current tab is highlighted in green and is also green if it supports the current environment. The commands are arranged by panels. The most common commands are larger in size while less used commands are smaller and positioned to the right of the larger command. Tools may also be available in the drop list of a command or in the name of the panel. For example there is a drop list available under the Circle icon and the Draw panel as shown in the following image that shows the Sketch tab active in a part file.

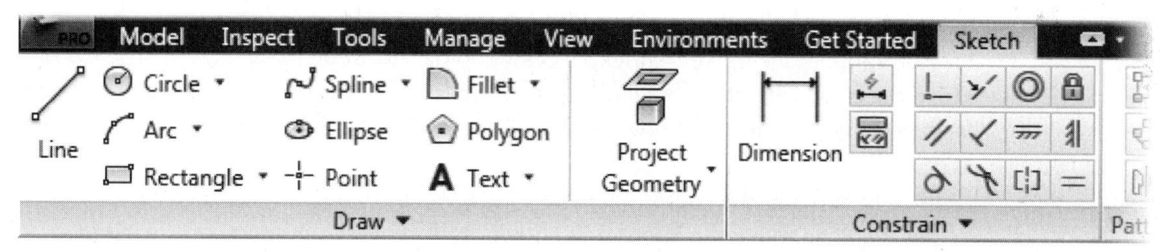

FIGURE 1.4

The ribbon can be modified by right clicking on the Panel; the following images show the available options.

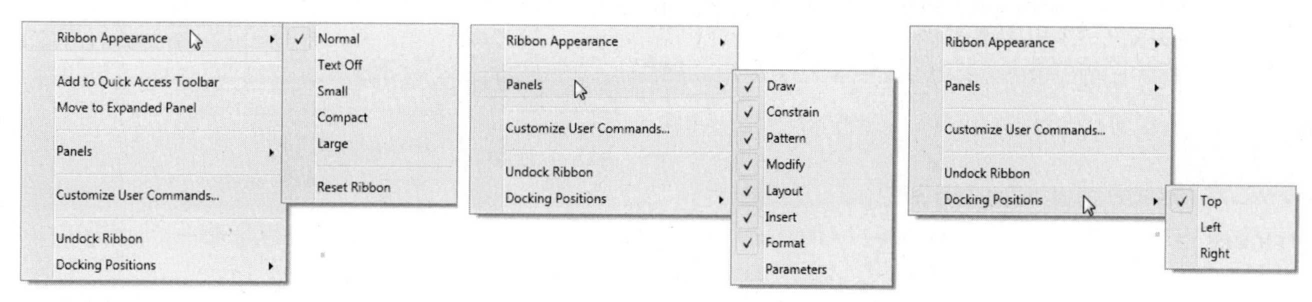

FIGURE 1.5

- Ribbon Appearance – Changes how the Ribbon looks, turns text off, and changes the size of the commands
- Panels – Adds and removes Panels
- Customize User Commands – Adds commands to a User Panel
- Undock Ribbon – Allows the ribbon to be freely moved
- Docking Position – Changes the location of the Ribbon

QUICK ACCESS TOOLBAR

Tools can be removed or added from the Quick Access Toolbar.

To add a command follow these steps

1. Move the cursor over a command to add and right-click.
2. Click Add to Quick Access Toolbar as shown in the following image on the left.

To remove a command follow these steps

3. Move the cursor over the command to remove and right-click.
4. Click Remove from Quick Access Toolbar as shown in the following image on the right.

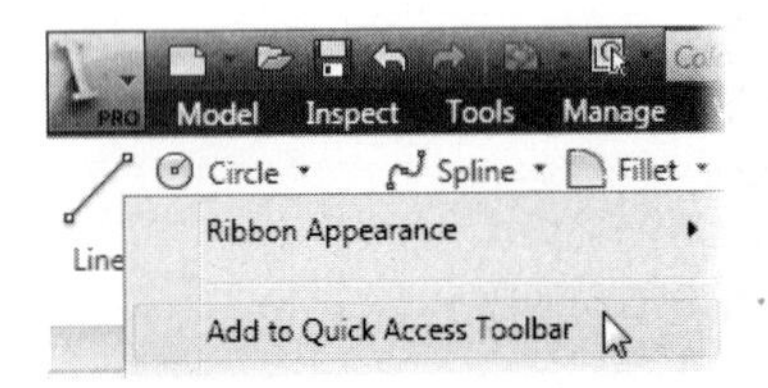

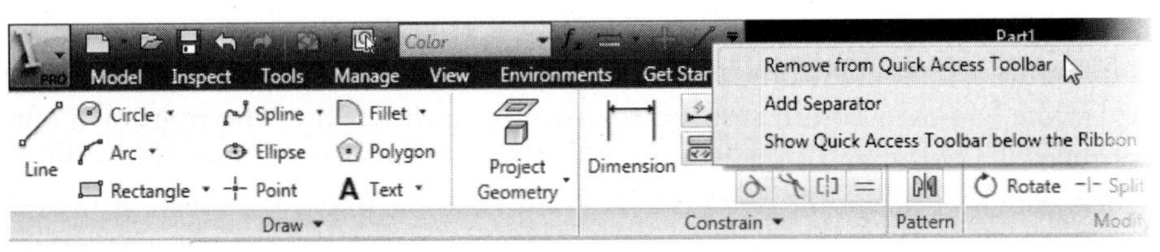

FIGURE 1.6

To gain more space for the Quick Access Toolbar it can be moved below the ribbon. Move the cursor over the Quick Access Toolbar and click Show Quick Access Toolbar below the Ribbon as shown in the following image on the left. Once the Quick Access Toolbar is placed below the ribbon, it can be moved back to its original location by moving the cursor over the Quick Access Toolbar and click Show Quick Access Toolbar above the Ribbon as shown in the following image on the right.

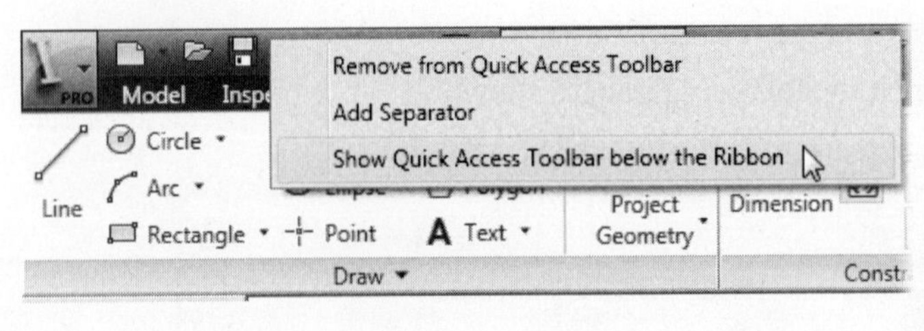

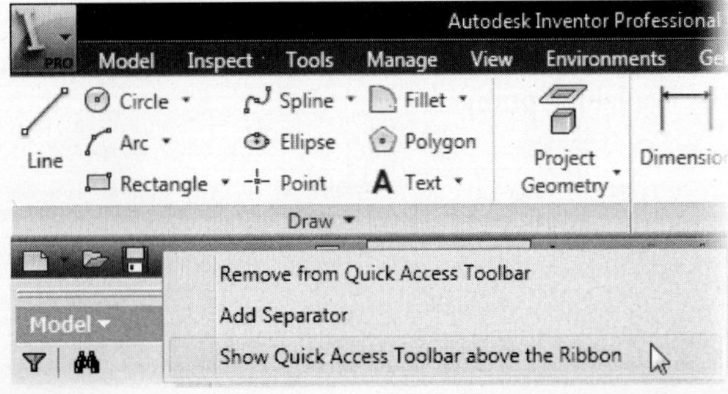

FIGURE 1.7

Open

To open files follow one of these techniques:

- Click the Open command in the Get Started tab in the Launch Panel as shown in the following image on the left.
- Click the Open command in the Quick Access Toolbar as shown in the following image on the left.
- Click the Inventor Application menu in the top-left corner and click Open as shown in the following image on the right.
- Press CTRL + O.

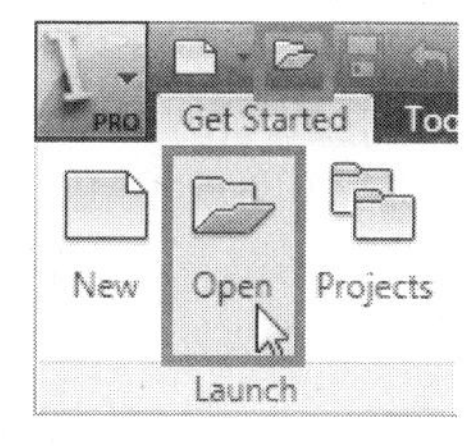

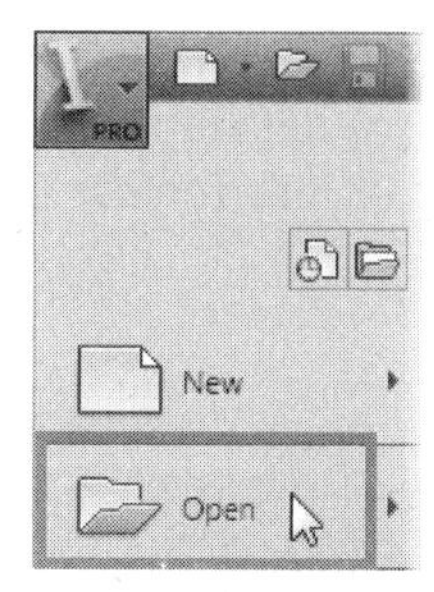

FIGURE 1.8

The Open dialog box will appear as shown in the following image. The directory that opens by default is set in the current project file. You can open files from other directories that are not defined in the current project file, but this is not recommended. Part, drawing, and assembly relationships may not be resolved when you reopen an assembly that contains components outside the locations defined in the current project file. Projects are covered later in this chapter.

FIGURE 1.9

New Files

Like the Open command there are many ways to create a new file.

To create a new Inventor file, follow one of these techniques:

- Click the New command in the Get Started tab in the Launch Panel as shown in the following image on the left.
- Click the New command in the Quick Access Toolbar as shown in the following image on the left.
- Click the Inventor Application button in the top-left corner and click New as shown in the following image in the middle.
- Press CTRL + N.
- To start a new file based on one of the default templates click the down arrow next to the New icon in the Quick Access Toolbar as shown in the following image on the right.

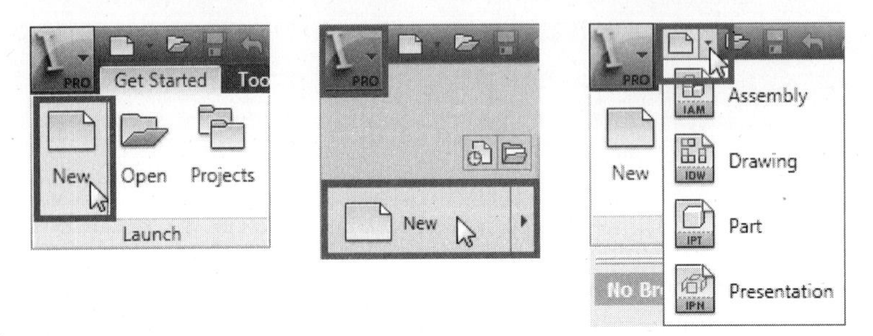

FIGURE 1.10

The New dialog box will appear as shown in the following image. Begin by selecting the type of file to create or one of the drafting standards named on the tabs, and then select a template for a new part, assembly, presentation file, sheet metal part, or drawing. If Autodesk Inventor Professional is installed, a Professional tab will exist.

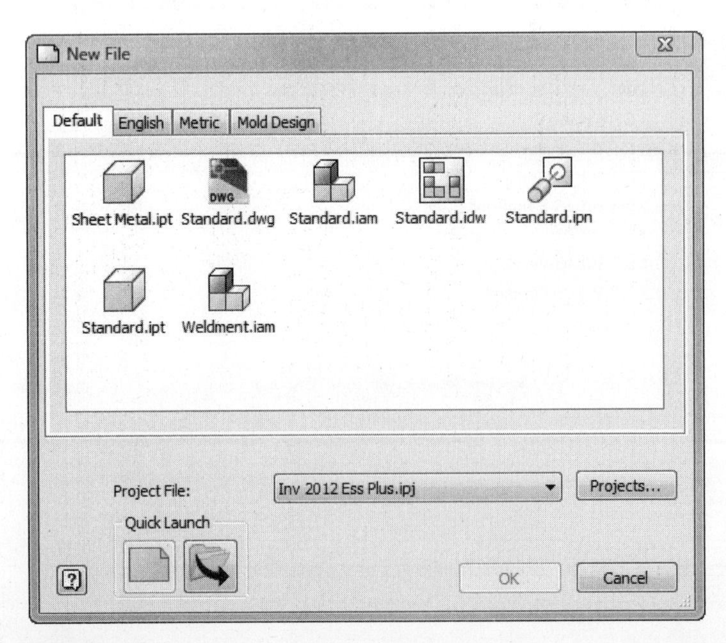

FIGURE 1.11

Quick Launch

While working you may accidentally click the New command when you wanted to click the Open command or vice versa. In the New and Open dialog boxes, there is a Quick Launch area in the lower left corner of the dialog box.

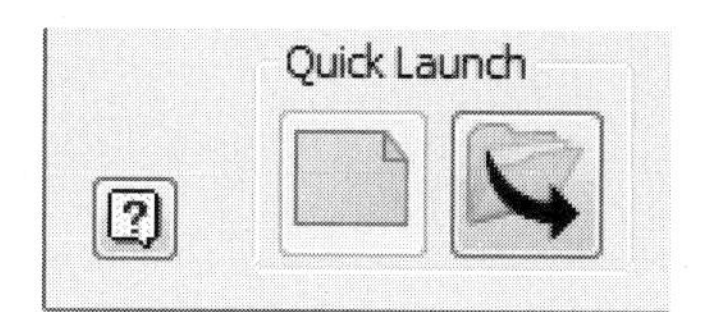

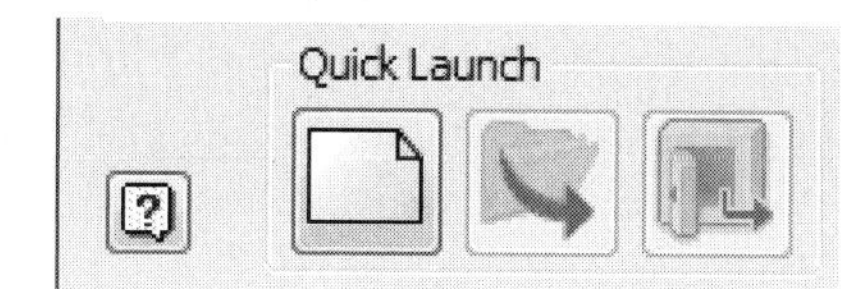

FIGURE 1.12

FILE INFORMATION

While creating parts, assemblies, presentation files, and drawing views, data is stored in separate files with different file extensions. This section describes the different file types and the options for creating them.

File Types

The following section describes the main file types that you can create in Autodesk Inventor, their file extensions, and descriptions of their uses.

Part (.ipt)

Part files contain only one part, which can be either 2D or 3D.

Assembly (.iam)

Assembly files can consist of a single part, multiple parts, or subassemblies. The parts themselves are saved to their own part file and are referenced (linked) in the assembly file. See Chapter 6 for more information about assemblies.

Presentation (.ipn)

Presentation files show parts of an assembly exploded in different states. A presentation file is associated with an assembly, and any changes made to the assembly will be updated in the presentation file. A presentation file can be animated, showing how parts are assembled or disassembled. The presentation file extension is *ipn*, but you save animations as AVI files. See Chapter 6 for more information about presentation files.

Sheet Metal (.ipt)

Sheet metal files are part files that have the sheet metal environment loaded. In the sheet metal environment, you can create sheet metal parts and flat patterns. You can create a sheet metal part while in a regular part. This requires that you load the sheet metal environment manually. See Chapter 10 for more information about creating sheet metal parts.

Drawing (.dwg and .idw)

Drawing files can contain 2D projected drawing views of parts, assemblies, and/or presentation files. You can add dimensions and annotations to drawing views. The parts and assemblies in drawing files are linked, like the parts and assemblies in assembly and presentation files. See Chapter 5 for more information about drawing views.

Project (.ipj)

Project files are structured XML files that contain search paths to locations of all the files in the project. The search paths are used to find the files in a project.

iFeature (.ide)

iFeature files can contain one or more 3D features or 2D sketches that can be inserted into a part file. You can place size limits and ranges on iFeatures to enhance their functionality. See Chapter 8 for more information about creating iFeatures.

Opening Multiple Documents

You can open multiple Autodesk Inventor files at the same time by holding down the CTRL key and selecting the files to open as shown in the following image. Each file will be opened in its own window in a single Autodesk Inventor session. To switch between the open documents, click the file on the Windows menu. The files can also be arranged to fit the screen or to appear cascaded. If the files are arranged or cascaded, click a file to activate it. Only one file can be active at a time.

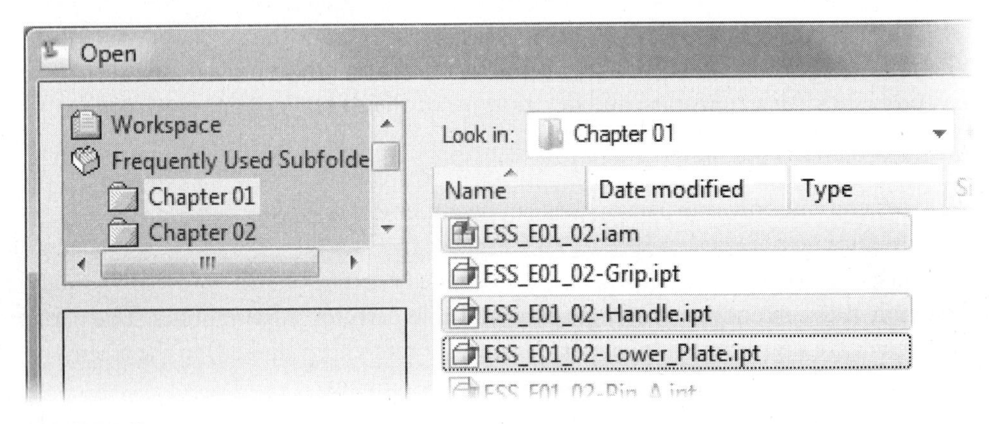

FIGURE 1.13

Document Tabs

When multiple documents are open in Inventor, each document appears in a tab in the lower left corner of the graphics window. The current document is represented with an "x" to the right of the file name and the tab's background is white. You can see a preview of an open document by hovering the cursor over the tab. In the same area, you can cascade, arrange, or list the open documents as shown in the following image.

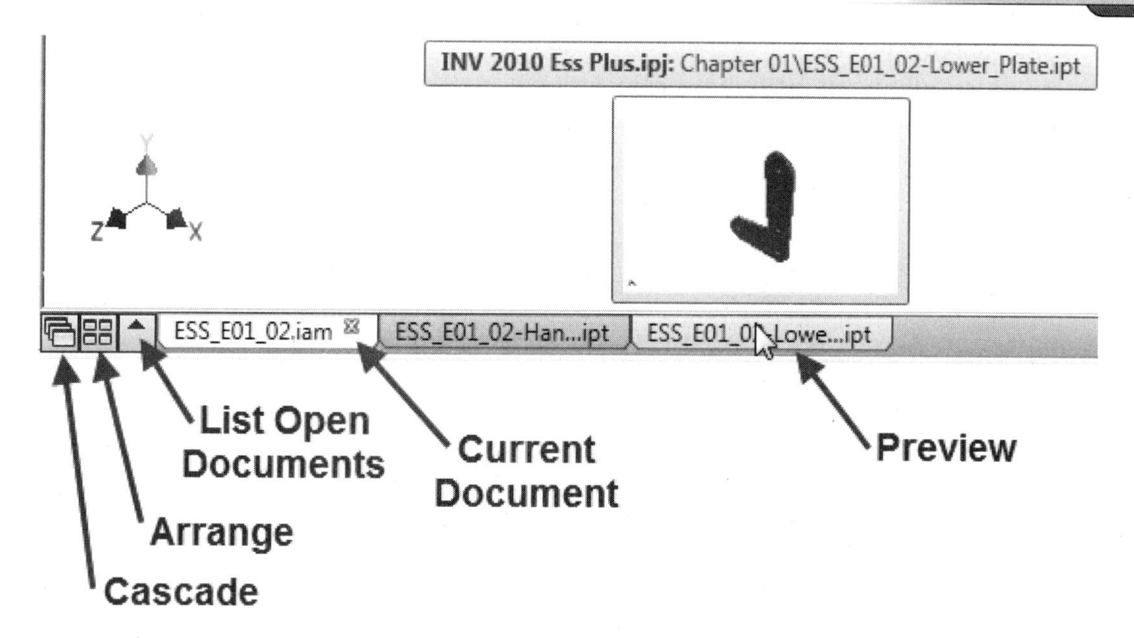

FIGURE 1.14

Save Options

There are three options on the Inventor Application button for saving your files: Save, Save Copy As, and Save All as shown in the following image.

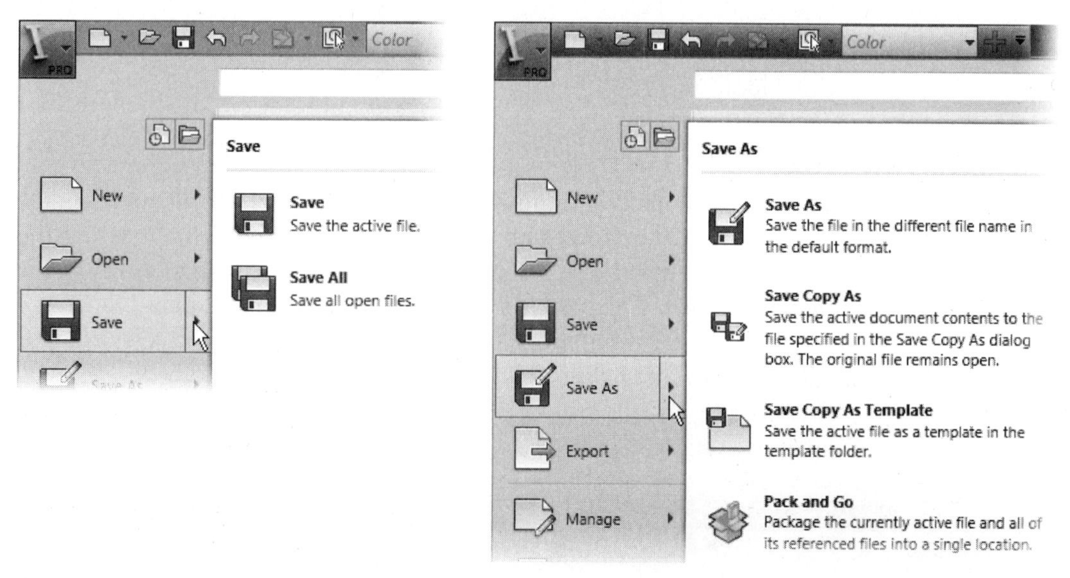

FIGURE 1.15

Save

The Save command saves the current document with the same name and to the location where you created it. If this is the first time that a new file is saved, you are prompted for a file name and file location.

TIP

To run the Save command, click the Save icon on the Quick Access Toolbar, use the shortcut keys CTRL + S, or click Save on the File menu.

Save All

Use the Save All command to save all open documents and their dependents. The files are saved with the same name to the location where you created them. The first time that a new file is saved, you will be prompted for a file name and file location.

Save As

Use the Save As command to save the active document with a new name and location, if required. A new file is created and is made active.

Save Copy As

Use the Save Copy As command to save the active document with a new name and location, if required. A new file is created but is not made active.

Save Copy As Template

Use the Save Copy As Template command to save the current file to the template folder. New files can be based on the template file. Templates can be saved in the existing folders, or you can create a subdirectory in the *Autodesk\Inventor (version number)\templates* directory, and add a file to it. A new template tab, with the same name as the subdirectory, is created automatically when a file is added to the new folder.

Pack and Go

Use the Pack and Go command to copy all the files that are used to create the current file to a specified location.

Save Reminder

You can have Inventor remind you to save a file. Inventor will NOT automatically save the file. After a predetermined amount of time has expired without saving the file, a notification bubble appears in the upper right corner of the screen as shown in the following image on the left. The time can be adjusted via the Application Options > Save tab as shown in the following image on the right. The time can be adjusted from 1 minute to 9999 minutes, or uncheck this option to turn off the notification.

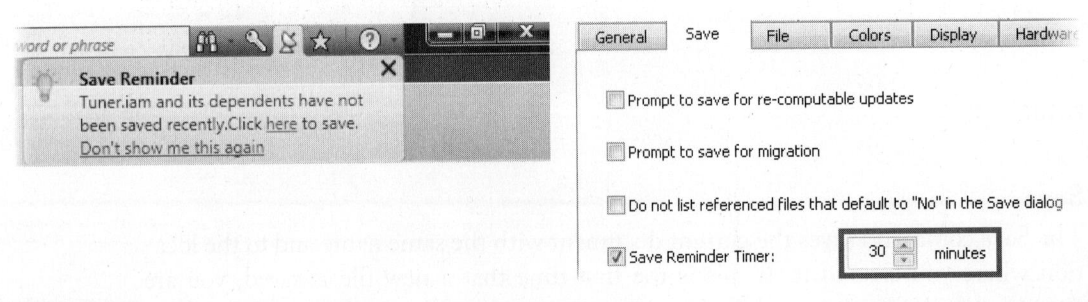

FIGURE 1.16

APPLICATION OPTIONS

Autodesk Inventor can be customized to your preferences. On the Inventor Application Menu, click Options, or from the Tools tab click Application Options, to open the Options dialog box as shown in the following image. You set options on each of the tabs to control specific actions in the Autodesk Inventor software. The application options affect all Inventor documents that are open or will be created. Each section is covered in more detail in the pertinent sections throughout this book. For more information about application options, see the Help system.

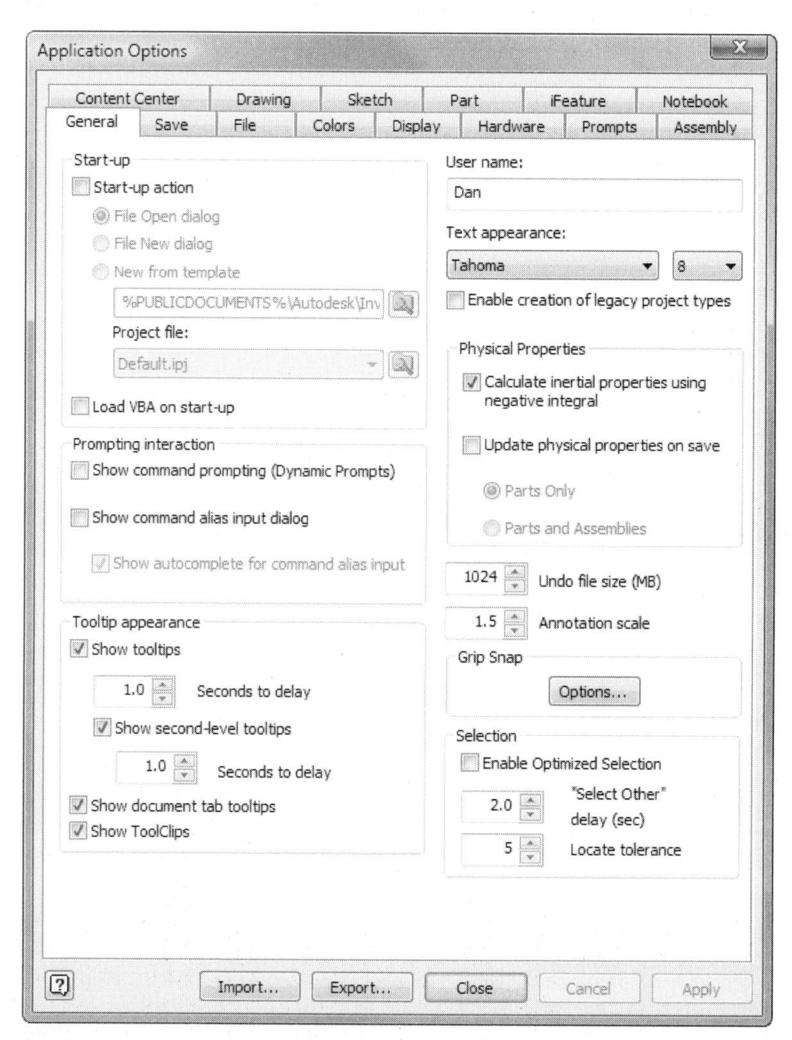

FIGURE 1.17

General

Set general options for how Autodesk Inventor operates.

Save

Set how files are saved.

File
Set where files are located.

Colors
Change the color scheme and color of the background on your screen. Determine if reflections and textures will be displayed.

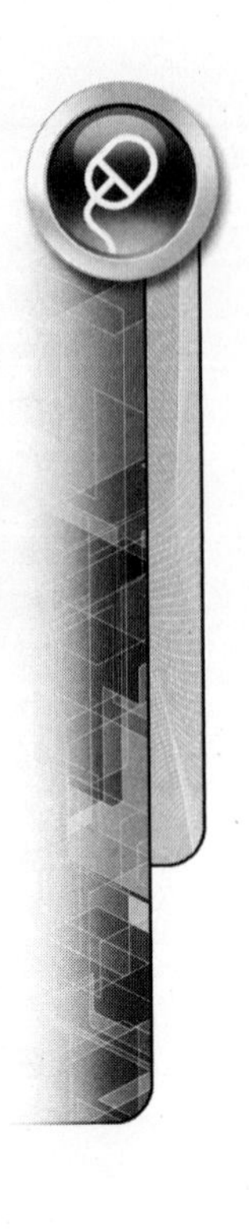

Display
Adjust how parts look. Your video card and your requirements affect the appearance of parts on your screen. Experiment with different settings to achieve maximum video performance.

Hardware
Adjust the interaction between your video card and the Autodesk Inventor software. The software is dependent upon your video card. Take time to make sure that you are running a supported video card and the recommended video drivers. If you experience video-related issues, experiment with the options on the Hardware tab. For more information about video drivers, click Graphics Drivers on the Help menu.

Prompts
Modify the response given to messages that are displayed.

Drawing
Specify the way that drawings are created and displayed.

Notebook
Specify how the Engineer's Notebook is displayed.

Sketch
Modify how sketch data is created and displayed.

Part
Change how parts are created.

iFeature
Adjust where iFeatures data is stored.

Assembly
Specify how assemblies are controlled and behave.

Content Center
Specify the preferences for using the Content Center.

EXERCISE 1-1: USER INTERFACE

In this exercise, you change the user interface by moving the Ribbon, Quick Access Toolbar, and add and remove commands to the Quick Access Toolbar.

1. Click the New command, click the English tab, and then double-click Standard (in).ipt.

2. Move the Ribbon to different locations. Move the cursor anywhere over the Ribbon and right-click, and from the menu click Docking Positions and click the different options.
3. Undock the Ribbon, right-click on the Ribbon, and from the menu click Undock Ribbon. Move the Ribbon to different locations.
4. Move the Ribbon back to its original top position, right-click on the Ribbon, and from the menu click Docking Positions > Top as shown in the following image.

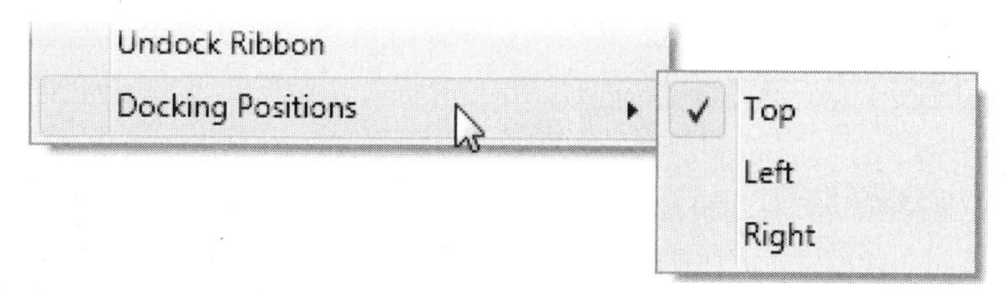

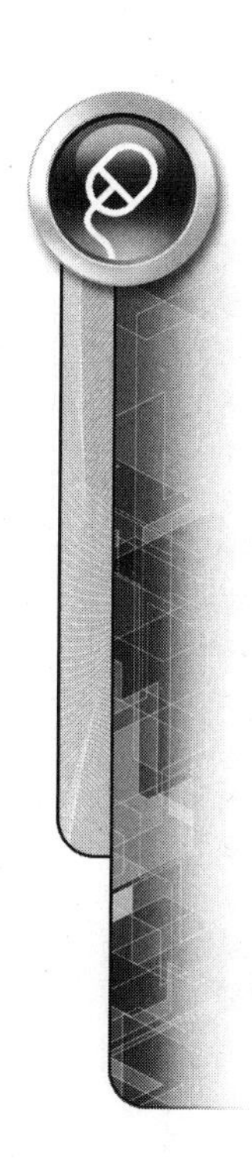

FIGURE 1.18

5. Change the appearance of the Ribbon, right-click on the Ribbon, and from the menu click Ribbon Appearance. Try the different options to change the Ribbon's appearance.
6. Reset the Ribbon back to its original state by clicking Reset Ribbon from the same menu.

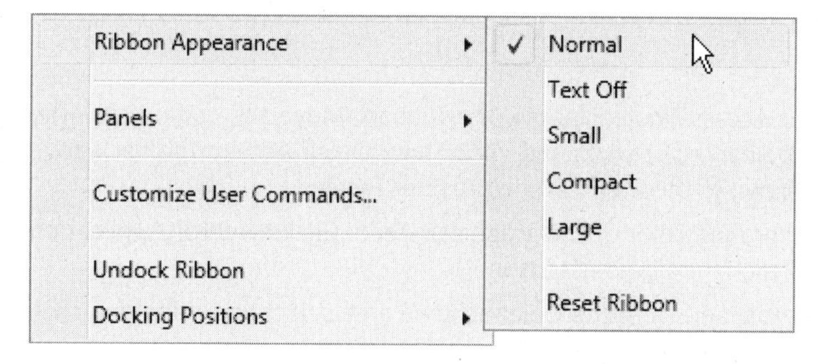

FIGURE 1.19

7. Add a command to the Quick Access Toolbar. Move the cursor over the Line command in the Draw panel, right-click, and click Add to Quick Access Toolbar.

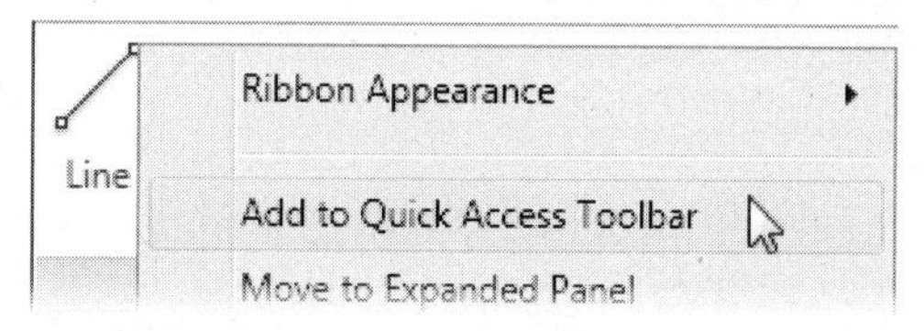

FIGURE 1.20

8. Move the Quick Access Toolbar below the Ribbon. Move the cursor over the Quick Access Toolbar, right-click, and click Show Quick Access Toolbar below the Ribbon.

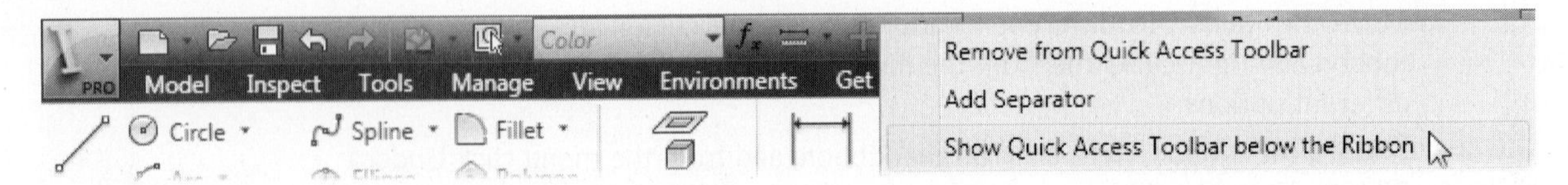

FIGURE 1.21

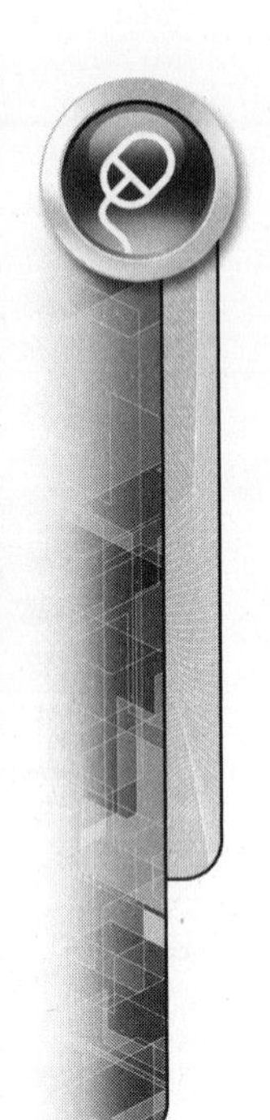

9. Remove the Line command from the Quick Access Toolbar. Move the cursor over the Line command in the Quick Access Toolbar, right-click, and click Remove from Quick Access Toolbar.

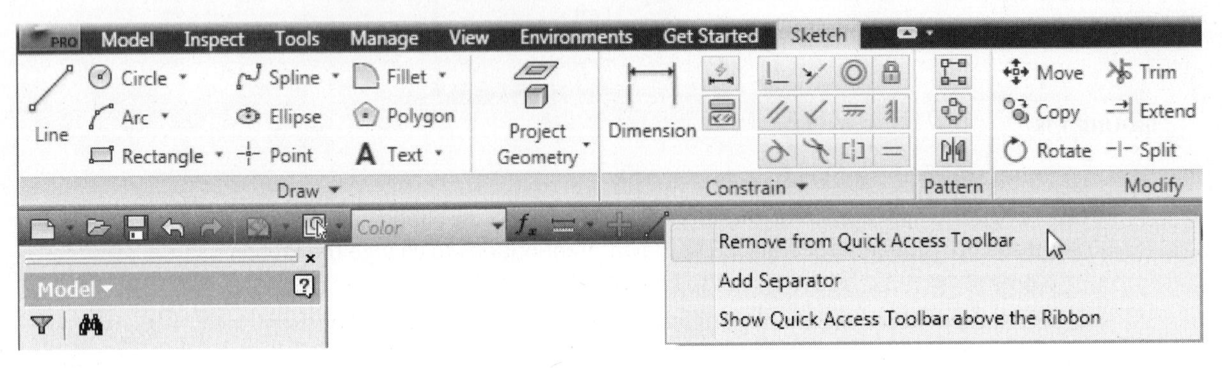

FIGURE 1.22

10. Move the Quick Access Toolbar above the Ribbon. Move the cursor over the Quick Access Toolbar, right-click, and click Show Quick Access Toolbar above the Ribbon as shown in the menu in the previous image.
11. Change the background color of the graphics screen. Click Inventor Application Menu, and then click the Options button.
12. Click the Colors tab, and from the Color scheme area select an option and click Apply to see the change.
13. Experiment with the background options.
14. Experiment changing the colors of the icons. In the Color Theme area, click the Amber option and click Apply. Notice the color of the icons change.
15. Change the icons color back to Cobalt as shown in the following image and click OK.

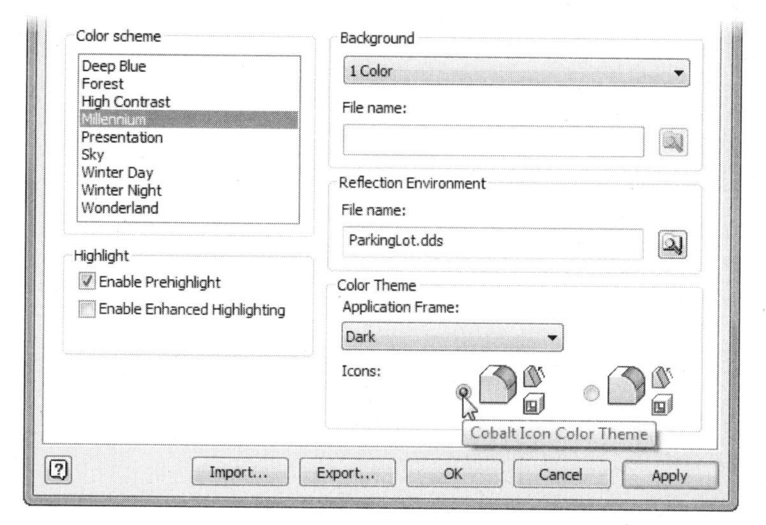

FIGURE 1.23

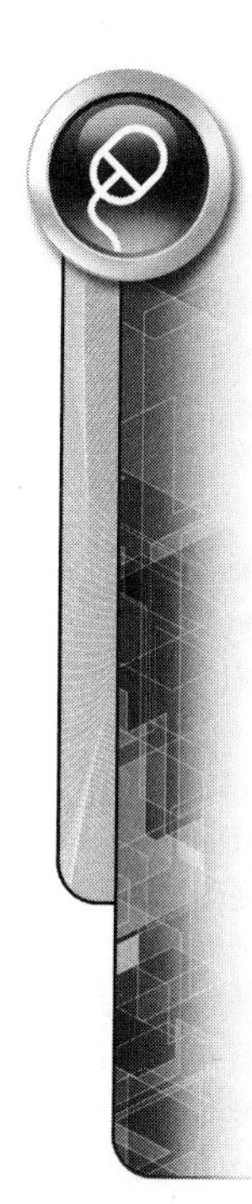

16. As you work with Inventor, adjust the user interface to meet your requirements.
17. Close the file. Do not save changes. End of exercise.

Command Entry

There are several methods to issue commands in Autodesk Inventor. In the following sections, you will learn how to start a command. There are no right or wrong methods for starting a command, and with experience, you will develop your own preference.

To stop a command, either press the Esc key, right-click, and click Done from the menu or select another icon.

Panels and Tooltips

In the last section, you learned how to control the appearance of the ribbon. The main function of the Ribbon is to hold the commands in a logical fashion, which is done by dividing the commands into panels. To start a command from a panel, move the cursor over the desired icon, and a command tip appears with the name of the command. You can control the tooltip from the Application Options under the General tab as shown in the following image on the left. The first level tooltip displays an abbreviated command description as shown in the following image in the middle. If the cursor hovers over the icon longer, a more detailed command description will appear as shown in the following image on the right.

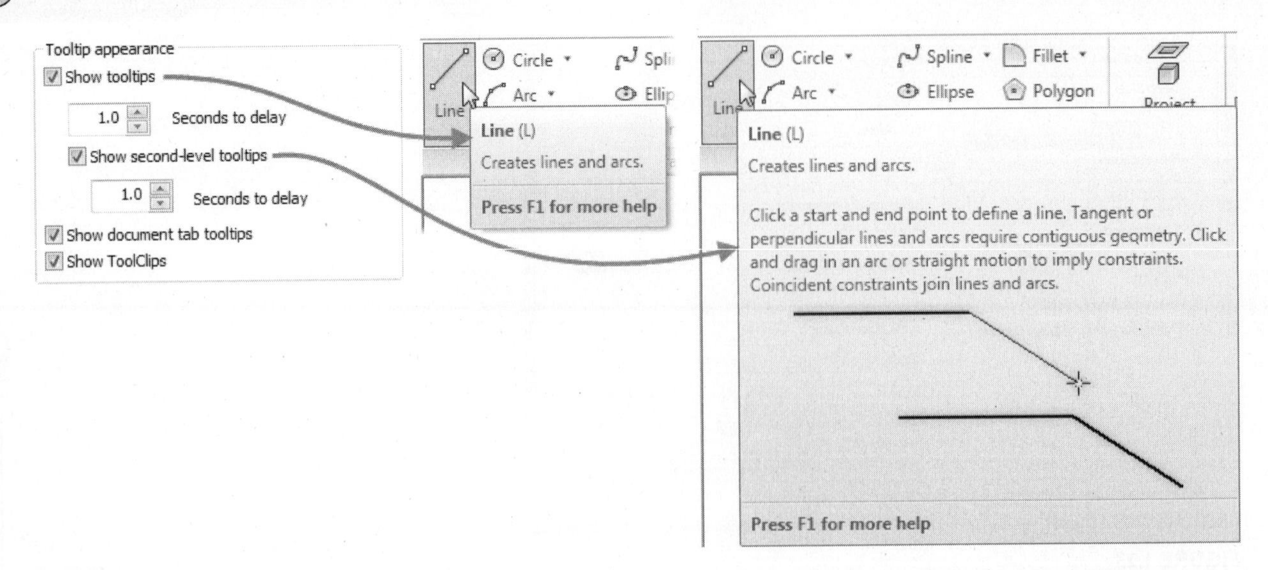

FIGURE 1.24

Some of the icons in the Panel have a small down arrow in the right side. Select the arrow to see additional commands. To activate a command, move the cursor over a command icon and click. The command that is selected will appear first in the list replacing the previous command.

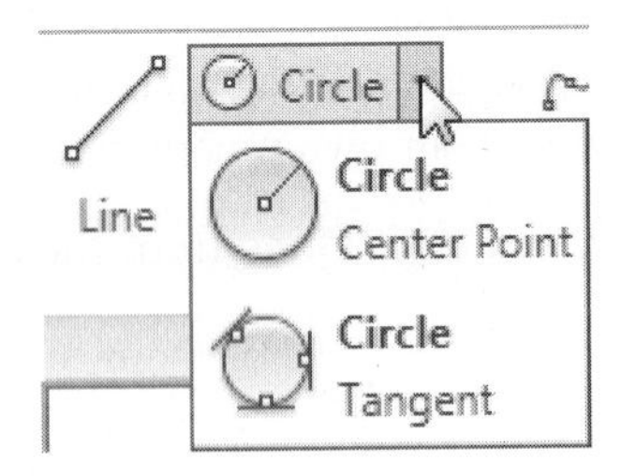

FIGURE 1.25

Marking Menus

Autodesk Inventor also uses marking menus. These are text menus that pop up when you press the right mouse button. The shortcut menus are context sensitive and contain options that are relevant to the current task. The following image on the left shows the marking menu that appears while in the Line command. The top section lists possible commands, while the bottom menu displays options and commands to use while in the command. As you gain experience with Inventor you can start a command from the top portion of the marking menu by right-clicking and moving the cursor (before the menu appears) in the direction of the command and release the mouse button as shown in the following image on the right. This technique is also referred to as gesture behavior.

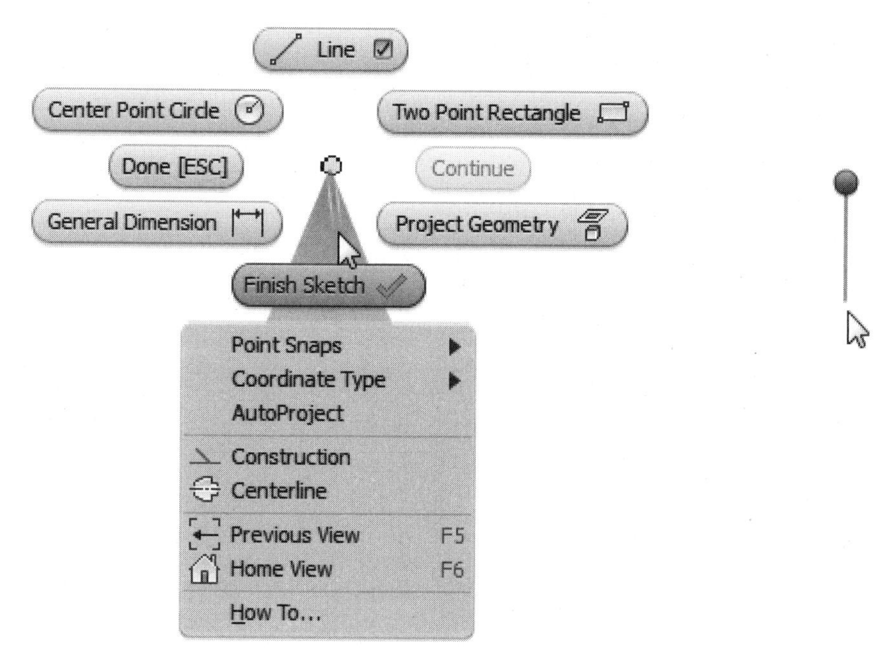

FIGURE 1.26

Autodesk Inventor Shortcut Keys

Autodesk Inventor has keystrokes called shortcut keys that are preprogrammed. While in a command, the tooltip displays the shortcut key in parenthesis if a shortcut key is available. To start a command via a shortcut key, press the desired preprogrammed key(s). The keys can be reprogrammed by clicking the Tools tab and click Customize.

REPEAT LAST COMMAND

To restart a command without reselecting the command in a panel bar, either press ENTER or the spacebar, or right-click and click the top entry in the menu, Repeat "the last command." The following image shows the Line command being restarted.

FIGURE 1.27

Minimize Dialog Boxes

You can control whether an individual dialog is minimized. To minimize a dialog box, click the up arrow near the bottom of the dialog box. The dialog box then displays only the horizontal title bar as shown in the image on the right. To maximize the dialog box, click the down arrow in the minimized dialog box as shown in the following image on the right. This option is set for each dialog box.

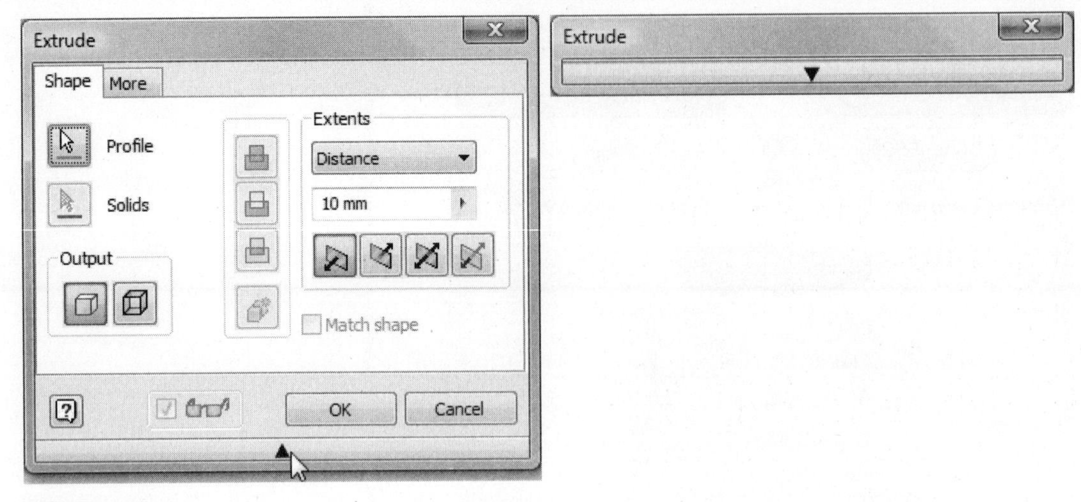

FIGURE 1.28

Undo and Redo

You may want to undo an action that you just performed, or undo an undo. The Undo command backs up Autodesk Inventor one function at a time. If you undo too far, you can use the Redo command to move forward one step at a time. The Zoom, Orbit, and Pan commands do not affect the Undo and Redo commands. To start the commands, select the command from the Quick Access Toolbar as shown in the following image. The Undo command is to the left, and the Redo command is to the right. The shortcut keys are CTRL + Z for Undo and CTRL + Y for Redo.

NOTE

To set the Undo file size allocation, click Tools tab > Application Options. On the General tab of the Options dialog box, change the Maximum size of Undo file (MB).

FIGURE 1.29

HELP SYSTEM

The Help system in Autodesk Inventor goes beyond basic command definition by offering assistance while you design. The commands in the Information Toolbar on the top-right corner of the screen will assist you while you design. To get help on a topic, enter a keyword in the area entitled "Type a keyword or phrase." To see the other help mechanisms that make up the Help System, click the drop arrow next to the question mark as shown in the following image.

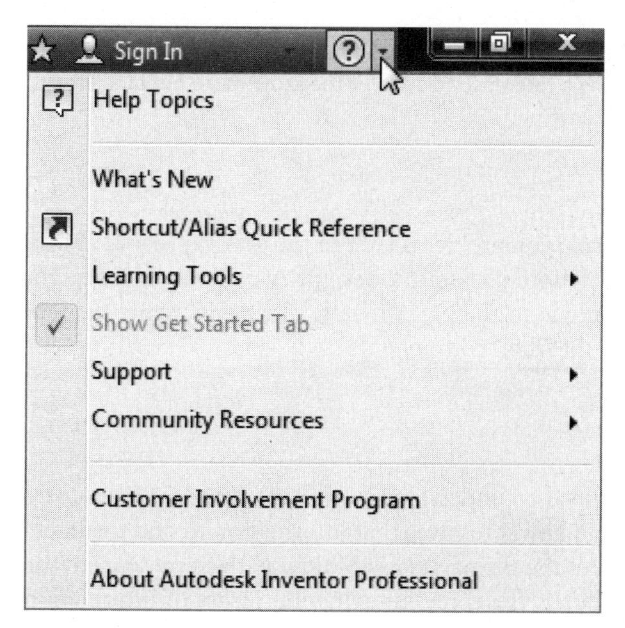

FIGURE 1.30

Other options to access the Help system include the following methods:

- Press the F1 key, and the Help system assists you with the active operation.
- Click an option on the Help menu.
- Click a Help option on the Information Toolbar.
- In any dialog box, click the ? icon.
- Click an option on the Get Started tab > Learn about Inventor panel.
- Click on How To on a shortcut menu initiated within an active command.

PROJECTS IN AUTODESK INVENTOR

Almost every design that you create in Autodesk Inventor involves more than a single file. Each part, assembly, presentation, and drawing created is stored in a separate file. Each of these files has its own unique file extension. There are many times when a design will reference other files. An assembly file, for example, will reference a number of individual part files and/or additional subassemblies. When you open the parent or top-level assembly, it must contain information that allows Autodesk Inventor to locate each of the referenced files. Autodesk Inventor uses a project file to organize and manage these file-location relationships. There is no limit to the number of projects you can create, but only one project can be active at any given time.

You can structure the file locations for a design project in many ways. A single-person design shop has different needs from a large manufacturing company or a design team with multiple designers working on the same project. In addition to project files, Autodesk Inventor includes a program called Autodesk Vault on the DVD that controls basic check-out and check-in file-reservation mechanisms; these control file access for multiuser design teams. Autodesk Inventor always has a project named Default. Specifically, if all the files defining a design are located in a single folder, or in a folder tree where each referenced part is located with its parent or in a subfolder underneath the parent, the Default project may be all that is required.

NOTE

It is recommended that files in different folders never have the same name to avoid the possibility of Autodesk Inventor resolving a reference to a file of the same name but in a different folder.

Project Setup

To reduce the possibility of file resolution problems later in the design process, always plan your project folder structure before you start a design. A typical project might consist of parts and assemblies unique to the project; standard components that are unique to your company; and off-the-shelf components such as fasteners, fittings, or electrical components.

Project File Search Options

Before you create a project, you need to understand how Autodesk Inventor stores cross-file reference information and how it resolves that information to find the referenced file. Autodesk Inventor stores the file name, a subfolder path (if present) to the file, and a library name (optionally) as the three fundamental pieces of information about the referenced file.

When you use the Default project file, the subfolder path is located relative to the folder containing the referencing file. It may be empty or may go deeper in the subfolder hierarchy, but it can never be located at a level above the parent folder.

When you create a project file, you do not need to add subfolder as search paths. The subfolder(s) path is automatically searched and does not need to be added to the project file.

Creating Projects

To create a new project or edit an existing project, use the Autodesk Inventor Project File Editor. The Project File Editor displays a list of shortcuts to previously active projects. A project file has an *.ipj* file extension and typically is stored in the home folder for the design-specific documents, while a shortcut to the project file is stored in the Projects Folder. The Projects Folder is specified on the Files tab of the Options dialog box as shown in the following image. All projects with a shortcut in the Projects Folder are listed in the top pane of the Project File Editor.

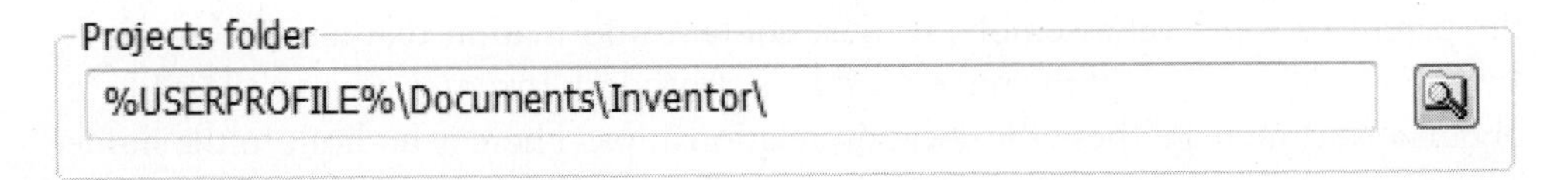

FIGURE 1.31

You create or edit a project file by clicking the Project button in the New or Open dialog box or by clicking Get Started tab > Launch panel > Projects as shown in the following image on the left, or from the Inventor Application Menu click Manage and click Projects as shown in the following image on the right.

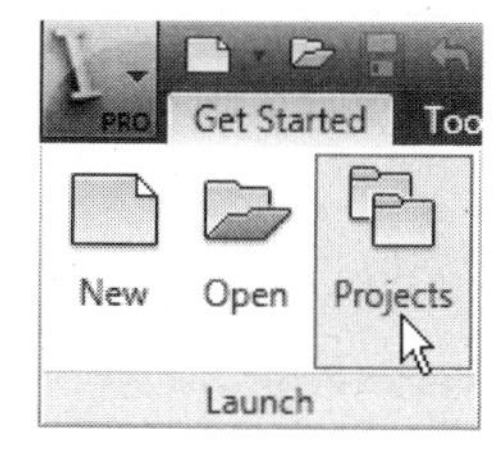

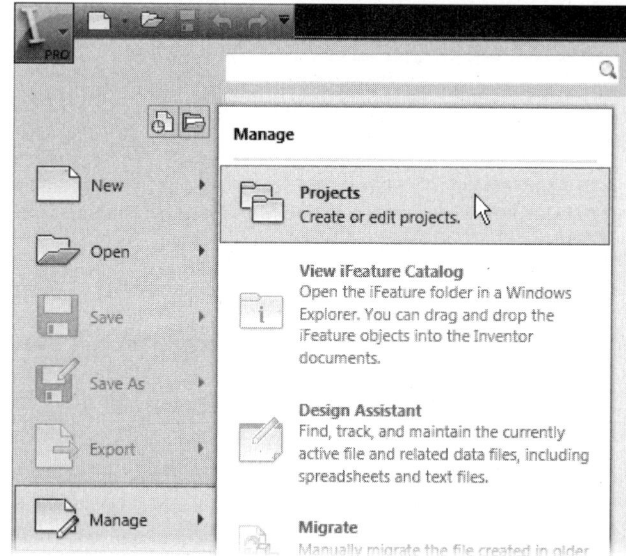

FIGURE 1.32

The Projects dialog box will appear as shown in the following image. The Projects dialog box is divided into two panes. The top pane lists shortcuts to the project files that have been active previously. Double-click on a project's name to make it the active project. All Inventor files must be closed before making a project current. Only one project file can be active in Autodesk Inventor at a time. The bottom portion reflects information about the project selected in the top pane. If a project file already exists, click on the Browse button on the bottom of the dialog box, then navigate to, and select the project file. The bottom pane of the dialog box lists information about the highlighted project.

NOTE

When defining a path to a folder on a network, it is better to define a Universal Naming Convention (UNC) path starting with the server name (\\Server\...) and not to use shared (mapped) network drives.

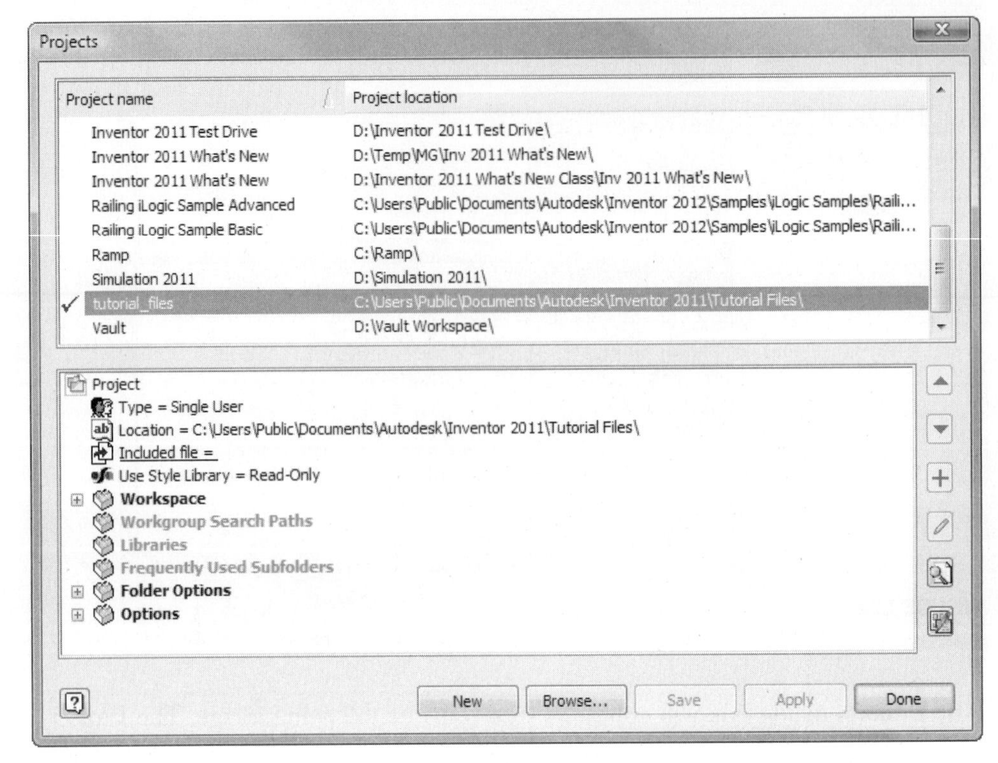

FIGURE 1.33

To create a new project, follow these steps:

1. In the Projects dialog box, click the New button at the bottom to initiate the Inventor project wizard.
2. In the Inventor project wizard, follow the prompts to the following questions.

What Type of Project Are You Creating?

If Autodesk Vault is installed, you will be prompted to create a New Vault project or a New Single-User project. If Autodesk Vault is not installed, only a New Single-User project type will appear in the list.

New Vault Project

This project type is used with Autodesk Vault and is not available until you install Autodesk Vault.

It creates a project with one workspace and any needed library location(s), and it sets the multiuser mode to Vault. More information about Autodesk Vault appears later in this section.

New Single-User Project

This is the default project type, which is used when only one user will reference Autodesk Inventor files. It creates one workspace where Autodesk Inventor files are stored and any needed library location(s), and it sets the Project Type to Single User. No workgroup is defined but can be defined later.

The next section covers the steps for creating a new single-user project. For more information on projects, consult the online Help system.

Creating a New Single-User Project

Click on New Single User Project as shown in the following image. If Autodesk Vault is not installed, New Single User Project will be selected for you.

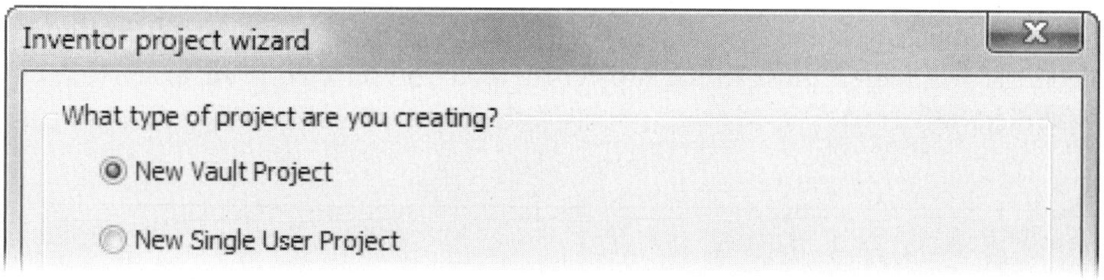

Inventor project wizard
What type of project are you creating?
New Vault Project
New Single User Project

FIGURE 1.34

Click the Next button, and specify the project file name and location on the second page of the Inventor project wizard as shown in the following image.

Name

Enter a descriptive project file name in the Name field. The project file will use this name with an *.ipj* file extension.

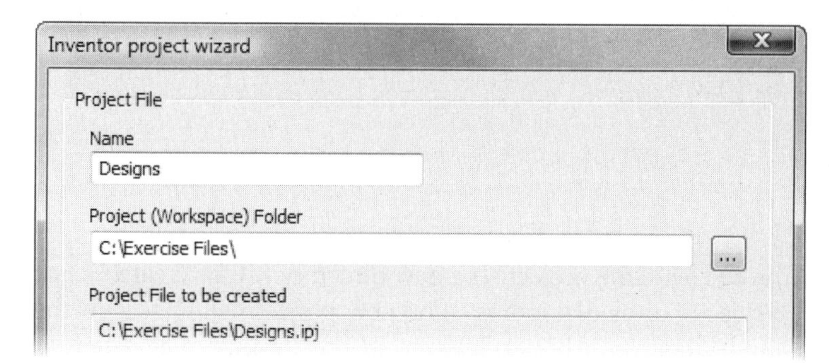

FIGURE 1.35

Project (Workspace) Folder

This specifies the path to the home or top-level folder for the project. You can accept the suggested path, enter a path, or click the Browse button (...) to manually locate the path. The default home folder is a subfolder under My Documents, or wherever you last browsed, that is named to match the project file name.

Project File to Be Created

The full path name of the project file is displayed below the Location field.

NOTE

You can specify the home or top-level folder as the location of your workspace, but it is preferable for the project file (*.ipj*) to be the only Autodesk Inventor file stored in the home folder. This makes it easier to create other project locations, such as library and workspace folders, as subfolders without creating nesting situations that can lead to confusion.

Click the Next button at the bottom of the Inventor project wizard, and specify the project library search paths.

You can add library search paths from existing project files to this new project file. The library search paths from every project with a shortcut in your Projects Folder are listed on the left in the Inventor project wizard dialog box as shown in the following image. You can add and remove libraries from the New Project area by clicking on their names in either the All Projects area or the New Project area and then clicking the arrows in the middle of the dialog box. The libraries listed by default in the All Projects list will match those in the project file that you selected prior to starting the New Project process.

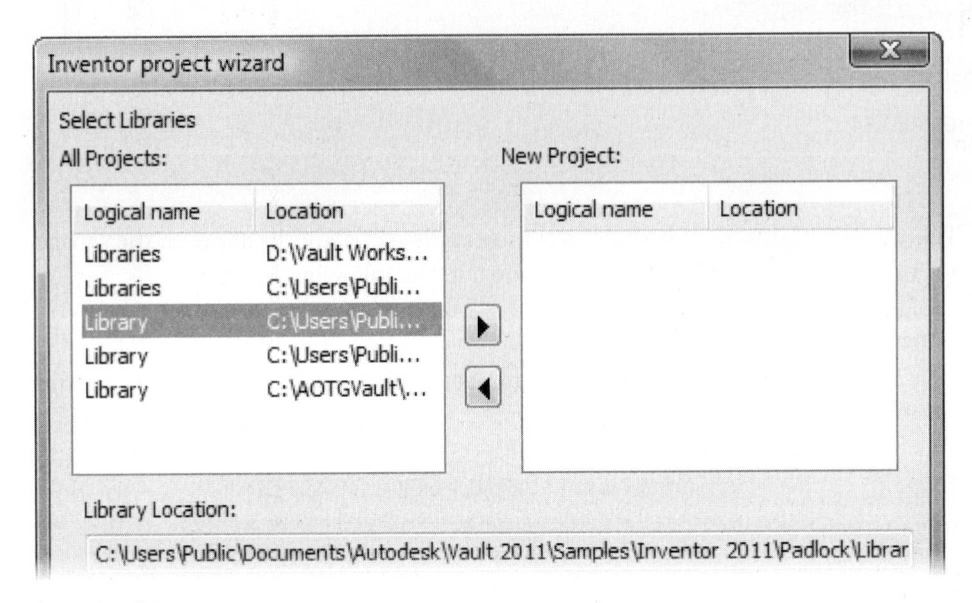

FIGURE 1.36

Click the Finish button to create the project. If a new directory will be created, click OK in the Inventor Project Editor dialog box. The new project will appear in the Open dialog box. Double-click on a project's name in the Projects dialog box to make it the active project. A checkmark will appear to the left of the active project as shown in the following image.

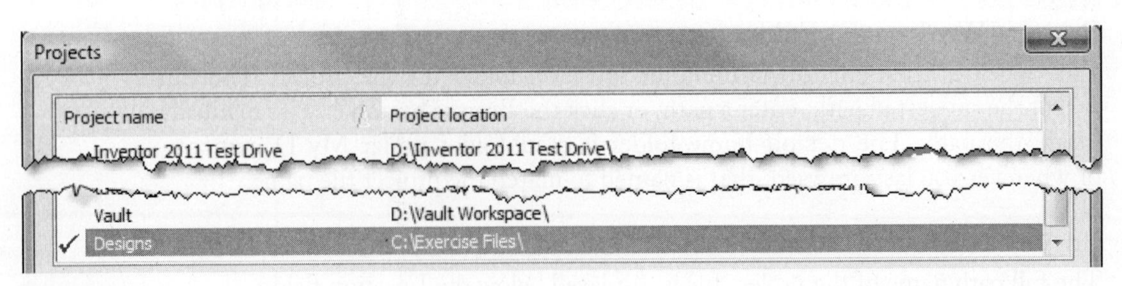

FIGURE 1.37

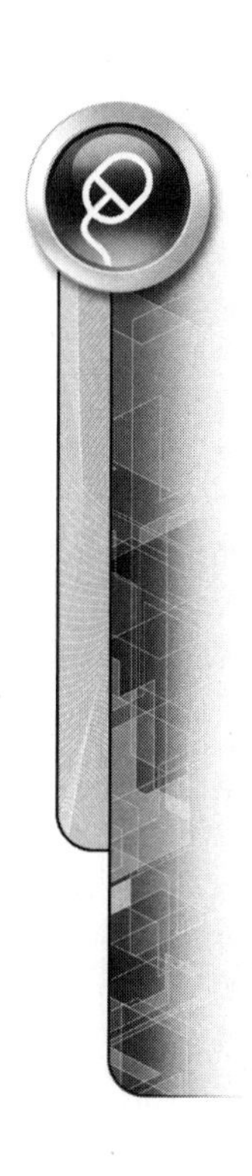

Autodesk Vault

Autodesk Vault is available on the Autodesk Inventor DVD. The Autodesk Vault enhances the data management process by managing more than just Autodesk Inventor files and by tracking file versions as well as team member access. Controlling access to data, tracking modifications, and communicating the design history are important aspects of managing collaborative data. When working with a vault project, your data files are stored in a central repository that records the entire development history of the design. The vault manages Inventor and non-Inventor files alike. In order to modify a file, it first must be checked out of the vault. When the file is checked back into the vault, the modifications are stored as the most recent version for the project, and the previous version is sequentially indexed as part of the living history of the design.

EXERCISE 1-2: PROJECTS

In this exercise, you create a project file for a single-user project, open existing files, delete the project file, and make an existing project file current.

1. Prior to creating a new project file, close all Inventor files and ensure that the exercise files have been installed.
2. From the Get Started tab > Launch panel click Projects.
3. You now create a new project file. Click the New button at the bottom of the Projects dialog box.
4. Click New Single User Project, and click Next.
5. In the next dialog box of the Inventor project wizard, enter **Essentials Plus Book** in the Name field.
6. Click the Browse button (...) to the right of the folder location and name of the Project (Workspace) Folder field.
7. Browse to and select the folder C:\Inv 2012 Ess Plus folder as shown in the following image. The project file (.ipj) will be placed in this folder. The selected folder is the top-level folder for the project. It is a good practice to place project component files such as parts, assemblies, and drawings in folders below the top-level folder, but not in the top-level folder.

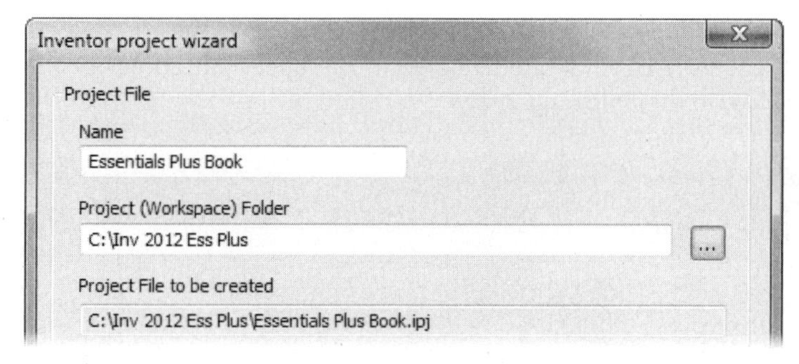

FIGURE 1.38

8. Click Next to add a library search path. The New Project list on the right side should be blank.
9. Click Finish. The new project file is highlighted in the upper pane of the Project File Editor. The search paths for the new project are listed in the lower pane.

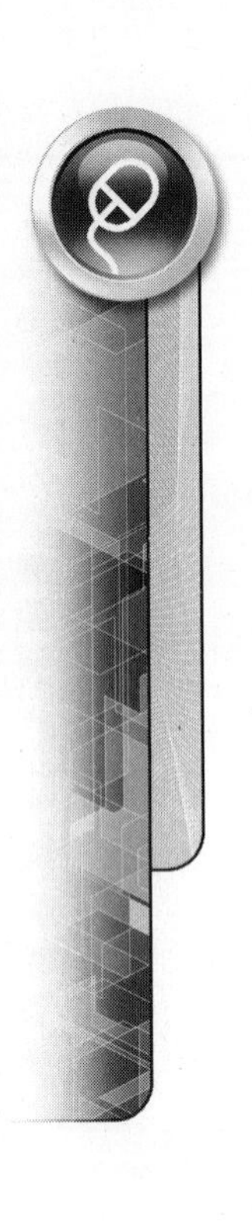

10. Activate the new project by double-clicking on the project named **Essentials Plus Book** in the upper pane of the Project File Editor.
11. Expand Workspace and notice that the Workspace search path is listed as a period "." as shown in the following image. The "." denotes that the workspace location is relative to the location where the project file is saved.

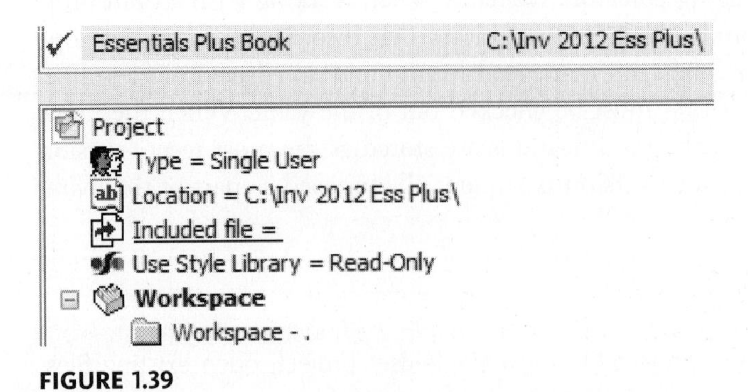

FIGURE 1.39

12. To make it easier to use subfolders below the workspace, you can add them to the Frequently Used Subfolders. Right-click on the Frequently Used Subfolders option, and click Add Paths from Directory as shown in the following image. Then navigate to and select the *C:\Inc 2012 Ess Plus* folder. Click OK to add the subfolders.

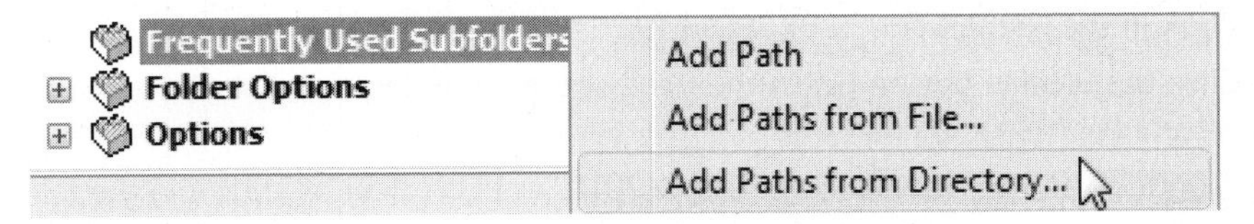

FIGURE 1.40

13. Click Save, and then click Done.
14. Click the Open icon on the Quick Access Toolbar. Notice that the 10 chapters appear in the Frequently Used Subfolders area in the upper-left corner of the dialog box as shown in the following image.

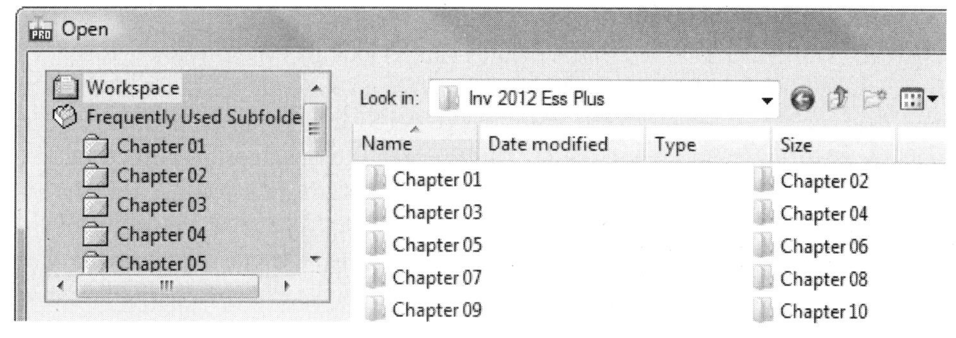

FIGURE 1.41

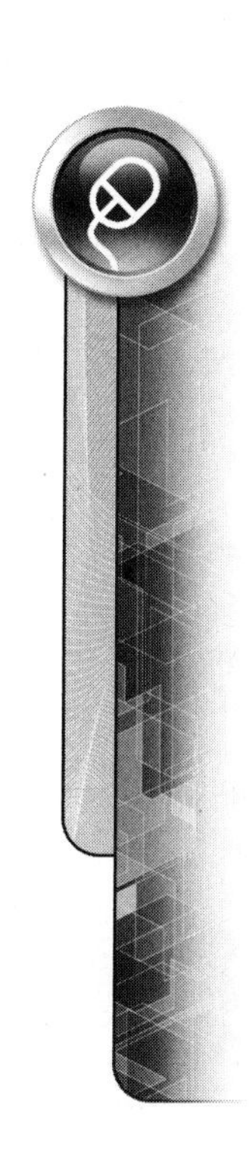

15. Click on any of the chapter folders under the Frequently Used Subfolders, and then open any of the Inventor files that exist in the folder. Notice that the Look in location changes to match the subfolder name.
16. **You must change the active project file to successfully complete the remaining exercises. Close all files currently open in Autodesk Inventor.**
17. Click Inventor Application Menu then click Manage > Projects.
18. To add an existing project file to the current list, click the Browse ... button at the bottom of the dialog box. Navigate to and select *C:\Inv 2012 Ess Plus\Inv 2012 ESS Plus.ipj*. This project should now be the current project. This project will be used for the remaining exercises. Expand the Frequently Used Subfolders section to verify that each chapter has its own subfolder.
19. Right-click on the project **Essentials Plus Book** in the upper pane and select delete as shown in the following image.

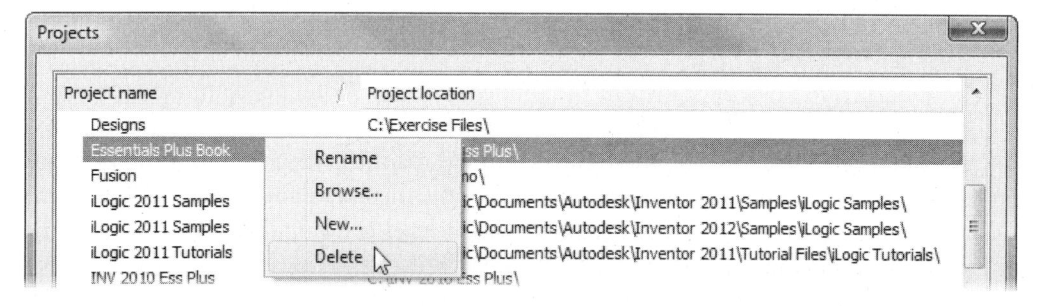

FIGURE 1.42

20. End of exercise.

VIEWPOINT OPTIONS

When you work on a 2D sketch, the default view is looking straight down at the XY plane, which is often referred to as a plan view. When you work in 3D, it is helpful to view objects from a different viewpoint and to zoom in and out or pan the objects on the screen. The next section guides you through the most common methods for viewing objects from different perspectives and viewpoints. As you use these commands, the physical objects remain unmoved. Your perspective or viewpoint of the objects is what creates the perceived movement of the part. If you are performing an operation while a viewing command is issued, the operation resumes after the transition to the new view is completed.

Home (Isometric) View

Change to an isometric viewpoint by pressing the F6 key, click the home icon above the ViewCube as shown in the following image on the left, or by right-clicking in the graphics window and then selecting Home View from the menu as shown in the following image on the right. The view on the screen transitions to a predetermined home view. You can redefine the home view with a ViewCube option. ViewCube is explained later in this section.

FIGURE 1.43

Navigation Bar

The Navigation bar, as shown in the following image, contains commands that will allow you to zoom, pan, and rotate the geometry on the screen. The default location for the Navigation bar is on the right side of the graphics window, but it can be repositioned. Commands are also available below the displayed commands. When commands are selected, they will become the top level command. Descriptions of the viewing commands follow.

NOTE

The navigation commands are also available on the View tab > Navigate panel.

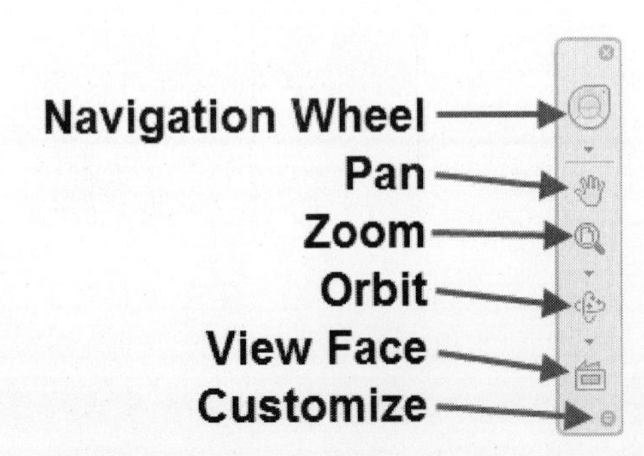

FIGURE 1.44

Navigation Wheel

Click the Navigation Wheel icon to turn on the steering wheel. Click one of the eight options and that option will be active.

Orbit – Rotates the viewpoint.

Zoom – Zooms in and zooms out.

Rewind – Changes to a previous view. A series of images appears representing the previous views. Move the cursor over the image that represents the view to restore.

Pan – Moves the view to a new location.

Center – Centers the view based on the position of the cursor over the wheel.

Walk – Swivels the viewpoint.

Up/Down – Changes the viewpoint vertically.

Look – Moves the view without rotating the viewpoint.

Click the down arrow below the Navigation Wheel icon to change the look of the Navigation wheel.

Pan

Moves the view to a new location. Issue the Pan View command, or select the F2 key. Press and hold the left mouse button, and the screen moves in the same direction that the cursor moves. If you have a mouse with a wheel, hold down the wheel, and the screen moves in the same direction that the cursor moves.

Zoom All

Maximizes the screen with all parts that are in the current file. The screen transitions to the new view. Other commands under Zoom All command are Zoom, Zoom Window, and Zoom Selected.

Zoom. Zooms in or out from the parts. Issue the Zoom In-Out command, or press the F3 key. Then, in the graphics window, press and hold the left mouse key. Move the mouse toward you to make the parts appear larger, and away from you to make the parts appear smaller. If you have a mouse with a wheel, roll the wheel toward you, and the parts appear larger; roll the wheel away from you, and the parts appear smaller.

Zoom Window. Zooms in on an area that is designated by two points. Issue the Zoom Window command and select the first point. With the mouse button depressed, move the cursor to the second point. A rectangle representing the window appears. When the correct window is displayed on the screen, release the mouse button, and the view transitions to it.

Zoom Selected. Fills the screen with the maximum size of a selected face, faces, or a part. Either select the face or faces and then issue the Zoom Selected command, or press the END key, or launch the Zoom Selected command, or press the END key. Then select the face or faces to which you wish to zoom. Use the Select Other command to select a part to zoom to.

Orbit

Rotates your viewpoint dynamically. Issue the Orbit command; a circular image with lines at the quadrants and center appears. To rotate your viewpoint, click a point inside the circle, and keep the mouse button pressed as you move the cursor. Your viewpoint rotates in the direction of the cursor movement. When you release the mouse button, the viewpoint stops rotating. To accept the view orientation, either press the ESC key or right-click and select Done from the menu. Click the outside of the circle to rotate the viewpoint about the center of the circle. To rotate the viewpoint about the vertical axis, click one of the horizontal lines on the circle and, with the mouse button pressed, move the cursor sideways. To rotate the viewpoint about the horizontal axis, click one of the vertical lines on the circle and, with the mouse button pressed, move the cursor upward or downward. The Constrained Orbit command is available below the Orbit command.

Constrained Orbit. Use the Constrained Orbit command to rotate the model about the horizontal screen axes like a turntable. Click one of the horizontal lines on the circle and, with the mouse button pressed, move the cursor sideways. The model is rotated about the model space center point set in the Navigation Wheel CENTER command.

View Face

Changes your viewpoint so that you are looking perpendicular to a plane or rotates the screen viewpoint to be horizontal to an edge. Issue the View Face command or press the PAGE UP key; then select a plane or edge. The View Face command can also be issued by selecting a plane or edge, right-clicking while the cursor is in the graphics window, and selecting Look At from the menu.

Customize the Navigation Bar

To modify the Navigation bar, click the down arrow on the bottom of the Navigation Toolbar as shown in the following image. Click a command to add or remove it from the Navigation bar. You can also reposition the toolbar by clicking the options under Docking position.

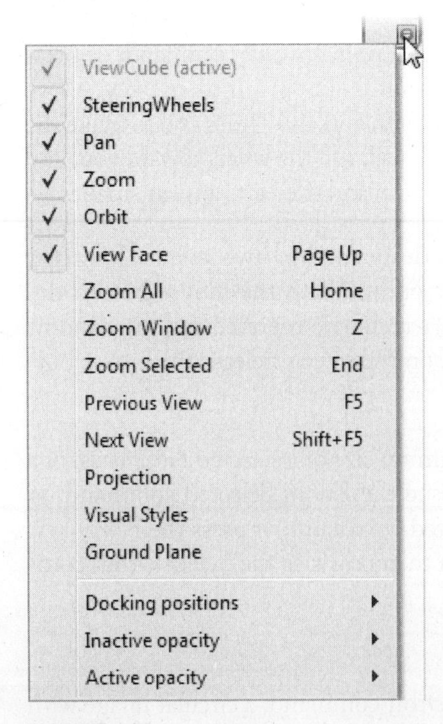

FIGURE 1.45

ViewCube

The ViewCube allows you to quickly change the viewpoint of the screen. With the ViewCube turned on, move the cursor over the ViewCube and the opacity becomes 100%. Move the cursor over the home in the ViewCube to return to the default home (isometric) view as shown in the following image on the left.

FIGURE 1.46

Change the viewpoint by using one of the following techniques.

Isometric

Change to a different isometric view by clicking a corner on the ViewCube as shown in the following image, second from the left.

Face

Change the viewpoint so it looks directly at a plane by clicking a plane on the ViewCube as shown in the following middle image.

Rotate in 90 Degree Increments

When looking at the plane of a ViewCube, you can rotate the view 90 degrees by clicking one of the arcs with arrows or one of the four inside facing triangles as shown in the following image second from the right.

Edge

You can also orientate the viewpoint to an edge. Click an edge on the ViewCube as shown in the following image on the right.

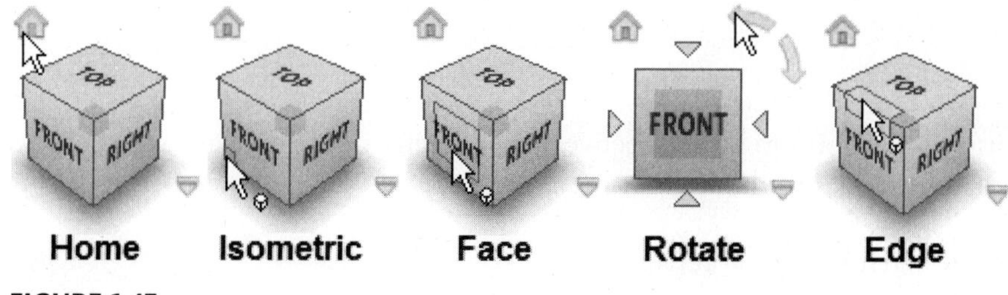

FIGURE 1.47

Dynamic Rotate

To dynamically rotate your viewpoint, click on the ViewCube and keep the mouse button pressed as you move the cursor. Your viewpoint rotates in the direction of the cursor movement. When you release the mouse button, the viewpoint stops rotating. When the viewpoint does not match a defined viewpoint, the ViewCube edges appear in dashed lines from one of the corners as shown in the following image. To smooth the rotation, right-click on the ViewCube and click Options > and uncheck Snap to closest view.

FIGURE 1.48

ViewCube Options

To change the ViewCube's options, move the cursor over the ViewCube and click the down arrow as shown in the following image. From the menu you can change the following:

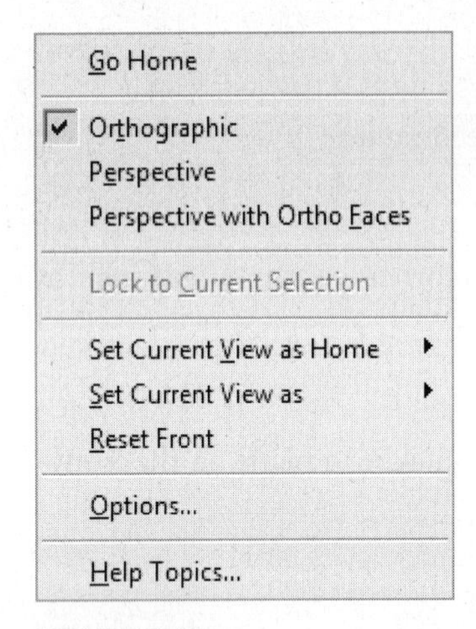

FIGURE 1.49

Go Home

Change the viewpoint to the default Home View.

Orthographic

Set the display viewpoint to be parallel; the lines are projected perpendicular to the plane of projection.

Perspective

Set the display viewpoint to be perspective; the geometry on the screen converges to a vanishing point similar to the way the human eye sees.

Perspective with Ortho Faces

Set the display to perspective except when looking directly at a plane of the ViewCube.

Lock to Current Selection

The center and distance does not change regardless of whether objects are selected or deselected.

Set Current View as Home – Fixed Distance

Defines the current view as the default home view. The view will return to the same size and orientation as when it was set.

Set Current View as Home – Fit to View

Defines the current view as the default home view. The view size will always display all of the parts, but have the same orientation as when it was set.

Set Current View as – Front or Top View

Defines the current view as the front or top view. The view does not need to be a plane on the ViewCube.

Reset Front

Resets the front view to the default setting.

Options

Opens the ViewCube Options dialog box.

Help Topics

Launches the Help system and displays the topic on the ViewCube.

Dynamic Rotation

Rotates viewpoint dynamically. While working, press and hold down the F4 key. The circular image appears with lines at the quadrants and center. With the F4 key still depressed, rotate the viewpoint. When you finish rotating the viewpoint, release the F4 key. If you are performing an operation while the F4 key is pressed, that operation will resume after you release the F4 key.

Another option to quickly rotate your viewpoint is to hold down the Shift + Wheel (middle mouse button). Once the rotate glyph appears on the screen, you can release the Shift key and, while holding down the wheel, move the mouse, and your viewpoint will rotate. With this option, no other options are available. The model(s) are rotated about the center of the model(s). Release the wheel to complete the operation.

Appearance Options

While working with a 3D model, you can adjust the perspective and shading option and control the shadows. To change the appearance, click the View tab > Appearance panel as shown in the following image, or you can add Visual Styles to the Navigation bar.

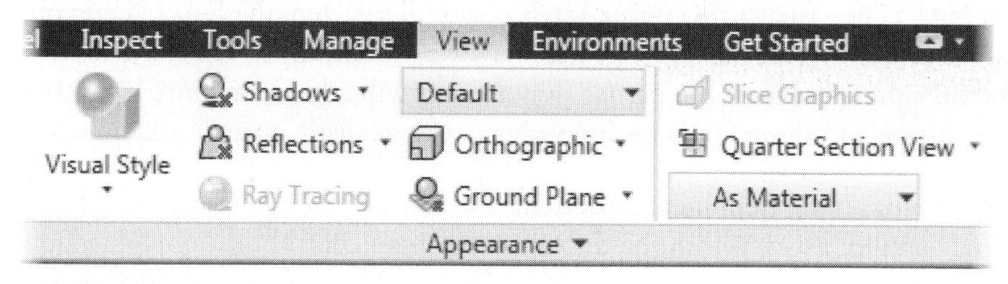

FIGURE 1.50

Visual Style

Set the appearance of model faces and edges in the graphics window. There are ten different styles to choose from as shown in the following image. The styles do not change the model properties; they just change how the model is displayed. Combine visual styles with the other appearance options to display your model(s) differently.

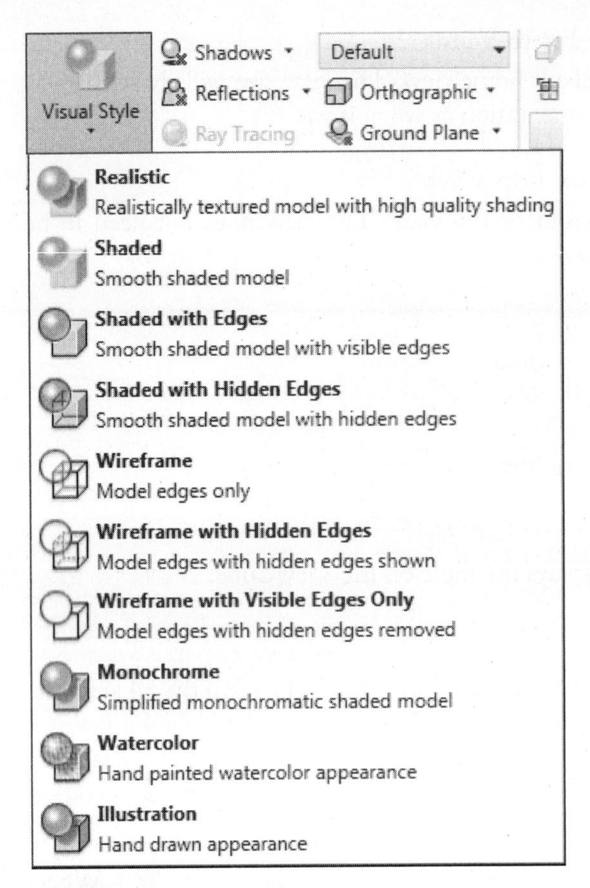

FIGURE 1.51

Shadows. To give your model a realistic look, you can choose different options for displaying shadows. By default shadows are turned off.

Reflections. Controls the color, grid spacing, and reflection of the ground plane. Click the Reflection command to toggle reflections on / off. The ground plane is parallel to the Bottom plane of the view cube, which can be adjusted by orienting the view so the view is parallel to the bottom of the part (that you want). Then right-click on the view cube > Set Current View as > Top.

Ray Tracing. When the Realistic visual style option is selected, the Ray Tracing option will be turned on. Ray tracing gives the images a more realistic effect by simulating how the light reflects off the part(s). Ray tracing will add to the time it takes to render the image and should only be done when a very realistic image is required.

Lights. Select or set different lighting options. From under the Lights command click the Settings option. From within the dialog box you can control the lights direction, color, add lights and control brightness and ambience. Also from this dialog box you can turn on Image Lighting, which allows you to set a background image and set the scale of the image. With a scene image set you can rotate your viewpoint 360 degrees. To move the image, select the Ground Plane Settings option, change the Position & Size option to Manual adjustment and a triad will appear. Select the arrows on the triad to move the image.

Orthographic. Sets the viewpoint to be orthographic or perspective. When set to orthographic the viewpoint is parallel, meaning that lines are projected perpendicular to the plane of projection. For clarity, this book shows all the images in the orthographic viewpoint. When set to perspective the geometry on the screen converges to a vanishing point similar to the way the human eye sees.

Color. Sets the color of the current part. The color can also be set from the Quick Access toolbar.

EXERCISE 1-3: VIEWING A MODEL

In this exercise, you use View Manipulation commands that make it easier to work on your designs.

1. Open C:\Inv 2012 Ess Plus\Chapter 01\ESS_E01_03.iam; the file contains a model of a clamp.

TIP

Click the Chapter 01 subfolder from the Frequently Used Subfolders area, and then click the file in the file area as shown in the following image.

FIGURE 1.52

2. Move the cursor over the bottom clamp grips and spin the wheel toward you. Inventor will zoom in/out to the location of the cursor. This will fill the screen as shown in the following image.

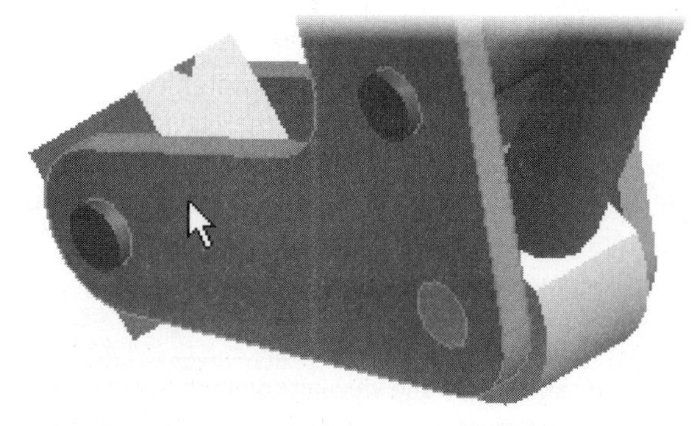

FIGURE 1.53

3. Press F5 to return to the previous view.
4. Use the Orbit command to rotate the viewpoint. From the Navigation bar, click the Orbit command. The 3D Rotate symbol appears in the window as shown in the following image.

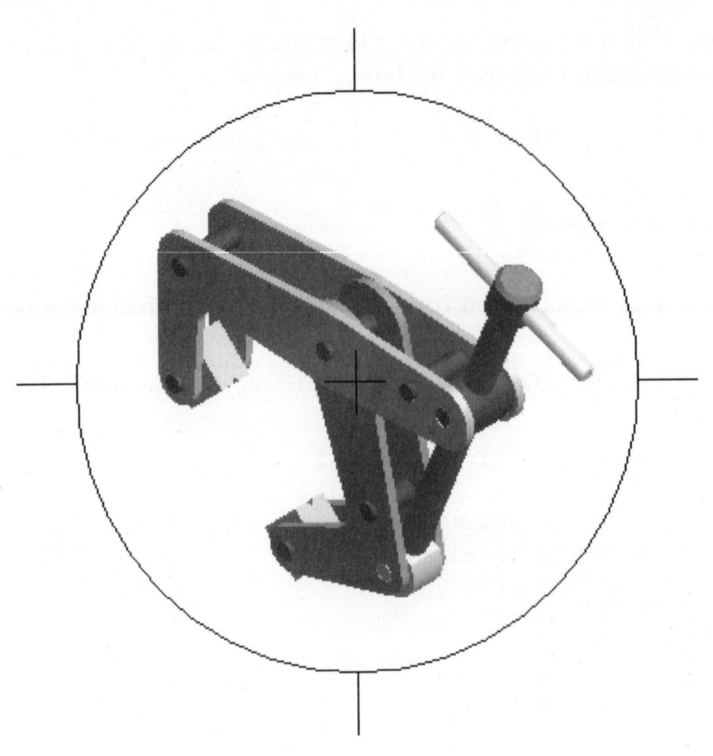

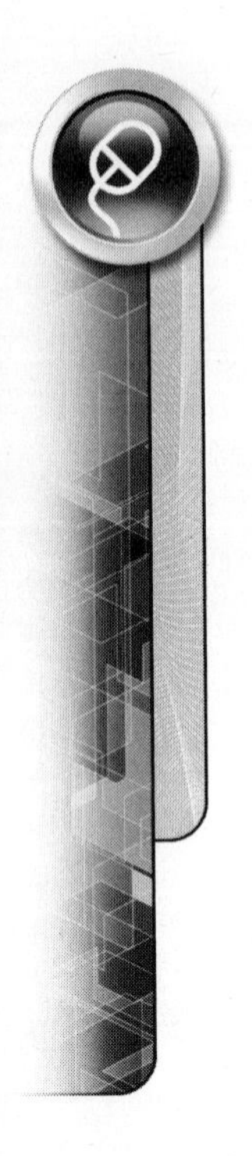

FIGURE 1.54

5. Move the cursor inside the circle of the 3D Rotate symbol, noting the cursor display.
6. Click and drag the cursor to rotate the model.
7. To return to the Home View, press F6.
8. Place the cursor on one of the horizontal handles of the 3D Rotate symbol, noting the cursor display.
9. Drag the cursor right or left to rotate the viewpoint about the Y axis. As you rotate the viewpoint, you will see the bottom of the assembly.
10. To again rotate the viewpoint about the vertical axis like a turntable, click the Constrained Orbit command below the Orbit command on the Navigation bar. Click one of the horizontal lines on the circle. With the mouse button pressed, move the cursor sideways. The view will rotate like a turntable; notice that you do not see the bottom of the assembly.
11. Return to the home view and click the Home symbol above the ViewCube.

TIP

While you are performing another process, you can click on the viewing commands, spin the wheel to zoom, press and hold the wheel to pan, or press and hold F4 to rotate the view.

12. To change the viewpoint to predetermined isometric views, or directly at a plane, use the ViewCube. Click the Top-Front left corner of the ViewCube as shown in the following image on the left. Continue clicking the other corners of the ViewCube as shown in the following middle two images.
13. Change to the front view by clicking the Front plane on the ViewCube as shown in the following image on the right.

FIGURE 1.55

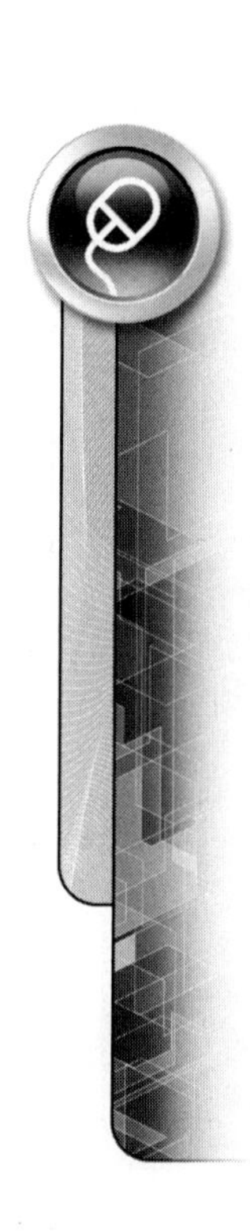

14. Rotate the view 90 degrees by clicking the arc with arrow in the ViewCube as shown in the following image on the left.
15. To look at the Top view that is also rotated by 90 degrees, click the left arrow as shown in the following image on the right.

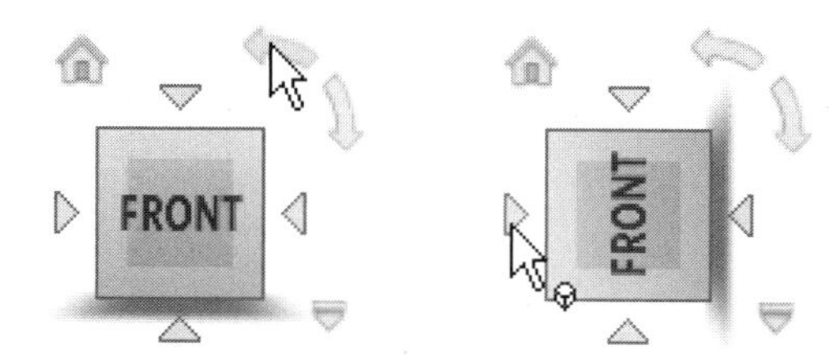

FIGURE 1.56

16. Practice changing views by clicking the corners, faces, and edges of the ViewCube.
17. Click the Navigation Wheel icon on the Navigation bar.
18. Click the different options shown in the following image on the Navigation Wheel and experiment with them.

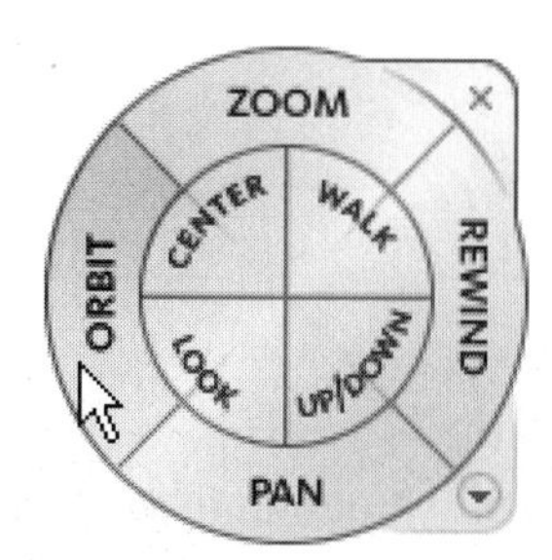

FIGURE 1.57

19. Move the cursor into a blank area in the graphics window, right-click, and click Home View from the menu.
20. Next you will apply different Visual Styles. Click the View tab > Appearance panel > under Visual Style; experiment by selecting different styles.
21. Change the visual style to Shaded.
22. Next you will apply different Shadows. Click the View tab > Appearance panel > under Shadows and experiment by turning the different shadows on and off.
23. Change the Shadow option to All Shadows.
24. Turn on reflections. From the View tab > Appearance panel click Reflections.

25. Next change to a perspective view. From the View tab > Appearance panel click the arrow beside Orthographic and click Perspective.
26. To change the perspective, press CTRL + Shift, and spin the wheel on the mouse.
27. Change back to an orthographic view. From the View tab > Appearance panel, click the arrow beside Perspective and click Orthographic, and your screen should resemble the following image.

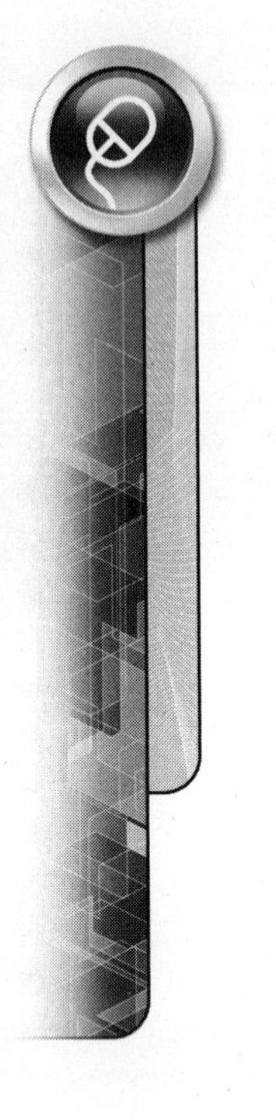

FIGURE 1.58

28. Apply different lights. From the View tab > Appearance panel, click the arrow beside Default and select different lighting configurations.
29. In the Style and Standard Editor dialog box from the Lighting option make the following changes as shown in the following image.
    - Click the Use Image Lighting option to turn it on.
    - Click the Display Scene Image option to turn it on.
    - From the Source drop list select Empty Lab.
    - Adjust the image Scale to 25% and move the slider under the Scale option.
    - To complete the operation click Done and then click Yes to save the changes. Note that you could have created a new Lighting Style.

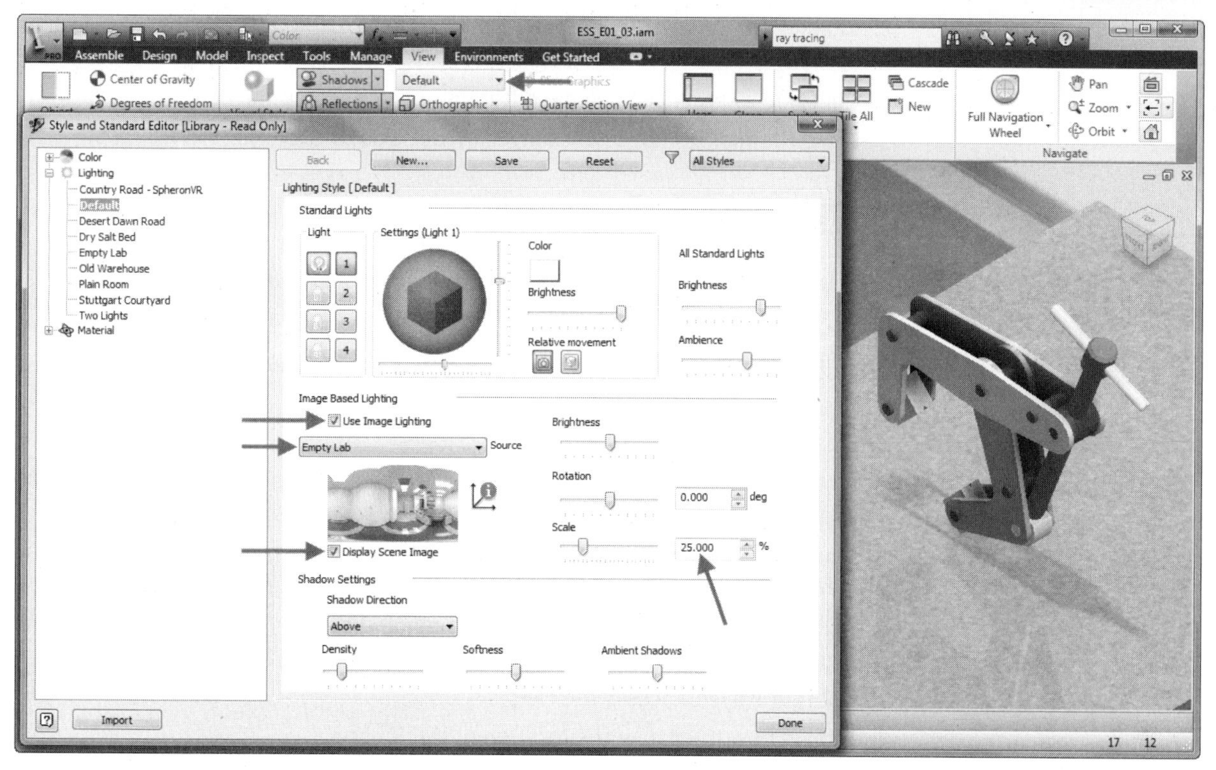

FIGURE 1.59

30. Adjust the location of the image by clicking the down arrow to the right of the Ground Plane command and click Settings. Change the Position & Size option to Manual Adjust, in the graphics window click and drag the arrows that appear. When done click OK.

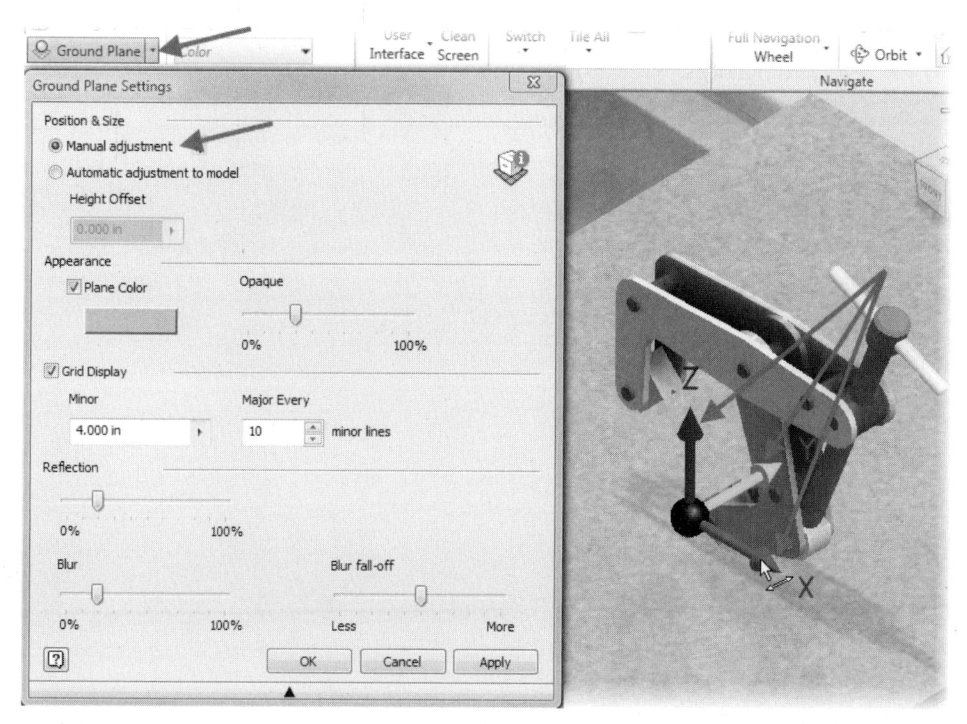

FIGURE 1.60

31. Change the Visual Style to Realistic and experiment with the Ray Tracing options.
32. Continue to practice applying different appearance options.
33. Close the file. Do not save changes model. End of exercise.

CHECKING YOUR SKILLS

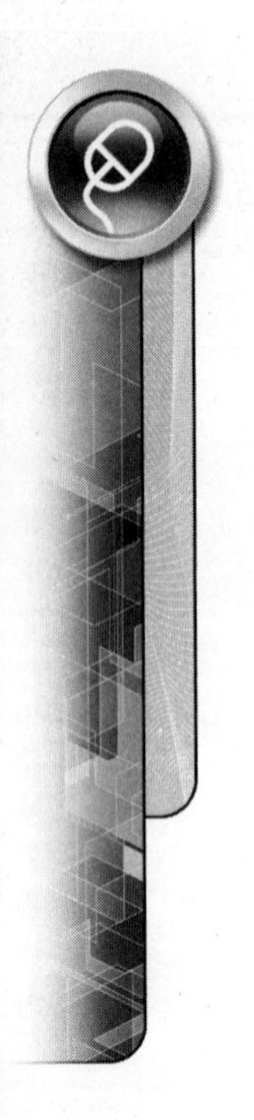

Use these questions to test your knowledge of the material covered in this chapter.

1. Explain the reasons why a project file is used.
2. True __ False __ Only one project can be active at any time.
3. True __ False __ Subfolders need to be added to the project file in order for the files to be found.
4. True __ False __ Autodesk Inventor stores the part, assembly information, and related drawing views in the same file.
5. True __ False __ Press and hold down the F4 key to dynamically rotate the viewpoint.
6. True __ False __ The Save Copy As command saves the active document with a new name, and then makes it current.
7. List four ways to access the Help system.
8. Explain how to change the location of the Ribbon.
9. True __ False __ Commands can be added to the Quick Access toolbar.
10. True __ False __ Visual styles change the physical properties of the model.
11. True __ False __ Ray Tracing is used to get an accurate mass property.
12. True __ False __ The settings in the Application Options are global and affect all open and new Inventor files.
13. Explain how to move the Quick Access toolbar below the ribbon.
14. Explain how to start a command from the marking menu without clicking a on a specific command.
15. True __ False __ By default Inventor will automatically save the current file every 10 minutes.

CHAPTER 2

Sketching, Constraining, and Dimensioning

INTRODUCTION

Most 3D parts in Autodesk Inventor start from a 2D sketch. This chapter first provides a look at the application options for sketching and part creation. It then covers the three steps in creating a 2D parametric sketch: sketching a rough 2D outline of a part, applying geometric constraints, and then adding parametric dimensions.

OBJECTIVES

After completing this chapter, you will be able to do the following:

- Change the part and sketch options as needed
- Sketch an outline of a part
- Create geometric constraints
- Use construction geometry to help constrain sketches
- Dimension a sketch
- Create dimensions using the automatic dimensioning command
- Change a dimension's value in a sketch
- Open and insert AutoCAD DWG data

PART AND SKETCH APPLICATION OPTIONS

Before you start a new part, examine the part and sketch options in Autodesk Inventor that will affect how the part file will be created and how the sketching environment will look and act. While learning Autodesk Inventor, refer back to these option settings to determine which ones work best for you—there are no right or wrong settings.

Part Options

You can customize Autodesk Inventor Part options to your preferences. Click Inventor Application Menu > Options button, and click on the Part tab, as shown in the following image. Descriptions of the Part options follow. These settings are global—they will affect all active and new Autodesk Inventor documents.

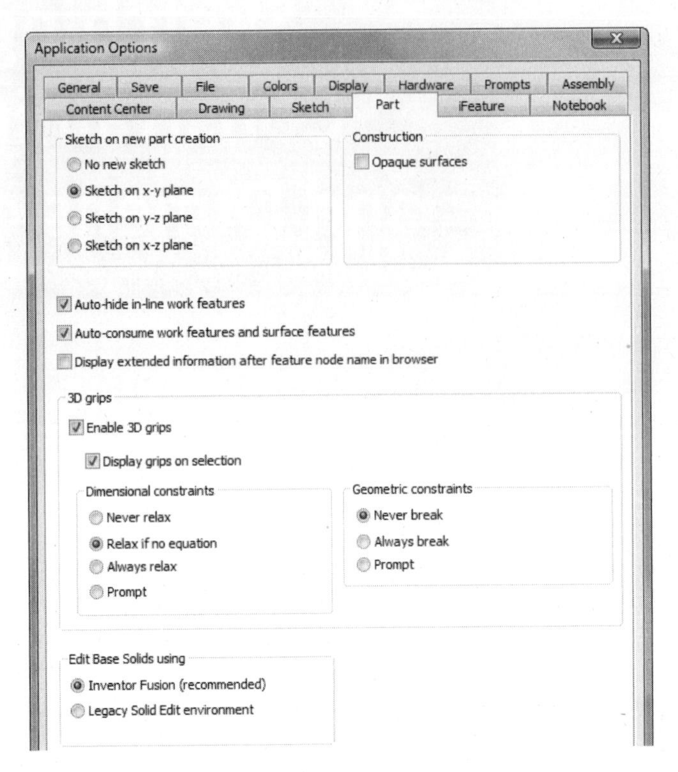

FIGURE 2.1

Sketch on New Part Creation

- No new sketch

When checked, does not set sketch plane when you create a new part.

- Sketch on x-y plane

When checked, sets the x-y plane as the current sketch plane when you create a new part.

- Sketch on y-z plane

When checked, sets the y-z plane as the current sketch plane when you create a new part.

- Sketch on x-z plane

When checked, sets the x-z plane as the current sketch plane when you create a new part.

Construction

- Opaque surfaces

When checked, surfaces you create are displayed as opaque; otherwise, they will be translucent. After you create a surface, its translucency can be controlled from a menu.

Auto-Hide In-Line Work Features

When checked, automatically hides work features that are consumed by another work feature.

Auto-Consume Work Features and Surface Features

When checked, work features and surface features will become children feature(s) of the feature that used them to be created. Display extended information after feature node in browser.

Display extended information after feature node in browser

When checked and in a part, information about the feature such as extrusion depth will be displayed after the name of the feature in the browser.

3D Grips

Sets the preferences for using 3D grips.

Enable 3D Grips. When checked, enables 3D grips. When unchecked, 3D grips will not be utilized.

Display Grips on Selection. When checked, displays the grip when selecting a face or an edge in a part (*.ipt*) or an assembly (*.iam*) file. The grip displays when the selection priority is set to faces and edges, and the face can be edited with 3D Grips. Clicking on a grip launches the 3D Grips command. Clear the check box to turn off the display of the grip.

Dimensional Constraints

- Never relax

When checked, a feature cannot be grip edited in a direction that has a linear or angular dimension.

- Relax if no equation

When checked, a feature cannot be grip edited in a direction that is defined by an equation that is based on linear or angular dimension. Dimensions without equations are not affected.

- Always relax

When checked, a feature can be grip edited, whether or not a linear, angular, or equation-based dimension is applied. Equation-based dimensions are converted to numeric values.

- Prompt

When checked, similar to Always relax, a dialog box is displayed warning if a grip edit will affect either dimensions or equation-driven dimensions. When accepted, dimensions and equations are relaxed, and both are updated as numerical values after grip editing.

Geometric Constraints

- Never break

When checked, a feature cannot be grip edited if a constraint exists.

- Always break

When checked, a feature can be grip edited, even if a constraint exists.

- Prompt

When checked, similar to Always break, a dialog box is displayed warning if a grip edit will break one or more constraints.

Edit Base Solids Using

- Inventor Fusion

When checked, Inventor Fusion will be used to edit base solids (solids that have no features).

- Legacy Solid Edit environment

When checked, Inventor commands from the Edit Base Solid tab are used to edit base solids (solids that have no features).

Sketch Options

Autodesk Inventor sketching options can be customized to your preferences. Click Inventor Application Menu > Options, and then click on the Sketch tab, as displayed in the following image. Descriptions of the Sketch options follow. These settings are global, and all of them affect currently active Autodesk documents and Autodesk Inventor documents you open in the future.

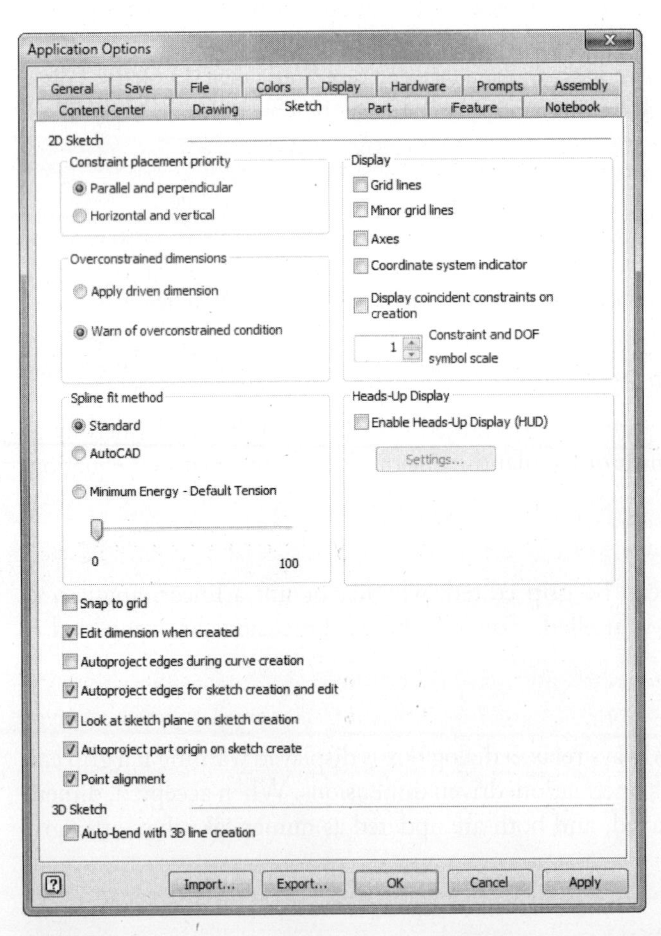

FIGURE 2.2

Constraint Placement Priority

- Parallel and perpendicular

When checked, and when a parallel or perpendicular condition exists while sketching, applies a parallel or perpendicular constraint before any other possible constraints that affect the geometry being created.

- Horizontal and vertical

When checked, and when a horizontal or vertical condition exists while sketching, applies a horizontal or vertical constraint before any other possible constraints that affect the geometry being created.

Display

- Grid lines

Toggles both minor and major grid lines on the screen on and off. To set the grid distance, click the Tools tab > Options panel > Document Settings, and on the Sketch tab of the Document Settings dialog box, change the Snap Spacing and Grid Display.

- Minor grid lines

Toggles the minor grid lines on the screen on and off.

- Axes

Toggles the lines that represent the X- and Y-axis of the current sketch on and off.

- Coordinate system indicator

Toggles the icon on and off that represents the X-, Y-, and Z-axes at the 0, 0, 0 coordinates of the current sketch.

- Display coincident constraints on creation

When checked, coincident constraints are represented with a dot after the constraint is created.

- Constraint and DOF Symbol Scale

Adjusts the scale of the sketch constraint toolbars and the sketch degree of freedom symbols to make them larger or smaller.

Overconstrained Dimensions

- Apply driven dimensions

When checked, and when you add dimensions that would overconstrain the sketch, adds the dimension as a driven (reference) dimension.

- Warn of overconstrained condition

When checked, and when you add dimensions that would overconstrain the sketch, a dialog box appears warning of the condition. This allows you to accept the placement of a driven dimension or cancel the dimension placement.

Spline Fit Method

- Standard

Sets the fit method to create a spline with smooth continuity (G3 minimum) between points. Suitable for Class A surfaces.

- AutoCAD

Sets the fit method to create a spline using the AutoCAD fit method (G2 minimum). Not suitable for Class A surfaces.

- Minimum Energy

Sets the fit method to create a spline with smooth continuity (G3 minimum) and good curvature distribution. Suitable for Class A surfaces. Takes the longest to calculate and creates the largest file.

- Minimum Energy – Default Tension

Set the slider bar to determined tightness or looseness for a 2D spline.

Snap to Grid

When checked, endpoints of sketched objects snap to the intersections of the grid as the cursor moves over them.

Edit Dimension When Created

When checked, edits the values of a dimension in the Edit Dimension dialog box immediately after you position the dimension. It is recommended to turn this setting on.

Autoproject Edges During Curve Creation

When checked, and while sketching, place the cursor over an object and it will be projected onto the current sketch. You can also toggle Autoproject on and off while sketching by right-clicking and selecting Autoproject from the menu.

Autoproject Edges for Sketch Creation and Edit

When checked, automatically projects all of the edges that define that plane onto the sketch plane as reference geometry when you create a new sketch.

Look at Sketch Plane on Sketch Creation

When checked, automatically changes the view orientation to look directly at the new or active sketch.

Autoproject Part Origin on Sketch Create

When checked, the part's origin point will automatically be projected when a new sketch is created. It is recommended to keep this setting on.

Point Alignment On

When checked, automatically infers alignment (horizontal and vertical) between endpoints of newly created geometry. No sketch constraint is applied. If this option is not checked, points can still be inferred; this technique is covered later in this chapter in the Inferred Points section.

Enable Heads-Up Display (HUD)

When checked Dynamic Input is turned on, which allows numeric and angular values to be entered into the value input boxes when creating sketch geometry. Click the Settings button to change the Heads-Up Display settings.

3D Sketch

- Auto-bend with 3D line creation

When checked, automatically places a tangent arc between two 3D lines as they are sketched. To set the radius of the arc, click the Tools tab > Options panel > Document Settings, and change the Auto-Bend Radius on the Sketch tab of the Document Settings dialog box.

UNITS

Autodesk Inventor uses a default unit of measurement for every part and assembly file. The default unit is set from the template file from which you created the part or assembly file. When specifying numbers in dialog boxes with no unit, the default unit will be used. You can change the default unit in the active part or assembly document by clicking Tools tab > Document Settings button and selecting the Units tab as shown in the following image. The unit system values change for all of the existing values in that file.

NOTE

In a *drawing* file, the appearance of dimensions is controlled by dimension styles. Drawing settings will be covered in Chapter 5.

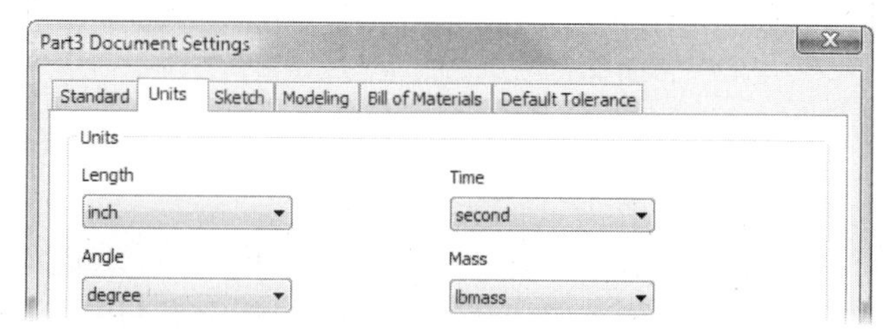

FIGURE 2.3

You can override the default unit for any value by entering the desired unit. If you were working in a metric file whose unit is set to mm, for example, and you placed a horizontal dimension whose default value was 50.8 mm, you could enter 2 in. Dimensions appear on the screen in the default units.

For the previous example, 50.8 mm would appear on the screen. When you edit a dimension, the overridden unit appears in the Edit Dimension dialog box as shown in the following image.

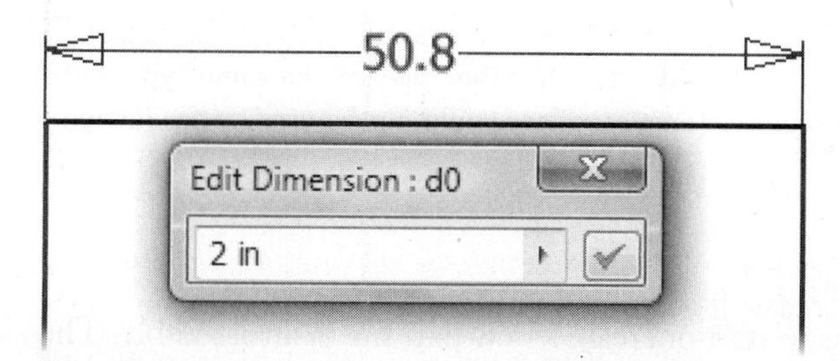

FIGURE 2.4

TEMPLATES

Each new file is created from a template. You can modify existing templates or add your own templates. As you work, make note of the changes that you make to each file. You then create a new template file or modify an existing file that contains all of the changes and save that file to your template directory, which by default in Windows XP is located at *C:\Program Files\Autodesk\Inventor 2012\templates* directory and in Windows Vista and Windows 7 is *C:\Users\Public\Public Documents\Autodesk\Inventor 2012\Templates* directory. You can also create a new subdirectory under the templates folder, and place any Autodesk Inventor file in this new directory. After adding an Inventor file the new tab will appear, and it will be available as a template.

You can use one of two methods to share template files among many users. You can modify the location of templates by clicking the Inventor Application Menu > Options button, clicking on the File tab, and modifying the Templates location as shown in the following image. The Templates location will need to be modified for each user who needs access to these common templates.

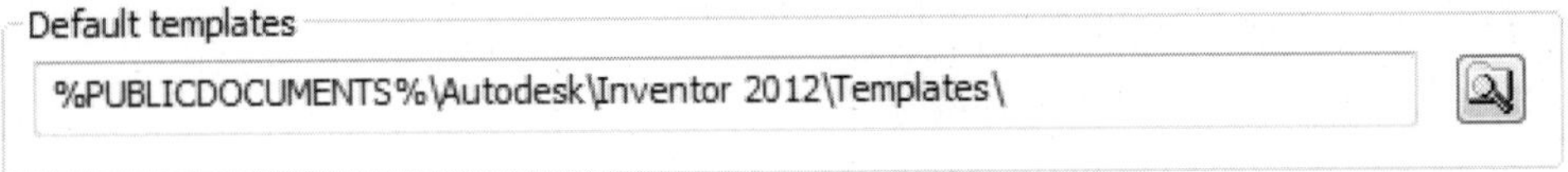

Default templates
%PUBLICDOCUMENTS%\Autodesk\Inventor 2012\Templates\

FIGURE 2.5

You can also set the Templates location in each project file. This method is useful for companies that need different templates files for each project. While editing a project file, change the Templates location in the Folder Options area. The following image shows the default location in Windows Vista. The Template location in the project file takes precedence over the Templates option in the Application Options, File tab.

Folder Options
Design Data (Styles, etc.) = [Default]
Templates = [Default]
Content Center C:\Users\Public\Documents\Autodesk\Inventor 2012\Templates\

FIGURE 2.6

NOTE

Template files have file extensions that are identical to other files of the same type, but they are located in the template directory. Template files should not be used as production files.

CREATING A PART

The first step in creating a part is to start or create a new part file in an assembly. The first four chapters in this book deal with creating parts in a new part file, and Chapter 6 covers creating and documenting assemblies. You can use the following methods to create a new part file:

- Click New on the Inventor Application menu, and then click the *Standard.ipt* icon on the Default tab, as shown in the following image on the left. You can also click *Standard (unit).ipt* on one of the other tabs.

- Click the down arrow of the New icon, and select Part from the left side of the Quick Access toolbar as shown in the following image on the right. This creates a new part file based on the units that were selected when Autodesk Inventor was installed or on the *Standard.ipt* that exists in the Templates folder.
- Click the New icon in the Quick Launch area of the Open dialog box, and then click the *Standard.ipt* icon on the Default tab, as shown in the following image on the left, or click *Standard (unit).ipt* on one of the other tabs.
- Use the shortcut key, CTRL-N, and then click the *Standard.ipt* icon on the Default tab, as shown in the following image. You can also click *Standard (unit).ipt* on one of the other tabs.

After starting a new part file using one of the previous methods, Autodesk Inventor's screen will change to reflect the part environment.

NOTE

The units for the files located in the Default tab are based upon the units you selected when you installed Autodesk Inventor.

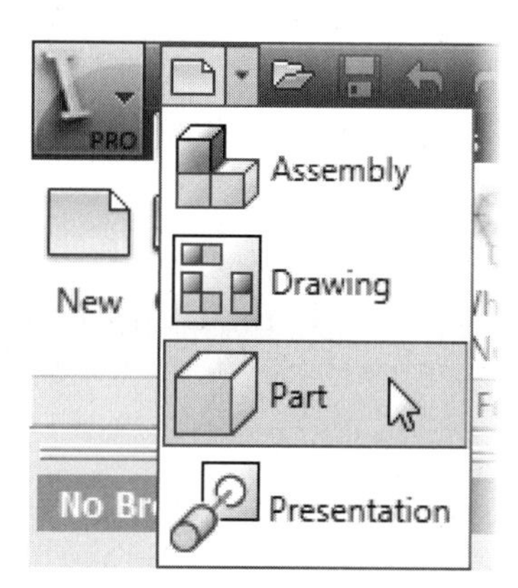

FIGURE 2.7

Sketches and Default Planes

Before you start sketching, you must have an active sketch on which to draw. A sketch is a plane on which 2D objects are sketched. You can use any planar part face or work plane to create a sketch. A sketch is automatically set by default when you create a new part file. You can change the default plane on which you will create the sketch by selecting the Inventor Application Menu > Options and clicking on the Part tab. Choose the sketch plane to which new parts should default.

Each time you create a new Autodesk Inventor, there are three planes (XY, YZ, and XZ), three axes (X, Y, and Z), and the center (origin) point at the intersection of the three planes. You can use these default planes to create an active sketch. By default, visibility is turned off to the planes, axes, and center point. To see the planes, axes, or center point, expand the Origin entry in the browser by clicking on the + to the left side of the text. You can then move the cursor over the names, and they will appear in the graphics area. The following image illustrates the default planes, axes, and center point in the graphics area shown in an isometric view with their visibility on. The browser shows the Origin menu expanded. To leave the visibility of the planes or axes on, right-click in the browser while the cursor is over the name, and select Visibility from the menu.

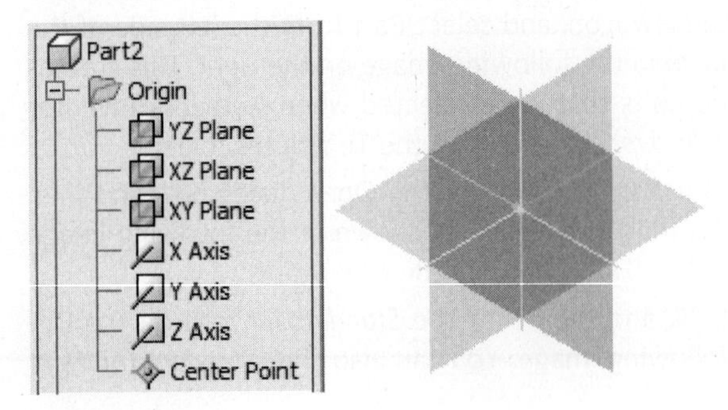

FIGURE 2.8

Origin 3D Indicator

When working in 3D, it is common to get your orientation turned around. By default in the lower left corner of the graphics screen, there is an XYZ axis indicator that shows the default (world) coordinate system as shown in the following image on the left. The direction of these planes and axes cannot be changed. The arrows are color coded:

Red arrow = X axis

Green arrow = Y axis

Blue arrow = Z axis

In the Application Options dialog box under the Display tab, you can turn the axis indicator and the axis labels on and off as shown in the following image on the right.

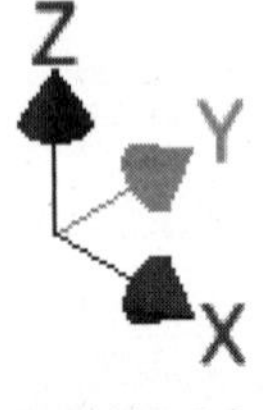

Origin 3D indicator

☑ Show Origin 3D indicator

☑ Show Origin XYZ axis labels

FIGURE 2.9

Autoproject Center (Origin) Point

To access another option that will automatically project the origin point for each new sketch of a part, click Tools tab > Application Options, click on the Sketch tab, and then check Autoproject part origin on sketch create as displayed in the following image. The origin point can be used to constrain a sketch to the 0, 0 point of the sketch. This option is on by default.

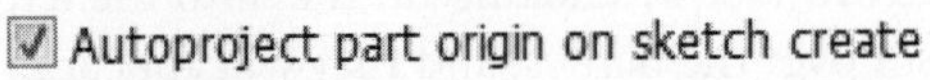

☑ Autoproject part origin on sketch create

FIGURE 2.10

New Sketch

Issue the Sketch command to create a new sketch on a planar part face or a work plane or to activate a nonactive sketch in the active part. To create a new sketch or make an existing sketch active, use one of these methods:

- Click the Model tab > Sketch panel > Create 2D Sketch as shown in the following image on the left. Then click a face, a work plane, or an existing sketch in the browser.
- Click a face, a work plane, or an existing sketch in the browser. Then click the Create 2D Sketch command on the Model tab.
- Press the S key (a keyboard shortcut) and click a face of a part, a work plane, or an existing sketch in the browser.
- Click a face of a part, a work plane, or an existing sketch in the browser, and then press the S key.
- While not in the middle of an operation, right-click in the graphics area, and select New Sketch from the marking menu. Then click a face, a work plane, or an existing sketch in the browser.
- While not in the middle of an operation, click a face of a part, a work plane, or an existing sketch in the browser. Then right-click in the graphics area, and select Create Sketch from the mini-toolbar as shown in the following image on the right.

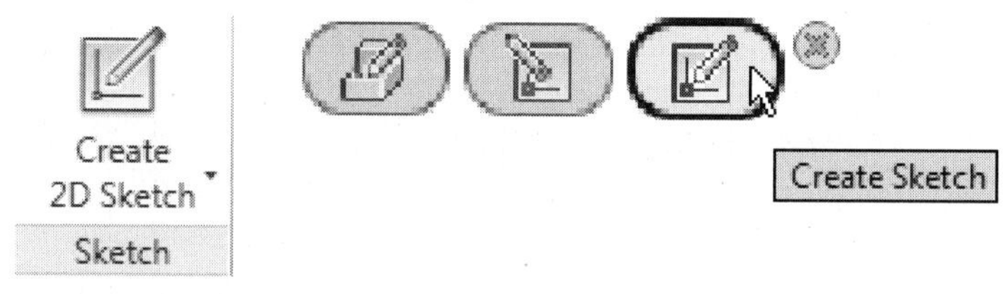

FIGURE 2.11

By default after you have activated the sketch, the X- and Y-axes will align automatically to this plane, and you can begin to sketch.

NOTE

This book assumes that when you installed Autodesk Inventor, you selected mm as the default unit. If you selected inch as the default unit, then select the *Standard (mm).ipt* template file from the Metric tab. This book will use the XY plane as the default sketch plane.

STEP 1—SKETCH THE OUTLINE OF THE PART

As stated at the beginning of this chapter, 3D parts usually start with a 2D sketch of the outline shape of the part. You can create a sketch with lines, arcs, circles, splines, or any combination of these elements. The next section will cover sketching strategies, commands, and techniques.

Sketching Overview

When deciding what outline to start with, analyze how the finished shape will look. Look for the 2-dimensional shape that best describes the part. When looking for this outline, try to look for a flat 2-dimensional shape that can be extruded or revolved to create a shape that other features can be added to, to create the finished part. It is usually easier to sketch a 2-dimensional geometry than 3-dimensional geometry. As you gain modeling experience, you can reflect on how you created the model and think about other ways that you could have built it. There is usually more than one way to generate a given part.

When sketching, draw the geometry so that it is close to the desired shape and size—you do not need to be concerned about exact dimensional values. Even though Autodesk Inventor allows islands in the sketch (closed objects that lie within another closed object) it is NOT recommended to sketch islands (when you extrude a sketch, island(s) may become voids in the solid). A better method is to place features, which make editing a part easier. For example, instead of sketching a circle inside a rectangle, place a hole feature. A sketch can consist of multiple closed objects that are coincident.

The following guidelines will help you to successfully generate sketches:

- Select a 2-dimensional outline that best represents the part. The 2D outline will be used to create the base feature.
- Draw the geometry close to the finished size. If you want a 20 inch square, for example, do not draw a 200 inch square. Use dynamic input to define the size of the geometry. Dynamic input is covered in a later section in this chapter.
- Create the sketch proportional in size to the finished shape. When drawing the first object verify its size in the lower-right corner of status bar. Use this information as a guide.
- Draw the sketch so that it does not overlap. The geometry should start and end at a single point, just as the start and end points of a rectangle share the same point.
- Do not allow the sketch to have a gap; all of the connecting endpoints should be coincident.
- Keep the sketches simple. Leave out fillets and chamfers when possible. You can easily place them as features after the sketch turns into a solid. The simpler the sketch, the fewer the number of constraints and dimensions that will be required to constrain the model.

If you want to create a solid, the sketch must form a closed shape. If it is open, it can only be turned into a surface.

Sketching Tools

Before you start sketching the outline of the part, examine the 2D sketching commands that are available. By default, the 2D sketch commands appear on the Panel Bar with Display Text with Icons turned on (text descriptions shown). As you become more proficient with Autodesk Inventor, you can turn off the text next to some of the commands, as shown in the following image. Do this by right-clicking on the ribbon, and from the Appearance menu click Text Off. The most frequently used commands will be explained throughout this chapter. Consult the help system for information about the remaining commands.

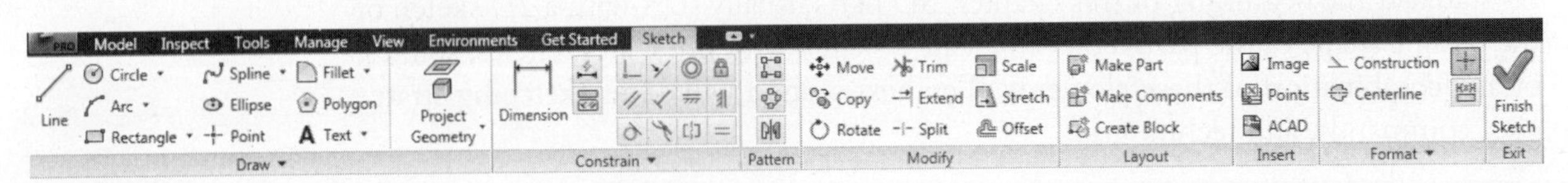

FIGURE 2.12

Using the Sketch Tools

After starting a new part, a sketch will automatically be active so that you can now use the sketch commands to draw the shape of the part. To start sketching, issue the sketch command that you need, click a point in the graphics area, and follow the prompt on the status bar. The sections that follow will introduce techniques that you can use to create a sketch.

Dynamic Input

Dynamic input in the sketch environment makes a Heads-Up Display (HUD), which displays information near the cursor for many sketching commands that helps you keep your eyes on the screen. While using the Line, Circle, Arc, Rectangle, or Point commands, you can enter values in the input fields. You can toggle between the value input fields by pressing the TAB key. The following image shows examples of entering Polar and Cartesian Coordinates.

NOTE

If no data is entered in the input fields and only points are picked, dimensions will NOT automatically be placed. You can manually place dimensions and constraints after the geometry is sketched.

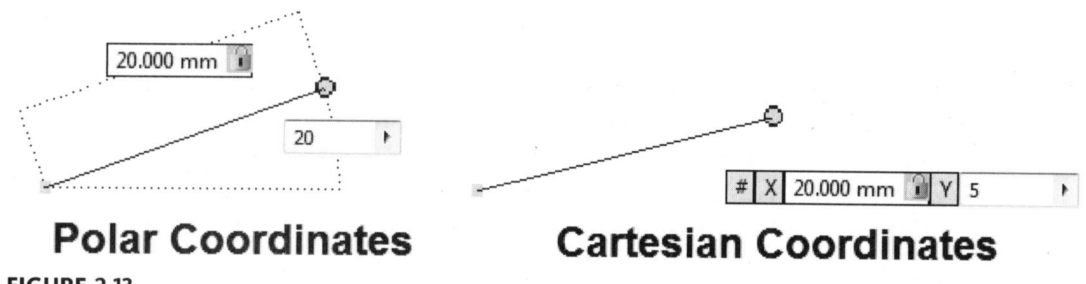

FIGURE 2.13

Coordinate Types

While sketching you can right-click and change the coordinate types shown in the following image. You must select one option from the top and bottom of the menu.

Absolute Coordinate #: Values are relative to the 0,0 coordinates on the sketch.

Relative Coordinate @: Values are relative to the last selected point.

Polar Coordinate <: #: When defining a second or subsequent point, enter X value and angle.

Cartesian Coordinate >: #: Enter X and Y coordinates.

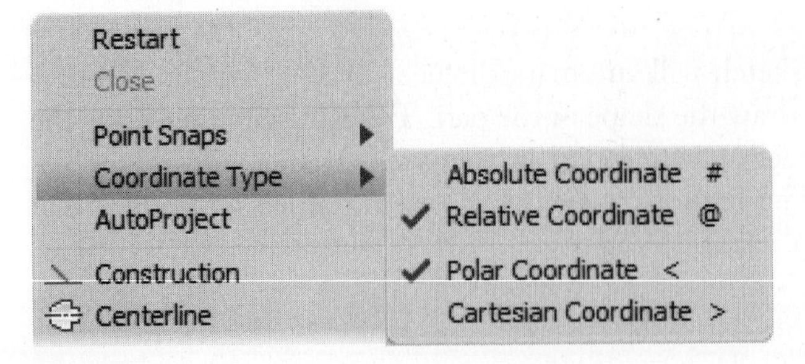

FIGURE 2.14

Polar Coordinates. By default you enter values using Polar coordinates by specifying the distance and angle of sketch elements. After entering the first value, press TAB to display a lock icon in the field, which constrains the cursor by the value that you entered and moves the cursor to the next field.

Cartesian Coordinates. The Rectangle and Point commands use Cartesian coordinates, where you specify X and Y coordinates. After entering the first value, press TAB to display a lock icon in the field, which constrains the cursor by the value that you entered and moves the cursor to the next field.

Persistent Dimensions

After entering values in the input fields, press ENTER to create the geometry, and automatically place dimension referred to as persistent dimensions. To disable persistent dimension and manually apply dimensions, turn off persistent dimensions by clicking Sketch tab > Constrain panel > click the drop arrow and click Persistent Dimensions as shown in the following image. Persistent dimensions are disabled until toggled back on. Persistent dimensions can also be turned on and off using the Heads-Up Display (HUD) settings on the Sketch tab in the Application Options dialog box. Also if you create geometry by clicking points, no dimensions will automatically be created.

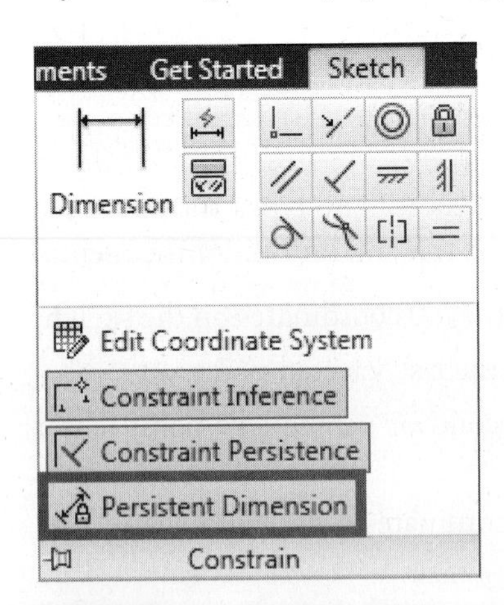

FIGURE 2.15

Heads-Up Display Settings

Pointer Input and Dimension Input can be turned on/off; change to Cartesian or Polar Coordinates and change the Heads-Up Display (HUD) settings. Click the Sketch tab in the Application Options dialog box as shown in the following image on the left. Click the Setting button and the Heads-Up Display Settings dialog box will appear as shown in the following image on the right. Select the options in the dialog box to control the heads-up display options when dimensioning a sketch.

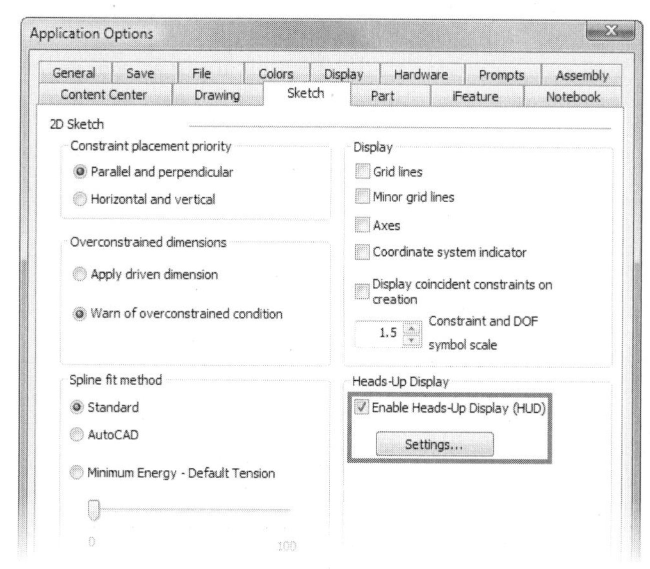

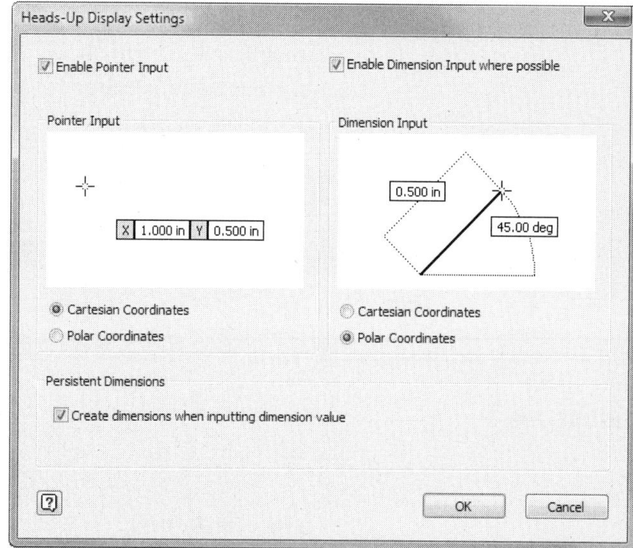

FIGURE 2.16

Pointer Input. When Pointer Input is on, the starting point for a sketch element is displayed as Cartesian coordinates (X and Y values) in a tooltip near the cursor. The default for second and subsequent points is relative Polar coordinates when defining lengths and angles.

When Pointer Input is off, the Cartesian coordinate displays the starting point of sketch geometry. Second and subsequent points are still displayed in the value input fields.

Dimension Input. When dimension input is on, the value input fields display polar coordinates when defining lengths and angles for a second point. The dimensional values change as you move the cursor. Press TAB to move to the next input field or click in another cell. After pressing the Tab key, the value will be locked and a lock icon will appear to the right of the value as shown in the following image. After a dimension's value is locked, the parametric dimension will be created after clicking a point or pressing the Enter key. You can change the value in an input filed by either clicking in the field or pressing the Tab key until the field is highlighted and then typing in a new value.

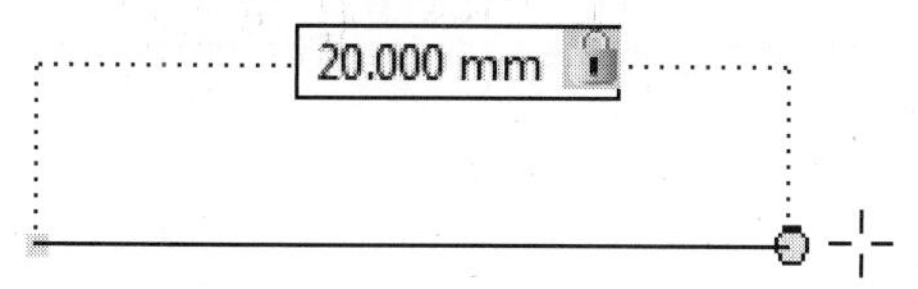

FIGURE 2.17

Line Command

The Line command, is one of the most powerful commands that you will use to sketch. Not only can you draw lines with it, but also you can draw an arc from the endpoint of a line segment. To start sketching lines, click the Line command from the Sketch tab > Draw panel as shown in the following image on the left, or right-click in a blank area in the graphics window and click Create Line from the marking menu as shown in the following image in the middle, or press the L key on the keyboard. After starting the Line command you will be prompted to click a first point, select a point in the graphics window, and then click a second point. The following image on the right shows the line being created with the dynamic input as well as the horizontal constraint.

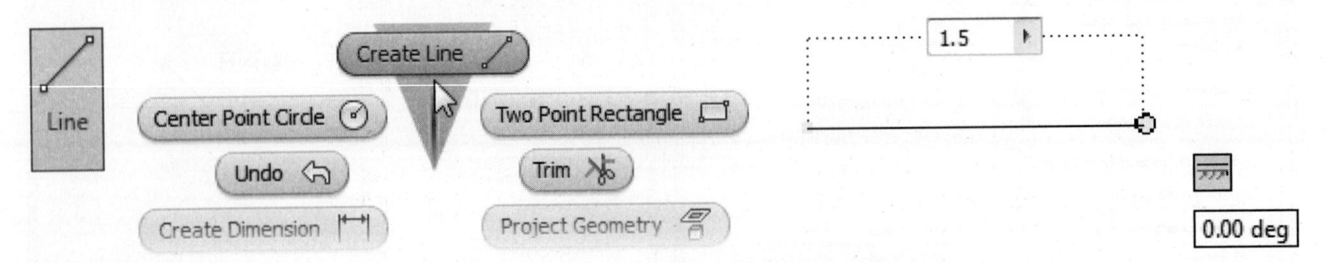

FIGURE 2.18

You can continue drawing line segments, or you can sketch an arc from the endpoint. Move the cursor over the endpoint of a line segment or arc, and a small gray circle will appear at that endpoint as shown in the following image on the left. Click on the small circle, and with the left mouse button pressed down, move the cursor in the direction that you want the arc to go. Up to eight different arcs can be drawn, depending upon how you move the cursor. The arc will be tangent to the horizontal or vertical edges that are displayed from the selected endpoint. The following image on the right shows an arc that is normal to the sketched line being drawn.

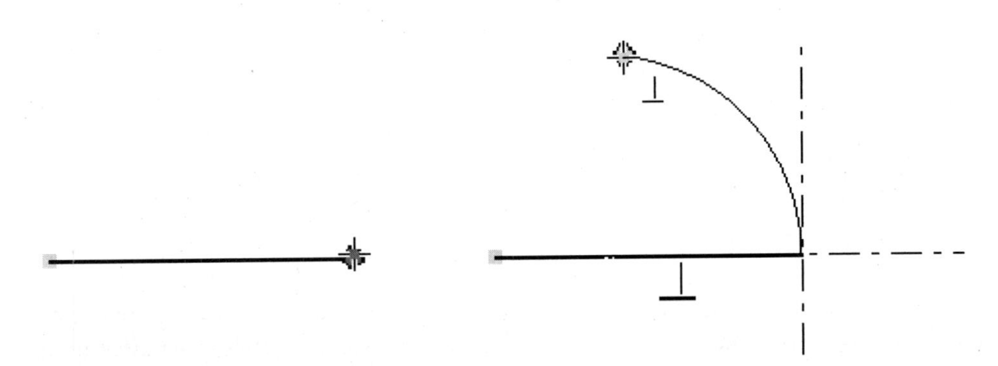

FIGURE 2.19

TIP

When sketching, look at the bottom-right corner of the status bar (bottom of the screen) to see the coordinates, length, and angle of the objects that you are drawing. The following image shows the status bar when a line is being drawn.

1.109 in, 0.330 in x=0.605 in y=0.000 in

FIGURE 2.20

Object Tracking – Inferred Points

If the Point Alignment On option is checked from the Sketch tab of the Application Options, as you sketch dashed lines will appear on the screen. These dashed lines represent the endpoints; midpoints; and theoretical intersections of lines, arcs, and center points of arcs and circles that represent their horizontal, vertical, or perpendicular positions. As the cursor gets close to these inferred points, it will snap to that location. If that is the point that you want, click that point; otherwise, continue to move the cursor until it reaches the desired location. When you select inferred points, no constraints (geometric rules such as horizontal, vertical, collinear, and so on) are applied from them. If the Point Alignment On option is unchecked, you can still infer points, move the cursor over a point, and move the cursor. Using inferred points helps create more accurate sketches. The following image shows the inferred points from two midpoints that represent their horizontal and vertical position.

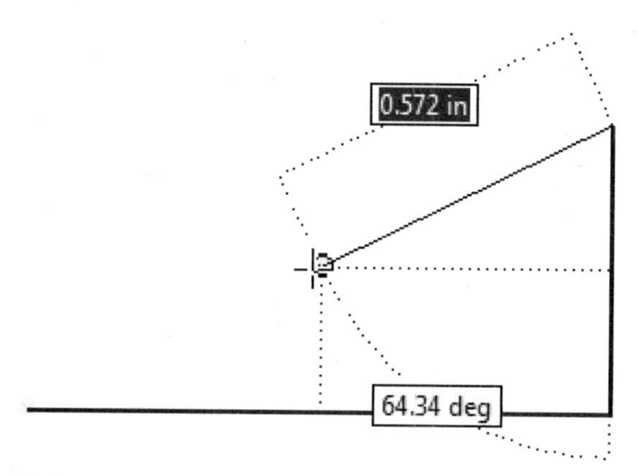

FIGURE 2.21

Automatic Constraints

As you sketch, a small constraint symbol appears that represents geometric constraint(s) that will be applied to the object. If you do not want a constraint to be applied, uncheck the Constraint Persistent option under the Constrain panel drop arrow or hold down the CTRL key when you select the point. The following image shows a line being drawn from the arc, tangent to the arc, and parallel to the angled line, and the dynamic input is also displayed. The symbol appears near the object from which the constraint is coming. Constraints will be covered in the next section.

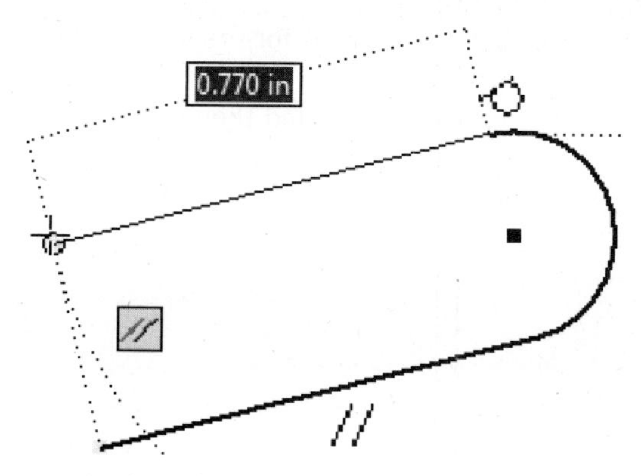

FIGURE 2.22

Scrubbing

As you sketch, you may prefer to apply a constraint different from the one that automatically appears on the screen. You may want a line to be perpendicular to a given line, for example, instead of being parallel to a different line. The technique to change the constraint is called *scrubbing*. To place a different constraint while sketching, move the cursor so it touches (scrubs) the other object to which the constraint should be related. Move the cursor back to its original location, and the constraint symbol changes to reflect the new constraint. The same constraint symbol will also appear near the scrubbed object, representing that it is the object to which the constraint is matched. Continue sketching as normal. The following image shows the top horizontal line being drawn with a perpendicular constraint that was scrubbed from the left vertical line. Without scrubbing the left vertical line, the applied constraint would have been parallel to the bottom line.

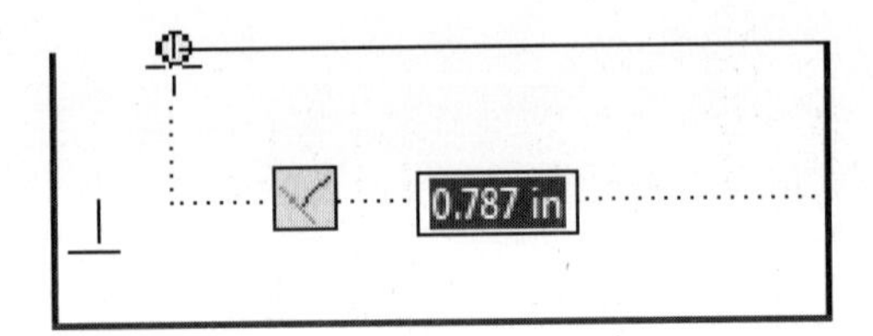

FIGURE 2.23

Common Sketch Tools

The following chart lists common 2D sketch commands that are not covered elsewhere in this chapter. Some commands are available by clicking the down arrow in the lower-right corner of the top command. Consult the help system for more information about these commands.

Tool	Function
Center-point Circle	Creates a circle by clicking a center point for the circle and then a point on the circumference of the circle.
Tangent Circle	Creates a circle that will be tangent to three lines or edges by clicking the lines or edges.
Three-Point Arc	Creates an arc by clicking a start and endpoint and then a point that will lie on the arc.
Tangent Arc	Creates an arc that is tangent to an existing line or arc by clicking the endpoint of a line or arc and then clicking a point for the other endpoint of the arc.
Center-Point Arc	Creates an arc by clicking a center point for the arc and then clicking a start and endpoint.
Two-Point Rectangle	Creates a rectangle by clicking a point and then clicking another point to define the opposite side of the rectangle. The edges of the rectangle will be horizontal and vertical.

(Continued)

Tool	Function
Three-Point Rectangle	Creates a rectangle by clicking two points that will define an edge and then clicking a point to define the third corner.
Fillet	Creates a fillet between two nonparallel lines, two arcs, or a line and an arc at a specified radius. If you select two parallel lines, a fillet is created between them without specifying a radius. When the first fillet is created, a dimension will be created. If many fillets are placed in the same operation, you choose to either apply or not apply an equal constraint.
Chamfer	Creates a chamfer between lines. There are three options to create a chamfer: both sides equal distances, two defined distances, or a distance and an angle.
Polygon	Creates an inscribed or a circumscribed polygon with the number of faces that you specify. The polygon's shape is maintained as dimensions are added.
Mirror	Mirrors the selected objects about a centerline. A symmetry constraint will be applied to the mirrored objects.
Rectangular Pattern	Creates a rectangular array of a sketch with a number of rows and columns that you specify.
Circular Pattern	Creates a circular array of a sketch with a number of copies and spacing that you specify.
Offset	Creates a duplicate of the selected objects that are a given distance away. By default, an equal-distance constraint is applied to the offset objects.
Trim	Trims the selected object to the next object it finds. Click near the end of the object that you want trimmed. While using the Trim command, hold down the SHIFT key to extend objects. If desired, hold down the CTRL key to select boundary objects.
Extend	Extends the selected object to the next object it finds. Click near the end of the object that you want extended. While using the Extend command, hold down the SHIFT key to trim objects. If desired, hold down the CTRL key to select boundary objects.

Selecting Objects

After sketching objects, you may need to move, rotate, or delete some or all of the objects. To edit an object, it must be part of a selection set. There are two methods that you can use to place objects into a selection set.

- You can select objects individually by clicking on them. To select multiple individual objects, hold down the CTRL or SHIFT key while clicking the objects. You can remove selected objects from a selection set by holding down the CTRL or SHIFT key and reselecting them. As you select objects, their color will change to show that they have been selected.
- You can select multiple objects by defining a selection window. Not all commands allow you to use the selection window technique and only allow single

selections. To define the window, click a starting point. With the left mouse button depressed, move the cursor to define the box. If you draw the window from left to right (solid lines), as shown in the following image on the left, only the objects that are fully enclosed in the window will be selected. If you draw the window from right to left (dashed lines), as shown in the following image on the right, all of the objects that are fully enclosed in the window *and* the objects that are touched by the window will be selected.

- You can use a combination of the methods to create a selection set.

When you select an object, its color will change according to the color style that you are using. To remove all of the objects from the selection set, click in a blank section of the graphics area.

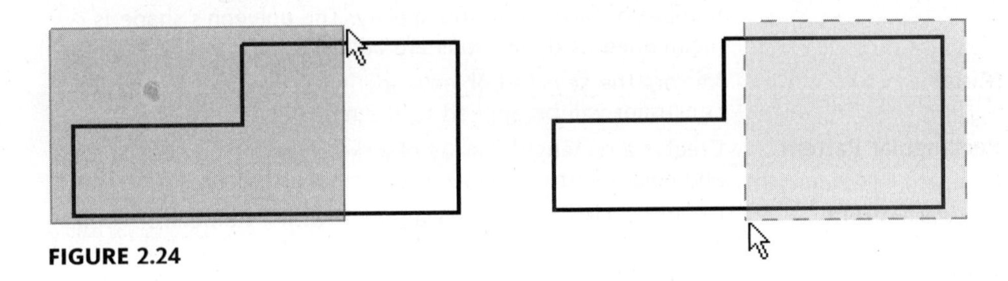

FIGURE 2.24

Deleting Objects

To delete objects first cancel the command that you are in by pressing the Esc key. Then select objects to delete, and then either press the DELETE key or right-click and select Delete from the menu as shown in the following image.

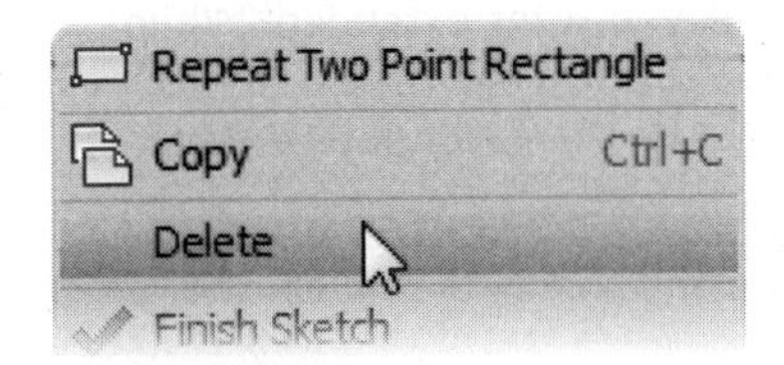

FIGURE 2.25

Measure Commands

Measure commands that assist in analyzing sketch, part, and assembly models are available. You can measure distances, angles, and loops, and you can perform area calculations. You can start the measure command first and then select the geometry, or select the geometry first and then start a measure command.

The Measure commands are on the Tools tab > Measure panel as shown in the following image. The following sections discuss these commands in greater detail.

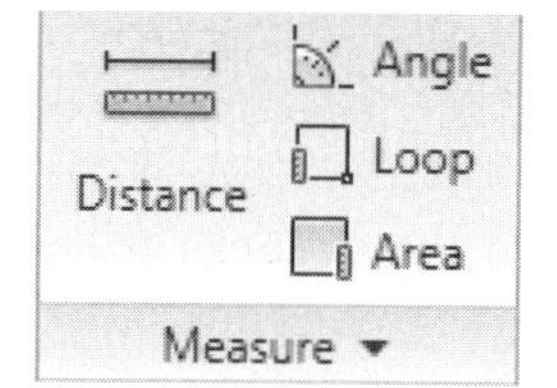

FIGURE 2.26

Measure Distance

Measures the length of a line or edge, length of an arc, distance between points, radius and diameter of a circle, or the position of elements relative to the active coordinate system. A temporary line designating the measured distance appears, and the Measure Distance dialog box displays the measurement for the selected length, as shown in the following image.

If two disjointed faces of a single part or two faces from different parts are selected, the minimum distance between the faces will be displayed.

Measure Angle

Measures the angle between two lines or edges. The measurement box displays the angle based on the selection of two lines or edges or two lines defined by the selection of three points.

Measure Loop

Measures the length of closed or open loops defined by face boundaries or other geometry. When moving your cursor over a part face, all edges of the face will become highlighted. Clicking on this face will calculate the closed loop or perimeter of the shape.

Measure Area

Measures the area of enclosed regions or faces. Moving your cursor inside the closed outer shape will cause the outer shape and all holes (referred to as "islands") to also become highlighted. Clicking inside this shape will calculate the area of the shape.

When you click the arrow beside the measurement box, a menu similar to the following image will appear.

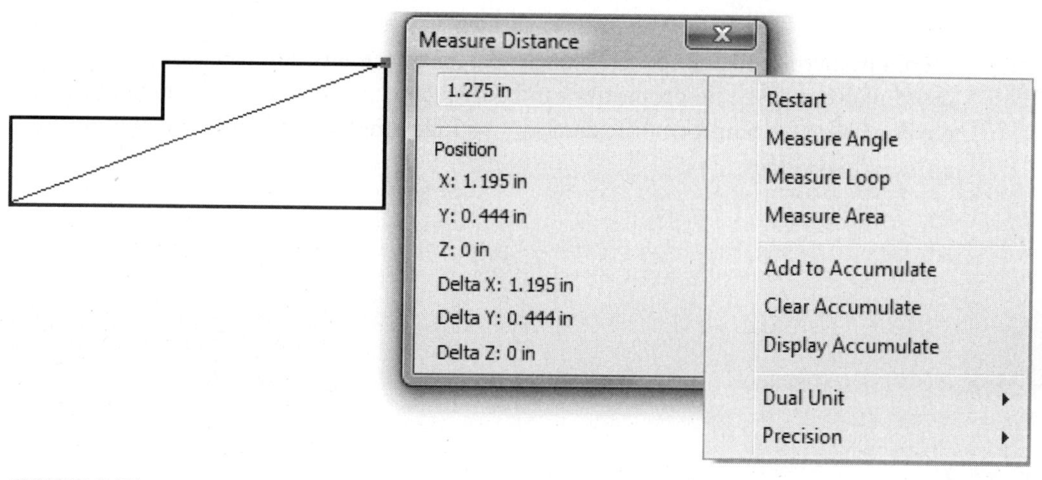

FIGURE 2.27

Brief explanations of each option follow:

Restart
Click to clear the measurement from the measurement box so that you can make another measurement.

Measure Angle
Click to change the measurement mode to Measure Angle.

Measure Loop
Click to change the measurement mode to Measure Loop.

Measure Area
Click to change the measurement mode to Measure Area.

Add to Accumulate
Click to add the measurement in the measurement box to accumulate a total measurement.

Clear Accumulate
Click to clear all measurements from the accumulated sum, resetting the sum to zero.

Display Accumulate
Click to display the sum of all measurements you have added to the accumulated sum.

Dual Unit
Click to display the measurement in another unit. The measurement in the second unit will be displayed below the first unit measurement.

Precision
Click to change the decimal display between showing all decimal places and showing the number of decimal places specified in the document settings for the active part or assembly.

Region Properties
Available by clicking the drop arrow on the bottom of the Measure panel. While in a sketch you can determine the properties such as the area, perimeter, and Moment of Inertia properties of a closed 2D sketch. All measurements are taken from the sketch coordinate system. The properties can be displayed in dual units, the default unit of the document, or a unit of your choice. The following image shows the region properties of two circles.

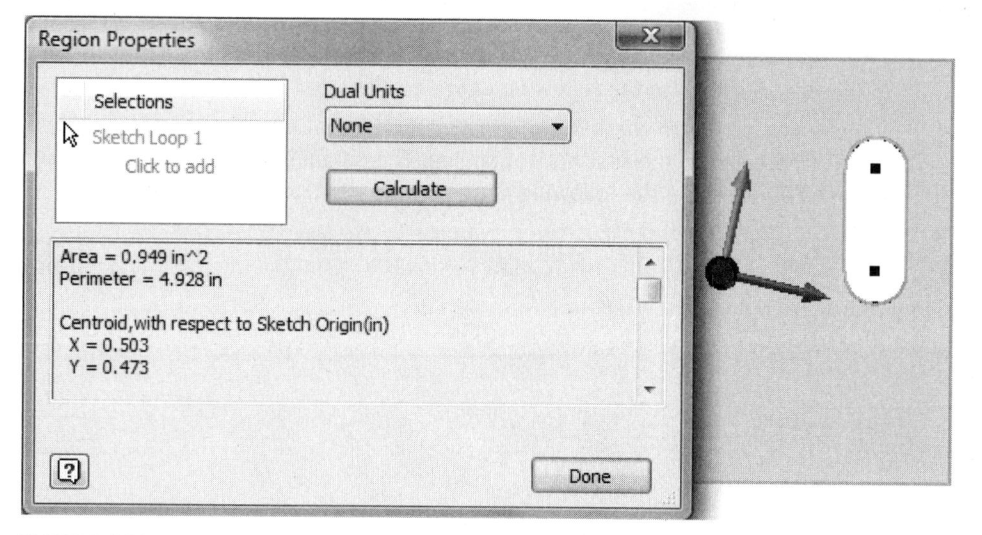

FIGURE 2.28

EXERCISE 2-1: CREATING A SKETCH WITH LINES

In this exercise, you create a new part file, and then create sketch geometry using basic construction techniques. In this exercise no dimensions will be created.

1. Click the New command, click the English tab, and then double-click *Standard (in).ipt,* or if inch is the default unit; from the left side of the Quick Access toolbar you can click the down arrow of the New icon, and select Part.
2. Click the Line command in the Draw panel.
3. Click on the origin point in the graphics window, move the cursor to the right approximately 4 inches, and, when the horizontal constraint symbol displays, click to specify a second point as shown in the following image. You may need to zoom back and pan the screen to draw the line.

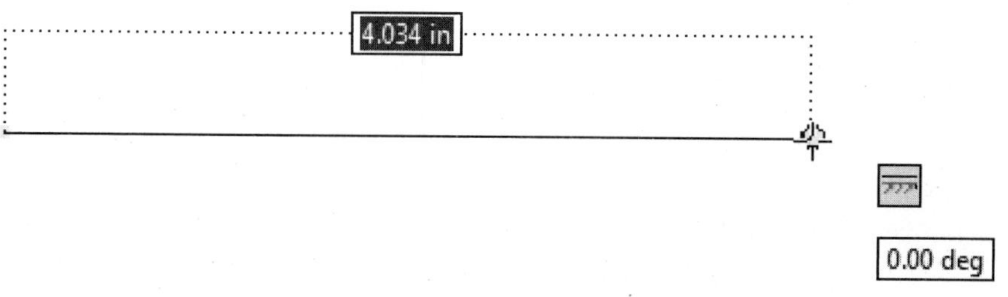

FIGURE 2.29

NOTE

Symbols indicate the geometric constraint. In the figure above, the symbol indicates that the line is horizontal. When you create the first entity in a sketch, make it close to final size. The length and angle of the line are displayed in the lower-right corner of the window to assist you.

4. Move the cursor up until the perpendicular constraint symbol displays beside the first line and then click to create a perpendicular line that is approximately 2 inches as shown in the following image on the left.
5. Move the cursor to the left and create a horizontal line approximately 1 inch, that is, parallel to the first horizontal line. The parallel constraint symbol is displayed as shown in the following image on the right.

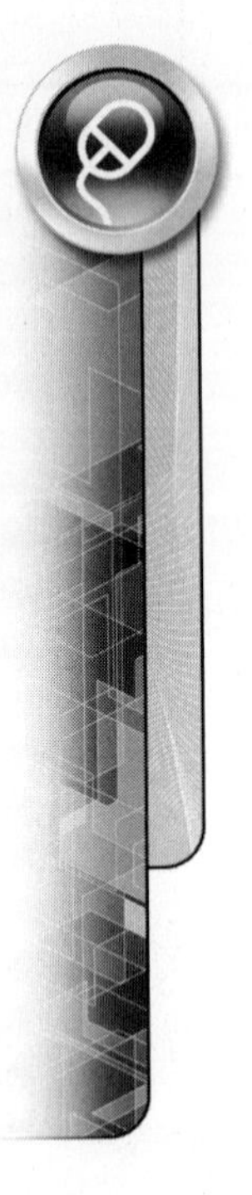

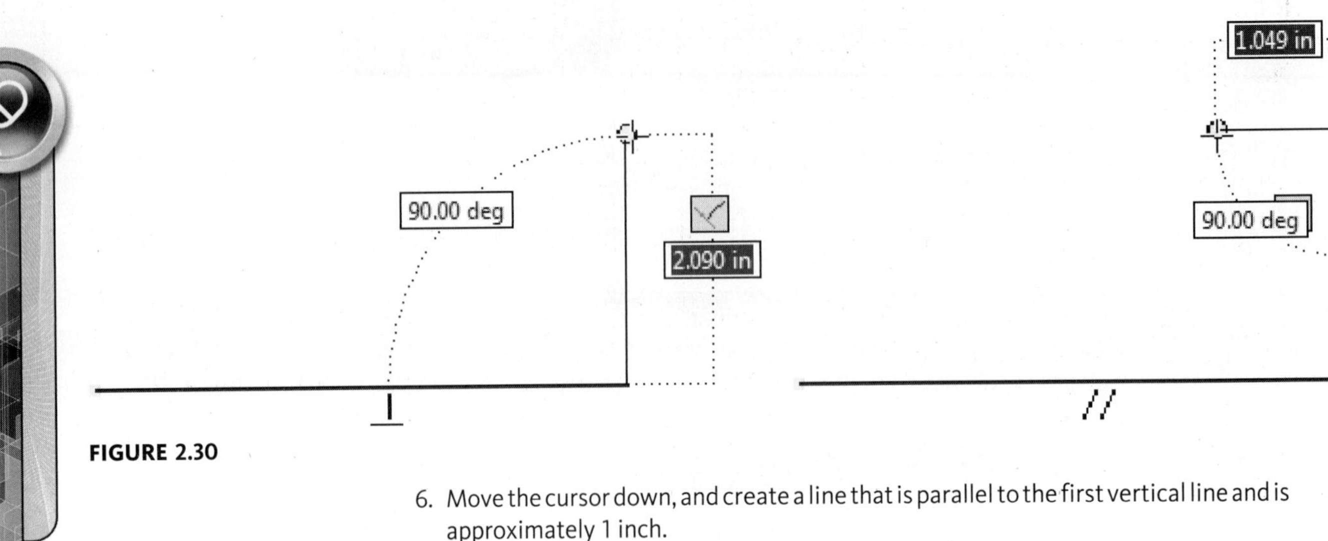

FIGURE 2.30

6. Move the cursor down, and create a line that is parallel to the first vertical line and is approximately 1 inch.
7. Move the cursor left to create a horizontal line of approximately 2 inches.
8. Move the cursor up until the parallel constraint symbol is displayed, and a dotted alignment line appears as shown in the following image on the left. If the parallel constraint does not appear, move (scrub) the cursor over the inside vertical line to create a relationship to it. Click to locate the point.
9. Move the cursor to the left until the parallel constraint symbol is displayed, and a dotted alignment line appears as shown in the following image on the right. Then click to locate the point.

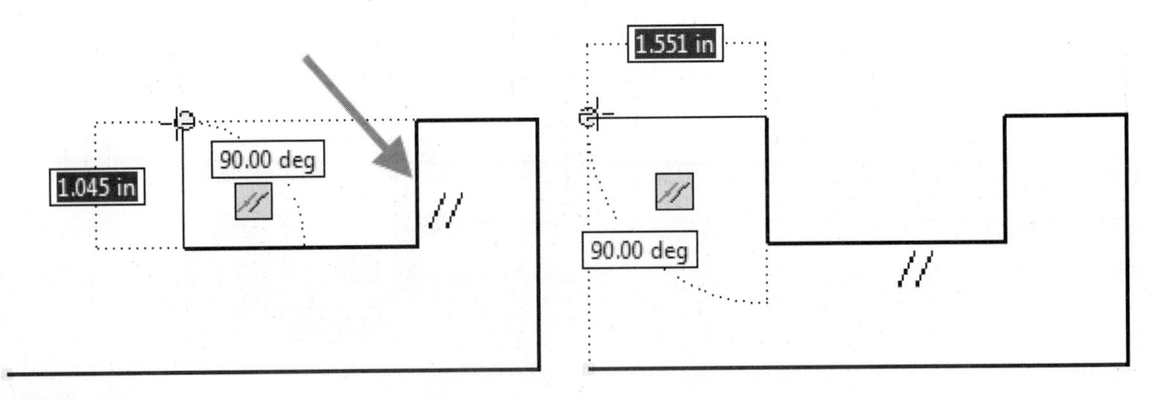

FIGURE 2.31

10. To close the profile right-click and click Close from the menu.
11. Your screen should resemble the following image.
12. Right-click in the graphics screen, and click Done to end the Line command.
13. Right-click in the graphics screen again, and click Finish Sketch.
14. Close the file. Do not save changes. End of exercise.

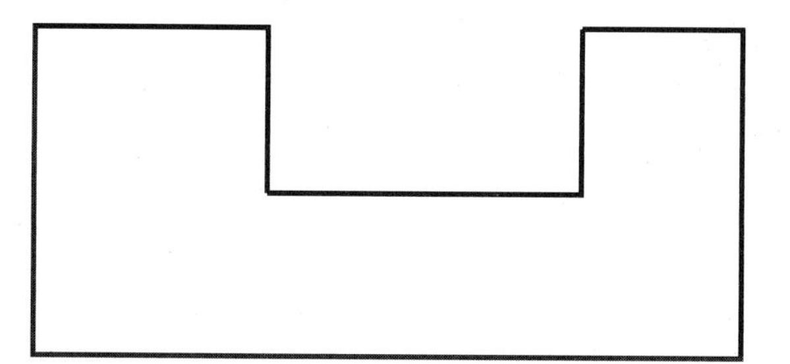

FIGURE 2.32

EXERCISE 2-2: CREATING A SKETCH WITH TANGENCIES

In this exercise, you create a new part file, and then you create a profile consisting of lines and tangent arcs.

1. Click the New command, click the English tab, and then double-click *Standard (inch).ipt*; or if inch is the default unit, from the left side of the Quick Access toolbar you can click the down arrow of the New icon, and select Part.
2. Click the Line command in the Draw panel.
3. Click on the projected origin point in the middle of the graphics window, and create a horizontal line to the right of the origin point and type 3 (inches will be assumed as the unit because the part file is based on the unit of inch) in the input field and then press enter.

If the second point of the line lies off the screen, roll the mouse wheel away from you to zoom out, hold down the mouse wheel, and drag to pan the view.

4. Create a perpendicular line, move the cursor up, type 2 in the input field as shown in the following image on the left, and press enter.
5. In this step, you infer points, meaning that no sketch constraint is applied. Move the cursor to the intersection of the midpoints of the right-vertical line and bottom horizontal line. Dashed lines (inferred points) appear as shown in the following image on the right, and then click to create the line. No dimension was created since a value was not entered.

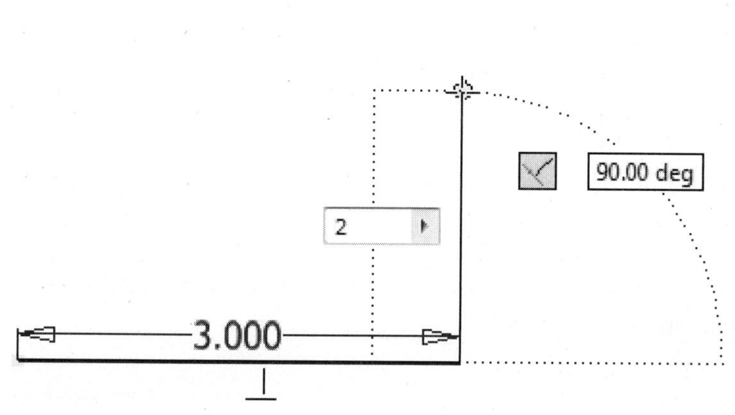

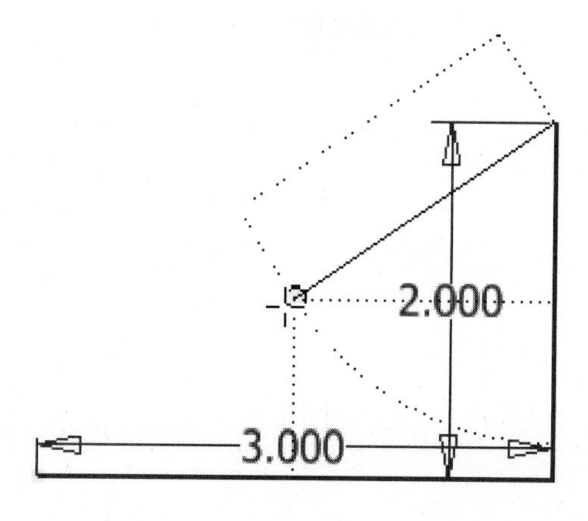

FIGURE 2.33

6. Next you create a line that is parallel to the bottom line. You scrub the bottom line by moving the cursor over the bottom line (do NOT click), and then move the cursor up and to the left until the vertical inferred line and the constraints are displayed as shown in the following image on the left, and then click to place the line.
7. Next you sketch an arc while in the line command. Click on the gray dot at the left end of the line, and hold and drag the endpoint to create a tangent arc. **Do not release the mouse button.**
8. Move the cursor over the endpoint of the first line segment until a coincident constraint (green circle) and the two tangency constraints at start and end points of the arc are displayed as shown in the following image on the right.

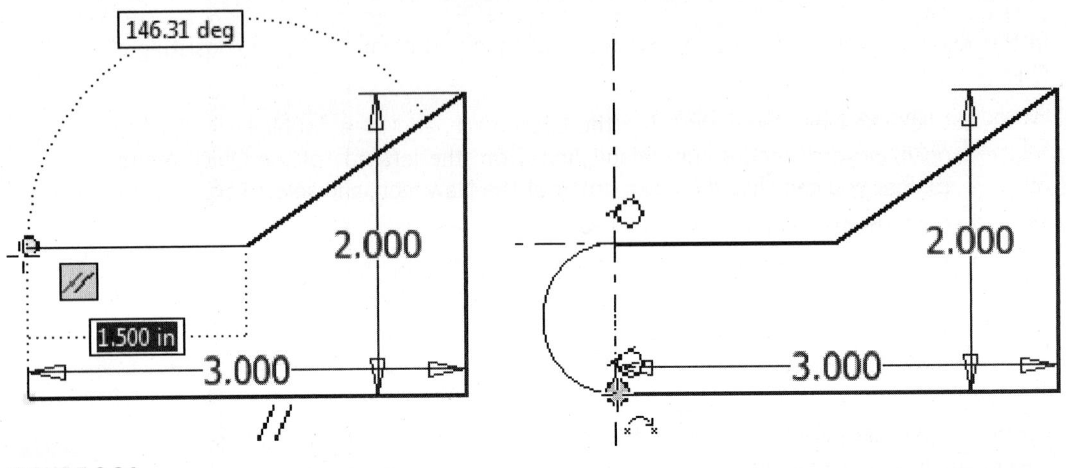

FIGURE 2.34

9. Release the mouse button to place the arc.
10. Right-click in the graphics window, and then click Done. Later in this chapter you will learn how to create dimensions.
11. Close the file. Do not save changes. End of exercise.

STEP 2—CONSTRAINING THE SKETCH

After you draw the sketch, you may want to add geometric constraints to it. Geometric constraints apply behavior to a specific object or create a relationship between two objects. An example of using a constraint is applying a vertical constraint to a line so that it will always be vertical. You could apply a parallel constraint between two lines to make them parallel to one another; then, as one of the line's angles changes, so will the other's. You can apply a tangent constraint to a line and an arc or to two arcs.

When you add a constraint, the number of constraints or dimensions that are required to fully constrain the sketch will decrease. On the bottom-right corner of Autodesk Inventor, the number of constraints or dimensions will be displayed similar to what is shown in the following image. A fully constrained sketch is a sketch whose objects cannot move or stretch.

4 dimensions needed 1 1

FIGURE 2.35

To help you see which objects are constrained, Autodesk Inventor will change the color of constrained objects if a point on the sketch has a coincident constraint applied to the origin point or another point of an existing sketch. If you have not sketched a point coincident to the projected origin point or another point of an existing sketch, the color of the sketch will not change. You could apply a fix constraint instead of using the origin point, but it is not recommended. If you have not sketched a point coincident to the origin point or applied a fix constraint to the sketch, objects are free to move in their sketch plane.

NOTE

Autodesk Inventor does not force you to fully constrain a sketch. However, it is recommended that you fully constrain a sketch, as this will allow you to better predict how a part will react when you change dimensions values.

Constraint Types

Autodesk Inventor has 12 geometric constraints that you can apply to a sketch. The following image shows the constraint types that can be applied from the marking menu and the symbols that represent them when the constrains are displayed in a sketch. Descriptions of the constraints follow.

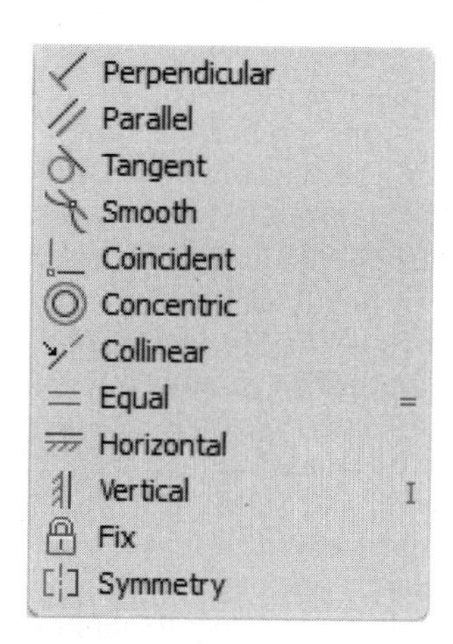

FIGURE 2.36

Button	Tool	Function
	Perpendicular	Lines will be repositioned at 90° angles to one another. The first line sketched will stay in its position, and the second will rotate until the angle between them is 90°.
	Parallel	Lines will be repositioned so that they are parallel to one another. The first line sketched will stay in its position, and the second will move to become parallel to the first.
	Tangent	An arc, circle, or line will become tangent to another arc or circle.
	Smooth (G2)	A spline and another spline, line, or arc that connect at an endpoint with a coincident constraint will represent a smooth G2 (continuous curvature) condition.
	Coincident	A point is constrained to lie on another point or curve (line, arc, etc.).
	Concentric	Arcs and/or circles will share the same center point.

Button	Tool	Function
	Collinear	Two selected lines will line up along a single line; if the first line moves, so will the second. The two lines do not have to be touching.
	Equal	If two arcs or circles are selected, they will have the same radius or diameter. If two lines are selected, they will become the same length. If one of the objects changes, so will the other object to which the Equal constraint command has been applied. If the Equal constraint command is applied after one of the arcs, circles, or lines has been dimensioned, the second arc, circle, or line will take on the size of the first one. If you select multiple similar objects (lines, arcs, etc.) before selecting this command, the constraint is applied to all of them. This rule applies to some of the other sketch constraints as well.
	Horizontal	Lines are positioned parallel to the X-axis, or a horizontal constraint can be applied between any two points in the sketch. The selected points will be aligned such that a line drawn between them will be parallel to the X-axis.
	Vertical	Lines are positioned parallel to the Y-axis, or a vertical constraint can be applied between any two points in the sketch. The selected points will be aligned such that a line drawn between them will be parallel to the Y-axis.
	Fix	Applying a fixed point or points will prevent the selected point from moving. The Fix constraint command overrides any other constraint. Any endpoint or segment of a line, arc, circle, spline segment, or ellipse can be fixed. Multiple points in a sketch can be fixed. If you select near the endpoint of an object, the endpoint will be locked from moving. If you select a line segment, the angle of the line will be fixed and only its length will be able to change. You can remove a fix constraint as needed. Deleting constraints will be covered later in this section.
	Symmetry	Selected points defining the selected geometry are made symmetric about the selected line.

NOTE

To fully constrain a base (first) sketch, constrain or dimension a point in the sketch to the center point or apply a fix constraint.

Adding Constraints

As stated previously in this chapter, you can apply constraints while you sketch objects. You can also apply additional constraints after the sketch is drawn. However, Autodesk Inventor will not allow you to overconstrain the sketch or add duplicate constraints. If you add a constraint that would conflict with another, you will be warned with the message, "Adding this constraint will overconstrain the sketch." For example, if you try to add a vertical constraint to a line that already has a horizontal constraint, you will be alerted. To apply a constraint, follow these steps:

1. Click a constraint from the Constrain panel, or right-click in the graphics window and click Create Constraint from the menu. Then click the specific constraint from the menu as shown in the previous image before the chart.
2. Click the object or objects to apply the constraints.

Showing and Deleting Constraints

To see the constraints that are applied to an object, use the Show Constraints command from the Constrain panel as shown on the left side of the following image. After issuing the Show Constraints command, select an object and a constraint icon and yellow squares that represent coincident constraints will appear with the constraints that are applied to the selected object—similar to what is shown in the middle of the following image. You can modify the size of the constraint toolbar by clicking Tools tab > Application Options, clicking on the Sketch tab, and modifying the size of the Constraint and DOF Symbol Scale. The following image on the right shows the scale increased from 1.0 to 1.5.

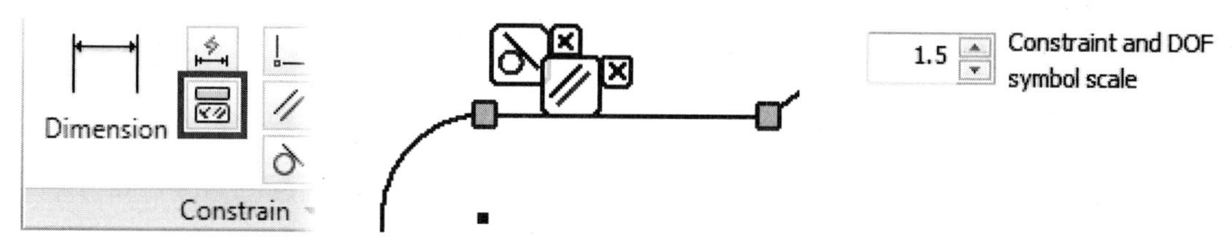

FIGURE 2.37

To show all the constraints for all of the objects in the sketch, do the following:

- While not in an operation, right-click in the graphics window, and from the menu, click Show All Constraints, or press the F8 key. To hide all the constraints, right-click in the graphics window, and click Hide All Constraints from the menu, or press the F9 key or right-click and click Hide All Constraints from the marking menu.
- If a constraint is added when sketch constraints are displayed you will need to exit out of the active command and then press the F8 key again.
- When the constraints are shown, all the constraints in the sketch will appear. You can control which constraints are visible by right-clicking in the graphics window while not in a command and then clicking Constraint Visibility from the menu as shown in the following image on the left. The Constraint Visibility dialog will appear; uncheck the constraint type that you do not want to see. These settings apply only to the current sketch.

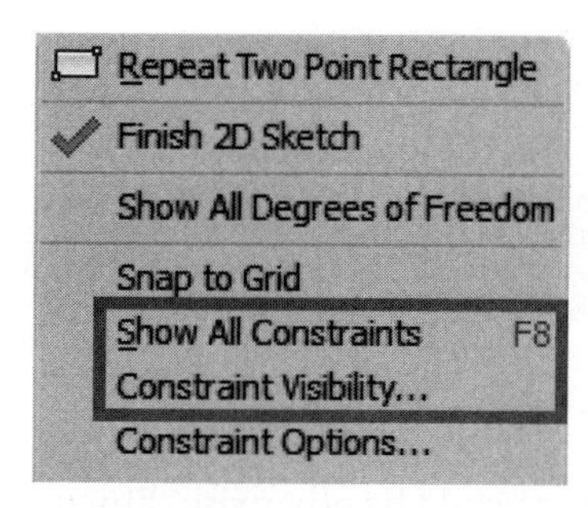

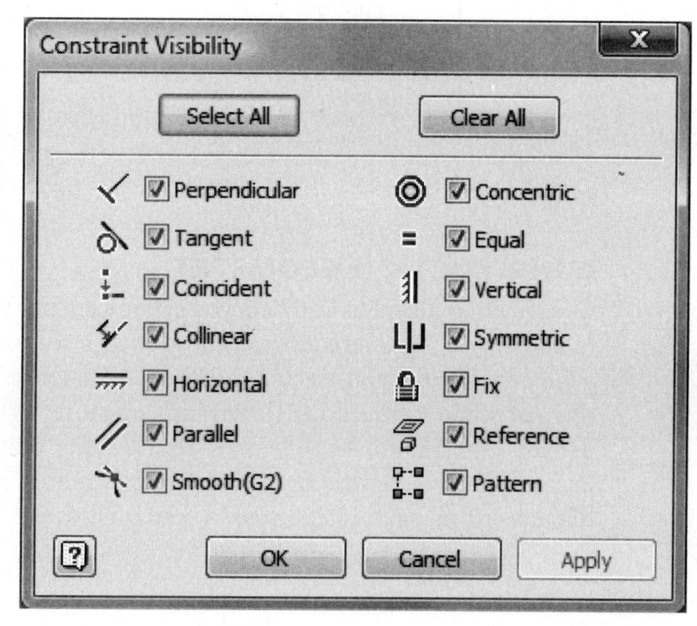

FIGURE 2.38

As you move the cursor over a constraint icon, the matching sketch constraint and the object that is linked to that constraint will be highlighted. The coincident constraint will appear as a yellow square at the point that the constraint exists. The perpendicular constraint will appear once at the location the constraint is applied. The following image on the left shows the cursor over the perpendicular constraint on the right vertical edge; the bottom horizontal and right vertical lines are highlighted. The following image on the right shows the cursor over the bottom horizontal line, and the constraints that are related to the line are highlighted.

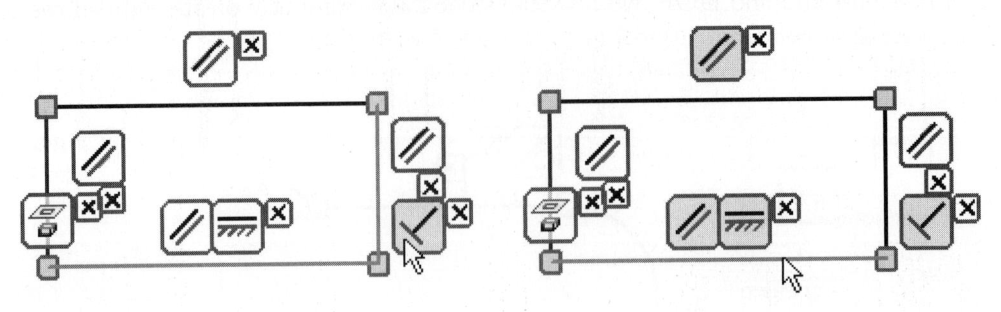

FIGURE 2.39

To delete the constraint, either click on it and then right-click, or right-click while the cursor is over the constraint, or select Delete from the menu. You can also click on it, and press the Delete key. The following image shows the parallel constraint being deleted.

To hide constraint icons, click the × on the right side of the constraint symbols toolbar.

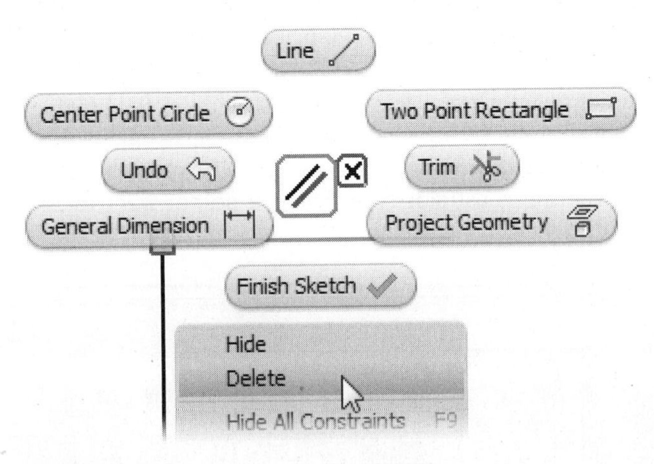

FIGURE 2.40

CONSTRUCTION GEOMETRY

Construction geometry can help you create sketches that would be otherwise difficult to constrain. You can constrain and dimension construction geometry like normal geometry, but the construction geometry will not be recognized as a profile edge in the part when you turn the sketch into a feature. When you sketch, the sketches by default have a normal geometry style, meaning that the sketch geometry is visible in the feature. Construction geometry can reduce the number of constraints and dimensions required to constrain a sketch fully, and it can help to define the sketch.

For example, a construction circle that is tangent to the inside a hexagon (drawn with individual lines and not the Polygon command) can drive the size of the hexagon. Without construction geometry, the hexagon would require six constraints and dimensions. With construction geometry, it would require only three constraints and dimensions; the circle would have tangent or coincident constraints applied to it and the hexagon. You create construction geometry by changing the line style before or after you sketch geometry in one of the following two ways:

- After creating the sketch, select the geometry that you want to change and click the Construction icon on the Format panel as shown in following image.
- Before sketching, click the Construction icon on the Format panel, as shown in the following image on the left. All geometry created will be construction until Construction is unselected.

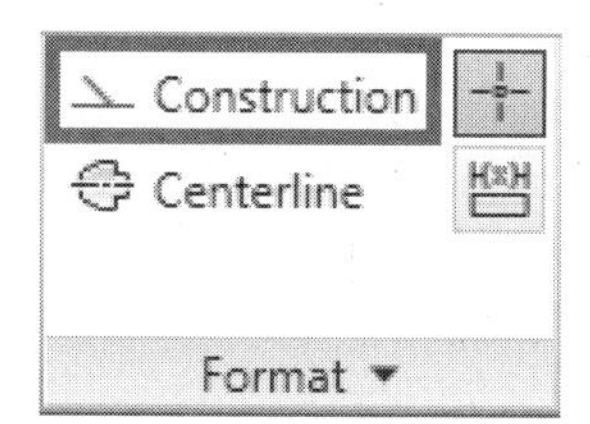

FIGURE 2.41

After turning the sketch into a feature, the construction geometry will be consumed with the sketch and is maintained in the sketch. When you edit a feature's sketch that you created with construction geometry, the construction geometry will reappear during editing and disappear when the part is updated. You can add or delete construction geometry to or from a sketch just like any geometry that has a normal style. In the graphics window, construction geometry will be displayed as a dashed line, lighter in color and thinner in width than normal geometry. The following image on the left shows a sketch with a construction line for the angled line. The angled line has a coincident constraint applied to it at every point that touches it. The image on the right shows the sketch after it has been extruded. Note that the construction line was not extruded.

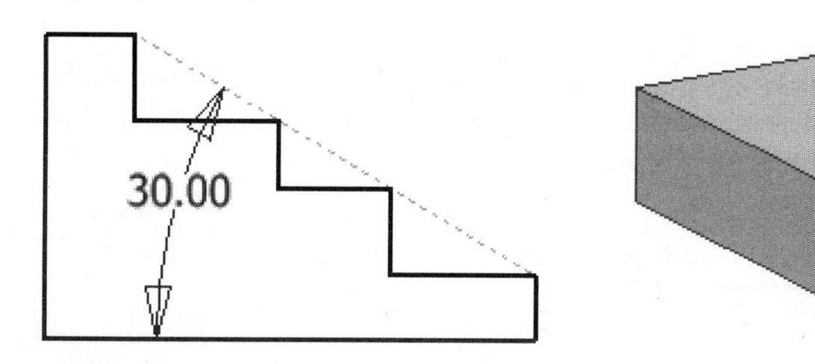

FIGURE 2.42

Snaps

Another method used to place geometry with a coincident constraint is snaps: midpoint, center, and intersection. After using a snap, a coincident constraint will be applied, and it will maintain the relationship that you define. For example, if you use a midpoint snap, the sketched point will always be in the middle of the selected object, even if the selected object's length changes.

To use snaps, follow these steps:

1. Start a sketch command, and right-click in the graphics window.
2. From the menu, select the desired snap.
3. Click on the object to which the sketched object will be constrained. For the intersection snap, select two objects.

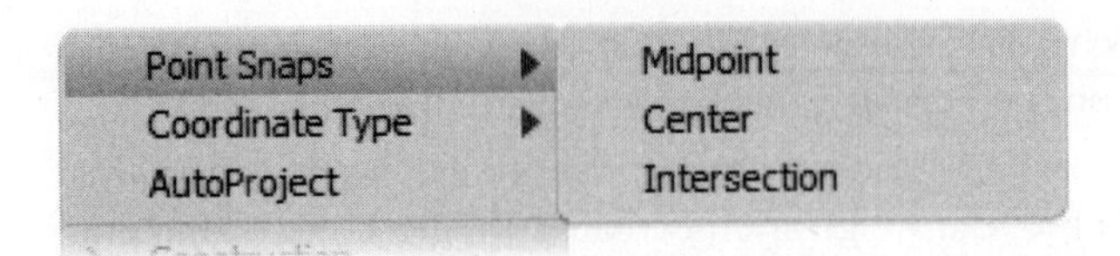

FIGURE 2.43

Sketch Degrees of Freedom

While constraining and dimensioning a sketch there are multiple ways to determine the open degrees of freedom. When you add a constraint or dimension the number of constraints or dimensions that are required to fully constrain, the sketch will decrease.

Status Bar

On the bottom-right corner of Autodesk Inventor, the number of constraints or dimensions will be displayed similar to what is shown in the following image on the left. A fully constrained sketch is a sketch whose objects cannot move or stretch and the message Fully Constrained will appear in the bottom-right corner of the screen as shown in the following image on the right.

FIGURE 2.44

Degrees of Freedom

To see the areas in the sketch that are NOT constrained, you can display the degrees of freedom. While in a sketch right-click and click Show Degree of Freedom as shown in the following image in the middle. Lines and arcs with arrows will appear as shown in the following image on the right. As constraints and dimensions are added to the sketch, degrees of freedom will disappear. To remove the degree of freedom symbols from the screen, right-click and click Hide All Degrees of Freedom from the menu as shown in the following image on the right.

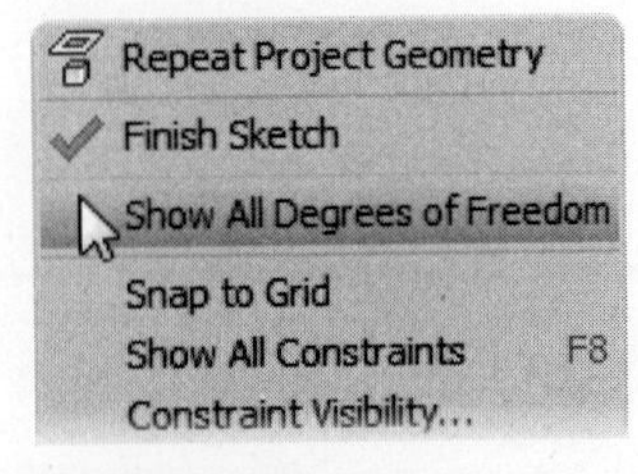

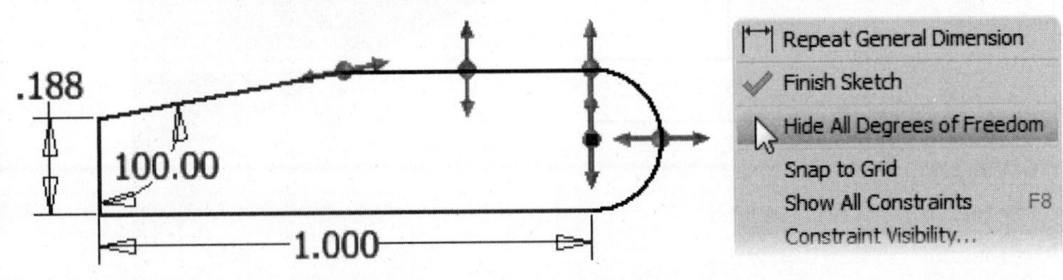

FIGURE 2.45

Dragging a Sketch

Another method to determine whether or not an object is constrained is to try to drag it to a new location. While not in a command, click a point or an edge, or select multiple objects on the sketch. With the left mouse button depressed, drag it to a new location. If the geometry stretches, it is underconstrained. For example, if you draw a rectangle that has two horizontal and two vertical constraints applied to it and you drag a point on one of the corners, the size of the rectangle will change, but the lines will maintain their horizontal and vertical behaviors. If dimensions are set on the object, they too will prevent the object from stretching.

EXERCISE 2-3: ADDING AND DISPLAYING CONSTRAINTS

In this exercise, you add geometric constraints to sketch geometry to control the shape of the sketch.

1. Click the New command, click the English tab, and double-click *Standard (in).ipt.*
2. Sketch the geometry as shown in the following image, with an approximate size of 2 inches in the X (horizontal) and 1.125 inches in the Y (vertical). Do not apply dimensions dynamically. Place the upper-left corner of the sketch on the projected origin point. Right-click in the graphics window, and then click Done. By starting the line at the origin point, that point is constrained to the origin.
3. Right-click in the graphics window, and click Show All Constraints, or press the F8 key. Your screen should resemble the following image.

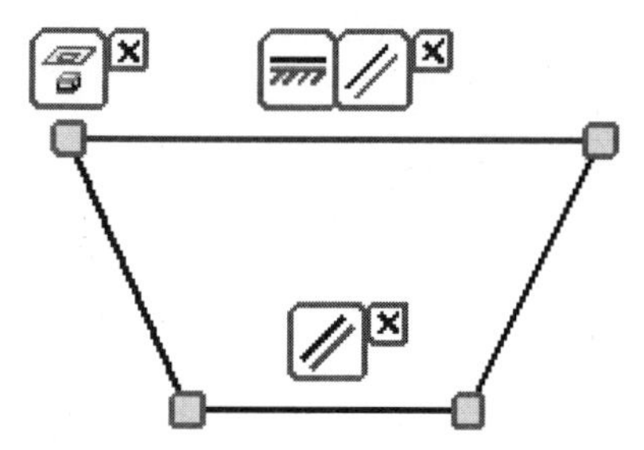

FIGURE 2.46

4. If another constraint appears, place the cursor over it, right-click, and then click Delete from the menu.
5. On the Constrain panel, click the Parallel constraint icon.
6. Click the two angled lines. Depending upon the order in which you sketched the lines, the angles may be opposite of the following image on the left.
7. Press the ESC key twice to stop adding constraints.
8. Press the F8 key to refresh the visible constraints. Your screen should resemble the following image on the right.

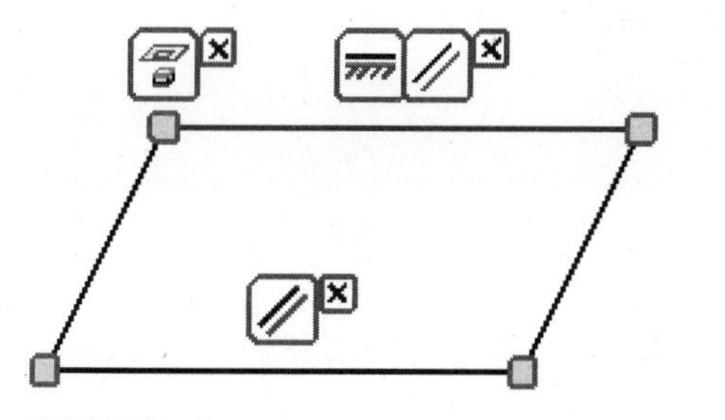

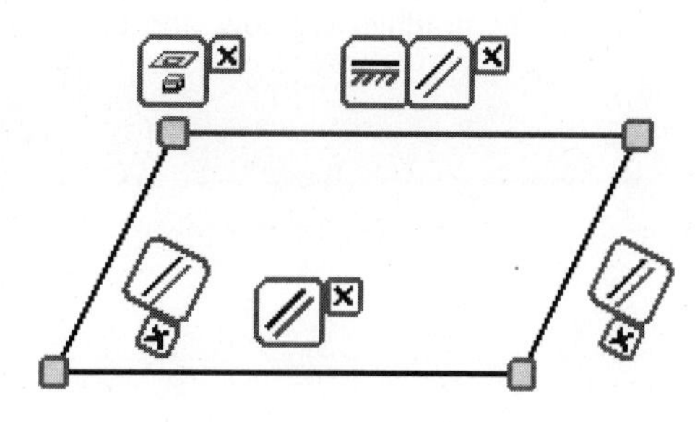

FIGURE 2.47

9. Click on the bottom horizontal line in the sketch and drag the line. Notice how the sketch changes its size, but not its general shape. Try to drag the top horizontal line. The line cannot be dragged as it is constrained.
10. Click the endpoint on the bottom-left horizontal line, and drag the endpoint. The lines remain parallel due to the parallel constraints.
11. Place the cursor over an icon for the parallel constraint for the angled lines, right-click, and click Delete from the menu as shown in the following image on the left. The parallel constraint that applies to both angled lines is deleted.
12. On the Constrain panel, click the Perpendicular constraint icon.
13. Click the bottom horizontal line and the angled line on the right side. Even though it may appear that the rectangle is fully constrained, the left vertical line is still unconstrained and can move.
14. Click the bottom horizontal line and the left angled line.
15. Press the F8 key to refresh the visible constraints. Your screen should resemble the following image on the right.
16. Right-click in the graphics window, and click Hide All Constraints, or press the F9 key.

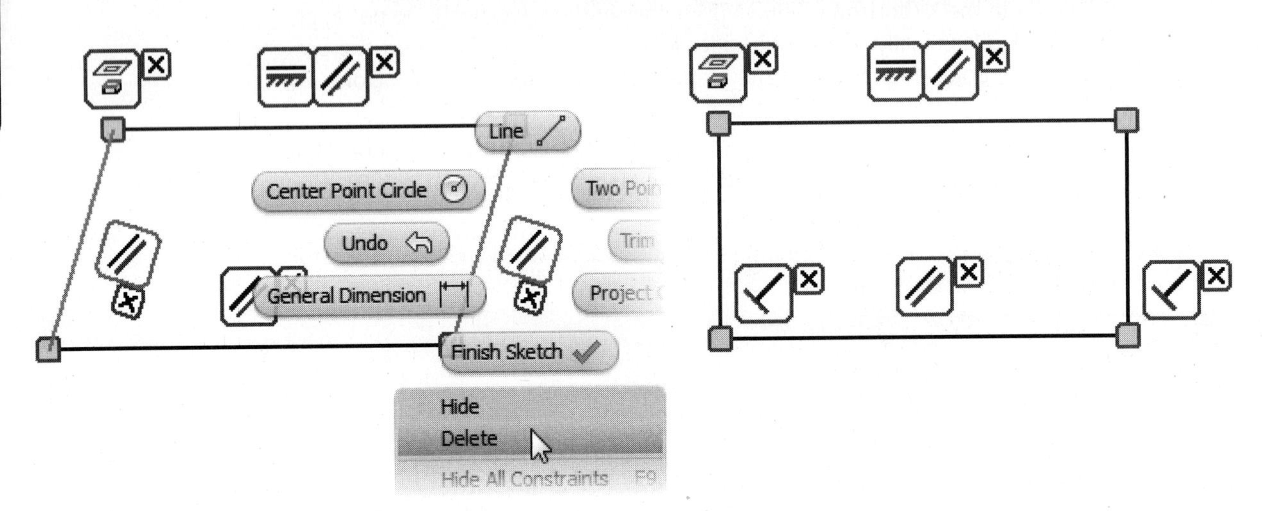

FIGURE 2.48

17. Drag the point at the lower-right corner of the sketch to verify that the rectangle can change size in both the horizontal and vertical directions, but its shape is maintained.
18. Close the file without saving the changes.
19. Click the New command, click the English tab, and then double-click *Standard (in).ipt*.
20. Sketch the geometry as shown with an approximate size of **2 inches** in the X and **1.375 inches** in the Y. Do not apply dimensions dynamically. Place the lower-left most point of the sketch on the projected center point. Right-click in the graphics window, and then click Done.

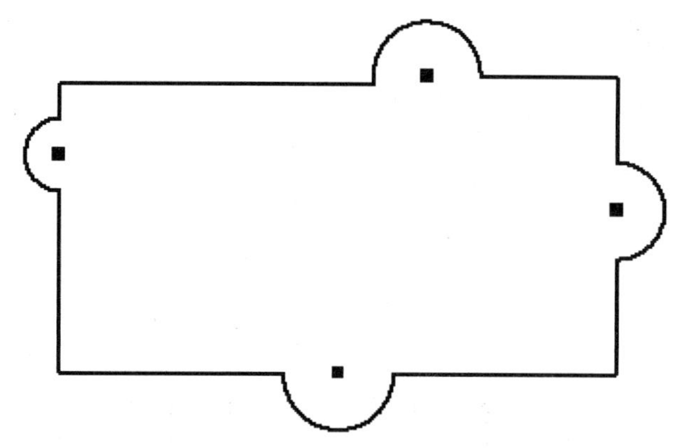

FIGURE 2.49

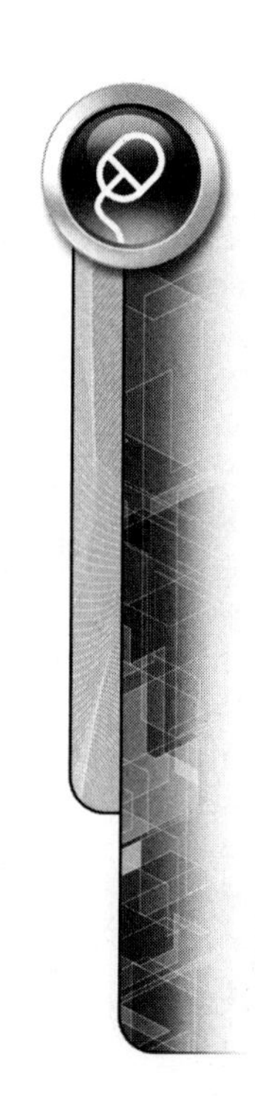

21. Inspect the constraints by dragging different points and edges.
22. Next you make the arcs equal in size. On the Constrain panel, click the Equal constraint icon or press the = key on the keyboard.
23. Click the arc on the left and the bottom arc.
24. Click the arc on the left and the arc on the right side.
25. Click the arc on the left and the arc on the top.
26. Next you align the line segments. On the Constrain panel, click the Collinear constraint icon or right-click in the graphics window, click Create Constraint, and then click Collinear. Note: In the following steps, if the constraint cannot be placed, you have a constraint that prevents it from being placed. Delete the constraint that is preventing the collinear constraint from being placed.
27. Click the two bottom horizontal lines.
28. Click the two top horizontal lines.
29. Click the two left vertical lines.
30. Click the two right vertical lines.
31. To stop applying the collinear constraint, either right-click and click Done from the marking menu or press the ESC key.
32. Next you will align the arcs. On the Constrain panel, click the Vertical constraint icon.
33. Click the center point of the bottom arc, and then click the center point of the top arc.
34. On the Constrain panel, click the Horizontal constraint icon.
35. Click the center point of the left arc, and then click the center point of the right arc.
36. To stop applying the constraints, right-click and click Done from the marking menu, or press the ESC key.
37. Display all of the constraints by pressing the F8 key. Your screen should resemble the following image.

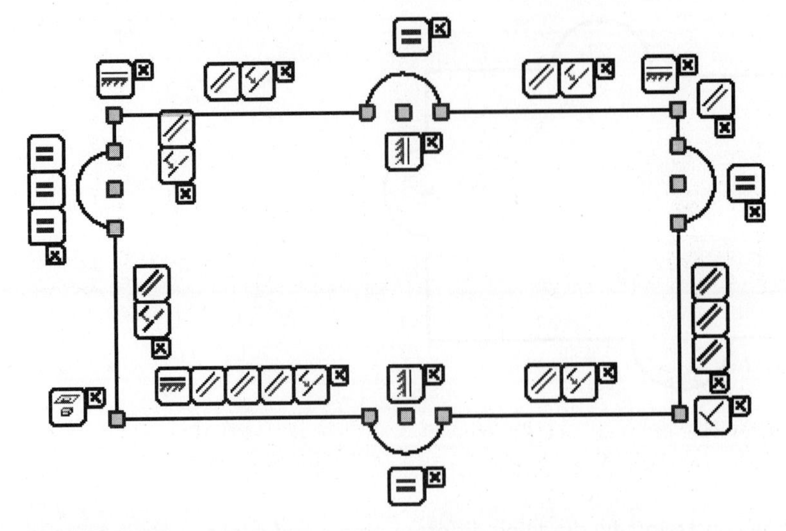

FIGURE 2.50

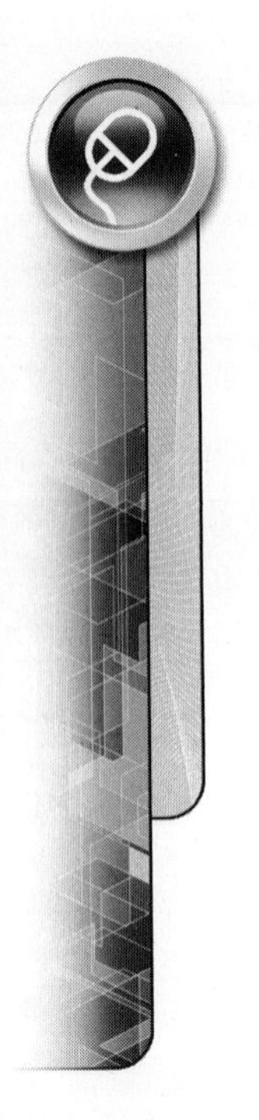

38. Hide all of the constraints by pressing the F9 key.
39. Click on an endpoint in the sketch and drag the endpoint. Try dragging different points, and notice how the sketch changes.
40. Delete the geometry as shown in the following image on the right. Press the Esc key twice to cancel any command, click a point above and to the left of the top arc, drag a window so it encompasses the arc on the right, release the mouse button, and press the Delete key on the key board.
41. Close the open line segments. Drag the open endpoints onto each other until your sketch resembles the following image on the right.

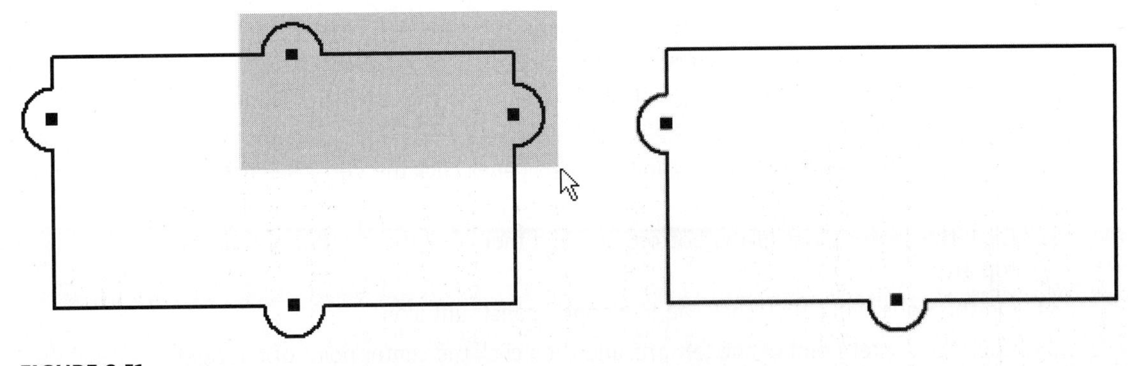

FIGURE 2.51

42. Next you center the arcs. On the Constrain panel, click the Vertical constraint icon.
43. Click the center point on the bottom arc and the midpoint of the top horizontal line as shown in the following image on the left.
44. On the Constrain panel, click the Horizontal constraint icon.
45. Click the center point on the left arc and the midpoint of the right vertical line as shown in the following image on the right.

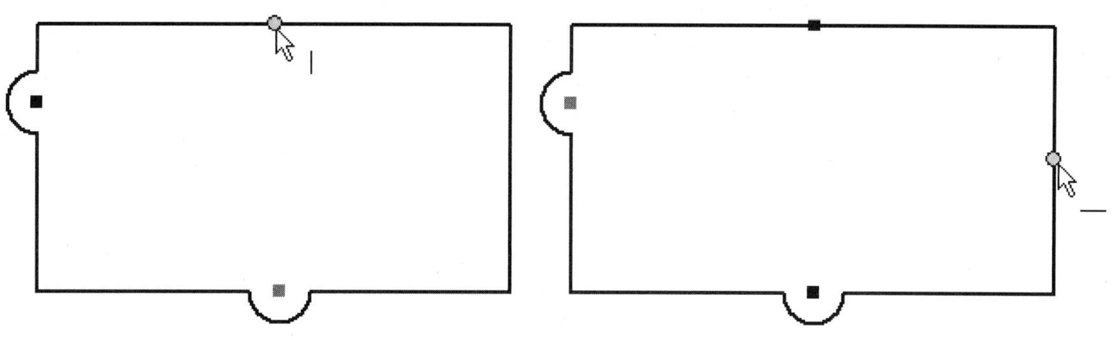

FIGURE 2.52

46. Click on different points and drag them, notice how the sketch changes shape, but the arcs are always centered as shown in the following image.

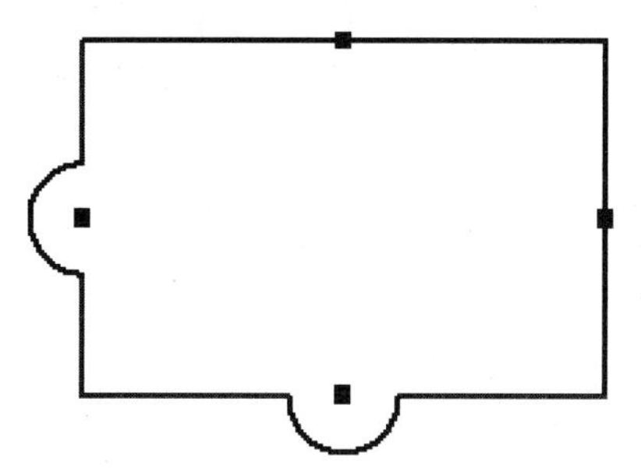

FIGURE 2.53

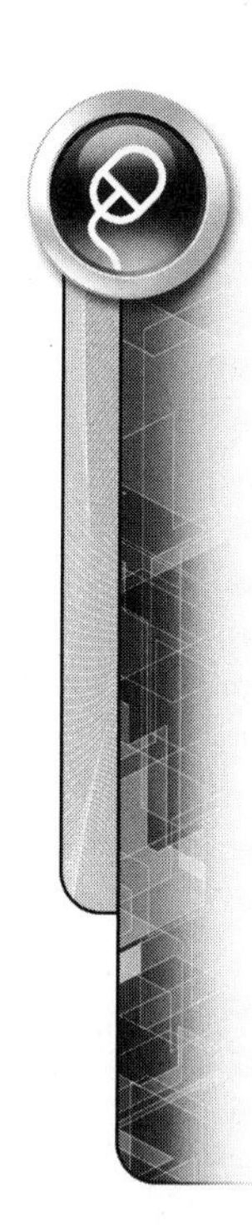

47. Close the file. Do not save changes. End of exercise. Note that dimensions would be added to completely define the sketch. Dimensions are covered in the next section.

STEP 3—ADDING DIMENSIONS MANUALLY

The last step to constraining a sketch is to add dimensions that were not added dynamically. The dimensions you place will control the size of the sketch and can also appear in the part drawing views when they are generated. When placing dimensions, try to avoid having extension lines go through the sketch, as this will require more cleanup when drawing views are generated. Click near the side from which you anticipate the dimensions will originate in the drawing views.

All dimensions that you create are parametric as well as the dynamic dimensions. Parametric means that they will change the size of the geometry. Parametric dimensions are manually created with either the General Dimension or Auto Dimension commands.

General Dimensioning

The General Dimension command can create linear, angle, radial, or diameter dimensions one at a time. The following image on the left shows an example of a dimensioned sketch. To start the General Dimension command, follow one of these techniques:

- Click the Create Dimension command from the Sketch tab > Constrain panel as shown in the following image on the right.

- Right-click in the graphics area and select Create Dimension from the marking menu.
- Press the shortcut key D.

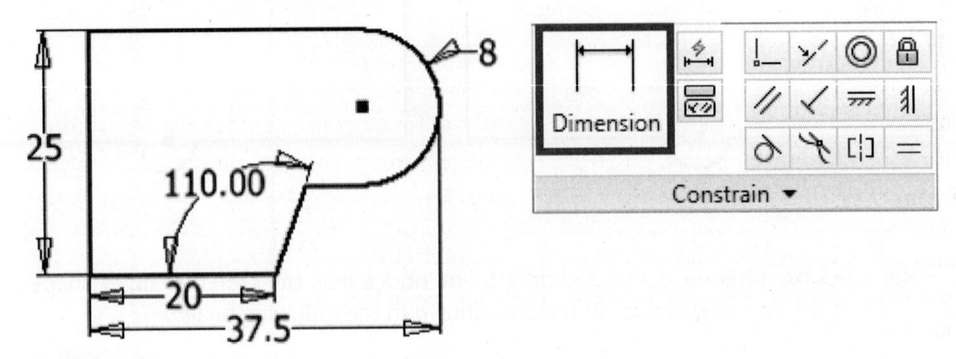

FIGURE 2.54

When you place a linear dimension, the extension line of the dimension will snap automatically to the nearest endpoint of a selected line; when an arc or circle is selected, it will snap to its center point. To dimension to a quadrant of an arc or circle, see "Dimensioning to a Tangent of an Arc or Circle," later in this chapter.

After you select the General Dimension command, follow these steps to place a dimension:

1. Click a point or points to locate where the dimension is to start and end.
2. After selecting the point(s) to dimension, a preview image will appear attached to your cursor showing the type of dimension. If the dimension type is not what you want, right-click, and then select the correct style from the menu. After changing the dimension type, the dimension preview will change to reflect the new style.
3. Click a point on the screen to place the dimension.

The next sections cover how to dimension specific objects and how to create specific types of dimensioning with the Dimension command.

Dimensioning Lines

There are two techniques for dimensioning a line. Issue the Dimension command and do one of the following:

- Click near two endpoints, move the cursor until the dimension is in the correct location, and click.
- To dimension the length of a line, click anywhere on the line; the two endpoints will be selected automatically. Move the cursor until the dimension is in the correct location and click.
- To dimension between two parallel lines, click one line and then the next, and then click a point to locate the dimension.
- To create a dimension whose extension lines are perpendicular to the line being dimensioned, click the line and then right-click. Click Aligned from the menu, and then click a point to place the dimension.

Dimensioning Angles

To create an angular dimension, issue the General Dimension command, click on two lines whose angle you want to define, move the cursor until the dimension is in the correct location, and place the dimension by clicking on a point.

Dimensioning Arcs and Circles

To dimension an arc or circle, issue the General Dimension command, click on the circle's circumference, move the cursor until the dimension is in the correct location, and click. To dimension the angle of an arc, click on the arc; click the arc center point or click an endpoint of the arc, the center point, and then the other endpoint; and then locate the dimension. By default, when you dimension an arc, the result is a radius dimension. When you dimension a circle, the default is a diameter dimension. To change the radial dimension to diameter or a diameter to radial, right-click before you place the dimension and select the other style from the menu.

You can dimension the angle of the arc. Starting the Dimension command, click on the arc's circumference, click the center point of the arc, and then place the dimension or click the center point and then click the circumference of the arc.

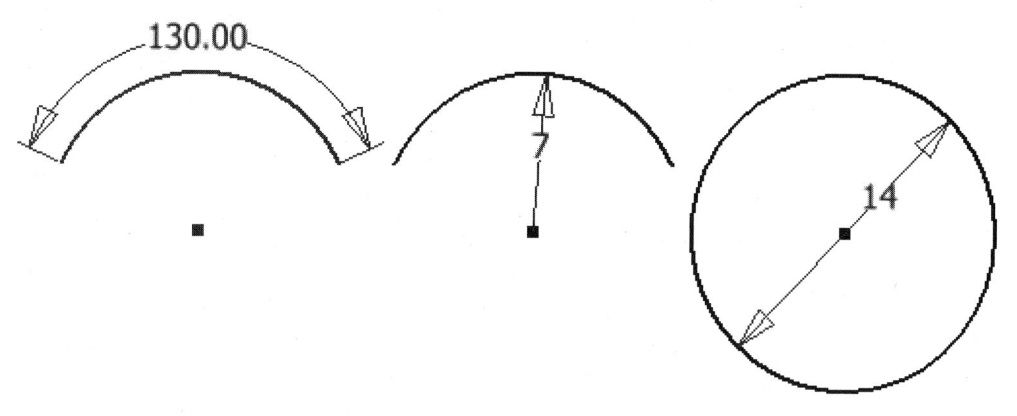

FIGURE 2.55

Dimensioning to a Tangent of an Arc or Circle

To dimension to a tangent of an arc or circle, follow these steps:

1. Issue the General Dimension command.
2. Click a line that is parallel to the tangent arc or circle that will be dimensioned.
3. Place the cursor over the tangent arc or circle that should be dimensioned.
4. Move the cursor over the tangent until the constraint symbol changes to reflect a tangent, as shown on the left side of the following image.
5. Click to accept the dimension type, and then move the cursor until the dimension is in the correct location. Click as shown on the right side of the following image.

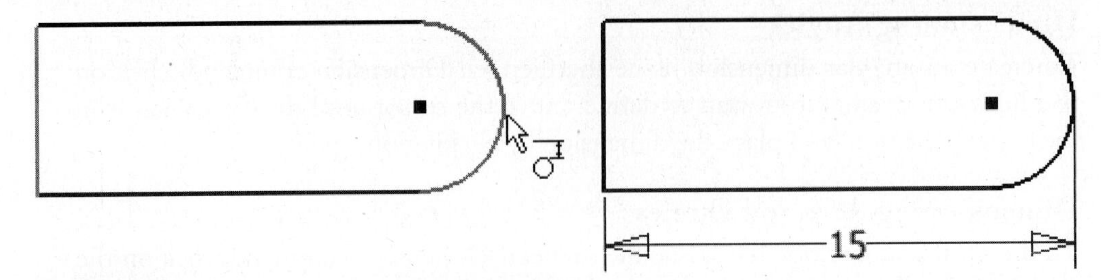

FIGURE 2.56

To dimension to two tangents, follow these steps:

1. Issue the General Dimension command.
2. Click an arc or circle that includes one of the tangents to which it will be dimensioned.
3. Place the cursor over the tangent edge of the second arc or circle to which it will be dimensioned.
4. Move the cursor over the area that is tangent until the constraint symbol appears, as shown on the left in the following image.
5. Click and then move the cursor until the dimension is in the correct location, and then click as shown in the following image on the right.

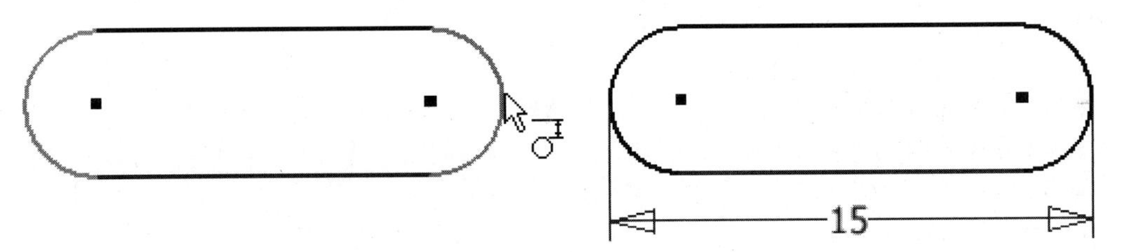

FIGURE 2.57

Entering and Editing a Dimension Value

After placing the dimension, you can change its value. Depending on your setting for editing dimensions when you created them, the Edit Dimension dialog box may or may not appear automatically after you place the dimension. To set the Edit Dimension option, do one of the following:

1. Click the Tools tab > Options panel > Application Options. On the Sketch tab of the Options dialog box, select Edit dimension when created as shown on the left side of the following image.
2. Or set this option by right-clicking in the graphics area while placing a dimension and selecting Edit Dimension from the menu as shown in the following image on the right. This method will change the Application Option sketch tab Edit Dimension.

If the Edit dimension when created option is checked, the Edit Dimension dialog box will appear automatically after you place the dimension. Otherwise, the dimension will be placed with the default value, and you will not be prompted to enter a different value.

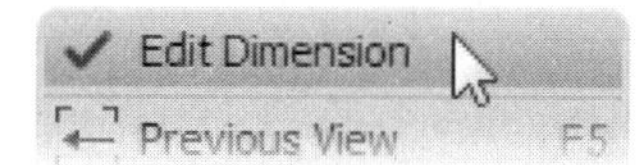

FIGURE 2.58

To edit a dimension that has already been created, double-click on the value of the dimension, and the Edit Dimension dialog box will appear, as shown in the following image. Enter the new value and unit for the dimension; then either press ENTER or click the checkmark in the Edit Dimension dialog box. If no unit is entered, the units that the file was created with will be used. When inputting values, enter the exact value—do not round up or down. The accuracy of the dimension that is displayed in a sketch is set in the Document Setting. For example, if you want to enter 4 1/16 decimally enter 4.0625 not 4.06.

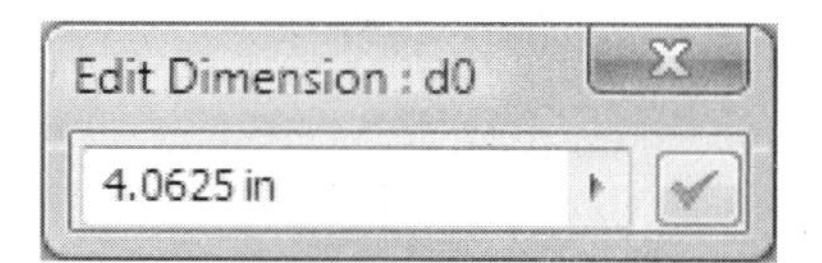

FIGURE 2.59

Fractions

Inventor also allows you to type in a fraction anywhere a value is required. When the length unit in the Tools tab > Options > Document Settings is set to a metric unit the decimal equivalent will be displayed in the graphics window but the fraction will be maintained in the edit dialog box. If the unit is set to any non-metric unit as shown in the following image on the left and a fraction is entered, a fraction will display in the graphics window. After inputting a fraction you can click on the right-faced arrow and set the type of dimension to display Decimal, Fractional, or Architectural as shown in the following middle image. For example, enter 4 1/16, not 4-1/16. Inventor would interpret the - as part of an equation and would return the value 3.9375. The following image on the right shows the fraction displayed in the graphics window.

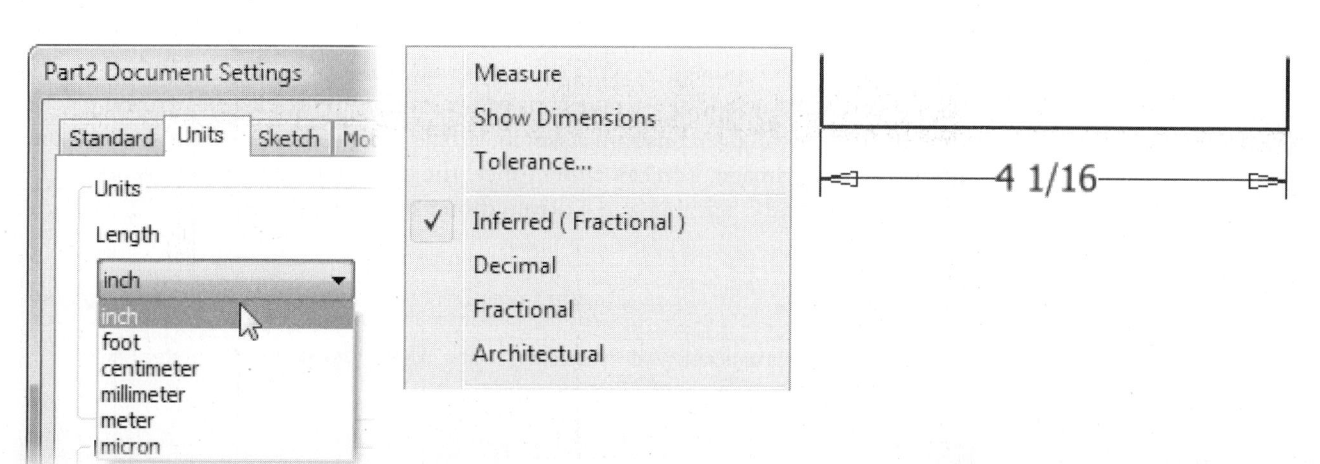

FIGURE 2.60

NOTE When placing dimensions, it is recommended that you place the smallest dimensions first. This will help prevent the geometry from flipping in the opposite direction.

Repositioning a Dimension

Once you place a dimension, you can reposition it, but the origin points cannot be moved. Follow these steps to reposition a dimension:

1. Exit the current operation either by pressing ESC twice or right-clicking and then selecting Done from the menu.
2. Move the cursor over the dimension until the move symbol appears as shown in the following image.
3. With the left mouse button depressed, move the dimension or value to a new location and release the button.

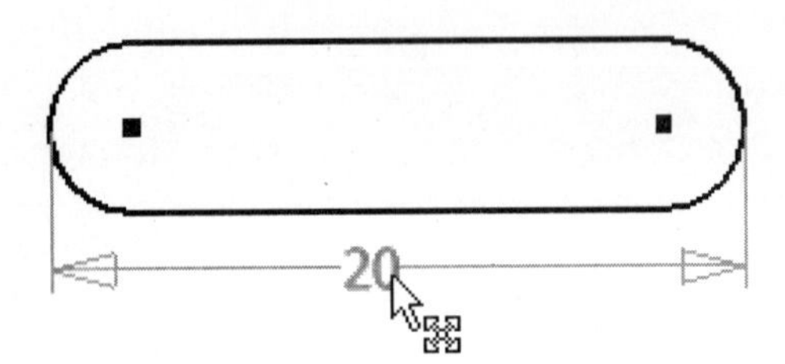

FIGURE 2.61

Overconstrained Sketches

As explained in the "Adding Constraints" section, Autodesk Inventor will not allow you to overconstrain a sketch or add duplicate constraints. The same is true when adding dimensions. If you add a dimension that will conflict with another constraint or dimension, you will be warned that this dimension will overconstrain the sketch or that it already exists. You will then have the option to either not place the dimension or place it as a driven dimension.

A driven dimension is a reference dimension. It is not a parametric dimension—it reflects the size of the points to which it is dimensioned, if the part changes the driven dimension will be updated to reflect the new value. A driven dimension will appear with parentheses around the dimensions value—for example, (30). When you place a dimension that will overconstrain a sketch, a dialog box will appear similar to the one in the following image. You can either cancel the operation and no dimension will be placed or accept the warning and a driven dimension will be created.

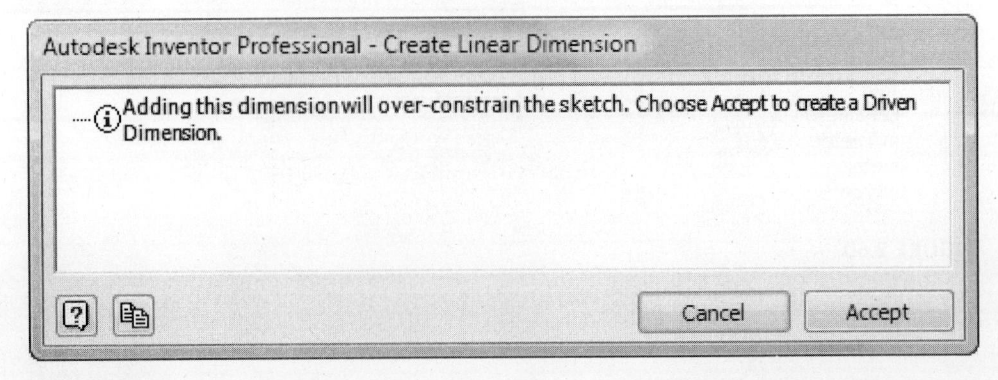

FIGURE 2.62

Autodesk Inventor gives you an option for handling overconstrained dimensions. To set the overconstrained dimensions option, click Tools tab > Options panel > Application Options, and on the Sketch tab of the Options dialog box, change the Overconstrained dimensions option as shown in the following image. You have the following two options:

- Apply driven dimension

When checked, this option automatically creates a driven dimension without warning you of the condition.

- Warn of overconstrained condition

When checked, this option causes a dialog box to appear stating that the dimension will overconstrain the sketch. Click Cancel not to place a driven dimension, or click Accept to place a driven dimension.

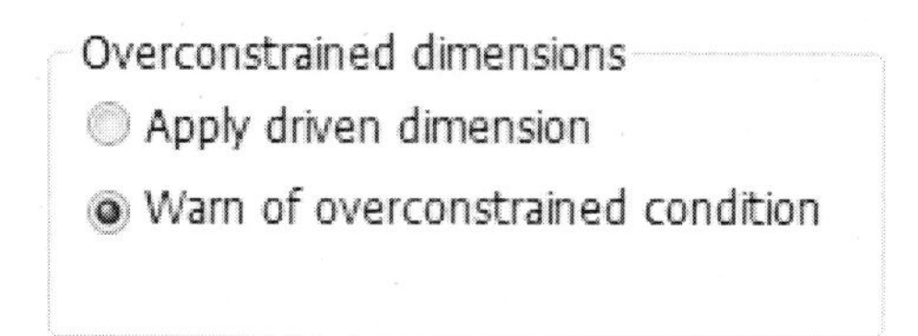

FIGURE 2.63

Another option for controlling the type of dimension that you create is to use the Driven Dimension command on the Format panel. If you issue the Driven Dimension command, any dimension you create will be a driven dimension. If you do not issue the command, you create a regular dimension. The following image shows the Driven Dimension command on the Format panel in its normal condition. The same Driven Dimension command can be used to change an existing dimension to either a normal or driven dimension by selecting the dimension and clicking the Driven Dimension command. Driven dimensions are represented in the sketch with parenthesis around the text as shown in the following image on the right, whereas parametric dimensions do not have parenthesis around the text.

TIP

Avoid over using driven dimensions as they do not parametrically control the size of the profile, and reference dimensions are used only in the manufacturing process to verify the size of the part.

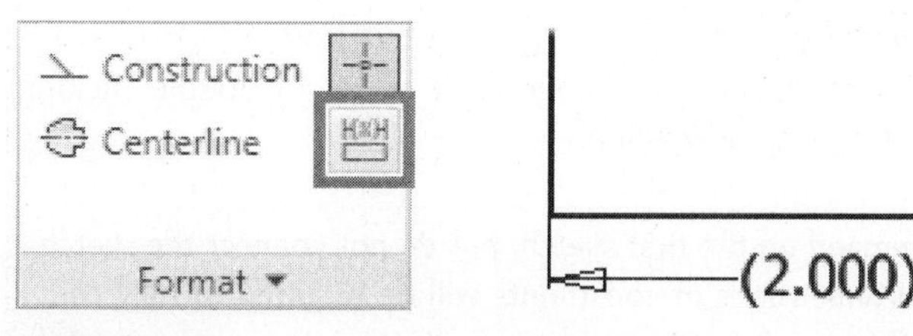

FIGURE 2.64

AUTO DIMENSION

Adding constraints and dimensions to a sketch or removing dimensions from a sketch can be a time-consuming task. To automate this process, you can use the Auto Dimension command to create or remove dimensions or to add constraints to selected geometry automatically. Before using the Auto Dimension command, you should apply critical constraints and dimensions using the appropriate sketch constraint or Dimension command. The Auto Dimension command will not override or replace any existing constraints or dimensions. Click the Auto Dimension command on the Sketch tab > Constrain panel as shown in the following image on the left. The Auto Dimension dialog box will appear as shown on the right.

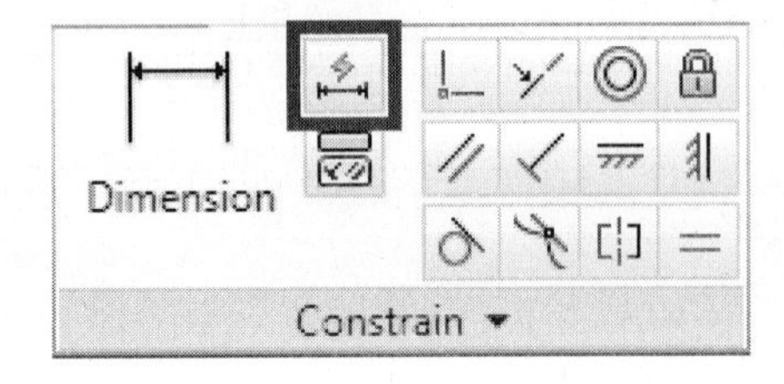

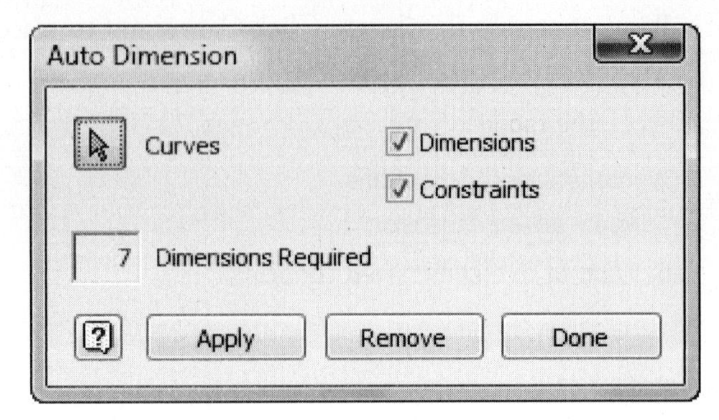

FIGURE 2.65

To use the Auto Dimension command, follow these steps:

1. Click the Auto Dimension command on the Constrain panel.
2. The number of constraints and dimensions required to fully constrain the sketch appear in the lower-left corner of the dialog box.
3. Determine if you want to create dimensions or constraints or remove the dimensions or constraints that you previously added using the Auto Dimension command.
4. Click the Dimensions box and/or the Constraints box.
5. Click the Curves option, and then select the objects with which to work in the graphics window.
6. Click the Apply button to create the dimensions and/or apply constraints to the selected curves, or click the Remove button to delete the selected dimensions and/or constraints.
7. If you clicked Apply, you can change the values of the dimensions that you placed by double-clicking on the dimension text and entering in a new value in the Edit Dimension dialog box.
8. After the dimensions are placed, you can change their values by double-clicking on the numbers and entering in a new value.

NOTE

If you use the Auto Dimension command on the first sketch, but do not connect the sketch to the center point, two additional dimensions or constraints will be required to fully constrain the sketch. Drag a point on the sketch to the projected center point, apply horizontal and vertical constraints between points on the sketch and the center point, or add dimensions between the center point and the sketch to fully constrain the sketch.

EXERCISE 2-4: DIMENSIONING A SKETCH

In this exercise, you add dimensional constraints to a sketch. Note: this exercise assumes that the "Edit dimension when created" and "Autoproject part origin on sketch create" options are checked in the Options dialog box under the Sketch tab. Experiment with Autodesk Inventor's color schemes to see how the sketch objects changes color to represent if they are constrained.

1. Click the New command and click the English tab, and then double-click *Standard (in).ipt*.
2. Start sketching by starting the line command and click the origin point, move the cursor to the right, type **5** in the distance input field, press the Tab key, and then click a point when the horizontal constraint is previewed above the input field for degrees as shown in the following image on the left.
3. Next you place an angle line and a dynamic dimension to define the angle. Press the tab key and type **150** for the angle input field, press the Tab key, and then click a point to the upper right as shown in the following image on the right. The distance should be about 2 inches but the dimension is not needed to define this sketch.

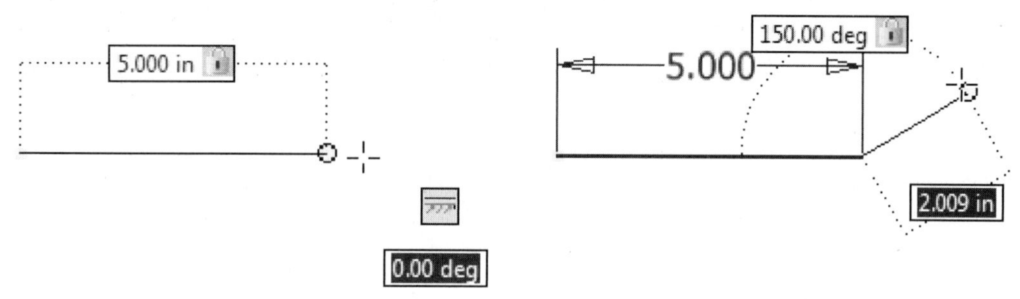

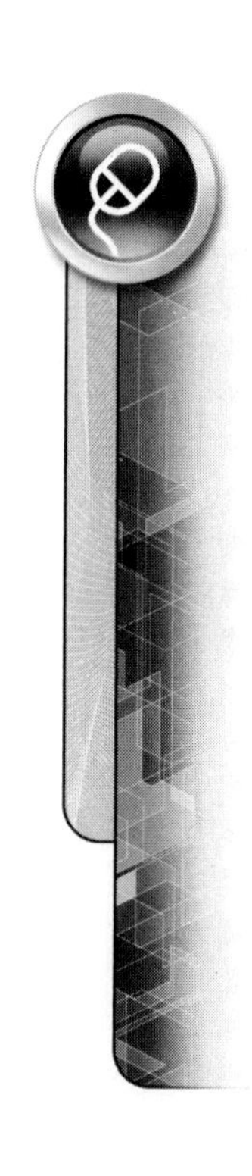

FIGURE 2.66

4. Sketch the geometry as shown in the following image. When sketching, ensure that a perpendicular constraint is not applied between the two angled lines. Hold down the CTRL key while sketching the top angle line to prevent sketch constraints from being applied. The arc should be tangent to both adjacent lines.

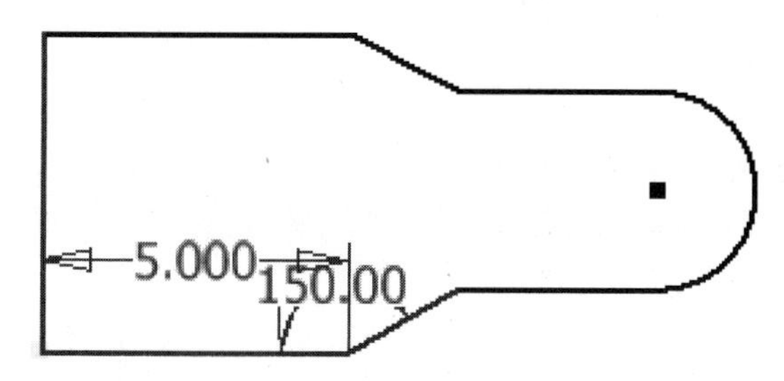

FIGURE 2.67

5. Add a horizontal constraint between the midpoint of the left vertical line and the center of the arc as shown in the following image on the left.
6. Add a vertical constraint between the endpoints of the angled lines nearest the right side of the sketch as shown in the following image on the right.

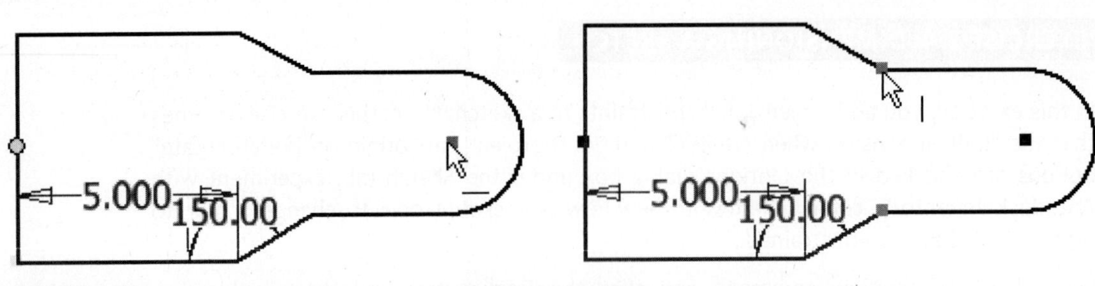

FIGURE 2.68

7. Add an equal constraint between the two angled lines by pressing the = key and then select the two angled lines.
8. Click the General Dimension command in the Constrain panel.
9. Add a radial dimension by clicking the arc. Drag the dimension to the left, click a point to locate it, enter **1.5** (if the Edit Dimension dialog box did not appear, double-click on the dimension and change the dimension to 1.5), and click the checkmark in the Edit Dimension dialog box.

To set the Edit Dimension option to appear when placing sketch dimensions, right-click in the graphics area while placing a dimension and select Edit Dimension from the menu.

10. Add a vertical dimension by clicking the vertical line. Drag the dimension to the left, click a point to locate it, enter **5**, and click the checkmark. When complete, your sketch should resemble the following image on the left. Notice on the bottom right of the status bar that the sketch requires 1 dimension as shown in the following image on the right.

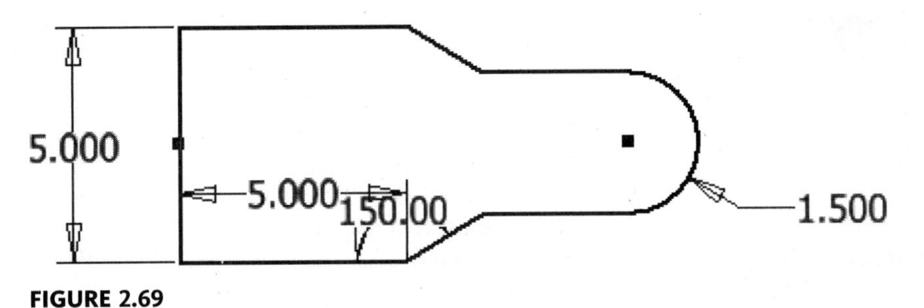

FIGURE 2.69

11. Add an overall horizontal dimension by clicking the vertical line (not an endpoint). Move the cursor near the right tangent point of the arc until the glyph of dimension to a circle appears, as shown in the following image on the left. Click, drag the dimension down, click a point to locate it, enter **12**, and click the checkmark. When complete, your sketch should be fully constrained and resemble the following image on the right.

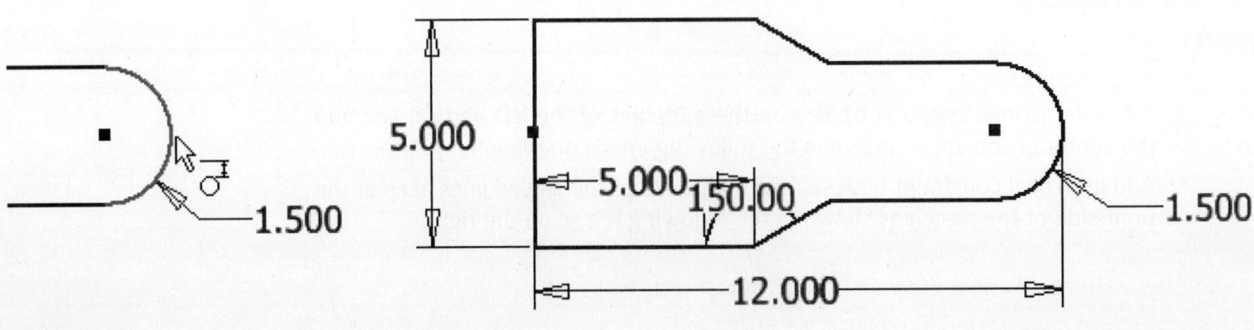

FIGURE 2.70

12. Edit the value of the dimensions, and examine how the sketch changes. The arc should always be in the middle of the vertical line. All of the dimensions are parametric, even the dimensions that were placed dynamically while sketching.
13. Delete the horizontal constraint between the center of the arc and the midpoint of the vertical line.
14. Delete the vertical constraint between the two angled lines.
15. Practice adding other types of constraints and dimensions.
16. Close the file. Do not save changes. End of exercise.

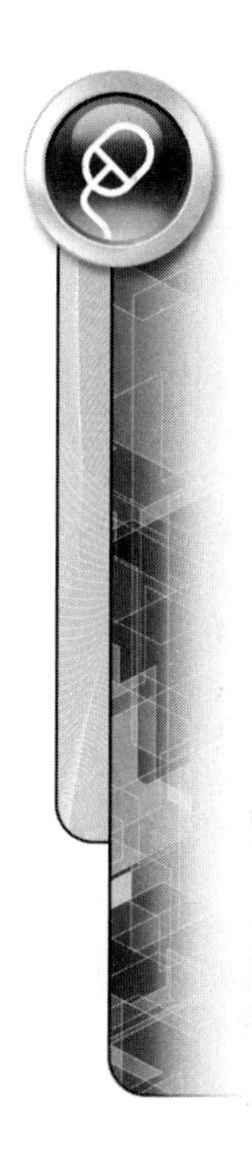

MOVE AND SCALE TOOLS

After importing data into Autodesk Inventor, you may need to move the object to another location. For example, you may wish to reposition it so that a point on the sketch is coincident to the origin point. You may need to scale the geometry if it was imported with incorrect units.

Move

To use the move command, follow these steps.

1. Click the Move command from the Modify panel.
2. In the graphics window, select the objects to move.
3. Click the Base Point button, and then select a point on the sketch that will be used as the starting point.
4. Click a point in the graphics window to move the objects to, such as the origin point.
   a. You can also use the Precise Input option to move the objects a specified distance.
   b. Click the Copy option to make a copy of the selected objects.
   c. If the move would not be possible because a dimension or constraint is not allowing it, you will be prompted to relax these dimensions or constraints. To set how dimensions or constraints will be handled, click the >> button, and select an option to relax the dimensions and/or constraints.
   d. Check the option Optimize for Single Selection if you want to use only one selection and then automatically be prompted to select the base point.

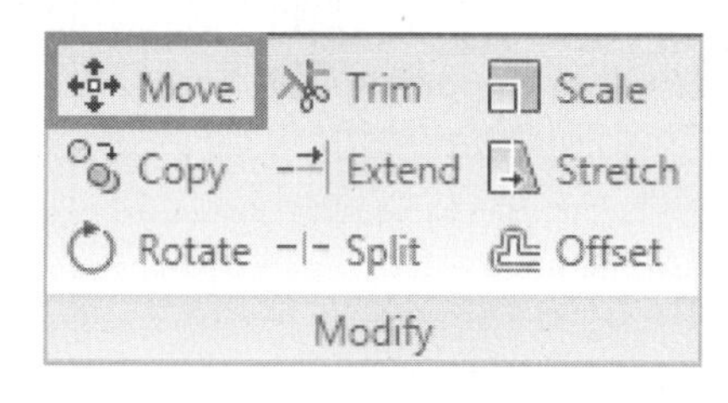

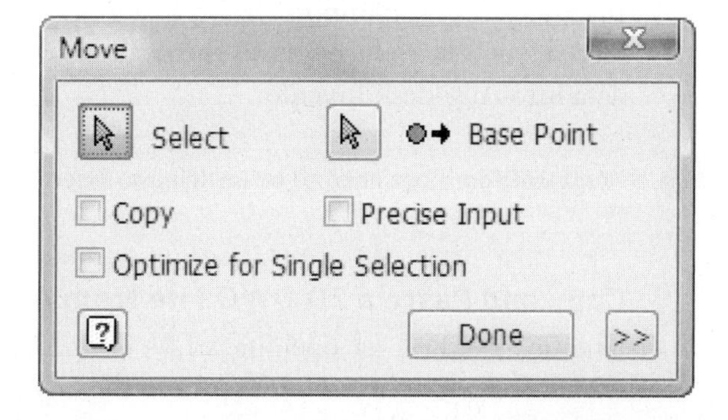

FIGURE 2.71

Scale

To use the scale command, follow these steps.

1. Click the Scale command from the Modify panel.

2. In the graphics window, select the objects to scale.
3. Click the Base Point button, and then select a point on the sketch from which the scale will be based.
    a. Type a Scale Factor or move the cursor and click a second point.
    b. You can also use the Precise Input option to scale the objects a specified distance by using the Precise Input toolbar; the distance between the two points determines the scale factor.
    c. If the scale would not be possible because a dimension or constraint is not allowing it, you will be prompted to relax these dimensions or constraints. To set how dimensions or constraints will be handled, click the >> button and select an option to relax the dimensions and/or constraints.
    d. Check the option Optimize for Single Selection if you want to use only one selection and then automatically be prompted to select the base point.

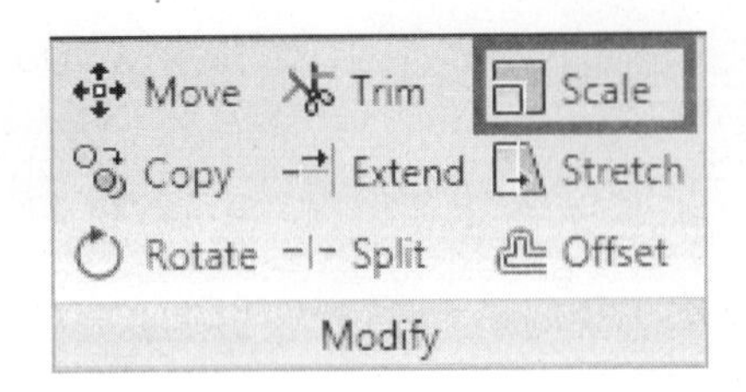

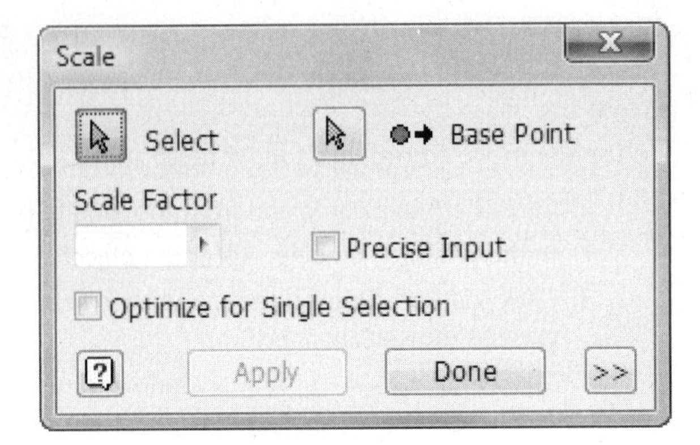

FIGURE 2.72

OPENING AND IMPORTING AUTOCAD FILES

Many Autodesk Inventor users store data in the AutoCAD DWG format. Instead of redrawing this data, you can import it into Autodesk Inventor drawings or into a part feature sketch. You can also import AutoCAD files containing 3D solids using the Open command. The AutoCAD solids will be opened in a new part or assembly file depending upon whether there are multiple AutoCAD solids. When importing a 2D DWG file into Autodesk Inventor, you can either copy the contents from the DWG file to the clipboard via Autodesk Inventor or AutoCAD and paste into Autodesk Inventor, or use an import wizard that guides you through the process. The following sections will introduce you to the options available when importing 2D AutoCAD data into Autodesk Inventor.

NOTE **AutoCAD does not need to be installed to insert AutoCAD geometry.**

Copy and Paste a 2D DWG File from AutoCAD or Autodesk Inventor

The first method for opening a DWG or DXF data is to open a drawing file from within Autodesk Inventor. Copy it to the clipboard and paste it into Autodesk Inventor by following these steps. This same procedure can be done by opening the file in AutoCAD and copying it to the clipboard.

1. From in Inventor Application Menu, click Open, or click Open on the Quick Access toolbar.
2. In the Open dialog box, navigate to and select the DWG file, as shown in the following image.

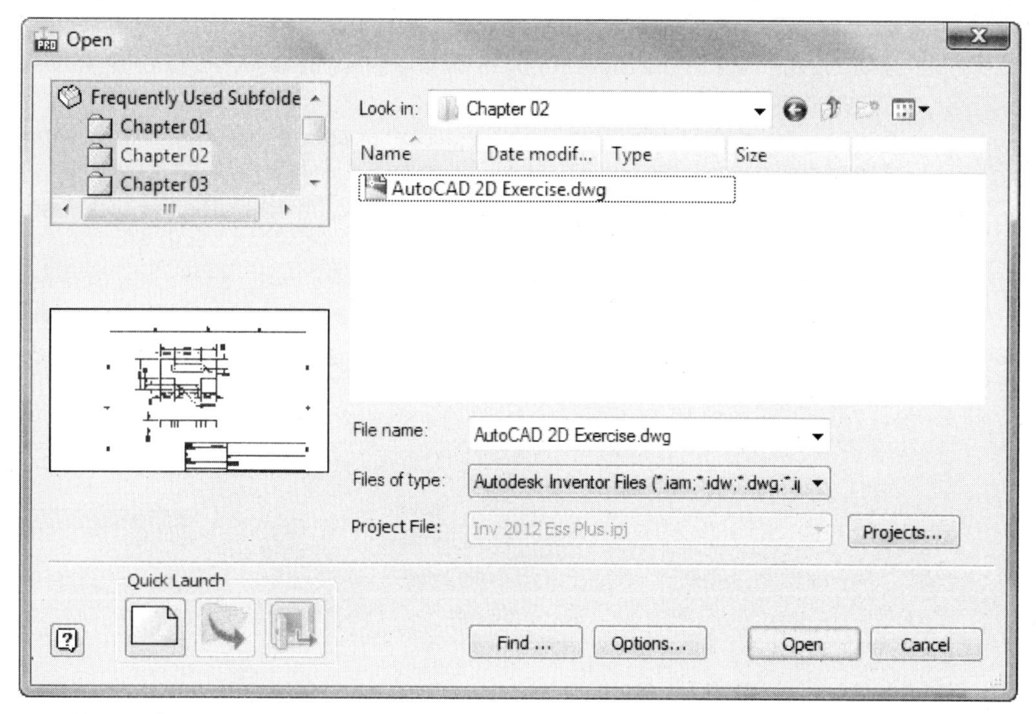

FIGURE 2.73

3. Either double-click on the file name or click Open from the dialog box, and the DWG file will open inside of Autodesk Inventor, as shown in the following image. The background color can be changed by right-clicking on Model (AutoCAD) entry in the browser and click Background Color from the menu.

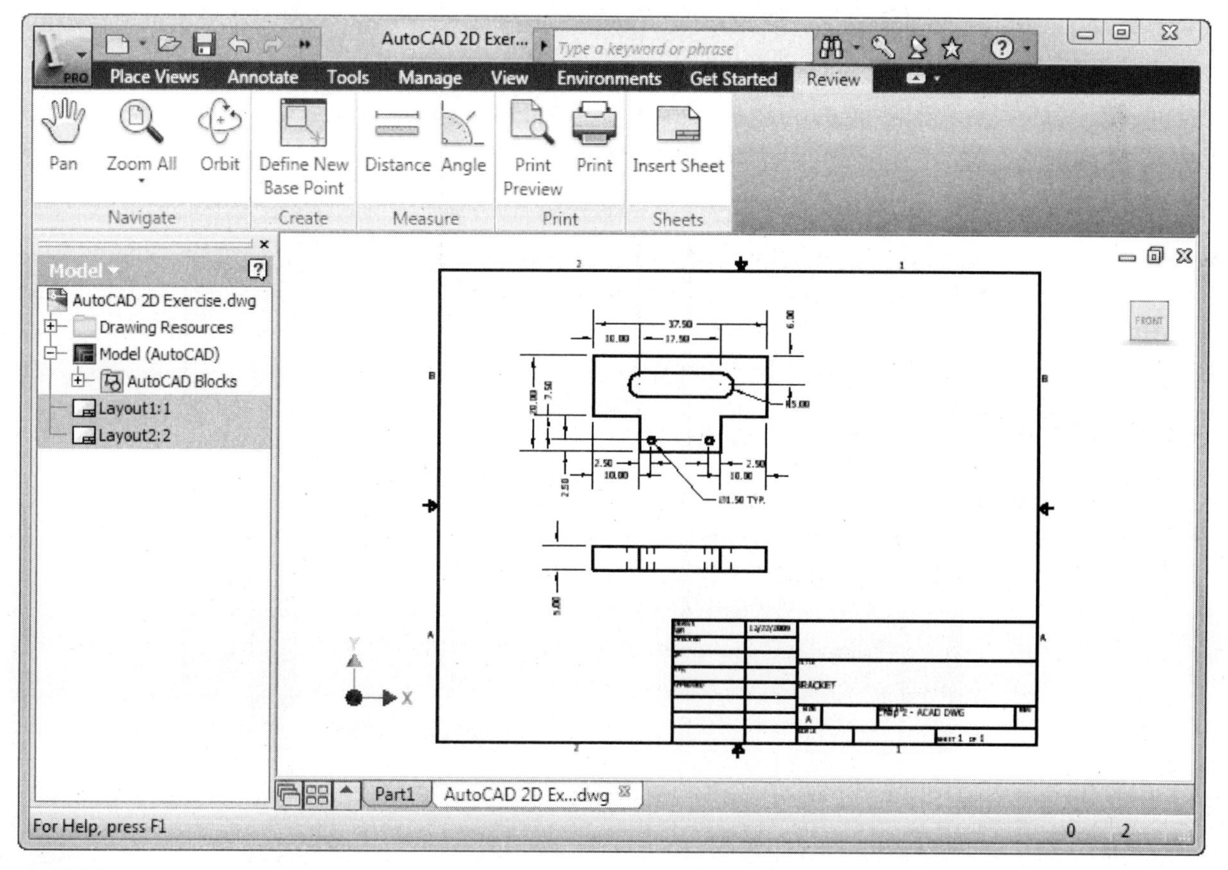

FIGURE 2.74

4. Select the geometry and dimensions you want to copy into Autodesk Inventor.
5. Copy the data to the clipboard by using the shortcut key CTRL-C, or right-click and click Copy from the menu.
6. Make the Autodesk Inventor part file active, or start a new Autodesk Inventor part file in which the DWG data will be placed. A sketch must be active.
7. Paste the data by using the shortcut key CTRL-V, or right-click and click Paste from the menu.
8. The bounding box appears as shown in the following image on the left. Depending upon the size of the copied objects, you may need to zoom out to see the bounding box. Then right-click, and click Paste Options from the menu, as shown in the following image on the right.

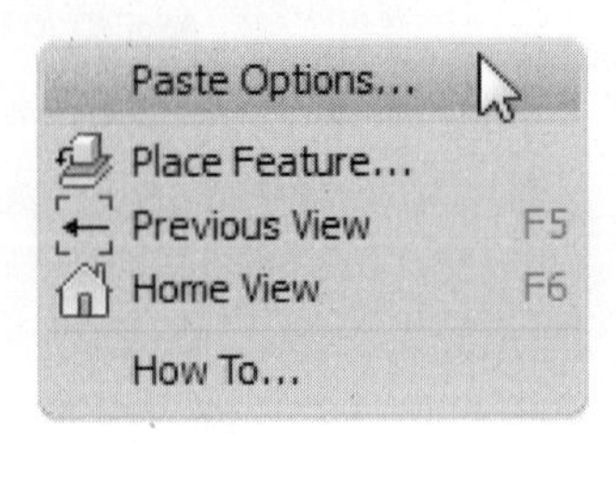

FIGURE 2.75

9. The Paste Options dialog box appears as shown in the following image. Select the unit in which the AutoCAD geometry was created, and check Constrain End Points if you want a coincident constraint to be applied to geometry where two endpoints touch. Select the Apply geometric constraints option to have 2D sketch constraints automatically applied to the sketch. The Import parametric constraints option will bring in geometric constraints that were created in AutoCAD. Check the AutoCAD Blocks to Inventor Blocks if you want the AutoCAD geometry to be an Inventor Sketch block.

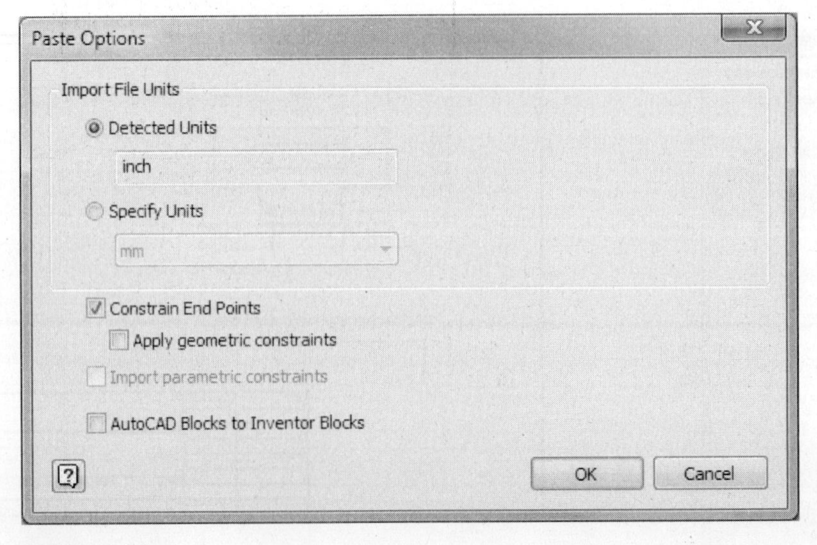

FIGURE 2.76

10. Click the OK button.
11. Click a point in the graphics window where you want the copied geometry to be placed. The DWG data is now in the active sketch.
12. Constrain the geometry as needed.

INSERTING 2D AUTOCAD DATA INTO A SKETCH

Another method that utilizes existing AutoCAD 2D data inserts the data into the active sketch in a part or drawing. To insert AutoCAD data into the active sketch, follow these steps:

1. Make a sketch active in a part file or a draft view active in a drawing file.
2. Click the Insert AutoCAD command on the Sketch tab > Insert panel as shown in the following image.

FIGURE 2.77

3. The Open dialog box will appear, browse to, and double-click the desired DWG file.

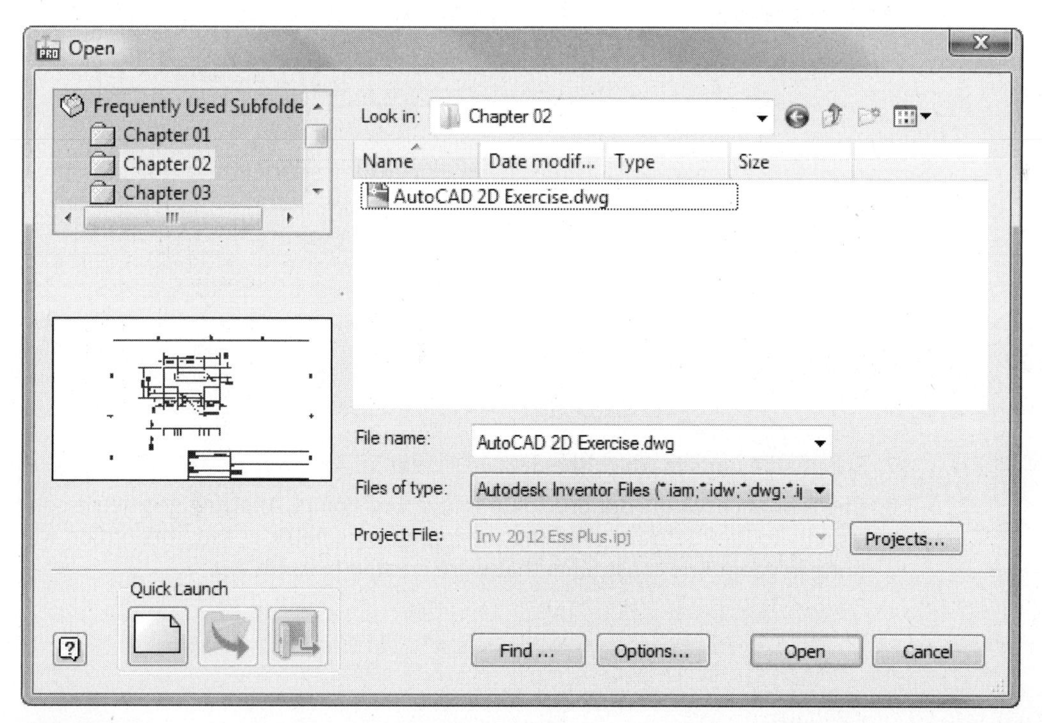

FIGURE 2.78

4. In the Selective import section, uncheck the layer names you do not want to see the data from.

5. Click the Open button on the bottom of the dialog box. To select specific objects to insert, uncheck the All option, and then select the desired data in the preview window. In the preview window you can zoom and pan as needed.
6. You can change the background color of the preview image by clicking the black or white icon at the top of the dialog box. Import objects from Model Space or from a layout within the DWG file by clicking the different tabs at the bottom of the screen. The names of the tabs are identical to the tab names in the AutoCAD file. Click the Next button to go to the next step.

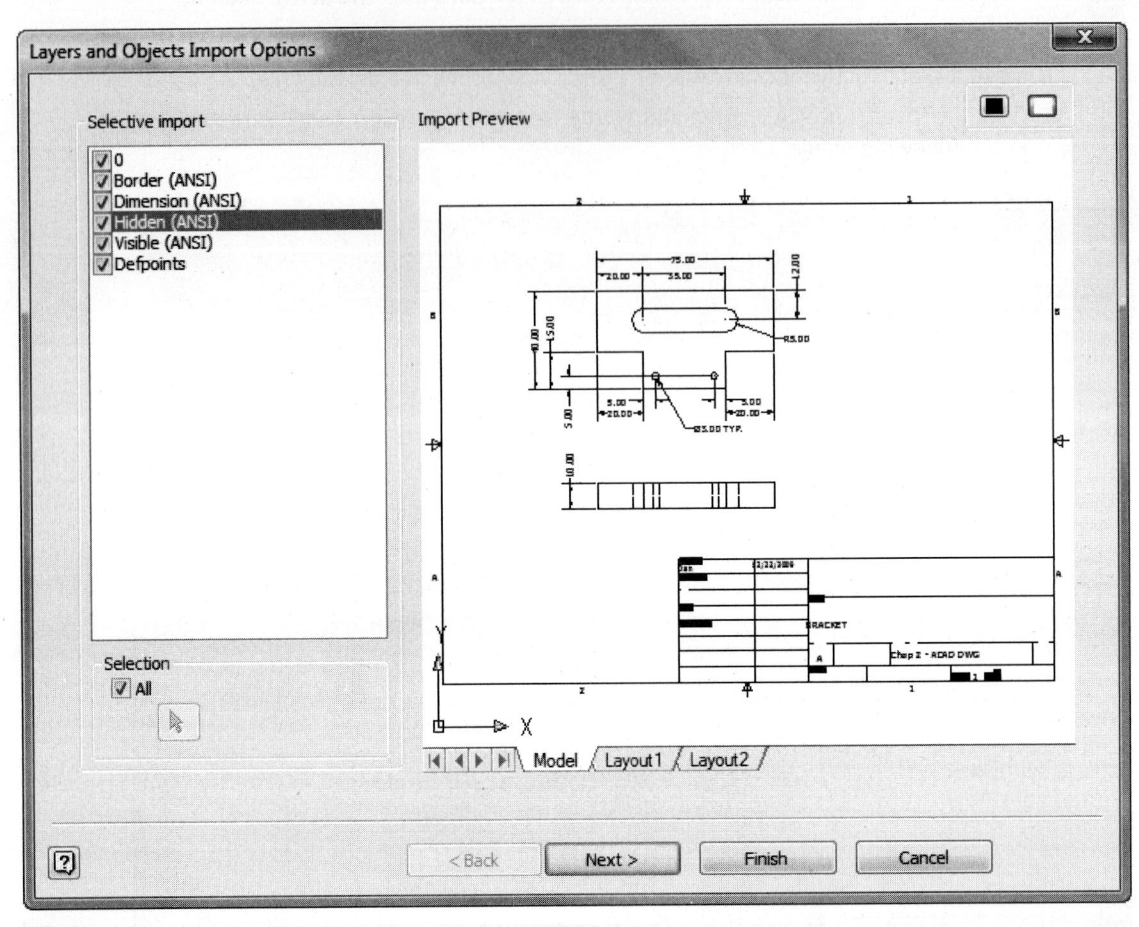

FIGURE 2.79

7. Specify the units in which the data was created.
8. Check or uncheck the options to Constrain End Points. Applying geometric constraints to the sketch or checking the Import parametric constraints option will bring in geometric constraints that were created in AutoCAD.

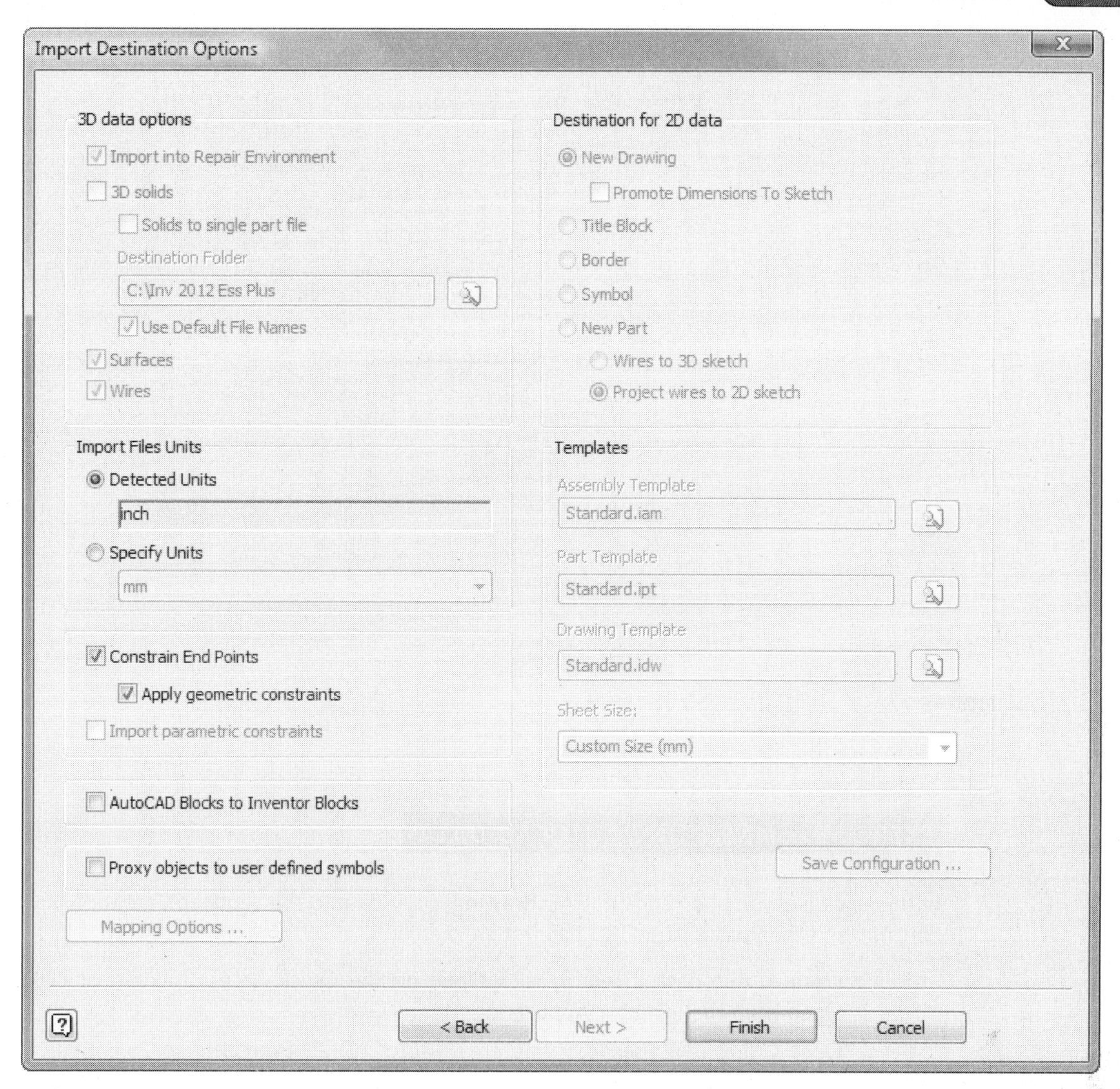

FIGURE 2.80

9. To complete the operation, click Finish.

OPEN OTHER FILE TYPES

Autodesk Inventor can also open parts and assemblies exported from other CAD systems. Autodesk Inventor models created from these formats are base solids or surface models, and no feature histories or assembly constraints are generated when you import a file in any of these formats. You can add features to imported parts, edit the base solids using Autodesk Inventor's solids editing commands or Inventor Fusion, and add assembly constraints to the imported components. To open file types such as DWF (markup files), DXF, Alias, Catia, IDF Board File IGES, JT, Parasolids, PRO/E, SAT, STEP, SolidWorks, and Unigraphics NX, click Inventor Application Menu > Open or click Open on the Quick Access toolbar.

In the Open dialog box, click the desired file format in the Files of type list. See the help system for more information about the different file types.

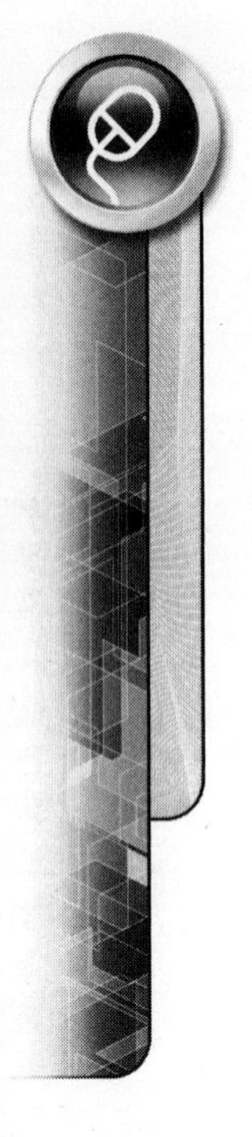

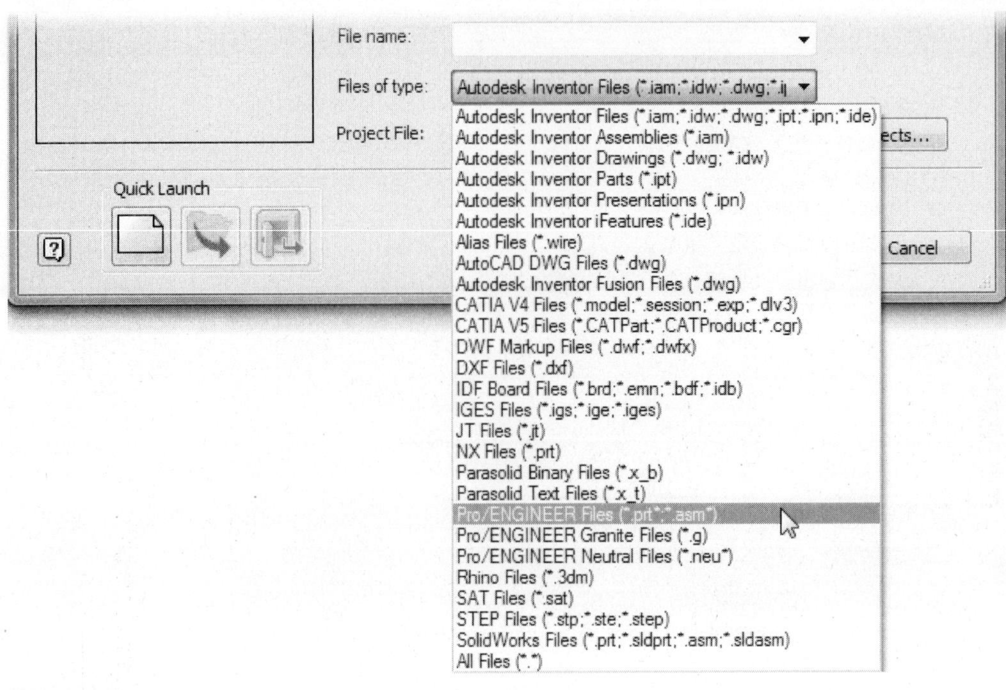

FIGURE 2.81

EXERCISE 2-5: INSERTING AUTOCAD DATA

In this exercise, you open an AutoCAD drawing, copy data to the clipboard, create a new part file, and then paste AutoCAD data into a sketch.

1. From in Autodesk Inventor, click Open on the Quick Access toolbar, or click Open from the Inventor Application Menu.
2. Open C:\Inv 2012 Ess Plus\Chapter 02\AutoCAD 2D Exercise.dwg.

TIP

Click the Chapter 02 subfolder from the Frequently Used Subfolder area and then click the file in the file area.

3. The DWG file will open in a new window; use the window selection technique to select the geometry and dimensions as shown in the following image on the left.
4. Right-click, and click Copy from the menu as shown in the following image on the right.

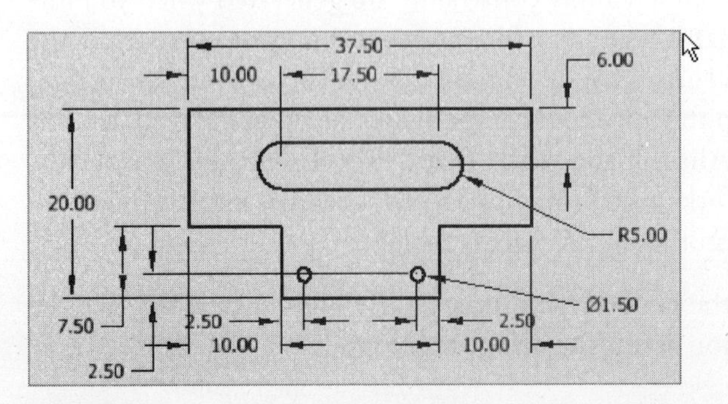

FIGURE 2.82

5. Click the New command, click the English tab, and then double-click *(in).ipt*.
6. Move the cursor into the graphics window (do not click), right-click, and click Paste from the menu.
7. Right-click, and click Paste Options as shown in the following image on the left.
8. In the Paste Options dialog box, if inch is not displayed as the default unit click Specify Units and verify that inch is selected. If needed, check both Constrain End Points and Apply geometric constraints, and then click OK. If the AutoCAD file had constraints applied, you could have used the Import parametric constraints option.

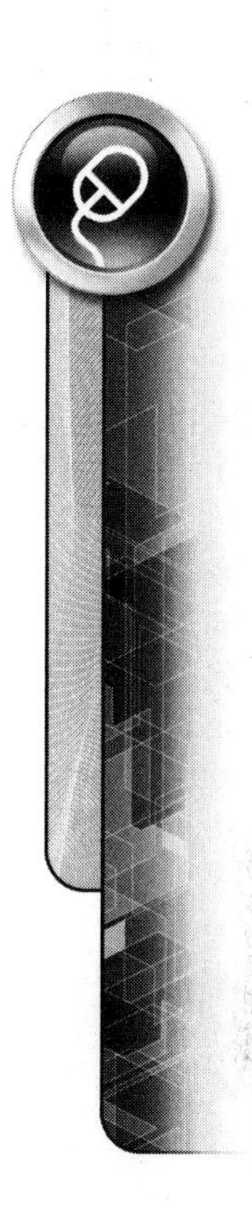

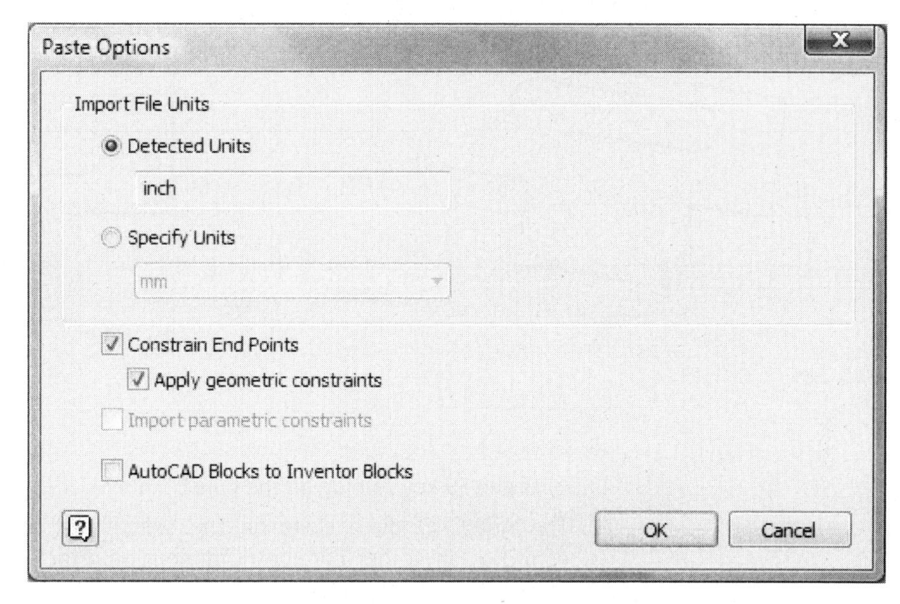

FIGURE 2.83

9. To paste the data, click a point in the graphics window.
10. Zoom out to see the geometry, click the Zoom All command from the Navigation bar.
11. Apply a horizontal constraint between the center points of the two circles.
12. Apply a collinear constraint between the two middle horizontal lines labeled (1) and (2) as shown in the following image on the left.
13. Press the Esc key twice to cancel the command.
14. The sketch is free to move. To constrain the sketch, drag the lower-left corner of the sketch labeled (3) in the following image to the center point.

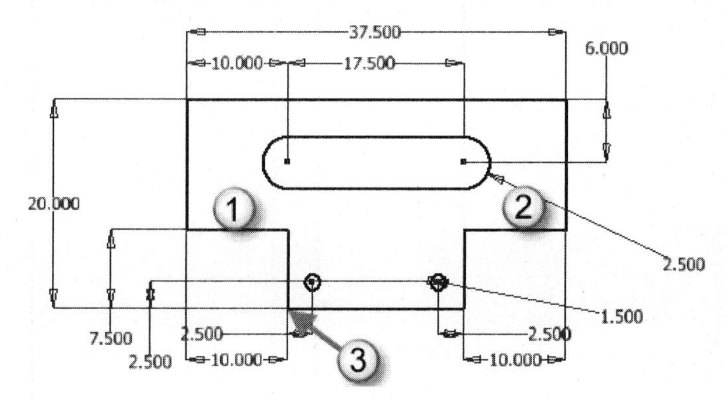

FIGURE 2.84

15. On the lower-right corner of Autodesk Inventor, the text should state that the number of constraints needed to constrain the sketch is 1.
16. Drag up the right-side endpoint on the lower horizontal line; the sketch will be rotated slightly as shown in the following image on the left.
17. Apply a horizontal constraint to the lower line, and this will fully constrain the sketch as shown in the following image on the right. The dimensions can be repositioned as needed.

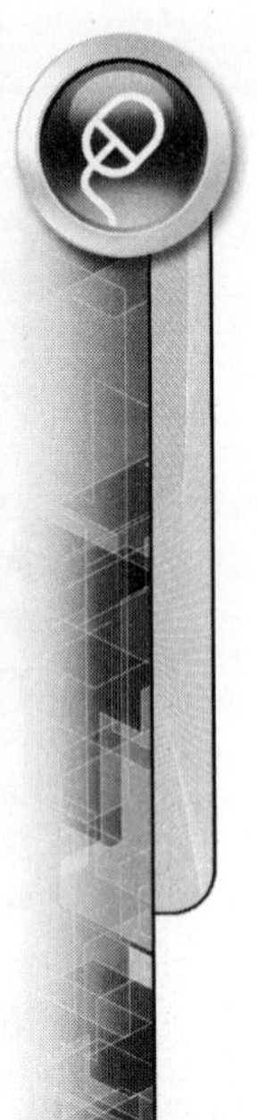

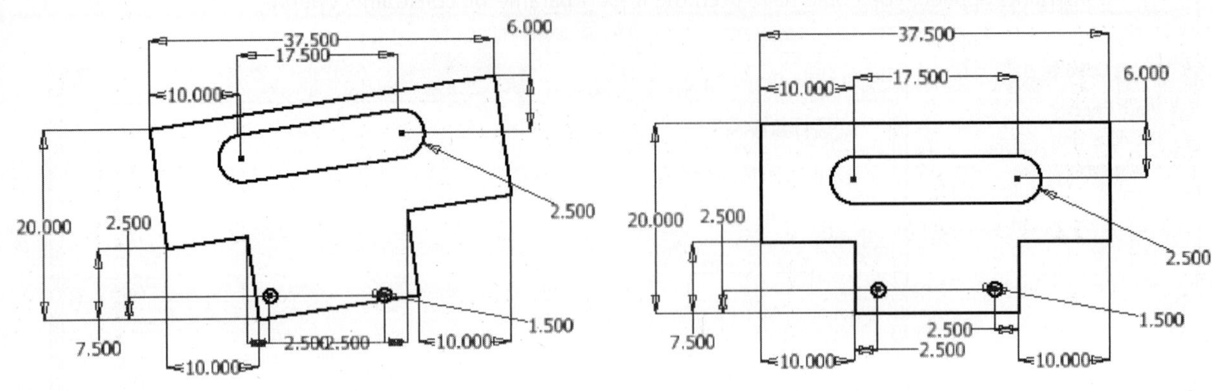

FIGURE 2.85

18. Click the F8 key to see all the constraints.
19. Click the F9 key to hide all the constraints.
20. The AutoCAD dimensions on the sketch are now parametric and can be edited as any other parametric dimension. Practice editing the values of the dimensions.
21. Close the file. Do not save changes. End of exercise.

APPLYING YOUR SKILLS

Skill Exercise 2-1

In this exercise, you create a sketch and then add geometric and dimensional constraints to control the size and shape of the sketch. Start a new part file based on the *Standard (in).ipt*, and create the fully constrained sketch as shown in the following image. Assume that the top and bottom horizontal lines are collinear, the center points of the arcs are aligned vertically, and the sketch is symmetric about the left and right sides. One of the bottom angled lines should be coincident with the center point of the lower arc (if the arc is drawn via the line command, the center point of the arc will automatically be coincident with the line it was drawn from). When done, close the file and do not save the changes.

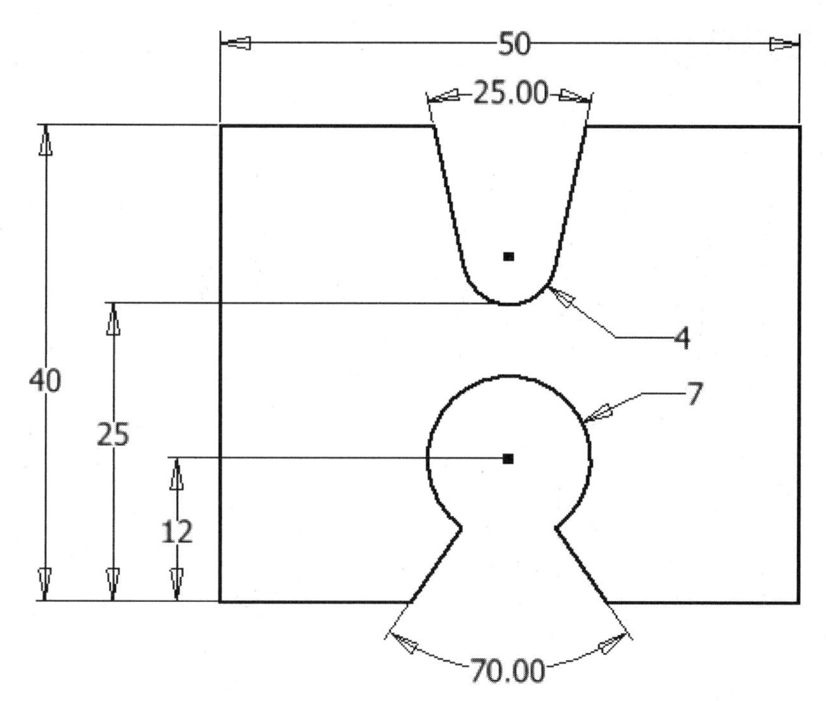

FIGURE 2.86

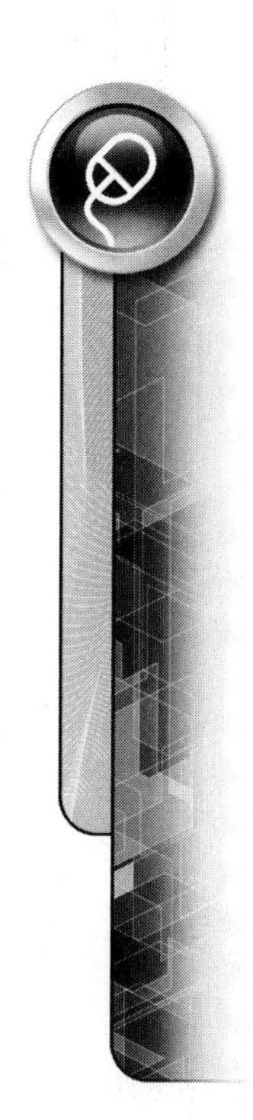

Skill Exercise 2-2

In this exercise, you create a sketch with linear and arc shapes, and then add geometric and dimensional constraints to fully constrain the sketch. Start a new part file based on the ***Standard (in).ipt***, and create the fully constrained sketch as shown in the following image. First create the two outer circles and align their center points horizontally. Then create the two lines, trim the two circles, and place a vertical constraint between the line endpoints on both ends. The dimension on the arcs define the arc's radius. When done, close the file and do not save the changes.

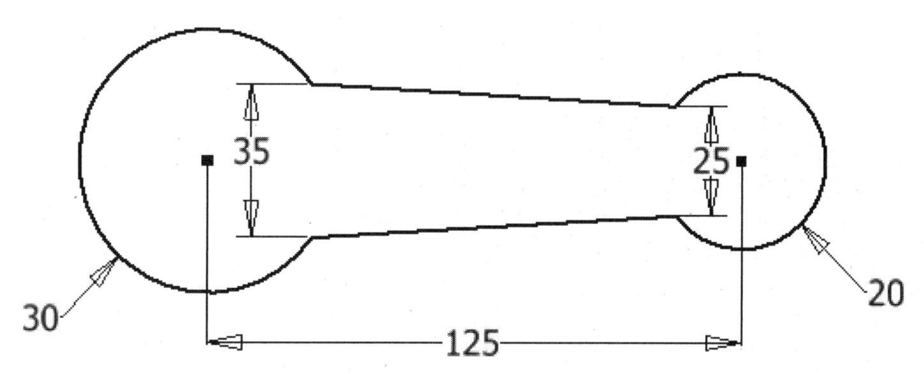

FIGURE 2.87

CHECKING YOUR SKILLS

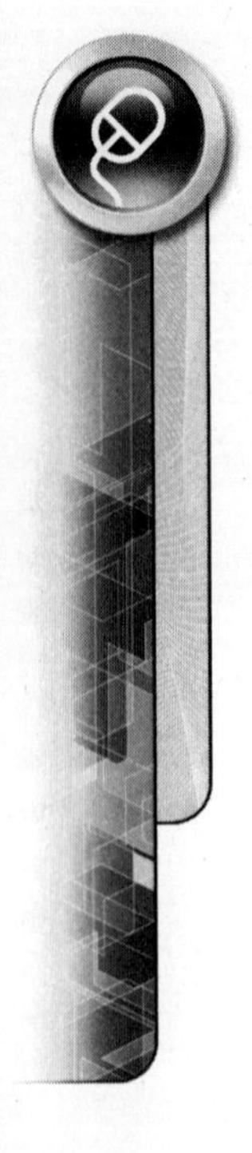

Use these questions to test your knowledge of the material in this chapter.

1. True ___ False___ When you sketch, constraints are not applied to the sketch by default.
2. True ___ False ___ When you sketch and a point is inferred, a constraint is applied to represent that relationship.
3. True ___ False ___ A sketch does not need to be fully constrained.
4. True ___ False ___ When working on an mm part, you cannot use English (inch) units.
5. True ___ False ___ After a sketch is constrained fully, you cannot change a dimension's value.
6. True ___ False ___ A driven dimension is another name for a parametric dimension.
7. True ___ False ___ Dimensions placed dynamically are not parametric.
8. True ___ False ___ You can import only 2D AutoCAD data into Autodesk Inventor.
9. Explain how to draw an arc while in the Line command.
10. Explain how to remove a geometric constraint from a sketch.
11. Explain how to change a vertical dimension to an aligned dimension while you create it.
12. Explain how to create a dimension between two quadrants of two arcs.
13. True ___ False ___ AutoCAD needs to be installed to insert AutoCAD geometry.
14. True ___ False ___ When a sketch is extruded that contains construction geometry, the construction geometry is deleted.
15. Explain how to change the unit type in a part file.

CHAPTER 3

Creating and Editing Sketched Features

INTRODUCTION

After you have drawn, constrained, and dimensioned a sketch, your next step is to turn the sketch into a 3D part. This chapter takes you through the process to create and edit sketched features.

OBJECTIVES

After completing this chapter, you will be able to perform the following:

- Understand what a feature is
- Understand Autodesk Inventor's browser
- Use direct manipulation techniques to create and edit a part
- Extrude a sketch into a part
- Revolve a sketch into a part
- Edit features of a part
- Edit the sketch of a feature
- Make an active sketch on a plane
- Create sketched features using one of three operations: cut, join, or intersect
- Project edges of a part

UNDERSTANDING FEATURES

After creating, constraining, and dimensioning a sketch, the next step in creating a model is to turn the sketch into a 3D feature. The first sketch of a part that is used to create a 3D feature is referred to as the base feature. In addition to the base feature, you can create sketched features in which you draw a sketch on a planar face or work plane, and you can either add or subtract material to or from existing features in a part. Use the Extrude, Revolve, Sweep, or Loft commands to create sketched features in a part. You can also create placed features such as fillets, chamfers, and holes by applying them to features that have been created. Placed features will be covered in Chapter 4. Features are the building blocks in creating parts.

A plate with a hole in it, for example, would have a base feature representing the plate and a hole feature representing the hole. As features are added to the part, they appear in the browser, and the history of the part or assembly, that is, the order in which the features are created or the parts are assembled, is shown. Features can be edited, deleted from, or reordered in the part as required.

Consumed and Unconsumed Sketches

You can use any sketch as a profile in feature creation. A sketch that has not yet been used in a feature is called an unconsumed sketch. When you turn a 2D sketched profile into a 3D feature, the feature consumes the sketch. The following image shows an unconsumed sketch in the browser on the left and a consumed sketch in the browser on the right. If desired, you can display more information in the browser about the features by clicking on the filter icon in the browser and click Show extended Names.

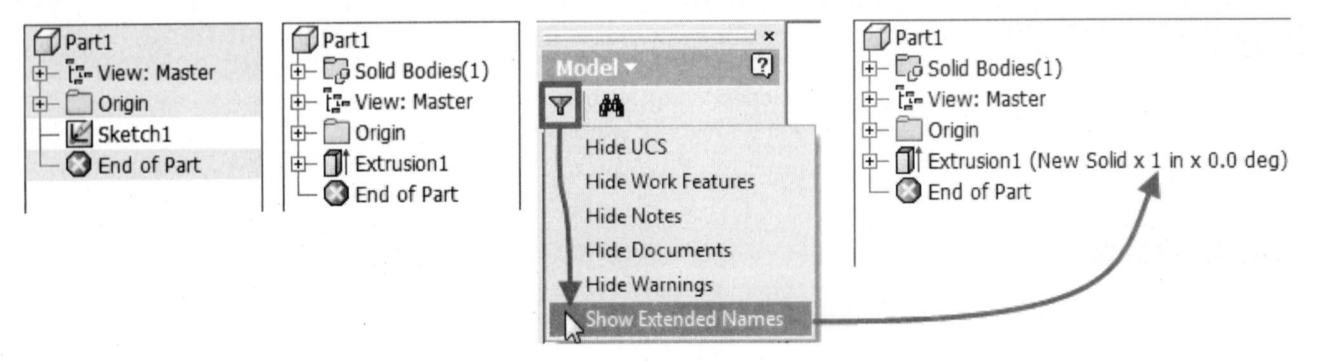

FIGURE 3.1

Although a consumed sketch is not visible as you view the 3D feature, you may need to access sketches and change their geometric or dimensional constraints in order to modify their associated features. A consumed sketch can be accessed from the browser by right-clicking and selecting Edit Sketch from the menu or by using the direct-manipulation mini-toolbar. The editing process will be covered later in this chapter. You may also access the sketch by starting the Sketch command and selecting the sketch from the browser. The following image on the left shows the unconsumed sketch, and the image on the right shows the extruded solid that consumes the sketch.

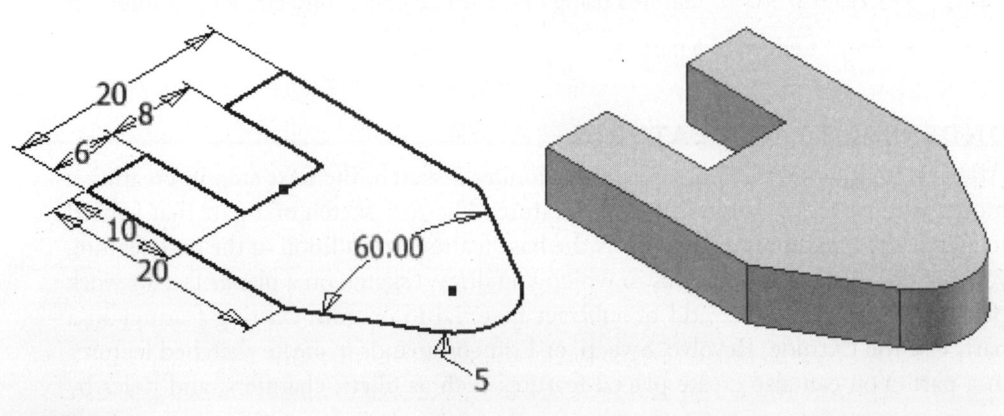

FIGURE 3.2

UNDERSTANDING THE BROWSER

The Autodesk Inventor browser, by default, is docked along the left side of the screen and displays the history of the file. In the browser you can create, edit, rename, copy, delete, and reorder features or parts. You can expand or collapse the browser to display the history of the features (the order in which the features were created) by clicking the + and – on the left side of the part or feature name in the browser. An alternate method of expanding the browser is to place your cursor on top of a feature icon but not click. The item in the browser automatically expands. To expand all the features, move the cursor into a blank area in the browser, right-click, and click Expand All from the menu.

The following image shows a browser with features of the part expanded.

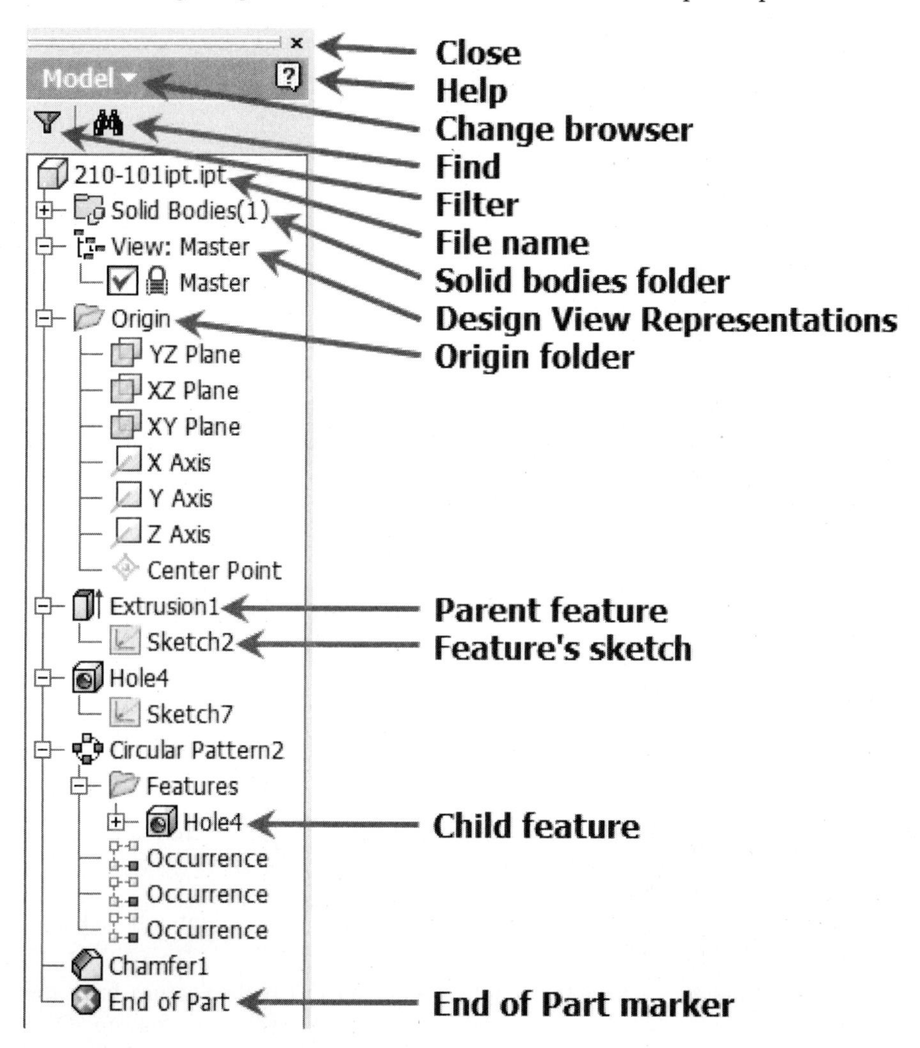

FIGURE 3.3

As parts grow in complexity, so will the information found in the browser. Dependent features are indented to show that they relate to the item listed above it. This is referred to as a *parent-child relationship*. If a hole is created in an extruded rectangle, for example, and the extrusion is deleted, the hole will also be deleted.

Each feature in the browser is given a default name. The first extrusion, for example, will be named Extrusion1, and the number in the name will sequence as you add similar features. The browser can also help you to locate features in the graphics area. To highlight a feature in the graphics window, simply move your cursor over the feature name in the browser.

To zoom in on a selected feature, right-click on the feature's name in the browser and select Find in Window on the menu or press the END key on your keyboard. The browser itself functions similarly to a toolbar except that you can resize it while it is docked. To close the browser, click the X in its upper-right corner. If the browser is not visible on the screen, you can display it by clicking the View tab > Windows panel > User Interface > Browser as shown in the following image.

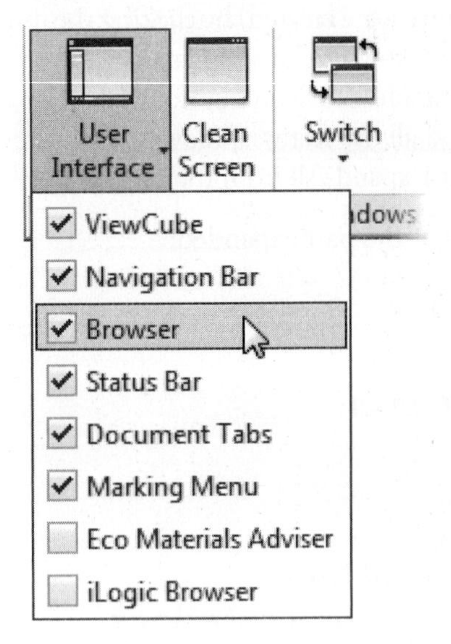

FIGURE 3.4

Specific functionality of the browser will be covered throughout this book in the pertinent sections. A basic rule is to either right-click or double-click the feature's name or icon to edit or perform a function on the feature.

SWITCHING ENVIRONMENTS

Up to this point, you have been working in the sketch environment where the work is done in 2D. The next step is to turn the sketch into a feature. To do so, you need to exit the *sketch environment* and enter the *part environment*. A number of methods can be used to accomplish this transition:

- Click the Finish Sketch command on the right side of the Sketch tab > Exit panel as shown in the following image on the left.
- In a blank area in the graphics window right-click and click Finish Sketch from the menu as shown in the middle of the following image.
- Press the S key on the keyboard.
- Click the Model tab as shown in the following image on the right and click on a command.
- You can also right-click in the graphics area, select Create Feature from the menu, and then click the command you need to create the feature. Only the commands that are applicable to the current situation will be available.
- Enter a shortcut key to initiate one of the feature commands.

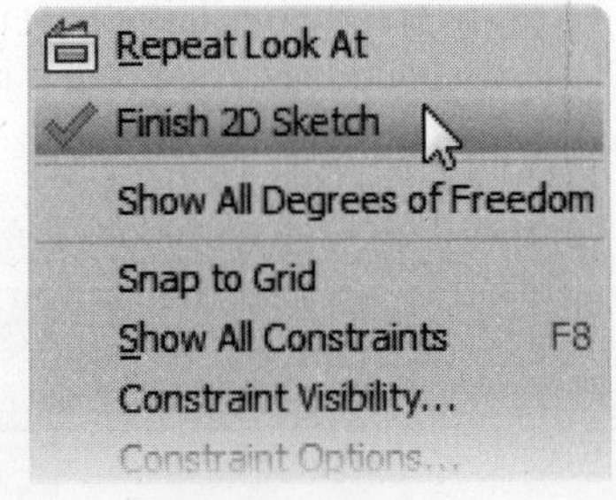

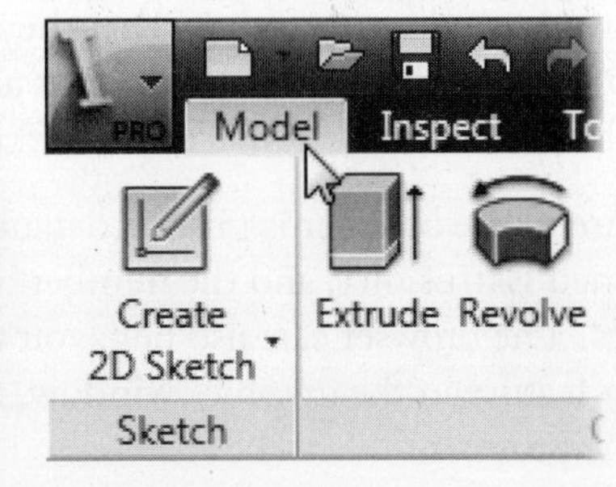

FIGURE 3.5

MODEL COMMANDS

When you exit the sketch environment, the Model Ribbon is active. The following image shows the Model tab. Many of these commands will be covered throughout this book.

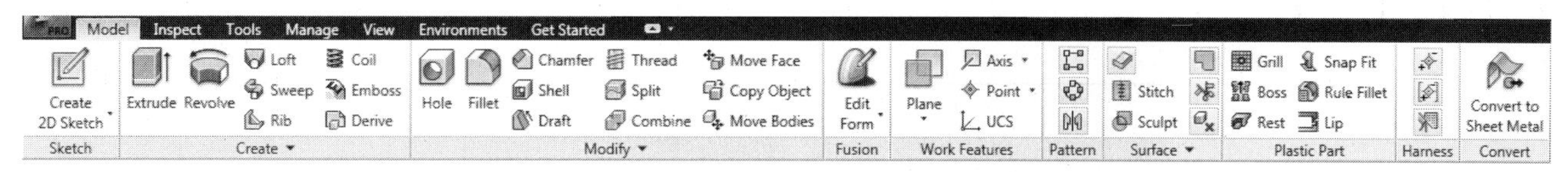

FIGURE 3.6

DIRECT MANIPULATION

Direct manipulation allows you to start common commands and operations by clicking directly on the geometry in the graphics window. The direct manipulation tools appear as mini-toolbars or in-canvas buttons. Each direct manipulation option is explained next.

Mini-Toolbars

While not in a command you have different options depending upon what geometry is selected. After selecting the geometry click on the desired option. The different options will be explained throughout the book as they pertain to the topic.

Sketch

While not in a command, not in the sketch environment, and you select a sketch, the mini-toolbar appears and provides you the option of performing Extrude, Revolve, Hole, or Edit Sketch operations.

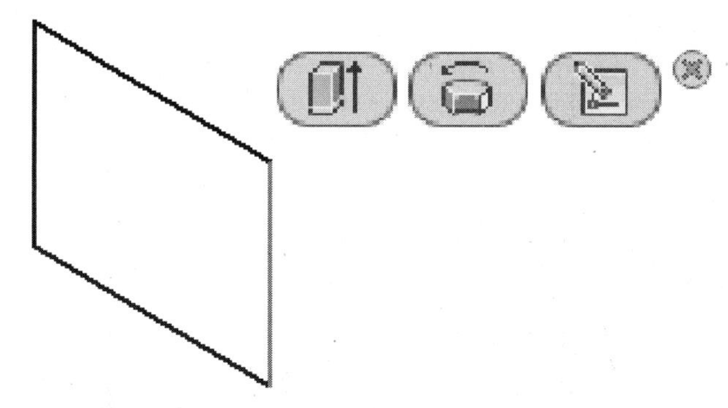

FIGURE 3.7

Face

While not in a command and you select a face, a mini-toolbar appears that allows you to quickly create a new sketch, edit an existing sketch, or edit a feature.

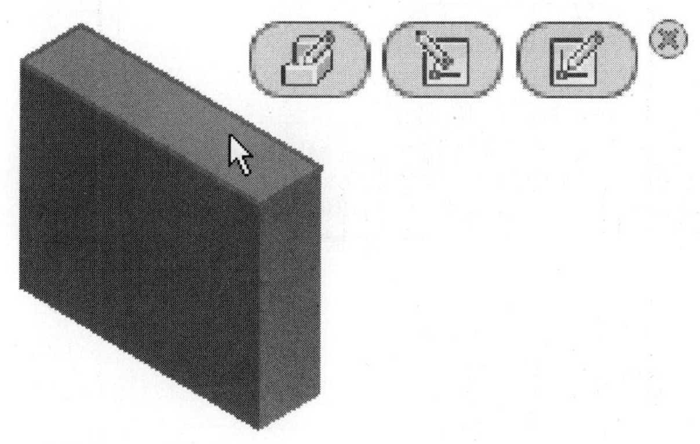

FIGURE 3.8

Edge

While not in a command and you select an edge, the mini-toolbar appears and provides you the option of performing a fillet or chamfer operation on the selected edge.

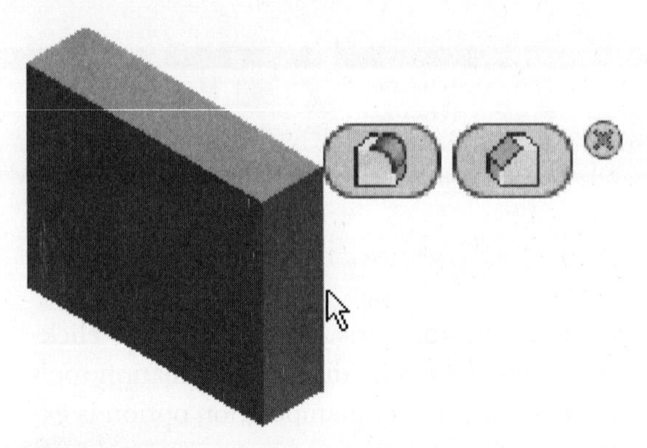

FIGURE 3.9

Mini-Toolbar–Command Options

After starting selected commands, the mini-toolbar will display command options in the graphics screen next to a selected object. The min-toolbar options are presented via buttons and selection tags. The in-canvas buttons and selection tags display command options and prompts such as profile, face, and axis. Min-toolbar display is enabled for Extrude, Revolve, Fillet, Chamfer, Hole and Face Draft commands. The following image on the left shows the min-toolbar buttons and the selection tag (profile) when extruding the first profile. You can move the mini-toolbar by clicking and dragging the vertical obround button in the upper left corner of the mini-toolbar. When the Pin Mini-Toolbar Position option is unchecked the mini-toolbars position will move as the arrow on the face is dragged. When checked the mini-toolbar will stay in its current position when the arrow on the face is dragged. The Fade option in the mini-toolbar will fade when the cursor moves away from the mini-toolbar. The option to pin and fade the mini-toolbar is controlled by clicking on the bottom-right button on the mini-toolbar as shown in the following image on the right. The min-toolbar display options are covered throughout the book where the command is covered.

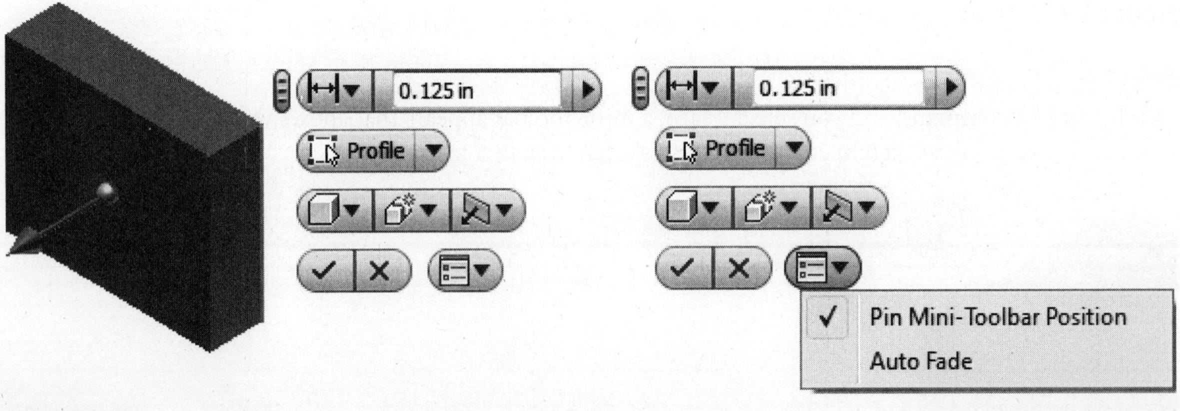

FIGURE 3.10

EXTRUDE A SKETCH

The most common method for creating a feature is to extrude a sketch and give it depth along the Z-axis. Before extruding, it is helpful to view the part in an isometric view. When you exit the first sketch the viewpoint will automatically change to the home view. Autodesk Inventor previews the extrusion depth and direction in the graphics window.

To extrude a sketch, click the Extrude command on the Create panel bar as shown in the following image on the left, or while not in a sketch click on the sketch and the mini-toolbar will appear, click the Create Extrude button as shown in the following image in the middle, or press the shortcut key E. Alternately, you can right-click in the graphics screen and select Extrude on the marking menu as shown in the following image on the right.

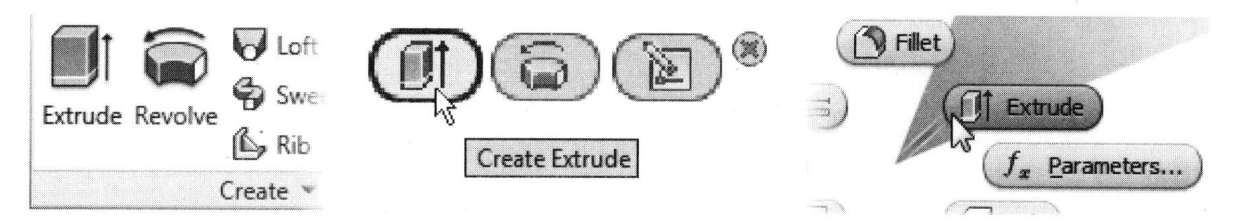

FIGURE 3.11

After starting the Extrude command, the in-canvas display options will appear in the graphics window and the Extrude dialog box will appear. The following image on the left shows the in-canvas extrude options that are available when extruding the first sketch and the image on the right shows the operation button that is available when creating additional extrude features. You can either enter data in the dialog box or via the in-canvas display. For the mini-toolbar buttons, click the down arrow next to each button to see more options and click and drag the arrow to change the distance or taper. The min-toolbar buttons will change depending upon what options are selected. If selection tags have a red arrow that command option needs to be satisfied before the command can be completed.

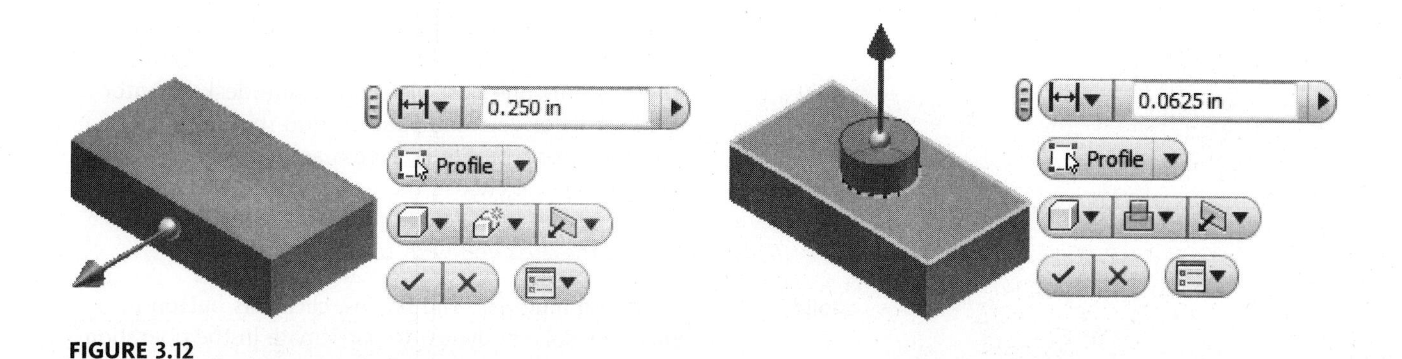

FIGURE 3.12

After you start the command, the Extrude dialog box also appears as shown in the following image. The Extrude dialog box has two tabs: Shape and More. When you make changes in the dialog box, the shape of the sketch will change in the graphics area to represent these values and options. When you have entered the values and the options you need, click the OK button to create the extruded feature. The extrude options will be explained in the next sections.

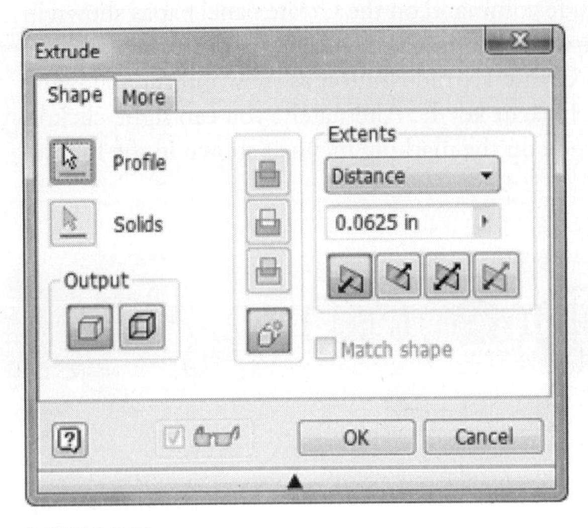

FIGURE 3.13

NOTE

When an arrow is red in a dialog box or in a selection tag in the mini-toolbar area that prompts for profiles, faces, and axes, the color indicates that Autodesk Inventor needs input from you to complete that function.

Shape

The Shape tab gives you options to specify the profile to use, operation, extents, and output type. The options are described below.

	Profile	Click this button to choose the sketch that you wish to extrude. If there are multiple closed profiles, you will need to select which sketch area you want to extrude. If there is only one possible profile, Autodesk Inventor will select it for you, and you can skip this step. If you select the wrong profile or sketch area, click the Profile button again and choose the desired profile or sketch area. To remove a selected profile, hold the CTRL key and click the area you wish to remove.
	Solid	If there are multiple solid bodies, click this button to choose the solid body(ies) to participate in the operation.

Operation

This is the unlabeled middle column of buttons. If this is the first sketch that you create a solid from, it is referred to as a base feature, and only the top button is available. The operation defaults to Join, which is the top button. Once the base feature has been established, you can extrude a sketch, adding or removing material from the part by using the Join or Cut options, or you can retain the common volume between the existing part and the newly defined extrude operation using the Intersect option.

	Join	Adds material to the part.
	Cut	Removes material from the part.
	Intersect	Removes material, keeping what is common to the existing part feature(s) and the new feature.
	New Solid	Creates a new solid body. The first solid feature created uses this option by default. Select to create a new body in a part file with an existing solid body.

Extents

The Extents option determines how the extruded sketch will be terminated. There are five options from which to choose as shown in the following image, but like the operation section, some options are not available until a base feature exists.

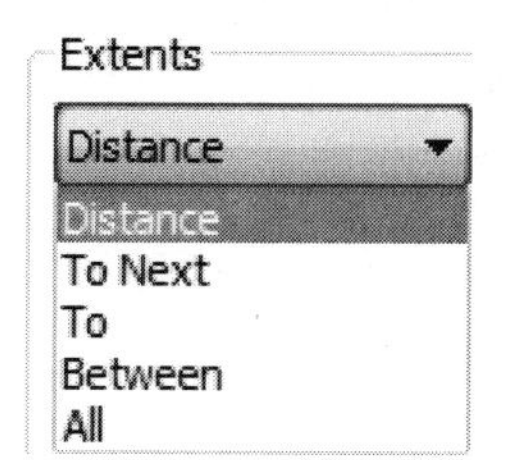

FIGURE 3.14

Distance

This option determines that the sketch will be extruded a specified distance.

To Next

This option determines that the sketch will be extruded until it reaches a plane or face. The sketch must be fully enclosed in the area to which it is projecting; if it is not fully enclosed, use the To termination with the Extend to Surface option. Click the Direction button to determine the extrusion direction.

To

This option determines that the sketch will be extruded until it reaches a selected face or plane. To select a point (midpoint or endpoint), plane, or face to end the extrusion, click the Select Surface to end the feature creation button, as shown in the following image, and then click a face or plane at which the extrusion should terminate.

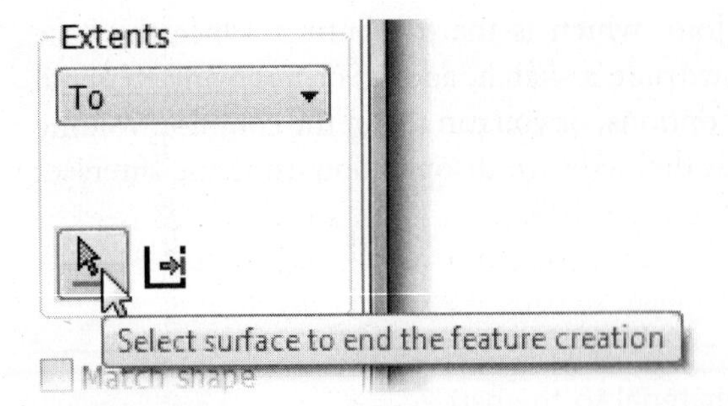

FIGURE 3.15

Between

This option determines that the extrusion will start at a selected plane or face and stop at another plane or face. Click the Select surface to start feature creation button as shown in the following image on the left, and then click the face or plane where the extrusion will start. Then click the Select surface to end the feature creation button, as shown in the following image on the right and then click the face or plane where the extrusion will terminate.

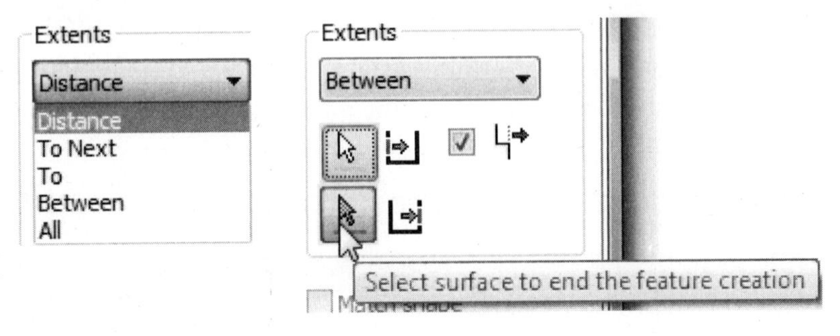

FIGURE 3.16

All

This option determines that the sketch will be extruded all the way through the part in one or both directions.

Distance Value

If the extent is set to Distance, you can either enter a value in the dialog box or in the in-canvas display where the sketch will be extruded or click and drag the arrow in the in-canvas area. Another option is to click the arrow to the right where the distance value is displayed, from here you have an option to click two points to determine a value, display the dimensions of previously created features to select from, or select from the list of the recent values used. After a value is determined, a preview image appears in the graphics area to show how the extrusion will look.

A preview image appears in the graphics area, and the corresponding value appears in the distance area in the dialog box and in the in-canvas area.

If values and units appear in red when you enter them, the defined distance is incorrect and should be corrected. For example, if you entered too many decimal places (e.g., 2.12.5) or an incorrect unit for the dimension value, the value will appear in red. You will need to correct the error before the extrusion can be created.

Direction

There are four buttons from which to choose for determining the direction. Choose from the first two to flip the extrusion direction, click the third button (symmetric) to have the extrusion go equal distances in the negative and positive directions, and click the asymmetric option to define different distances for the negative and positive direction. With the midplane option if the extrusion distance is 2 inches, for example, the extrusion will go 1 inch in both the negative and positive Z directions.

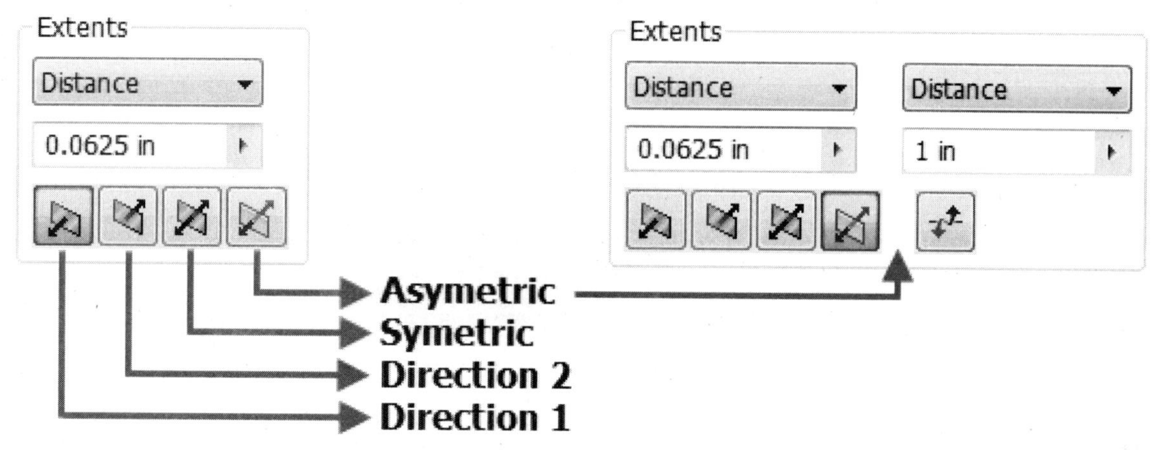

FIGURE 3.17

Output

Two options are available to define the type of output that the Extrude command will generate:

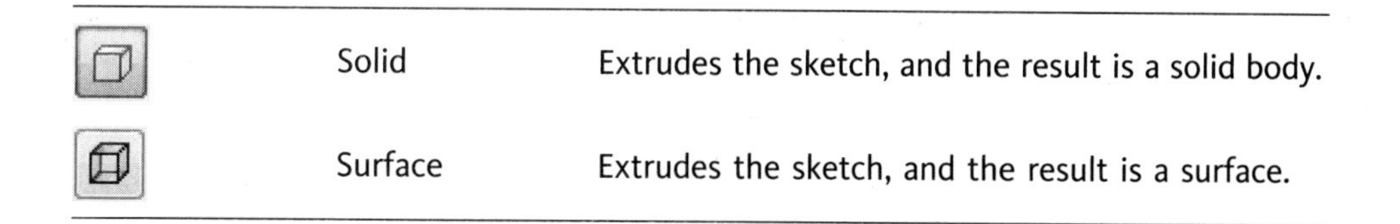

	Solid	Extrudes the sketch, and the result is a solid body.
	Surface	Extrudes the sketch, and the result is a surface.

Match Shape

You can use the match shape option when working with an open profile that you want to extrude. The edges of the open profile are extended until they intersect geometry of the model. This provides a "flood-fill"–type effect for the extrude feature.

When you convert a sketch into a part or feature, the dimensions on the sketch are consumed (they disappear). When you edit the feature, the dimensions reappear. The dimensions can also be displayed when drawing views are made. For more information on editing parts or features, see the "Editing 3D Parts" section later in this chapter.

More

The More tab, as shown in the following image, contains additional options to refine the feature being created:

Alternate Solution

Alternate Solution terminates the feature on the most distant solution for the selected surface. An example is shown in the following image.

Minimum Solution

Minimum Solution option when checked terminates the feature on the first possible solution for the selected surface. An example is shown in the following image.

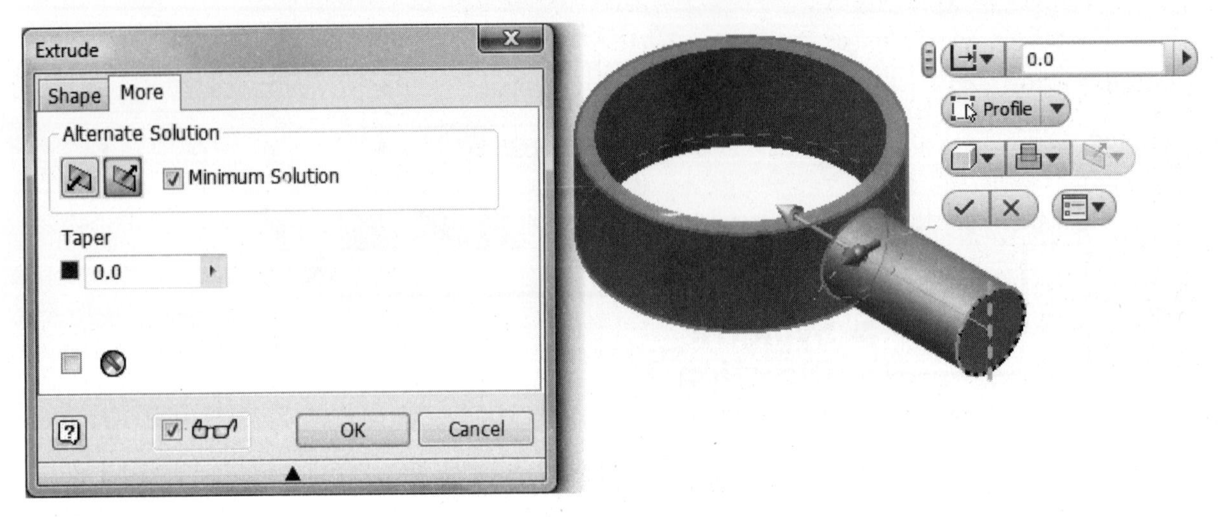

FIGURE 3.18

Taper

Taper extrudes the sketch and applies a taper angle to the feature. To extend the taper angle out from the part, give the taper angle a positive or negative number. This increases or decreases the volume of the resulting extruded feature. You can also click on the sphere in the mini-toolbar area and an arrow will appear as shown in the following image. Drag the arrow or enter a value.

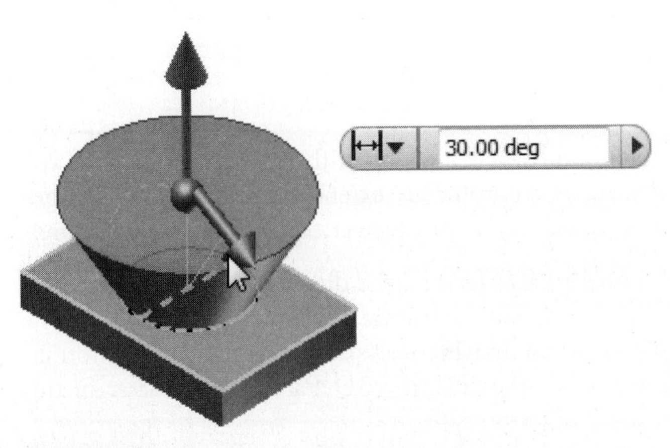

FIGURE 3.19

Infer iMates

Check this box to automatically create an iMate on a full circular edge. Autodesk Inventor attempts to place the iMate on the closed loop most likely to be used.

EXERCISE 3-1: EXTRUDING A SKETCH

In this exercise, you will create a base feature by extruding an existing profile. You will examine the direction options available in the Extrude dialog box.

1. Open *ESS_E03_01.ipt* from the Chapter 03 subfolder.
2. From the Model tab > Create panel click the Extrude command or click on the sketch and in the mini-toolbar click the Create Extrude button. Since there is only one closed profile, the profile is automatically selected.
3. In the Extrude dialog box, set the Distance to **.5 in**.
4. In the graphics window select the arrow on the extrusion in the graphics window and drag the arrow until a distance of **.75** in is displayed in the value field of the Extrude dialog box and in mini-toolbar display as shown in the following image, and then release the mouse button.

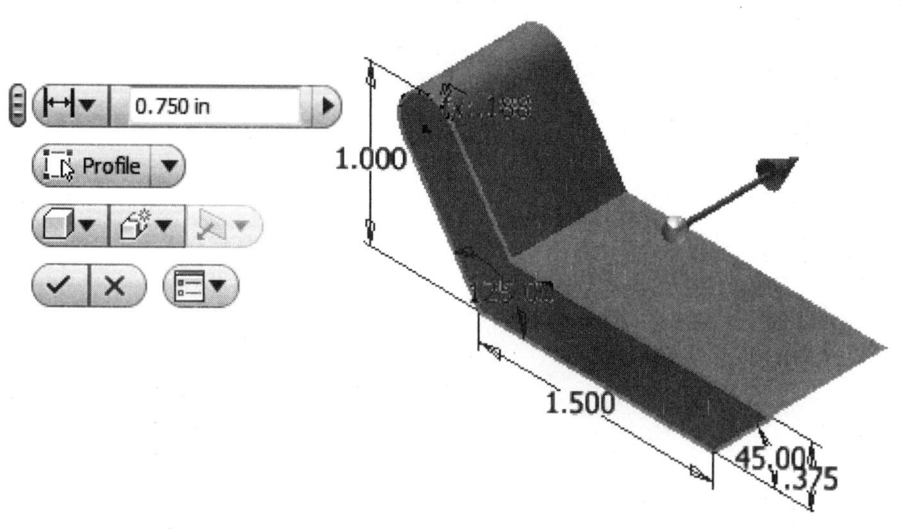

FIGURE 3.20

5. Click the More tab in the Extrude dialog box or click the sphere at the start of the arrow in the graphics window.
6. Adjust the value of the Taper to **20**, by entering 20 in the Extrude dialog box. In the graphics window you could also have applied taper by clicking on the sphere on the back of the extrusion and drag the arrow.

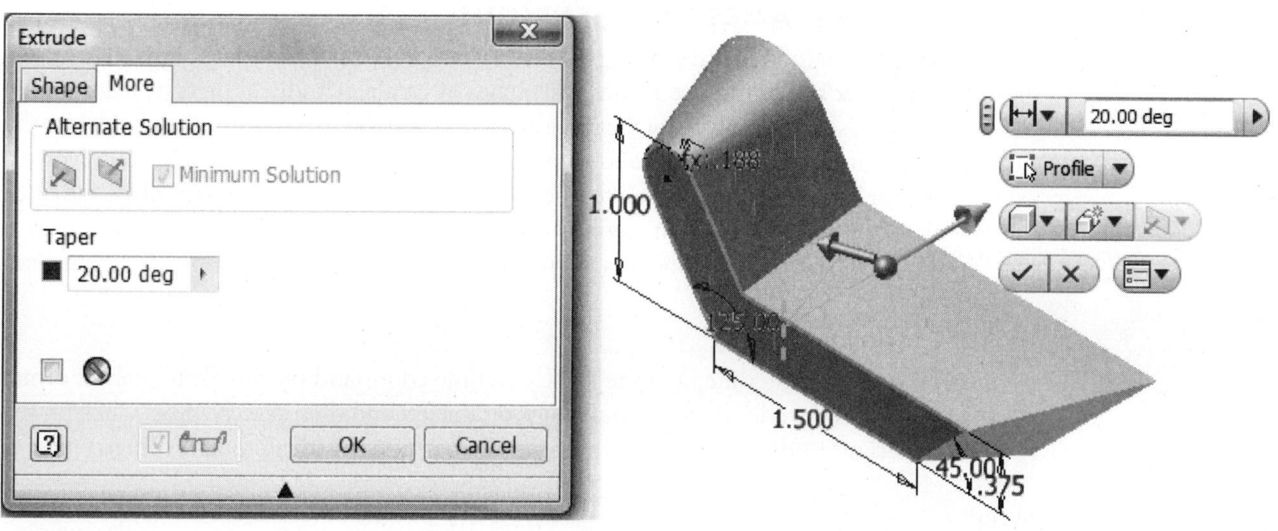

FIGURE 3.21

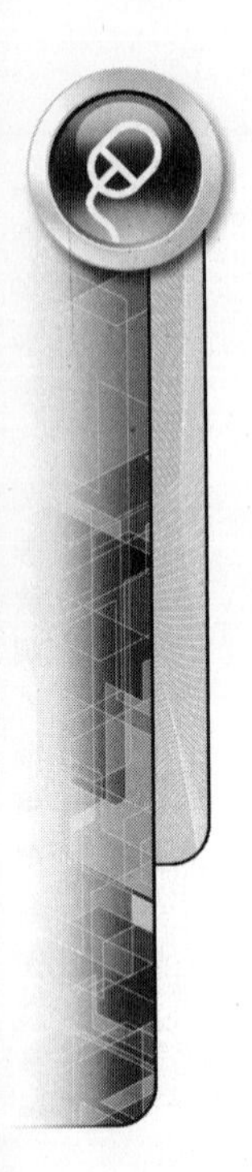

7. In the graphics window click and drag the active arrow (for the taper) to adjust the taper to 15° and then change it back to 10°.
8. Click the Shape tab in the Extrude dialog box.
9. Change the direction of the extrusion to go in the negative Y direction by selecting the Direction 2 button (second from the left button) on the direction area in the dialog box. This will flip the direction of the extrusion.
10. In the dialog box change the direction to symmetric (go evenly in both directions) by selecting the second from the right button.
11. In the min-toolbar display area change the extent to asymmetric as shown in the following image on the left.
12. In the graphics window click and drag the arrows to different values as shown in the following image on the right.

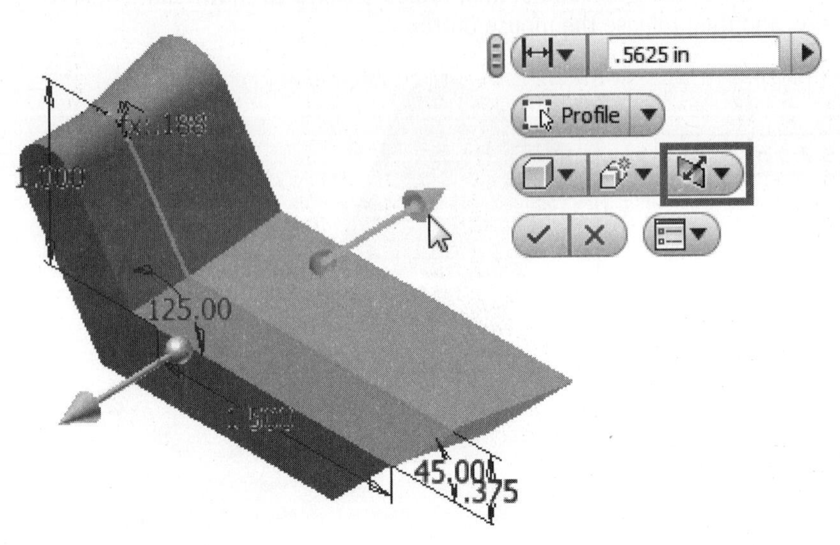

FIGURE 3.22

13. Practice changing the values and directions to see the results. When done, click OK or click the green check mark in the mini-toolbar display area and the extrusion has been created. Later in this chapter you will learn how to edit features.
14. Close the file. Do not save changes. End of exercise.

LINEAR DIAMETER DIMENSIONS

Another method for creating a part is to revolve a sketch around a straight edge or axis (centerline). You can use revolved sketches to create cylindrical parts or features. To revolve a sketch, you follow the same steps that you did to extrude a sketch. Create the sketch and add constraints and dimensions. When revolving a sketch, you revolve a quarter section of the completed part. To define the sketch you can place radial or diameter dimensions that define the sketch. To place diameter dimensions on a sketch you use a centerline or a normal line to define the linear diameter dimension.

Centerline

To create a centerline, activate the Centerline command on the Sketch tab >Format panel, as shown in the following image on the left and then sketch a line. Click on the centerline to deactivate it. If the command is not deactivated, all geometry that is sketched will be a centerline. Or you can change an existing sketch entity to a centerline by selecting an existing sketched entity and then select the Centerline command. To create a linear diameter dimension by using a centerline, follow these steps:

1. Draw a sketch that represents a quarter section of the finished part.
2. Draw a centerline or change an existing line into a centerline that the sketch will be revolved around.
3. Start the Dimension command.
4. Click the centerline, or a point or edge to be dimensioned.
5. Click the centerline, or a point or edge to be dimensioned. It does not matter the order that the entities are selected; however, one of the selections needs to be the centerline not just an endpoint of the centerline.
6. Move the cursor until the diameter dimension is in the correct location and click.
7. The following image on the right shows the linear diameter dimension created using the centerline to define the axis of revolution.

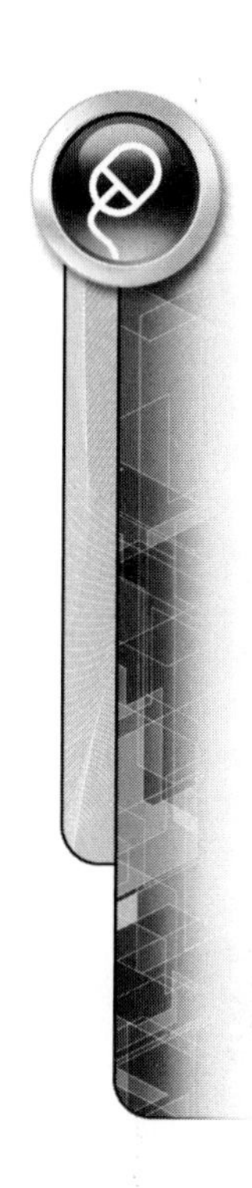

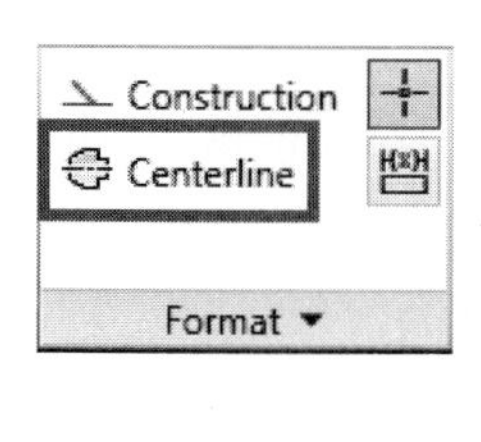

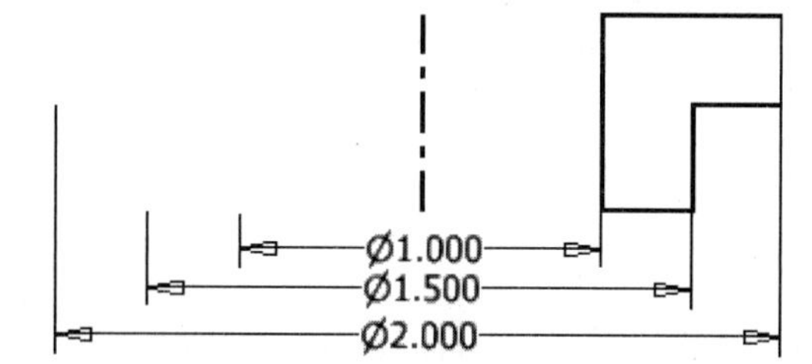

FIGURE 3.23

Normal Line

When a normal line will be used to revolve the sketch around, you can create linear diameter (diametric) dimensions for sketches that represent a quarter outline of a revolved part. To create a linear diameter dimension, follow these steps:

1. Draw a sketch that represents a quarter section of the finished part.
2. Draw a line, if needed, or use a line in the sketch that the sketch will be revolved around.
3. Start the Dimension command.
4. Click the line (not an endpoint) that will be the axis of rotation.
5. Click the other point or line to be dimensioned.
6. Right-click and select Linear Diameter from the menu as shown in the following image on the left.
7. Move the cursor until the diameter dimension is in the correct location and click. The following image on the right shows a sketch that represents a quarter section of a part with a linear diameter dimension placed.

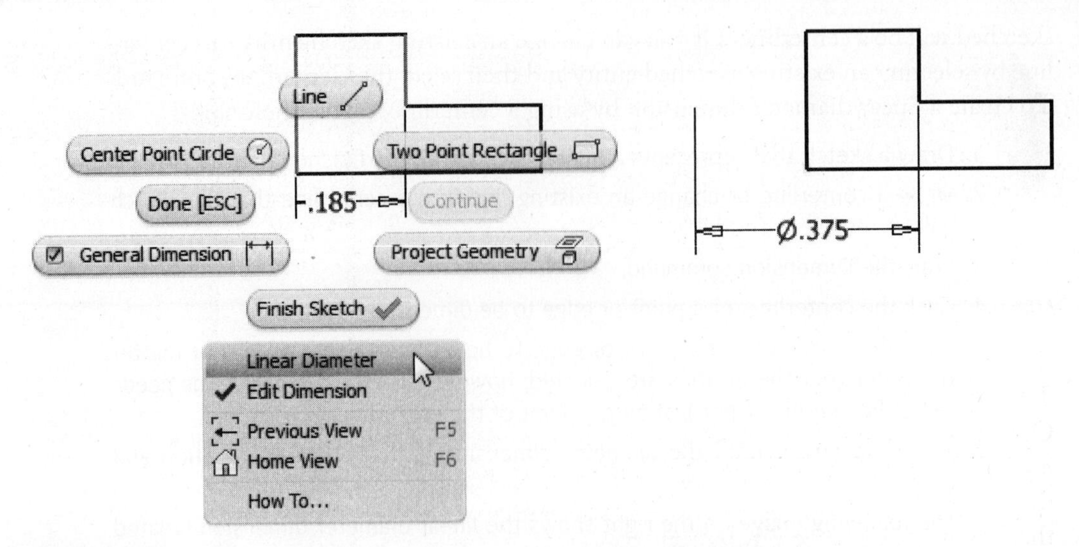

FIGURE 3.24

REVOLVE A SKETCH

After defining the sketch that will be revolved, click the Revolve command on the Create panel as shown in the following image on the left; or while not in a sketch, in the graphics window click on the sketch geometry and the mini-toolbar will appear, click the Create Revolve button or press R or right-click in the graphics window and select Create Feature > Revolve from the menu. The Revolve dialog box appears, as shown in the following image on the right.

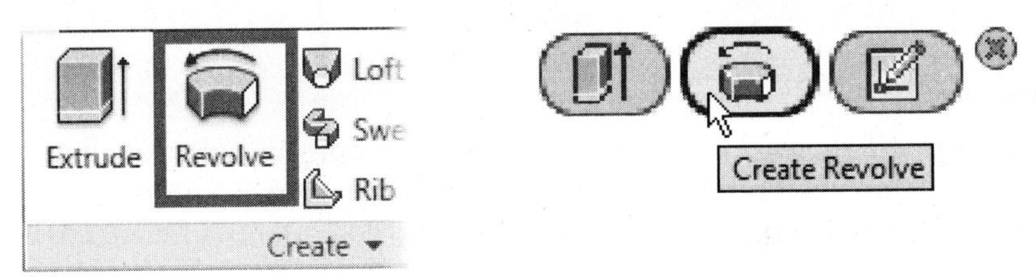

FIGURE 3.25

After starting the Revolve command, the in-canvas display options will appear in the graphics window and the Revolve dialog box will appear. The following image on the left shows the mini-toolbar revolve options that are available when revolving the first sketch and the image on the right shows the operation button that is available when creating additional revolve features. You can either enter data in the dialog box or via the in-canvas display. For the in-canvas buttons click the down arrow next to each button to see more options and click and drag the arrow to change the distance or taper. The in-canvas buttons will change depending upon what options are selected. If selection tags have a red arrow, that command option needs to be satisfied before the command can be completed.

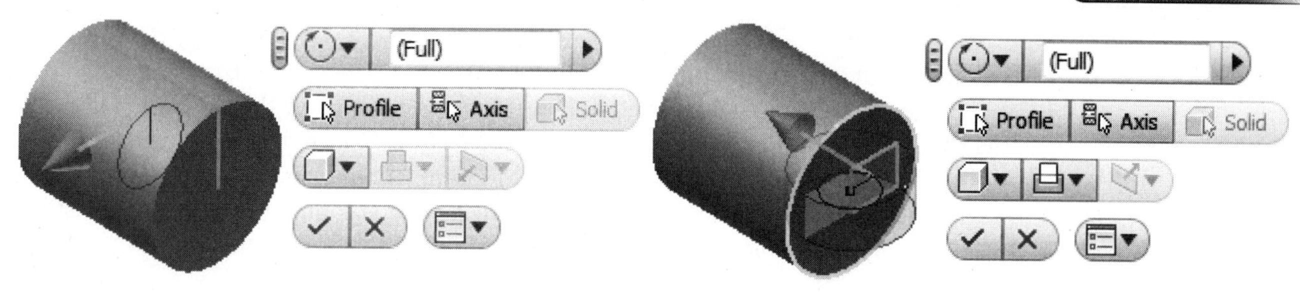

FIGURE 3.26

The Revolve dialog box as shown in the following image has five sections: Shape, Operation, Extents, Output, and Match Shape. When you make changes in the dialog box, the preview for the revolved feature changes in the graphics area to represent the values and options selected. When you have entered the values and the options that you need, click the OK button.

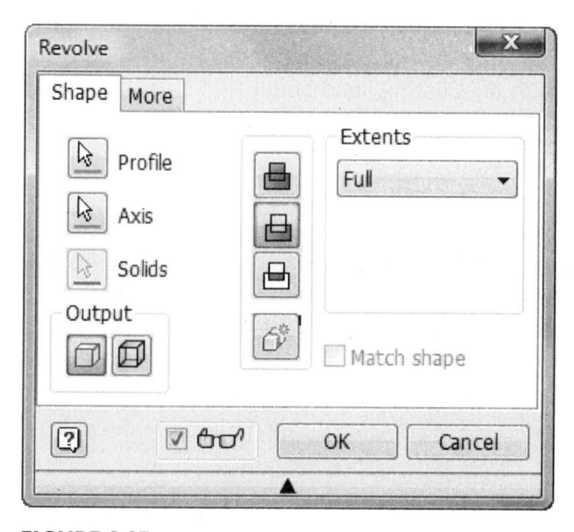

FIGURE 3.27

Shape

This section has two options: Profile and Axis.

	Profile	Click this button to choose the profile to revolve. If the Profile button is shown depressed, this is telling you that a profile or sketch needs to be selected. If there are multiple closed profiles, you will need to select the profile that you want to revolve. If there is only one possible profile, Autodesk Inventor will select it for you, and you can skip this step. If the wrong profile or sketch area is selected, click the Profile button, and choose the new profile or sketch area. To remove a selected profile, hold the CTRL key and click the profile to remove.
	Axis	Click a straight edge, centerline, work axis, or origin axis about which the profile(s) should be revolved. See the section below on how to create a centerline and create diametric dimensions.
	Solid	If there are multiple solid bodies, click this button to choose the solid body(ies) to participate in the operation.

Operation

This is the unlabeled middle column of buttons. If this is the first sketch that you create a solid from, it is referred to as a base feature, and only the top button is available. The operation defaults to Join (the top button). Once the base feature has been established, you can then revolve a sketch, adding or removing material from the part using the Join or Cut options, or you can keep what is common between the existing part and the completed revolve operation using the Intersect option.

	Join	Adds material to the part.
	Cut	Removes material from the part.
	Intersect	Removes material, keeping what is common to the existing part feature(s) and the new feature.
	New Solid	Creates a new solid body. The first solid feature created uses this option by default. Select to create a new body in a part file with an existing solid body.

Extents

The Extents determines if the sketch will be revolved 360°, a specified angle, stop at a specific plane, or start and stop at specific planes.

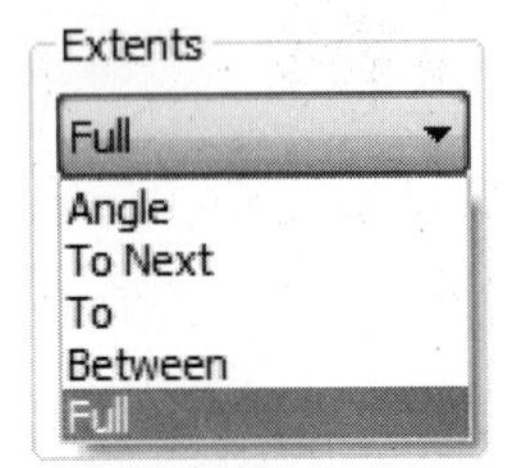

FIGURE 3.28

Full

Full is the default option; it will revolve the sketch 360° about a specified edge or axis.

Angle

Click this option from the drop-down list, and the Revolve dialog box displays additional options, as shown in the following image. Enter an angle for the sketch to be revolved. The three buttons below the degree area will determine the direction of the revolution. Choose the first two to flip the revolution direction, or click the right-side button to have the revolution go an equal distance in the negative and positive directions. If the angle was set to 90°, for example, the revolution will go 45° in both the negative and positive directions.

To Next

This option determines that the sketch will be revolved until it reaches a plane or face. The sketch must be fully enclosed in the area to which it is projecting; if it is not fully enclosed, use the To termination with the Extend to Surface option. Click the Direction button to determine the revolve direction.

To

This option determines that the sketch will be revolved until it reaches a selected face, plane, or point. To select a plane, face, or point to end the revolve, click the Select Surface button, as shown in the following image, and then click a face, plane, or point at which the revolve should terminate.

Between

This option determines that the revolve will start at a selected plane or face and stop at another plane or face. Click the Select surface to start feature creation button, and then click the face or plane where the revolve will start. Click the Select surface to end the feature creation button, and then click the face or plane where the extrusion or revolve will terminate.

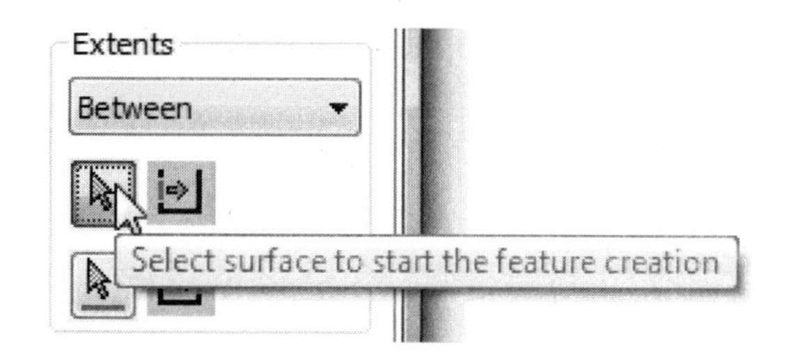

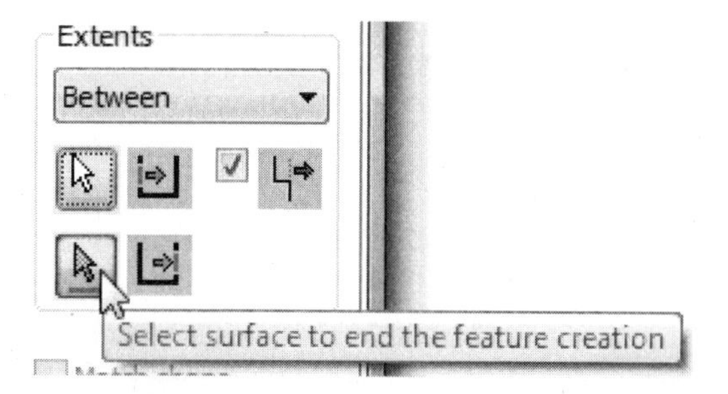

FIGURE 3.29

Direction

When the Extents is set to Angle there are four buttons from which to choose for determining the direction. Choose from the first two to flip the angle direction, click the third button (symmetric) to have the revolution go equal distances in the negative and positive directions, and click the asymmetric option to define different angles for the negative and positive direction. With the symmetric option, if the angle is 90 degrees, for example, the angle will go 45 degrees in both the negative and positive Z directions.

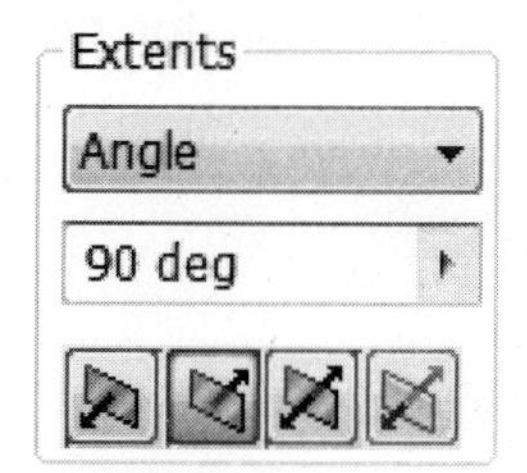

FIGURE 3.30

Output

Two options are available to select the type of output that the Revolve command will generate.

	Solid	Revolves the sketch, and the resulting feature is a solid body.
	Surface	Revolves the sketch, and the resulting feature is a surface.

Match Shape

You can use the Match shape option when working with an open profile. The edges of the open profile are extended until they intersect the geometry of the model. This provides a "flood-fill"–type effect for the revolve feature.

More

The More tab, as shown in the following image, contains additional options to refine the feature being created.

Alternate Solution

Alternate Solution terminates the feature on the most distant solution for the selected surface. An example is shown in the following image.

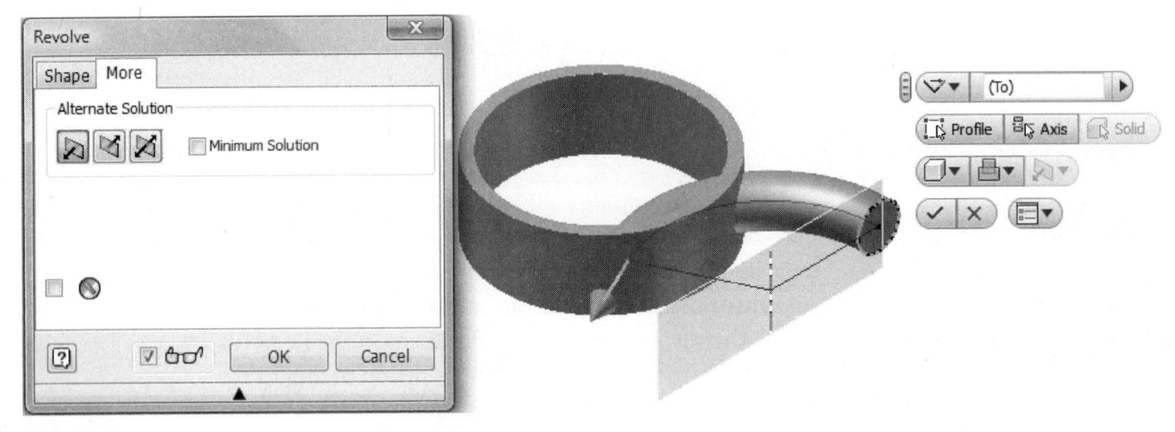

FIGURE 3.31

Minimum Solution

Minimum Solution terminates the feature on the first possible solution for the selected surface. An example is shown in the following image.

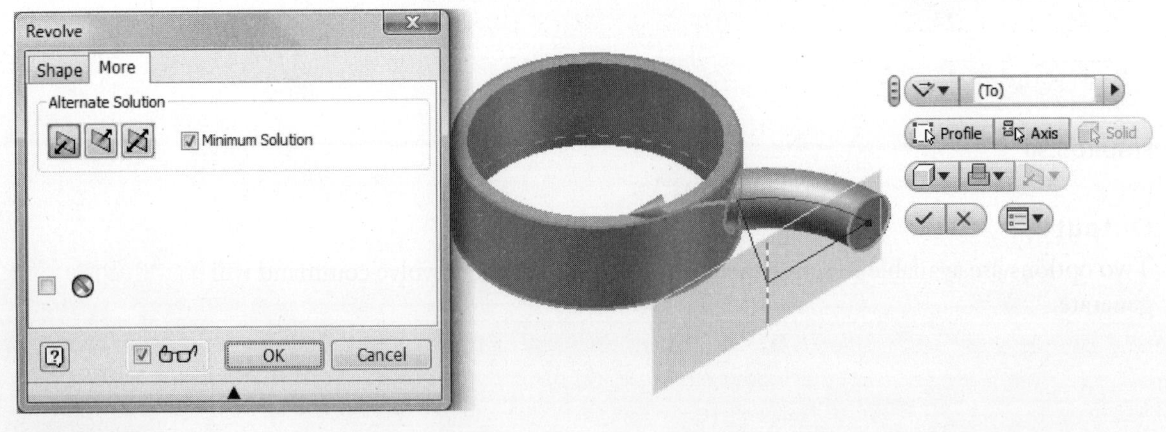

FIGURE 3.32

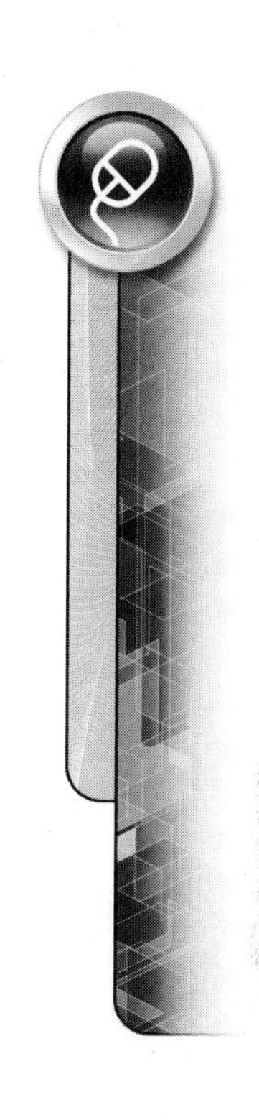

Infer iMates

Check this box to automatically create an iMate on a full circular edge. Autodesk Inventor attempts to place the iMate on the closed loop most likely to be used.

EXERCISE 3-2: REVOLVING A SKETCH

In this exercise, you will create a sketch and then create a revolved feature to complete a part. This exercise demonstrates how to revolve sketched geometry about an axis to create a revolved feature.

1. Click the New command.
2. Select the English tab, and then double-click Standard (in).ipt.
3. Create the sketch geometry as shown. Place the lower endpoint of the centerline at the projected origin point. Then use the Centerline command on the Format panel to change the left vertical line to a centerline as shown in the following image on the left.
4. Click the Dimension command.
5. Add the linear diameter dimensions shown in the following image on the right by selecting the centerline and selecting an endpoint on the sketch. Then click a point to place the dimension.

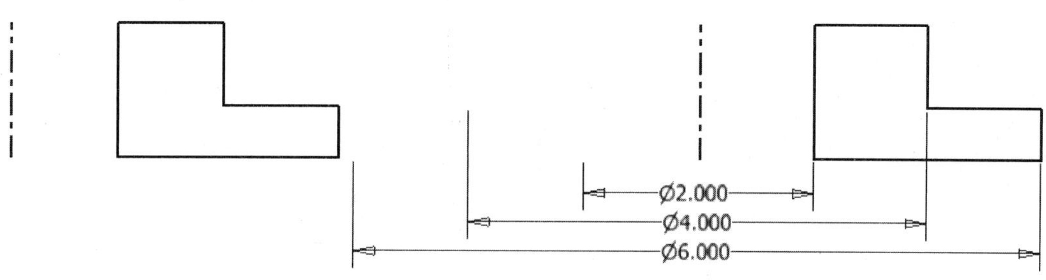

FIGURE 3.33

6. Right-click in the graphics window, and click Finish Sketch from the menu.
7. Change to an isometric view, on the View Cube click the corner of the Top, Front, and Right.
8. From the Create panel bar, click the Revolve command or click on the sketch mini-toolbar and click the Create Revolve button. Since there is only one possible closed profile, the profile is selected for you. Since there is only one centerline the centerline is also automatically selected as the axis.
9. Change the extents to Angle.
10. Enter **45 deg** or drag the arrow in the mini-toolbar display area until **45.00 degrees** is displayed. The preview of the model updates to reflect the change as shown in the following image.

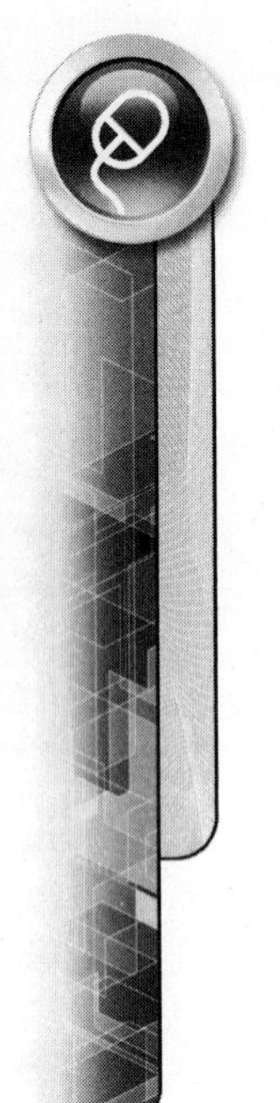

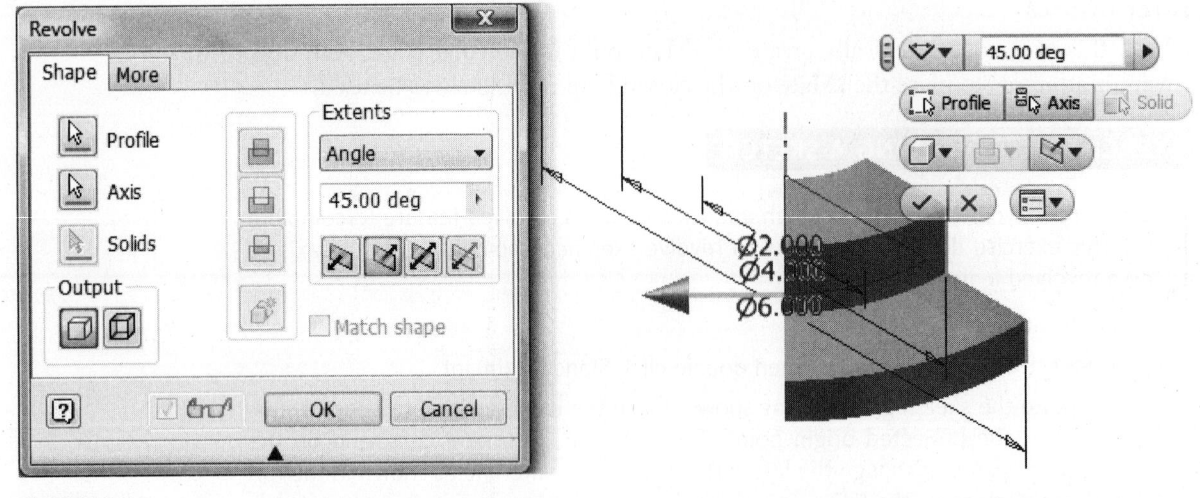

FIGURE 3.34

11. Change the direction of the revolve to go counter clockwise by selecting the left flip button in the dialog box or click and drag the arrow in the graphics screen to go **45.00 degrees** in the opposite direction. The preview image will reverse the direction counter clockwise.
12. Set the revolve direction to symmetric.
13. Enter **90 deg** for the angle. The preview should resemble the following image.

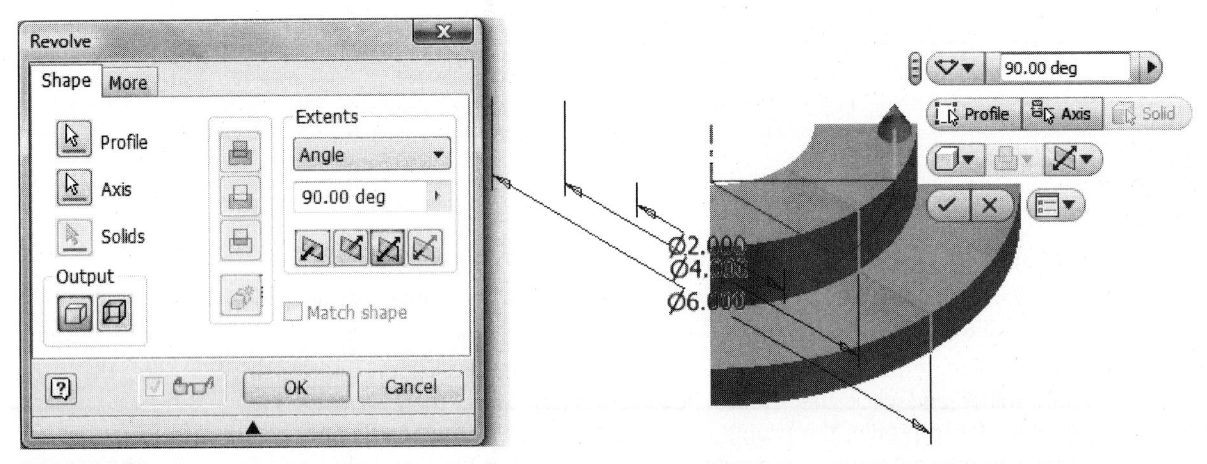

FIGURE 3.35

14. In the mini-toolbar display area change the extent to asymmetric as shown in the following image.
15. In the graphics window click and drag the arrows to different values as shown in the following image.

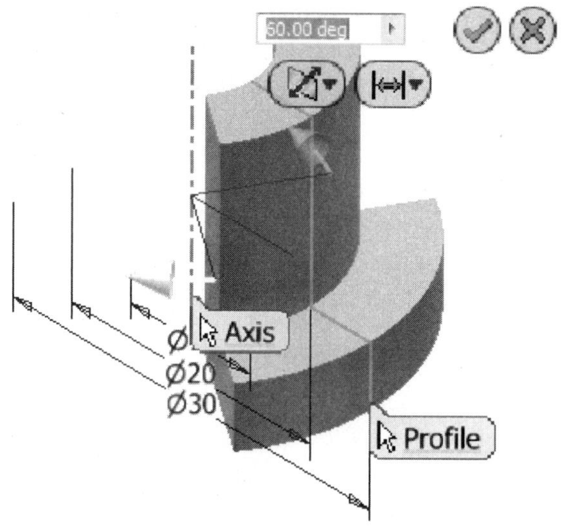

FIGURE 3.36

16. Change the extents type to Full.
17. Click the OK button or click the green check mark in the mini-toolbar to create the feature.
18. Your part should resemble the following image.

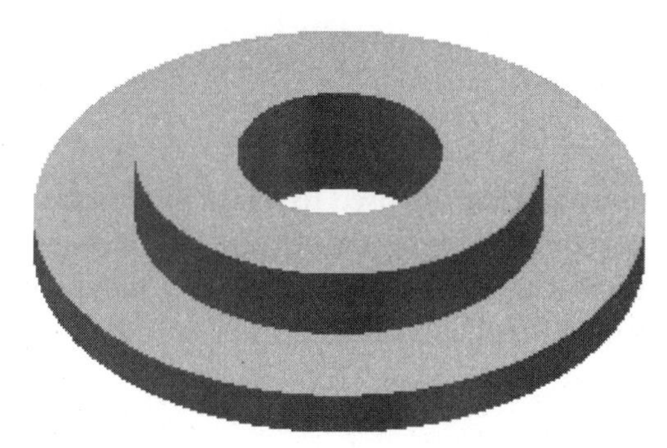

FIGURE 3.37

19. Close the file. Do not save changes. End of exercise.

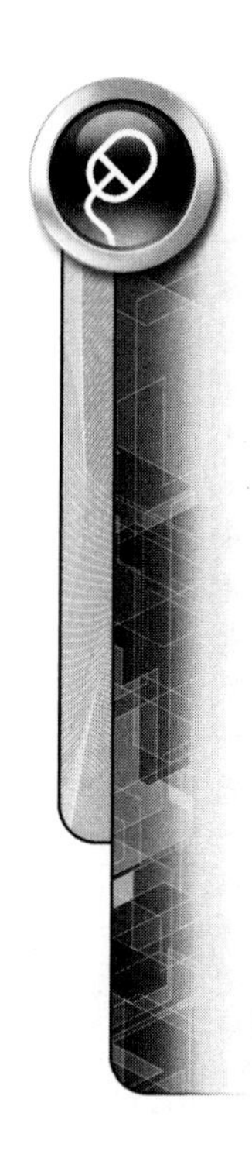

EDITING A FEATURE

After you create a feature, the feature consumes all of the dimensions that were visible in the sketch. If you need to change a feature's options and values such as operation, extents, distance, taper etc. you edit the feature. You can also edit the feature's sketch and change the values of the dimensions or add or remove geometry. Editing a feature and feature's sketch are covered in the next sections.

Editing Feature Options

To change a feature's options, you need to edit the feature. There are multiple methods that you can use to edit the feature. There is no preferred method; use the method that works best for your workflow.

- While not in a sketch, move the cursor over the feature in the graphics window and click. Click the Edit "Feature Name" button from the mini-toolbar. The following image on the left shows the mini-toolbar button for editing an extrusion.

- Double-click the features name in the browser; the dialog box that was used to create the feature will appear.
- In the browser, right-click the feature's name and select Edit Feature from the menu as shown in the following image on the right.

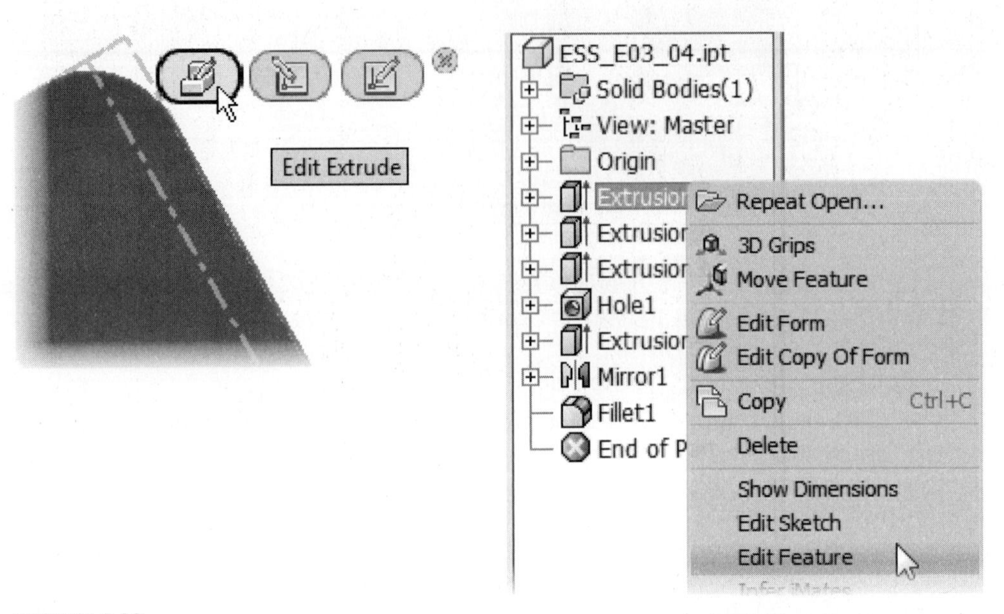

FIGURE 3.38

When editing a feature, the dialog box will appear that was used to create the feature. While editing you can change the feature's values and its options except the join operation on a base feature (first feature in the browser) and the output (solid or surface). The following image on the left shows the menu that appears after right-clicking on Extrusion1.

Editing a Feature's Sketch

In the last section, you learned how to edit the dimensions and the settings in which the feature was created. In this section, you will learn how to add and delete constraints, dimensions, and geometry in the original 2D sketch. To edit the 2D sketch of a feature, do the following:

- While not in a sketch, move the cursor over the feature in the graphics window whose sketch you want to edit and click. Then click the Edit Sketch button from the mini-toolbar as shown in the following image on the left.
- In the browser, expand the feature, right-click the name of the feature or sketch, and click Edit Sketch from the menu as shown in the following image on the right.
- In the browser, expand the feature and double-click on the sketch icon.
- Another method to change the value of dimensions on a sketch is to right-click the feature's name in the browser, and select Show Dimensions from the menu.

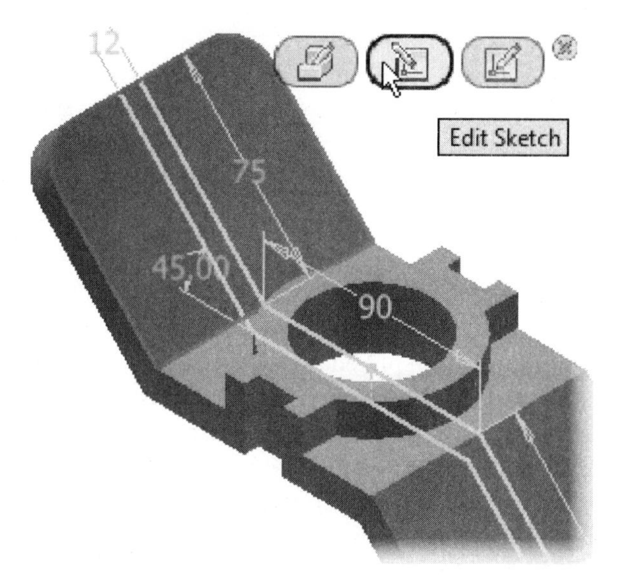

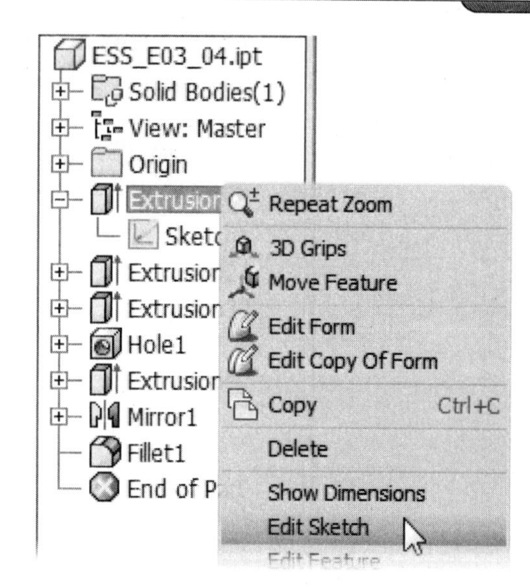

FIGURE 3.39

While editing the sketch, you can both add and remove objects. You can add geometry lines, arcs, circles, and splines to the sketch. To delete an object, right-click it and select Delete from the menu, or click it and press the DELETE key. If you delete an object from the sketch that has dimensions associated with it, the dimensions are no longer valid for the sketch, and they will also be deleted. You can also delete the entire sketch and replace it with an entirely new sketch. When replacing entire sketches, you should first delete other features that would be consumed by the new objects and re-create them.

You can also add or delete constraints and dimensions. When the dimensions are visible on the screen, double-click the dimension text that you want to edit. The Edit Dimension dialog box appears. Enter a new value, and then click the checkmark in the dialog box or press the ENTER key. Continue to edit the dimensions and, when finished click Finish Sketch from the ribbon or click the Local Update button on the Quick Access toolbar as shown in the following image. The sketch will be consumed by the feature and the new values will be used to regenerate the part.

FIGURE 3.40

NOTE

If you receive an error after updating the part, make sure that the sketch forms a closed profile. If the appended or edited sketch forms multiple closed profiles, you will need to reselect the profile area.

Renaming Features and Sketches

By default, each feature is given a name. These feature names may not help you when trying to locate a specific feature of a complex part, as they will not be descriptive to your design intent. The first extrusion, for example, is given the name Extrusion1 by default, whereas the design intent may be that the extrusion is the thickness of a plate. To rename a feature, slowly double-click the feature name and enter a new name. Spaces are allowed.

Deleting a Feature

You may choose to delete a feature after it has been placed. To delete a feature, right-click the feature name in the browser, and select Delete from the menu, as shown in the following image. The Delete Features dialog box will then appear, and you should choose what you want to delete from the list. You can delete multiple features by holding down the CTRL or SHIFT key, clicking their names in the browser, right-clicking one of the names, and then selecting Delete from the menu. The Delete Features dialog box appears; it allows you to delete consumed or dependent sketches and features.

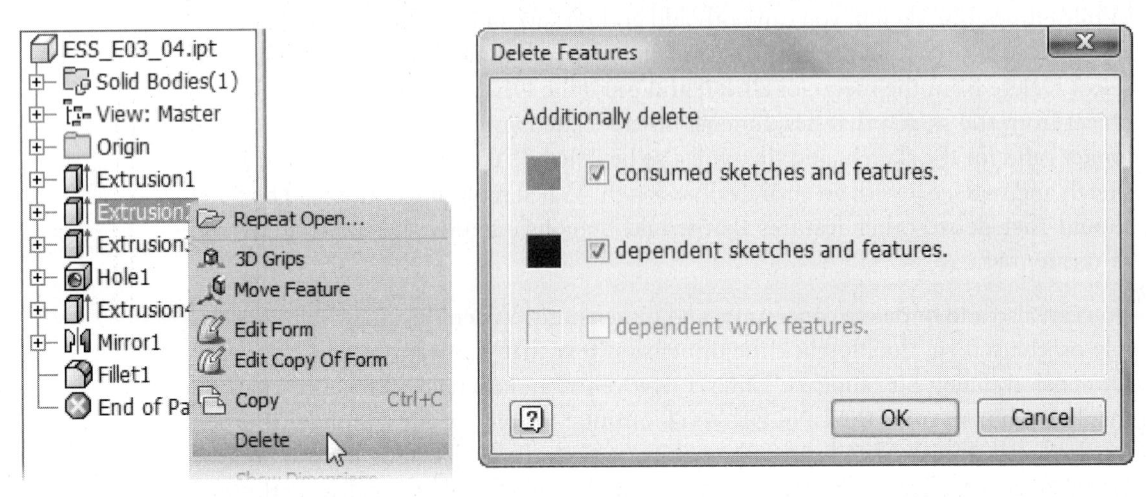

FIGURE 3.41

Failed Features

If the feature in the browser turns red and have a yellow triangle with an explanation point before the features icon as shown in the following image. After updating the part, this is an alert that the new values or settings were not regenerated successfully. To see what the feature looked like in its last successful state, move the cursor over the features name in the browser and it will highlight in the graphics window like what is shown in the following image on the right of the last good fillet feature. You can then edit the values, enter new values, or select different settings to define a valid solution. Once you define a valid solution, the feature should regenerate without error.

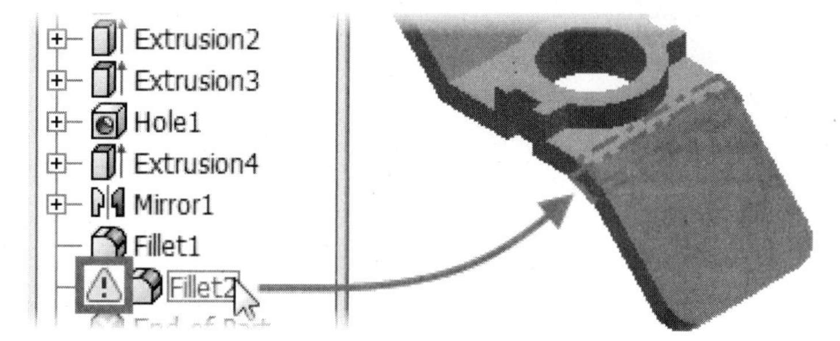

FIGURE 3.42

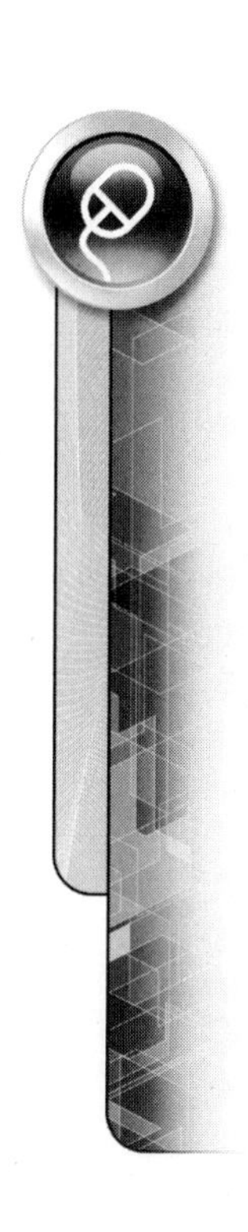

EXERCISE 3-3: EDITING FEATURES AND SKETCHES

In this exercise, you will edit a consumed sketch in an extrusion and update the part. In the next section, you learn how to create sketched features that were created in this model.

1. Open *ESS_E03_03.ipt* from the Chapter 03 folder.
2. In the browser, right-click Extrusion1, and Click Edit Sketch from the menu. The feature's sketch is displayed.
3. Double-click the .375 radius dimension. In the field, enter **.25**, and then press Enter or click the checkmark in the dialog box.
4. Double-click the 2.000 dimension, and change the value to **3** and then press Enter or click the checkmark in the dialog box. The sketch should resemble the following image on the left.
5. Click the Finish Sketch command on the Sketch tab > Exit panel. The feature is updated with the new values as shown in the following image on the right.

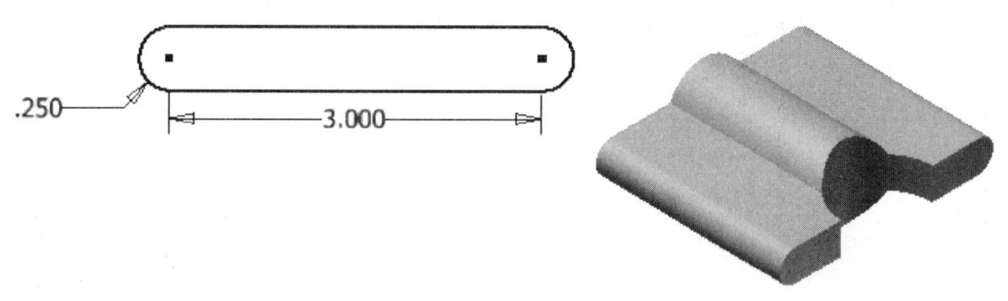

FIGURE 3.43

6. You now edit the extents method that Extrusion2 was created with. In the graphics window click on the inside circular face and then click on the Edit Extrude mini-toolbar button as shown in the following image on the left. The Extrude dialog box and mini-toolbar buttons are displayed.
7. Change the Operation to Join as shown in the following image on the right or use the mini-toolbar buttons.
8. Change the Extents option to To and select the bottom face as shown in the following image. The extrusion will stop at this face no matter what dimensional changes occur to the part.

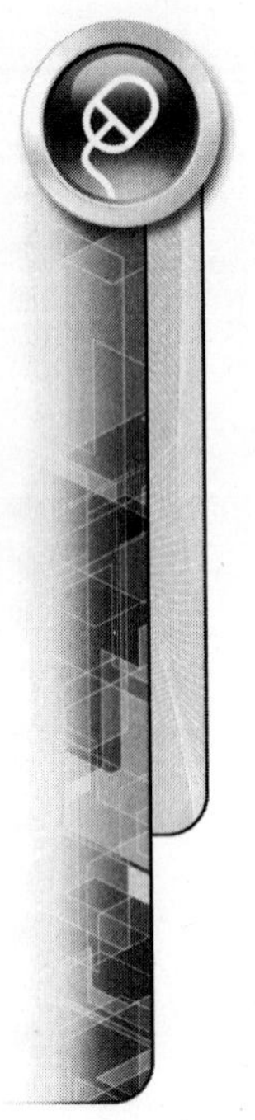

FIGURE 3.44

9. Click OK or the green check mark in the mini-toolbar display. The feature is updated, as shown in the following image on the left.
10. In the browser, right-click Extrusion3. Click Delete and then click OK in the dialog box to delete consumed sketches and features. When done, your part should resemble the following image on the right.

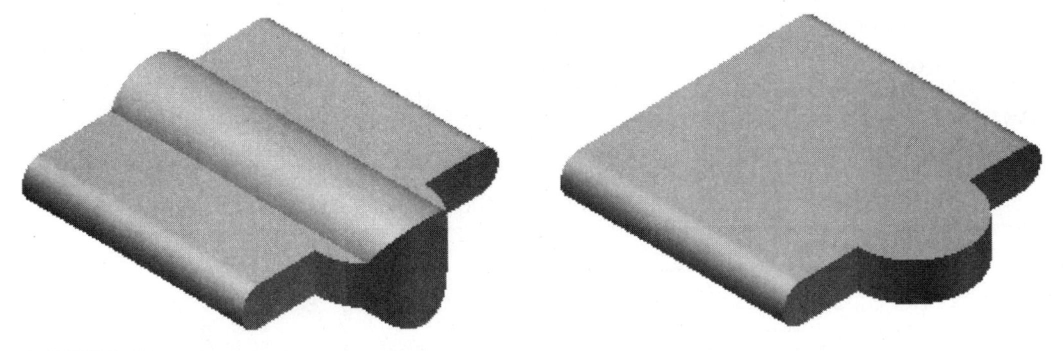

FIGURE 3.45

11. Practice editing the sketch dimensions and features.
12. Close the file. Do not save changes. End of exercise.

SKETCHED FEATURES

A sketched feature is created from a sketch that you draw on a planar face or work plane and then add or remove material from the part. The basic steps to create a sketched feature are as follows:

1. Create a 2D sketch or make an existing sketch active.
2. Draw the geometry that defines the sketch.
3. Add constraints and dimensions.
4. Perform a boolean operation that will either add material to the part, remove material from the part, or keep whatever is common between the part and the feature that is being created. The following image shows a sketch on the left and the affect the three operations would have on the part.

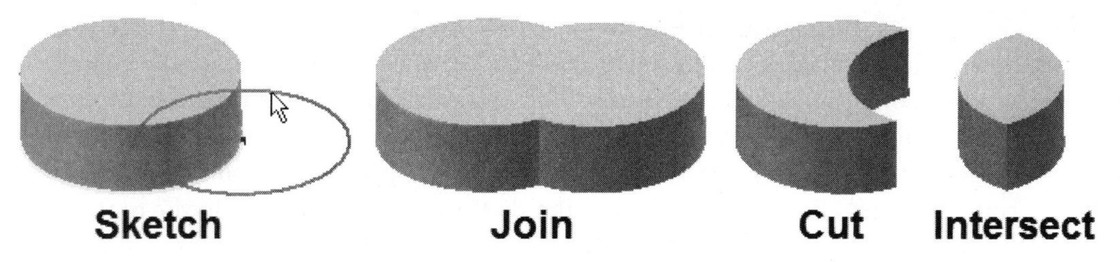

FIGURE 3.46

There are no limits to the number of sketched features that can be added to a part. Each sketched feature is created on its own plane, and multiple features can reference the same plane. You can sketch curves, apply constraints, and apply dimensions exactly as you did with the first sketch. In addition to constraining and dimensioning the new sketch, you can also constrain the new sketch to the existing part. You can place dimensions to geometry that does not lie on the current plane; the dimensions, however, will be placed on the current plane. After the sketch has been constrained and dimensioned, it can be extruded, revolved, swept, or lofted. Exit the sketch environment by right-clicking in the graphics area and selecting Finish Sketch from the menu or by clicking the Finish Sketch command on the Exit panel. Only one sketch can be active at a time.

In the following section, you will learn how to assign a plane to the active sketch and then how to work with sketched features.

DEFINE THE ACTIVE SKETCH PLANE

As stated previously, each sketch must exist on its own plane. The active sketch has a plane on which the sketch is drawn. To assign a plane to the active sketch, the plane on which the sketch will be created must be a planar face, a work plane, or an origin plane. The planar face does not need to have a straight edge. A cylinder has two planar faces, one on the top and the other on the bottom of the part. Neither has a straight edge, but a sketch can be placed on either face.

To make a sketch active, use one of the following methods:

- Click the 2D Sketch command from the Model tab > Sketch panel as shown in the following image on the left, and then click the plane where you want to place the sketch.
- While not in a sketch, click on the planar face or workplane that you want to create the active sketch on. Then click the Create Sketch button from the mini-toolbar buttons as shown in the following image on the right.
- Press the key S, and then click the plane where you want to place the sketch.
- Click a plane that will contain the active sketch, and press the hot key S.
- Click a plane that will contain the active sketch, right-click in the graphics area, and select New Sketch from the menu.
- Issue the Sketch command, expand the Origin folder in the browser, and click one of the origin planes.
- Expand the Origin folder in the browser, right-click on one of the default planes, and select New Sketch on the menu.
- To make a previously created sketch active, issue the Create 2D Sketch command and click the sketch name in the browser.

Once you have created a sketch, it appears in the browser with the name Sketch#, and sketch commands appear in the 2D Sketch panel bar. The number will sequence for each new sketch that is created. To rename a sketch, go to the browser, slowly double-click the existing name, and enter the new name. When you create (PE) a new sketch or make a sketch active, by default the view will automatically change so you are looking straight at the sketch. This is controlled by the Application Option > Sketch tab > Look at sketch plane on sketch creation plan view, that is, looking straight at, or normal to, the sketch plane. This can be done by using the View Face command and click any geometry on the sketch or click the sketch in the browser.

FIGURE 3.47

Select Other-Face Cycling

Autodesk Inventor has dynamic face highlighting that helps you to select the correct face, edge, part etc. to activate and to select objects. As you move the cursor over a given face, the edges of the face are highlighted. If you continue to move the cursor, different faces are highlighted as the cursor passes over them.

To cycle to a face that is behind another one, start a command and then move the cursor over the face that is in front of one that you want to select and hold the cursor still for two seconds. The Select Other tool appears, as shown in the following image on the left. Select the drop down arrow and move the cursor over the available objects in the list until the correct face is highlighted and then click. The number of objects that appear in the list will depend upon the geometry and the location of the cursor, when the Select Other tool appears. You can also access the Select Other command by right-clicking the desired location in the graphics area and clicking Select Other from the marking menu.

You can specify the amount of time before the Select Other command will appear automatically. Click Application Options on the Tools tab > Options panel and click the General tab. In the Selection area as shown in the following image on the right, the "Select Other" delay (sec) feature can be specified in tenths of a second. If you do not want the Select Other command to open automatically, specify OFF in the field. The default value is 1.0 second.

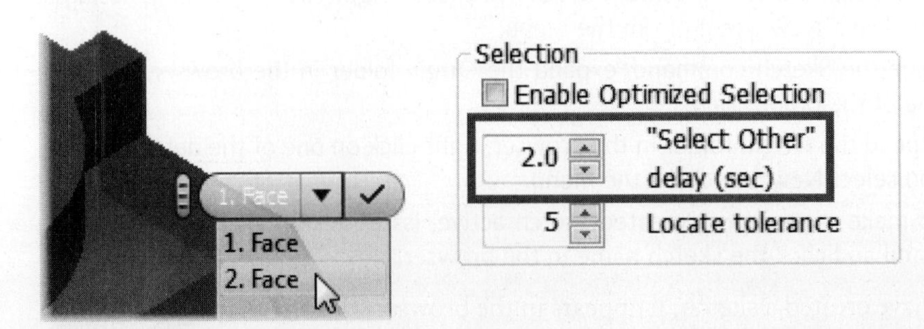

FIGURE 3.48

SLICE GRAPHICS

While creating parts, you may need to sketch on a plane that is difficult to see because features are obscuring the view. The Slice Graphics option will temporarily slice away the portion of the model that obscures the active sketch plane on which you want to sketch. The following image on the left shows a revolved part with an origin plane visible and the Slice Graphics menu. The image on the right shows the graphic sliced and the origin plane's visibility turned off. To temporarily slice the graphics screen, follow these steps:

1. Make a planar face or work plane that the graphics of the active sketch will be sliced through.
2. Rotate the model so the correct side will be sliced, that is, the side of the model that faces up will be sliced away.
3. While editing the sketch, right-click and select Slice Graphics from the menu as shown in the following image, press the F7 key or click Slice Graphics on the View tab > Appearance panel. The part will be sliced on the active sketch plane.
4. Use sketch commands from the Sketch tab to create geometry on the active sketch.
5. To exit the sliced graphics environment, right-click and select Slice Graphics, press the F7 key or click the Finish Sketch button on the Exit panel.

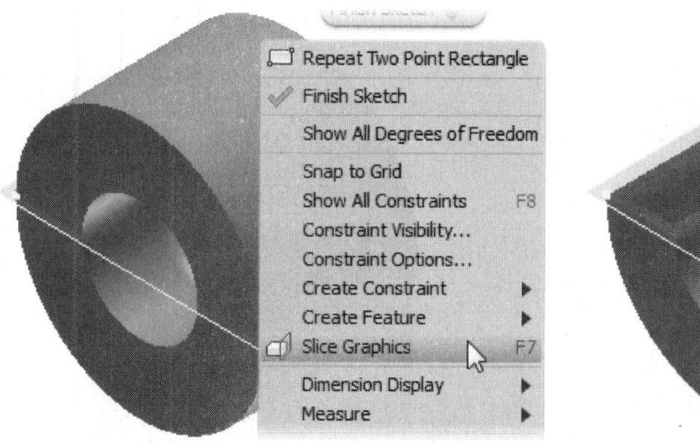

FIGURE 3.49

NOTE

When working in an assembly, you have additional Assembly Section commands available from the View tab > Appearance panel.

EXERCISE 3-4: SKETCH FEATURES

In this exercise, you will create a sketch on the angled face of a part and then create an oblong on the new sketch plane.

1. Open *ESS_E03_04.ipt* from the Chapter 03 folder.
2. Click the Create 2D Sketch command on the Sketch panel, and then click the top-inside angled face as shown in the following image on the left. You could have clicked on the face and click the Create Sketch mini-toolbar button.
3. The view should change so you are looking straight at the sketch, if not click the View Face command from the Navigation bar, and then select Sketch6 in the browser.

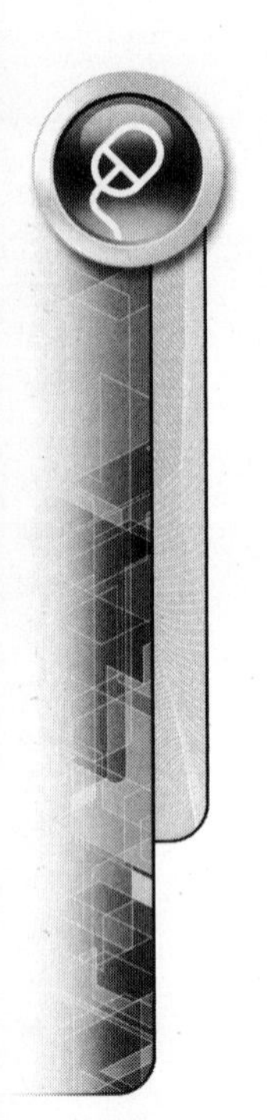

4. Next you create a sketch for an obround in the part and center the slot vertically. Sketch a slot using dynamic input to create a **1.5 inch** high obround as shown in the following image on the right. Both arcs should be tangent to the adjacent lines.

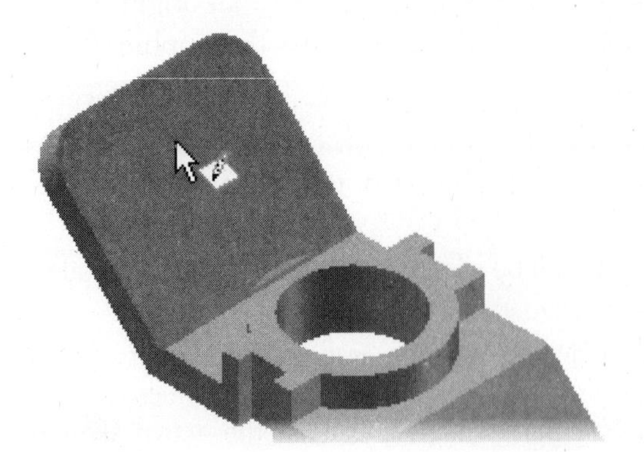

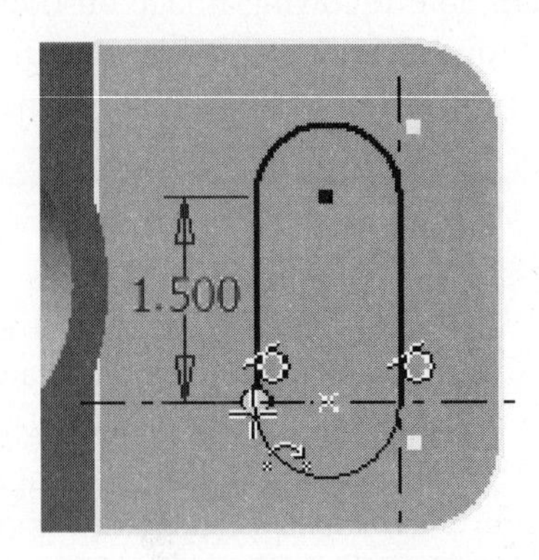

FIGURE 3.50

5. Click the Horizontal constraint command on the Constrain panel as shown in the following image on the left.
6. Select the midpoint of the right vertical edge of the oblong and the midpoint of the right vertical edge of the part as shown in the following image on the right. The obround is centered vertically using the horizontal constraint.

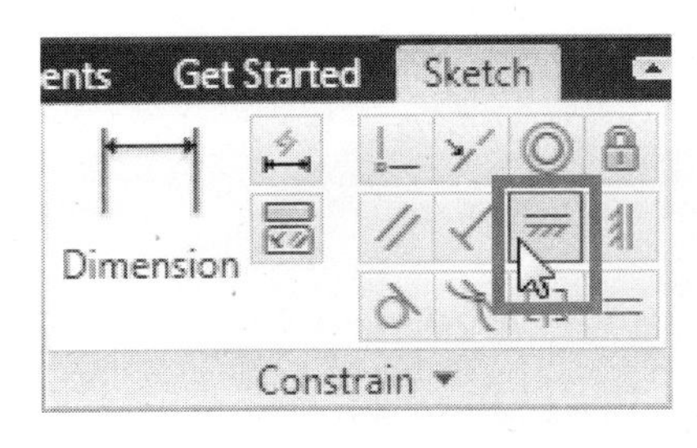

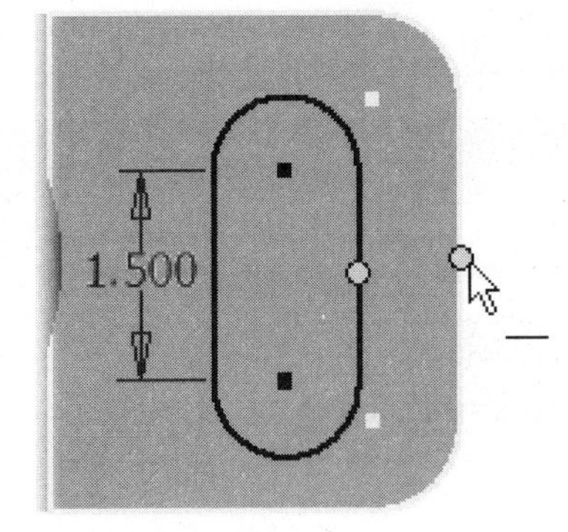

FIGURE 3.51

7. Click the Dimension command on the Constrain panel as shown in the following image on the left. To fully constrain the sketch place the two dimensions as shown in the following image on the right.

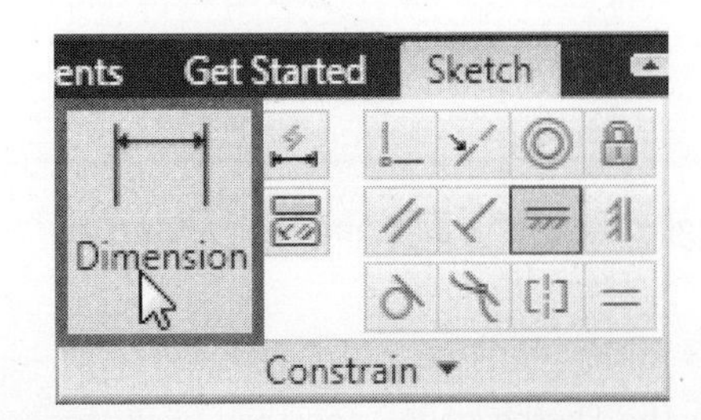

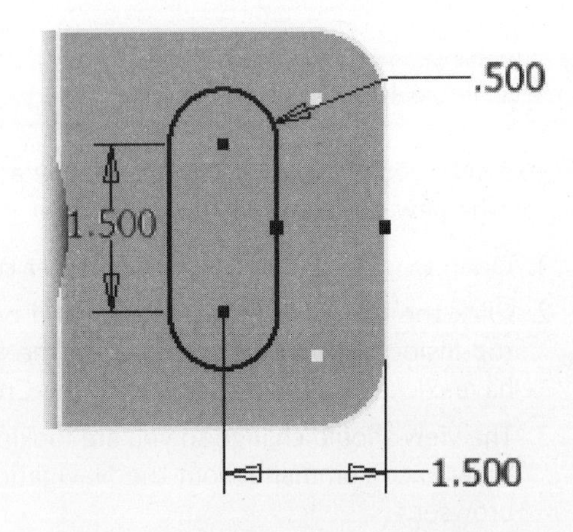

FIGURE 3.52

8. Click the Finish Sketch command on the Exit panel.
9. If the view did not automatically change to the home view press the F6 key to change to the Home View.
10. In the graphics window click on an entity on the obround and click Create Extrude from the mini-toolbar buttons.
11. For the profile, click inside the oblong.
12. In the mini-tool click the Cut operation, and change the Extents to All. Ensure that the direction is pointing into the part as shown in the following image on the left.
13. Click OK. The completed part is shown in the following image on the right.

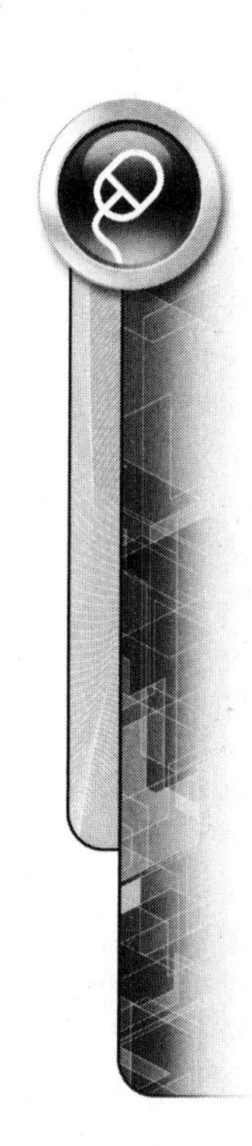

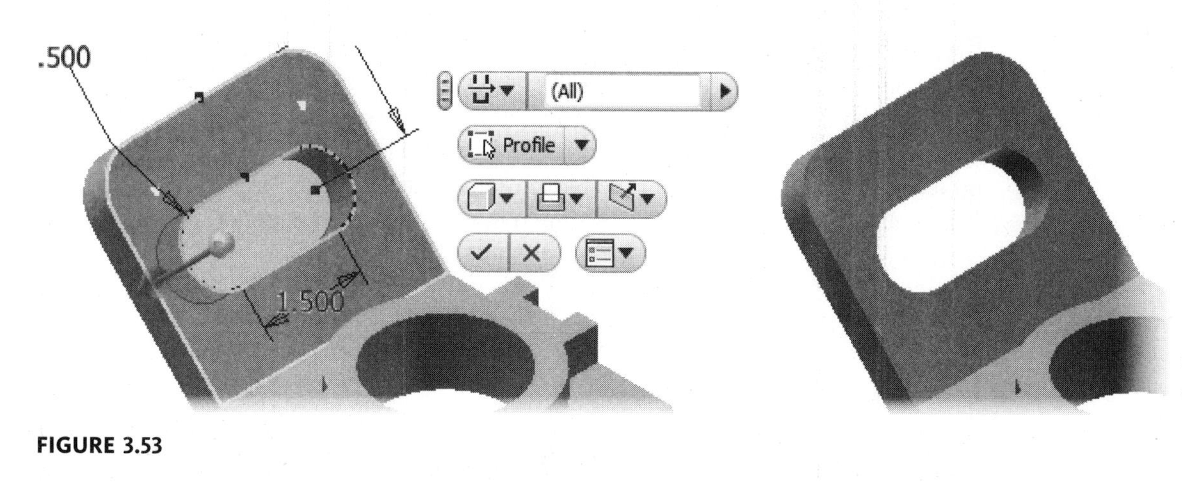

FIGURE 3.53

14. Practice creating sketched features trying different options.
15. Close the file. Do not save changes. End of exercise.

PROJECTING PART EDGES

Building parts based partially on existing geometry is done often, and you will frequently need to reference faces, edges, or loops from features that have been created.

While in a part file, you can project an edge, face, point, or loop onto a sketch. Projected geometry can maintain an associative link to the original geometry that is projected. If you project the face of a feature onto another sketch, for example, and the parent sketch is modified, the projected geometry will update to reflect the changes.

Direct Model Edge Referencing

While you sketch, you can use direct model edge referencing to

- Automatically project edges of the part to the sketch plane as you sketch a curve.
- Create dimensions and constraints to edges of the part that do not lie on the sketch plane.
- Control the automatic projection of part edges to the sketch plane.

Creating Reference Geometry

There are two ways to automatically project part edges to the sketch plane:

- Click an edge of the part while creating a dimension or constraint.

- On the Sketch tab in the Application Options dialog box, you can control two options.
  - Autoproject edges during curve creation, which controls the ability to touch and project edges while sketching a curve;
  - Autoproject edges for sketch creation and edit, which controls the automatic projection of the edges of the selected face when you start a sketch on a planar face of the part.

NOTE

Neither of these options disables the ability to reference part edges when creating dimensions and constraints.

PROJECT EDGES

In this section, you learn about using the Project Geometry command that can project selected edges, vertices, work features, curves, the silhouette edges of another part in an assembly, or other features in the same part to the active sketch. There are three project commands available on the Sketch Panel Bar: Project Geometry, Project Cut Edges, and Project Flat Pattern, as shown in the following image.

Project Geometry

Use to project geometry from a sketch or feature onto the active sketch.

Project Cut Edges

Use to project part edges that touch the active sketch. The geometry is only projected if the uncut part would intersect the sketch plane. For example, if a sphere has a sketch plane in the center of the part and the Project Cut Edges command is initialized, a circle will be projected onto the active sketch.

Project Flat Pattern

Use to project a selected face or faces of a sheet metal part flat pattern onto the active sheet metal part sketch plane. Creation of sheet metal parts is covered in Chapter 10.

Project to 3D Sketch

Use to project geometry from the active 2D sketch onto selected faces to create a 3D sketch. 3D Sketches are covered in Chapter 7.

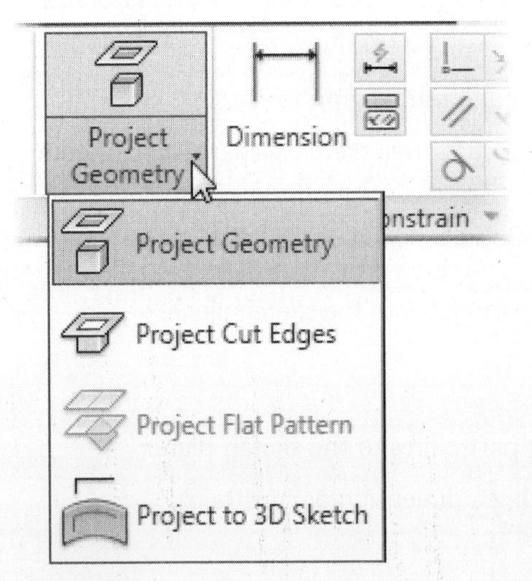

FIGURE 3.54

To project geometry, follow these steps:

1. Make a plane in the active sketch to which the geometry will be projected.
2. Click the Project Geometry command on the 2D Sketch Panel Bar.
3. Select the geometry to be projected onto the active sketch. Click a point in the middle of a face, and all edges of the face will be projected onto the active sketch plane. If you want to project all edges that are tangent to an edge (a loop), use the Select Other command to cycle through until they all appear highlighted, as shown in the following image on the left. The following image on the right shows the projected geometry.
4. To exit the operation, press the ESC key or click another command.

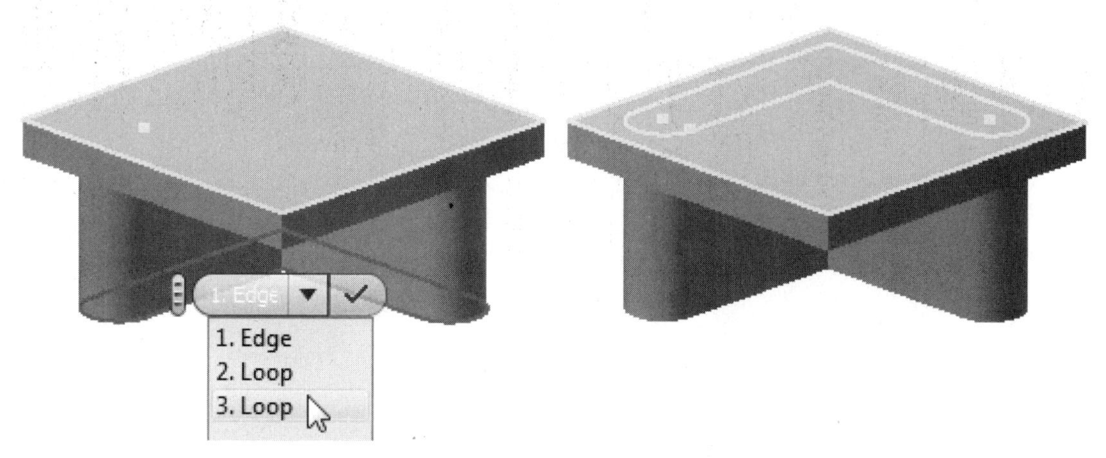

FIGURE 3.55

If you clicked a loop for the projection, the sketch is updated to reflect the modification when any part of the profile changes. If a face is projected, the internal islands that are defined on the face are also projected and will update accordingly. For example, if the loop of edges of the following image on the left is projected, the four lines that define the cutout will be projected and updated if the original feature is modified. To disassociate projected geometry from the original geometry, show the sketch constraints and delete the reference constraint that is created for each projected edge/curve.

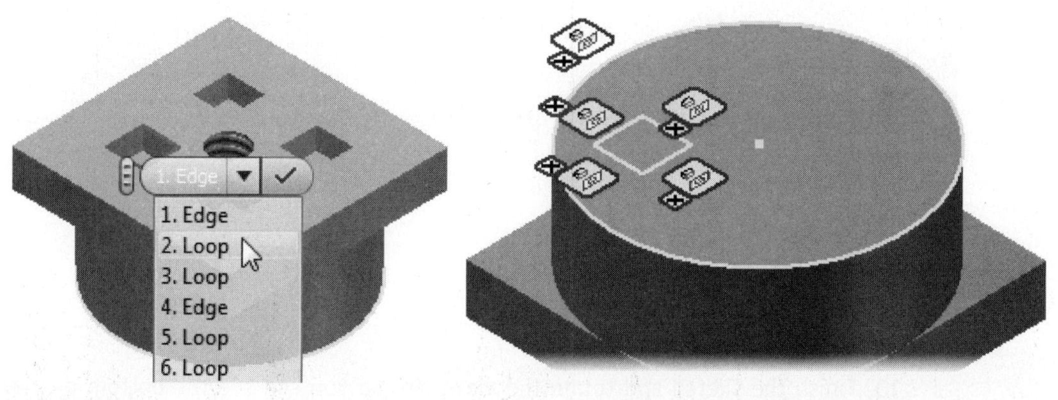

FIGURE 3.56

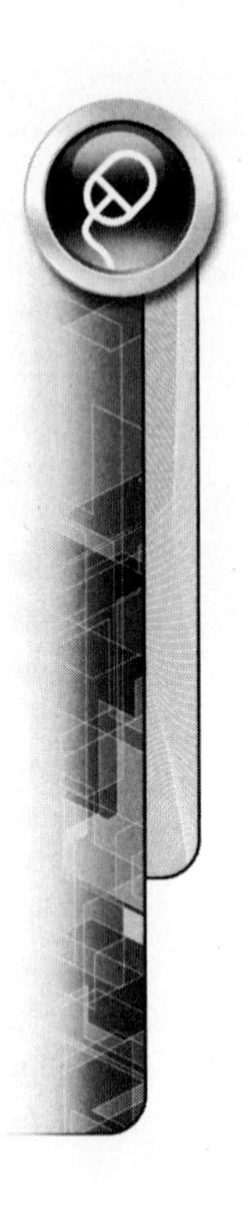

EXERCISE 3-5: PROJECT GEOMETRY

In this exercise, you will project geometry to create a feature, edit the original sketch and update the linked feature and then delete the sketch constraints that were created from the projected geometry.

1. Open ESS_E03_05.ipt from the Chapter 03 folder.
2. Create a sketch on the bottom face of the cylinder (opposite side of Extrusion2) as shown in the following image on the left.
3. Rotate your viewpoint so you can see the front of the part.
4. Click the Project Geometry command on the Draw panel and select the front face the Extrusion2 as shown in the following image on the right.

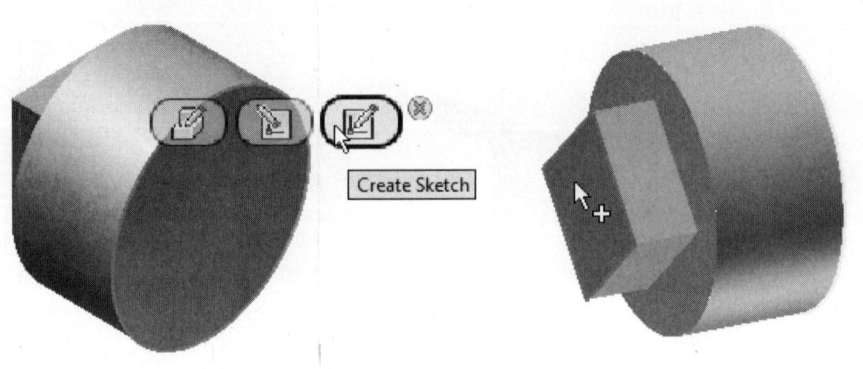

FIGURE 3.57

5. Again rotate your viewpoint so you can see the bottom of the part. Notice that the square has been projected onto the active sketch and is fully constrained.
6. Finish the sketch.
7. Extrude the sketch **.5** inches, cutting material into the part as shown in the following image on the left.
8. Edit the sketch of Extrusion2 (the original extruded square).
9. Change the 1 in dimension to **.5** in as shown in the following image on the right.

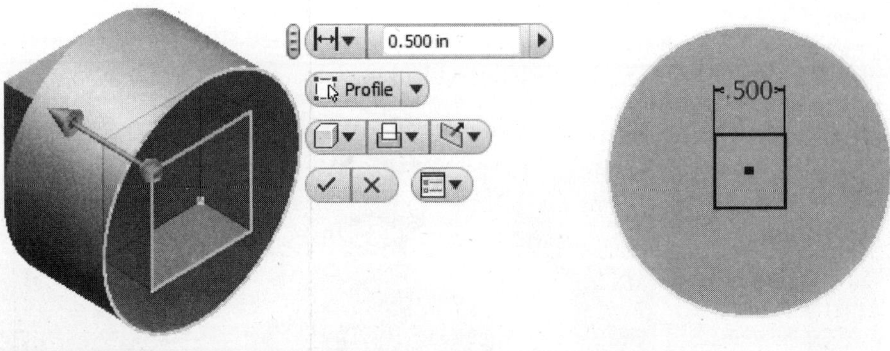

FIGURE 3.58

10. Finish the sketch and rotate the viewpoint so you can see the bottom of the part. Notice that Extrusion3's sketch and feature have been updated to reflect the change to the geometry that was projected.
11. Edit the sketch of Extrusion3.

12. Next you will delete the reference constraints that were created when the geometry was projected. Press the F8 key to display the constraints as shown in the following image on the left.
13. Delete the four reference constraints that surround the square (these constraints were automatically created when the geometry was projected onto this sketch). The sketch should now have 8 dimensions/constraints required to fully constrain the sketch.
14. Press the F8 key to refresh the constraints that are displayed, the only constrains that remain on the sketch are the four coincident constraints.
15. Click the Automatic Dimension command from the Constrain panel, select the four inside lines for the curves to Auto Dimension, uncheck the Dimension option (only constraints will be applied) as shown in the following image on the right and then click Apply in the dialog box.

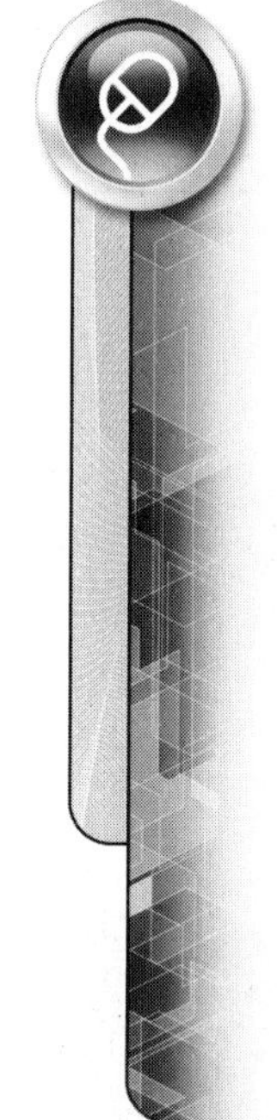

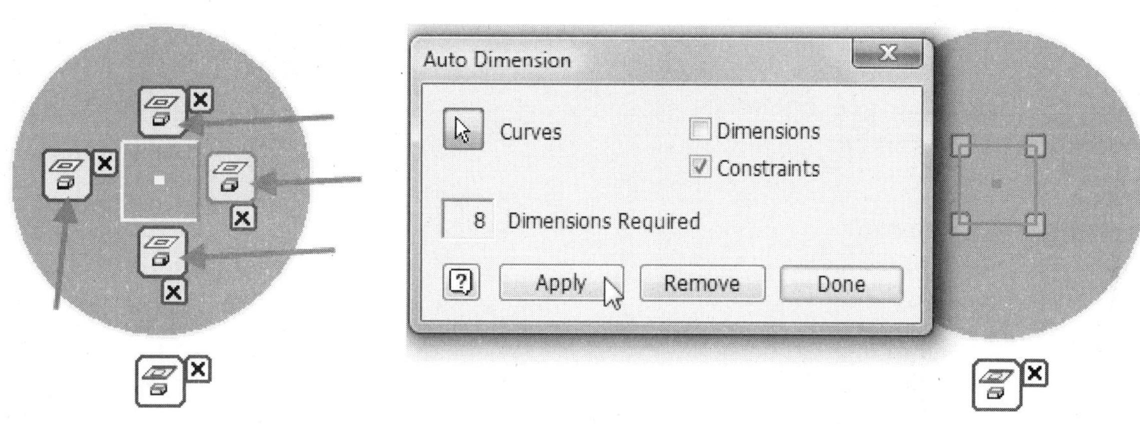

FIGURE 3.59

16. Click the Done button in the dialog box.
17. To display the constraints that were added in the last step press the F8 key.
18. Place a horizontal constraint to the bottom line.
19. Press the F9 key to hide the constraints.
20. To fully constrain the sketch, add the four dimensions as shown in the following image in the middle.
21. Finish the sketch and rotate the part so you see that Extrusion3 is a different size than Extrusion2 as shown in the following image on the right.

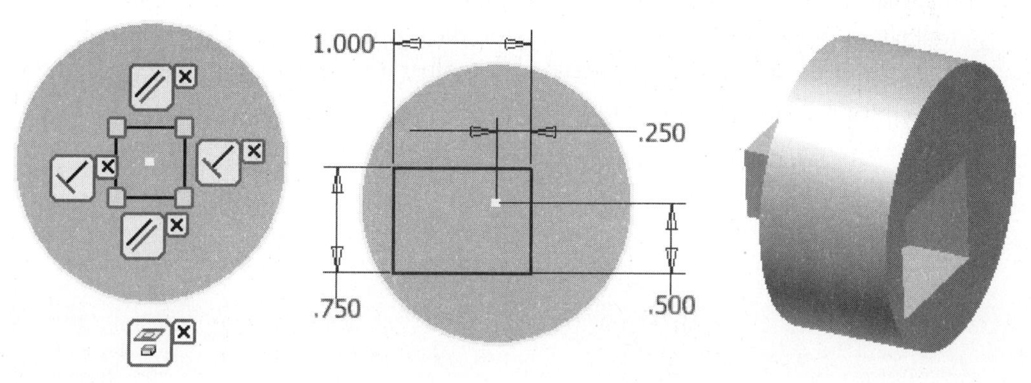

FIGURE 3.60

22. Close the file. Do not save changes. End of exercise.

PART MATERIAL, PROPERTIES AND COLOR

To change a part's physical properties and appearance to a specific material, click the Inventor Application Menu and click iProperties from the menu. You can also right-click on the part's name in the browser and click iProperties from the menu or click Physical Properties from the Color list on the Quick Access toolbar. Then in the iProperties dialog box click the Physical tab and select a material from the material drop-down list as shown in the following image on the left. Click the Update button in the dialog box to update the properties of the file as shown in the following image on the right.

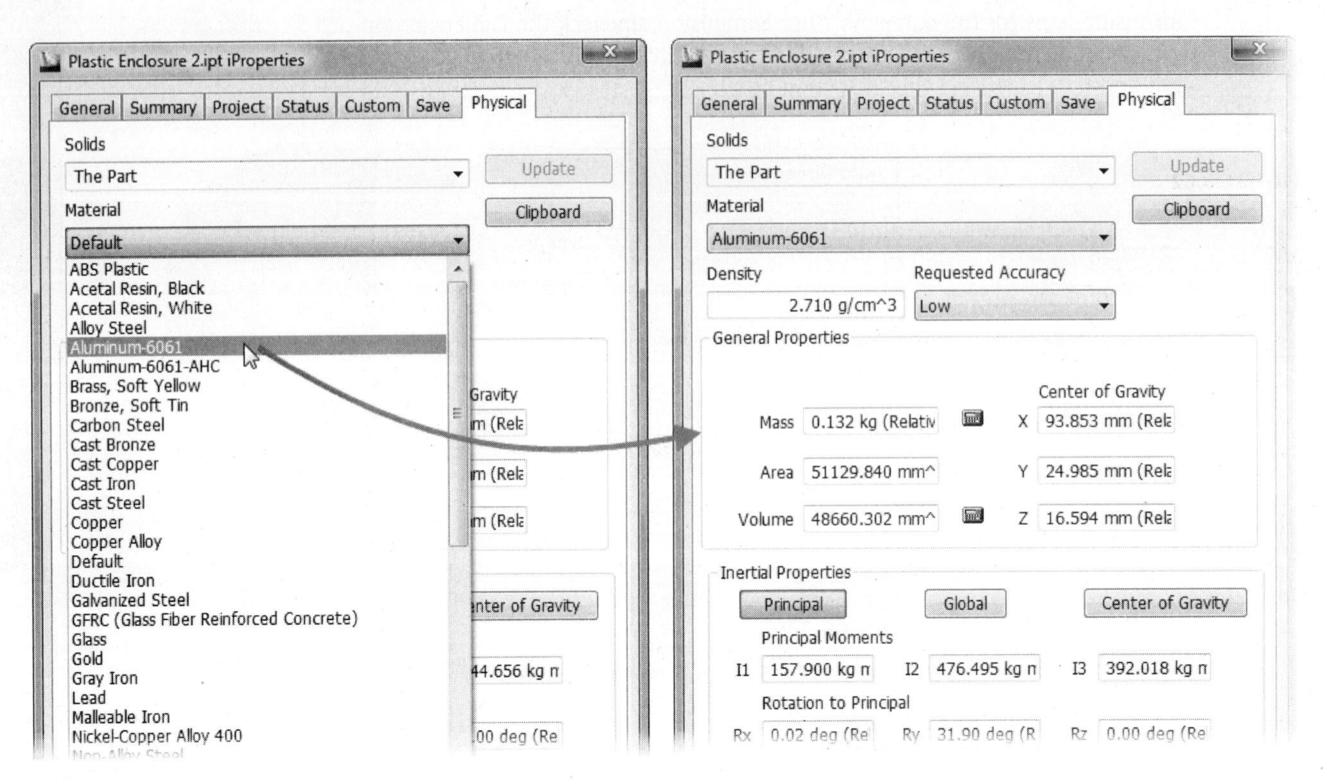

FIGURE 3.61

Click the OK button in the dialog box to complete the operation, and the color of the part change to reflect the selected material. If the part color does not update when the material is edited, click the As Material option in the Color area of the Quick Access toolbar, as shown in the following image on the right. If you selected a color from the Color drop-down list on the Quick Access toolbar it will change only the color of the part, not the physical material or properties of the part.

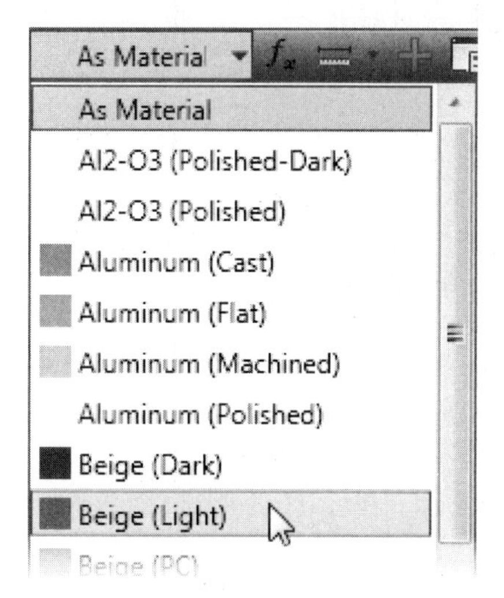

FIGURE 3.62

APPLYING YOUR SKILLS

Skill Exercise 3-1

In this exercise, you will create a bracket from a number of extruded features. Assume that the part is symmetrical about the center of the horizontal slot.

1. Start a new part based on the English Standard (in).ipt template.
2. Create the sketch geometry for the base feature.
3. Add geometric constraints and dimensions. Make sure that the sketch is fully constrained.
4. Extrude the base feature.
5. Create the three remaining features to complete the part.

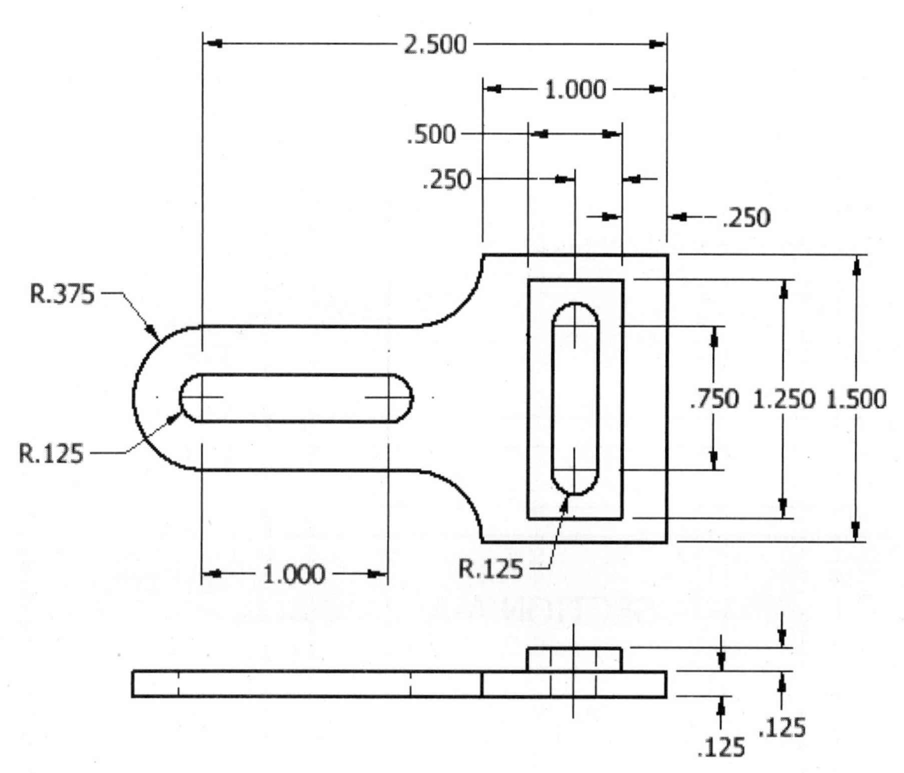

FIGURE 3.63

The completed part should resemble the following image. When done close the file. Do not save changes. End of exercise.

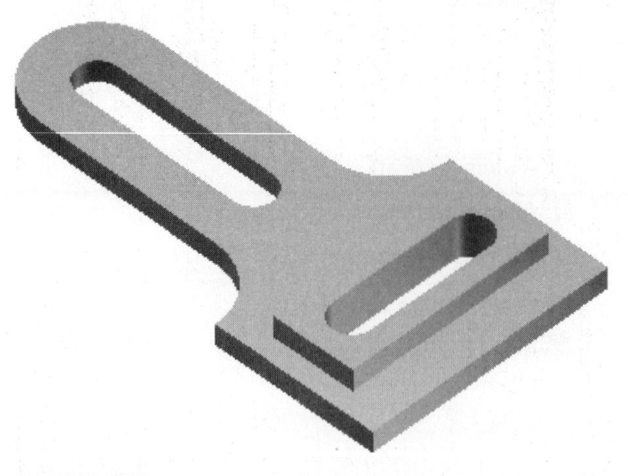

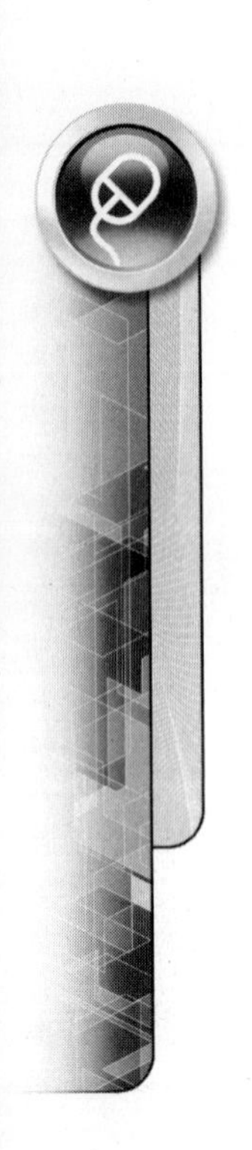

FIGURE 3.64

Skill Exercise 3-2

In this exercise, you will create a connecting rod and add draft to extrusions during feature creation.

1. Start a new part based on the English Standard (in).ipt template.
2. Create the sketch geometry for the outside of the connecting rod.
3. Add geometric constraints and dimensions to fully constrain the sketch.
4. Extrude the base feature using the midplane option, adding a −10° taper.
5. Create a separate feature for each pocket by projecting geometry (don't mirror the feature, the mirror command will be covered in Chapter 7). The sides of the pocket are parallel to the sides of the connecting rod.

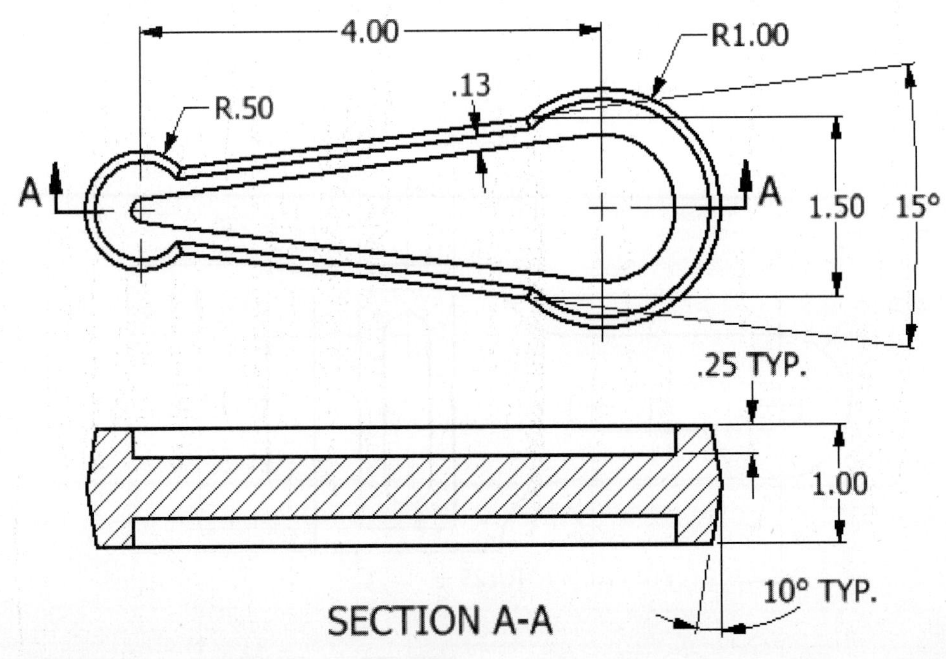

FIGURE 3.65

The completed part should resemble the following image. When done close the file. Do not save changes. End of exercise.

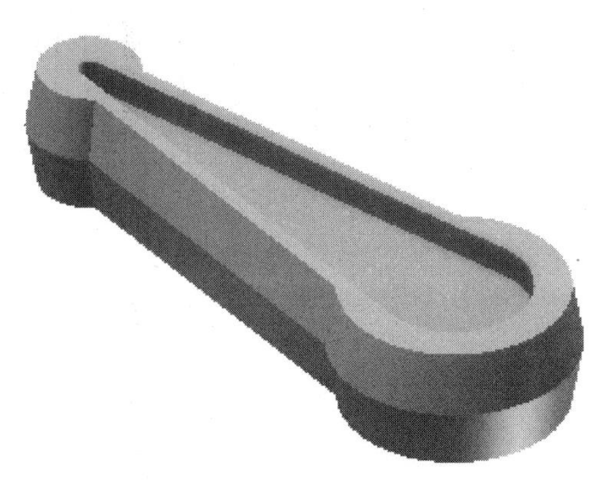

FIGURE 3.66

Skill Exercise 3-3

In this exercise, you will create a pulley using a revolved feature. Assume that the part is symmetric about the middle of the part vertically.

1. Start a new part based on the English Standard (in).ipt template.
2. Create the sketch geometry for the cross-section of the pulley.

TIP

To create a centerline, draw a line, select it, and then click the Centerline command on the Sketch tab > Format panel.

3. Apply appropriate sketch constraints.
4. Add dimensions.
5. Revolve the sketch.

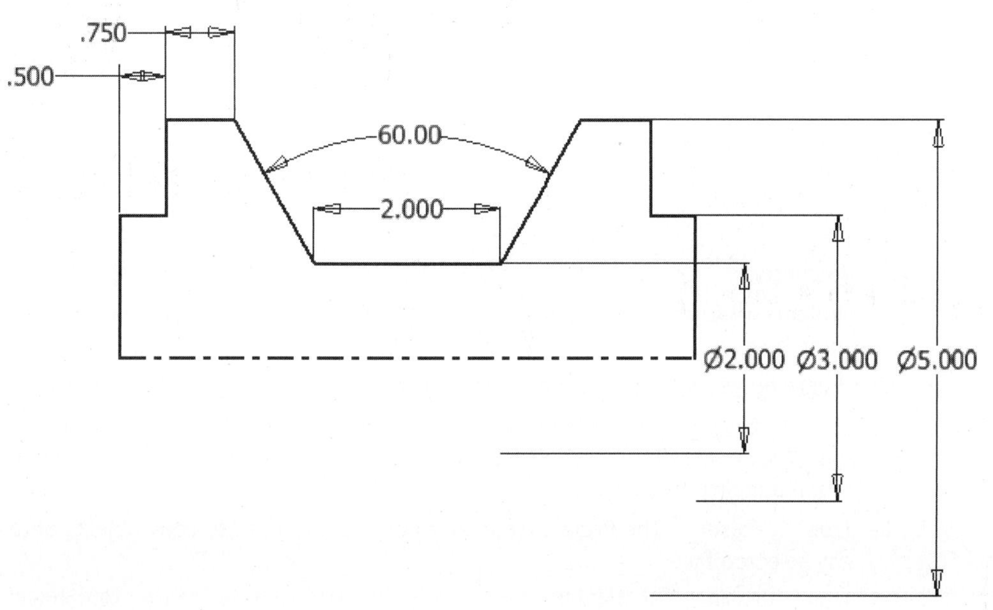

FIGURE 3.67

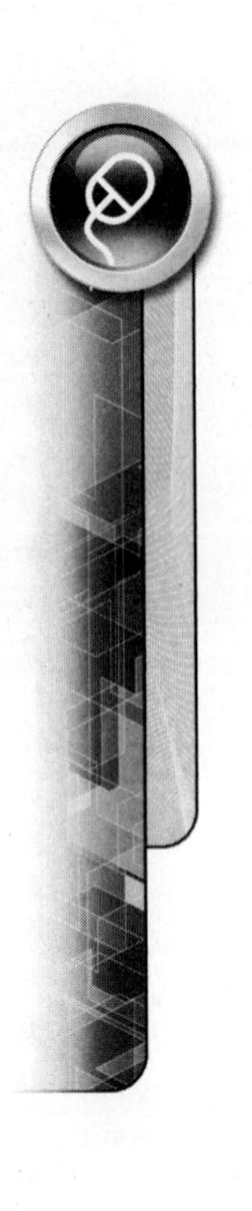

The completed part should resemble the following image. When done, close the file. Do not save changes. End of exercise.

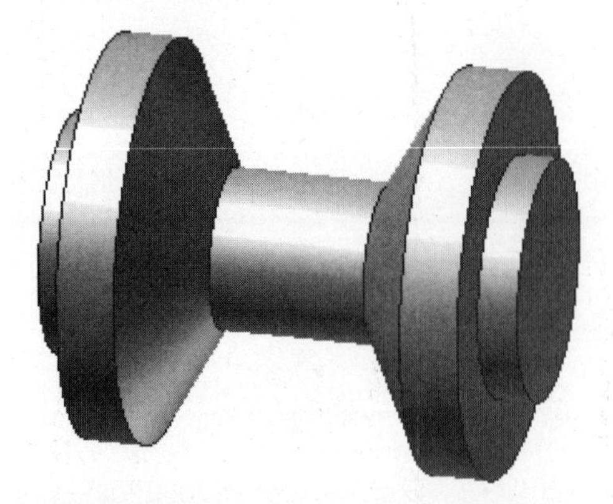

FIGURE 3.68

CHECKING YOUR SKILLS

Use these questions to test your knowledge of the material covered in this chapter.

1. What is a base feature?
2. True ___ False ___ When creating a feature with the Extrude or Revolve command, the mini-toolbar buttons only allows you to define the distance or angle.
3. When creating a revolve feature, which objects can be used as an axis of revolution?
4. Explain how to create a linear diameter (diametric) dimension on a sketch.
5. Name two ways to edit an existing feature.
6. True ___ False ___ Once a sketch becomes a base feature, you cannot delete or add constraints, dimensions, or objects to the sketch.
7. Name three operation types used to create sketched features.
8. True ___ False ___ A direct manipulation technique can only be started by clicking on a face of a part.
9. True ___ False ___ Once a sketched feature exists, its extents type cannot be changed.
10. True ___ False ___ By default geometry that is projected from one feature to a sketch will update automatically based on changes to the original projected geometry.
11. Explain what the asymmetric option does for the extrude and revolve commands.
12. Where do you set the physical material property of a part?
13. True ___ False ___ After setting a part's material properties, the color setting in the Quick Access toolbar must be set to Color Matching for the color to match the material of the part.
14. True ___ False ___ The Project Geometry command is used to copy objects onto any selected face.
15. True ___ False ___ By default, when a feature is deleted, the feature's sketch will be maintained.

CHAPTER 4

Creating Placed Features

INTRODUCTION

In Chapter 3, you learned how to create and edit base and sketched features. In this chapter, you will learn how to create *placed* features. These are features that are predefined except for specific values and only need to be located. You can edit placed features in the browser like sketched features. When you edit a placed feature, either the dialog box that you used to create it will open or feature values will appear on the part.

When creating a part, it is usually better to use placed features instead of sketched features wherever possible. To make a through hole as a sketched feature, for example, you can draw a circle profile, dimension it, and then extrude it with the cut operation, using the All extension. You can also create a hole as a placed feature—you can select the type of hole, size it, and then place it using a dialog box. When drawing views are generated, the type and size of the hole are easy to annotate, and they automatically update if the hole type or values change.

OBJECTIVES

After completing this chapter, you will be able to perform the following:

- Create fillets
- Create chamfers
- Create holes
- Shell a part
- Create work axes
- Create work points
- Create work planes
- Create a UCS
- Pattern features

FILLETS

Fillet features consist of fillets and rounds. Fillets add material to interior edges to create a smooth transition from one face to another. Rounds remove material from exterior edges. The following image shows a part without fillets on the left and the part with fillets on the right.

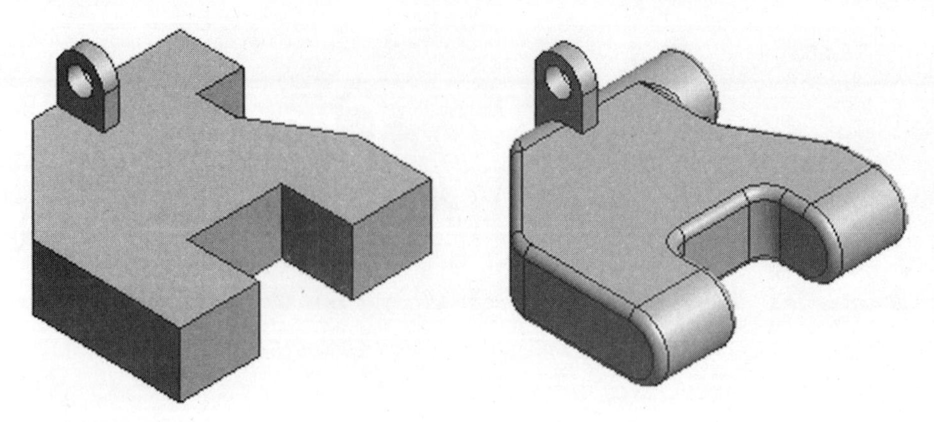

FIGURE 4.1

To create fillets, you select the edge that needs to be filleted; the fillet is created between the two faces that share the edge, or you can select two or three faces that a fillet will go between. When placing a fillet between two faces, the faces do not need to share a common edge. When creating a part, it is good practice to create fillets and chamfers as some of the last features in the part. Fillets add complexity to the part, which in turn adds to the size of the file. They also remove edges that you may need to place other features.

To create a fillet feature, click the Fillet command from the Model tab > Modify panel, as shown in the following image on the left, while not in a sketch click on an edge and the mini-toolbar will appear click the Create Fillet button as shown in the following image on the right or press the shortcut key F.

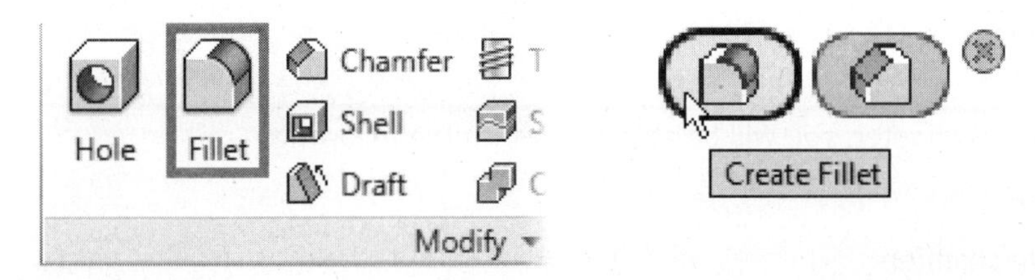

FIGURE 4.2

After you click the command, the Fillet dialog box appears as shown in the following image on the left and the mini-toolbar will appear as shown in the following image on the right. Along the left side of the Fillet dialog box, there are three types of fillets: Edge, Face, and Full Round. When the Edge option is selected, you see three tabs: Constant, Variable, and Setbacks. Each tab creates a fillet along an edge(s) with different options. The options for each of the tabs and fillet types are described in the following sections. A preview of a fillet will appear on the part when the preview option on the bottom of the dialog box is checked and a valid fillet can be created from the data you input into the Fillet dialog box or the mini-toolbar.

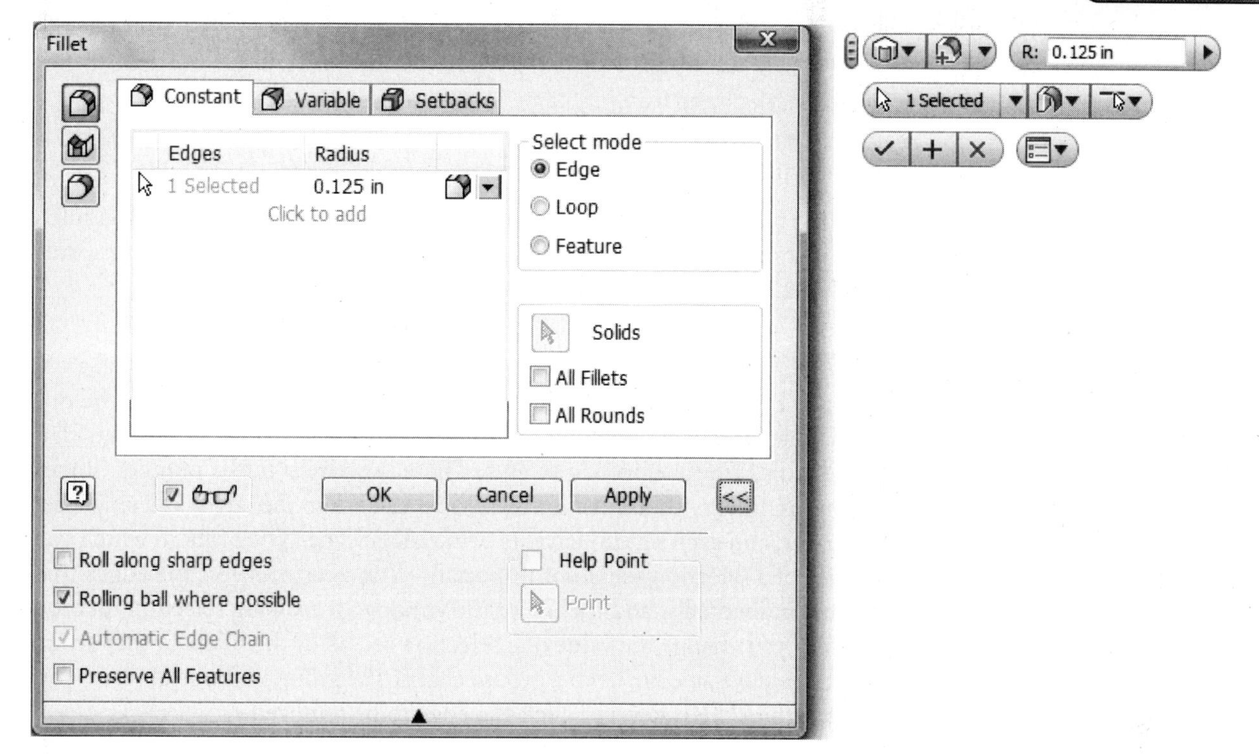

FIGURE 4.3

Before we look at the tabs, let's discuss the methodology that you will use to create fillets. You can click an edge or edges, or faces to fillet, or select the type of fillet to create. If you click an edge or face before you issue the Fillet feature command, it is placed in the first selection set. Selection sets contain the edges or faces that will be filleted when the OK button is clicked.

Each fillet feature can contain multiple selection sets, each having its own unique fillet value. There is no limit to the number of selection sets that can exist in a single instance of the feature. An edge, however, can only exist in one selection set. All of the selection sets included in an individual fillet command appear as a single fillet feature in the browser. Click the type of fillet that you want to create. To add edges to the first selection set, select the edges that you want to have the same radius. To create another selection set, select Click to add, and then click the edges that will be part of the next selection set. To remove an edge or face that you have selected, click the selection set that includes the edge or face, and the edges or faces will be highlighted. Hold down the CTRL key, and click the edge(s) or face(s) to be removed from the selection set. Enter the desired values for the fillet. As changes are made in the dialog box, a representation of the fillet is previewed in the graphics area. When the fillet type and value are correct, click the OK button to create the fillet.

To start editing a fillet's type and radius, use one of the following methods to start the editing process:

- In the graphics window click on the fillet to edit and click the Edit Fillet button that appears in the graphics window.

- Start the Edit Feature command by right-clicking on the fillet's name in the browser and selecting Edit Feature from the menu.
- Double-click on the feature's name in the browser.
- Change the select priority to Select Feature and double-click on the fillet that is on the part.

The Fillet dialog box appears with all of the settings that you used to create the fillet. Change the fillet settings as needed. Then if needed click the Local Update command on the Quick Access toolbar to update the part.

Edge Fillet

With the Edge fillet type selected, the Constant tab will be activated. With the options on the Constant tab, as shown in the previous image, you can create fillets that have the same radius from beginning to end. There is no limit to the number of part edges that you can fillet with a constant fillet. You can select the edges as a single set or as multiple sets, and each set can have its own radius value. The order in which you select the edges of a selection set is not important. You need to select the edges that are to be filleted individually, and the use of the window or crossing selection method is not allowed. If you change the value of a selection set, all of the fillets in that group will change. To remove an edge from a group, choose the group in the select area, and then hold down the CTRL key and click the edge to be removed.

Constant Tab

The following section describes the options that are available on the Constant tab.

Select Edge, Radius, and Continuity

Edge

By default, after issuing the Fillet command, you can click edges, and they appear in the first selection set. You can continue to select multiple edges. To remove an edge from the set, hold down the CTRL key and select the edge.

Radius

Enter a size for the fillet. The size of the fillet will be previewed on the selected edges.

Continuity

To adjust the continuity of the fillet, select either a tangent or smooth (G2) condition (continuous curvature), as shown in the following image in both the dialog box and in the mini-toolbar.

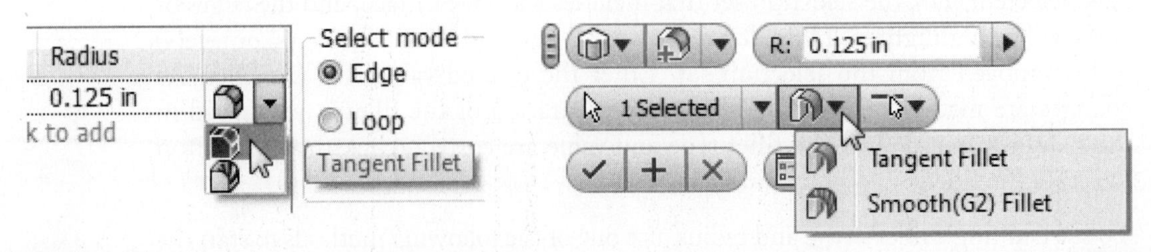

FIGURE 4.4

To create another selection set, select Click to add, and then click the edges that will be part of the next selection set. After clicking an edge, a preview image of the fillet appears on the edge that reflects the current values that are set in the dialog box or mini-toolbar, as shown in the following image.

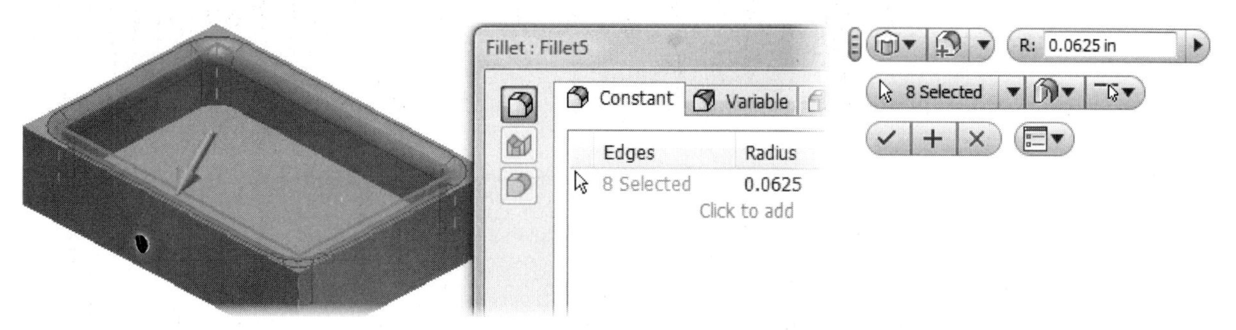

FIGURE 4.5

Select Mode

Edge

Click the Edge mode to select individual edges to fillet. By default, any edge that is tangent to the clicked edges is also selected. If you do not want to have tangent edges automatically selected, uncheck the Automatic Edge Chain option in the More (>>) section.

Loop

Click the Loop mode to have all of the edges that form a closed loop with the selected edge filleted.

Feature

Click the Feature mode to select all of the edges of a selected feature.

Solids

If multiple solid bodies exist, select the solid body to apply All Fillets or All Rounds.

All Fillets

Click the All Fillets option to select all concave edges of a part that you have not filleted already—see the following image on the left. The All Fillets option adds material to the part, and it requires a separate edge selection set to remove material from the remaining edges using the All Rounds option.

All Rounds

Click the All Rounds option to select all convex edges of a part that you have not filleted already, as seen in the following image on the right. The All Rounds option removes material from the part, and it requires a separate edge selection set from All Fillets.

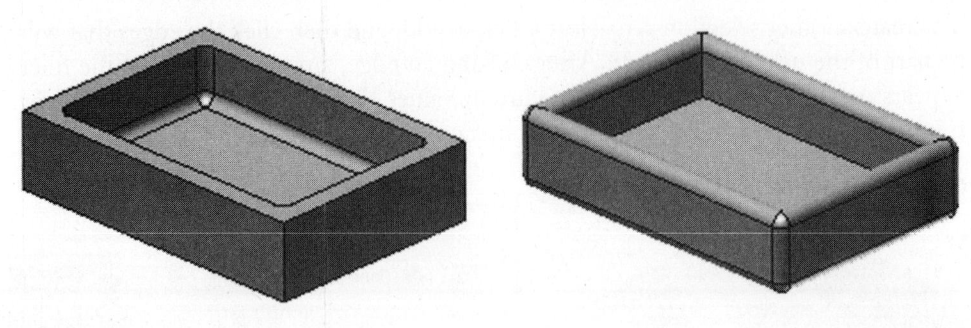

FIGURE 4.6

More (>>) Options. Click the More (>>) button to access other options, as shown in the following image.

Roll along Sharp Edges

Click this option to adjust the specified radius when necessary to preserve the edges of adjacent faces.

Rolling Ball Where Possible

Click this option to create a fillet around a corner that looks like a ball has been rolled along the edges that define a corner, as shown in the following (middle) image. When the Rolling ball where possible solution is possible, but it is not selected, a blended solution is used, as shown in the following image on the right.

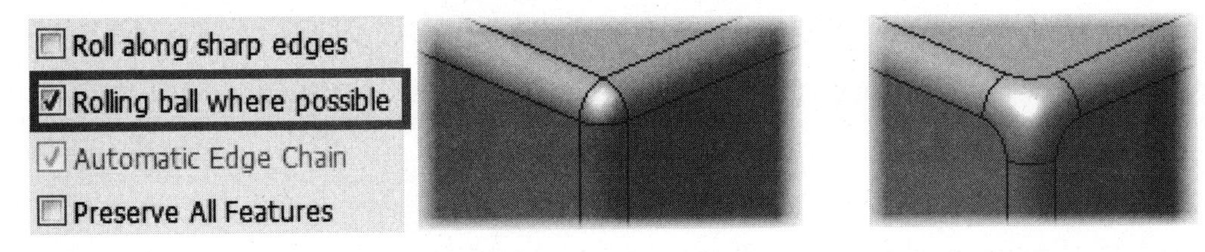

FIGURE 4.7

Automatic Edge Chain

Click this option to select tangent edges automatically when you click an edge.

Preserve All Features

Click to check all features that intersect with the fillet and to calculate their intersections during the fillet operation. If the option's checkbox is clear, only the edges that are part of the fillet operation are calculated during the operation.

For example, if Preserve All Features is checked, a fillet is placed on the outside edge of a shelled box. Since the fillet is larger than the shell thickness, a gap will exist in the fillet, as shown in the image on the left. If the Preserve All Features is unchecked, the inside edges for the shell will appear in the fillet, as shown in the following image on the right. The default option is NOT to preserve the features.

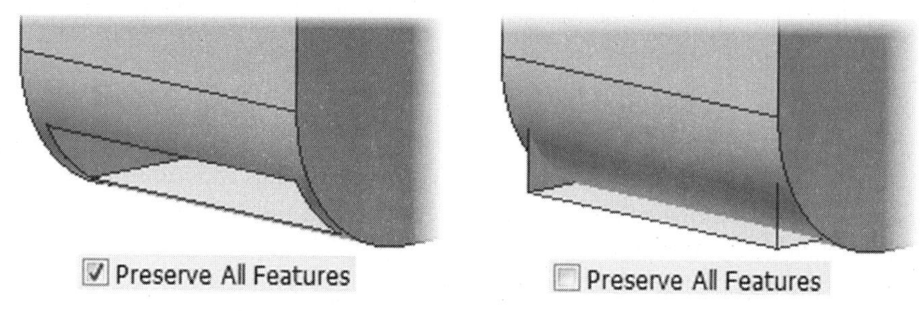

FIGURE 4.8

Variable Tab

You can also create a variable radius fillet that has a different starting and ending radius and/or a different radius between the starting and ending radius. To create a variable radius fillet, click the Edge Fillet option in the upper-left corner of the dialog box, and click the Variable tab, as shown in the following image on the left or click the Add Variable Set option from the mini-toolbar on the right.

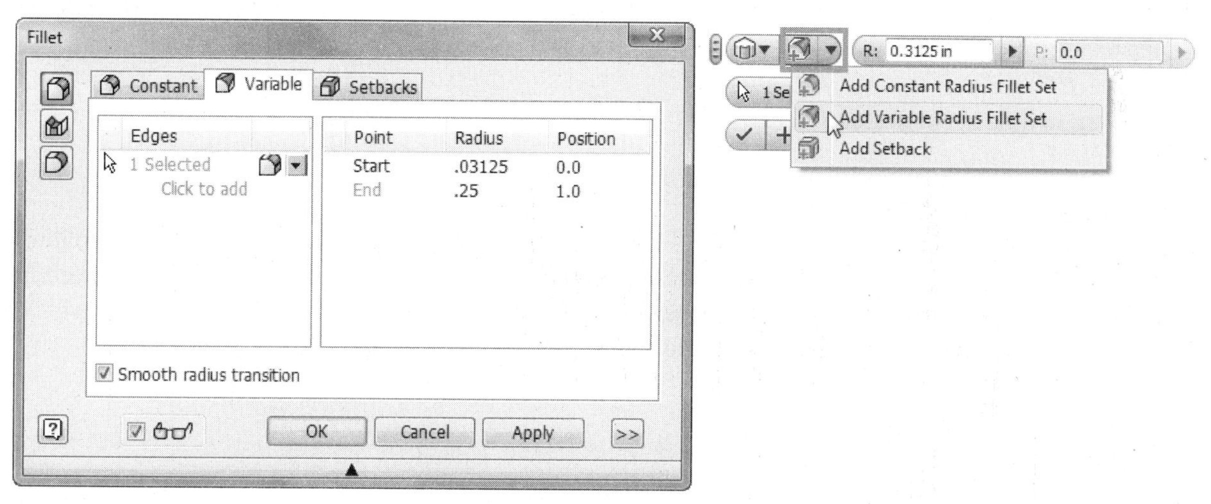

FIGURE 4.9

The following image on the left shows a variable fillet with the smooth option, and the image on the right shows a variable fillet blending in a straight line.

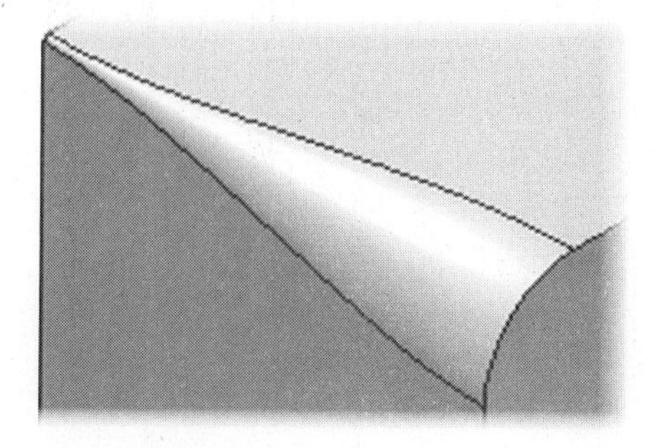

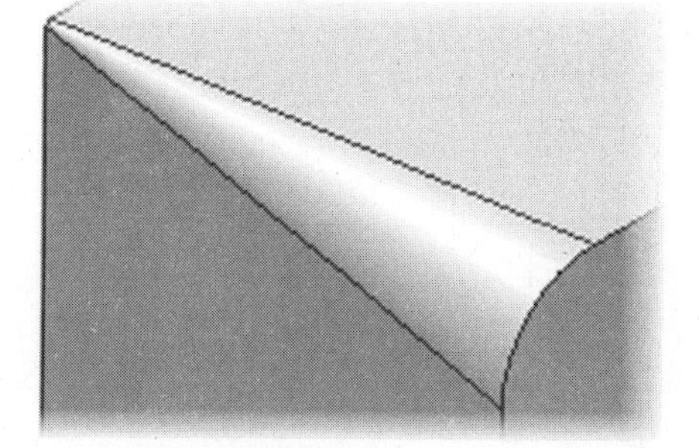

FIGURE 4.10

Setbacks Tab

You can specify the distance at which a fillet starts its transition from a vertex with the options on the Setbacks tab. Using these options, you can model special fillet applications where three or more edges converge, as shown in the following image. You can choose a different radius for each converging edge if needed. Click the minimal option to create a setback with the smallest possible fillet. You can only use setbacks where three or more filleted edges form a vertex. To setback fillets, follow these steps:

1. In the graphics window, select three or more edges to fillet. The fillets must converge at a point.
2. Click the Setbacks tab.
3. In the graphics window, click the vertex point.
4. In the Fillet dialog box, set the setback distance for each edge.

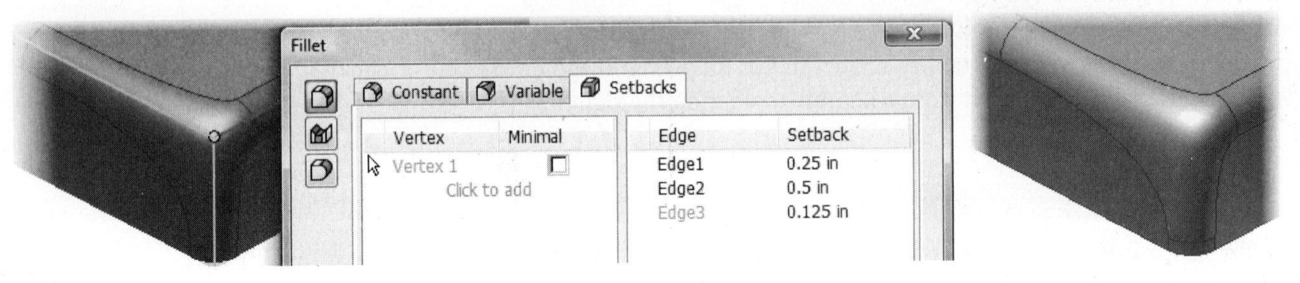

FIGURE 4.11

Face Fillet

With the Face fillet type selected, the dialog box will change, as shown in the following image on the left and the mini-toolbar on the right. Create a face fillet by selecting two or more faces; the faces do not need to be adjacent. If a feature exists that will be consumed by the fillet, the volume of the feature will be filled in by the fillet.

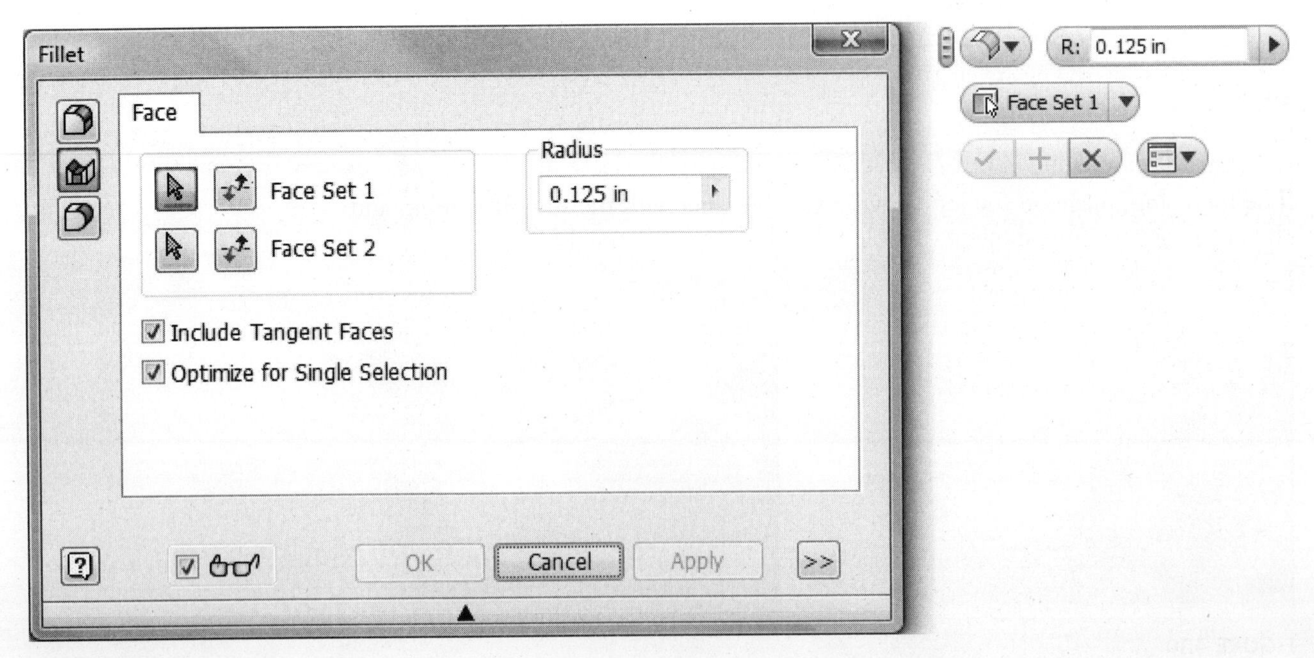

FIGURE 4.12

To create a face fillet, follow these steps:

1. With the Face Set 1 button active, select one or more tangent contiguous faces on the part to which the fillet will be tangent.
2. With the Face Set 2 button active, select one or more tangent contiguous faces on the part to which the fillet will be tangent.

Check the Include Tangent Faces option to automatically chain all faces that are tangent to faces in the selection set.

Check the Optimize for Single Selection to automatically make the next selection set button active after selecting a face.

The following image on the left shows a part that has a gap between the bottom and top extrusion, the middle image shows the preview of the face fillet, and the image on the right shows the completed face fillet. Notice the rectangular extrusion on the top face is consumed by the fillet.

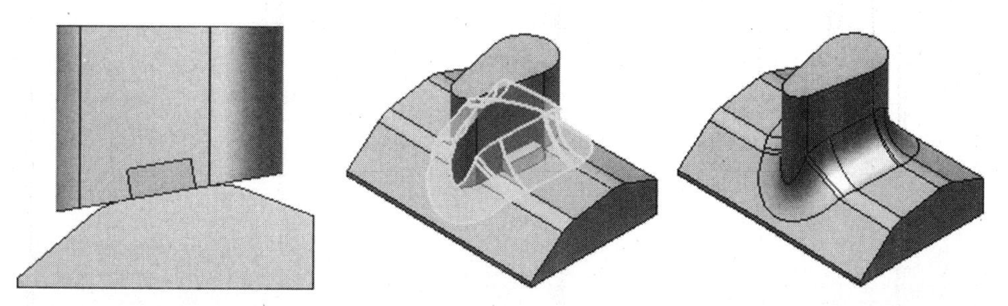

FIGURE 4.13

FullRound Fillet

With the FullRound fillet type selected, the dialog box will change, as shown in the following image on the left and the mini-toolbar on the right. The full round fillet option creates a fillet that is tangent to three faces; the faces do not need to be adjacent.

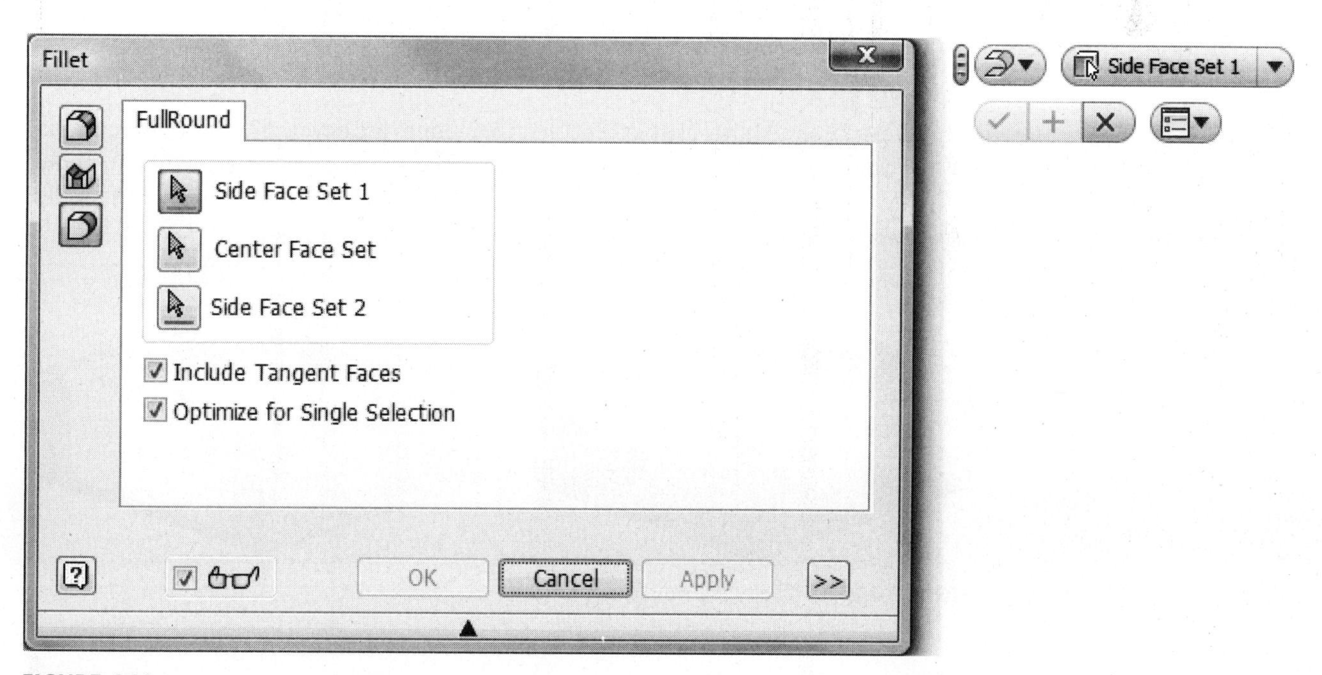

FIGURE 4.14

To create a FullRound fillet, follow these steps:

1. With the Side Face Set 1 button active, select a face on the part that the fillet will start at and to which it will be tangent.
2. With the Center Face Set button active, select a face on the part that the middle of the fillet will be tangent.
3. With the Side Face Set 2 button active, select a face on the part that the fillet will end at and to which it will be tangent.

Check the Include Tangent Faces option to automatically chain all faces that are tangent to faces in the selection set.

Check the Optimize for Single Selection to automatically make the next selection set button active after selecting a face.

The following image shows three faces selected on the left and the resulting fullround fillet on the right.

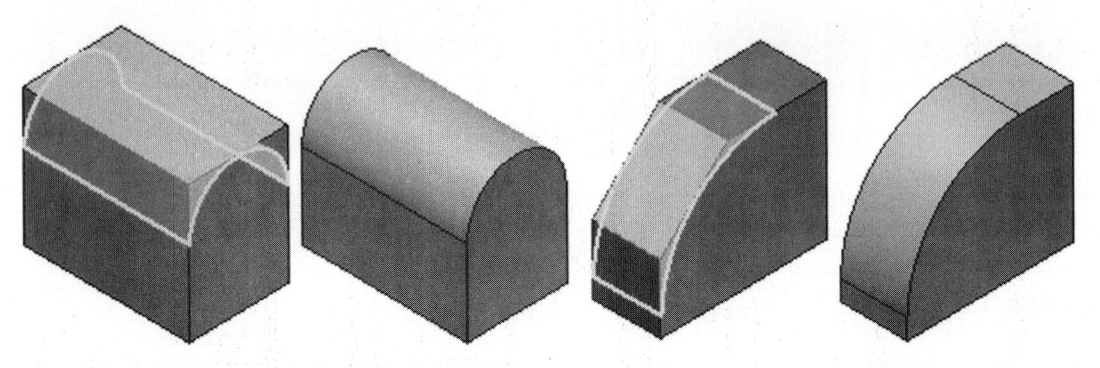

FIGURE 4.15

TIP

If you get an error when creating or editing a fillet, try to create it with a smaller radius. If you still get an error, try to create the fillet in a different sequence or create multiple fillets in the same operation.

CHAMFERS

Chamfers are similar to fillets except that their edges are beveled rather than rounded. When you create a chamfer on an interior edge, material is added to your model. When you create a chamfer on an exterior edge, material is cut away from your model, as shown in the following image.

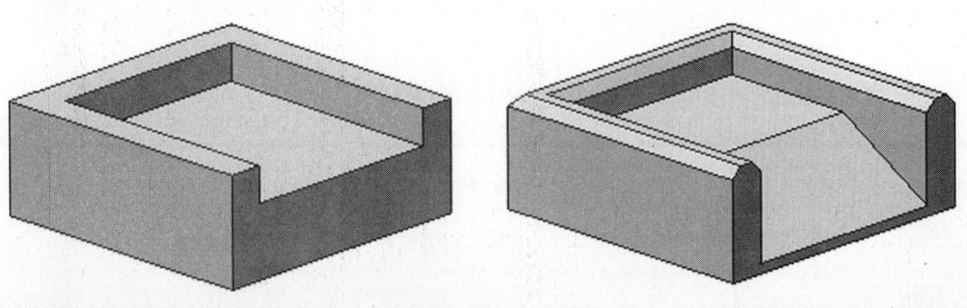

FIGURE 4.16

To create a chamfer feature, follow the same steps that you used to create fillet features. Click the common edge, and the chamfer is created between the two faces sharing the edge. To create a chamfer feature, click on the Chamfer command on the Model tab > Modify panel, as shown in the following image on the left, or press the keys CTRL SHIFT K. Another method, while not in a sketch, click on an edge and the mini–toolbar will appear, then click the Create Chamfer button in the mini-toolbar, as shown in the following image on the right.

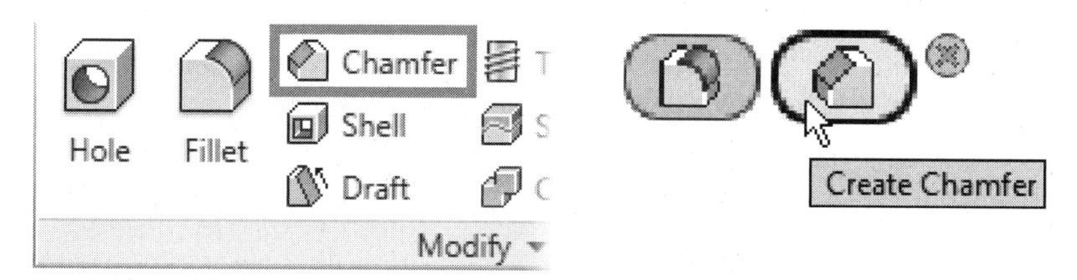

FIGURE 4.17

After you start the command, the Chamfer dialog box appears, as shown in the following image on the left and the mini-toolbar appears as shown in the following image on the right. As with fillet features, you can select multiple edges to be included in a single chamfer feature. From the dialog box or mini-toolbar, click a method, click the edge or edges to chamfer, enter a distance and/or angle, and then click OK.

FIGURE 4.18

To edit the type of chamfer feature or distances, use one of the following methods to start the editing process:

- In the graphics window click on the chamfer to edit and click the Edit Chamfer button that appears in the mini-toolbar.
- Double-click the feature's name or icon in the browser to edit a distance.
- Right-click the chamfer's name in the browser, and select Edit Feature from the menu. The Chamfer dialog box appears with all the settings that you used to create the feature. Adjust the settings as desired.
- Change the select priority to Feature Priority, and then double-click the chamfer on the part. The Chamfer dialog box appears with all the settings that you used to create the feature. Adjust the settings as desired.

- The Chamfer dialog box appears with all of the settings that you used to create the fillet. Change the chamfer settings as needed. Then if needed click the Local Update command on the Quick Access toolbar to update the part.

Method

Distance

Click the Distance option to create a 45° chamfer on the selected edge. You determine the size of the chamfer by typing a distance in the dialog box. The value is the offset from the common edge of the two adjacent faces. You can select a single edge, multiple edges, or a chain of edges. A preview image of the chamfer appears on the part. If you select the wrong edge, hold down the CTRL key and select the edge to remove. The following image illustrates the use of the Distance option on the left.

Distance and Angle

Click the Distance and Angle option to create a chamfer offset from a selected edge on a specified face, at the defined angle. In the dialog box, enter an angle and distance for the chamfer, then click the face to which the angle is applied and specify an edge to be chamfered. You can select one edge or multiple edges. The edges must lie on the selected face. A preview image of the chamfer appears on the part. If you have selected the wrong face or edge, click on the Edge or Face button, and choose a new face or edge. The following image in the middle illustrates the use of the Distance and Angle option.

Two Distances

Click the Two Distances option to create a chamfer offset from two faces, each being the amount that you specify. Click an edge first, and then enter values for Distance 1 and Distance 2. A preview image of the chamfer appears. To reverse the direction of the distances, click the Flip button. When the correct information about the chamfer is in the dialog box, click the OK button. You can only use a single edge or chained edges with the Two Distances option. The following image illustrates the use of the Two Distances option on the right.

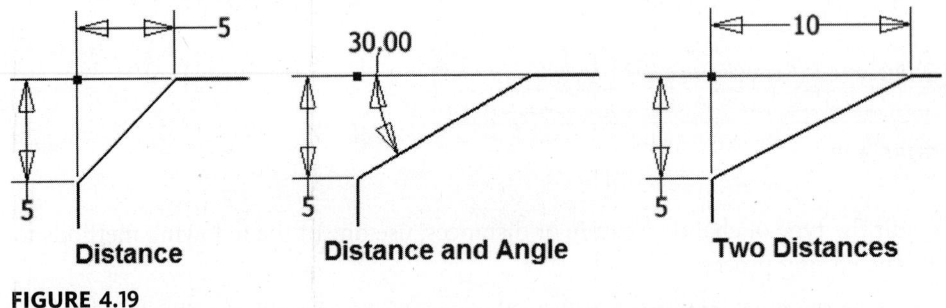

FIGURE 4.19

Edge and Face

Edges

Click an edge or edges to be chamfered.

Face

Click a face on which the chamfer will be based.

Flip

Click the button to reverse the direction of the distances for a Two Distances chamfer.

Distance and Angle

Distance

Enter a distance to be used for the offset.

Angle

Enter a value that will be used for the angle if creating the Distance and Angle chamfer type.

Edge Chain and Setback

Edge Chain

Click this option to include tangent edges in the selection set automatically, as shown in the following image.

Setback

When the Distance method is used and three chamfers meet at a vertex, click this option to have the intersection of the three chamfers form a flat edge (left button) or to have the intersection meet at a point as though the edges were milled (right button), as shown in the following image.

Preserve All Features

Click this option to check all features that intersect with the chamfer and to calculate their intersections during the chamfer operation, as shown in the following image. If the option's checkbox is clear, only the edges that are part of the fillet operation are calculated during the operation.

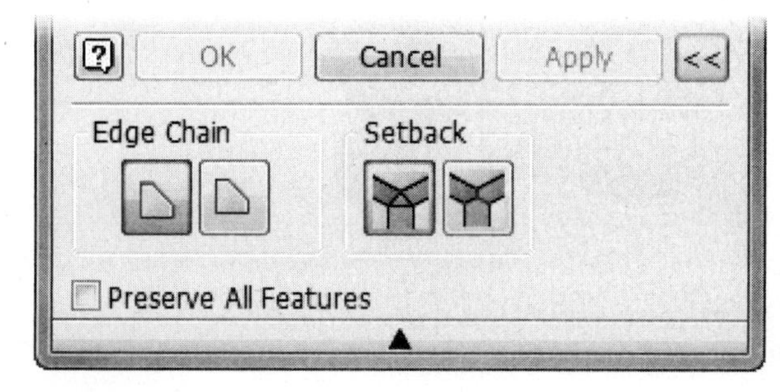

FIGURE 4.20

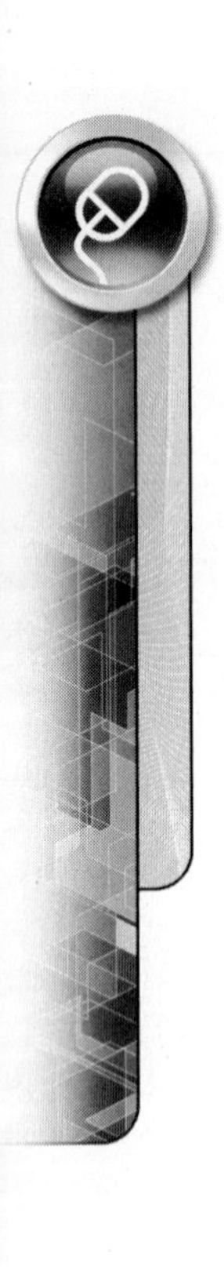

EXERCISE 4-1: CREATING FILLETS AND CHAMFERS

In this exercise, you will create constant radius fillets, full round fillets, face fillet, and chamfers.

1. Open *ESS_E04_01.ipt* in the Chapter 04 folder.
2. Click the Fillet command in the Model tab > Modify panel or click the inside edge of the cutout and click Create Fillet from the mini-toolbar.
3. In the Fillet dialog box, click on the first entry in the Radius column, and type **.25** or in the mini-toolbar enter **.25** as shown in the following image.

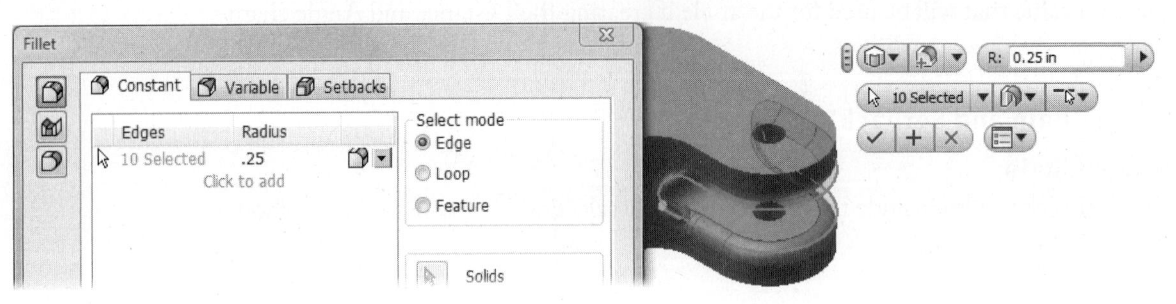

FIGURE 4.21

4. Click Apply to create the fillet.
5. Next, create a Full Round Fillet that will be tangent to the front three faces. In the Fillet dialog box, click the Full Round Fillet option in the left column.
6. Click the front-inside face, then the left-front face, and then the back-vertical face of the part, as shown in the following image.

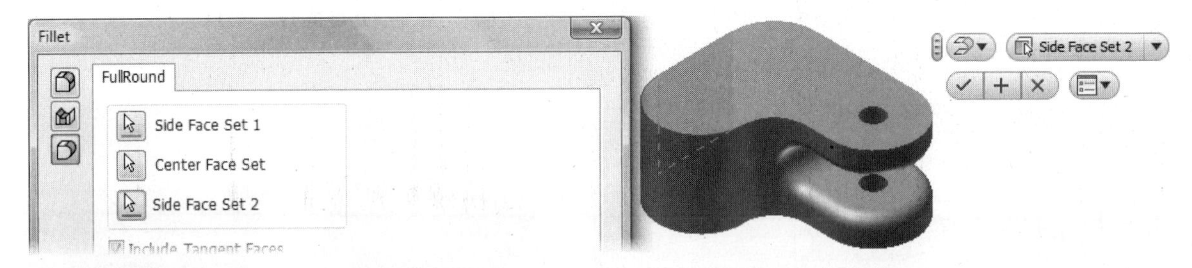

FIGURE 4.22

7. Click OK to create the fillet.
8. In the browser, right-click on Extrusion4, and click Unsuppress Features.
9. Next, create a Face Fillet. Click the Fillet command in the Modify panel. In the Fillet dialog box, click the Face Fillet option in the left column.
10. Click the cylindrical face of Extrusion4 and then the full round fillet in the model you created in the last step, as shown in the following image.

11. In the Fillet dialog box, type **.375 in** in the Radius field, as shown in the following image.

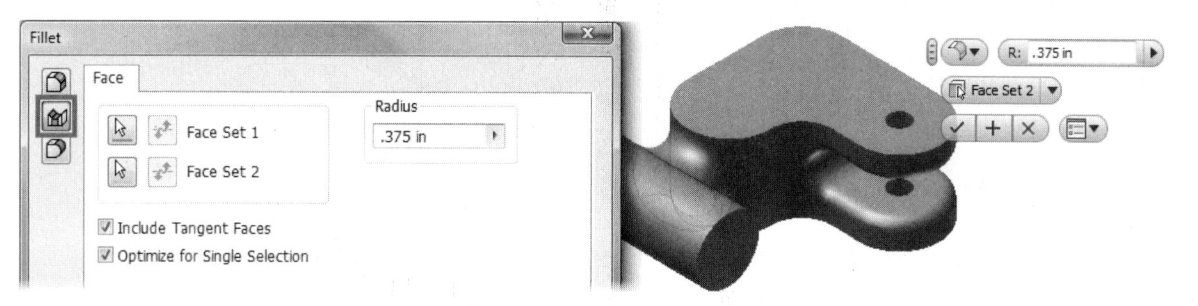

FIGURE 4.23

12. Click OK to create the fillet and rotate the viewpoint to examine the fillet features.
13. Move the cursor into a blank area in the graphics window, right-click, and click Repeat Fillet from the menu.
14. Next you fillet two edges. Click the top and bottom edges of the model, as shown in the following image.
15. In the mini-toolbar input radius field, type **.125**, as shown in the following image.

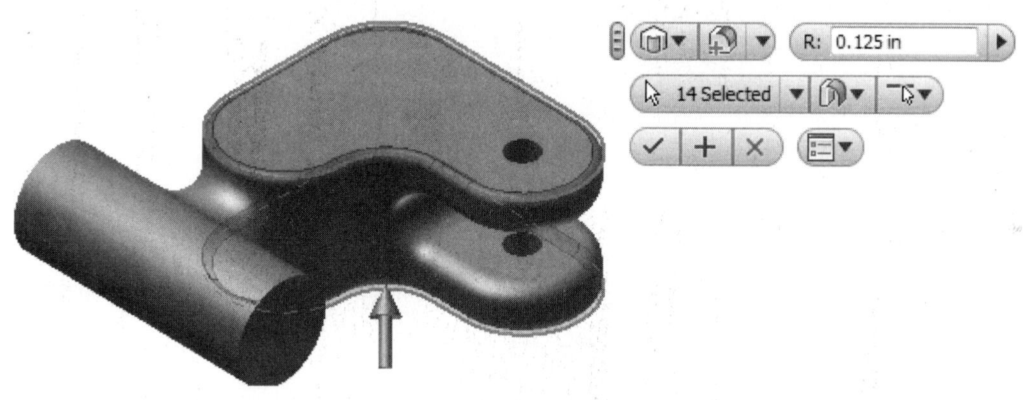

FIGURE 4.24

16. Click the green check mark (OK button) in the mini-toolbar to create the fillets.
17. Click the back cylindrical edge on Extrusion4 and click Create Chamfer from the mini-toolbar as shown in the following image on the left.
18. In the mini-toolbar's distance field, type **.25**, as shown in the following image on the right.
19. To create the chamfer and keep the dialog box open, click green plus button in the mini-toolbar. The green plus button is the same as the Apply button in the dialog box.

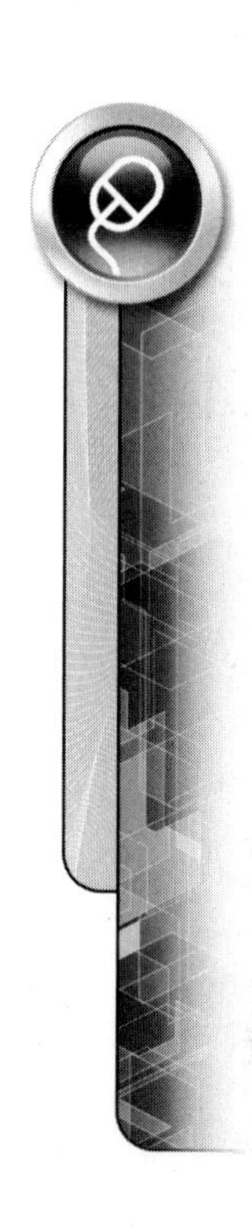

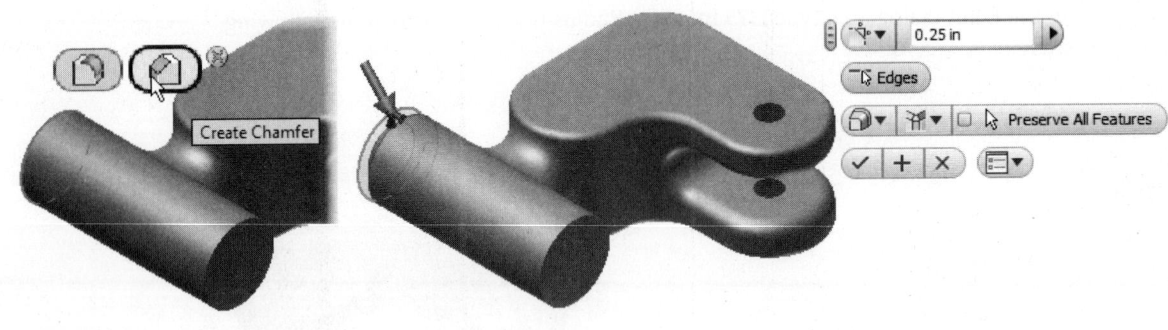

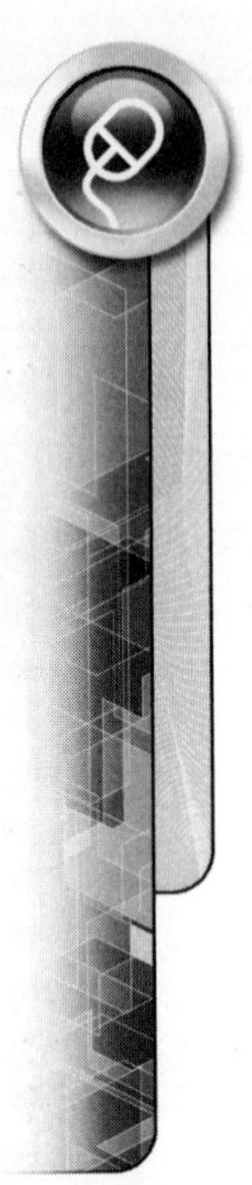

FIGURE 4.25

20. In the Chamfer dialog box or in the mini-toolbar, click the Distance and Angle option.
21. Click the front-circular face on Extrusion4.
22. Click the front-circular edge of the same face.
23. In the Chamfer dialog box or in the mini-toolbar, enter **.25** in the Distance field and **60 deg** in the Angle field, as shown in the following image.

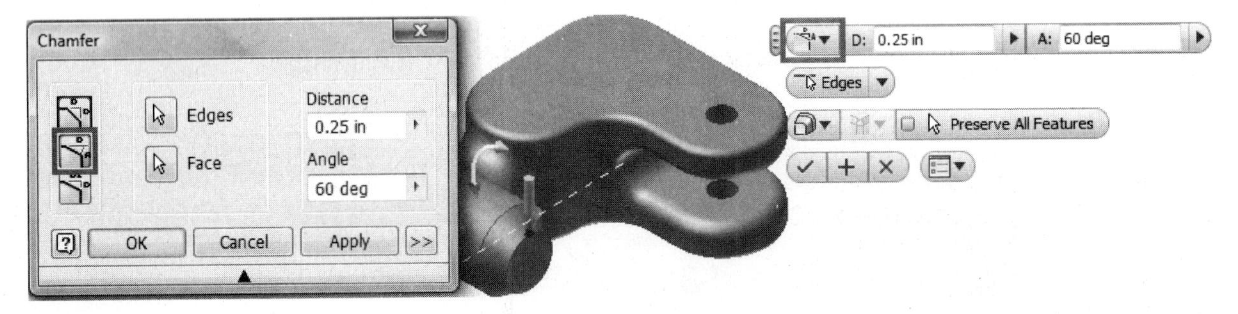

FIGURE 4.26

24. Click OK to create the chamfer. When done, your model should resemble the following image.

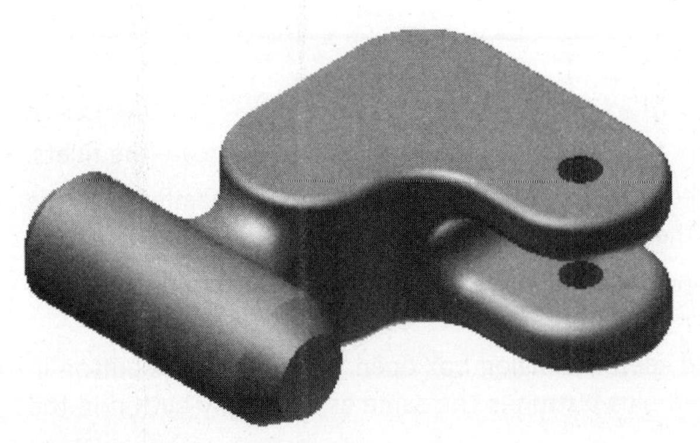

FIGURE 4.27

25. Practice editing and placing fillets and chamfers.
26. Close the file. Do not save changes. End of exercise.

HOLES

The Hole command lets you create drilled, counterbored, spotface, countersunk, clearance, tapped, and taper tapped holes, as shown in the following image. You can place holes using sketch geometry or existing planes, points, or edges of a part. You can also specify the type of drill point and thread parameters.

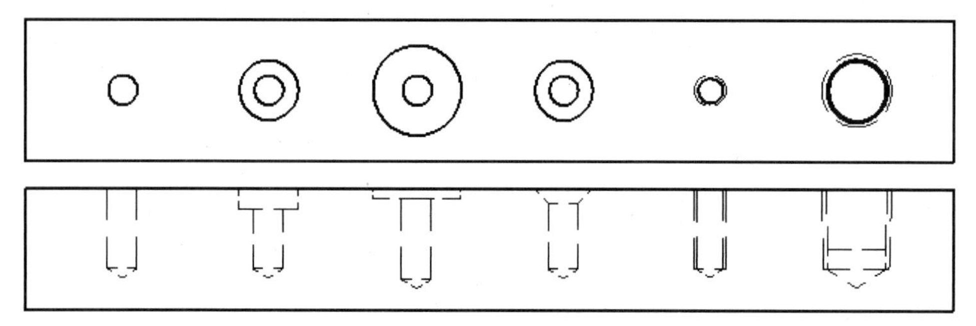

FIGURE 4.28

To create a hole feature, follow these steps:

1. Create a part that you want to place a hole on.
2. Click the Hole command from the Model tab > Modify panel, as shown in the following image on the left, or press the hot key H. You could also right-click and click Hole from the marking menu.
3. The Holes dialog box appears, as shown in the following image on the right. Four placement options are available: From Sketch, Linear, Concentric, and On Point. The Placement options are covered in the next section. After you have chosen the hole placement options, select the desired hole style options from the Holes dialog box. As you change the options, the preview image of the hole(s) is updated. When you are done making changes, click the OK button to create the hole(s).

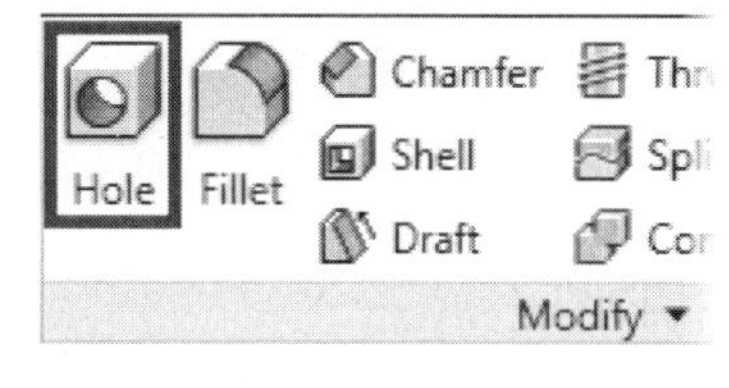

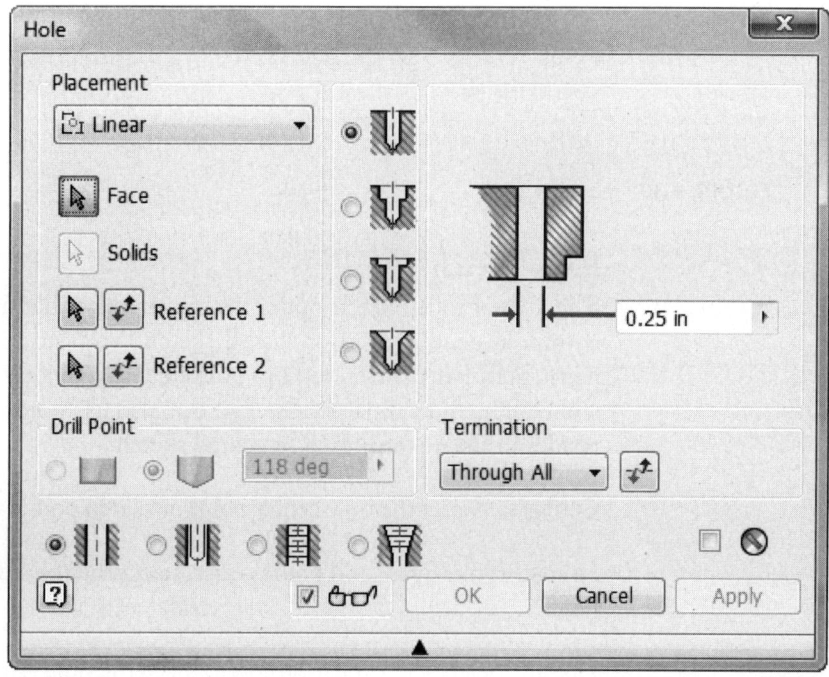

FIGURE 4.29

Editing Hole Features

To edit the type of hole feature or distances, use one of the following methods:

- Double-click the feature's name or icon in the browser to display the Holes dialog box with the dimensions and option you used to create the hole feature.
- Change the select priority to Feature Priority, and then double-click the hole feature on the part in the graphics area. This will display the Holes dialog box with the dimensions and option you used to create the hole feature.
- Right-click the hole's name or icon in the browser, and select Edit Feature from the menu to display the Holes dialog box with the dimensions and option you used to create the hole feature.

Holes Dialog Box

In the Holes dialog box, you establish the placement method, type of hole, its termination, and additional options such as type of drill point, angle, and tapped properties.

Placement

Select the appropriate placement method. If you select From Sketch, a sketch containing hole centers or any point, such as an endpoint of a line, must exist on the part. Hole centers are described in the next section. The Linear, Concentric, and On Point options do not require an unconsumed or shared sketch to exist in the model and are based on previously created features. Depending on the placement option you select, the input parameters will change, as shown in the following images and as described below.

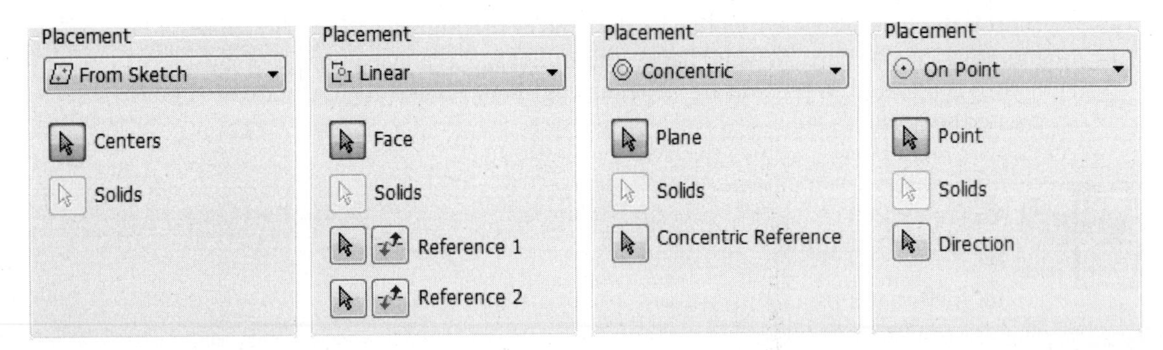

FIGURE 4.30

From Sketch

Select the From Sketch option to create holes that are based on a location defined within an unconsumed or shared sketch. You can base the center of the hole on a point/hole center or endpoints of sketched geometry like endpoints, centers of arcs and circles, and spline points. You can also use points from projected geometry that resides in the unconsumed or shared sketch.

Centers. Select the hole center point or sketch points where you want to create a hole.

Solids. If multiple solid bodies exist, select the solid body where you want to create a hole.

Linear. Select the Linear option to place the hole relative to two selected face edges.

Face. Select the face on the part where the hole will be created.

Solids. If multiple solid bodies exist, select the solid body where you want to create a hole.

Reference 1. Select a face edge as a positional reference for the center of the hole. When you select the edge, a dimension appears that can be edited to constrain the center of the hole dimensionally.

Reference 2. Select a face edge as a positional reference for the center of the hole. When you select the edge, a dimension appears that can be edited to constrain the center of the hole dimensionally.

Flip Side. Click this button to position the hole on the opposite side of the selected edge.

Concentric. Select the Concentric option to place the hole on a planar face and concentric to a circular or arc edge or a cylindrical face.

Plane. Select a planar face or plane where you want to create the hole.

Solids. If multiple solid bodies exist, select the solid body where you want to create the hole.

Concentric Reference. Select a circular or arc model edge or cylindrical face to constrain the center of the hole to be concentric with the selected entity.

On Point

Select the On Point option to place the center of the hole on a work point. The work point must exist on the model prior to selecting this option.

Point. Select a work point to position the center of the hole.

Solids. If multiple solid bodies exist, select the solid body where you want to create the hole.

Direction. Select a plane, face, work axis, or model edge to specify the direction of the hole. When selecting a plane or face, the hole direction will be normal to the face or plane.

Hole Option

Click the type of hole that you want to create: drilled, counterbore, spotface, or countersink, and enter the appropriate dimensions. The following image shows the counterbore option.

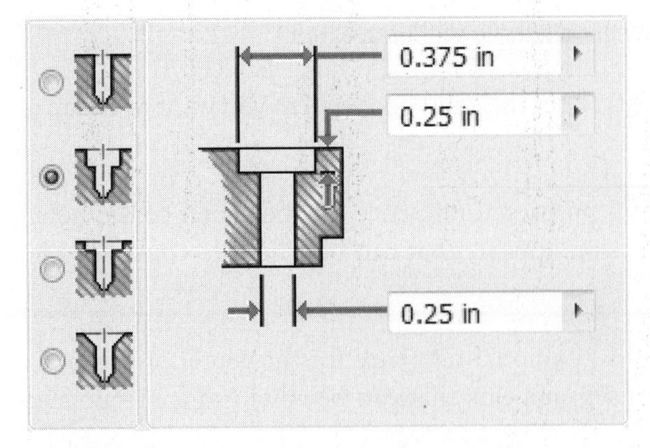

FIGURE 4.31

Termination

Select how the hole will terminate.

- *Distance:* Specify a distance for the depth of the hole.
- *Through All:* Choose to extend the hole through the entire part in one direction.
- *To:* Select a plane at which the hole will terminate.
- *Flip:* Reverse the direction in which the hole will travel.

Drill Point

Select either a flat or angle drill point. If you select an angle drill point, you can specify the angle of the drill point.

Infer iMates

Check this box to automatically create an iMate on a full circular edge. Autodesk Inventor attempts to place the iMate on the closed loop most likely to be used.

Dimensions

To change the diameter, depth, countersink, counterbore diameter, countersink angle, or counterbore depth of the hole, click the dimension in the dialog box, and enter a desired value.

Hole Type

Click the type of hole you want to create. There are four options: Simple Hole, Clearance Hole, Tapped Hole, and Taper Tapped Hole. The following image shows the tapped hole selected. After selecting the hole type, fill in the dialog box with the specific data for the hole you need to create.

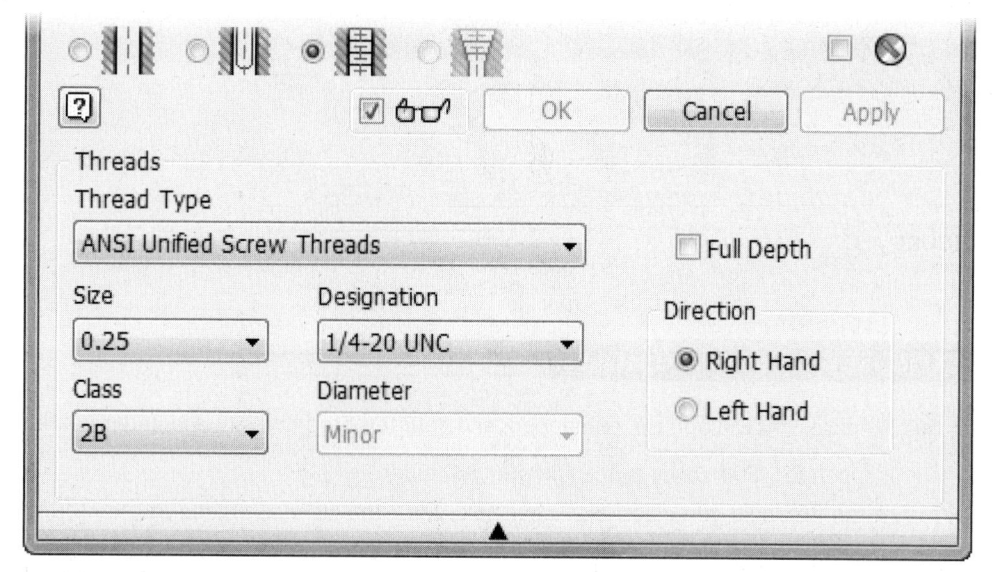

FIGURE 4.32

Simple Hole. Click the Simple Hole option to create a (drilled) hole feature with no thread features or properties.

Clearance Hole. Click the Clearance Hole option to create a simple hole feature that does not have thread.

Tapped Hole. Click the Tapped Hole option if the hole is threaded. Thread information appears in the dialog box area so that you can specify the thread properties, as shown in the above image.

Taper Tapped Hole. Click the Taper Tapped Hole option if the hole is tapered thread. The taper tapped hole information appears in the dialog box area so you can specify the thread properties.

Center Points

Center points are sketched entities that can be used to locate hole features. To create a hole center, follow these steps:

1. Make a sketch active.
2. Click the Point, Center Point command in the Sketch tab > Draw panel as shown in the following image on the left.
3. Click a point to locate the hole center where you want to place the hole, and constrain and dimension the point as desired.

You can also switch between a hole center and a sketch point by using the Center Point command on the Sketch tab > Format panel, as shown in the following image on the right, when the sketch environment is active. When you press the button, you create a center point. If you use a center point style when you activate the From Sketch option of the Hole command, all center points that reside in the sketch are selected as centers for the hole feature automatically. This can expedite the process of creating multiple holes in a single hole feature. You can also select endpoints of lines, arcs, splines, center points of arcs and circles, or spline control points as hole centers. Points can be deselected by holding down the CTRL or SHIFT key and then selecting the points.

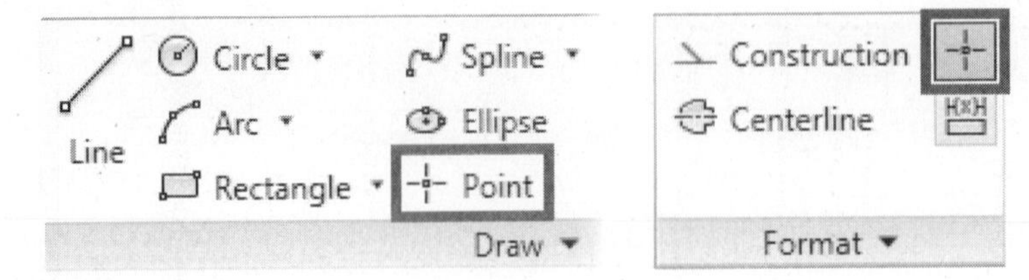

FIGURE 4.33

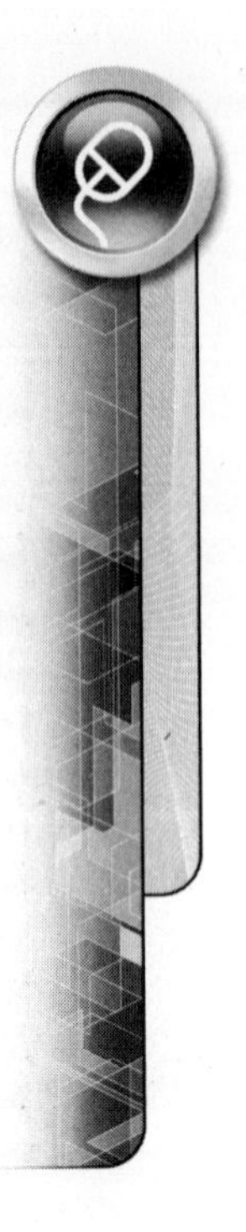

EXERCISE 4-2: CREATING HOLES

In this exercise, you will add drilled, tapped, and counterbored holes to a cylinder head.

1. Open *ESS_E04_02.ipt* in the Chapter 04 folder.
2. The first hole you place is a linear hole. Click the Hole command in the Modify panel.
   - For the Face, click the top planar face of Extrusion1.
   - For Reference 1 and 2, select the bottom left and left-angled edge, and place the point **.5 in** from each edge.
   - Verify that the Drilled hole option is selected.
   - Change the Termination to Through All.
   - Change the hole's diameter to **.25 in**, as shown in the following image.
   - Click OK to create the hole.

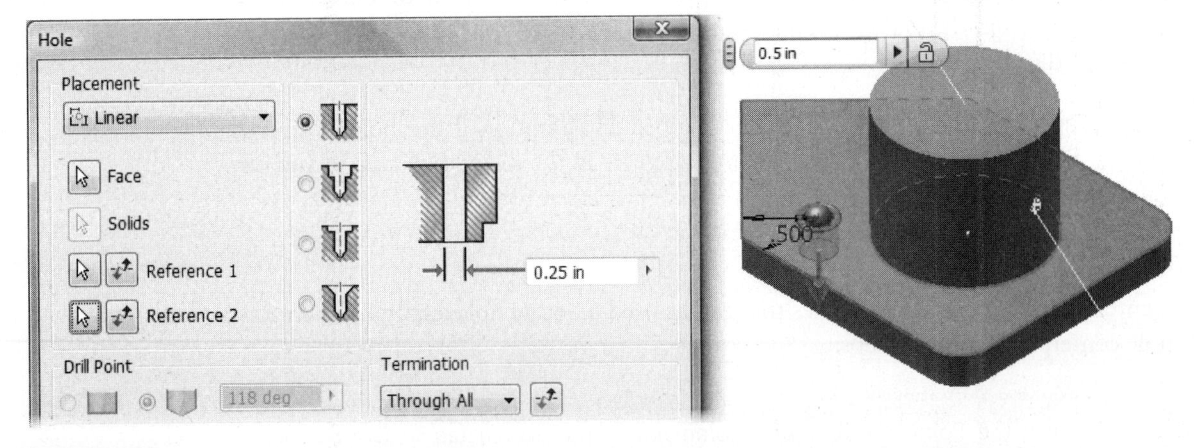

FIGURE 4.34

3. Next, you will create a hole based on the Sketch option. Click the Create 2D Sketch command on the Sketch panel, and then click the top planar face of Extrusion1. The center point of the arc is projected onto the sketch.
4. Move the cursor to a blank area in the graphics window and right-click, and then click Finish Sketch from the menu.
5. Click the Hole command in the Modify panel or press the H key.
   - Click the center points that were projected from the two filleted edges.
   - Change the hole option to counterbore.

- Ensure the Termination is set to Through All.
- Change the hole's values as shown in the following image.
- Click OK to create the holes.

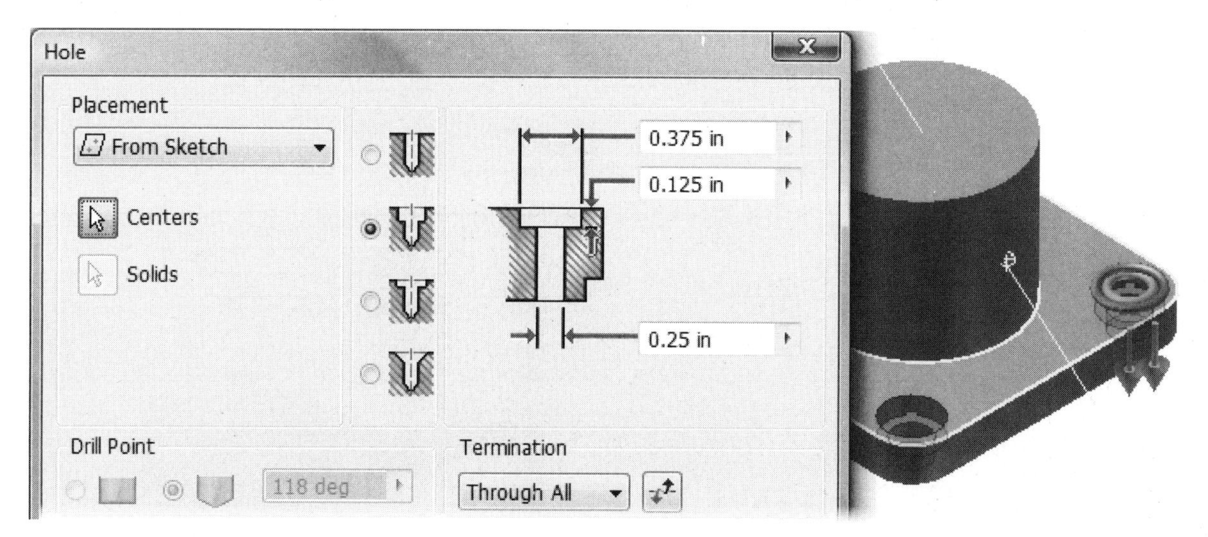

FIGURE 4.35

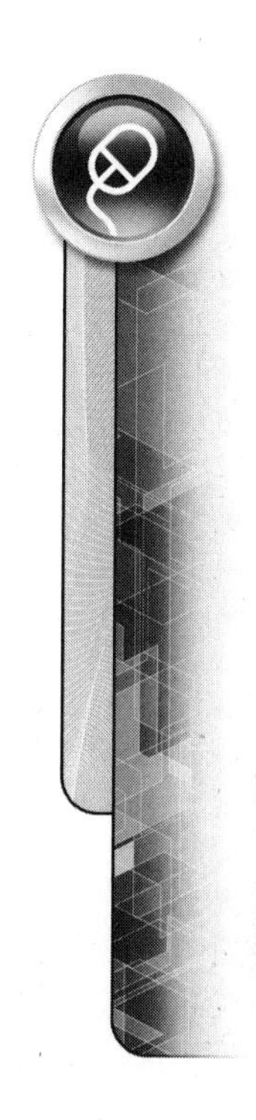

6. Next, you create a tapped hole based on the Sketch option, but place it by first creating a center point. Click the Create 2D Sketch command and then click the top face of Extrusion1.
7. Click the View Face command from the Navigation bar, and click Sketch4 in the browser.
8. Click the Point, Center Point command in the 2D Sketch Panel, and place a point to the right of the angled edges.
9. Add a horizontal constraint between the left point where the angled edges converge.
10. Add a **.5 inch** dimension between the point that the angled edges converge and the center point as shown in the following image.

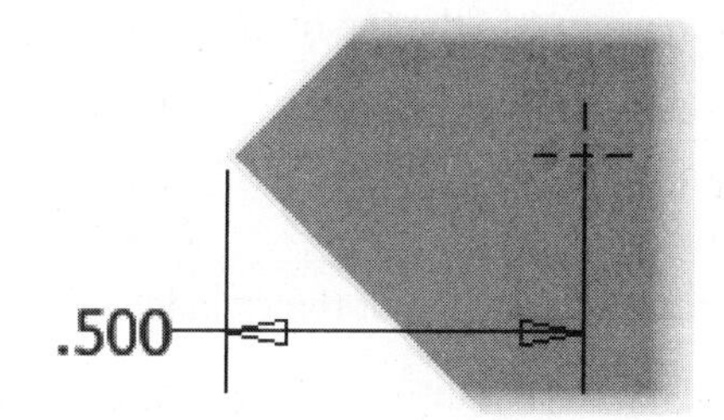

FIGURE 4.36

11. Finish the dimension command, right-click, and click Done.
12. Right-click in the graphics window, click Finish Sketch, the view should change to the home view. If not press the F6 key to change to the Home View.
13. Click the Hole command in the Modify panel bar or press the H key.
    - The Center Point is automatically selected as the Center.
    - Change the hole option to Drilled.

- Change the hole type to Tapped.
- Ensure the Termination is set to Through All.
- Change the Thread Type to **ANSI Metric M Profile**.
- Set the size to **.375** and the Designation is **3/8-16 UNC**, as shown in the following image.
- Click OK to create the hole.

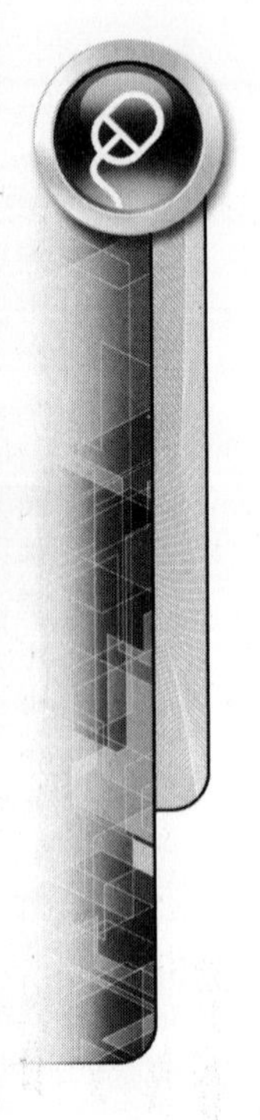

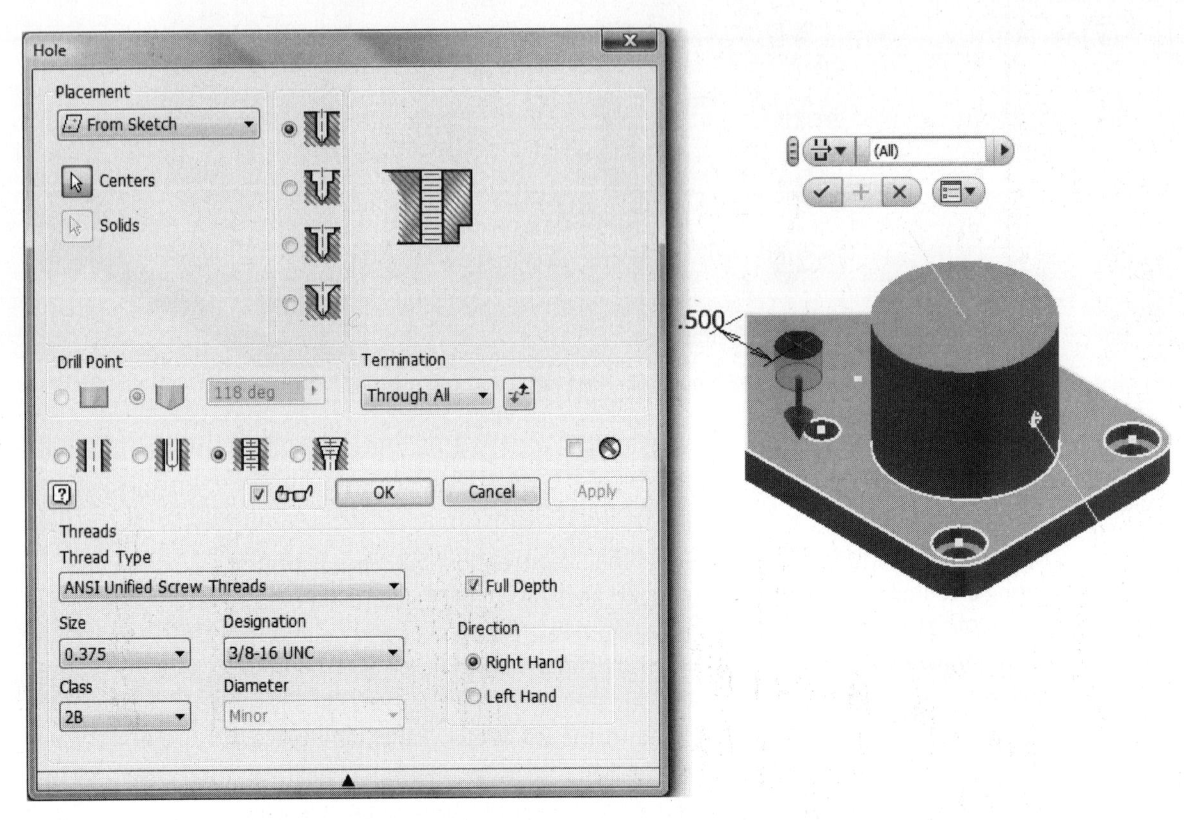

FIGURE 4.37

14. Next, you create a Taper Tapped hole that is concentric to the top of the cylinder.
    - Press the ENTER key to restart the Hole command.
    - Change the Placement option to Concentric.
    - For the Plane, select the top face of the cylinder.
    - For the Concentric Reference, select the top circular edge of the cylinder.
    - Ensure the hole option is set to Drilled.
    - Click the Taper Tapped Hole type.
    - Change the Thread Type to NPT.
    - Change the size to **1/2**.
    - Change the Termination to Through All, as shown in the following image.
    - Click Apply to create the hole.

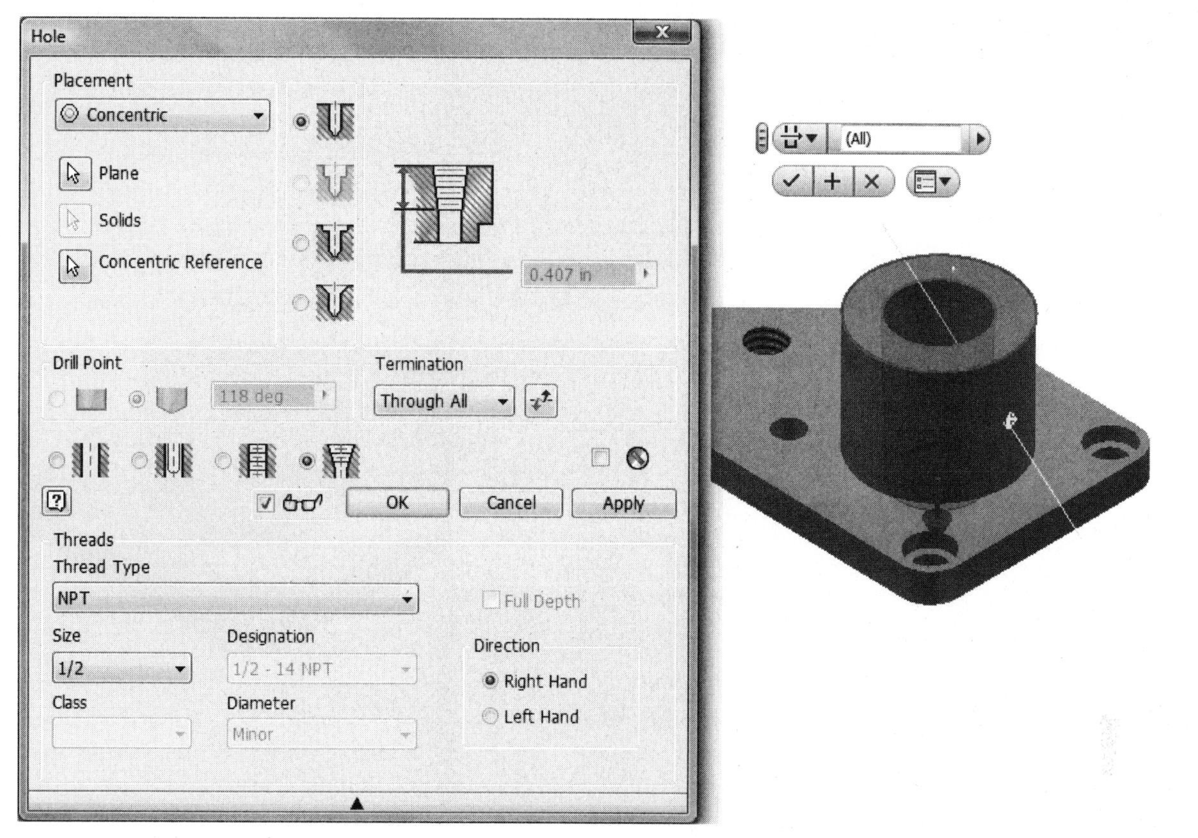

FIGURE 4.38

15. For the last step, you create a tapped hole that uses the work point and work axis on the cylinder. Work point and axis will be covered later in this chapter.
    - Change the Placement option to On Point.
    - For the Point, select the work point on the cylinder.
    - For the Direction, select the work axis that goes through the cylinder.
    - Click the Tapped Hole type.
    - Change the Thread Type to Tapped Hole.
    - Change the size to **.25**.
    - Change the Termination to To and select the inside circular face of the tapered hole, as shown in the following image.
    - Click OK to create the hole.

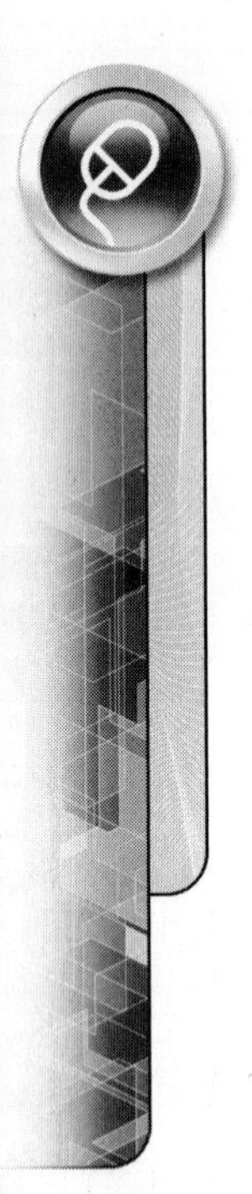

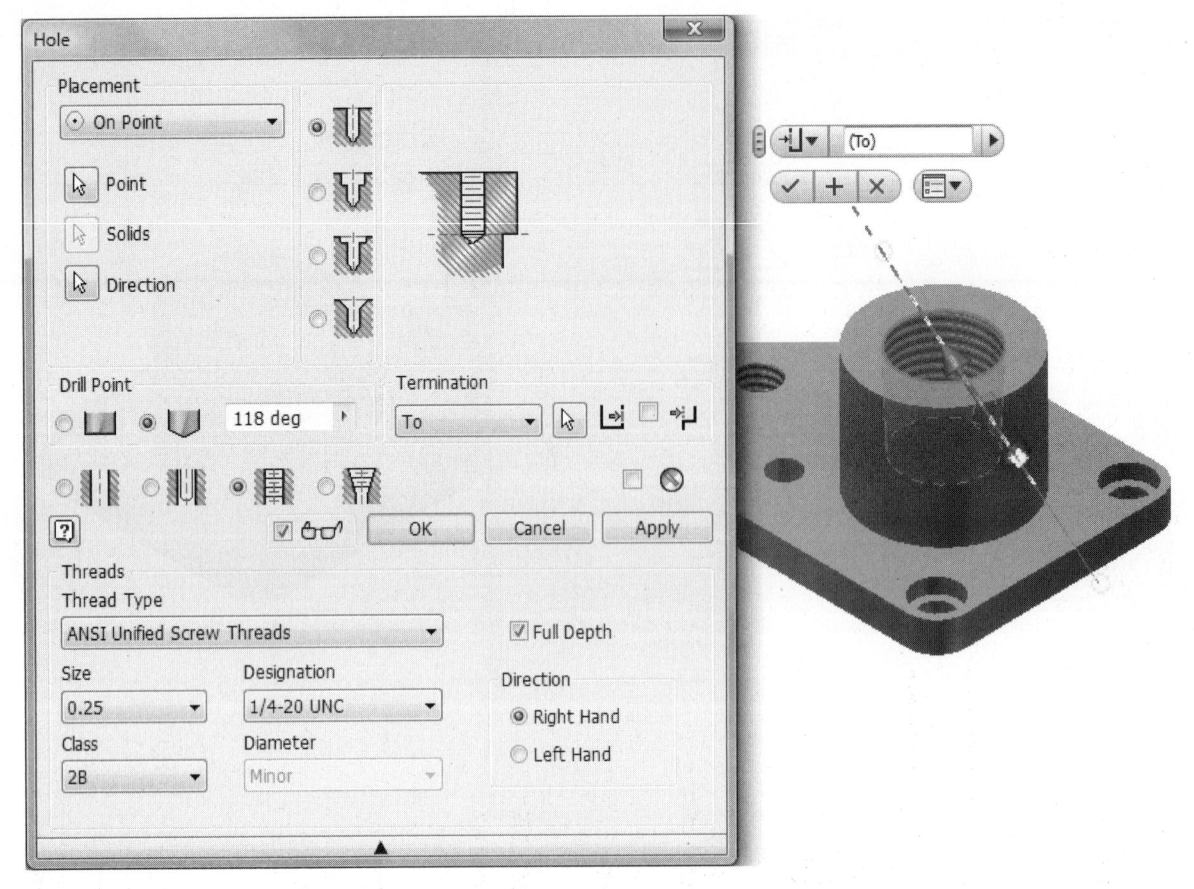

FIGURE 4.39

16. The completed part is shown in the following image (the visibility of the work axis and work plane have been turned off for better visibility). Rotate the part and examine the holes.

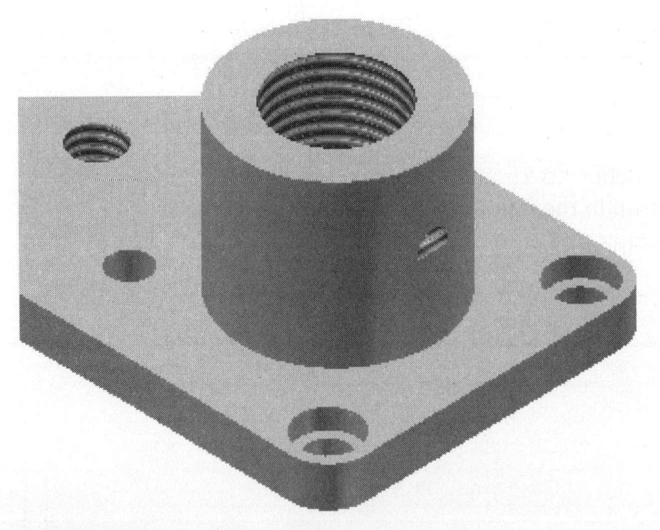

FIGURE 4.40

17. Practice editing the holes and placing new holes.
18. Close the file. Do not save changes. End of exercise.

SHELLING

As you design parts, you may need to create a model that is made of thin walls. The easiest way to create a thin-walled part is to create the main shape and then use the Shell command to remove material.

The term *shell* refers to giving a wall thickness to the outside shape of a part and removing the remaining material. Essentially, you are scooping out the inside of a part and leaving the walls a specified thickness, as shown in the following image. You can offset the wall thickness in, out, or evenly in both directions. If the part you shell contains a void, such as a hole, the feature will have the thickness built around it.

A part may contain more than one shell feature, and individual faces of the part can have different thicknesses. If a wall has a different thickness than the shell thickness, it is referred to as a unique face thickness. If a face that you select for a unique face thickness has faces that are tangent to it, those faces will also have the same thickness. You can remove faces from being shelled, and these faces remain open. If no face is removed, the part is hollow on the inside.

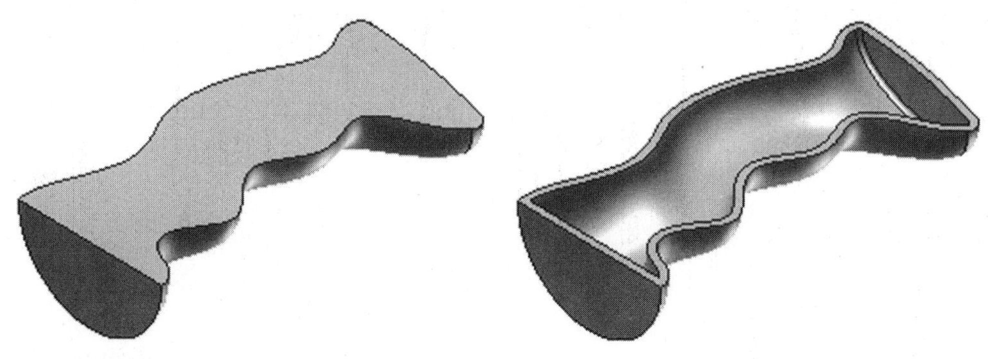

FIGURE 4.41

To create a shell feature, follow these steps:

1. Create a part that will be shelled.
2. Issue the Shell command from the Model tab > Modify panel, as shown in the following image on the left.
3. The Shell dialog box appears, as shown in the following image on the right. Enter the data as needed.

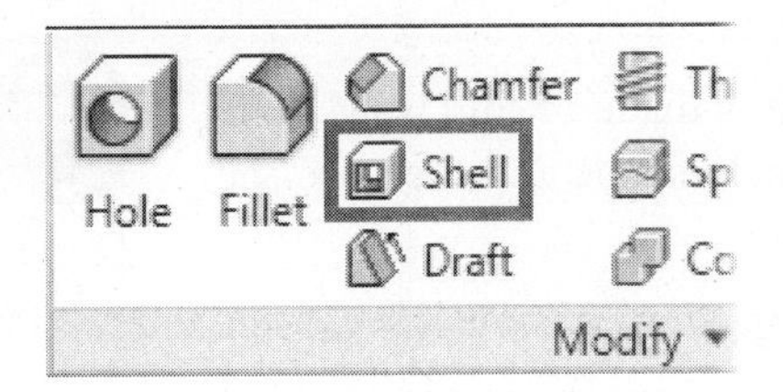

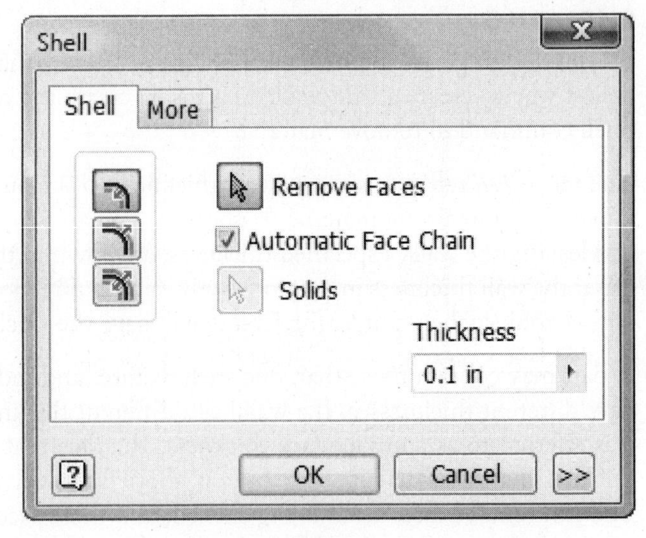

FIGURE 4.42

4. After filling in the information in the dialog box, click OK, and the part is shelled.

To edit a shell feature, use one of the following methods:

- Right-click the name of the shell feature in the browser, and select Edit Feature from the menu. Alternately, you can double-click the feature's name or icon in the browser, and the Shell dialog box appears with all of the settings you used to create the feature. Change the settings as needed.
- Change the select priority to Feature Priority, double-click the shell feature, and the Shell dialog box appears with all of the settings you used to create the feature.

The following sections explain the options that are available for the Shell command.

Direction

Inside

Click this button to offset the wall thickness into the part by the given value.

Outside

Click this button to offset the wall thickness out of the part by the given value.

Both

Click this button to offset the wall thickness evenly into and out of the part by the given value.

Remove Faces

Click the Remove Faces button, and then click the face or faces to be left open. To deselect a face, click the Remove Faces button, and hold down the CTRL key while you click the face.

Automatic Face Chain

When you are removing faces and this option is checked, faces that are tangent to the selected face are automatically selected. Uncheck this option to select only the selected face.

Solids

If multiple solid bodies exist, select the solid body to shell.

Thickness

Enter a value or select a previously used value from the drop-down list to be used for the shell thickness.

Unique Face Thickness

Unique face thickness is available by clicking the More (>>) button that is located on the lower-right corner of the dialog box, as shown in the following image on the left.

To give a specific face a thickness, click on Click to add, select the face, and enter a value. A part may contain multiple faces that have a unique thickness, as shown in the following image on the right.

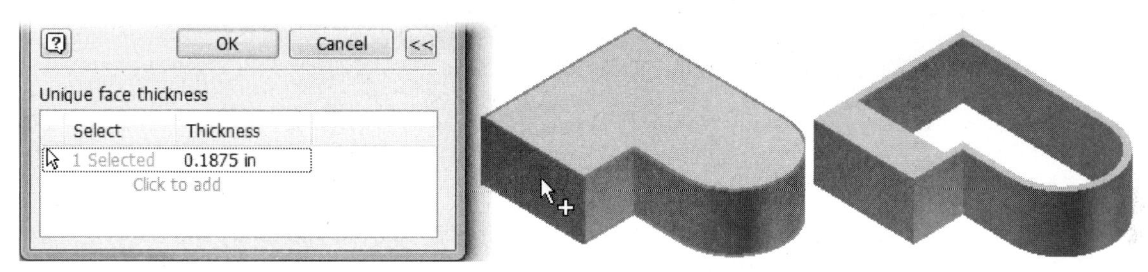

FIGURE 4.43

EXERCISE 4-3: SHELLING A PART

In this exercise, you will use the Shell command to create a shell on a part.

1. Open *ESS_E04_03.ipt* in the Chapter 04 folder.
2. Click the Shell command on the Modify panel.
3. Remove the top face by selecting the top face of the part.
4. Type a thickness of **.0625 in**, as shown in the following image.

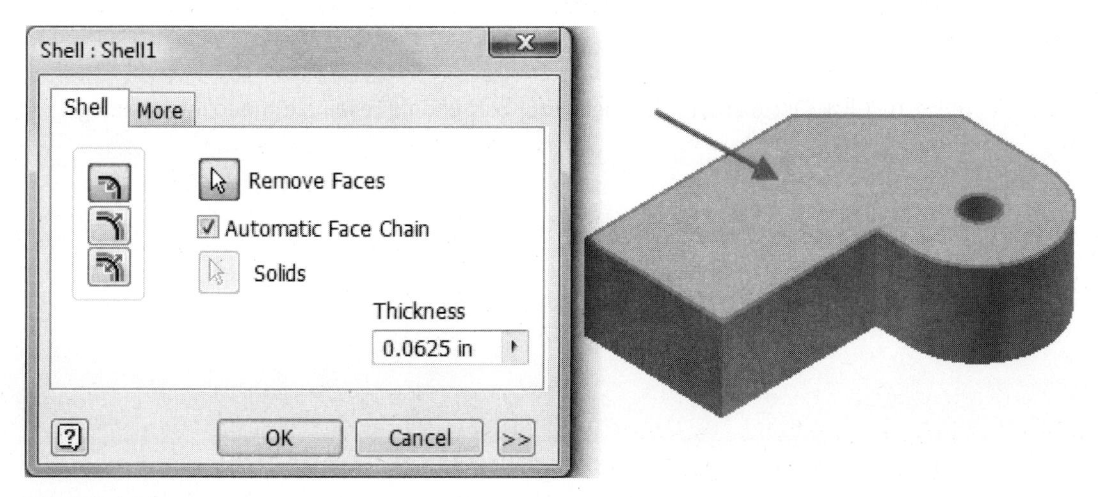

FIGURE 4.44

5. Click OK to create the shell feature.
6. Rotate the part and examine the shell. The bottom face of the part should be closed.
7. Click the Home View.
8. Edit the Shell feature that you just created. Click on one of the inside faces of the shell and from the mini-toolbar click Edit Shell as shown in the following image.

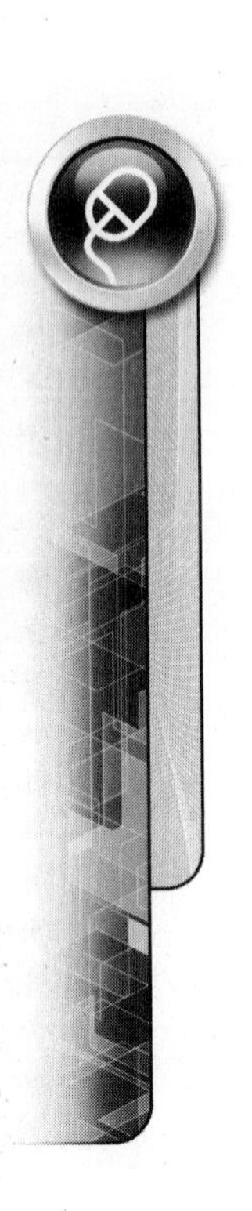

FIGURE 4.45

9. Select the More (>>) button, click in the "Click to add" area, and select the left outside-vertical face.
10. Enter a value of **.5 in**, as shown in the following image.

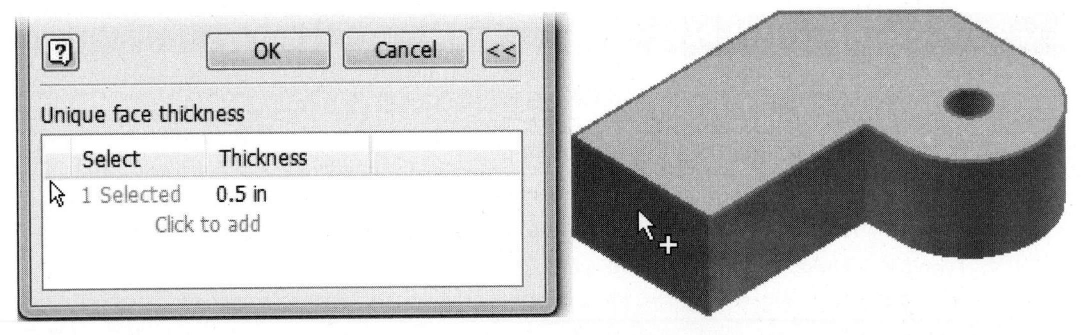

FIGURE 4.46

11. Click OK to create the shell. Your part should resemble the following image.

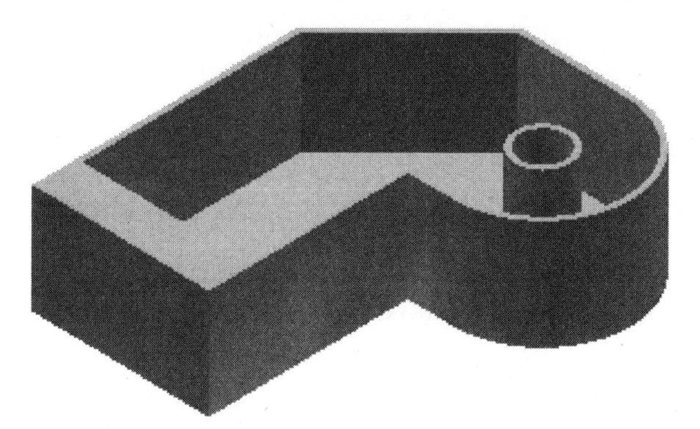

FIGURE 4.47

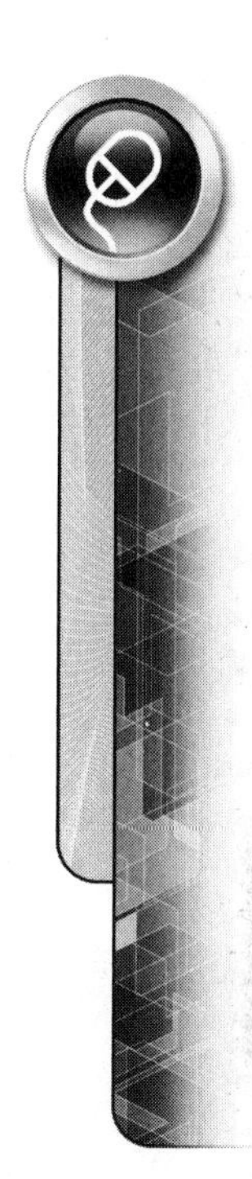

12. Practice editing the shell feature; change the thickness direction, thickness and delete the unique thickness by selecting the unique thickness entry in the dialog box and then press the Delete key.
13. Close the file without saving changes. End of exercise.

WORK FEATURES

When you create a parametric part, you define how the features of the part relate to one another; a change in one feature results in appropriate changes in all related features. Work features are special construction features that are attached parametrically to part geometry or other work features. You typically use work features to help you position and define new features in your model. There are three types of work features: work planes, work axes, and work points.

Use work features in the following situations:

- To position a sketch for new features when a planar part face is not available.
- To establish an intermediate position that is required to define other work features. You can create a work plane at an angle to an existing face, for example, and then create another work plane at an offset value from that plane.
- To establish a plane or edge from which you can place parametric dimensions and constraints.
- To provide an axis or point of rotation for revolved features and patterns.
- To provide an external feature termination plane off the part, such as a beveled extrusion edge, or an internal feature termination plane in cases where there are no existing surfaces.

Creating a Work Axis

A work axis is a feature that acts like a construction line. In the database, it is infinite in length but displayed a little larger than the model, and you can use it to help create work planes, work points, and subsequent part features. You can also use work axes as axes of rotation for polar arrays or to constrain parts in an assembly using assembly constraints. Their length always extends beyond the part—as the part changes size, the work axis also changes size. A work axis is tied parametrically to the part. As changes occur to the part, the work axis will maintain its relationship to the points, edge, or cylindrical face from which you created it. To create a work axis, use the Work Axis command on the Model tab > Work Features panel or press the key / (forward slash). The Axis command will allow you to select all geometry in the graphics window to create an axis. Another option is to click the down arrow in the lower right corner of the Axis button and select an option for creating an axis using only certain types of geometry as shown in the following image. These options will only select the predetermined geometry and will filter out all other types of geometry. For example, if the Through Two Points option is selected, only points can be selected in the graphics window.

You can also create a work axis when using the Work Plane or Work Point commands; right-click and select Create Axis when one of these work feature commands is active. Then, after the work axis is created, the command you were running will be active. The created axis will be indented as a child of the work axis or the work plane in the browser.

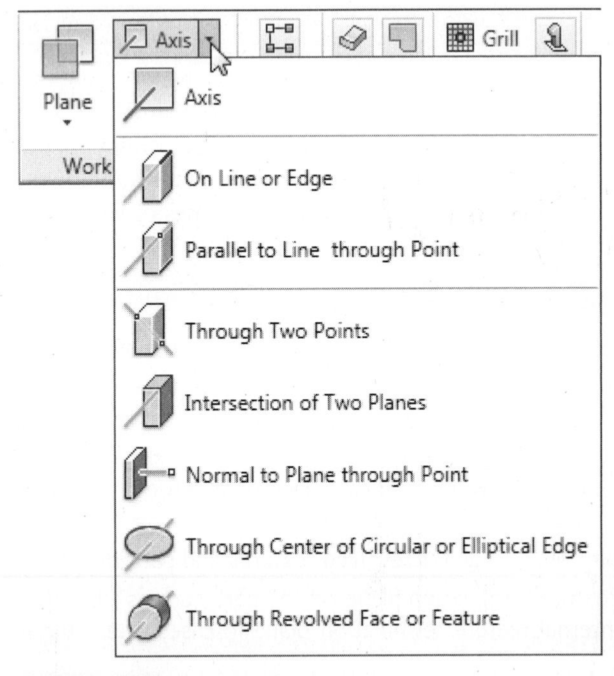

FIGURE 4.48

Use one of the following methods to create a work axis:

Axis Type	Selectable Geometry	Process
Axis	All geometry if available	Depending upon selected geometry.
On Line or Edge	Lines and linear edges	The work axis will be collinear with the selected line or linear edge.
Parallel to Line through Point	Linear edge or sketch line and an endpoint, midpoint, sketch point, or work point	The work axis will be parallel to the selected line or linear edge and through the selected point.
Through Two Points	Two endpoints, intersections, midpoints, sketch points, or work points	The work axis will go through the two selected points.
Intersection of Two Planes	Two nonparallel work planes or planar faces	The work axis will be created coincident with the intersection of the two planes.
Normal to Plane through Point	A planar face or work plane and an endpoint, midpoint, sketch point, or work point	The work axis is created perpendicular to the selected plane and through the point.
Through Center of Circular or Elliptical Edge	A circular edge, elliptical edge or an edge of a fillet	The work axis will create through the center point and perpendicular to the selected edge.
Through Revolved Face or Feature	A circular face	The work axis will create through the center of the selected circular face.

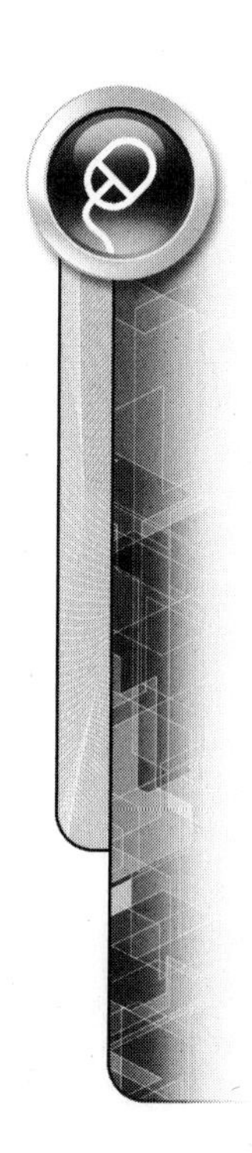

EXERCISE 4-4: CREATING WORK AXES

In this exercise, you will create a work axis to position a circular pattern. Circular patterns are covered later in this chapter.

1. Open *ESS_E04_04.ipt* in the Chapter 04 folder.
2. Click the Create 2D Sketch command and select the inside planar face as shown in the following image on the left.
3. If the view did not change, you are looking directly at the face, click the View Face command on the Navigation bar and then in the browser select the sketch you just created.
4. Click the Point, Center Point command in the Draw panel, and place a point near the middle of the sketch.
5. Place a vertical constraint between the center point you just created and the midpoint on the top line. Apply a horizontal constraint between the point and the midpoint of the edge on the right, as shown in the following image on the right.

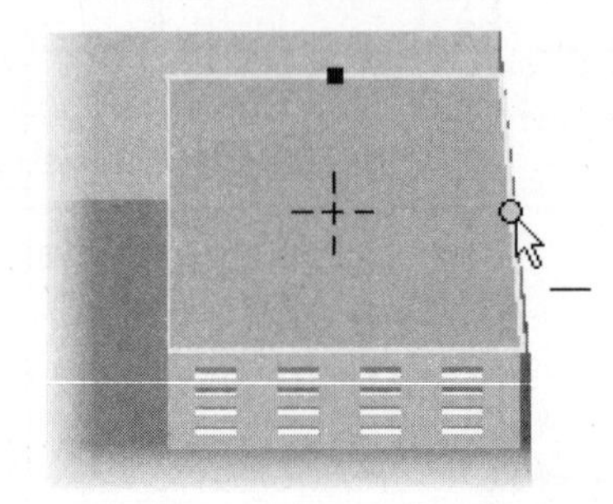

FIGURE 4.49

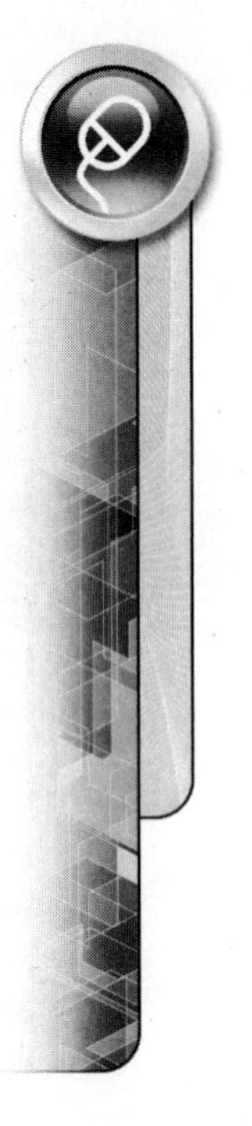

6. Right-click in the graphics window, and click Done to finish the constraint command.
7. Right-click in the graphics window, and then select Finish Sketch.
8. If the view did not automatically change to the home view, right-click in the graphics window, and then click Home View from the menu.
9. From the Model tab > Work Features panel click the down arrow on the right of the Work Axis command and from the list click Normal to Plane through Point.
10. Select the angled planar face, and then select the center point. The work axis is created through this point and normal to the plane, as shown in the following image on the left.

NOTE

One use of a work axis is to use it as a Rotation Axis when creating a circular pattern. The following image on the right shows a hole that was patterned around the work axis. Patterns are covered later in this chapter.

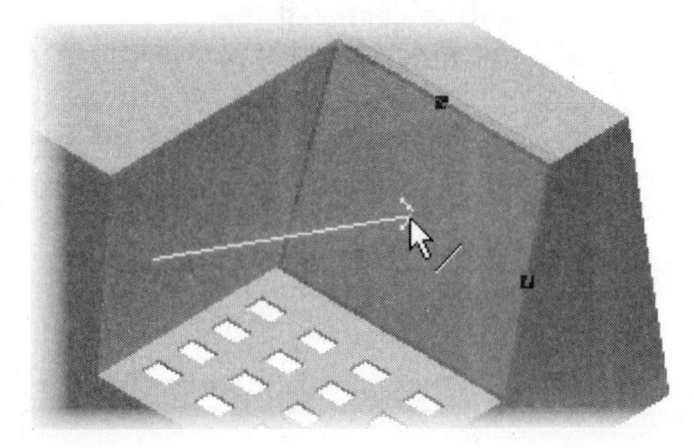

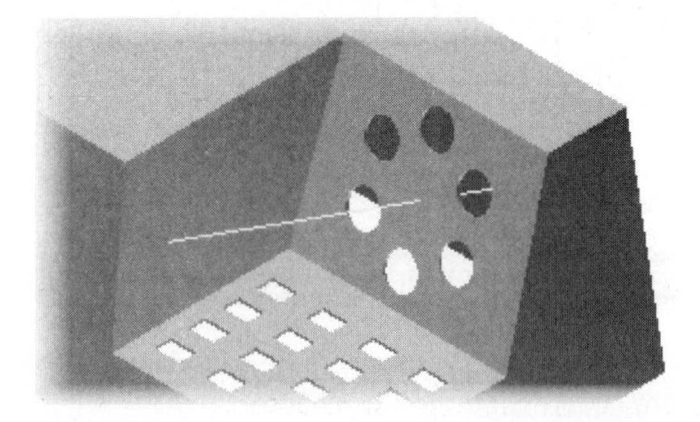

FIGURE 4.50

11. From the Model tab > Work Features panel click the down arrow on the right of the Work Axis command and from the list click Through Center of Circular or Elliptical Edge.
12. Select the top circular edge on the back of the part as shown in the following image on the left. The axis is created through the center of the circular edge.
13. Undo the last work axis you just created. Click the Undo command from the Quick Access toolbar and the work axis will removed.
14. From the Model tab > Work Features panel click the down arrow on the right of the Work Axis command and from the list click Through Revolved Face or Feature.
15. Select the back circular face on the part as shown in the following image on the right. The axis is created through the center of the circular face.

NOTE

Note that the results for the work axis are the same if the Through Center of Circular or Elliptical Edge or the option Through Revolved Face or Feature are used since both sets of geometry share the same centers.

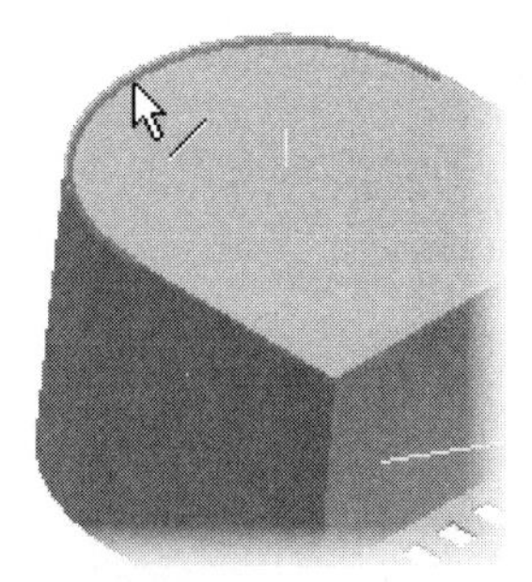

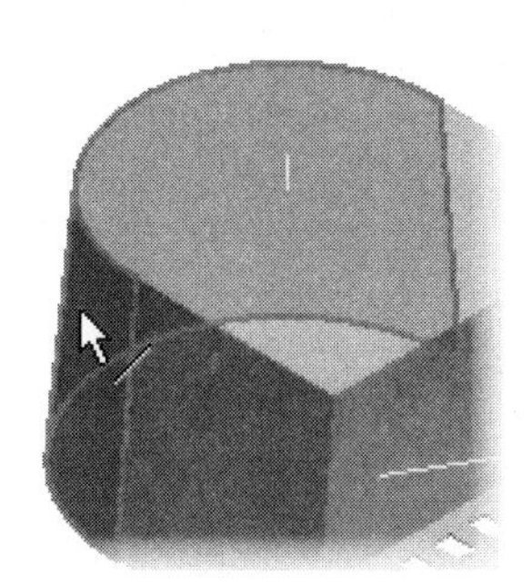

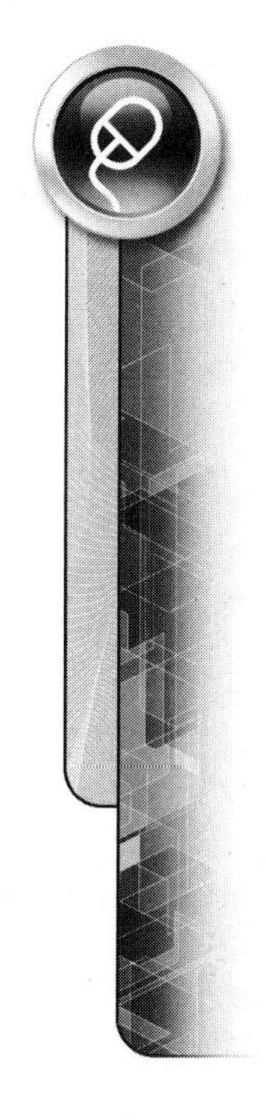

FIGURE 4.51

16. Practice creating work axes using different options.
17. Close the file. Do not save changes. End of exercise.

Creating Work Points

A work point is a feature that you can create on the active part or in 3D space and can be used to create other work features such as work axis and work planes, place a hole using the On Point option and place 3D lines on the work points. Work points are created relative to selected geometry, if the selected geometry changes location, the work point will move to keep the relationship to the new location of the geometry. To create a work point, use the Point command on the Model tab > Work Features panel, as shown in the following image, or press the key (.) (period). The point command will allow you to select all geometry in the graphics window to create a work point. Another option is to click the down arrow in the lower right corner of the Point button and select an option for creating a work point using only a filtered geometry type(s) as shown in the following image. These options will only select the predetermined geometry and will filter out all other types of geometry. For example, if the Intersection of Three Planes option is selected, only planes can be selected in the graphics window.

You can also create a work point when using the Work Plane or Work Axis commands; right-click and select Create Point when one of these work feature commands is active. Then, after the work point is created, the command you were running will be active. The created point will be indented as a child of the work axis or the work plane in the browser.

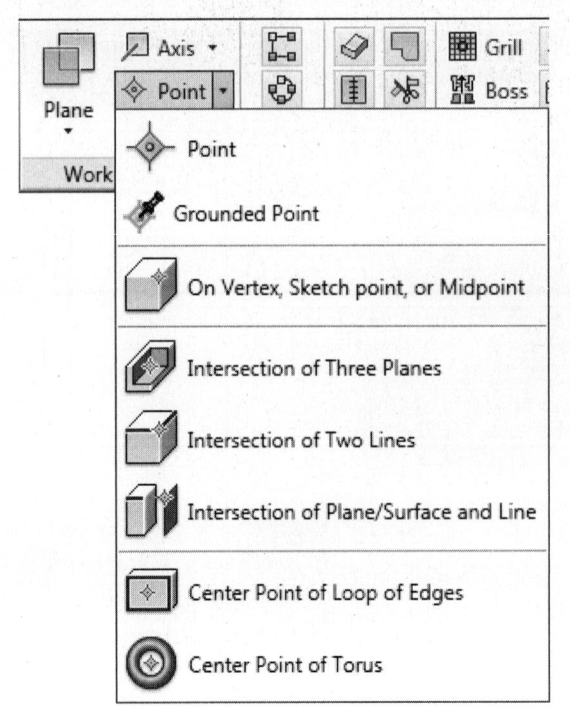

FIGURE 4.52

Use one of the following methods to create a work point:

Work Point Type	Process
Point	Select desired geometry, no filters are applied.
Grounded Point	Select a work point, midpoint, or vertex. Grounded work points will be covered in a following section.
On Vertex, Sketch point, or Midpoint	Select a 2D or 3D sketch point, vertex, or the endpoint or midpoint of a line or linear edge.
Intersection of Three Planes	Select three work planes or planar faces.
Intersection of Two Lines	Select any two lines including linear edges, 2D or 3D sketch lines, or work axes.
Intersection of Plane/Surface and Line	Select a planar face or work plane and a work axis or line. Or select a surface and a sketch line, straight edge, or work axis.
Center Point of Loop of Edges	Select an edge of a closed loop of edges.
Center Point of Torus	Select a torus.

Grounded Work Points

You can create grounded work points that are positioned in 3D space. Grounded work points are not associated with the part or any other work features, including the original locating geometry. When you modify surrounding geometry, the grounded work point remains in the specified location. To create a grounded work point, use the Grounded Work Point command on the Model tab > Work Features panel and click the arrow next to the Point command, as shown in the previous image, or press the hot key (;) (semicolon).

After starting the Grounded Point command, select a vertex, midpoint, sketch point, or work point on the model. When you have selected the vertex or point, the 3D Move/Rotate dialog box and a triad appear, as shown in the following image. The initial orientation of the triad matches the principle axes of the part. These colors represent the three axes: red = X, green = Y, and blue = Z.

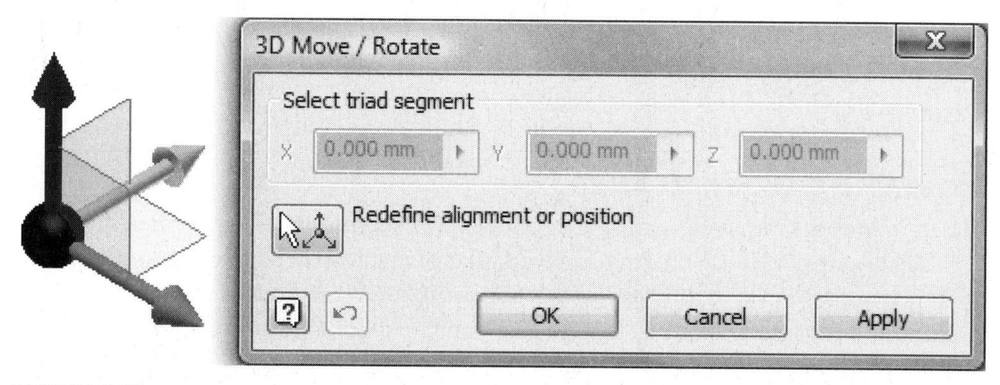

FIGURE 4.53

Enter values in the 3D Move/Rotate dialog box to precisely position the grounded work point relative to the selected point. You can also select areas of the triad to move the triad and locate the grounded work point in the desired direction, as shown in the following image and as described in the following sections.

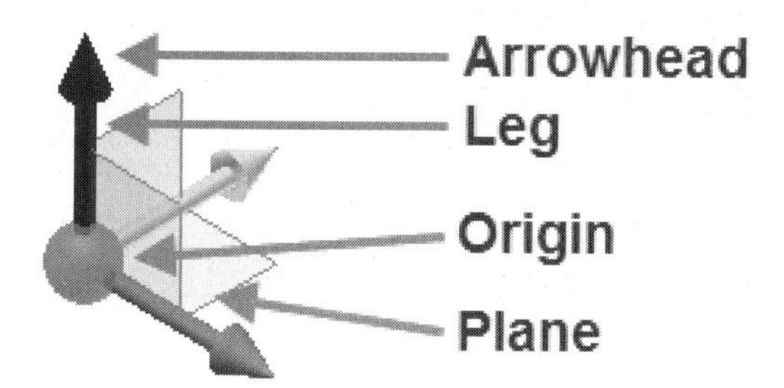

FIGURE 4.54

Arrowheads

Select an arrowhead to specify a position along a particular axis, and move the cursor or enter a value.

Legs

Select a leg to rotate about that axis, and move the cursor or enter an angle.

Origin

Click to move the triad freely in 3D space, move it to a selected vertex or point, or to enter X, Y, or Z coordinates.

Planes

Select a plane to restrict movement to the selected plane.

Once you position the triad properly, click Apply or OK in the 3D Move/Rotate dialog box to create the grounded work point. By clicking Apply the dialog box will remain open, allowing you to create another grounded work point. You can identify a grounded work point in the browser by the thumbtack icon that is placed on the work point. The following image shows a regular work point (Work Point1) and a grounded Work Point (Work Point2) represented in the browser.

FIGURE 4.55

CREATING WORK PLANES

Before introducing work planes, it is important that you understand when you need to create a work plane. You can use a work plane when you need to create a sketch and no planar face exists at the desired location, or if you want a feature to terminate at a plane and no face exists to select. If you want to apply an assembly constraint to a plane on a part and no part face exists, you will need to create a work plane. If a face exists in any of these scenarios, you should use it and not create a work plane. A new sketch can be created on a work plane.

A work plane looks like a rectangular plane. It is tied parametrically to the part. Though extents of the plane will appear slightly larger than the part, the plane is in fact infinite. If the part or related feature moves or resizes, the work plane will also move or resize. For example, if a work plane is tangent to the outside face of a 10 mm diameter cylinder and the cylinder diameter changes to 20 mm, the work plane moves with the outside face of the cylinder. You can create as many work planes on a part as needed, and you can use any work plane to create a new sketch. A work plane is a feature and is modified like any other feature.

Before creating a work plane, ask yourself where this work plane needs to exist and what you know about its location. You might want a plane to be tangent to a given face and parallel to another plane, for example, or to go through the center of two arcs. Once you know what you want, select the appropriate options and create a work plane. There are times when you may need to create an intermediate work plane before creating the final work plane. You may need to create a work plane, for example, that is at 30° and tangent to a cylindrical face. You should first create a work plane that is at a 30° angle and located at the center of the cylinder; then create a work plane parallel to the angled work plane that is also tangent to the cylinder.

To create a work plane, click the Plane command on the Model tab > Work Features panel, or press the hot key (]) (end bracket). The Plane command allows you to select all geometry types in the graphics window to create a work plane. Another option is to click the down arrow in the lower right corner of the Plane button and select an option for creating a work plane using only a filtered geometry type as shown in the following image. These options will only select the predetermined geometry and will filter out all other types of geometry. For example if the Three Points option is selected, only points can be selected in the graphics window.

You can also create a work plane when using the Work axis or Work Point commands; right-click and select Create Plane when one of these work feature commands is active. Then, after the work plane is created, the command you were running will be active. The created plane will be indented as a child of the work axis or the work point in the browser.

Work planes can also be associated to the default reference (origin) planes, axis and points that exist in every part. These origin entities initially have their visibility turned off, but you can make them visible by expanding the Origin folder in the browser, right-clicking on a plane or planes, and selecting Visibility from the menu. You can use the origin planes to create a new sketch or to create other work planes.

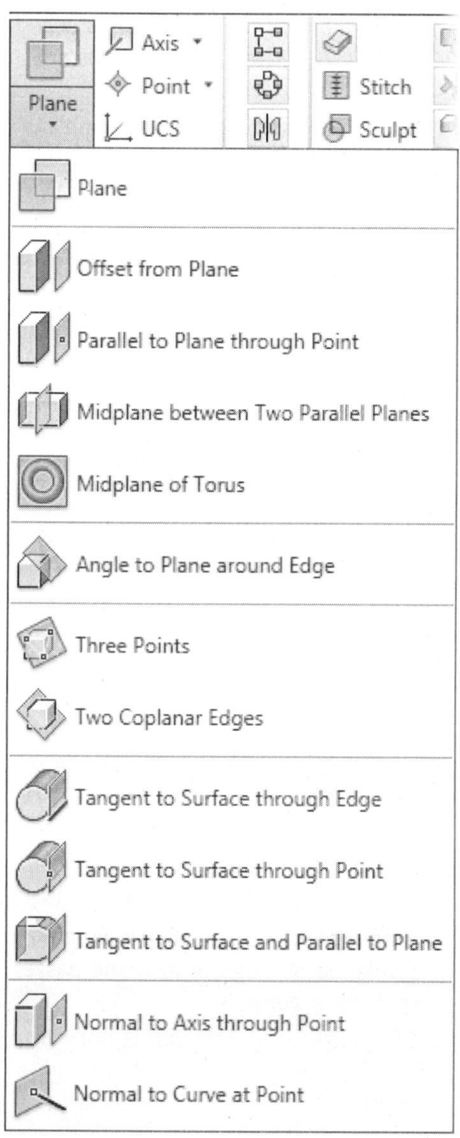

FIGURE 4.56

Use one of the following methods to create a work plane:

Work Plane Type	Selectable Geometry	Process
Plane	Select vertices, edges, or faces	Click the desired geometry.
Offset from Plane	A planar face	Click a plane, drag the new work plane to a selected location. To specify the offset distance, enter a value in the input field.
Parallel to Plane through Point	A planar face, work plane and a point	Select a planar face or work plane and a point; order of selection does not matter.
Midplane between Two Parallel Planes	Planar faces or work planes	Create a work plane at the midplane defined by selecting two parallel planes.
Midplane of Torus	A torus	Create a work plane that goes through the center, or midplane, of a torus.
Angle to Plane around Edge	A face or plane and an edge or line parallel to the face	Create a work plane about an edge or line whose angle is determined from a planar face or plane. Enter the desired angle in the input field.
Three Points	Endpoints, intersections, midpoints, work points	Create a work plane based on three points.
Two Coplanar Edges	Work axes, edges, or lines	Select two coplanar work axes, edges, or lines.
Tangent to Surface through Edge	Circular face and a linear edge	Create a work plane that passes through the linear edge and is tangent to the circular face.
Tangent to Surface through Point	Circular face and an endpoint, midpoint, or work point	Create a work plane that passes through a point and is tangent to the circular face.
Tangent to Surface and Parallel to Plane	Circular face and a planar face or work plane, in either order	Create a work plane that is parallel to a plane and tangent to the circular face.
Normal to Axis through Point	Linear edge or axis and a point	Create a work plane that is perpendicular to an edge or axis and passes through a point.
Normal to Curve at Point	Nonlinear edge or sketch arc, circle, ellipse, or spline and a vertex, edge midpoint, sketch point, or work point on the curve	Create a work plane that is normal to a circular edge and passes through a point.

When you are creating a work plane, and more than one solution is possible, the Select Other command appears. Click the forward or reverse arrows from the Select Other command until you see the desired solution displayed. Click the checkmark in the selection box. If you clicked a midpoint on an edge, the resulting work plane links to the midpoint. If the selected edge's length changes, the location of the work plane will adjust to the new midpoint.

NOTE

The order in which points or planes are selected is irrelevant.

Use the Show Me animations located in the Visual Syllabus to view animations that display how to create certain types of work planes.

UCS—User Coordinate System

A UCS is similar to the data in the origin folder; it contains three work planes, three axes, and a center point. Unlike the data in the origin folder you can create as many UCSs as required and position them as needed. The following list shows the common uses for a UCS.

- Create a UCS where it would be difficult to create a work plane, for example a compound angle.
- Locate a sketch on a UCS plane.
- Start and terminate features on UCS planes.
- A UCS axis can be used as a rotation axis to pattern features or parts.
- A UCS axis can be used to rotate parts.
- In an assembly, you can constrain UCSs of two parts.
- In a part or an assembly you can measure to the planes, axis or point in a UCS.
- Measure to the planes, axis or origin.

To start the UCS command, click Model tab > Work Features panel > UCS as shown in the following image.

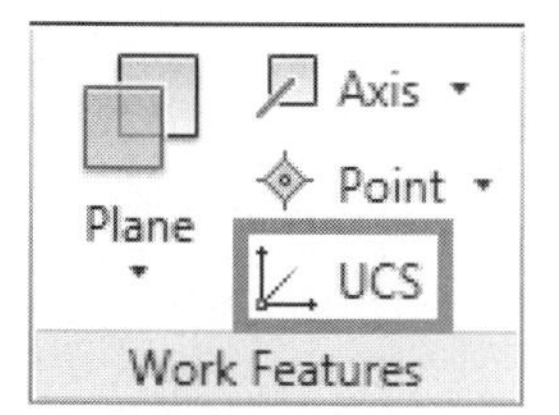

FIGURE 4.57

There are multiple methods to place a UCS:

- In a part file, a UCS can also be placed on existing geometry.
- A UCS can be positioned using absolute coordinates in a part or an assembly file.

The first method is to place a UCS by selecting geometry:

- Start the UCS command, click Model tab > Work Features panel > UCS.
- Select a point to locate the origin.
- Select a point to define the direction of the X axis.
- Select a point to define the direction of the Y axis.

When selecting points to locate or position a UCS valid inputs are: vertex of an edge, midpoint of an edge, sketch, work point origin, solid circular edge, or solid elliptical edge.

The following image shows a UCS being placed at the top left vertex in the left image. The X axis is positioned by selecting a vertex as shown in the middle image, and the Y axis is also positioned by selecting a vertex as shown in the following image on the right. The midpoints of the edges could have also been used to align the UCS.

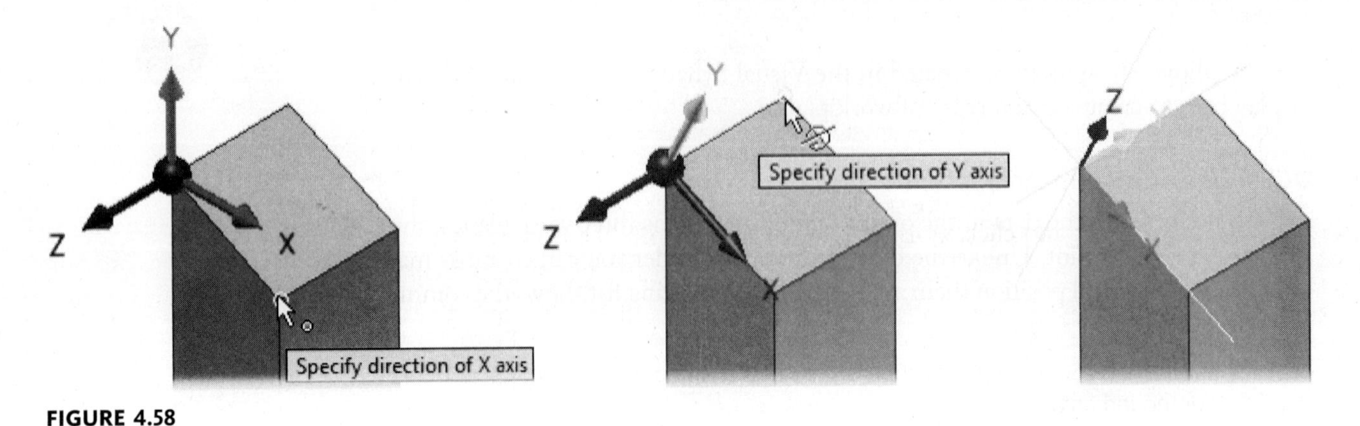

FIGURE 4.58

The second method to place a UCS is to enter absolute coordinates:

- Start the UCS command, click Model tab > Work Features panel > UCS.
- Enter a value for the X, Y, and Z location. Press the Tab key to switch between the cells as shown in the following image.
- After the three points are defined, press Enter or click in the graphics window.
- To define the direction of the UCS, click on an arrow and enter a value.
- To rotate the UCS about an axis on the UCS, click on the shaft of an arrow and enter a value.

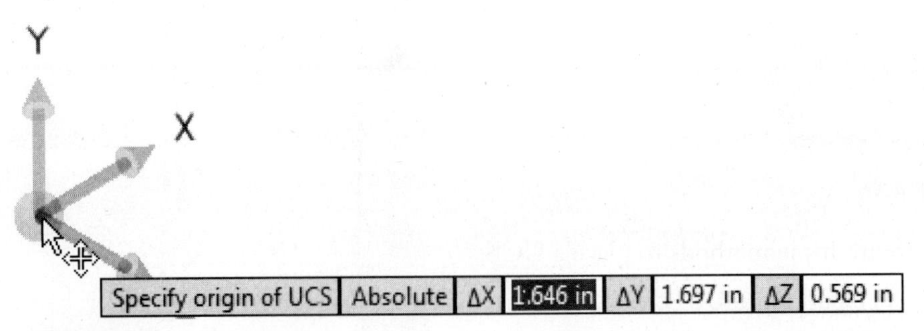

FIGURE 4.59

After creating a UCS, it will appear in the browser as shown in the following image. The UCS in the browser can be treated like the origin folder, turn the visibility on and off and measure to the planes, axis and center point.

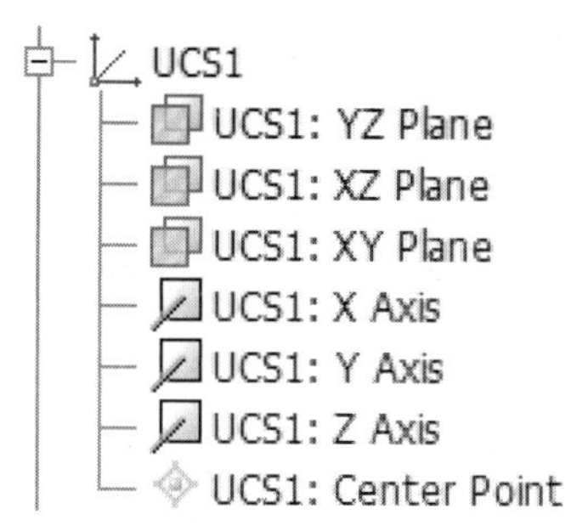

FIGURE 4.60

To edit a UCS, move the cursor over the UCS in the graphics window or in the browser and double-click or right-click, and then click Redefine Feature as shown in the following image on the left. Select on the desired UCS segment: arrow, leg, or origin, and enter a new value or drag to a new location. When done relocating the UCS, right-click and click Finish from the menu. Do NOT click Done as this will cancel the operation. The following image on the right shows the prompt to select a segment to edit.

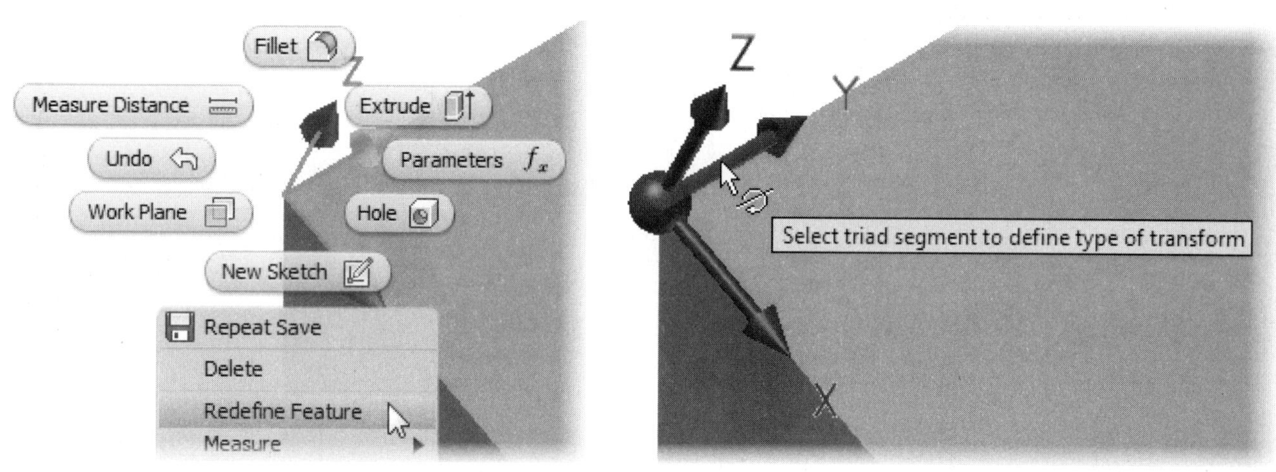

FIGURE 4.61

Feature Visibility

You can control the visibility of the origin planes, origin axes, origin point, user work planes, user work axes, user work points, and sketches by either right-clicking on them in the graphics area or on their name in the browser and selecting Visibility from the menu. You can also control the visibility for all origin planes, origin axes, origin point, user work planes, user work axes, user work points, sketches, solids, and UCS triad, planes, axis, and points from the View tab > Visibility panel > Object Visibility. Visibility can be checked to turn visibility on or cleared to turn it off, as shown in the following image.

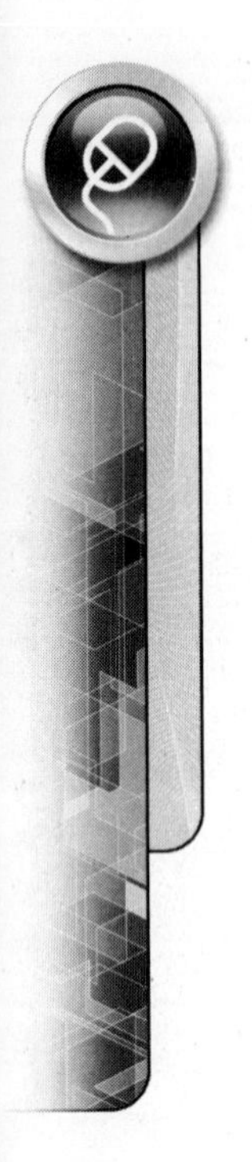

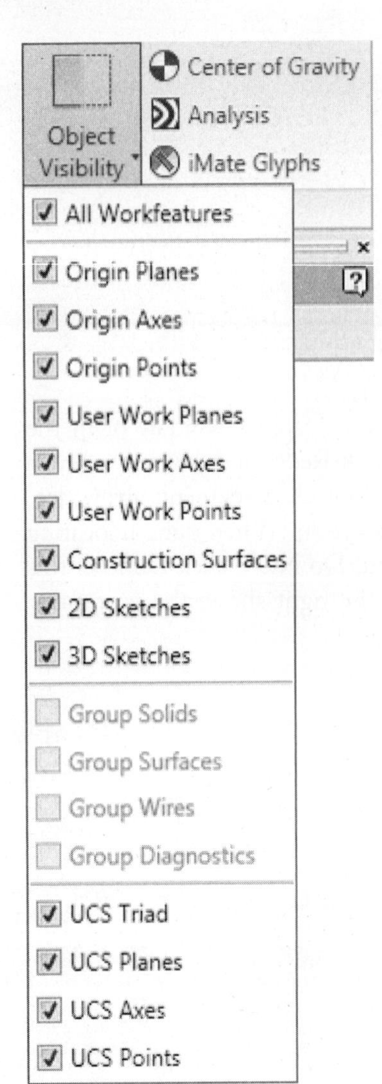

FIGURE 4.62

EXERCISE 4-5: CREATING WORK PLANES AND A UCS

In this exercise, you will create work planes in order to create a boss on a cylinder head and a slot in a shaft.

1. Open *ESS_E04_05_1.ipt* in the Chapter 04 folder.
2. From the Model tab > Work Features panel click the down arrow on the lower right corner of the Plane command and from the list click Angle to Plane around Edge.
3. Click the top rectangular face and the top-back left edge, and then enter a value of **-30** as shown in the following image on the left. Click the green checkmark in the mini-toolbar to create the work plane.
4. Make the top rectangular face the active sketch. Click on the top face of Extrusion1 and click Create Sketch command on the mini-toolbar.
5. Create and dimension a circle and apply a horizontal constraint between the center of the circle and the midpoint of the right vertical line as shown in the following image on the right. Finish the sketch.

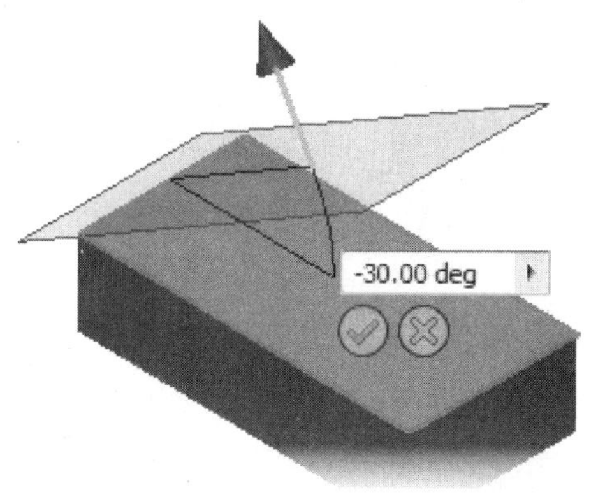

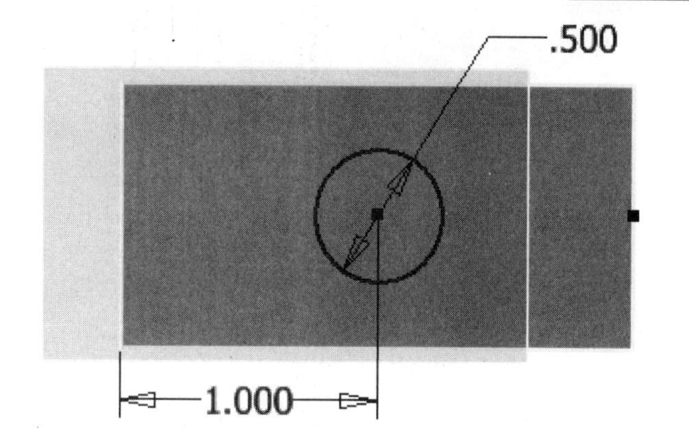

FIGURE 4.63

6. Extrude the circle with the Extents set to To, and select the work plane, as shown in the following image.

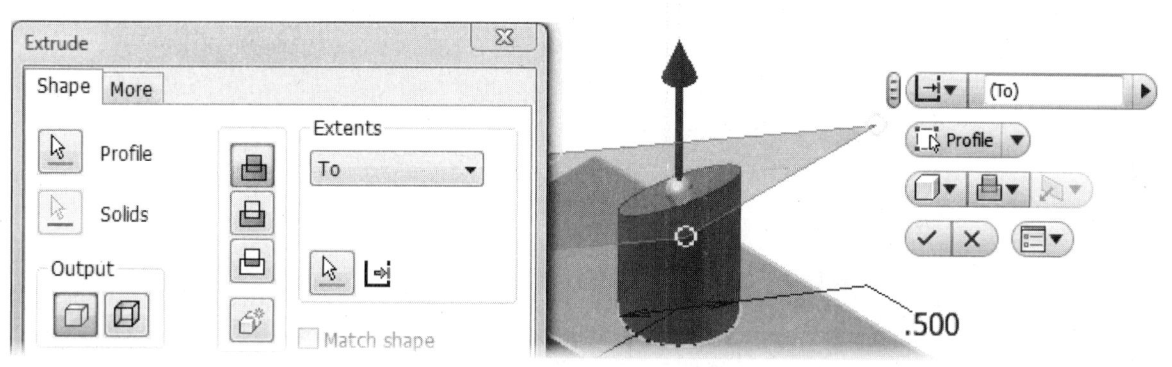

FIGURE 4.64

7. In the browser, double-click on Work Plane1 the angled work plane you created in step 3 and enter different values for the angle; click the local Update command on the Quick Access toolbar to update the model. Change the angle of the work plane back to **-30 degrees**.
8. Turn off the visibility of the work plane by moving the cursor over an edge of the work plane in the graphics window, right-click, and click Visibility from the marking menu.
9. Another option to define a plane is to create a UCS. Click the UCS command on the Work Features panel.
10. Align the UCS on the top face of Extrusion1 by selecting the three vertices in the order as shown in the following image on the left.
11. Edit the UCS by double-clicking on one the arrowheads on the UCS in the graphics window.
12. Click the X leg of the UCS and enter a value of -**40** as shown in the following image on the right. To complete the edit, right-click and click Finish from the menu.

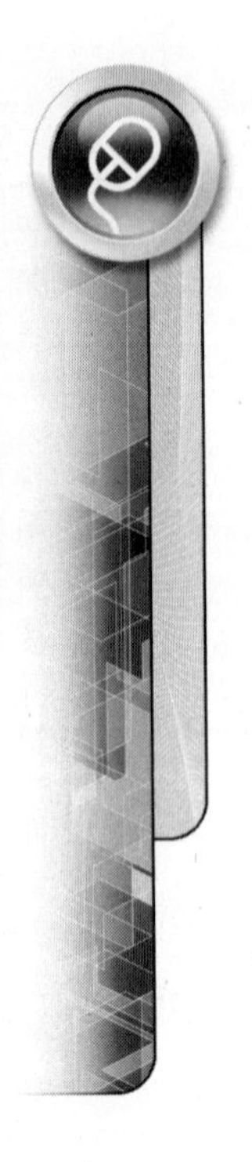

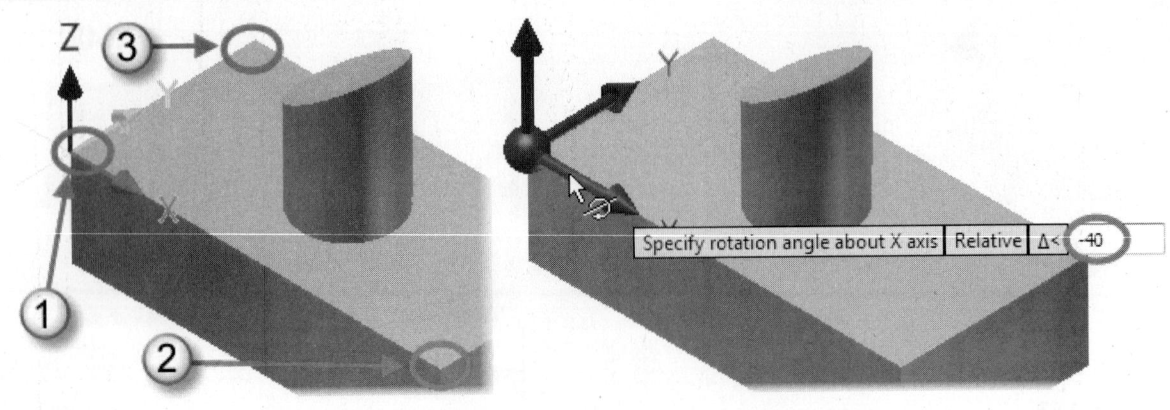

FIGURE 4.65

13. Reorder the UCS to it exists in the browser before Extrusion2. In the browser, drag the UCS entry so it is above Extrusion2 in the browser as shown in the following image. Reordering features is covered in Chapter 7.

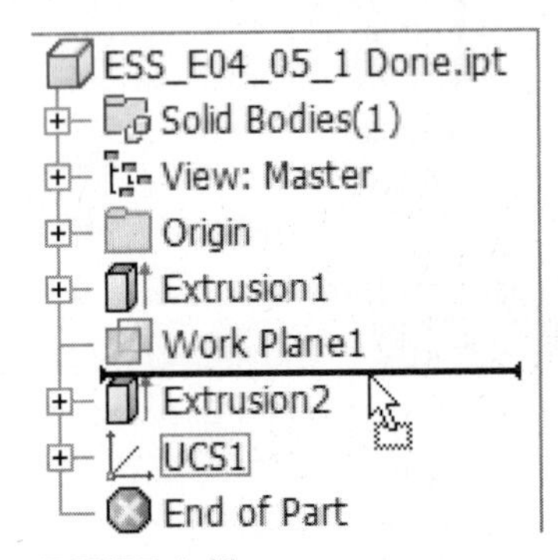

FIGURE 4.66

14. Edit Extrusion2 and click the arrow button for the option Select surface to end the feature creation and in the browser, click UCS1: XY Plane in the UCS as shown in the following image.

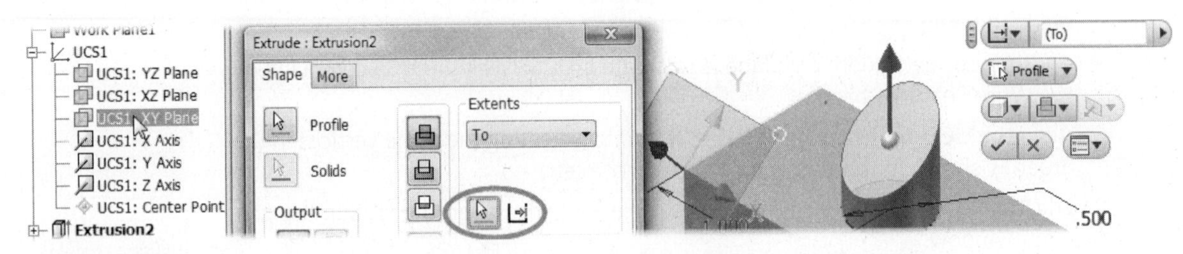

FIGURE 4.67

15. Click OK to complete the edit.
16. Create a work plane that is centered between two parallel planes. From the Model tab > Work Features panel click the down arrow on the right side of the Plane command and from the list click Midplane between Two Parallel Planes.
17. Click the front-left face and the back-right face, as shown in the following image on the left.

18. Click one of the edges of the work plane you just created and click Create Sketch from the menu.
19. Slice the graphics by pressing the F7 key or right-click and click Slice Graphics from the menu.
20. To project the edges of the part, click the Project Cut Edges command from the Draw panel (the Project Cut Edges command may be under the Project Geometry command).
21. Create and dimension a circle on the top edge of the projected geometry, as shown in the following image on the right.

NOTE

If the circle is placed at the midpoint of the projected edge, the horizontal dimension cannot be placed.

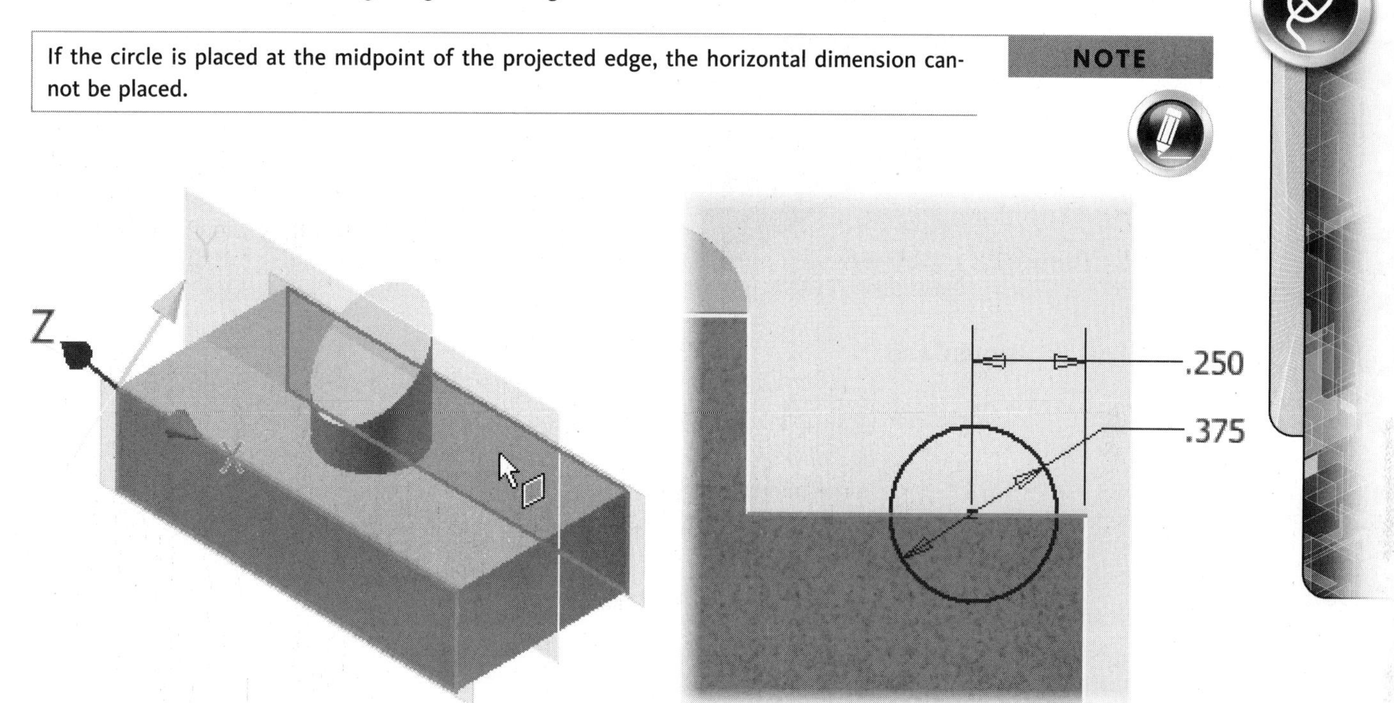

FIGURE 4.68

22. Finish the sketch, click Finish Sketch from the Exit panel.
23. Extrude the circle **.5** with the symmetric option, as shown in the following image.

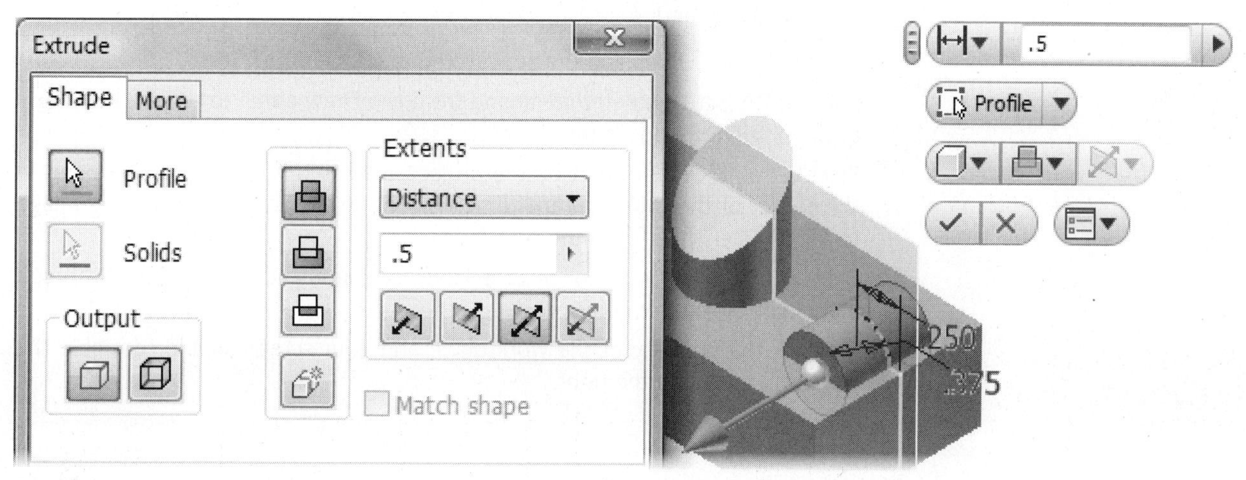

FIGURE 4.69

24. Turn off the visibility of the work plane and the UCS. Your screen should resemble the following image on the left.

25. To verify that the symmetric extrusion will be maintained in the center of the part even if the width changes, edit Sketch1 of Extrusion1, and change the 1 inch dimension to **1.5 inches**.
26. If needed click the Local Update command on the Quick Access toolbar. Your screen should resemble the following image on the right.

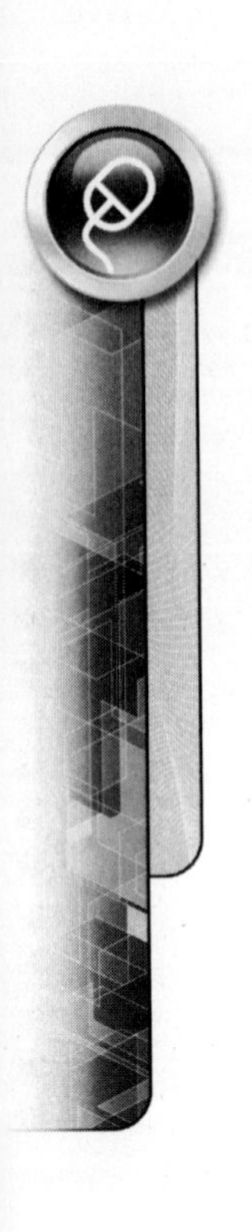

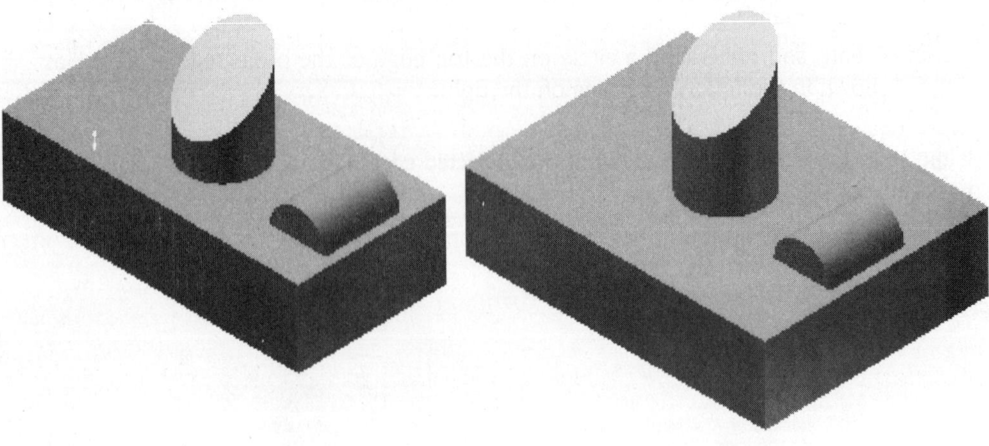

FIGURE 4.70

27. Use the Free Orbit command to view the part from different perspectives. Notice how Extrusion2 still terminates at the UCS XY plane and Extrusion3 is still in the middle of the part.
28. Close the file. Do not save changes.
29. In this portion of the exercise, you will place two holes on a cylinder. Open *ESS_E04_05_2.ipt* from the Chapter 04 folder.
30. Create a work plane that is parallel to an origin plane and tangent to the cylinder.
    - In the browser, expand the Origin folder.
    - Create a work plane on the outside of the cylinder. From the Model tab > Work Features panel click the down arrow on the right side of the Work Plane command and from the list click Tangent to Surface and Parallel to Plane.
    - In the browser click the YZ plane under the Origin folder.
    - Click a point to the front right of the cylinder, as shown in the following image on the left.
31. Make the new work plane the active sketch.
32. Use the Project Geometry command from the Draw panel to project the Z axis of the origin folder onto the sketch.
33. Place a Point, Center Point on the projected axis—this will center the point in the center of the cylinder—and dimension it, as shown in the following image in the middle. Then finish the sketch.
34. Press H on the keyboard to start the Hole command and place a **.5 inch** through all hole at the center point.
35. Turn off the visibility of the work plane and your screen should resemble the following image on the right.

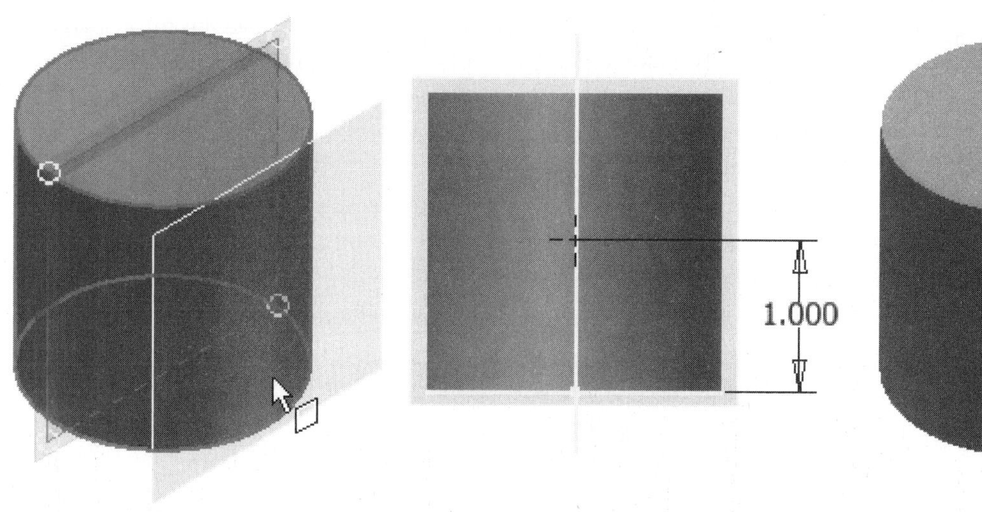

FIGURE 4.71

36. Next, you place a hole on the cylinder at an angle. First you create an angled work plane.

    - Create a work plane in the middle of the cylinder. From the Model tab > Work Features panel click the down arrow on the right side of the Plane command and from the list click Angle to Plane around Edge.
    - In the browser, click the YZ plane under the Origin folder.
    - In the browser click, the Z Axis under the Origin folder.
    - In the Angle dialog box, enter **315** as shown in the following image on the left.
    - Create the work plane, click the green check mark in the mini-toolbar.

37. Next, create a work plane that is parallel to an angled plane and tangent to the cylinder.

    - Create a work plane on the outside of the cylinder. From the Model tab > Work Features panel click the arrow on the right side of the Plane command and from the list click Tangent to Surface and Parallel to Plane.
    - Click the angle work plane that you just created.
    - Click a point to the front face of the cylinder, as shown in the following image on the right.

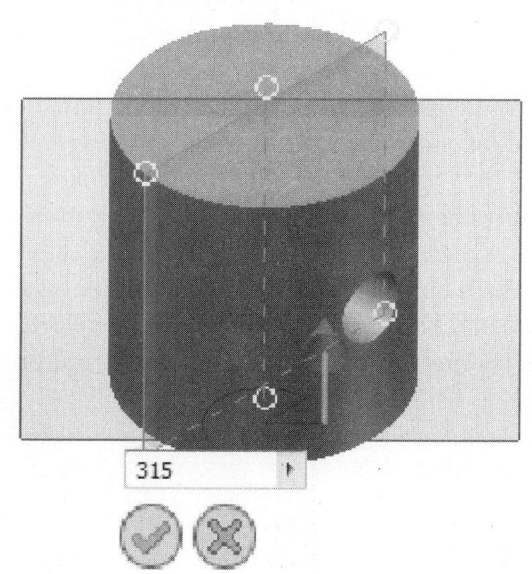

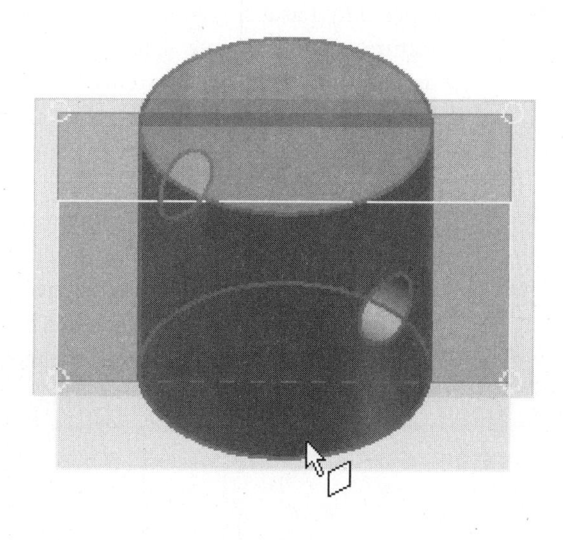

FIGURE 4.72

38. Turn off the visibility of the inside angled work planes. Move the cursor over the work plane. Right-click, and click Visibility from the menu.
39. Create a new sketch on the new work plane.
40. Use the Project Geometry command to project the Z axis of the origin folder onto the sketch.
41. Place a Point, Center Point on the projected axis, and dimension it, as shown in the following image on the left.
42. Finish the sketch.
43. Start the Hole command, and place a counterbore hole with the size of your choice at the center point.
44. Turn off the visibility of the work plane and your screen should resemble the following image on the right.

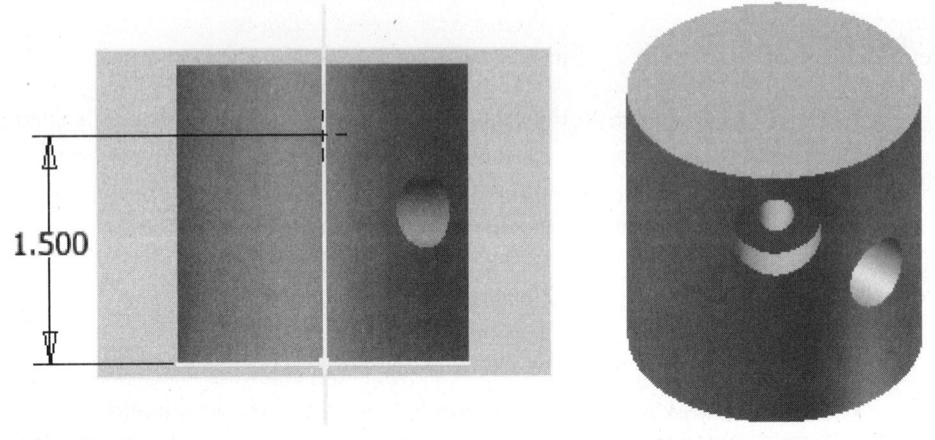

FIGURE 4.73

45. Expand Work Plane3 in the browser, double-click on Work Plane2, and enter a new value. Update the model to see the change.
46. Close the file. Do not save changes. End of exercise.

PATTERNS

There are two types of pattern methods: rectangular and circular. The pattern is represented as a single feature in the browser, but the original feature and individual feature occurrences are listed under the pattern feature. You can suppress the entire pattern or individual occurrences except for the first occurrence. Both rectangular and circular patterns have a child relationship to the parent feature(s) that you patterned. If the size of the parent feature changes, all of the child features will also change. If you patterned a hole, and the parent hole type changes, the child holes also change. Because a pattern is a feature, you can edit it like any other feature. You can pattern the base part or feature, as well as patterns. A rectangular pattern repeats the selected feature(s) along the direction set by two edges on the part or edges that reside in a sketch. These edges do not need to be horizontal or vertical, as shown in the following image. A circular pattern repeats the feature(s) around an axis, a cylindrical or conical face, or an edge.

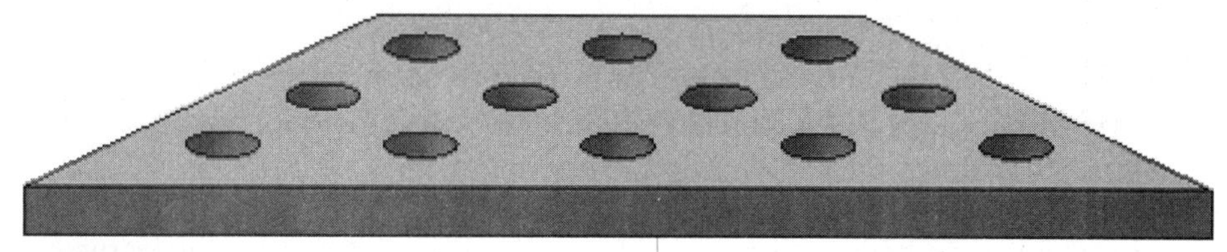

FIGURE 4.74

Rectangular Patterns

When creating a rectangular pattern, you define two directions by clicking an edge or line segment in a sketch to define alignment. After you click the Rectangular Pattern on the Model tab > Pattern panel, as shown in the following image on the left, or press the hot key CTRL + SHIFT + R (rectangular pattern), the Rectangular Pattern dialog box appears, as shown in the following image on the right. Click one or more features to pattern, enter the values, and click the edges or axes as needed. A preview image of the pattern appears in the graphics area.

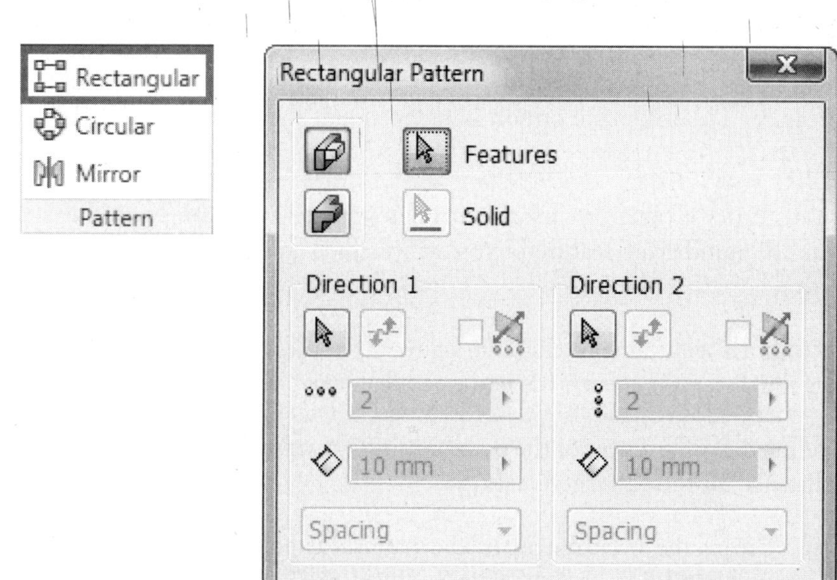

FIGURE 4.75

The following options are available for rectangular patterns.

Pattern Individual Features

Click this button to pattern a feature or features. When you select this option, you activate a features button, as described below.

Pattern the Entire Solid

Click this button to pattern a solid body. When you select this option, you select the entire part as the item to pattern. You also have the Include Work Features option when patterning an entire solid.

Features

Click this button, and then click a feature or features to be patterned from either the graphics window or the browser. You can add or remove features to or from the selection set by holding down the CTRL key and clicking them.

Solid. If multiple solid bodies exist, select the solid body that you want the feature(s) patterned on.

Direction 1

In Direction 1, you define the first direction for the alignment of the pattern. It can be an edge, an axis, or a path.

Path. Click this arrow button, and then click an edge or sketch that defines the alignment along which you will pattern the feature.

Flip. If the preview image shows the pattern going in the wrong direction, click this button to reverse its direction.

Midplane. Check this option to have the occurrences patterned on both sides of the selected feature. The midplane option is independent for both Direction 1 and Direction 2.

Column Count. Enter a value or click the arrow to choose a previously used value that represents the number of feature(s) you will include in the pattern along the selected direction or path.

Column Spacing. Enter a value or click the arrow to choose a previously used value that represents the distance between the patterned features.

Distance. Define the occurrences of the pattern using the provided dimension as the total overall distance for the patterned features.

Curve Length. Create the occurrences of the pattern at equal spacing along the length of the selected curve.

Direction 2

In Direction 2, you can define a second direction for the alignment of the pattern. It can be an edge, an axis, or a path, but it cannot be parallel to Direction 1.

Path. Click this button, and then click an edge or sketch that defines the alignment along which you will pattern the feature.

Flip. If the preview image shows the pattern going in the wrong direction, click this button to reverse its direction.

Midplane. Check this option to have the occurrences patterned on both sides of the selected feature. The midplane option is independent for both Direction 1 and Direction 2.

Row Count. Enter a value or click the arrow to choose a previously used value that represents the number of feature(s) that you will include in the pattern in the second direction.

Row Spacing. Enter a value or click the arrow to choose a previously used value that represents the distance between the patterned features.

Distance. Define the occurrences of the pattern using the provided dimension as the total overall distance for the patterned features.

Curve Length. Create the occurrences of the pattern at equal spacing along the length of the selected curve.

Options for the start point of the direction, compute type, and orientation method, as shown in the following image, are available by clicking the More (>>) button located in the bottom-right corner of the Rectangular Pattern dialog box.

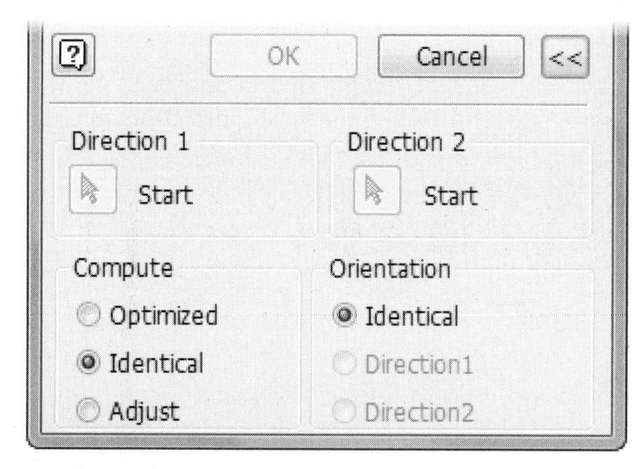

FIGURE 4.76

Start

Click the Start button to specify where the start point for the first occurrence of the pattern will be placed. The pattern can begin at any selectable point on the part. You can select the start points for both Direction 1 and Direction 2.

Compute

Optimized. Click this option to pattern the feature's faces instead of the feature(s) to calculate all of the occurrences in the pattern. This option is ideal when the occurrences you are creating do not intersect and are all identical. It can improve the performance of pattern creation.

Identical. Click this option to use the same termination as that of the parent feature(s) for all of the occurrences in the pattern. This is the default option.

Adjust. Click this option to calculate the termination of each occurrence individually. Since each occurrence is calculated separately, the processing time can increase. You must use this option if a parent feature terminates to a face or plane.

Orientation

Identical. Click this option to orient all of the occurrences in the pattern the same as the parent feature(s). This is the default option.

Direction 1. Click this option to control the position of the patterned features by the selected direction. Each occurrence of the pattern is rotated to maintain proper orientation with the 2D tangent vector of the path.

Direction 2. Click this option to control the position of the patterned features by the selected direction. Each occurrence of the pattern is rotated to maintain proper orientation with the 2D tangent vector of the path.

Circular Patterns

When creating a circular pattern, you must have a work axis, a part edge, or a cylindrical face about which the features will rotate. After you click the Circular Pattern command on the Model tab > Pattern panel, as shown in the following image on the left, or press the hot key CTRL + SHIFT + O, the Circular Pattern dialog box appears, as shown in the following image on the right. Click a feature or features to pattern, enter the values, and click the edges and axis as needed. A preview image of the pattern appears in the graphics area.

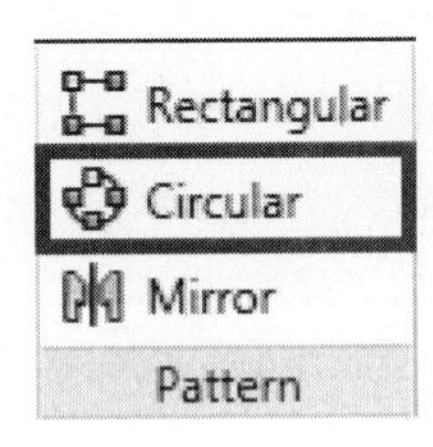

FIGURE 4.77

The following options are available for circular patterns.

Pattern Individual Features

Click this button to pattern a feature or features. When you select this option, the features button is available and operates as described below.

Pattern the Entire Solid

Click this button to pattern a solid body. When you select this option, you select the entire part as the item to pattern. You also have the Include Work Features option when patterning an entire solid.

Features

Click this button, and then click a feature or features to be patterned. You can add or remove features to or from the selection set by holding down the CTRL key and clicking them.

Rotation Axis

Click the button and then click an edge, axis, or cylindrical face (center) that defines the axis about which the feature(s) will rotate.

Flip. If the preview image shows the pattern going in the wrong direction, click this button to reverse its direction.

Solid. If multiple solid bodies exist, select the solid body you want the feature(s) patterned on.

Placement

Occurrence Count. Enter a value or click the arrow to choose a previously used value that represents the number of feature(s) that you will include in the pattern. A positive number will pattern the feature(s) in the clockwise direction; a negative number will pattern the feature in the counterclockwise direction.

Occurrence Angle. Enter a value or click the arrow to choose a previously used value that represents the angle that you will use to calculate the spacing of the patterned features.

Midplane. Check this option to have the occurrences patterned evenly on both sides of the selected feature.

By clicking the More (>>) button, located in the bottom-right corner of the Circular Pattern dialog box, you can access options for the creation method and positioning method of the feature, as shown in the following image.

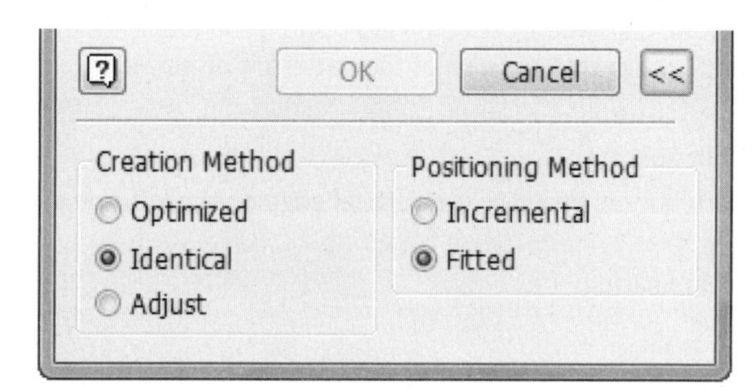

FIGURE 4.78

Creation Method

Optimized. Click this option to pattern the feature's faces instead of the feature(s) to calculate all of the occurrences in the pattern. This option is ideal when the occurrences you are creating do not intersect and are all identical. It can improve the performance of pattern creation. This method is recommended when there are 50 or more occurrences.

Identical. Click this option to use the same termination as that of the parent feature(s) for all of the occurrences in the pattern. This is the default option.

Adjust. Click this option to calculate each occurrence termination individually. Because each occurrence is calculated separately, the processing time can increase. You must use this option if a parent feature terminates to a face or plane.

Positioning Method

Incremental. Click this option to separate each occurrence by the number of degrees specified in Angle in the dialog box.

Fitted. Click this option to space each occurrence evenly within the angle specified in Angle in the dialog box.

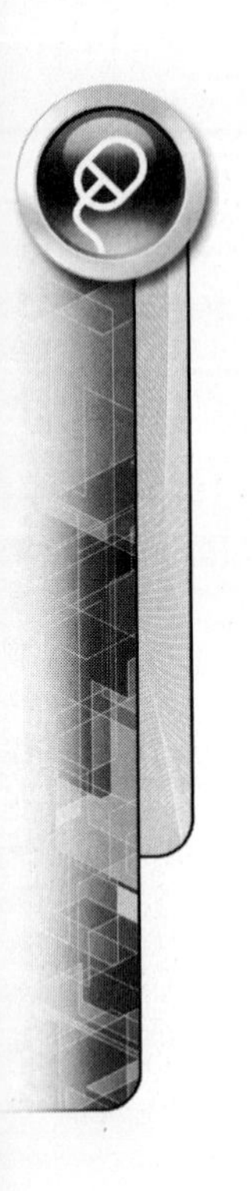

TIP A work axis, an edge, or a cylindrical face (the centerline of the circular face will be used) about which the feature will rotate must exist before you create a circular pattern.

Linear Patterns—Pattern along a Path

You can also use the Rectangular Pattern command to pattern a feature about a path. You can define a path by a complete or partial ellipse, an open or closed spline, or a series of curves (lines, arcs, splines, etc.).

To pattern along a path, click the Path button, and use the options described above for rectangular patterns. The path you use can be either 2D or 3D.

EXERCISE 4-6: CREATING RECTANGULAR PATTERNS

In this exercise, you will create a rectangular pattern of holes to a cover plate.

1. Open *ESS_E04_06.ipt* in the Chapter 04 folder.
2. You pattern the hole. Click the Rectangular Pattern command in the Pattern panel.
3. Click the small hole feature as the feature to be patterned.
4. In the Direction 1 area of the dialog box, click the Path button, and then click the bottom horizontal edge of the part. A preview of the pattern is displayed.
5. Click the Flip button.
6. Enter **5** Count and **.625 in** Spacing.
7. Click the Direction 2 Path button, and click the vertical edge on the left side of the part.
8. Enter **4** Count and **.625 in** Spacing.

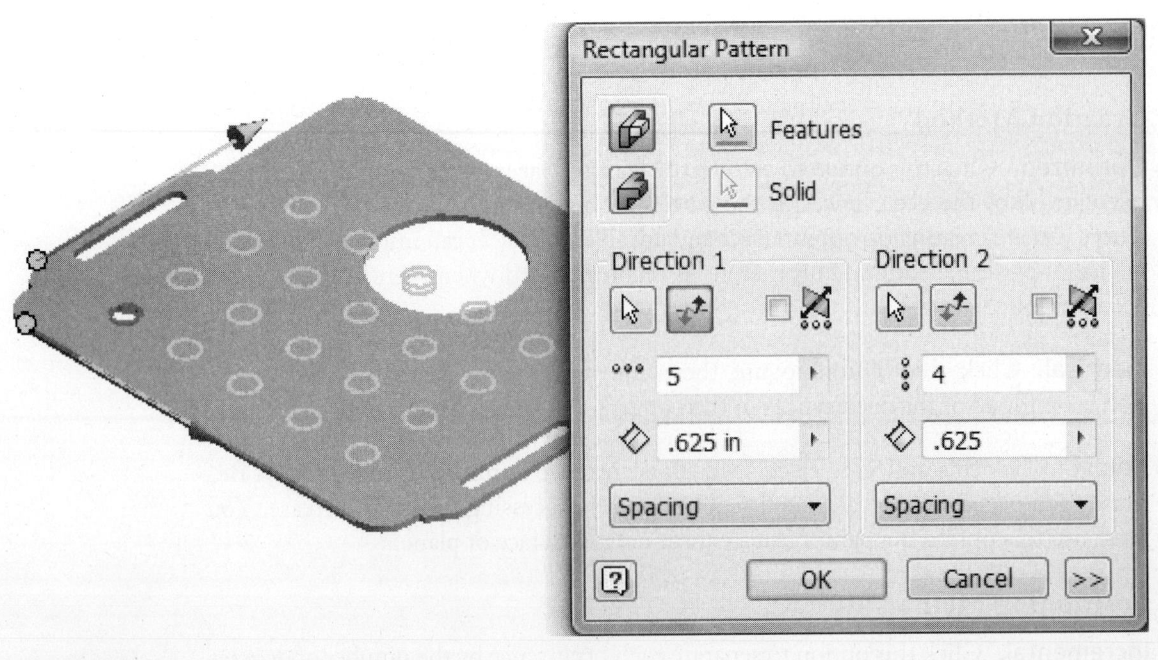

FIGURE 4.79

9. Click OK to create the pattern.

   You now suppress three of the holes that are not required in the design.

10. Expand the rectangular pattern feature entry in the browser to display the occurrences.
11. In the browser, move the cursor over the occurrences. Each occurrence highlights in the graphics window as you point to it in the browser.
12. Hold the CTRL key down, and click the three occurrences the holes will highlight on the model, as shown in the following image on the left.
13. Right-click on any one of the highlighted occurrences in the browser, and click Suppress, as shown in the following image.

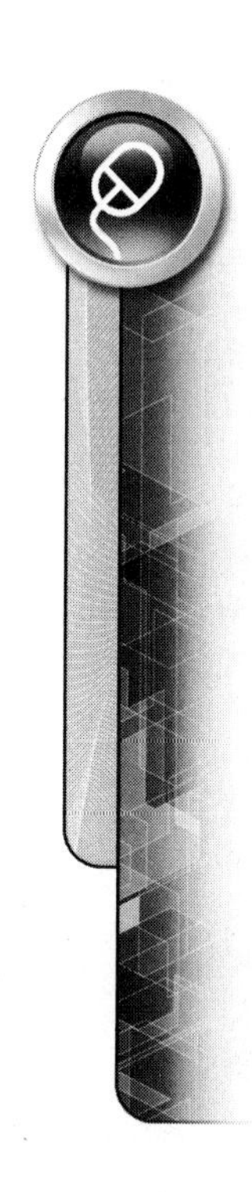

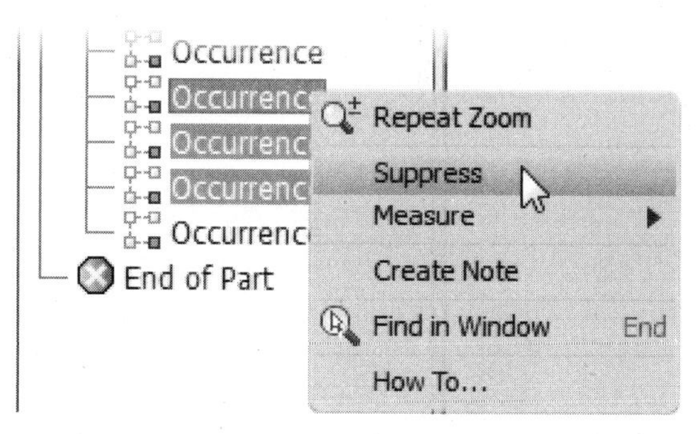

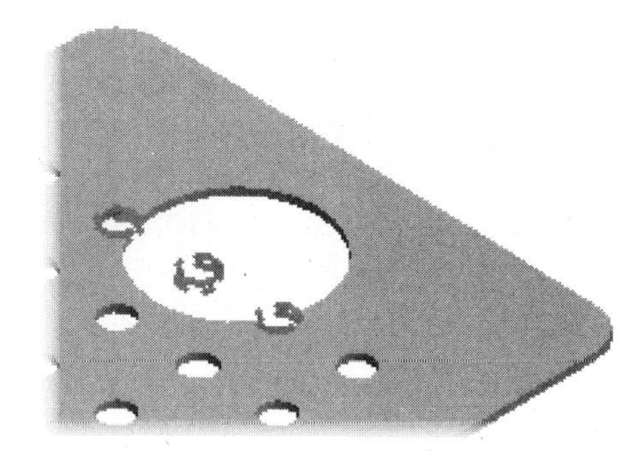

FIGURE 4.80

14. The holes are suppressed in the model, as shown in the following image.

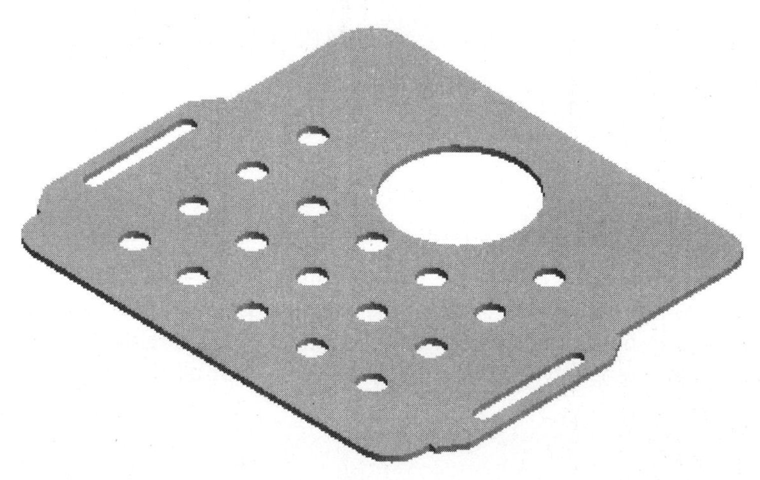

FIGURE 4.81

15. Practice editing the feature pattern, and change the count and spacing for each direction. The suppressed occurrences are still suppressed.
16. Close the file. Do not save changes. End of exercise.

EXERCISE 4-7: CREATING CIRCULAR PATTERNS

In this exercise, you will create a circular pattern of a counterbore hole.

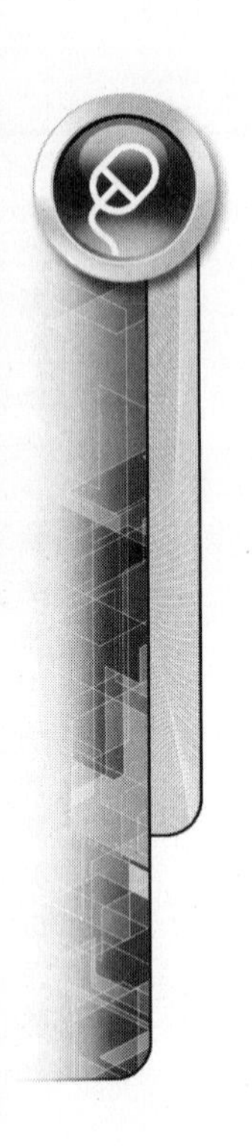

1. Open *ESS_E04_07.ipt* in the Chapter 04 folder.
2. Edit the Hole1 feature to verify that the termination for the hole is Through All.
3. Click the Cancel button to close the Hole dialog box.
4. Click the Circular Pattern command in the Pattern panel.
5. Click the hole feature as the feature to pattern.
6. Click the Rotation Axis button in the Circular Pattern dialog box.
7. In the graphics window click the work axis to specify the rotation axis. A preview of the pattern is displayed in the graphics window.
8. Type **8** in the Count field as shown in the following image.

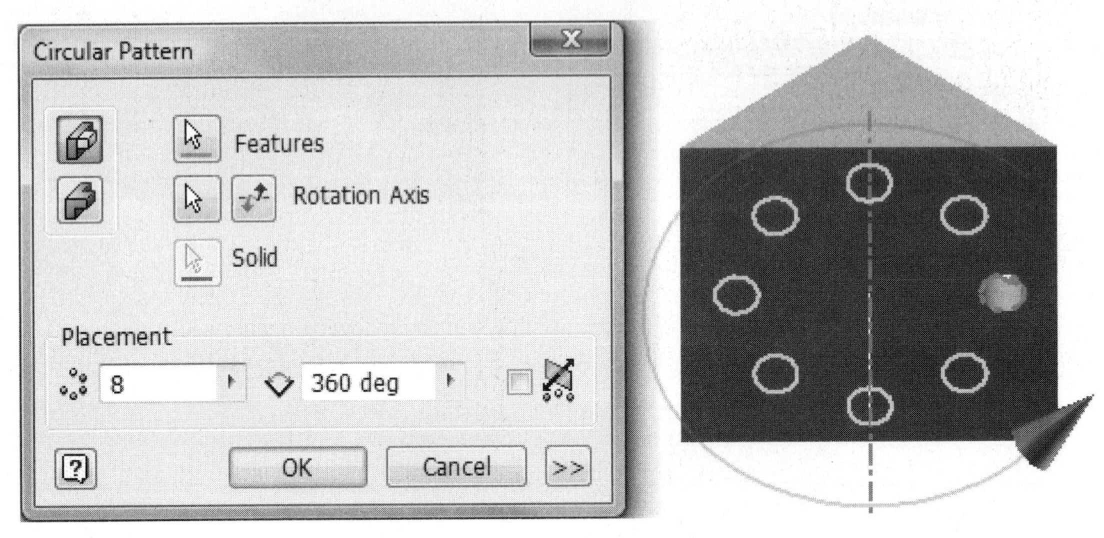

FIGURE 4.82

9. Click the OK button to create the pattern.
10. Rotate the model to see the back side, as shown in the following image. The other six holes do not go through because the Identical Creation Method was selected; that is, the hole that is patterned is identical to the original hole. If desired, change the visual style to Wireframe to verify that all the holes are identical.

FIGURE 4.83

11. In the browser right-click on Circular Pattern1 and click Edit Feature from the menu.
12. Click the More (>>) button.
13. Under Creation Method, select the Adjust option, as shown in the following image on the left.
14. Click OK to create the circular pattern. When done, your model should resemble the following image on the right.

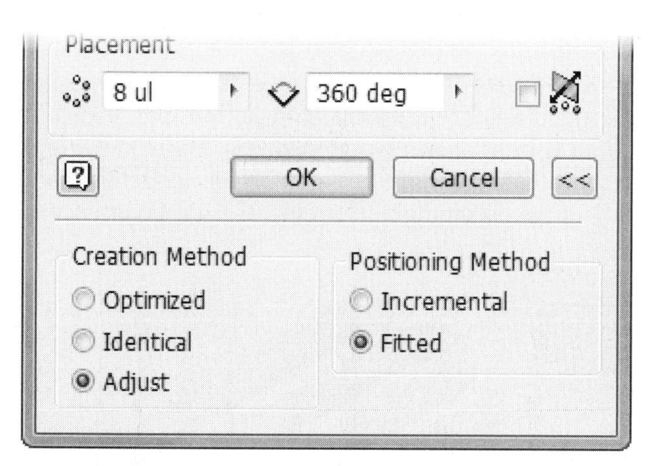

FIGURE 4.84

15. Edit the circular pattern, and try different combinations of count, angle, creation method, and positioning method.
16. Close the file. Do not save changes. End of exercise.

EXERCISE 4-8: CREATING A PATTERN ALONG A NONLINEAR PATH

In this exercise, you will pattern a boss and hole along a nonlinear path.

1. Open *ESS_E04_08.ipt* in the Chapter 04 folder. Note that the visibility of the sketch that the pattern will follow is on.
2. Click the Rectangular Pattern command in the Pattern panel.
3. For the features to pattern, in the browser, click both Extrusion2 and Hole2 feature.
4. In the Direction 1 area in the Rectangular Pattern dialog box, click the Path arrow button. To select the entire path in the graphics window click the sketch line near the extrusion feature. A preview of the pattern is displayed, as shown in the following image.

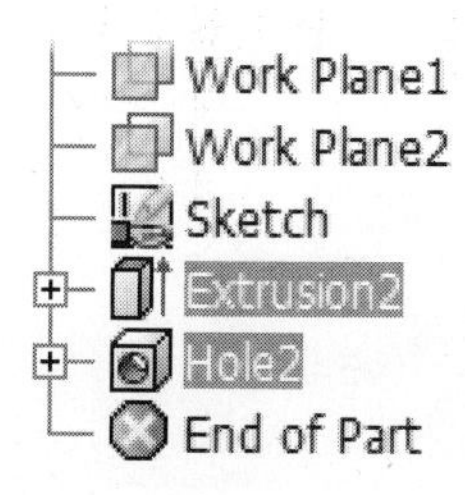

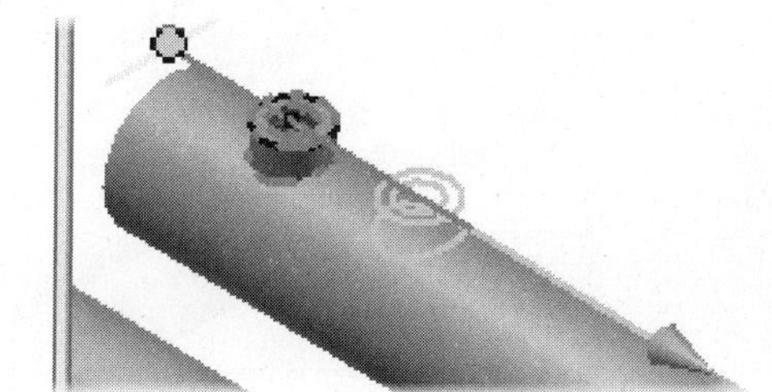

FIGURE 4.85

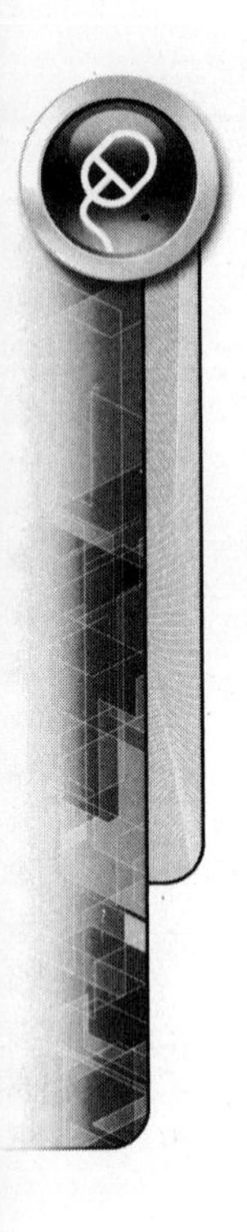

5. In Count, enter **40**, and in Spacing, enter **1.25**. The preview shows 1.25 inches between each occurrence.
6. From the Spacing drop-down list, select Distance. The preview shows 40 occurrences fit within 1.25 inches.
7. From the Distance drop-down list, select Curve Length. The preview shows the 40 occurrences fitting within the entire length of the path.
8. Rotate the model to verify that the occurrences on the right side are hanging off the model; this is because the first occurrence is spaced 1 inch away from the start of the path.
9. To solve the starting point issue, click the More (>>) button.
10. In the Direction 1 area in the dialog box, click the Start button, and then click the center point of the first hole, as shown in the following image on the left. The preview updates to show that all occurrences are now located on the part, as shown in the following image on the right. However, the last occurrence is on the edge of the part.

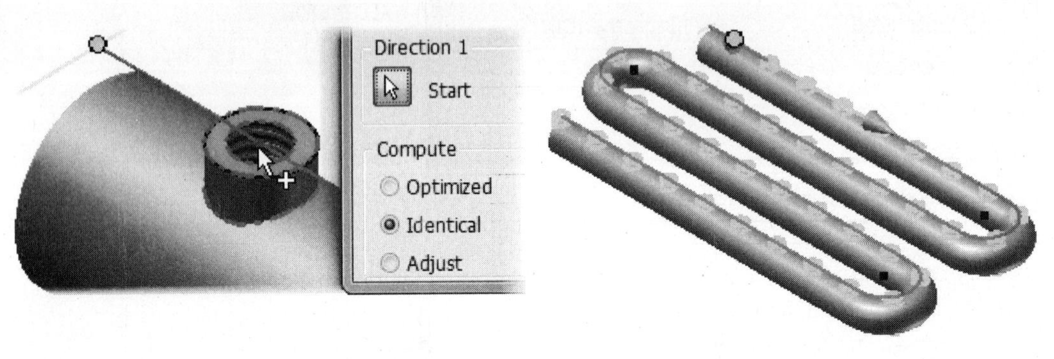

FIGURE 4.86

11. From the Curve Length drop-down list, select Distance. The curve length remains in the distance area but the value can be edited.
12. Click in the distance area and arrow to the right of the value of the four lines and three arcs. Subtract –**1 in** from the distance, as shown in the following image on the left.
13. Click OK to create the pattern.
14. In the browser, right-click on the entry Sketch, and uncheck Visibility from the menu. When done, your model should resemble the following image on the right.

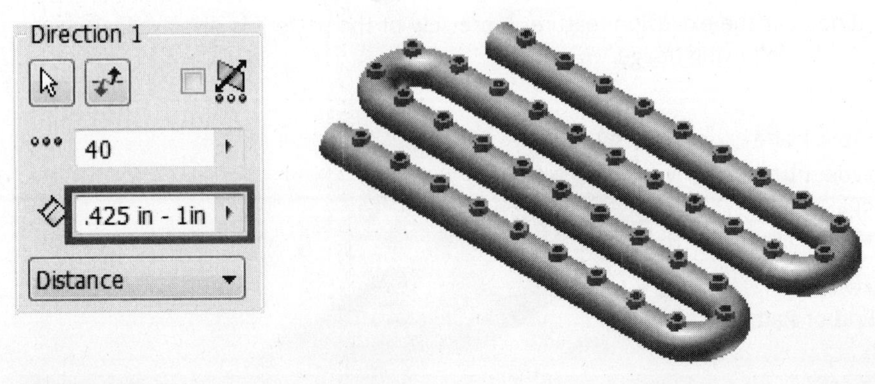

FIGURE 4.87

15. Edit the pattern trying different combinations.
16. Close the file. Do not save changes.

APPLYING YOUR SKILLS

Skill Exercise 4-1

In this exercise, you will create a drain plate cover.

1. Start a new part based on the English *Standard (in).ipt* template.
2. Use the extrude, shell, hole, and rectangular pattern commands to create the part.

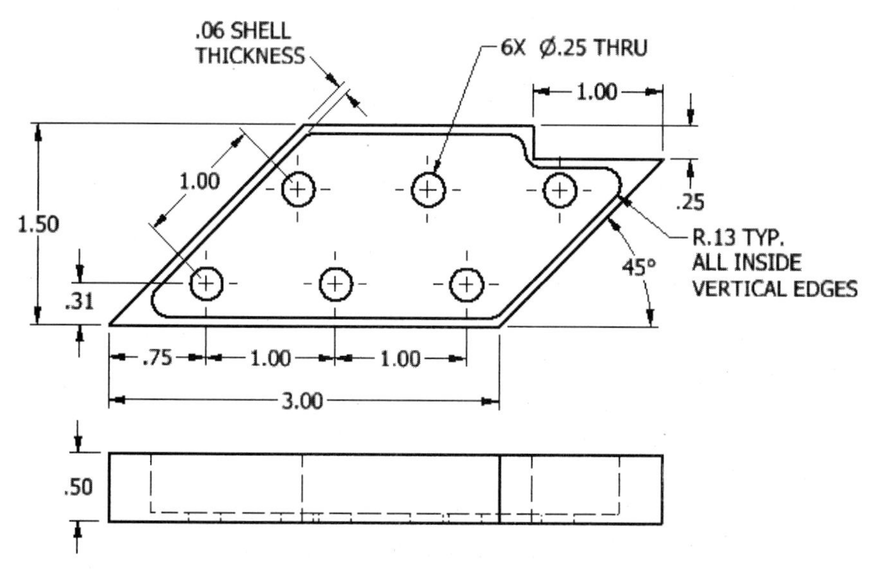

FIGURE 4.88

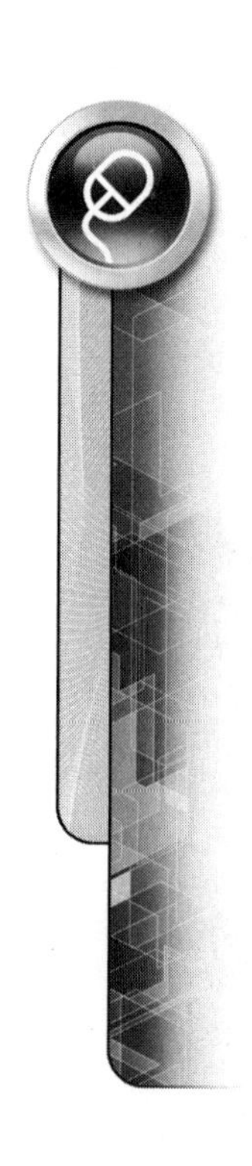

Skill Exercise 4-2

In this exercise, you will create a connector.

1. Start a new part based on the English *Standard (in).ipt* template.
2. Use the revolve, work plane, hole, chamfer, fillet, and circular pattern commands to complete the part.

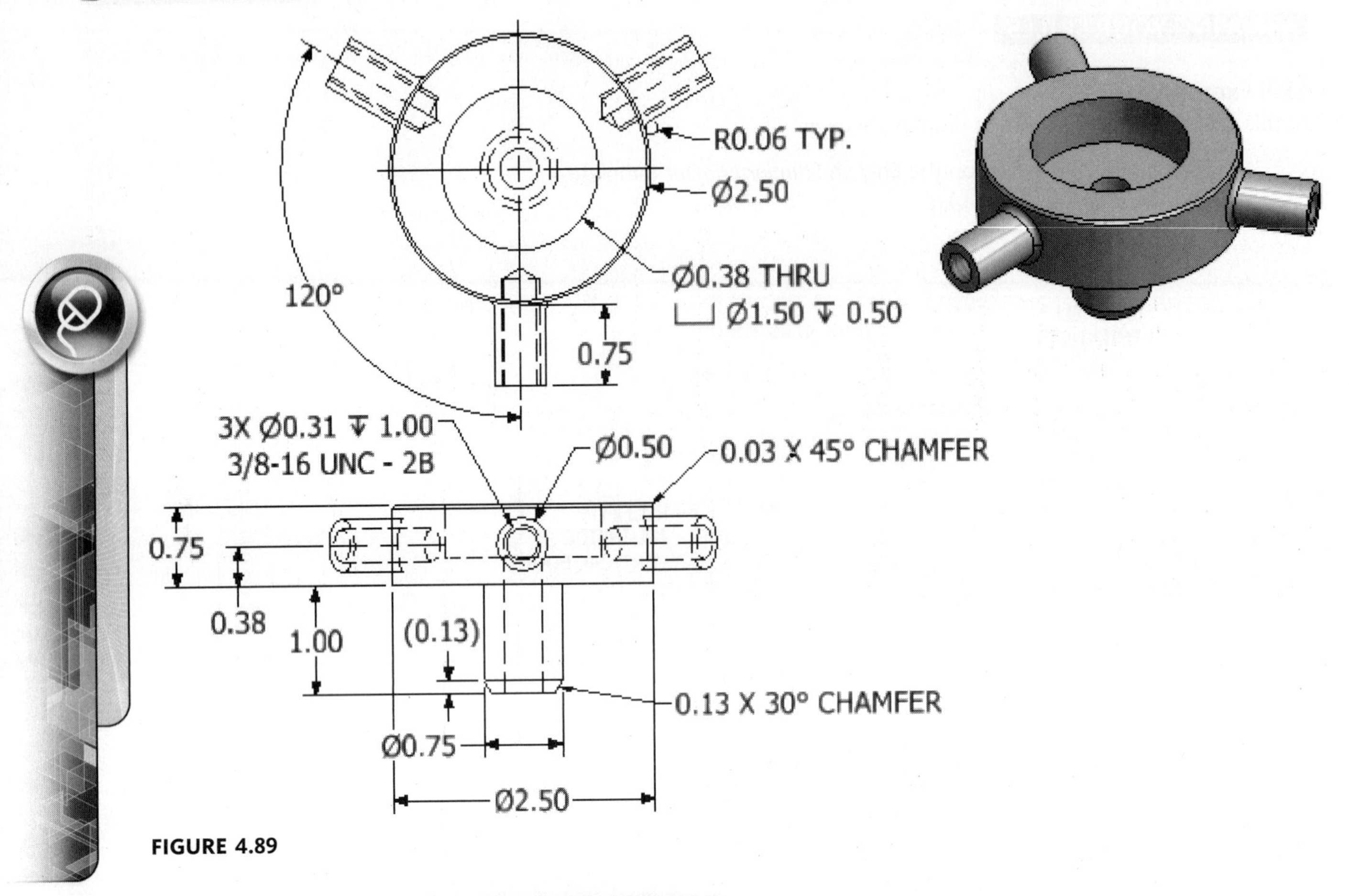

FIGURE 4.89

CHECKING YOUR SKILLS

Use these questions to test your knowledge of the material covered in this chapter.

1. True ___ False ___ When creating a fillet feature that has more than one selection set, each selection set appears as an individual feature in the browser.
2. In regard to creating a fillet feature, what is a smooth radius transition?
3. True ___ False ___ When you are creating a fillet feature with the All Fillets option, material is removed from all concave edges.
4. True ___ False ___ When you are creating a chamfer feature with the Distance and Angle option, you can only chamfer one edge at a time.
5. True ___ False ___ When you are creating a hole feature, you do not need to have an active sketch.
6. What is a Point, Center Point used for?
7. True ___ False ___ A part may contain only one shell feature.
8. True ___ False ___ The only method to create a work axis is by clicking a cylindrical face.
9. True ___ False ___ You need to derive every new sketch from a work plane feature.
10. Explain the steps to create an offset work plane.
11. True ___ False ___ The options for creating work features filter out geometry that is not valid for the selected option.
12. True ___ False ___ You cannot create work planes from the default work planes.
13. True ___ False ___ A UCS can only be placed on existing geometry.

14. True ___ False ___ When you are creating a rectangular pattern, the directions along which the features are duplicated must be horizontal or vertical.
15. True ___ False ___ When you are creating a circular pattern, you can only use a work axis as the axis of rotation.
16. True ___ False ___ To start the Fillet command from the mini-toolbar, select the two planes that define the location of the fillet.
17. True ___ False ___ Use the Filet command with the Face option to create a fillet that is tangent to three faces.
18. While in the shell command, explain how to create a unique face thickness within the shell command.
19. Explain three reasons why you would create a UCS.
20. True ___ False ___ Use the Linear Pattern command to pattern a feature along a selected path, consisting of multiple lines and arcs.

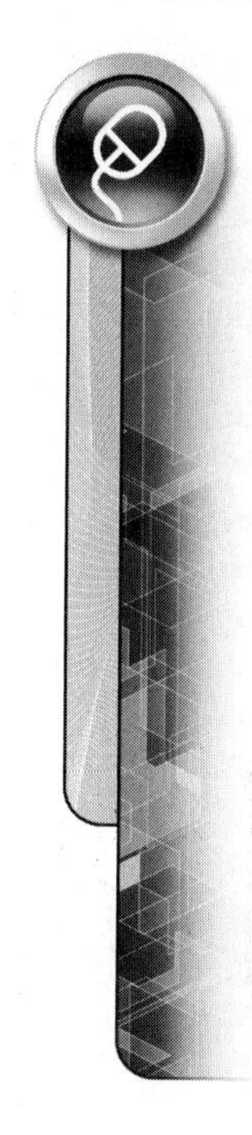

CHAPTER 5

Creating and Editing Drawing Views

INTRODUCTION

After creating a part or assembly, the next step is to create 2D drawing views that represent that part or assembly. To create drawing views, start a new drawing file, select a 3D part or assembly on which to base the drawing views, insert or create a drawing sheet with a border and title block, project orthographic views from the part or assembly, and then add annotations to the views. You can create drawing views at any point after a part or assembly exists. The part or assembly does not need to be complete because the part and drawing views are associative in both directions (bidirectional). This means that if the part or assembly changes, the drawing views will automatically be updated. If a parametric dimension changes in a drawing view, the part will be updated before the drawing views get updated. This chapter will guide you through the steps for setting up styles, creating drawing views of a single part, editing dimensions, and adding annotations.

When creating drawings you adhere to specific drawing standards that communicate information about designs in a consistent manner. There are many different drawing standards, such as ANSI and ISO. Each company usually adopts a drawing standard and modifies it to meet their requirements. This book follows the ANSI drawing standard. More information about drawing standards and good practices are covered throughout this chapter.

OBJECTIVES

After completing this chapter, you will be able to perform the following:

- Make changes under the Drawing tab of the Application Options dialog box
- Create base and projected drawing views from a part
- Create auxiliary, section, detail, broken, break out, and cropped views
- Edit the properties and location of drawing views
- Retrieve and arrange model dimensions for use in drawing views
- Edit, move, and hide dimensions
- Add automated centerlines

- Add general dimensions, baseline, chain, and ordinate dimensions
- Add annotations such as text, leaders, Geometric Dimensioning & Tolerancing (GD&T), surface finish symbols, weld symbols, and datum identifiers
- Create hole notes
- Open a model from a drawing
- Open a drawing from a model
- Create hole, general, and revision tables

Drawing Options

Before drawing views are created, the drawing options should be set to your preferences. To set the drawing options, click Application Options on the Tools panel. The Options dialog box will appear. Click the Drawing tab, your screen should resemble the following image. Make any changes to the options before creating the drawing views, otherwise the changes may not affect drawing views that you have already created. The following sections describe the drawing options.

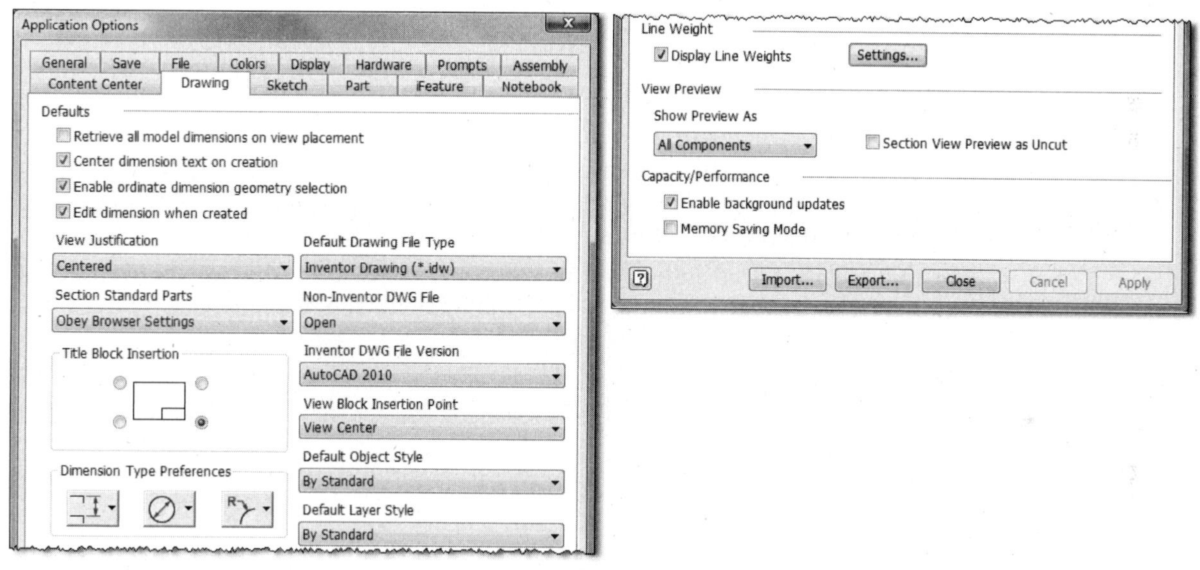

FIGURE 5.1

Retrieve All Model Dimensions on View Placement

Click this box to add applicable model dimensions to drawing views when they are placed. If the box is clear, no model dimensions will be placed automatically. You can override this setting by manually selecting the All Model Dimensions in the Drawing View dialog box when creating base views.

Center Dimension Text on Creation

Click this option to have dimension text centered as you create the dimension.

Enable Ordinate Dimension Geometry Selection

Click this option to allow geometry to be selected with the window / crossing selection technique when placing ordinate dimensions.

Edit Dimension When Created

Click this option to see the Edit Dimension dialog box everytime a dimension is placed in a drawing.

View Justification

Use this option to set the default justification for drawing views. Two modes are available: Centered and Fixed.

Section Standard Parts

Use this option to control whether standard parts placed into an assembly from the supplied Inventor parts library, such as nuts, bolts, and washers, are sectioned. Three options are available in this area: Always, Never, and Obey Browser Settings.

Title Block Insertion

Click to select the title block insertion point for the first and subsequent drawing sheets.

Dimension Type Preferences

Use this area to set the preferred type of dimensions. Three buttons are available to control the type of dimension being placed. The first button controls linear dimensions, Linear diametric, and Linear symmetric. The second button controls whether circles utilize diametric or radial dimensions. The last button controls the arc preference and includes radial, diametric, angle, arc length, and chord length.

Default Drawing File Type

In this section select either Inventor drawing IDW or DWG as the default file format. Also set how Inventor should handle DWG and drawing data.

Non-Inventor DWG File

When a non-Inventor DWG file is selected, you set the default behavior to either open or import the DWG file.

Inventor DWG File Version

Set the Inventor DWG file version to save.

View Block Insertion Point

Set the default insertion point of a view block to be View Center or Model Origin.

Default Object Style

Set the default object style to be By Standard or Last Used.

Default Layer Style

Set the default layer style to be By Standard or Last Used.

Line Weight Display Options

Use this option to control the display of line weights in a drawing.

Display Line Weights. Click this option to allow line weights to appear in drawings. Visible lines will appear in drawings with the line weights defined in the active drafting standard. Clear the box to display all lines without weights.

View Preview Display

Controls how components are previewed when drawing view are created.

Show Preview As. Sets how components are previewed when creating drawing views; All Components. Partial or Bounding Box. Partial and Bounding Box options reduce memory consumption. The preview has no effect on the computed drawing view.

Section View Preview as Uncut. Set this option to preview the model as uncut or uncheck this option to preview as cut. The preview has no effect on the computed drawing view.

Capacity/Performance

Controls if Inventor should conserve memory at the cost of performance when creating and editing drawing views.

Enable Background Updates

When checked, precise drawing views will be calculated while you continue to work. A drawing view is being updated when a small horizonal and a vertical line appear in the corners of a drawing view.

Memory Saving Mode. When checked Inventor reduces the amount of memory while creating drawing views at the expense of performance. Memory is conserved by changing the way components are loaded and unloaded.

Creating a Drawing

The first step in creating a drawing from an existing part or assembly is to create a new drawing IDW or DWG file by one of the following methods. You can use the following methods to create a new part file:

- Click New on the Inventor Application menu, and then click the *Standard.idw or dwg* icon on the Default tab. You can also click *an idw or dwg* template on one of the other tabs.
- Click the New icon in the Quick Launch area of the Open dialog box, and then click the *Standard.idw or dwg* icon on the Default tab, as shown in the following image on the left, or click *a idw or dwg* template on one of the other tabs.
- From the Quick Access toolbar click the down arrow of the New icon and select Drawing from the menu as shown in the following image on the right. This creates a new drawing file based on the units that were selected when Autodesk Inventor was installed or on the *Standard.idw or dwg* that exists in the Templates folder.
- Use the shortcut key, CTRL-N, and then click the *Standard.idw or dwg* icon on the Default tab, as shown in the following image on the left. You can also click *a idw or dwg* template on one of the other tabs.

After starting a new drawing file using one of the previous methods, Autodesk Inventor's screen will change to reflect the drawing environment.

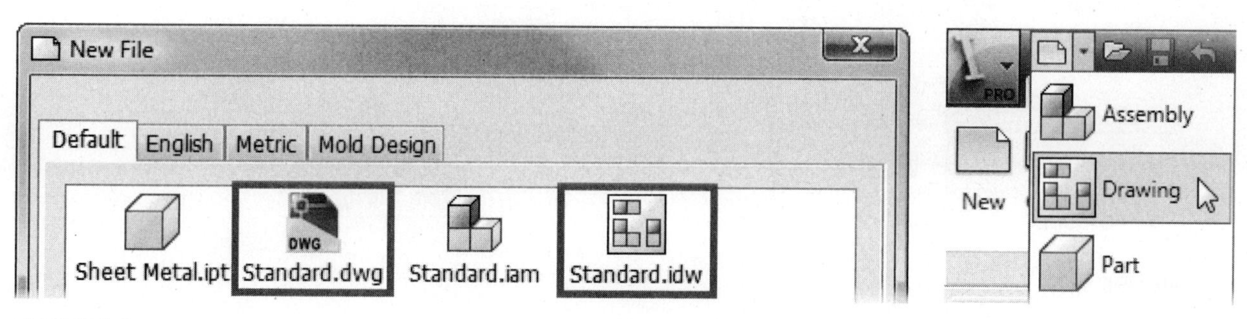

FIGURE 5.2

Seven drafting standard/templates, included with Autodesk Inventor, are presented under the Metric tab. If needed select one of these templates to start a drawing. They include the following:

- ANSI (American National Standards Institute)
- BSI (British Standards Institute)
- DIN (The German Institute for Standardization)
- GB (The Chinese National Standard)
- GOST (The Russian Standard)
- ISO (International Organization for Standardization)
- JIS (Japan Industrial Standard)

DWG TrueConnect

DWG TrueConnect allows you to work directly with AutoCAD DWG files without the need for translations. This feature allows you to view, plot, and measure AutoCAD drawing data while using Autodesk Inventor and perform the same actions to Autodesk Inventor drawing data while using AutoCAD, with the exception of editing the drawing views. The drawing view data in an Inventor DWG file is controlled by Inventor. In this way, you can supply AutoCAD customers with DWG drawings that contain familiar named items such as layers and object types such as blocks and title blocks. As was discussed in the Drawing Options section, from the Drawing tab of the Application Option dialog box set the Default Drawing File Type to idw or dwg. This sets the default file format that is used when a new drawing is created from the Quick Access toolbar > down arrow of the New icon > Drawing from the menu.

DRAWING SHEET PREPARATION

When you create a new drawing file using one of the provided template files or a template file that you created. Inventor displays a default drawing sheet with a default title block and border. The template DWG or IDW file that is selected determines the default drawing sheet, title block, and border. The drawing sheet represents a blank piece of paper on which you can alter the border, title block, and drawing views. There is no limit to the number of sheets that can exist in the same drawing, but you must have at least one drawing sheet. To create a new sheet, click the New Sheet command from the Place Views tab > Sheets panel as shown in the following image on the left. Alternately, you can right-click in the browser and click New Sheet from the menu as shown in the following image in the middle, or on the current sheet in the graphics window right-click and click New Sheet from the marking menu as shown in the following image on the right.

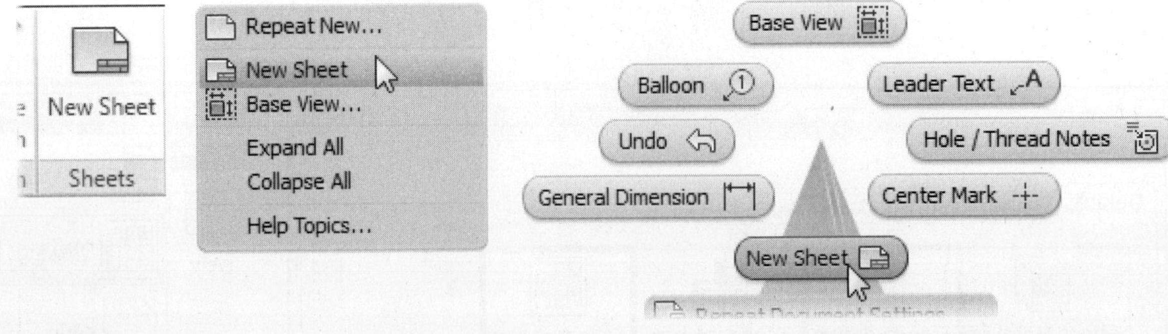

FIGURE 5.3

A new sheet will appear in the browser, and the new sheet will appear in the graphics window with the same size, border and title block of the active sheet. To edit the sheet's size right-click on the sheet name in the browser, and select Edit Sheet from the menu, as shown in the following image on the left. Then select a size from the list, as shown in the following image on the right. To use your own values, select Custom Size from the list, and enter values for height and width. From the dialog box you can also change the sheet's name or slowly double-click on its name in the browser and then enter a new name.

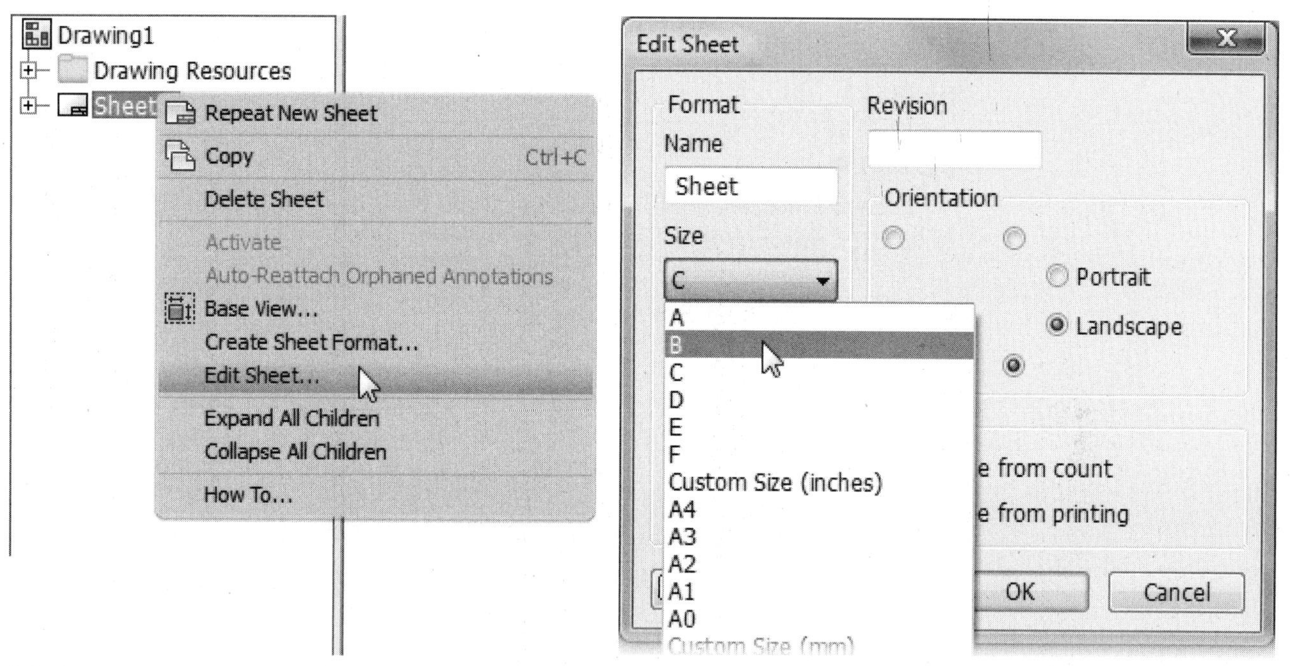

FIGURE 5.4

NOTE

The sheet size is inserted full scale (1:1) and should be plotted at 1:1. The drawing views will be scaled to fit the sheet size.

TITLE BLOCKS

To change a title block on a drawing sheet, you can either insert a default title block or construct a customized title block and insert it into a drawing sheet.

Inserting a Default Title Block

To insert a default title block, follow these steps:

1. If a title block exists in the sheet, it must be deleted before a new title block can be inserted. Make the sheet active, from the browser right-click on the title block entry and click Delete from the menu as shown in the following image on the left.
2. Insert a title block by expanding Drawing Resources > Title Blocks in the browser, as shown in the following image. Either double-click on the title block's name, as shown in the following image on the right, or right-click on the title block's name and select Insert from the menu.

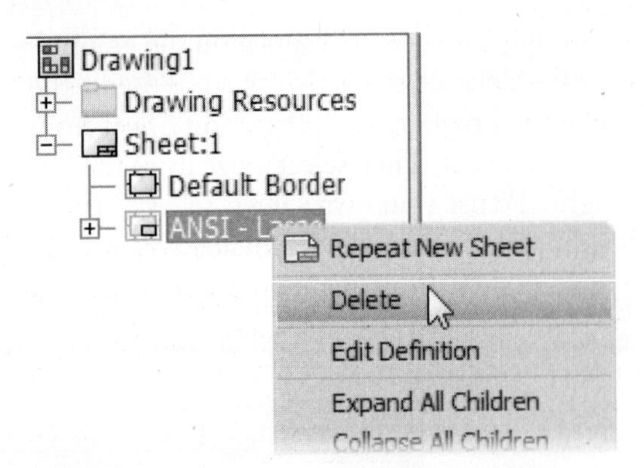

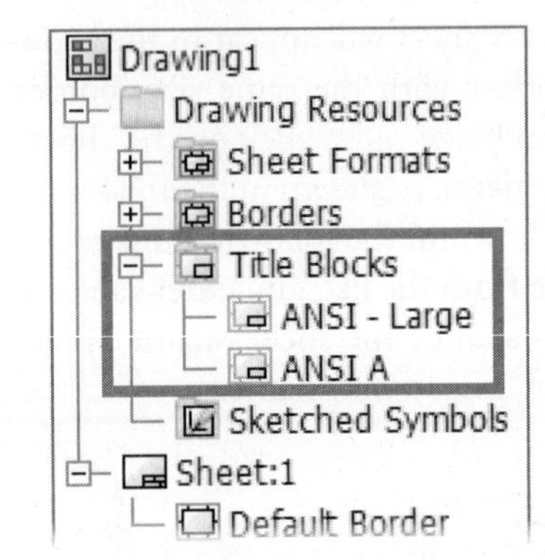

FIGURE 5.5

BORDER

To change a border on a drawing sheet, you can either insert a default border or construct a customized border and insert it into a drawing sheet.

Inserting a Default Border

To insert a default border, follow these steps:

1. If a border exists in the sheet, it must be deleted before a new border can be inserted. Make the sheet active, from the browser right-click on the Default Border entry and click Delete from the menu as shown in the following image on the left.
2. Insert a border by expanding Drawing Resources > Borders in the browser.
   - To insert a generic border double-click on the border's name and a generic border will be inserted.
   - To control the border's appearance right-click on Borders > Default Border and click Insert Drawing Border, as shown in the following image in the middle. The Default Drawing Border Parameters dialog box will appear, modify the options as needed, click the More (>>) button to control text and layer data as well as the sheet's margin as shown in the following image on the right.
   - Consult the help system to manually create a border.

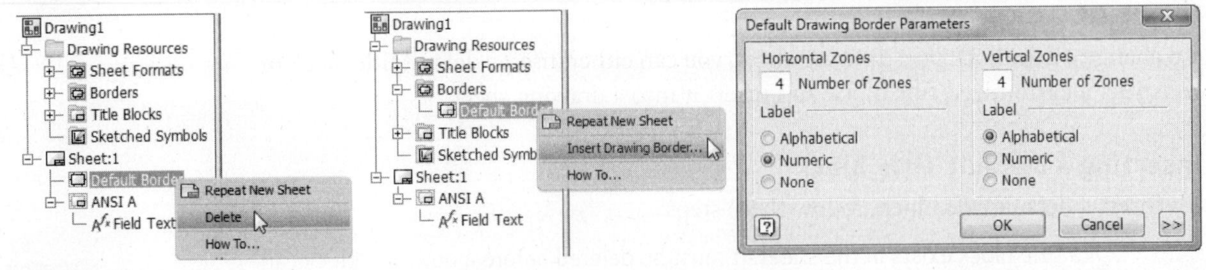

FIGURE 5.6

Edit Property Fields Dialog Box

The time will come when you will need to fill in title block information. Expanding the default title block in the browser and double-click or right-click on the Field Text entry and click Edit Field Text from the menu as shown in the following image on the left will display the Edit Property Fields dialog box. By default, the following information will already be filled in: Sheet Number, Number of Sheets, Author, Creation Date, and Sheet Size. To fill in other title block information such as Part Number, Company Name, Checked By, and so on, select the iProperties command button as shown in the following image in the middle. The Drawing Properties dialog box will appear, as shown in the following image on the right, and you can fill in the information as needed. You can find most title block information under the Summary, Project, and Status tabs.

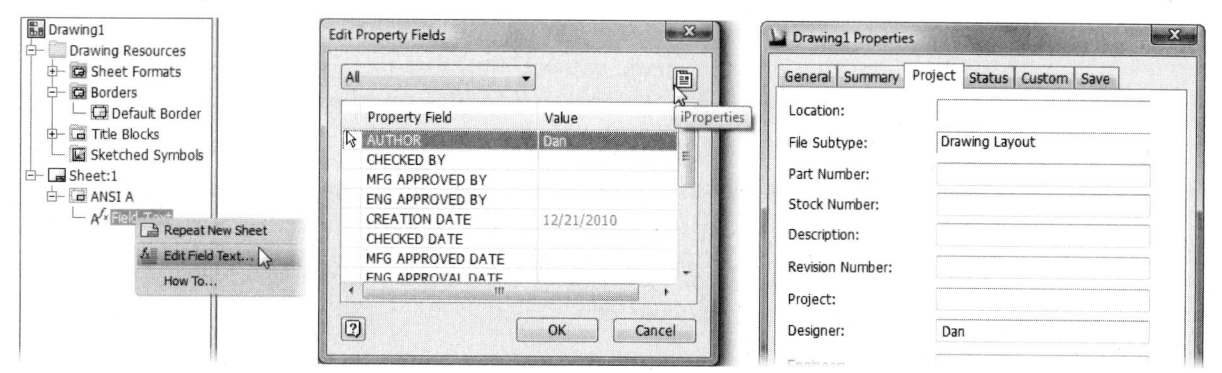

FIGURE 5.7

SAVE DRAWING DATA TO A TEMPLATE

After getting a drawing setup, you can save the drawing to the template folder so you can create a new drawing that contains the changes. From the Application Menu click Save As > Save Copy As Template as shown in the following image. Save the file to the correct folder; Templates (the default location), English and Metric or create a new folder. The location of the template files is set in the Application Options > File tab > Default templates or overridden in the active project's Folder Options > Templates.

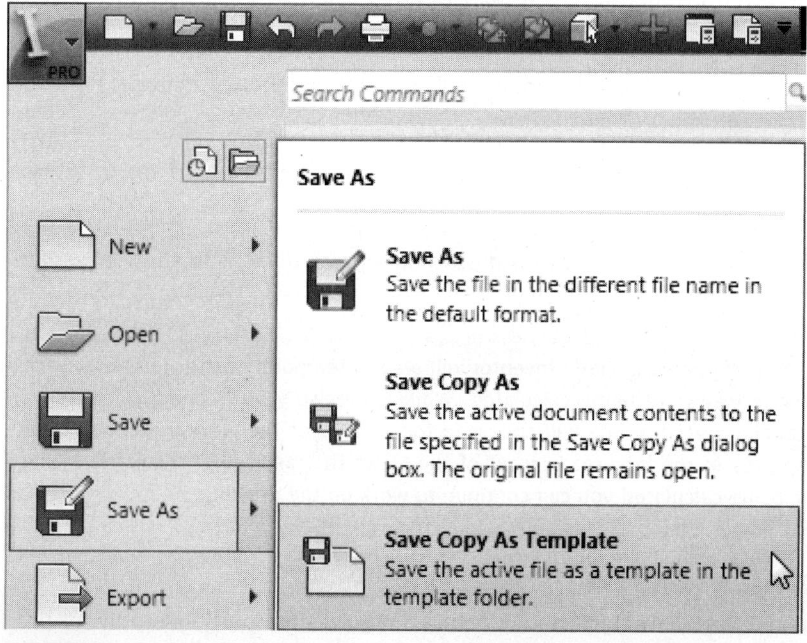

FIGURE 5.8

CREATING DRAWING VIEWS

After you have set the drawing sheet format, border, and title block you can create drawing views from an existing part, assembly, or presentation file. The file from which you will create the views does not need to be open when a drawing view is created. It is suggested, however, that both the file and the associated drawing file be stored in the same directory and that the directory be referenced in the project file. When creating drawing views, you will find that there are many different types of views you can create. The following sections describe these view types.

Base View. This is the first drawing view of an existing part, assembly, or presentation file. It is typically used as a basis for generating the following dependent view types. You can create many base views in a given drawing file.

Projected View. This is a dependent orthographic or isometric view that is generated from an existing drawing view.

- Orthographic (Ortho View): A drawing view that is projected horizontally or vertically from another view.
- Iso View: A drawing view that is projected at a 30° angle from a given view. An isometric view can be projected to any of four quadrants.

Auxiliary View. This is a dependent drawing view that is perpendicular to a selected edge of another view.

Section View. This is a dependent drawing view that represents the area defined by a slicing plane or planes through a part or assembly.

Detail View. This is a dependent drawing view in which a selected area of an existing view will be generated at a specified scale.

Broken View. This is a dependent drawing view that shows a section of the part removed while the ends remain. Any dimension that spans over the break will reflect the actual object length.

Break Out View. This is a drawing view that has a defined area of material removed in order to expose internal parts or features.

Crop View. This drawing view allows a view to be clipped based on a defined boundary.

Overlay View. This drawing view uses positional representations to show an assembly in multiple positions in a single view.

NOTE

When you are creating drawing views Inventor will place a temporary raster image for each view while precise views are being calculated. While a precise view is being calculated a small horizontal and a vertical line will appear in the corners of the view and two circular arrows are displayed in the browser in front of the views that are being calculated. While the view(s) are being calculated you can continue to work on the drawing.

CREATING A BASE VIEW

A base view is the first view that you create from the selected part, assembly, or presentation file. When you create a base view, the scale is set in the dialog box, and from this view, you can project other drawing views. There is no limit to the number of base views you can create in a drawing based on different parts, assemblies, or presentation files. As you create a base view, you can select the orientation of that view from the Orientation list on the Component tab of the Drawing View dialog box. By default there is an option to create projected views immediately after placing a base view.

To create a base view, follow these steps:

1. Click the Base View command on the Place Views tab as shown in the following image on the left, or right-click in the graphics window and select Base View.
2. The Drawing View dialog box will appear, also shown in the following image. On the Component tab, click the Open an existing file icon to navigate to and select the part, assembly, or presentation file from which to create the base drawing view. After making the selection, a preview image will appear attached to your cursor in the graphics window. Do not place the view until the desired view options have been set.
3. Select the type of view to generate from the Orientation list.
4. Select the scale for the view.
5. Select the style for the view.
6. By default the option is on that allows you to create projected views immediately after placing a base view. Uncheck this option if you want to only place a base view.
7. Locate the view by selecting a point in the graphics window.

As shown in the following image on the right, the Drawing View dialog box has three tabs: Component, Model State, and Display Options. The following sections describe these tabs.

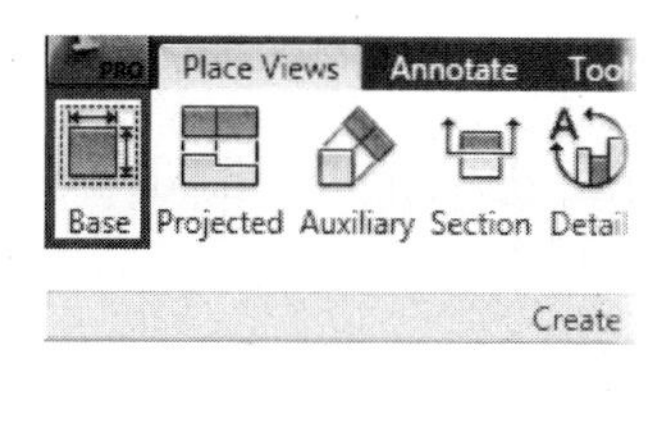

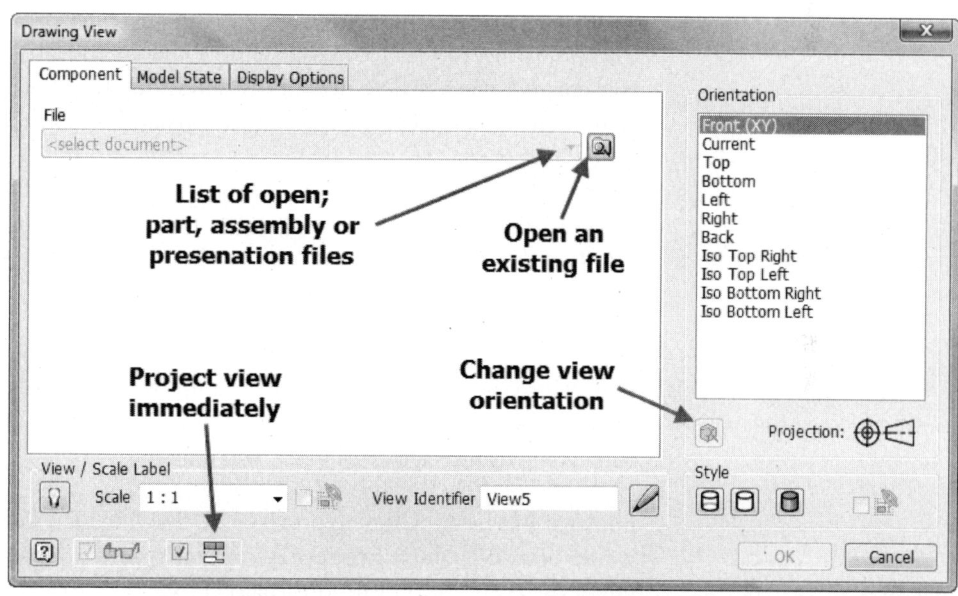

FIGURE 5.9

The Component Tab

The following categories are explained in detail under the Drawing View Component tab.

File. Any open part, assembly, or presentation files will appear in the drop-down list. You can also click the Explore directories icon and navigate to and select a part, assembly, or presentation file.

Orientation. After selecting the file, choose the orientation in which to create the view. After selecting an orientation, a preview image will appear in the graphics window attached to your cursor. If the preview image does not show the view orientation that you want, select a different orientation view by selecting another option. A symbol that identifies the type of drawing projection is also displayed under the Orientation area and is a symbol that identifies the type of drawing projection, such as First Angle or Third Angle Projection.

The following view orientations are available:

- Front, Current (current orientation of the part, assembly, or presentation in its file), Top, Bottom, Left, Right, Back, Iso Top Right, Iso Top Left, Iso Bottom Right, and Iso Bottom Left. A Change view orientation button is also present that allows you to create a custom view.

View/Scale Label (Light Bulb). Click to display or hide the view scale label. Clicking the checkbox displays the scale label, and leaving the box unchecked hides the scale label.

Scale. Enter a number for the scale in which to create the view. Note that the drawing sheet will be plotted at full scale (1:1), and the drawing views are scaled as needed to fit the sheet. You can edit the scale of the views after the views have been generated.

Scale from Base. Click to set the scale of a dependent view to match that of its base view, as shown in the following image on the left. To change the scale of a dependent view, clear the option's box, and modify the scale value for that view. This feature activates when you edit existing views.

View Identifier. Use to include and/or change the label for the selected view. When you create a view, a default label is determined by the active drafting standard. To change the label, select the label in the box and enter the new label.

Style. Choose how the view will appear. There are three choices: Hidden Line, Hidden Line Removed, and Shaded. The preview image will not update to reflect the style choice. When the view is created, the chosen style will be applied. The style can be edited after the view has been generated.

Style from Base. Click to set the display style of a dependent view to be the same as that of its base view, as shown in the following image on the right. To change the display style of a dependent view, clear the checkbox and modify the style for that view. This feature activates when you edit existing views.

Feature Preview. Click to preview the drawing view before it is created. This option is unavailable if Show Preview As option is set to All Components on the Drawing tab of the Application Options dialog box.

Create Projected Views Immediately after Base View Creation. By default the option is on that allows you to create projected views immediately after placing a base view. Uncheck this option if you want to only place a base view.

The Model State Tab

Use this tab to specify the weldment state and member states of an iAssembly or iPart to use in a drawing view, as shown in the image below. Other items are used to control line style and hidden line calculations.

Weldment. Enable this area when you select a document that is a weldment. Weldments have four states: Assembly, Preparations, Welds, and Machining.

Member. Allows you to select the member of an iAssembly or iPart to represent in a drawing view.

Line Style. Three options are presented: As Reference Parts, As Parts, and Off. When you select As Reference Parts for shaded view types, the reference data will appear transparently shaded with tangent edges turned off and all part edges as phantom lines. The reference data will appear on top of the product data that is shaded. When you select As Parts, reference data will have no special display characteristics. The reference data will appear on top of product data, and the reference data will appear with tangent edges turned off and all other edges as the phantom line type. When you select Off, reference data will not appear.

Hidden Line Calculation. Specifies if hidden lines are calculated for Reference Data Separately or for All Bodies.

Margin. To see more reference data, set the value to expand the view boundaries by a specified value on all sides.

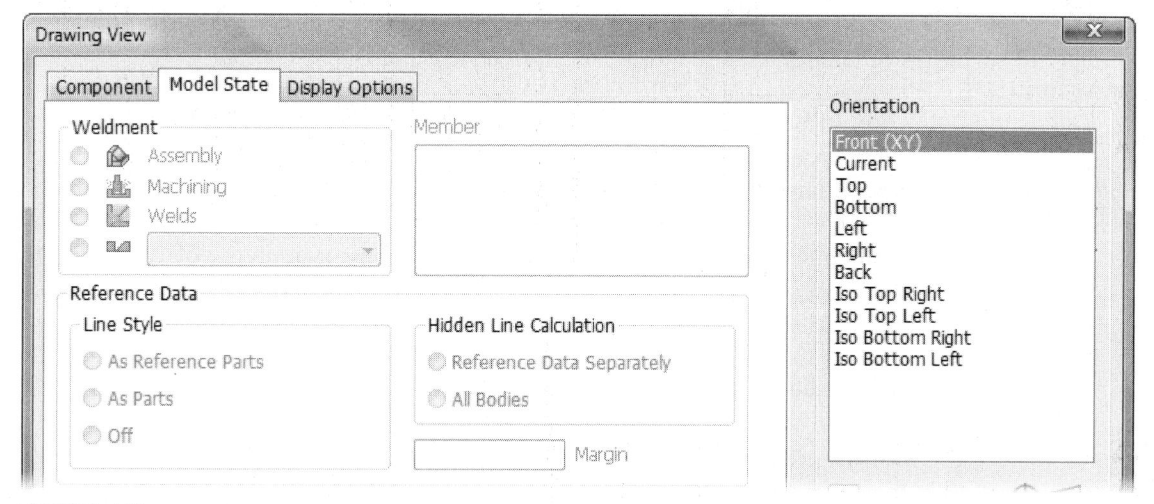

FIGURE 5.10

The Display Options Tab

The following image shows the Display Options tab of the Drawing View dialog box. The following sections describe the tab's options.

All Model Dimensions. Click to see model dimensions in the view, which is only active upon base view creation. If the option's box is clear, model dimensions will not be placed automatically upon view creation. When checked, only the dimensions that are parallel to the view and have not been retrieved in existing views on the sheet will appear.

Model Welding Symbols. This box is active only if you are creating a drawing view of a weldment. Click to retrieve welding symbols placed in the model in the drawing.

Bend Extents. This box is active only if you are creating a drawing view of a sheet metal part flat pattern. Click to control the visibility of bend extent lines or edges.

Thread Feature. This box is active only if you are creating a view of an assembly model. Click to set the visibility of thread features in the view.

Weld Annotations. This box is active only if creating a view of a weldment. Click to control the display of weld annotations.

User Work Features. Click to display work features in the drawing view.

Interference Edges. When selected, a drawing view will display both hidden and visible edges due to an interference condition such as a press or interference fit.

Tangent Edges. Click to set the visibility of tangent edges in a selected view. Checking the box displays tangent edges, and leaving the box clear hides them. If enabled, tangent edges can also be adjusted to display Foreshortened.

Show Trails. Click to control the display of trails in drawing views based on presentation files.

Hatching. Click to set the visibility of the hatch lines in the selected section view. Checking the box displays hatch lines, and leaving the box clear hides them.

Align to Base. Click to remove the alignment constraint of a selected view to its base view. When the box is checked, alignment of views exists. Leaving the box clear breaks the alignment and labels the selected view and the base view.

Definition in Base View. Click to display or hide the section view projection line or detail view boundary. Checking the box displays the line, circle, or rectangle and text label; leaving the box clear hides the line, circle, or rectangle and text label.

Cut Inheritance. Use this option to turn on and off the inheritance of a Break Out, Break, Section, and Slice cut for the edited view. Selecting the appropriate checkbox will inherit the corresponding cut from the parent view.

Section Standard Parts. Use this option to control whether or not standard parts placed into an assembly from the supplied Inventor parts library, such as nuts, bolts, and washers, are sectioned. Three options are available in this area: Always, Never, and Obey Browser Settings.

View Justification. Use to control the position of drawing views when the size or position of the model changes. This area is especially helpful with creating drawing views from assembly models. This area contains two modes: Center and Fixed. The Center mode keeps the model image centered in the drawing view. If, however, the model's view increases or decreases in size, the drawing view will shift on the drawing sheet. In some cases, this shift could overlap the drawing border or title block. The Fixed mode keeps the drawing view anchored on the drawing sheet. In the event that the model image increases, an edge of the drawing view will remain fixed.

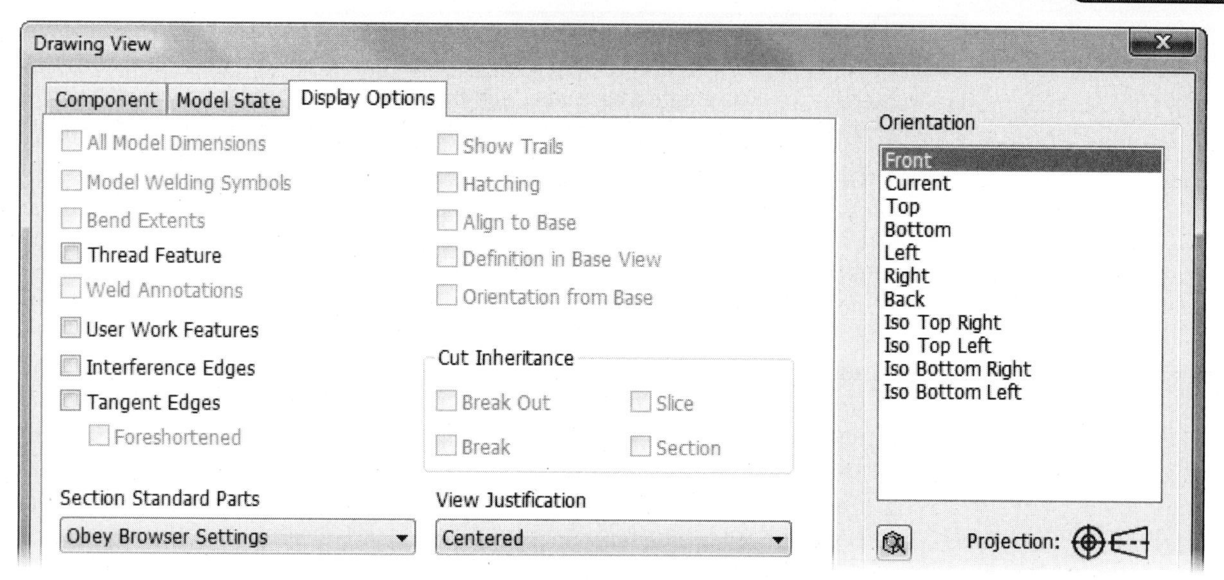

FIGURE 5.11

Creating Projected Views

A projected view can be an orthographic or isometric view that you project from a base view or any other existing view. When you create a projected view, a preview image will appear, showing the orientation of the view you will create as the cursor moves to a location on the drawing. There is no limit to the number of projected views you can create. To create a projected drawing view, follow these steps:

1. Click the Projected command on the Place Views tab > Create panel as shown in the following image. You could also right-click inside the bounding area of an existing view box, displayed as dashed lines when the cursor moves into the view. Select Create View > Projected View from the menu.

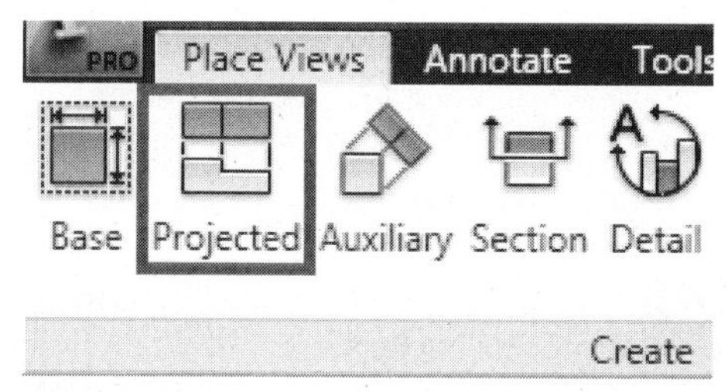

FIGURE 5.12

2. If you selected the Projected command, click inside the desired view to start the projection.
3. Move the cursor horizontally, vertically, or at an angle (to create an isometric view) to get a preview image of the view you will generate. Keep moving the cursor until the preview matches the view that you want to create, and then press the left mouse button. Continue placing projected views.
4. When finished, right-click, and select Create from the menu.

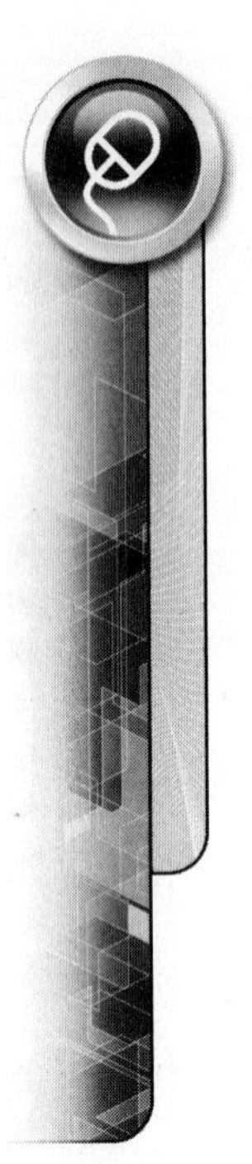

EXERCISE 5-1: CREATING A MULTIVIEW DRAWING

In this exercise, you will create a base view, and then you will add projected views to create a multiview orthographic drawing. Finally, you will add an isometric view to the drawing.

1. Open the file *ESS_E05_01.idw.* in the Chapter 05 folder. This drawing file contains a single sheet with a border and title block. In addition to the title block information, notice also the numbers and letters present on the outside of the border. Use these as reference points to locate a specific detail, view, or dimension on a large sheet. Numbers are located horizontally along the top and bottom of the border. Letters are located vertically along the left and right border edges. A typical references example would be C2. Here, you would identify the C along the vertical portion of the border and the number 2 along the horizontal portion of the border. Where both of these two references intersect, your search item should be easily found.
2. Create a base view by clicking on the Base View command; this will display the Drawing View dialog box.
3. Under the File area of the Component tab, click the Open an existing file button (1), and double-click the file *ESS_E05_01.ipt.* in the Chapter 05 folder to use it as the view source.
4. In the Orientation area, verify that Front (XY) is selected (2).
5. In the Scale list, verify that 1:1 is current (3).
6. Verify that the Create projected views immediately after base view creation option is checked (4).
7. In Style, verify that the Hidden Line button in the lower right corner of the Drawing View dialog box is selected (5). The following image shows all the settings in the Drawing View dialog box.

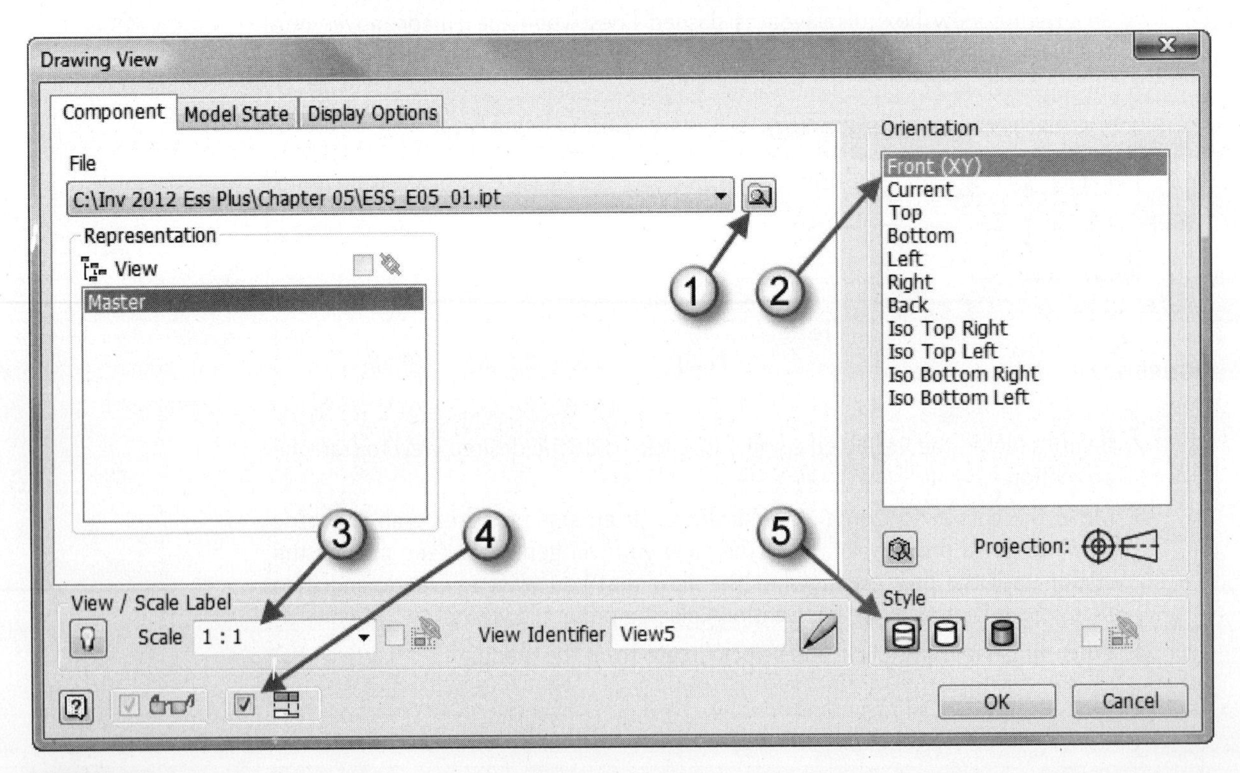

FIGURE 5.13

8. Click the Display Options tab, and ensure that All Model Dimensions is not checked.
9. Position the view preview in the lower-left corner of the sheet (in Zone C6), and then click to place the view as shown in the following image.

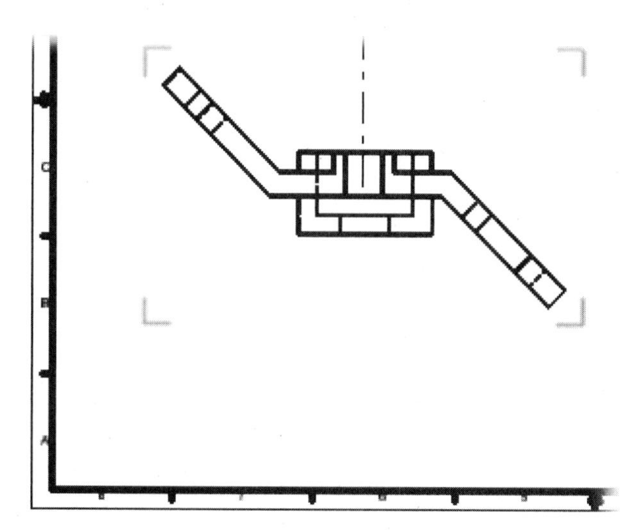

FIGURE 5.14

10. Next you place projected views. While still in the Base View command move the cursor to the right of the base view. Click in Zone C2 to place the right-side view.
11. Move the cursor to the top of the base view. Click in Zone E6 to place the top view.
12. Right-click, and click Create from the menu to create the views, as shown in the following image.

You could have created the right side and top view using the Projected view command.

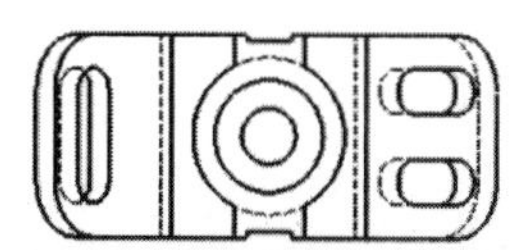

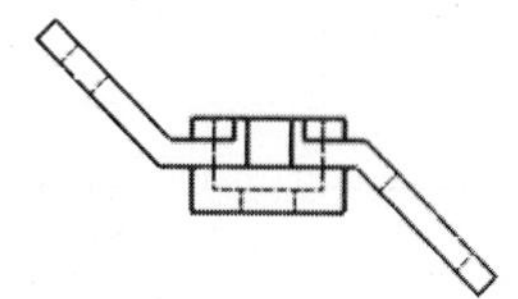

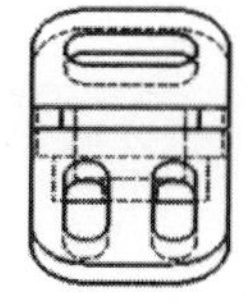

FIGURE 5.15

13. The views are crowded with a scale of 1:1. In the steps that follow, you will reduce the size of all drawing views by setting the base view scale to 1:2. The dependent views will update automatically. Edit the base view (front view) by either double-clicking in the base view or right-click in the base view and click Edit View from the menu.

To edit the view, right-click on the view border or inside the view. Do not right-click on the geometry.

NOTE

14. When the Drawing View dialog box displays, select 1 / 2 from the Scale list, and click OK.
15. The scale of all views updates, as shown in the following image.

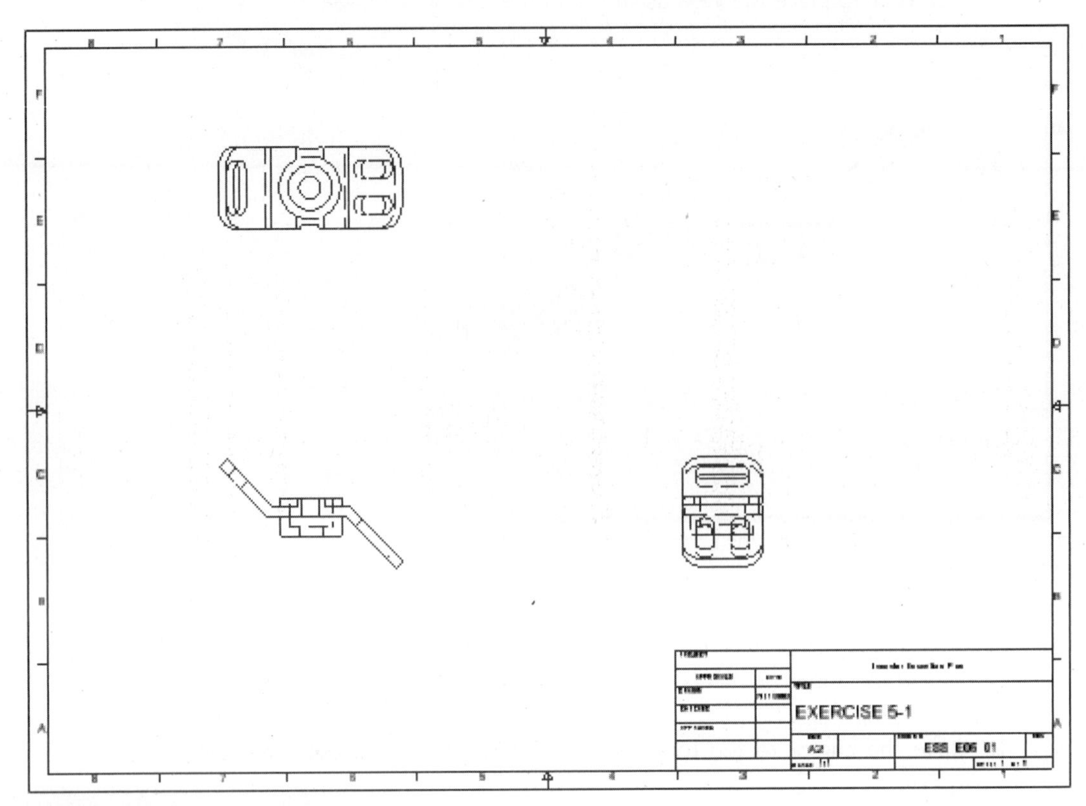

FIGURE 5.16

16. Now change the drawing scale in the title block. Begin this process by expanding Sheet:1 in the browser.
17. From the browser, also expand the title block ANSI-Large under Sheet:1.
18. Right-click on the Field Text icon, and choose Edit Field Text from the menu.
19. The Edit Property Fields dialog box will display. In the SCALE cell, click on 1:1 and enter a new scale of 1:2 as shown in the following image on the left.
20. Click OK. The title block updates to the new scale, as shown in the following image on the right.

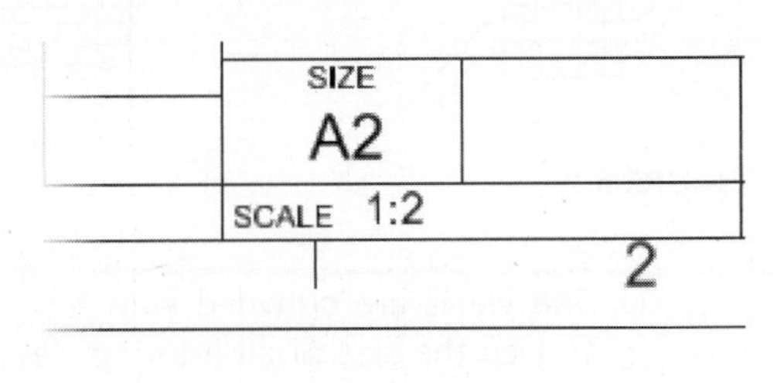

FIGURE 5.17

21. To complete the drawing views, an isometric view will be created. Begin by clicking the Projected view command on the Place Views tab > Create panel.
22. Select the base view (front view) and move your cursor to a point above and to the right of the base view.
23. Click in Zone E3 to place the isometric view. Right-click the sheet and choose Create. Your drawing should appear similar to the following image.

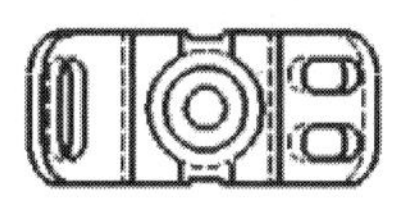

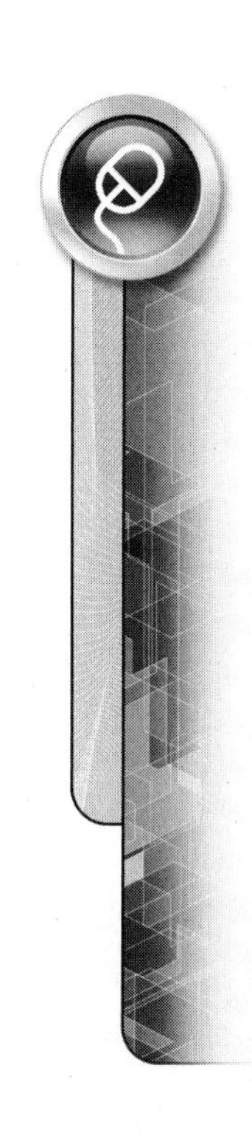

FIGURE 5.18

24. Double-click the isometric view and click the Display Options tab.
25. In the Display tab clear the check box for Tangent Edges, and then click OK.
26. Complete this exercise by moving drawing views to better locations. Select the right-side view, and drag it to Zone C4. Select the isometric view, and drag it to Zone E4. Your drawing should appear similar to the following image.

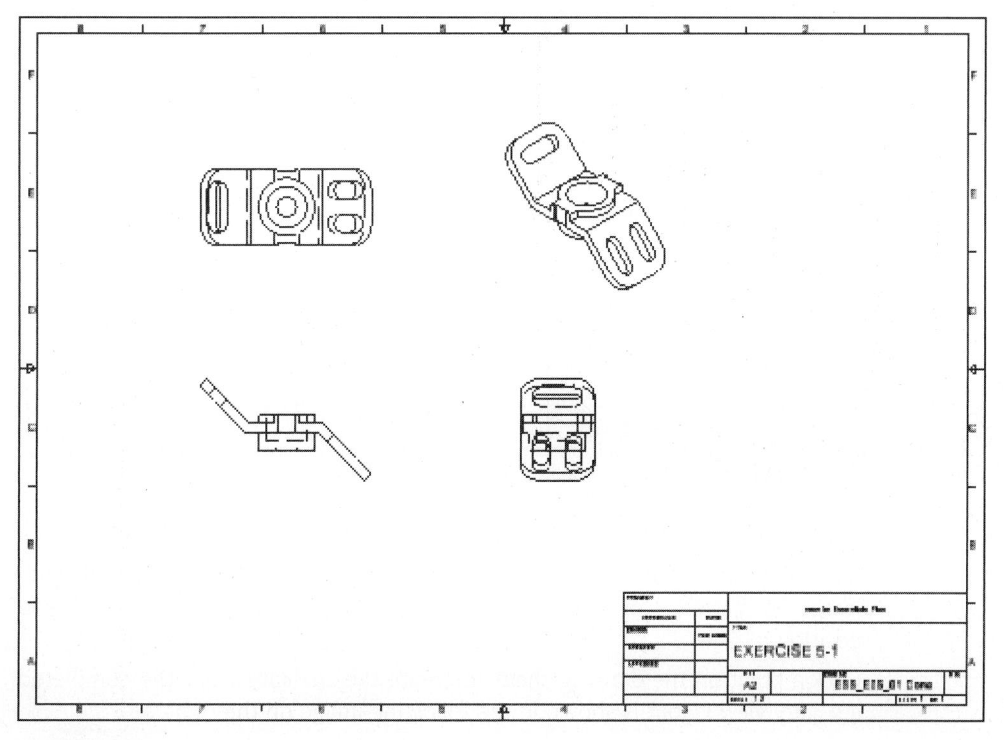

FIGURE 5.19

27. Close all open files. Do not save changes. End of exercise.

Creating Auxiliary Views

An auxiliary view is a view that is projected perpendicular to a selected edge or line in a base view. It is designed primarily to view the true size and shape of a surface that appears foreshortened in other views.

To create an auxiliary drawing view, follow these steps:

1. Click the Auxiliary command on the Place Views tab, as shown in the following image. You can also right-click inside the bounding area of an existing view box, shown as a dotted box when the cursor moves into the view, and then select Create View > Auxiliary View from the menu.

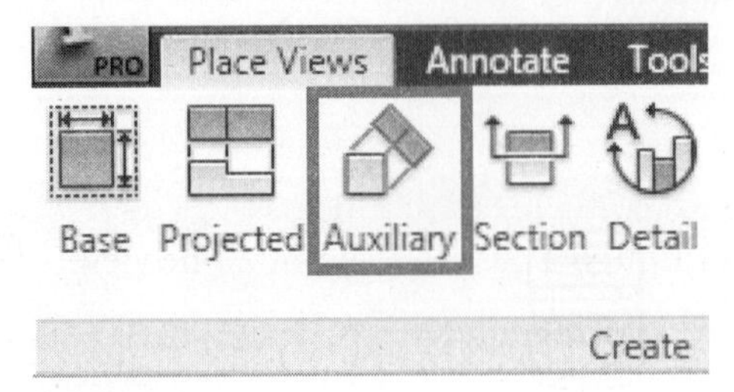

FIGURE 5.20

2. If you selected the Auxiliary command, click inside the view from which the auxiliary view will be projected. The Auxiliary View dialog box will appear, as shown in the following image on the left. Type in a name for Label and a value for Scale, and then select one of the Style options: Hidden, Hidden Line Removed, or Shaded.
3. In the selected drawing view, select a linear edge or line from which the auxiliary view will be perpendicularly projected, as shown in the following image on the right.

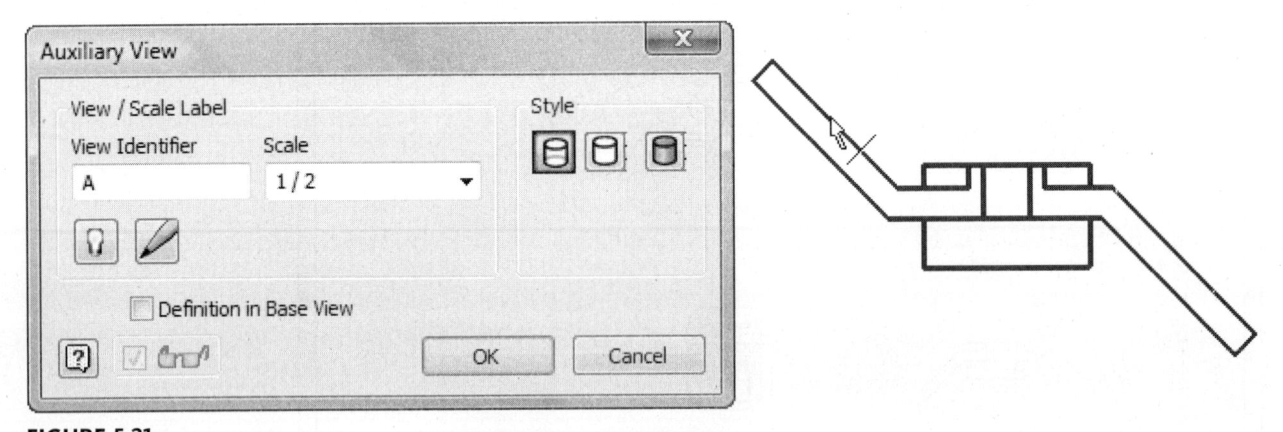

FIGURE 5.21

4. Move the cursor to position the auxiliary view, as shown in the following image on the left. Notice that the view takes on a shaded appearance as you position it.
5. Click a point on the drawing sheet to create the auxiliary view. The completed auxiliary view layout is shown in the following image on the right.

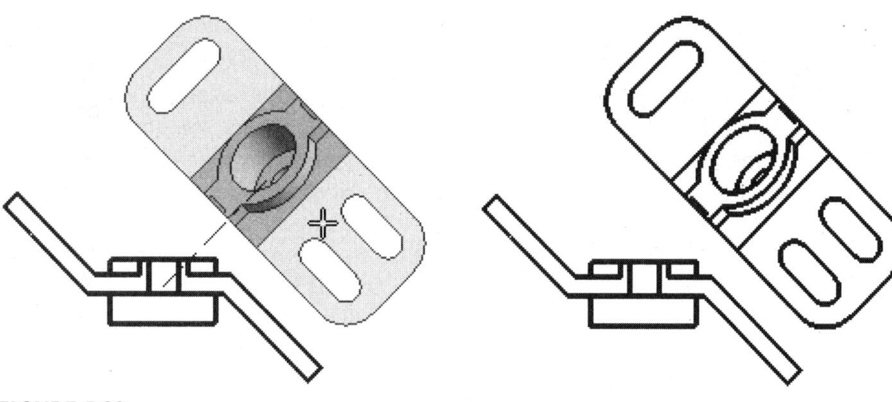

FIGURE 5.22

Creating Section Views

A section view is a view you create by defining a section line or multiple section lines that will represent the plane(s) that will be cut through a part or assembly. The view will represent the surface of the cut area and any geometry shown behind the cut face from the direction being viewed. When defining a section, you sketch line segments that are horizontal, vertical, or at an angle. You cannot use arcs, splines, or circles to define section lines. When you sketch the section line(s), geometric constraints will automatically be applied between the line being sketched and the geometry in the drawing view. You can also infer points by moving the cursor over (or scrubbing) certain geometry locations, such as centers of arcs, endpoints of lines, and so on, and then moving the cursor away to display a dotted line showing that you are inferring or tracking that point. To place a geometric constraint between the drawing view geometry and the section line, click in the drawing when a green circle appears; the glyph for the constraint will appear. If you do not want the section lines to have constraint(s) applied to them automatically when they are created, hold down the CTRL key when sketching the line(s). Because the area in the section view that is solid material appears with a hatch pattern by default, you may want to set the hatching style before creating a section view.

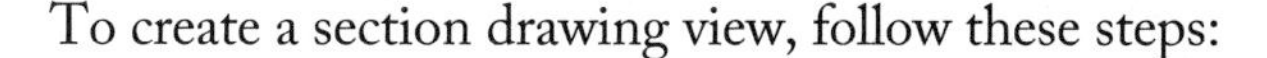

To create a section drawing view, follow these steps:

1. Click the Section command on the Place Views tab > Create panel, as shown in the following image. You can also right-click inside the bounding area of an existing view, displayed as a dotted box when the cursor moves into the view. You can then select Create View > Section from the menu.
2. If you selected the Section command, click inside the view from which to create the section view.
3. Sketch the line or lines that define where and how you want the view to be cut. In the following image, a vertical line is sketched through the center of the object. When sketching and a green dot appears a coincident constraint will be applied.

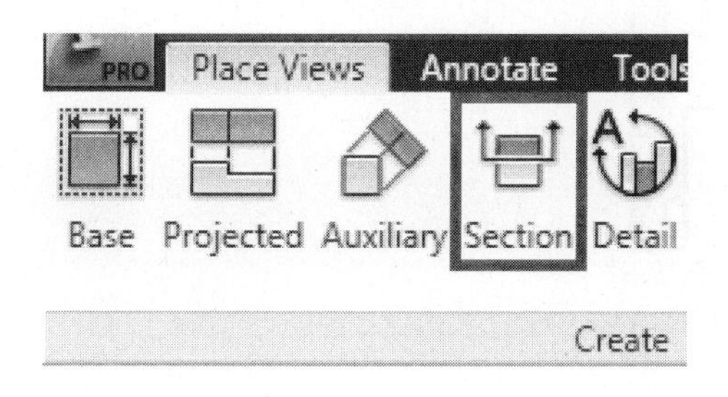

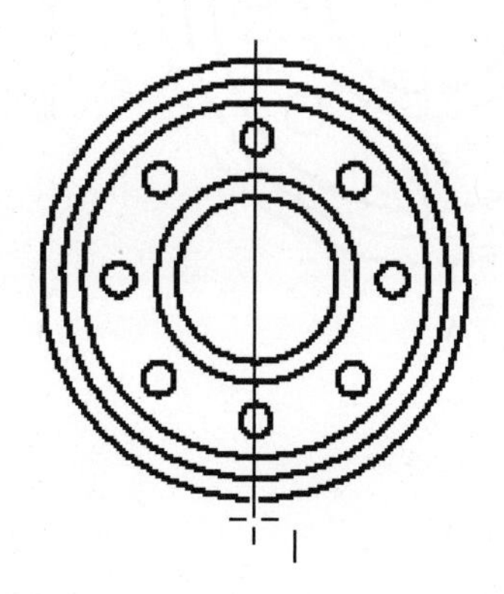

FIGURE 5.23

4. When you finish sketching the section line(s), right-click and select Continue from the menu as shown in the following image on the left.
5. The Section View dialog box will appear, as shown in the following image on the right. Fill in the information for how you want the label, scale, and style to appear in the drawing view. When the Include Slice option is checked, a section view will be created with some components sliced and some components sectioned, depending on their browser attribute settings. Placing a check in the box next to Slice All parts will override any browser component settings and will slice all parts in the view according to the Section line geometry. Components that are not crossed by the Section Line will not be included in the slicing operation.

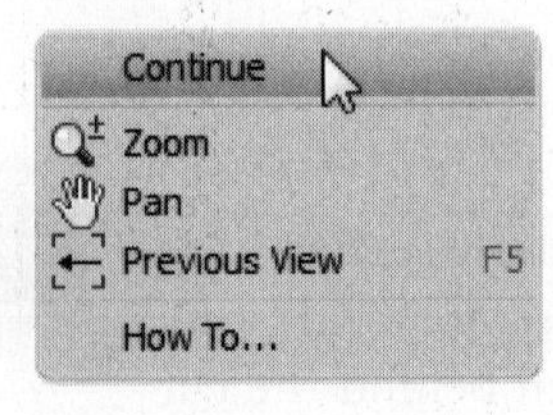

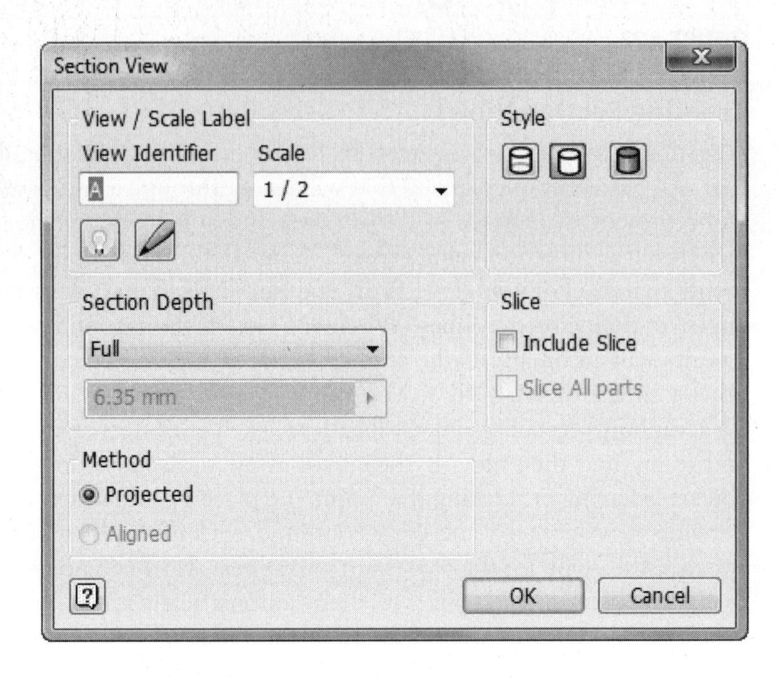

FIGURE 5.24

6. Move the cursor to position the section view (the section arrows will flip direction depending upon the location of the cursor), as shown in the following image, and select a point to place the view.

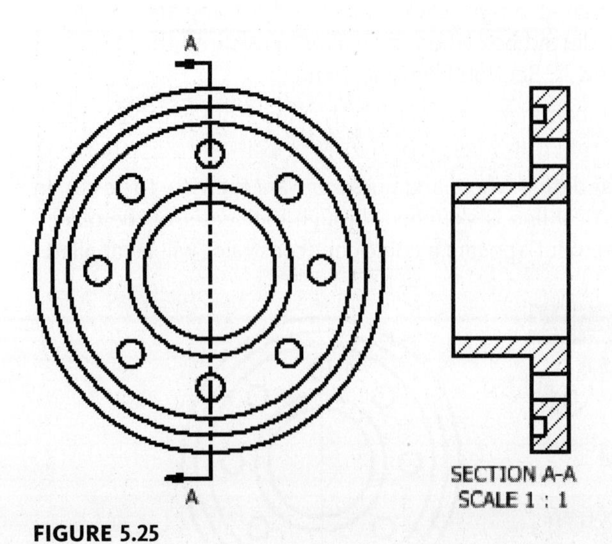

FIGURE 5.25

Creating Aligned Sections

Aligned sections take into consideration the angular position of details or feature of a drawing instead of projecting the section view perpendicular to the view, as shown in the following image on the left. As illustrated on the left in the following image, it is difficult to obtain the true size of the angled elements of the bottom area in the section view. In the Side view, they appear foreshortened or not to scale. Hidden lines were added as an attempt to better clarify the view.

An aligned section view creates a section view that is perpendicular to its section lines. This prevents objects in the section view from being distorted.

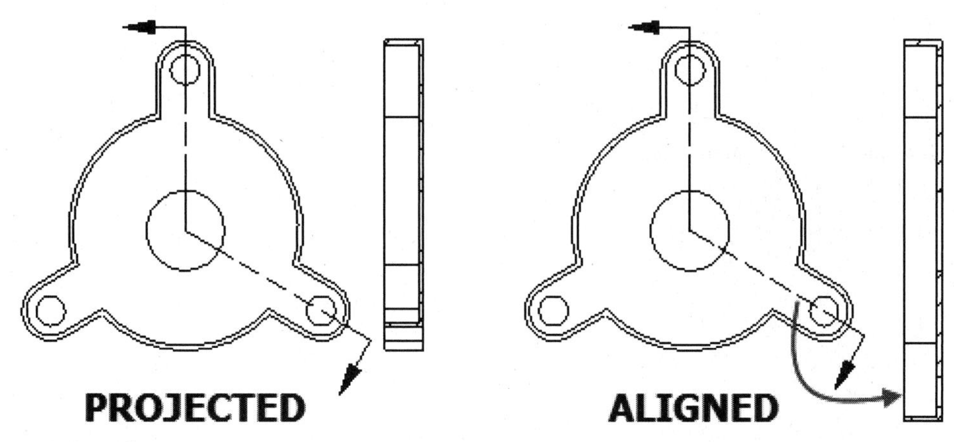

FIGURE 5.26

To create an aligned section view, follow these steps:

1. Click the Section command on the Place Views tab. You can also right-click inside the bounding area of an existing view box, shown as a dotted box when the cursor moves into the view. You can then select Create View > Section from the menu.
2. If you selected the Section command, click inside the view from which to create the aligned section view.
3. Sketch the line or lines that define where and how you want the view to be cut. In the following image, a vertical line is sketched up to the center of the object. Then, the line changes direction to catch the counterbore feature as shown in the following image on the left. When sketching and a green dot appears a coincident constraint will be applied.
4. When you finish sketching the section line(s), right-click and select Continue.
5. The Section View dialog box will appear, as shown in the following image on the right. Fill in the information for how you want the label, scale, and style to appear in the drawing view. Verify that the type of section being created is Aligned. When the Include Slice option is checked, a section view will be created with some components sliced and some components sectioned, depending on their browser attribute settings. Placing a check in the box next to Slice All parts will override any browser component settings and will slice all parts in the view according to the Section line geometry. Components that are not crossed by the Section Line will not be included in the slicing operation.

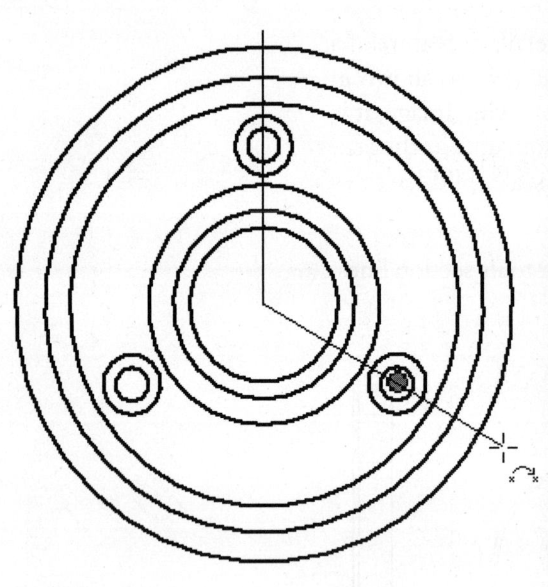

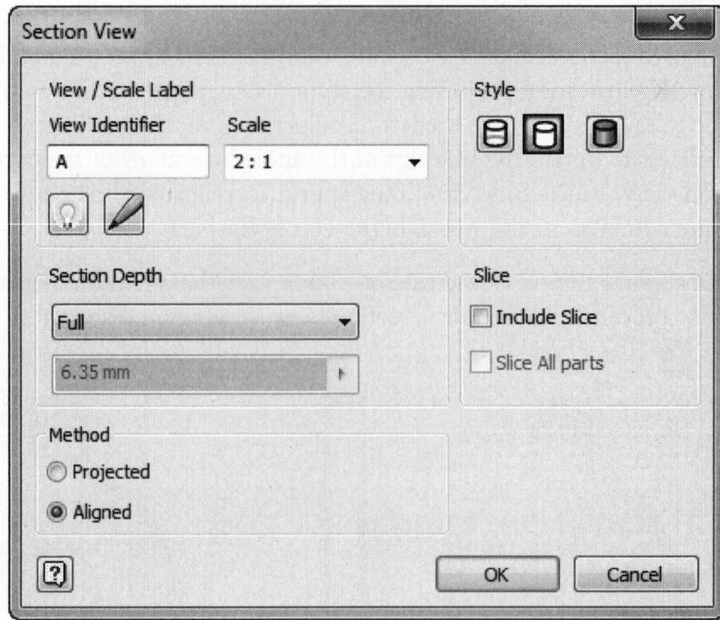

FIGURE 5.27

6. The completed aligned section is displayed in the following image. The bottom counterbore hole was rotated until it appears as a full section.
7. To edit the location of the section line(s) right-click on the section line and click Edit from the menu. The sketch environment will be current, add or delete constraints as needed. Dimensions can also be added between a section line and existing geometry. Before adding dimensions use the Project Geometry command to project the existing geometry onto the active sketch.

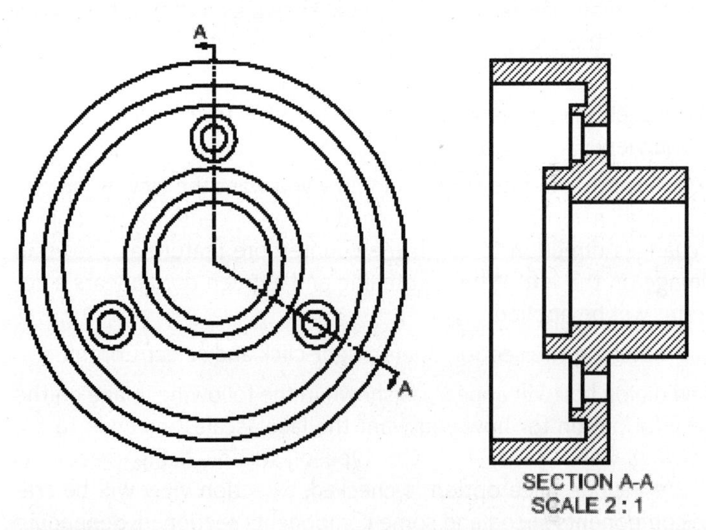

FIGURE 5.28

Modifying Section View and a Hatch Pattern

You can edit the section view by right-clicking in the section view and clicking Edit Section Properties. In the Edit Section dialog box you can change the section depth, to include a slice and change the method to Projected or Aligned if the view supports it. You can also edit the hatch pattern by right-clicking on the hatch pattern in the section view and selecting Edit from the menu. This will launch the Edit Hatch Pattern dialog box, as shown in the following image. Make the desired changes in the

Pattern, Angle, Scale, Line Weight, Shift (shifts the hatch pattern a specified distance, but stills stays within the section boundary), Double, and Color areas. Click the OK button to reflect the changes in the drawing.

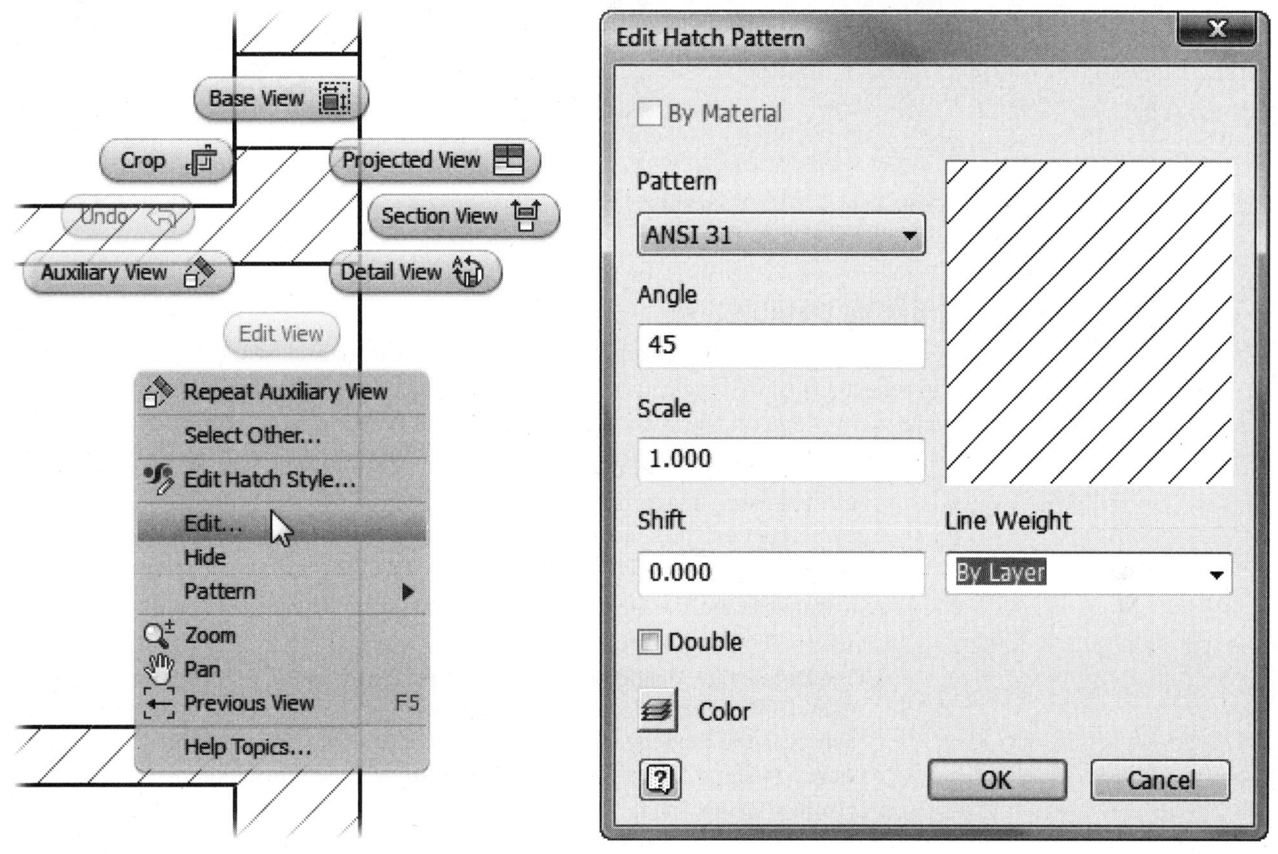

FIGURE 5.29

Hatching Isometric Views

In addition to applying crosshatching to orthographic view, as shown in the following image on the left, you can also apply crosshatching to an isometric view. To perform this operation, double-click or right-click inside of the bounding box that holds the isometric view, and pick Edit View from the menu, as shown in the following image.

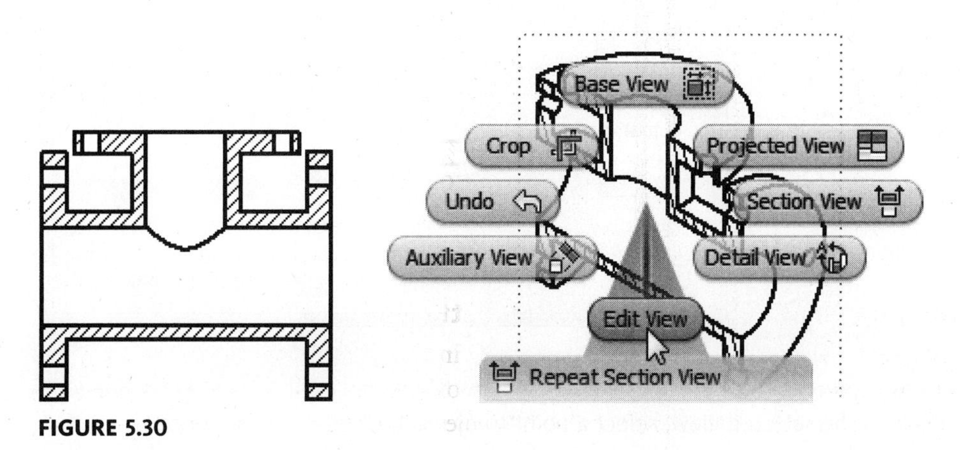

FIGURE 5.30

When the Drawing View dialog box appears, click on the Display Options tab, and place a check in the box next to Hatching, as shown in the following image on the left. The results are illustrated in the following image on the right with the hatching applied to the isometric view.

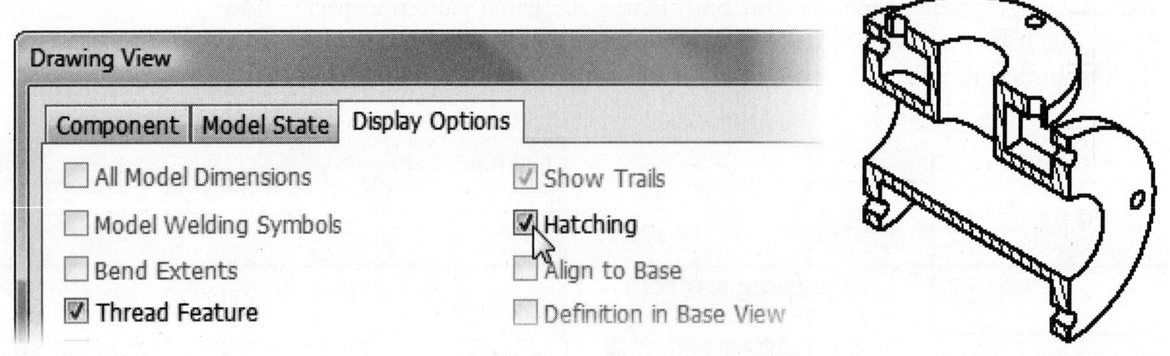

FIGURE 5.31

Creating Detail Views

A detail view is a drawing view that isolates an area of an existing drawing view and can reflect a specified scale. You define a detailed area by a circle or rectangle and can place it anywhere on the sheet. To create a detail drawing view, follow these steps:

1. Click the Detail command on the Place Views tab > Create panel as shown in the following image on the left. You can also right-click inside the bounding area of an existing view box, shown as dashed lines when the cursor moves into the view, and select Create View > Detail View from the menu.
2. If you selected the Detail command, click inside the view from which you will create the detail view.
3. The Detail View dialog box will appear, as shown in the following image on the right. Fill in information according to how you want the label, scale, and style to appear in the drawing view, and pick the desired view fence shape to define the view boundary. Do not click the OK button at this time, as doing so will generate an error message.

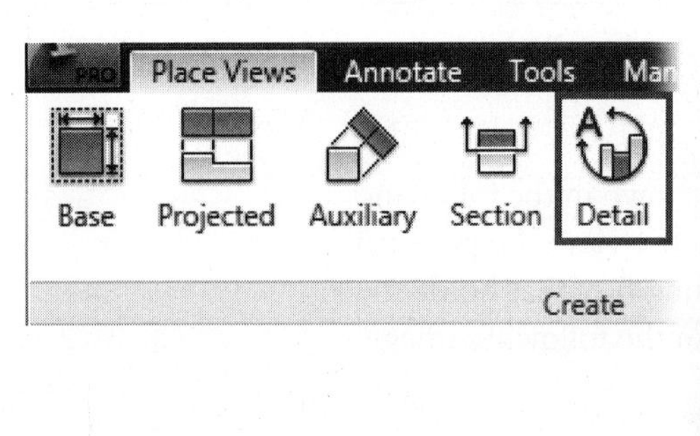

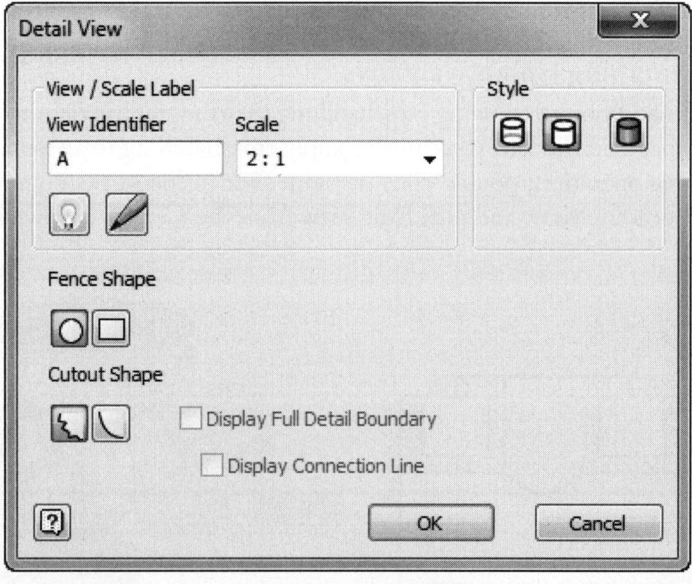

FIGURE 5.32

4. In the selected view, select a point to use as the center of the fence shape that will describe the detail area. The following image on the left shows the icon that appears when creating a circular fence. The image on the right shows the icon used for creating a rectangular fence.

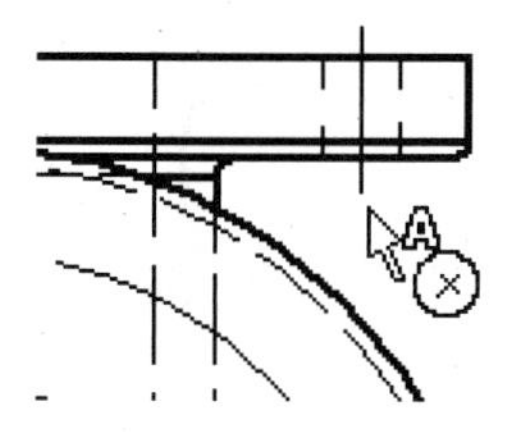
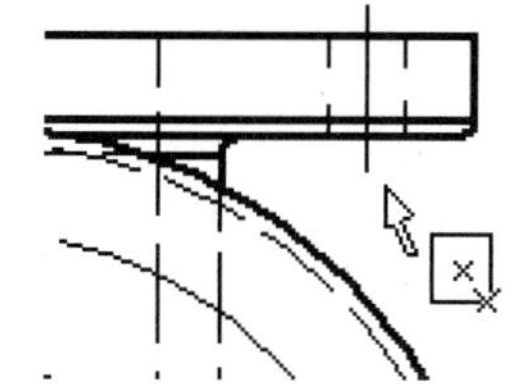

FIGURE 5.33

NOTE

You can right-click at this point and select Rectangular Fence to change the bounding area of the detail area definition.

5. Select another point that will define the radius of the detail circle, as shown in the following image on the left, or corner of the rectangle, as shown in the following image on the right. As you move the cursor, a preview of the boundary will appear.

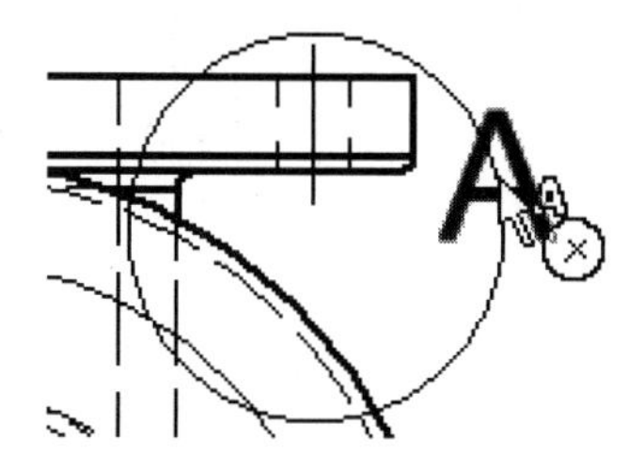

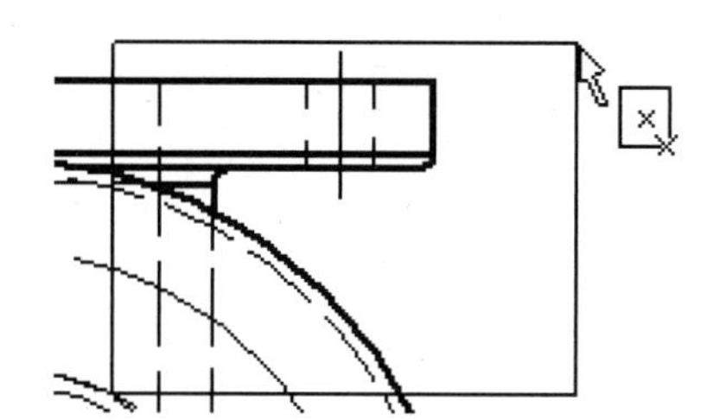

FIGURE 5.34

6. Select a point on the sheet where you want to place the view. There are no restrictions on where you can place the view. The following image on the left shows the completed detail view based on a circular fence; a similar detail based on a rectangular fence is shown on the right.

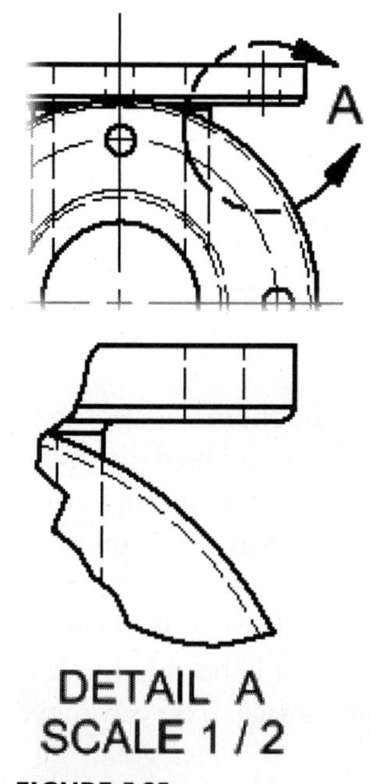

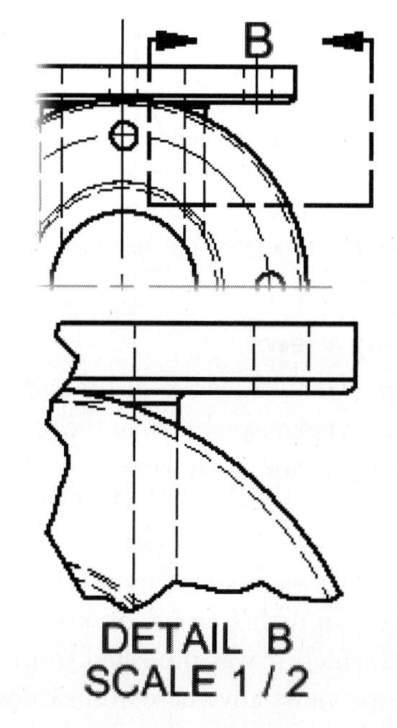

FIGURE 5.35

Modifying a Detail View

After a detail view is created, numerous options are available to fine-tune the detail. To access these controls, right-click on the edge of the detail circle and choose options, as shown in the following image on the right. The three detail view options are explained below.

Smooth Cutout Shape: Placing a check in this box will affect actual detail view. In the following image, instead of the cutout being ragged, a smooth cutout shape is present.

Full Detail Boundary: This option displays a full detail boundary in the detail view when checked, as shown in the following image.

Connection Line: This option will create a connection line between the main drawing and the detail view, as shown in the following image.

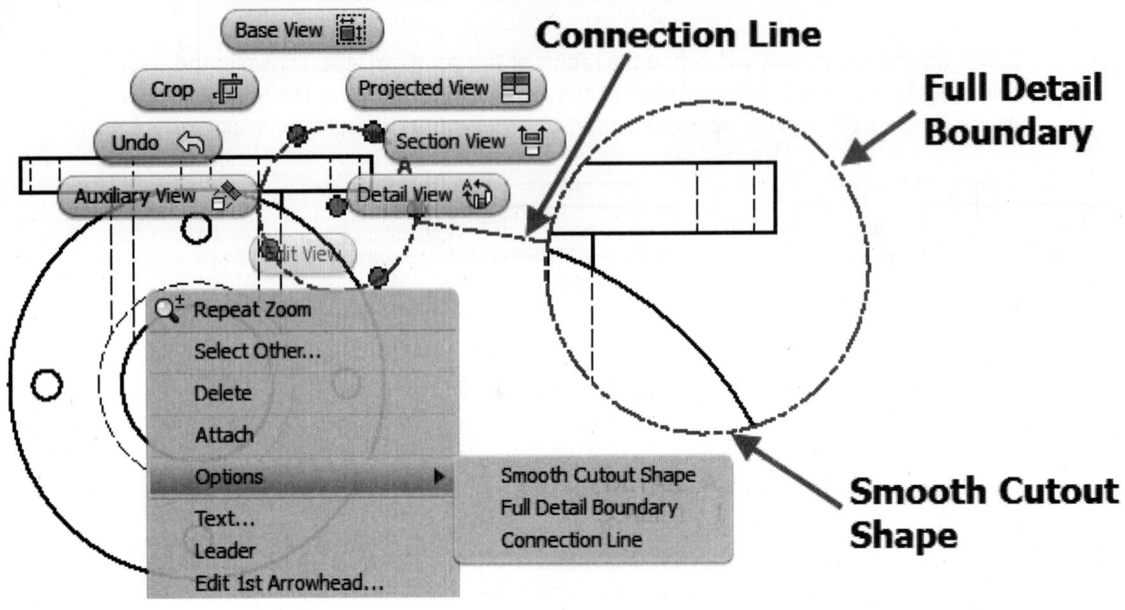

FIGURE 5.36

Once a connection line has been created in a detail view, you can add a new vertex by right-clicking the detail boundary or connection line and clicking Add Vertex from the menu. Picking a point on the connection line will allow you to add the vertex and then move the new vertex to the new position. You can even remove a vertex by right-clicking a vertex and choosing Delete Vertex from the menu.

Moving Drawing Views

To move a drawing view, move the cursor over the view until a bounding box consisting of dotted lines appears, as shown in the following image. Press and hold down the left mouse button, and move the view to its new location. Release the mouse button when you are finished. As you move the view, a rectangle will appear that represents the bounding box of the drawing view. If you move a base view, any projected children or dependent views will also move with it as required to maintain view alignment. If you move an orthographic or auxiliary view, you will only be able move it along the axis in which it was projected from the part edge or face. You can move detail and isometric views anywhere in the drawing sheet.

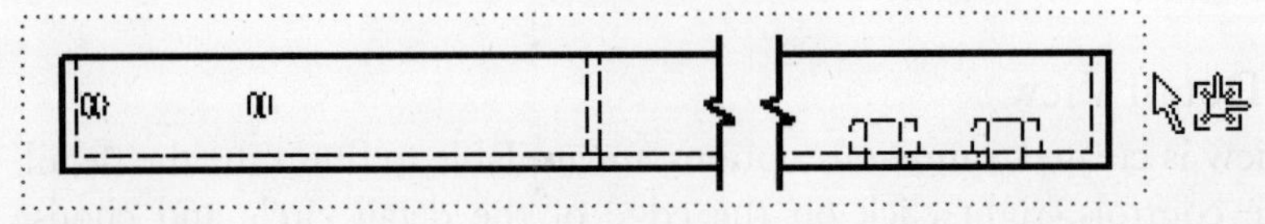

FIGURE 5.37

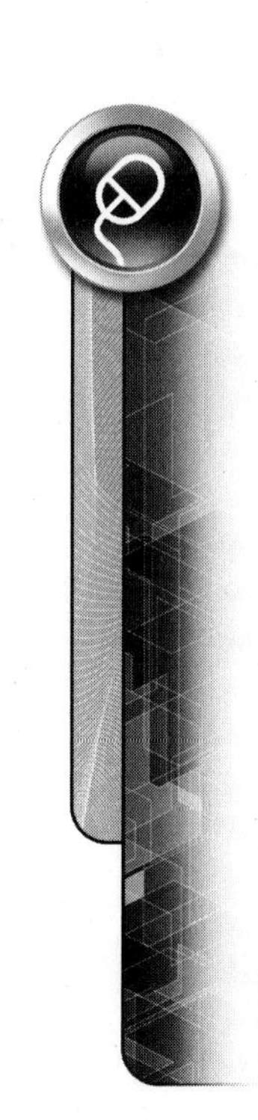

EXERCISE 5-2: CREATING AUXILIARY, SECTION, AND DETAIL VIEWS

In this exercise, you create a variety of drawing views from a model of a cover.

1. Open the file *ESS_E05_02.idw*. Use the Zoom command to zoom in on the top view.
2. First you create an auxiliary view. Click the Auxiliary command on the Place Views tab > Create panel.
3. In the graphics window, click in the top view.
4. Select the left-outside angled line, as shown in the following image on the left, to define the projection direction.
5. Place the view to the left as shown in the following image on the right.

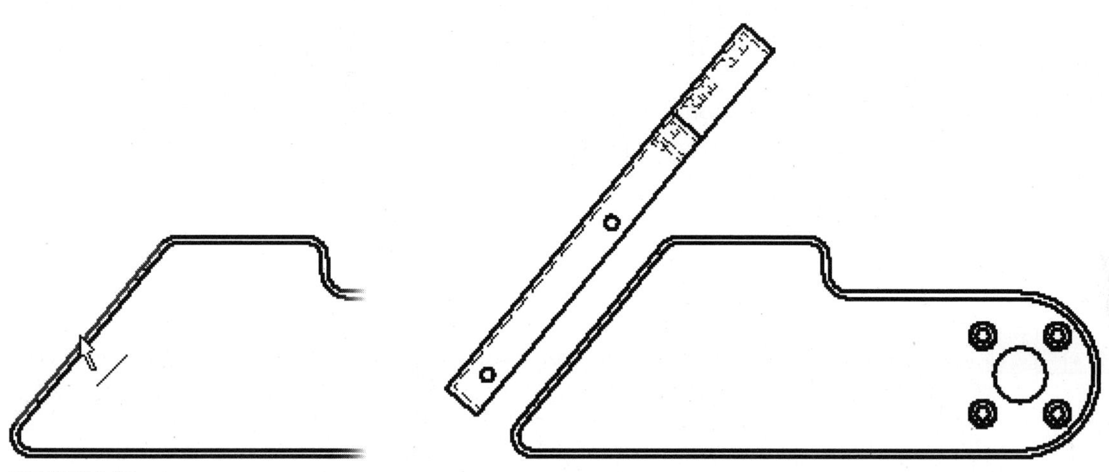

FIGURE 5.38

6. In the graphics window, select and drag the border of the views to fit them in the drawing.
7. You now create a section view. Click the Section command on the Place Views tab > Create panel.
8. In the graphics window, click in the top view.
9. To define the first point of the section line, hover the cursor over the center hole, and then click a point directly above the hole (the dotted lines represent the inferred point), as shown in the following image.
10. Click a point below the geometry to create a vertical section line.
11. Right-click and click Continue from the menu.
12. Place the section view with the default settings to the right of the top view, as shown in the following image on the right.
13. If the section view does not show the entire part, drag the endpoints of the section line in the top view so it is above the view.

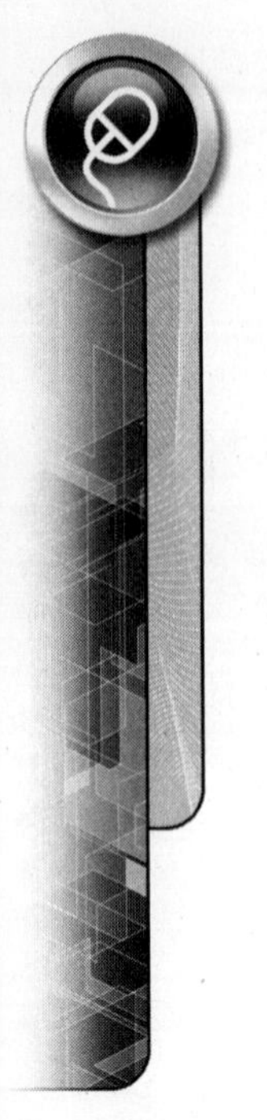

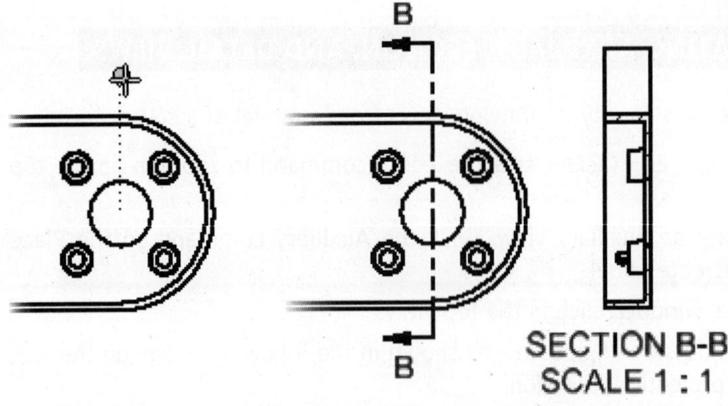

FIGURE 5.39

14. Try to drag the section line in the top view. If the section line can move follow these steps.
    - Right-click on the section line and click Edit from the menu.
    - Use the project geometry command and project the circle that is in the center of the part.
    - Apply a coincident constraint between the section line and the center point of the projected circle.
    - Click the Finish Sketch command from the Exit panel.

NOTE

If the sketch line cannot move and you need to reposition it, edit the section line as described in step 14, but remove a constraint that is holding it in place.

15. If desired, project an isometric view from the sectioned view.
16. Delete the section view by right-clicking in the section view and click Delete from the menu.
17. Next you create an aligned section view. Click the Section command on the Place Views tab > Create panel.
18. In the graphics window, click in the top view.
19. To define the first point of the section line, hover the cursor over the center hole, and then click a point directly above the hole (the dotted lines represent the inferred point), as you did in step 9.
20. Click the center point of the center circle.
21. Move the cursor over the center point of the lower-right circle (a green dot will appear) and then click a point outside of the geometry as shown in the following image on the left.
22. Right-click and click Continue from the menu.
23. Place the section view with the default settings to the right of the top view, as shown in the following image on the right.

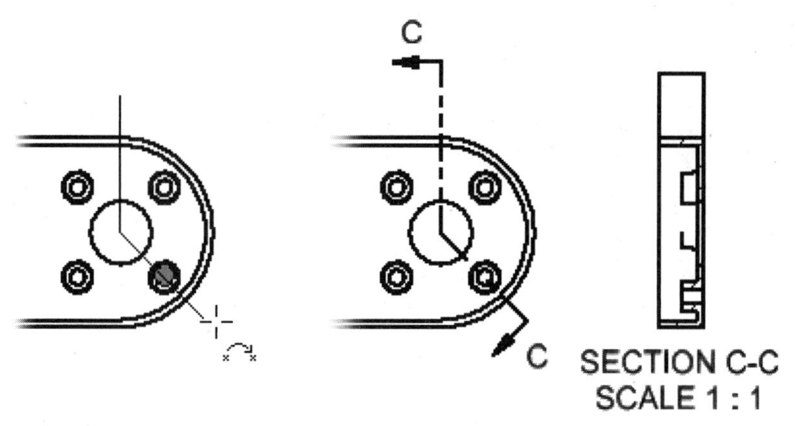

FIGURE 5.40

24. If the section view does not show the entire part, drag the endpoints of the section line in the top view so the entire part is sectioned.
25. Finally, create a detail view. Click the Detail command from the Place Views tab > Create panel.
26. In the graphics window, click in the top view.
27. Select a point near the left corner of the view to define the center of the circular boundary of the detail view.
28. Drag the circular boundary to include the entire lower-left corner of the view geometry, as shown in the following image on the left.
29. Click to position the detail view below the break view. The following image in the middle shows the completed detail in the top view and the image on the right shows the completed detail view.

Your view identifier (letter) may be different than what is shown.

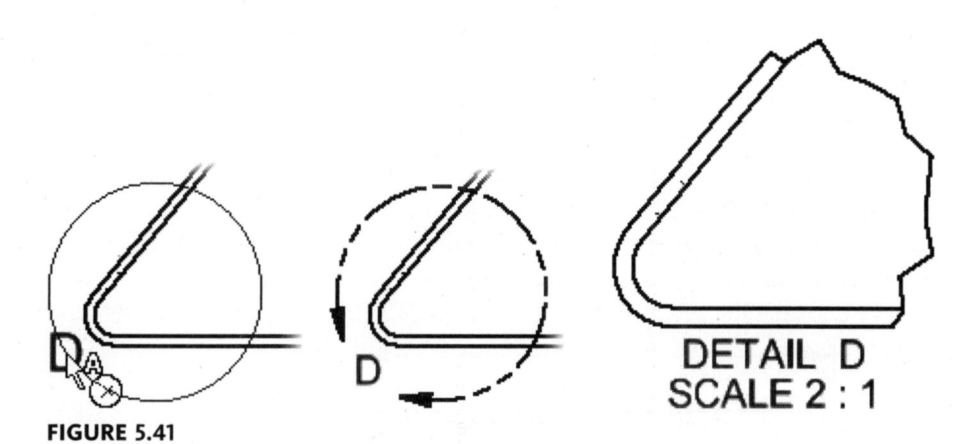

FIGURE 5.41

30. Reposition the drawing views as shown.

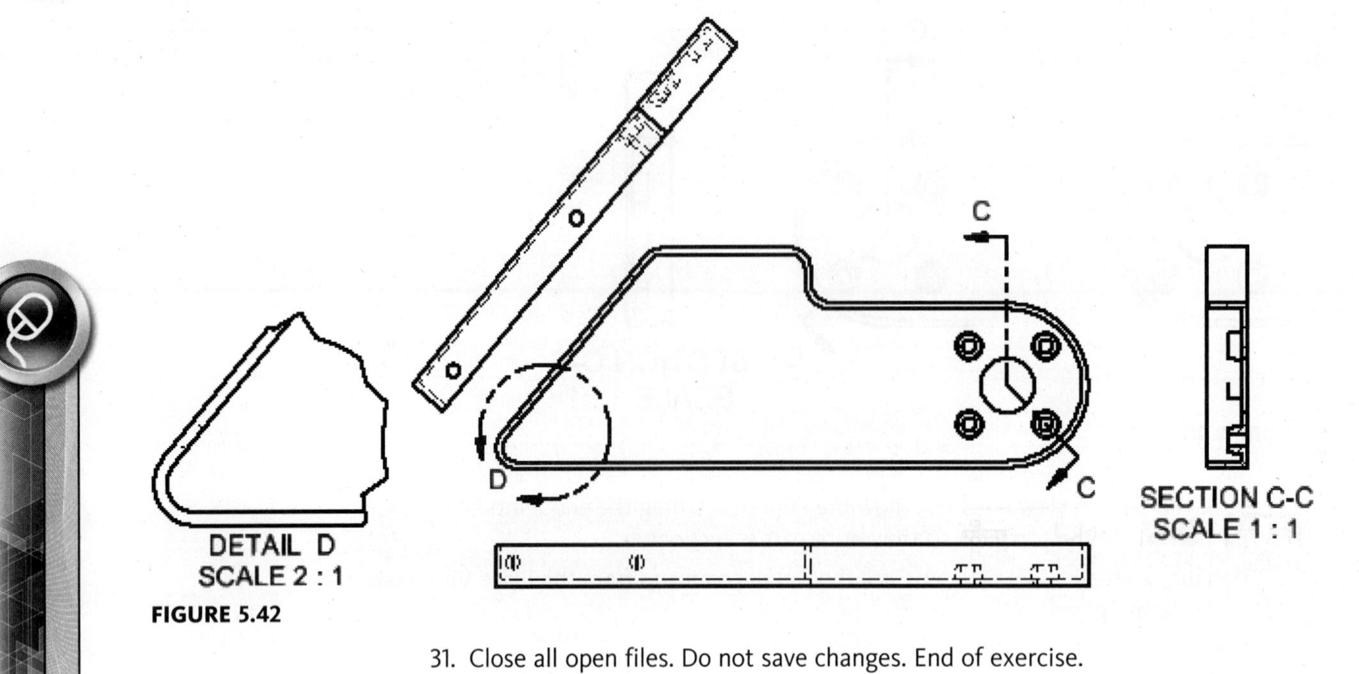

FIGURE 5.42

31. Close all open files. Do not save changes. End of exercise.

Creating Break Views

When creating drawing views of long parts, you may want to remove a section or multiple sections from the middle of the part in a drawing view and show just the ends. This type of view is referred to as a "break view." You may, for example, want to create a drawing view of a 2 × 2 × 1/4 angle, as shown in the following image, which is 48 inches long and has only the ends chamfered. When you create a drawing view, the detail of the ends is small and difficult to see because the part is so long.

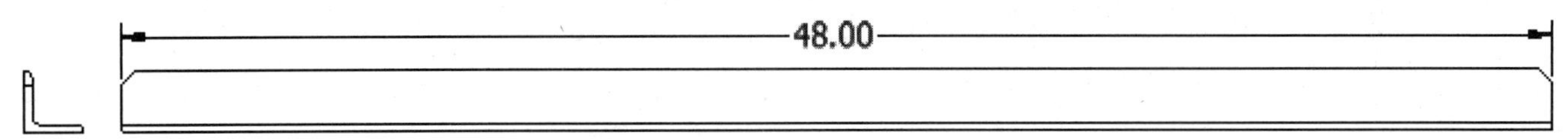

FIGURE 5.43

In this case, you can create a break view that removes the middle of the angle and leaves a small section on each end. When you place an overall length dimension that spans the break, it appears as 48 inches, and the dimension line shows a break symbol to note that it was based from a break view, as shown in the following image.

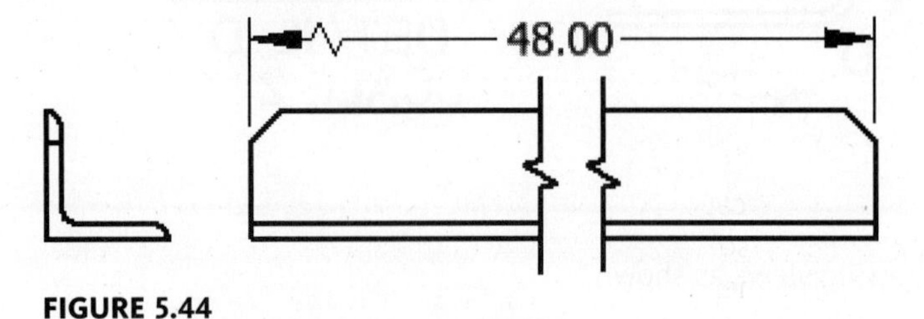

FIGURE 5.44

You create a break view by adding breaks to an existing drawing view. The view types that can be changed into break views are as follows: part views, assembly views, projected views, isometric views, auxiliary views, section views, break out views, and detail views. After creating a break view, you can move the breaks dynamically to change what you see in the broken view.

To create a break view, follow these steps:

1. Create a base or projected view or one that will eventually be shown as broken.
2. Click the Break command on the Place Views tab > Modify panel, as shown in the following image on the left. You can also right-click inside the bounding area of an existing view box, shown as a dotted rectangle when the cursor is moved into the view. Select Create View > Break from the menu.
3. If you selected the Break command on the Place Views tab > Modify panel, click inside the view from which to create the break view.
4. The Break dialog box will appear, as shown in the following image on the right. Do not click the OK button at this time, as this will end the command.

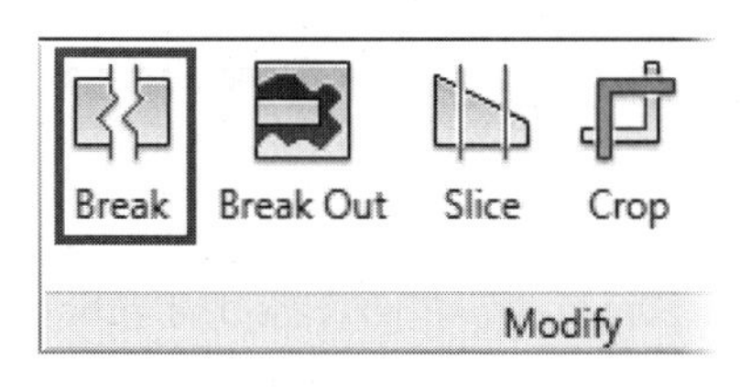

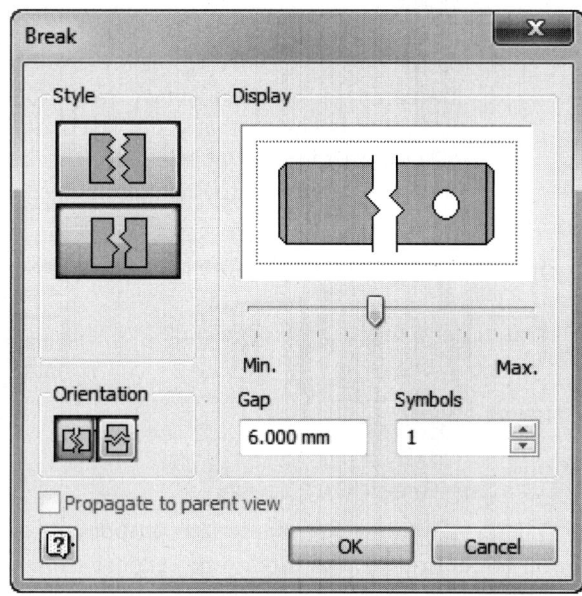

FIGURE 5.45

5. In the drawing view that will be broken, select a point where the break will begin, as shown in the following image.

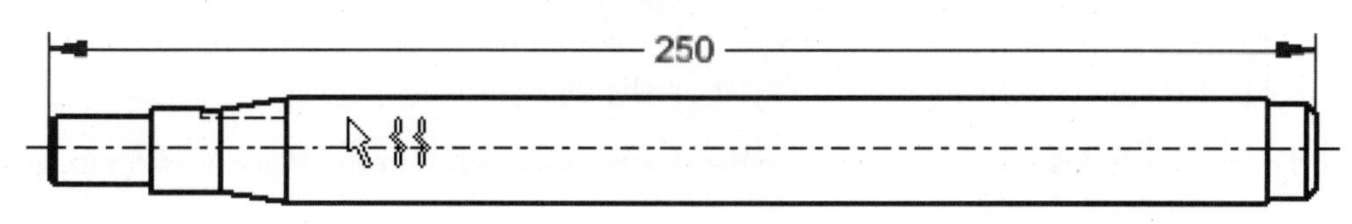

FIGURE 5.46

6. Select a second point to locate the second break, as shown in the following image. As you move the cursor, a preview image will appear to show the placement of the second break line.

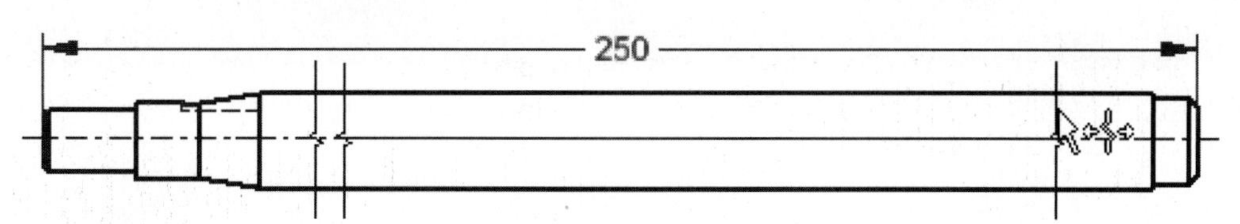

FIGURE 5.47

7. The following image illustrates the results of creating a broken view. Notice that the dimension line appears with a break symbol to signify that the view is broken.

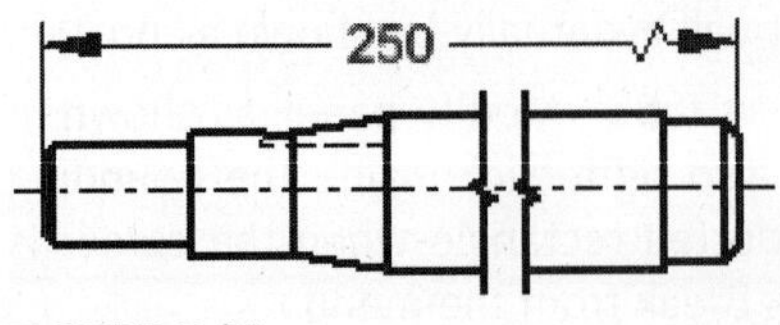

FIGURE 5.48

8. To edit the properties of the break view, move the cursor over the break lines, and a green circle will appear in the middle of the break. Right-click and select Edit Break from the menu. The same Break dialog box will appear. Edit the data as needed, and click the OK button to complete the edit.
9. To move the break lines, click on one of the break lines and, with the left mouse button pressed down, drag the break line to a new location, as shown in the following image. Drag the cursor away from the other break line to reduce the amount of geometry that is displayed, drag the cursor into the other break line to increase the amount geometry that is displayed. In either case the other break line will follow to maintain the gap size.

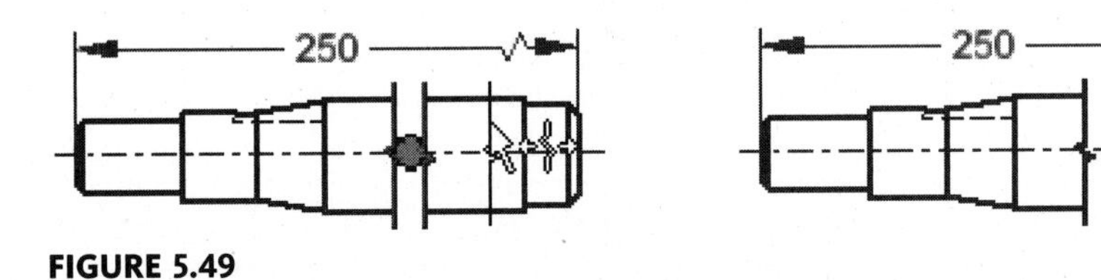

FIGURE 5.49

Creating Break Out Views

When you want to expose internal components or features, Autodesk Inventor allows you to cut or peel away a body and expose those internal parts; this is called a "break out view." It is not unusual for assemblies to have housings or covers that hide internal components. Break out views make it possible to expose these components. You can create break out views on assemblies as well as on part files.

To use break out views, you will need to be able to define two items: a closed profile boundary over the area to break and the depth of material to remove or the portion of a component to remove in order to see other components. As with broken views, the break out view is defined and displayed on the same view.

To create a break out view, click the Break Out command on the Place Views tab, as shown in the following image on the left.

After you select the view in which the break out will occur, the Break Out dialog box will appear, as shown in the following image. The termination options, From Point, To Sketch, To Hole, and Through Part, are available to give you control over how the break out section is created. The following sections explain these options.

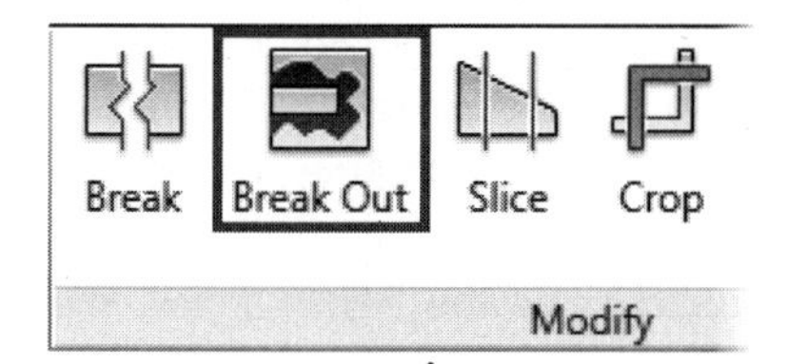

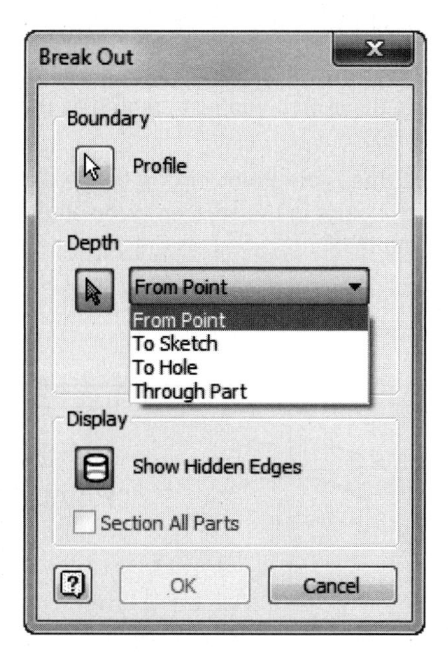

FIGURE 5.50

From Point. Select this to make the break out occur at a specified distance from a point located in a view. The point could be located in the base view or in a projected view.

To Sketch. Select this to use sketched geometry associated with another view to define the depth of the break out.

To Hole. Select this to base the break out depth on a hole feature whose axis determines the termination of the break.

Through Part. Select this to remove drawing content from inside a closed profile through selected components located in the browser. This termination option is especially useful when you want to reveal internal components or features.

Creating a Break Out View Using the From Point Option

To create a break out view using the From Point option, follow these steps:

1. Click on the view in which to create the break out section.
2. Click on the Create Sketch command on the Place Views tab > Sketch tab, and sketch a closed profile over the area you want broken out, as shown in the following image. When done, right-click, and select Finish Sketch from the menu.

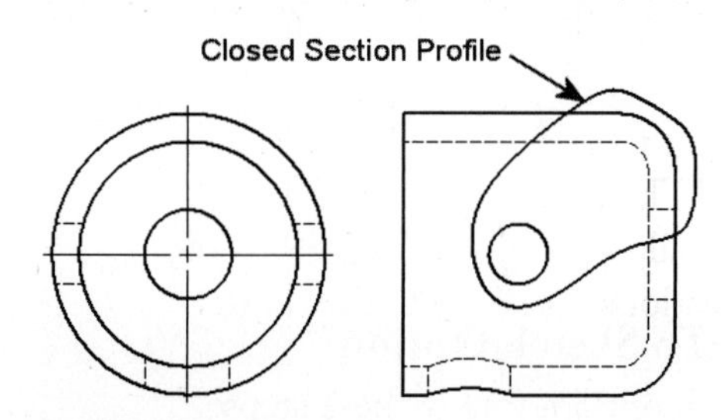

FIGURE 5.51

3. Click the Break Out command; this will activate the Break Out dialog box. Select the view in which the break out will occur, and select the defined boundary. Select the sketch you just created as the boundary in the view that will contain the break out.
4. Select the From Point option in the dialog box, click on the Depth arrow, and select a point in the adjacent projected view. This point will be used to calculate the depth of cut based on the distance, as shown in the following image.

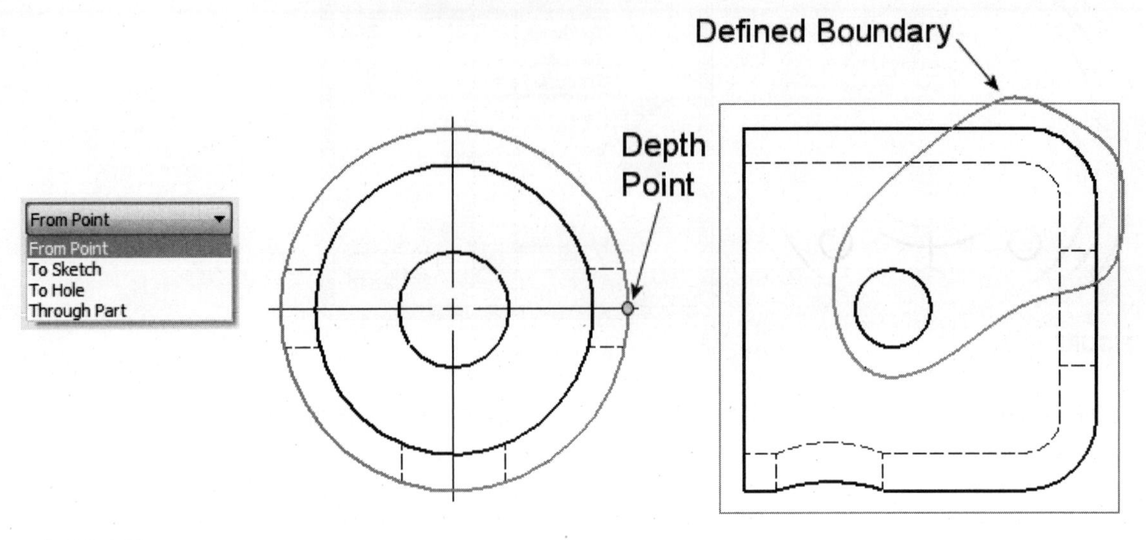

FIGURE 5.52

5. The following image shows the results of using the From Point option for creating a break out. The depth of 1 inch from the selected point cuts the view at the middle of the circular features, thus displaying the wall thickness of the part.

You can also select a point at the center of the circle for the depth of the cut. In this case, the distance value would change to 0, because the depth of the cut was specified by a point.

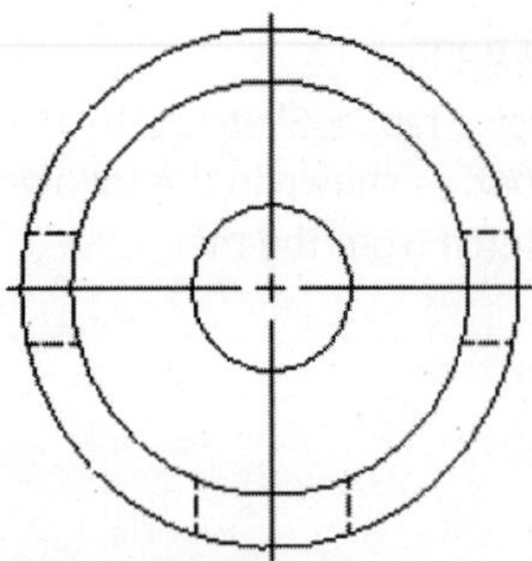

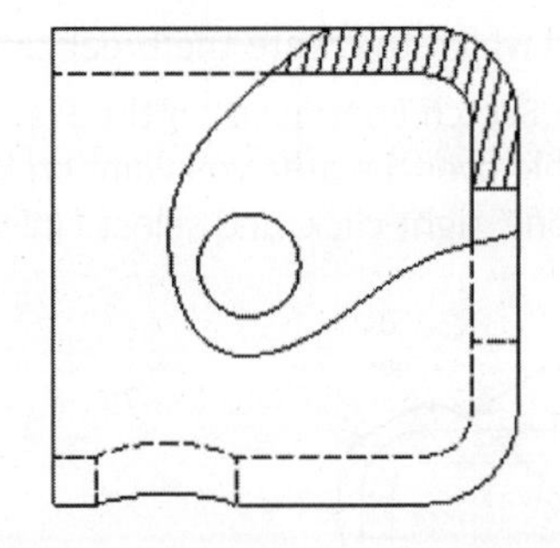

FIGURE 5.53

Creating a Break Out View Using the To Sketch Option

To create a break out view using the To Sketch option, follow these steps:

1. Activate the view in which the break out will occur, and sketch a closed profile over the area you want broken out. In the following image, two closed profiles will be used to create the break out view. When finished with this operation, right-click, and select Finish Sketch.

2. Activate the adjacent projected view and create another sketch. This sketch will be used to determine the depth of the cut for the break out, as shown in the following image. When finished with this operation, right-click, and select Finish Sketch.

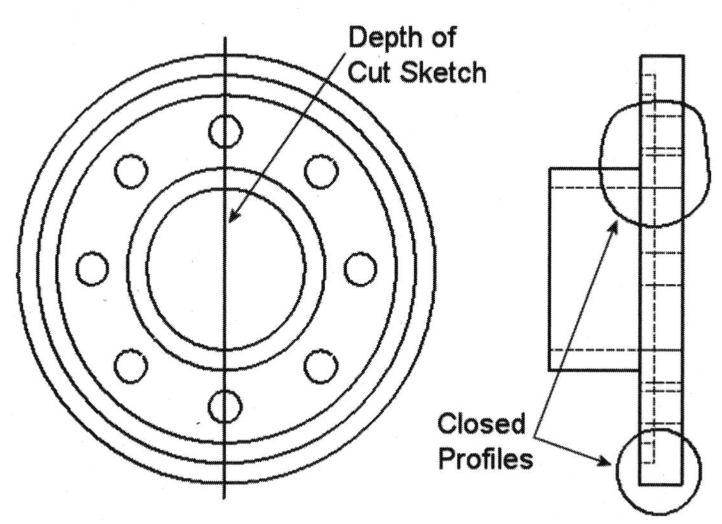

FIGURE 5.54

3. Click the Break Out command, and select a view as the base for the break out view.
4. When the Break Out dialog box appears, select the initial sketch or sketches to use as the boundary profile. In the Depth area of the dialog box, change the option to Sketch. In the adjacent projected view, select the sketch that will determine the depth of the cut for creating the break out view, as shown in the following image on the left.
5. The following image on the right shows the break out view created based on the first group of sketched boundaries as well as the second sketch, which determined the depth of the cut.

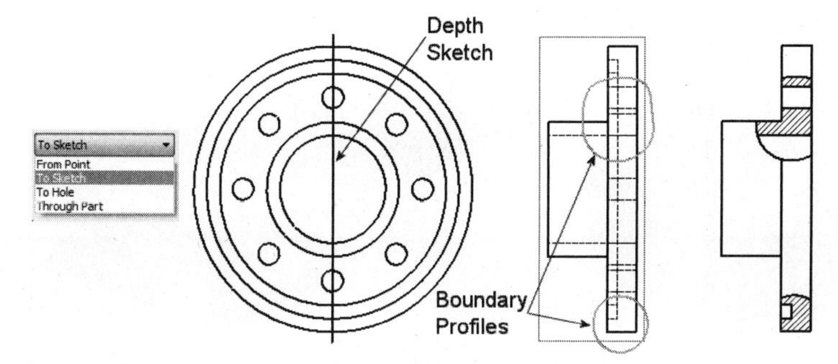

FIGURE 5.55

Creating a Break Out View Using the To Hole Option

The To Hole option for creating break out views is similar to the To Sketch option. Instead of basing the break out on a sketch, it will base it on a hole feature. The axis of the hole feature determines the termination depth.

To create a break out view using the To Hole option, follow these steps:

1. Click the Break Out command on the Place Views tab.
2. Activate the view in which the break out will occur, and sketch a closed profile around the hole you want broken out.

3. In the graphics area, click to select the view. When the Break Out dialog box appears, click to select the defined boundary. The boundary profile must be on a sketch associated to the selected view. In the Break Out dialog box, click the arrow next to the Depth type box, and select the To Hole option, as shown in the following image on the left.
4. Click the select arrow, and then select the hole feature in the graphic window, as shown in the middle in the following image. The depth is defined by the axis of the hole.

If the hole feature is hidden, edit the view, and click the Show Hidden Edges button to temporarily display hidden lines. You can also select the hole in another view.

5. When the view is defined fully, click OK to create the view, as shown in the following image on the right.

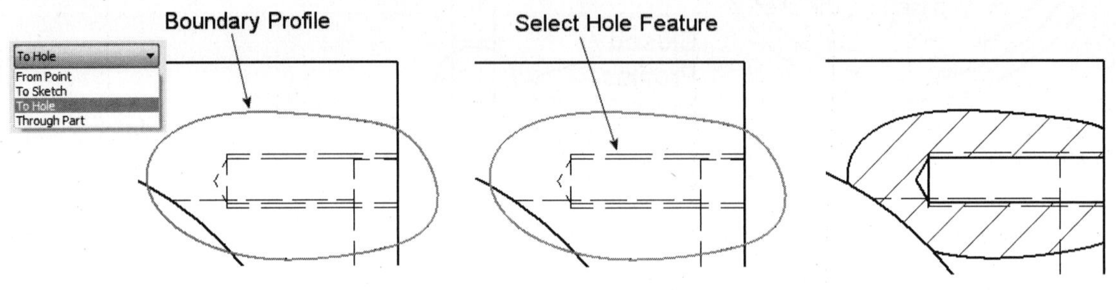

FIGURE 5.56

Creating a Break Out View Using the Through Part Option

In the following image, a housing hides internal details of an assembly. You can cut away a segment of the housing to expose obscured parts or features using the Through Part option of the Break Out command.

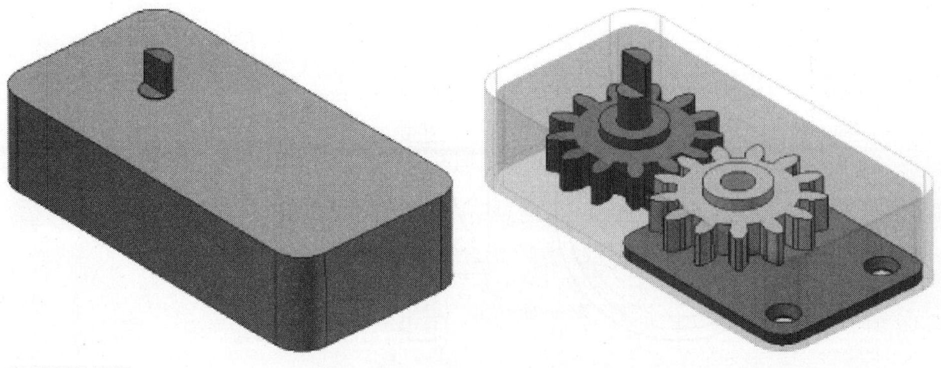

FIGURE 5.57

To create a break out view using the Through Part option, follow these steps:

1. Activate the view through which you wish to cut, and click on the Create Sketch command on the Place Views tab.
2. Next sketch a closed profile, as shown in the following image on the left, as the area through which you wish to cut. When finished, right-click, and select Finish Sketch.
3. Click on the Break Out command. Click in the view in which to perform the operation will activate the Break Out dialog box. The boundary profile consisting of the previous sketch will be selected automatically. In the Depth area of the dialog box, click on the Through Part option, as shown in the following image in the middle.

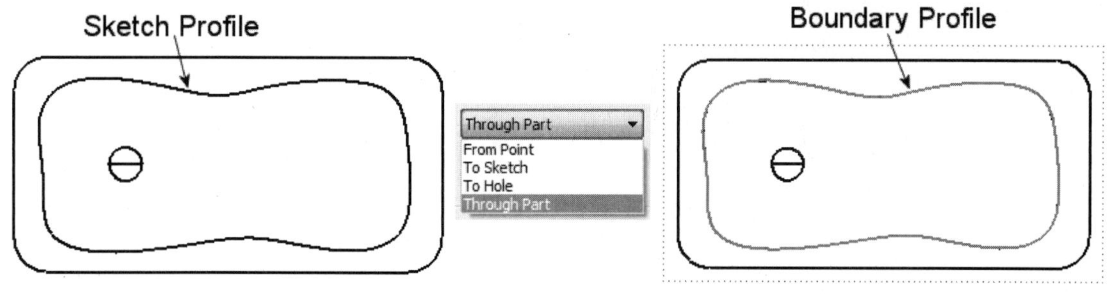

FIGURE 5.58

4. Move your cursor over the edge of the part or parts (you can also select the parts in the browser) and select this edge, as shown in the following image on the left.
5. When finished, click the OK button. The following image on the right illustrates the results of using the Through Part option. The gear cover has been sliced away, based on the sketched boundary, to expose the internal workings of the gear mechanism.

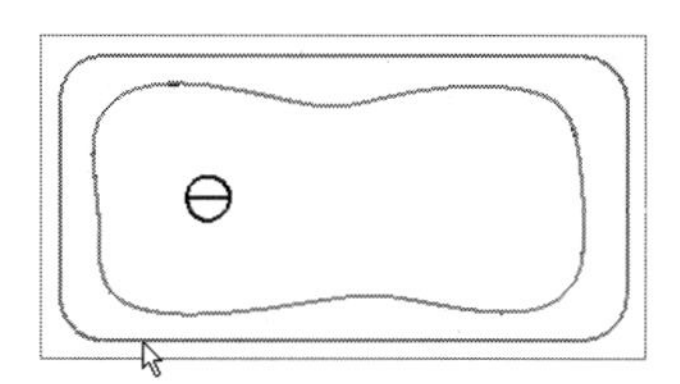

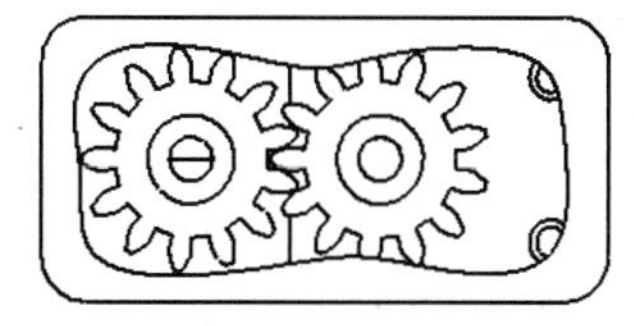

FIGURE 5.59

The Through Part option can be very effective when applied to an isometric view. As shown in the following image, a spline sketch was created on an isometric view. Notice how the sketch covers the top and sides of the isometric. The image on the right in the figure illustrates the completed break out view. All sides surrounded by the spline sketch break out to display the internal components.

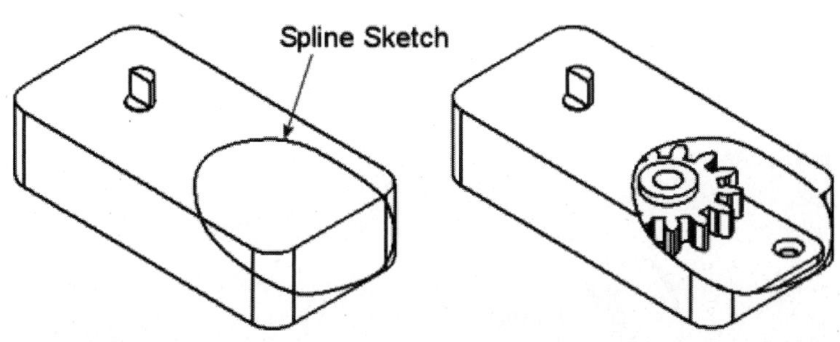

FIGURE 5.60

Creating Slice Views

A sliced view is a different type of section cut that has no depth and allows you to create slices at key locations based on a sketch located in the source view. The target view is the view displaying the various slices. The Slice command can be activated from the Place Views tab > Modify panel, as shown in the following image on the left. You can also right-click inside the bounding area of an existing view box, shown as dashed lines when the cursor moves into the view, and select Create View > Slice from the menu.

To create a slice view, follow these steps:

1. Select the bounding box of the view of the Source view (where the sketch geometry will be drawn).

2. Click the Create Sketch command from the Sketch panel to open a drawing sketch associated with the view.
3. Create sketch geometry to define an open slice sketch. When finished, exit the sketch environment.
4. Click the Slice command, and select the Target view that will be sliced.
5. When the Slice dialog appears, select a point in the target view to be sliced, then click on sketch geometry that was drawn in step 3.

When dealing with multiple parts, the Slice All Parts checkbox will activate. Use this feature to override any component settings made in the browser and slice all parts in the view, depending on the sketch geometry. Components that are not crossed by this geometry will not be sliced.

6. Click OK to complete the Slice operation.

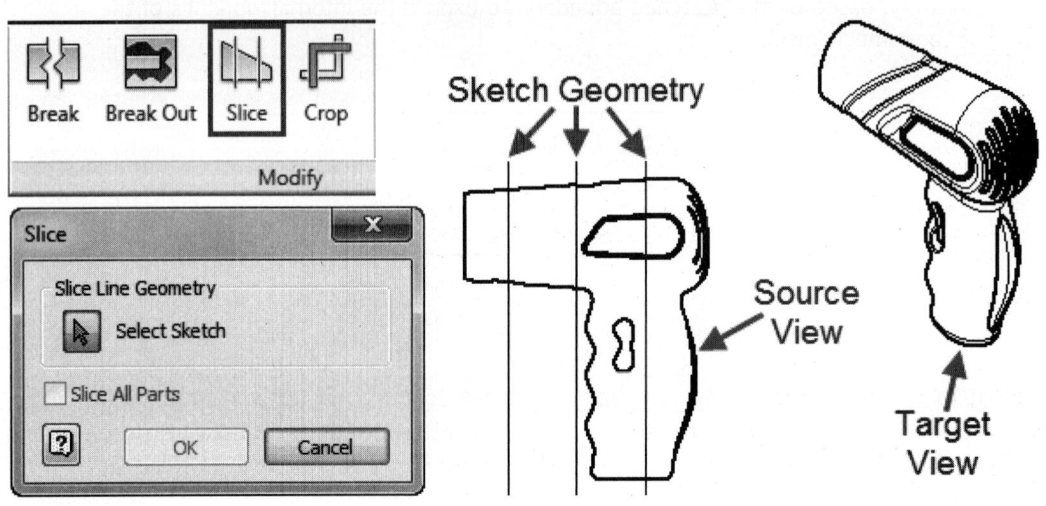

FIGURE 5.61

The results of the slice operation are illustrated in the following image on the right. A cross-section was created from every location of the hair dryer where vertical sketch geometry was sketched.

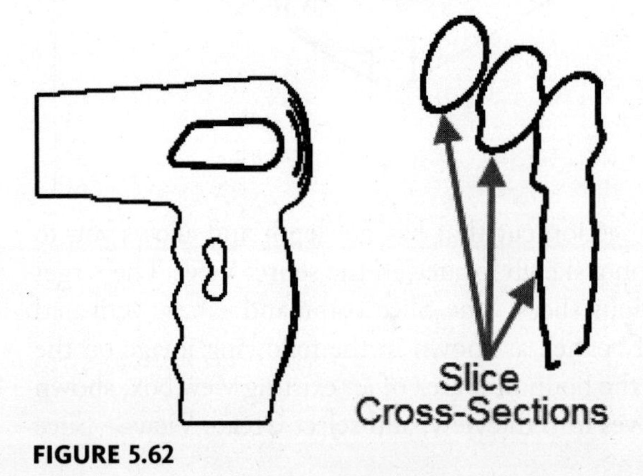

FIGURE 5.62

Creating Cropped Views

Cropped views are used to show a portion of a part compared to a larger portion. The following image on the left illustrates a complete view showing object and hidden lines. After cropping this view, only a portion of the view is shown. This is different from a detail view, which was explained previously. A detail view is generated by a parent view and references this parent, complete with callout information. The cropped view will not create a callout and is meant for stand-alone use.

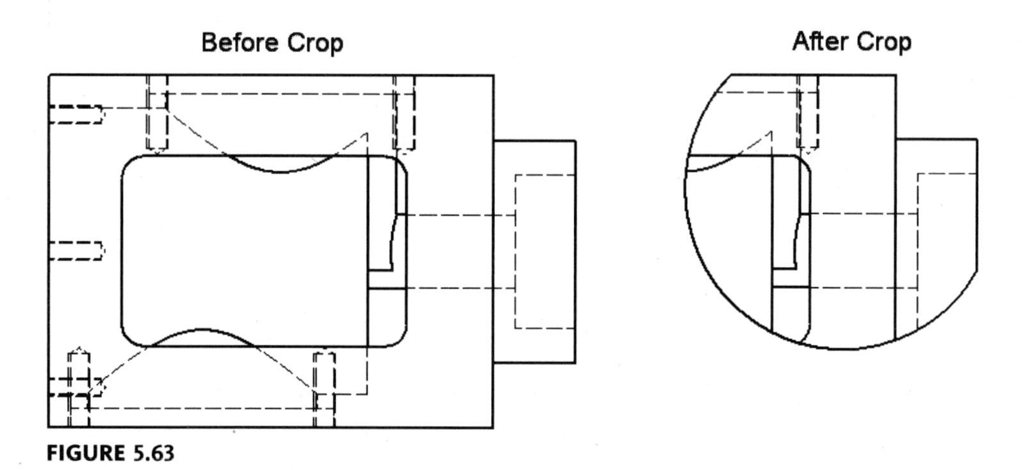

FIGURE 5.63

The steps used for creating a cropped view are as follows.

1. On the Place Views tab, click the Crop command from the Modify panel as shown in the following image.

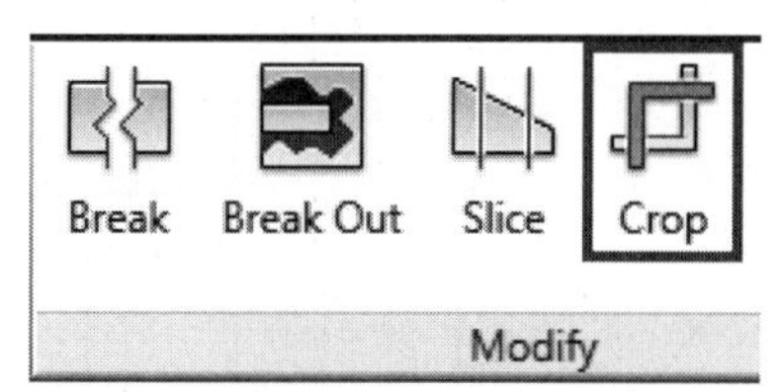

FIGURE 5.64

2. Select the view to crop by clicking the edge of the view.
3. Right-click the menu to select the boundary type that can be either circular or rectangular. When specifying the circular boundary, a center point acts as the boundary center; for rectangular boundaries, two diagonal points are used to create the rectangle as shown in the following image.

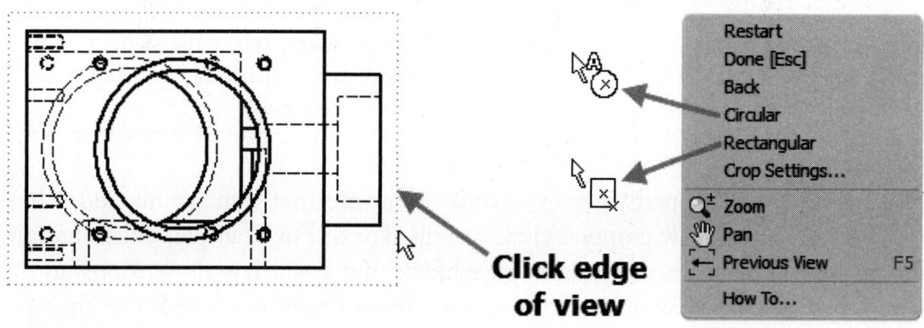

FIGURE 5.65

While creating the crop view you can also control the display of the crop cut lines in the cropped view. After starting the Crop command, right-click and click Crop Settings from the menu to display the Crop Settings dialog box as shown in the following image on the left. In addition to the circular and rectangular boundary type already mentioned, you can turn on or off the display of crop cut lines as shown in the following image on the left. After cropping a view you can turn off the crop cut lines by right-clicking on the Crop entry in the browser and uncheck the Display Crop Cut Lines option from the menu as shown in the following image on the right.

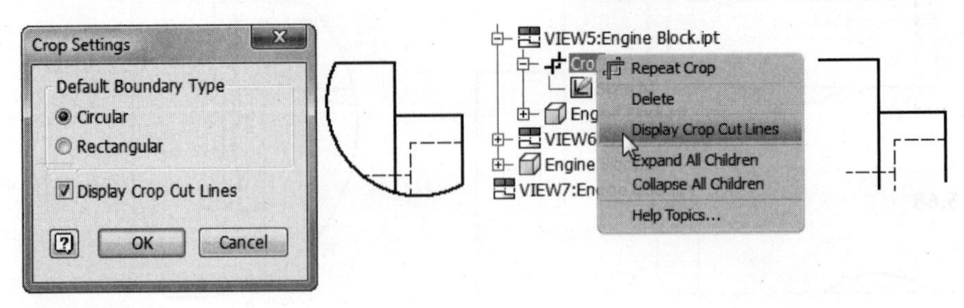

FIGURE 5.66

You can also create a cropped view that is based on an existing sketch. A few rules to follow are be sure that the sketch does not contain any non–self-intersecting loops and be sure the sketch is associated with the view that is currently being cropped. The illustration on the left in the following image shows a spline that was sketched. As long as this sketch is associated with this view, you will be able to create a cropped view by clicking the sketch as shown in the following image on the right.

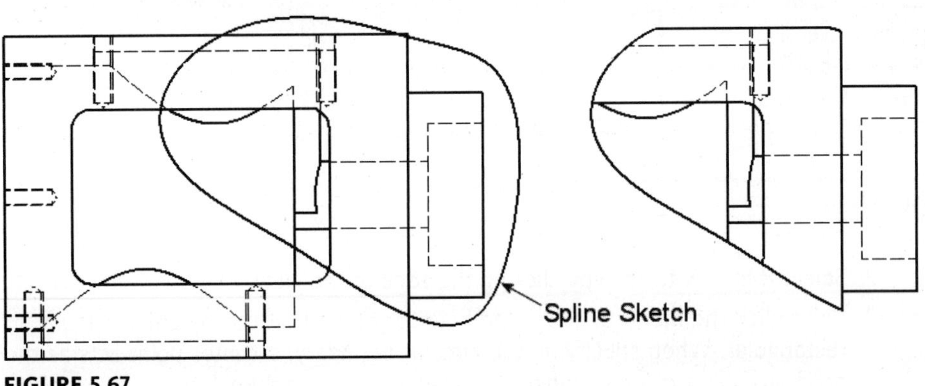

FIGURE 5.67

NOTE

To edit a cropped sketch, open the cropped view in the browser, select the sketch, right-click, and select edit. Make the necessary changes to the sketch and click the Return button. This will leave the sketch environment and update the cropped view.

When working with cropped views, you must be aware that dimensions and other annotations may affect how cropped views are displayed. For example, illustrated on the left in the following image is an engine block that has a few dimensions and a centerline through the view. However, once a cropped view is created, the anchors used as endpoints for the dimensions will no longer use these anchors, resulting in an orphaned dimension as shown in the following image on the right. Either the anchor would need to be reestablished by editing the cropped view sketch or the dimension should be deleted.

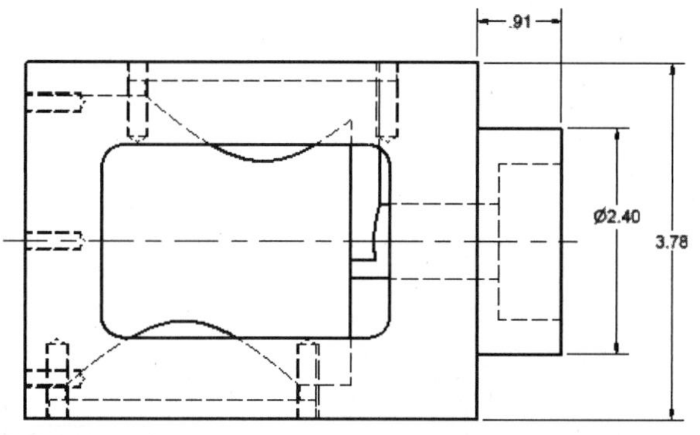

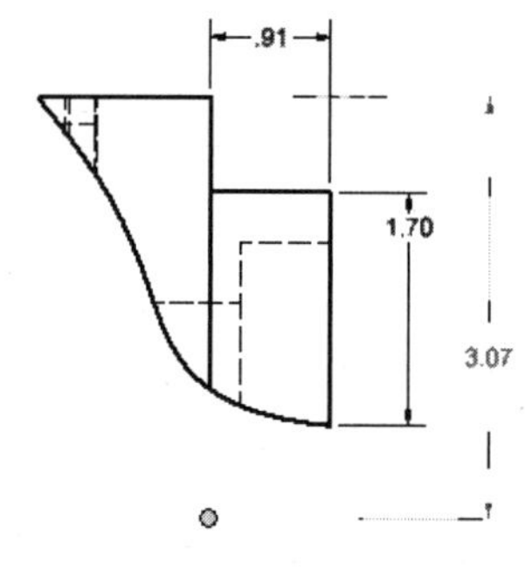

FIGURE 5.68

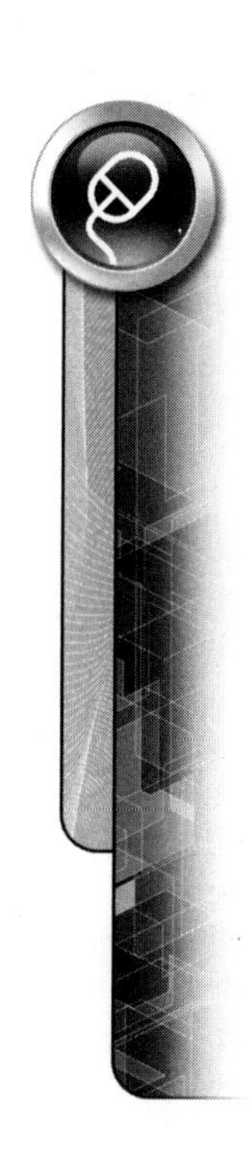

EXERCISE 5-3: CREATING BREAK, AND BREAK OUT VIEWS

In this exercise, you create a break view and a break out view.

1. Open the file *ESS_E05_03_1.idw*. This file has a base view and a side view that is too long to fit in the drawing.
2. You now create a break view. First click the Break command on the Place Views tab > Modify panel.
3. In the graphics window, click in the side view to break. Select the start and end points for the material to be removed, as shown in the following image.

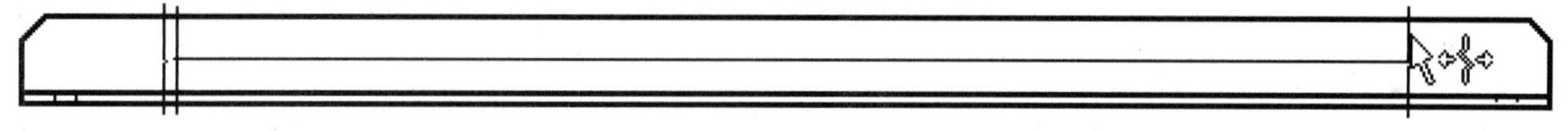

FIGURE 5.69

4. The side view will now be broken as shown in the following image on the left.
5. To edit the properties of the break view, move the cursor over the break lines, and a green circle will appear in the middle of the break. Right-click and select Edit Break from the menu.
6. In the Break dialog box change the Display to Max and the Gap to **0.50 inches** as shown in the following image in the middle. Click the OK button to complete the edit.
7. To move the break lines, click on one of the break lines and, with the left mouse button pressed down, drag the break line to a new location. Drag the cursor away from the other break line to reduce the amount of geometry that is displayed, drag the cursor into the other break line to increase the amount of geometry that is displayed. In either case the other break line will follow to maintain the gap size.
8. Next you will create a dimension. Press the D key on the keyboard and place the dimension on the two outside points as shown in the following image on the left. Notice the break symbol in the dimension. Dimensioning a drawing will be covered later in this chapter.
9. When done, your screen should resemble the following image on the right.

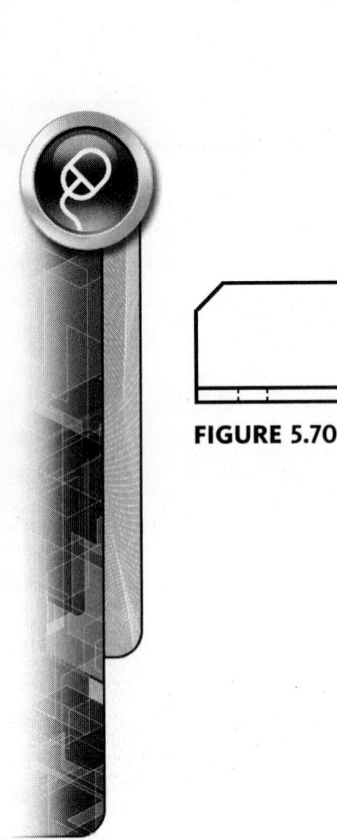

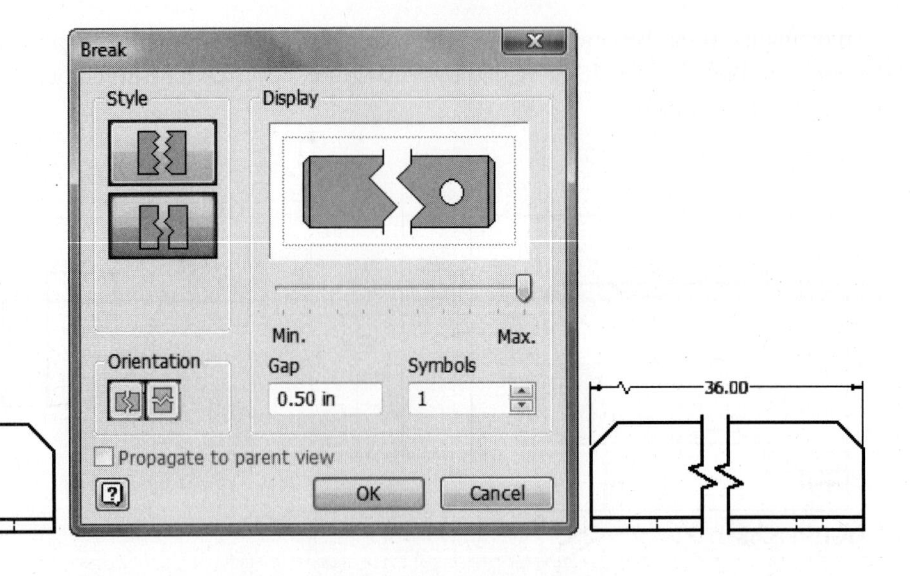

FIGURE 5.70

10. Close all open files. Do not save changes.
11. Open the file *ESS_E05_03_2.idw*. In this portion of the exercise you create a break out view of a part. This drawing contains a front and top view. Note that a break out view can also be created for assemblies.
12. First you create a sketch based on the top view. Select the bounding edge of the top view as shown in the following image on the left and click Create Sketch from the Place Views tab > Create panel. If the edge of the top view is not selected, the sketch will not be associated to the view and the break out view will not be able to be created.
13. Create a closed profile of your choice. This closed profile will be used to cut away a section in the view. The following image on the right shows a closed spline.

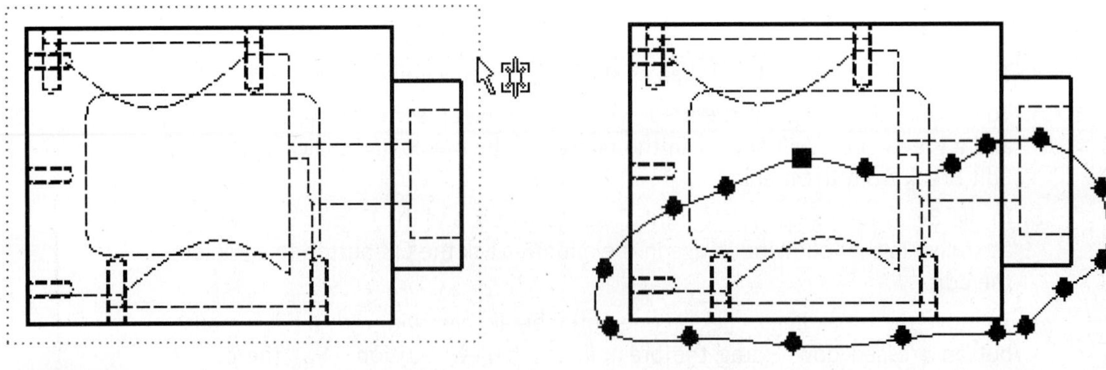

FIGURE 5.71

14. Exit the sketch by clicking Finish Sketch from the Place Views tab > Exit panel.
15. You now create a break out view. First click the Break Out command on the Place Views tab > Modify panel.
16. In the graphics window click in the top view, this is the view that will have the closed profile break out a section.

17. The Break Out dialog box will appear. The default Depth option is From Point, in the front view click on the point as shown in the following image.

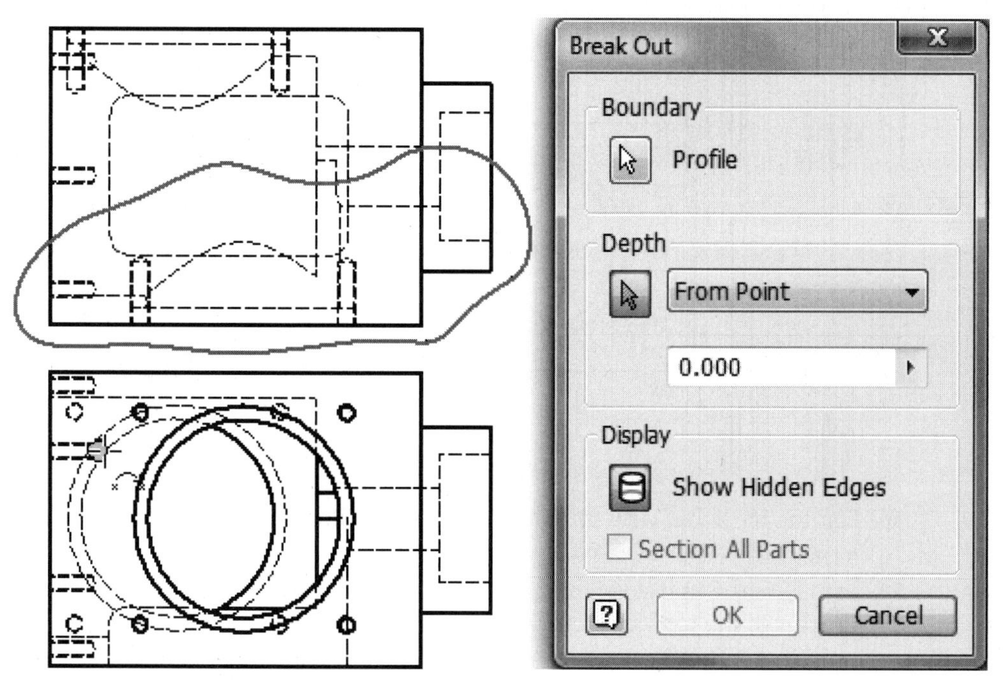

FIGURE 5.72

18. Click OK in the Break Out dialog box, the closed profile will break out a section in the top view as shown in the following image on the left.
19. Next you project an isometric view of the top broken out view. Click the Projected View command on the Place Views tab > Create panel.
20. Click in the top view and then locate the isometric view by clicking a point to the lower left.
21. Edit the isometric view and from the Display Options tab check the Hatching option and click OK to complete the edit, and the isometric view should resemble the following image on the right side. Move the isometric view to the right side of the drawing.

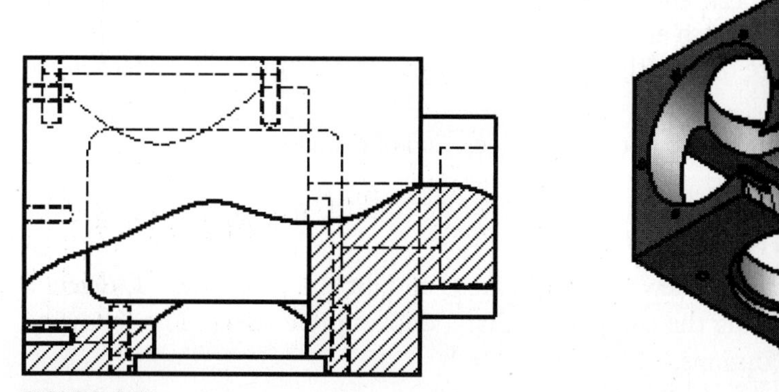

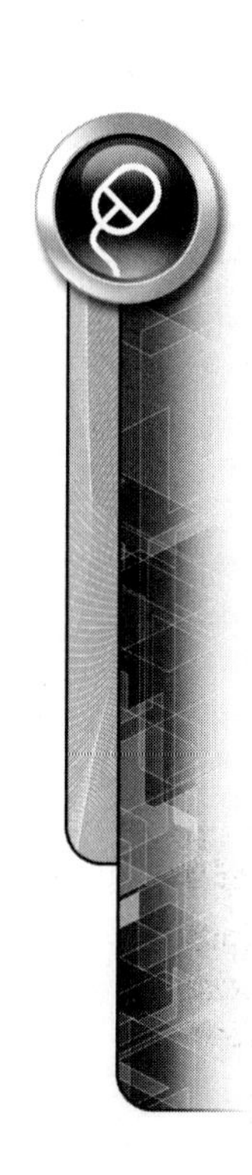

FIGURE 5.73

22. Edit the break out view by right-clicking on the Break Out View in the browser and click Edit Break Out from the menu as shown in the following image on the left.

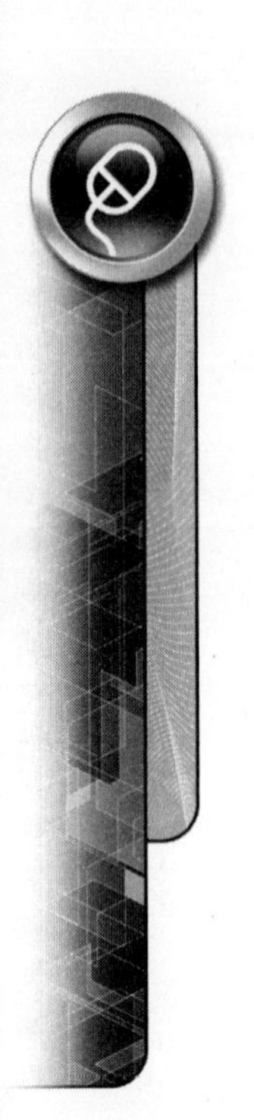

23. In the Break Out dialog box change the Depth type to To Hole and select the lower left hole in the front view as shown in the following image in the middle.
24. Click OK in the Break Out dialog box and the top and isometric view will be updated. The following image on the right shows the updated isometric view.

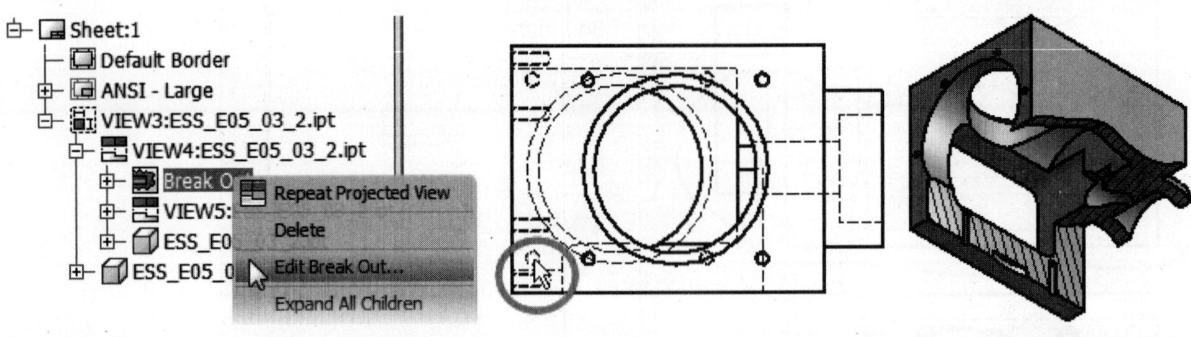

FIGURE 5.74

25. Edit the break out view by right-clicking on the Break Out View in the browser as you did in step 22.
26. In the Break Out dialog box change the Depth type to Through Part; an edge of the part in the top view is shown in the following image on the left.
27. Click OK in the Break Out dialog box and the top and isometric view will be updated. The following image on the right shows the updated isometric view.

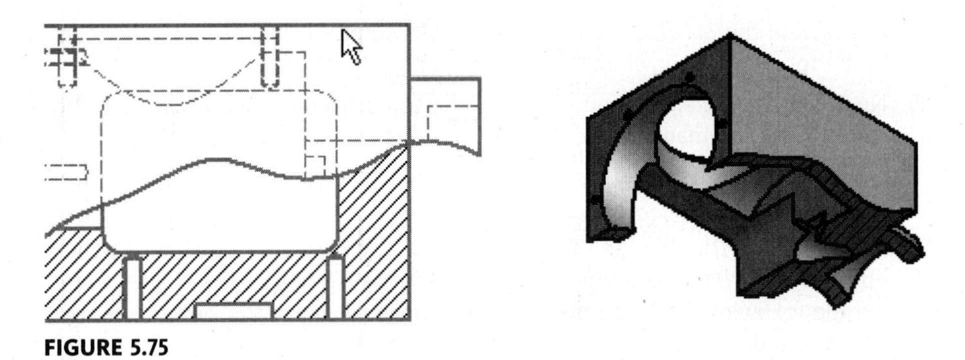

FIGURE 5.75

28. If desired, you can edit the sketch that the breakout view used by double-clicking on the Sketch entry under the Break Out view in the browser. Then reposition / resize the sketch as needed. When done, click the Finish Sketch in the Exit panel.
29. Close all open files. Do not save changes. End of exercise.

EDITING DRAWING VIEWS

After creating the drawing views, you may need to edit the properties of, delete a drawing view, or replacing the component that a drawing references. The following sections discuss these options.

Editing Drawing View Properties

After creating a drawing view, you may need to change the label, scale, style, or hatching visibility, break the alignment constraint to its base view, or control the visibility of the view projection lines for a section or auxiliary view. To edit a drawing view, follow one of these steps:

- Double-click in the bounding area of the view.
- Double-click on the icon of the view you want to edit in the browser.
- Right-click in the drawing view's bounding area or on its name, and select Edit View from the marking menu, as shown in the following image.

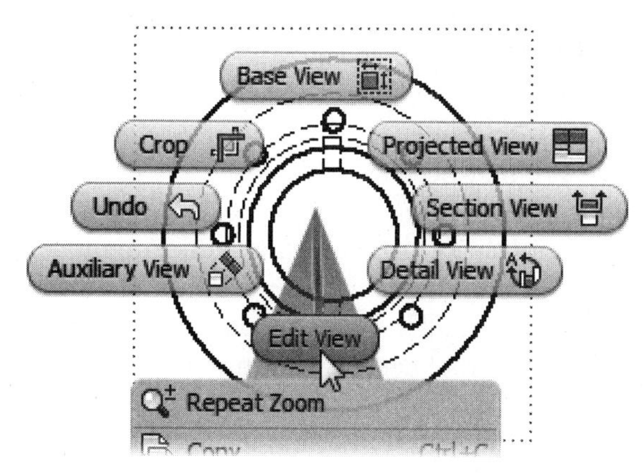

FIGURE 5.76

When you perform the operations listed above, the Drawing View dialog box will appear, as shown in the following image. This is the same dialog box used to create drawing views. Depending on the view that you selected, certain options may be grayed out from the dialog box. Make the necessary changes, and click the OK button to complete the edit. When you change the scale in the base view, all the dependent views will be scaled as well. You can also change the orientation of the base view; dependent views will then be updated to reflect the new orientation.

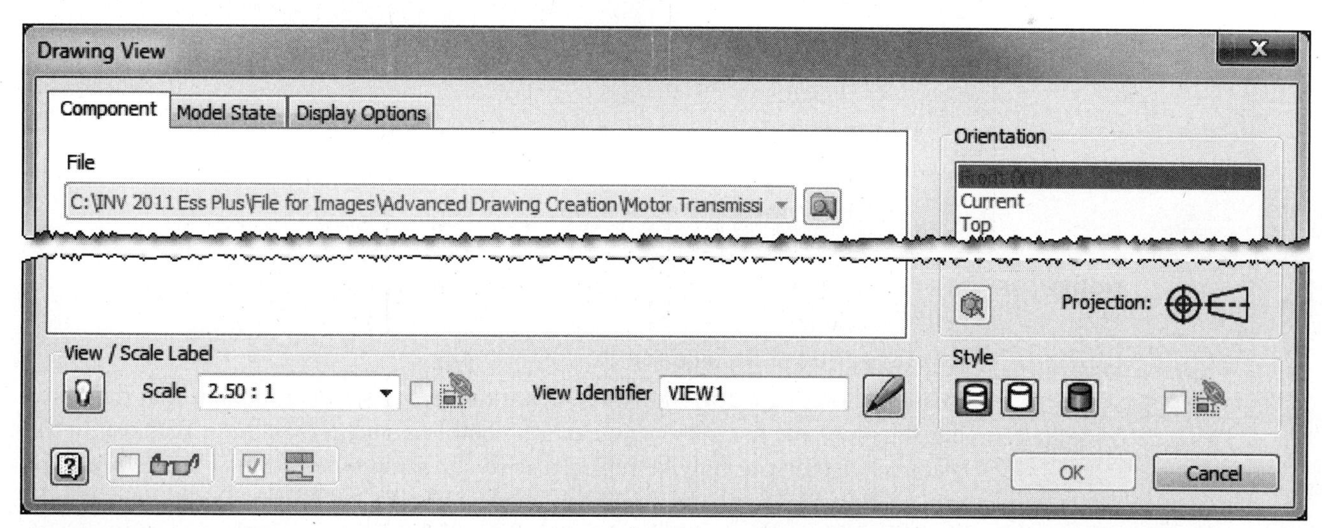

FIGURE 5.77

See the Using the Drawing View Dialog Box section in this chapter for detailed descriptions of Component and Options tab elements.

Deleting Drawing Views

To delete a drawing view, either right-click in the bounding area of the drawing view or on its name in the browser, and select Delete from the menu. You can also click in the bounding area of the drawing view, and press the Delete key on the keyboard. A dialog box will appear, asking you to confirm the deletion of the view. If the selected view has a view that is dependent on it, you will be asked if the dependent views should also be deleted. By default, the dependent view will be deleted. To exclude a dependent view from the delete operation, expand the dialog box, and click on the Delete cell to change the selection to No.

Break and Align Views

After creating or deleting drawing views you may need to move a view to a projected view to a different location on the drawing but the view can only move orthographically to its parent view. To break this relationship either click the Break Alignment command on the Place Views tab > Modify panel as shown in the following image on the left and click in the view to break alignment. The other option is to right-click in the view and click Alignment > Break from the menu as shown in the following image on the right. To align a view to another view follow the same process but click Horizontal to align a view horizontally to a selected view, Vertical to align a view vertically to a selected view and In Position to keep a view near its current position in relation to a selected view.

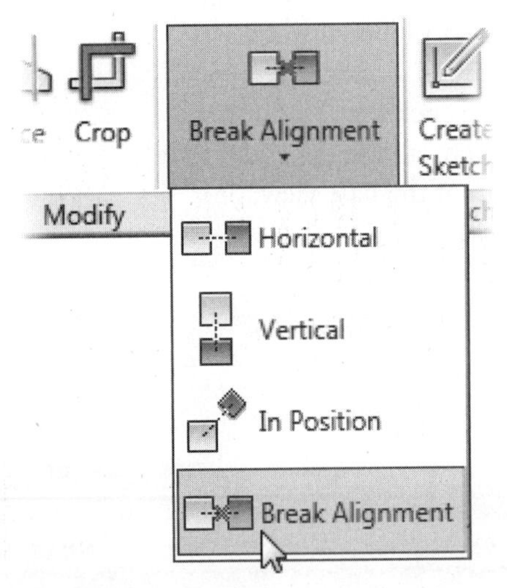

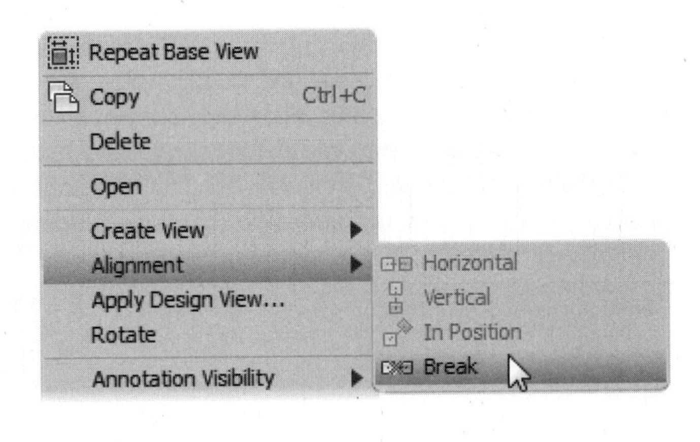

FIGURE 5.78

Replace Model Reference

While annotating a drawing you may want to change the component that the drawing reference for the geometry. Parts, assemblies, and presentation files can be replaced, but they can only be replaced with the same type of file. The file replacing the existing file should be similar in shape and size, otherwise dimensions may be orphaned. To replace a model that a drawing references, follow these steps.

1. Click the Replace Model Reference command on the Manage tab > Modify panel as shown in the following image on the left.
2. The Replace Model Reference dialog box will appear as shown in the following image on the right. All the files that are referenced in the drawing will appear in the list. In the dialog box click on the file to replace.

3. In the dialog box click the Navigate button.
4. Browse to the new file and select it.
5. In the Open dialog box click Open.
6. Click Yes in the warning dialog box to replace the model.
7. In the Replace Model Reference dialog box click OK to complete the command.
8. Cleanup the drawing annotations and the views orientation as needed.

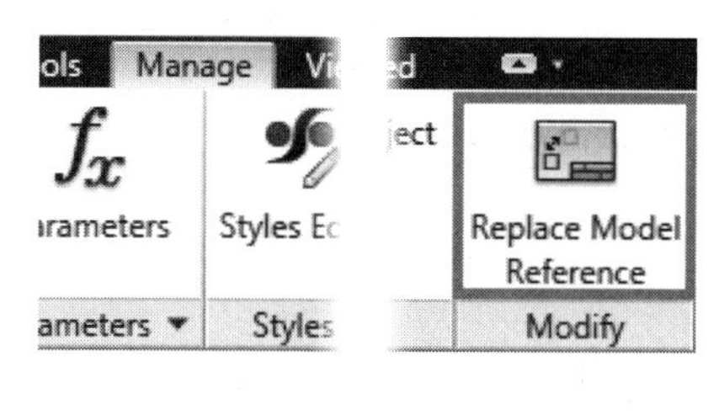

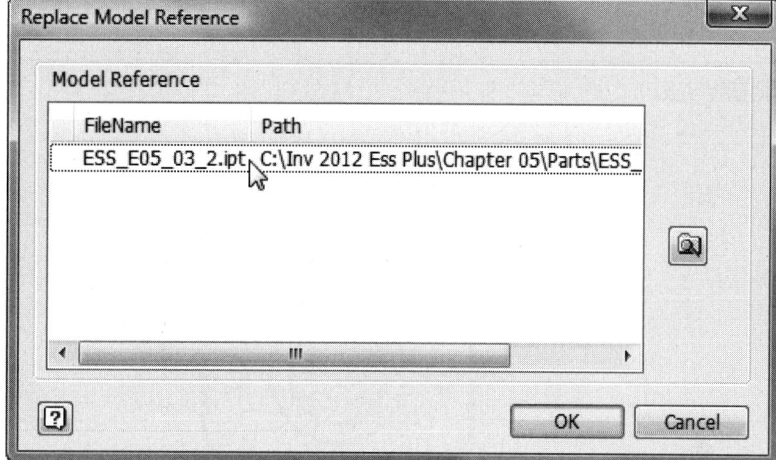

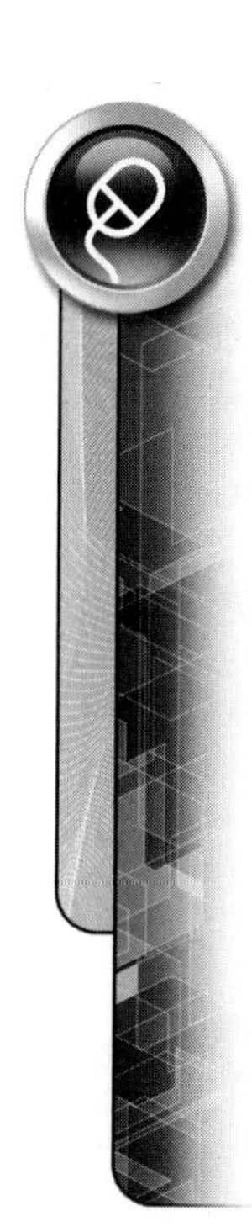

FIGURE 5.79

EXERCISE 5-4: EDITING DRAWING VIEWS

In this exercise, you will delete the base view while retaining its dependent views. The base view is not required to document the part, but its dependent views are. Next you align the section view with the right-side view to maintain the proper orthographic relationship between the views. You also modify the hatch pattern of the section to better represent the material.

1. Open the drawing file *ESS_E05_04.idw*. This drawing contains three orthographic views, an isometric view, and a section view.
2. Right-click on the base view (the middle view on the left side of the drawing), as shown in the following image, and choose Delete from the menu as shown in the following image on the left.
3. In the Delete View dialog box, notice that the projected views are highlighted on the drawing sheet. Select the More (>>) button in the dialog box.
4. Click on the word Yes in the Delete column for each dependent view to toggle each to No. This will delete the base view but leave the dependent views present on the drawing sheet as shown in the following image on the right.

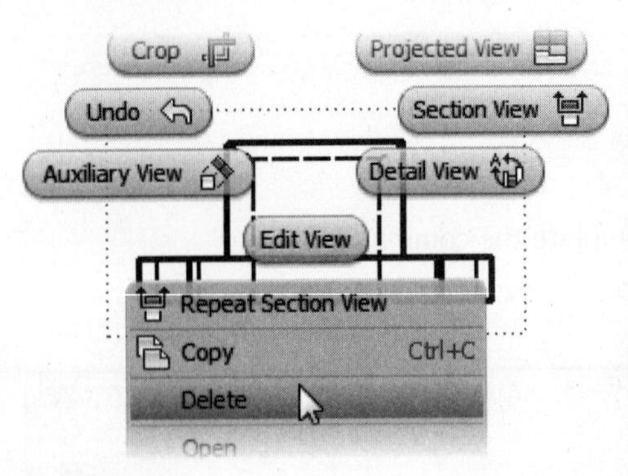

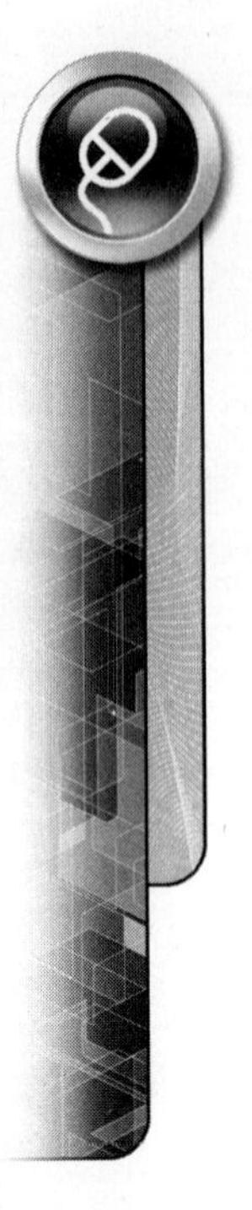

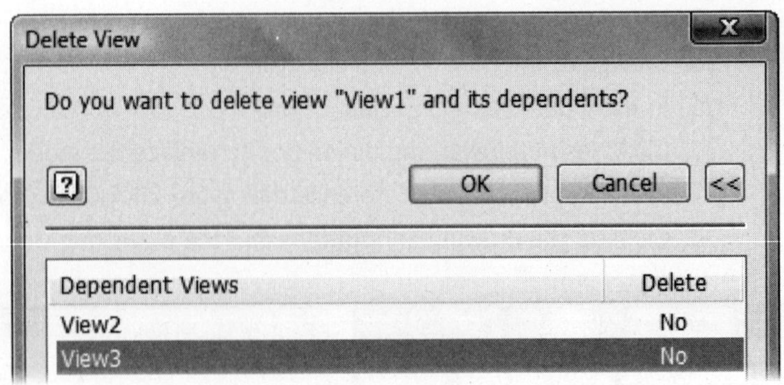

FIGURE 5.80

5. In the dialog box click OK to delete the base view and retain the two dependent views. Your drawing views should appear similar to the following image.

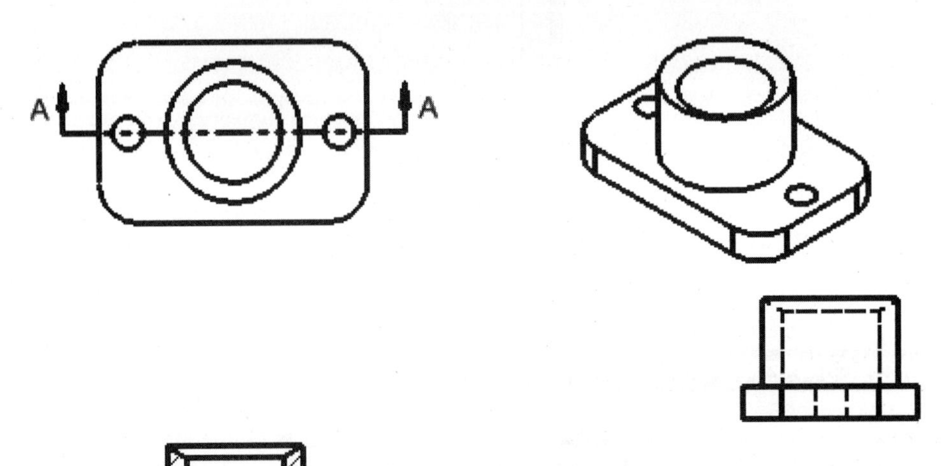

FIGURE 5.81

With the front view deleted, the section view needs to be identified as the new base view and aligned with the right-side view.

6. Right-click in the right-side view, and click Alignment > Horizontal from the menu.
7. Select the section view as the base view.
8. Select the section view, and drag the view vertically up to the area previously occupied by the front view. Notice how the right-side view remains aligned to the section view.
9. Right-click the isometric view, and select Alignment > In Position.
10. Select the section view as the base view. Move the section view, and notice that the isometric view now moves with the section view. Your drawing should appear similar to the following image.

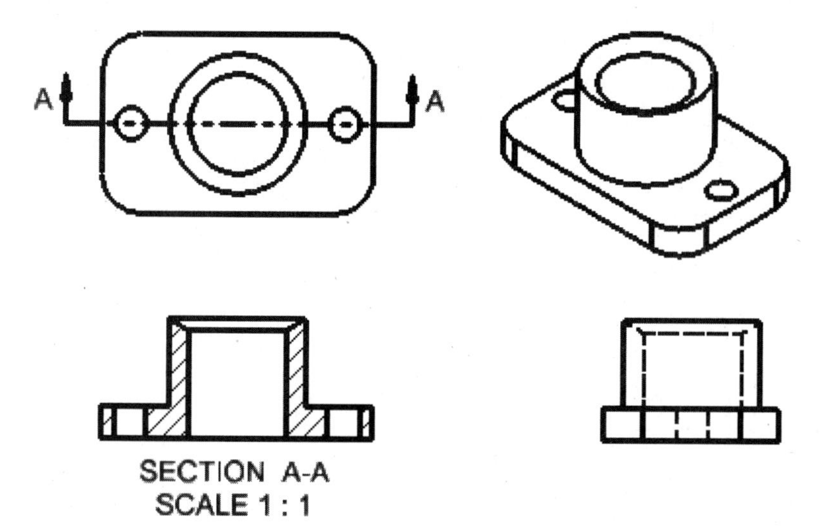

FIGURE 5.82

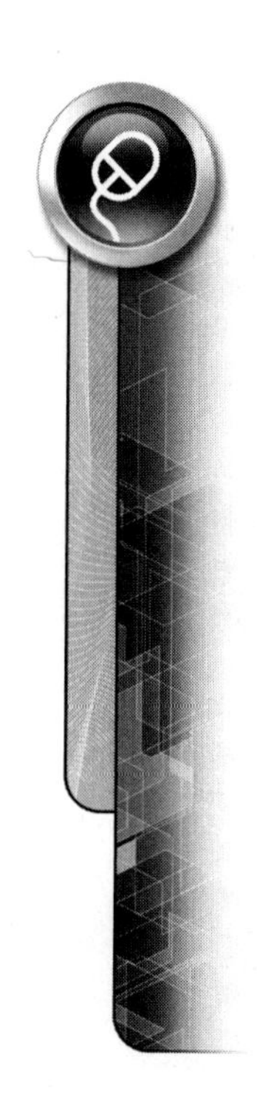

You now edit the section view hatch pattern to represent the material as bronze using the ANSI 33 hatch pattern.

11. Right-click on the hatch pattern in the section view, and choose Edit from the menu.
12. The Edit Hatch Pattern dialog box is displayed. Select ANSI 33 from the Pattern list, and click OK. The hatch pattern changes, as shown in the following image.

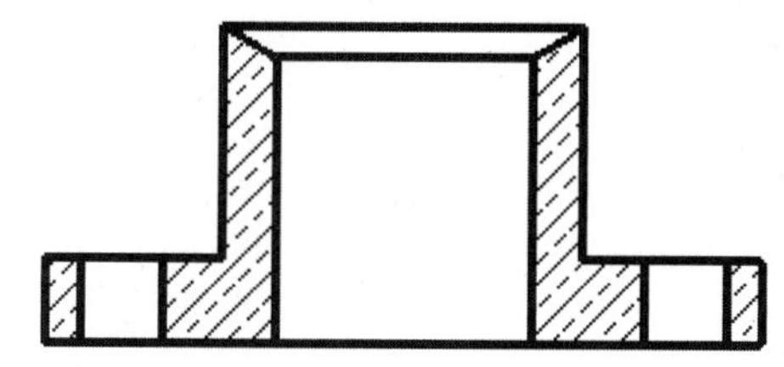

FIGURE 5.83

Lastly you replace the part with a similar part but the holes are in different locations.

13. Click the Replace Model Reference command on the Manage tab > Modify panel.
14. In the Replace Model Reference dialog box click on the only file in the list.
15. In the dialog box click the Navigate button and browse to the Chapter 05 folder and select the file *ESS_E05_04_Replace.ipt* and then click Open in the dialog box.
16. In the warning dialog box click Yes to replace the model.
17. In the Replace Model Reference dialog box click OK to complete the command.
18. Notice from the new part there are four holes centered on the fillets but the hatch pattern and view alignment was maintained.

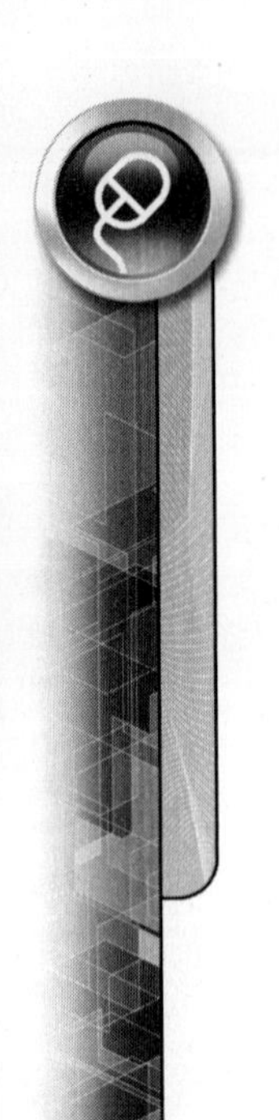

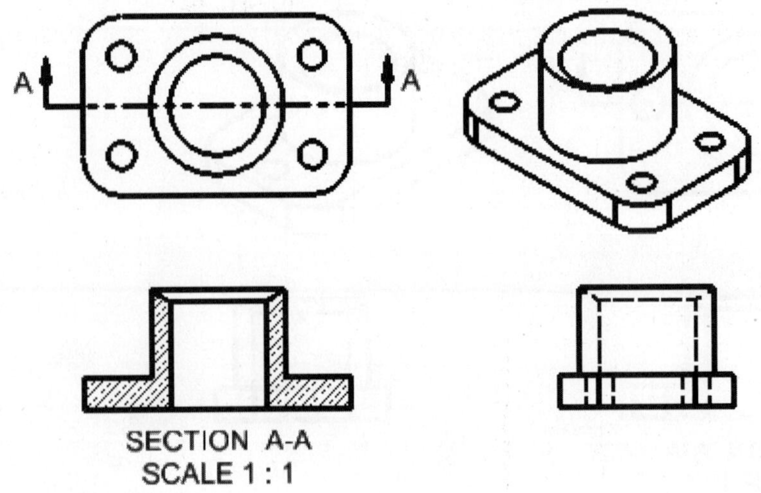

FIGURE 5.84

19. Close all open files. Do not save changes. End of exercise.

ANNOTATIONS

To complete an engineering drawing, you must add annotations such as centerlines, dimensions, surface texture symbols, welding symbols, geometric tolerance symbols, text, bills of materials, and balloons. How these objects appear is defined by styles. Styles will be covered as the first topic in this section.

DRAWING STANDARD AND STYLES

Each drawing is controlled by a drawing standard; the drawing standard controls overall drafting standards and which styles are available in a drawing. Only one standard can be active. Autodesk Inventor uses styles to control how objects appear. Styles control objects such as dimensions, centerlines, hole tables, and text. Styles are saved within an Autodesk Inventor file or to a project library location, and they can be saved to a network location so that many users can access the same styles. This section introduces you to styles. Once a style is used in a document, it is stored in the document. Changes made to a style are only saved in that document. The updated style library can be saved to the main style library but requires permission set in the local project file and rights of the folder that the styles are saved in. To help maintain standards, it is recommended that only a limited number of people have access to write to the style library. For more information on how to save styles to a style library consult the help system topic "Save styles to a style library".

Style Name/Value

Autodesk Inventor uses the style's name as the unique identifier of that style: only one name for the same style type can exist. For example, in a drawing, only one dimension style with a specific name "Default ANSI" can exist. An object can only be associated to one style.

Editing a Style

After dimensions have been placed, you may want to change how all of the dimension appear. Instead of changing each dimension, you can alter the dimension style or create a new style. The easiest way to edit a dimension style is to move the cursor over a dimension whose style you want to change and right-click and click Edit Dimension Style from the menu as shown in the following image.

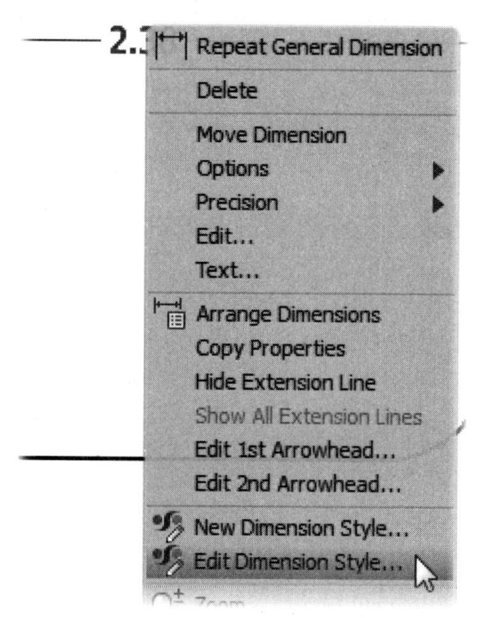

FIGURE 5.85

Then the Style and Standard Editor dialog box will appear as shown in the following image. Make changes to the style as needed and then click the Save button on the top of the dialog box.

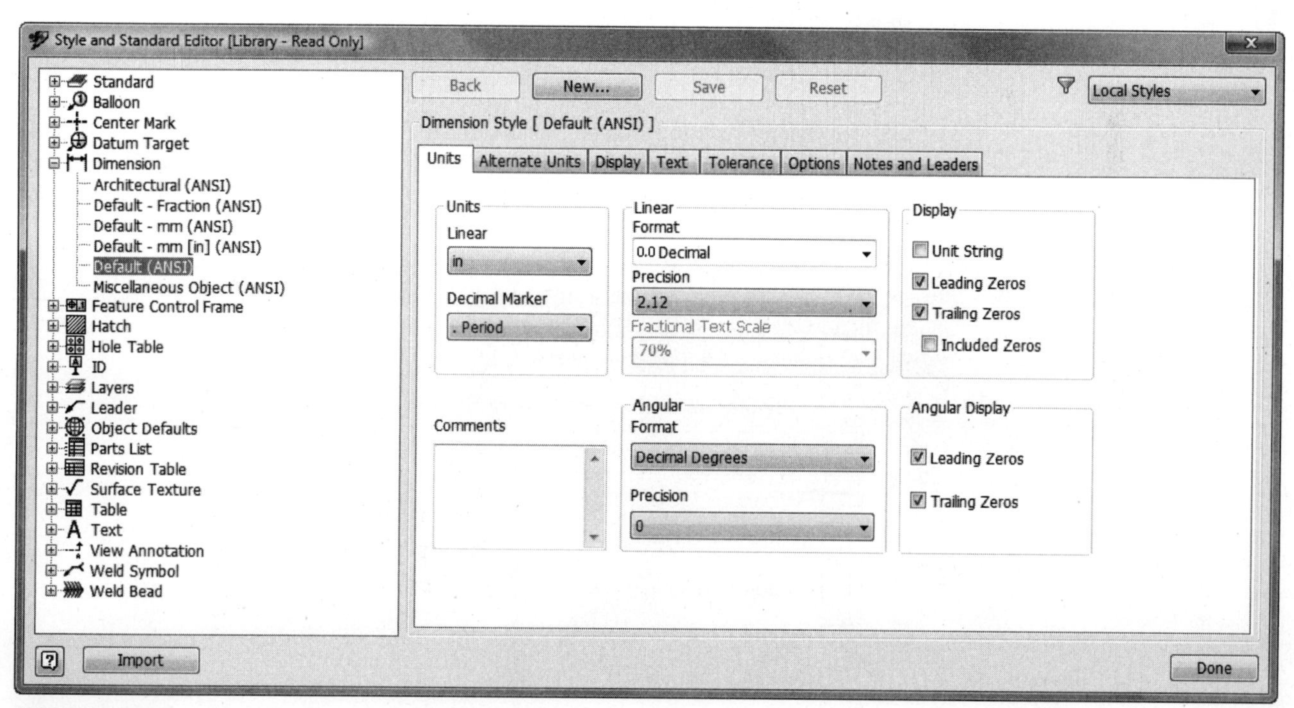

FIGURE 5.86

Creating a Style

Instead of editing a current dimension style, you can create a new style. To create a new style, follow these steps:

1. Click the Styles Editor command on the Manage tab > Styles and Standards panel.
2. Click and expand the style section for which you want to create a new style.
3. Right-click the style on which the new style will be based, and select New Style, as shown in the following image on the left. This image shows a new style being created from the Default (ANSI) dimension style.
4. The New Style Name dialog box will appear, as shown in the following image on the right. Enter a style name, and if you do not want the style to be used in the standard, uncheck Add to standard.
5. Make changes to the style, and save the changes.

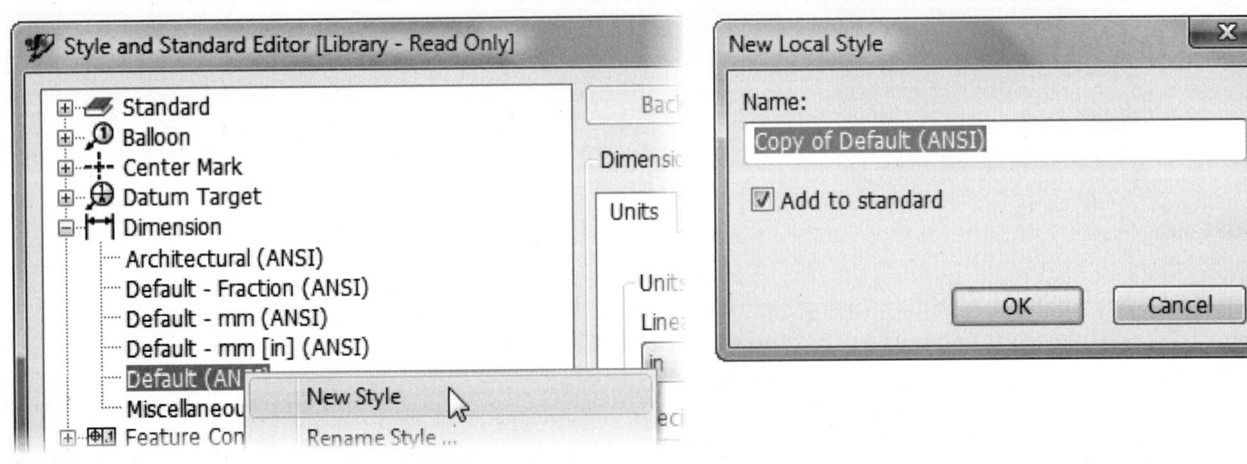

FIGURE 5.87

6. To change existing objects to a different style, select the object or objects whose style you want to change (selected objects need to be the same object type, i.e., all dimensions), and then select a style from the Style area on the Annotate tab > Format panel as shown in the following image.

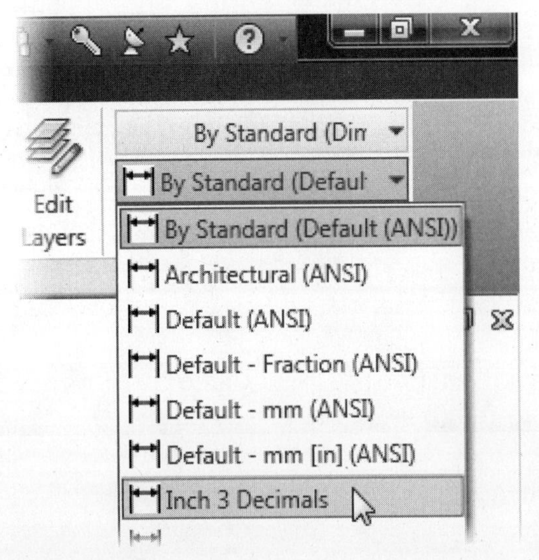

FIGURE 5.88

Style will be covered through the exercises in this chapter.

NOTE

Center Marks and Centerlines

When you need to annotate the centers of holes, circular edges, or the middle (center axis) of two lines, four methods allow you to construct the needed centerlines. Use the Center Mark, Centerline, Centerline Bisector, and Centered Pattern commands under the Annotate tab, as shown in the following image. The centerlines are associated to the geometry that you select when you create them. If the geometry changes or moves, the centerlines update automatically to reflect the change. This section outlines the steps for creating the different types of centerlines.

FIGURE 5.89

Centerline

To add a centerline, follow these steps and refer to the following image:

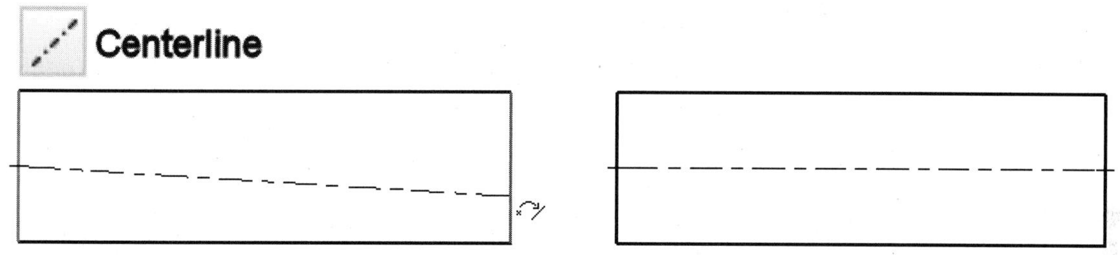

FIGURE 5.90

1. Click the Centerline command found on the Annotate tab > Symbols panel.
2. In the graphics window, select a piece of geometry for the start of the centerline.
3. Click a second piece of geometry for the ending location. You can also select multiple circles or points. If multiple points or circles are selected whose center points fall on a circle, a circle with the centerline line type (often referred to as a bolt circle) will be created.
4. Continue placing centerlines by selecting geometry.
5. Right-click, and select Create from the menu to create the centerline. The centerline will be attached to the midpoints of selected edges and center point of selected arcs and circles.
6. To complete the operation, right-click, and select Done from the menu or press the Esc key.

Centerline Bisector

To add a centerline bisector, follow the steps and refer to the following image:

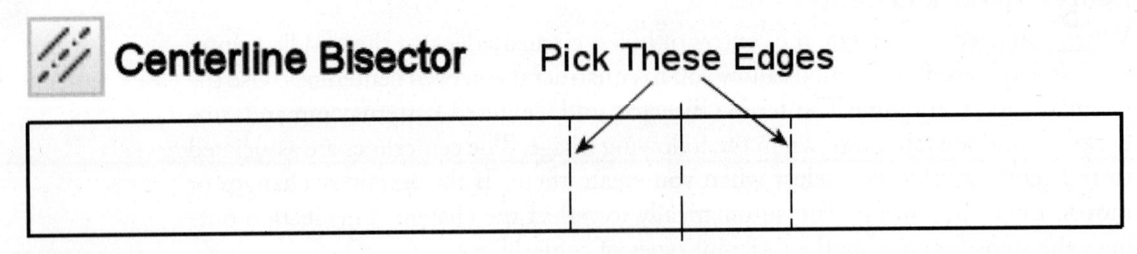

FIGURE 5.91

1. Click the Centerline Bisector command found on the Annotate tab > Symbols panel.
2. In the graphics window, select two lines between which you want to place the centerline bisector. The lines do not need to be parallel to each other.
3. Continue placing centerline bisectors by selecting geometry.
4. To complete the operation, right-click, and click Done from the menu.

Center Mark

To add a center mark, follow the steps and refer to the following image:

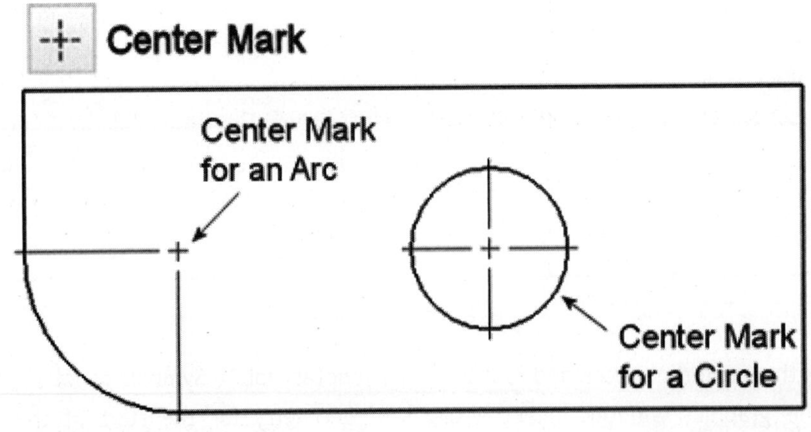

FIGURE 5.92

5. Click the Center Mark command found on the Annotate tab > Symbols panel.
6. In the graphics window, select a center point of a circle or arc or select a circle or arc geometry in which you want to place a center mark.
7. Continue placing center marks by selecting arcs and circles.
8. To complete the operation, right-click, and click Done from the menu.

Centered Pattern

To add a centered pattern, follow the steps and refer to the following image:

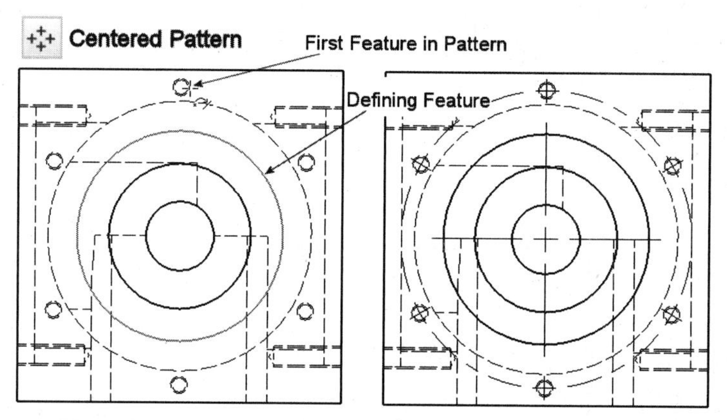

FIGURE 5.93

1. Click the Centered Pattern command found on the Annotate tab > Symbols panel.
2. In the graphics window, select the circular face whose center point will be used as the center of the circle that a set of center marks will fall on, commonly referred to a bolt circle.
3. Click the first feature of the pattern.
4. Continue selecting features in a clockwise or counter clockwise direction until all of the features are added to the selection set.
5. Right-click, and select Create from the menu to create the centered pattern.
6. To complete the operation, right-click, and select Done from the menu.

Automated Centerlines from Models

Creating centerlines automatically can eliminate a considerable amount of work. You can control which features automatically get centerlines and marks and in which views these occur.

You can create automated centerlines for drawing views by right-clicking on the view boundary or hold down the Ctrl key and select multiple view boundaries and right-click to display the menu and clicking Automated Centerlines from the menu as shown in the following image on the left. The Automated Centerlines dialog box will appear as shown in the following image on the right. Use this dialog box to set the type of feature(s) to which you will apply automated centerlines, such as holes, fillets, cylindrical features, etc., and to choose the projection type (plan or profile). The following sections discuss these actions in greater detail.

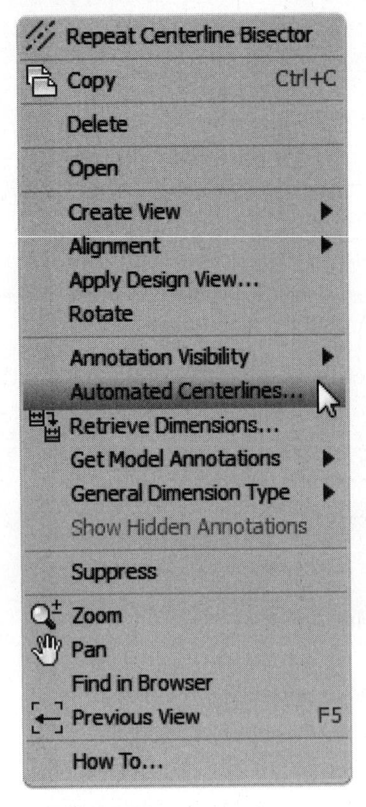

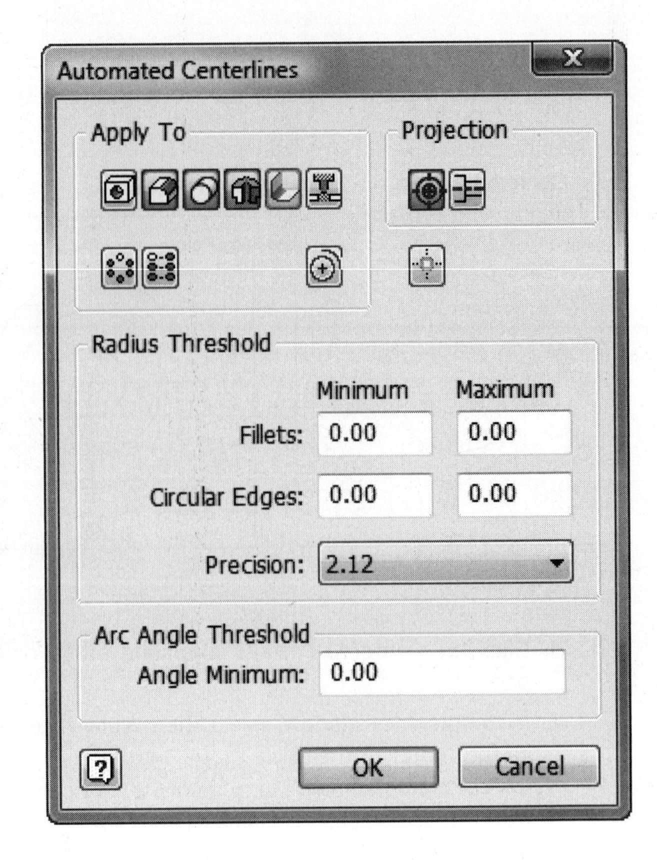

FIGURE 5.94

Apply To

This area controls the feature type to which you want to apply automated centerlines. Feature types include holes, fillets, cylindrical features, revolved features, circular patterns, rectangular patterns, sheet metal bends, punches, and circular sketched geometry. Click on the appropriate button to activate it, and automated centerlines will be applied to all features of that type in the drawing. You can click on multiple buttons to apply automated centerlines to multiple features. To disable centerlines in a feature, click on the feature button a second time.

Projection

Click on the projection buttons to apply automated centerlines to plan (axis normal) and/or profile (axis parallel) views.

Radius Threshold

Thresholds are minimum and maximum value settings and are provided for fillet features, arcs, and circles. Any object residing within a range should get the appropriate center mark. The values are based upon the model values, not the drawing values. This allows you to know what will or will not receive a centerline regardless of the view scale in the document. For example, if you set a minimum value of 0.50 for the fillet feature, a fillet that has a radius of 0.495 will not receive a center mark. A zero value on both threshold settings (min/max) denotes no restriction. This means that center marks will be placed on all fillets regardless of size.

Arc Angle Threshold

This option sets the minimum angle value for creating a center mark or centerline on circles, arcs, or ellipses.

NOTE

Automated centerlines can easily be created by preselecting the edges of multiple views and right-clicking the menu to display Automated Centerlines as shown in the following image.

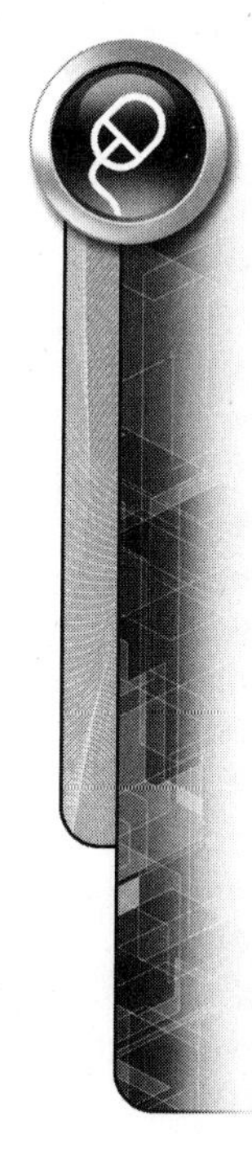

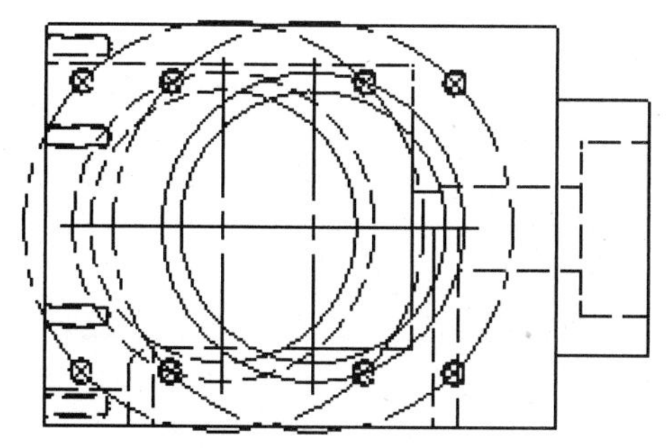

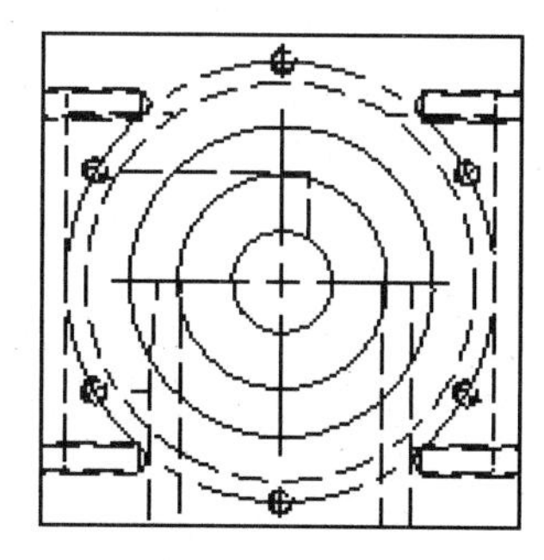

FIGURE 5.95

EXERCISE 5-5: ADDING CENTERLINES

In this exercise, you add centerlines, notes, and move dimensions between views to a drawing that you work on in Exercise 5-5.

Both model dimensions and drawing dimensions are used to document feature size.

1. Open the drawing *ESS_E05_05.dwg*. The drawing file contains four drawing views.
2. Centerlines and center marks need to be added to the drawing. Zoom in on the top view.
3. Click the Center Mark command on the Annotate tab > Symbols panel.
4. Select the left outside arc as shown in the following image on the left. Then right-click and click Done from the menu.
5. Notice that the centerline does not extend beyond the center of the circle, this is because the arc is only on the left side. Click on the centerline on the right side and drag it so it extends beyond the hidden line as shown in the following image in the middle. A center mark could have also been placed on the hole on the left side and the centerlines extended as needed.
6. Click the Centerline command on the Annotate tab > Symbols panel.
7. Select the two arcs that define the obround in the middle of the part as shown in the following image on the right. Then right-click and click Create from the menu.

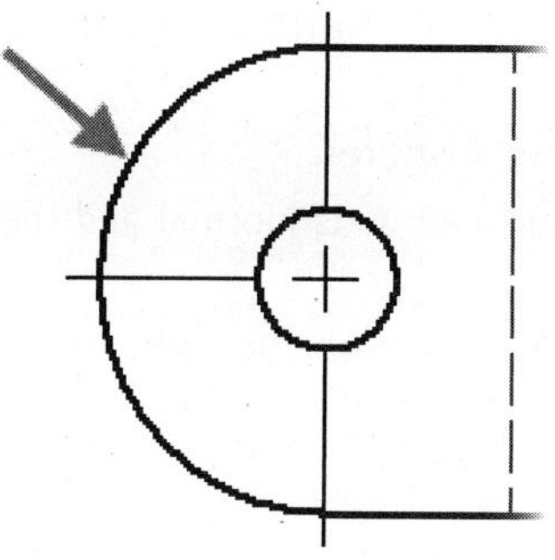

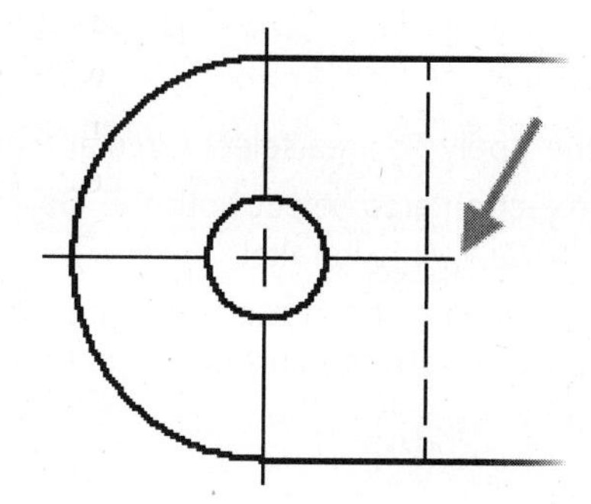

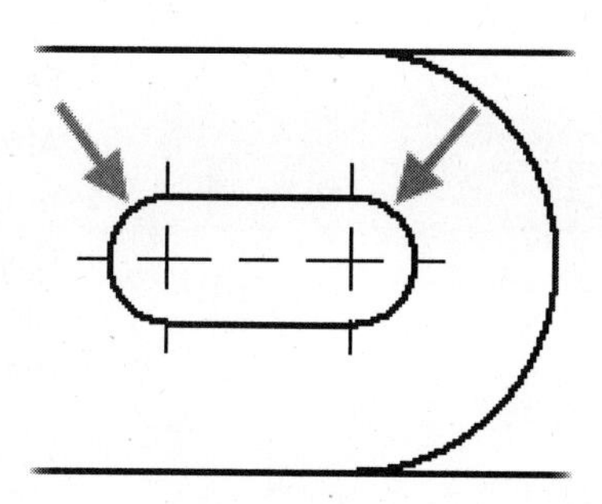

FIGURE 5.96

8. On the right-side of the top view you create a centered pattern (bolt circle). Click the Centered Pattern command on the Annotate tab > Symbols panel.
9. First select the larger center circle labeled in the following image on the left. The center of this circle will be the center of the centered pattern. Then select the six holes that surround the circle (you can click on the circle or on the circle's center point) and then right-click and click Create from the menu. The following image on the right shows all the center lines on the top view.

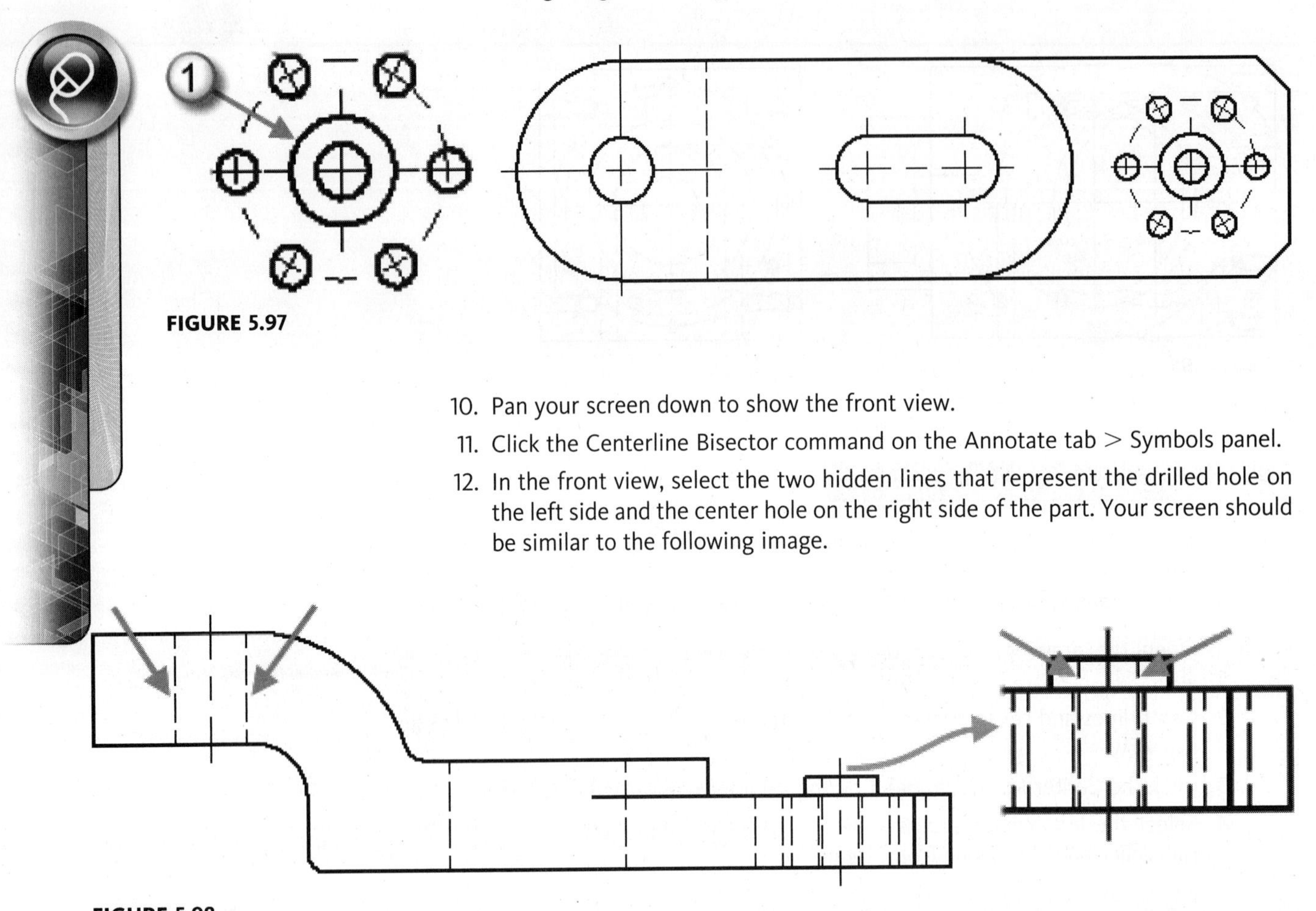

FIGURE 5.97

10. Pan your screen down to show the front view.
11. Click the Centerline Bisector command on the Annotate tab > Symbols panel.
12. In the front view, select the two hidden lines that represent the drilled hole on the left side and the center hole on the right side of the part. Your screen should be similar to the following image.

FIGURE 5.98

13. In the last portion of this exercise you create centerlines using the automated centerlines command. Close the drawing and reopen it or delete all the centerlines in the drawing.
14. Hold down the Ctrl key and click in the top, front, and side view and then right-click Automated Centerlines from the menu.
15. In the Automated Centerlines dialog box, make the following selections as shown in the following image on the left.
    - In the Apply To area select; Hole Features, Fillet Features, Cylindrical Features.
    - Also in the Apply To area select Circular Patterned Features.
    - In the Projection area select both the Objects in View, Axis Normal and the Objects in View, Axis Parallel.
    - Click OK to create the centerlines. The following image on the right shows the three views with the automatic centerlines.

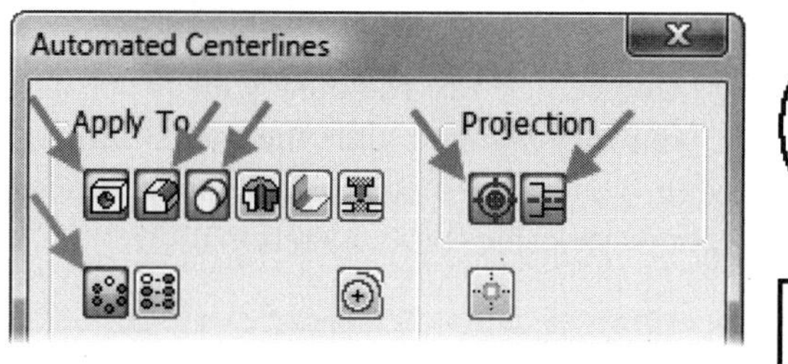

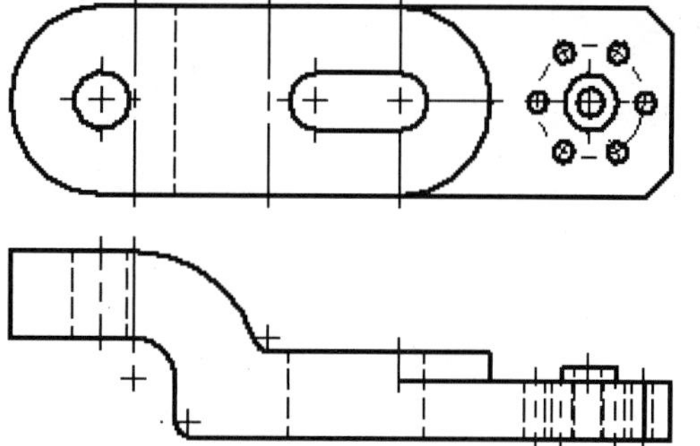

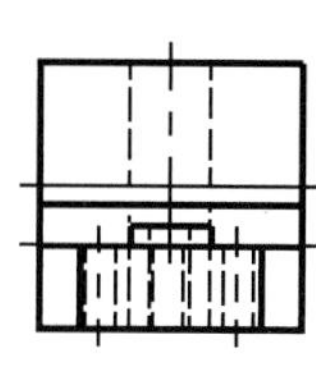

FIGURE 5.99

16. Zoom in and examine the centerlines that were created. If desired you can resize the centerlines by dragging their endpoints.
17. If the centerline and hidden lines do not display correctly you can adjust the global line scale. To adjust the global line scale, click the Styles Editor command from the Manage tab > Styles and Standards panel.
18. In the Styles and Standard Editor on the left side expand the Standard entry and click Default Standard (ANSI) (the bold text denotes that this is the active standard).
19. On the right side of the dialog box, enter a new value in the Global Line Scale area. To see the change in the centerlines and hidden lines in the graphics window, press Enter on the keyboard and then click the Save button on the top of the dialog box. Try different values, when done click Done on the lower right corner of the dialog box. Styles will be covered in more detail later in this chapter.

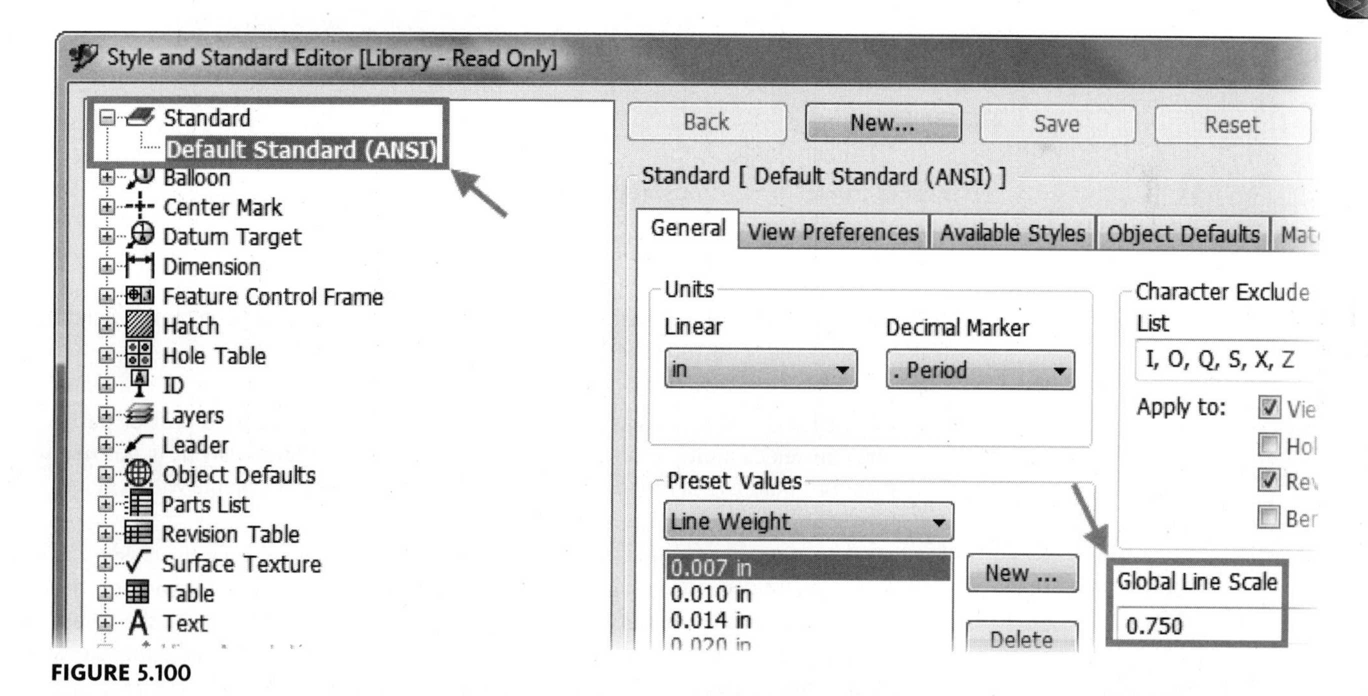

FIGURE 5.100

20. Close all open files. Do not save changes. End of exercise.

ADDING DIMENSIONS TO A VIEW

Once you have created the drawing view(s), and centerlines you mostly likely will add dimensional annotations to the drawing. You may want to alter the model dimensions, hide certain dimensions, add drawing (general) dimensions, or move dimensions to a new location. The following sections describe these operations.

Retrieving Model Dimensions

Model dimensions may not appear automatically when you create a drawing view. You can use the Retrieve Dimensions command to select valid model dimensions for display in a drawing view. Only those dimensions that you placed in the model on a plane parallel to the view will appear.

To activate this command, use one of the following three methods:

- Click on the Retrieve command located on the Annotate tab > Dimension panel as shown in the following image on the left.
- Right-click in the bounding area of a drawing view, and select Retrieve Dimensions as shown in the following image on the right.

The appropriate dimensions for the view that were used to create the part will appear.

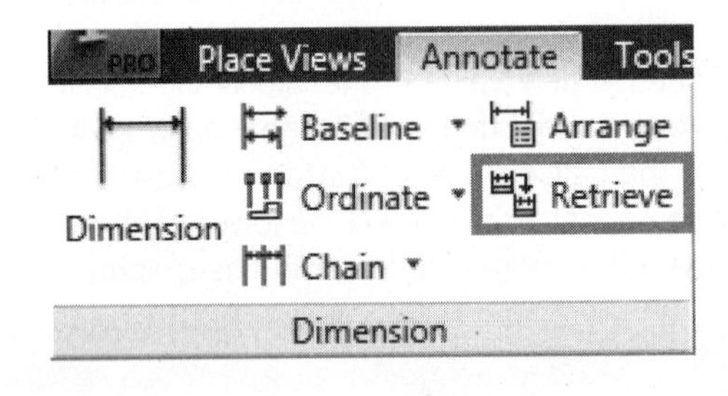

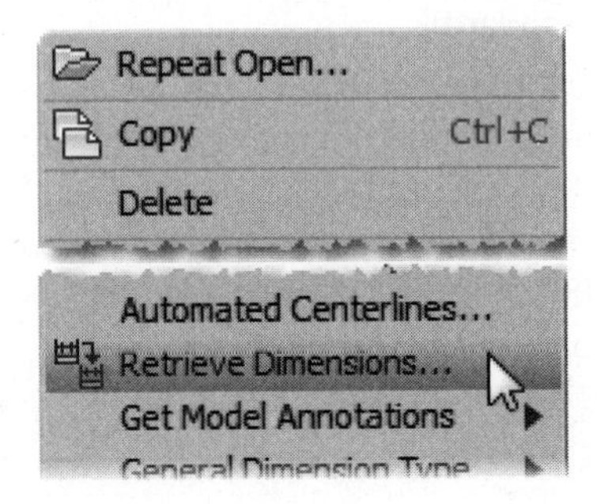

FIGURE 5.101

To retrieve model dimensions, follow these steps:

1. Click Retrieve on the Annotate tab > Dimension panel.
2. Select the drawing view in which to retrieve model dimensions.
3. To retrieve model dimensions based on one or more features, click the Select Features radio button. Then select the desired features.
4. If you want to retrieve model dimensions based on the entire part in a drawing view, click the Select Parts radio button. Then select the desired part or parts. You can select multiple objects by selecting them simultaneously. It is not necessary to hold the SHIFT or CTRL keys. You can also select objects with a selection window or crossing selection box.
5. Or to see all the dimensions that are available in the view, click the Select Dimensions button. If the dimensions were selected with the Feature or the model dimensions appear in the drawing view in preview mode, click the Select Dimensions button, and select the desired dimensions. The selected dimensions will be highlighted.
6. Click the Apply button to retrieve the model dimensions, and leave the dialog box active for further retrieval operations.
7. Click the OK button to retrieve the selected dimensions, and dismiss the Retrieve Dimensions dialog box.

The following image shows the effects of retrieving model dimensions using the Select Features mode. In this example, the large rectangle of the engine block was selected. This resulted in the horizontal and vertical model dimensions being retrieved.

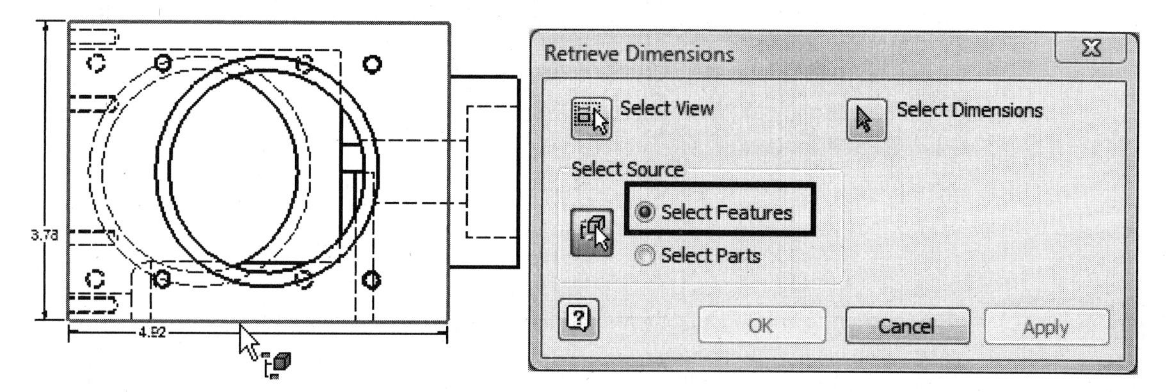

FIGURE 5.102

The following image shows the effects of retrieving model dimensions using the Select Parts mode. In this example, the entire engine block was selected. This resulted in all model dimensions parallel to this drawing view being retrieved.

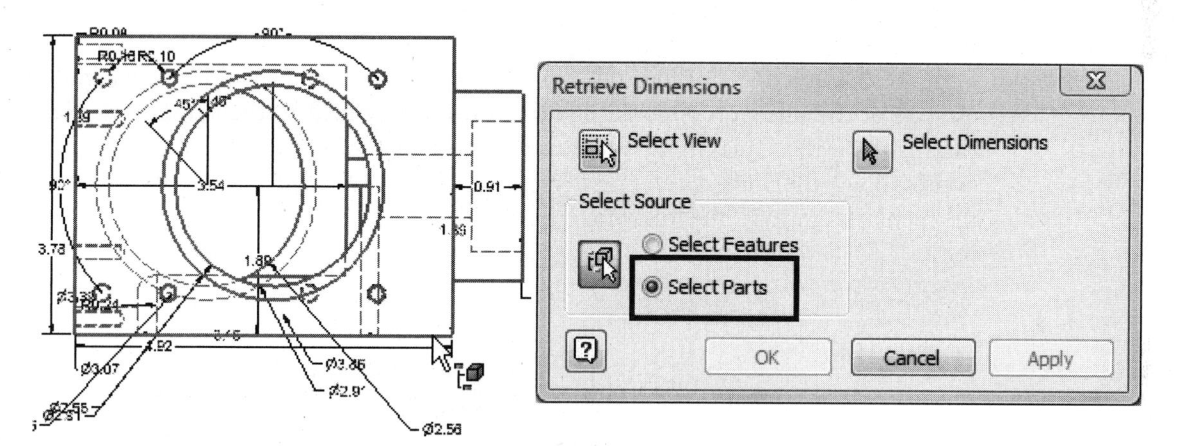

FIGURE 5.103

Changing Model Dimension Values

While working in a drawing view, you may find it necessary to change the model (parametric) dimensions of a part. You can open the part file, change a dimension's value, and save the part, and the change will then be reflected in the drawing views. You can also change a model dimension's value in the drawing view by right-clicking on the dimension and selecting Edit Model Dimension from the menu, as shown in the following image on the left.

The Edit Dimension text box will appear. Enter a new value and click the checkmark, as shown in the following image on the right, or press ENTER. The associated part will be updated and saved, and the associated drawing view(s) will be updated automatically to reflect the new value.

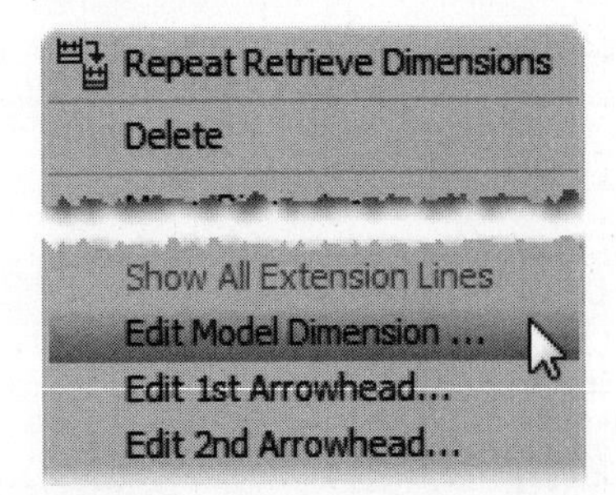

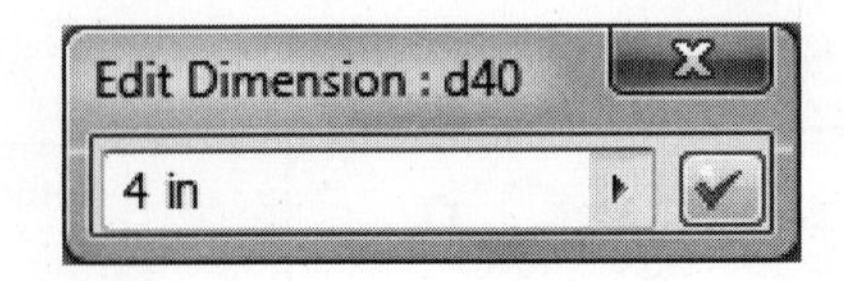

FIGURE 5.104

NOTE

The function to edit model dimensions in a drawing is an option that can be turned on or off when you install Autodesk Inventor. However, it is considered poor practice to make major dimension changes from the drawing. It is highly recommended that you make all dimension changes through the part model.

Selecting Dimensions

As you annotate a drawing you will need to select multiple dimensions. For example you would select multiple dimensions to arrange them, or change their dimension styles. To select multiple dimensions you can use the same selection techniques window, crossing, etc. that you learned about in the Selecting Objects section in Chapter 2. There are also four selection options that are available in a drawing. To use one of the selection options click the down arrow next to the current selection command on the Quick Access toolbar as shown in the following image on the left or press down the Shift key and right-click and click the desired selection option from the menu as shown in the following image on the right.

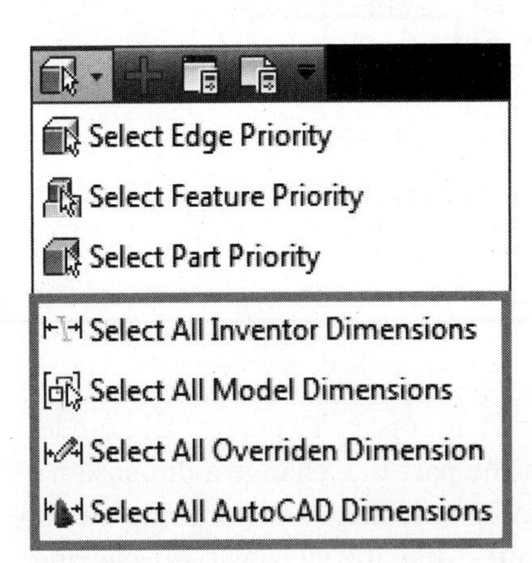

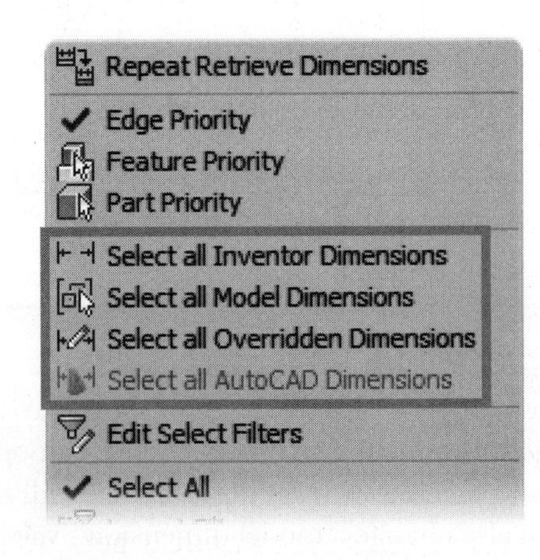

FIGURE 5.105

Below are explanations for the selection options.

- **Select All Inventor Dimensions:** This option selects all dimensions that are in the current sheet.
- **Select All Model Dimensions:** This option selects all model dimensions that were retrieved in the current sheet.

- **Select All Overridden Dimensions:** This option selects all dimensions whose values have been changed or values are hidden.
- **Select All AutoCAD Dimensions:** This option is only available for DWG files and selects all AutoCAD dimensions in the current sheet.

Auto-Arrange Model Dimensions

After retrieving model dimensions, the results are usually the display of dimensions that need to be rearranged in order to conform to standard dimensioning practices. You can manually reposition the dimensions or a better technique would be to have Inventor perform an automatic arrangement of the linear dimensions. To begin this process select the dimensions using any selection technique or selection option. With all of the dimensions selected, click the Arrange command from the Annotate tab > Dimension panel as shown in the following image on the left or right-click in the graphics window and click Arrange Dimensions from the menu as shown in the following image on the right. The linear dimensions will be arranged according to the current dimension style. If needed you will have to manually arrange dimensions like radial and diameter dimensions.

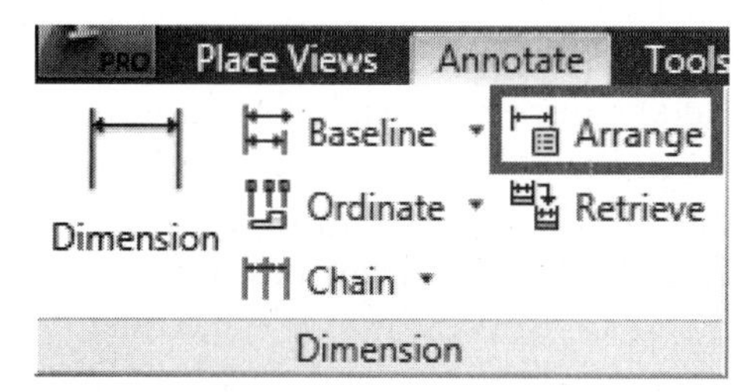

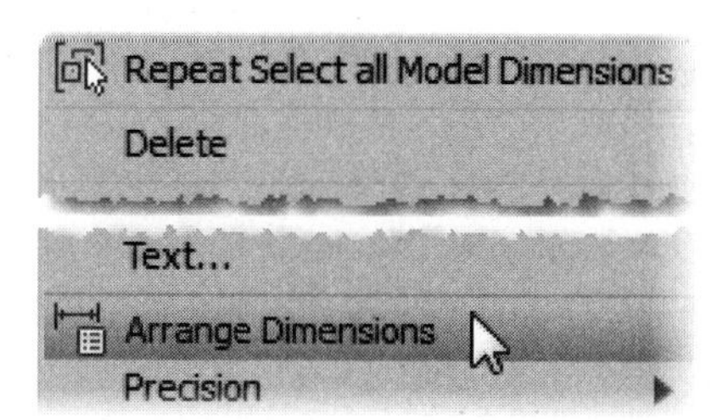

FIGURE 5.106

Moving and Centering Dimension Text

To move a dimension by either lengthening or shortening the extension lines or moving the text, position the cursor over the dimension until it becomes highlighted, as shown in the following image on the left. An icon consisting of four diagonal arrows will appear attached to your cursor. Notice the appearance of a centerline. This represents the center of the dimension text. You can drag the dimension text up, down, right, or left in order to better position it.

If you want to re-center the dimension text, slide the text toward the centerline. The dimension will snap to the centerline. In the following image on the right, the centerline changes to a dotted line to signify that the dimension text is centered. This centerline action will work on any type of linear dimension text including horizontal, vertical, and aligned. This feature is not active when repositioning the text of radius, diameter, or hole note dimensions.

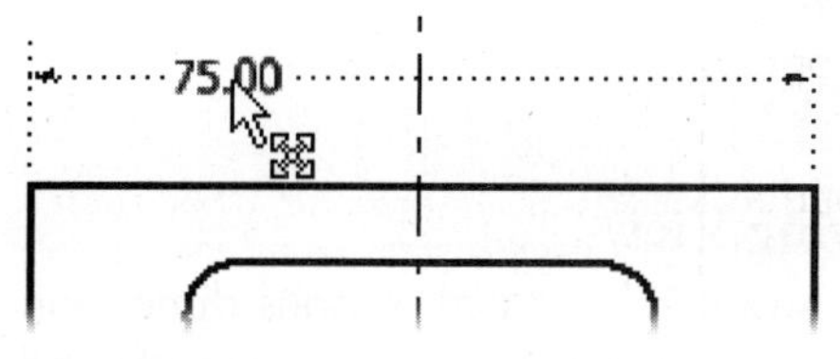

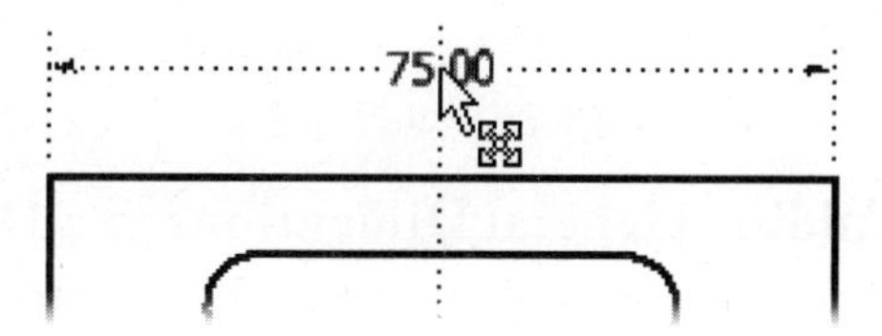

FIGURE 5.107

Creating General Dimensions

After laying out the drawing views, you may find that a dimension other than an existing model dimension or an additional dimension is required to better define the part. You can go back and add a parametric dimension to the sketch if the part was underconstrained, add a driven dimension if the sketch was fully constrained, or add a general dimension to the drawing view. A general dimension is not a parametric dimension; it is associative to the geometry to which it is referenced. The general dimension reflects the size of the geometry being dimensioned. After you create a general dimension, and the value of the geometry that you dimensioned changes, the general dimension will be updated to reflect the change. You add a general dimension by using the General Dimension command found on the Annotate tab > Dimension panel as shown in the following image on the left, right-click in a blank area in the graphics screen and click the General Dimension command from the marking as shown in the following image in the middle, press the D key. You create a general dimension in the same way you would create a parametric dimension on a part's sketch. The dimensions you create in a drawing follow a similar process that you did when creating a dimension in a part's sketch, except when a drawing dimension is placed, the Edit Dimension dialog box will appear as shown in the following image on the right. In the Edit Dimension dialog box the <<>> represents that value of the geometry being dimensioned. You can add text or symbols before or after the <<>>. You can also add a tolerance or add an inspection symbol to the dimension from the two other tabs. By default, when a drawing dimension is placed the Edit Dimension dialog box will appear. You can turn off this behavior by unchecking the Edit dimension when creating an option at the bottom of the dialog box or from the Application Options > Drawing tab. You can edit an existing drawing dimension by double-clicking on a dimension or right-clicking on a dimension and click Edit from the menu. The Edit Dimension dialog box will appear. Edit the dimension as needed.

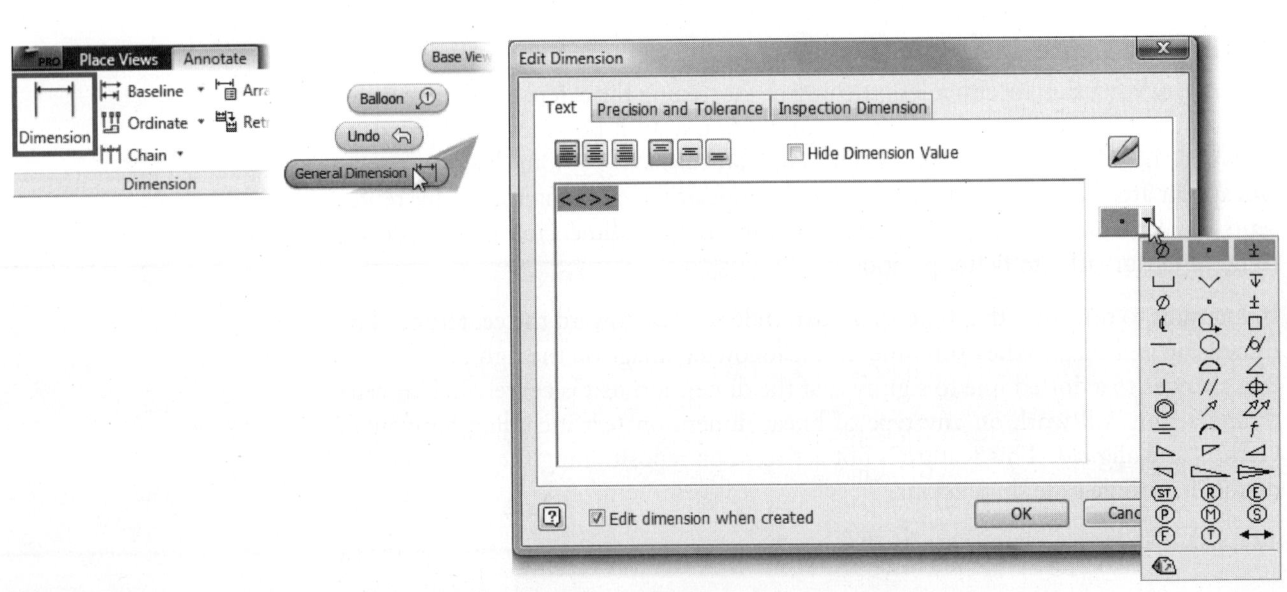

FIGURE 5.108

Adding General Dimensions to a Drawing View

The following image on the right illustrates a simple object with various dimension types. General dimensions can take the form of linear dimensions as shown at (A) and (B), diameter dimensions as shown at (C), radius dimensions as shown at (D), and angular dimensions as shown at (E). When using the General Dimension command, Autodesk Inventor automatically chooses the dimension type depending on the object you chose. In this example, the counterbored hole is dimensioned using a special command called a hole note. This command will be described later in this chapter.

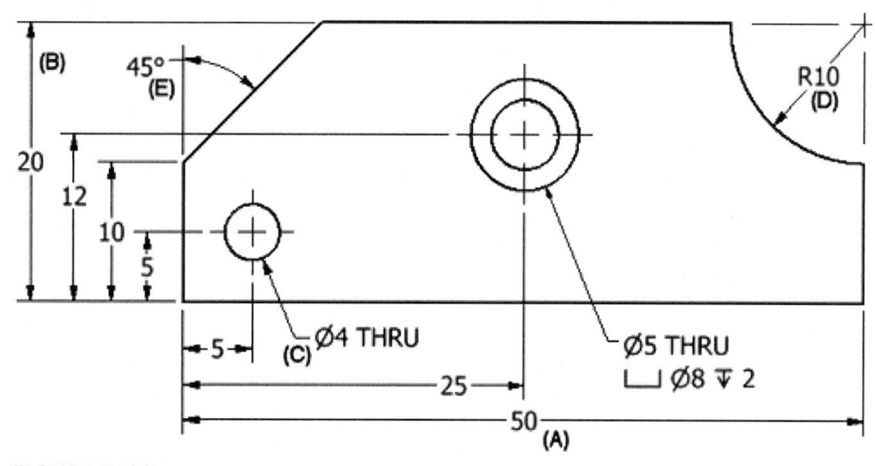

FIGURE 5.109

Adding Dimensions to Isometric Views

You can retrieve model dimensions or use of the General Dimension command to add dimensions to isometric drawing views as well as to standard orthographic views. When placing a dimension on an isometric view, the dimension text, dimension lines, extension lines, and arrows are all displayed as oblique and are aligned to the geometry being dimensioned. The following image shows a typical isometric view complete with oblique dimensions.

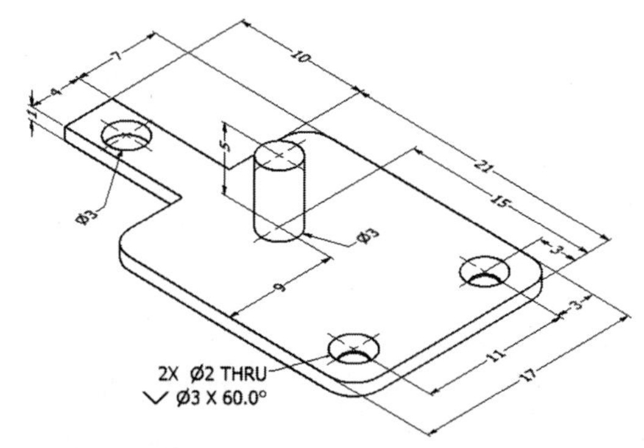

FIGURE 5.110

Depending on the object being dimensioned, additional controls are available to manipulate the isometric dimension being created. Once edges or points in a drawing are selected, the dimension displays based on an annotation plane, as shown in the following image on the left. If more than one annotation plane exists, you can toggle between these planes by pressing the spacebar. The results are shown in the following image on the right.

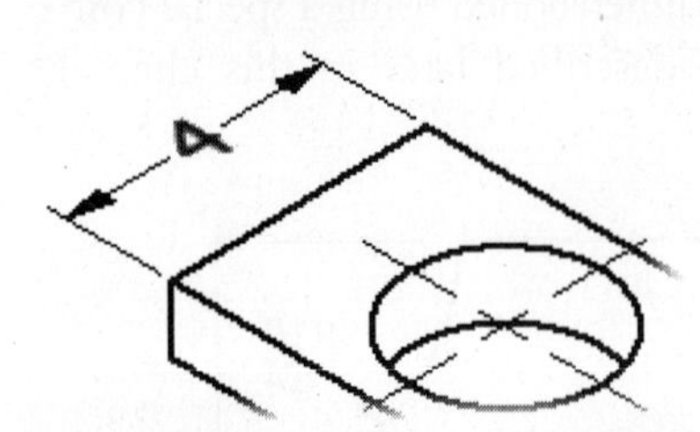

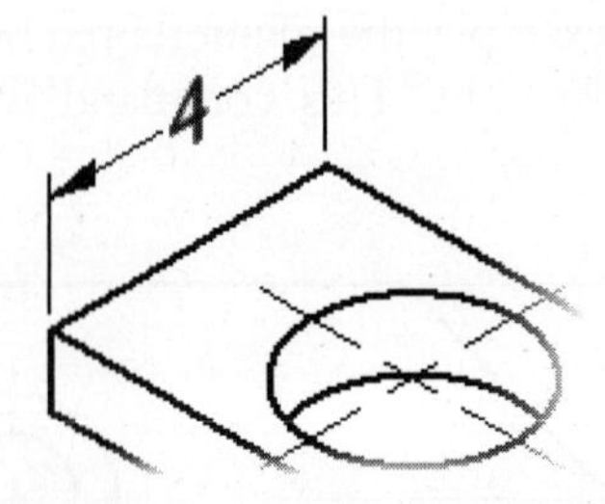

FIGURE 5.111

You can select other annotation planes by right-clicking when placing the dimension, selecting Annotation Plane, and selecting one of the following options:

- **Show All Part Work Planes**
- **Displays all default and user-defined work planes. These planes can then be selected as new annotation planes.**
- **Show All Visible Work Planes**
- **Displays only those work planes that are currently visible in the model. These planes can then be selected as new annotation planes.**
- **Use Sheet Plane**
- **Uses the current drawing sheet as a plane for placing the isometric dimension. Linear dimensions placed on the sheet plane are not true values.**

EXERCISE 5-6: ADDING DIMENSIONS

In this exercise, you add dimensions to a drawing of a clamp and edit and create a dimensions style.

1. Open the drawing *ESS_E05_06.dwg*. The drawing file contains four drawing views with centerlines.
2. In this step, you retrieve model dimensions for use in the drawing view. Begin by zooming in on the front view.
3. Right-click in the front view, and click Retrieve Dimensions from the menu. This will launch the Retrieve Dimensions dialog box.
4. In this dialog box, click the Select Dimensions button. The model dimensions will appear on the view.
5. Drag a selection window around all dimensions. This will select the dimensions to retrieve. When finished, click the OK button. The model dimensions that are planar to the view are displayed, as shown in the following image.

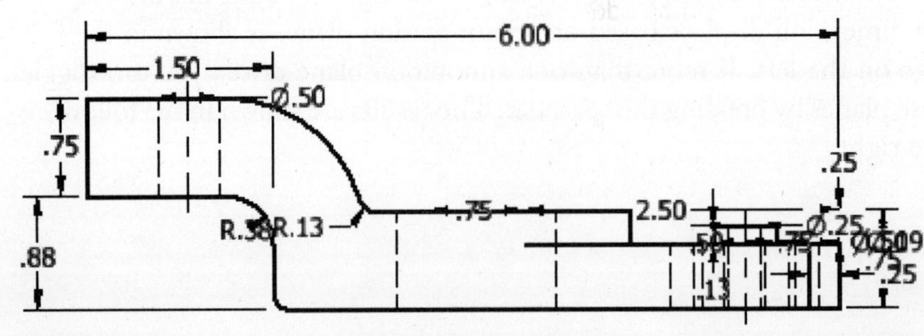

FIGURE 5.112

6. Instead of manually dragging dimensions to new locations, click the Arrange command on the Annotate tab > Dimension panel.
7. Drag a selection window around all dimensions. This will select the dimensions to arrange. When finished, right-click and click Done from the menu. The dimensions are arranged according to the settings in the current dimension style.
8. There are nine dimensions that were retrieved that are not deleted. Hold down the Ctrl key and select the nine dimensions that are highlighted in the following image and press the Delete key on the keyboard or right-click and click Delete from the menu. The dimensions that were deleted can be retrieved again if needed.

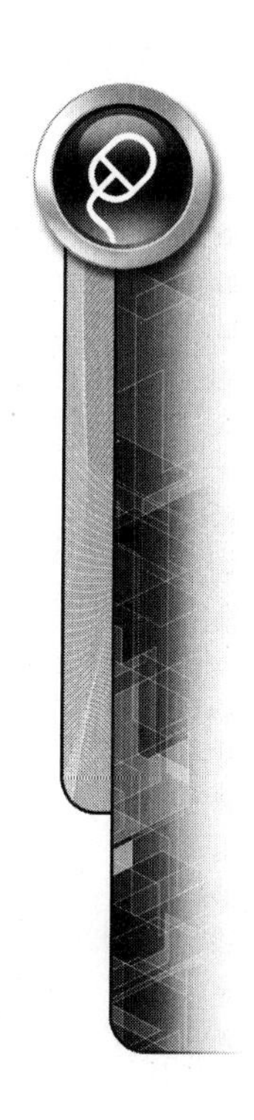

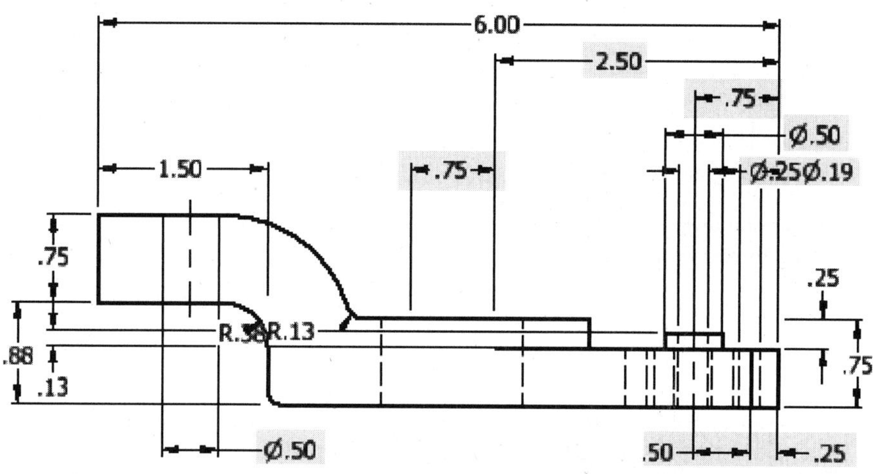

FIGURE 5.113

9. Use the arrange dimension command again to arrange the dimensions.

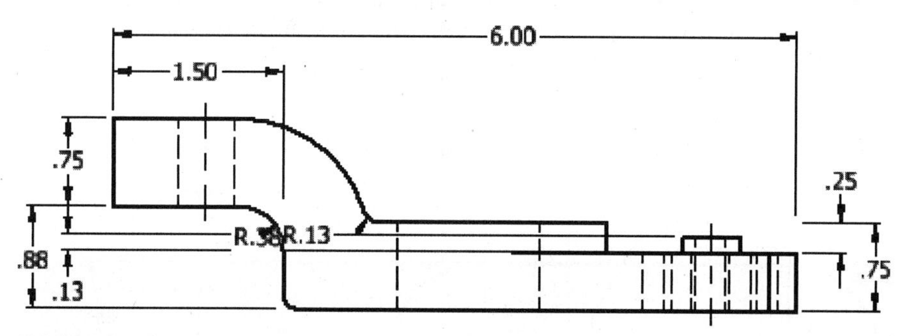

FIGURE 5.114

10. A few of the dimensions need to be repositioned. Click and drag the dimensions to locations similar to what is shown in the following image. Note that additional dimensions could be added to the view as will be described in the next few steps.

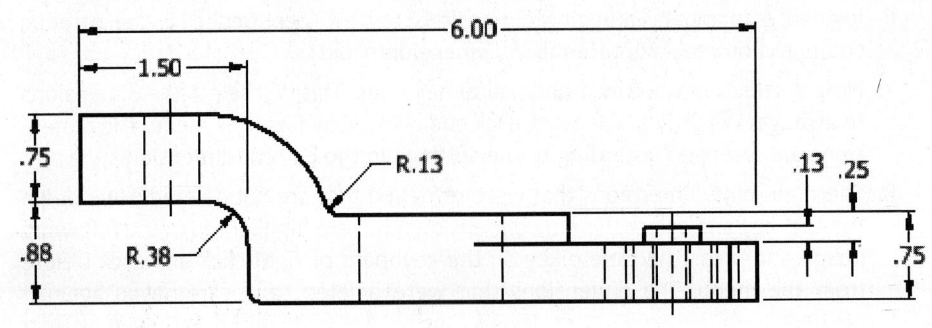

FIGURE 5.115

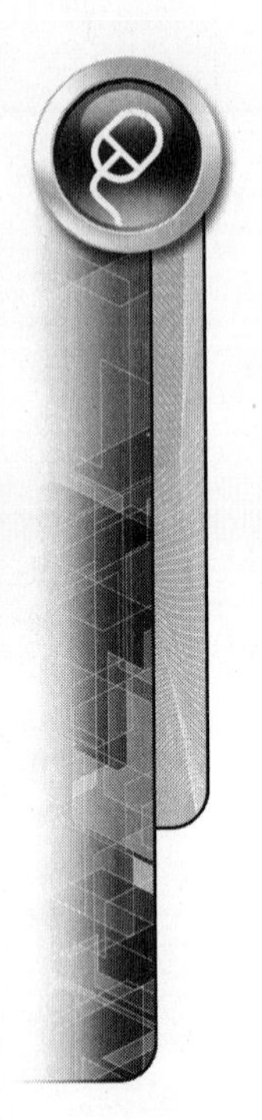

NOTE

To reposition dimension text, make sure that no command is active.

Click a dimension text object, and drag it into position. The dimension will be highlighted when it is a preset distance from the model. This distance is set in the dimension style.

You can also click and drag an endpoint of an extension line; you want to do this if an extension line is going through existing lines.

Radial dimensions can be repositioned by selecting the handle at the annotation end of the leader.

11. In the top view you could retrieve dimensions. Instead you will manually add dimensions to the top view. Pan to display the top view.
12. Click the General Dimension command on the Annotate tab > Dimension panel.
13. Place the nine dimensions as shown in the following image. If needed, you can double-click on a dimension to edit a dimension's text. To change a radius dimension to a diameter dimension or vice versa you must right-click and select the dimension type before the dimension is placed.

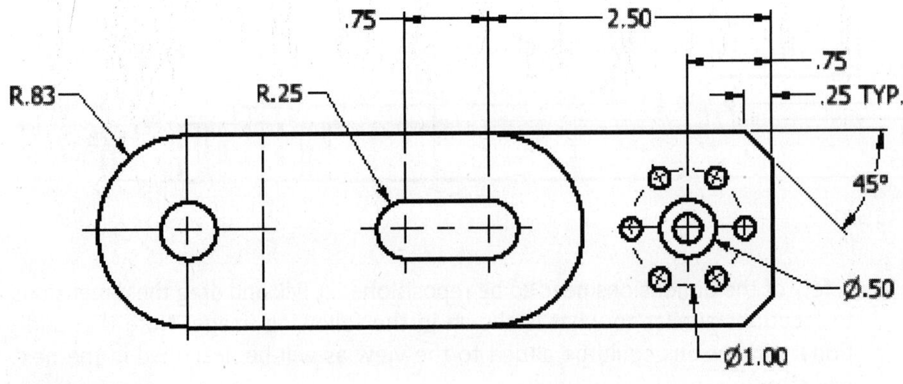

FIGURE 5.116

14. When the .25 TYP. horizontal dimension is created, type in **TYP**. after the <<>> as shown in the following image.

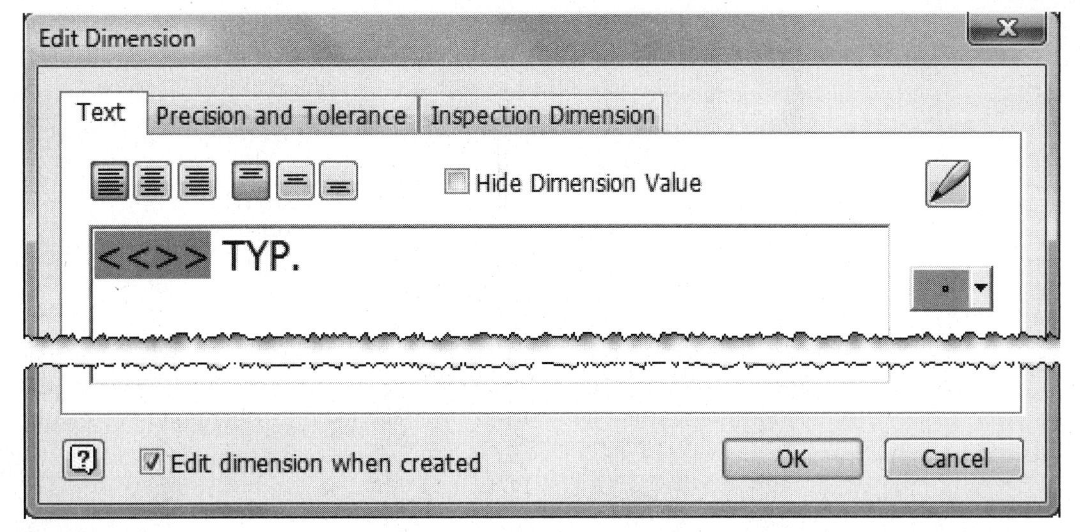

FIGURE 5.117

15. Next you change a model dimension's value via the drawing. In the front view, right-click on the 6.00 dimension and click Edit Model Dimension from the menu. Enter a value of 7 and then click on the green check mark in the Edit Dimension dialog box.

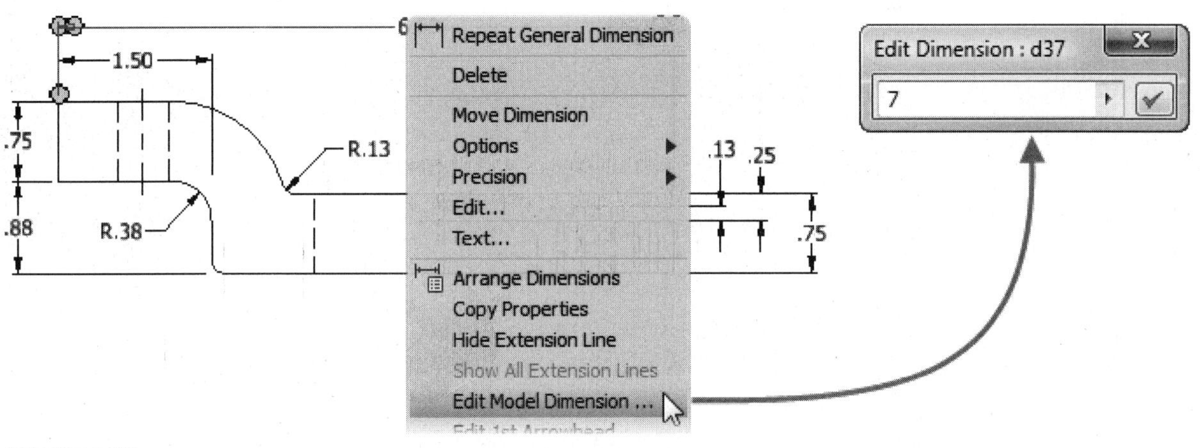

FIGURE 5.118

16. Notice how the drawing views and dimensions update to reflect the new value.
17. Next you will open the part file and change the horizontal dimension back to its original value. Move the cursor into a blank area in one of the drawing views and click Open from the menu as shown in the following image on the left.
18. Edit Sketch1 under Extrusion1 as shown in the following image on the right.

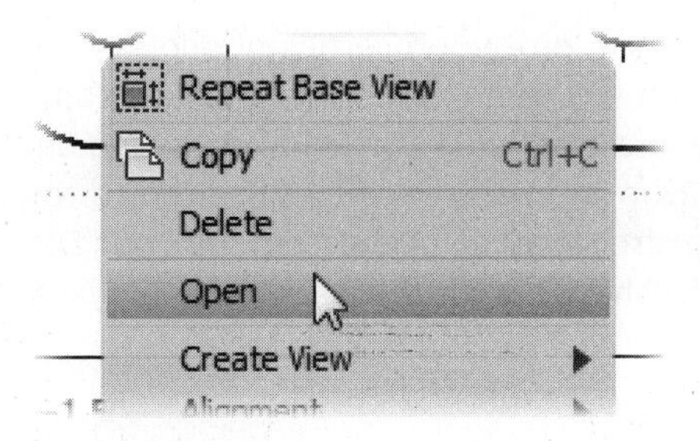

FIGURE 5.119

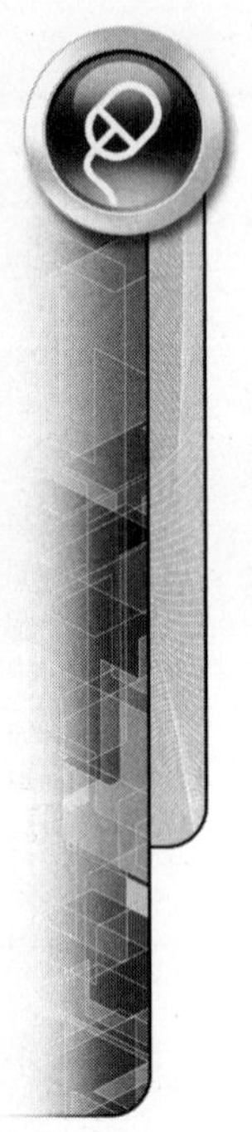

19. Double-click on the 7.000 horizontal dimension and enter a value **6** and then press ENTER on the keyboard.
20. Finish the sketch and then save and close the part file.
21. The drawing should be the current file; if not, make it the active file by clicking on the *ESS_E05_06.dwg* tab.
22. The drawing views should update to reflect the updated model.
23. Next you will change the dimension style. Move the cursor over any dimension and right-click and click Edit Dimension Style from the menu.
24. In the Style and Standard Editor the Default (ANSI) style should be current. From the Units tab, change the Precision to 1.1 (represents one decimal place) and in the Display area, check the Leading Zeros as shown in the following image and then click Save on the top of the dialog box.

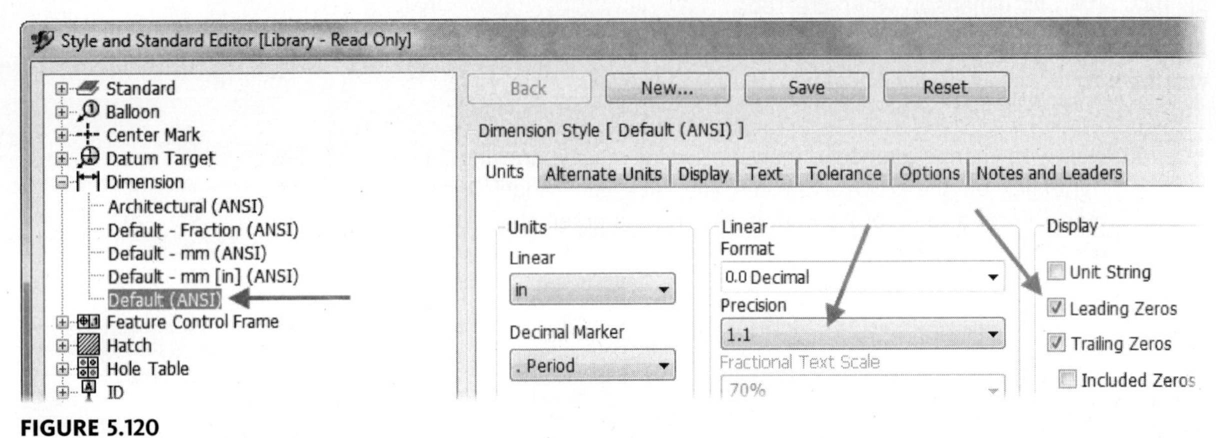

FIGURE 5.120

25. To close the dialog box, click Done on the bottom-right of the dialog box. All the dimensions should update to a single decimal place and a leading zero on the dimensions smaller than one inch in length or radius.
26. Next you create a new dimension style. Right-click on the 6.0 horizontal dimension and click New Dimension Style from the menu. The new dimension style will take on the properties of the dimension style of the selected dimension.
27. In the New Dimension Style dialog box in the upper left corner, enter a name for the new style as **3 Decimal (ANSI)**; from the Units tab change the Precision to 3.123. Click OK in the dialog box to create the dimension style.
28. The 6.0 dimension should now be 6.000 but the other dimensions are still at a single decimal place. Only 6.000 dimension is using the new 3 Decimal place dimension style.
29. Next you change a few dimensions to the new style. In the drawing, hold down the CTRL key and select a few dimensions and from the Annotate tab > Format panel select the 3 Decimal (ANSI) dimension style from the list as shown in the following image on the left.
30. Another option to change dimension to take on the properties of another dimension is to right-click on the dimension whose properties you want to copy and click Copy Properties from the menu as shown in the following image on the right. Then select the dimension(s) who you want the properties copied to. You can also right-click and click Settings from the menu and from the Copy Dimension Properties dialog box you can uncheck the properties you do not want copied.

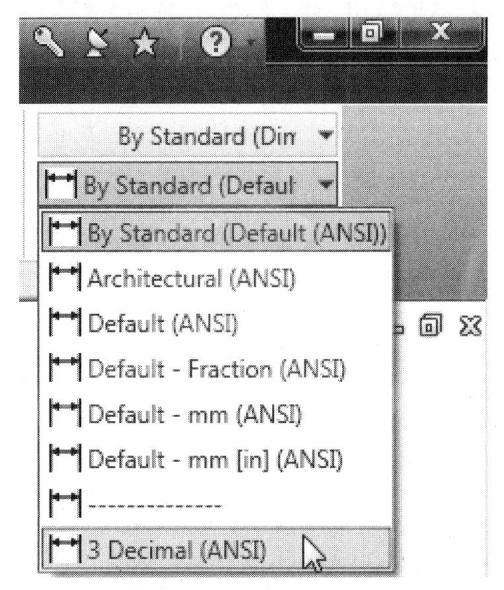

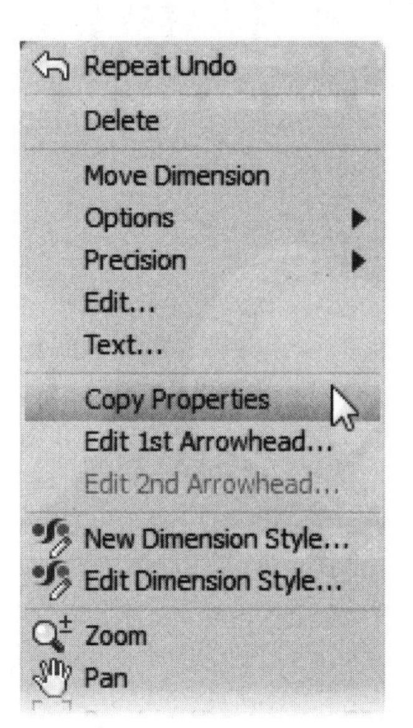

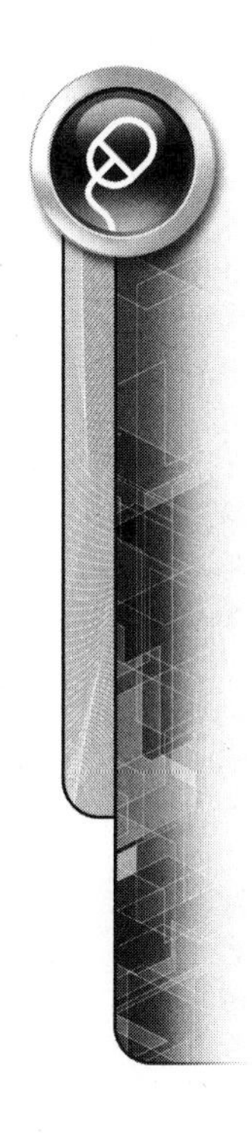

FIGURE 5.121

31. Practice editing and creating new dimension styles.
32. In Exercise 5-8, you will continue to work on a similar drawing where you will add: hole notes, texture symbol, and a text note.
33. Close all open files. Do not save changes. End of exercise.

CREATING BASELINE DIMENSIONS

To add multiple drawing general dimensions to a drawing view in a single operation, use the Baseline Dimension command, which will place a baseline dimension about the selected geometry. The dimensions can be either horizontal or vertical. Two commands are available to assist with the creation of these dimensions: Baseline Dimension and Baseline Dimension Set.

The Baseline Dimension command allows you to create a group of horizontal or vertical dimensions from a common origin. However, the group of dimensions are not considered one group of objects; rather, each dimension that makes up a Baseline Dimension is considered a single dimension.

To create and edit baseline dimensions, follow these steps:

1. Select the Baseline Dimension command from the Annotate tab > Dimension panel as shown in the following image.

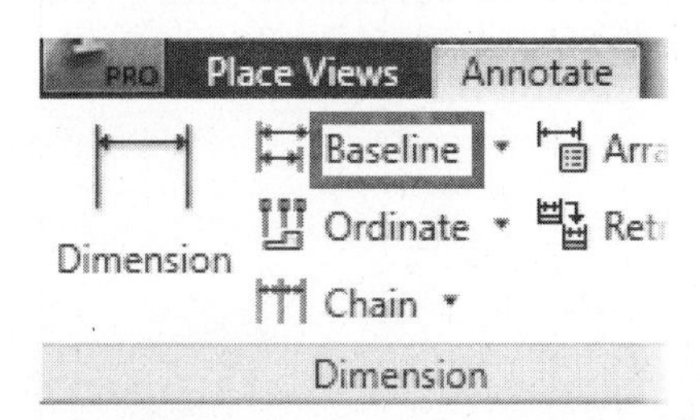

FIGURE 5.122

2. Individually select or drag a selection window around the geometry that you want to dimension. To window select, move the cursor into position where the

first point of the box will be. Press and hold down the left mouse button, and move the cursor so the preview box encompasses the geometry that you want to dimension, as shown in the following image on the left, and then release the mouse button.

3. When you are finished selecting geometry, right-click, and select Continue from the menu.
4. Move the cursor to a position where the dimensions will be placed, as shown in the following image on the right. When the dimensions are in the correct place, click the point with the left mouse button to anchor the dimensions.
5. To change the origin of the dimensions, move the cursor over the extension line that will be the origin, right-click, and click Make origin from the menu.
6. To complete the operation, right-click and select Create from the menu.
7. After creating dimensions using the Baseline Dimension command, the same editing operations can be performed on these dimensions as general dimensions. To perform these edits, move the cursor over the baseline dimension, and small green circles will appear on the dimension. Right-clicking will display the menu, as shown in the following image on the right.

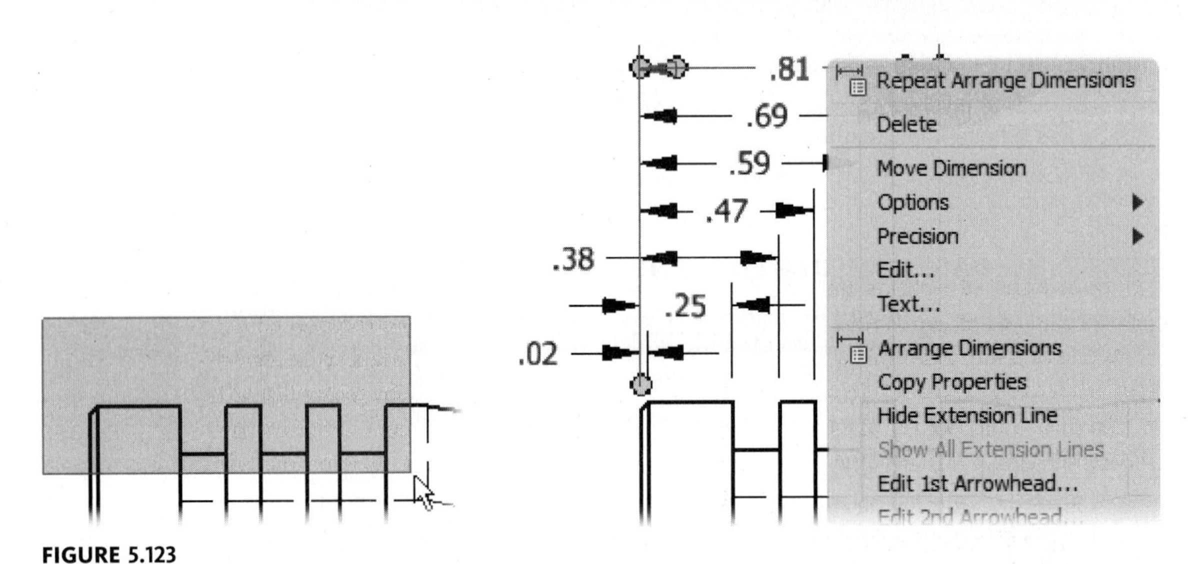

FIGURE 5.123

Creating a Baseline Dimension Set

Creating a Baseline Dimension Set is similar to the Baseline Dimension as explained in the previous section. With the Baseline Dimension Set command, however, dimensions are grouped together into a set. Select the Baseline Dimension Set command from the Annotate tab > Dimension panel > click the down arrow next to the Baseline Dimension command, as shown in the following image.

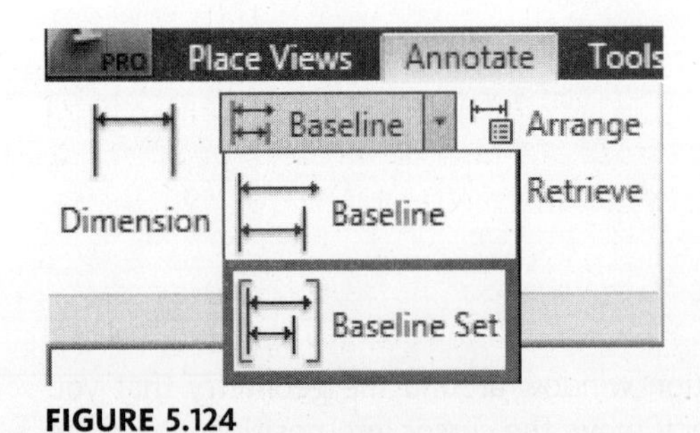

FIGURE 5.124

Members can be added and deleted from the set, a new origin can be placed, and the dimensions can be automatically rearranged to reflect the change. Right-clicking on any dimension in the set will highlight all dimensions and display the menu, as shown in the following image.

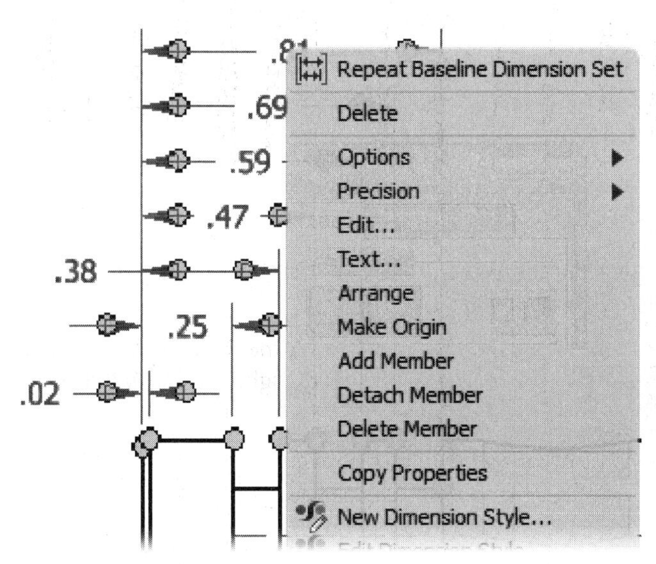

FIGURE 5.125

The main options for editing a baseline dimension set are described as follows:

Delete. This option will delete all baseline dimensions in the set. Move your cursor over any dimension in the set. When the green circles appear on each dimension, right-click, and select Delete from the menu.

Arrange. This command will rearrange the dimensions to the spacing defined in the Dimension Style.

Make Origin. This option will change the origin of the baseline dimensions. Move the cursor over an extension line of the baseline dimension set that you want to define as the new baseline, right-click, and click Make Origin from the menu.

Add Member. This option will add a drawing dimension to the baseline dimensions. Move the cursor over the baseline dimension set, when the green circles appear on the dimensions, right-click and click Add Member from the menu and then click a point that you want to add a dimension to. The dimension will be added to the set of baseline dimensions and rearranged in the proper order.

Detach Member. This option will remove a dimension from the set of baseline dimensions. Move your cursor over any dimension in the set. When the green circles appear on each dimension, right-click on the green circle on the dimension that you want to detach, right-click, and select Detach Member from the menu. The detached dimension can be moved and edited like a normal drawing dimension.

Delete Member. This option will erase a dimension from the set of baseline dimensions. Move your cursor over the dimension in the set that you want to erase, and then select Delete Member from the menu. The dimension will be deleted.

CREATING CHAIN DIMENSIONS

Another option for placing dimensions is to create chain dimensions, dimensions whose extensions lines are shared between two dimensions. Chain dimensions can be arranged horizontal or vertical. Like the baseline dimensions commands there are two Chain dimension commands, Chain and Chain Set.

The Chain command allows you to create a group of horizontal or vertical dimensions. However, the group of dimensions are not considered one group of objects; rather, each dimension that makes up the chain dimension are considered single dimensions.

To create and chain dimensions, follow these steps:

1. Select the Chain dimension command from the Annotate tab > Dimension panel as shown in the following image.

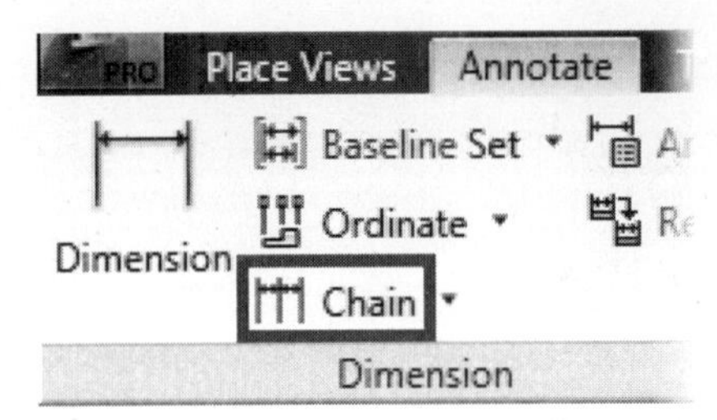

FIGURE 5.126

2. Individually select or drag a selection window around the geometry that you want to dimension. To window select, move the cursor into position where the first point of the box will be. Press and hold down the left mouse button, and move the cursor so the preview box encompasses the geometry that you want to dimension. In the following image on the left, four individual objects were selected.
3. When you are finished selecting geometry, right-click, and select Continue from the menu.
4. Move the cursor to a position where the dimensions will be placed, as shown in the following image on the right. When the dimensions are in the correct place, click the point with the left mouse button to anchor the dimensions.

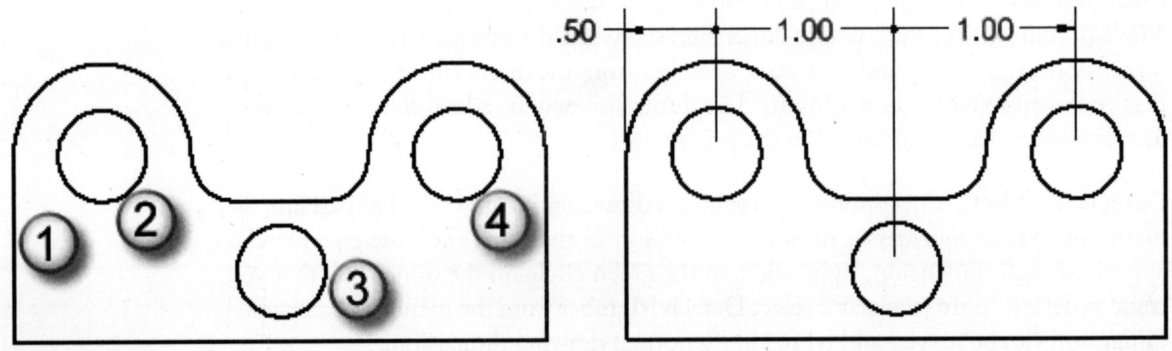

FIGURE 5.127

5. To complete the operation, right-click and select Create from the menu.
6. After creating dimensions using the Chain dimension command, the same editing operations can be performed on these dimensions as general dimensions.

Creating a Chain Dimension Set

Creating a Chain dimension Set is similar to Chain dimensions as explained in the previous section. With the Chain Set command, however, dimensions are grouped together into a set. Select the Chain Set command from the Annotate tab > Dimension panel > click the down arrow next to the Chain dimension command, as shown in the following image.

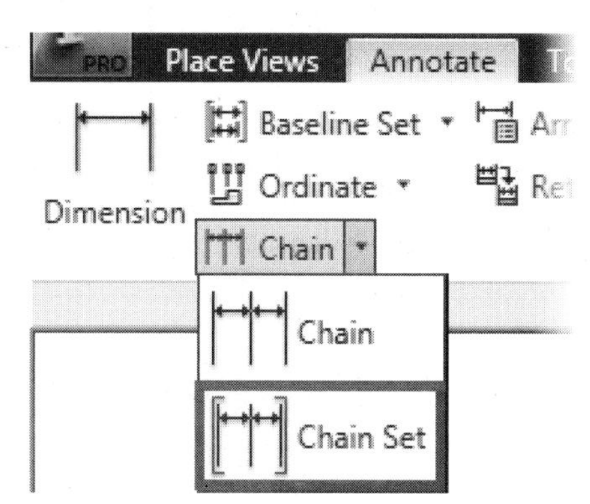

FIGURE 5.128

Members can be added and deleted from the set. Right-click on any dimension in the set will highlight all dimensions and display the menu, as shown in the following image.

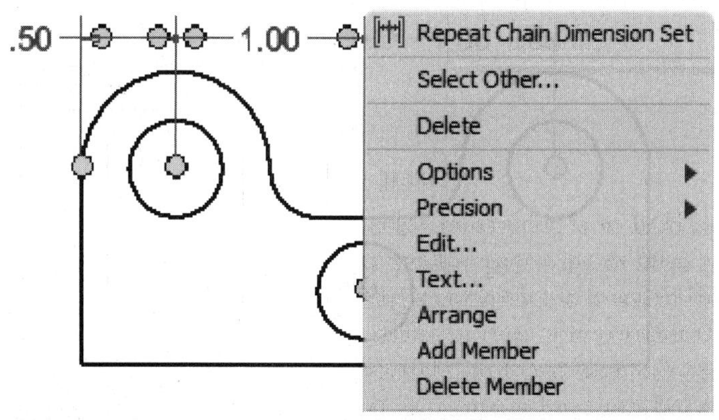

FIGURE 5.129

The main options for editing a chain dimension set are described as follows:

Delete. This option will delete all chain dimensions in the set. Move your cursor over any dimension in the set. When the green circles appear on each dimension, right-click, and select Delete from the menu.

Options. From the Options area you can lock the chain, control the arrowheads and add a leader to a dimension that the cursor was over when you right-click.

Add Member. This option will add a drawing dimension to the chain dimensions. Move the cursor over the chain dimension set, when the green circles appear on the dimensions, right-click and click Add Member from the menu and then click a point that you want to add a dimension to. The dimension will be added to the set of chain dimensions and rearranged in the proper order.

Delete Member. This option will erase a dimension from the set of chain dimensions. Move your cursor over the dimension in the set that you want to delete. When the green circles appear on each dimension, right-click, and then select Delete Member from the menu. The dimension will be deleted.

EXERCISE 5-7: CREATING BASELINE AND CHAIN DIMENSIONS

In this exercise, you will add annotations using baseline and chain dimensions.

1. Open *ESS_E05_07.dwg.*
2. You will now add a baseline dimension set. Select the Baseline Dimension Set command from the Annotate tab > Dimension panel > click the down arrow next to the Baseline Dimension command.
3. Select the three part edges as shown in the following image on the left.
4. Right-click, and click Continue from the menu.
5. Click a point above the geometry to position the dimension set as shown in the following middle image.
6. Right-click, and click Create from the menu. Notice the creation of the baseline dimension set.
7. However, all baseline dimensions need to reference the opposite edge of the object. To perform this task, move the cursor over the extension line on the right, Right-click, and click Make Origin from the menu. Notice how the dimensions are regenerated from the new origin, as shown in the following image on the right.

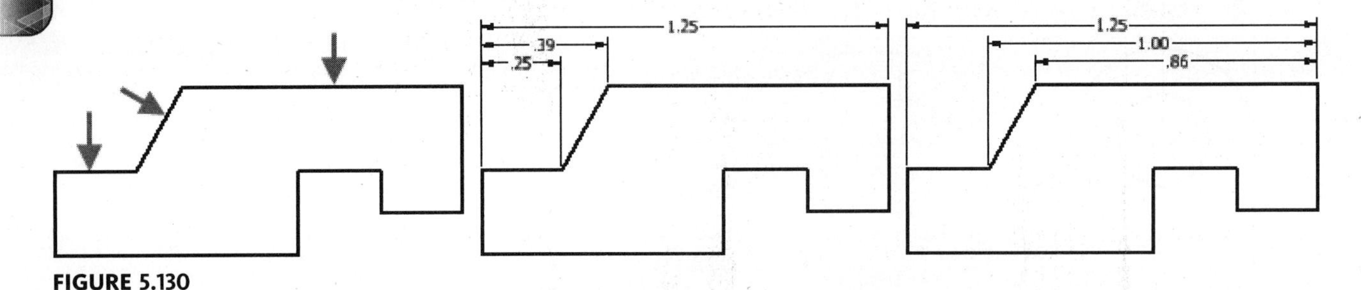

FIGURE 5.130

8. To add a dimension to the set move the cursor over the baseline dimension set, when the green circles appear on the dimensions, right-click and click Add Member from the menu and then click the point as shown in the following image on the left. Right-click and click Done from the menu and the dimension will be added to the set and the dimensions will be rearranged as shown in the following image on the right.

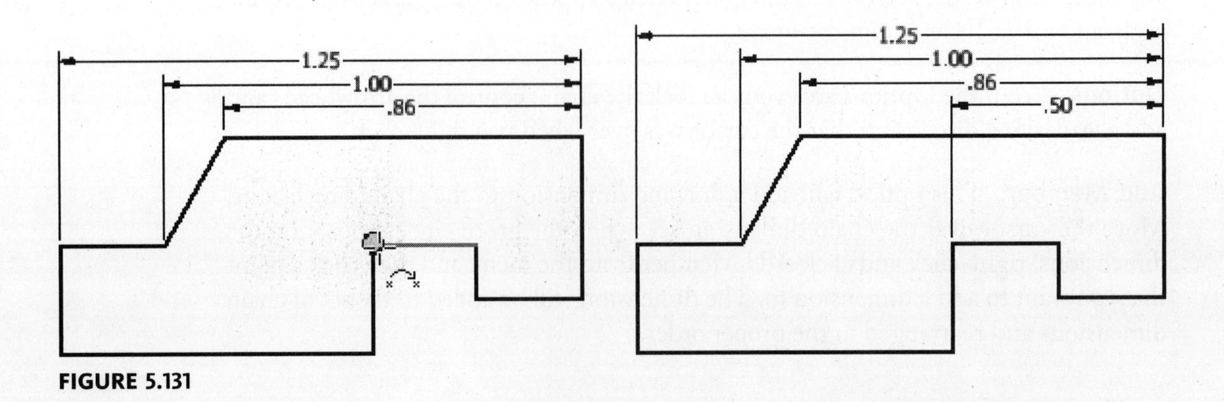

FIGURE 5.131

9. Detach a dimension, move the cursor over the **.86** dimension, right-click and click Delete Member from the menu. The dimension will be deleted as shown in the following image on the left. Notice the dimensions were not automatically arranged.
10. Next you arrange the dimensions, move the cursor over the dimension set, when the green circles appear on the dimension right-click and click Arrange from the menu. The dimension will be rearranged as shown in the following image on the right.

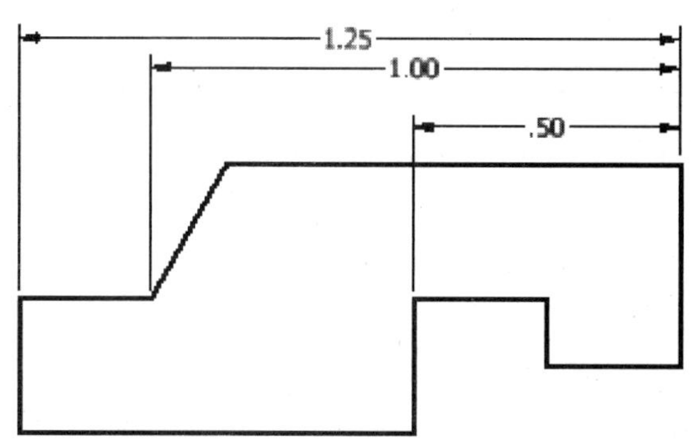

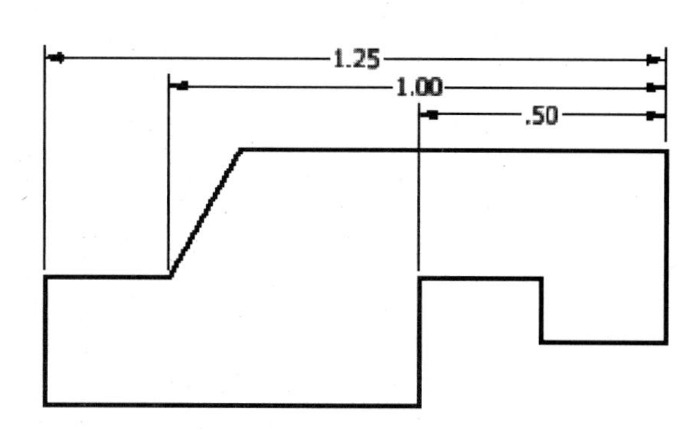

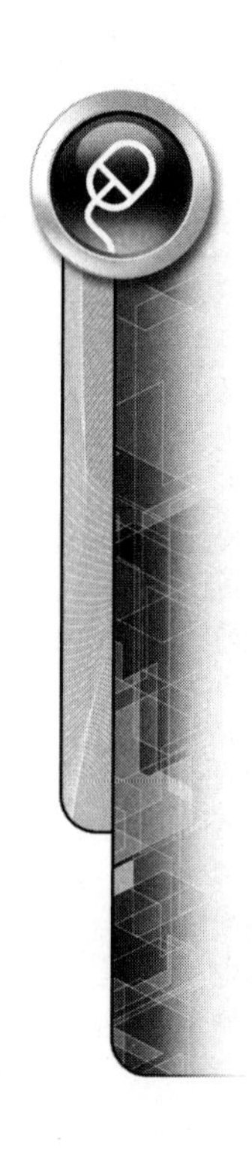

FIGURE 5.132

11. Next delete the baseline dimension set. Move the cursor over the dimension set, when the green circles appear on the dimension right-click and click Delete from the menu. The dimension set will be deleted.
12. Next you create a chain dimension set. Select the Chain Set command from the Annotate tab > Dimension panel > click the down arrow next to the Chain Set dimension command.
13. Window select the part edges as shown in the following image on the left.
14. Right-click, and click Continue from the menu.

    Click a point below the geometry to position the dimension set as shown in the following image on the right.
15. Right-click, and click Create from the menu and the chain dimension set will be created.

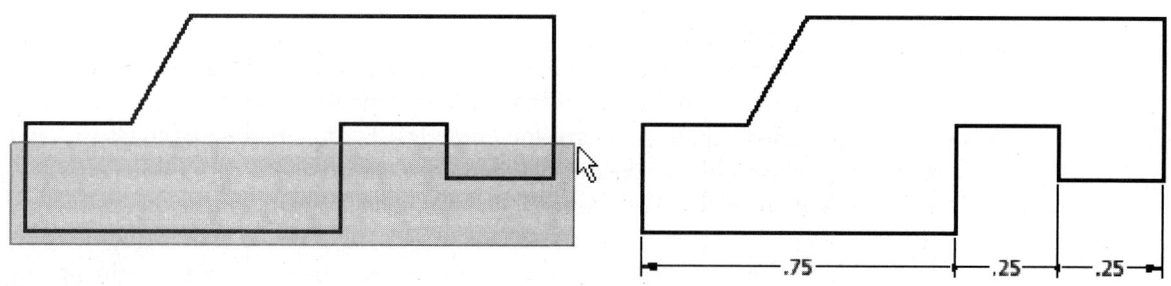

FIGURE 5.133

16. To add a dimension to the set move the cursor over the chain dimension set, when the green circles appear on the dimensions, right-click and click Add Member from the menu and then click the point as shown in the following image on the left. Right-click and click Done from the menu. The dimension will be added to the set and the dimensions will be rearranged as shown in the following image on the right.

NOTE

If a dimension does not fit within its extension lines you can right-click on the dimension and from the menu click Options > Leader and then drag the dimension and leader as needed.

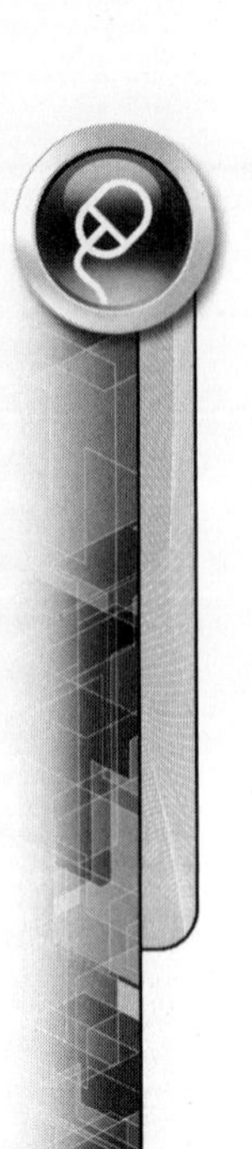

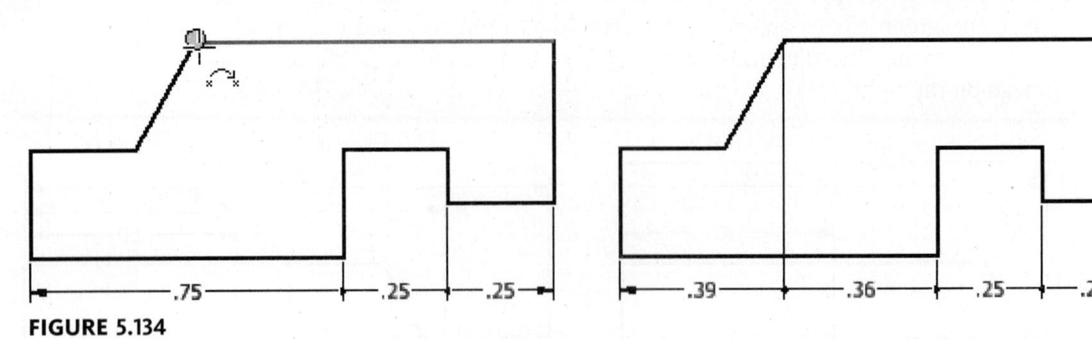

FIGURE 5.134

17. Detach a dimension, move the cursor over the **.36** dimension, right-click, and click Delete Member from the menu. To center the text, right-click on the dimension set and click Arrange from the menu. The dimension will be deleted as shown in the following image. Notice the dimensions were automatically arranged.

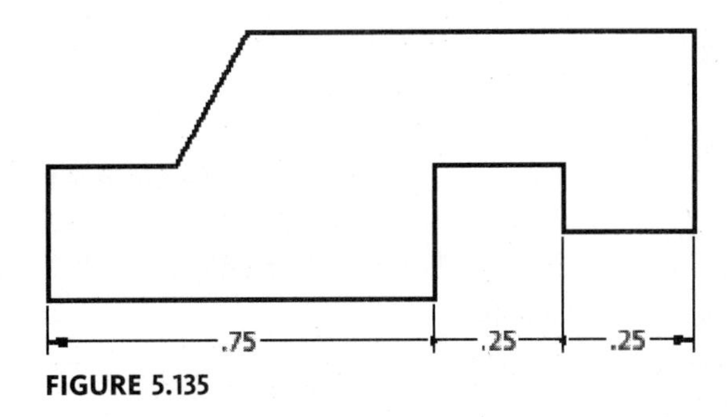

FIGURE 5.135

18. Close all open files. Do not save changes. End of exercise.

CREATING ORDINATE DIMENSIONS

Ordinate dimensions are used to indicate the location of a particular point along the X or Y axis from a common origin point. This type of dimensioning is especially suited for describing part geometry for numerical control tooling operations. Two commands are available for creating ordinate dimensions: Ordinate Dimension and Ordinate Dimension Set. Both commands will create drawing dimensions that reference the geometry and will be updated to reflect any changes in the geometry to which they are dimensioned. Ordinate dimensions can be placed on circular or straight edges.

Ordinate Dimensions

Ordinate dimensions created with the Ordinate Dimension command are recognized as individual objects, and an origin indicator will be created as part of the operation. If the origin indicator location is moved, the other ordinate dimensions will be updated to reflect the change.

To create an ordinate dimension using the Ordinate Dimension command, follow these steps:

1. Create a drawing view.
2. Click the Ordinate Dimension command on the Annotate tab > Dimension panel, as shown in the following image on the left.
3. In the graphics window, select in the view to dimension.
4. Select a point to set the origin indicator for the dimensions, as shown in the following image.

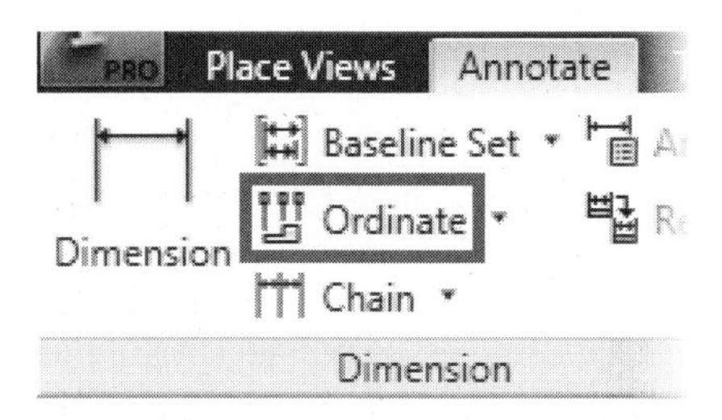

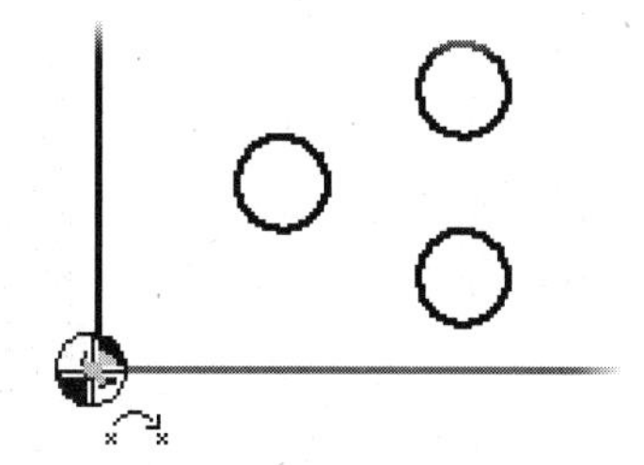

FIGURE 5.136

5. Select geometry that the ordinate dimensions will be applied to. When done selecting geometry, right-click and click Continue from the menu.
6. Locate the dimensions horizontally or vertically by moving the cursor and then click a point when the dimensions are previewed in the correct orientation, as shown in the following image on the right.
7. To create the dimensions and end the operation, right-click, and click Done from the menu.
8. To edit a dimension, right-click on a dimension and select the desired option from the menu, as shown in the following image.

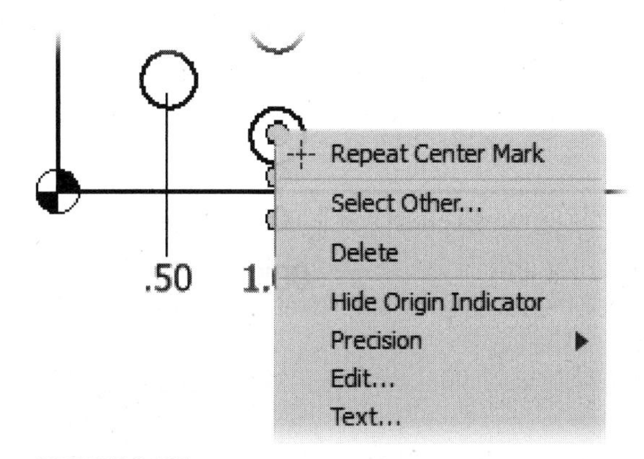

FIGURE 5.137

9. To edit an ordinate dimension leader, grab an anchor point (a green circle), and drag it to the desired location.
10. To edit the origin indicator, do one of the following:
    - Move the cursor over the origin indicator, and drag it to the desired location.
    - Double-click the origin indicator, and enter the precise location in the Origin Indicator dialog box.

Ordinate Dimension Set

In an ordinate dimension set, all the ordinate dimensions that are created in a single operation will be grouped together and can be edited individually or as a set. When creating an ordinate dimension set, the first dimension created will be used as the origin. The origin dimension needs to be a member of the set and can later be changed to a different dimension. If the location of the origin or the origin member changes, the other members will be updated to reflect the new location.

When you place ordinate dimensions, they will automatically be aligned to avoid interfering with other ordinate dimensions.

To create an ordinate dimension set, follow these steps:

1. Create a drawing view.
2. Select the Ordinate Dimension Set command on the Annotate tab > Dimension panel > click the down arrow next to the Ordinate Dimension command, as shown in the following image on the left.
3. Select a point to set the origin for the dimensions.
4. Select geometry that the ordinate dimensions will be applied to. When done selecting geometry, right-click and click Continue from the menu.
5. Locate the dimensions horizontally or vertically by moving the cursor and then click a point when the dimensions are previewed in the correct orientation, as shown in the following image on the right.

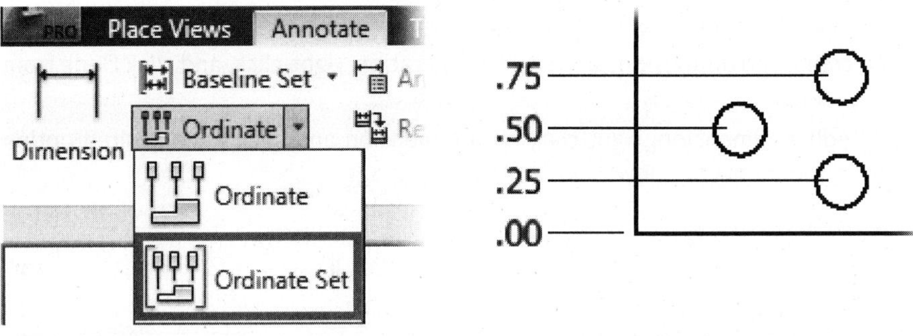

FIGURE 5.138

6. To change options for the dimension set, right-click, and then click Options, as shown in the following image on the left. This can also be done after creating the dimensions by moving the cursor over the dimension set, right-click and click Options from the menu.
7. To create the dimensions set and end the command, right-click, and click Create from the menu.
8. To edit the origin, right-click on the dimension that will be the origin and click Make Origin from the menu as shown in the following image on the right. The other dimension values will be updated to reflect the new origin.
9. To edit a dimension set, right-click on a dimension in the set and select the desired option from the menu as shown in the following image on the right.

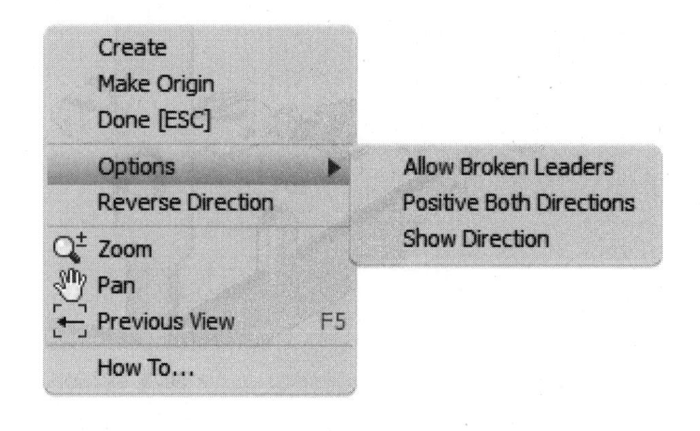

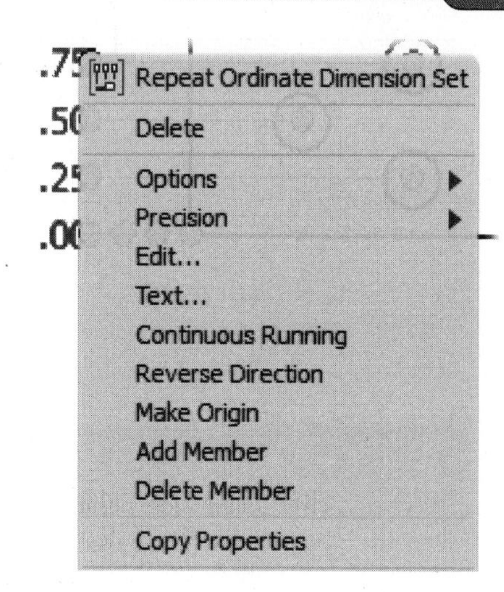

FIGURE 5.139

Text and Leaders

To add text to the drawing, click either the Text or the Leader Text command on the Annotate tab > Text panel, as shown in the following image.

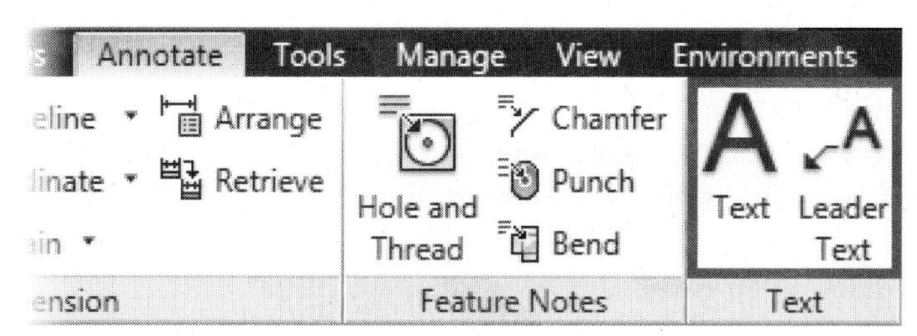

FIGURE 5.140

The Text command will add text while the Leader Text command will add a leader and text. Select the desired text command, and define the leader points and/or text location. Once you have chosen the location in the graphics window, the Format Text dialog box will appear. When placing text through the Format Text dialog box, select the orientation and text style as needed and type in the text, as shown in the following image. Click the OK button to place the text in the drawing. To edit the text or text leader position, move your cursor over it, click one of the green circles that appear, and drag to the desired location. To edit the text or text leader content, right-click on the text, and select Edit Leader Text or Edit Text from the menu.

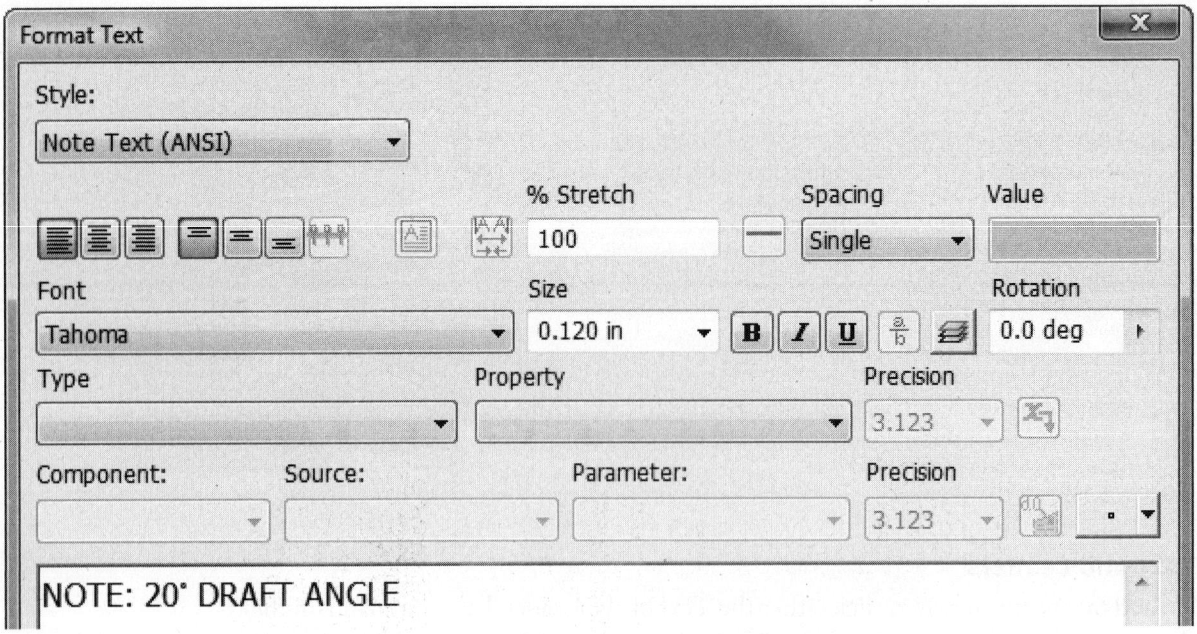

FIGURE 5.141

Additional Annotation Commands

To add more detail annotations to your drawing, you can add surface texture symbols, welding symbols, feature control frames, feature identifier symbols, datum identifier symbols, and datum targets by clicking the corresponding command on the Annotate tab > Symbols panel, as shown in the following image. The Import command is only available when you are working in an Inventor DWG file.

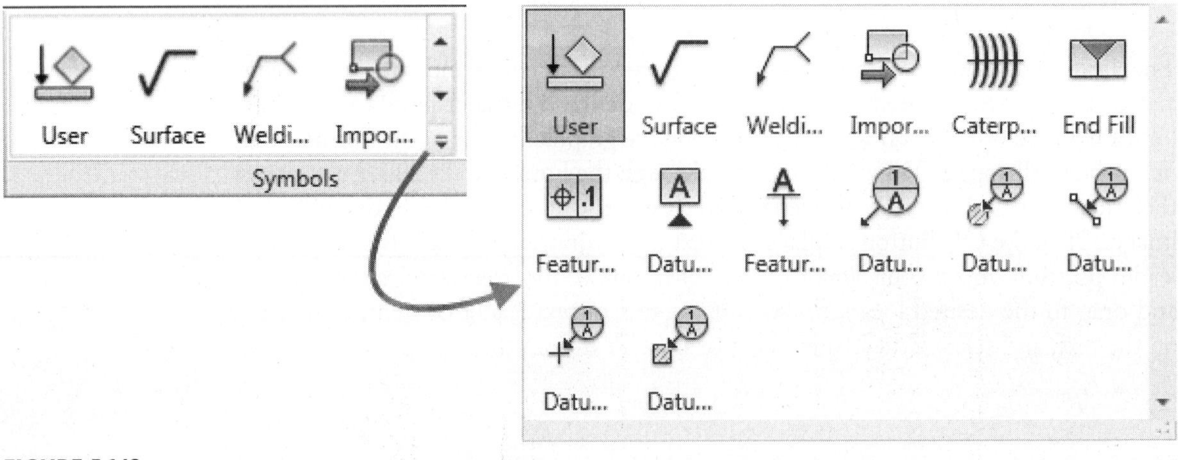

FIGURE 5.142

Follow these steps for placing symbols:

1. Click the appropriate symbol command on the Annotate tab > Symbols panel.
2. Select a point at which the leader will start. If you don't want a leader, click a point that the symbol will be placed and proceed to step 4.
3. Continue selecting points to position the leader lines.
4. Right-click, and select Continue from the menu.
5. Fill in the information as needed in the dialog box.

6. When done, click the OK button in the dialog box.
7. To complete the operation, right-click, and select Done from the menu.
8. To edit a symbol, position the cursor over it. When the green circles appear, right-click, and click the corresponding Edit option from the menu.

Hole and Thread Notes

Another annotation you can add is a hole or thread note. Before you can place a hole or thread note in a drawing, a hole or thread feature must exist. You can also annotate extruded circles using the Hole or Thread Note command. If the hole or thread feature changes, the note will be updated automatically to reflect the change. The following image shows examples of counterbore, countersink, and threaded/tapped hole notes.

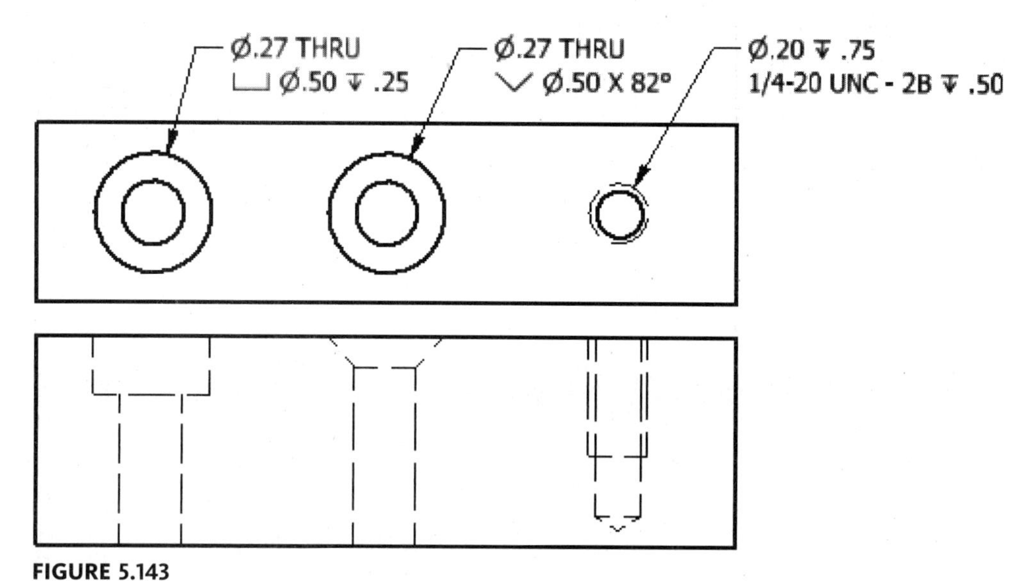

FIGURE 5.143

To create a hole or thread note, follow these steps:

1. Click the Hole/Thread Notes command on the Annotate tab > Feature Notes panel, as shown in the following image.

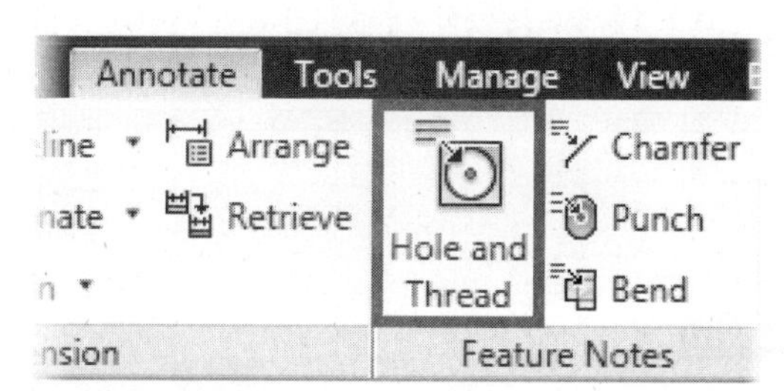

FIGURE 5.144

2. Select the hole or thread feature to annotate.
3. Select a second point to locate the leader and the note.
4. To complete the operation, right-click, and select Done from the menu.
5. To edit a note, position the cursor over it. When the green circles appear, right-click, and select the corresponding Edit option from the menu.

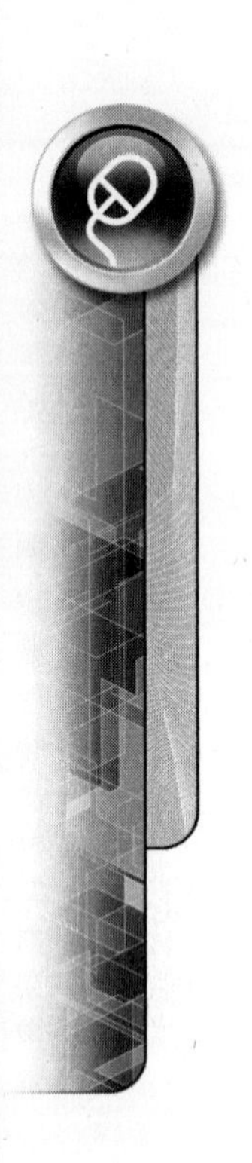

Editing Hole Notes

A hole note edit is invoked by right-clicking on an existing hole note to display the menu, as shown in the following image.

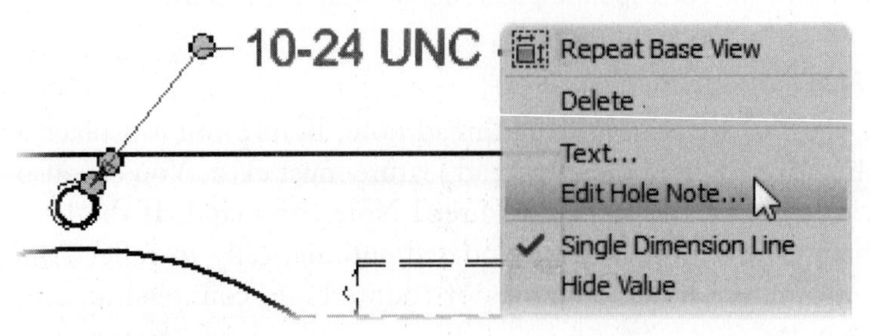

FIGURE 5.145

Selecting Edit Hole Note from this menu will display the Edit Hole Note dialog box, as shown in the following image. You can perform edits on a single hole note by unchecking the Use Default option and adding options to the hole note from the Values and Symbols area in the Edit Hole Note dialog box as shown in the following image. The following image illustrates the use of adding the <QTYNOTE> (quantity note) modifier to the hole note. This modifier will calculate the number of occurrences of the hole being noted, which is considered a highly valuable productivity command.

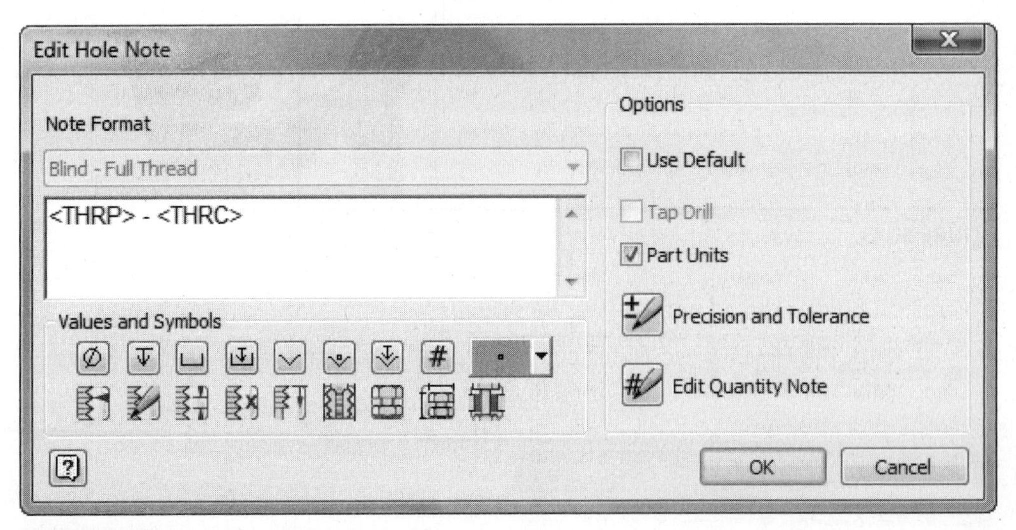

FIGURE 5.146

EXERCISE 5-8: ADDING ANNOTATIONS

In this exercise, you add hole notes, a surface texture symbol, and a text note.

1. Open the drawing *ESS_E05_08.dwg.* The drawing file contains four drawing views with centerlines and dimensions.
2. First you add three hole notes. Begin by zooming in on the top view.
3. Click the Hole and Thread command on the Annotate tab > Feature Notes panel.
4. Select the tapped hole on the left side and click a point to place the hole note as shown in the following image on the left.

5. Notice in the hole note there is no information about the tap drill. To add tap drill information, double-click on the hole note and in the Options area in the Edit Hole Note dialog box check the Tap Drill option as shown in the following image on the right.

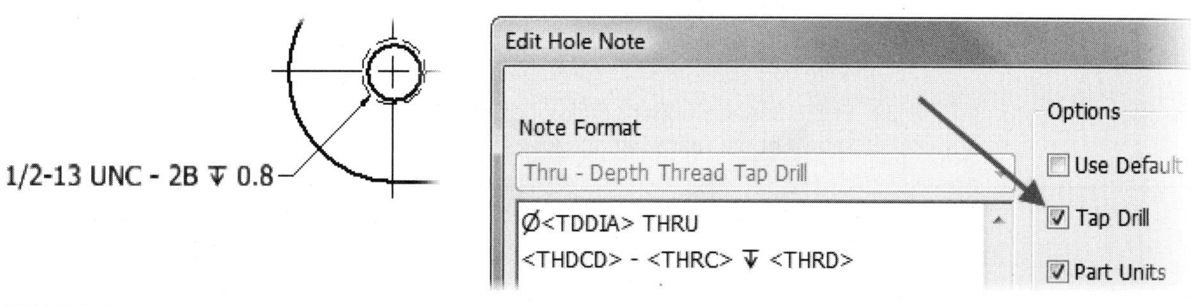

FIGURE 5.147

6. Click OK to finish the edit. The following image on the left shows the updated hole note. On the right side of the top view, add two hole notes as shown in the following image on the right.

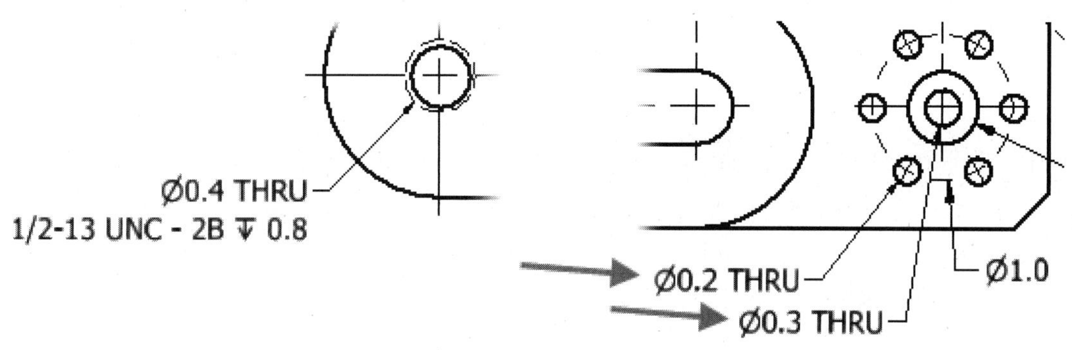

FIGURE 5.148

7. The Ø0.2 THRU hole note defines the holes that fall on the bolt circle you need to add a quantity. To add a quantity note, double-click on the Ø0.2 THRU hole note and in the Note Format area add a space after THRU and then click on Quantity Note as shown in the following image on the left.
8. Click OK to finish the edit. The following image on the right shows the updated hole note.

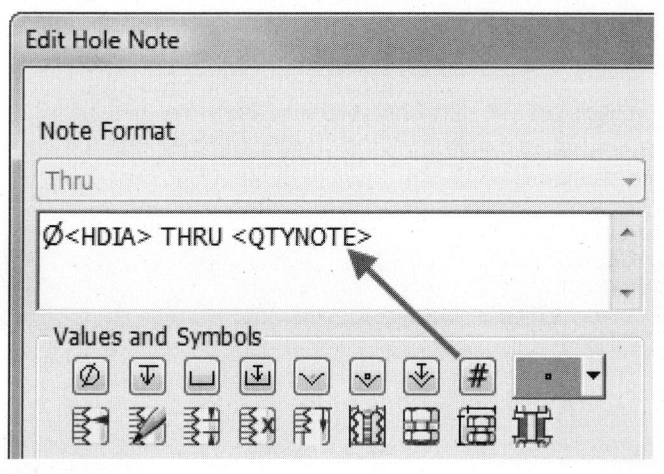

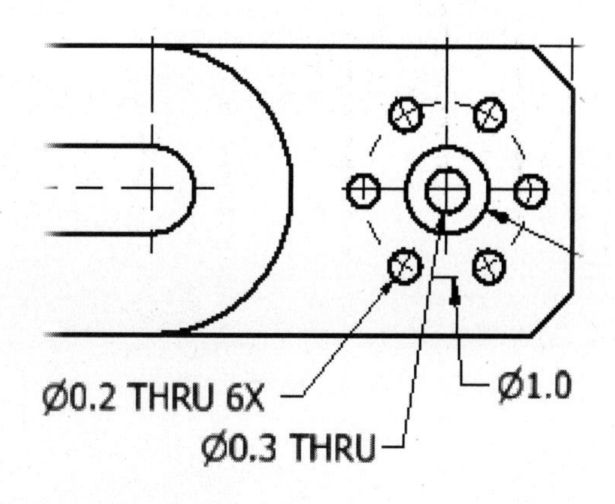

FIGURE 5.149

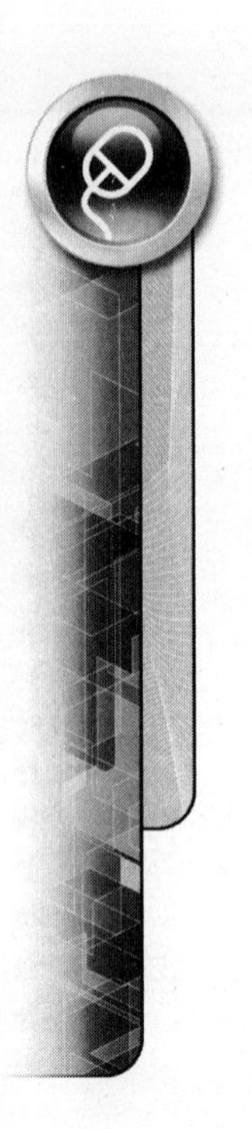

9. Next you add a surface finish note. Pan the screen so you can see the front view.
10. Click the Surface Texture Symbol command on the Annotate tab > Symbols panel.
11. Click on the top edge of the part as shown in the following image and then right-click and click Continue from the menu. Since the symbol will sit on the edge you don't need to click again to create an arrow.
12. In the Surface Texture dialog box enter **32** in the B Production method area as shown in the following image.

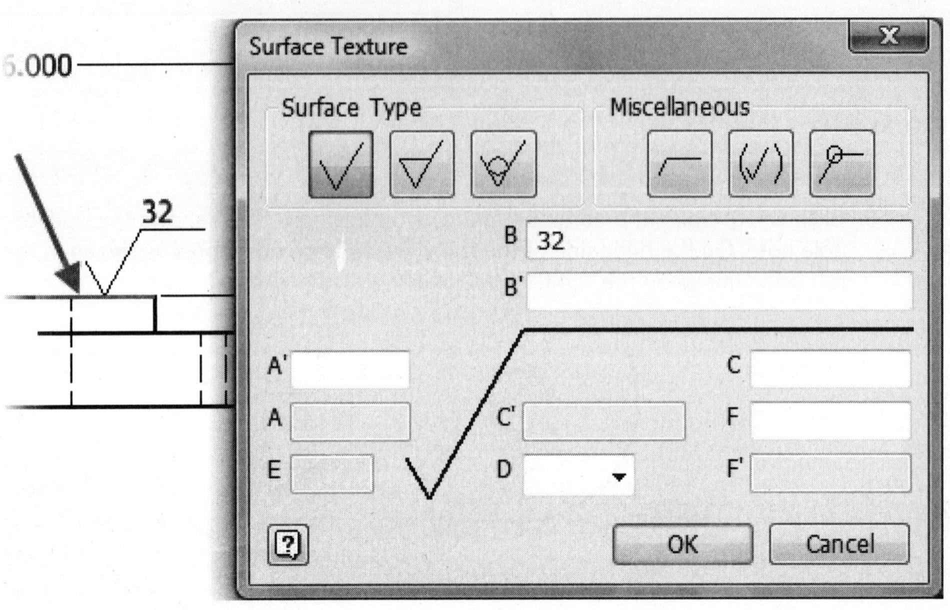

FIGURE 5.150

13. Click OK to create the surface texture symbol.
14. Next you add a text note. Click the Text command on the Annotate tab > Text panel.
15. To locate the position where the text will start, click on a point below the front view and in the Format Text dialog box enter the text as shown in the following image. To add the ± symbol, select the symbol from Insert symbol area in the dialog box.

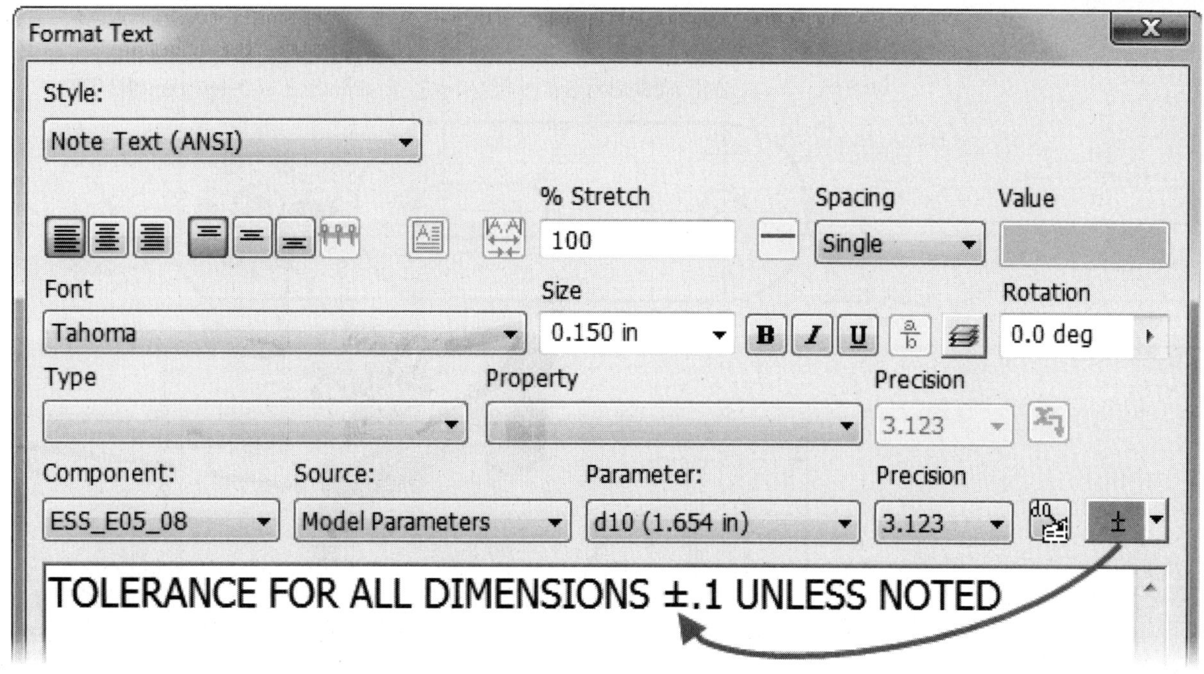

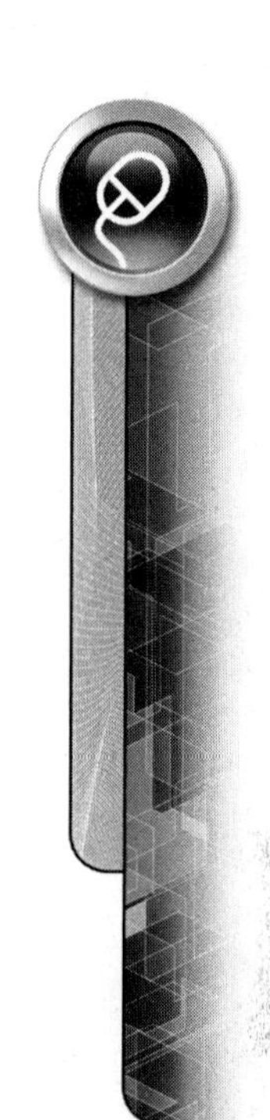

FIGURE 5.151

16. To create the text click OK in the dialog box.
17. The text is displayed on a single line. To change the text so it appears on multiple lines, click on the text in the graphics window and then click and drag the lower-right green circle of the text's bounding area and drag it to the left as shown in the following image on the left. Release the mouse button and the text will be fitted with the new defined area as shown in the following image on the right.

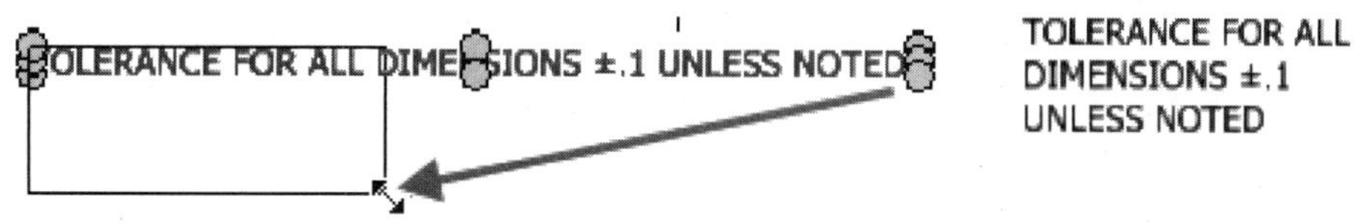

FIGURE 5.152

18. Drag a selection window around all dimensions. This will select the dimensions to arrange. When finished, right-click and click Done from the menu. The dimensions are arranged according to the settings in the current dimension style.
19. When done, your screen should resemble the following image.

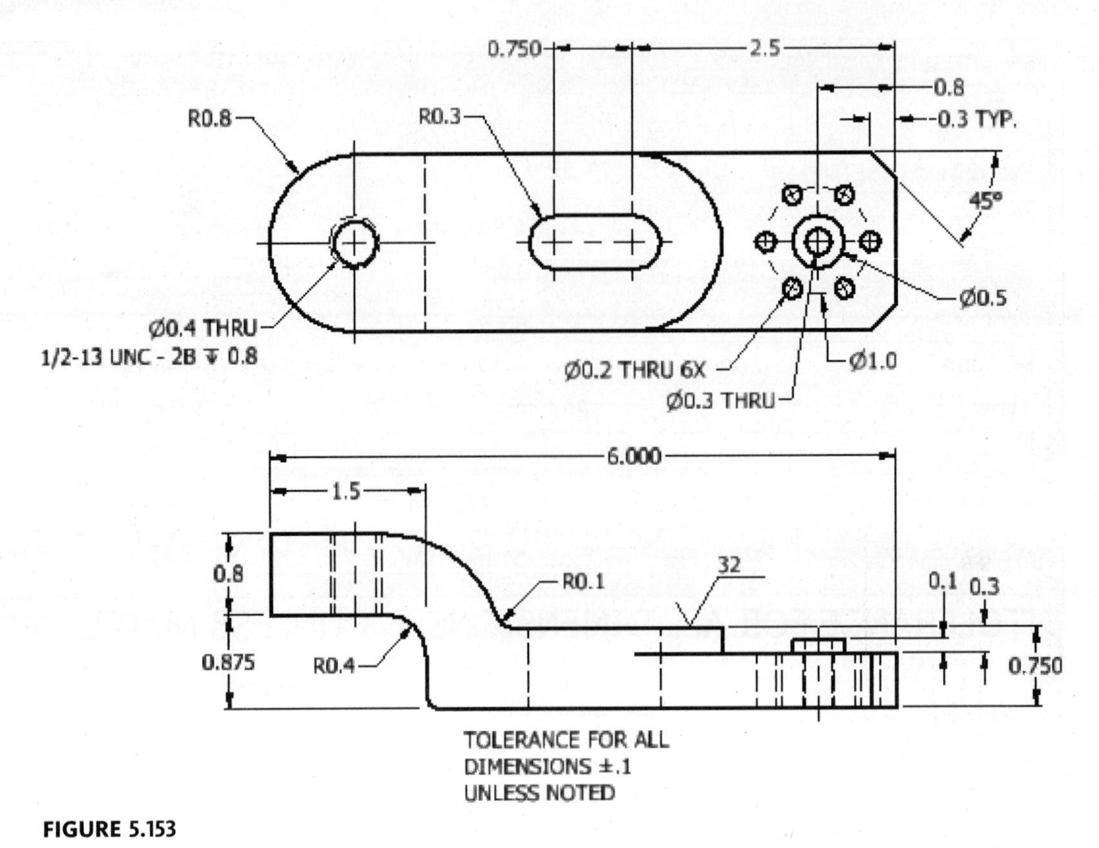

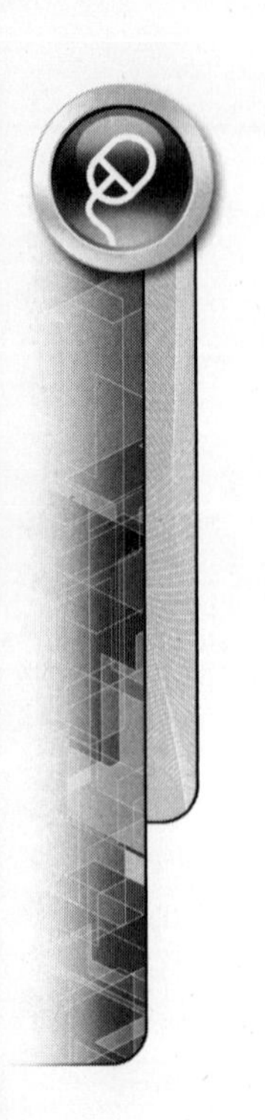

FIGURE 5.153

20. Close all open files. Do not save changes. End of exercise.

OPENING A MODEL FROM A DRAWING

While working inside of a complex drawing view, you may need to make changes to a part, assembly, or presentation file. Rather than close down the drawing and open the individual file, you can open a part, assembly, or presentation file directly from the drawing using a number of techniques.

One technique is to right-click in a drawing view to display the menu in the following image on the left, and then click Open to open the model file. In this example, an assembly file was opened from the drawing view.

Another technique involves expanding the drawing view in the browser to expose the assembly. Then continue expanding the browser until the file you want to open is visible in the list. Right-click the desired file in the browser and select Open from the menu to display the part model as shown in the following image on the right.

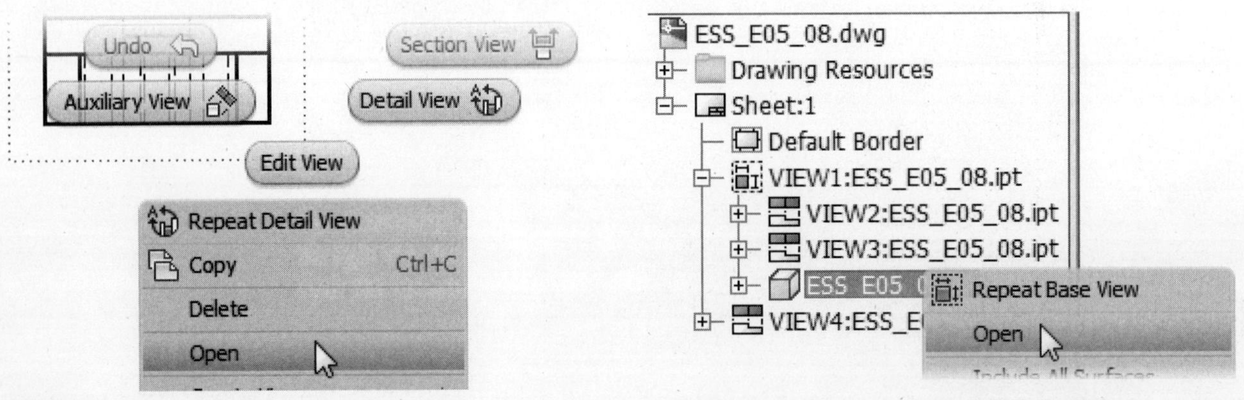

FIGURE 5.154

A third technique used to open a part file from a drawing is to first change the selection filter to Part Priority. After doing so, right-click on the specific part in the drawing view and select Open from the menu. This will open the source file.

OPENING A DRAWING FROM A MODEL

You can open a drawing from a part or assembly model by right-clicking on the part, assembly presentation file name and click Open Drawing from the menu as shown in the following image. This will open the drawing file associated with the part, assembly or presentation file. The drawing file must have the same name as the part, assembly or presentation file. Use this technique to prevent laborious searches for drawing files.

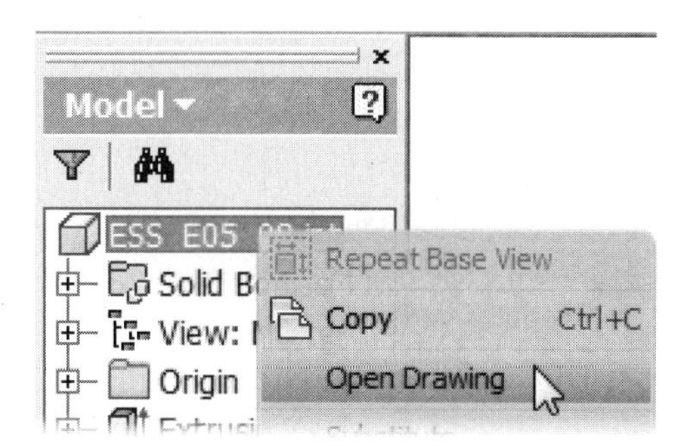

FIGURE 5.155

CREATING HOLE TABLES

If the drawing view that you are dimensioning contains holes, you can locate them by placing individual dimensions or by creating a hole table that will list the location and size of all the holes, or just the selected holes, in a view. The hole locations will be listed in both X- and Y-axis coordinates with respect to a hole datum that will be placed before creating the hole table.

After placing a hole table, if you add, delete, or move a hole, the hole table will automatically be updated to reflect the change after the part is saved and the drawing is the active document. Each hole in the table is automatically given an alphanumeric tag as its name. It can be edited by double-clicking on the tag in the drawing view or hole table. Alternately, you can right-click on the tag in the drawing view or hole table, and select Edit Tag from the menu. Type in a new tag name, and the change will appear in the drawing view and hole table. There are three commands for creating a hole table:

- Hole Table—Selection
- Hole Table—View
- Hole Table—Selected Feature

The commands are located on the Annotate tab > Table panel, as shown in the following image.

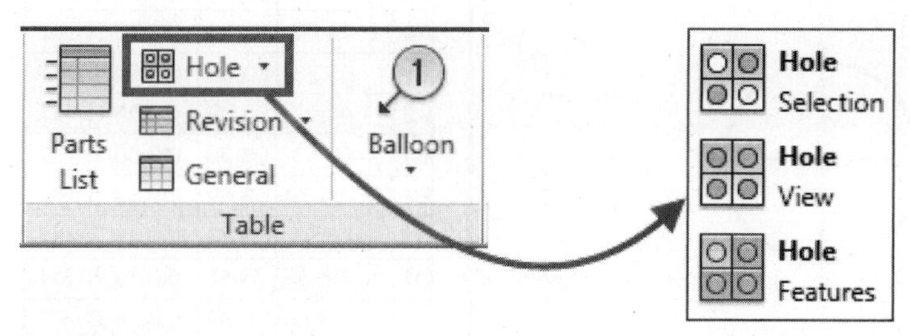

FIGURE 5.156

To create a hole table based on an existing drawing view that shows holes in a plan view, follow these steps:

1. Select the desired Hole Table command from the Table menu on the Annotate tab > Table panel.
2. Select the view on which the hole table will be based.
3. Select a point to locate the origin. Typical origins include the corners of rectangular objects and the centers of drill holes used for data.
4. **Hole Selection:** Select the holes to include in the hole table. The holes can be selected individually, or you can drag a selection window around the holes. When done selecting holes, right-click and click Create from the menu.

   Hole View: All the holes in the selected view will be annotated.

   Hole Features: Select the individual hole or holes that will act as references for all other holes of the same type. When done selecting holes, right-click and click Create from the menu.
5. Select a point in the drawing to locate the table.
6. The contents of the hole table can be edited by either right-clicking on the hole table, on the tag name in the hole table, or on the tag name in the drawing view. Next choose the desired option from the menu.

When using the Hole Selection command, a hole table is created from only the selected holes, as shown in the following image.

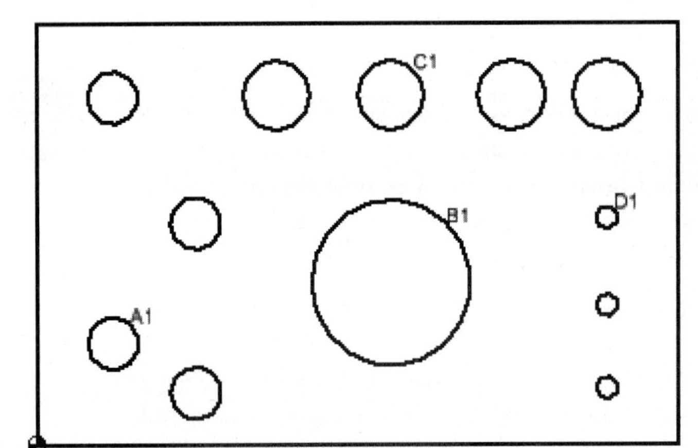

Hole Table			
HOLE	XDIM	YDIM	DESCRIPTION
A1	14.28	18.40	Ø0.38 THRU
B1	66.53	29.23	Ø1.18 THRU
C1	66.53	63.09	Ø0.50 THRU
D1	106.89	40.89	Ø0.16 THRU

FIGURE 5.157

When using the Hole View command, a hole table is created based on all the holes in a selected view, as shown in the following image. Notice that all holes are given an alphanumeric identifier, which locates each hole by X- and Y-axis coordinates.

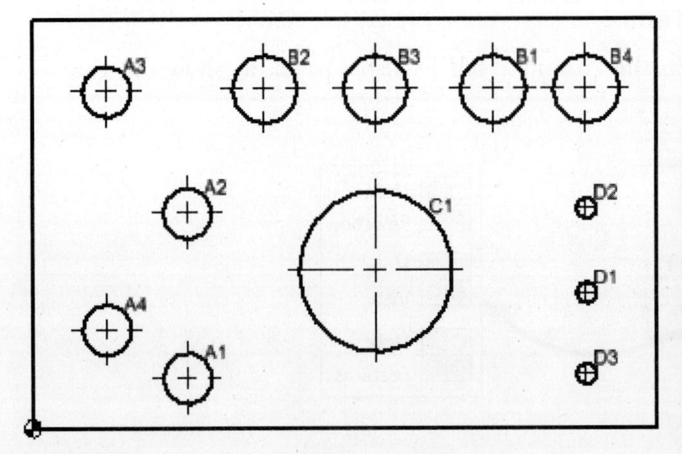

Hole Table			
LOC	XDIM	YDIM	SIZE
A1	29.77	9.39	Ø0.38 THRU
A2	29.77	40.02	Ø0.38 THRU
A3	14.28	62.73	Ø0.38 THRU
A4	14.28	18.40	Ø0.38 THRU
B1	89.23	63.09	Ø0.50 THRU
B2	44.91	63.09	Ø0.50 THRU
B3	66.53	63.09	Ø0.50 THRU
B4	106.89	63.09	Ø0.50 THRU
C1	66.53	29.23	Ø1.18 THRU
D1	106.89	25.12	Ø0.16 THRU
D2	106.89	40.89	Ø0.16 THRU
D3	106.89	10.11	Ø0.16 THRU

FIGURE 5.158

When using the Hole Features command, a hole table is created based on holes that are identical to selected hole. In the following image, one of the larger holes located in the upper part of the object (A4) is selected using the Hole Features command.

When the hole table is created, all holes that share the same type and size will be added to the hole table, as shown in the following image.

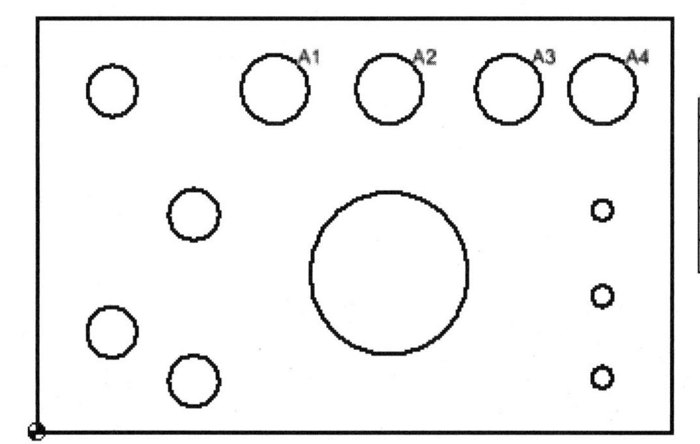

Hole Table			
HOLE	XDIM	YDIM	DESCRIPTION
A1	44.91	63.09	Ø0.50 THRU
A2	66.53	63.09	Ø0.50 THRU
A3	89.23	63.09	Ø0.50 THRU
A4	106.89	63.09	Ø0.50 THRU

FIGURE 5.159

EXERCISE 5-9: CREATING HOLE TABLES

In this exercise, you will create a hole chart for a shim plate. After creating the hole table, you will modify the plate model to add new holes and update the hole table accordingly.

1. In preparation for creating the hole table, first open the existing drawing, *ESS_E05_09.idw*.
2. To create the hole table, Click the Annotate tab > Table panel > click the down arrow next to Hole and click the Hole View command.
3. Select the front view of the plate as the view to which the hole table will be associated. To place the hole table datum, aquire the theoretical intersection of the lower left corner of the front view, as shown in the following image on the left. Then place the hole table on the right side of the sheet, as shown in the following image on the right.

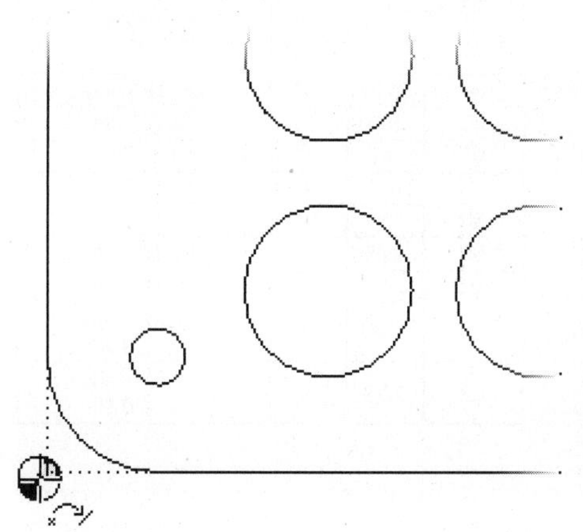

HOLE TABLE			
HOLE	XDIM	YDIM	DESCRIPTION
A1	.25	.25	Ø0.13 THRU
B1	.63	.39	Ø0.38 THRU
B2	1.10	.39	Ø0.38 THRU
B3	1.57	.39	Ø0.38 THRU
B4	2.05	.39	Ø0.38 THRU
B5	2.52	.39	Ø0.38 THRU
B6	.63	.91	Ø0.38 THRU
B7	1.10	.91	Ø0.38 THRU
B8	1.57	.91	Ø0.38 THRU
B9	2.05	.91	Ø0.38 THRU
B10	2.52	.91	Ø0.38 THRU
C1	.38	2.00	Ø0.25 THRU
C2	2.75	2.00	Ø0.25 THRU

FIGURE 5.160

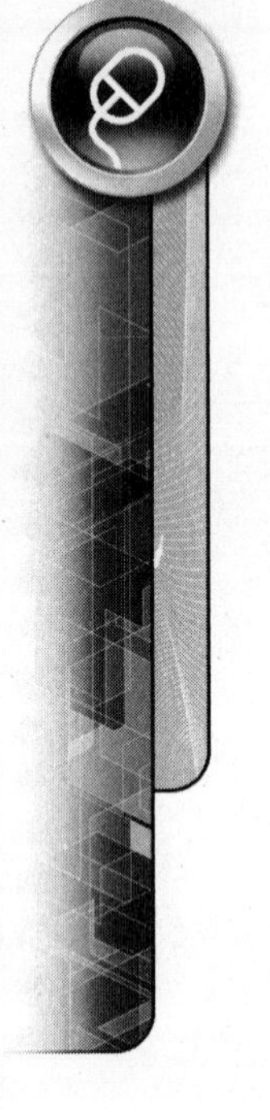

4. The plate lug holes need to be changed to counterbore holes. You will now switch from the drawing to the part file by expanding View3 in the browser and right-click on *ESS_E05_09.ipt* and click Open from the menu as shown in the following image. This will open the part file associated with the drawing in a new window.

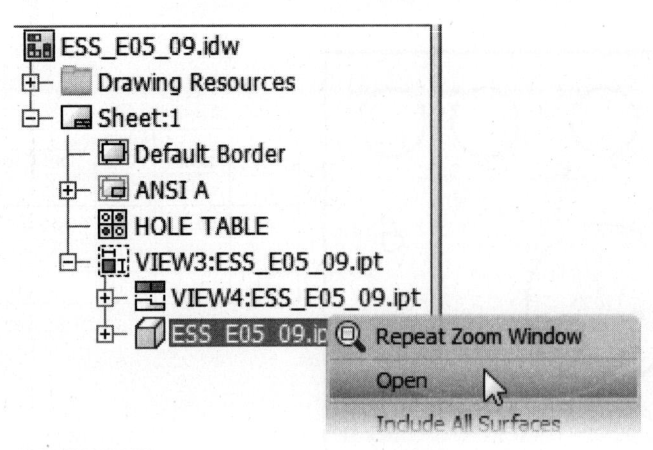

FIGURE 5.161

5. In the browser double-click the Hole1 feature.
6. While in the Hole dialog box, change the hole type to Spotface and change the hole specifications, as shown in the following image on the left. Click the OK button to update the holes.
7. Save the part file and then switch back to the drawing, click the tab near the bottom-left corner of the graphics window for the file *ESS_E05_09.idw*. Note the updated hole table; the holes labeled with the letter C reflect the change to the holes, as shown in the following image on the right.

Another method to switch between windows in an application is to hold down the CTRL key and press the TAB key.

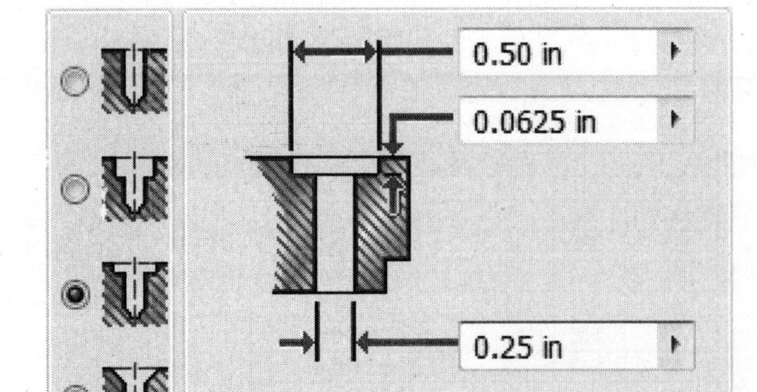

HOLE TABLE			
HOLE	XDIM	YDIM	DESCRIPTION
A1	.25	.25	Ø0.13 THRU
B1	.63	.39	Ø0.38 THRU
B2	1.10	.39	Ø0.38 THRU
B3	1.57	.39	Ø0.38 THRU
B4	2.05	.39	Ø0.38 THRU
B5	2.52	.39	Ø0.38 THRU
B6	.63	.91	Ø0.38 THRU
B7	1.10	.91	Ø0.38 THRU
B8	1.57	.91	Ø0.38 THRU
B9	2.05	.91	Ø0.38 THRU
B10	2.52	.91	Ø0.38 THRU
C1	.38	2.00	Ø0.25 THRU ⌴ Ø0.50
C2	2.75	2.00	Ø0.25 THRU ⌴ Ø0.50

FIGURE 5.162

8. A series of mounting holes needs to be created in the model consisting of countersink holes. Switch back to *ESS_E05_09.ipt* to view the part. Add two

countersink mounting holes using the following specifications, as shown in the following image. Both countersink holes are aligned horizontally.

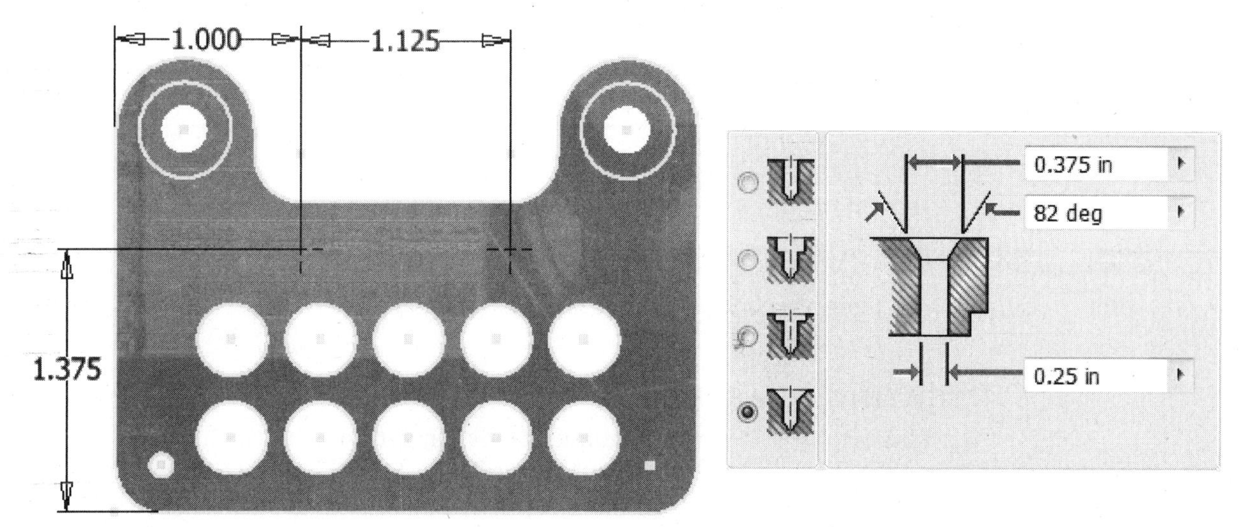

FIGURE 5.163

9. Save the part file and then make the file *ESS_E05_09.idw* current to view the drawing. Note in the following image that the new holes have been automatically added to the hole table.

C2	2.75	2.00	Ø0.25 THRU ⌴ Ø0.50
D1	1.00	1.38	Ø0.25 THRU ⌵ Ø0.38 X 82°
D2	2.13	1.38	Ø0.25 THRU ⌵ Ø0.38 X 82°

FIGURE 5.164

10. A hole tag will now be modified. Click on A1, then click on the green circle at the location where the tag is attached to the geometry and drag and position the A1 hole tag, as shown in the following image on the left. Next double-click on the A1 tag to open the Format Text dialog box, add the text-**TOOLING HOLE** to the hole tag, and click OK. The hole table updates to include the Tooling Hole label, as shown in the following image on the right.

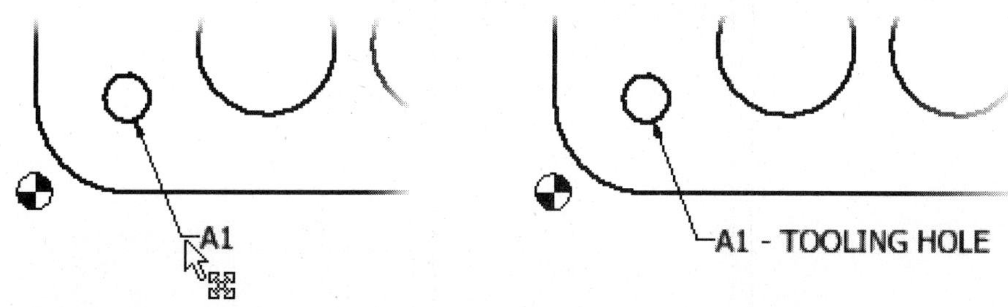

FIGURE 5.165

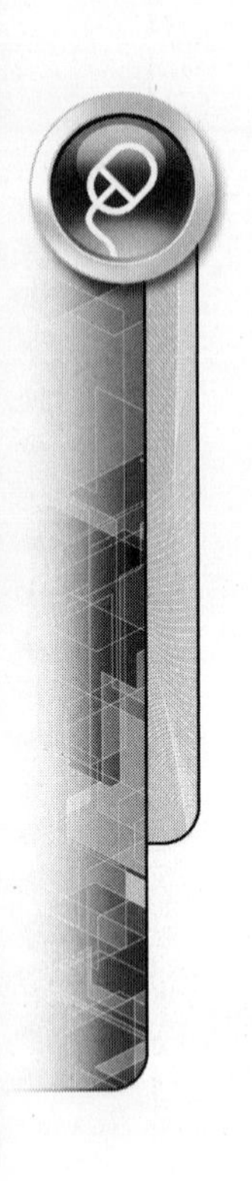

11. In the hole table notice that the information is updated to include the -Tooling Hole label. You can also resize the columns by clicking and dragging on a cell's border, as shown in the following image.

HOLE TABLE			
HOLE	XDIM	YDIM	DESCRIPTION
A1 - TOOLING HOLE	.25	.25	Ø0.13 THRU
B1	.63	.39	Ø0.38 THRU

FIGURE 5.166

12. Close all open files. Do not save changes. End of exercise.

CREATING A TABLE

In addition to organizing hole patterns, tables can also contain data from an iPart or iAssembly. They can be generated manually or created from a Microsoft Excel spreadsheet.

Creating a General Table

To create a generic table that is blank, use the following steps:

To activate the Table command click the General button on the Annotate tab > Table panel, shown in the following image.

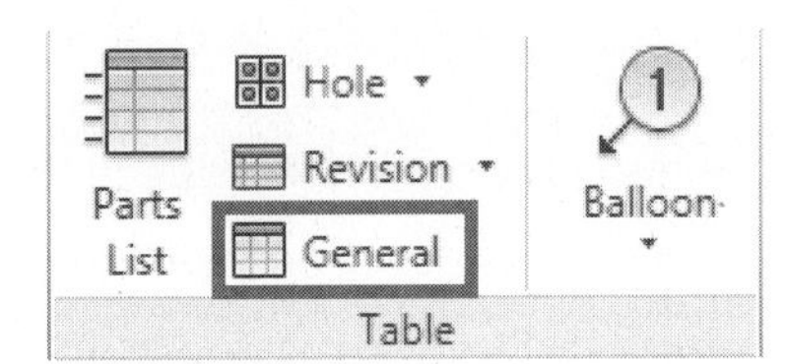

FIGURE 5.167

1. When the Table dialog box appears, verify that <Empty table> is selected from the source list. Then specify the number of rows and columns, as shown in the following image on the left. On the Table dialog box, select <Empty table> from the source list. Clicking the OK button will create the table, as shown in the following image on the right. Notice that the title of the table is blank. Notice also the column headers labeled Column 1 and so on. Both of these items are derived from the current table style.

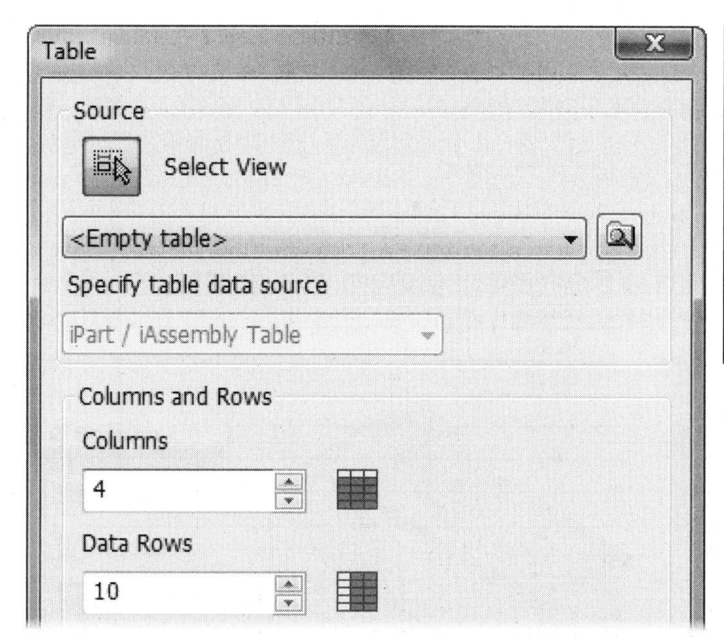

Parameters			
Parameter	Column 2	Column 3	Column 4

FIGURE 5.168

2. To edit the title to the table, double-click the table, and from the Table dialog box click the Table Layout button as shown in the following image on the left and in the Table Layout dialog box type in a new title name, as shown in the following image on the right. Click OK to complete the edit.

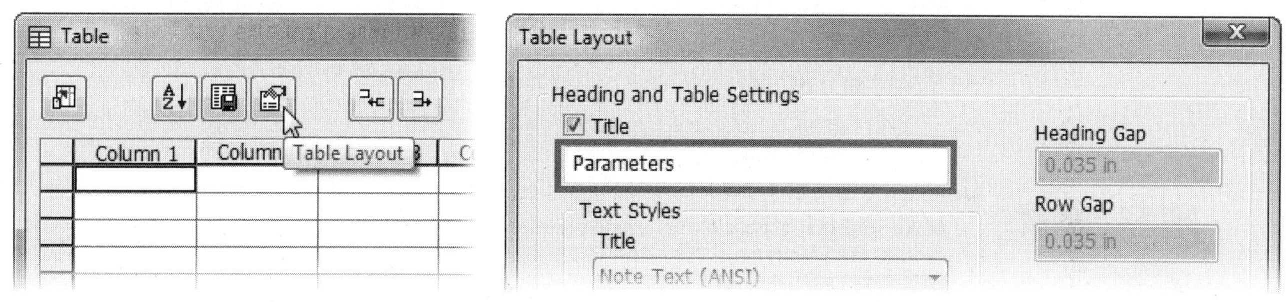

FIGURE 5.169

3. While still in the Table dialog box you can change the column headers by right-clicking on a column header and click Format Column from the menu as shown in the following image on the left. Then in the Format Column dialog box type in a new column heading name as shown in the following image on the right. You can also adjust column width from this dialog box. Click OK to complete the edit.

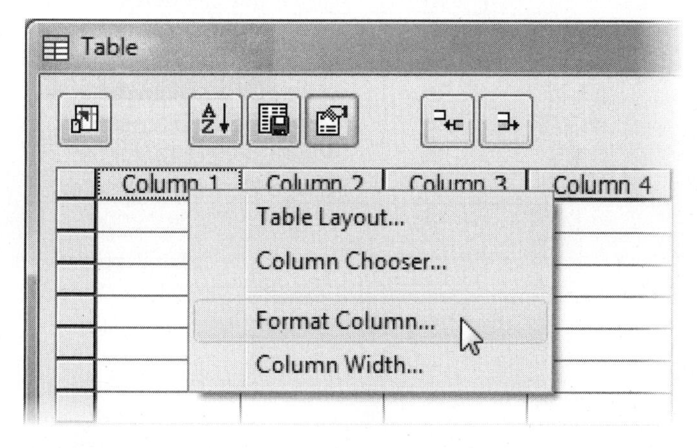

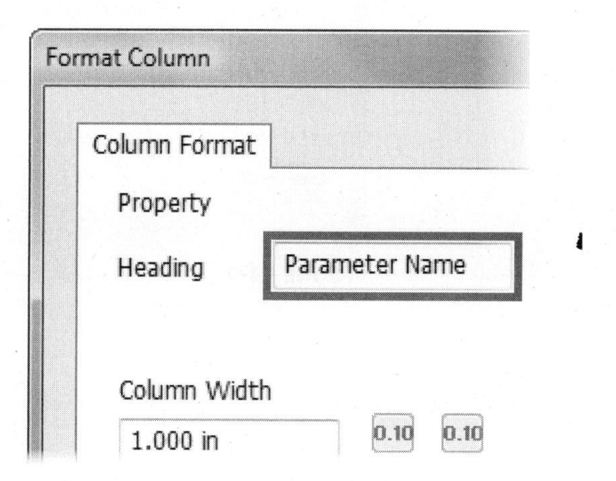

FIGURE 5.170

4. While still in the Table dialog box you can add information to the individual table cells by double-clicking on a cell and type in new information, like what is shown in the following image on the left. Click OK to complete the edit. The following image on the right shows the edited table. You can change the column width by dragging the edges of the table.

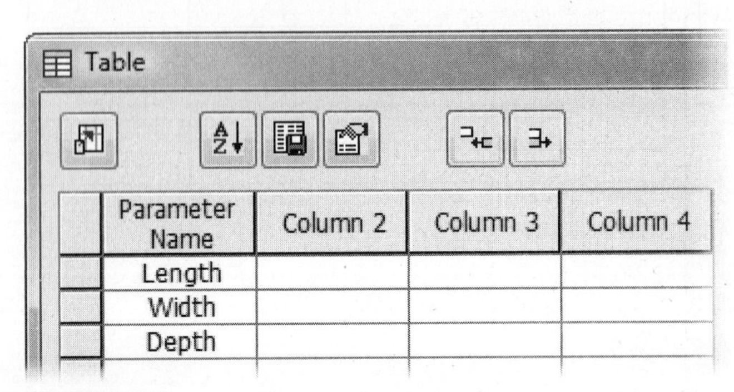

Parameters			
Parameter	Column 2	Column 3	Colu
Length			
Width			
Depth			

FIGURE 5.171

Creating a Table from an Excel Spreadsheet

Tables can also be created from an external Excel spreadsheet using the following steps:

1. To activate the Table command click the General button on the Annotate tab > Table panel.
2. When the Table dialog box appears, click on the Browse for file button to locate the Excel file as the source.
3. In the Table dialog box, specify the start cell and column header row, as shown in the following image on the left.
4. Click OK to place the table on the drawing sheet, as shown in the following image on the right.

NOTE **Make sure that the Excel file is saved in the current project path(s).**

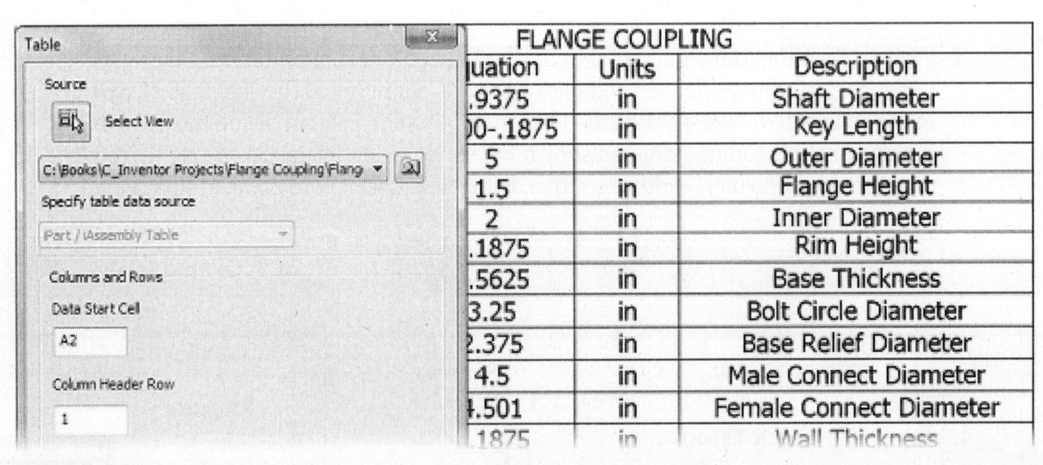

FLANGE COUPLING		
uation	Units	Description
.9375	in	Shaft Diameter
)0-.1875	in	Key Length
5	in	Outer Diameter
1.5	in	Flange Height
2	in	Inner Diameter
.1875	in	Rim Height
.5625	in	Base Thickness
3.25	in	Bolt Circle Diameter
2.375	in	Base Relief Diameter
4.5	in	Male Connect Diameter
4.501	in	Female Connect Diameter
.1875	in	Wall Thickness

FIGURE 5.172

CREATING A REVISION TABLE

During the course of documenting mechanical designs, the drawing document typically undergoes numerous changes. For legal, technical, and other reasons, companies have adopted the practice of maintaining records of these changes, also known as revisions. Revision records come in the form of Engineering Change Orders (ECO) and Engineering Change Notices (ECN). To formally display and keep track of these changes on the drawing, you should insert a revision table. In the revision table add a new row for each revision. Two commands are available for tracking revisions in a drawing: the Revision Table command and the Revision Tag command. You can find the Revision Table command on the Annotate tab, as shown in the following image on the left.

Clicking on the Revision Table command will launch the Revision Table dialog box, as shown in the following image on the right. Use this dialog box to define the scope and revision indexing methods. These items are described as follows:

Two settings are contained under the Table Scope heading. The Entire Drawing setting will create a revision table for the entire drawing no matter how many sheets are created. The Active Sheet setting will create a revision table for the active sheet. In this manner, different revision tables can be applied to the different drawing sheets.

Under the Revision Index heading, Auto-Index is checked by default. This setting automatically indexes the revisions. If this feature is not checked, the revision cell for any new revision rows will remain empty or blank. The Alpha setting uses alphabetic indexing, and the Numeric setting uses numeric indexing. The Start Value sets the initial revision number or letter.

When the Update property on revision number edit box is selected, the revision number in the active row of the revision table is connected with the revision number property saved in drawing iProperties or sheet properties.

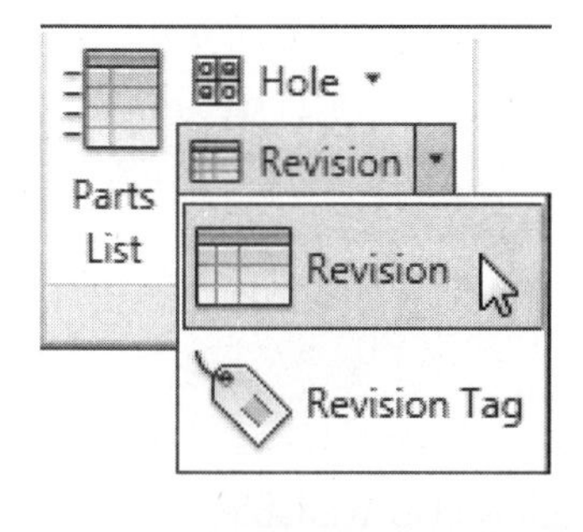

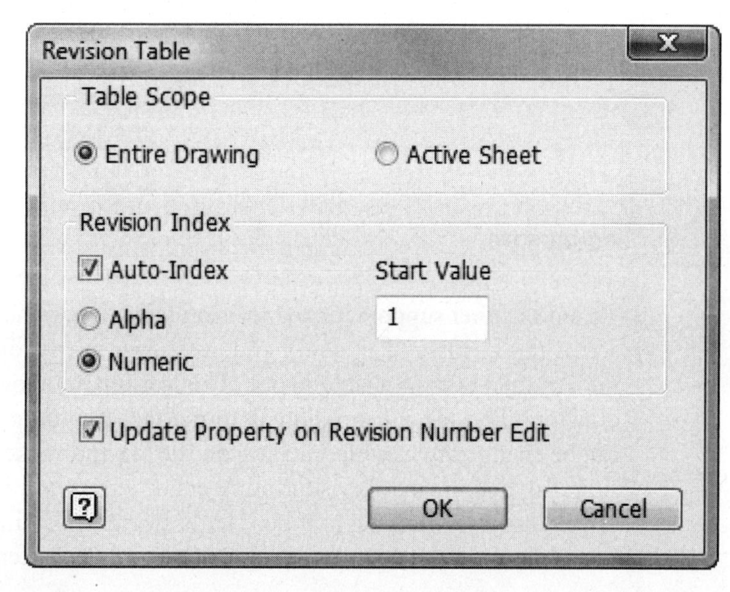

FIGURE 5.173

After making changes in the Revision Table dialog box, the revision table attaches to your cursor for placement in a drawing. The following image shows the default revision table.

REVISION HISTORY				
ZONE	REV	DESCRIPTION	DATE	APPROVED
	1			

FIGURE 5.174

When the revision table is placed in the desired location, you edit the contents of the various fields in order to reflect the revision change in the drawing. To do this, double-click on the text in the revision table. This will launch the Revision Table: Drawing Scope dialog box, as shown in the following image. Add the appropriate text to the dialog box, and click the OK button when finished. The cells can also be linked to iProperty fields, after typing in a value and clicking in a different cell and if the previous cell turns blue, this cell's value is different than what is in the iProperty field. To change the value to the iProperty value right-click in the cell and click Static Value from the menu.

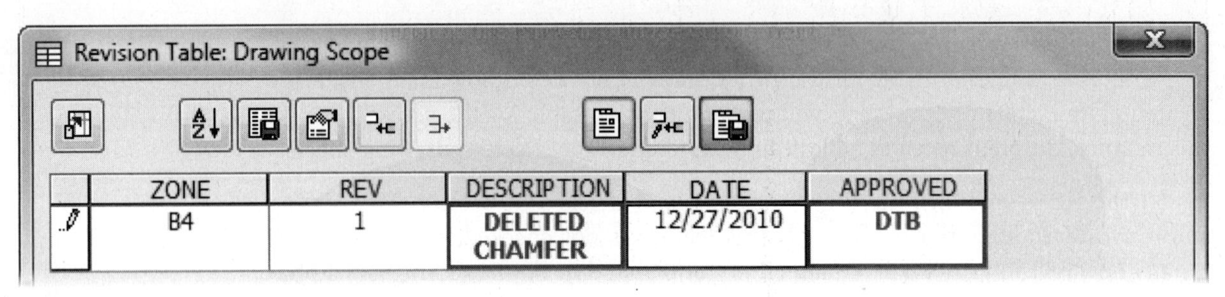

FIGURE 5.175

Extra rows can be added to the revision table by right-clicking on the revision table and picking Add Revision Row from the menu, as shown in the following image.

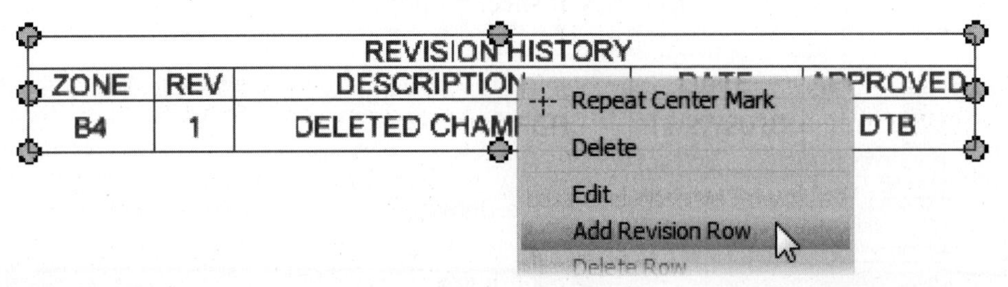

FIGURE 5.176

To lend further support for the revision table, a revision tag is also available. You can access the Revision Tag command, as shown in the following image on the left, on the Annotate tab > Table panel. This command allows you to tag an item on the drawing. The tag information is formatted according to the revision tag style but can be changed by double-clicking on the tag and make changes from the Edit Revision Tag dialog box. Revision numbers or letters will increment automatically. You can change a revision tag number / letter by right-clicking on a tag and click Tag from the menu and select the desired revision number / letter.

The following image illustrates a revision table and tag applied to an object in a drawing file.

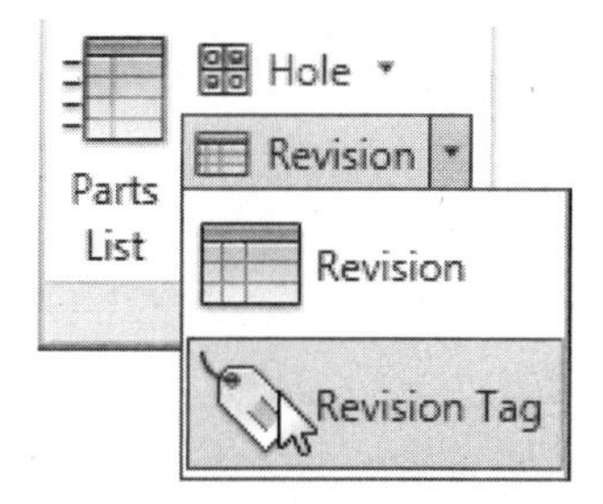

REVISION HISTORY				
ZONE	REV	DESCRIPTION	DATE	APPROVED
B4	1	DELETED CHAMFER	12/27/2010	DTB

1

0.65

FIGURE 5.177

APPLYING YOUR SKILLS

Skill Exercise 5-1

In this exercise, you will create a drawing of a part.

1. Create a new drawing based on the *English ANSI (in).idw* template.
2. Change the sheet size to A.
3. Create drawing views from the file *ESS_Skills_5-1.ipt* from the Chapter 05 folder. Use a scale of 1:1 for all views.
4. Add centerlines, dimensions, and hole notes.

When done your drawing should resemble the following image.

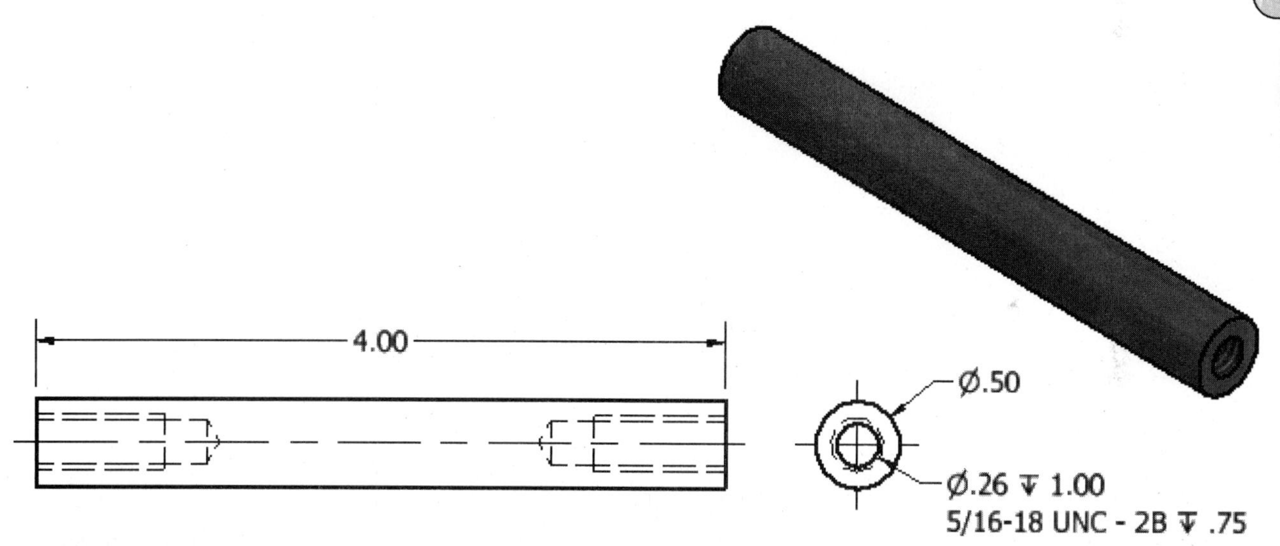

FIGURE 5.178

Skill Exercise 5-2

In this exercise, you will create a drawing for a drain plate cover.

1. Begin by starting a new drawing using a *English ANSI (in).dwg template.*
2. Change the sheet size to A.
3. Create three views of the part *ESS_Skills_5-2.ipt* in the Chapter 05 folder. Use a scale of 1:1 for all views.
4. Add center marks to the views.
5. Change the Default (ANSI) dimension style:
   - Change the Extension length that the extension line goes beyond the dimension leader line to **.0625 in**.
   - Set the precision to **3.123**.
   - Set text height to **.100 in**.

6. Add dimensions and annotations. Your display should resemble the following image.

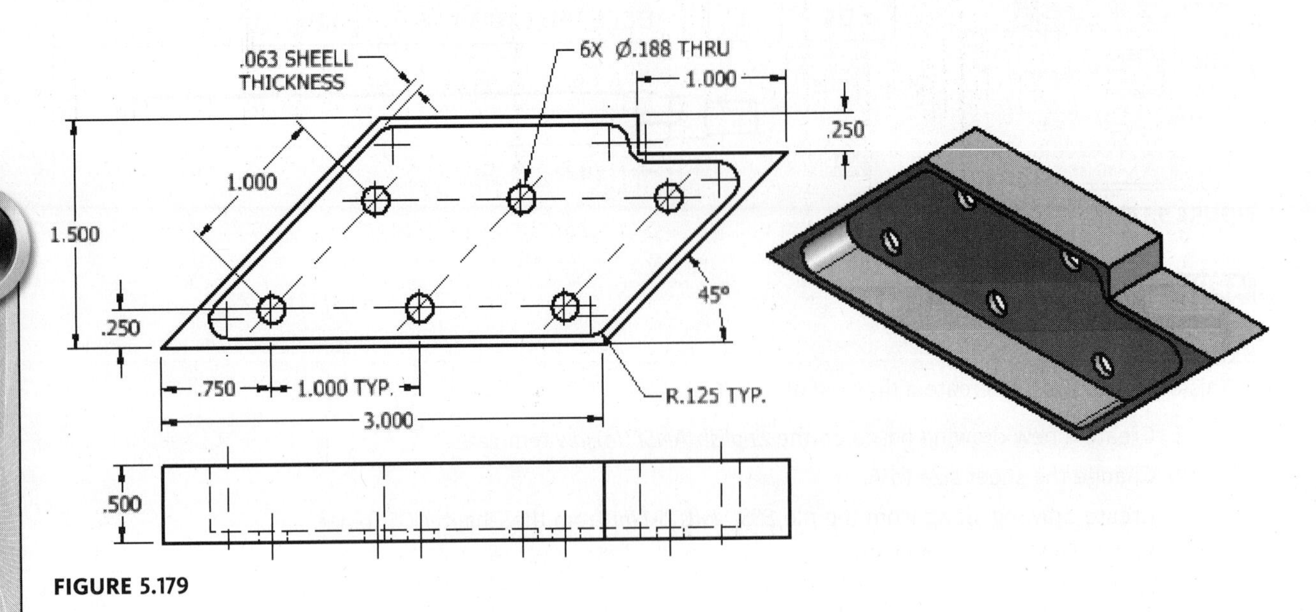

FIGURE 5.179

CHECKING YOUR SKILLS

Use these questions to test your knowledge of the material covered in this chapter.

1. True ___ False ___ A drawing can have an unlimited number of sheets.
2. Explain how to change a sheet's size.
3. True ___ False ___ There can only be one base view per sheet.
4. True ___ False ___ An inclined view is a view that is projected perpendicular to a selected edge or line in a base view.
5. True ___ False ___ An isometric view can only be projected from a base view.
6. True ___ False ___ A section view is a view created by sketching a line or multiple lines that will define the plane(s) that will be cut through a part or assembly.
7. True ___ False ___ Drawing dimensions can drive dimensional changes parametrically back to the part.
8. Explain how to shade an isometric drawing view.
9. True ___ False ___ When creating a hole note using the Hole/Thread Notes command, circles that are extruded to create a hole can be annotated.
10. True ___ False ___ An assembly model can open an individual part from its browser; however, it cannot open a drawing file from the browser.
11. True ___ False ___ When you create a title block in a drawing, the new title block will automatically be written back to the template file.
12. True ___ False ___ When creating a section view that goes through holes at different angles, you use the projected option to prevent the holes from being distorted.
13. True ___ False ___ A hatch pattern in a section view is changed via the Document Settings.
14. True ___ False ___ Create a Segment View to create a section view that has no depth.

15. True ___ False ___ When creating a Detail View you can use a circular or rectangular fence.
16. True ___ False ___ Use the Break Out command to remove a middle section from a drawing view of long part.
17. True ___ False ___ The Styles and Standards Editor only controls dimensions styles.
18. True ___ False ___ The Centerline Bisector command can only be used to create a centerline between two parallel lines.
19. True ___ False ___ Use the table command to insert data from an Excel file into a drawing.
20. True ___ False ___ A revision table data is only filled in by entering data into the revision table cells.

CHAPTER 6

Creating and Documenting Assemblies

INTRODUCTION

In the first four chapters, you learned how to create a component in its own file. In this chapter, you will learn how to place individual component files into an assembly file. You will also learn to create components in the context of the assembly file. After creating components, you will learn how to constrain the components to one another using assembly constraints, edit the assembly constraints, check for interference, and create presentation files that show how the components are assembled or disassembled. Manipulating and editing a Bill of Materials (BOM) is also discussed, including the placement of a Parts List and identifying balloons.

OBJECTIVES

After completing this chapter, you will be able to perform the following:

- Understand the assembly options
- Place components into an assembly
- Create components and assemblies
- Create subassemblies
- Constrain components together using assembly constraints
- Edit assembly constraints
- Pattern components in an assembly
- Check parts in an assembly for interference
- Drive constraints
- Create a presentation file
- Manipulate and edit the Bill of Materials (BOM)
- Create individual and automatic balloons
- Create and perform edits on a Parts List of an assembly

CREATING ASSEMBLIES

As you have already learned, component files have the *.ipt* extension, and they can only have one component each. In this chapter, you will learn how to create assembly files (*.iam* file extension). An assembly file holds the information needed to assemble the components together. All of the components in an assembly are *referenced in*, meaning that each component exists in its own component IPT file, and its definition is linked into the assembly. You can edit the components while in the assembly, or you can open the component file and edit it. When you have made changes to a component and saved the component, the changes will be reflected in the assembly after you open or update it. There are three methods for creating assemblies: bottom-up, in-place, and a combination of both. *Bottom-up* refers to an assembly in which all of the components were created in individual component files and are referenced into the assembly. The *in-place* approach refers to an assembly in which the components are created from within the context of the assembly. In other words, the user creates each component from within the top-level assembly. Each component in the assembly is saved to its own IPT file. The following sections describe the bottom-up and in-place assembly techniques.

Note that assemblies are not made up only of individual parts. Complex assemblies are typically made up of subassemblies, and therefore you can better manage the large amount of data that is created when building assemblies.

To create a new assembly, click New on the Get Started tab, and then click the Standard.iam icon, as shown in the following image. Alternately, you can click the down arrow of the Inventor icon located in the upper-left corner and click the New icon. Using either method, select Assembly from the list of templates. After issuing the new assembly operation, Autodesk Inventor's commands will change to reflect the new assembly environment. The assembly commands appear on the ribbon, and these commands are covered throughout this chapter.

NOTE

There are a number of valid approaches to creating an assembly. You will determine which method works best for the assembly that you are creating based on experience. You can create an assembly using the bottom-up technique, the in-place technique, or a combination of both. Whether you place or create the components in the assembly, all of the components will be saved to their own individual IPT files, and the assembly will be saved as an IAM file.

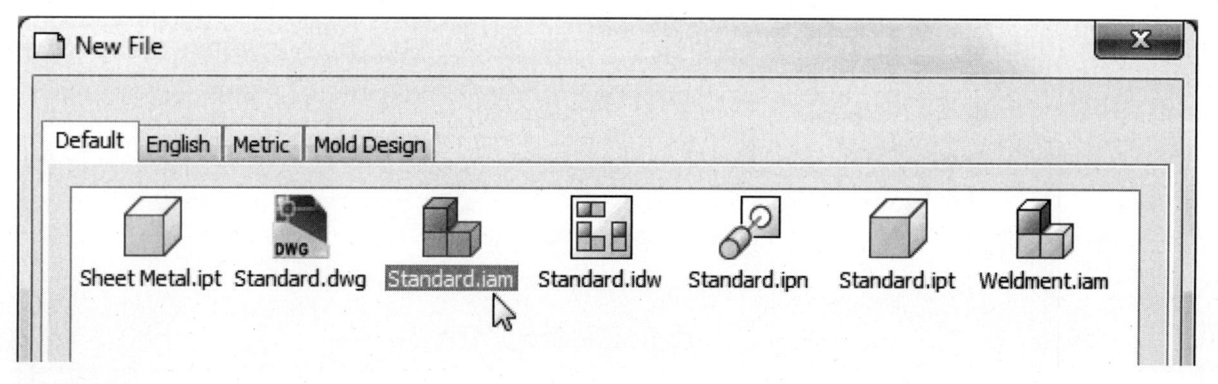

FIGURE 6.1

Assembly Options

Before creating an assembly, review the assembly option settings. On the Tools tab, click Application Options, and the Application Options dialog box will appear. Click on the Assembly tab, as shown in the following image. The following sections describe the various assembly settings. These settings are global and will affect how new components are created, referenced, analyzed, or placed in the assembly.

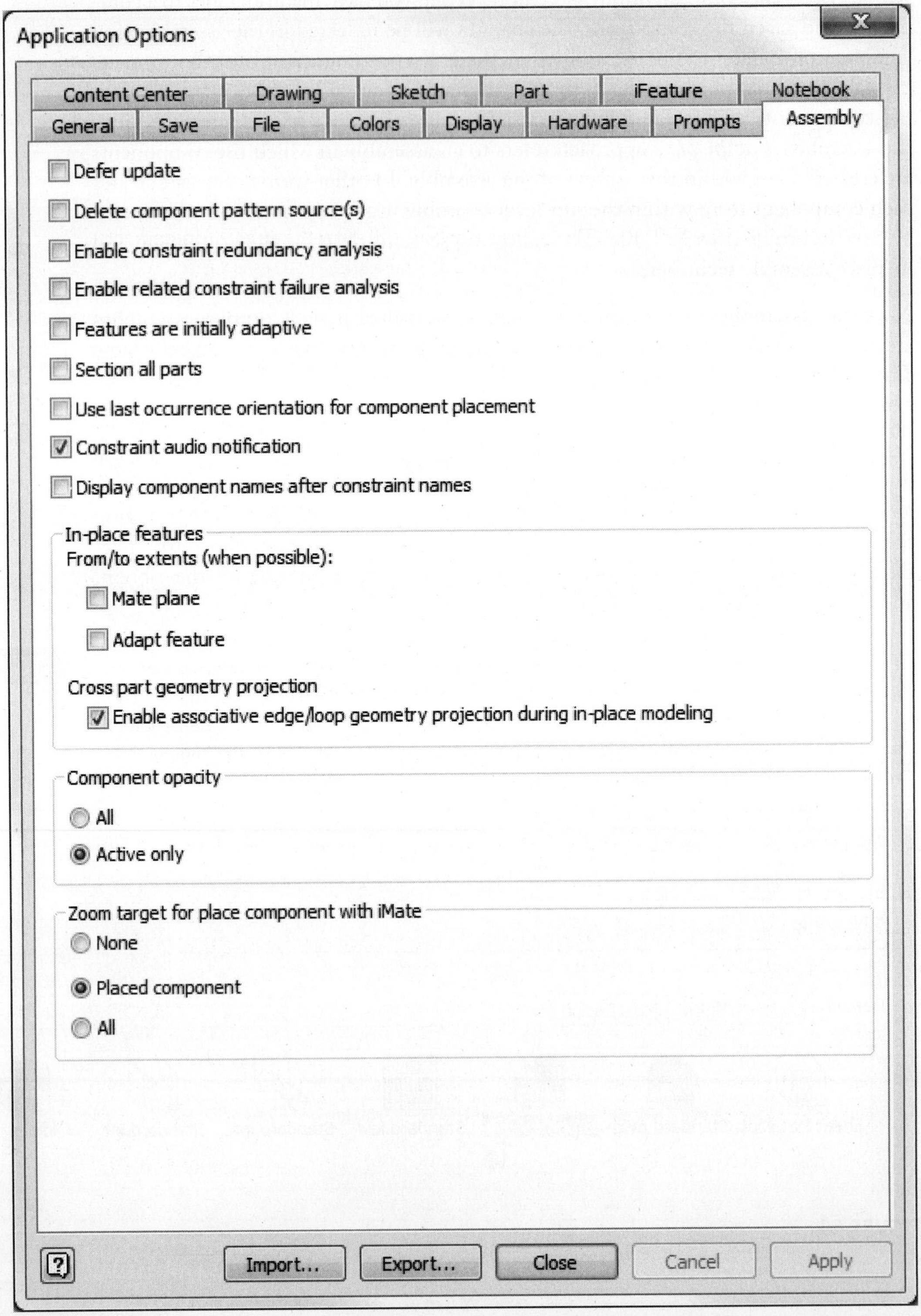

Application Options

Content Center | Drawing | Sketch | Part | iFeature | Notebook
General | Save | File | Colors | Display | Hardware | Prompts | Assembly

Defer update
Delete component pattern source(s)
Enable constraint redundancy analysis
Enable related constraint failure analysis
Features are initially adaptive
Section all parts
Use last occurrence orientation for component placement
Constraint audio notification
Display component names after constraint names

In-place features
From/to extents (when possible):
Mate plane
Adapt feature
Cross part geometry projection
Enable associative edge/loop geometry projection during in-place modeling

Component opacity
All
Active only

Zoom target for place component with iMate
None
Placed component
All

Import... | Export... | Close | Cancel | Apply

FIGURE 6.2

Defer Update. Click this option so that when you change a component that will affect the assembly, you will have to click the Update button manually to update the assembly. If you leave the box clear, the assembly will update automatically to reflect the change to a component.

Delete Component Pattern Source(s). Click to delete the source component of a component pattern when you delete the pattern. If you leave the box clear, the source component will not be deleted when you delete the pattern.

Enable Constraint Redundancy Analysis. Click to perform a secondary analysis of all assembled component constraints. You are notified when redundant constraints exist. The component degrees of freedom (DOF) are also updated but not displayed.

Enable Related Constraint Failure Analysis. This setting enables you to perform an analysis to identify the constraints and components if constraints fail. By default, this setting is turned off. Placing a check in the box will turn the setting on and activate an option in the Constraint Doctor dialog box to enable the analysis.

Features Are Initially Adaptive. Check this option box to make features adaptive when they are created. This is useful when you are creating a large number of adaptive features. However, having too many adaptive features may result in an unstable assembly, and for this reason, leave this mode unchecked.

Section All Parts. Click to section standard parts placed in an assembly from the Content Center libraries when you create a section drawing view through them. To prevent the sectioning of standard parts, leave this box unchecked.

Use Last Occurrence Orientation for Component Placement. This setting controls the orientation of multiple instances of the same component placed into an assembly. When checked, the orientation of the last occurrence in the browser is used to determine the orientation of new instances.

Constraint Audio Notification. This setting controls whether a sound is made once a constraint is created in an assembly. By default, this option is turned on.

Display Component Names After Constraint Names. When checked the component's instance name is appended to the constraints name in the browser.

In-Place Features

From/To Extents (When Possible). This command determines whether or not a feature will be adaptive when the To or From/To option is selected for the extrusion extent. If both options are selected, Autodesk Inventor will try to make the feature adaptive. If it cannot, it will terminate at the selected face.

Mate plane. If checked, when you create a new component it will have a mate constraint applied to the plane on which it was constructed. It will not be adaptive.

Adapt feature. If checked, a new component will adapt to the plane on which it was constructed.

Cross Part Geometry Projection

Enable Associative Edge/Loop Geometry Projection during In-Place Modeling. If checked, when geometry is projected from another part onto the active sketch, the projected geometry will be associative and it will update when changes are made to the parent part. You can use projected geometry to create a sketched feature.

Component Opacity. When a component is edited inside an existing assembly, the remainder of the assembly takes on a faded appearance; this action will be discussed later in this chapter. In this section, you determine if all components or only the active component will be opaque when it is edited in place in the assembly.

All. Click to make all components in the assembly opaque when the active component is edited in place in the assembly.

Active Only. Click to make only the active component in the assembly opaque when the active component is edited in place in the assembly.

Zoom Target for Place Component with iMate. Set the default zoom behavior for the graphics window when placing components with iMates.

None. Click to perform no zooming and leave the graphics display as is.

Placed Component. Click to zoom in on the placed part so that it fills the graphics window.

All. Click to zoom in on the assembly so that all elements in the model fill the graphics window.

Placing Components

To place a component(s) you create the components in their individual files and then place them into an assembly. If you place an assembly file into another assembly, it will be brought in as a subassembly. Be sure to include the path(s) for the file location(s) of the placed components in the project file; otherwise, Autodesk Inventor may not be able to locate the referenced component when you reopen the assembly.

To insert a component into the current assembly, click the Place Component command on the Assemble tab, as shown in the following image on the left, press the shortcut key P, or right-click in the graphics window, and select Place Component from the marking menu.

The Place Component dialog box will appear, as shown in the following image. Select the component to place or select multiple components by holding down the CTRL or SHIFT key and select the components from the list. If the component you are placing is the first in the assembly, an instance of it will be placed into the assembly automatically. If the component is not the first, you pick a point in the assembly where you want to place the component. If you need multiple occurrences of the component in the assembly, continue selecting placement points. When done, press the ESC key, or right-click and select Done from the menu.

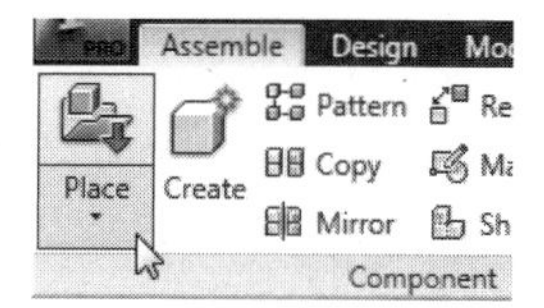

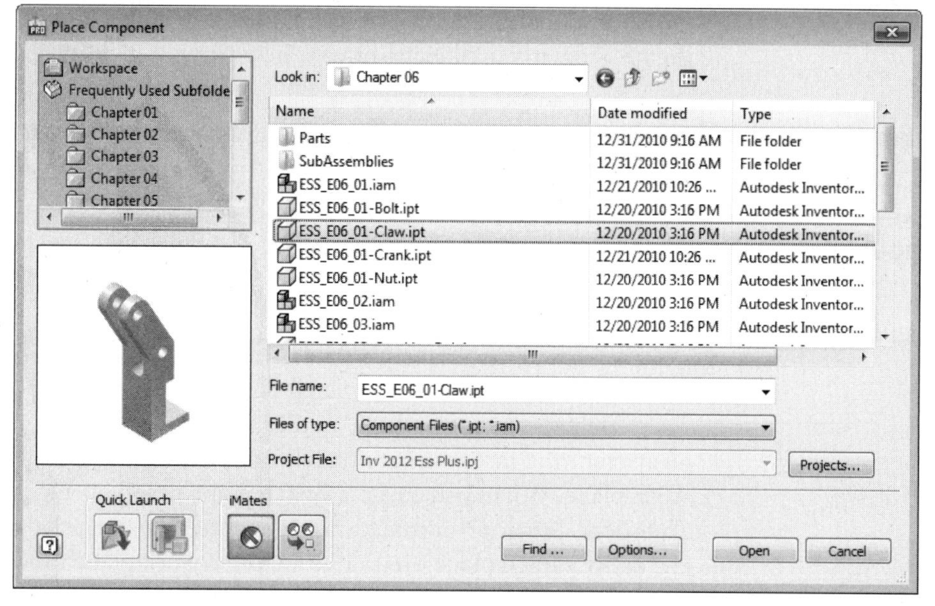

FIGURE 6.3

NOTE

When placing or creating components in an assembly, it is recommended to list them in the order in which they are assembled. The order is important when placing assembly constraints. To reorder a component, in the browser click and drag on a component to move it up or down.

CREATING PARTS IN PLACE

Most components in the assembly environment are created in relation to existing components in the assembly. When creating an in-place component, you can sketch on the face of an existing assembly component or a work plane. You can also click the graphics window background to define the current view orientation as the XY plane of the assembly in which you are presently working. If the YZ or XZ plane is the default sketch plane, you must reorient the view to see the sketch geometry. Click Application Options on the Tools tab, and then click on the Part tab to set the default sketch plane.

When you create a new component, you can select an option in the Create In-Place Component dialog box to constrain the sketch plane to the selected face or work plane automatically, as shown in the following image. After you specify the location for the sketch, the new part immediately becomes active, and the browser and ribbon switch to the part environment.

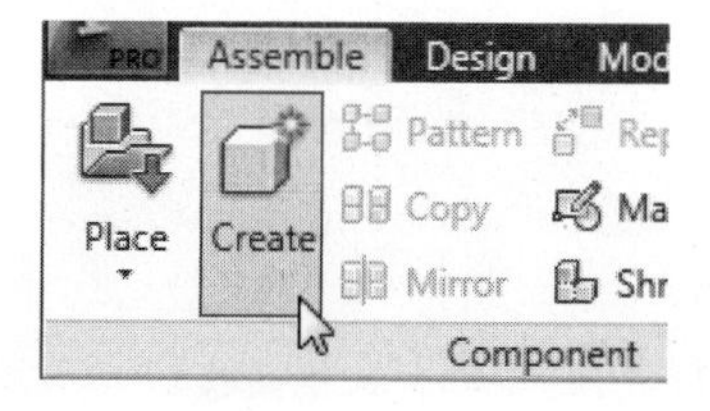

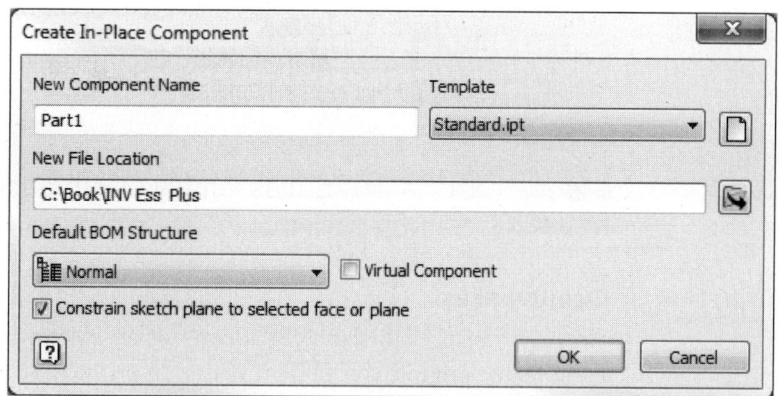

FIGURE 6.4

Notice also that the Sketch tab, as shown in the following image, is available to create sketch geometry in Sketch1 of your new part.

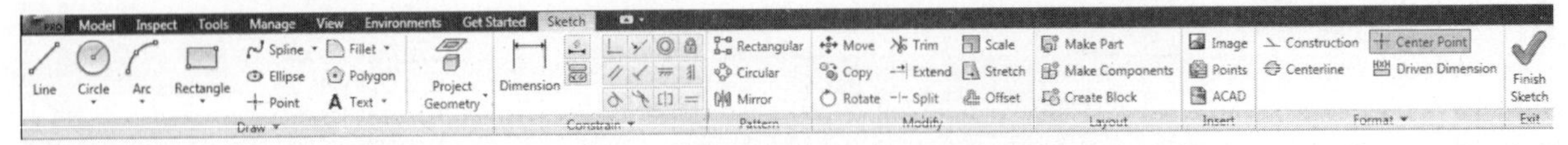

FIGURE 6.5

After you create the base feature of your new part, you can define additional sketches based on the active part or other parts in the assembly. When defining a new sketch, you can click a planar face of the active part or another part to define the sketch plane on that face. You can also click a planar face and drag the sketch away from the face to create the sketch plane automatically on the resulting offset work plane. When you create a sketch plane based on a face of another component, Autodesk Inventor automatically generates an adaptive work plane and places the active sketch plane on it. The adaptive work plane moves as necessary to reflect any changes in the component on which it is based. When the work plane adapts, your sketch moves with it. Features based on the sketch then adapt to match its new position.

After you finish creating a new part, you can return to assembly mode by double-clicking the assembly name in the browser. In assembly mode, assembly constraints become visible in the browser. If you selected the Constrain Sketch Plane to Selected Face option when you created your new part, a flush constraint will appear in the Assembly browser. As with all constraints, you can delete or edit this constraint at any time. No flush constraint is generated if you create a sketch by clicking in the graphics window or if the box is clear when selecting an existing part face.

The Assembly Browser

The Assembly browser displays the hierarchy of all parts and subassembly occurrences and assembly constraints in the assembly, as shown the following image. Each occurrence of a component is represented by a unique name. In the browser, you can select a component for editing, move components between assembly levels, reorder assembly components, control component status, rename components, edit assembly constraints, and manage design views and representations.

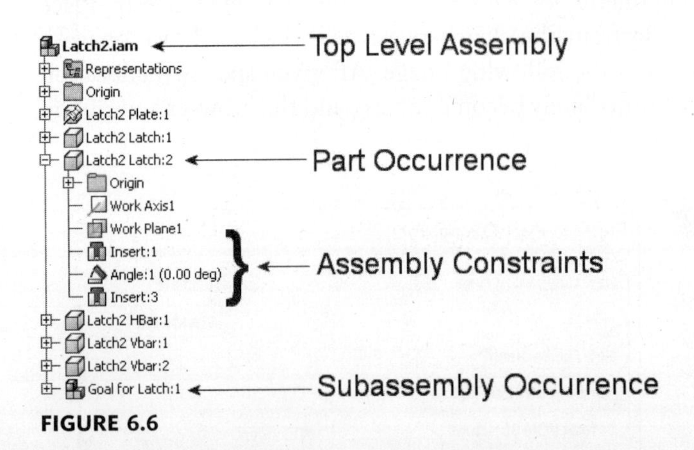

FIGURE 6.6

Occurrences

An occurrence, or instance, is a copy of an existing component and has the same name as the original component with a colon and a sequenced number. If the original component is named Bracket, for example, the next occurrence in the assembly will be Bracket:1 and a subsequent occurrence will be Bracket:2, as shown in the following image. If the original component changes, all of the component instances will reflect the change.

To create an occurrence, place the component. If the component already exists in the assembly, you can click the component's icon in the browser, and drag an additional occurrence into the assembly. You can also use the copy-and-paste method to place additional components by right-clicking on the component name in the browser, or on the component itself in the graphics window, and then right-clicking and selecting Copy from the menu. Then right-click and select Paste from the menu. You can also use the Windows shortcuts CTRL-C and CTRL-V to copy and paste the selected component. If you want an occurrence of the original component to have no relationship with its source component, make the original component active, use the Save Copy As command. This command is found by clicking the Inventor Application button located in the upper left corner of the display screen. Then click the arrow next to Save As and click Save Copy As, where you will enter a new name. The new component will have no relationship to the original, and you can place it in the assembly using the Place Component command.

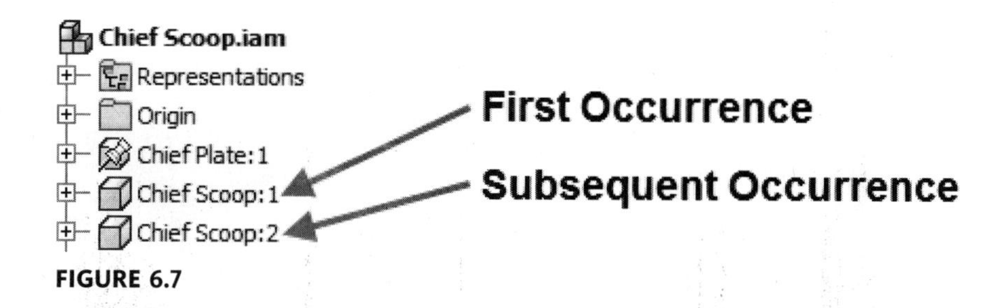

FIGURE 6.7

The Assembly Capacity Meter

To provide feedback on the resources used by an assembly model, use the Assembly Capacity Meter. This meter is located in the lower-right corner of the display screen and is present only in the assembly environment. Three different types of information are displayed in the status bar of this meter. The first numbered block deals with the number of occurrences or instances in the active assembly. The second numbered block displays the total number of files open in order to display the assembly. The third block is a graphical display of the total amount of memory being used to open and work in this assembly.

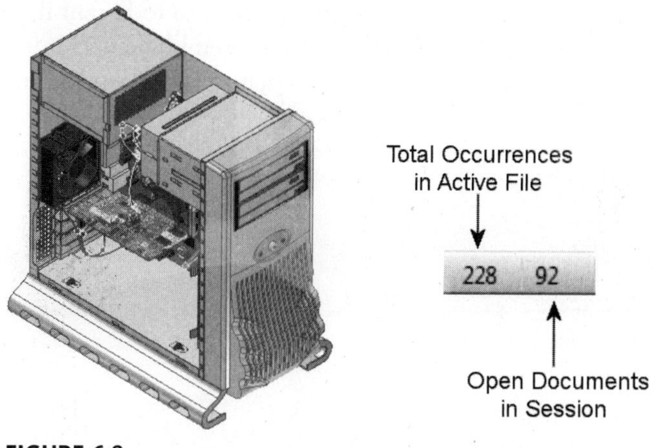

FIGURE 6.8

EDITING A COMPONENT IN PLACE

To edit a component while in an assembly, activate the component. Only one component in the assembly can be active at a time. To make a component active, double-click on the component in the graphics window, or double-click on the file name or icon in the browser. Alternately, right-click on the component name in the browser or graphics window, and select Edit from the menu. Once the component is active, the other component names in the browser will appear shaded, as shown in the following image. If Component Opacity, in Application Options on the Assembly tab, is set to Active Only, the other components in the assembly will take on a faded appearance in the graphics window.

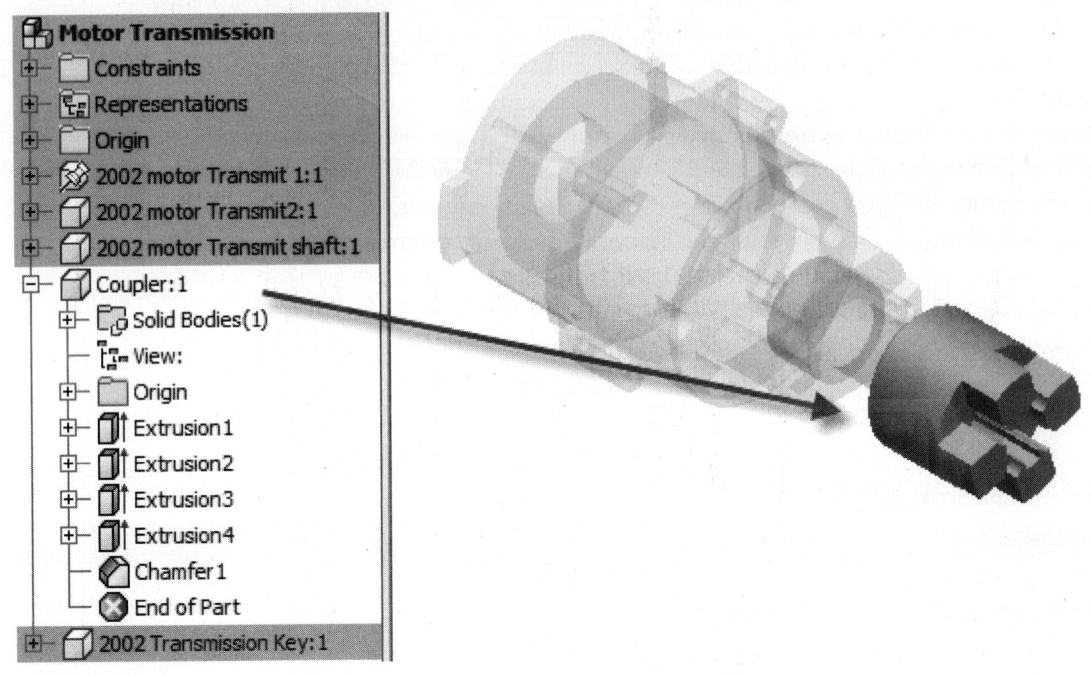

FIGURE 6.9

You can edit the component and save the changes by using the Save command. Only the active component will be saved. To make the assembly active, click the Return button on the Return panel, as shown in the following image on the left which will take you up one level, double-click on the assembly name in the browser, or right-click in the graphics window and select Finish Edit from the marking menu, as shown in the following image on the right. Under the Return button there are two other buttons, as shown in the middle of the following image; Return to Parent that will return you up one level (from the part level to the part subassembly) and Return to Top that will return you to the top level in the browser no matter how deep you are in the browser tree.

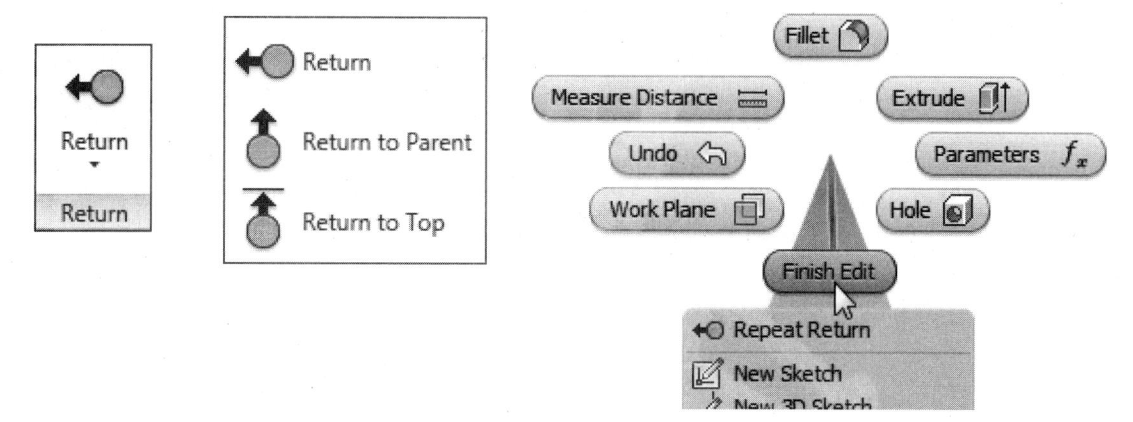

FIGURE 6.10

OPENING AND EDITING ASSEMBLY COMPONENTS

Another way to edit a component in the assembly is to open the component in another window. Click the Open command on the Inventor icon, or right-click on the component's name in the browser or on the component in the graphics window. Select Open from the menu, as shown in the following image. The component will appear in a new window. Edit the component as needed, save the changes, activate the assembly file, and the changes will appear in the assembly.

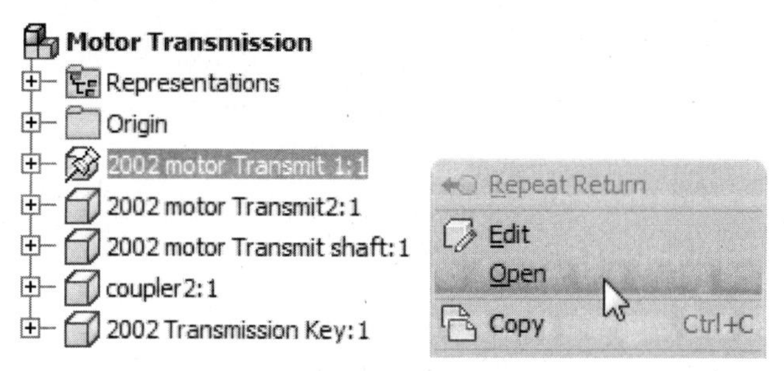

FIGURE 6.11

Grounded Components

When assembling components, you may want to make a component or multiple components grounded or stationary, meaning that they will not move. When applying assembly constraints, the unconstrained components will be moved to the grounded component(s). By default, the first component placed in an assembly is grounded. There is no limit to how many components can be grounded. It is strongly recommended that at least one component in the assembly be grounded; otherwise, the assembly can move. A grounded component is represented with a pushpin superimposed on its icon in the browser, as shown in the following image on the left. To ground or unground a component, right-click on the component's name in the browser, and select or deselect Grounded from the menu, as shown in the following image on the right.

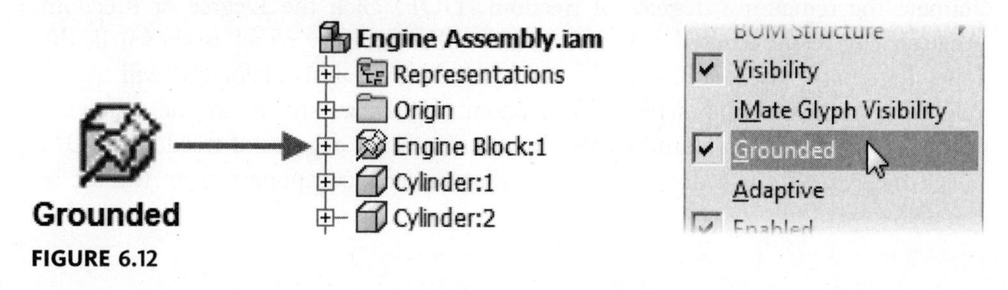

FIGURE 6.12

DEGREES OF FREEDOM (DOF)

You have now learned how to create assembly files, but the components had no relationship to one another except for the relationship that was defined when you created a component from the context of an assembly and in reference to a face on another component. For example, if you placed a bolt in a hole and the hole moved, the bolt would not move to the new hole position. Use assembly constraints to create relationships between components. With the correct constraint(s) applied, if a hole moves, the bolt will move to the new hole location.

In a previous chapter, you learned about geometric constraints. When you apply geometric constraints to sketches, they reduce the number of dimensions or constraints required to fully constrain a profile. When you apply assembly constraints, they reduce the degrees of freedom (DOF) that allow the components to move freely in space.

There are six degrees of freedom: three are translational and three are rotational. Translational means that a component can move along an axis X, Y, or Z. Rotational means that a component can rotate about an axis X, Y, or Z. As you apply assembly constraints, the number of the DOF decreases.

To see a graphical display of the DOF remaining on all of the components in an assembly, select Degrees of Freedom on the View tab, as shown in the following image.

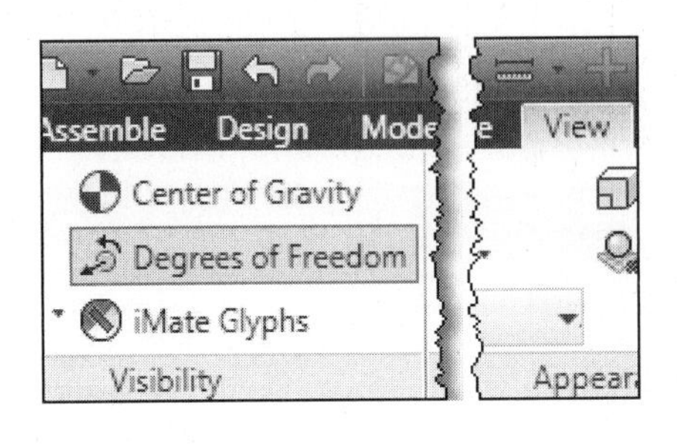

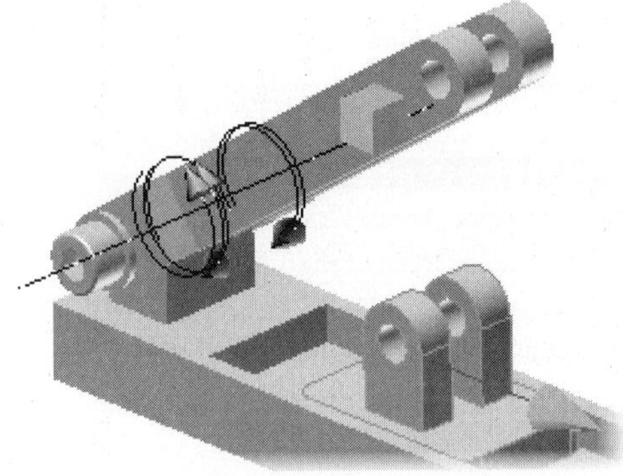

FIGURE 6.13

An icon will appear in the center of the component that shows the DOF remaining on the component. The line and arrows represent translational freedom, and the arc and arrows represent rotational freedom. To turn off the DOF icons, again click Degrees of Freedom on the View menu.

TIP

You can turn on the DOF symbols for single or multiple component(s) by right-clicking on the component's name in the canvas or in the browser, clicking iProperties from the menu, and clicking Degrees of Freedom on the Occurrence tab. If the symbols are turned on, following the same steps will toggle them off.

To better see the remaining degrees of freedom (DOF) you can animate them. To animate the remaining degrees of freedom (DOF) click the Degree of Freedom Analysis command from the Assemble tab > Productivity panel as shown in the following image on the left. The Degree of freedom Analysis dialog box will appear that lists the remaining degrees of freedom (DOF) as shown in the following image on the right. To animate components remaining degrees of freedom (DOF) check the Animate Freedom option and then select a component from the list in the dialog box.

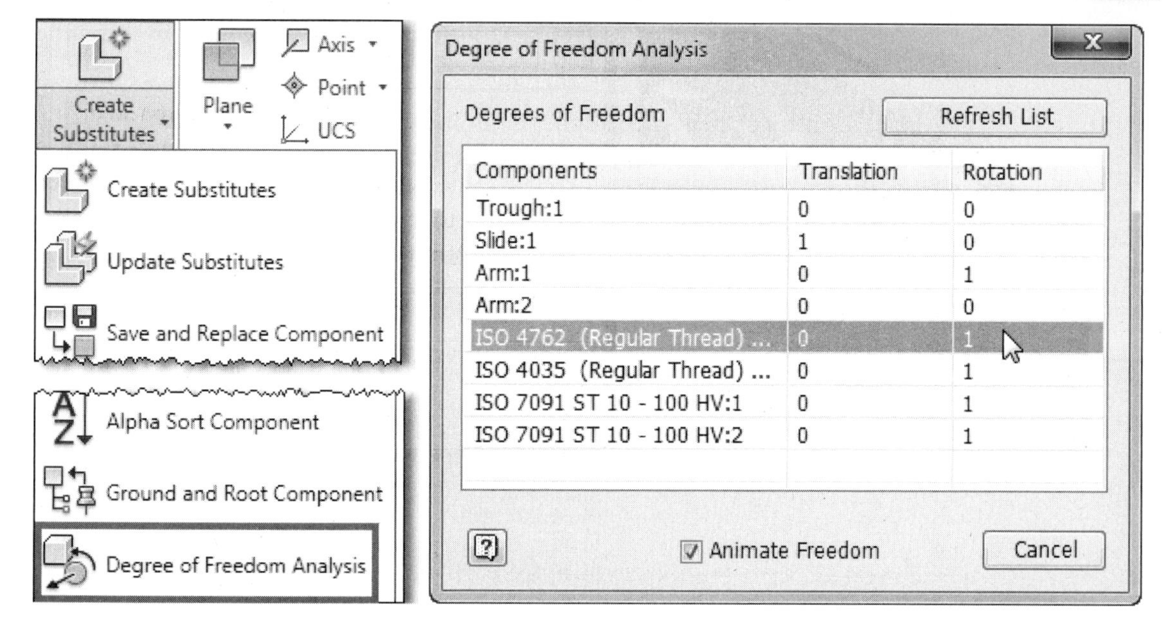

FIGURE 6.14

ASSEMBLY CONSTRAINTS

When constraining components to one another, you will need to understand the terminology. The following terminology is used with assembly constraints:

Line. This can be the centerline of an arc, a circular edge, a cylindrical surface, a selected edge, a work axis, or a sketched line.

Normal. This is a vector that is perpendicular to a planar face.

Plane. This can be defined by the selection of a plane or face to include the following: two noncollinear but coplanar lines or axes, three points, or one line or axis and a point that does not lie on the line or axis. When you use edges and points to select a plane, this creates a work plane, and it is referred to as a construction plane.

Point. This can be an endpoint or midpoint of a line, the center or end of an arc or circular edge, or a vertex created by the intersection of an axis and a plane or face.

Offset. This is the distance between two selected lines, planes, or points, or any combination of the three.

Autodesk Inventor does not require components to be fully constrained. By default, the first component created or added to the assembly will be grounded and will have zero DOF. As discussed earlier, more than one component can be grounded. Other components will move in relation to the grounded component(s).

Assembly Constraint Types

Autodesk Inventor uses four types of assembly constraints (mate, angle, tangent, and insert), two types of motion constraints (rotation and rotation-translation), a transitional constraint, and a constraint set. You can access the constraints through the Constraint command found on the Assemble tab, as shown in the following image on the left, by right-clicking and selecting Constraint from the menu, or by using the hot key C.

The Place Constraint dialog box appears, as shown in the following image on the right. The dialog box is divided into four areas, which are described in the following sections. Depending upon the constraint type, the option titles may change.

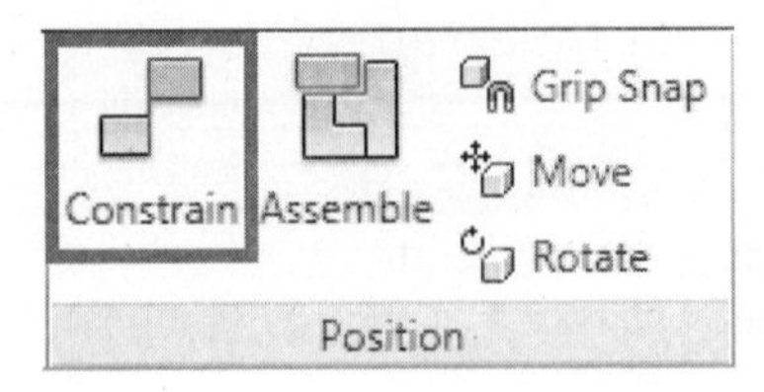

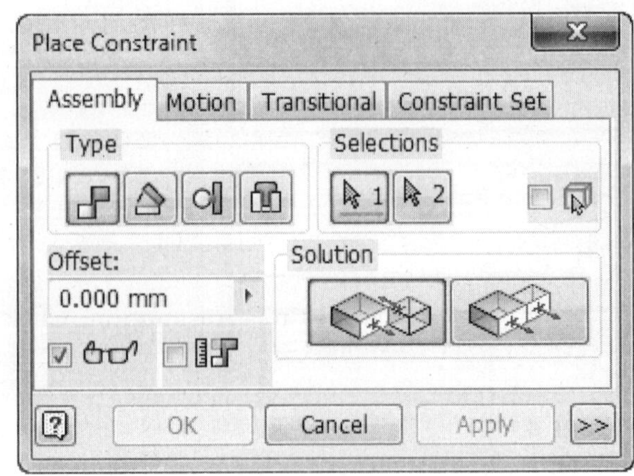

FIGURE 6.15

Assembly Tab

Type. Select the type of assembly constraint to apply: mate, angle, tangent, or insert.

Selections. Click the button with the number 1, and select a component's edge, face, point, and so on, on which to base the constraint type. Then click the button with the number 2, and select a component's edge, face, point, and so on, on which to base the constraint type. By default, the second arrow will become active after you have selected the first input.

Color coding is also available to assist with the assembly process. For example, when picking a face with the number 1 button, the color blue is associated with this selection. In the same way, the color green is associated with the number 2 button selection. This schema allows you to better recognize the selections, especially if they need to be edited.

You can edit an edge, face, point, and so on, of an assembly constraint that has already been applied by clicking the number button that corresponds to the constraint and then selecting a new edge, face, point, and so on. While working on complex assemblies, you can click the box on the right side of the Selections section called Pick part first. If the box has a check, select the component before selecting a component's edge, face, point, and so on.

Offset/Angle. Enter or select a value for the offset or angle. The Offset option changes to Angle when the Angle constraint is being applied.

Solution. Select how the constraint will be applied; the normals will be pointing in the same or opposite directions.

Show Preview. Click and when constraints are applied to two components, you will see the under-constrained components previewed in their constrained positions. If you leave the box clear, you will not see the components assembled until you click the Apply button.

Predict Offset and Orientation. Click to display the existing offset distance between two components. This allows you to accept this offset distance or enter a new offset distance in the edit box.

Motion Tab

Type. Select the type of assembly constraint to apply: rotation or rotation-translation.

Selections. Click the button with the number 1 and select a component's face or axis on which to base the constraint type. You will see a glyph in the graphics window previewing the direction of rotation motion. Click the button with the number 2, and select the component's axis or face on which to base the constraint type. A second glyph appears showing the direction of rotation.

Ratio. Enter or select a value for the ratio.

Solution. Select how the constraint type will be applied. The components will rotate in the same or opposite directions as previewed by the graphics window glyphs.

Transitional Tab

A transitional constraint will maintain contact between the two selected faces. You can use a transitional constraint between a cylindrical face and a set of tangent faces on another part.

Type. Select transitional as the type of assembly constraint to apply.

Selections. Click the First Selection button, and select the first face on the part that will be moving. Click the Second Selection button, and select a face around which the first part will be moving. If there are tangent faces, they will become chained automatically as part of the selected face.

Constraint Set Tab

This option will constrain two UCSs together.

Type. Select the type of assembly constraint to apply: UCS to UCS.

Selections. Select two UCSs on different parts.

Constraint Limits

While placing assembly constraints the geometry may not perfectly fit with the value of the constraint but may fit within a tolerance that is OK. By setting constraint limits you specify the distance or rotation that the geometry can deviate from an exact location without having the constraint fail. To set a limit for a constraint click the More button >> on the lower-right corner of the Place Constraint dialog box and the Limits section will appear as shown in the following image on the left. An assembly constraint that has limits will appear in the browser with a +/– symbol appended to its name as shown in the following image on the right.

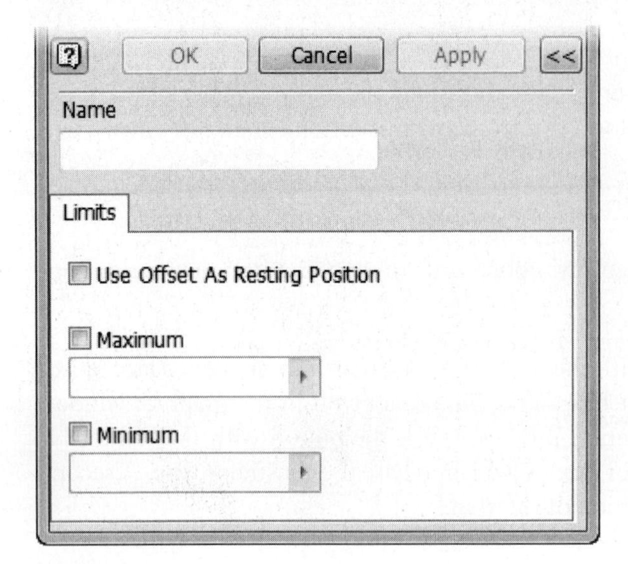

Mate:8 +/-

FIGURE 6.16

Name. Enter a unique name for the constraint in the browser. If left blank a default name is used.

Use Offset or Angle as Resting Position. This option sets a value as the default position of a constraint with limits. When checked, the value entered in either the Offset or Angle field is used as the resting position.

Maximum. This option sets the maximum value that a constraint can move or rotate.

Minimum. This option sets the minimum value that a constraint can move or rotate.

Assembly Constraint Types

This section explains each of the assembly constraint types.

Mate

There are three types of mate constraints: plane, line, and point.

Mate Plane. The mate plane constraint assembles two components so that the faces on the selected planes will be planar to and opposing one another. In the following image, the mate condition is being applied; it is selected in the Solution area of the Place Constraint dialog box.

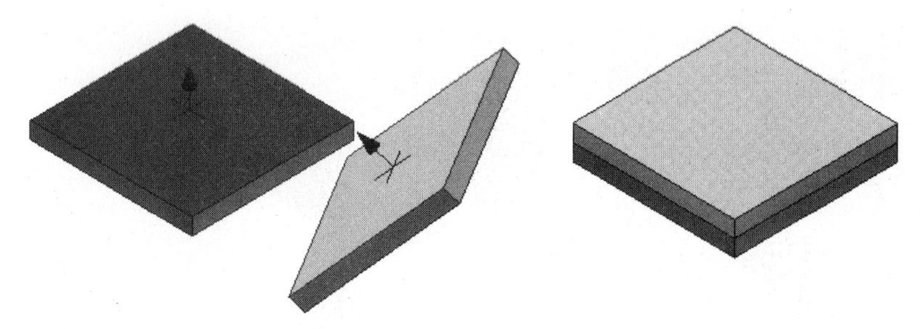

FIGURE 6.17

Mate Line. The mate line constraint assembles the edges of lines to be collinear, as shown in the following image.

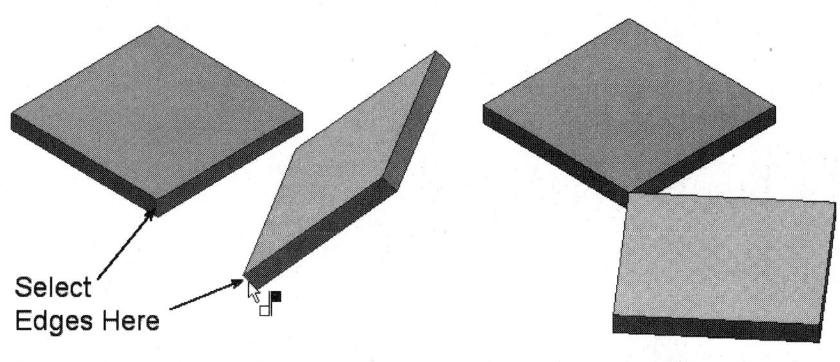

FIGURE 6.18

You can also use the mate line constraint to assemble the center axis of a cylinder with a matching hole feature or axis, as shown in the following image.

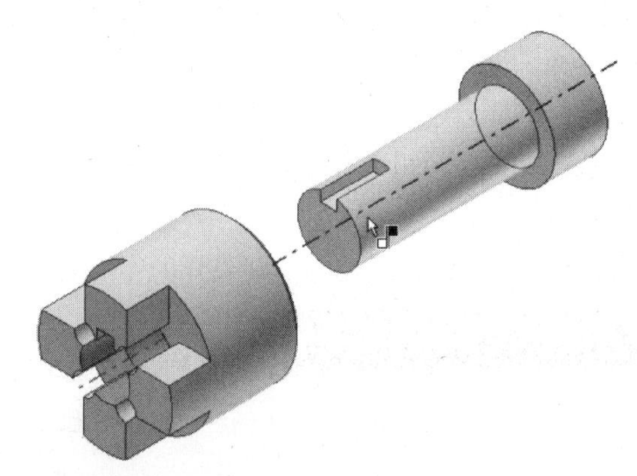

FIGURE 6.19

Mate Point. The mate point constraint assembles two points, such as centers of arcs and circular edges, endpoints, and midpoints, to be coincident, as shown in the following image.

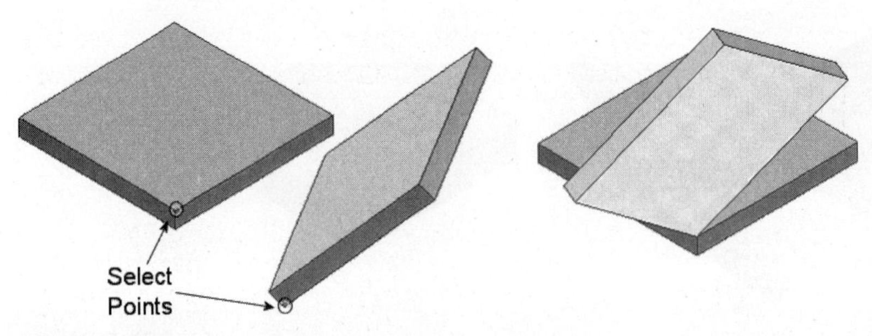

FIGURE 6.20

Mate Flush Solution

The mate flush solution constraint aligns two components so that the selected planar faces or work planes face the same direction or have their surface normals pointing in the same direction, as shown in the following image. Planar faces are the only geometry that can be selected for this constraint.

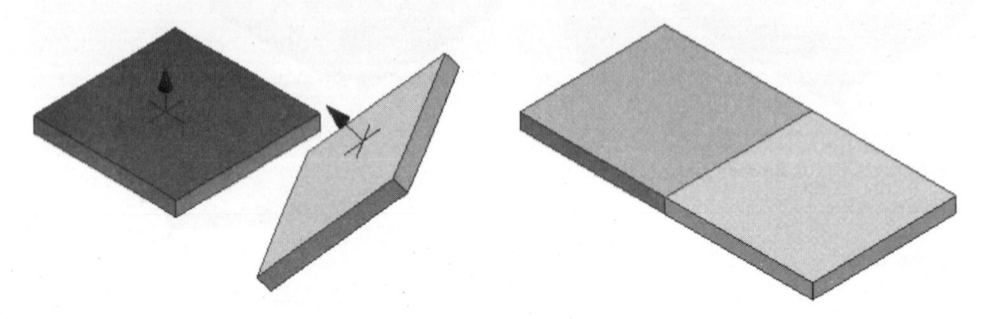

FIGURE 6.21

Angle

The angle constraint specifies the degrees between selected planes or faces or axes. The following image shows the angle constraint with two planes selected and a 30° angle applied.

Three solutions are available when placing an angle constraint: directed angle, undirected angle, and Explicit Reference Vector. The Explicit Reference Vector option requires a third selection that defines the Z axis. You can experiment with both solutions, especially when driving the angle constraint and observing the behavior of the assembly.

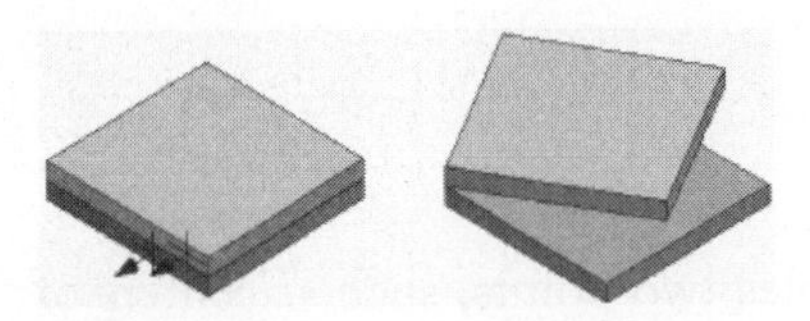

FIGURE 6.22

Tangent

The tangent constraint defines a tangent relationship between planes, cylinders, spheres, cones, and ruled splines. At least one of the faces selected needs to be a curve, and you can apply the tangency to the inside or outside of the curve.

The following image shows the tangent constraint applied to one outside curved face and a selected planar face, as well as the piece with the outer and inner solutions applied.

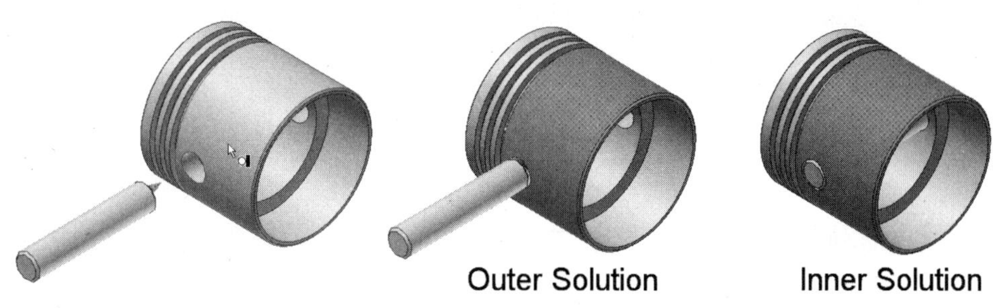

FIGURE 6.23

Insert

The insert constraint takes away five degrees of freedom (DOF) with one constraint, but it only works with components that have circular edges. Select the circular edges of two different components. The centerlines of the selected circles or arcs will be aligned, and a mate constraint will be applied to the planes defined by the circular edges. Circular edges define a centerline/axis and a plane. The following image shows the insert constraint with two circular edges selected and the opposed solution applied.

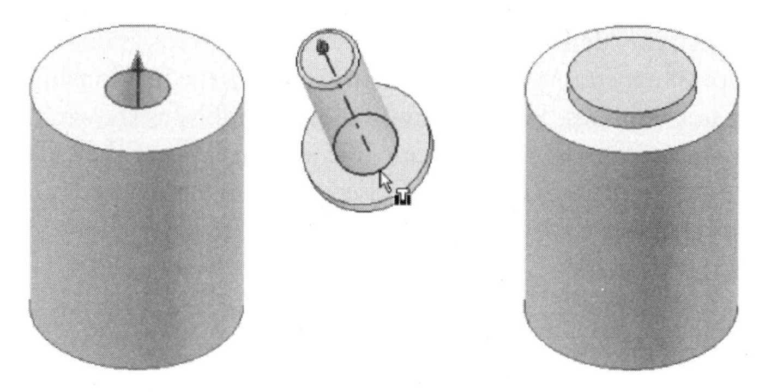

FIGURE 6.24

Motion Constraint Types

There are two types of motion constraints: rotation and rotation-translation, as shown in the Type section of the following image. Motion constraints allow you to simulate the motion relationships of gears, pulleys, rack and pinions, and other devices. By applying motion constraints between two or more components, you can drive one component and cause the others to move accordingly.

Both types of motion constraints are secondary constraints, which means that they define motion but do not maintain positional relationships between components. Constrain your components fully before you apply motion constraints. You can then suppress constraints that restrict the motion of the components you want to animate.

Rotation

The rotation constraint defines a component that will rotate in relation to another component by specifying a ratio for the rotation between the two components. Use this constraint for showing the relationship between gears and pulleys. Selecting the tops of the gear faces displays the rotation glyph, as shown in the following image. You may also have to change the solution type from Forward to Backward, depending on the desired results.

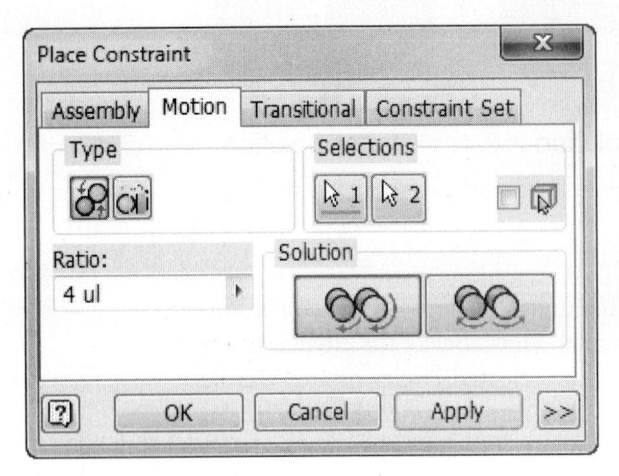

FIGURE 6.25

Rotation-Translation

The rotation-translation constraint defines the rotation relative to translation between components. This type of constraint is well suited for showing the relationship between rack and pinion gear assemblies. In a rack and pinion assembly, as shown in the following image, the top face of the pinion and one of the front faces of the rack are selected. You supply a distance the rack will travel based on the pitch diameter of the pinion gear, and then you can drive the constraints and test the travel distance of the mechanism.

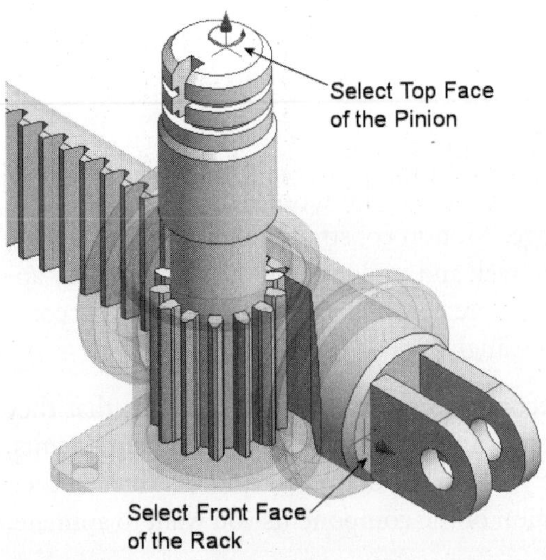

FIGURE 6.26

Creating a Transitional Constraint

The transitional constraint specifies the intended relationship between, typically, a cylindrical part face and a contiguous set of faces on another part, such as a cam follower in a cam slot. The transitional constraint maintains contact between the faces as you slide the component along open DOF. Access this constraint type through the Transitional tab of the Place Constraint dialog box, as shown in the following image.

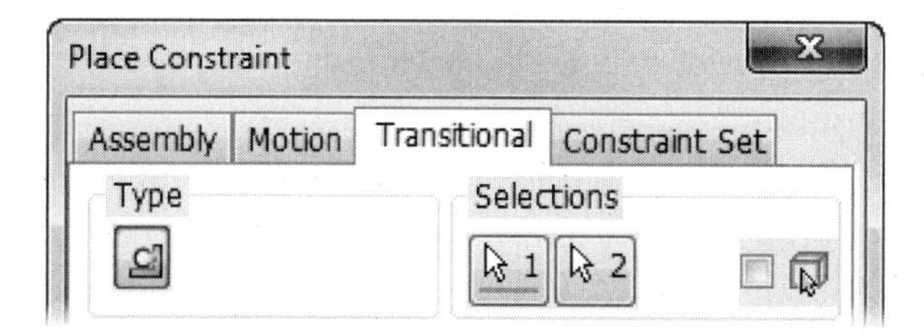

FIGURE 6.27

Select the moving face first on the cam as shown in the following image on the left. Next, select the transition face, as shown in the following image in the middle. The transitional face will now contact and follow the cam rotation, as shown on the right in the following image.

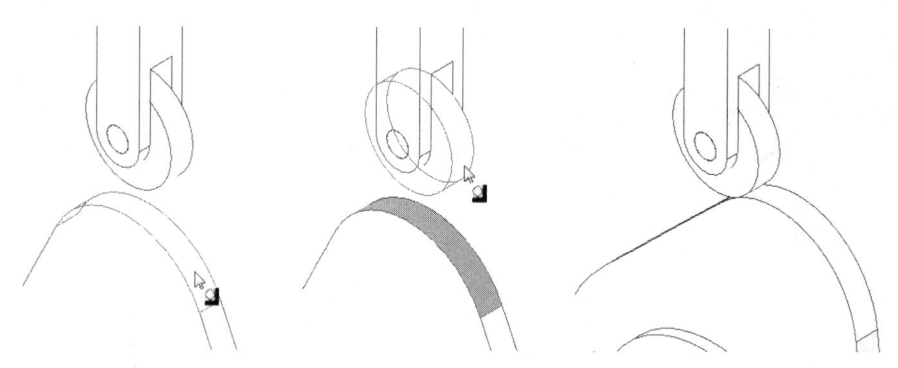

FIGURE 6.28

Creating a Constraint Set

If User Coordinate Systems were defined in individual part or assembly files, these UCSs can be constrained together. The buttons found under the Type and Selections areas of the dialog box allow for the selecting of individual UCSs and having the constraints be applied to the selections.

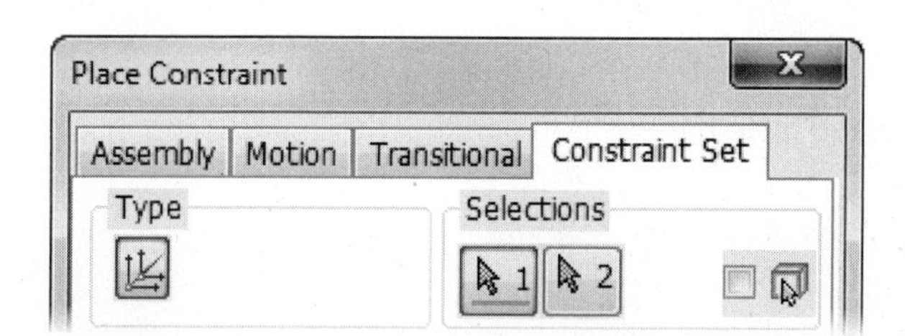

FIGURE 6.29

Selecting Geometry and the Select Other Command

After selecting the type of assembly constraint that you want to apply, the Selections button with the number 1 will become active; if it does not automatically become active, click the button. Position the cursor over the face, edge, point, and so on, to apply the first assembly constraint.

You may need to cycle through the selection set using the Select Other command, as shown in the following image, until the correct location is highlighted. Cycle by clicking on the down arrow of the command and select an option from the menu until you see the desired constraint condition, and then press the left mouse button.

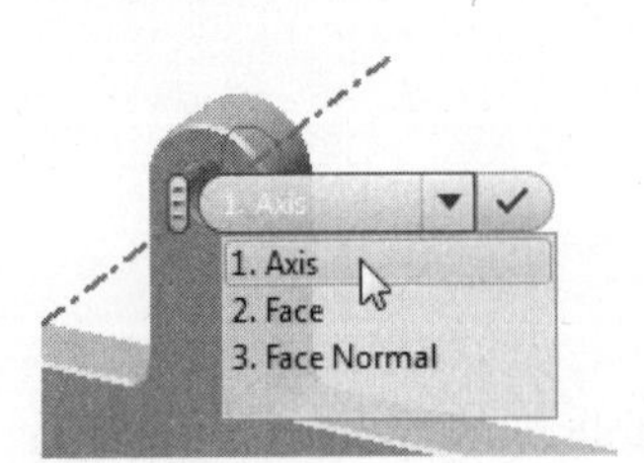

FIGURE 6.30

The next step is to position the cursor over the face, edge, point, and so on, and select the second geometry input for the assembly constraint. Again, you may need to cycle through the selection set until the correct location is highlighted. If the Show Preview option is selected in the dialog box, the components will move to show how the assembly constraint will affect the components, and you will hear a snapping sound when you preview the constraint. To change either selection, click on the button with the number 1 or 2, and select the new input. Enter a value as needed for the offset or angle, and select the correct Solution option until the desired outcome appears. Click the Apply button to complete the operation. Leave the Constraint dialog box active to define subsequent constraint relationships.

Direct Select - Assembly Constraints

Instead of placing assembly constraint via a dialog box you can use the Assemble command from the Assemble tab > Position panel as shown in the following image on the left. The Assemble command can be used to place Mate, Flush, Directed Angle, Tangent, Insert, and UCS constraints by selecting geometry and the constraint types are inferred based on the selections. After starting the command a mini toolbar displays in the graphics window as shown in the center of the following image. It allows you to choose a specific constraint option and only geometry that matches that option is selectable. For example, the Insert Opposed option will only select circular edges. To create a constraint with the Assemble command, follow these steps

1. Click the Assemble command from the Position panel.

2. If needed select a constraint option from the mini toolbar. The Automatic option, which is also the top level command, allows you to select all types of geometry.
3. Select geometry for the first selection set (this part will move).
4. Move the cursor over your second selection. Keep moving the cursor over the geometry until the solution type is previewed then click.
5. The mini toolbar will change. The new state of the mini toolbar will allow you to change the solution type and enter a value for the offset or angle.
6. When the options are set either press the ENTER key, press the Apply, or the OK button on the mini toolbar as shown in the following image on the right.

NOTE

The Assemble command always moves the first component to the second one. If the first component is grounded, it will move to the new location and maintain its grounded status.

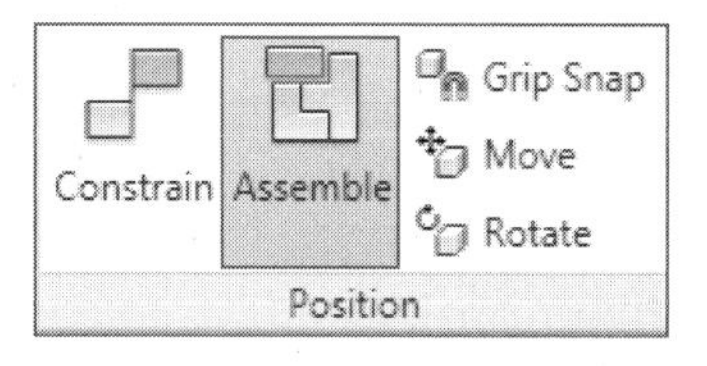

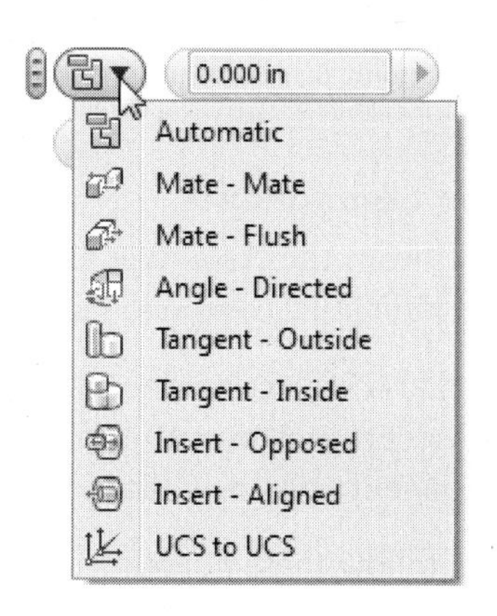

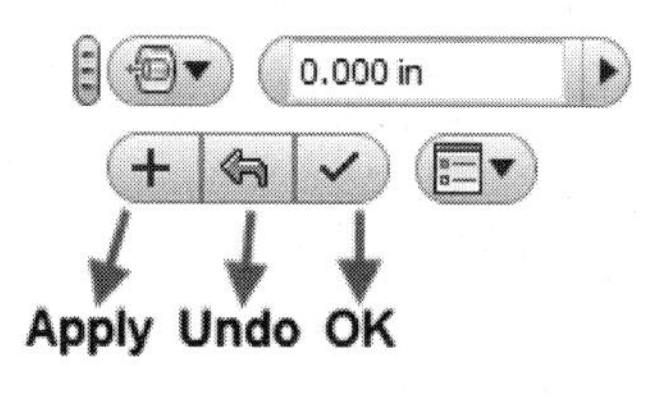

FIGURE 6.31

ALT + DRAG CONSTRAINING

Another way to apply an assembly constraint is to hold down the ALT key while dragging a part edge or face to another part edge or face; no dialog box will appear. The key to dragging and applying a constraint is to select the correct area on the part. Selecting an edge will create a different type of constraint than if a face is selected. If you select a circular edge, for example, an insert constraint will be applied. To apply a constraint while dragging a part, you cannot have another command active.

To apply an assembly constraint, follow these steps:

1. While holding down the ALT key, select the face, edge, or other part on the part that will be constrained.
2. Select a planar face, linear edge, or axis to place a mate or flush constraint. Select a cylindrical face to place a tangent constraint. Select a circular edge to place an insert constraint.
3. Drag the part into position. As you drag the part over features on other parts, you will preview the constraint type. If the face you need to constrain to is behind another face, pause until the Select Other command appears. Cycle through the possible selection options, and then click to accept the selection.

To change the constraint type previewed while you drag the part, release the ALT key and press the space bar to flip to the solution direction and if needed press one of the following shortcut keys:

Key	Constraint Type
M or 1	Mate
A or 2	Angle
T or 3	Tangent
I or 4	Insert
R or 5	Rotation-Motion
S or 6	Rotation-Translation
X or 8	Transitional

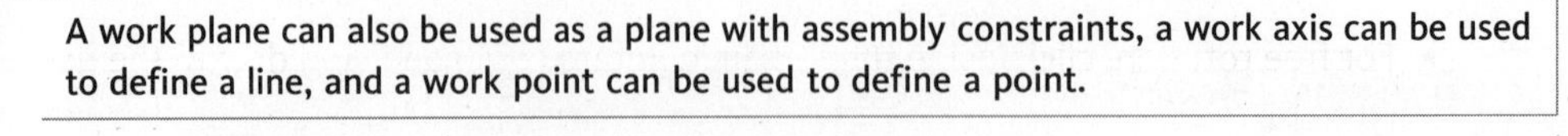

NOTE

A work plane can also be used as a plane with assembly constraints, a work axis can be used to define a line, and a work point can be used to define a point.

Moving and Rotating Components

Use the Move command from the Assemble tab > Position panel, as shown in the following image, to drag individual components in any direction in the viewing plane.

To perform a move operation on a component, activate the Move Component command. Click and hold the left mouse button on the component to drag it to a new location. Drop the component at its new location by releasing the button.

Moved components will follow these guidelines:

- An unconstrained component remains in the new location when moved until you constrain it to another component.
- A full or partially constrained component initially remains in the new location. When the assembly is updated, the component adjusts its location to comply with the constraints that you have already applied.

Use the Rotate command on the Assemble tab > Position panel, as shown in the following image, to rotate an individual component. This command is very useful when constraining faces that are hidden from your view.

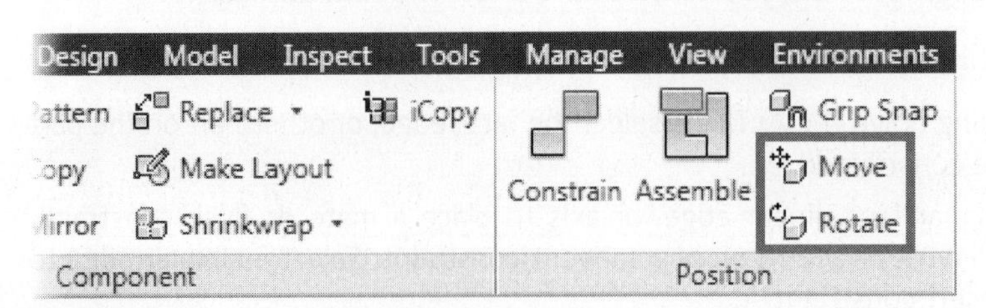

FIGURE 6.32

Follow these steps for rotating a component in an assembly:

1. Activate the Rotate command, and select the component to rotate. Notice the appearance of the 3D rotate symbol on the selected component in the following image.

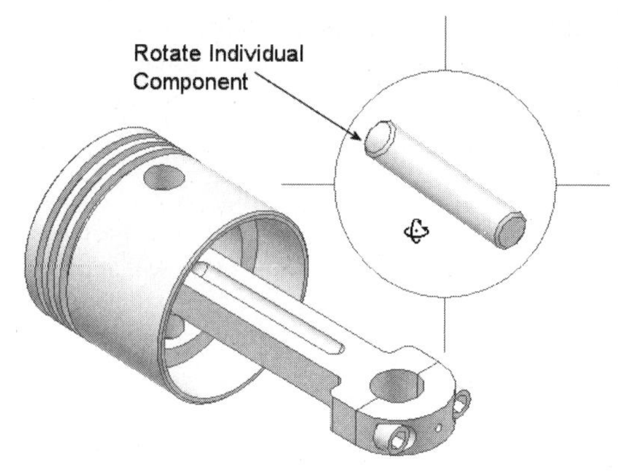

FIGURE 6.33

2. Drag your cursor until you see the desired view of the component.

- For free rotation, click inside the Dynamic Rotate command, and drag in the desired location.
- To rotate about the horizontal axis, click the top or bottom handle of the Dynamic Rotate command, and drag your cursor vertically.
- To rotate about the vertical axis, click the left or right handle of the Dynamic Rotate command, and drag your cursor horizontally.
- To rotate planar to the screen, hover over the rim until the symbol changes to a circle, click the rim, and drag in a circular direction.
- To change the center of rotation, click inside or outside the rim to set the new center.

Release the mouse button to drop the component into the rotated position.

If you click the Update button after moving or rotating components in an assembly, any components constrained to a grounded component will snap to their constrained positions in the new location. A fully constrained component can be moved or rotated temporarily. Once the assembly constraints are updated, the component will resolve the constraints and return to a fully constrained location/orientation.

NOTE

EDITING ASSEMBLY CONSTRAINTS

After you have placed an assembly constraint, you may want to edit, suppress, or delete it to reposition the components. There are two ways to edit assembly constraints.

Both methods are executed through the browser. In the browser, activate the assembly or subassembly that contains the component that you want to edit. Expand the component name, and you will see the assembly constraints, as shown in the following image on the left. Double-click on the constraint name, and an Edit Dimension dialog box will appear, allowing you to edit the constraint offset value. You can also right-click on the assembly constraint's name in the browser and select Delete, Edit, or Suppress from the menu, as shown in the following image.

If you select Edit, the Edit Constraint dialog box will appear. If you select Suppress, the assembly constraint will not be applied. Select Drive Constraint to drive a constraint through a sequence of steps, simulating mechanical motion. Select Delete, and the assembly constraint will be deleted from the component. If you try to place or edit an assembly constraint and it cannot be applied, a warning window that explains the problem will appear. You will have to either select new options for the operation or suppress or delete another assembly constraint that conflicts with it.

NOTE

When editing offset dimension values, click once with the left mouse button on the constraint with the offset value. Then change the offset value from the edit box that will appear at the bottom of the Assembly browser.

If an assembly constraint is conflicting with another, a triangular, yellow icon with an exclamation point will appear in the browser, as shown in the following image on the right. To edit a conflicting constraint in the browser, either double-click on its name or right-click on its name and select Recover from the menu, as shown in the following image on the right. The Design Doctor will appear and guide you through the steps to fix the problem.

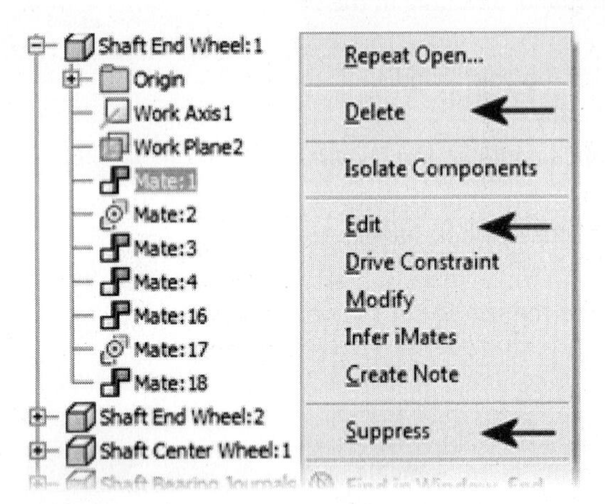

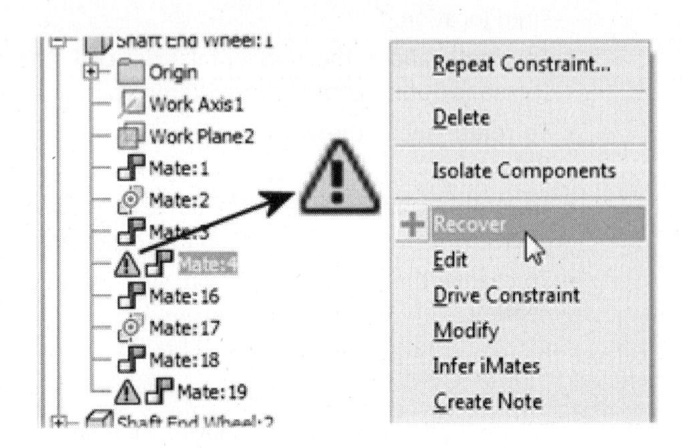

FIGURE 6.34

Constraint Offset Value Modification

When editing a constraint offset value, use the standard value edit control. This process is similar to editing work plane offsets and sketch dimensions, and it will allow you to measure while editing constraint offset values. Either double-click or right-click on the constraint to edit in the browser, and select Modify from the menu, as shown in the following image on the left. The Edit Dimension dialog box will appear; edit the offset value.

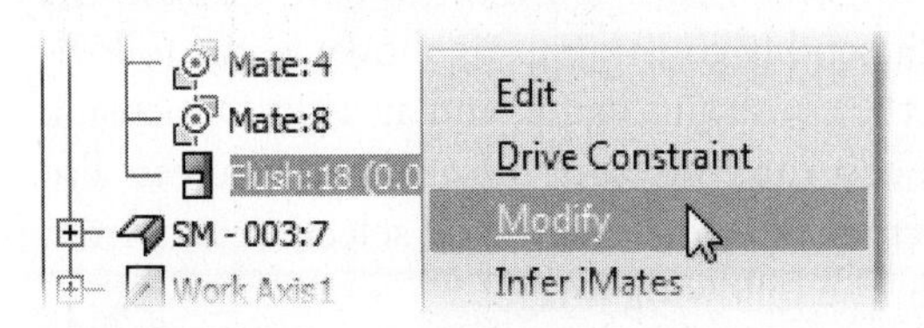

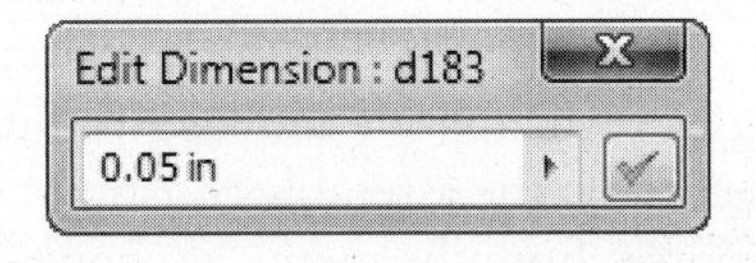

FIGURE 6.35

NOTE

You could also click on the constraint in the browser and a dialog box appears at the bottom of the Assembly browser, enabling you to make changes to the constraint's value.

EXERCISE 6-1: ASSEMBLING PARTS

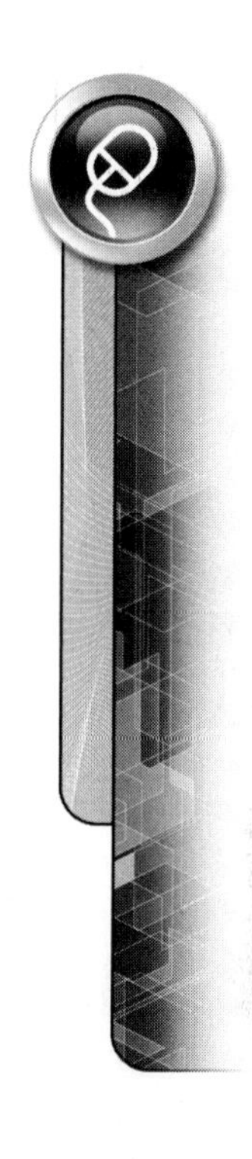

1. Open *ESS_E06_01.iam*.

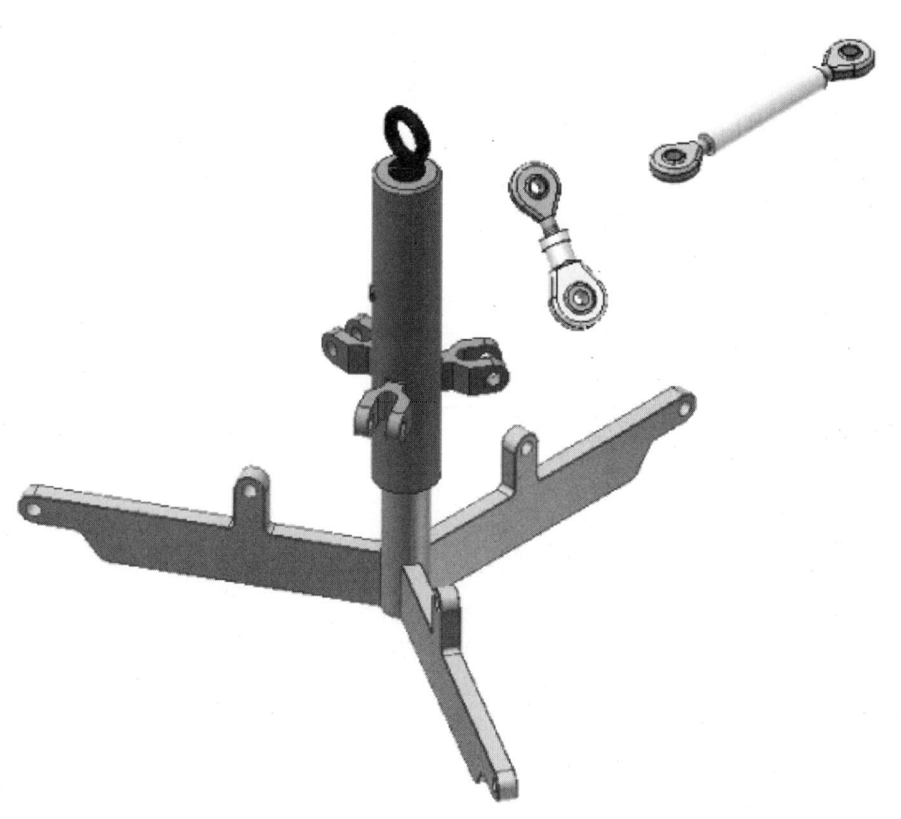

FIGURE 6.36

2. Begin assembling the connector and sleeve:

   a. Right-click in the graphics window, select Home View (F6 function key).

   b. Zoom in on the small connector and sleeve.

   c. Drag the connector so that the small end is near the sleeve, as shown in the following image.

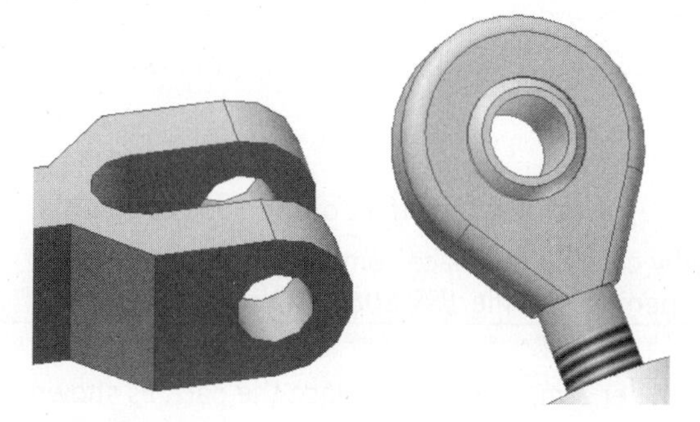

FIGURE 6.37

3. Next add a mate constraint between the centerlines of both components. Click the Constrain command from the Assemble tab > Position panel. Mate is the default constraint.
4. Move the cursor over the hole in the arm on the sleeve. Click when the centerline displays, as shown in the following image.
5. Move the cursor over the hole in the link. Click when the centerline displays, as shown in the following image.

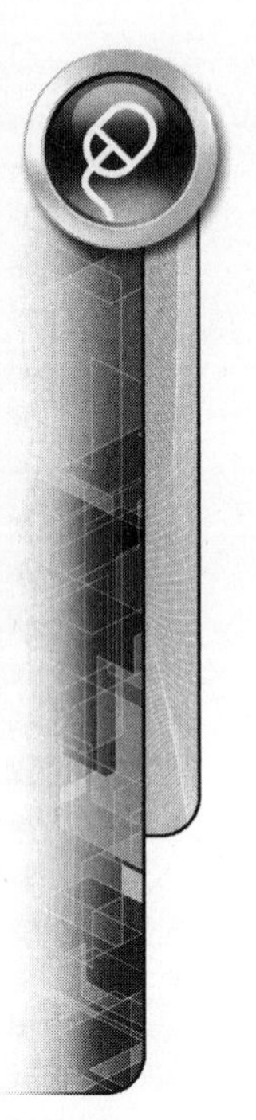

NOTE

If the green dot displays, move the cursor until the centerline displays, or use Select Other to cycle through the available choices to select the axis.

6. Click Apply to accept this constraint.

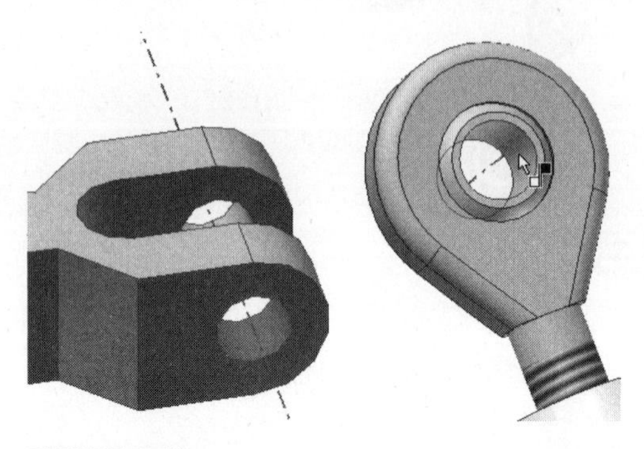

FIGURE 6.38

7. Now add a mate between the faces of both parts. Click the small flat face on the ball in the link end, as shown in the following image.
8. Click the inner face of the slot on the sleeve, as shown in the following image.
9. Click Apply to accept this constraint. Click Cancel to close the Place Constraint dialog box.

FIGURE 6.39

10. Click on the small link arm, and drag it to view the effect of the two constraints.
11. Place the crank in the assembly by clicking the Place command on the Assemble tab > Component panel and opening the file *ESS_E06_01-Crank.ipt.* from the Chapter 06 folder.
12. Move the component near the spyder arm, and click to place the part, as shown in the following image. Right-click, and select Done.

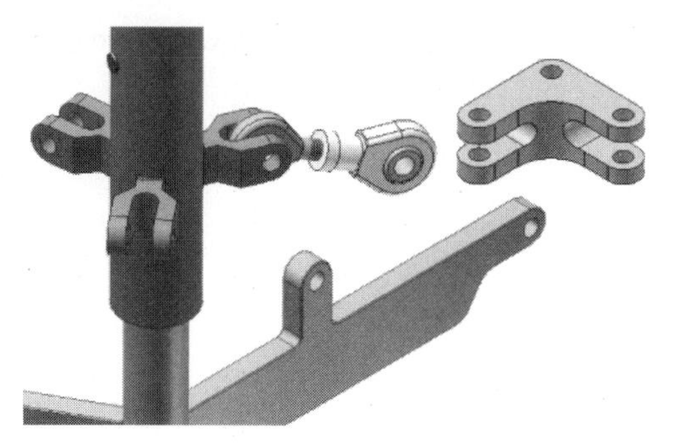

FIGURE 6.40

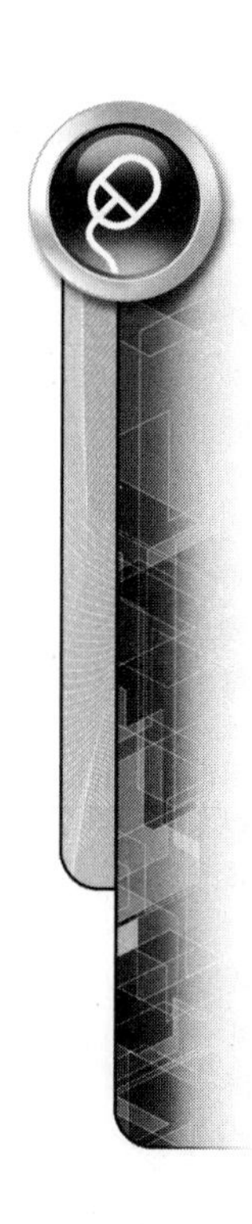

13. To assemble the crank to the spyder you will use the Assemble command with the Automatic option. Begin by rotating your view of the model so you can see the inner edge of a hole on the crank and the spyder as shown in the following image.

FIGURE 6.41

14. Click the Assemble command from the Assemble tab > Position panel. By default, the Automatic option is shown in the heads-up display.

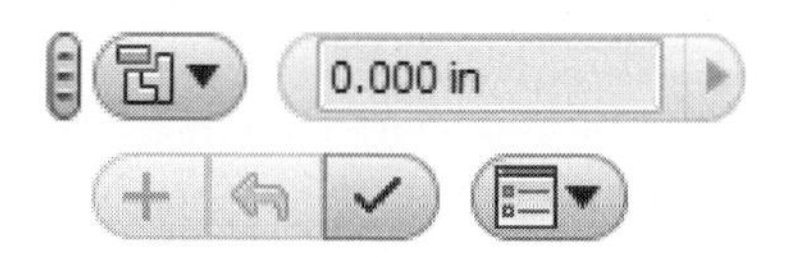

FIGURE 6.42

15. Move the cursor over the hole in the arm on the crank, and click when the centerline displays.

FIGURE 6.43

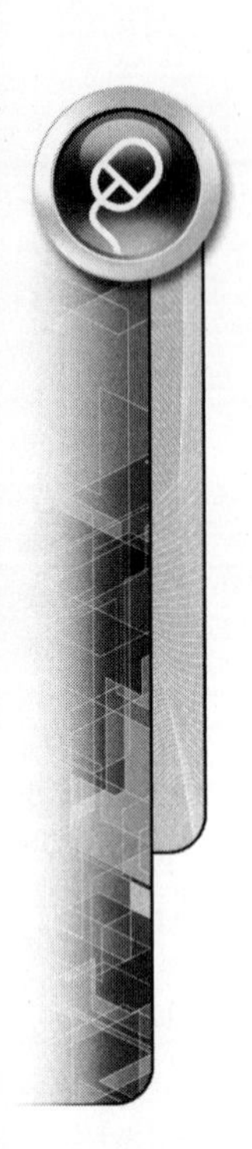

16. Move the cursor over the hole on the spyder and click when the centerline is displayed as shown in the following image.

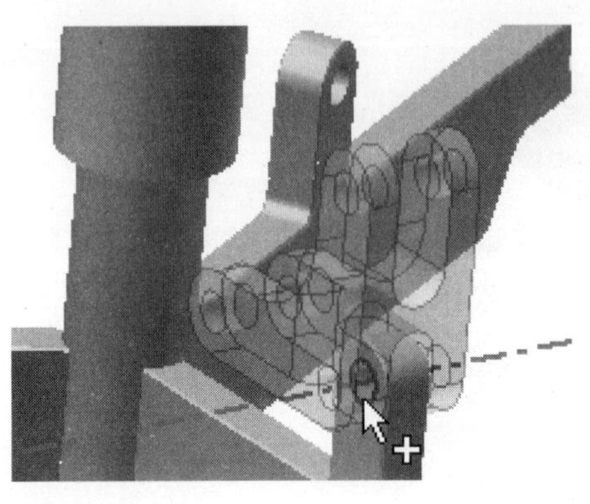

FIGURE 6.44

17. Click OK on the heads-up display or press ENTER to add the constraint.
18. Next, you will assemble the crank and link. Return to the Home View.
19. Drag the crank so you can see the affect of the constraint and to position it approximately as shown in the following image.

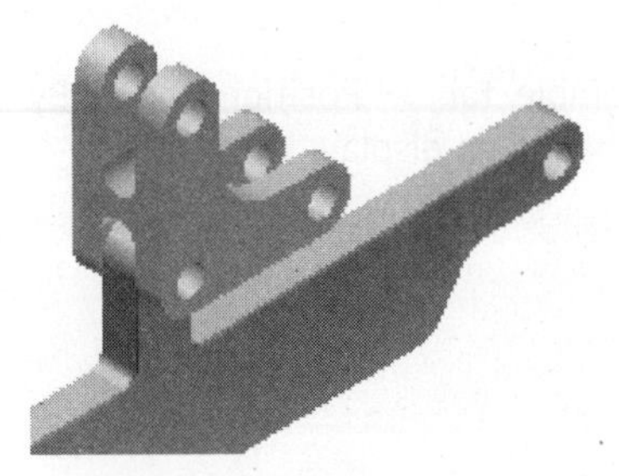

FIGURE 6.45

20. Drag the end of the small link arm close to the crank, as shown in the following image.
21. Zoom in to the crank and link.

22. Use the Constrain command to place a mate constraint between the centerlines of the two holes, as shown in the following image.

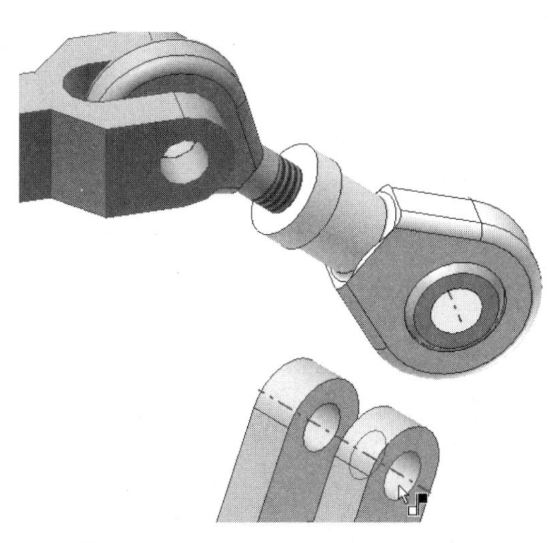

FIGURE 6.46

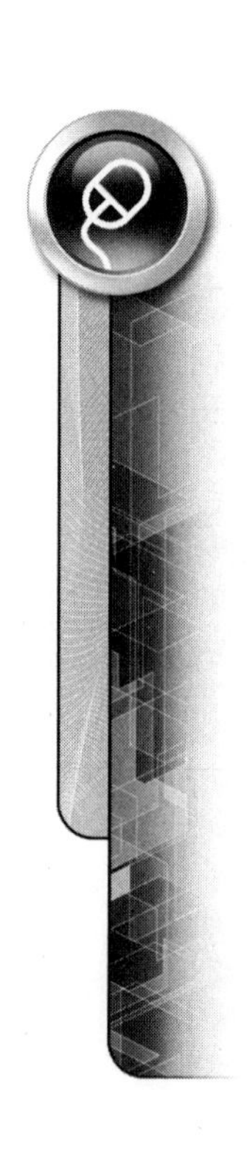

23. You will now place a claw in the assembly. Click the Place command from the Assemble tab > Component panel, and open the file *ESS_E06_01-Claw.ipt*.
24. Move the component near the end of the spyder arm, and click to place the part. Right-click, and select Done. Your display should appear similar to the following image.

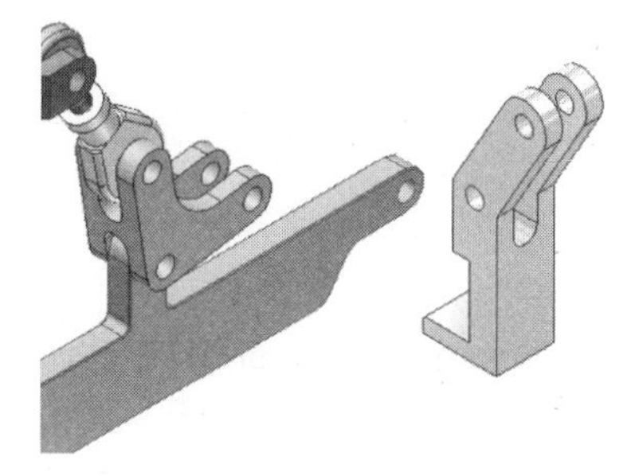

FIGURE 6.47

25. Constrain the claw to the spyder. Click Constrain from the Assembly tab > Position panel.
26. Click Insert from the type area of the Place Constraint dialog box.
27. Select the hole on the end of the spyder arm. Then select the inside edge of the hole on the claw as shown in the following image.

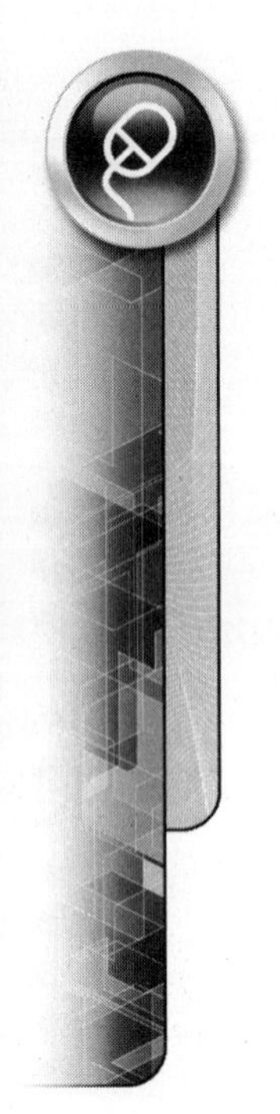

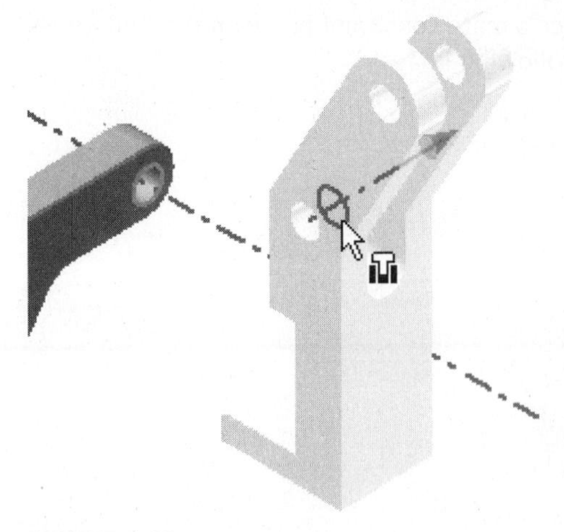

FIGURE 6.48

28. Click OK to place the constraint and close the Place Constraint dialog box.
29. Assemble the link rod to the crank and claw. First drag the claw and link rod into the position as shown in the following image.

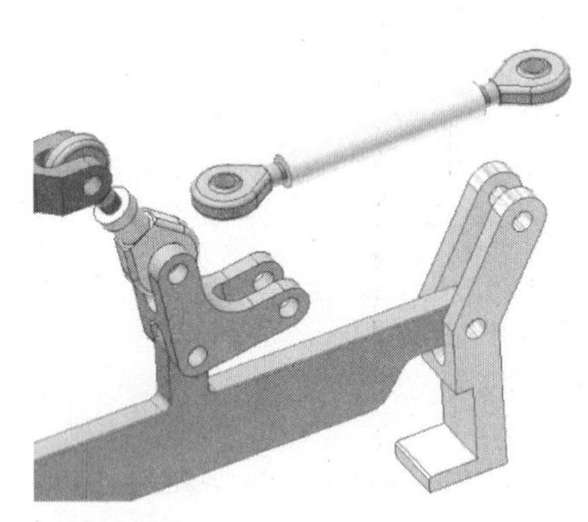

FIGURE 6.49

30. Next you use the Assemble command as an alternate method of placing an insert constraint between the link rod and the crank. Click the Assemble command from the Assemble tab > Position panel.
31. Select Insert – Opposed from the heads-up display drop-down list.

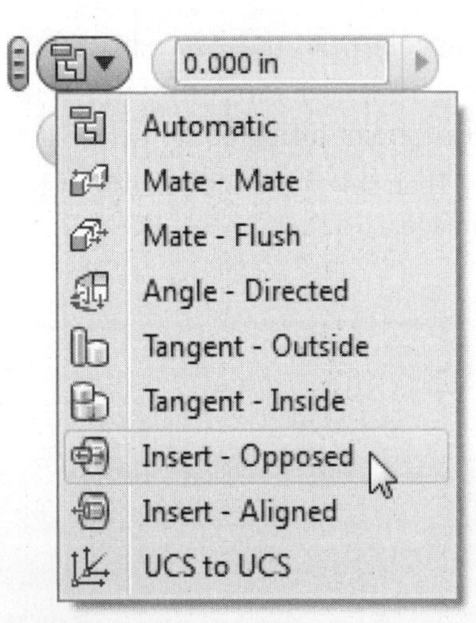

FIGURE 6.50

32. Select the top edge of the ball in the end of the link rod as shown in the following image on the left.
33. Select the inner edge of the hole in the crank as shown in the following image on the right.

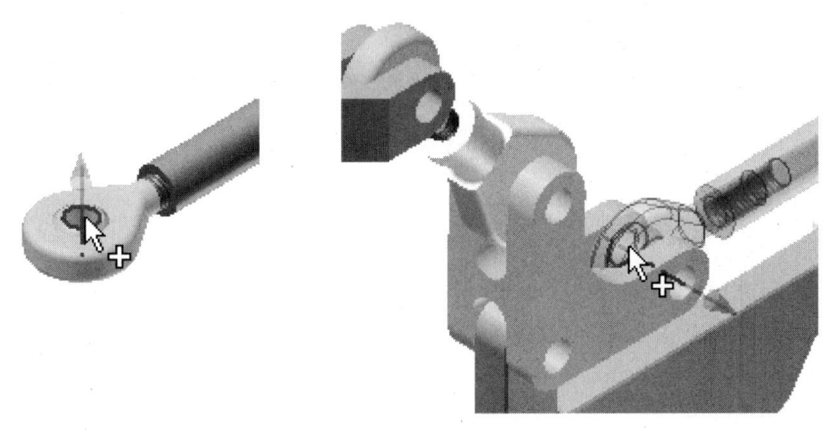

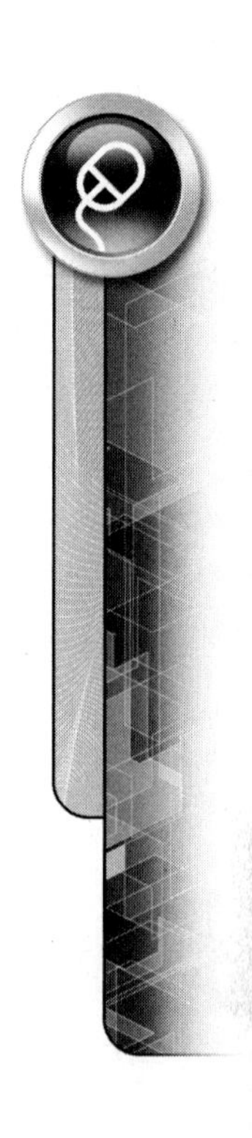

FIGURE 6.51

34. Press ENTER or click OK to place the constraint.
35. Click the Constrain command from the Assemble tab > Position panel.
36. Add a mate constraint between the centerlines of the claw and end of the link rod.
37. Click OK to apply the constraint and close the Place Constraint dialog box.
38. Drag the claw to view the effect of the assembly constraints, as shown in the following image.

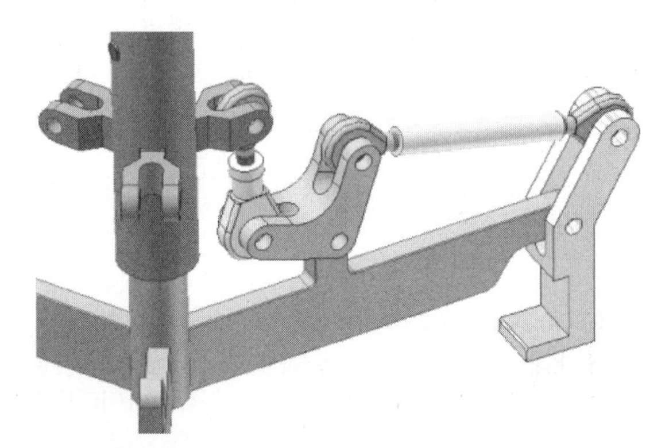

FIGURE 6.52

39. You will now add bolts and nuts to the spyder assembly. Begin by clicking the Place command on the Assemble tab > Component panel. Hold down the CTRL key and select ESS_E06_01-Bolt.ipt and ESS_E06_01-Nut.ipt. Click Open.
40. Place six nuts and bolts near their final position in the assembly, as shown in the following image. Right-click and select Done.

FIGURE 6.53

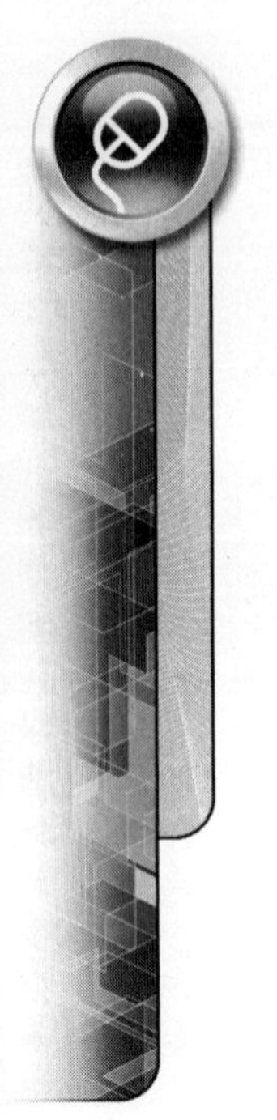

41. Insert the bolts into the assembly by clicking the Constrain command from the Assemble Tab > Position panel and in the Type area, click Insert to select an insert constraint.
42. Insert this bolt by selecting the top of the bolt's shank and then the edge of the hole, as shown in the following image.
43. Click OK to place the constraint and close the dialog box.

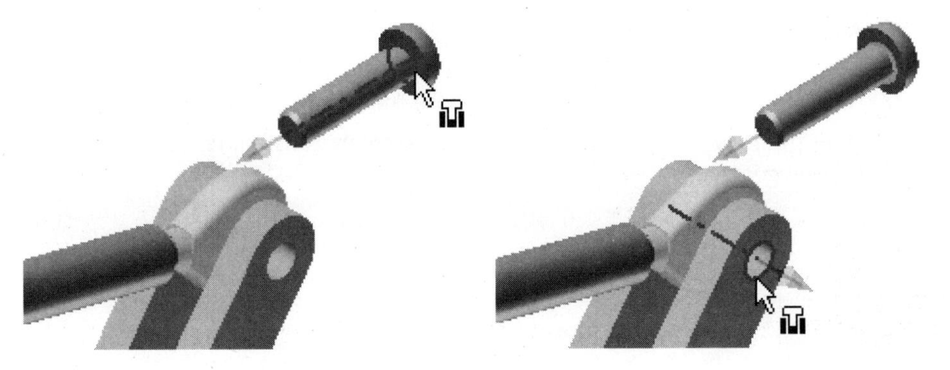

FIGURE 6.54

44. Use the Assemble command with either the Automatic or Insert – Opposed option to add the appropriate insert constraints to the remainder of the bolts.
45. Now add the nuts to the end of the bolts. Rotate the model so that you can see the back of the claw. Drag two nuts near their final location, as shown in the following image.
46. Use either the Constrain or Assemble command to place an insert constraint between the edge of the hole in the nut and the edge of the hole on the claw as shown in the following image on the right.

NOTE You must select an edge on the face of the nut. Do not select the inner chamfered edge.

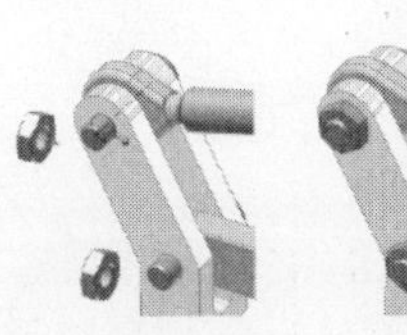

FIGURE 6.55

47. Follow the same process to add the appropriate Insert constraints to the remainder of the nuts.
48. The completed assembly is displayed in the following image.

FIGURE 6.56

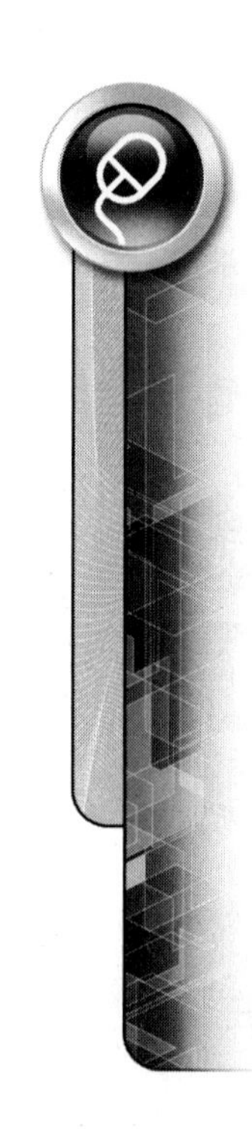

49. Close all open files. Do not save changes. End of exercise.

ADDITIONAL CONSTRAINT COMMANDS

Additional commands available to manipulate, navigate, and edit assembly constraints include browser views, Other Half, Constraint Offset Value Modification, Isolating Assembly Components, and Isolating Constraint Errors. The following sections describe these commands.

Browser Views

Two modes for viewing assembly information are located in the top of the browser toolbar: Assembly View and Modeling View. Assembly View, the default mode, displays assembly constraint symbols nested below both constrained components, as shown in the following image on the left. In this mode, the features used to create the part are hidden.

When Modeling View is set, assembly constraints are located in a Constraints folder at the top of the assembly tree, as shown in the following image and on the right. In this mode, the features used to create the part are displayed just as they are in the original part file.

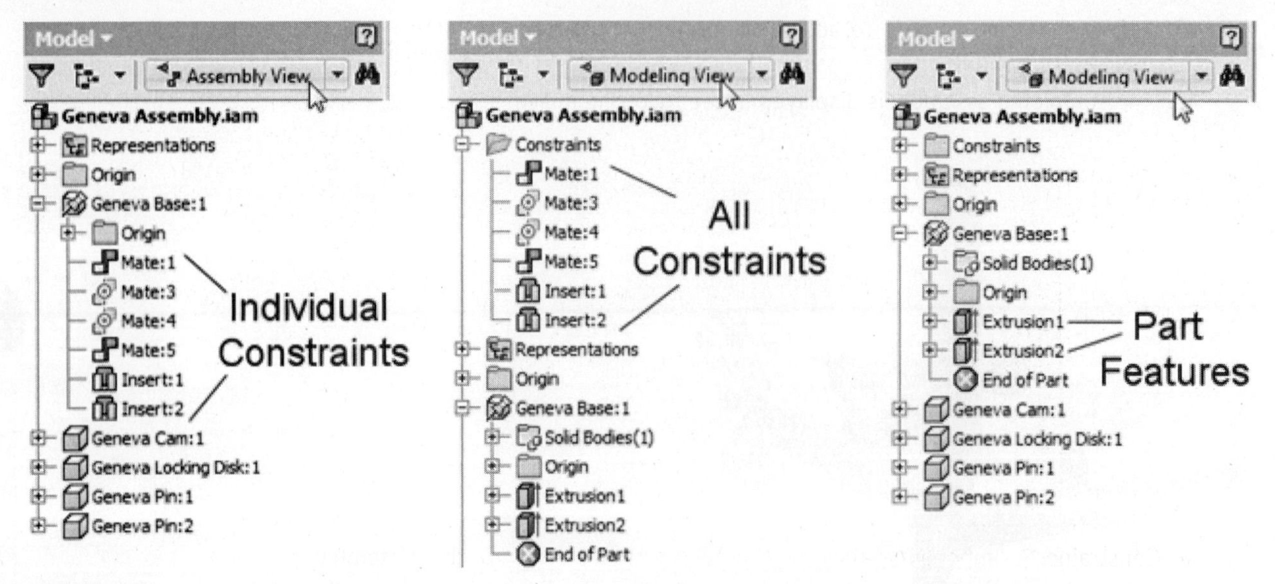

FIGURE 6.57

Other Half

You can use the Other Half command to find the matching part that participates in a constraint placed in an assembly. As you add parts over time, you may wish to highlight an assembly constraint and find the part(s) to which it is constrained. As shown in the following image on the left, a mate constraint has been highlighted. Half of this constraint has been applied to a part called Engine Block:1. To view the part sharing a common constraint, right-click on the constraint in the browser, and select Other Half from the menu, as shown in the following image on the left. The browser will expand and highlight the second half of the constraint, as shown in the following image in the middle. In this example, the other half of the mate constraint is a part called Cylinder:1.

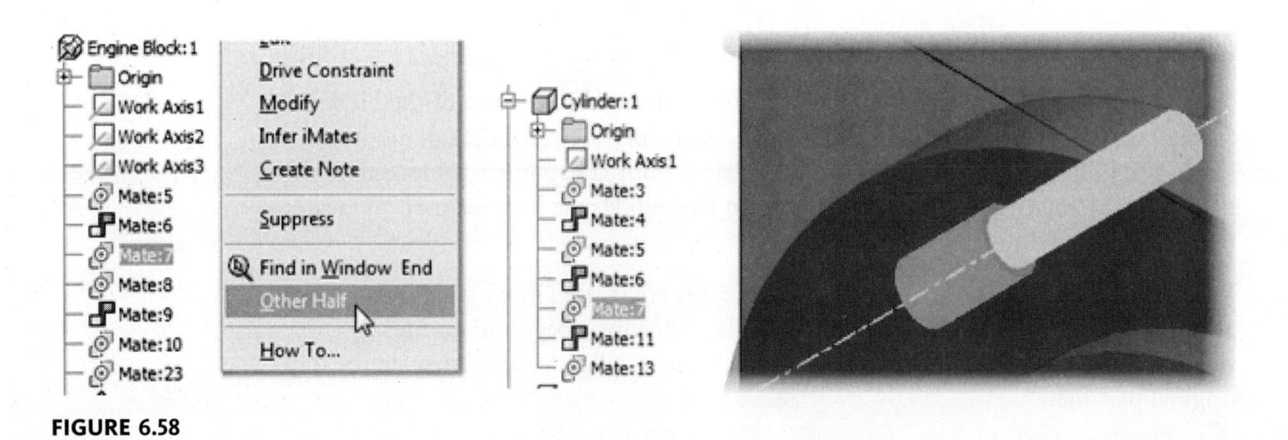

FIGURE 6.58

Constraint Tooltip

To display all property information for a specific constraint, move your cursor over the constraint icon, and a tooltip will appear, as shown in the following image. Although the constraint name is highlighted in the image, you must hover your cursor over the constraint icon to view the tooltip information.

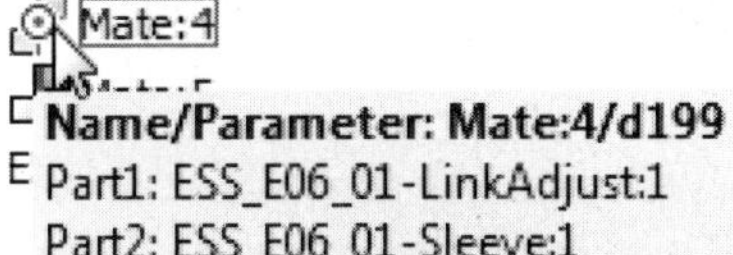

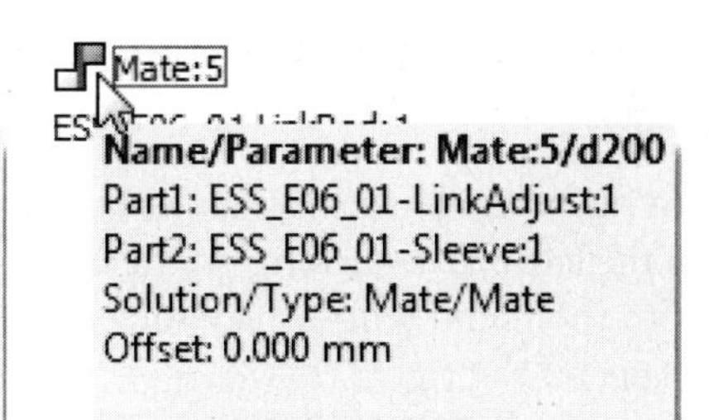

FIGURE 6.59

The following information is displayed in the tooltip:

- Constraint name and parameter name, applicable to offset and angle parameters
- Constrained components, that is, the two part names from the Assembly browser
- Constraint solution and type
- Constraint offset or angle value

User-Defined Assembly Folders

User-defined folders in an assembly allow you to organize your work by grouping assembly components under a single folder name. This method allows you to simplify the appearance of an assembly in the browser. To create a user-defined assembly folder, select those components that you want to group, right-click and pick Add to new folder, as shown in the following image on the left. You will be prompted in the browser to rename the default folder to something more meaningful, such as Fasteners, as shown in the following image on the right. Notice also a unique icon that identifies the user-defined assembly folder.

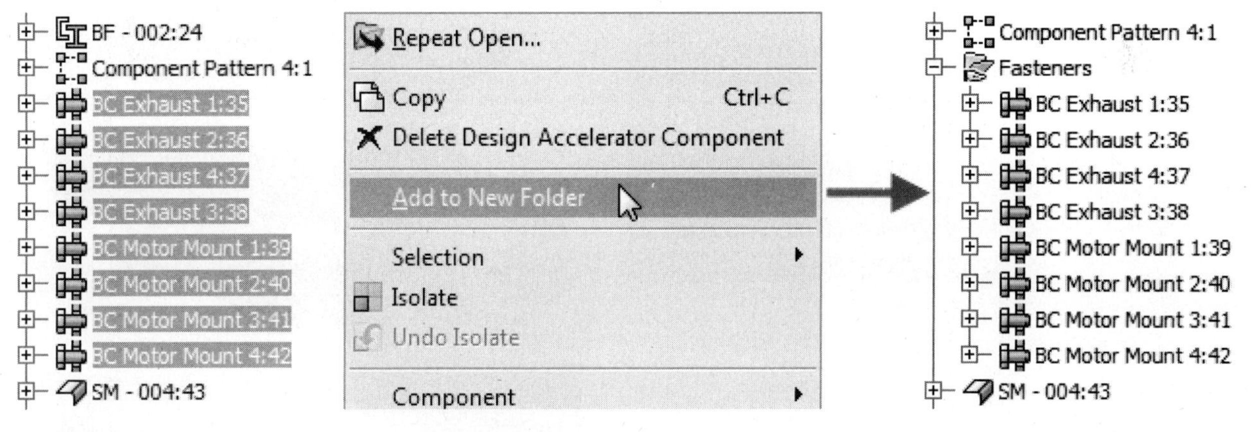

FIGURE 6.60

Isolating Assembly Components

Components can be isolated as a means of viewing smaller sets of components, especially in a large assembly. In the assembly browser, click on the components to isolate, right-click to display a menu, and select Isolate. All unselected components will be set to an invisible status.

Components can be isolated based on an assembly constraint. As shown in the following image on the left, right-clicking on the Mate:7 constraint and selecting Isolate Components from the menu will display the engine block and cylinder as shown in the following image on the right.

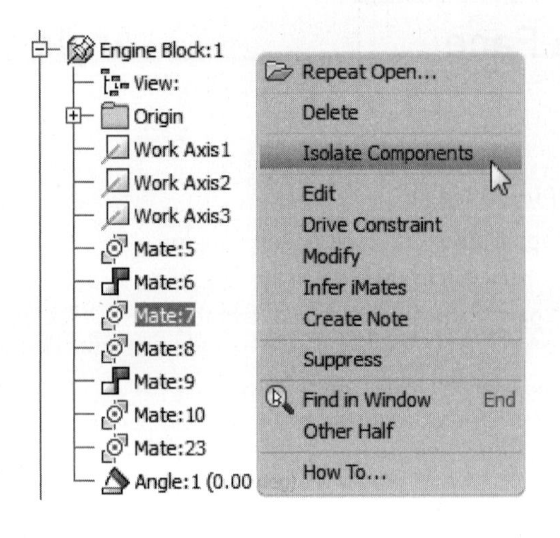

FIGURE 6.61

To return the assembly to its previous assembled state, right-click inside the browser, and select Undo Isolate from the menu, as shown in the following image.

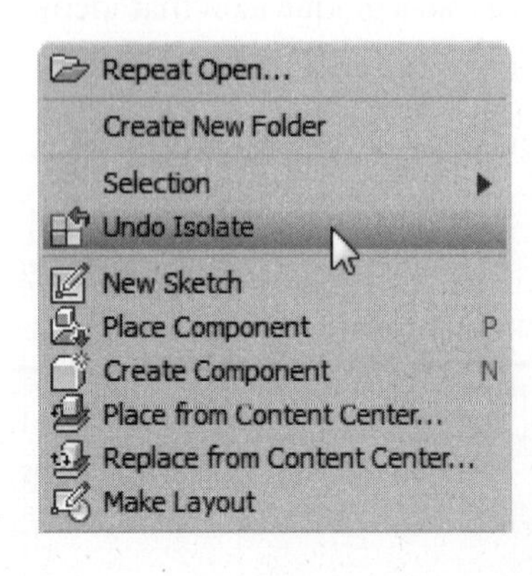

FIGURE 6.62

Isolating Constraint Errors

Assembly constraints can be isolated when errors in their placement occur. When editing a constraint through the Design Doctor, use the Isolate Related Failed Constraint option, as shown in the following image on the right. This action will turn off all components except those that participate in the common constraint, and it will display the Constraint dialog box, allowing you to edit the constraint with the errors.

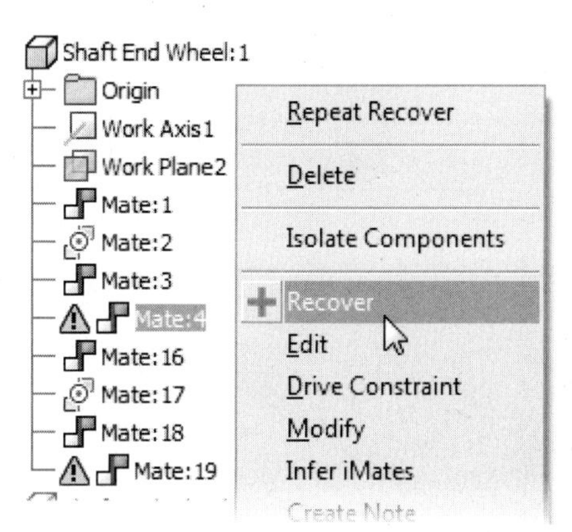

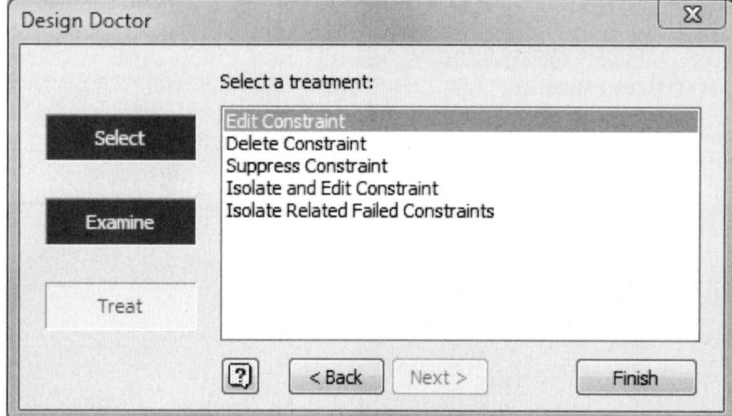

FIGURE 6.63

ASSEMBLY BROWSER COMMANDS

Additional commands are available through the Assembly browser as a means of better controlling and managing data in an assembly file. These commands include In-Place Activation, Visibility Control, Assembly Reorder, Restructuring an Assembly, demoting and promoting assembly components, and using browser filters. A few of these commands are explained as follows.

In-Place Activation

The level of the assembly that is currently active determines whether or not you can edit components or features. You can take some actions only in the active assembly and its first-level children, while other operations are valid at all levels of the active assembly.

Double-click any subassembly or part occurrence in the browser to activate it, or right-click the occurrence in the browser and select Edit. All components not associated with the active component appear shaded in the browser, as shown in the following image on the left.

NOTE

Double-clicking directly on a component in the graphics window will also activate it for editing.

FIGURE 6.64

If you are working with a shaded display, the active component appears shaded in the graphics window, and all other components appear translucent.

You can perform the following actions on the first-level children of the active assembly:

- Delete a component
- Display the DOF of a component
- Designate a component as adaptive
- Designate a component as grounded
- Edit or delete the assembly constraints between first-level components

You can edit the features of an activated part in the assembly environment. The ribbon changes to reflect the part environment when a part is activated.

NOTE

Double-click a parent or top-level assembly in the browser to reactivate it.

Visibility Control

Controlling the visibility of components is critical to managing large assemblies. You may need some components only for context, or the part you need may be obscured by other components. You can change the visibility of any component in the active assembly even if the component is nested many layers deep in the assembly hierarchy. To change the visibility of a component, expand the browser until the component occurrence is visible, right-click the occurrence, and select Visibility, as shown in the following image.

NOTE

You can also right-click on a component in the graphics window and select Visibility.

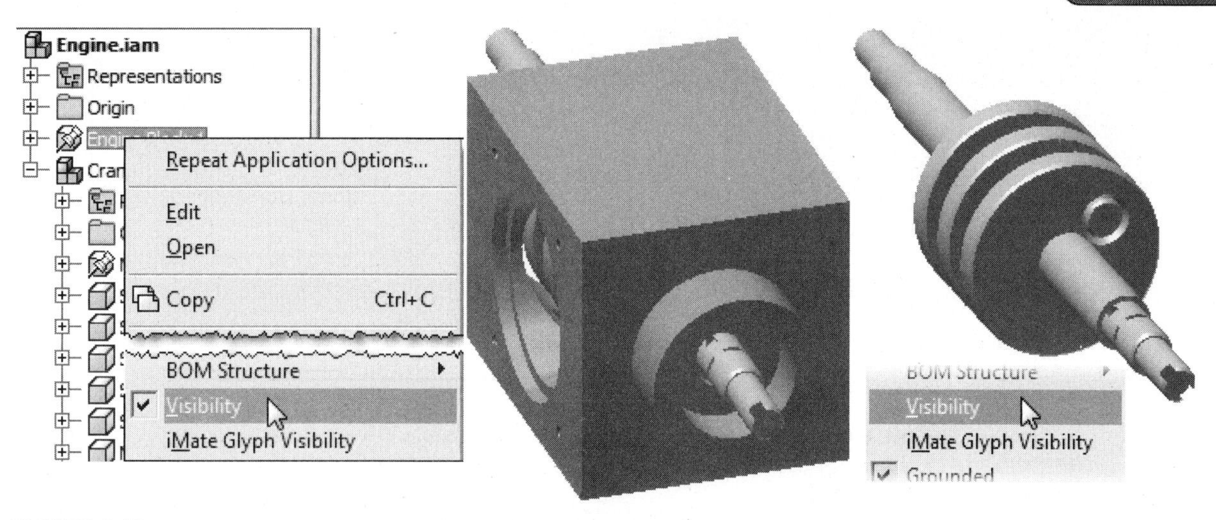

FIGURE 6.65

ADAPTIVITY

This section will introduce you to the concept of adaptivity. To learn more about adaptivity see the Help system. Adaptivity is the Autodesk Inventor function that allows the size of a part to be determined by setting up a relationship between two parts in an assembly. Adaptivity allows under-constrained sketches—features that have undefined angles or extents, hole features, and subassemblies, which contain parts that have adaptive sketches or features—to adapt to changes. The adaptivity relationship is defined by applying assembly constraints between an adaptive sketch or feature and another part. If a sketch is fully constrained, it cannot be made adaptive. However, the extruded length or revolved angle of the part can be. A part can only be adaptive in one assembly at a time. In an assembly that has multiple placements of the same part, only one occurrence can be adaptive. The other occurrences will reflect the size of the adaptive part.

An example of adaptivity would be determining the diameter of a pin from the size of a hole. You could determine the diameter of the hole from the size of the pin, and you can turn adaptivity on and off as needed. Once a part's size is determined through adaptivity and adaptivity is no longer useful you may want to turn its adaptivity off. If you want to create adaptive features, Autodesk Inventor includes features that will speed up the process of creating them. These features are not covered in detail in this book, but can be found on the Assembly tab of the Application Options dialog box.

Under-constrained Adaptive Features

The next series of images show how parts adapt when you apply assembly constraints. In the following image, a rectangular sketch for a small plate is not dimensioned (unconstrained) along its length.

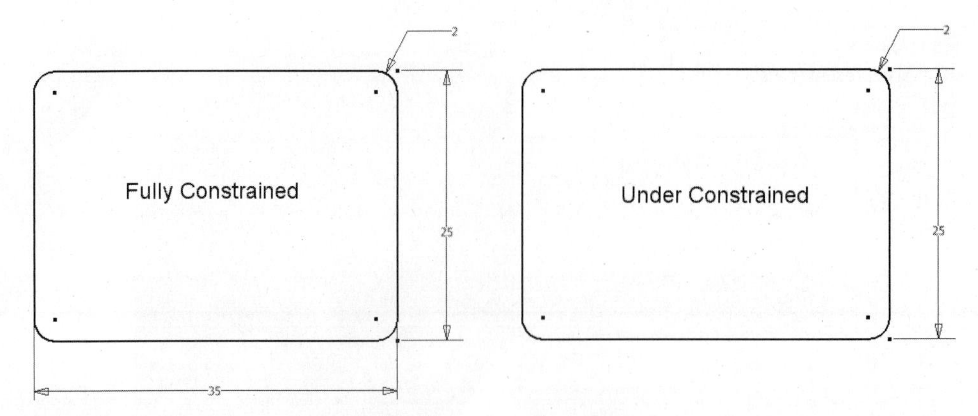

FIGURE 6.66

The extruded feature is then defined as adaptive by right-clicking on the extrusion in the browser as shown in the following image. Selecting Adaptive from the menu applies the adaptive property to the sketch and the extrusion.

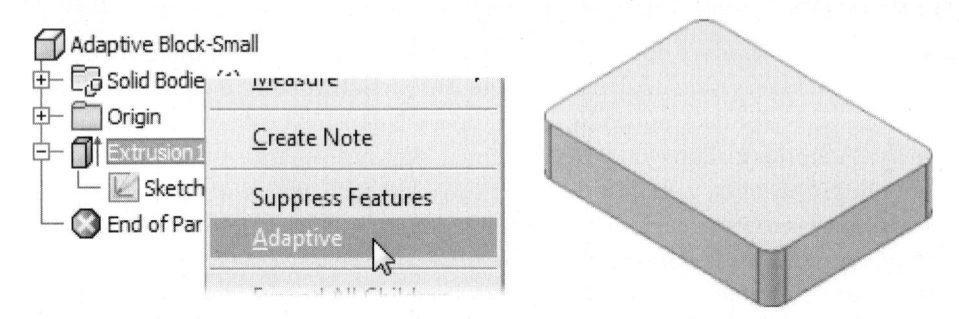

FIGURE 6.67

Once you have placed the parts in an assembly, right-clicking on the component in the browser activates a menu. Selecting Adaptive from the menu adds an icon consisting of two arcs with arrows next to the component, as shown in the following image on the left. The presence of this icon in the browser means that part is now adaptive. As flush constraints are placed along the edge faces of the plates, the small plate will adapt its length to meet the length of the large plate.

The results are shown in the following image on the right.

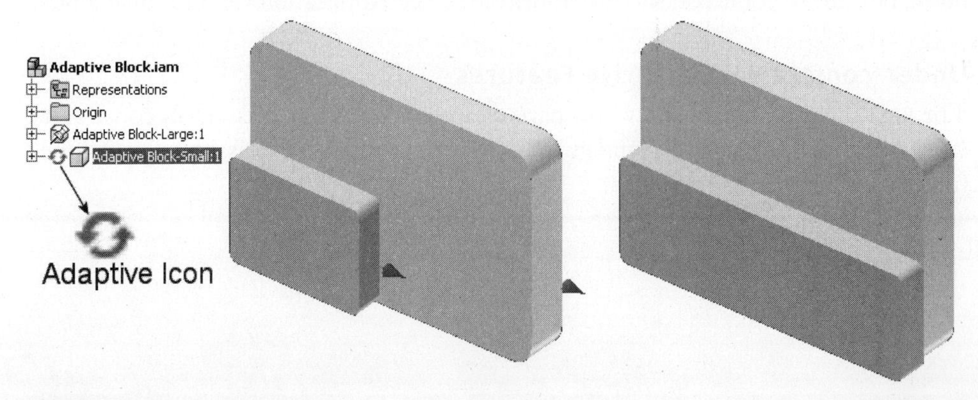

FIGURE 6.68

ASSEMBLY SECTIONS

Assembly section views can be used to visualize portions of an assembly, especially where components are hidden or obscured within chambers. While the assembly is sectioned, part and assembly tools can be used to create or modify parts within the assembly. You can use the exposed cut edges, along with the Project Cut Edges command, to assist in sketch construction for additional features that need to be created in a part. There are four assembly section tools available on the View tab > Appearance panel as shown in the following image on the left. The other two images show a before and after view of an assembly that is using the Half Section View tool.

FIGURE 6.69

The general process for using the section tools are:

1. Select the desired assembly section tool.
2. Select any existing planar face(s) or work plane(s).
3. Specify which side of the selected section remains visible by right clicking and using the Flip Section option, shown in the following image.
4. Right-click and select Done.

You can change the position of section planes in a section view using the Virtual Movement option from the context menu, as shown in the following image. When defining the section plane, you can enter a distance in the Offset dialog box, drag the section plane in the canvas, or use the mouse wheel while your cursor is on top of the offset field in the Offset dialog box, to move the section plane. The Scroll Step Size option can be used to adjust the distance of the mouse wheel scroll.

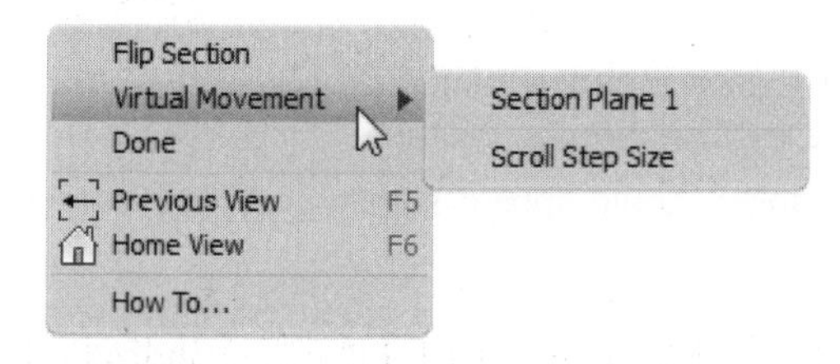

FIGURE 6.70

EXERCISE 6-2: DESIGNING PARTS IN THE ASSEMBLY CONTEXT

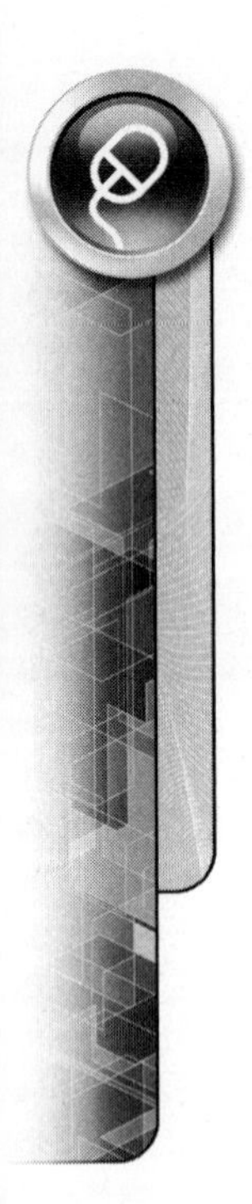

In this exercise, you create a lid for a container based on the geometry of the container. This cross part sketch geometry is adaptive, and it automatically updates to reflect design changes in the container.

1. Start with an existing assembly. You will then create a new part based on cross part sketch geometry. Begin by opening the file *ESS_E06_02.iam,* as shown in the following image on the left.
2. Begin the process of creating a new component in the context of an assembly by clicking the Create command from the Assemble tab > Component panel.
3. Under New File Name, enter *ESS_E06_02-Lid.*
4. Click the Browse Templates button, and in the English tab, click *Standard (in).ipt.* Click OK. There should be a checkmark beside Constrain Sketch Plane to Selected Face or Plane at the bottom of the Create In-Place Component dialog box.
5. Click the OK button to exit the dialog box, and then select the top face of the container, as shown in the following image on the right. This face becomes the new sketch plane for the new part.

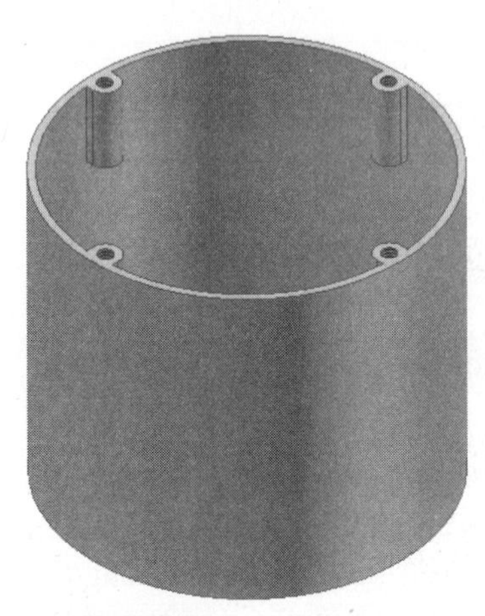

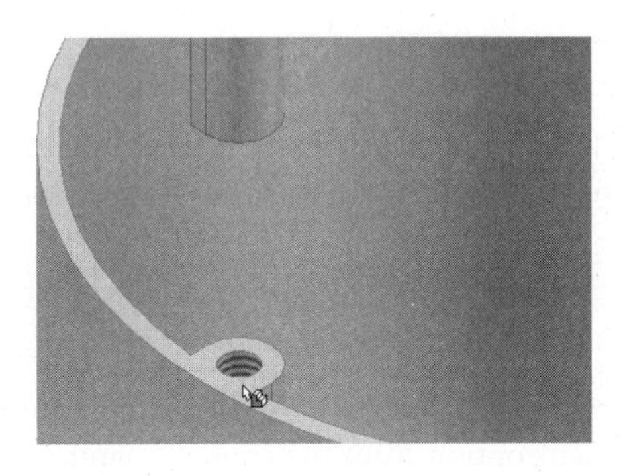

FIGURE 6.71

6. You will now project all geometry contained in this face. First click the Project Geometry command from the Sketch tab > Draw panel.
7. Move the cursor onto the face of the base part until the profile of the entire face is highlighted, as shown in the following image on the left.
8. Click to project the edges. Your display should appear similar to the following image on the right.

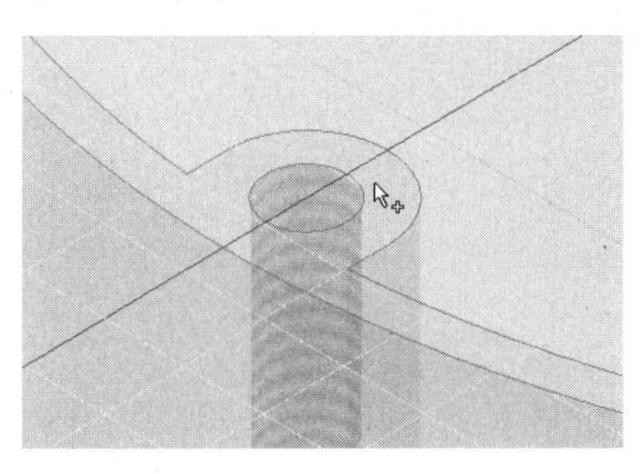

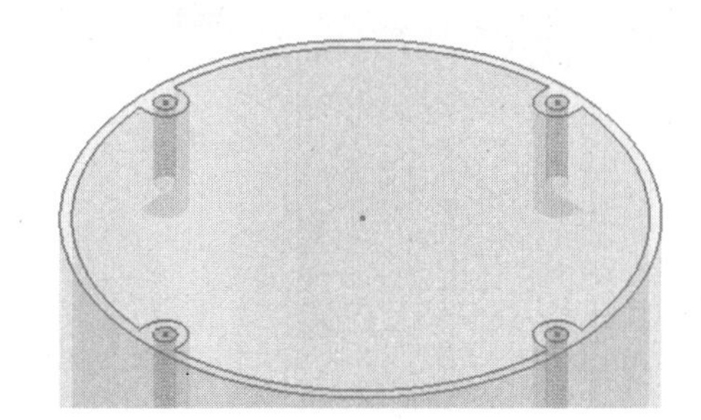

FIGURE 6.72

9. You will now create a series of clearance holes. First zoom into a tapped hole.
10. Click the Center Point Circle command, and click the projected circle's center point.
11. Move the cursor, and click to create a circle larger than the projected circle, as shown in the following image on the left.
12. Create three additional circles centered on the remaining projected holes, as shown in the following image on the right.

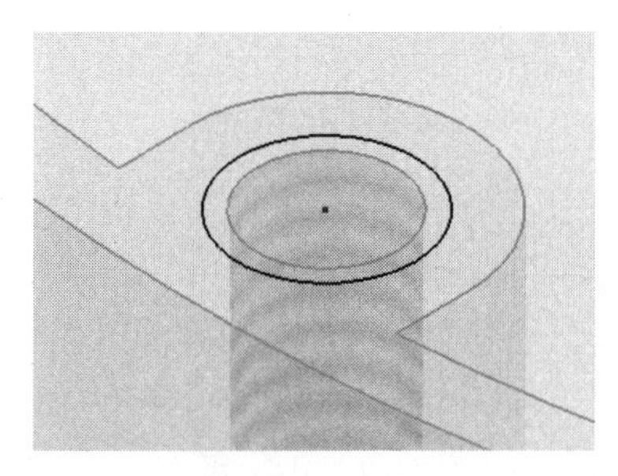

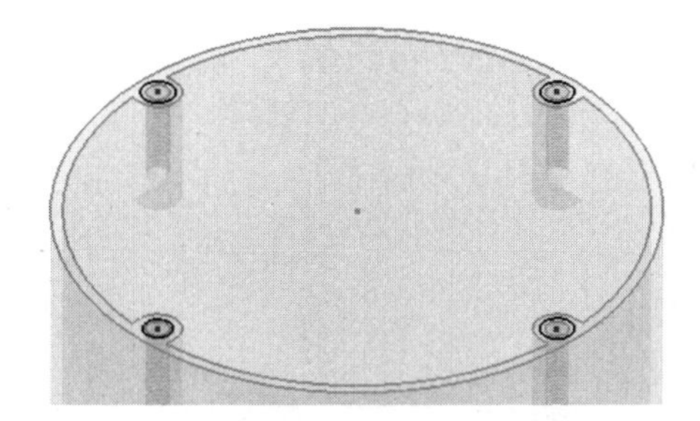

FIGURE 6.73

13. All circle diameters need to be made equal. Perform this operation by clicking the Equal constraint command from the Sketch tab > Constrain panel.
14. Click one circle, and then click one of the other circles. Click the first circle again, and click a third circle. Click the first circle again, and then click the fourth circle. These steps will make all circles equal to each other in diameter.

NOTE

The first entity selected is the source size that the next selection will adapt and become equal to.

15. Now add a dimension to one of the circles to fully constrain the sketch. First click the Dimension command from the Sketch tab > Constrain panel.
16. Click the edge of one of the circles, and click outside the circle to place the dimension.
17. Enter **.15 in** in the dimension edit box, and press ENTER to apply the dimension. Your display should appear similar to the following image on the left.
18. Next click the Finish Sketch button to exit the Sketch environment. Click the Extrude command from the Model tab > Create panel, and select the two profiles as shown in the following image on the right.

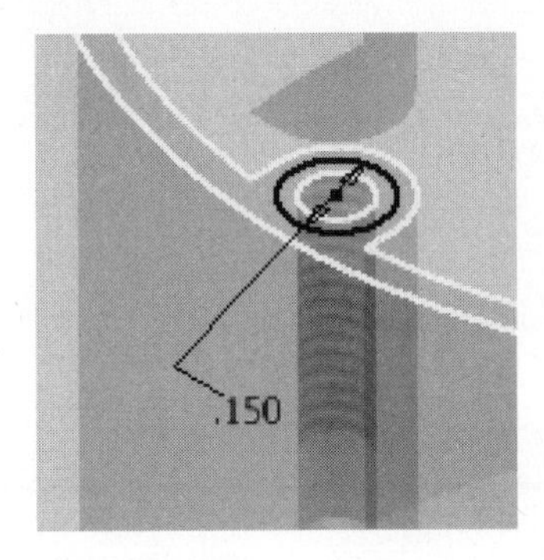

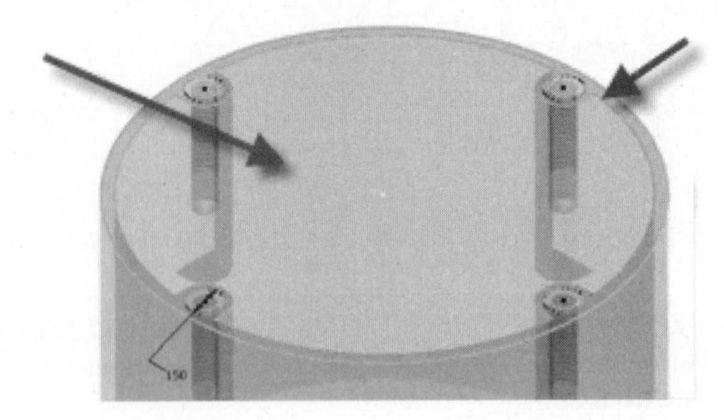

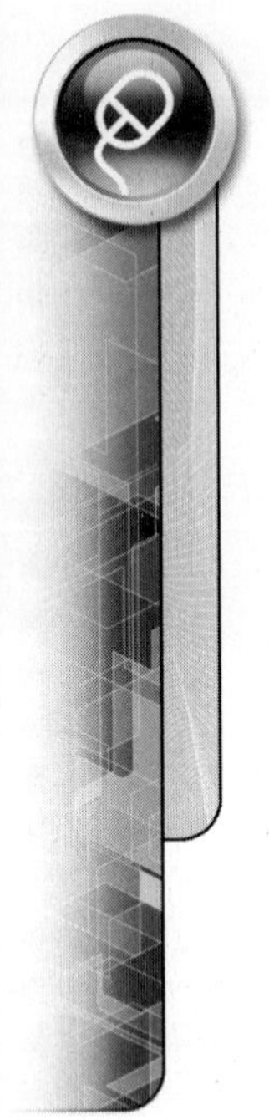

FIGURE 6.74

19. In either the Extrude dialog box or the heads-up display, enter a distance of **.25 in**, and then click OK.
20. Click the Return command to activate the assembly. Your display should appear similar to the following image on the right.

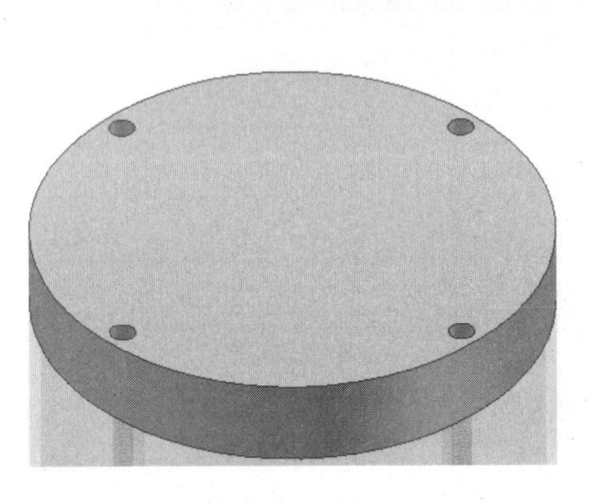

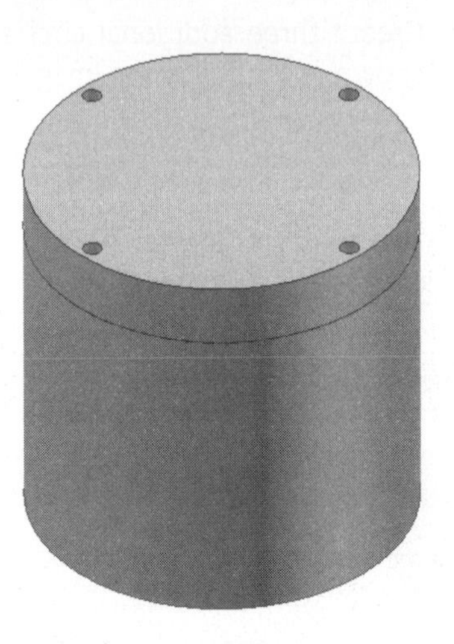

FIGURE 6.75

21. You will now modify the container base and observe how this affects the lid. In the browser, right-click ESS_E06_02-Container:1, and then select Edit. Your display should appear similar to the following image.

FIGURE 6.76

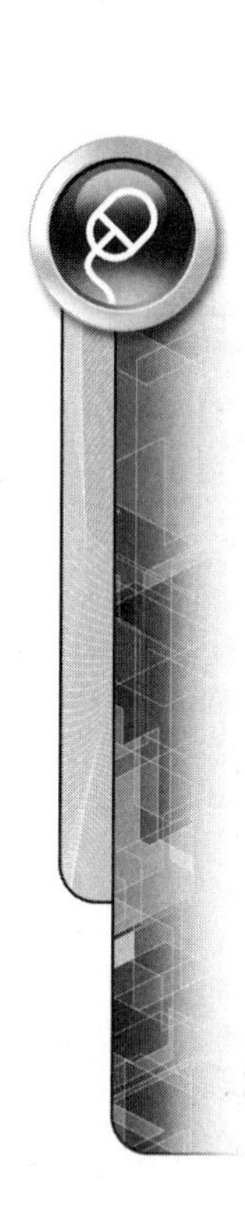

22. In the browser, right-click Extrusion1, and then select Edit Sketch.
23. Double-click the 2 in dimension, and change the value to **2.5 in**.
24. Click the Finish Sketch command. The model should appear as shown in the following image on the left.
25. Click the Return command again to return to the assembly context. Notice that the lid adapts to the modified dimensions of the base, as shown in the following image on the right.

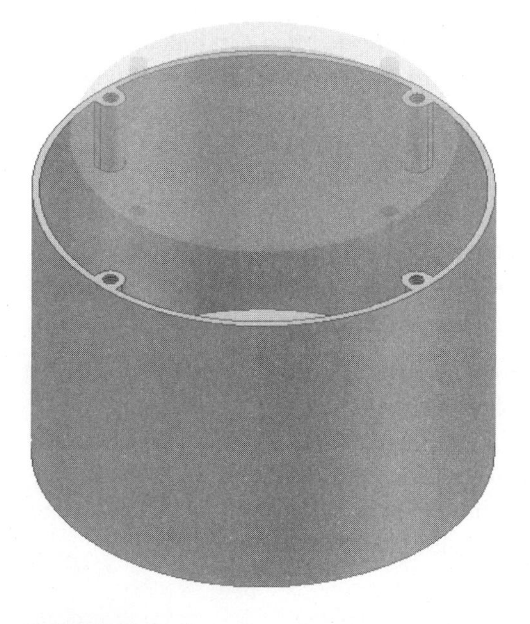

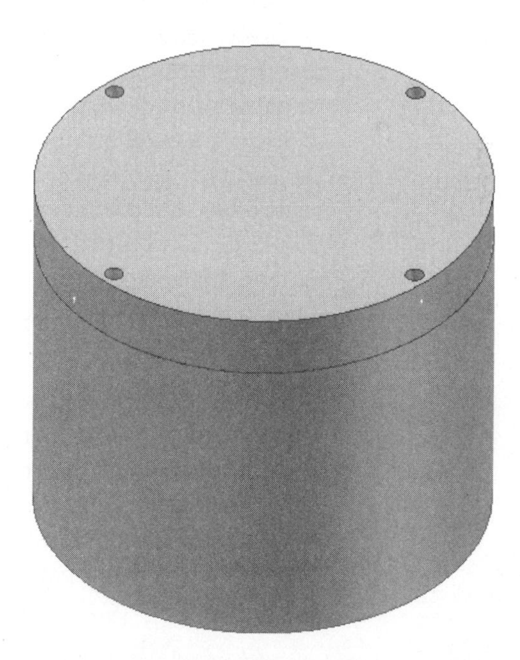

FIGURE 6.77

26. Close all open files. Do not save changes. End of exercise.

PATTERNING COMPONENTS

You can use the Pattern command on the Assemble tab > Component panel, as shown in the following image, when you place multiple occurrences of selected parts and subassemblies that match a feature pattern on another part (a component pattern) or that have a set of circular or rectangular part patterns in an assembly (an assembly pattern).

Three tabs are available in the Pattern Component dialog box: Associate, Rectangular, and Circular, as shown in the following image. The Associate tab is the default.

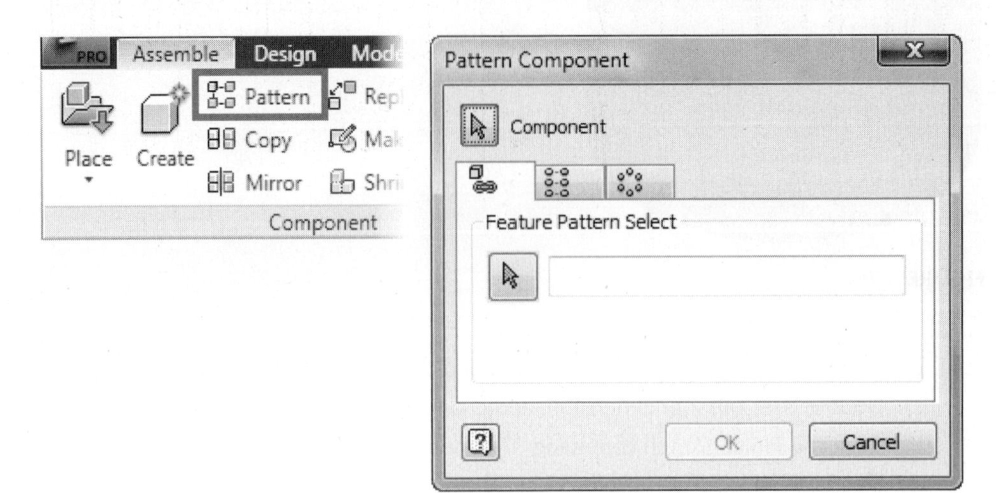

FIGURE 6.78

Associated Patterns

An associated component pattern will maintain a relationship to the feature pattern that you select. For example, a bolt is component-patterned to a part bolt-hole circular pattern that consists of four holes. If the feature pattern, the bolt-hole, changes to six holes, the bolts will move to the new locations, and two new bolts will be added for the two new holes. To create an associative component pattern, there must be a feature-based rectangular or circular pattern, and the part that will be patterned should be constrained to the parent feature, that is, the original feature that was patterned in the component. Issue the Pattern command from the Assemble tab > Component panel, and the Pattern Component dialog box will appear, as shown in the following image on the left.

By default, the Component selection option is active. Select the component, such as the cap screw, or components to pattern. Next click the Feature Pattern Select button in the dialog box and select a feature, such as a hole, that is part of the feature pattern. Do not select the parent feature. After selecting the pattern, it will highlight on the part and the pattern name will appear in the dialog box. When done, click OK to create the component pattern, as shown in the following image on the right.

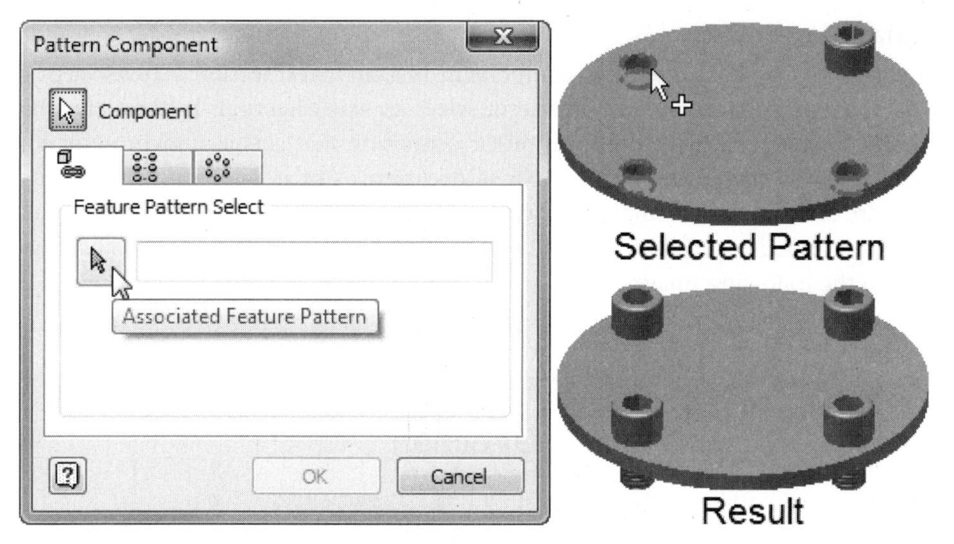

FIGURE 6.79

Rectangular Patterns

A rectangular pattern operation is illustrated in the following image. When performing this operation, two edges are chosen that define the direction of the columns and rows of the pattern. Enter the number of columns and rows and then the spacing or distance between these rows and columns. The resulting pattern acts like a feature pattern. After creating it, you can edit it to change its numbers, spacing, and so on.

In the following image, because one of the part edges is inclined, work axes are turned on and used for defining the X and Y directions of the pattern.

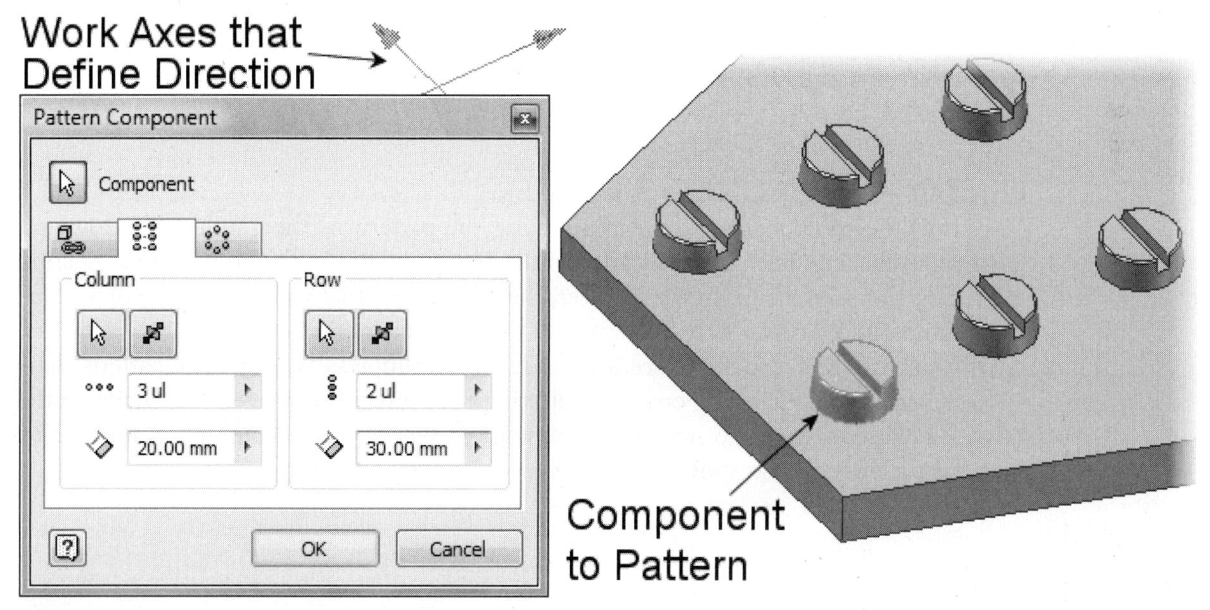

FIGURE 6.80

Circular Patterns

Circular patterns will copy selected components in a circular direction. After selecting the component or components to pattern, select an axis direction. In the following example, this element takes the form of the centerline that acts as a pivot point for the pattern. You then enter the number of occurrences or items that will make up the pattern and the circular angle. In the following example, since the bolt component is being patterned in a full circle, enter 360° as the value for the circular angle, as shown in the following image.

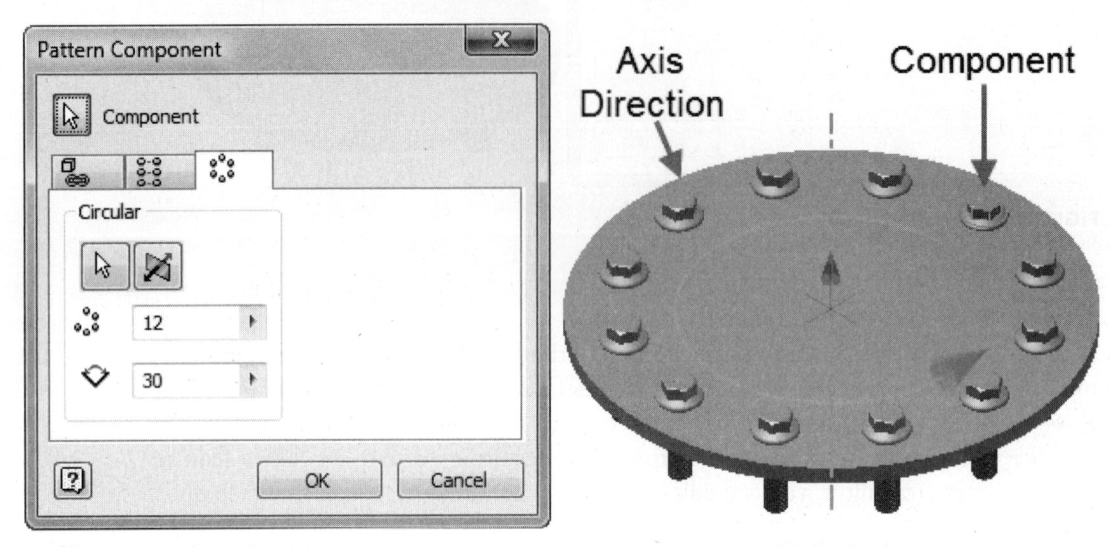

FIGURE 6.81

The completed pattern then acts as a single part. If one part moves, all parts move. The patterned component will be consumed into a component pattern in the browser, and each of the part occurrences will also appear as an element that you can expand.

Pattern Editing

Edit the component pattern by selecting the pattern in the graphics window and right-clicking, or by right-clicking on the pattern's name in the browser and selecting Edit from the menu. In the browser, the component that you patterned will be consumed into a component pattern, and the part occurrences will appear as elements, which can be expanded to see the part. You can suppress an individual pattern component by right-clicking on it and selecting Suppress from the menu, as shown in the following image. The suppressed component will be grayed out and a line will be struck through it as shown in the following image on the right.

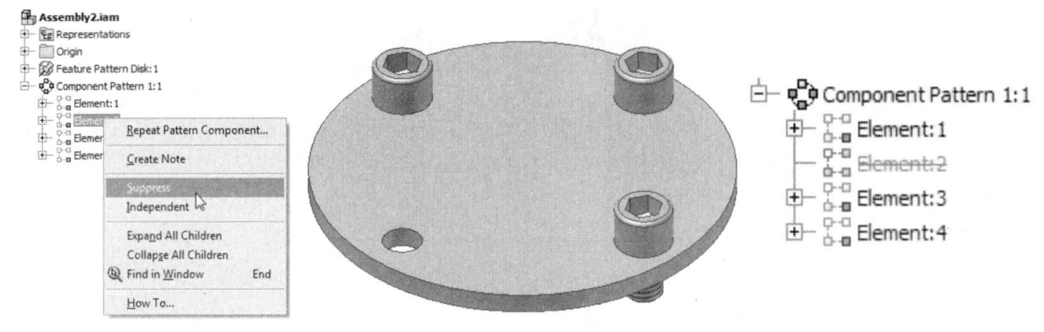

FIGURE 6.82

You can break an individual part out of the pattern by right-clicking on it and selecting Independent from the menu, as shown in the following image. Once a part is independent, it no longer has a relationship with the pattern.

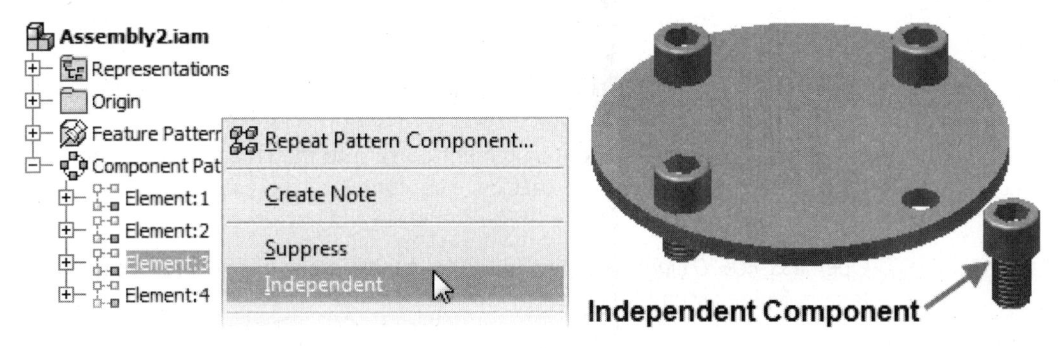

FIGURE 6.83

Replacing a Component Pattern

When replacing a component in a component pattern, you replace all occurrences in the selected component pattern with a newly selected component. Expand one of the elements in the browser, and select the component to replace. With this item highlighted in the browser, right-click and select Replace from the menu, as shown in the following image on the left. The Open dialog box will appear and will enable you to select the replacement component.

The following image on the right shows the result of performing the operation to replace a pattern component. Since the use of component patterns allows for better capture of design intent, replacement of all occurrences maintains this design intent without manually replacing each component in the pattern. Overall ease of assembly use is improved, and component patterns can be completed more quickly.

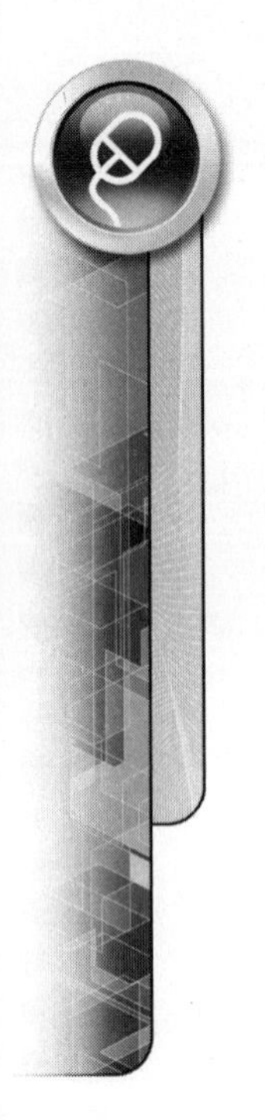

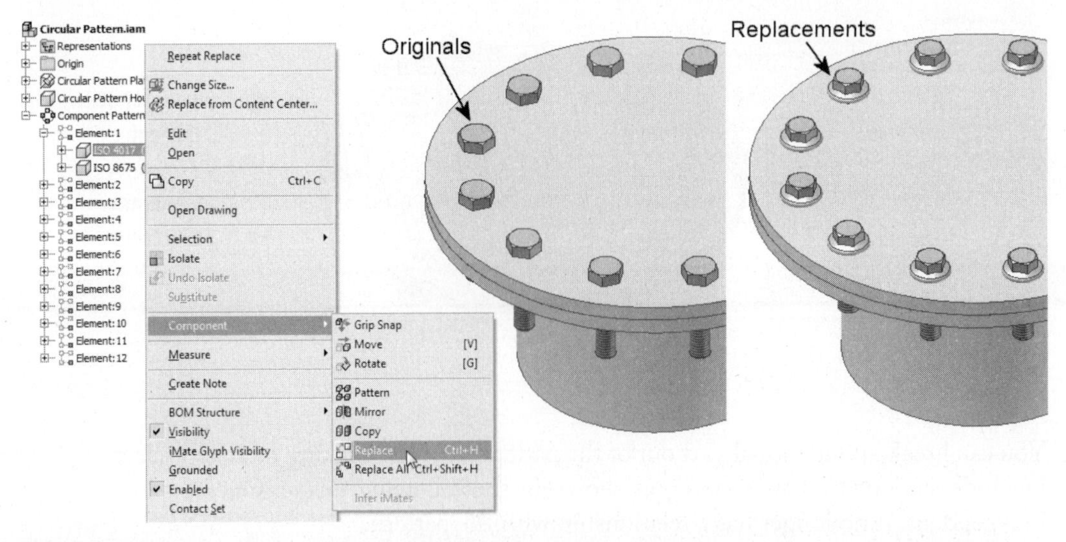

FIGURE 6.84

EXERCISE 6-3: PATTERNING COMPONENTS

In this exercise, you create a fastener component pattern to match an existing hole pattern and then replace the bolt in the pattern. You complete the exercise by demoting components to create a subassembly with the cap plate and patterned fasteners.

1. Open *ESS_E06_03.iam*, as shown in the following image on the left. The file contains a T pipe assembly.
2. In the browser, expand the parts *ESS_E06_03-Cap_Bolt.ipt:1* and *ESS_E06_03-Cap_Nut.ipt:1*. Notice that the parts already have insert constraints applied to them, as shown in the following image on the right.

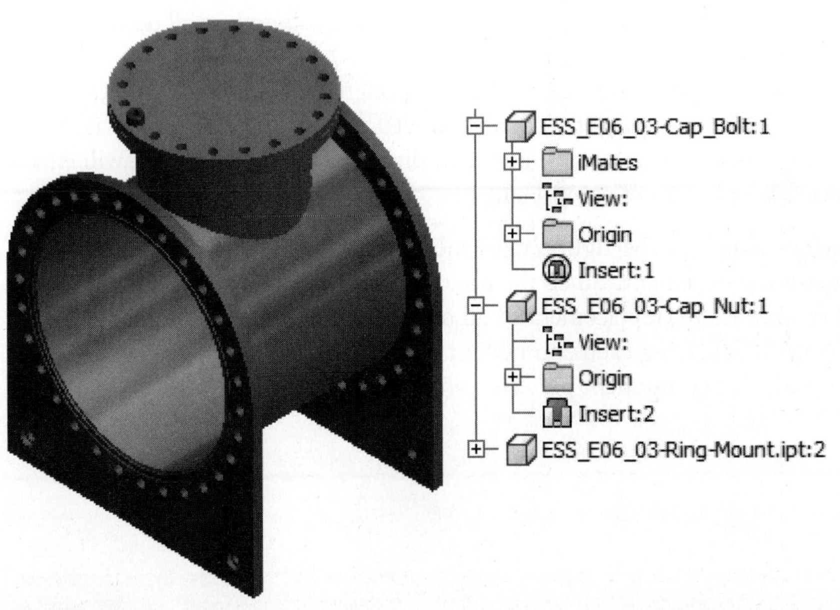

FIGURE 6.85

3. Now create a circular pattern consisting of a collection of nuts and bolts. Click the Pattern command from the Assemble tab > Component panel.
4. Press and hold the CTRL key and select the *ESS_E06_03-Cap_Bolt.ipt:1* and *ESS_E06_03-Cap_Nut.ipt:1* parts in the browser.
5. Click the Associated Features Pattern button found under the Feature Pattern Select area of the Pattern Component dialog box. In the graphics window, point to any hole in the Cap_Plate part. When the circular pattern of holes in the cap plate is highlighted, pick the edge of one of the holes, as shown in the following image on the left.
6. You should see a preview of the component pattern, as shown in the following image on the right.

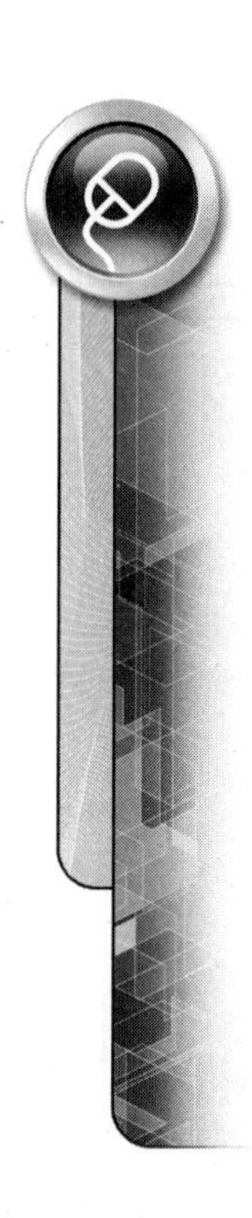

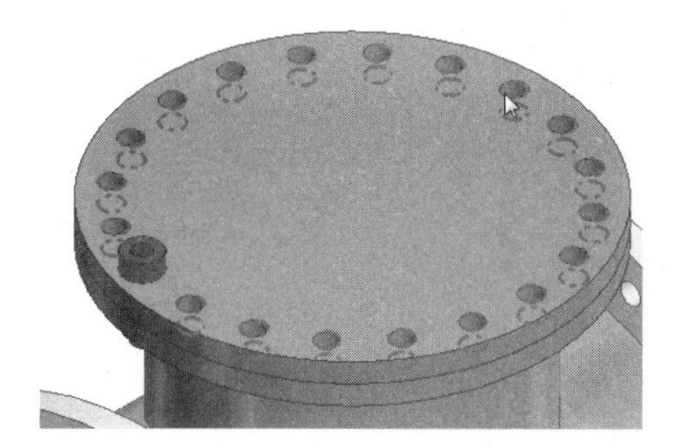

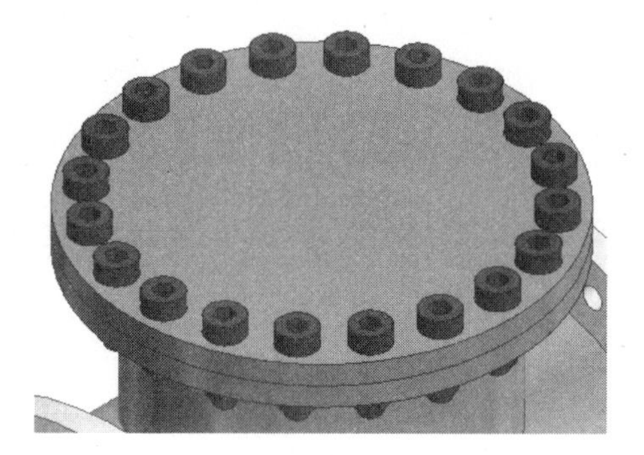

FIGURE 6.86

7. Click the OK button to create the component pattern. Expand the display of the Component Pattern 1 feature in the browser.
8. In the browser, expand Element:1 and the parts beneath it. Notice that the constraints on the original components were retained, as shown in the following image.

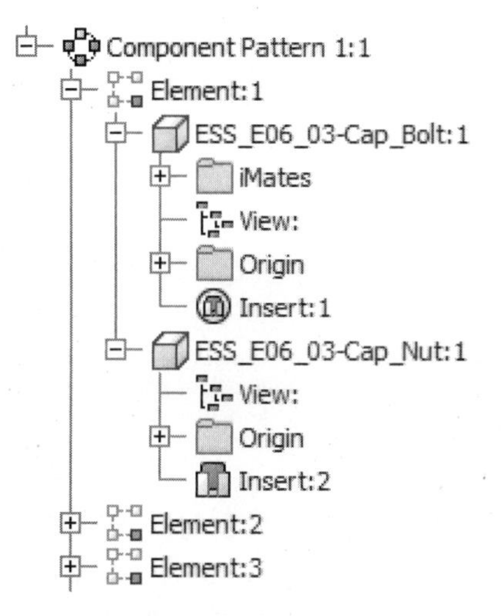

FIGURE 6.87

9. Now replace the bolt in the pattern with a similar but different fastener. Under Element:1 in the browser, right-click the *ESS_E06_03-Cap_Bolt,* select Component, and click Replace from the menu.

10. In the Place Component dialog box, select *ESS_E06_03-Cap_Hex_Bolt.ipt*, from the Chapter 06 folder and click Open. A warning message box will appear to notify you that some assembly constraints may be lost.
11. Click the OK button. The pattern is updated to reflect the new bolt, as shown in the following image.

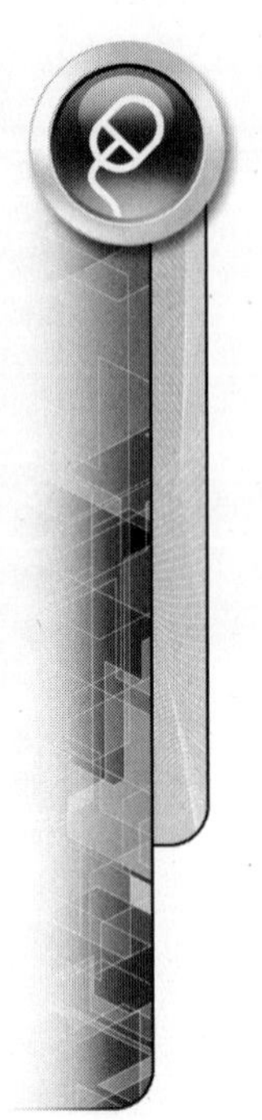

NOTE

The fastener in this exercise is constrained to the plate with an Insert iMate constraint. Since the replacement fastener has a similar Insert iMate, the constraint to the plate is retained.

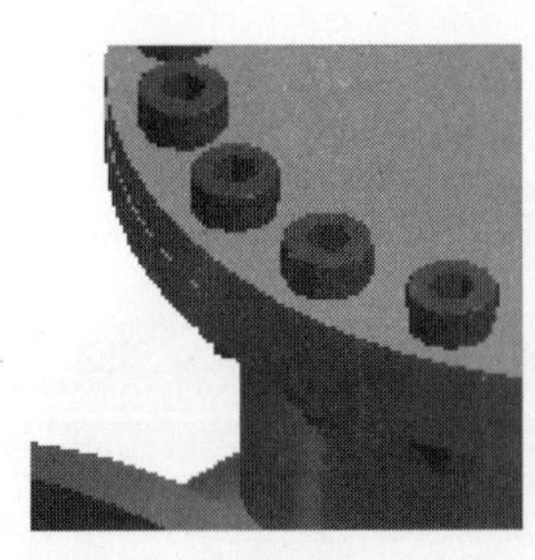

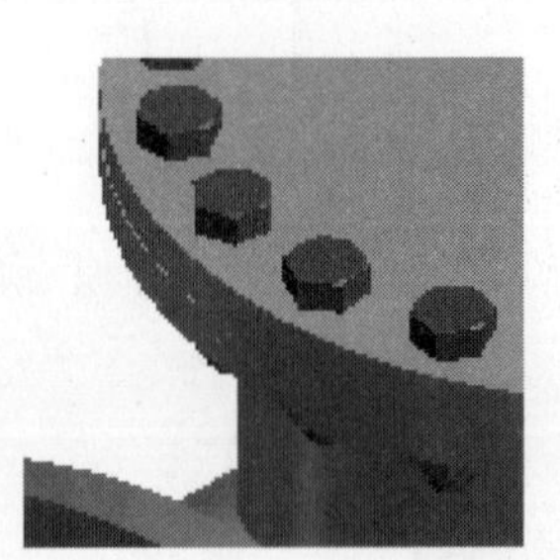

FIGURE 6.88

12. Next, you demote a number of components; this will create a subassembly with the cap plate and component pattern.
13. Hold down CTRL while you click *ESS_E06_03-Cap_Plate:1* and Component Pattern 1 in the browser.
14. Right-click, and select Component followed by Demote from the menu. In the Create In-Place Component dialog box, enter *Cap_Kit_Assembly* in the New Component Name field. When finished, click the OK button. A warning message box appears to notify you that some assembly constraints may be lost. Click Yes.
15. In the browser, expand Cap_Kit_Assembly:1. Notice that the cap plate and component pattern are now part of the new subassembly. Notice also that the constraints on the original components were retained. Your display should appear similar to the following image.

FIGURE 6.89

16. Close all open files. Do not save changes. End of exercise.

ANALYSIS COMMANDS

Various commands are available that assist in analyzing sketch, part, and assembly models. You can calculate minimum distance between components, calculate the center of gravity of parts and assemblies, and perform interference detection.

The Minimum Distance Command

You can easily obtain the minimum distance between any components, parts, or faces in an assembly. While in the assembly, select the Measure Distance command from the Inspect tab > Measure panel as shown in the following image on the left. Then change the selection priority, depending on what you are measuring. The three available selection modes are shown in the following image on the right.

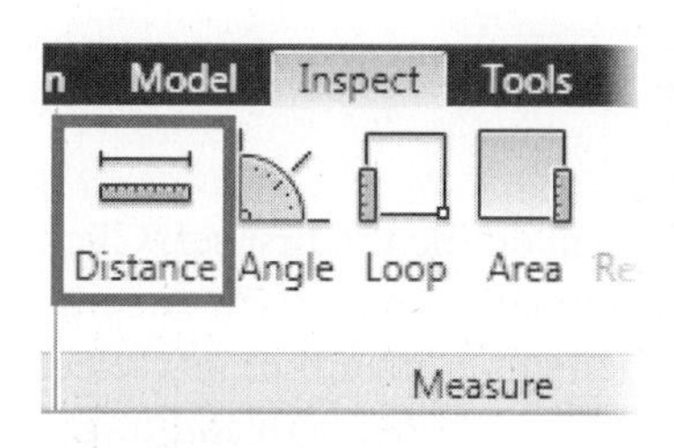

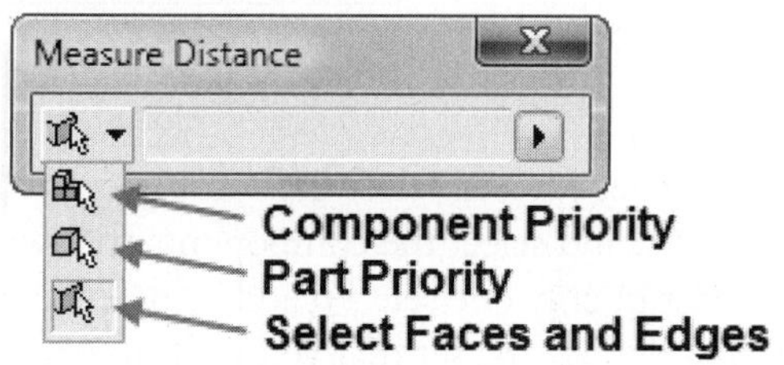

FIGURE 6.90

Identifying the Center of Gravity of an Assembly

A center of gravity placed in an assembly could be critical to the overall design process of that assembly, whether it is used in the next assembly or in the main assembly. On the View tab > Visibility panel, click the Center of Gravity command as shown in the following image on the left. The image on the right shows the center of gravity icon applied to an assembly model. This icon actually consists of a triad displaying the X, Y, and Z directions. Three selectable work planes and a selectable work point area are also available for the purpose of measuring distances and angles.

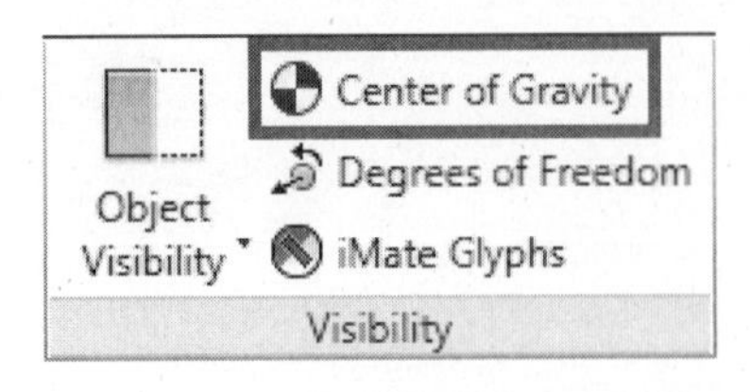

FIGURE 6.91

Interference Checking

You can check for interference in an assembly using one or two sets of objects. To check the interference among sets of stationary components, make the assembly or subassembly in question active. Then click the Analyze Interference command on the Inspect tab > Interference panel as shown in the following image on the left. The Interference Analysis dialog box will appear, as shown in the following image on the right.

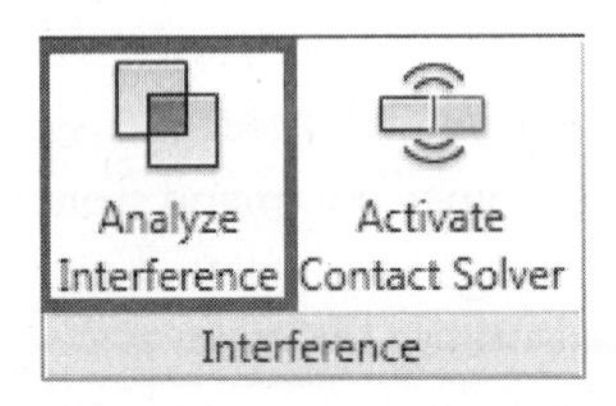

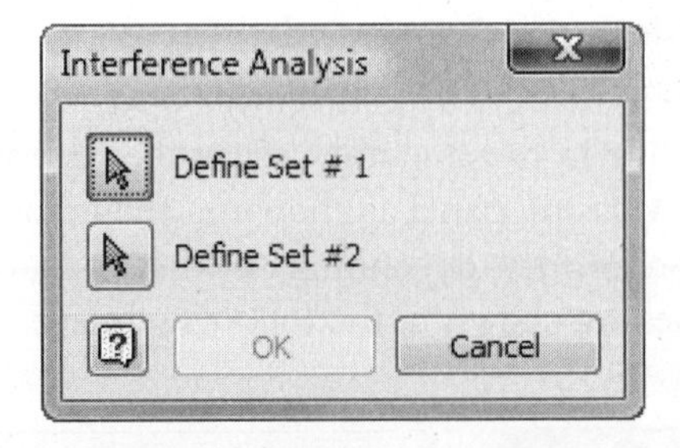

FIGURE 6.92

Click on Define Set #1, and select the components that will define the first set. Click on Define Set #2, and select the components that will define the second set. A component can exist in only one set. To add or delete components from either set, select the button that defines the set that you want to edit. Click components to add to the set, or press the CTRL key while selecting components to remove from the set.

NOTE

Use only Define Set #1 if you want to check for interference in a single group of objects.

Once you have defined the sets, click the OK button. The order in which you selected the components has no significance. If interference is found, the Interference Detected dialog box will appear, as shown in the following image.

FIGURE 6.93

The information in the dialog box defines the X, Y, and Z coordinates of the centroid of the interfering volume. It also lists the volume of the interference and the components that interfere with one another. A temporary solid will also be created in the graphics window that represents the interference. You can copy the interference report to the clipboard or print it from the commands in the Interference Detected dialog box. When the operation is complete, click the OK button, and the interfering solid will be removed from the screen.

Performing an interference detection does not fix the interfering problem; it only presents a graphical representation of the problem. After analyzing and finding an interference, edit the assembly or components to remove the interference. You can also detect interference when driving constraints.

EXERCISE 6-4: ANALYZING AN ASSEMBLY

In this exercise, you analyze a partially completed assembly for interference between parts and then check the physical properties to verify design intent.

1. Open the assembly *ESS_E06_04.iam*. The linkage is displayed, as shown in the following image on the left.
2. Begin the process of checking for interferences in the assembly. First zoom into the linkages, as shown in the following image on the right.

FIGURE 6.94

3. Click Analyze Interference from the Inspect tab > Interference panel.
4. When the Analyze Interference dialog box displays, select the two components for Set #1, as shown in the following image on the left.
5. For Set #2, select the spyder, as shown in the following image on the right.

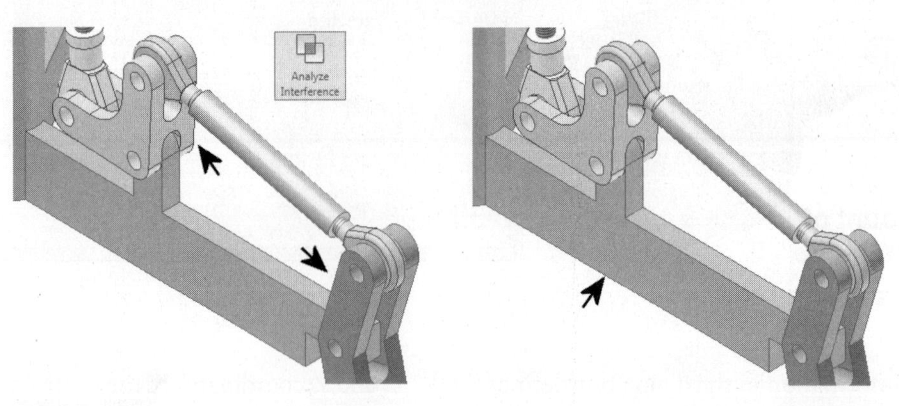

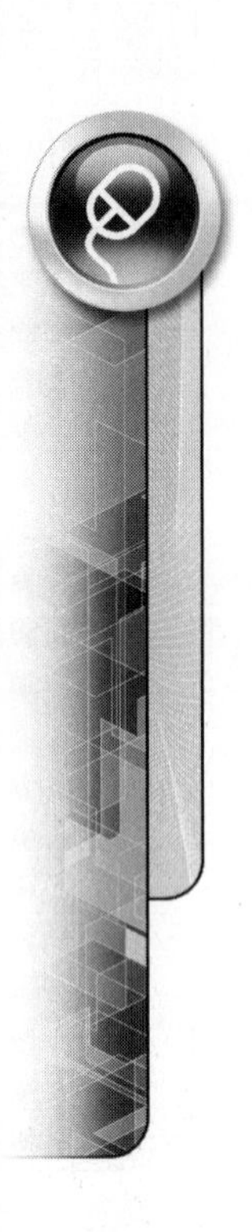

FIGURE 6.95

6. Click the OK button. Notice that the Interference Detected dialog box is displayed, and the amount of interference displays on the parts of the assembly, as shown in the following image.

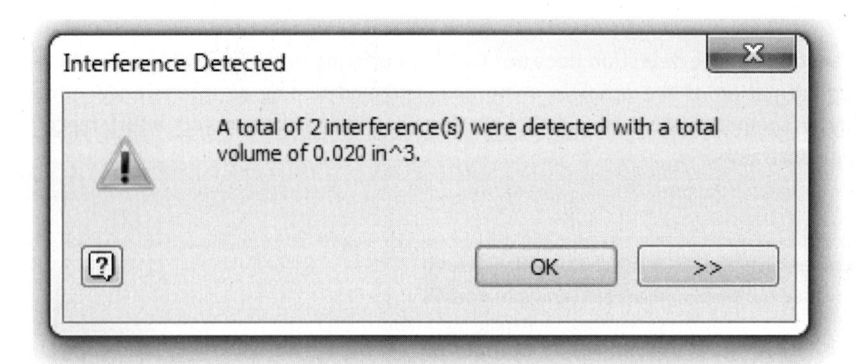

FIGURE 6.96

7. Click the More (>>) button to expand the dialog box and display additional information.
8. Click the OK button to close the Interference Detected dialog box.
9. The design intent of the lifting mechanism relies on the correct selection of materials. This impacts the strength and mass of the assembly. You will now display the physical properties of the entire assembly. Return to the Home View of the assembly.
10. In the browser, right-click *ESS_E06_04.iam*, and then select iProperties, as shown following image.

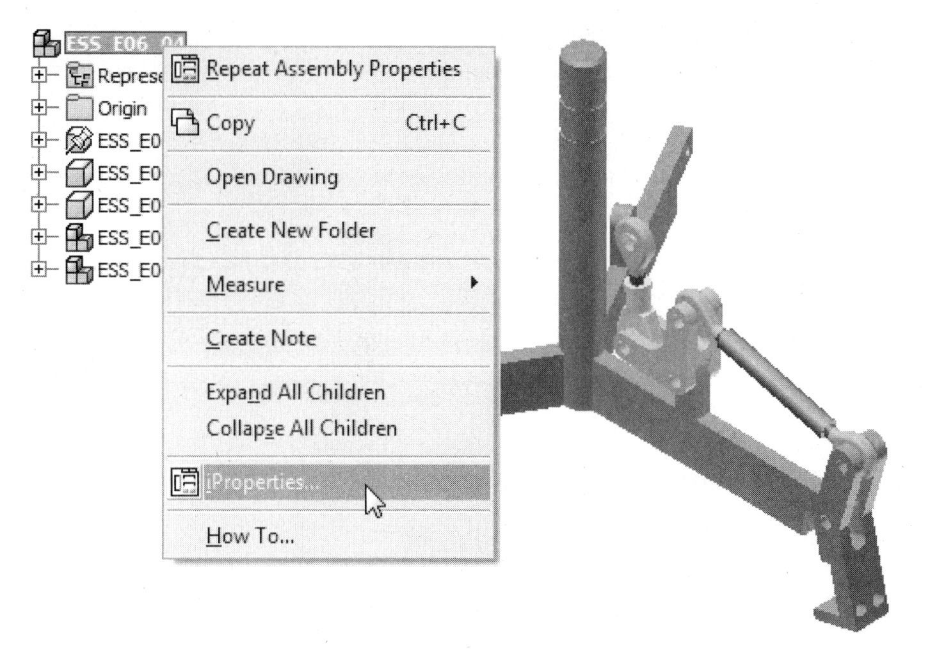

FIGURE 6.97

11. When the iProperties dialog box displays, click the Physical tab.
12. Click the Update button. Notice that the physical properties of the assembly are displayed.

NOTE

To get a more accurate account of the assembly properties, you can change the Requested Accuracy setting to Very High. The updating action may take some time, depending on the size of the assembly being analyzed.

13. Click the Close button to dismiss this dialog box.
14. Close all open files. Do not save changes. End of exercise.

DRIVING CONSTRAINTS

You can animate or drive mechanical motion using the Drive Constraint command. To simulate motion, an angle, mate, tangent, or insert assembly constraint must exist. You can only drive one assembly constraint at a time, but you can use equations to create relationships to drive multiple assembly constraints simultaneously. To drive a constraint, right-click on the desired constraint in the browser, and select Drive Constraint from the menu, as shown in the following image on the left.

The Drive Constraint dialog box will appear, as shown in the following image on the right. Depending on the constraint that you are driving, the units may be different. Enter a Start value; the default value is the angle or offset for the constraint. Enter a value for End and a value for Pause Delay if you want a dwell time between the steps. In the dialog box, you can also choose to create an animation (AVI or WMV) file that will record the assembly motion. You can replay the recorded file without having Autodesk Inventor installed.

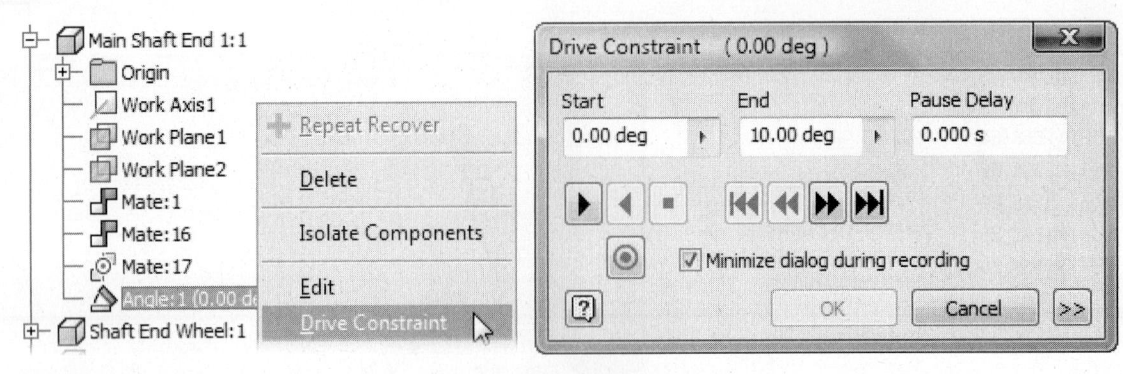

FIGURE 6.98

The following sections describe the Start, End, and Pause Delay controls.

Start. Sets the start position of the offset or angle.

End. Sets the end position of the offset or angle.

Pause Delay. Sets the delay between steps. The default delay units are seconds.

Use the following Motion and video buttons to control the motion and to create a (PE) recorded video file.

Button	Description
Forward	Drives the constraint forward, from the start position to the end position.
Reverse	Drives the constraint in reverse, from the end position to the start position.
Pause	Temporarily stops the playback of the constraint drive sequence.
Minimum	Returns the constraint to the starting value and resets the constraint driver. This button is not available unless the constraint driver has been run.
Reverse Step	Reverses the constraint driver one step in the sequence. This button is not available unless the drive sequence has been paused.
Forward Step	Advances the constraint driver one step in the sequence. This button is not available unless the drive sequence has been paused.
Maximum	Advances the constraint sequence to the end value.
Record	Begins capturing frames at the specified rate for inclusion in an animation file named and stored to a location that you define.

To set more conditions on how the motion will behave, click the More (<<) button. This action will display the expanded Drive Constraint dialog box, as shown in the following image. The following sections describe the options in the dialog box.

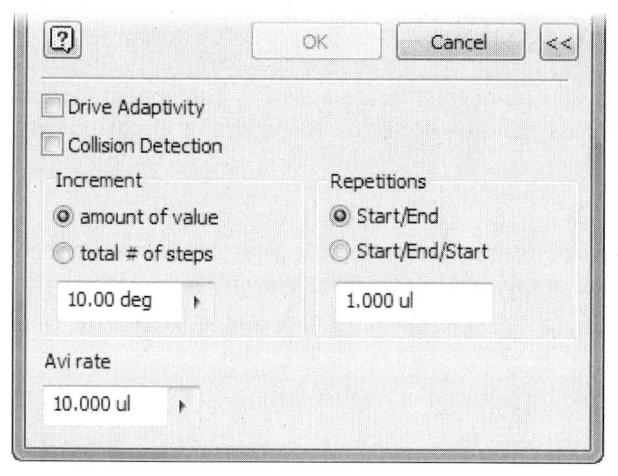

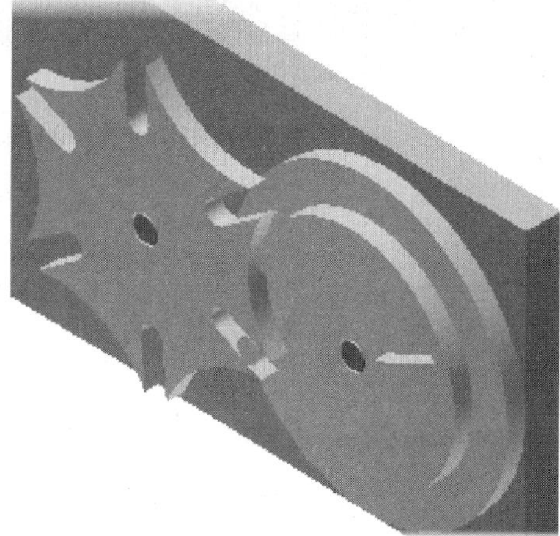

FIGURE 6.99

Drive Adaptivity. Click this option to adapt the component while the constraint is driven. It only applies to assembly components for which adaptivity has been defined and enabled.

Collision Detection. Click this option to drive the constraint until a collision is detected. When interference is detected, the drive constraint will stop and the components where the collision occurs will be highlighted. It also shows the constraint value at which the collision occurred.

Increment. Determines the value that the constraint will be incremented during the animation.

Amount of Value. Increments the offset or angle value by this value for each step.

Total # of Steps. The constraint offset or angle is incremented by the same value per step based on the number of steps entered and the difference between the start and stop values.

Repetitions. Sets how the driven constraint will act when it completes a cycle and how many cycles will occur.

Start/End. Drives the constraint from the start value to the end value and resets at the start value.

Start/End/Start. Drives the constraint from the start value to the end value and then drives it in reverse from the end value to the start value.

AVI Rate. Specifies how many frames are skipped before a screen capture is taken of the motion that will become a frame in the completed AVI file.

NOTE

If you try to drive a constraint and it fails, you may need to suppress or delete another assembly constraint to allow the required component degrees of freedom. To reduce the size of a WMV or AVI movie file, reduce the screen size before creating the file and use a solid background color in the graphics window.

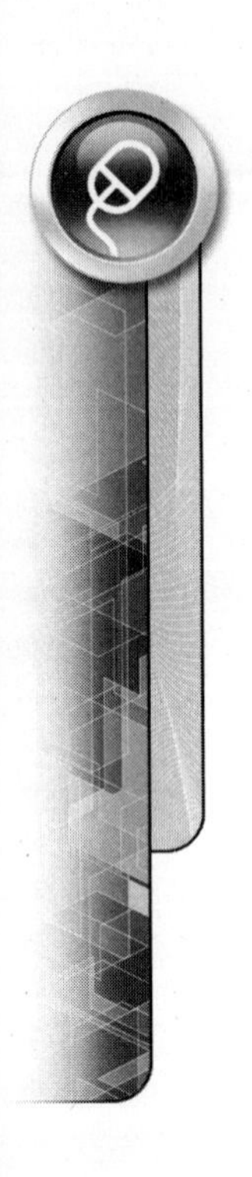

EXERCISE 6-5: DRIVING CONSTRAINTS

In this exercise, you drive an angle constraint to simulate assembly component motion. Then you drive the constraint and use collision detection to determine if components interfere.

1. Open *ESS_E06_05.iam.*
2. Use the Rotate and Zoom commands to examine the assembly. When finished, right-click in the graphics window, and select Home View from the menu.
3. Now create an angle constraint between the pivot base and arm. Click the Constrain command.
4. In the Place Constraint dialog box, click the Angle button.
5. Select the Base face and then arm face, select the Directed Angle solution as shown in the following image.
6. Enter an angle of **45°**. Click the OK button.

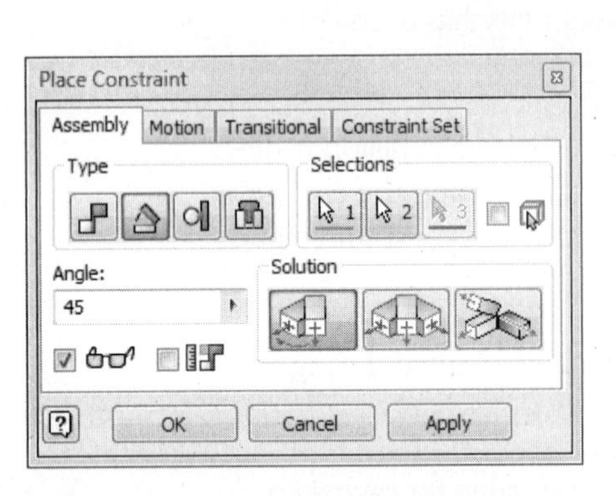

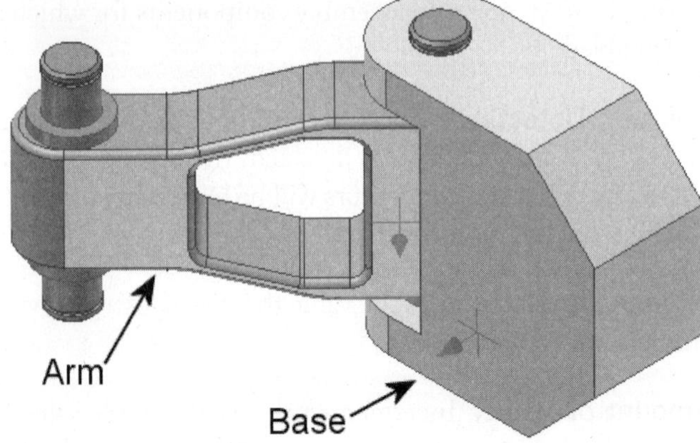

FIGURE 6.100

7. Now drive the angle constraint you just placed. In the browser, expand *ESS_E06_05-Pivot_Base:1.*
8. Right-click the Angle constraint, and then select Drive Constraint from the menu.
9. In the Drive Constraint dialog box, enter **45.00°** as the Start value and **120.00°** as the End value.
10. Click the More (>>) button.
11. In the Increment section, enter **2.00 Deg**, as shown in the following image.
12. Click the Forward button. Notice that the arm assembly interferes with the pivot base between 45.00° and 120.00° but Inventor continues to drive the constraint.
13. Click the Reverse button.

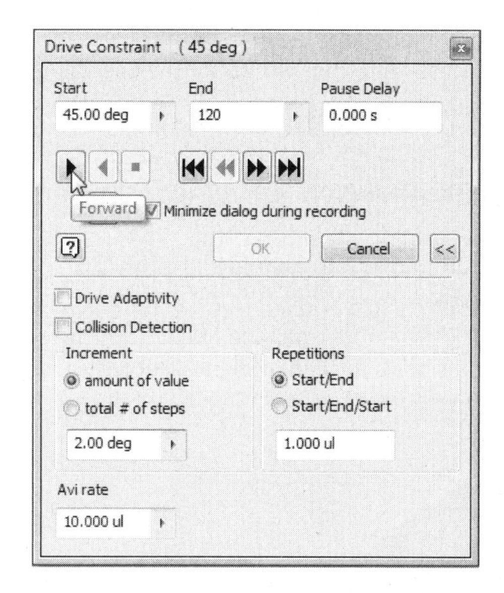

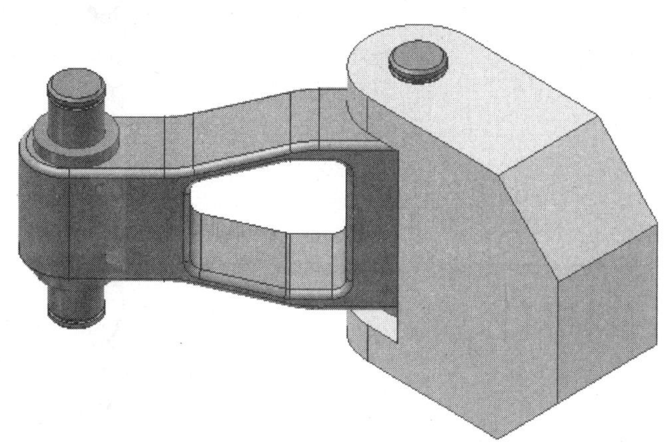

FIGURE 6.101

14. Now perform a collision detection operation. In the Drive Constraint dialog box place a check in the Collision Detection box.
15. Change the Increment value to **.1**.
16. Click the Forward button. Notice that a collision is detected at 80.7° and that the interfering parts are highlighted, as shown in the following image. Click OK to return to the Drive Constraint dialog box.

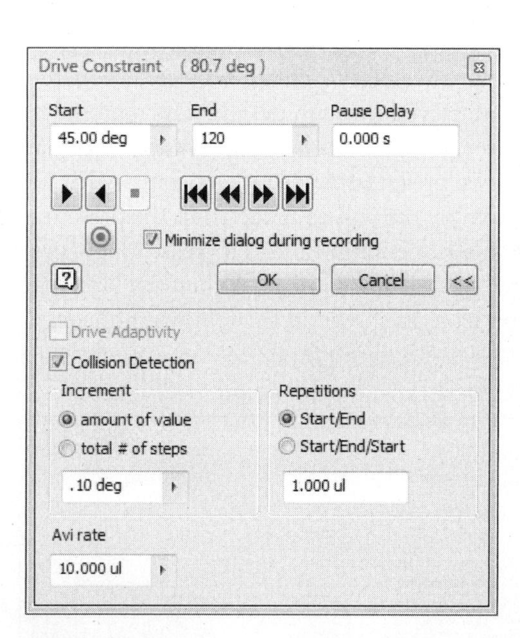

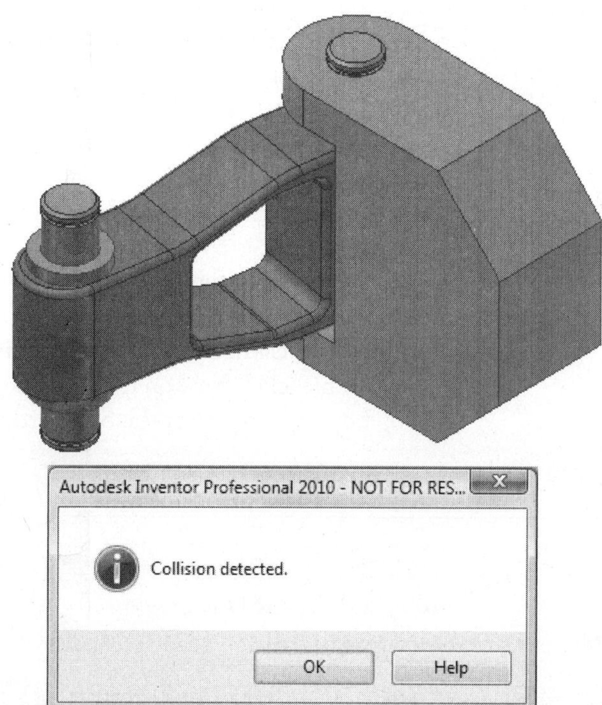

FIGURE 6.102

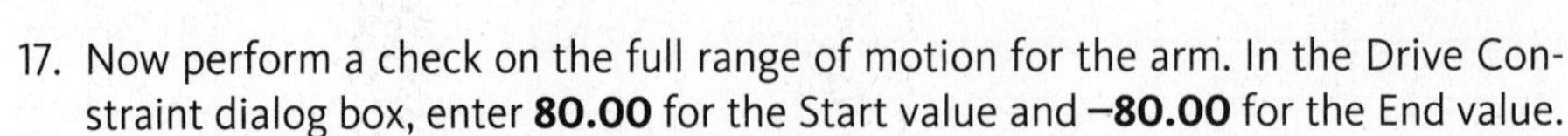

17. Now perform a check on the full range of motion for the arm. In the Drive Constraint dialog box, enter **80.00** for the Start value and **–80.00** for the End value.
18. Ensure that the Collision Detection option is checked.
19. Change the Increment value to **2.00**.
20. Select Start/End/Start and enter 4 in the Repetitions area of the Drive Constraint dialog box.
21. Click the Reverse button to test the full range of motion for interference. No collision is detected in this range of motion.
22. In the Drive Constraint dialog box, click the OK button. The 80.00°, which represents the maximum angle before collision, is applied to the Angle constraint, as shown in the following image.

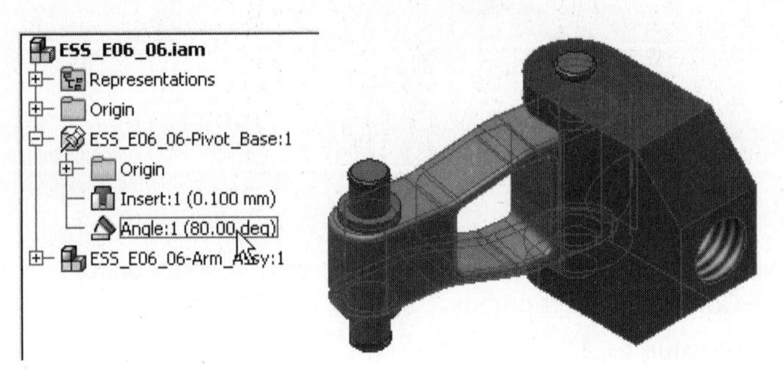

FIGURE 6.103

23. Close all open files. Do not save changes. End of exercise.

CREATING PRESENTATION FILES

After creating an assembly, you can create drawing views based on how the parts are assembled. If you need to show the components in different positions similar to an exploded view, hide specific components, or create an animation that shows how to assemble and disassemble the components, you need to create a presentation file. Components can also be hidden and later retrieved through design view representations. This topic is covered in a later chapter. A presentation file is separate from an assembly and has a file extension of *.ipn*. The presentation file is associated with the assembly file, and changes made to the assembly file will be reflected in the presentation file. You cannot create components in a presentation file. To create a new presentation file, click New on the Get Started tab > Launch panel. When the New File dialog box appears, select the Standard.ipn icon, as shown in the following image.

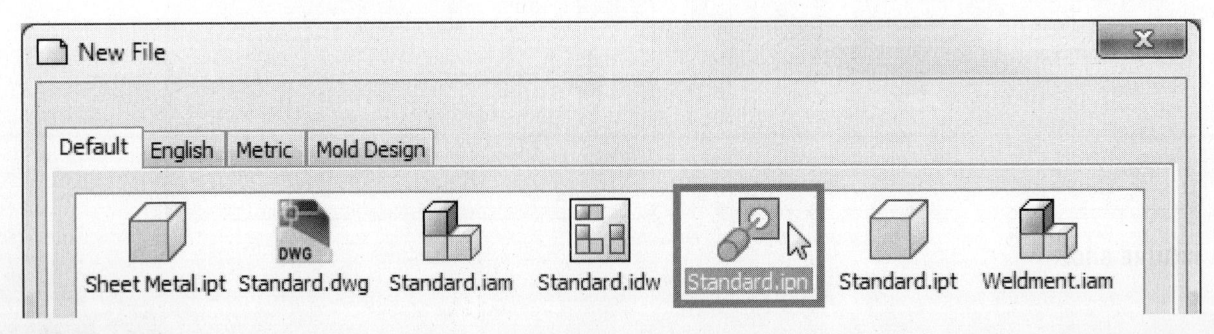

FIGURE 6.104

By default, the ribbon will show the presentation commands available to you, as shown in the following image. The following sections explain these commands.

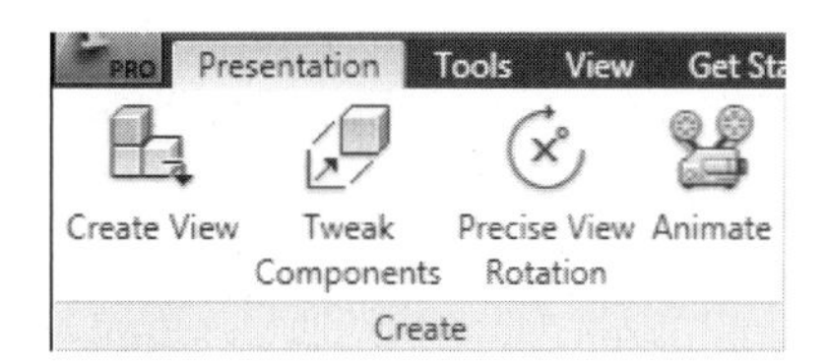

FIGURE 6.105

Create View. Click to create views of the assembly. Created views are based on the same assembly model. Once created, a view will be listed as an explosion in the browser.

Tweak Components. Click to move and/or rotate parts in the view.

Precise View Rotation. Click to rotate the view by a specified angle and direction using a dialog box.

Animate. Click to create an animation of the assembly tweaks and saved camera views; an AVI or WMV file can be output.

The basic steps for creating a presentation view are to reposition (tweak) the parts in specific directions and then create an animation, if needed. The following sections discuss these steps.

Creating Presentation Views

The first step in creating a presentation is to create a presentation view. Issue the Create View command on the Presentation tab > Create panel as shown in the following image on the left, or right-click and select Create View from the menu. The Select Assembly dialog box appears, as shown in the following image on the right. The dialog box is divided into two sections: Assembly and Explosion Method.

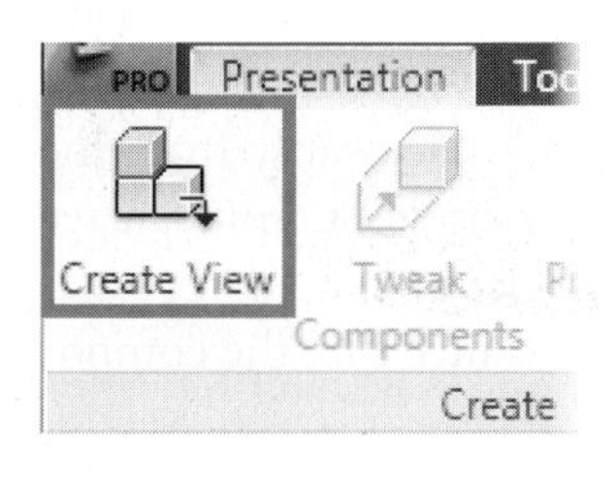

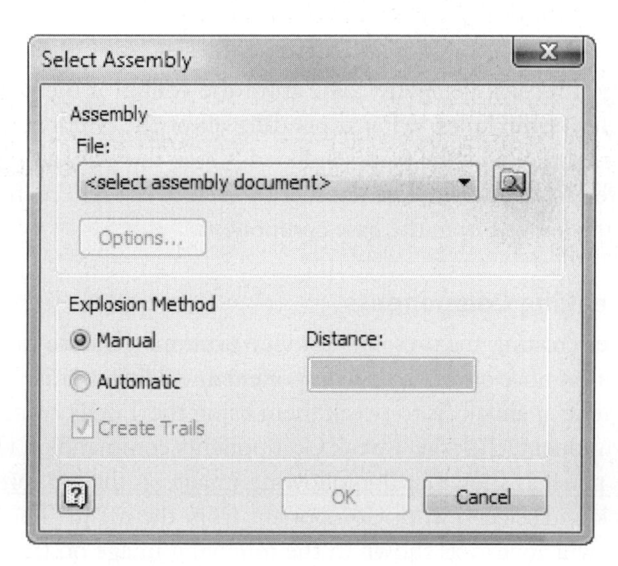

FIGURE 6.106

Assembly

In this section, you determine on which assembly and design view to base the presentation view.

File. If an assembly file or files are open they will appear in the drop-down list. Select an assembly file or navigate to and select an assembly file on which to base the presentation view.

Options. Click to display the File Open Options dialog box. Use this dialog box to select design view representations, positional representations, and level of detail representations, which are covered in Chapter 9.

Explosion Method

In this section, you can choose to manually or automatically explode the parts, meaning that you separate the parts a given distance.

Manual. Click so that the components will not be exploded automatically. After the presentation view is created, you can add tweaks that will move or rotate the parts.

Automatic. Click so that the parts will be exploded automatically with a given value.

Create Trails. If you clicked Automatic, click to have visible trails. Trails are lines that show how the parts are exploded.

Distance. If Automatic is selected, enter a value that the parts will be exploded.

After making your selection, click the OK button to create a presentation view. The components will appear in the graphics window and a presentation view will appear in the browser. If you clicked the Automatic option, the parts will be exploded automatically. To determine how the parts will be exploded, Autodesk Inventor analyzes the assembly constraints. If you constrained parts using the mate plane option and the arrows perpendicular to the plane (normals) for both parts were pointing outward, they will be exploded away from each other the defined distance. Think of the mating parts as two magnets that want to push themselves in opposite directions. The grounded component in the browser will be the component that stays stationary, and the other components will move away from it. If you are creating trails automatically, they will generally come from the center of the part, not necessarily from the center of the holes. After expanding all of the children in the browser, you can see numerically how the parts exploded. Once the parts are exploded, the distance is referred to as a tweak. The number by each tweak reflects the distance that the component is moved from the base component.

Tweaking Components

After creating the presentation view, you may choose to edit the tweak of a component or move or rotate the component an additional distance. To reposition the components manually is to tweak them using the Tweak command. To manually tweak a component, click the Tweak Components command on the Presentation tab > Create panel as shown in the following image on the left, press the hot key T, or right-click and select Tweak Component from the menu. The Tweak Component dialog box will appear, as shown in the following image on the right. The Tweak Component dialog box has two sections: Create Tweak and Transformations.

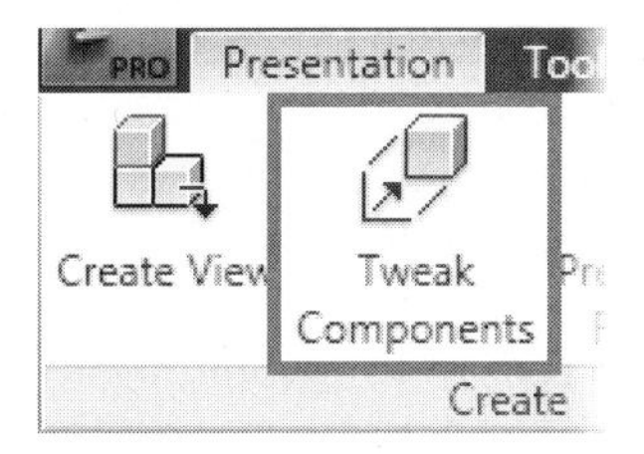

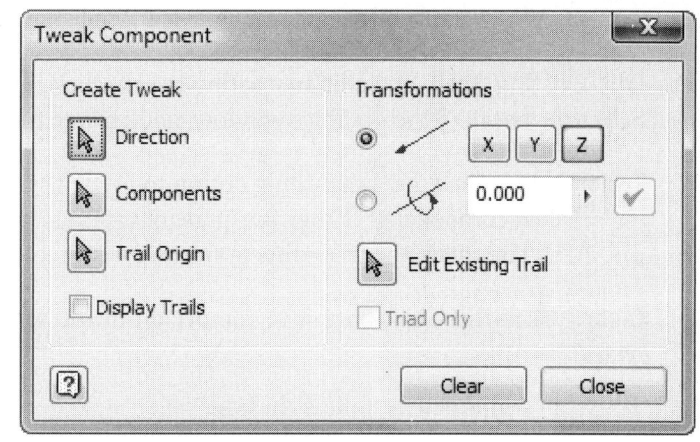

FIGURE 6.107

Create Tweak

In this section, you select the components to tweak, set the direction and origin of the tweak, and control trail visibility.

Direction. Determine the direction or axis of rotation for the tweak. After clicking the Direction button, select an edge, face, or feature of any component in the graphics window to set the direction triad (X, Y, and Z) for the tweak. The edge, face, or feature that you selected does not need to be on the components you are tweaking.

Components. Select the components to tweak. Click the Components button, and then click the components in the graphics window or browser to tweak. If you selected a component when you started the Create Tweak operation, it will be included in the components automatically. To remove a component from the group, press and hold the CTRL key and click the component.

Trail Origin. Set the origin for the trail. Click the Trail Origin button, and then click in the graphics window to set the origin point. If you do not specify the trail origin, it will be placed at the center of mass for the part.

Display Trails. Click if you want to see the tweak trails for the selected components. Clear the checkbox to hide the trails.

Transformations. In this section, you set the type and value of a tweak.

Linear. Click the button next to the arrow and line to move the selected components in a linear fashion.

Rotation. Click the button next to the arrow and arc to rotate the selected components.

X, Y, Z. Click the X, Y, or Z coordinate button to determine the direction for a linear tweak or the axis for a rotational tweak. Alternately, you can select the arrow on the triad that represents the X, Y, or Z direction.

Tweak Distance/Tweak Angle Field. Enter a positive or negative value for the tweak distance or rotation angle, or click a point in the graphics window and move the cursor with the mouse button depressed to set the distance.

Apply. After making all selections, click the Apply button to complete the tweak.

Edit Existing Trail. To edit an existing tweak, click the Edit Existing Trail button. Select the tweak in the graphics window, and change the desired settings.

Triad Only. Click the Triad Only option to rotate the direction triad without rotating selected components. Enter the angle of rotation, and then click the Apply button. After you rotate the triad direction, you can use it to define tweaks.

Clear. Click the Clear button to remove all of the settings and set up for another tweak.

To tweak a component, follow these steps:

1. Issue the Tweak Component command.
2. Determine the direction or axis of rotation for the tweak by clicking the Direction button and selecting an edge, face, or feature.
3. Select the components to tweak by clicking the Components button and clicking the components in the graphics window or browser that will be tweaked.
4. Select any additional trail origin points.
5. Determine whether or not you want trails to be visible.
6. Set the type of tweak to linear or rotation.
7. Click the X, Y, or Z coordinate button to determine the direction for a linear tweak or the axis for a rotational tweak.
8. Enter a value for the tweak in the text box, or select a point on the screen and drag the part into its new position.
9. Click the Apply button in the Tweak Component dialog box.

To edit an individual tweak, click on its value in the browser, as shown in the following image. Enter a new value in the text box in the lower-left corner of the browser, and press ENTER to use the new value.

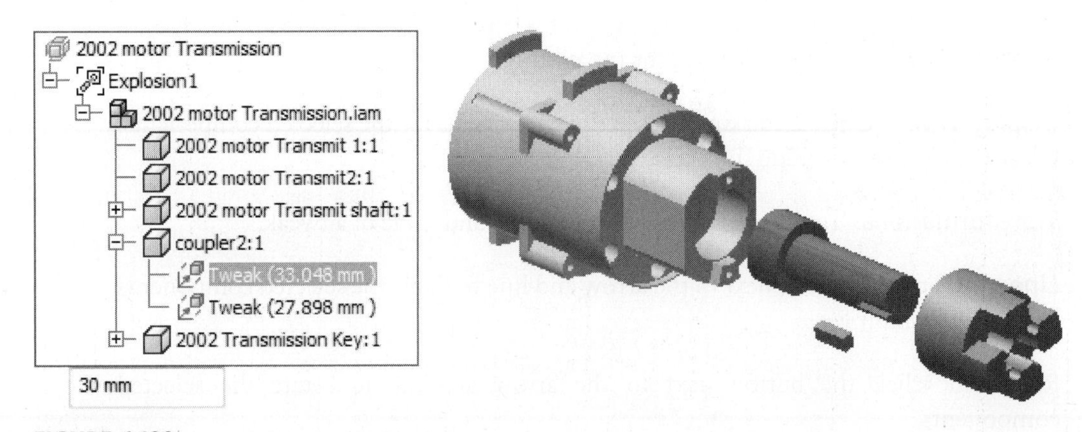

FIGURE 6.108

To extend all of the tweaks the same distance, right-click on the assembly's name in the browser, and select Auto Explode from the menu, as shown in the following image on the left. Enter a value, and click the OK button, as shown in the following image on the right, and all of the tweaks will be extended the same distance. You cannot use negative values with the Auto Explode method.

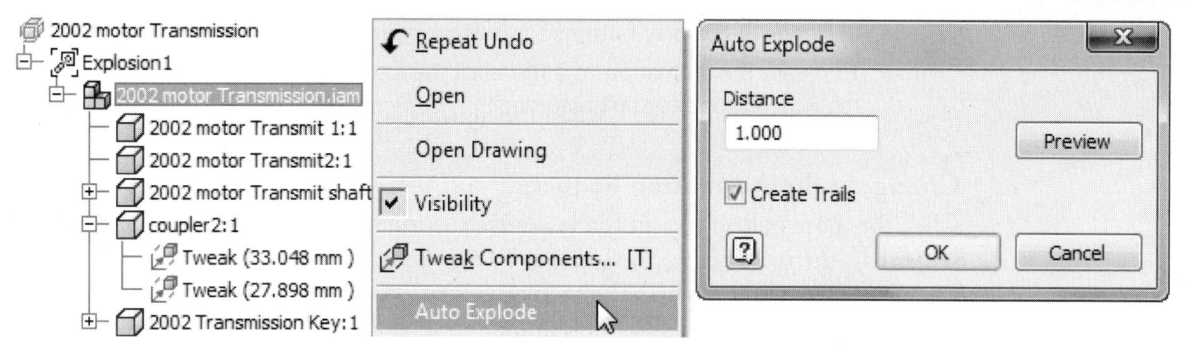

FIGURE 6.109

Animation

After you have repositioned the components, you can animate the components to show how they assemble or disassemble. To create an animation, click the Animate command on the Presentation tab > Create panel as shown in the following image on the left. The Animation dialog box will appear as shown in the following image. The Animation dialog box has three sections: Parameters, Motion, and Animation Sequence, which appears under the More (<<) button.

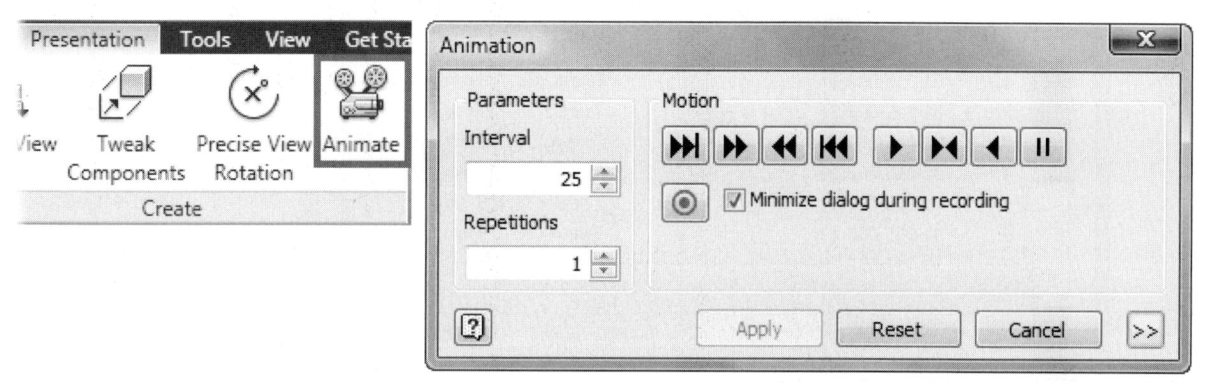

FIGURE 6.110

Parameters

In this section, you specify the playback speed and the number of repetitions for the animation.

Interval. Set this value for the playback speed of the animation in frames. The higher the number, the greater the number of steps in the tweak and the slower the animation. A smaller number will speed up the animation.

Repetitions. Set the number of times to repeat the playback. Enter the desired number of repetitions, or use the up or down arrow to select the number.

NOTE

To change the number of repetitions, click the Reset button on the dialog box, and then enter a new value.

To animate components, follow these steps:

1. Issue the Animation command.
2. Set the number of repetitions.
3. Adjust the tweaks as needed.

4. Click one of the play buttons to view the animation in the graphics window.
5. To record the animation to a file, click the Record button, and then click one of the Play buttons to start recording.

Changing the Animation Sequence

Click the more button >> in the lower-right corner of the Animation dialog box to see or edit the sequence of the tweaks. In the Animation Sequence section, you can change the sequence in which the tweaks happen, select the tweak, and then select the needed operation. Expanding the Animation dialog box displays the following image. For example, in the image on the left, each component will move until its complete cycle is over before another component starts to move. In the image on the right, notice all sequence numbers have changed to a value of 1. This means when the animation begins, all components will move at the same time at the same rate of speed. Experiment with the sequencing of an exploded assembly to gain more dramatic animation results.

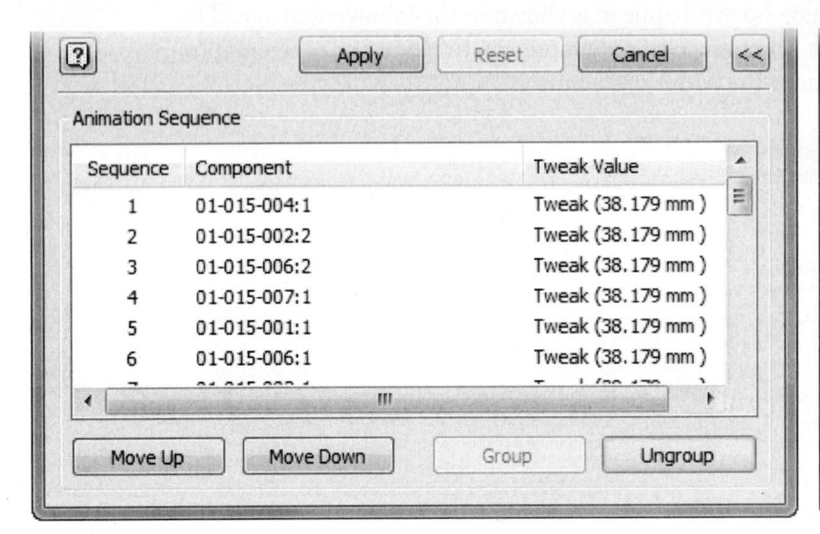

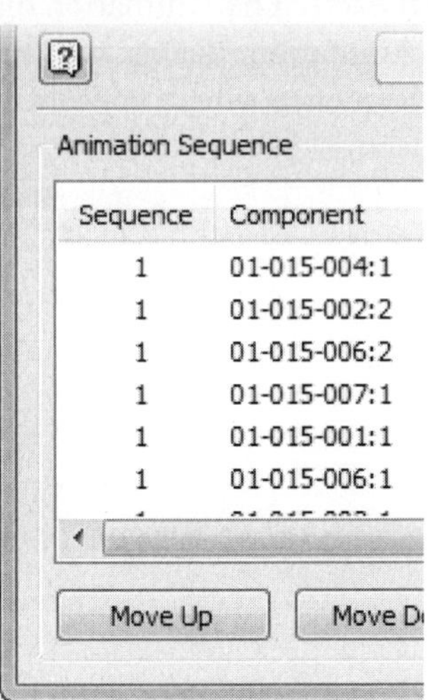

FIGURE 6.111

Move Up. Click to move the selected tweak up one place in the list.

Move Down. Click to move the selected tweak down one place in the list.

Group. Select a number of tweaks, and then click the Group button. When tweaks are grouped, all of the tweaks in the group will move together as you change the sequence. The group assumes the sequence order of the lowest tweak number.

Ungroup. After selecting a tweak that belongs to a group, you can click the Ungroup button, and the tweak can then be moved individually in the list. The first tweak in the group assumes a number that is one higher than the group. The remaining tweaks are numbered sequentially following the first.

TIP You can search the Help system for *Sequence View* or *Edit Tasks and Sequences* to discover additional browser views and editing capabilities when working with Presentation files.

AUTODESK INVENTOR PUBLISHER

Although beyond the scope of this book, another product, Autodesk Inventor Publisher may be useful for you to know about as it relates to illustrating or documenting your assemblies. Publisher is an application that leverages CAD models to easily create product documentation—maintaining an associative link so that updates are easily incorporated as your designs change.

After importing your CAD data into Inventor Publisher, you will find tools allowing you to easily reposition and change the appearance of components in different process steps and views, automatically explode 3D assemblies, add annotations, and more, in order to quickly and easily communicate detailed process steps from any viewpoint.

Some of the authoring tools include:

- Component move/pull reposition tools make it easy to position individual parts and restore them to their home position.
- Storyboarding a sequence of instructions that can be animated or viewed at specified speed and transition.
- Automated and Manual tools will explode all parts or just sub-assemblies.
- Edit multiple snapshots in the storyboard at the same time, re-order, or one-step reverse the storyboard.
- Text annotations can be manually entered or inherited from the CAD model.
- Imported images and labels can be used to annotate and supplement the instructions.
- Pre-defined styles for the appearance of the assembly or individual parts can be used to obtain the required appearance.

Designs can be published for use with a wide range of applications, viewers, and media players so your documentation can be accessed by anyone. An entire document, particular storyboard or a snapshot can be selected to be published. Snapshots are published as views and storyboards are published as animations or a sequence of views. The following image shows an example of a file opened and being edited in Inventor Publisher.

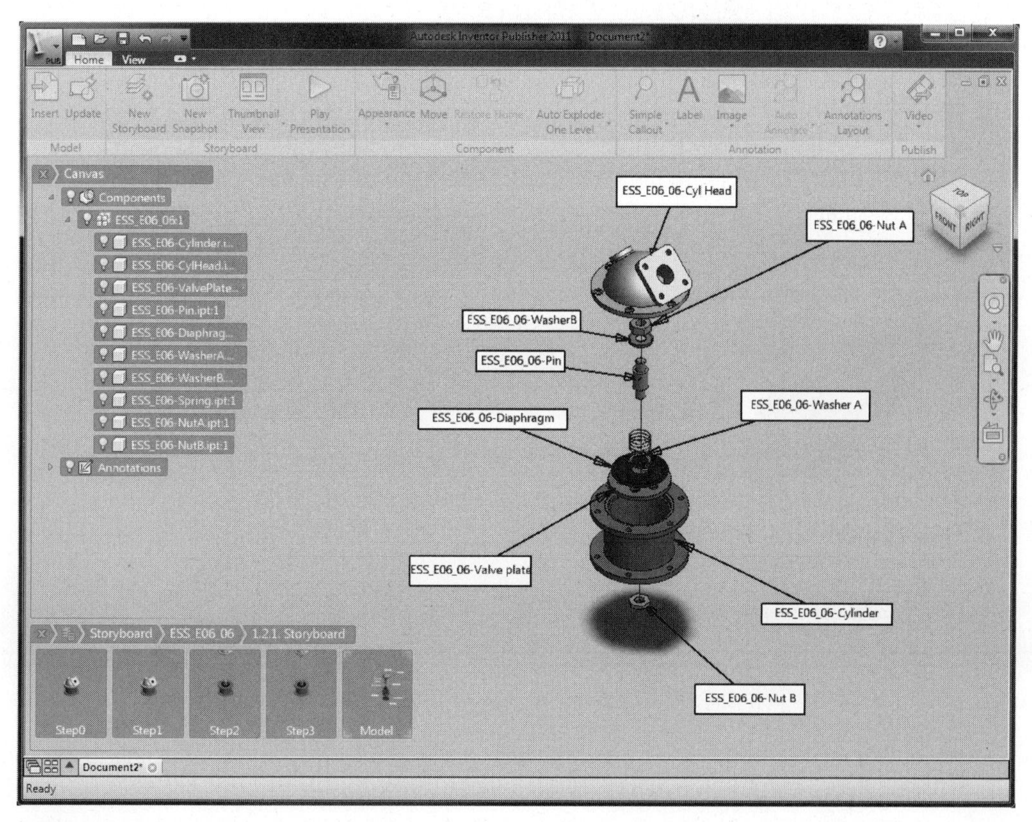

FIGURE 6.112

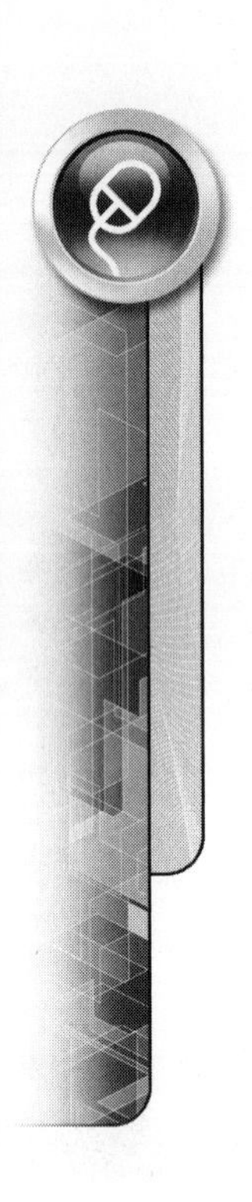

EXERCISE 6-6: CREATING PRESENTATION VIEWS

In this exercise, you create a presentation view of an existing assembly, and you add tweaks to the assembly components to create an exploded view.

1. Begin creating a presentation by clicking the New command.
2. Select the English tab, and then double-click Standard (in).ipn.
3. Click the Create View command on the Presentation tab > Create panel.
4. In the Select Assembly dialog box, verify that the Explosion Method is set to Manual. Then click the Open an Existing File button next to the File field. Open *ESS_E06_06.iam*, and click the Options button in the Select Assembly dialog box.
5. In the Design View Representation field, select Internal Components, as shown in the following image and click OK.

FIGURE 6.113

6. Click OK to create the view. Your display should appear similar to the following image.

FIGURE 6.114

7. Now set the tweak direction used to create an exploded presentation. In the browser, click the plus sign in front of Explosion1 and the plus sign in front of *ESS_E06_06.iam* to expand their displays. The assembly components are displayed so you can select components in the graphics screen or in the browser.
8. Click the Tweak Components command on the Presentation tab > Create panel.
9. In the Tweak Component dialog box, verify that Direction is selected and Display Trails is checked.
10. Position your cursor over a cylindrical face until the temporary Z axis aligns with the axis of the assembly, as shown in the following image.
11. Click to accept the axis orientation.

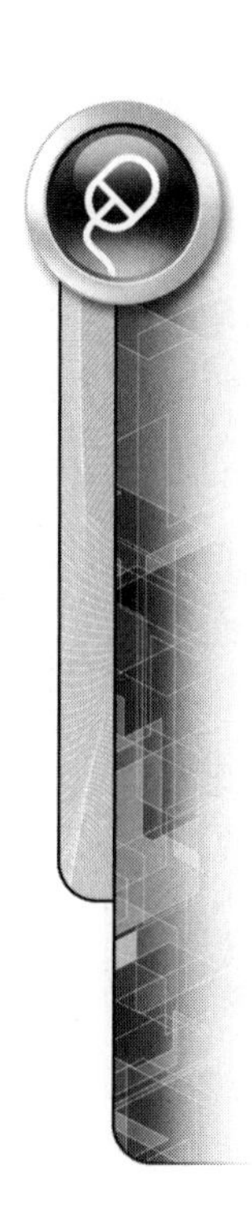

FIGURE 6.115

12. Click on the nut at the base of the assembly to select it. In the browser, observe that *ESS_E06-NutB* is highlighted.
13. In the Tweak Components dialog box, verify that the Z button is selected, and then enter **–1.75** in the Tweak Distance field.
14. Click Apply, the green checkmark, to create the tweak. Your display should appear similar to the following image.

FIGURE 6.116

TIP

After you define the tweak axes, you can select a component or group of components and then drag them to create a tweak. This way, you can quickly arrange your components visually. You can later edit the values of the tweaks from the browser to define their final positions.

15. Next click the valve plate to select it. Hold down the CTRL key, and click the nut to deselect it. Observe that the valve plate is the only entry highlighted in the browser.
16. Enter **–0.875** in the Tweak Distance field of the dialog box.
17. Click Apply to create the tweak. Your display should appear similar to the following image.

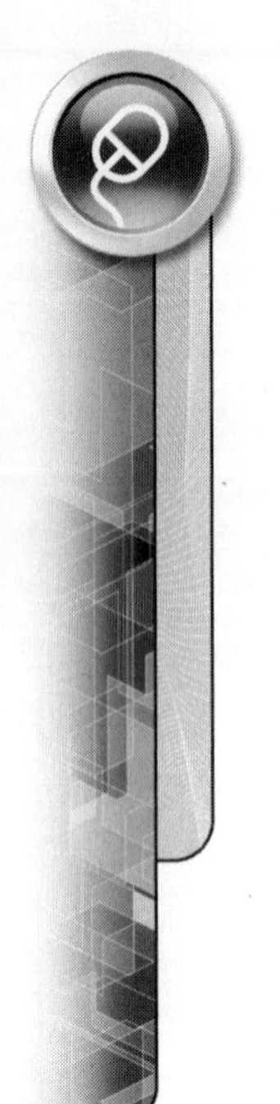

FIGURE 6.117

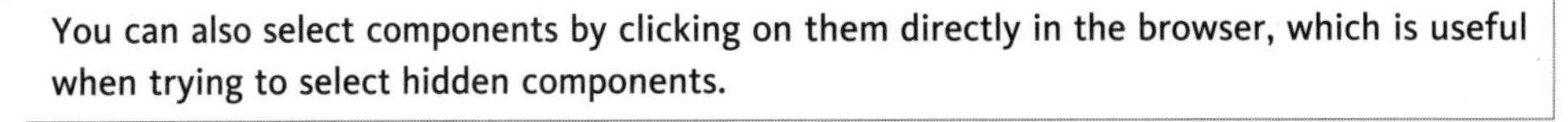

TIP

You can also select components by clicking on them directly in the browser, which is useful when trying to select hidden components.

18. Hold down the CTRL key and click the valve plate to deselect it. The next component to tweak is the top nut. In the browser, click *ESS_E06-NutA* to select it.
19. Enter **4.5** in the Tweak Distance field, and click Apply to create the tweak. Your display should appear similar to the following image.

FIGURE 6.118

TIP

Examine the browser. Expand the components that have a plus sign in front of them to view the tweaks you have applied.

20. Complete this phase of the exercise by adding the following tweaks:

Part Name	Tweak
ESS_E06-WasherB.ipt:1	4
ESS_E06-Pin.ipt:1	3.125
ESS_E06-Spring.ipt:1	1.5
ESS_E06-WasherA.ipt:1	1
ESS_E06-Diaphragm.ipt:1	0.5

21. Close the Tweak Component dialog box when you are finished. Your display should appear similar to the following image.

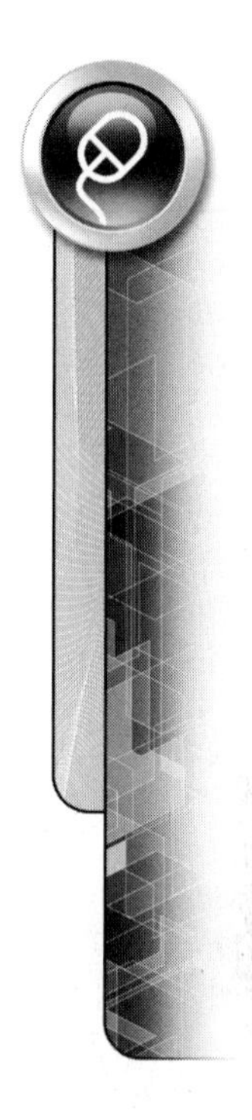

FIGURE 6.119

22. Now adjust an existing tweak. In the browser, expand *ESS_E06-Diaphragm.ipt:1*, *ESS_E06-Spring.ipt:1*, and *ESS_E06-WasherA.ipt:1*. The hierarchy displays tweaks nested below each component definition.
23. Right-click each of the tweaks, and select Visibility to turn off the trail display for the tweak.
24. Click Tweak (0.500 in) under *ESS_E06-Diaphragm.ipt:1*.
25. Below the browser, enter a new value of **0.125** in the Offset field, and then press ENTER. Observe that the name of the tweak reflects the change and the diaphragm moves to its new position.
26. Close all open files. Do not save changes. End of exercise.
27. Click the Animate command from the Presentation tab > Create panel.
28. Click the Play Forward button to view the animation. When finished, click Cancel in the Animation dialog box.
29. Close all open files. Do not save changes. End of exercise.

CREATING DRAWING VIEWS FROM ASSEMBLIES AND PRESENTATION FILES

After creating an assembly or presentation, you may want to create drawings that use the data. You can create drawing views from assemblies using information from an assembly or a presentation file. You can select from them a specific design view or presentation view. To create drawing views based on assembly data, start a new drawing file or open an existing drawing file. Use the Base View command and select the IAM or IPN file from which to create the drawing. If needed, specify the design view or presentation view in the dialog box. The following image shows a presentation view being selected after a presentation file was selected.

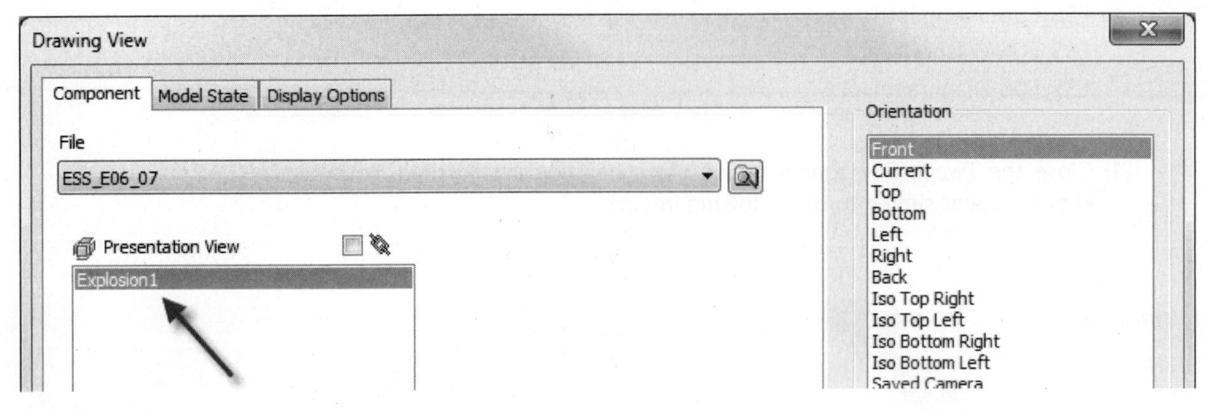

FIGURE 6.120

THE BILL OF MATERIAL (BOM) EDITOR

A Bill of Materials, usually referred to as a BOM, is a table that contains information about the components inside an assembly. It can include item number, quantity, part number, description, vendor, and other information needed to describe the assembly. The BOM is also considered associative; when a component is added or removed from the assembly, the quantity field in the BOM will update to reflect this change in the assembly.

To facilitate managing item numbers, every item is automatically assigned a number, and the item number can be changed or reordered if necessary. Whenever these types of changes are made in the BOM, the same changes are updated automatically in the Parts List and balloons.

Values in the BOM can be changed in several ways. The first way is to add information while inside each individual part through the Properties dialog box. Another way is to add the part information directly into the BOM with the BOM Editor; then, when the BOM is saved, the information supplied inside of the BOM is also saved to each individual part. This is an efficient way to edit the properties of multiple parts at the same time.

In order to activate the BOM Editor, you must be inside an assembly model or editing a Parts List in a drawing. In an assembly click Bill of Materials on the Assemble Tab > Manage panel as shown in the following image on the left and the Bill of Materials dialog box appears as shown in the following image on the right.

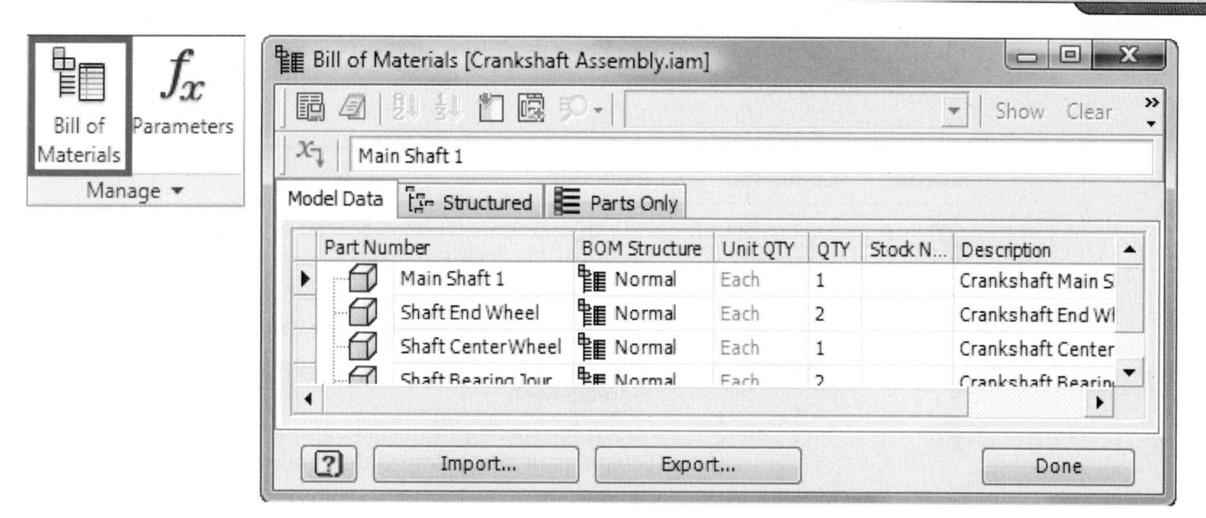

FIGURE 6.121

There are three tabs in the BOM Editor: Model Data, Structured, and Parts Only. The Model Data tab is shown in the following image on the left. Here the components that make up the assembly are arranged in a format that resembles the assembly browser. Crankshaft Assembly can be expanded to display its individual parts. The information under this tab is not meant to be reported to a Parts List.

The Structured tab for this example is shown in the following image in the middle. This is the actual BOM data that is reported to the Parts List. When arranging an assembly in Structured mode, Crankshaft Assembly is considered an individual item along with the other parts listed.

When the Parts Only tab is selected, Crankshaft Assembly is not listed. However, the individual parts that make up Crankshaft Assembly are listed along with the other parts, as shown in the following image on the right. This arrangement of parts is referred to as a flat list.

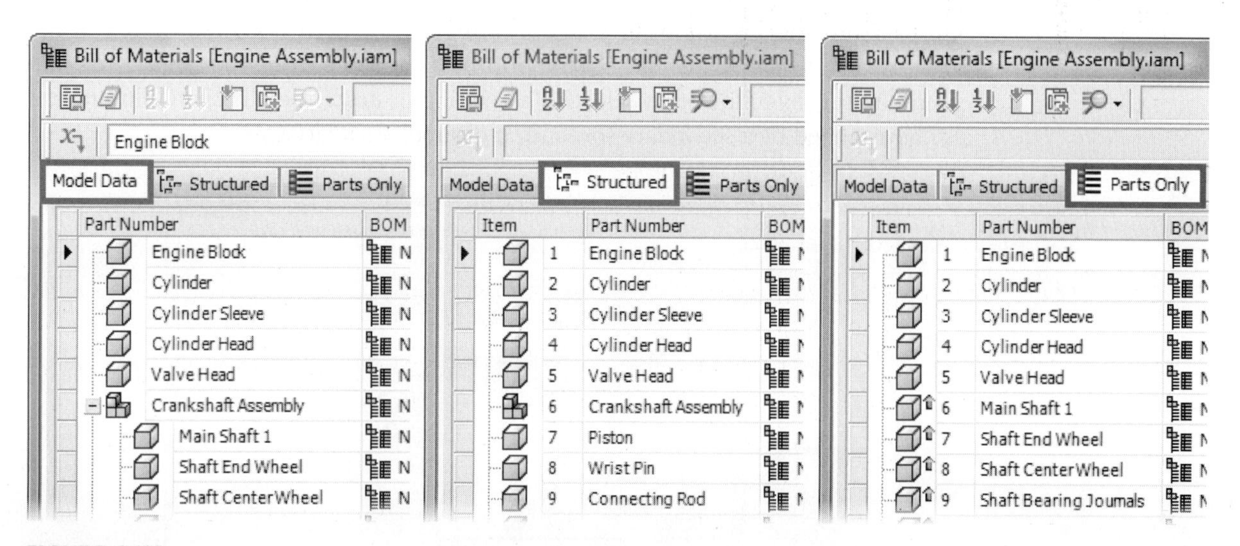

FIGURE 6.122

The BOM Editor also contains several buttons that provide further control over the information supplied in the BOM. These buttons are described as follows:

Button	Command	Function
	Export Bill of Materials	Bill of material information can be exported out to an external file. The following file formats are supported: mdb, xls, dbf, txt, and csv.
	Engineer's Notebook	The BOM can be added to the Engineer's Notebook in the form of a table as a note that is linked to an Excel spreadsheet.
	Sort Items	Used for sorting the values in one or more BOM columns in ascending or descending order.
	Renumber Items	Allows items to be sorted based on the current sort order of the BOM.
	Choose Columns	Allows extra informational columns to be added or removed from the BOM.
	Add Custom iProperties Column	Adds the new values to the Custom tab in the corresponding iProperties dialog box.
	View Options	Enables or disables the BOM information from being displayed in the Model Data, Structured, and Parts Only tabs.
	Part Number Row Merge Settings	Allows different components with the same part number to be treated as a single component.
	Update Mass Properties	Calculates the mass and volume of all items in the BOM.

Sorting Items in the BOM

Clicking on the Sort button located in the BOM Editor will display the Sort dialog box, as shown in the following image. Sorting can be performed only when in the Structured and Parts Only tabs. In the Sort by area of the dialog box, select the first column to sort by followed by selecting ascending or descending order. Columns can also be arranged by a secondary sort and, if desired, a tertiary sort.

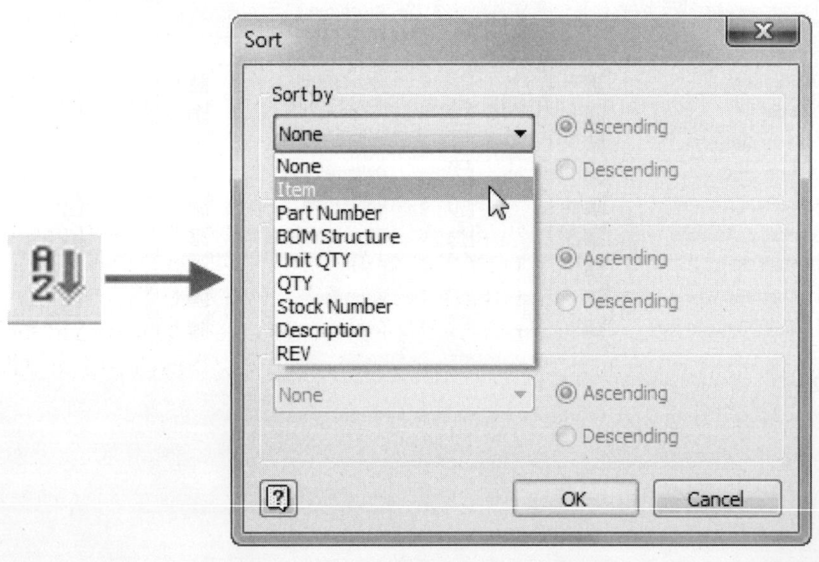

FIGURE 6.123

Renumbering Items in the BOM

Clicking on the Renumber Items button will display the Item Renumber dialog box, as shown in the following image. This operation is usually performed after sorting information in the BOM Editor.

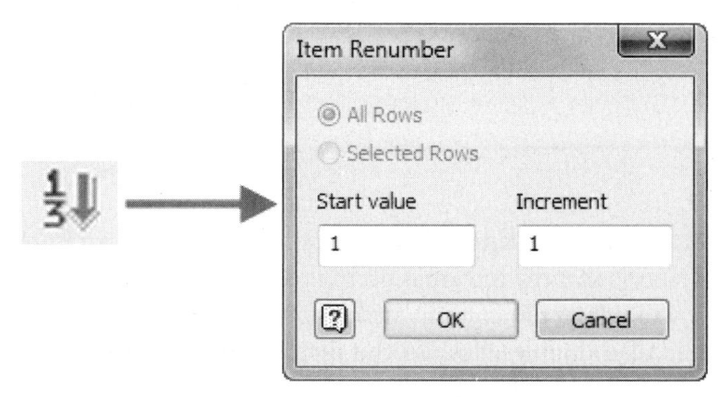

FIGURE 6.124

Choosing Extra Columns for the BOM

Clicking on the Choose Columns button will display a Customization dialog box in the lower-right corner of the BOM Editor, as shown in the following image. Choose the desired column from the list, and then drag and drop this listing onto an existing column in the BOM Editor. This action will add the custom column to the BOM. You can drag existing column headers to remove them from the BOM Editor; drop the column when a large X appears at your cursor position. Removed headers are added to the Customization dialog so that you can restore them to the BOM Editor.

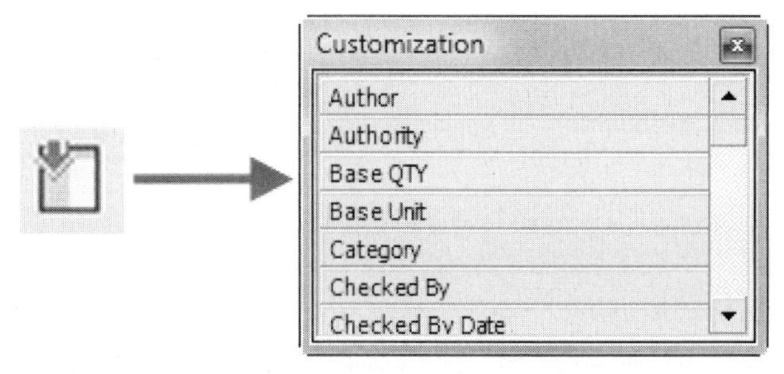

FIGURE 6.125

Changing the View Options of the BOM

Clicking on the View Options button activates a pull-down menu, as shown in the following image on the left. You can enable or disable the BOM view based on the current tab. In this example, the current tab is Structured. Clicking on View Properties from the menu will activate the Structured Properties dialog box, as shown in the following image. Use this dialog to switch the BOM view from First Level to All Levels.

FIGURE 6.126

Editing a BOM Column

Right-clicking on a column in the BOM Editor will display a menu, as shown in the following image. Use this menu to sort the information in the column by ascending or descending order. Best Fit will resize the column width to fit the information that occupies each cell; you could also double-click on the divider between the column headers. In this case, the column to the left of the divider is resized to fit the contents of the cells in that column. Clicking on Best Fit (all columns) will resize the information in all columns.

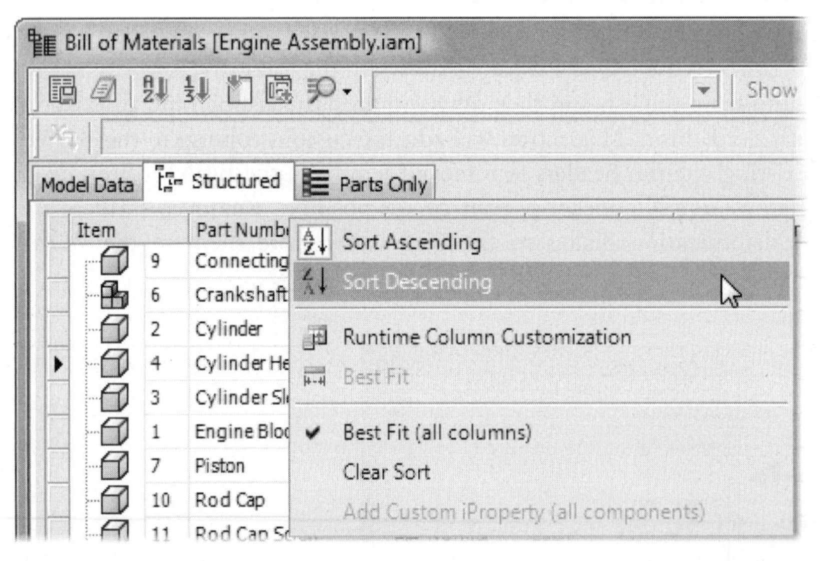

FIGURE 6.127

Highlighting Components in the Browser and BOM

Selecting a cell or group of cells in the BOM editor will highlight all corresponding components located in the assembly browser and in the BOM editor, as shown in the following image.

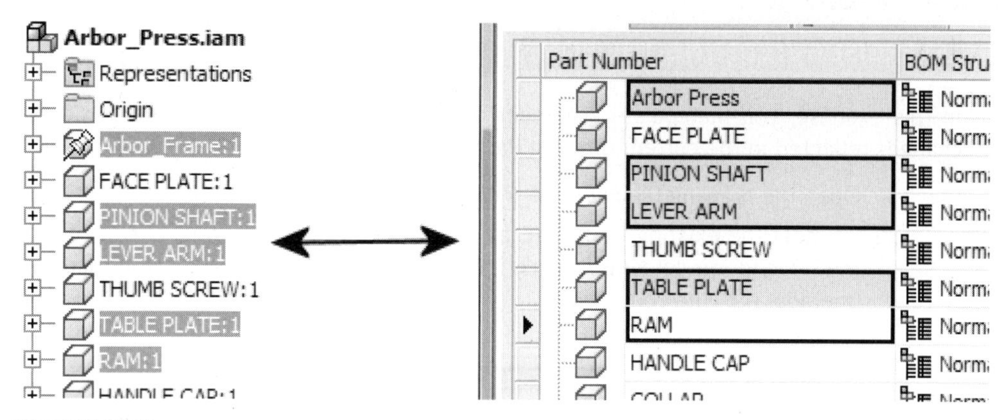

FIGURE 6.128

Opening a Part from the BOM Editor

The BOM editor can be used to open a part or assembly file. To perform this task, open the BOM editor, select the part from the list, right-click and select Open from the menu as shown in the following image.

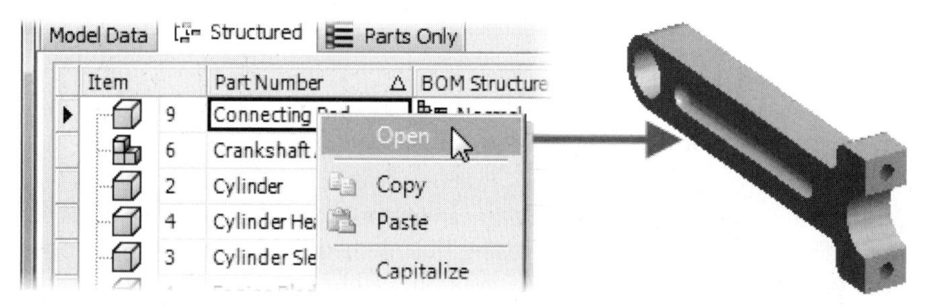

FIGURE 6.129

NOTE

While in the BOM editor, the following editing tools work similarly when in Microsoft Excel:

- Use the SHIFT or CTRL keys to make selection sets of information while in the BOM editor.
- Use the grip point located in the lower-right corner of a cell to expand a cell or blocks of cells.
- BOM cells can be copied from the BOM Editor to another BOM view or to an external editor.
- Copy and paste information from the BOM editor into Microsoft Office Excel.
- Additional command options are available through a menu. These options include Open, Copy, Paste, Capitalize, Find, and Replace.

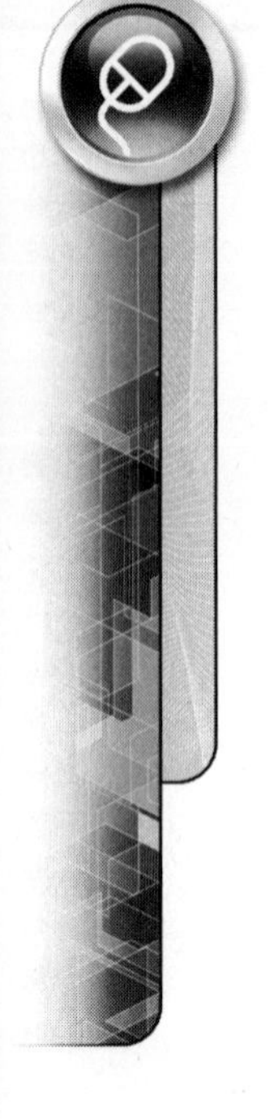

EXERCISE 6-7: EDITING THE BOM

In this exercise, you manipulate a number of items in the BOM Editor and see these changes reflected in the current drawing Parts List.

1. Open the drawing file *ESS_ E06_07.idw*.
2. Examine the existing Parts List, as shown in the following image. Notice that the Description column is blank. The fields are populated from the iProperties (metadata) of the components.

Parts List			
ITEM	QTY	PART NUMBER	DESCRIPTION
1	2	ESS_E06_07-003	
2	1	ESS_E06_07-001	
3	2	ESS_E06_07-002	
4	2	ESS_E06_07-006	
5	1	ESS_E06_07-007	
6	1	ESS_E06_07-004	
7	6	ESS_E06_07-008	
8	1	ESS_E06_07-005	

FIGURE 6.130

(Courtesy US FIRST Team 342—Robert Bosch Corp., Trident Technical College, Fort Dorchester High School, Summerville High School; Charleston, South Carolina)

3. The individual descriptions of each part can be easily added through the BOM Editor while working in the *.idw* file. Using this method, you will not have to open each individual part file to add a description. Right-click on the Parts List in the browser or in the graphics window and click Bill of Materials from the menu.
4. The BOM Editor will be displayed, as shown in the following image. Of the three tabs available, (Model Data, Structured, and Parts Only), the Parts Only tab will be made active for the remainder of this exercise. Notice the appearance of the Description column.

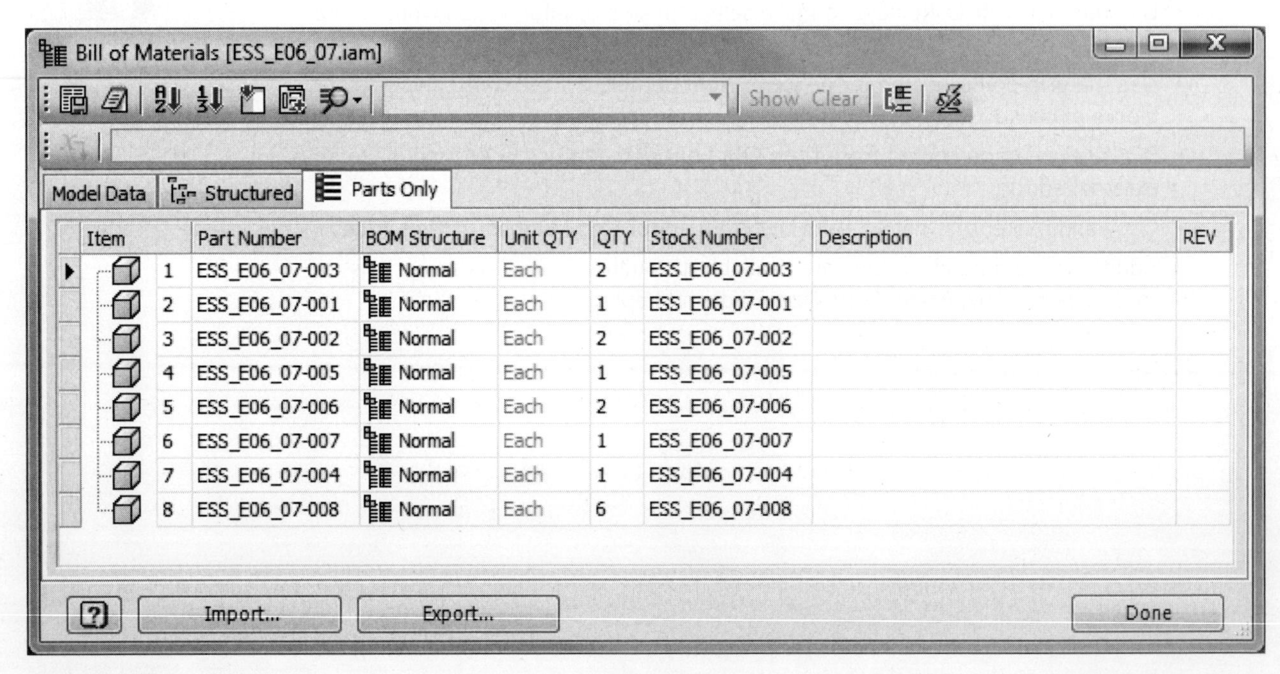

FIGURE 6.131

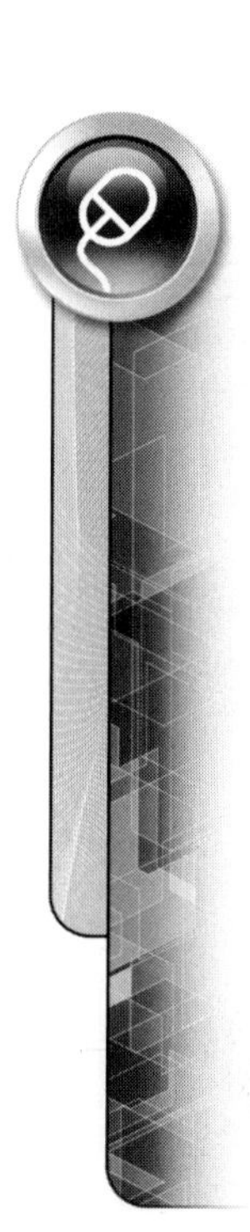

5. Complete the Description column by clicking in each cell and filling in all needed information as shown in the following image. When finished, click the Done button to close the BOM Editor.

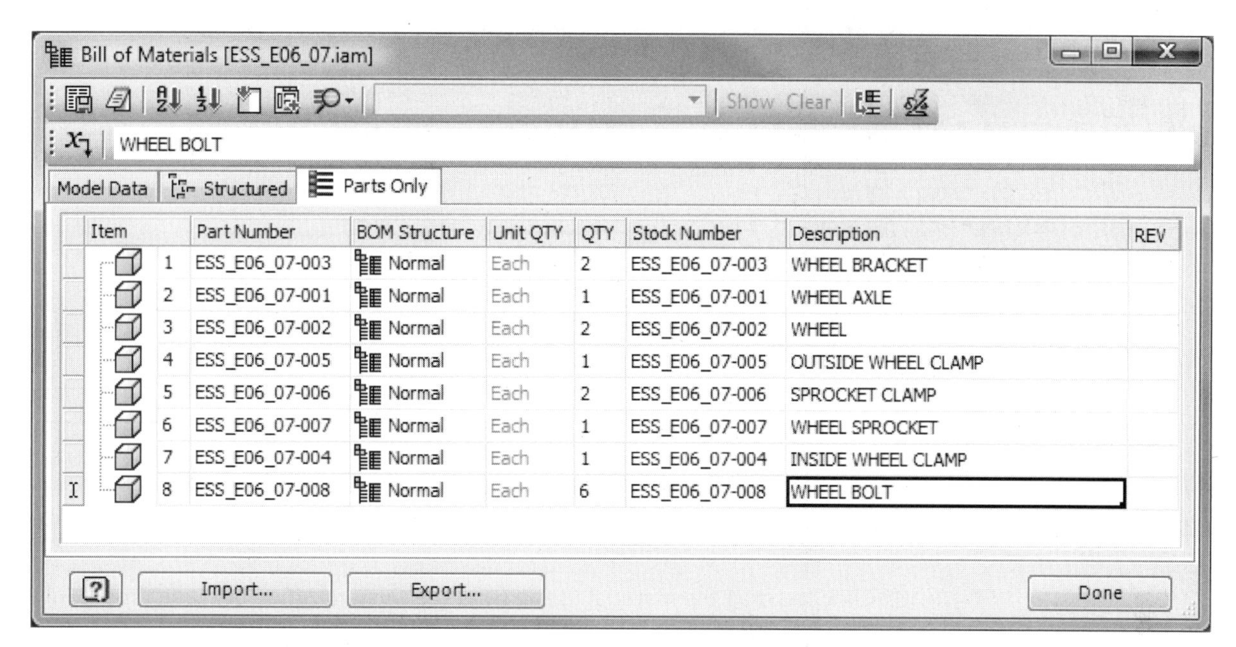

FIGURE 6.132

6. When you return to the drawing file, notice that the Description column of the Parts List updates to reflect the new information, as shown in the following image.

Parts List			
ITEM	QTY	PART NUMBER	DESCRIPTION
1	2	ESS_E06_07-003	WHEEL BRACKET
2	1	ESS_E06_07-001	WHEEL AXLE
3	2	ESS_E06_07-002	WHEEL
4	2	ESS_E06_07-006	SPROCKET CLAMP
5	1	ESS_E06_07-007	WHEEL SPROCKET
6	1	ESS_E06_07-004	INSIDE WHEEL CLAMP
7	6	ESS_E06_07-008	WHEEL BOLT
8	1	ESS_E06_07-005	OUTSIDE WHEEL CLAMP

FIGURE 6.133

7. Next add a Material column to the BOM Editor to capture the material type for each part. Activate the BOM Editor again from the Parts List in the browser.
8. To add a column called Material, click the Choose Columns button as shown in the following image. This will display the Customization box of properties in the lower-right corner of the dialog box.
9. Scroll down in the Customization box and locate Material as shown in the following image.

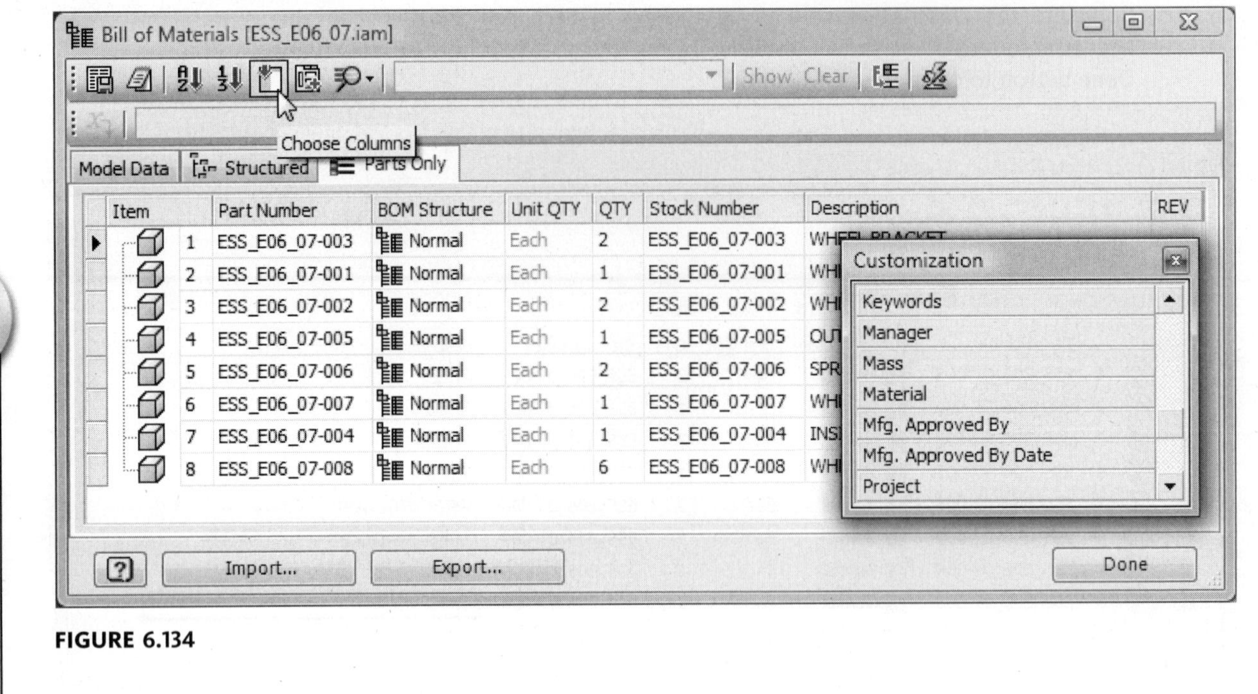

FIGURE 6.134

10. Click and drag the Material listing from the Customization box, and drop it into the Description Column heading as shown in the following image.

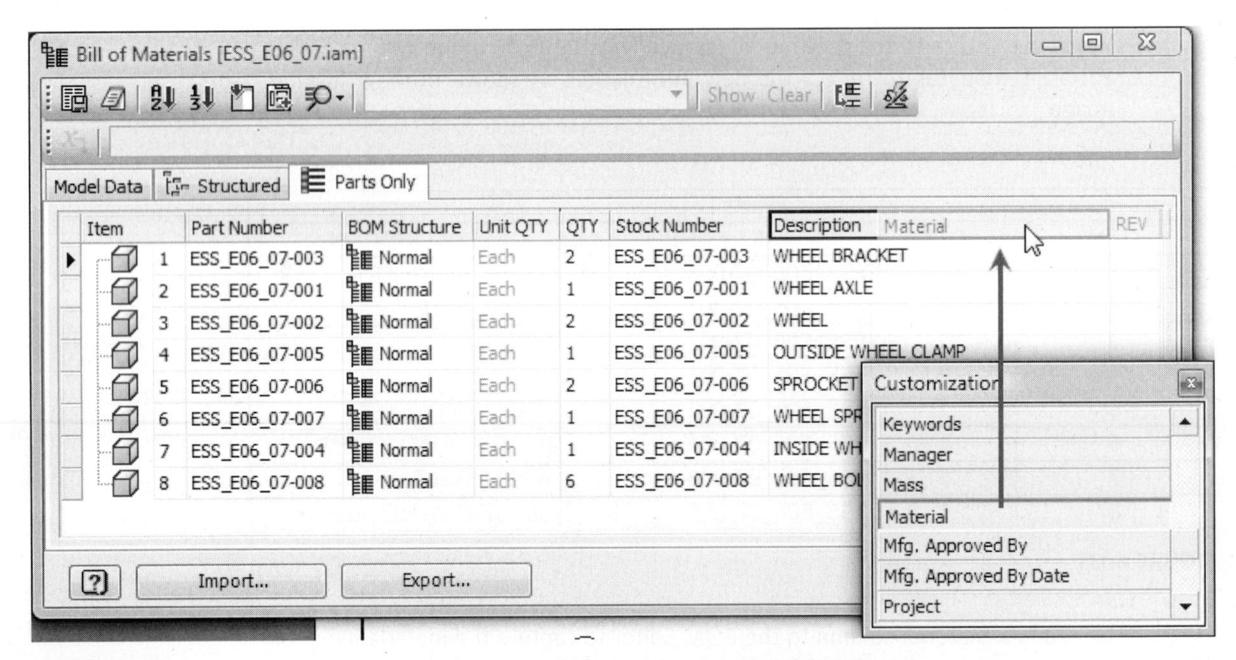

FIGURE 6.135

11. When finished, dismiss the Customization box by clicking on the Close button (X).

12. With the addition of the Material column, your display should appear similar to the following image. Notice that all Materials are listed as default. You will now add a material to each part.

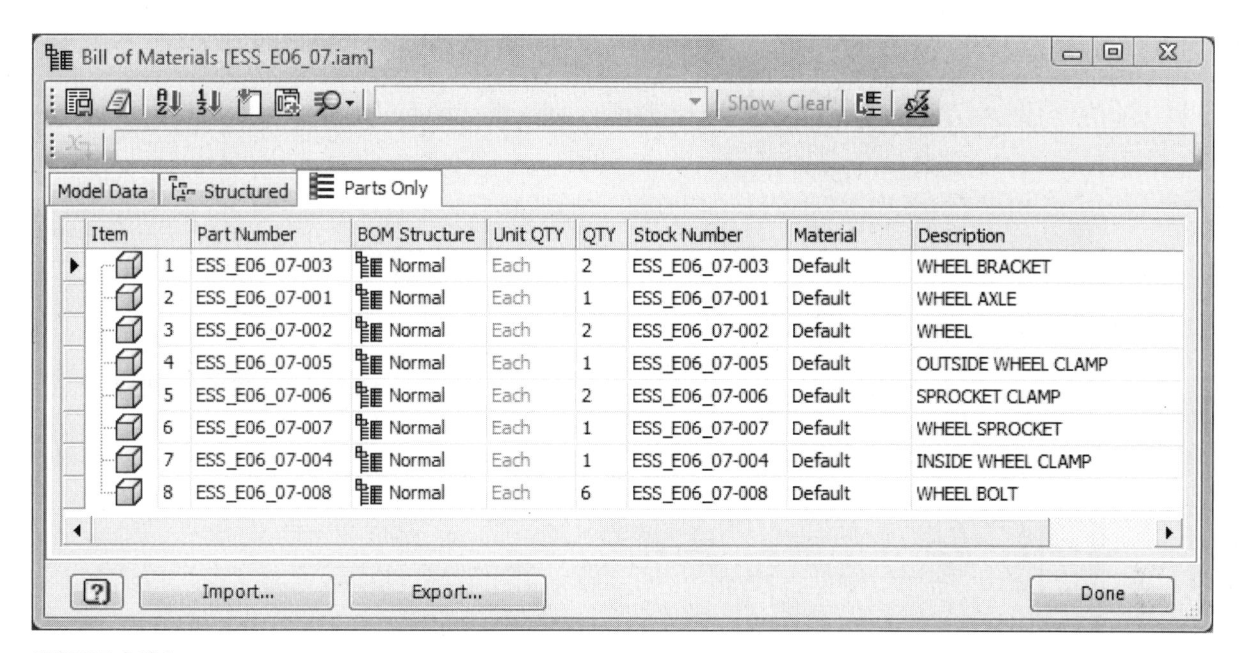

Item	Part Number	BOM Structure	Unit QTY	QTY	Stock Number	Material	Description
1	ESS_E06_07-003	Normal	Each	2	ESS_E06_07-003	Default	WHEEL BRACKET
2	ESS_E06_07-001	Normal	Each	1	ESS_E06_07-001	Default	WHEEL AXLE
3	ESS_E06_07-002	Normal	Each	2	ESS_E06_07-002	Default	WHEEL
4	ESS_E06_07-005	Normal	Each	1	ESS_E06_07-005	Default	OUTSIDE WHEEL CLAMP
5	ESS_E06_07-006	Normal	Each	2	ESS_E06_07-006	Default	SPROCKET CLAMP
6	ESS_E06_07-007	Normal	Each	1	ESS_E06_07-007	Default	WHEEL SPROCKET
7	ESS_E06_07-004	Normal	Each	1	ESS_E06_07-004	Default	INSIDE WHEEL CLAMP
8	ESS_E06_07-008	Normal	Each	6	ESS_E06_07-008	Default	WHEEL BOLT

FIGURE 6.136

13. Begin filling in the material type by clicking on Default for Item 1. Double-click the cell to activate the material drop-down list. Select the proper material from the list supplied. Make material assignments that correspond to the part numbers, as shown in the following image.

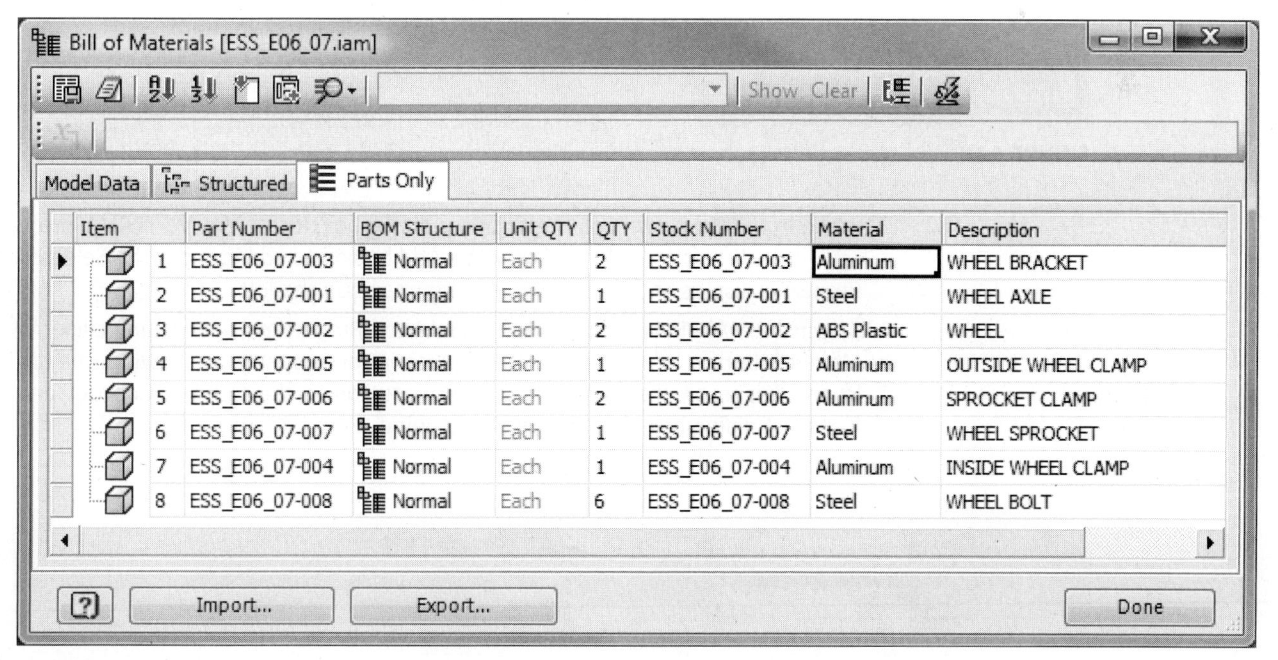

Item	Part Number	BOM Structure	Unit QTY	QTY	Stock Number	Material	Description
1	ESS_E06_07-003	Normal	Each	2	ESS_E06_07-003	Aluminum	WHEEL BRACKET
2	ESS_E06_07-001	Normal	Each	1	ESS_E06_07-001	Steel	WHEEL AXLE
3	ESS_E06_07-002	Normal	Each	2	ESS_E06_07-002	ABS Plastic	WHEEL
4	ESS_E06_07-005	Normal	Each	1	ESS_E06_07-005	Aluminum	OUTSIDE WHEEL CLAMP
5	ESS_E06_07-006	Normal	Each	2	ESS_E06_07-006	Aluminum	SPROCKET CLAMP
6	ESS_E06_07-007	Normal	Each	1	ESS_E06_07-007	Steel	WHEEL SPROCKET
7	ESS_E06_07-004	Normal	Each	1	ESS_E06_07-004	Aluminum	INSIDE WHEEL CLAMP
8	ESS_E06_07-008	Normal	Each	6	ESS_E06_07-008	Steel	WHEEL BOLT

FIGURE 6.137

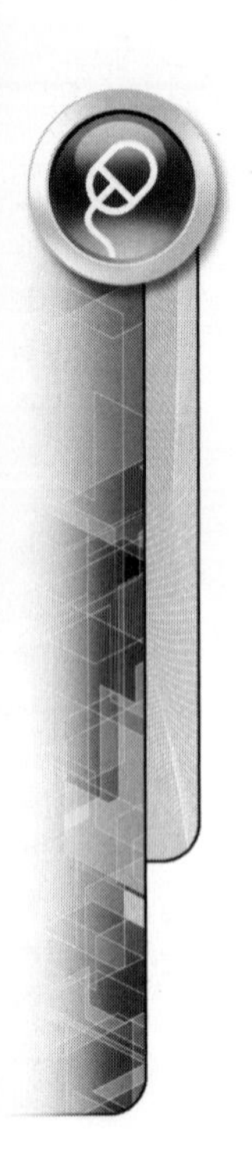

14. When finished, click the Done button to return to the drawing file. Note: Upon leaving the BOM Editor, the iProperty data is written to the individual files.
15. When you return to the drawing, the Material column will not automatically display in the Parts List. To display this column of information, right-click on the Parts List in the browser or in the graphics window, and click Edit Parts List from the menu.
16. When the Parts List displays, click the Column Chooser button, as shown in the following image.

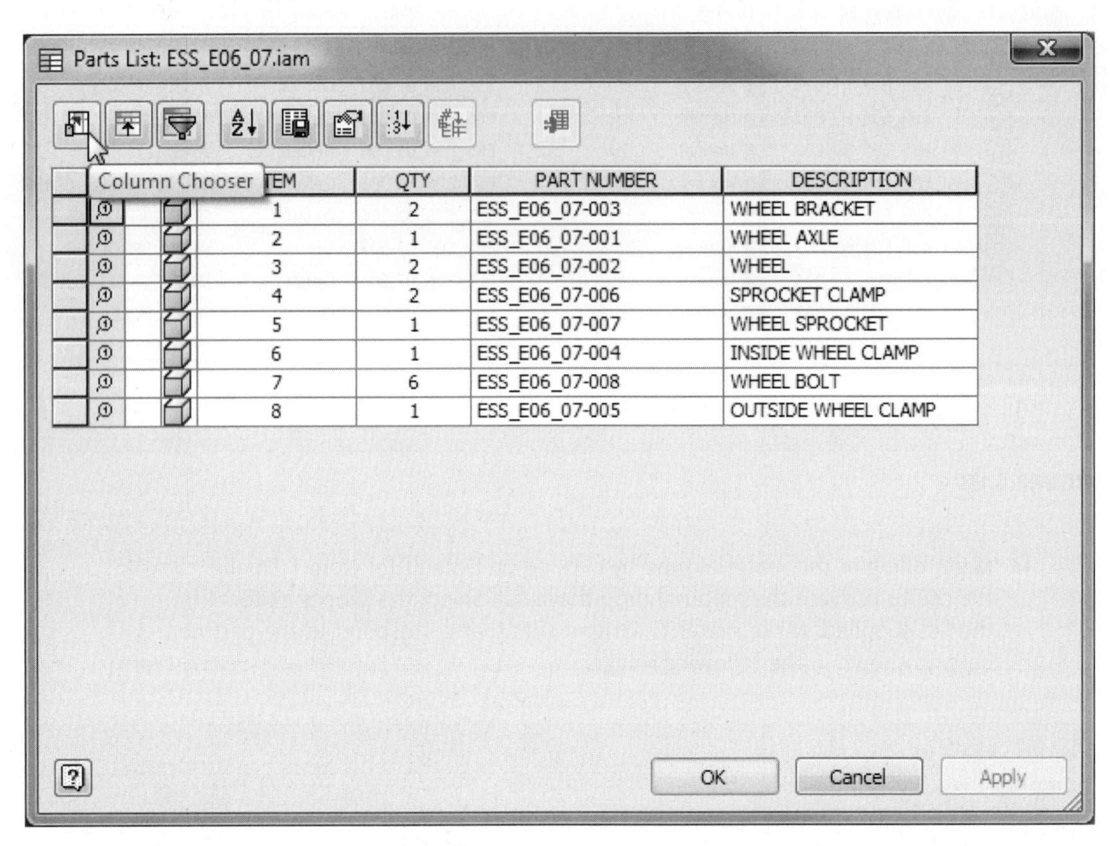

FIGURE 6.138

17. While in the Parts List Column Chooser dialog box, locate the MATERIAL property in the list on the left, and click the Add button to add it under the Selected Properties list, as shown in the following image on the left.
18. Select the MATERIAL property in the Selected Properties window, and click the Move Up button to place it above the Description property, as shown in the following image on the right. Click the OK button in the Column Chooser and Parts List dialog box to return to the drawing.

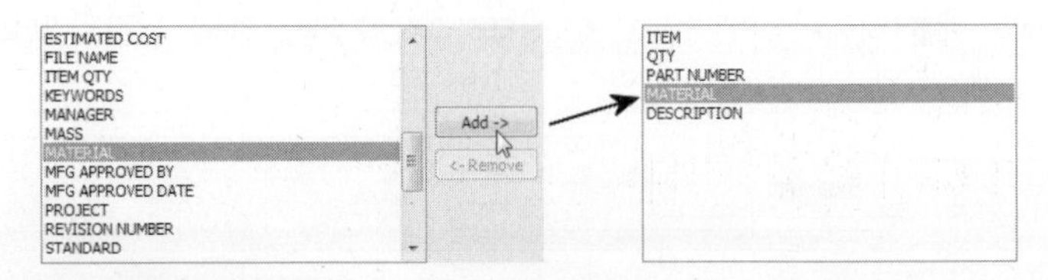

FIGURE 6.139

19. Notice that the Parts List now displays the MATERIAL column and that all information is filled in, as shown in the following image.

Parts List				
ITEM	QTY	PART NUMBER	MATERIAL	DESCRIPTION
1	2	ESS_E06_07-003	Aluminum	WHEEL BRACKET
2	1	ESS_E06_07-001	Steel	WHEEL AXLE
3	2	ESS_E06_07-002	ABS Plastic	WHEEL
4	2	ESS_E06_07-006	Aluminum	SPROCKET CLAMP
5	1	ESS_E06_07-007	Steel	WHEEL SPROCKET
6	1	ESS_E06_07-004	Aluminum	INSIDE WHEEL CLAMP
7	6	ESS_E06_07-008	Steel	WHEEL BOLT
8	1	ESS_E06_07-005	Aluminum	OUTSIDE WHEEL CLAMP

FIGURE 6.140

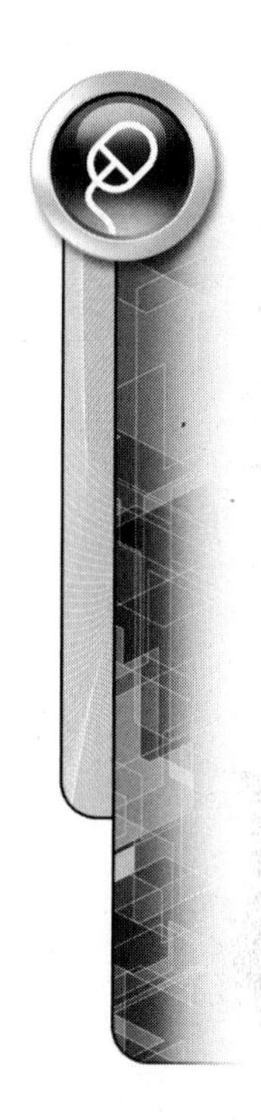

20. Next adjust the width of all columns by the amount of information contained under the heading. To perform this task, right-click the Parts List located in the browser and click Edit Parts List.
21. When the Parts List dialog box displays, right-click the column and pick Column Width from the menu. When the Column Width dialog box appears, enter a value. Change the column headings to the following widths:
    - ITEM = **25**
    - QTY = **20**
    - PART NUMBER = **40**
    - MATERIAL = **30**
    - DESCRIPTION = **55**
22. Click OK to close the Parts List dialog box. The Parts List now reflects the new column widths, as shown in the following image.

Parts List				
ITEM	QTY	PART NUMBER	MATERIAL	DESCRIPTION
1	2	ESS_E06_07-003	Aluminum	WHEEL BRACKET
2	1	ESS_E06_07-001	Steel	WHEEL AXLE
3	2	ESS_E06_07-002	ABS Plastic	WHEEL
4	2	ESS_E06_07-006	Aluminum	SPROCKET CLAMP
5	1	ESS_E06_07-007	Steel	WHEEL SPROCKET
6	1	ESS_E06_07-004	Aluminum	INSIDE WHEEL CLAMP
7	6	ESS_E06_07-008	Steel	WHEEL BOLT
8	1	ESS_E06_07-005	Aluminum	OUTSIDE WHEEL CLAMP

FIGURE 6.141

23. Close the file. Do not save changes. End of exercise.

CREATING BALLOONS

After you have created a drawing view of an assembly, you can add balloons to the parts and/or subassemblies. You can add balloons individually to components in a drawing, or you can use the Auto Balloon command to automatically add balloons to all components.

To add individual balloons to a drawing, follow these steps:

1. On the Annotate tab > Table panel, click the arrow below the Balloon command, as shown in the following image. Two icons will appear: the first is used for ballooning components one at a time, and the second is used for automatic ballooning of components in a view via a single operation.
2. To balloon a single component, click the Balloon command or use the hot key B.
3. Select a component to balloon.
4. The BOM Properties dialog box will appear, as shown in the following image.

NOTE

This dialog box will not appear if a Parts List or balloons already exist in the drawing.

5. Select which Source and BOM Settings to use, as shown in the following image, and select OK. A preview image of the balloon will appear attached to the cursor.

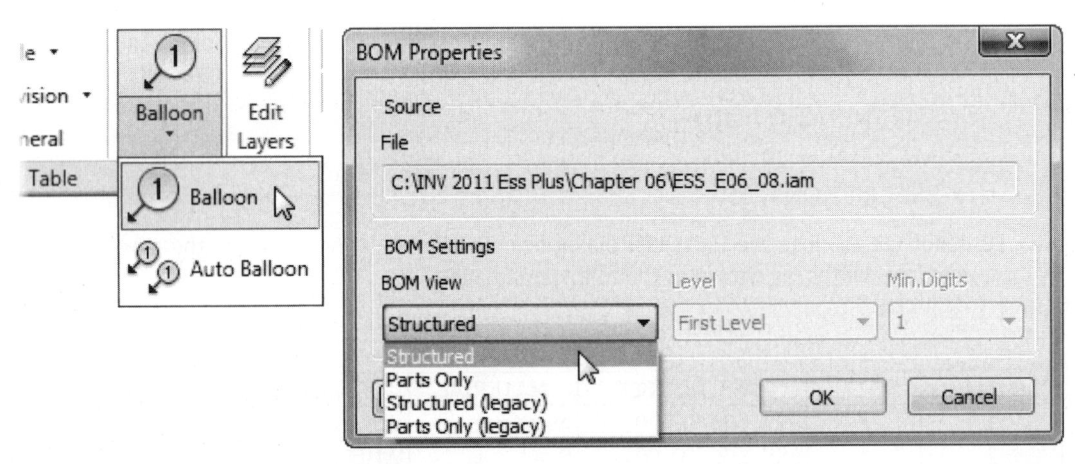

FIGURE 6.142

6. Position the cursor to place the second point for the leader, and then press the left mouse button.
7. Continue to select points to add segments to the balloon's leader, if desired.
8. When finished adding segments to the leader, right-click, and select Continue from the menu to create the balloon.
9. Select Done from the menu to cancel the operation without creating an additional balloon, or select the Back option to undo the last step that was created for the balloon.

NOTE

When placing balloons, the direction of the balloon leader line snaps to 15° increments. The center point of a new balloon can be aligned horizontally or vertically with the center point of a balloon that already exists in a drawing.

BOM Properties Dialog Box Options

Source

Specify the source file on which the BOM will be based.

BOM Settings and BOM View

Structured. A Structured list refers to the top-level components of an assembly in the selected view. Subassemblies and parts that belong to the subassembly will be ballooned, but parts that are in a subassembly will not be.

Parts Only View. Click to balloon only the parts of the assembly in the selected view. Subassemblies will not be ballooned, but the parts in the subassemblies will be.

NOTE

In a Parts Only list, components that are assemblies are not presented in the list unless they are considered inseparable or purchased.

Level

First Level or All Levels determines the level of detail that the Parts List and balloons will display. For example, if an assembly consists of a number of subassemblies and you choose First Level, each subassembly will be considered a single item. If you choose All Levels, the individual parts that make up the subassembly will also be treated as individual components when ballooned.

Min.Digits

Use this command to set the minimum amount of digits displayed for item numbering. The range is fixed from 1 to 6 digits.

Auto Ballooning

In complex assembly drawings, it will become necessary to balloon a number of components. Rather than manually add a balloon to each component, you can use the Auto Balloon command to perform this operation on several components in a single operation automatically. Clicking on the Auto Balloon button under the Annotation tab will display the Auto Balloon dialog box, as shown in the following image. The areas in this dialog box allow you to control how you place balloons.

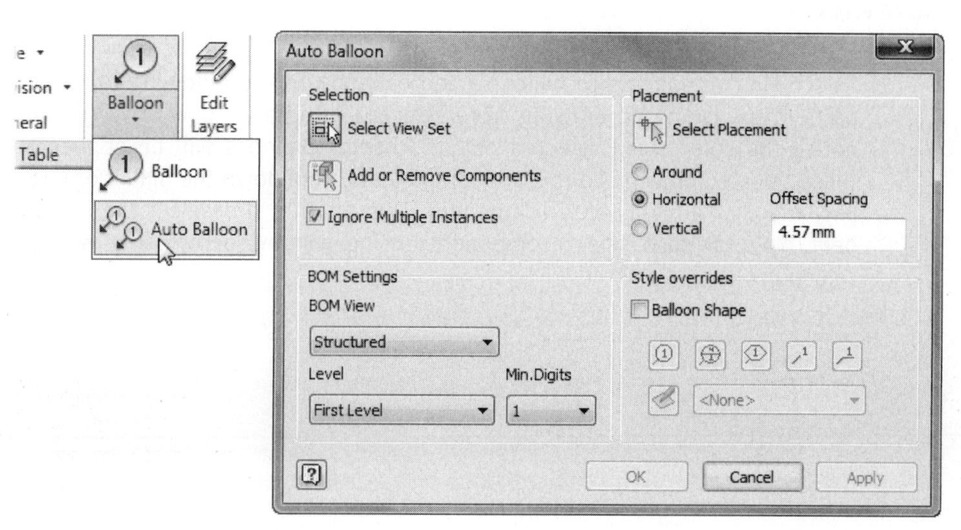

FIGURE 6.143

Selection

This area requires you to select where to apply the balloons. With the view selected, you then add or remove components to the balloon. The Ignore Multiple Instances option appears in this area, and when the box is checked, multiple instances of the same component will not be ballooned. This action greatly reduces the number of balloon callouts in the drawing. If your application requires multiple instances to each have a balloon, remove the check from this box.

Placement

This area allows you to display the balloons along a horizontal axis, along a vertical axis, or around the view being ballooned.

BOM Settings

This area allows you to control if balloons are applied to structured, or first-level components or to parts only.

Style Overrides

The Style overrides area allows you to change the shape of the balloon and assign a user-defined balloon shape. Click Balloon Shape to enable balloon shape style overrides.

To create an auto balloon, follow these steps:

1. Select the view to which to add the balloons, as shown in the following image.

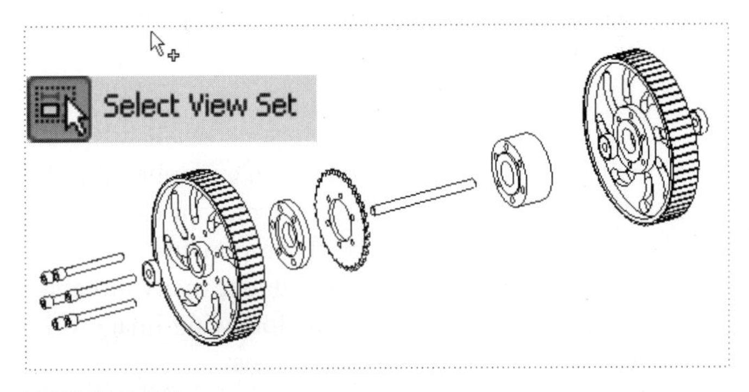

FIGURE 6.144

2. Select the components to which to add balloons. In the following image, all parts in the isometric drawing were selected using a window box. You will notice the color of all selected objects change. Balloons will be applied to these selected objects. To remove components, hold down the SHIFT or CTRL key, and click a highlighted component. This action will deselect the component. You can also select components using window or crossing window selections.

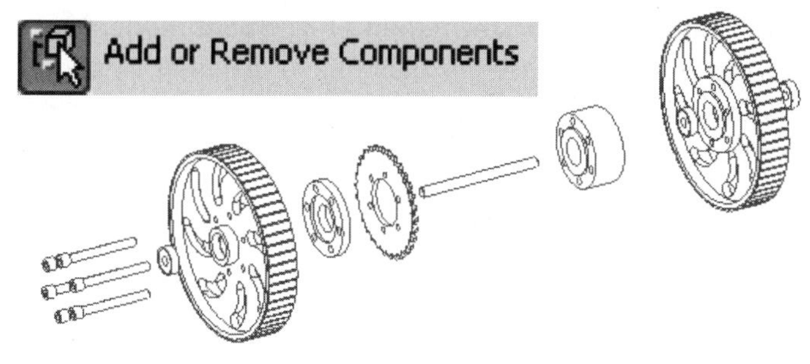

FIGURE 6.145

3. Select one of the three placement modes for the balloons. The following image shows an example of using the Horizontal Placement mode. You can further locate this balloon group by moving your cursor around the display screen. This balloon arrangement will retain its horizontal order.

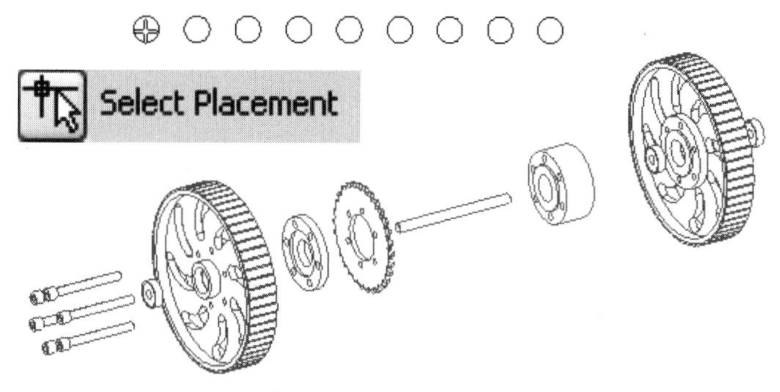

FIGURE 6.146

The following image shows an example of using the Vertical Placement mode. You can further locate this balloon group by moving your cursor around the display screen. This balloon arrangement will retain its vertical order.

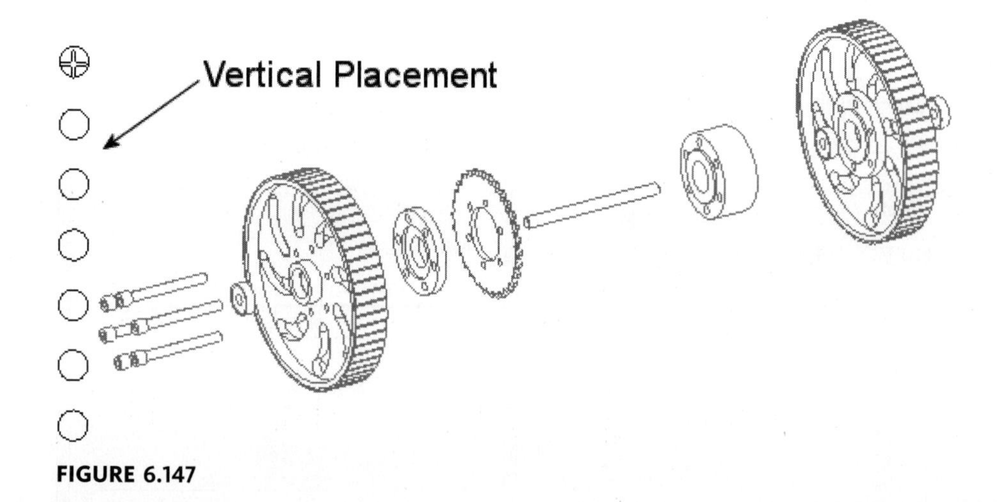

FIGURE 6.147

The following image shows an example of using the Around Placement mode. Move the cursor from the center of the view to readjust the balloons to different locations. In this example, the balloons are located in various positions outside of the view box. After you have placed these types of balloons, you may want to drag the balloons and arrow terminators to fine-tune individual balloon locations.

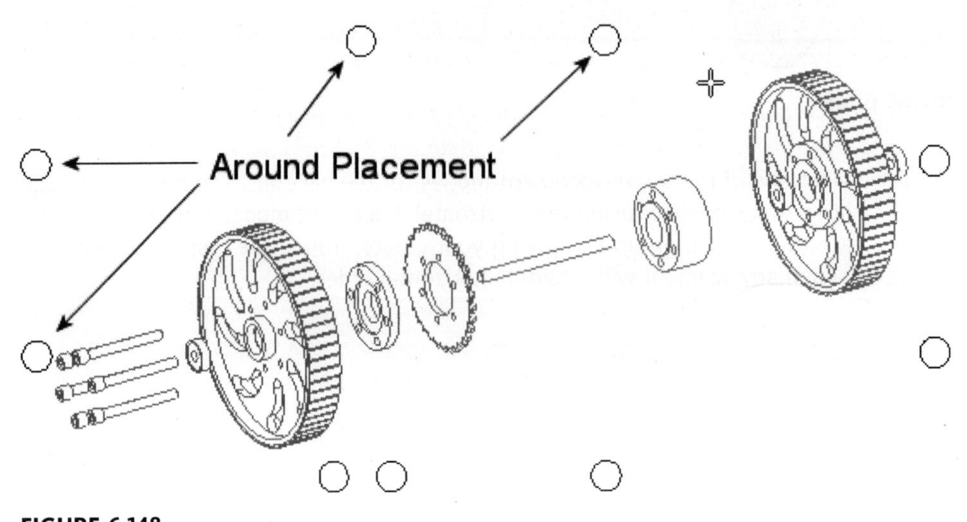

FIGURE 6.148

Editing Balloons

Edit a balloon by right-clicking on it. This action will activate a menu, as shown in the following image on the left.

Click Edit Balloon from the menu to activate the Edit Balloon dialog box, as shown in the following image on the right.

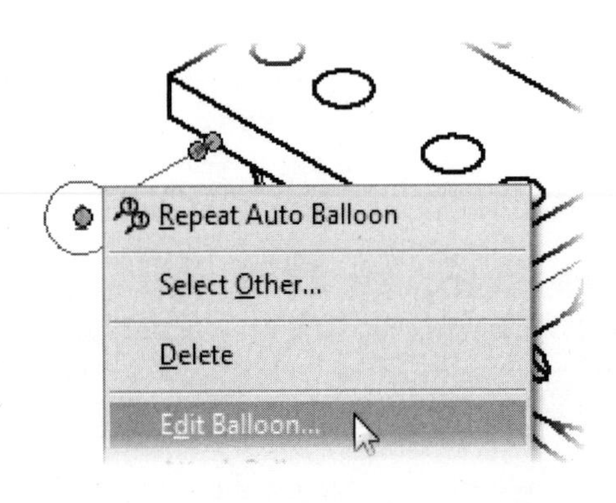

FIGURE 6.149

Use this dialog box to make changes to a balloon on an individual basis. The following features may be edited: Balloon Type and Balloon Value.

Editing the Balloon Value

The columns in the Balloon Value area of the Edit Balloon dialog box in the previous illustration specify Item and Override. When you edit the Item value, changes will be made to the balloon and will be reflected in the Parts List. In the following image, an Item was changed to 2A. Notice in the following image, on the left, that both the balloon and the Parts List have updated to the new value.

When you make a change in the Override column, the balloon will update to reflect this change, but the Parts List remains unchanged. The following image, on the right, shows this. The balloon was overridden with a value of 2A, but the Parts List has the item Rod Cap listed as 2.

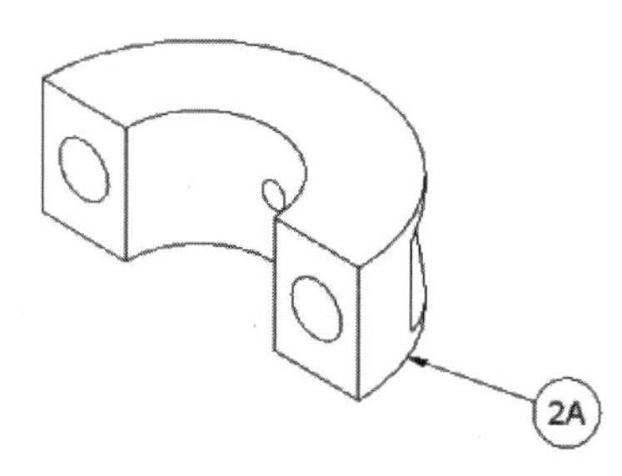

Parts			
ITEM	QTY	PART NUMBER	
1	1	Connecting Rod	
2A	1	Rod Cap	
3	2	Rod Cap Screw	
4	1	Piston	
5	1	Wrist Pin	

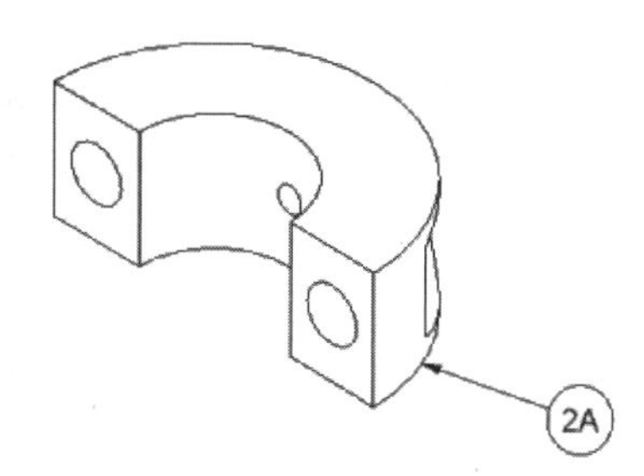

Parts			
ITEM	QTY	PART NUMBER	
1	1	Connecting Rod	
2	1	Rod Cap	
3	2	Rod Cap Screw	
4	1	Piston	
5	1	Wrist Pin	

FIGURE 6.150

NOTE

Any changes performed with the Balloon Edit dialog box will be reflected in all Parts Lists associated with that view.

Attaching a Custom Balloon

If you have predefined a custom part, you can select it from a list of custom parts for custom balloon content. To attach a custom balloon, right-click on the balloon, and select Attach Balloon From List, as shown in the following image on the left. If no custom part is available, this menu item will be grayed out.

The Attach Balloon dialog box, as shown in the following image on the right, will prompt you for your selection of custom balloons. You can select only one custom item at a time.

After selecting the custom part, click the OK button to return to the drawing. Drag the cursor around for placement of the attached balloon. Click to place the balloon in the desired location, as shown in the following image.

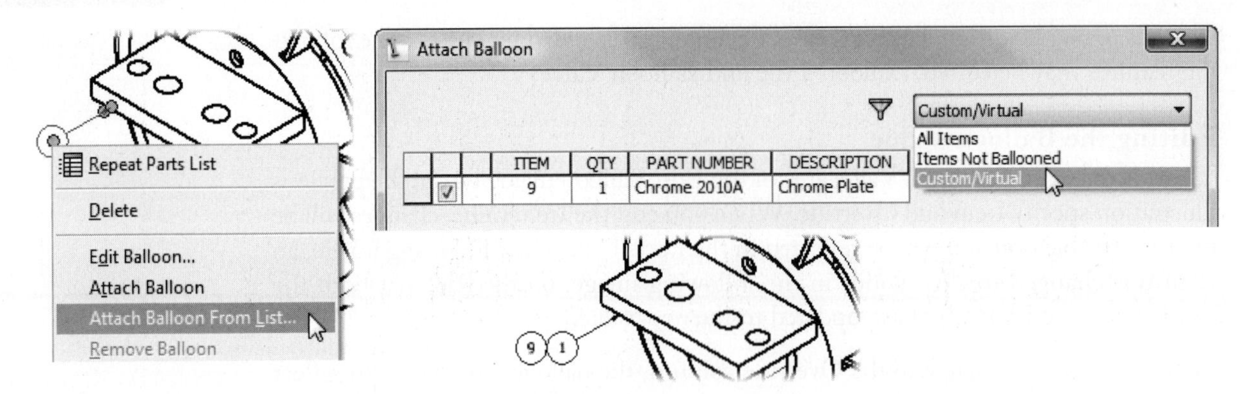

FIGURE 6.151

PARTS LISTS

After you have created the drawing views and placed balloons, you can also create a Parts List, as shown in the following image. As noted earlier in this section, components do not need to be ballooned before creating a Parts List.

Parts List				
ITEM	QTY	PART NUMBER	DESCRIPTION	MATERIAL
1	1	Arbor Press	ARBOR PRESS	Cast Steel
2	1	FACE PLATE	FACE PLATE	Steel, Mild
3	1	PINION SHAFT	PINION SHAFT	Steel, High Strength Low Alloy
4	1	LEVER ARM	LEVER ARM	Steel, Mild
5	1	THUMB SCREW	THUMB SCREW	Steel, Mild
6	1	TABLE PLATE	TABLE PLATE	Steel, High Strength Low Alloy
7	1	RAM	RAM	Steel, High Strength Low Alloy
8	2	HANDLE CAP	HANDLE CAP	Rubber
9	1	COLLAR	COLLAR	Steel, Mild
10	1	GIB PLATE	GIB PLATE	Steel, Mild
11	1	GROOVE PIN	GROOVE PIN	Steel, Mild

FIGURE 6.152

To create a Parts List, select the Annotate tab > Table panel and click the Parts List command, as shown in the following image on the left. The Parts List dialog box will appear, as shown in the following image on the right. Select a view on which to base the Parts List. After specifying your options, click the OK button, and a Parts List frame preview will appear attached to your cursor. If the components were already ballooned, the list preview will appear without displaying the Parts List dialog box. Click a point in the graphics window to place the Parts List. The information in the list is extracted from the properties of each component.

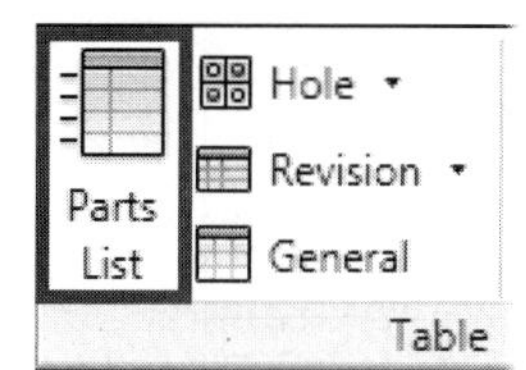

FIGURE 6.153

You can split or wrap a long Parts List into multiple columns through the Parts List dialog box. Notice the Table Wrapping area, as shown in the previous image on the right. Place a check in the Enable Automatic Wrap box to activate this feature, and then enter the number of sections to break the Parts List into. In this example, 2 is used.

The following image shows the results of performing the wrapping operation on a Parts List. Notice the two distinct sections that have been created.

Parts List

ITEM	QTY	PART NUMBER	DESCRIPTION
8	2	HANDLE CAP	HANDLE CAP
9	1	COLLAR	COLLAR
10	1	GIB PLATE	GIB PLATE
11	1	GROOVE PIN	GROOVE PIN
14	4	ANSI B18.3 - 1/4 - 20 UNC - 7/8	Hex Cap Screw
15	4	BS 4168 - M5 x 16	Hex Cap Screws
16	1	ISO 4766 - M5 x 5	Hex Cap Screws

Parts List

ITEM	QTY	PART NUMBER	DESCRIPTION
1	1	Arbor Press	ARBOR PRESS
2	1	FACE PLATE	FACE PLATE
3	1	PINION SHAFT	PINION SHAFT
4	1	LEVER ARM	LEVER ARM
5	1	THUMB SCREW	THUMB SCREW
6	1	TABLE PLATE	TABLE PLATE
7	1	RAM	RAM

FIGURE 6.154

Editing Parts List BOM Data

Once a Parts List is created, you can open the BOM Editor directly from the drawing environment and edit the assembly BOM. All changes are saved in the assembly and the corresponding component files. Right-click on the Parts List in the graphics screen or Drawing browser, and then click Bill of Materials from the menu, as shown in the following image.

ITEM	QTY		MATERIAL
1	1	Arbor	Cast Steel
2	1	FACE	Steel, Mild
3	1	PINIO	Steel, High Strength Low Alloy
4	1	LEVEF	Steel, Mild
5	1	THUM	Steel, Mild
6	1	TABLE	Steel, High Strength Low Alloy
7	1	RAM	Steel, High Strength Low Alloy
8	2	HAND	Rubber
9	1	COLL	Steel, Mild

Repeat Parts List
Copy Ctrl+C
Delete
Bill of Materials...
Save Item Overrides to BOM

FIGURE 6.155

This action will launch the Bill of Materials editor based on a specific assembly file. The dialog box provides a convenient location to edit the iProperties and Bill of Material properties for all components in the assembly. You can even perform edits on multiple components at once. Notice in the following image that the Bill of Materials editor displays the Bill of Materials Structure. In the Bill of Materials editor, perform the desired edits. When finished, click Done to close the dialog box. All changes will be updated in the Parts List and balloons.

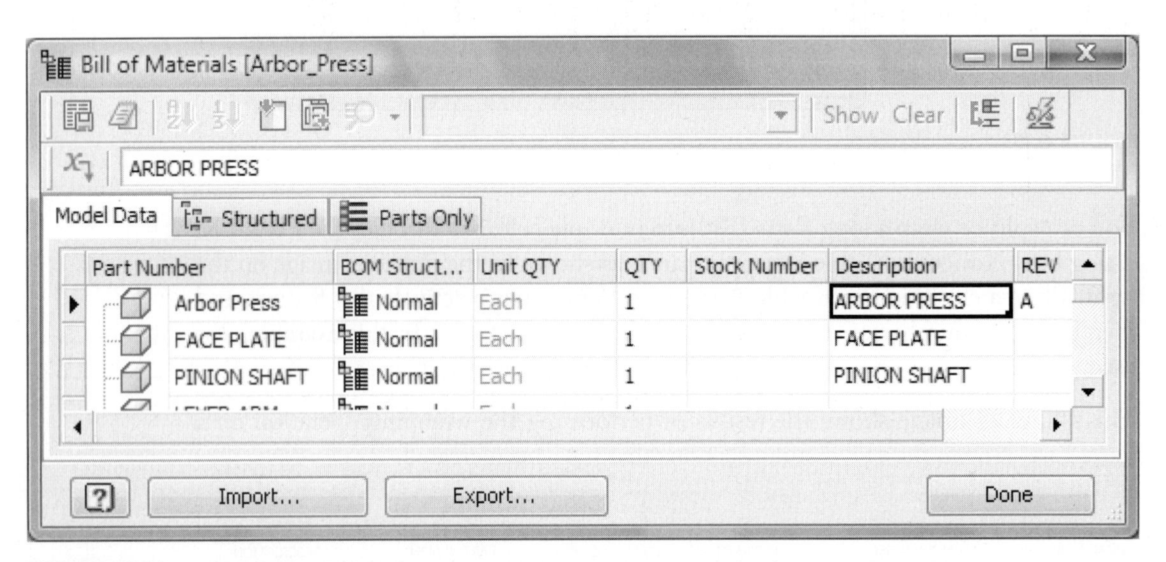

FIGURE 6.156

PARTS LIST COMMANDS

The information supplied earlier in this chapter covered the basics of inserting and manipulating a Parts List. This section will expand on the capabilities of the Parts List and will cover the following topics:

- A detailed account of the Parts List Operations commands available at the top of the Edit Parts List dialog box
- The function of the spreadsheet view
- Additional sections of the Edit Parts List dialog box
- Expanded capabilities of the Edit Parts List dialog box
- Nested Parts Lists

Parts List Operations (Top Icons)

Once you have placed a Parts List in the drawing, controls are available to help you modify the Parts List. Right-click on the Parts List in a drawing and select Edit Parts List from the menu, as shown in the following image, or double-click on the Parts List in the drawing.

ITEM	QTY	PART NUM	MATERIAL
1	1	Arbor Press	Cast Steel
2	1	FACE PLATE	Steel, Mild
3	1	PINION SHAFT	Steel, High Strength Low Alloy
4	1	LEVER ARM	Steel, Mild
5	1	THUMB SCREW	Steel, Mild
6	1	TABLE PLATE	Steel, High Strength Low Alloy
7	1	RAM	Steel, High Strength Low Alloy
8	2	HANDLE CAP	Rubber
9	1	COLLAR	Steel, Mild
10	1	GIB PLATE	Steel, Mild
11	1	GROOVE PIN	Steel, Mild
14	4	ANSI B18.3 - 1/4 - 7/8 HS HCS	Steel, Mild
15	4	BS 4168 - M5 x 16	Steel, Mild

FIGURE 6.157

This action displays the Edit Parts List dialog box, as shown in the following image. The row of icons along the top of the dialog box consist of numerous features that allow you to change the Parts List.

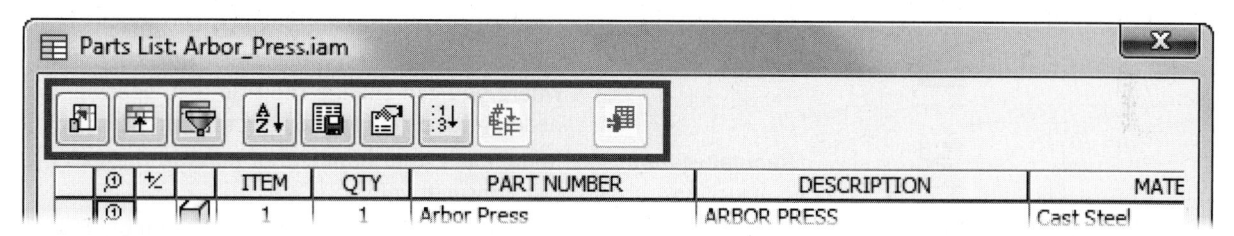

FIGURE 6.158

The Parts List table gives detailed explanations of all Parts List icons:

Parts List Button	Function	Explanation
	Column Chooser	Opens the Parts List Column Chooser dialog box, where you can add, remove, or change the order of the columns for the selected Parts List. Data for these columns is populated from the properties of the component files.
	Group Settings	Opens the Group Settings dialog box, where you select Parts List columns to be used as a grouping key and group different components into one Parts List row. This button is active only when generating a Parts List using the Structured mode.
	Filter Settings	Opens the Filter Settings dialog box, where you can define such filter settings as Assembly View Representations, Ballooned Items Only, Item Number Range, Purchased Items, and Standard Content from which to filter the information. Once a filter is selected, you then filter out rows without changing the data in the Parts List. You can also add Parts List filters to a Parts List style.
	Sort	Opens the Sort Parts List dialog box, where you can change the sort order for items in the selected Parts List.
	Export	Opens the Export Parts List dialog box so that you can save the selected Parts List to an external file of the file type you choose.
	Table Layout	Opens the Parts List Table Layout dialog box where you can change the title text, spacing, or heading location for the selected Parts List.
	Renumber Items	Renumbers item numbers of parts in the Parts List consecutively.
	Save Item Overrides to BOM	Saves item overrides back to the assembly Bill of Materials.
	Member Selection	Used with iAssemblies. Clicking this button opens a dialog box where you can select which members of the iAssembly to include in the Parts List.

Creating Custom Parts

Custom parts are useful for displaying parts in the Parts List that are not components or graphical data, such as paint or a finishing process. To add a custom part, right-click on an existing part in the Edit Parts List dialog box, as shown in the following image, and click Insert Custom Part. When a custom part no longer needs to be documented in the Parts List, you can right-click in the row of the custom part, and click Remove Custom Part from the menu.

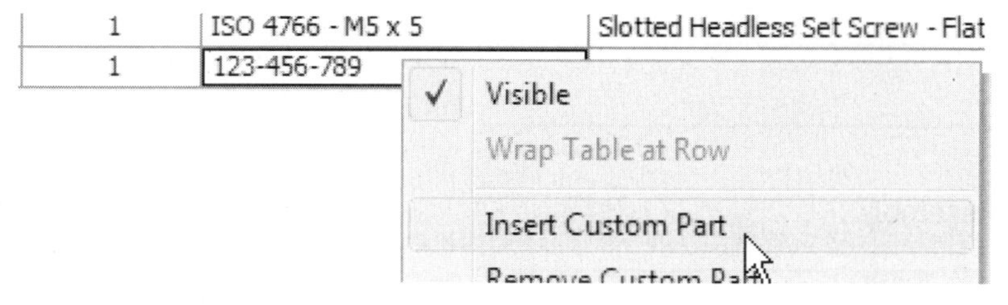

FIGURE 6.159

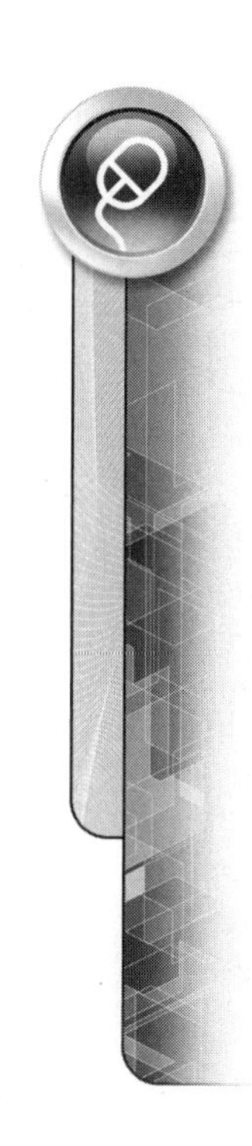

EXERCISE 6-8: CREATING ASSEMBLY DRAWINGS

In this exercise, you create drawing views of an assembly and then add balloons and a Parts List.

1. This exercise begins with formatting a drawing sheet. Open the drawing file *ESS_E06_08.idw* from the Chapter 06 folder. This drawing consists of one blank D sheet with a border and title block.
2. In the browser, right-click Sheet:1 and select Edit Sheet.
3. In the Edit Sheet dialog box, rename the sheet by typing **Assembly** in the Name field.
4. In the Size list, select C. When finished, click the OK button.
5. The next phase of this exercise involves updating the file properties in order to update the title block fields. Select the Inventor Application Menu click iProperties.
6. Click on the Summary tab and make the following changes:
   - In Title, enter **Pump Assembly**.
   - In Author, enter your name or initials.
7. Next click on the Project tab and make the following changes:
   - In Part number, enter **123-456-789**.
   - In Creation Date, click the down arrow, and then select today's date.
8. When finished, click the OK button. Zoom in to the title block and notice that the entries have updated, as shown in the following image on the left.
9. Now create drawing views. Zoom out to view the entire sheet.
10. Click the Base command on the Place Views tab > Create panel to display the Drawing View dialog box.
11. Click the Open an Existing File button next to the File list, select Assembly Files from the Files of Type drop-down list, and double-click *ESS_E06_08.iam* from the Chapter 06 folder as the view source.
12. From the Orientation list, select Top.
13. Set the Style to Hidden Line. Do not select Hidden Line Removed.
14. Set the Scale to 1:1
15. Move the view to the upper-left corner of the sheet, and then click to place the view, right-click and select Done. The drawing view should appear similar to the following image on the right.

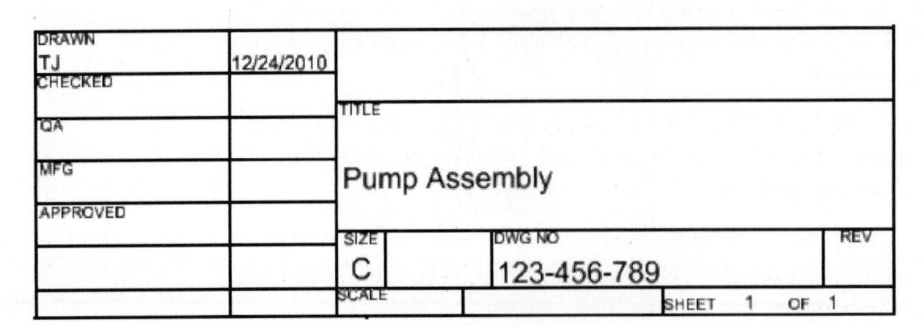

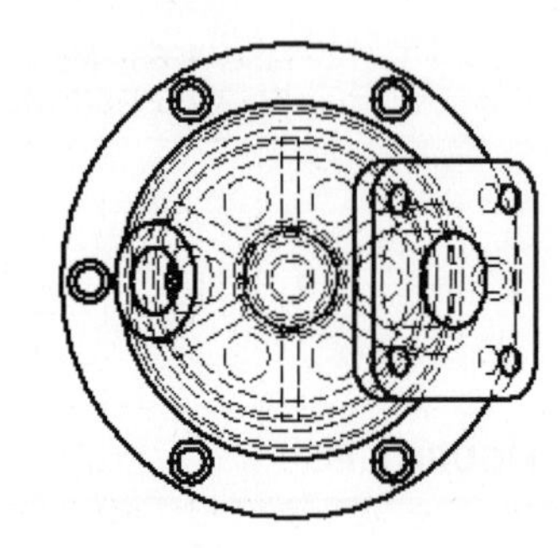

FIGURE 6.160

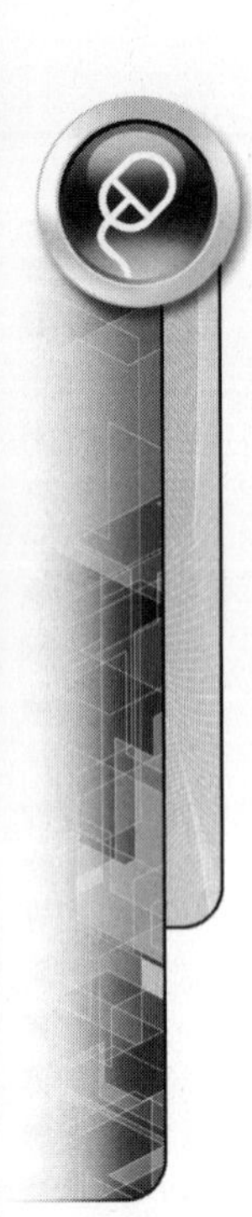

16. Now begin to generate a section view from the existing top view.
17. Before you create the section view, you will need to turn off the Section property for some components so that they are not sectioned by following the next series of directions:
    - In the browser, expand the Assembly sheet.
    - In the browser, expand View1: *ESS_E06_08.iam*. Your view number may be different.
    - Expand *ESS_E06_08.iam*.
    - Select *ESS_E06-Pin.ipt*, *ESS_E06-Spring.ipt*, *ESS_E06-NutA.ipt*, and *ESS_E06-NutB.ipt*.

TIP	**To select multiple items, hold down the CTRL key while clicking the items from the browser.**

Right-click any of the highlighted parts in the browser, select Section Participation > None from the menu.

18. Now create the section view by clicking the Section command on the Place Views tab > Create panel.
19. Select the top view of the assembly, and draw a horizontal section line through the center of the top view, as shown in the image below.
20. Right-click, and select Continue.
21. Move the preview below the top view, and click to place the view.

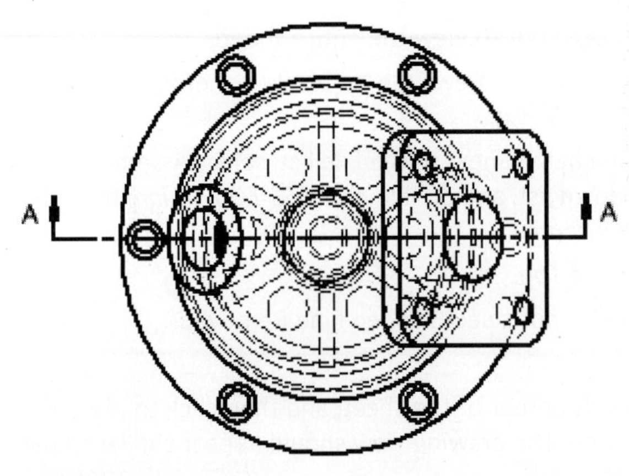

FIGURE 6.161

22. Notice that the nuts, spring, and pin are visible in the view but not sectioned, as shown in the following image on the left. Notice also the presence of the section label and drawing view scale.
23. Now edit the section view, and turn off the label and scale. On the sheet, select the sectioned assembly view, right-click, and select Edit View from the menu to display the Drawing View dialog box.
24. Click the light bulb button (Toggle Label Visibility) to turn off the scale and label in the section view. Click the OK button. The label and scale no longer display at the bottom of the view, as shown in the following image on the right.

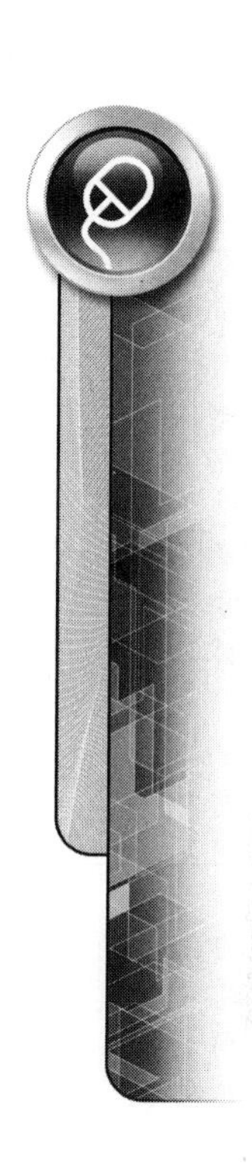

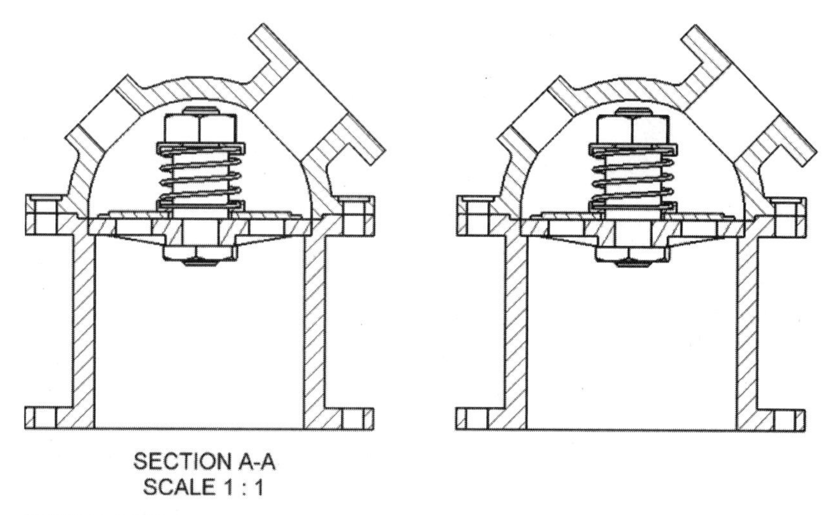

FIGURE 6.162

25. Now hide certain components from displaying in the top view. In the browser, under the expanded component listing for *ESS_E06_08.iam*, select both *ESS_E06-CylHead:1* and *ESS_E06-Spring:1*.
26. Right-click, and clear the checkmark from Visibility. The cylinder head and the spring do not display in the top view, as shown in the following image.

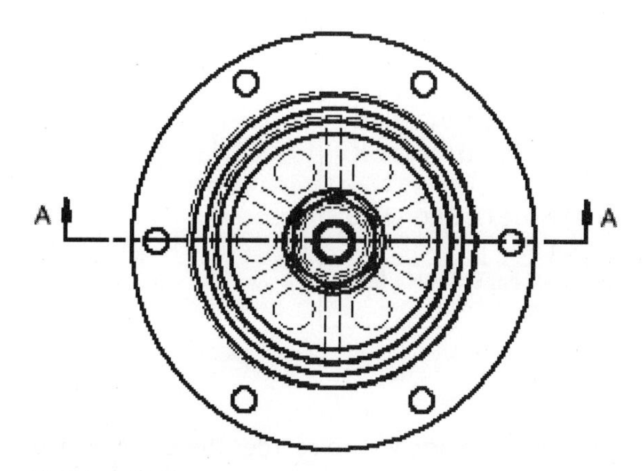

FIGURE 6.163

27. Next add a text note to the top view stating that the cylinder head and spring are not visible. Click the Text command in the Annotate tab > Text panel.
28. Select a point to the left of the top assembly view.

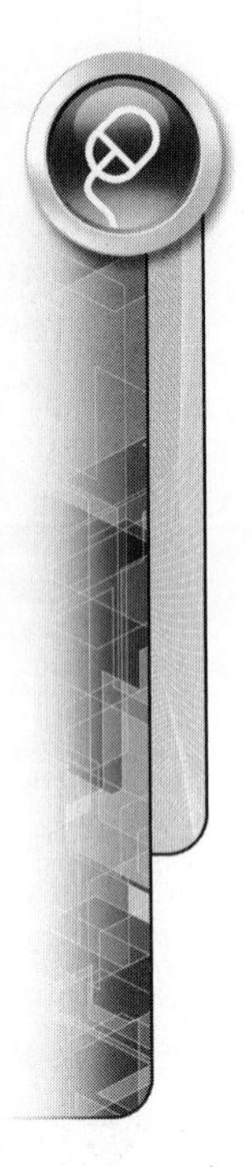

29. In the text entry area, type **CYLINDER HEAD AND**, press ENTER, and type **SPRING REMOVED**. Click OK. Right-click and select Done. The text will be added to the top view, as shown in the following image. If necessary, select and drag the text to a better location.

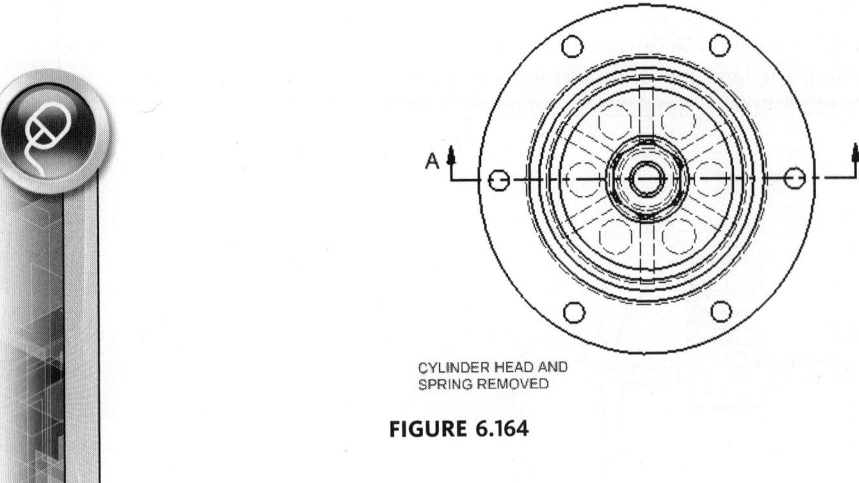

FIGURE 6.164

30. Next, you create a Parts List to identify the item number, quantity, part number, and description. Begin by clicking the Parts List command on the Annotate tab > Table panel.
31. The Parts List dialog box will display. Select the sectioned assembly view.
32. Verify that the BOM View is set to Structured, and click the OK button.
33. Move the Parts List until it joins the title block and border, and then click to place it. Zoom in to display the Parts List, as shown in the following image.

Parts List			
ITEM	QTY	PART NUMBER	DESCRIPTION
1	1	Cylinder	CYLINDER
2	1	Cyl Head	CYLINDER HEAD
3	1	Valve plate	VALVE PLATE
4	1	Pin	THREADED PIN
5	1	Diaphragm	DIAPHRAGM
6	1	Washer A	RETAINING WASHER
7	1	ESS_E06_05-WasherB	RETAINING WASHER
8	1	Spring	SPRING - Ø1 X Ø22 X 25
9	1	Nut A	FLAT NUT, REG - M16 X 1.5
10	1	Nut B	FLAT NUT, THIN - M16 X 1.5

FIGURE 6.165

34. In the following steps, you will edit the Parts List to reorder the columns and adjust the column widths. Begin by selecting the Parts List in the graphics window or from the browser.
35. Right-click and select Edit Parts List from the menu to display the Parts List dialog box.

36. Click the Column Chooser button to display the Parts List Column Chooser dialog box and make the following changes:
    - In the Available Properties list, select MATERIAL.
    - Click Add.
    - In the Selected Properties list, select DESCRIPTION.
    - Click Move Up two times, and then click OK.
37. Now edit the column widths. While still in the Parts List dialog box, right-click on the DESCRIPTION header, and choose Column Width from the menu. Make the following column width changes:
    - Enter **2.25** in the Column Width dialog box.
    - Change the PART NUMBER field's width to **2**.
    - Verify the ITEM field's width is set to **1**.
    - Change the QTY field's width to **1**.
    - Change the MATERIAL field's width to **1.5**.
38. Click the OK button to close the dialog box.
39. Move the Parts List so that it is flush with the border and the title block. Your display should appear similar to the following image.

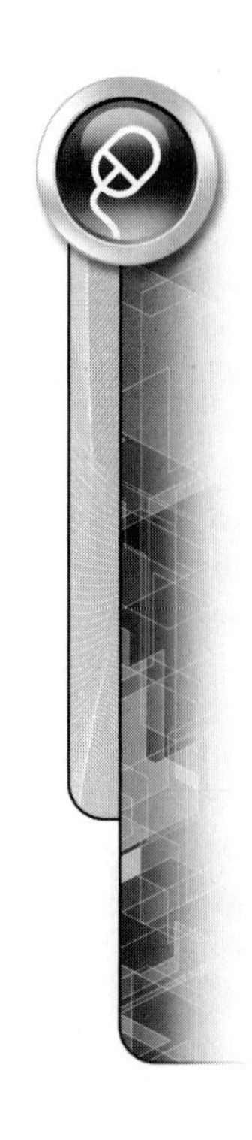

Parts List				
ITEM	DESCRIPTION	QTY	PART NUMBER	MATERIAL
1	CYLINDER	1	Cylinder	Cast Iron
2	CYLINDER HEAD	1	Cyl Head	Cast Iron
3	VALVE PLATE	1	Valve plate	Steel, Mild
4	THREADED PIN	1	Pin	Steel, Mild
5	DIAPHRAGM	1	Diaphragm	Stainless Steel
6	RETAINING WASHER	1	Washer A	Steel, Mild
7	RETAINING WASHER	1	ESS_E06_05-WasherB	Steel, Mild
8	SPRING - Ø1 X Ø22 X 25	1	Spring	Stainless Steel
9	FLAT NUT, REG - M16 X 1.5	1	Nut A	Steel, Mild
10	FLAT NUT, THIN - M16 X 1.5	1	Nut B	Steel, Mild

FIGURE 6.166

40. Now place balloons in the assembly drawing. The item number in the balloon corresponds to the item number in the Parts List.
41. Pan and zoom to display the sectioned assembly view.
42. Click the Balloon command on the Annotate tab > Table panel.
43. Select the edge of the component as the start of the leader as shown in the following image.
44. Click a point on the sheet to define the end of the first leader segment.
45. Right-click, and select Continue from the menu to place the balloon, as shown in the following image.

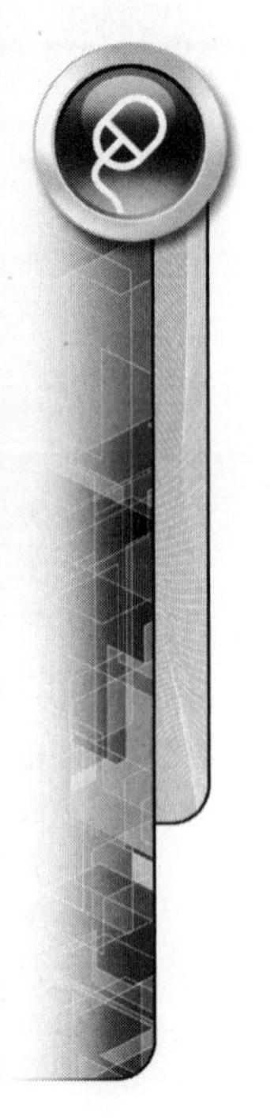

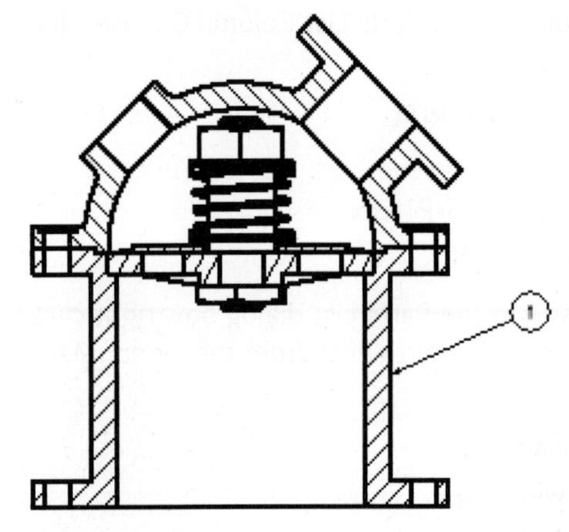

FIGURE 6.167

46. Select the Auto Balloon command from the Annotate tab > Table panel. Select the sectioned assembly view.
47. Window select all the components in the view. The cylinder will not be selected because it has already been ballooned.
48. Click the Select Placement button and select a location above the section view.
49. Select Horizontal for the placement and enter **0.5 in** for the Offset Spacing.
50. Click OK to place the balloons as shown in the following image.
51. Close all open files. Do not save changes. End of exercise.

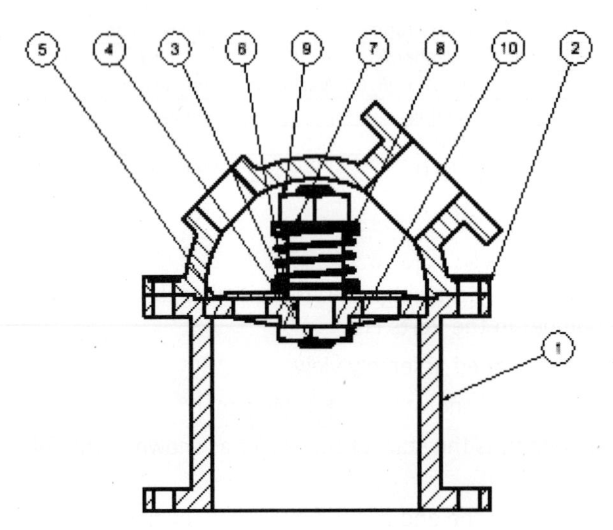

FIGURE 6.168

APPLYING YOUR SKILLS

Skill Exercise 6-1

In this exercise, you create a new component for a charge pump and then assemble the pump.

1. This exercise uses the skills you have learned in previous exercises to assemble a pump and create a new part in place. Open *ESS_E06_09.iam*.

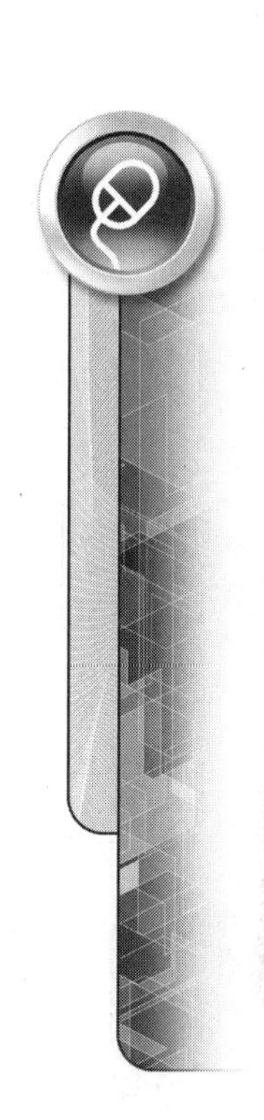

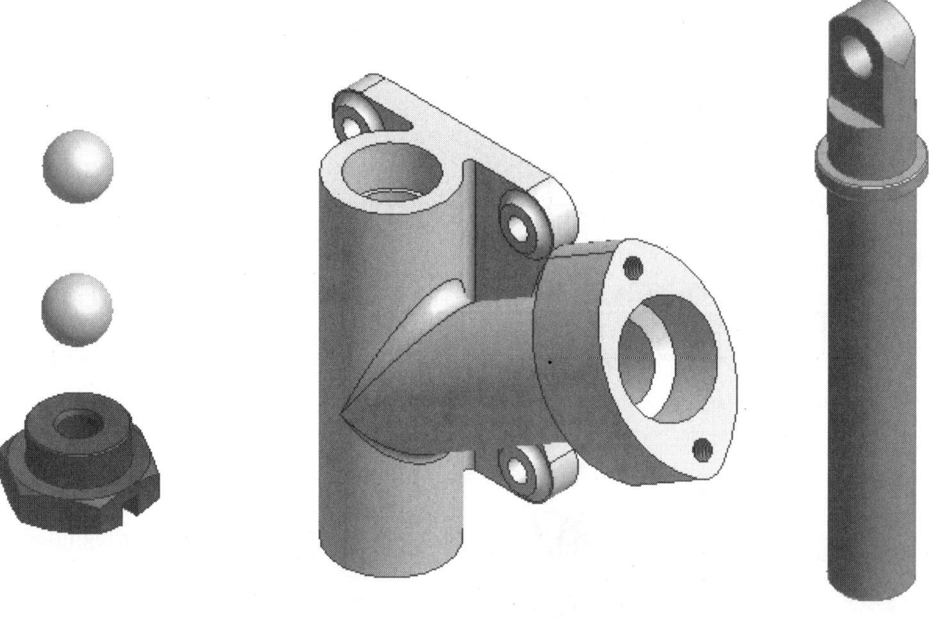

FIGURE 6.169

2. Place the following predefined components into the assembly:
   - 1 occurrence of *ESS_E06_09-Union.ipt*
   - 1 occurrence of *ESS_E06_09-Seal.ipt*
   - 2 occurrences of *ESS_E06_09-M8x30.ipt*
3. Next create the gland in place. Project edges from the pump body to define the flange. See the following image for the dimensions needed to complete the gland.

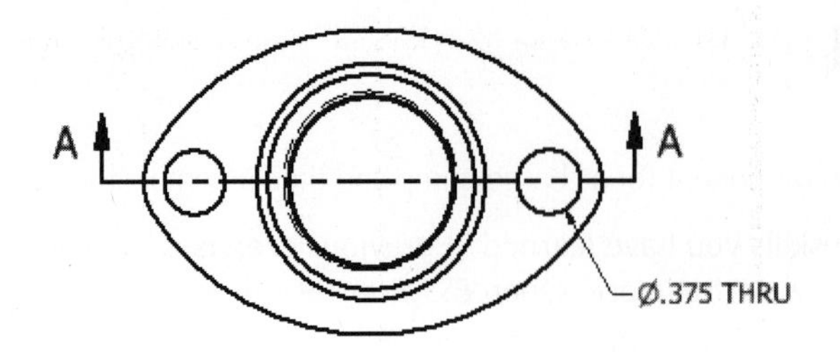

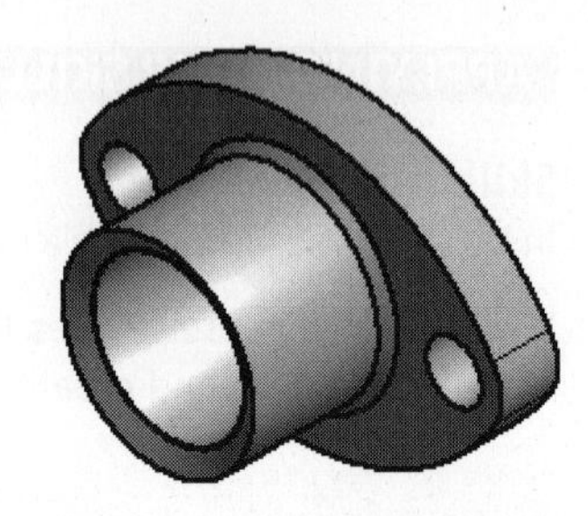

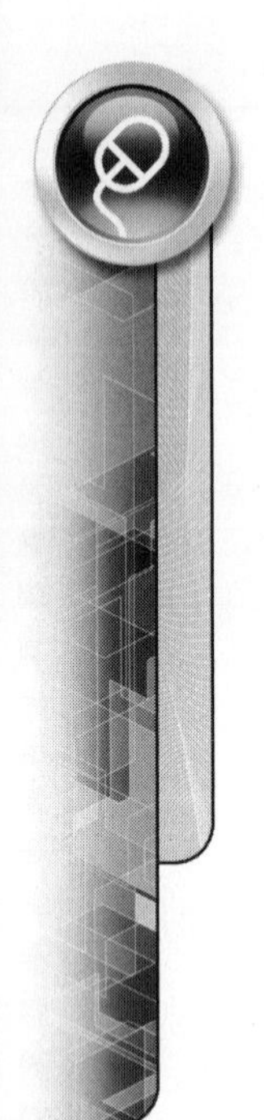

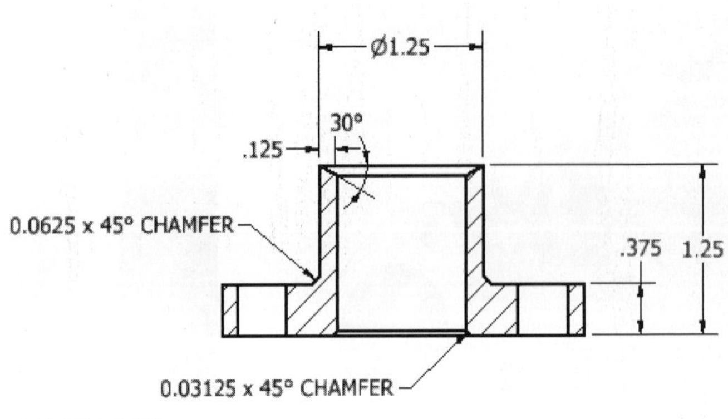

FIGURE 6.170

4. Use assembly constraints to build the model, as shown in the following image.

TIP

To assemble the balls, mate the center point of the ball with the centerline of the body and then place a tangent constraint between the ball's surface and the sloped surface of the seat.

To assemble the seal into the body, first apply a mate constraint between the conical surface of the seal and the conical surface of the seal's seat in the body. You must use the Select Other command to select the first conical surface. Apply a similar constraint between the gland and the seal.

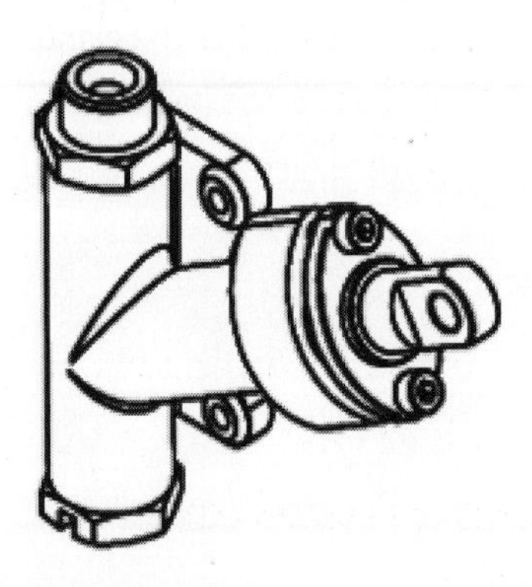

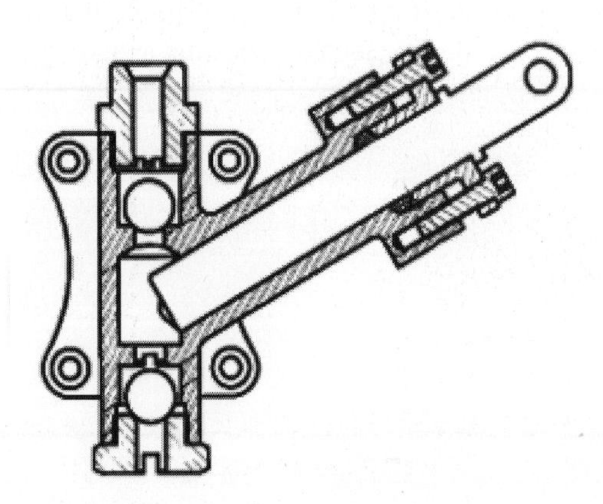

FIGURE 6.171

5. The completed assembly model should appear similar to the following image.

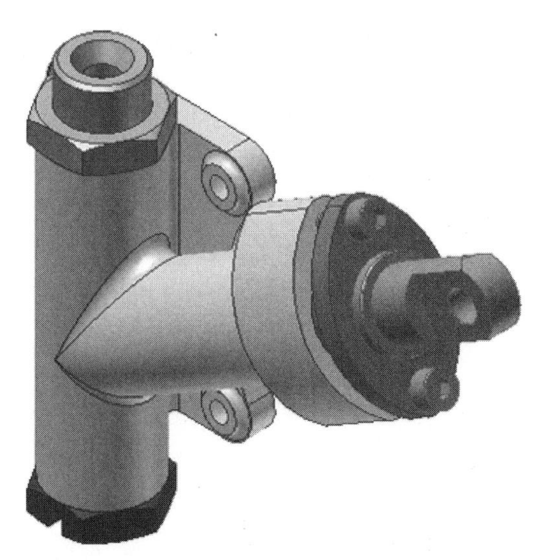

FIGURE 6.172

6. Close all open files. Do not save changes. End of exercise.

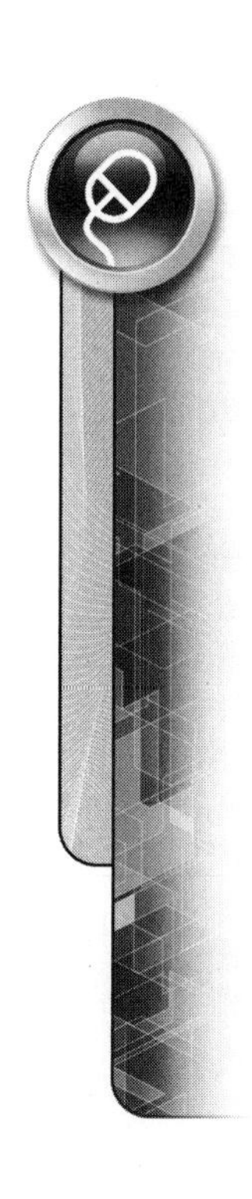

Skill Exercise 6-2

In this exercise, you create an assembly drawing, a Parts List, and balloons for a charge pump. The charge pump assembly is shown in the following image.

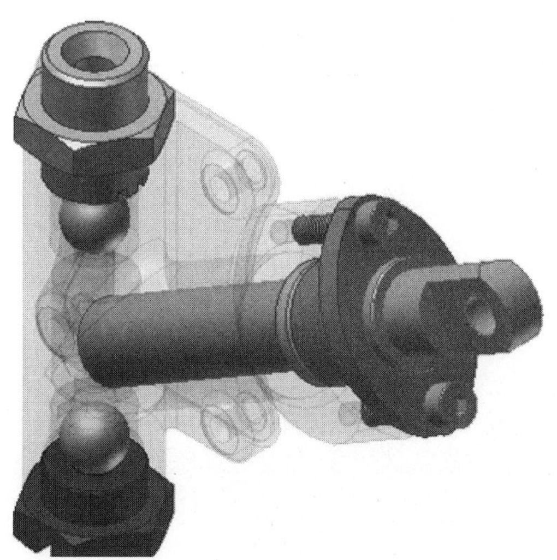

FIGURE 6.173

1. Create a new drawing using the ANSI (in) DWG or IDW template with a single D sheet named Assembly.
2. Insert a top view of *ESS_E06_10.iam* with a scale of 1:1.
3. Using the top view as the base view, create a sectioned front view. Exclude the ram, valve balls, and machine screws from sectioning.
4. Use the Base command and the Change View Orientation button to create an independent top right isometric view of the pump assembly as shown in the following image.

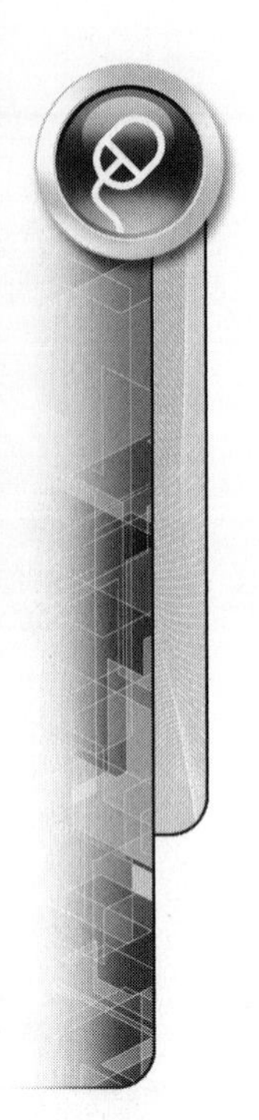

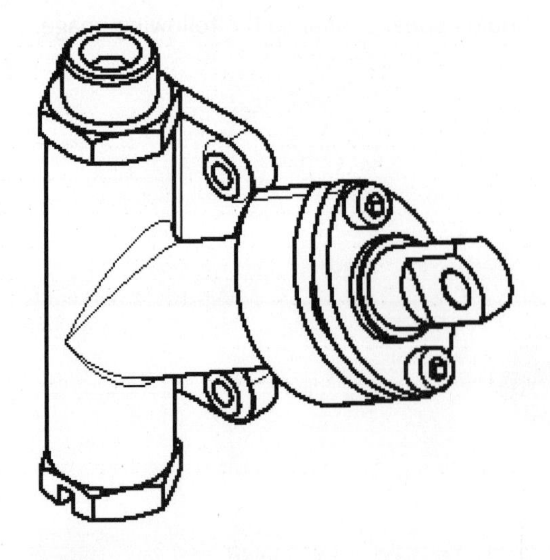

FIGURE 6.174

5. Insert the Parts List, add a Material column and then modify its format according to the column widths shown in the following table.

Column	Width
ITEM	1
QTY	1
DESCRIPTION	1.5
PART NUMBER	2
MATERIAL	2.25

6. Modify iProperties and Parts List parameters so that the Parts List and title block match those shown in the following image. Note how the item numbers are arranged in ascending order and the header row is at the bottom of the table.

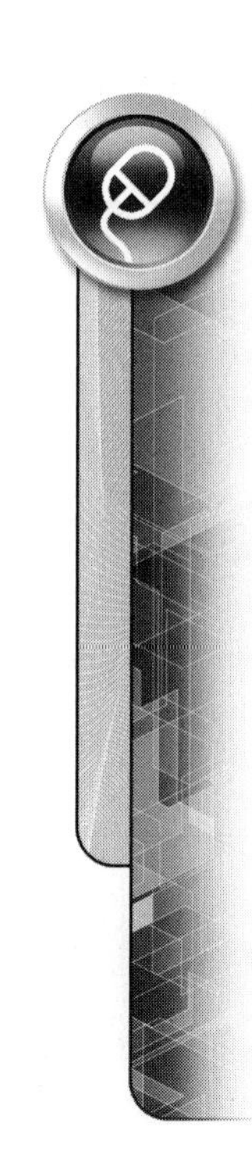

8	1	GLAND	ESS_E06_10-Gland	Brass, Soft Yellow Brass
7	2	PURCHASED	ESS_E06_09-M8x30	Default
6	1	SEAL	ESS_E06_09-Seal	Rubiconium
5	1	UNION	ESS_E06_09-Union	Rubiconium
4	1	SEAT	ESS_E06_09-Seat	Rubiconium
3	1	RAM	ESS_E06_09-Ram	Rubiconium
2	2	BALL	ESS_E06_09-Ball	Rubiconium
1	1	BODY	ESS_E06_09-Body	Rubiconium
ITEM	QTY	DESCRIPTION	PART NUMBER	MATERIAL
PARTS LIST				

DRAWN
TJ 12/24/2010
CHECKED
QA
MFG
APPROVED
TITLE
CHARGE PUMP ASSEMBLY
SIZE D
DWG NO
REV
SCALE
SHEET 1 OF 1

FIGURE 6.175

7. Add balloons to identify the components as shown in the following image.

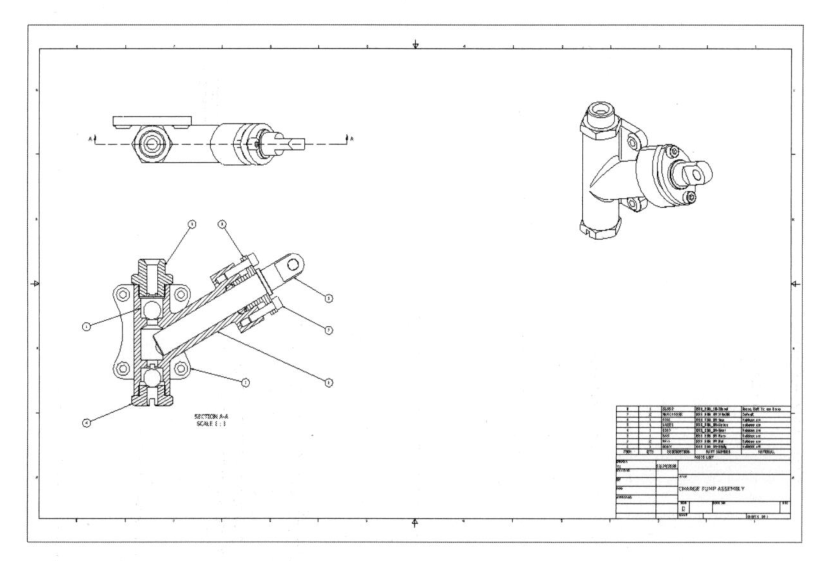

FIGURE 6.176

8. Close all open files. Do not save changes. End of exercise.

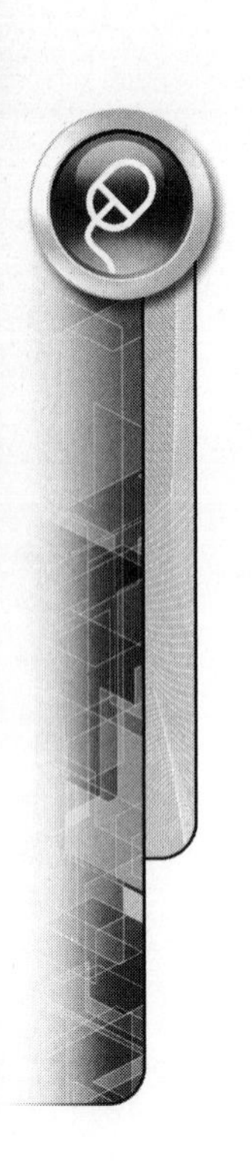

CHECK YOUR SKILLS

Use these questions to test your knowledge of the material covered in this chapter.

1. True ___ False ___ The only way to create an assembly is by placing existing parts into it.
2. Explain how to create a component in the context of an assembly.
3. True ___ False ___ An occurrence is a new copy of an existing component.
4. True ___ False ___ Only one component can be grounded in an assembly.
5. True ___ False ___ Autodesk Inventor does not require components in an assembly to be fully constrained.
6. True ___ False ___ A sketch must be fully constrained to adapt.
7. What is the purpose of creating a presentation file?
8. True ___ False ___ Balloons can only be placed in a drawing after placing a Parts List.
9. True ___ False ___ When creating drawing views from an assembly, you can create views from multiple presentation views or design views.
10. True ___ False ___ A Bill of Materials only retrieves its data from a Parts List.
11. True ___ False ___ A presentation file can be saved as a Flash animation.
12. True ___ False ___ A BOM Structured view shows all subassemblies and individual parts at the same assembly level.
13. True ___ False ___ An Associated Component Pattern will maintain a relationship to a feature pattern.
14. True ___ False ___ User-defined folders allow you to organize an assembly browser by group assembly constraints in a single folder.
15. True ___ False ___ In a Parts List, you can display custom parts that are not components or graphical data, e.g. paint.

CHAPTER 7

Advanced Part Modeling Techniques

INTRODUCTION

In this chapter, you will learn how to use advanced modeling techniques. Using advanced modeling techniques, you can create transitions between parts that would otherwise be difficult to create. You can edit advanced features such as other sketched and placed features but typically their creation requires more than one unconsumed sketch. This chapter also introduces you to techniques that will help you be more productive in your modeling.

OBJECTIVES

After completing this chapter, you will be able to perform the following:

- Section a part
- Create a design view representation in a part file
- Create ribs, and rib networks
- Emboss text and profiles
- Create sweep features
- Create coil features
- Create loft features
- Split a part or split faces of a part
- Bend a part
- Reorder part features
- Mirror model features
- Suppress features of a part
- Create a derived part based on another part or an assembly
- Shrinkwrap an assembly
- Create plastic and cast features

ADJUSTABLE SECTIONS VIEWS IN A PART FILE

In chapter 6 you learned how to section an assembly to better see inside it. The same functionality is available while working in a part file. As in an assembly, there are four section commands available while working in a part file. To access the section commands, click on the View tab > Appearance panel as shown in the following image on the left. The following image in the middle shows a part before being sectioned and the view on the right shows the part sectioned with the Three Quarter Section View command. Note that the sections are temporary and sketches cannot be actively placed on a section. As always, you can create a sketch on a plane that was selected to create the section.

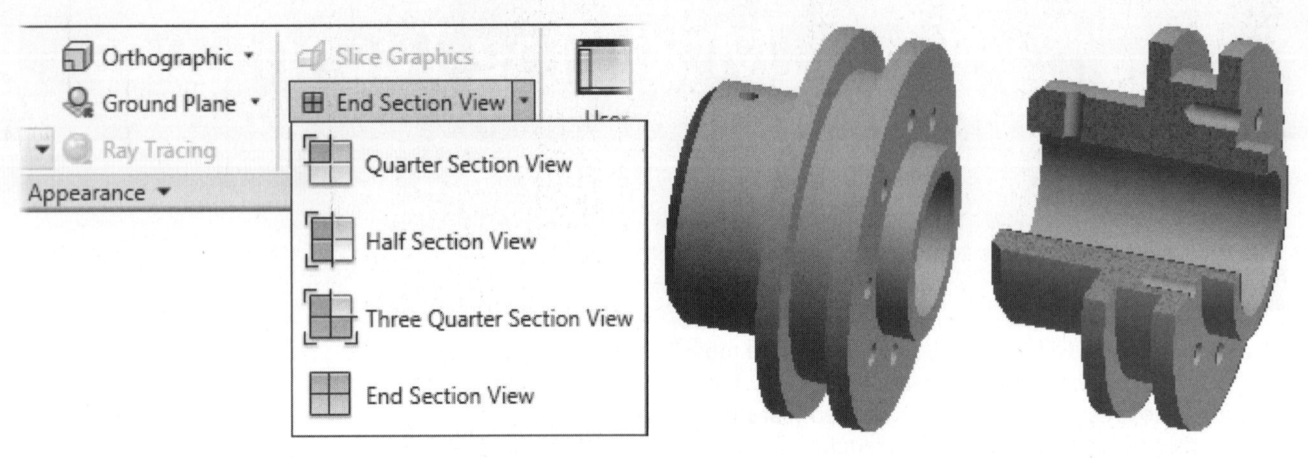

FIGURE 7.1

To section a part, follow these steps.

1. Select the desired section command on the View tab > Appearance panel.
2. Select any existing planar face(s), work plane(s) or origin plane(s).
3. Adjust the section depth by clicking and dragging the visible arrow in the graphics window; enter a value in the Offset cell in the graphics window as shown in the following image on the left. Another option to control the offset is to move the cursor over the Offset cell and scroll the wheel on the mouse. You can control the distance that each step takes when you scroll the wheel by right-clicking and click Virtual Movement > Scroll Step Size.
4. To accept the current location click the green check mark as shown in the following image in the middle, press the ENTER key or right-click and click OK in the menu.
5. If you are creating a quarter or three quarter section view, follow the same process to select a second plane.
6. While still in the command, you can specify which side of the section remains visible by right clicking and click the Flip Section or change the section type as shown in the following image on the right.
7. To finish the command, right-click and click Done from the menu.
8. To clear the section, click the End Section View command on the View tab > Appearance panel.

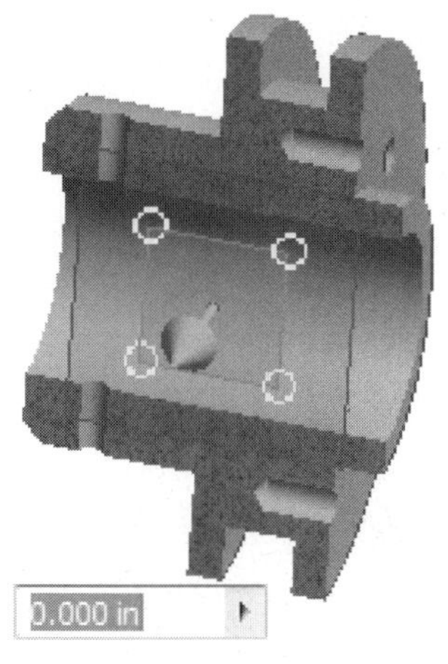

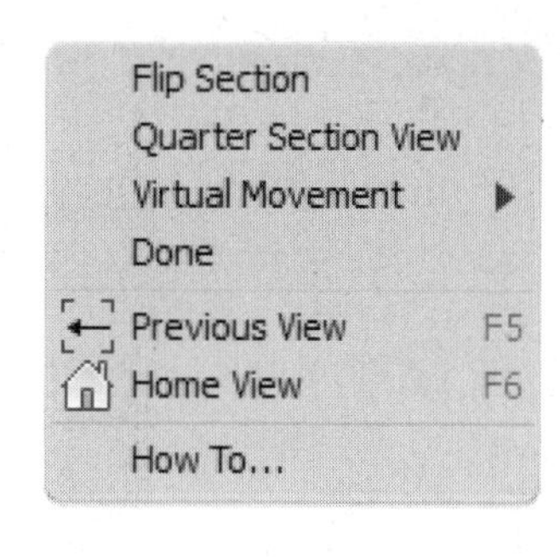

FIGURE 7.2

DESIGN VIEW REPRESENTATION IN A PART FILE

While working on a part file, you can save a design view representation that stores' information related to the current viewpoint, part color, and work feature visibility. You can save as many design view representations as needed, but only one can be current. By default, there is a design view representation named Master, this master design view cannot be changed, deleted, or locked. To create a design view representation in a part file, follow these steps.

1. In the browser, right-click on View: Master and click New from the menu as shown in the following image on the left.
2. A generic design view will be created; you can rename the design view by slowly double-clicking on its name and type in a new name.
3. Once a design view is created, change the view point, part color, and work feature visibility and those changes are captured in the current design view.
4. Since these changes are actively captured to the current design view, you may want to lock a design view so any change to the current viewpoint, part color, or work feature visibility are not saved to the current design view. To lock a design view, right-click on its name in the browser and click Lock from the menu as shown in the following image on the right.
5. Create as many design views as needed.
6. Lock and unlock the current design view to capture the settings as needed.
7. To make a design view current, in the browser double-click on its name or right-click on its name and click Activate from the menu. The last settings in the design view or the locked settings in the design view will be displayed in the graphics window.

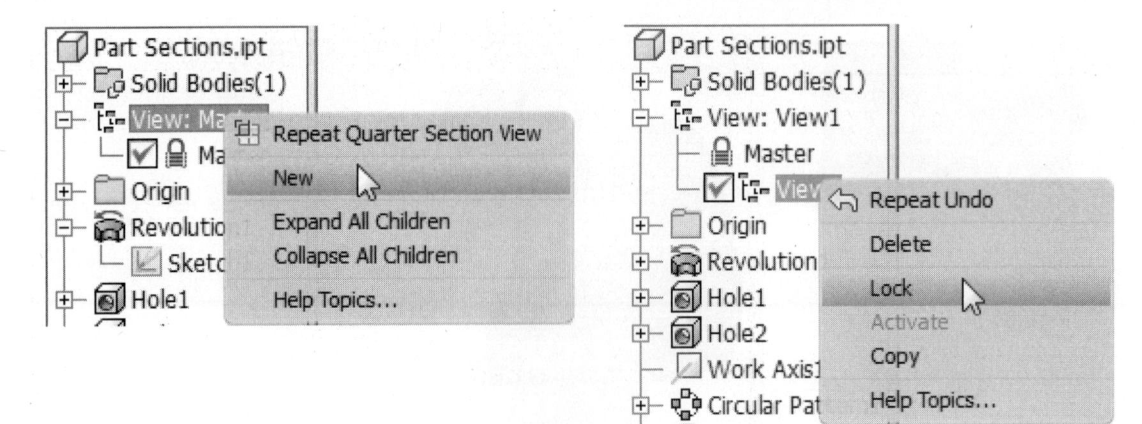

FIGURE 7.3

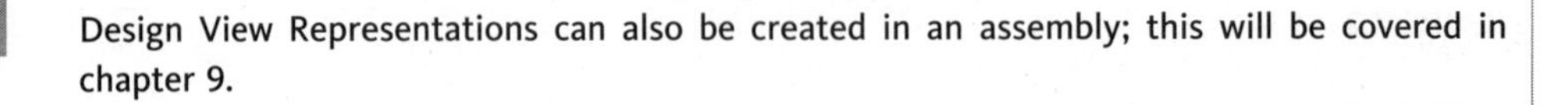

Design View Representations can also be created in an assembly; this will be covered in chapter 9.

RIB FEATURES

Ribs are used primarily to reinforce or strengthen features in mold and cast parts, but you can also use them in machined parts and in other cases where additional support and minimal weight are required. The following image shows a part with a sketch and with the sketch used to create a rib that uses the To Next and the Finite thickness options.

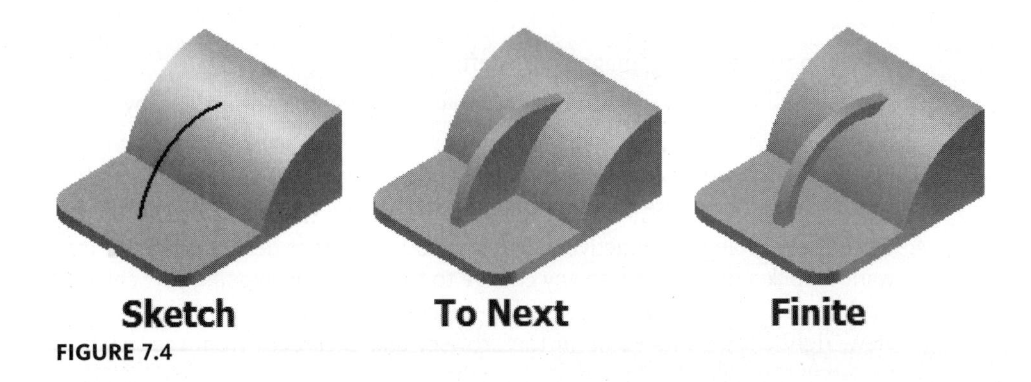

FIGURE 7.4

Using the Rib command on the Model tab > Create panel, as shown in the following image on the left, you can create ribs, and rib networks.

- A rib is a thin-walled feature that is typically closed.
- A rib network consists of a series of thin-walled support features.

A rib feature is defined by a single, open, unconsumed profile that is refined using the options in the Rib dialog box. If no unconsumed sketch exists in the part file, Autodesk Inventor will warn you with the message: “No visible unadaptive sketches.” After starting the Rib command, the Rib dialog box will appear, as shown in the following image on the right. The following sections describe the options in the dialog box.

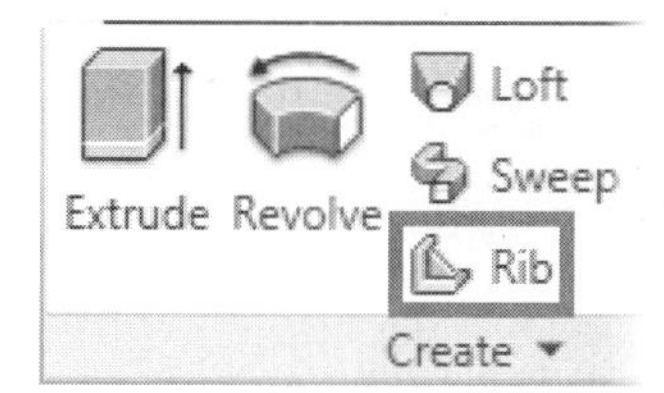

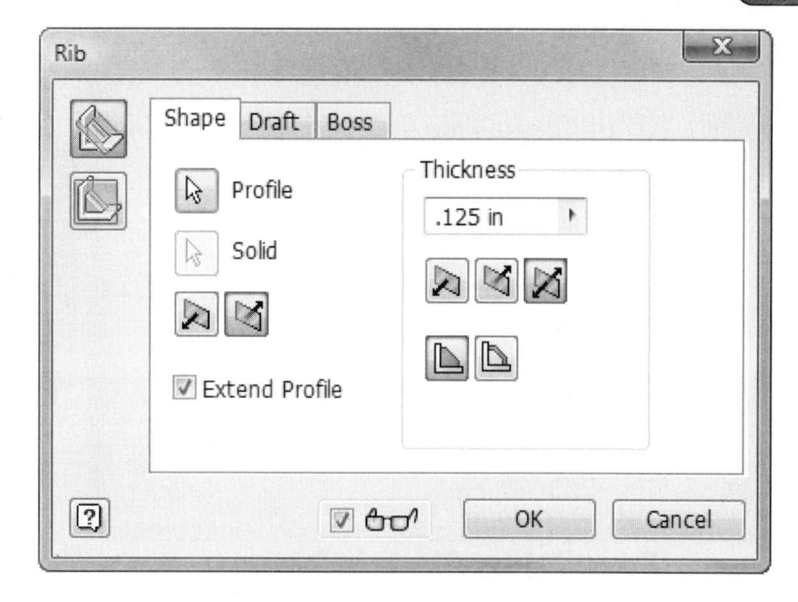

FIGURE 7.5

Shape

Profile

Select this button to choose the sketch to extrude. The sketch can be an open profile or you can select multiple intersecting or nonintersecting profiles to define a rib or web network.

Rib Type—Normal or Parallel to Sketch Plane

The two options on the left side are Normal to Sketch Plane and Parallel to Sketch Plane. The Normal to Sketch Plane type extrudes the profile normal to the sketch plane and its thickness is parallel to the sketch plane. The Parallel to Sketch Plane type extrudes the profile parallel to the sketch plane and its thickness is normal to the sketch plane. The following two images show a Normal to Sketch Plane and a Parallel to Sketch Plane type created from the same sketch that is drawn on the visible work plane. Notice how the direction of the thickness changes. The following images show web type and a rip type.

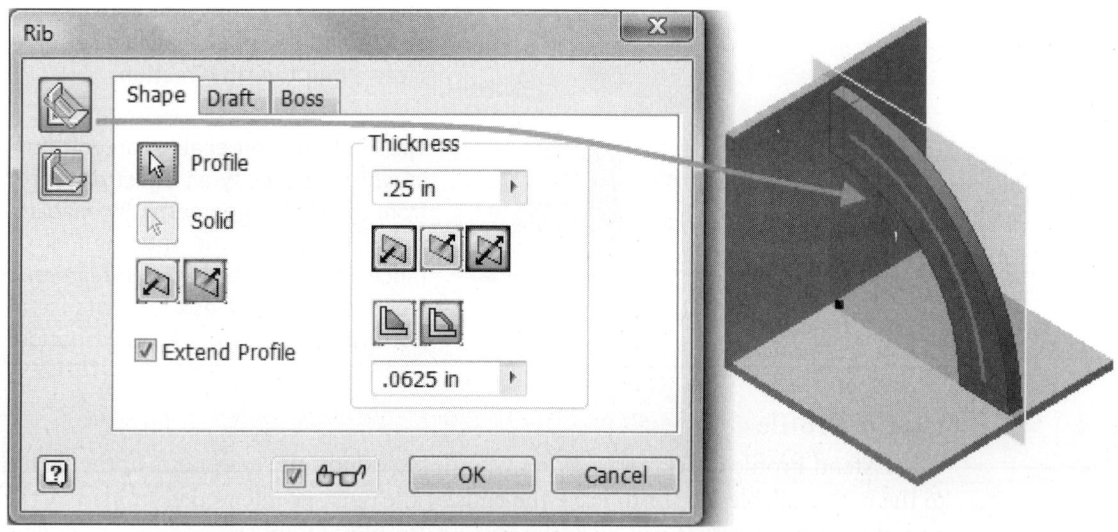

Normal to Sketch Plane

FIGURE 7.6A

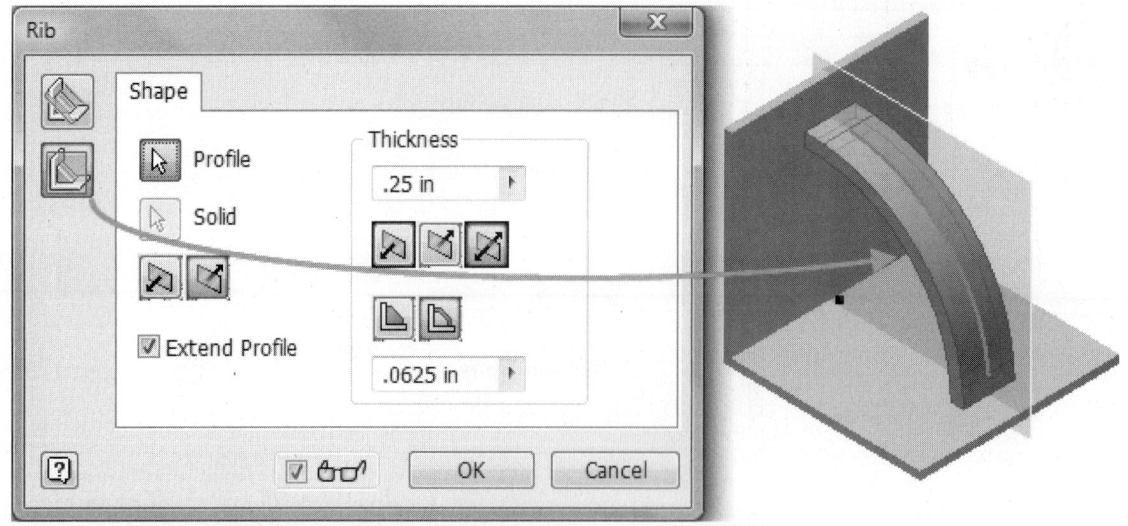

FIGURE 7.6B

Thickness

In this section, you specify the thickness and thickness direction of the rib.

Value

Enter the width of the rib feature using this edit box.

Flip Buttons

Select the flip buttons to specify which side of the profile to apply the thickness value to or to add the same amount of material to both sides of the profile.

Extents

In this section, you specify a rib (To Next) or a web (Finite).

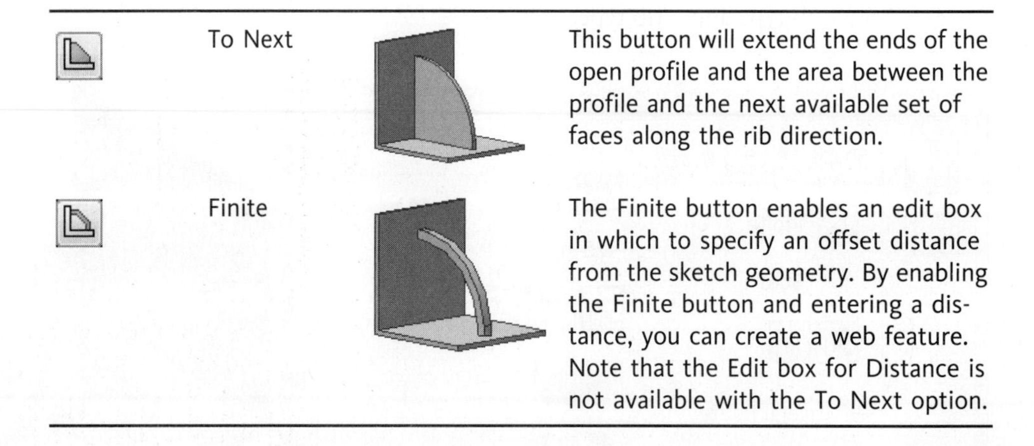

	To Next		This button will extend the ends of the open profile and the area between the profile and the next available set of faces along the rib direction.
	Finite		The Finite button enables an edit box in which to specify an offset distance from the sketch geometry. By enabling the Finite button and entering a distance, you can create a web feature. Note that the Edit box for Distance is not available with the To Next option.

Extend Profile

The Extend Profile checkbox specifies whether to extend the endpoints of the sketch to the next available face or to leave the ends of the open profile as determined by the end of the rib feature. If you click the Extend Profile option, the ends are extended; if you leave the box clear, the ends cap at the end of the sketched profile. The Extend

Profile option is always available for a finite web, but it is only active for the To Next option when the direction and face(s) on the model meet appropriate conditions.

Draft

The Draft tab is only available when the Web Type is selected. In the Draft tab you choose which thickness to hold, the top or root, and define the draft angle as shown in the following image.

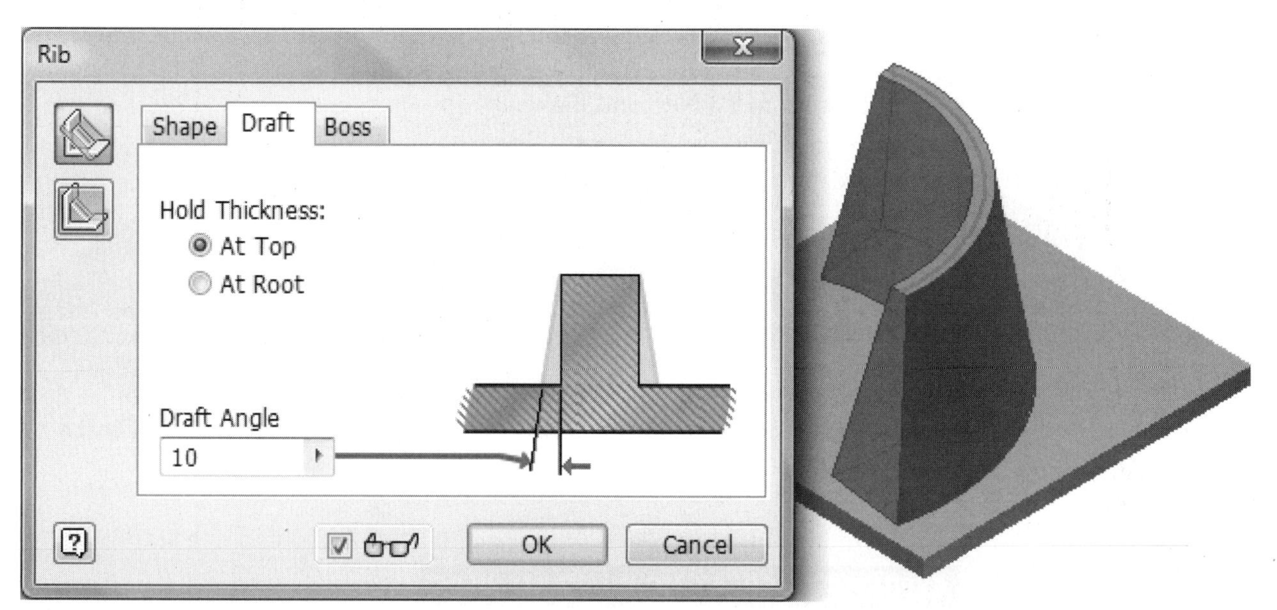

FIGURE 7.7

The Boss tab is only available when the Web Type is selected. In the Boss tab you can add a boss to a rib. You place a point, center point(s), in which the boss will be defined. The following image shows a boss created on the arc.

Creating Ribs

To create a rib follow these steps:

1. Create an active sketch in the location where you will place the rib.
2. Sketch an open profile that defines the basic shape of one edge of the rib.
3. Add constraints and dimensions to the sketch as needed.
4. Click the Rib command on the Create panel.
5. Specify the rib type, normal or parallel to the sketch.
6. Enter a value for the rib or web in the Thickness edit box.
7. Also in the Thickness section, use the Flip buttons to choose which side of the profile to apply the thickness value or to add the same amount of material to both sides of the profile.
8. Specify the depth of the profile by clicking either the To Next or the Finite buttons in the Extents area of the Rib dialog box.
9. If you use the Finite option, enter an offset distance, and click Extend Profile if the endpoints are to be extended to the next available face.
10. If the web type is selected you can also define draft and a boss(es).
11. To complete the operation, click the OK button.

Rib Network

You can also create a rib network using the Rib command. You can use multiple intersecting or nonintersecting sketch objects within a single profile to create a rib network. The creation process is the same as for creating a single web, except that the thickness is applied to all objects within the profile. When you select the profile objects, you will need to select them individually. If the rib network is to have equal spacing between the objects, use the 2D Rectangular Pattern or the 2D Circular Pattern command in the Sketch tab > Pattern panel to create the profile. The following image shows a part with multiple intersecting lines defined in a single profile, the same profile used to create a rib network using the Normal to Sketch Plane type and the thickness set to To Next and Finite.

FIGURE 7.8

EXERCISE 7-1: CREATING RIBS AND WEBS

In this exercise, you sketch an open profile, use the Rib command to create a rib, and then edit the rib feature to change the rib to a web. You complete the exercise by sketching overlapping lines and creating a rib network.

1. Open *ESS_E07_01.ipt* in the Chapter 07 folder.
2. Use the Free Orbit command to examine the part.
3. Create a new sketch on Work Plane1.
4. Turn off the visibility of the work plane.
5. Sketch the line and arc, as shown in the following image. Make sure that the arc you sketch, if extended to the right, would intersect the top of the part. Constraints and dimensions could be added as required.

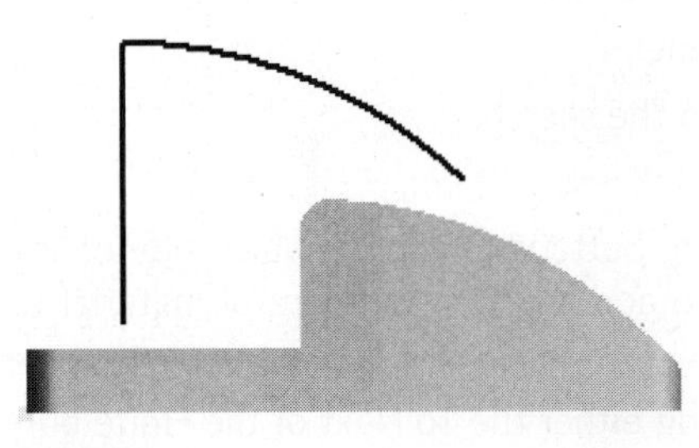

FIGURE 7.9

6. Click the Finish Sketch command on the Exit panel to finish the sketch.
7. Click the Rib command on the Model tab > Create panel. Make the Web Type current, the Thickness **.25 in**, change the Extents to Finite with a distance of **2 in**, as shown in the following image.
8. Click OK to create the rib.

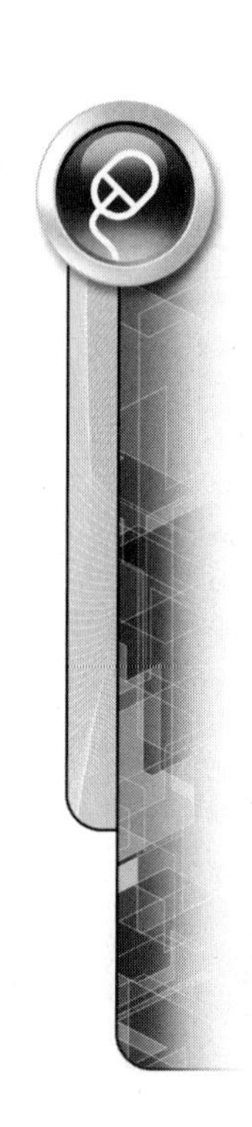

FIGURE 7.10

9. Edit the rib, in the browser double-click the Rib feature.
10. In the Rib dialog box, make the Rib Type (Parallel to Sketch Plane) current, the Thickness **2 in**, ensure the Extents is set to Finite with a distance of **.25 in** as shown in the following image. Notice how the distances are going in the opposite direction from the Web type.

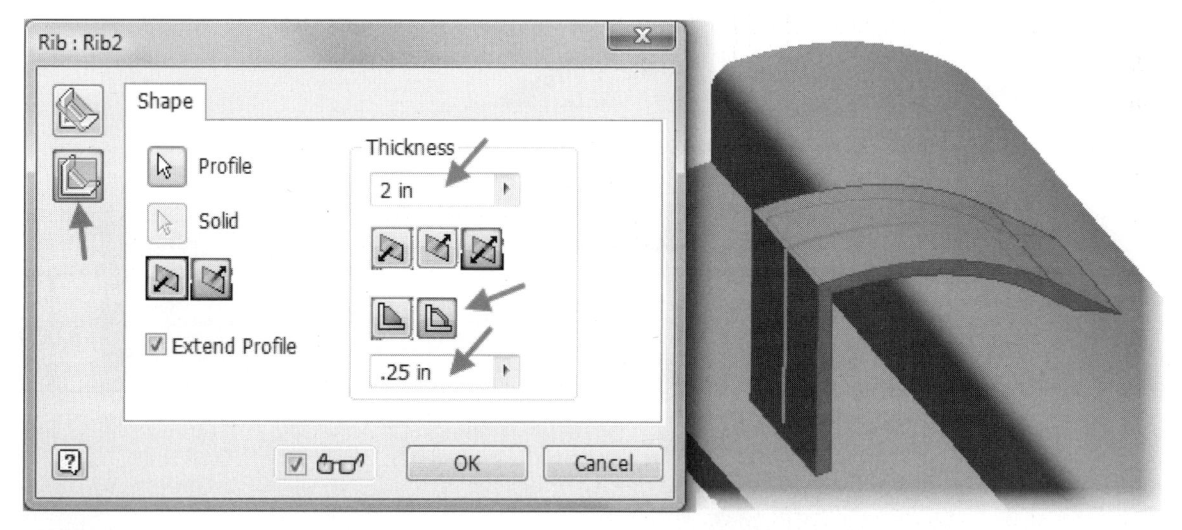

FIGURE 7.11

11. While still editing the rib, change the Extents to To Next as shown in the following image. Notice how the rib extends the area between the profile and the next available set of faces. Click OK to complete the edit.

FIGURE 7.12

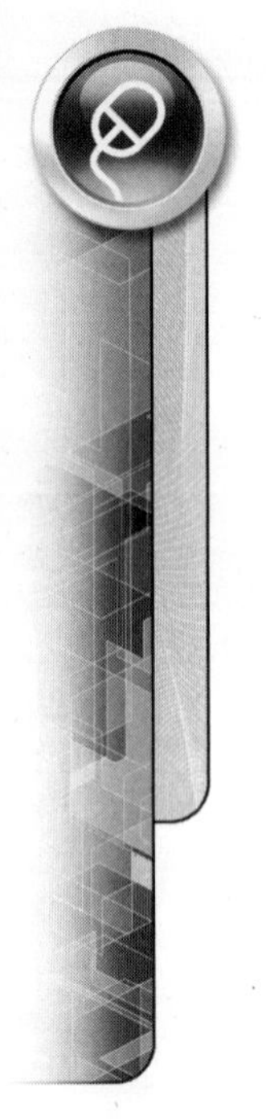

12. Use the Free Orbit command to view the bottom of the part, as shown in the following image on the left.
13. Create a new sketch on the inside flat face, as shown in the following image on the left.
14. Sketch and dimension three, overlapping line segments, as shown in the following image on the right. The line should be parallel and perpendicular to the existing edges but not touch the edges. The edges will be extended when the rib is created.

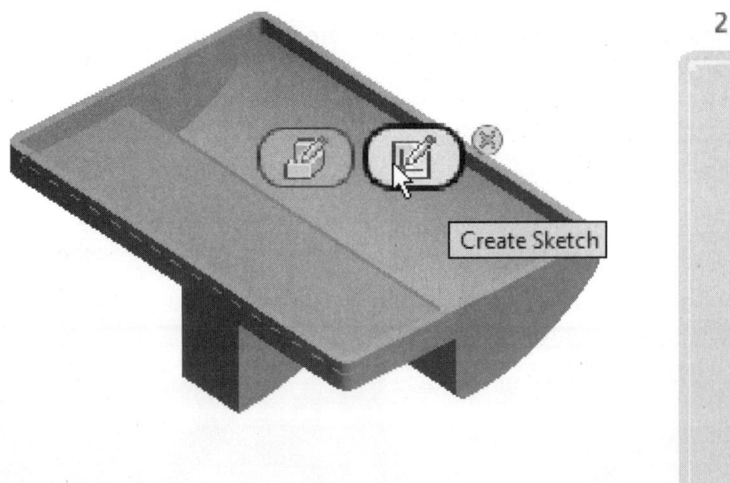

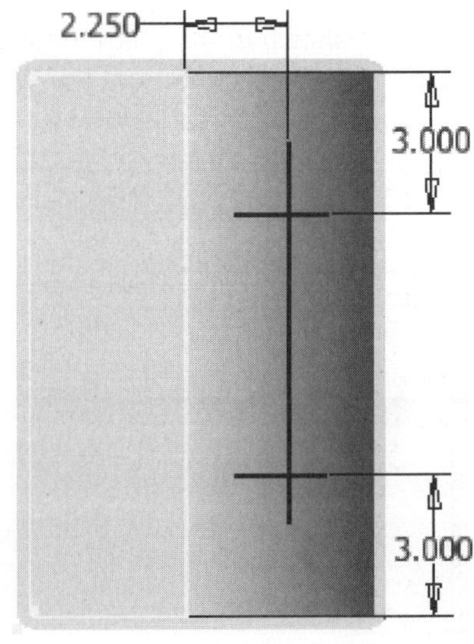

FIGURE 7.13

15. Click the Finish Sketch command on the Exit panel.
16. Click the Rib command on the Create panel.
17. Select the three sketched line segments for the profile.
18. In the Rib dialog box, make the Web Type current, the Thickness **.5 in**, change the Extents to Finite with a distance of **.25 in**, as shown in the following image.

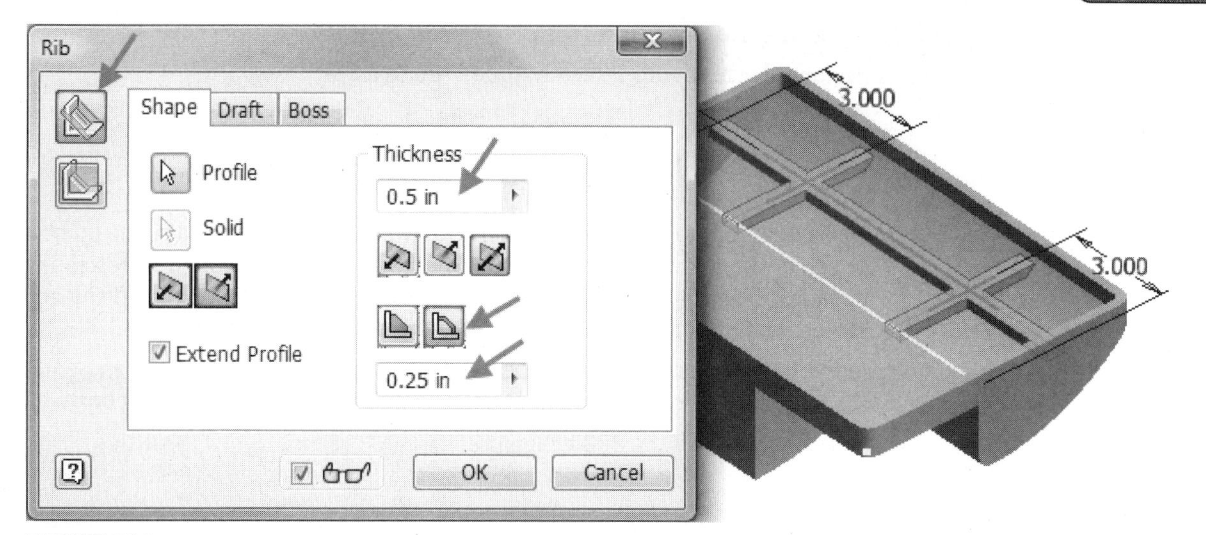

FIGURE 7.14

19. Click OK in the Rib dialog box, and your rib network should resemble the following image.

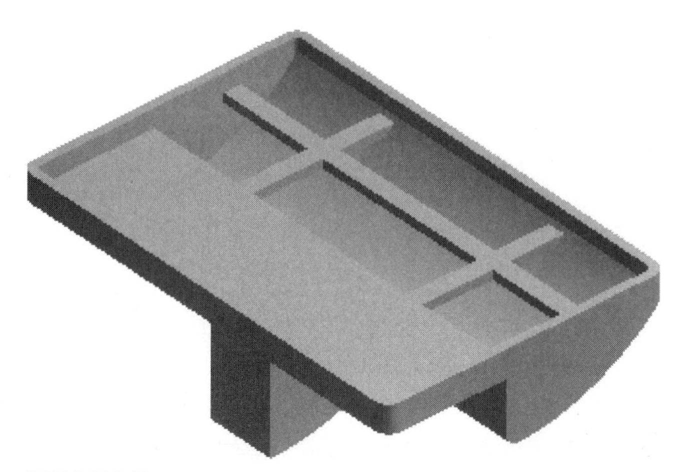

FIGURE 7.15

20. Use the Zoom and Free Orbit commands to examine the rib network.
21. Close the file. Do not save changes. End of exercise.

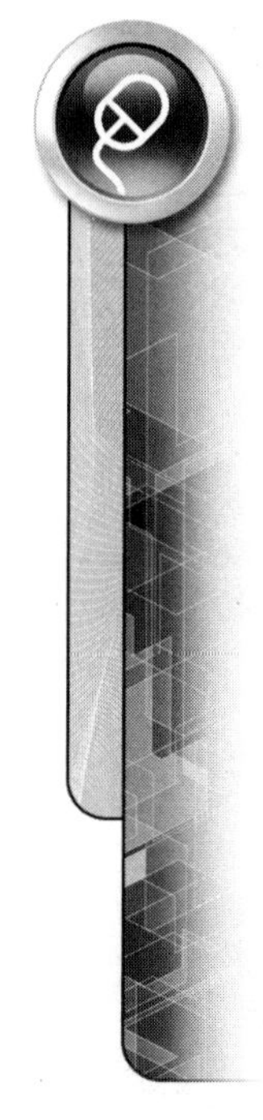

EMBOSSED TEXT AND CLOSED PROFILES

To better define a part, you may need to have a shape or text either embossed (raised) or engraved (cut) into a model. In this section, you will learn how you can emboss or engrave a closed shape or text onto a planar or curved face. You can define a shape using the sketch commands on the 2D Sketch panel bar. The shape needs to be closed. There are two steps to emboss or engrave: first create the shape or text. Second, you emboss the shape or text onto the part. The text can be oriented in a rectangle or about an arc, circle, or line. The following sections describe the steps.

Step 1: Creating Rectangular Text

To place rectangular text onto a sketch, follow these steps:

1. Create a sketch.
2. Click the Text command on the Sketch tab > Draw panel, as shown in the following image on the left.
3. Pick a point or drag a rectangle where you will place the text. If a single point is selected, the text will fit on a single line and can be grip edited to resize the bounding box. If a rectangle is used, it defines the width for text wrapping and can be modified. The Format Text dialog box will appear.
4. In the Format Text dialog box, specify the text font and format style, and enter the text to place on the sketch. The following image shows the dialog box with text entered in the bottom pane.

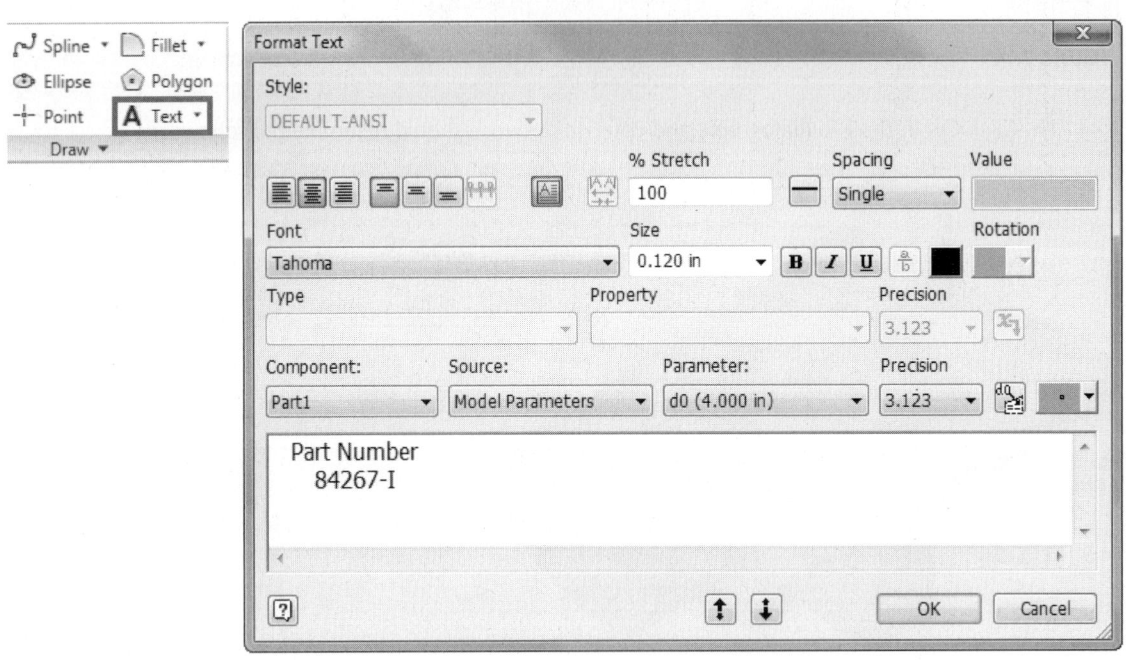

FIGURE 7.16

5. When done typing text, click OK.
6. When the text is placed, a rectangular set of construction lines defines the perimeter of the text. You can click and drag a point on the rectangle and change the size of the rectangle. Dimensions or constraints can be added to these construction lines to refine the text's location and orientation, as shown in the following image.
7. To edit the text, move the cursor over the text in the graphics window, right-click, and select Edit Text from the menu. The same Format Text dialog box that was used to create the text will appear.

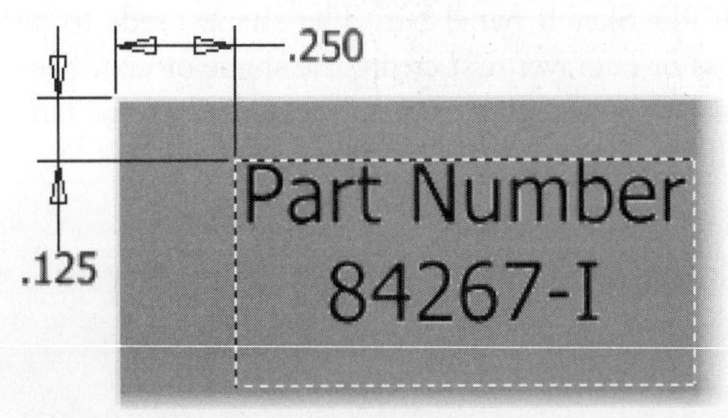

FIGURE 7.17

Step 2: Creating Text about Geometry

To place text about an arc, circle, or line in a sketch, follow these steps:

1. Make a sketch active.
2. Sketch and constrain an arc, circle, or line that the text will follow; it is recommended to make this geometry construction geometry as it will not be used as the geometry for the profile.
3. Click the Geometry Text command on the Sketch tab > Draw panel > Geometry Text command as shown in the following image on the left. The command may be under the Text command.
4. In the sketch, select the arc, circle, or line and the Geometry-Text dialog box will appear.
5. In the Geometry-Text dialog box, specify the text font and format style, direction, position, and start angle and enter the text to place on the sketch. The start angle is relative to the left quadrant point of a circle or the start point of the arc. The following image on the right shows the dialog box with text entered in the bottom pane.

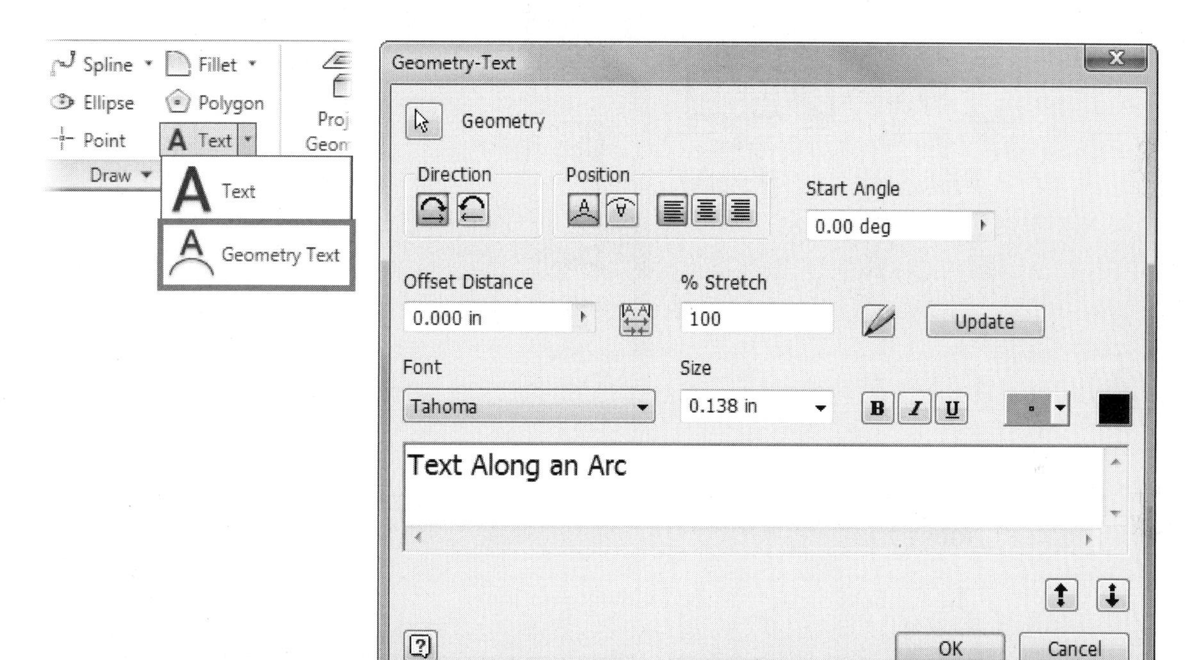

FIGURE 7.18

6. When done entering text, click the Update button in the dialog box to see the preview in the graphics window.
7. Click OK to complete the operation. The following image shows text placed along an arc.
8. To edit the text, move the cursor over the text in the graphics window, right-click, and click Edit Geometry Text from the menu. The same Format Text dialog box that was used to create the text will appear.

FIGURE 7.19

Step 3: Embossing Text

To emboss text or a closed profile, click the Emboss command on the Model tab > Create panel, as shown in the following image on the left. The Emboss dialog box will appear, as shown in the following image on the right.

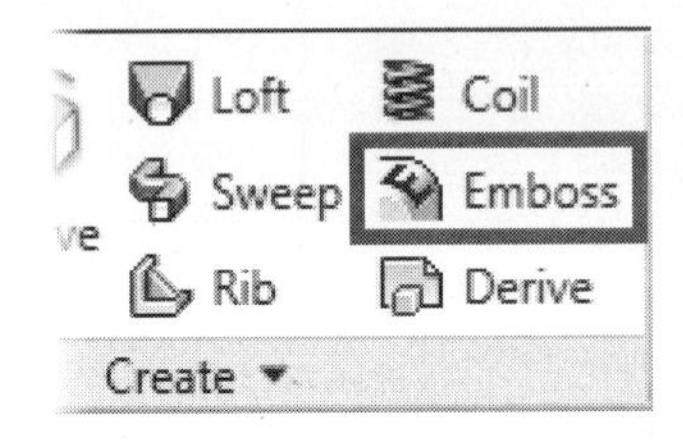

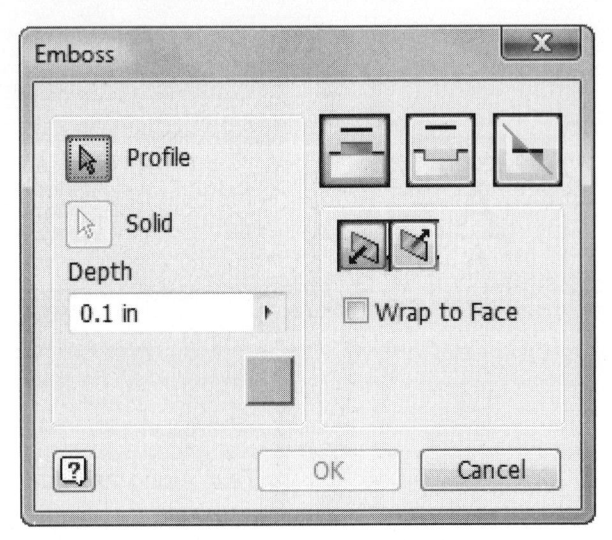

FIGURE 7.20

The Emboss dialog has the following options:

	Profile	Select a profile, meaning closed shape or text, to emboss. You may need to use the Select Other command to select the text.
1 mm	Depth	Enter an offset depth to emboss or engrave the profile if you selected the Emboss from Face or Engrave from Face types.
	Top Face Color	Select a color from the drop-down list to define the color of the top face of the embossed area, not its lateral sides.
	Emboss from Face	Select this option to add material to the part.
	Engrave from Face	Select this option to remove material to the part.

(Continued)

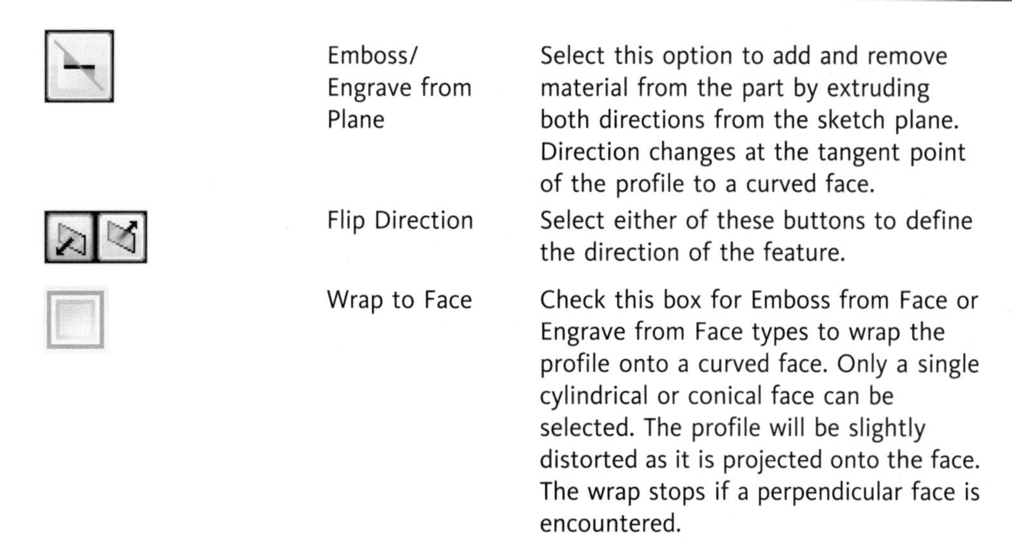

	Emboss/ Engrave from Plane	Select this option to add and remove material from the part by extruding both directions from the sketch plane. Direction changes at the tangent point of the profile to a curved face.
	Flip Direction	Select either of these buttons to define the direction of the feature.
	Wrap to Face	Check this box for Emboss from Face or Engrave from Face types to wrap the profile onto a curved face. Only a single cylindrical or conical face can be selected. The profile will be slightly distorted as it is projected onto the face. The wrap stops if a perpendicular face is encountered.

To emboss a closed shape or text, follow these steps:

1. Click the Emboss command on the Create panel.
2. Define the profile by selecting a closed shape or text. If needed, use the Select Other command.
3. Select the type of emboss: Emboss from Face, Engrave from Face, or Emboss/ Engrave from Plane.
4. Specify the depth, color, direction, and face as needed. If you selected Emboss/ Engrave from plane option, you can also add a taper angle to the created emboss/engrave feature.

NOTE

You edit the embossed feature like any other feature.

EXERCISE 7-2: CREATING TEXT AND EMBOSS FEATURES

In this exercise, you emboss and engrave sketched profile objects on faces of a razor handle model. You then create a sketch text object and engrave it on the handle.

1. Open *ESS_E07_02.ipt* in the Chapter 07 folder.
2. Use the Free Orbit and Zoom commands to examine the part.
3. From the browser turn on visibility of Sketch11 by right-clicking on Sketch11 and click Visibility from the menu and rotate the view to display the top triangular face, as shown in the following image.

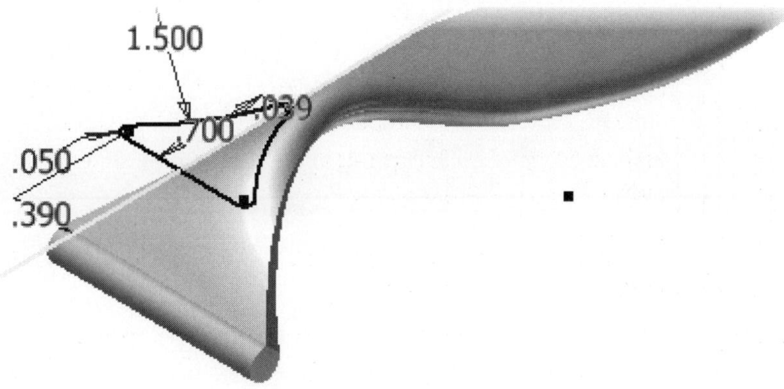

FIGURE 7.21

4. Click the Emboss command on the Create panel to engrave a closed sketch profile. Notice that the only visible closed profile is automatically selected.
   - In the Emboss dialog box, click the Engrave from Face option (middle button).
   - Change the Depth to **.03125 in**.
   - Click the Top Face Color button, and choose Aluminum (Flat) from the Color dialog box drop-down list, as shown in the following image on the left.
5. Click OK twice to close both dialog boxes and create the engraved feature, as shown in the following image on the right.

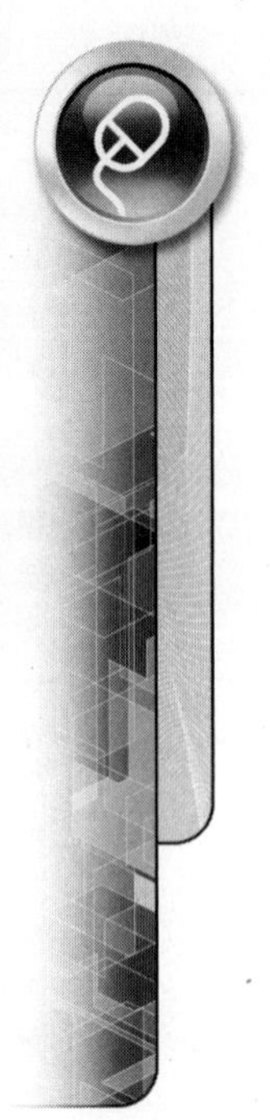

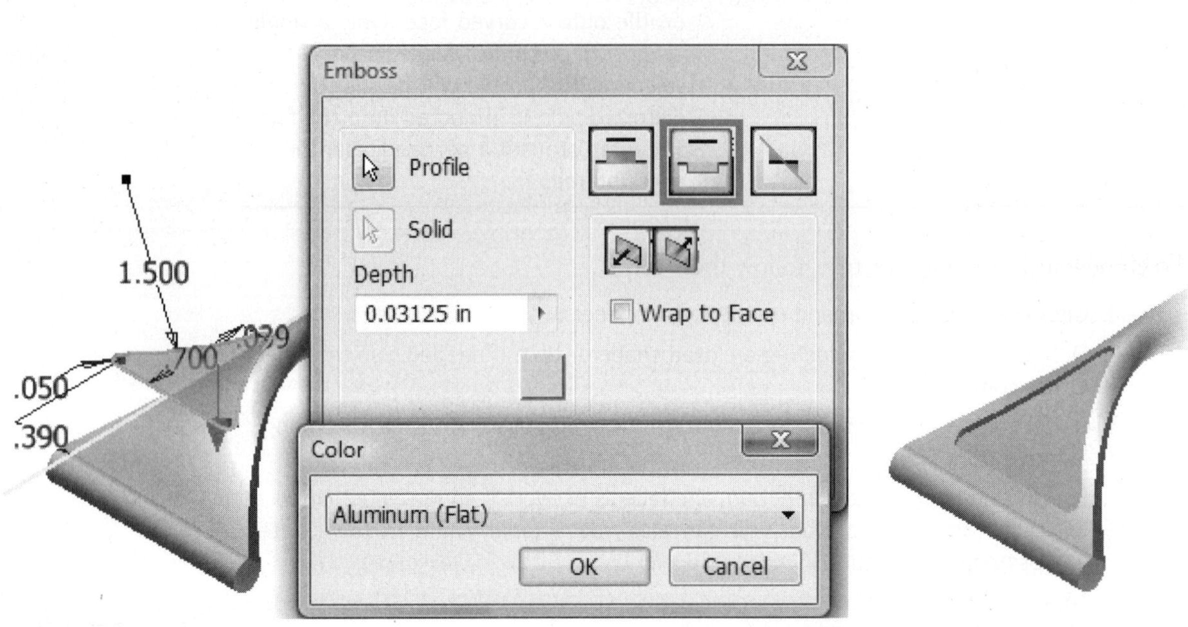

FIGURE 7.22

6. Change to the Home View, and from the browser turn on the visibility of Sketch10.
7. Click the Emboss command from the Create panel.
   - For the profile, click the oblong and the six closed profiles. Be sure to select the left and right halves of the herringbone profiles, as shown in the following image on the left.
   - In the Emboss dialog box, click the Emboss from Face option (left button).
   - Change the Depth to **.015625 in**.
   - Click the Top Face Color button, and choose Black from the Color dialog box drop-down list, as shown in the following image on the right.

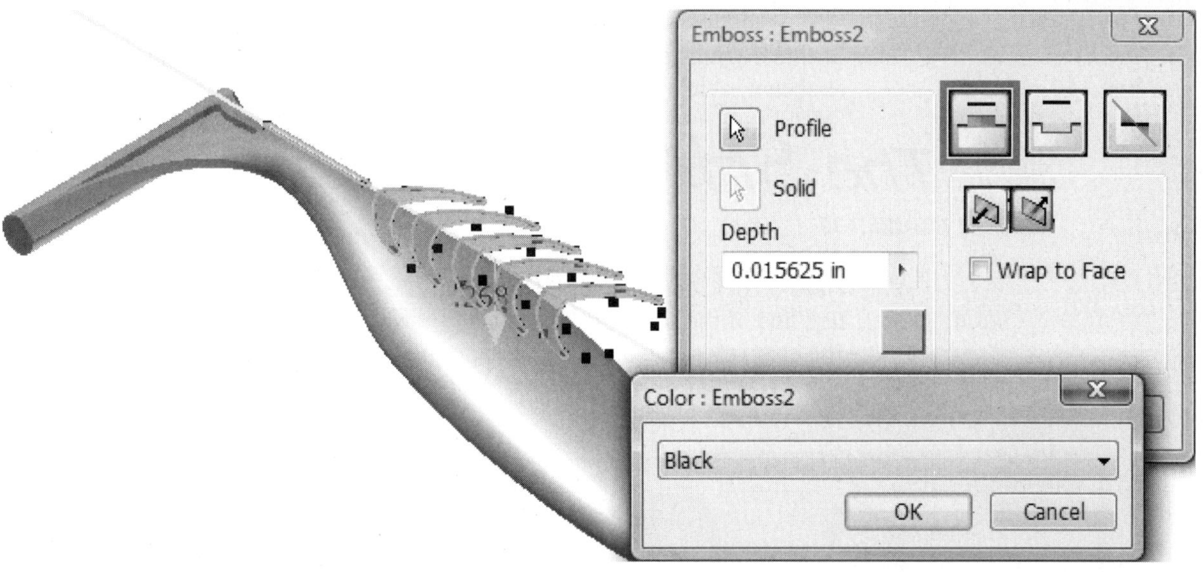

FIGURE 7.23

8. Click OK twice to close both dialog boxes and create the emboss feature.
9. Use the Orbit command to examine the engraved and embossed features.
10. Next you create a text sketch object. From the browser turn on the visibility of Sketch9.
11. In the browser, double-click Sketch9 to make it the active sketch.
12. Click the Text command in the Draw panel.
    - Click in an open area below the part and under the left edge of the construction rectangle in the sketch to specify the insertion point and display the Format Text dialog box.
    - Click the Center and Middle Justification buttons.
    - Click the Italic option.
    - Change the % Stretch value to **120**.
    - Click in the text field, and type The SHARP EDGE.
    - In the text field, double-click on the word "The," and change the text size from 1.20 in to **.09 in**, as shown in the following image.

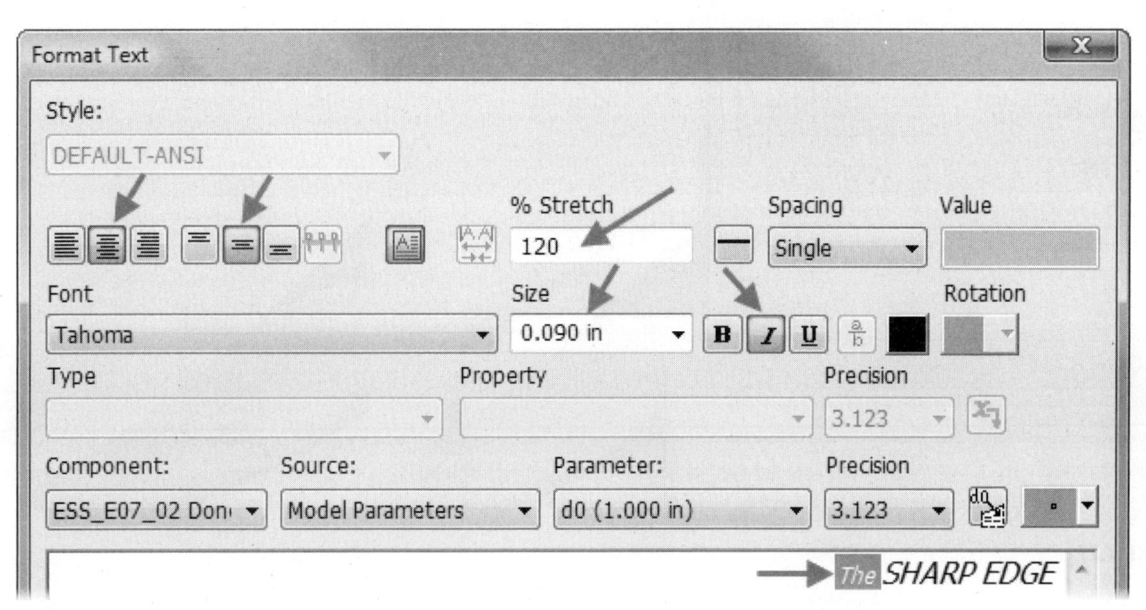

FIGURE 7.24

13. Click OK to create the text.
14. Draw a diagonal construction line coincident with the text bounding box corners, as shown in the following image.

The SHARP EDGE

FIGURE 7.25

15. Place a coincident constraint between the midpoint of the diagonal construction line of the text and Sketch9, as shown in the following image on the left. The completed operation is shown in the following image on the right.

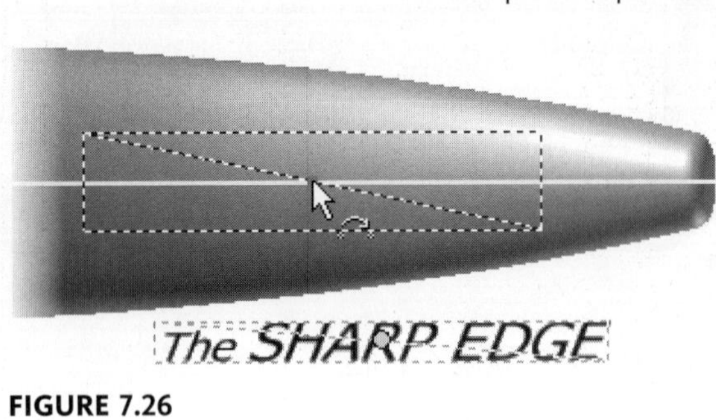

FIGURE 7.26

16. Click the Finish Sketch command to finish editing the sketch.
17. Click the Emboss command.
    - For the profile, select the text object.
    - In the Emboss dialog box, click the Engrave from Face option.
    - Change the Depth to **.015625 in**.
    - For the direction, click the right button to change the direction down.
    - Click the Top Face Color button, and click Black from the Color drop-down list as shown in the following image on the left.
18. Click OK twice to close both dialog boxes and create the emboss feature.
19. Use the Free Orbit command to examine the engraved text feature as shown in the following image on the right.

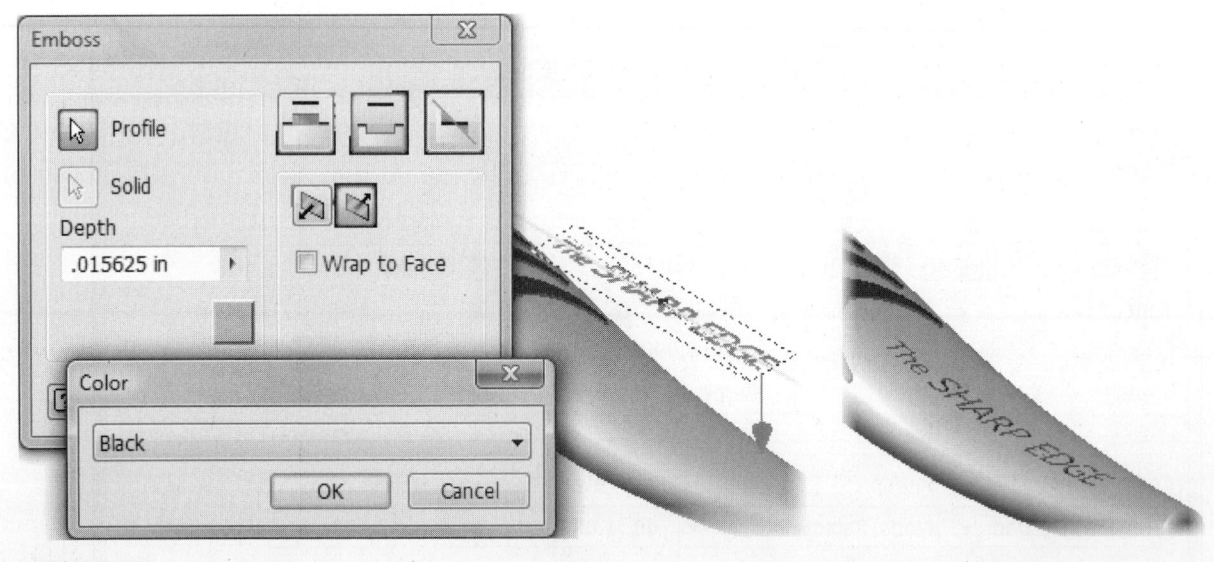

FIGURE 7.27

20. Delete the Emboss3 feature you just created.
21. Edit Sketch9 by double-clicking on it in the browser.
22. In the sketch, delete the existing text, rectangle and angled lines; do NOT delete the yellow horizontal line in the middle of the part (your color may be different depending upon your color scheme).
23. Sketch and dimension a construction circle as shown.

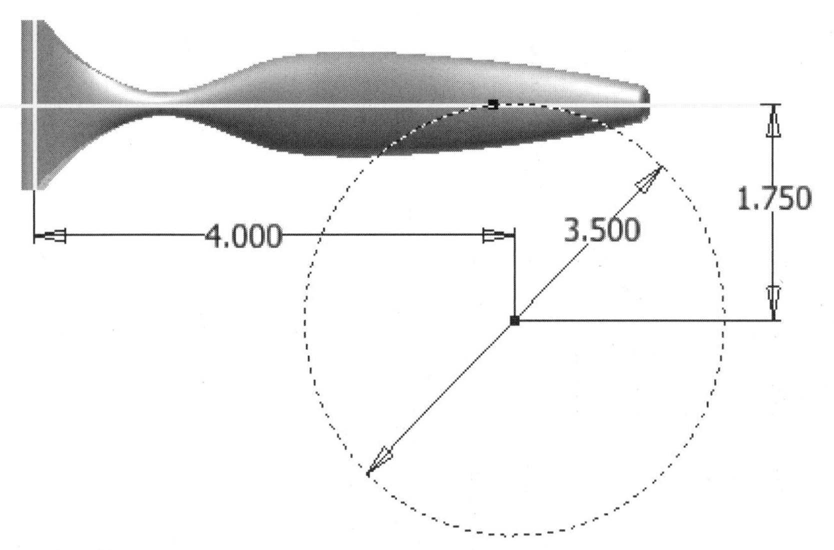

FIGURE 7.28

24. Click the Geometry Text command and select the circle for the Geometry selection.
    - Click the Center Justification button (middle button).
    - Change the Start Angle to **90.00 deg**.
    - Check the Bold button.
    - Change the text height to **.125 in**.
    - Click in the text field, and enter **The SHARP EDGE** as shown in the following image on the left (do not press Enter, the words in your dialog box may be on a single line).
    - To preview the text on the geometry, click the Update button in the Geometry-Text dialog box as shown in the following image on the right.

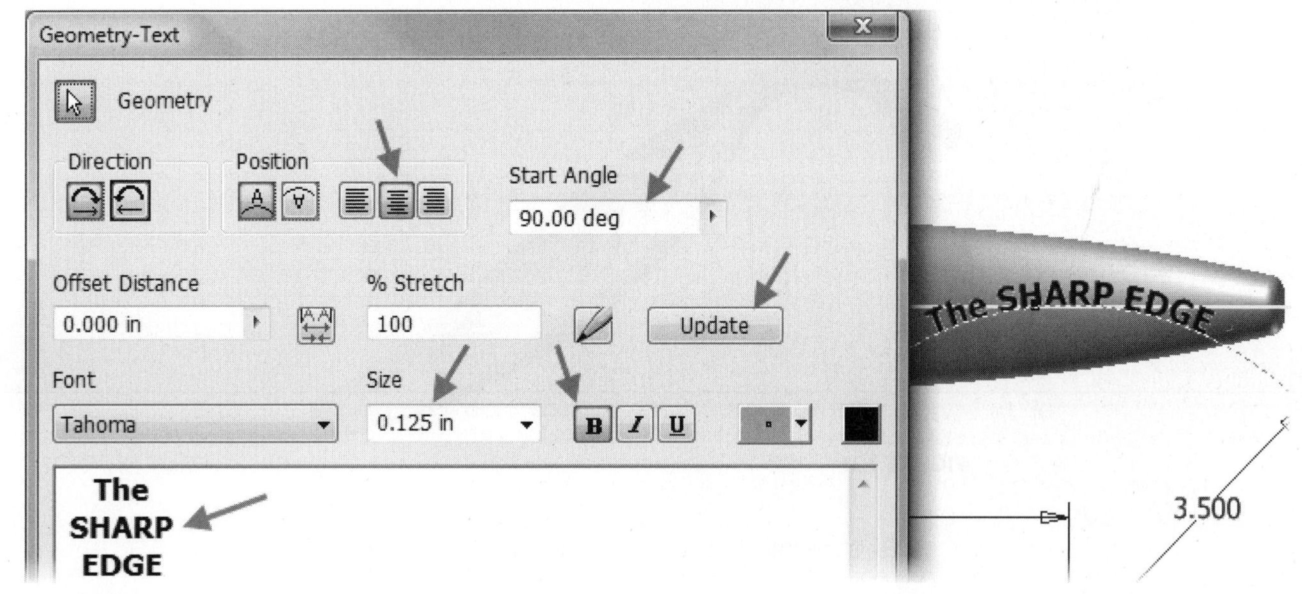

FIGURE 7.29

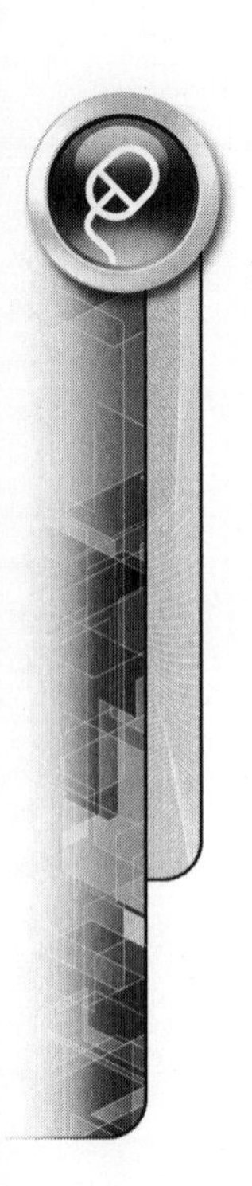

- Click OK to complete the command.

25. Click the Finish Sketch command to finish editing the sketch, and then change to the Home View.
26. Click the Emboss command.
    - For the profile, select the text object.
    - In the Emboss dialog box, click the Engrave from Face option.
    - Flip the direction so it points down.
    - Change the Depth to **.015625 in**.
    - Click the Top Face Color button, and click Black from the Color drop-down list.
27. Click OK twice to close both dialog boxes and create the emboss feature.

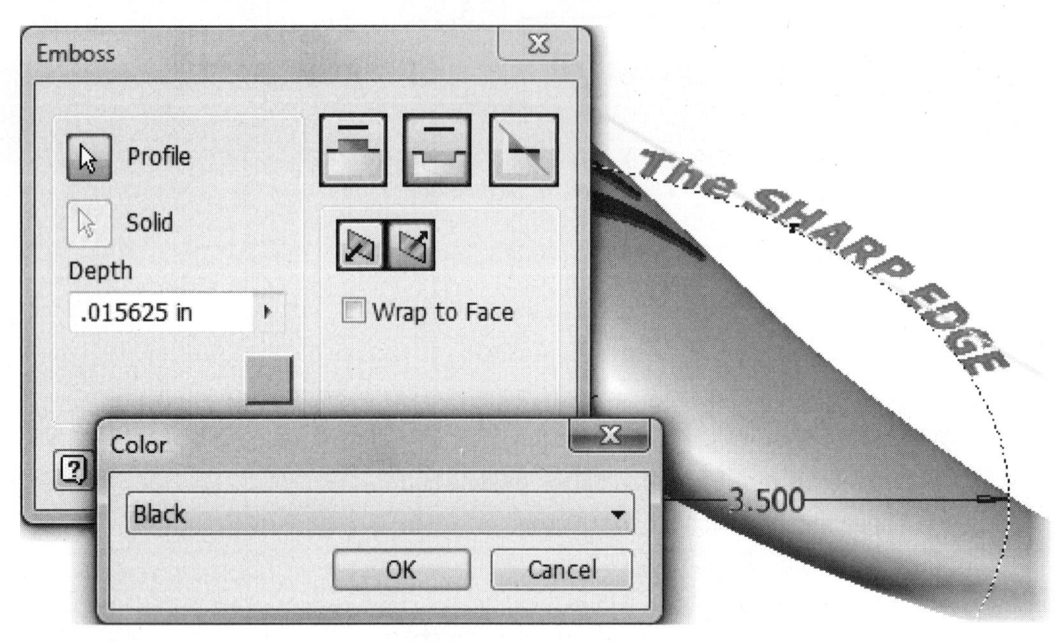

FIGURE 7.30

28. Use the Free Orbit command to examine the emboss features as shown in the following image.

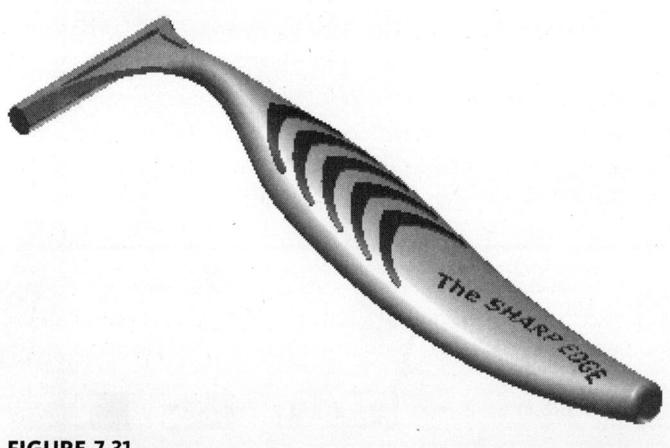

FIGURE 7.31

29. Close the file. Do not save changes. End of exercise.

SWEEP FEATURES

Unlike other sketched features, a sweep feature requires two unconsumed sketches: a profile to be swept and a path that the profile will follow. Additional profiles can be used as guide rails or a surface can also be used to help shape the feature. The profile sketch and the path sketch cannot lie on the same or parallel planes. The path can be an irregular shape or can be based on a part edge by projecting and including the edges onto the active sketch. The path can be either an open or closed profile and can lie in a plane or lie in multiple planes (3D Sketch). Handles, cabling, and piping are examples of sweep features. A sweep feature can be a base or a secondary feature. To create a sweep feature, use the Sweep command on the Model tab > Create panel, as shown in the following image on the left. The Sweep dialog will appear, as shown in the following image on the right. The following list includes descriptions of the options in the Sweep dialog box.

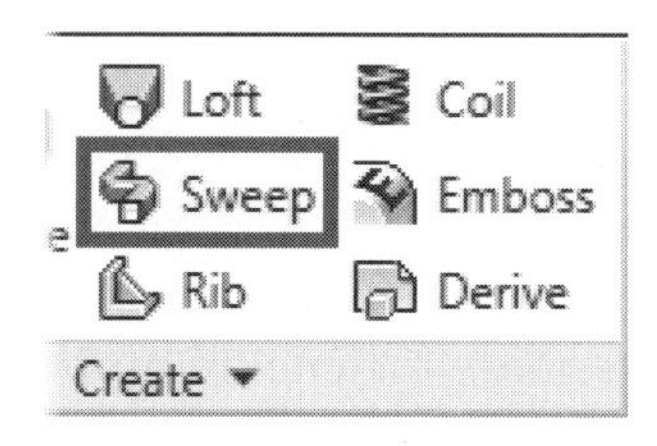

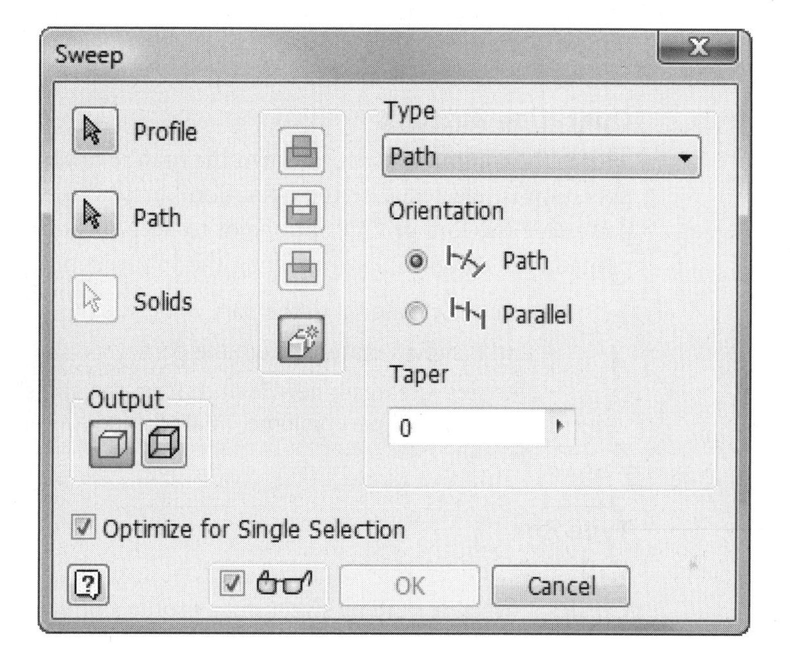

FIGURE 7.32

Profile

Click this button to select the sketch profile to sweep. If the Profile button is depressed and red, it means that you need to select a sketch or sketch area. If there are multiple closed profiles, you will need to select the profile that you want to sweep. If there is only one possible profile, Autodesk Inventor will select it for you and you can skip this step. If you selected the wrong profile or sketch area, depress the Profile button, and deselect the incorrect sketch by clicking it while holding down the CTRL key. Release the CTRL key, and select the desired sketch profile.

Path

Click this button to select the path along which to sweep the profile. The path can be an open or a closed profile but must pierce the sketch that the profile was drawn on. You can also use edge(s) of a part as a path. The profile is typically perpendicular to and intersects with the start point of the path. The start point of the path is often projected into the profile sketch to provide a reference point.

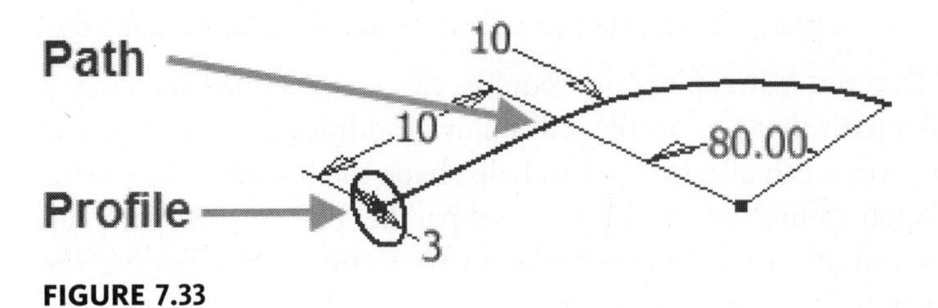

FIGURE 7.33

Solids

If there are multiple solid bodies, click this button to choose the solid body(ies) to participate in the operation.

Output Buttons

In the Output section, click Solid to create a solid feature or Surface to create the feature as a surface.

Operation Buttons

This is the column of buttons down the middle of the dialog box. By default, the Join operation is selected. Use the operation buttons to add or remove material from the part using the Join or Cut options or to keep what is common between the existing part and the completed sweep using the Intersect option.

- *Join:* Adds material to the part.
- *Cut:* Removes material from the part.
- *Intersect:* Creates a new feature from the shared volume of the sweep feature and existing part volume. Material not included in the shared volume is deleted.

Type Box

Path

Creates a sweep feature by sweeping a profile along a path.

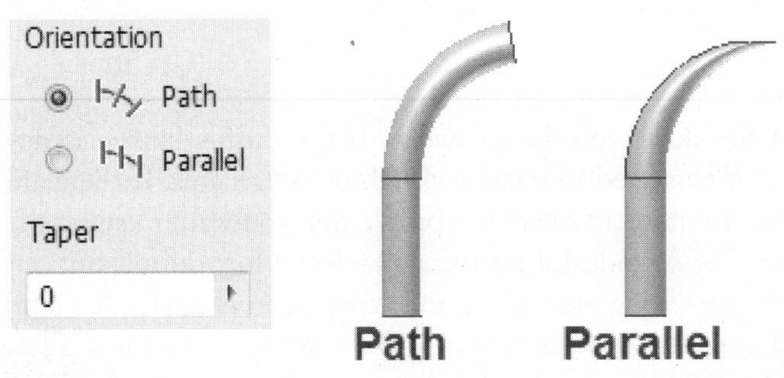

FIGURE 7.34

Orientation

Path

Holds the swept profile constant to the sweep path. All sweep sections maintain the original profile relationship to the path, as shown in the previous image.

Parallel

Holds the swept profile parallel to the original profile as shown in the previous image.

Taper Enter a value for the angle you want the profile to be drafted. By default, the taper angle is 0, as shown in the previous image.

Path & Guide Rail

Creates a sweep feature by sweeping a profile along a path and uses a guide rail to control scale and twist of the swept profile. The following image shows the options for the path and guide rail type.

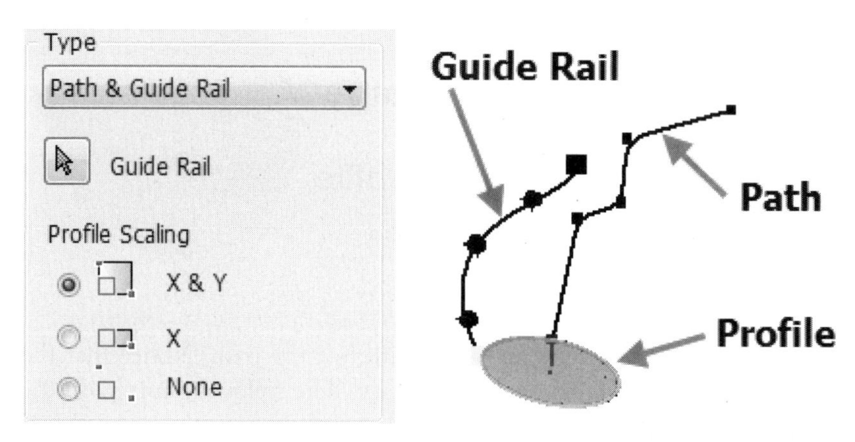

FIGURE 7.35

Guide Rail. Select a guide curve or rail that controls the scaling and twist of the swept profile. The guide rail must touch the profile plane. If you project an edge to position the guide rail and the projected edge is not the path, change the projected edge to a construction line.

Profile Scaling. Specify how the swept section scales to meet the guide rail. The following image shows the path and the profile and guide rail from the previous image with the three different profile scaling options.

X and Y. Scales the profile in both the X and Y directions as the sweep progresses.

X. Scales the profile in the X direction as the sweep progresses. The profile is not scaled in the Y direction.

None. Keeps the profile at a constant shape and size as the sweep progresses. Using this option, the rail controls only the profile twist and is not scaled in the X or Y direction.

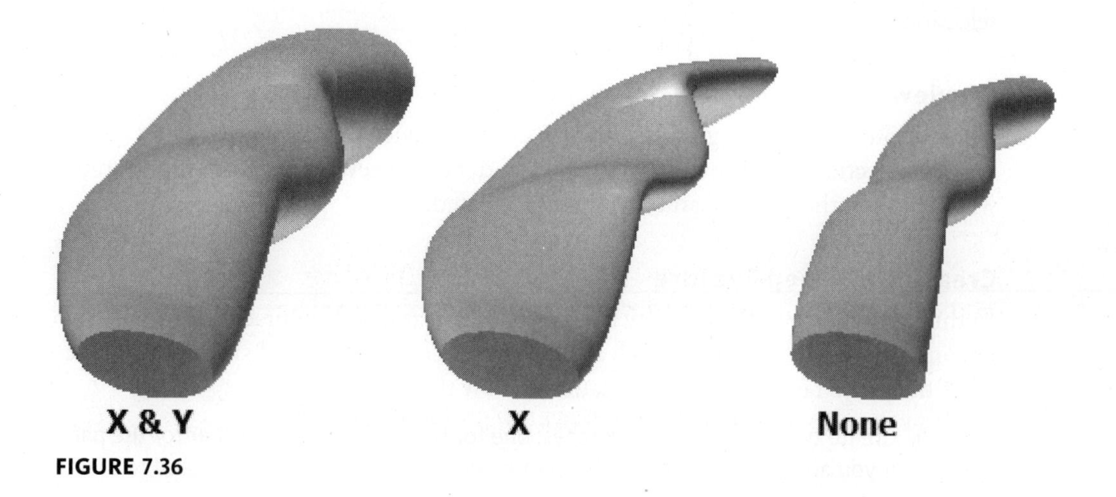

FIGURE 7.36

Path & Guide Surface

Creates a sweep feature by sweeping a profile along a path and a guide surface. The guide surface controls the twist of the swept profile. For best results, the path should touch or be near the guide surface.

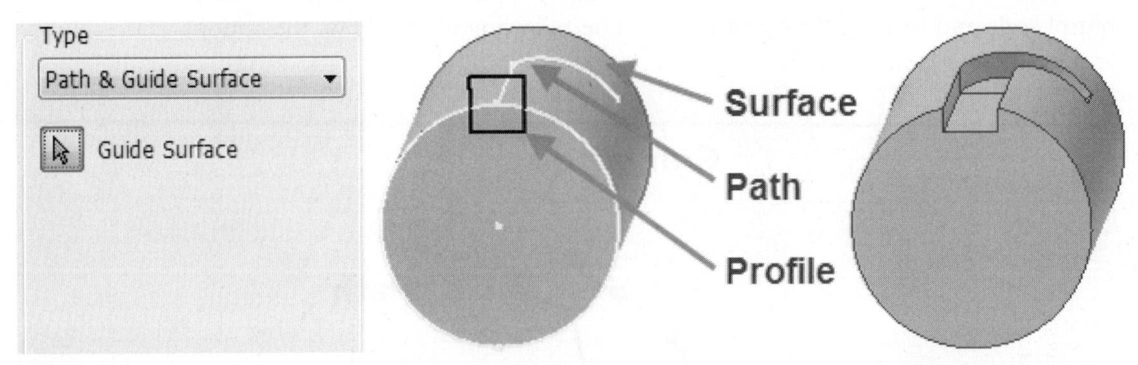

FIGURE 7.37

The following image on the left shows a sweep with just the path. Notice that the back inside edge of the cut is angled from the surface. The following image on the right shows the same sweep with a guide surface. Notice that the back inside edge is parallel to the top of the surface.

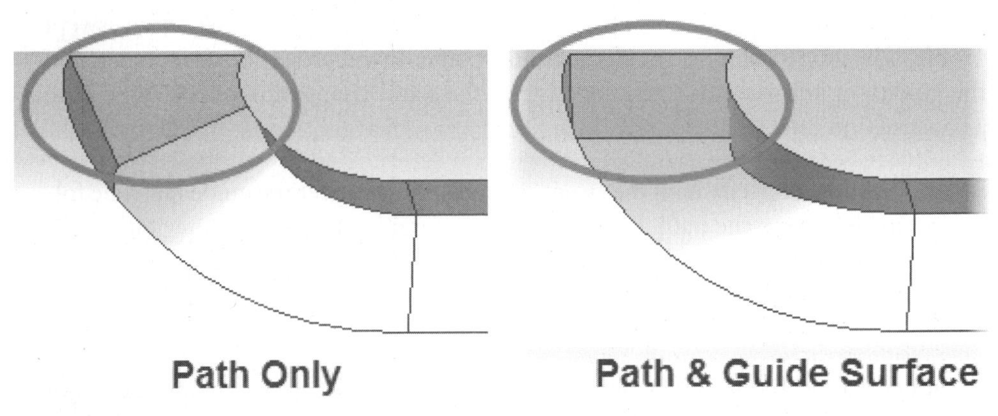

FIGURE 7.38

Optimize for Single Selection

When this option is checked, Autodesk Inventor automatically advances to the next selection option once a single selection is made. Clear the checkbox when multiple selections are required.

Preview

The button with the eyeglasses provides a solid preview of the sweep based on the current selections. If Preview is enabled and no preview appears in the graphics window, then the sweep feature will not be created.

Creating a Sweep Feature

In this section, you will learn how to create sweep features. You first need to have two unconsumed sketches. One sketch will be swept along the second sketch that represents the path. To create a sweep feature, follow these steps:

1. Create two unconsumed sketches: one for the profile and the other for the path. If you are going to use existing part edges for the path you don't need to create

a sketch for the path. The profile and path must lie on separate nonparallel planes. It is recommended that the profile intersect the path. Use work planes to place the location of the sketches, if required. The sketch that you use for the path can be open or closed. Add dimensions and constraints to both sketches as needed. If needed, create a sketch that will be used as a rail, or create a surface to be used as a guide surface.

2. Click the Sweep command on the Model tab > Create panel.
3. The Sweep dialog box will appear. If two unconsumed sketches do not exist, Autodesk Inventor will notify you that two unconsumed sketches are required.
4. Click the Profile button, and then select the sketch that will be swept in the graphics window. If only one closed profile exists, this step is automated for you.
5. If it is not already depressed, click the Path button, and then select the sketch to be used as the path in the graphics window.
6. Select the type of sweep you are creating: Path, Path & Guide Rail, or Path & Guide Surface.
7. Define whether or not the resulting sweep will create a solid or a surface by clicking either the Solid or the Surface button in the Output area. Select the required options, as outlined in the previous descriptions.
8. If this is a secondary feature, click the operation that will specify whether material will be added or removed or if what is common between the existing part and the new sweep feature will be kept.
9. If you want the sweep feature to have a taper, click on the More tab, and enter a value for the taper angle.
10. Click the OK button to complete the operation.

EXERCISE 7-3: CREATING SWEEP FEATURES

In the first part of this exercise, you create a component with the sweep command. Three sketches will be created, one each for the profile, the path, and the guide rail. In the second part of the exercise you sweep a profile using edges on a part as the path.

1. Open *ESS_E07_03_1.ipt* from the Chapter 07 folder.
2. Click the Sweep command on the Create panel. The profile and path are automatically selected since only one closed profile (ellipse) and one open sketch (arc) are visible. Note that the profile or path can be open or closed, but if it is closed, you will need to manually select it; if the profile is open, a surface will be created.
3. Change the Orientation from Path to Parallel, as shown in the following image. Notice how the profile changes orientation.

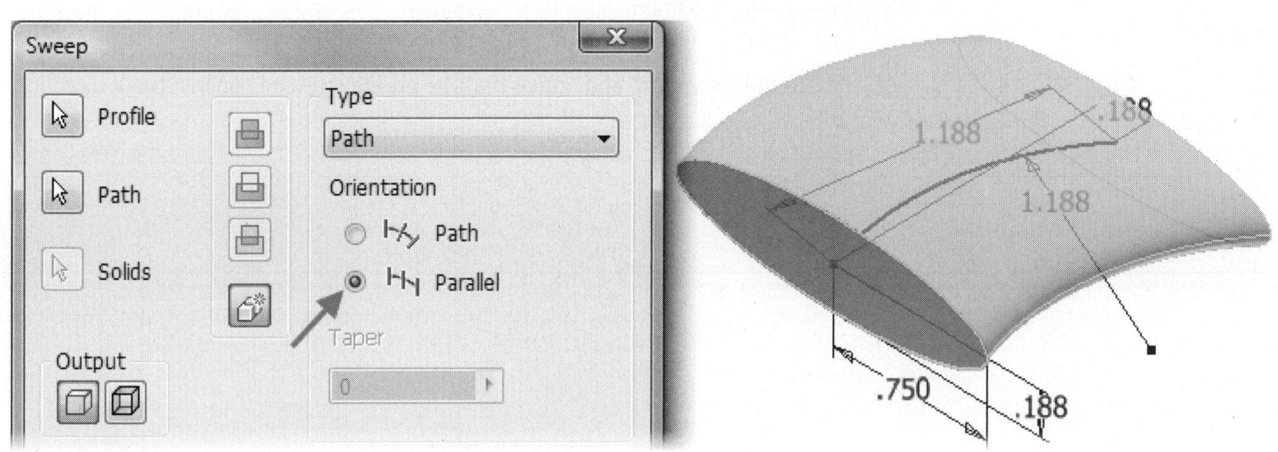

FIGURE 7.39

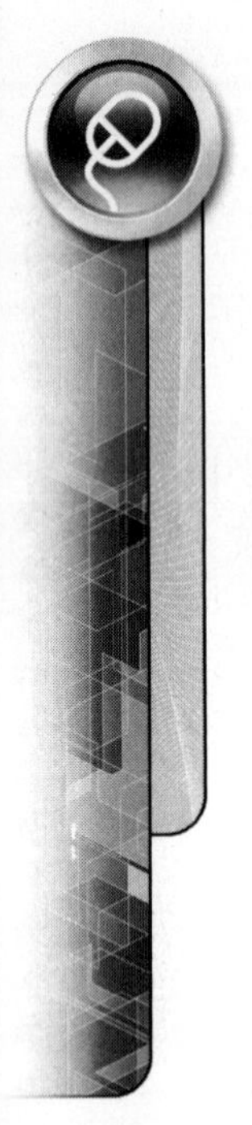

4. Click OK to create the sweep feature.
5. Turn on the visibility of the guide rail by moving the cursor over the entry Sketch Guide Rail in the browser. Right-click, and click Visibility from the menu.
6. Rotate the viewpoint, and notice that the front end of the guide rail is coincident to the profile. To better see the geometry, change the Visual Style to Wireframe from the Navigation Bar or from the View tab > Appearance panel > Visual Style.
7. Right-click in the graphics window, and click Home View from the menu.
8. Edit the sweep feature by moving the cursor over the "Sweep1" entry in the browser and double-click.
9. Change the type to Path & Guide Rail and, in the graphics window, select the spline for the Guide Rail, as shown in the following image.

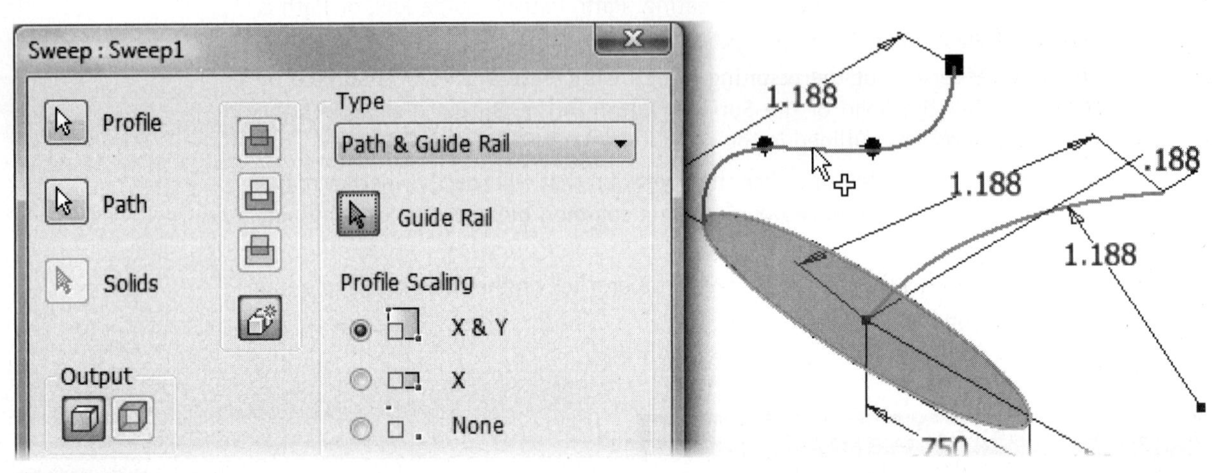

FIGURE 7.40

10. Notice that the Profile Scaling is set to X & Y. Click OK to create the part. Note that the sweep could have been completed as a Path & Guide Rail immediately.
11. Change the Visual Style back to Shaded. The following image on the left and in the middle shows the Profile Scaling set to X & Y.
12. Rotate the model, and notice that the profile stretches in both X and Y directions.
13. Right-click in the graphics window, and click Home View from the menu.
14. Edit the sweep feature by clicking on the model in the graphics window and click the Edit Sweep button from the mini-toolbar that is displayed in the graphics window.
15. Change the Profile Scaling to X, as shown in the following image on the right, and click OK to update the sweep feature.
16. Rotate the viewpoint, and notice that the profile stretches only in the X direction.

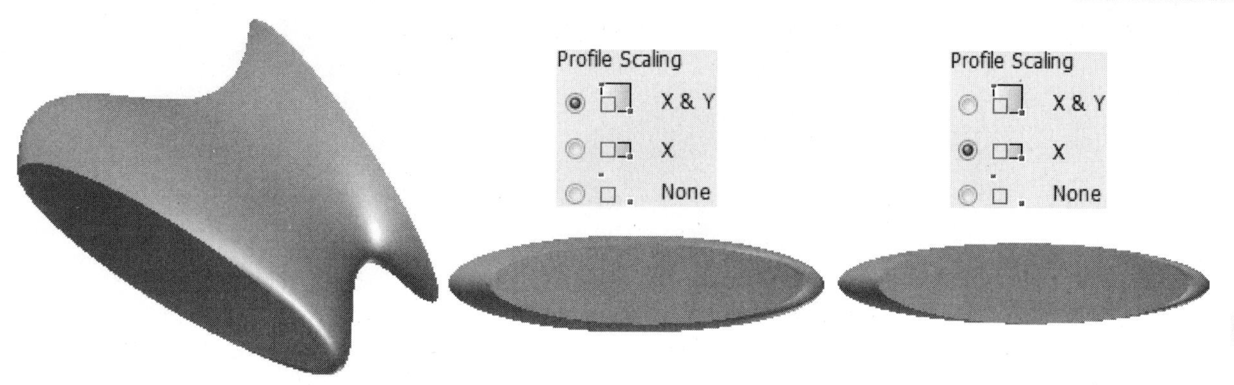

FIGURE 7.41

17. Close the file.
18. Open *ESS_E07_03_2.ipt* in the Chapter 07 folder.
19. Zoom in on the lower right-side of the part, as shown in the following image on the left.
20. Click on the front face of the part and click Create Sketch from the mini-toolbar as shown in the following image on the left.
21. The viewpoint will change so you are looking directly at the sketch. However, it is difficult to see where you are sketching. To return to your previous view, press the F5 key.
22. Sketch a circle on the projected point and add a **.0625 in** dimension, as shown in the following image on the right.

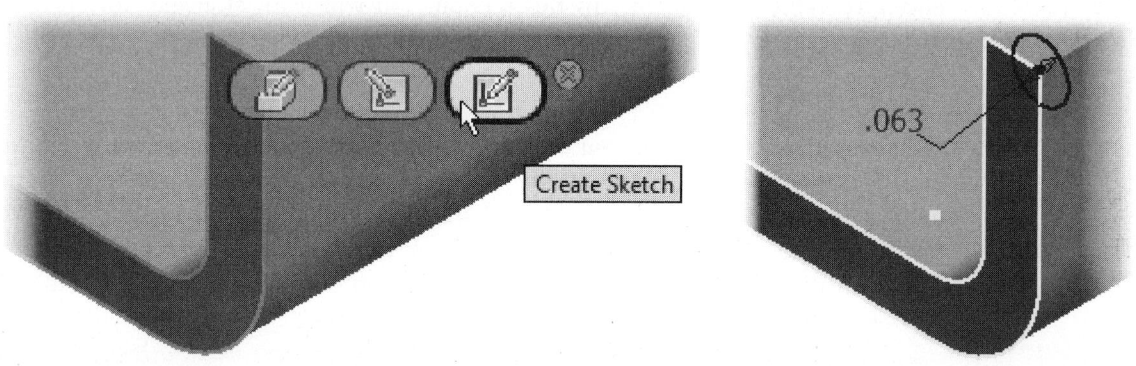

FIGURE 7.42

23. Click the Finish Sketch command on the Exit panel.
24. Next, you sweep the circle along edges of the part using the Sweep command. Click the Sweep command on the Create panel.
25. For the profile select inside the circle and the Path button will then be active. Click on a top-outside edge of the part, Inventor will automatically select all of the top-outside edges that go around the part, as shown in the following image on the left.
26. In the Sweep dialog box, click the Cut operation to remove material.
27. Click the OK button in the Sweep dialog box. The following image on the right shows the completed part.

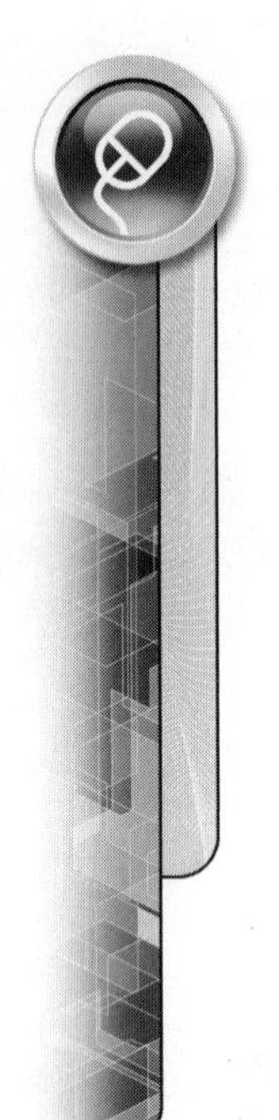

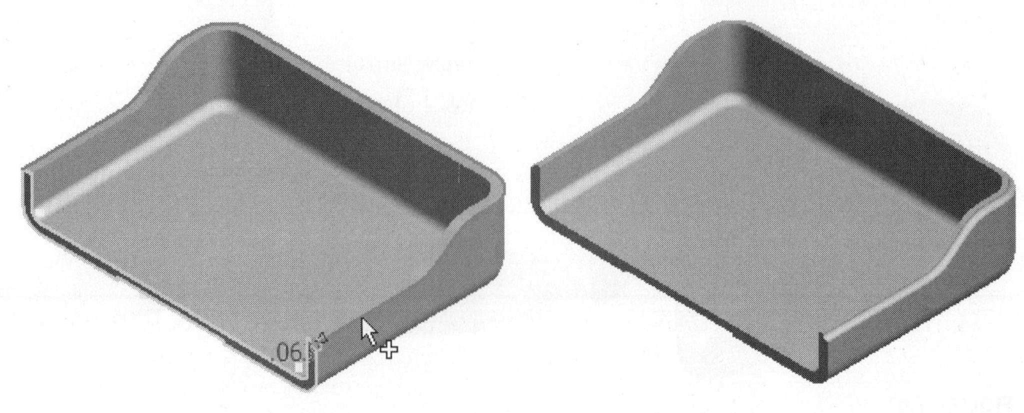

FIGURE 7.43

28. Use the Zoom and Free Orbit commands to examine the Sweep feature.
29. Close the file. Do not save changes. End of exercise.

3D Sketching

To create a sweep feature whose path does not lie on a single plane, you need to create a 3D sketch that will be used for the path. You can use a 3D sketch to define the path for a lip or to define the routing path for an assembly component, such as a pipe or duct work that crosses multiple faces on different planes. You need to define a 3D sketch in the part environment, and you can do this within an assembly or in its own part file. You can use the Autodesk Inventor adaptive technology during 3D sketch creation to create a path that updates automatically to reflect changes to referenced assembly components. In this section, you will learn strategies for creating 3D sketches.

3D Sketch Overview

When creating a 3D sketch, you use many of the sketching techniques that you have already learned with the addition of a few commands. 3D sketches use work points and model edges or vertices to define the shape of the 3D sketch by creating line or spline segments between them. You can also create bends between line segments. When creating a 3D sketch, you use a combination of lines, splines, fillet features, work features, constraints, and existing edges and vertices.

3D Sketch Environment

The 3D sketch environment is used to create 3D or a combination of both 2D and 3D curves. Before creating a 3D sketch, change the environment to the 3D sketch environment by clicking the Create 3D Sketch command on the Model tab > Sketch panel beneath the Create 2D Sketch command, as shown in the following image on the left. The commands on the ribbon will change to the 3D Sketch commands, as shown in the following image on the right. The most common 3D sketch commands are explained throughout this chapter. While in the 3D Sketch environment, all features appear in the browser with a 3D sketch name. Once a feature uses the 3D sketch, it will be consumed under the new feature in the browser.

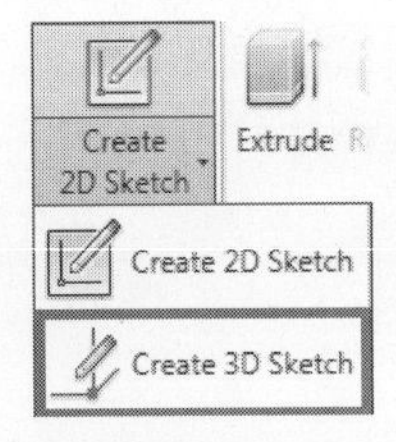

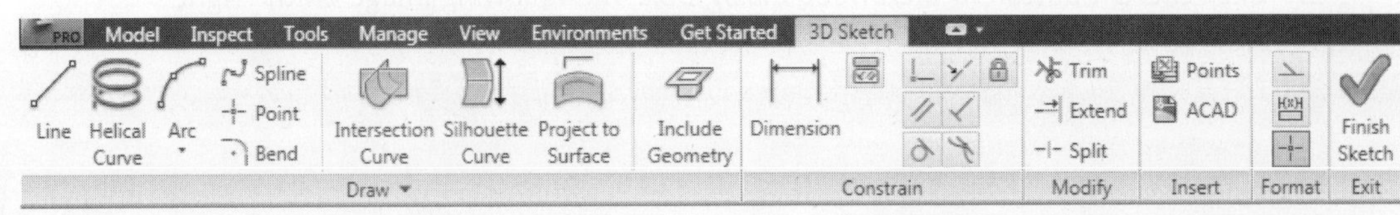

FIGURE 7.44

3D Path from Existing Geometry

One way to create a 3D path is to use existing geometry. If you want to create a lip on an existing part, you can use the Include Geometry command to select specific edges to define the path. You can include existing geometry by projecting part edges, vertices, and geometry from visible sketches into a 3D sketch. To use existing geometry to create a 3D path, follow these steps:

1. Create a 3D Sketch by clicking the Create 3D Sketch command on the Sketch panel under the Create 2D Sketch command. No dialog box will appear.
2. Click the Include Geometry command on the Draw panel, as shown in the following image on the left. This command allows you to project existing sketch geometry and existing part edges to a 3D sketch. If the original geometry or sketch changes, the projected geometry is updated to reflect changes to the original geometry.
3. Click each of the model edges that you want to use for the 3D sketch. When finished, right-click and select Done from the menu, or press the ESC key on the keyboard. If you click an incorrect edge, you can delete it from the sketch manually. The following image on the right shows an example of including outside edges of a part but not the top three edges in a 3D sketch.

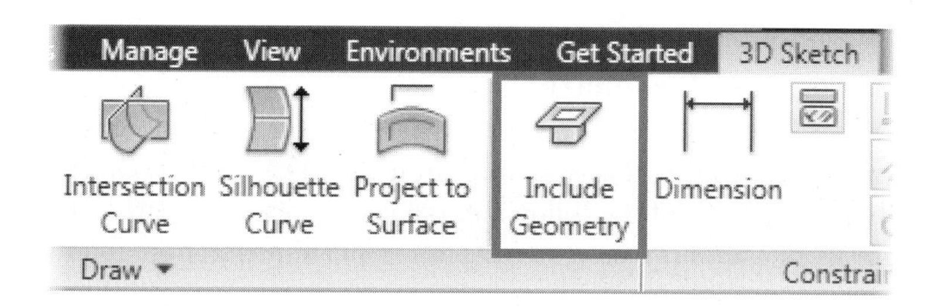

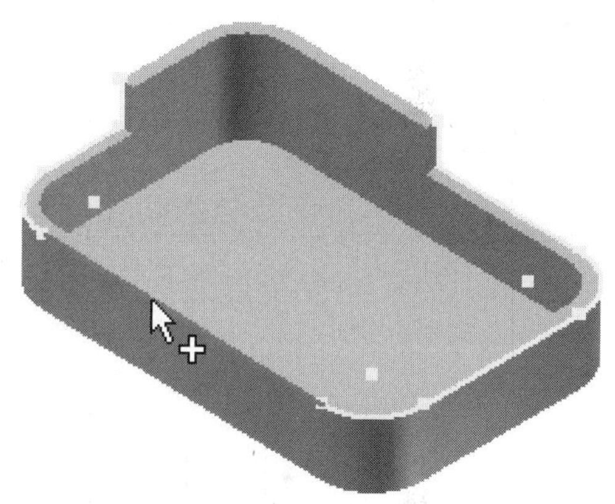

FIGURE 7.45

4. Exit the 3D sketch by clicking Finish Sketch on the Exit panel.
5. If a plane does not exist where you want to place the profile to be swept, create a work plane that defines the plane.
6. Use the Create 2D Sketch command to create a 2D sketch on an existing plane or work plane.
7. Sketch, constrain, and dimension the profile that will be swept. The following image on the left shows a sketch created, constrained, and dimensioned on the top planar face.
8. Click the Sweep command on the Create panel.
9. If the Profile button is not already depressed in the Sweep dialog box, click it, and then select the sketch that will be swept in the graphics window.
10. Click the Path button, and then select the 3D sketch to use as the path in the graphics window, you may need to use the Select Other tool to select the 3D Sketch.
11. Select the operation that will specify whether or not material will be added or removed or if what is common between the existing part and the new sweep feature will be kept.

12. Click the OK button to complete the operation. The following image on the right shows the completed part.

TIP

If you need to create a rectangular lip or a groove on a plastic part try the Lip command that is covered later in this chapter.

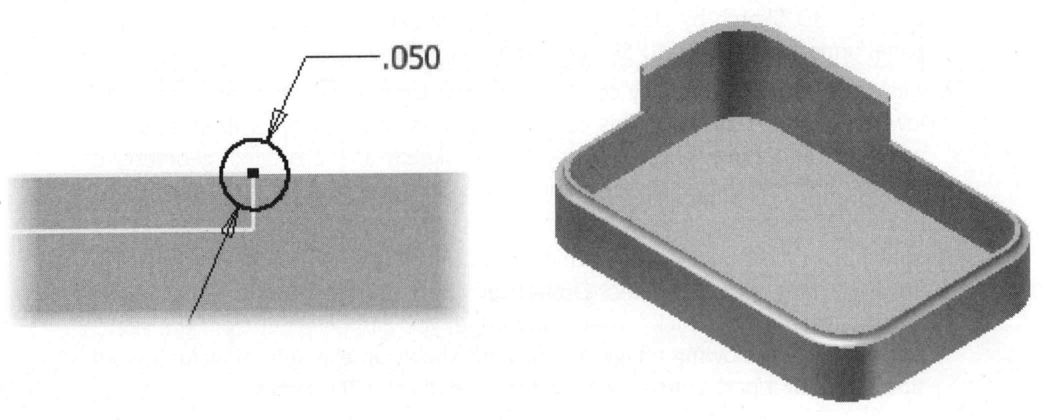

FIGURE 7.46

3D Sketch from Intersection Geometry

Another way to create a 3D path is to use geometry that intersects with the part. If the intersecting geometry defines the 3D path, you can use it. The intersection can be defined by a combination of any of the following: planar or nonplanar part faces, surface faces, a quilt, or work planes. To create a 3D path from an intersection, follow these steps:

1. Create the intersecting features that describe the desired path.
2. Change to the 3D Sketch environment by clicking the Create 3D Sketch command on the Sketch panel under the 2D Sketch command.
3. Click the Intersection Curve command, as shown in the following image on the left on the Draw panel.
4. The 3D Intersection Curve dialog box appears, as shown in the following image in the middle.
5. Select the two intersecting geometry.
6. To complete the operation, click OK.

The following image on the right shows a 3D path being created on a cylindrical face at the edge defined by a work plane that intersects it at an angle.

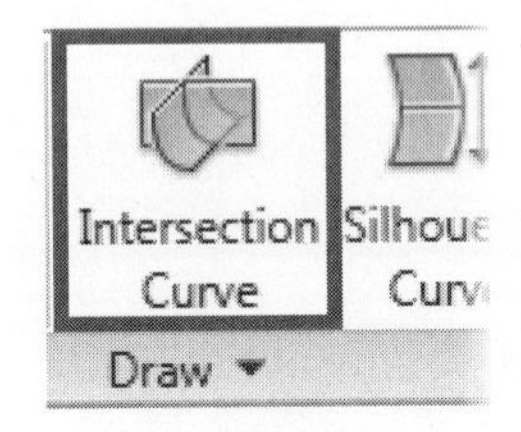

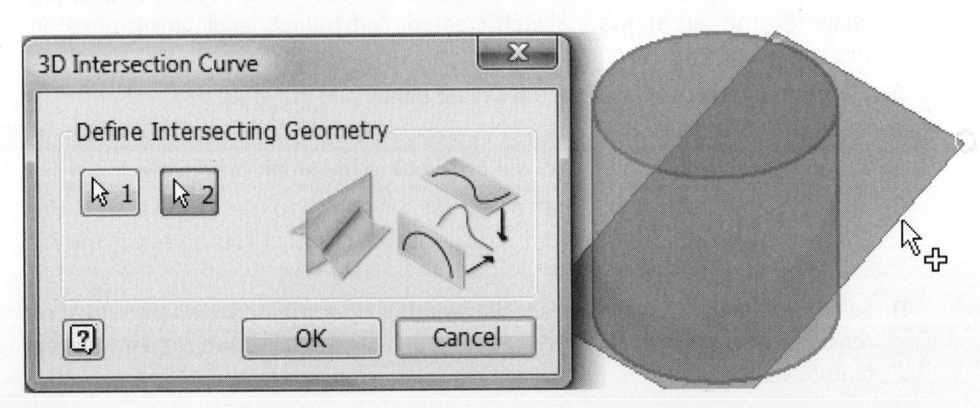

FIGURE 7.47

Project to Surface

While in a 3D sketch, you can project curves, 2D or 3D geometry, part edges, and points onto a face or onto selected faces of a surface or solid. To project curves onto a face, follow these steps:

1. Create the solid or surface onto which the curves will be projected.
2. Create the curves that will be projected onto the face(s).
3. Click the 3D Sketch command on the Sketch panel.
4. Click the Project to Surface command on the Draw panel, as shown in the following image on the left.
5. The Project Curve to Surface dialog box will appear, as shown in the following image in the middle. The Faces button will be active. In the graphics window, select the face(s) onto which the curves will be projected.
6. Click the Curves Button, and then in the graphics window, select the individual objects to project. The following image on the right shows the face and curves selected.

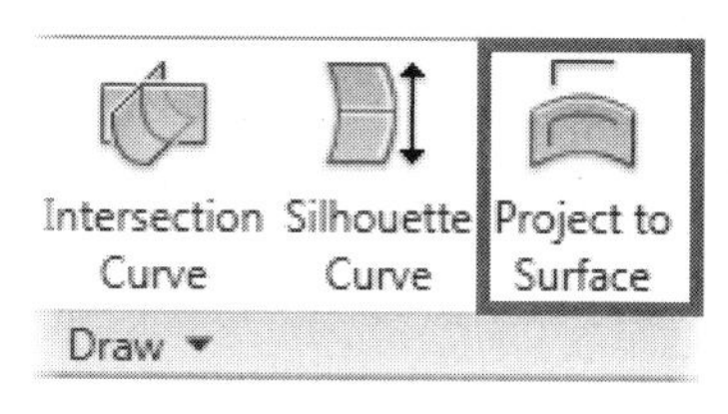

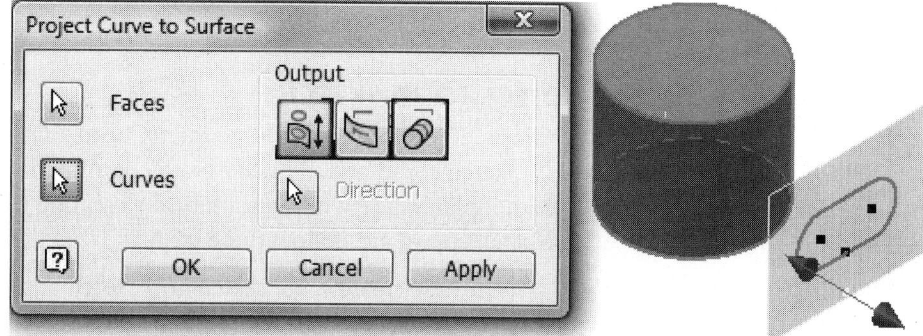

FIGURE 7.48

7. In the Output area, select one of the following options.

	Project along Vector	Specify the vector by clicking the Direction button and selecting a plane, edge, or axis. If a plane is selected, the vector will be normal (90°) from the plane. The curves will be projected in the direction of the vector.
	Project to Closest Point	Projects the curves onto the surface normal to the closest point.
	Wrap to Surface	The curves are wrapped around the curvature of the selected face or faces.

8. Click OK.

By default, the projected curves are linked to the original curve. If the original curves change size, they will be updated. To break the link, move the cursor into the browser over the name of the Projected to Surface entry, right-click, and select Break Link in the menu, as shown in the following image. You could also display the sketch constraints and delete the reference constraints. You can also change the way the curves were projected by right-clicking on the Project to Surface entry in the browser and click Edit Projection Curve from the menu as shown in the following image, and the Project Curve to Surface dialog box will appear. Make the changes as required.

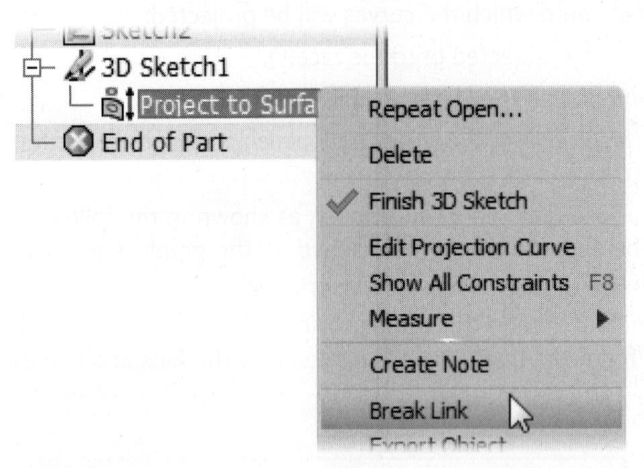

FIGURE 7.49

PROJECT TO 3D SKETCH

Another method to project 2D geometry onto a non-planar face is to use the Project to 3D Sketch command. The 3D Sketch command will project 2D sketch geometry onto a nonplanar face while automatically creating a 3D sketch. To use the Project to 3D Sketch command, follow these steps.

1. To First create a 2D sketch and sketch and constrain geometry as needed or make an existing sketch active that contains the geometry that you want to project.
2. Click the Project to 3D Sketch command from the Sketch tab > Draw panel as shown in the following image on the left.
3. In the Project to 3D Sketch dialog box check the Project option.
4. Select a face or faces that the geometry will be projected onto. All of the geometry on the active sketch will preview how it will be projected on the selected face(s) as shown in the following image on the right.

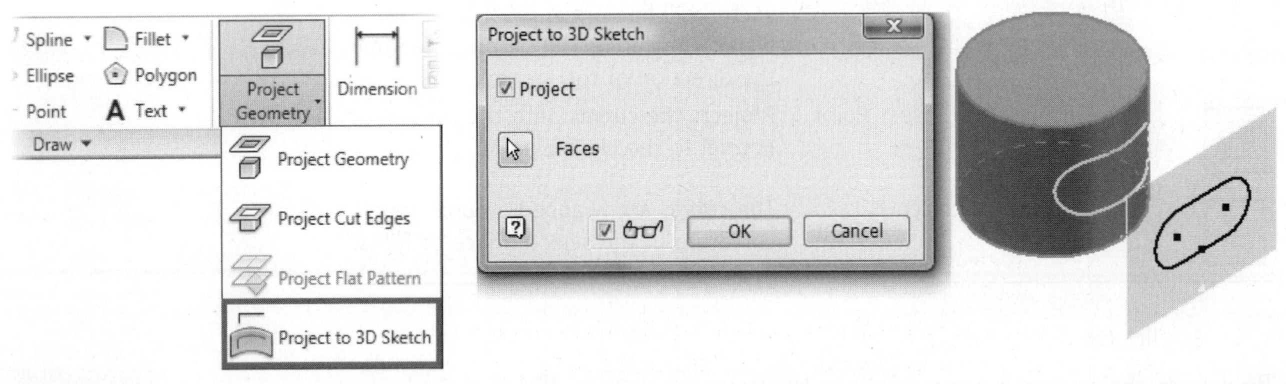

FIGURE 7.50

5. Click OK to create a 3D sketch and project all the geometry on the active sketch to the selected face(s).

Constructed 3D Paths

Another option for creating 3D paths is to define a path by creating work points at locations where the 3D path will intersect and then connecting the points using a 3D line or spline. Because this method of constructing a 3D path depends on work points, you should review the Creating Work Points section in Chapter 4 to become comfortable creating and editing work points. The following is a brief review of the methods used to create work points:

- Click an endpoint or midpoint of an edge or sketch line. A work point is also generated automatically if you select a vertex while creating a 3D line.
- Click two edges or sketch lines that lie on the same plane. A work point is created at the intersection, or theoretical intersection, of the two.
- Click an edge or sketch line and plane. A work point is created at the intersection, or theoretical intersection, of the two.
- Click three nonparallel faces or planes, and a work point is created at their intersection or theoretical intersection.

You can also use grounded work points, but they are not associated dynamically with the part or any other work features including the original locating geometry. When you modify surrounding geometry, the grounded work point remains in the specified location.

3D Lines. There are two ways to create a 3D path. The first method is to use the line command with the Inventor Precise Input dialog box. Follow these steps:

1. While in a 3D Sketch, click the Line command, and enter data into the Inventor Precise Input dialog box, as shown in the following image. Consult the help system "Create a 3D Line" for more information about using Precise Input.

FIGURE 7.51

2. Parametric dimensions can be applied to the lines.
3. By default, a bend is not applied between 3D line segments, but this option can be toggled on and off by right-clicking while in the 3D line command and selecting or deselecting Auto-Bend on the menu, as shown in the following image on the left.
4. To set the default radius of the bend, Click the Document Settings command on Tools tab > Options panel, and then change the 3D Sketch Auto-Bend Radius setting on the Sketch tab.
5. To manually add a bend between two 3D lines, use the Bend command on the Draw panel, as shown in the following image on the right, or right-click in a blank area in the graphics window and click Bend from the marking menu as shown in the following image on the left. In the 3D Sketch Bend dialog box, enter a value for the bend, and then select two 3D lines or the endpoint where they meet. Once the bend is placed, you can edit it by double-clicking on the dimension and entering a new bend radius value.

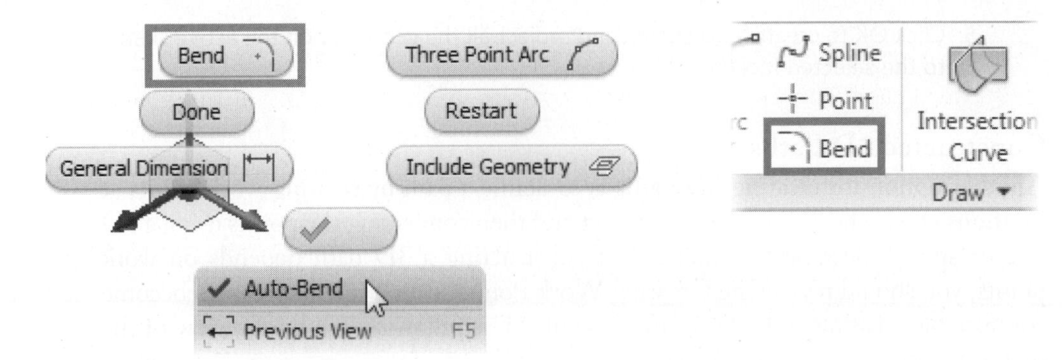

FIGURE 7.52

The second method to create a 3D path is to use work points and the line command. The lines will be linked to the work points, and if a location changes, so will the line. Follow these steps:

1. Create work points and grounded work points, as needed, to define the 3D path.
2. Change to the 3D Sketch environment by clicking the 3D Sketch command on the Standard toolbar under the 2D Sketch command.
3. If you want the 3D path to place a bend automatically, you can set the size of the bend as described in Step 4 in the last procedure.
4. Click the Line or Spline command on the Draw panel.
5. Select the work points in the order that the path will follow. By default, there is no bend between the line segments. To create a bend between line segments as they are created, right-click while in the 3D line command, and select Auto-Bend from the menu.
6. To manually add a bend between two 3D lines, use the Bend command, as described in Step 5 of the previous procedure.
7. When you are done selecting points, right-click, and select Done from the menu.
8. Next create a new sketch that defines the desired profile that will be swept along the 3D path.
9. Click the Sweep command, and create the 3D sweep by selecting the profile and path sketches.

The following image on the left shows a part with 3D lines and dimensions. The image on the right shows the completed sweep.

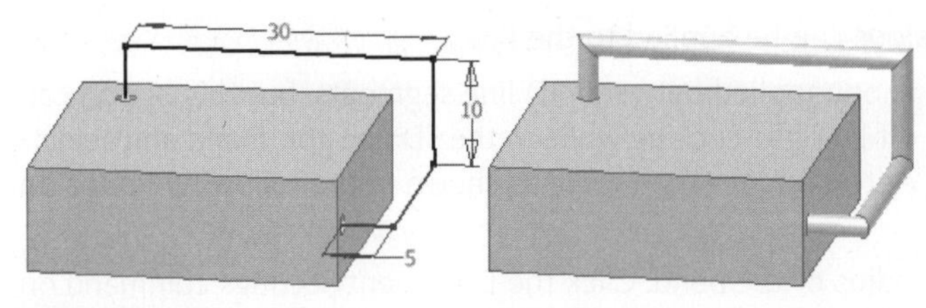

FIGURE 7.53

3D Splines. You can also create a spline between work points or vertices on a part, or you can use the precise input method.

To create a 3D path using work points and a 3D spline, follow these steps:

1. Create work points and/or grounded work points, as needed, to define the 3D path or input the spline's control point via the Inventor Precise Input dialog box as was previously explained in this chapter, in the 3D Lines section.

2. Change to the 3D Sketch environment by clicking the Create 3D Sketch command on the Sketch panel under the Create 2D Sketch command.
3. Click the Spline command on the Draw panel, as shown in the following image on the left.
4. Select the work points in the order that the path will follow as shown in the following image in the middle.
5. To exit the spline command, right-click, and click Create from the menu. Then either press the ESC key or right-click and select Done from the menu.
6. You can add constraints between the spline and a part edge by clicking the desired constraint command on the Constrain panel. The following image on the right shows the available 3D sketch constraints.

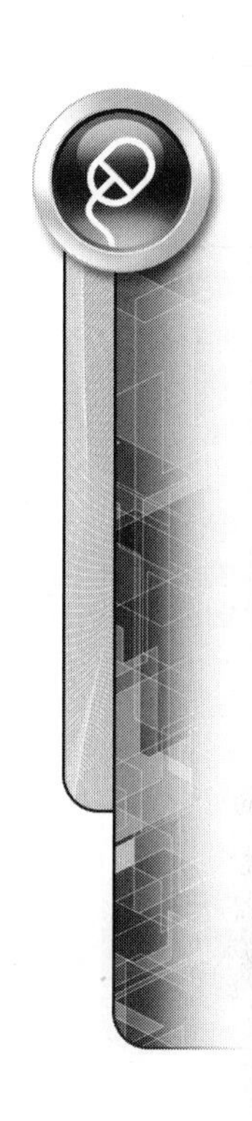

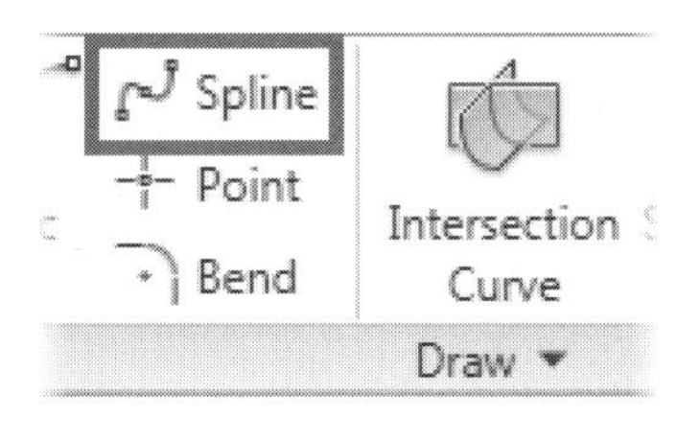

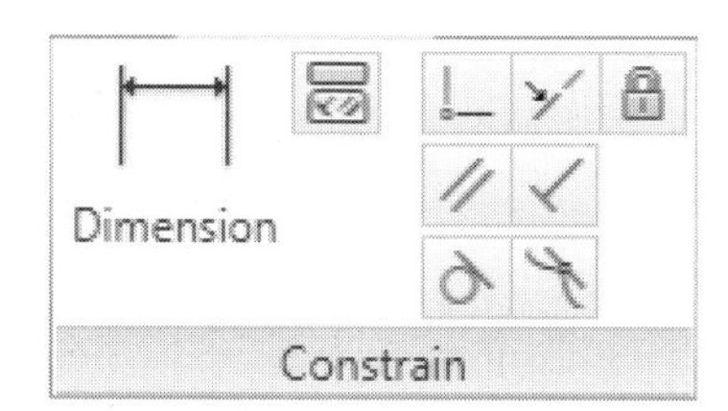

FIGURE 7.54

EXERCISE 7-4: 3D SKETCH—SWEEP FEATURES

In this exercise, you use the Include Geometry, 3D Sketch from Intersection Geometry, Project to Surface, Project to 3D Sketch and the Sweep commands.

1. In this portion of this exercise you will use the Include Geometry command to create a path that the sweep command will use. Open *ESS_E07_04_1.ipt* from the Chapter 07 folder. The part contains a sketch that has a constrained circle.
2. Create a 3D Sketch by clicking the Create 3D Sketch command on the Sketch panel under the Create 2D Sketch command.
3. Click the Include Geometry command from the 3D Sketch tab › Draw panel.
4. Select the nine edges that go around the part as shown in the following image on the left.
5. Finish the 3D Sketch by pressing the Esc key or right-click and click Done from the marking menu.
6. Finish the 3D sketch by clicking on Finish Sketch from the Exit panel.
7. Use the sweep command with the following options.
    - Use the circle as the profile.
    - Use the included geometry as the path (use the Select Other command to select the Sweep Path as shown in the following image in the middle).
    - Change the operation to cut.
    - Click OK to create the sweep. When done, your screen should resemble the following image on the right.

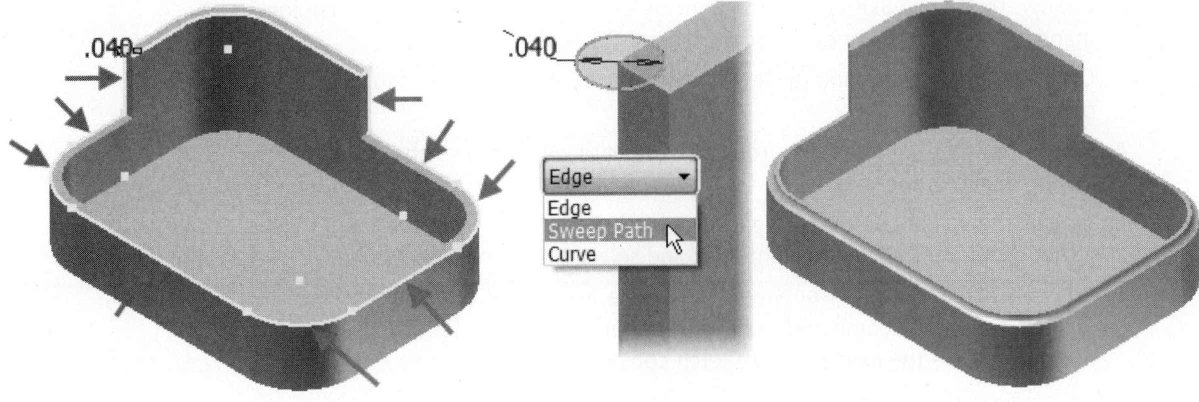

FIGURE 7.55

Note that you could have used the Sweep command without creating a 3D sketch and just select the edges on the part for the path.

8. Close the file. Do not save changes.
9. In this portion of this exercise you will use the 3D Sketch from Intersection Geometry command to create a path that the sweep command will use. Open *ESS_E07_04_2.ipt* from the Chapter 07 folder. The part contains an angled work plane.
10. Create a 3D Sketch by clicking the Create 3D Sketch command on the Sketch panel under the Create 2D Sketch command.
11. Click the Intersection Curve command from the 3D Sketch tab > Draw panel.
12. Select the work plane and the circular face of the part as shown in the following image on the left.
13. Click OK to create the curve.
14. Turn off the visibility of the work plane and, when done, your screen should resemble the following image on the right.

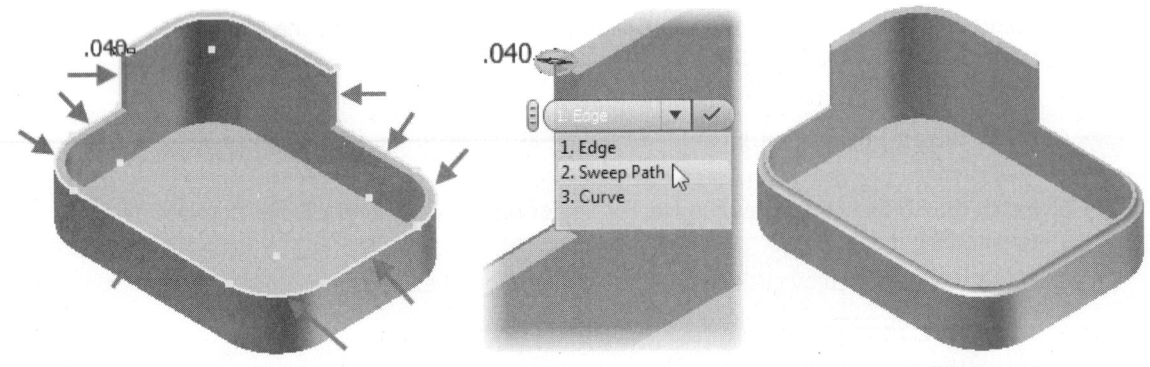

FIGURE 7.56

15. Finish the 3D sketch by clicking on Finish Sketch from the Exit panel.
16. Next you create a 2D sketch that the sweep command will use as the profile. Click the Create 2D Sketch command on the Sketch panel under the Create 3D Sketch command, and from the Origin folder in the browser click the YZ Plane, as shown in the following image on the left.
17. Change to the Home view.

18. Press the F7 key to slick the graphics.
19. Use the Project Geometry command to project the geometry that was created from the 3D Sketch from Intersection Geometry command.
20. Change the projected line to a construction line. This will prevent the projected line from being used when you create the sweep.
21. Create a **.5 inch** diameter circle on the bottom left corner of the projected geometry as shown in the following image in the middle.
22. Finish the Sketch.
23. Use the sweep command with the following options.
    - Use the circle as the profile.
    - Use the geometry that was created from the intersecting plane and face for the sweep path.
    - Change the operation to cut.
    - Click OK to create the sweep. When done, your screen should resemble the following image on the right.

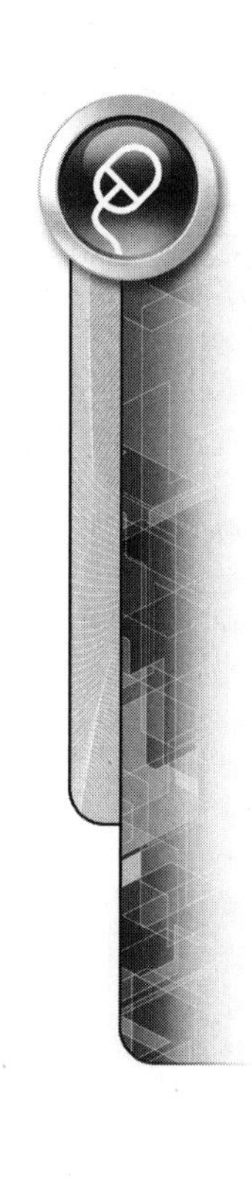

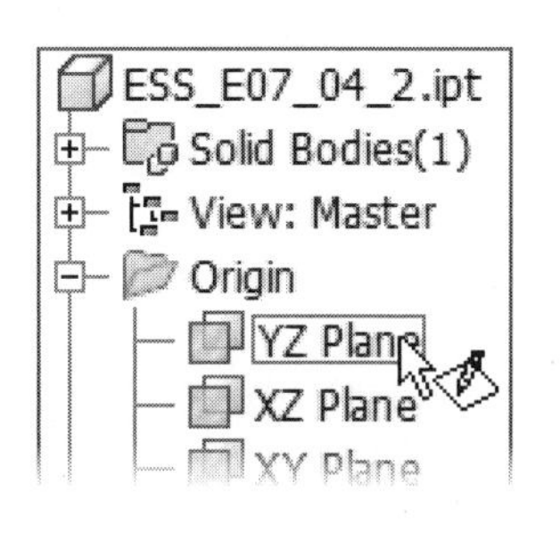

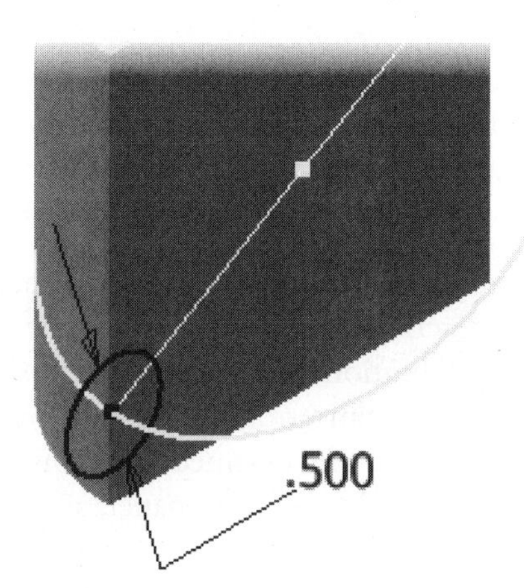

FIGURE 7.57

24. Close the file. Do not save changes.
25. In this portion of this exercise you will use the Project to 3D Sketch command to project geometry onto a cylindrical face. Open *ESS_E07_04_3.ipt* from the Chapter 07 folder. The part contains a constrained sketch.
26. Make Sketch2 active by double-clicking on Sketch2 in the browser.
27. To better see the part, change to the Home view.
28. Click the Project to 3D Sketch command on the Sketch tab > Draw panel.
29. In the Project to 3D Sketch dialog box check the Project option and click the inside circular face on the part as shown in the following image on the left.
30. Click OK to complete the operation.
31. Finish the sketch.
32. Turn off the visibility of Sketch2. When done, your screen should resemble the following image on the right.

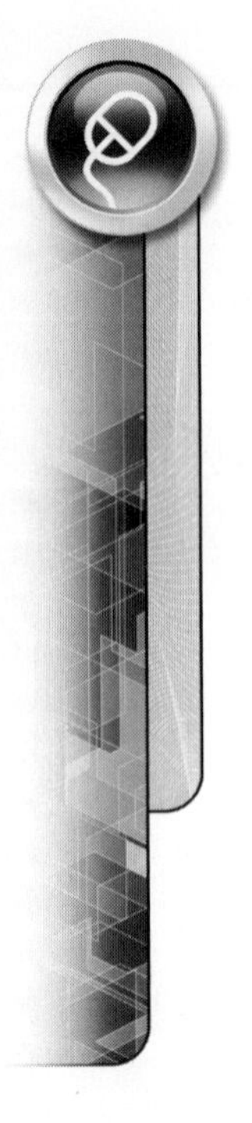

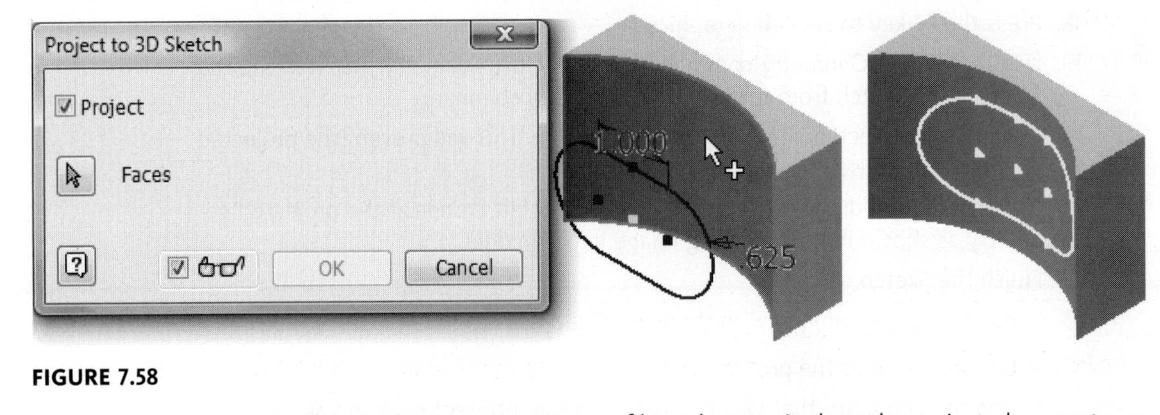

FIGURE 7.58

33. If desired, you can create a profile and sweep it along the projected geometry as you did earlier in this exercise. Note that you could have also used the Project to Surface command in the 3D sketch tab to project the geometry onto the face.
34. Close the file. Do not save changes. End of exercise.

COIL FEATURES

Using the Coil feature, you can easily create many types of helical, coil, or spiral geometry. You can create various types of springs by selecting different settings in the Coil dialog box. You can also use the Coil feature to remove or add a helical shape around the outside of a cylindrical part to represent a thread profile.

To create a coil, you need to have at least one unconsumed or shared sketch available in the part. This sketch describes the profile or shape of the coil feature and can also describe the coil's axis of revolution. If no unconsumed sketch is available, Autodesk Inventor will prompt you with an error message stating, "No unconsumed visible sketches on the part." After an unconsumed sketch is available, you can click the Coil command on the Model tab > Create panel, as shown in the following image on the left. The Coil dialog box appears as shown in the following image on the right. The following sections explain the tabs.

Coil Shape Tab

The Coil Shape tab allows you to specify the geometry and orientation of the coil.

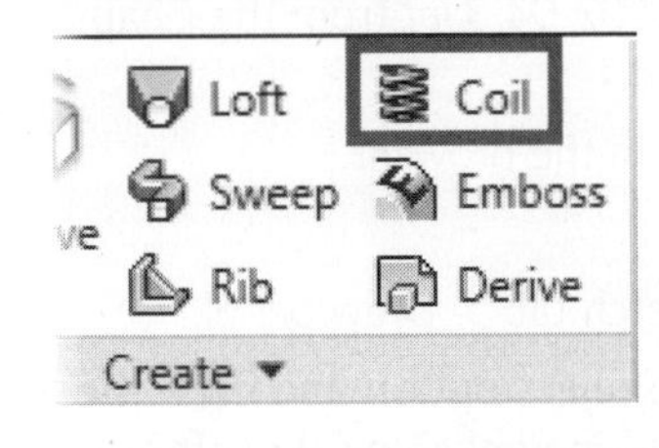

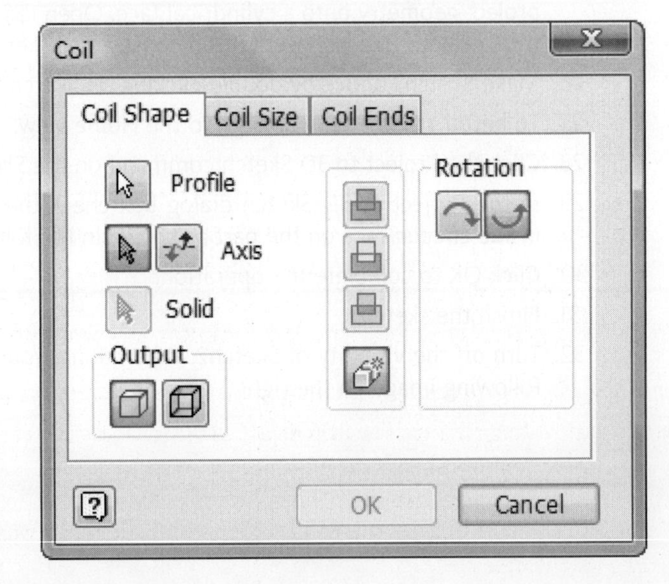

FIGURE 7.59

Profile. Click to select the sketch that you will use as the profile shape of the coil feature. By default, the Profile button is shown depressed; this tells you that you need to select a sketch or sketch area. If there are multiple closed profiles, you will need to select the profile that you want to revolve. If there is only one possible profile, Autodesk Inventor will select it for you, and you can skip this step. If you select the wrong profile or sketch area, click the Profile button and select a new profile or sketch area. You can only use one closed profile to create the coil feature.

Axis. Click to select a sketched line or centerline, a projected straight edge, or an axis about which to revolve the profile sketch. If selecting an edge or sketched centerline, it must be part of the sketch. If selecting a work axis, it cannot intersect the profile.

Flip. Click to change the direction in which the coil will be created along the axis. The direction will be changed on either the positive or negative X or Y axis, depending upon the edge or axis that you selected. You will see a preview of the direction in which the coil will be created.

Solid. If there are multiple solid bodies, click this button to choose the solid body to participate in the operation.

Rotation. Click to specify the direction in which the coil will rotate, either clockwise or counterclockwise.

Operation. The operation buttons are the column of buttons along the center of the dialog box that will appear if a base feature exists, as shown in the following image. The operation buttons are only available if the Coil feature is not the first feature in the part. By default, the Join operation is selected. You can select the other operations to either add or remove material from the part using the Join or Cut options or to keep what is common between the existing part volume and the completed coil feature using the Intersect option.

- *Join:* Adds material to the part.
- *Cut:* Removes material from the part.
- *Intersect:* Keeps what is common to the part and the coil feature.
- *New solid*: Creates a new solid body. The first solid feature that is created uses this option by default. Select to create a new body in a part file with an existing solid body.

Output. Click to create a solid or a surface.

Coil Size Tab

The Coil Size tab, as shown in the following image on the left, allows you to specify how the coil will be created. You have various options for the type of coil that you can create. Based on the type of coil that you select, the other parameters for Pitch, Height, Revolution, and Taper will become active or inactive. Specify two of the three available parameters, and Autodesk Inventor will calculate the last field for you.

Type. Select the parameters that you want to specify: Pitch and Revolution, Revolution and Height, Pitch and Height, or Spiral. If you select Spiral as the Coil Type, only the Pitch and Revolution values are required.

Pitch. Type in the value for the height to which you want the helix to elevate with each revolution.

Revolution. Specify the number of revolutions for the coil. A coil cannot have zero revolutions, and fractions can be used in this field. For example, you can create a coil that contains 2.5 turns. If end conditions are specified, as mentioned in the Coil Ends tab section, the end conditions are included in the number of revolutions.

Height. Specify the height of the coil. This is the total coil height as measured from the center of the profile at the start to the center of the profile at the end.

Taper. Type an angle at which you want the coil to be tapered along its axis.

NOTE

A spiral coil type cannot be tapered.

Coil Ends Tab

The Coil Ends tab, as shown in the following image on the right, lets you specify the end conditions for the start and end of the coil. When selecting the Flat option, the helix, not the profile that you selected for the coil, is flattened. The ends of a coil feature can have unique end conditions that are not consistent between the start and the end of the coil.

Start. Select either Natural or Flat for the start of the helix. Click the down arrow to change between the two options.

End. Select either Natural or Flat for the end of the helix. Click the down arrow to change between the two options.

Transition Angle. This is the rotational angle, specified in degrees, in which the coil achieves the coil start or end transition. It normally occurs in less than one revolution.

Flat Angle. This is the rotational angle, specified in degrees, that describes the amount of flat coil that extends after the transition. It specifies the transition from the end of the revolved profile into a flattened end.

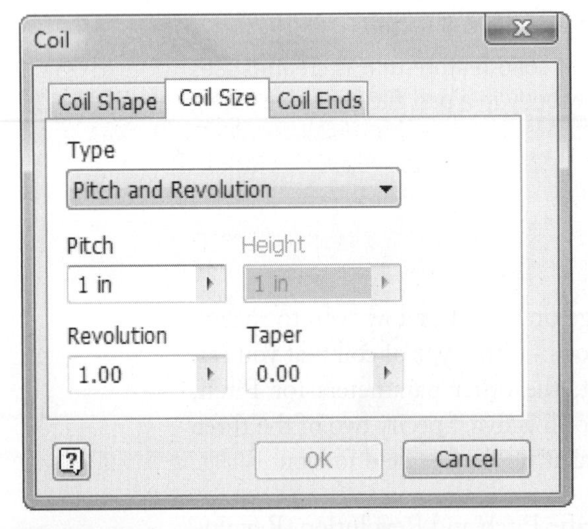

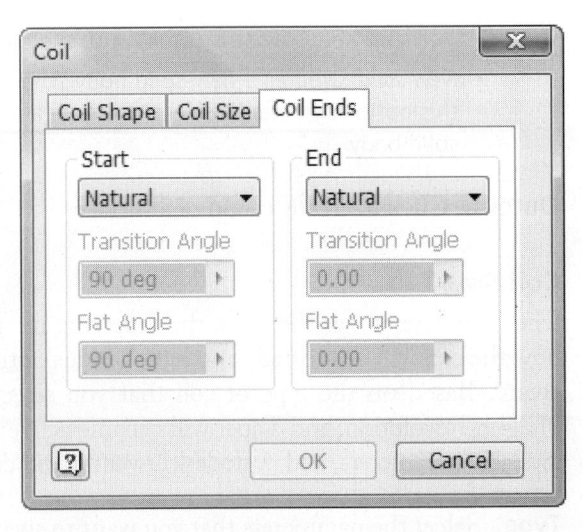

FIGURE 7.60

The following image shows a coil created as the base feature. The image on the left shows the coil in its sketch stage; the rectangle will be used as the profile, and the centerline will be used as the axis of rotation. The finished part, as shown on the right, shows the coil with flat ends.

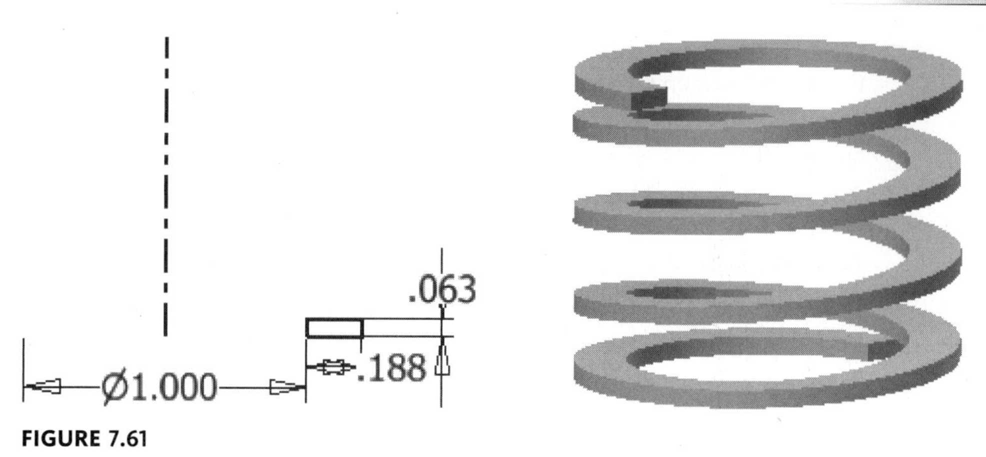

FIGURE 7.61

The following image shows a coil created as a secondary feature. The image on the left shows the coil in its sketch stage with the Coil dialog box displayed. The sketch, a rectangle, is drawn tangent to the cylinder. The rectangle will be used as the profile, and the work axis will be used as the axis of rotation. The finished part, as shown on the right, shows the coil with the flat end on the top.

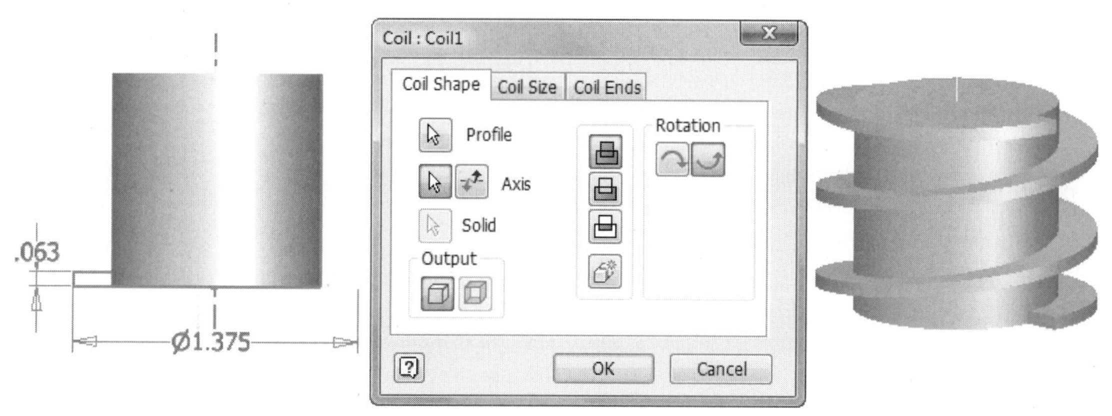

FIGURE 7.62

LOFT FEATURES

The Loft command creates a feature that blends a shape between two or more different sections or profiles. A point can also be used to define the beginning and ending section of the loft. Loft features are used frequently when creating plastic or molded parts. A loft is similar to a sweep, but it can have multiple sections and rails. Many of these types of parts have complex shapes that would be difficult to create using standard modeling techniques. You can create loft features that blend between two or more cross-section profiles that reside on different planes. You can also control the area of a specific section in the loft. You can use a rail, multiple rails, or a centerline to define a path(s) that the loft will follow. There is no limit to the number of sections or rails that you can include in a loft feature. Four types of geometry are used to create a loft: sections, rails, centerlines, and points. The following sections describe these types of geometry.

Create a Loft

To create a loft feature, follow these steps:

1. Create the profiles or points that will be used as the sections of the loft. If required, use work features, sketches, or projected geometry to position the profiles.

2. Create rails or a centerline that will be used to define the direction or control of the shape between sections.
3. Click the Loft command on the Create panel, as shown in the following image on the left.
4. On the Curves tab of the Loft dialog box, the Sections option will be the default. In the graphics window, click the sketches, face loops, or points in the order in which the loft sections will blend.
5. If rails are to be used in the loft, click Click to add in the Rails section of the Curves tab, and then click the rail or rails.
6. If needed, change the options for the loft on the Conditions and Transition tabs.

NOTE **The loft options are also available by right-clicking in a blank area in the graphics window and clicking an option on the menu.**

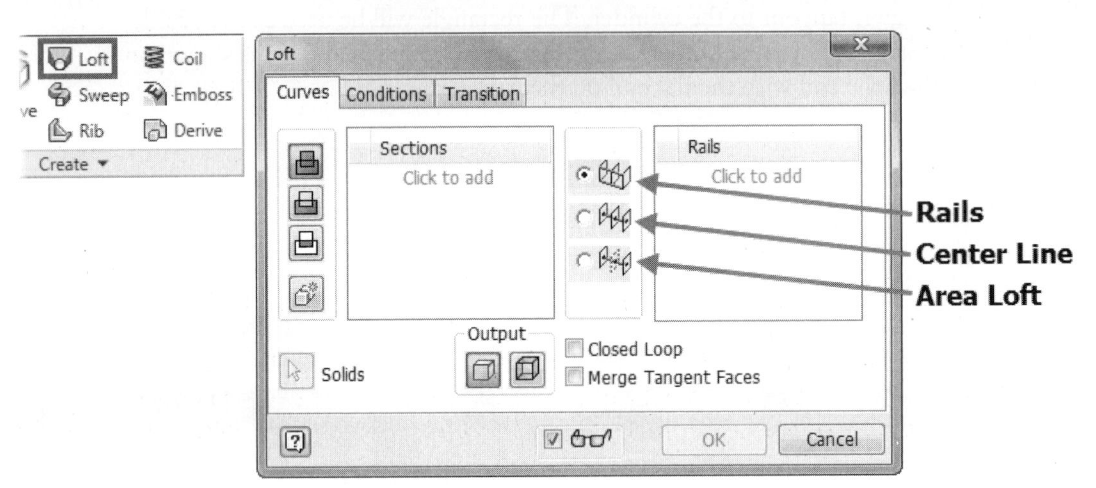

FIGURE 7.63

The following image shows a loft created from two sections and a centerline rail. The image on the left shows two sections and a centerline in their sketch stages, shown in the top view. The image on the right shows the completed loft.

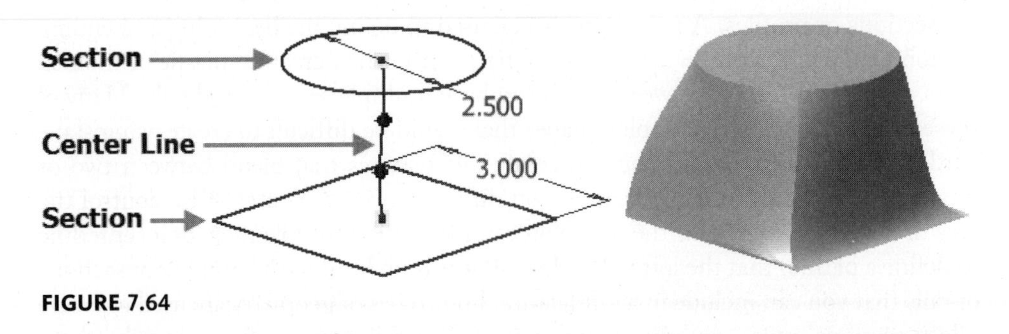

FIGURE 7.64

The following sections explain the options in the Loft dialog box and on its tabs.

Curves Tab

The Curves tab, as shown in the first image in this section, allows you to select which sketches, part edges, part faces, or points will be used as sections to select whether or not a rail or centerline will be used and to determine the output condition.

Sections

You can define the shape(s) between which the loft will blend. The following rules apply to sections:

1. There is no limit to the number of sections that you can include in the loft feature.
2. Sections do not have to be sketched on parallel planes.
3. You can define sections with 2D sketches (planar), 3D sketches (nonplanar), and planar or nonplanar faces, edges on a part, or points.
4. All sections must be either open or closed. You cannot mix open and closed profile types within the same loft operation. Open profiles result in a lofted surface.

Points

A point can be used as a section to help define the loft. The following rules apply to points used for a loft profile:

1. A point can be used to define the start or end of the loft.
2. An origin point, sketch point, center point, edge point, or work point can be used.

Rails

You can define rails by the following elements: 2D sketches (planar), 3D sketches (nonplanar), or part faces and edges. The following rules apply to rails:

1. There is no limit to the number of rails that you can create or include in the loft feature.
2. Rails must not cross each other and must not cross mapping curves.
3. Rails affect all of the sections not just faces or sections that they intersect. Section vertices without defined rails are influenced by neighboring rails.
4. All rail curves must be open or closed.
5. Closed rail curves define a closed loft, meaning that the first section is also the last section.
6. No two rails can have identical guide points, even though the curves themselves may be different.
7. Rails can extend beyond the first and last sections. Any part of a rail that comes before the first section or after the last is ignored.
8. You can apply a 2D or 3D sketch tangency or smooth constraint between the rail and the existing geometry on the model.

Centerline

A centerline is treated like a rail, and the loft sections are held normal to the centerline. When a centerline is used, it acts like a path used in the sweep feature, and it maintains a consistent transition between sections. The same rules apply to centerlines as to rails except that the centerline does not need to intersect sections and only one centerline can be used.

Area Loft

Select a sketch to be used as the centerline, and then click on the centerline to define the area of the profile at the selected point. After picking a point, the Section Dimensions dialog box appears, as shown in the following image. You define its position either as a proportional or absolute distance, and you can define the section's size by area or scale factor related to the area of the original profile.

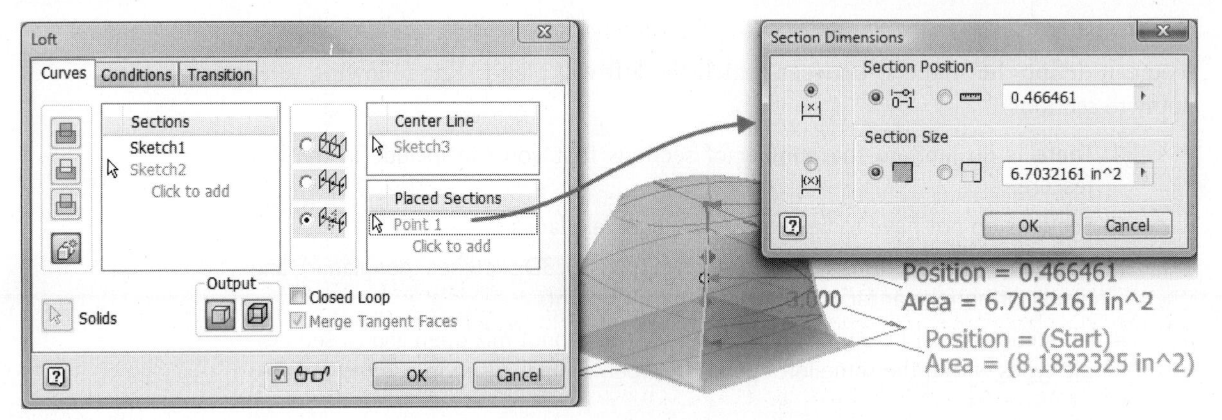

FIGURE 7.65

Output. Select whether the loft feature will be a solid or a surface. Open sketch profiles selected as loft sections will define a lofted surface.

Operation. Select an operations button to add or remove material from the part, using the Join or Cut options, or to keep what is common between the existing part and the completed loft using the Intersect option. By default, the Join operation is selected.

Closed Loop. Click to join the first and last sections of the loft feature to create a closed loop.

Merge Tangent Faces. Click to join tangent faces on the completed loft into a single face.

Conditions Tab

The Conditions tab, as shown in the following image, allows you to control the tangency condition, boundary angle, and weight condition of the loft feature. These settings affect how the faces on the loft feature relate to geometry at the start and end profiles of the loft. This may be existing part geometry or the plane or work plane containing the loft section sketch.

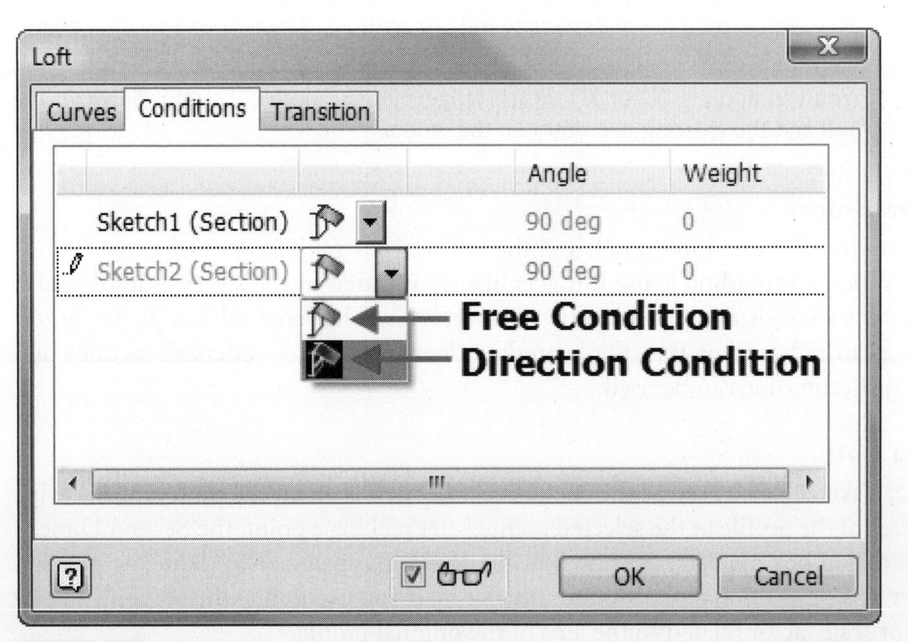

FIGURE 7.66

Conditions. The column on the left lists the sketches and points specified for the sections. To change a sketch's condition, click on its name, and then select a condition option.

Condition Boundaries Sketches

Two boundary conditions are available when the first or last section is a point, as shown in the previous image.

- Free Condition (top button): With this option, there is no boundary condition, and the loft will blend between the sections in the most direct fashion.
- Direction Condition (bottom button): This option is only available when the curve is a 2D sketch. When selected, you specify the angle at which the loft will intersect or transition from the section.

Condition Boundaries Face Loop. When a face loop or edges from a part are used to form a section, as shown in the following image on the left, three conditions are available, as shown in the following image on the right.

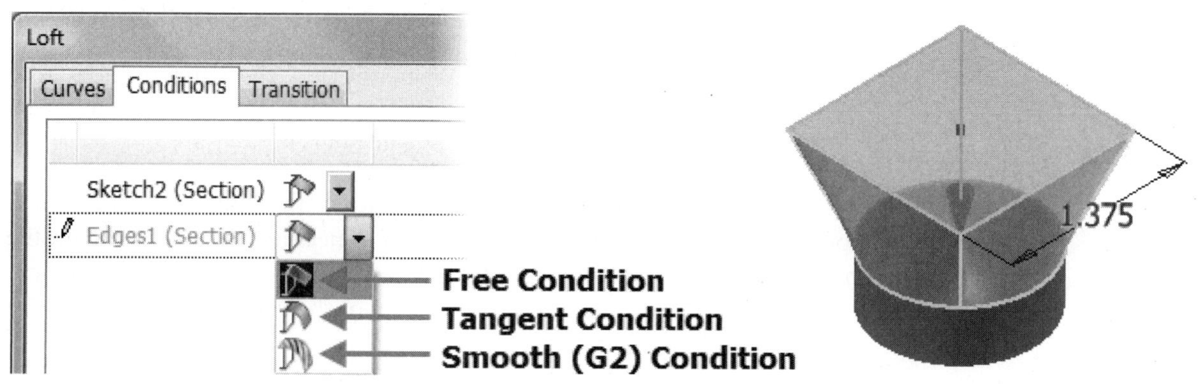

FIGURE 7.67

- Free Condition (top button): With this option, there is no boundary condition, and the loft will blend between the sections in the most direct fashion.
- Tangent Condition (middle button): With this option, the loft will be tangent to the adjacent section of face.
- Smooth (G2) Condition (bottom button): With this option, the loft will have curvature continuity to the adjacent section of face.

Condition Boundaries Point. Three conditions are available when a point is selected for the loft profile, as shown in the following image.

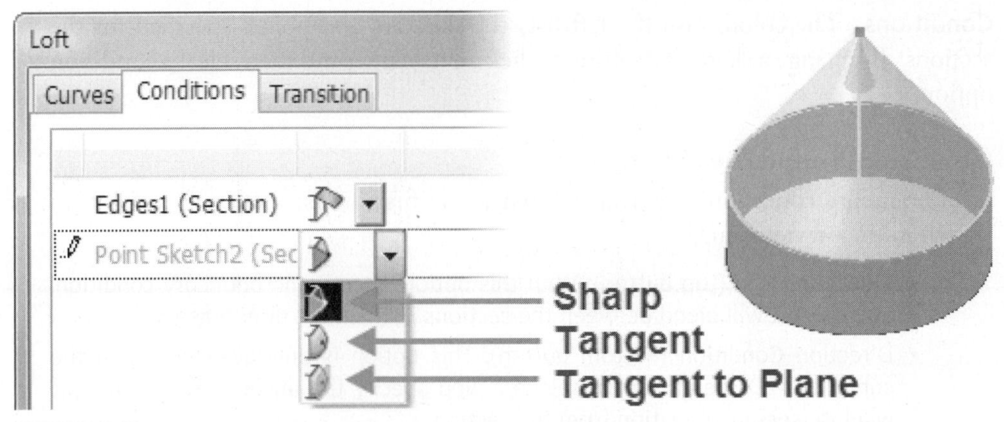

FIGURE 7.68

- Sharp Point (top button): With this option, there is no boundary condition, and the loft will blend from the previous section to the point in the most direct fashion.
- Tangent (middle button): When selected, the loft transitions to a rounded or domed shape at the point.
- Tangent to Plane (bottom button): When selected, the loft transitions to a rounded dome shape. You select a planar face or work plane to be tangent to. This option is not available when a centerline loft is used.

Angle. This option is enabled for a section only when the boundary condition is Tangent or Direction. The default is set to 90° and is measured relative to the profile plane. The option sets the value for an angle formed between the plane that the profile is on and the direction to the next cross-section of the loft feature. Valid entries range from 0.0000001° to 179.99999°.

Weight. The default is set to 0. The weight value controls how much the angle influences the tangency of the loft shape to the normal of the starting and ending profile. A small value will create an abrupt transition, and a large value creates a more gradual transition. High weight values could result in twisting the loft and may cause a self-intersecting shape.

Transition Tab

The Transition tab, as shown in the following image with the Automatic mapping box unchecked, allows you to specify point sets. A point set is used to define section point relationships and control how segments blend from one section to the segments of the adjacent sections. Points are reoriented or added on two adjacent sections.

Map Points

You can map points to help define how the sections will blend into each other. The following rules apply to mapping points:

Point Set. The name of the point set appears here.

Map Point. The corresponding sketch for the selected point set appears here.

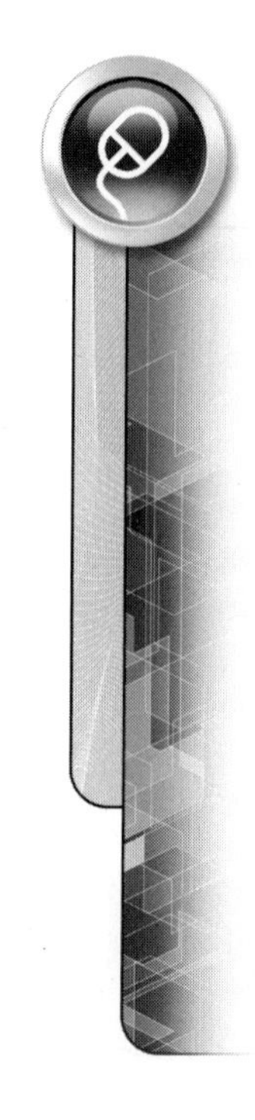

Position. The location of the selected map point appears here. You can modify the position by entering a new value or by dragging the point to a new location in the graphics window.

To modify the default point sets or to add a point set, follow these steps:

1. Click on the Transition tab, and uncheck the Automatic Mapping box. The dialog box will populate the automatic point data for each section. The list is sorted in the order in which the sections were specified on the Curve tab as shown in the following image on the left.
2. To modify a point's position, select the point set in which the point is specified. When you select its name, it will become highlighted in the graphics window.
3. Click in the Position section of the map point that you want to modify, and enter a new value or drag the point to a new location in the graphics window.
4. To add a point set, select Click to add in the point set area.
5. Click a point on the profile of two adjacent sections. As you move the cursor over a valid region of the active section, a green point appears. As the points are placed, they are previewed in the graphics window, as shown in the following image on the right.
6. You can modify the new point set in a similar way to the default point set.

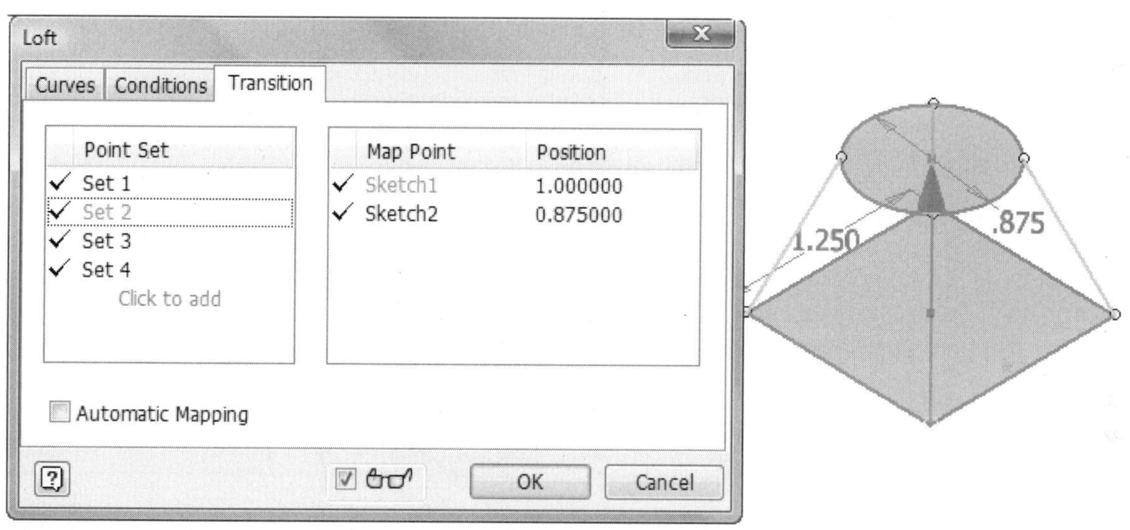

FIGURE 7.69

EXERCISE 7-5: CREATING LOFT FEATURES

In this exercise, you use the loft command to define the shape of a razor handle.

1. Open *ESS_E07_05.ipt* in the Chapter 07 folder.
2. In the browser, double-click Sketch1 to edit the sketch.
3. Click the Point, Center Point command on the Draw panel.
4. Click a point coincident with the spline near the bottom of the curve, making sure that the coincident glyph is displayed as you place the point, as shown in the following image on the left.
5. Draw two construction lines coincident with the sketched point and the nearest spline points on both sides of the Point that you just created, as shown in the following image on the right.

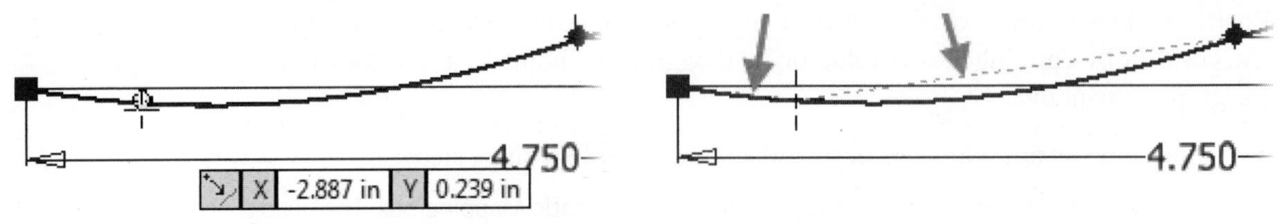

FIGURE 7.70

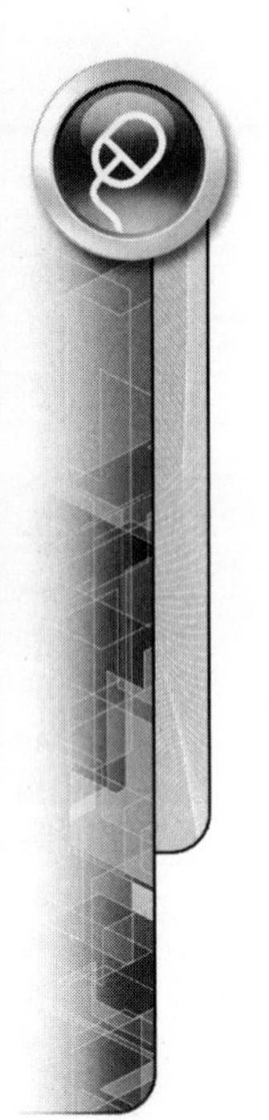

6. To parametrically position the sketched point midway between the spline points, place an equal constraint between the two construction lines.
7. Change the 4.750 overall horizontal dimension to **5 in**, and verify that the sketched point moves along the spline to maintain its position on the spline.
8. Click Finish Sketch on the Exit panel to finish editing the sketch.
9. Click the Work Plane command on the Work Features panel. To create a work plane perpendicular to the spline, click the Center Point you just created, and then click the spline, not the construction lines as shown in the following image on the left.
10. Next create a profile for a loft section on the new work plane. Create a new sketch on the new work plane.
11. Change to the Home View.
12. Project the Point, Center Point that you created in an earlier step onto the sketch.
13. Draw an ellipse with its center coincident with the projected point and the second point so the ellipse's axis is horizontally constrained to the Point. Click a third point, as shown in the following image.
14. Place **.1875 in** and **.375 in** dimensions to control the ellipse size, as shown in the following image on the right.

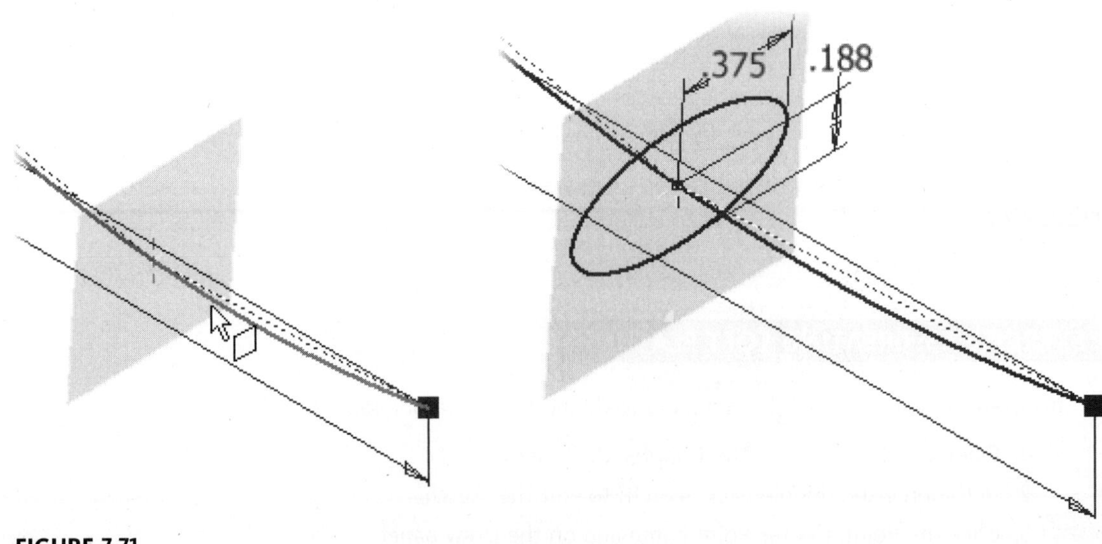

FIGURE 7.71

15. Finish the sketch.
16. Click the Loft command on the Create panel.
17. For the first section, click the concave 3D face, as shown in the following image.

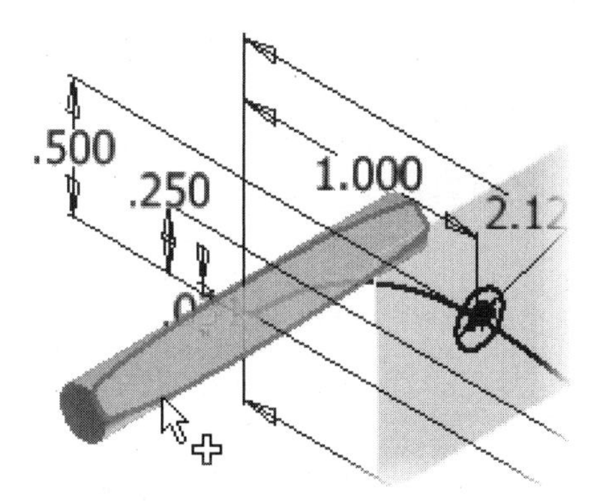

FIGURE 7.72

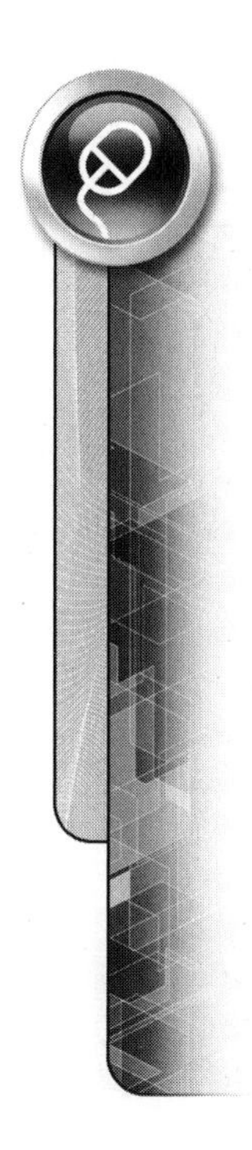

18. Click the other three profile sections from left to right, and then click the point on the right end of the spline.
19. Click the Center Line option in the dialog box, and then select the spline. When finished, the preview looks like the following image on the left.

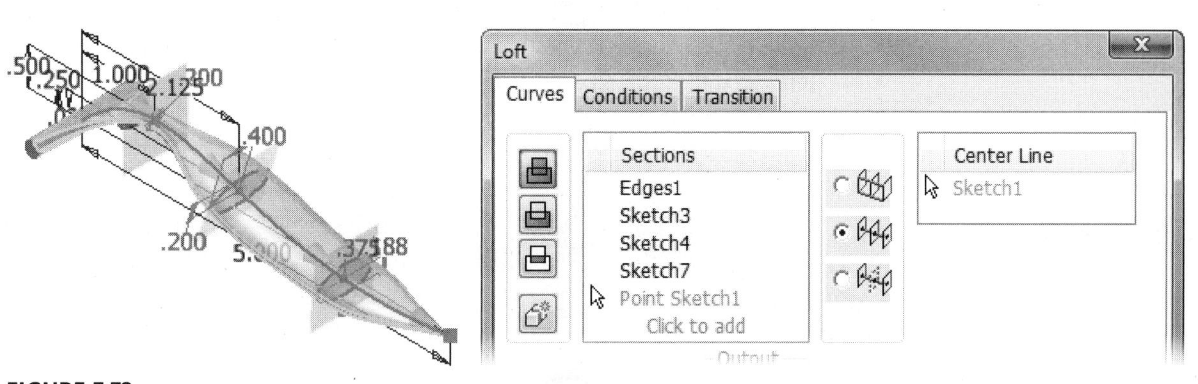

FIGURE 7.73

20. Click OK to create the loft.
21. Turn off the visibility of all the work planes by pressing the ALT and] keys, and then use the Orbit command to examine the loft. Notice that the end is sharp at the point.

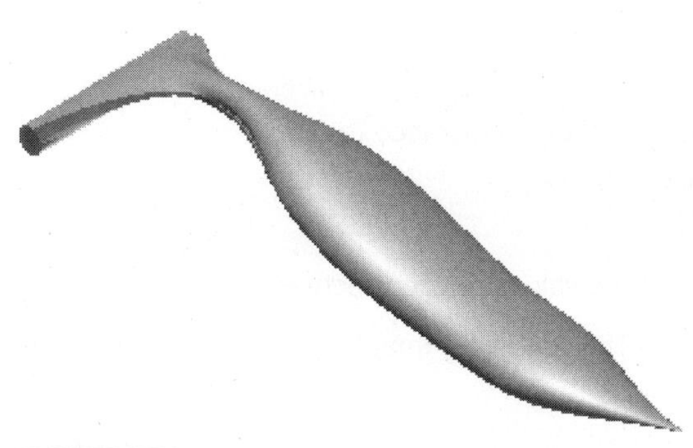

FIGURE 7.74

22. To edit the loft move the cursor over the Loft in the graphics window, and click the Edit Loft button from the mini-toolbar that is displayed. Click the Conditions tab, and change the Point entry to a Tangent condition, as shown in the following image on the left.
23. Click OK to update the loft. The right side of the part should resemble the following image on the right.

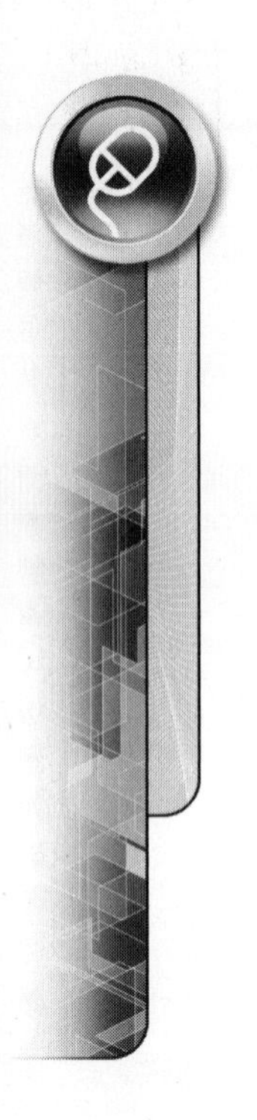

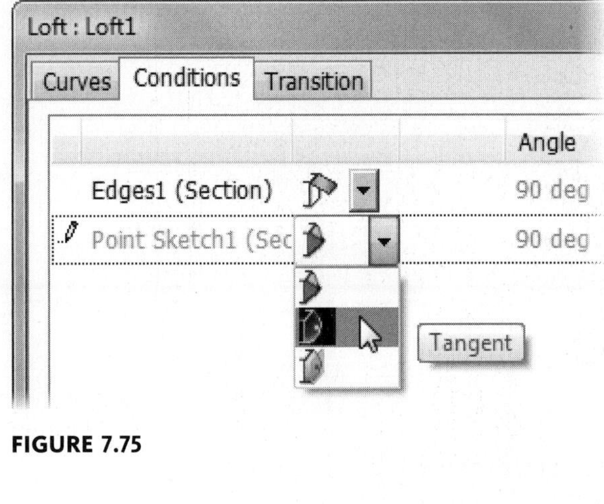

FIGURE 7.75

24. Next add another section that defines the section's area. In the browser, move the cursor over the entry Loft1 and double-click.
    - From the Curves tab, click Area Loft option.
    - In the Placed Sections area, click Click to add.
    - In the graphics window, click a point near the middle of the spline, as shown in the following image.

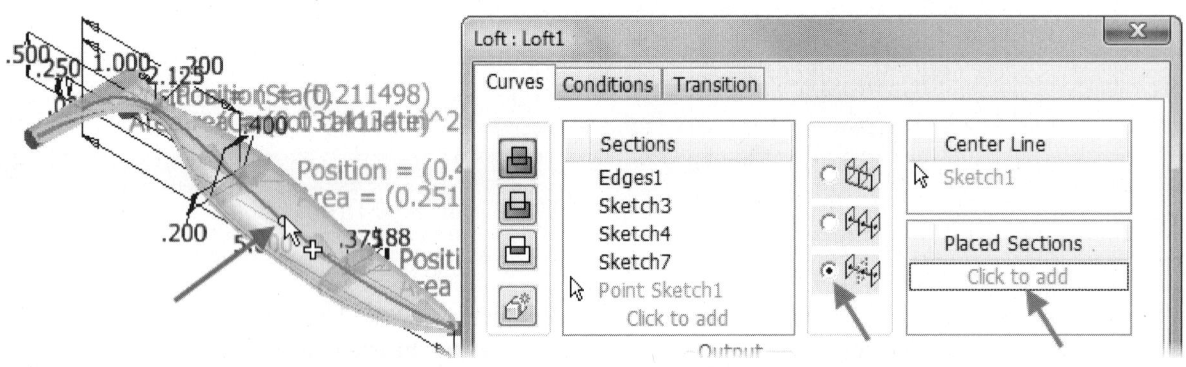

FIGURE 7.76

25. In the Section Dimensions dialog box:
    - Change the Section Position proportional distance to **0.5**.
    - Change the Section Size area to **0.375 in^2.**
    - Click OK.
    - In the graphics window, notice the values of each section.
26. In the Loft dialog box, click OK. The following image on the right shows the updated loft.

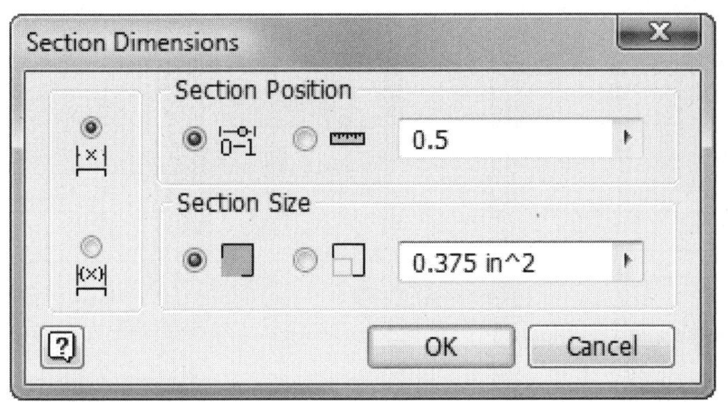

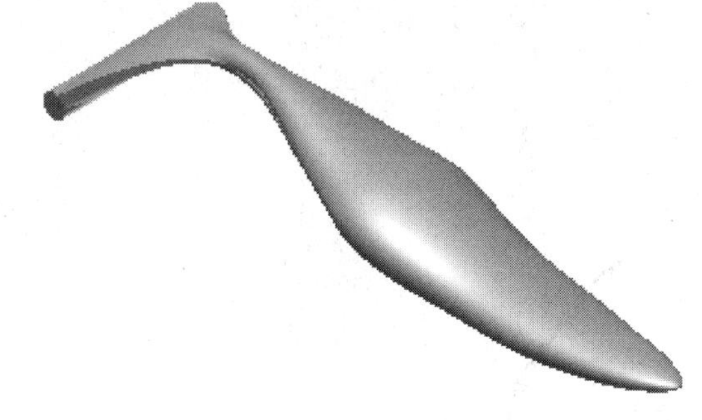

FIGURE 7.77

27. Close the file. Do not save changes. End of exercise.

MULTI-BODY PARTS

As was explained throughout the previous chapters commands such as extrude, revolve, sweep, and loft have a New Solid option that allows you to create a solid body in a part file. There is no limit to the number of solid bodies that can be created in a part file. You can also create a single part and then split it into multiple parts and then, if required, export the solid bodies to their own part files. This modeling method is helpful for designing parts that have complex relationships between the parts.

There are three main methods for creating solid bodies in a part: create within the part, derive an existing part into a part file or split a part into multiple solid bodies.

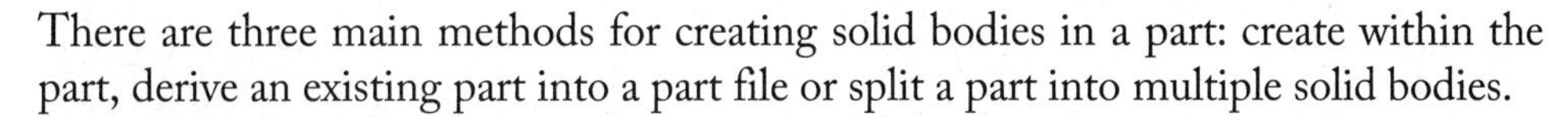

Create a Solid Body in a Part File

This method creates solid bodies within a part file. To create a solid body within a part file, follow these steps:

1. If needed create a new sketch.
2. Issue the extrude, revolve, sweep, and loft commands.
3. In the dialog box select the New Solid option, as shown in the extrude tool in the following image on the left or from the in-canvas operation button, as shown in the following image on the right.
4. Define the operation. Only Join and Intersection will create a new solid body.
5. Define the extents of the feature.
6. Create the solid body.

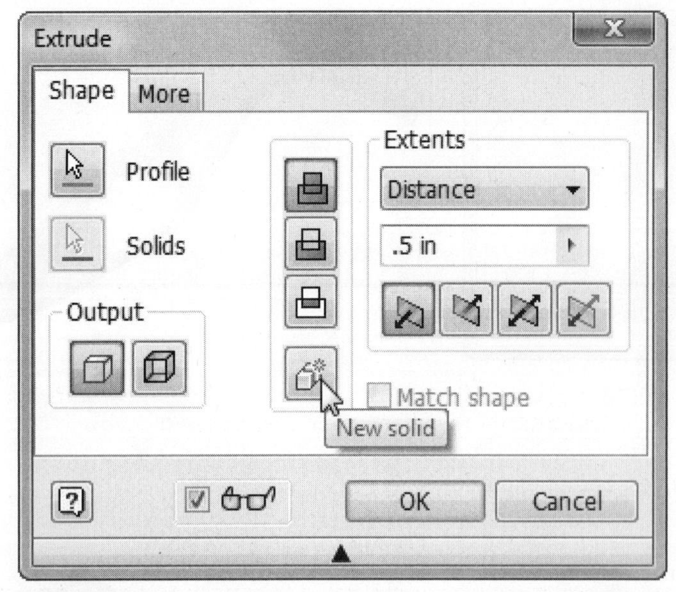

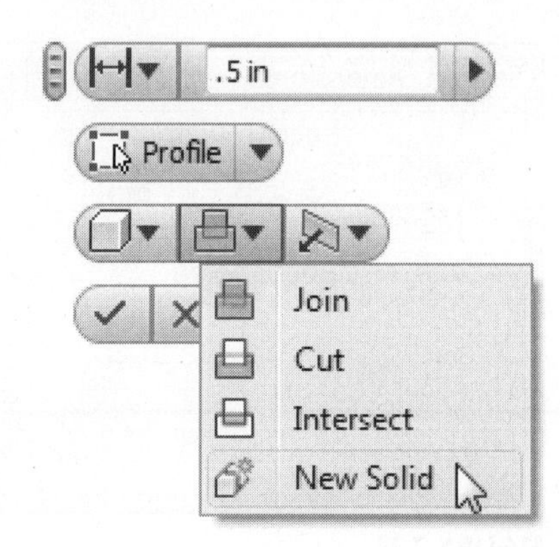

FIGURE 7.78

Derive a Solid Body into a Part File

This method links an existing part file into a part file. To derive a solid body into a part file, follow these steps:

1. While in a part file, start the Derive command from the Model tab > Create panel or from the Manage tab > Insert panel.
2. Select a part or an assembly file.
3. In the Derive Part dialog box, select the geometry that you want to derive.

NOTE The Derive command will be covered later in this chapter.

Split a Part into Solid Bodies

This method will take a solid body and split it into multiple solid bodies. To split a part into multiple solid bodies, follow these steps:

1. Create the part that will be split.
2. Create a work plane or surface that will be used to split the part.
3. Start the split command and select the Split Solid option.

NOTE The Split command will be covered in the next section.

Edit Solid Bodies

After a solid body is created it will appear in the browser under the Solid Bodies folder as shown in the following image.

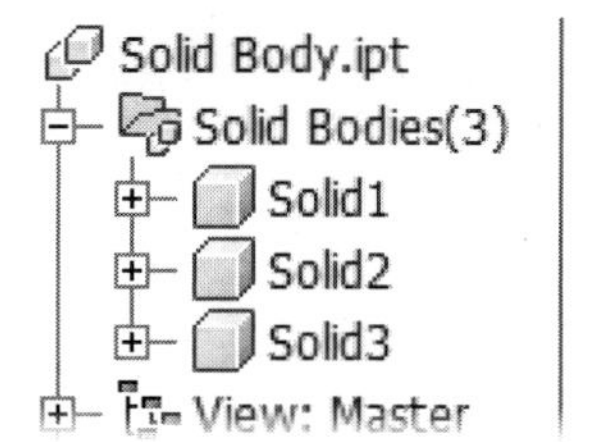

FIGURE 7.79

Autodesk Inventor has tools that can be used to move and combine solid bodies.

Move Bodies

Use the Move Bodies command to move or rotate the location of solid bodies in a part file. The following image shows the Move Bodies dialog box. The Move Bodies command is located on the Model tab > Modify panel, as shown in the following image on the left.

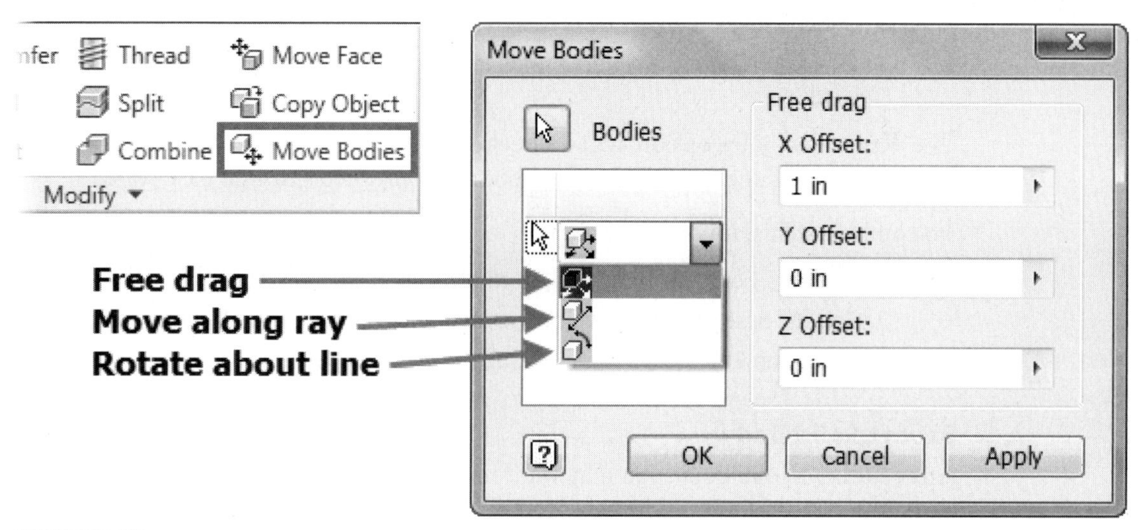

FIGURE 7.80

To move a solid body, you have three Move Type options:

- Free drag: Enter an X, Y, or Z value or select the solid body and free drag the solid by moving the mouse.
- Move along ray: Enter a linear offset value and then select an edge or axis to define the vector.
- Rotate about line: Select an edge or axis to rotate and enter a value for the angle.

To move a solid body(ies), follow these steps:

- Click the Move Bodies command from the Modify panel.
- Select a solid body(ies) to move or rotate.
- Select the Move Type and enter the required information.

Combine

The Combine command will join two or more solids together, remove material from a base solid, or keep the common volume of the selected solids. The Combine command is located on the Model tab > Modify panel, as shown in the following image on the left. The following image on the right shows the Combine dialog box.

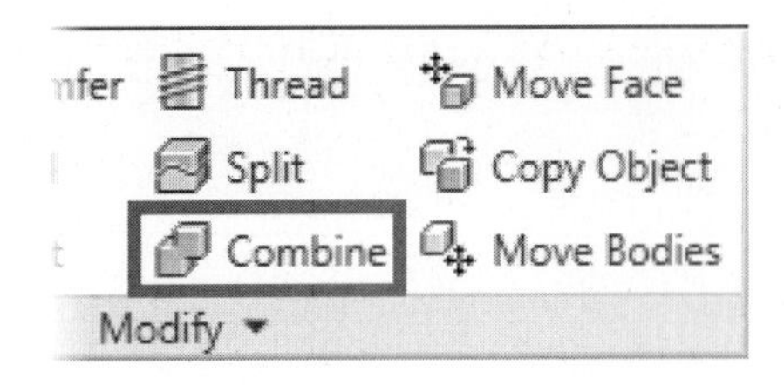

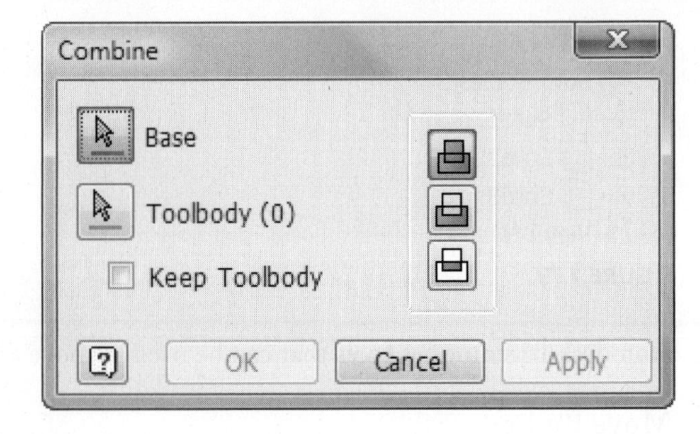

FIGURE 7.81

When combining solids, there are three operations to choose from that are similar to the operations in the feature creation commands.

- *Join:* Combines two or more solid bodies.
- *Cut:* Removes the volume of a solid(s) from the base solid.
- *Intersect:* Keeps the volume that is common between the base and selected solids.

The Keep Toolbody option when checked will keep the solid body in the Solid Bodies folder. If unchecked, the solid body is consumed into the base solid.

To combine solids, follow these steps:

1. Click the Combine command on the Modify panel.
2. Select a base solid from which the other solids will be joined, cut, or intersected.
3. Select the Cut, Join, or Intersect option.

Export Solid Body

After editing a solid body, you may want to export the solid(s) to its own file so it can be used in assemblies and documented. There are two commands to export a solid: Make Part and Make Components. The exported part(s) is linked to the original solid body via the Derived functionality that will be covered later in this chapter.

Make Part

The Make Part command will export a single solid to a new part file. To start the Make Part command, click the Manage tab > Layout panel, as shown in the following image on the left and its dialog box is on the right. Via the dialog box, select the objects to be exported in the Status area. Set the scale factor and check the Mirror part option if you want to mirror the part. Then select a template, file name location, and determine if the part should be placed in a new assembly. When done, click OK and the solid will be exported.

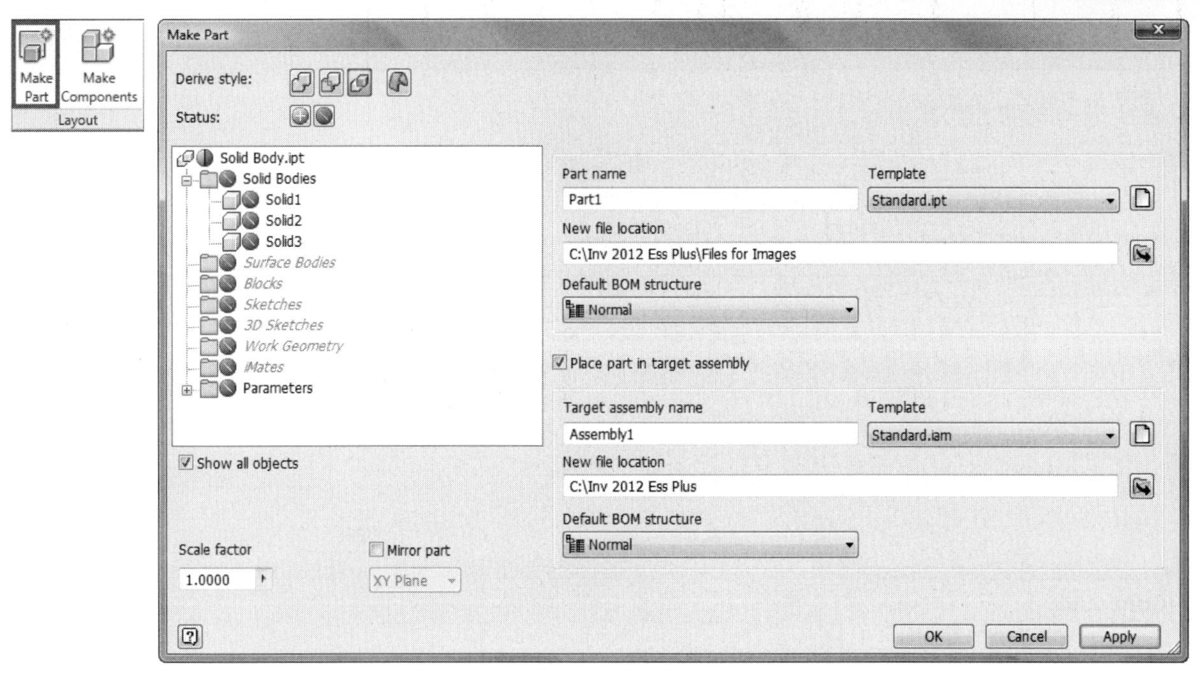

FIGURE 7.82

To Make Part follow these steps.

Make Components

The Make Components command is similar to the Make Part command except it allows multiple solid bodies to be exported to individual files in one operation, to start name of the target assembly, or to uncheck the Insert components in target assembly if you just want the solids to go to a part file. When done, click Next.

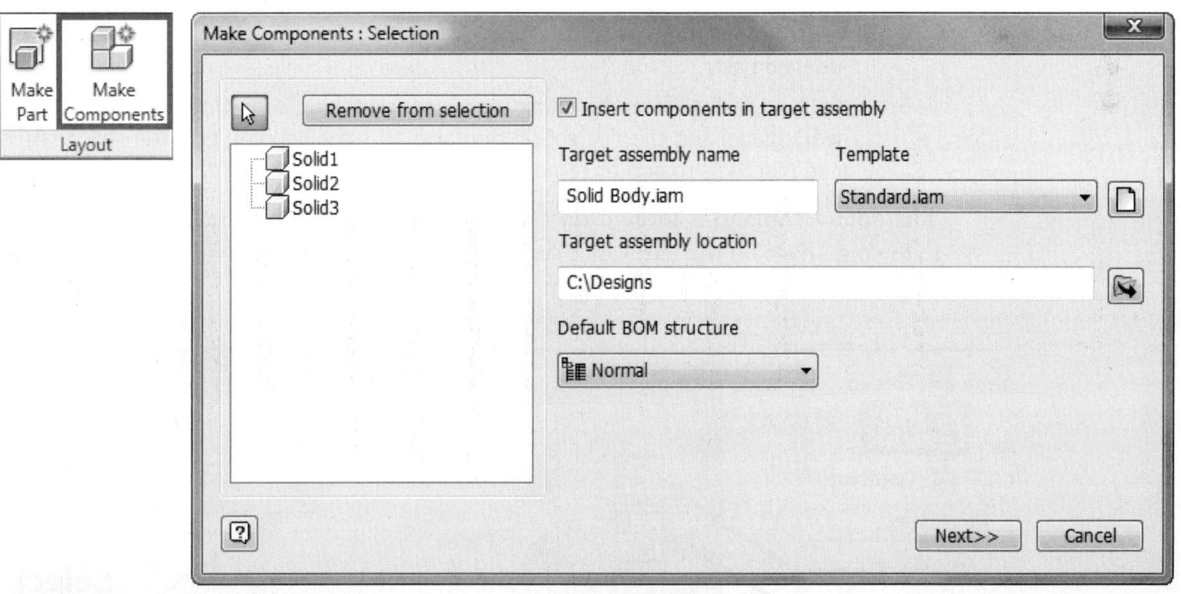

FIGURE 7.83

The Make Components: Bodies dialog box will appear. If needed, change the components name, template, BOM structure, file location, scale factor, determine if the part will be mirrored, as shown in the following image, and then click OK and the solids will be exported.

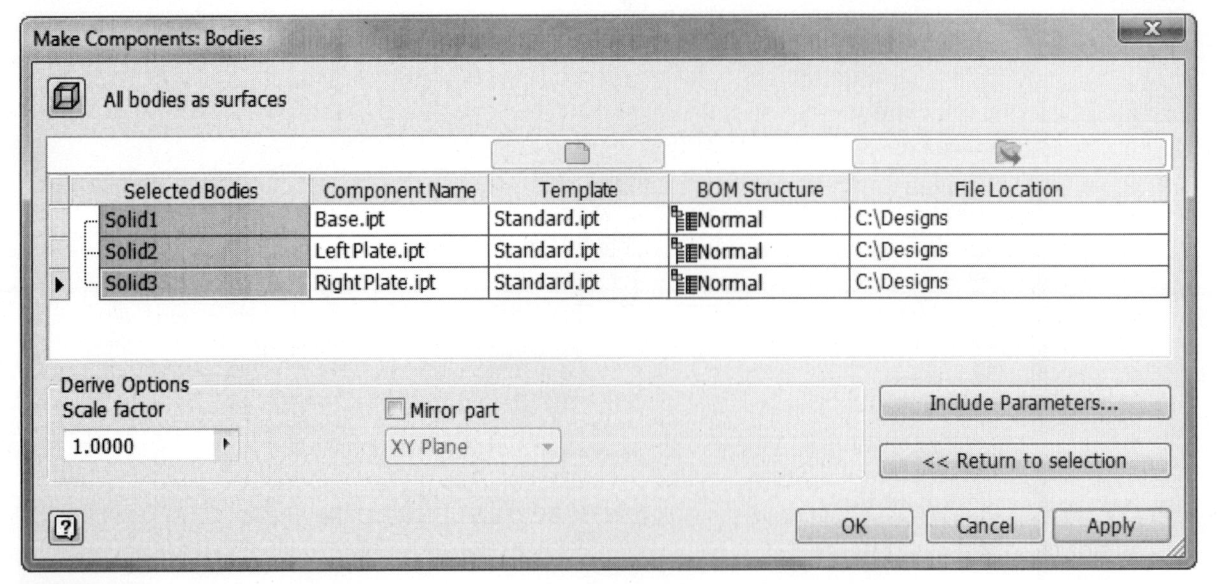

FIGURE 7.84

SPLIT A SOLID, PART, OR FACE

The Split command allows you to split a part into two solids, split the solid by removing one portion of the part, to split individual faces, or to split all faces. A typical application to split a face is to allow the creation of face drafts to the split faces of a part. You can use the Split command to perform the following:

- Split a solid into multiple solid bodies.
- Remove a section of the part by using a surface, planar face, or a work plane and to cut material from the part in the direction you specify. The side that is removed is suppressed rather than deleted. To create a part with the other side removed, edit the split feature, redefine it to keep the other side, save the other half of the part to its own file using the Save Copy As option, or create a derived part.
- Split individual faces by using a surface, sketching a parting line, or placing a work plane, and then selecting the faces to split. You can edit the split feature and modify it to add or remove the desired part faces to be split.

The Split command is located on the Model tab > Modify panel, as shown in the following image on the left. Once selected, the Split dialog box will appear, as shown in the following image on the right.

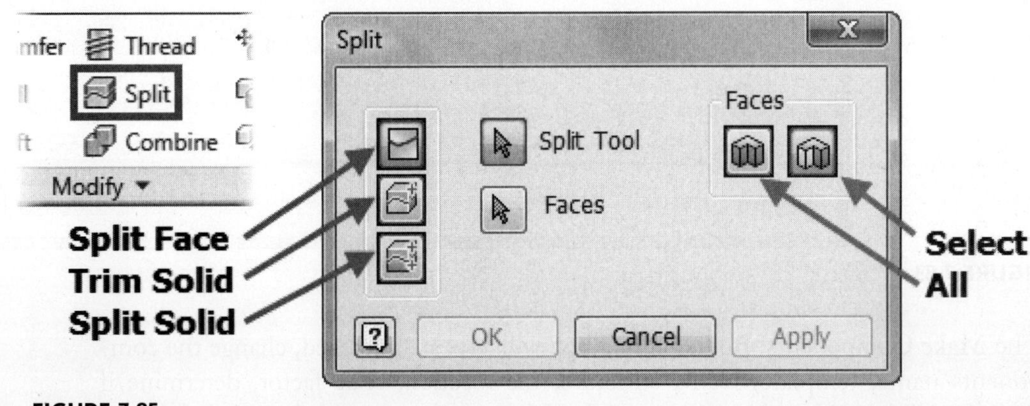

FIGURE 7.85

The Split dialog box contains the following sections:

Method

Split Face. Click this button to split individual faces of a part by selecting a work plane, surface, or sketched geometry and then selecting the faces to split. The split face method can split individual faces or all the faces on the part. When you select the split face method, the Remove area in the dialog box will be replaced with the Faces area, which allows you to select all or individual faces.

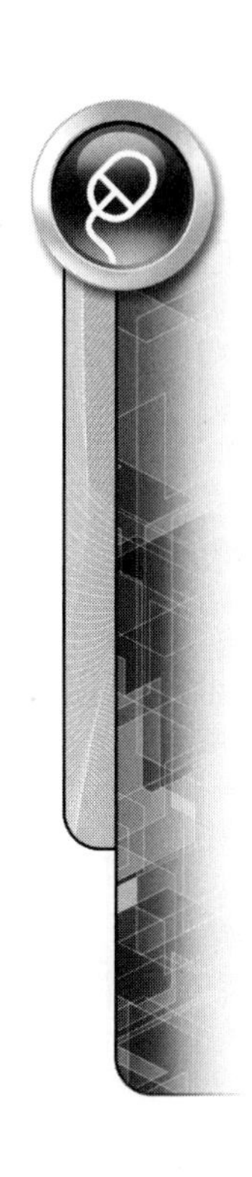

Trim Solid. Click this button to split a part using a selected work plane, surface, or sketched geometry to cut or remove material. If you select this option, you are prompted to choose the direction of the material that you want to remove.

Split Solid. Click this button to split a solid into two solid bodies. Use a surface, plane, or work plane that at least touches the outside edges of the solid; it can exceed the exterior faces of the part.

Remove. The option to remove material is only available when you use the Trim Solid method. After splitting a part, you can retrieve the cut material by editing the split feature and clicking to remove the opposite side or by deleting the split feature.

Faces

The Faces option is available only when you select the split face method.

All. Click this button to split all faces of the part that intersect the Split command.

Select. Click this button to enable the selection of specific faces that you want to split. After clicking the Select button, the Faces to Split command becomes active.

Faces. Click this button, and select a surface, work plane, or sketch that you want to use to split the part.

EXERCISE 7-6: SPLITTING A PART INTO MULTIPLE SOLID BODIES

In this exercise, you split a part into multiple solid bodies and export them to an assembly.

1. Open *ESS_E07_06.ipt* in the Chapter 07 folder.
2. In this step you use an origin plane to split the part into two solids. Click the Split command in the Modify panel.
   - Click the Split Solid method button, as shown in the following image.
   - For the Split tool, click XY Plane from the origin folder in the browser.

FIGURE 7.86

3. Click OK to split the part into two solids.
4. In the browser, expand the Solid Bodies folder and click on the two solids to verify that the solids have been created.
5. Next you turn off the visibility of a solid. In the browser, under the Solid Bodies folder right-click on Solid2 and on the menu click Visibility.
6. Use the Zoom and Free Orbit commands to examine the solid.
7. Next you switch the visibility of the solids. In the browser, under the Solid Bodies folder right-click on Solid2 and on the menu click Hide Others.
8. Use the Zoom and Orbit commands to examine the solid.
9. Turn the visibility of the second solid back on, under the Solid Bodies folder right-click on Solid3 and on the menu click Visibility or Show All.

NOTE

You can rename a solid by slowly double-clicking on the solid name in the browser or right-clicking on the solid in the browser and clicking Properties from the menu.

10. Next you create a hole feature. From the Model tab > Modify panel click Hole. Add a Through All **1/4 in** concentric hole to the top right side of the part, as shown in the following image.

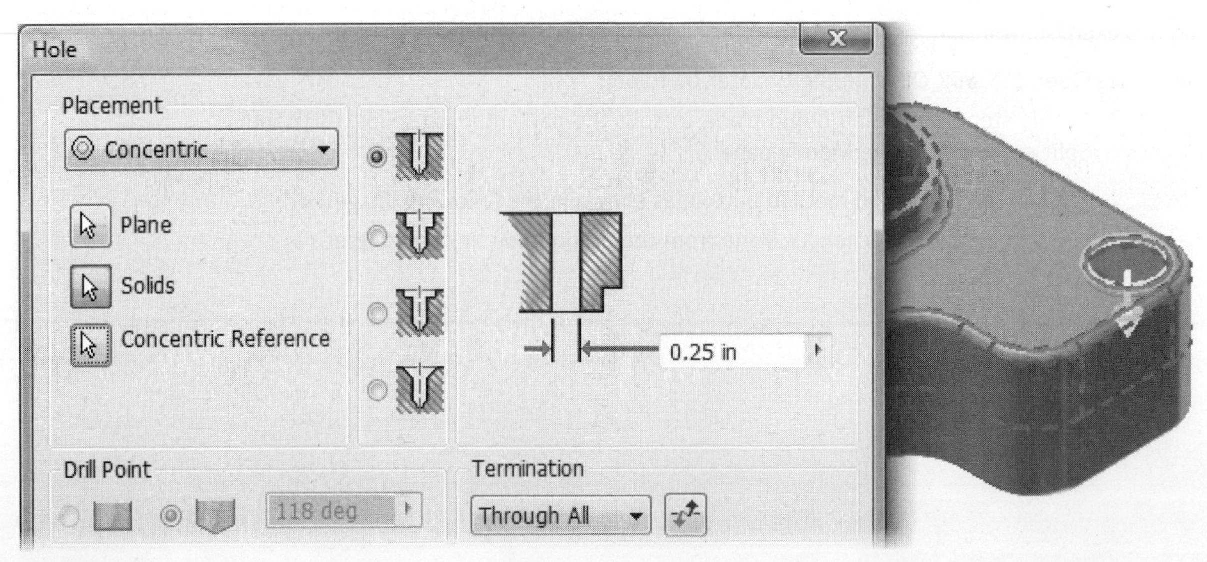

FIGURE 7.87

11. Use the Free Orbit command to examine the bottom of the part. Notice that the hole does not go through the bottom solid.
12. Change to the Home View.
13. In the browser expand the Solid Bodies folder and the Solid2 and Solid3 entry. Notice that Hole1 is only under Solid2, as shown in the following image on the left.
14. Edit the hole feature and click the Solids button, and then select the lower solid (Solid3) in the graphics window. Click OK to complete the edit.
15. Use the Free Orbit command to ensure that the hole goes through both solids.
16. In the browser also notice that the Hole1 is now under both Solid2 and Solid3.

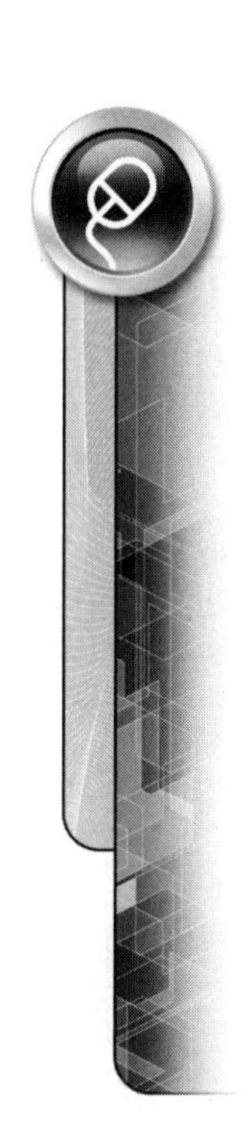

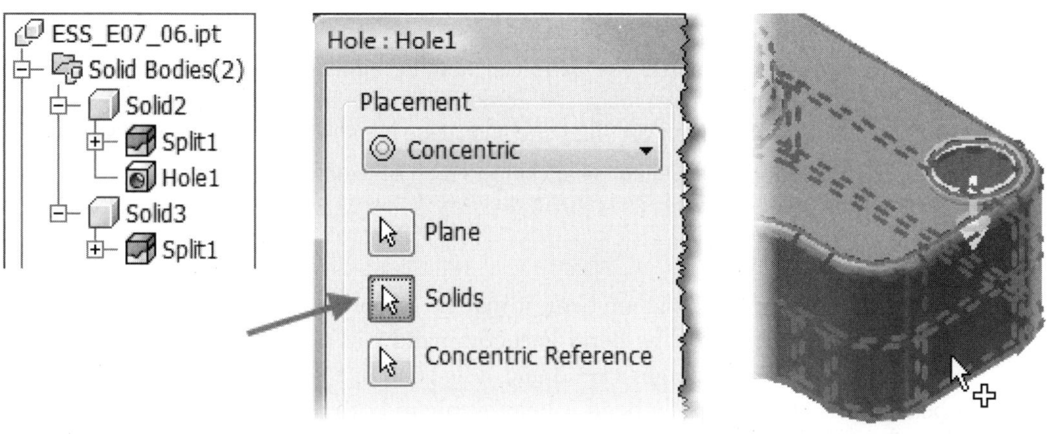

FIGURE 7.88

17. Next you export the solids into an assembly. Click the Make Components command on the Manage tab > Layout panel.
    - Select the two solids.
    - Change the Target assembly location to **C:\Inv 2012 Ess Plus\Chapter 07** as shown in the following image.

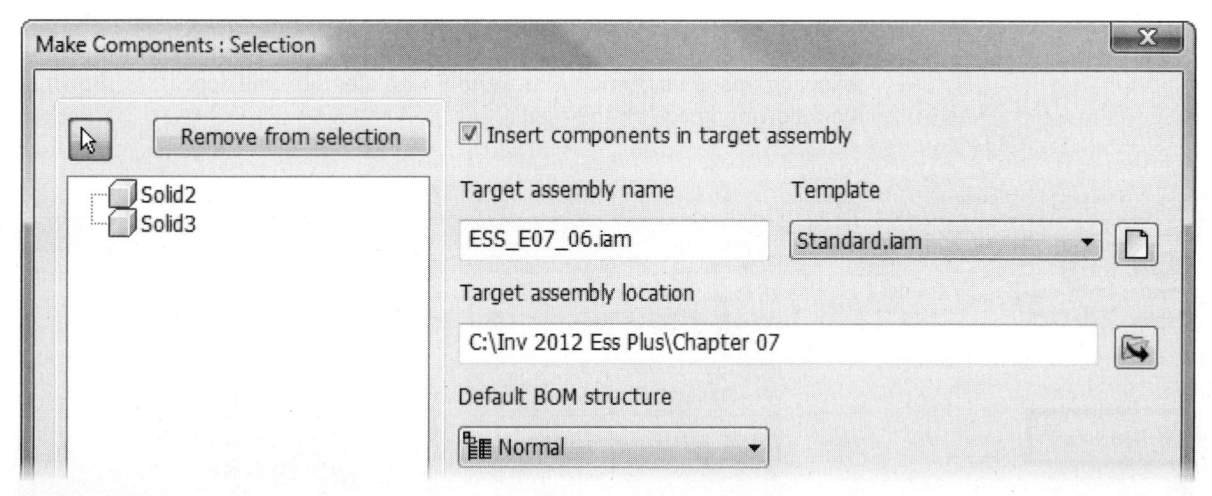

FIGURE 7.89

18. Click Next and change the components names, as shown in the following image.

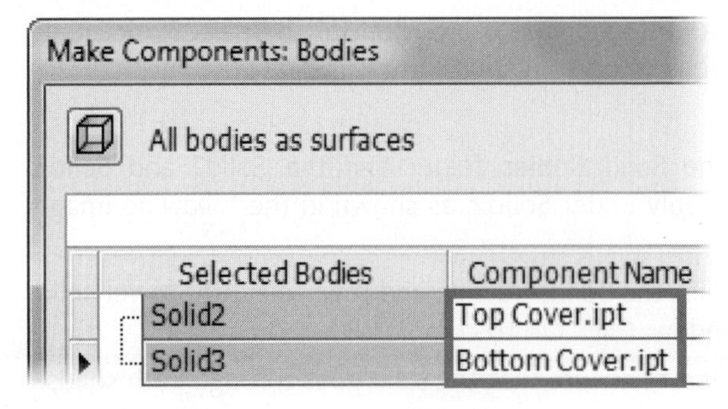

FIGURE 7.90

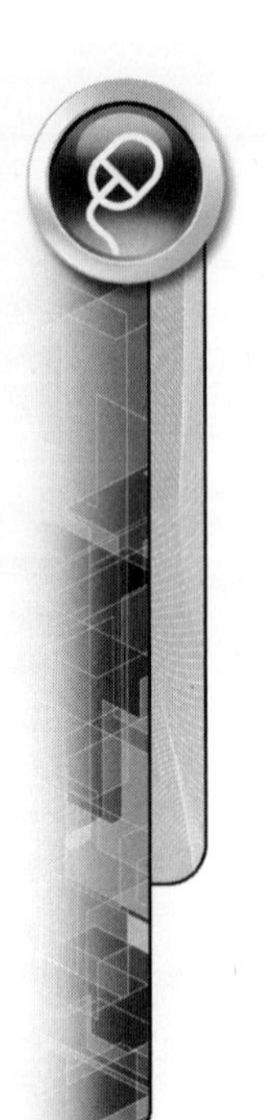

19. Click OK to create the parts and assembly.
20. Examine the parts. The parts can now be edited, for example, plastic feature placed. In the assembly, the components can be ungrounded and assembled. Note that the new part files size and shape are still driven by the original part file.
21. Close the files. Do not save changes. End of exercise.

BEND PART

While designing parts that are bent, it may be easier to create the part in a flattened state and then bend it. The Bend Part command will bend a side or both sides of a part based on the location of a bend line. The bend line will be the location where the bend or the centerline of a bend is bent evenly in both directions. After creating the bend, it will appear in the browser as a feature and can be edited like any other feature. To bend a part, follow these steps.

1. Create or open a part that will be bent.
2. Create a sketch where the bend line will be placed. The plane must intersect the part or lie on a planar face of the part.
3. Draw a line that will be used as the bend line, the line does not need to extend to the limits of the part.
4. Exit the sketch.
5. Click the Bend Part command from the Model tab > Modify panel and click the drop arrow next to Modify to see the Bend Part command, as shown in the following image on the left. The Bend Part dialog box will appear, as shown in the following image on the right.

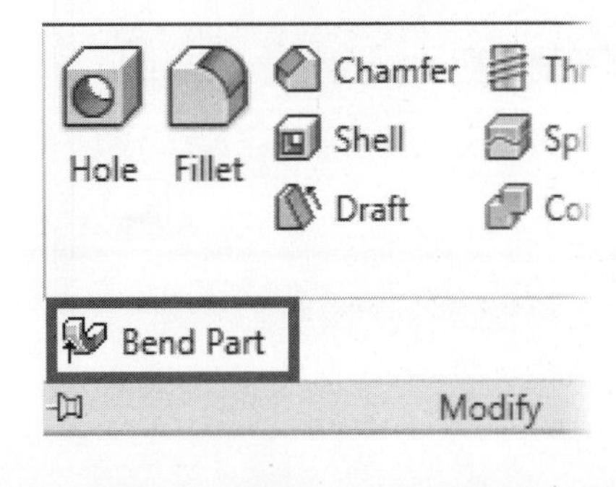

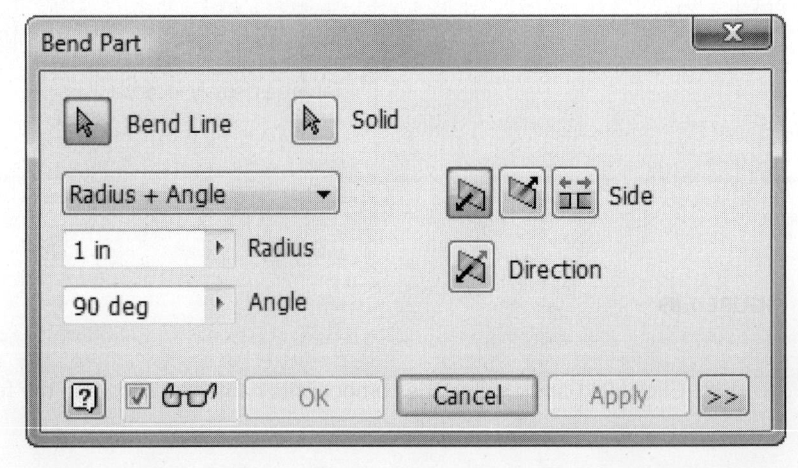

FIGURE 7.91

6. In the dialog box, specify how you want to calculate the bend: Radius + Angle, Radius + Arc Length, or Arc Length = Angle.
7. Enter the data to create the bend.
8. If needed, change the side and direction that the bend occurs.
9. The following image on the left shows the part with a bend line in the middle of the inside face and the image on the right shows the bent part.

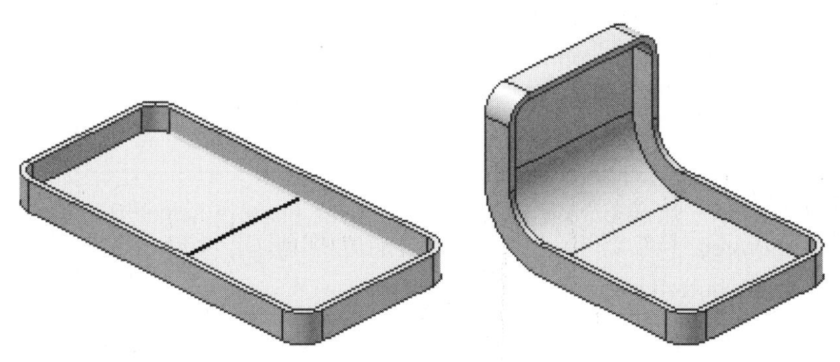

FIGURE 7.92

MIRROR FEATURES

When creating a part that has features that are mirror images of one another, you can use the Mirror Feature command to mirror a feature(s) about a planar face or work plane instead of recreating the features from scratch. Before mirroring a feature, a work plane or planar face that will be used as the mirror plane must exist. The feature(s) will be mirrored about the existing plane; it can be a planar face, a work plane on the part, or an origin plane. The mirrored feature(s) will be dependent on the parent feature—if the parent feature(s) change, the resulting mirror feature will also update to reflect the change. To mirror a feature or features, use the Mirror command on the Model tab > Pattern panel, as shown in the following image on the left. The Mirror dialog box will appear, as shown on the right. The following sections explain the options in the Mirror dialog box.

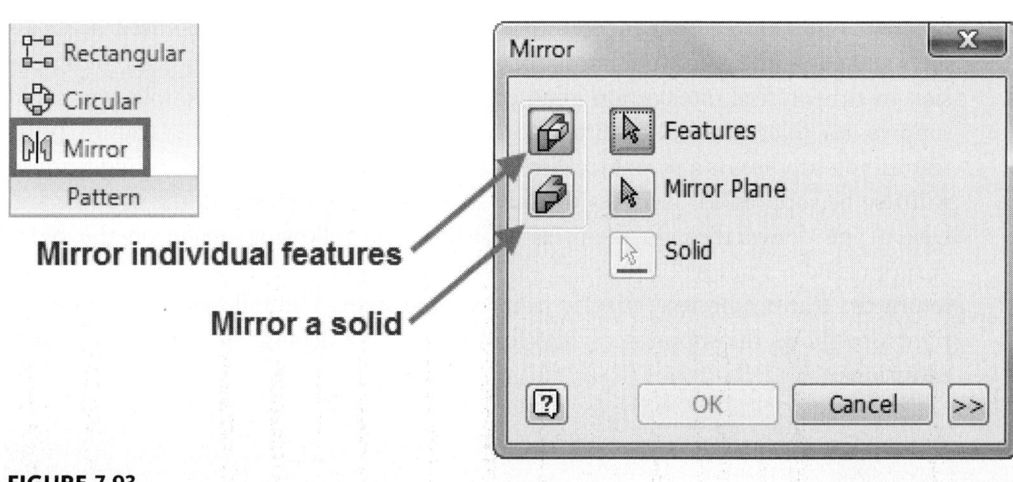

FIGURE 7.93

Mirror Individual Features. Click this option to mirror a feature or features.

Mirror the Entire Solid. Click this option to mirror the solid body.

Mirror Plane. Click and choose a planar face or work plane on which to mirror the feature(s).

Solid. If there are multiple solid bodies, click this button to choose the solid body(ies) to receive the mirrored feature.

To mirror a feature or features, follow these steps:

1. Click the Mirror Feature command on the Pattern panel. The Mirror dialog box will appear.
2. Click the Mirror individual features option or the Mirror the entire solid option.
3. Select the feature(s) or solid body to mirror.
4. Click the Mirror Plane button, and then select the plane on which the feature will be mirrored.
5. Click the OK button to complete the operation.

The following image on the left shows a part with the features that will be re-created, the middle image shows the part with a work plane that the features will be mirrored about, and the image on the right shows the features mirrored.

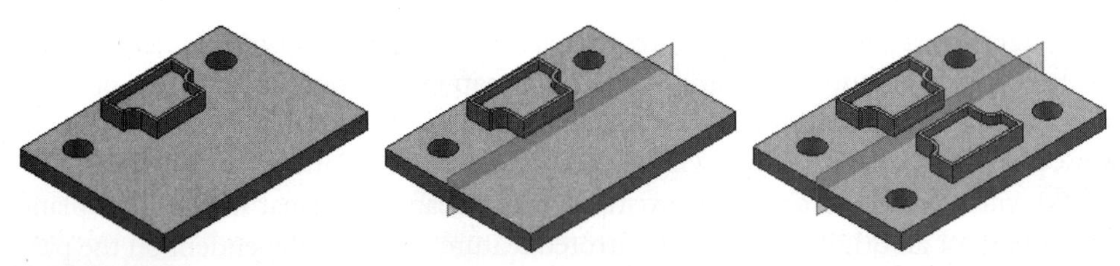

FIGURE 7.94

SUPPRESSING FEATURES

You can suppress a model's feature or features to temporarily turn off their display, as shown in the following image on the left. Feature suppression can be used to simplify parts, which may increase system performance. This capability also shows the part in different states throughout the manufacturing process, and it can be used to access faces and edges that you would not otherwise be able to access. If you need to dimension to a theoretical intersection of an edge that was filleted, for example, you could suppress the fillet and add the dimension and then unsuppress the fillet feature. If the feature you suppress is a parent feature for other dependent features, the child features will also be suppressed. Features that are suppressed appear gray in the browser and have a line drawn through them, as shown in the following image on the right. A suppressed feature will remain suppressed until it is unsuppressed, which will also return the features browser display to its normal state. The following image on the right also shows the suppressed child features that are dependent on the suppressed extrusions.

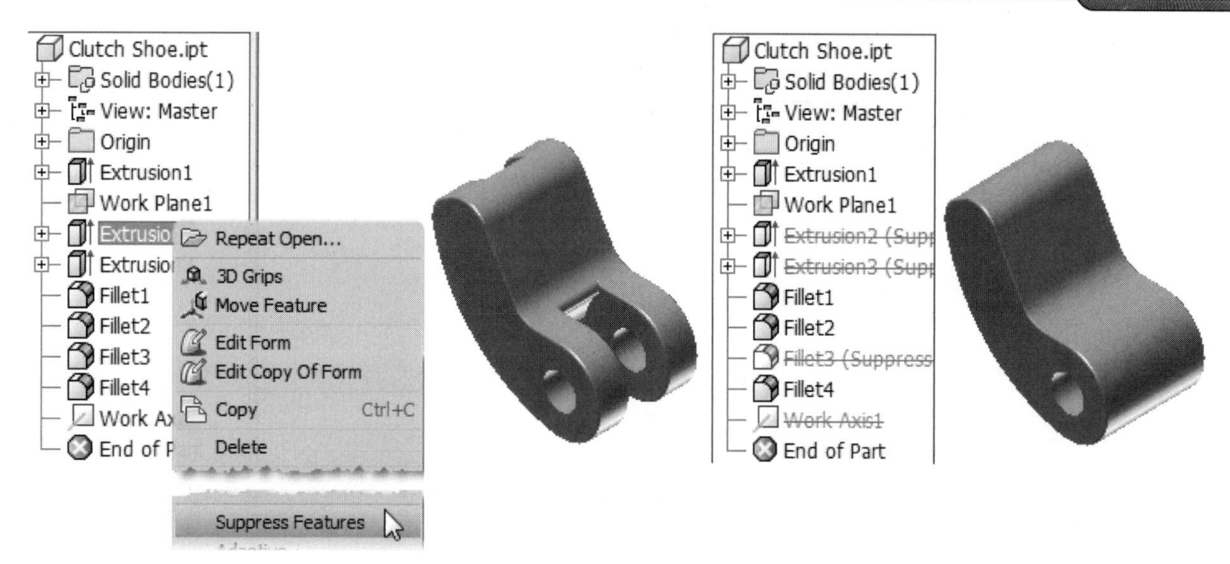

FIGURE 7.95

To suppress a feature on a part, use one of these methods:

1. Right-click on the feature in the browser, and select Suppress Features from the menu, as shown in the previous image on the left, or
2. Click the Select Feature option from the Quick Access toolbar. This option allows you to select features on the parts in the graphics window.
3. Right-click on the feature in the graphics window, and click the Suppress Features option from the menu.

To unsuppress a feature, right-click on the suppressed feature's name in the browser, and select Unsuppress Features from the menu.

REORDERING A FEATURE

You can reorder a feature in the browser. If you created a fillet feature using the All Fillets or All Rounds option and then created an additional extruded feature, such as a boss, you can move the fillet feature below the boss and include the edges of the new feature in the selected edges of the fillet. To reorder features, follow these steps:

1. Click the feature's name or icon in the browser. Hold down the left mouse button and click and drag the feature to the desired location in the browser. A horizontal line will appear in the browser to show you the feature's relative location while reordering the feature. The following image on the left shows a hole feature being reordered in the browser.
2. Release the mouse button and the model features will be recalculated in their new sequence. The following image on the right shows the browser and the reordered hole features.

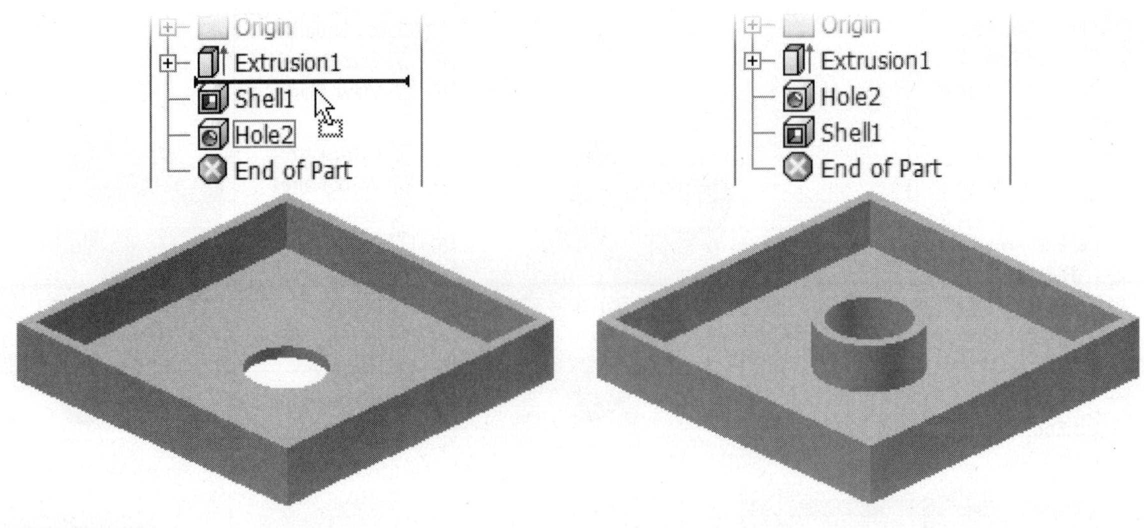

FIGURE 7.96

If you cannot move the feature due to parent-child relationships with other features, Autodesk Inventor will not allow you to drag the feature to the new position. In the browser, the cursor will change to a No symbol instead of a horizontal line, as shown in the following image.

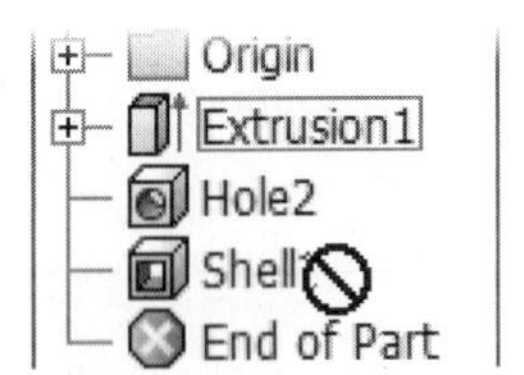

FIGURE 7.97

FEATURE ROLLBACK

While designing, you may not always place features in the order that your design later needs. Earlier in this chapter, you learned how to reorder features, but reordering features will not always allow you to create the desired results. To solve this problem, Autodesk Inventor allows you to roll back the design to an earlier state and then place the additional new features. To roll back a design, drag the End of Part marker in the browser to the location where the new feature will be placed. To move the End of Part marker, follow these steps:

1. Move the cursor over the End of Part marker in the browser.
2. With the left mouse button depressed, drag the End of Part marker to the new location in the browser. While dragging the marker, a line will appear, as shown in the following image on the left.
3. Release the mouse button, and the features below the End of Part marker are removed temporarily from calculation of the part. The End of Part marker will be moved to its new location in the browser. The following image in the middle shows the End of Part marker in its new location.

4. Create new features as needed. The new features will appear in the browser above the End of Part marker.
5. To return the part to its original state, including the new features, drag the End of Part marker below the last feature in the browser.
6. If needed, you can delete all features below the End of Part marker by right-clicking on the End of Part marker. Then click Delete all features below EOP, as shown in the following image on the right.

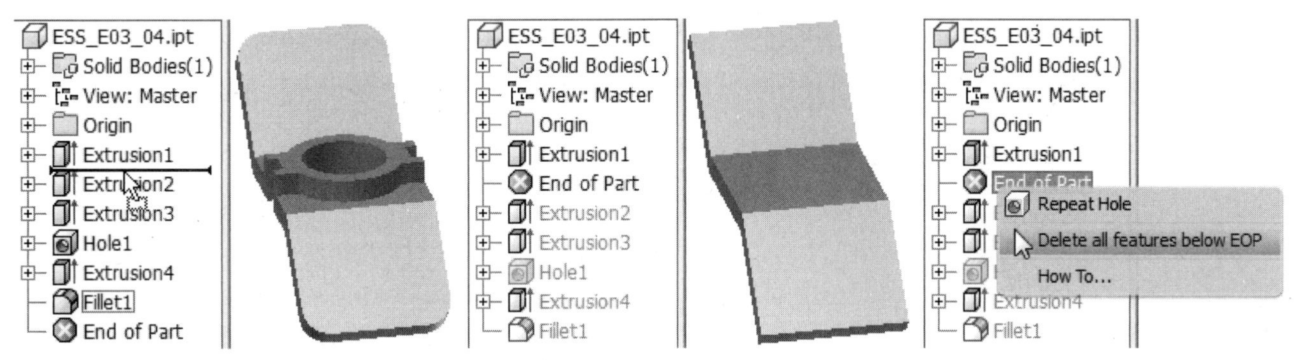

FIGURE 7.98

DERIVED PARTS

Derived parts are used to capture the design intent of a part, assembly, specific sketch, work geometry, surface, parameter, or iMate by either creating a new part file or by importing any of the aforementioned selections. The derived technique can be used in two ways: You can push a derived part out, which was covered in the Make Part and Make Components sections of this chapter. The second method is to pull the information into a part file. A derived part is a part file that links in a selected part, assembly, or sketch from another file. The derived operation is also used in the skeletal modeling technique. Additional features can then be added to the derived part or solid. If the original part changes, the derived part will update to include any changes that are made to the original file. This associativity can be broken if you do not want changes made to the original file to be included in the linked part file. You can derive in as many parts or assemblies into a part as required.

Some possible uses for a derived part or assembly are as follows:

- Use existing parts as solid bodies.
- Create different parts that are machined from a single casting blank.
- Scale or mirror a derived part upon creation: note that this does not apply to derived assemblies. You can derive sketches to be used within an assembly as a layout.
- Derive a part as a work surface that can be used to simplify the representation of a part.

You create a derived part using the Derive command on the Manage tab > Insert panel, as shown in the following image on the left.

Navigate to and select the Part or Assembly file that you want to derive. If you select a part file, the Derived Part dialog box opens, as shown in the following image in the middle. If you select an assembly file, the Derived Assembly dialog box opens to provide you with different options, as shown in the following image on the right.

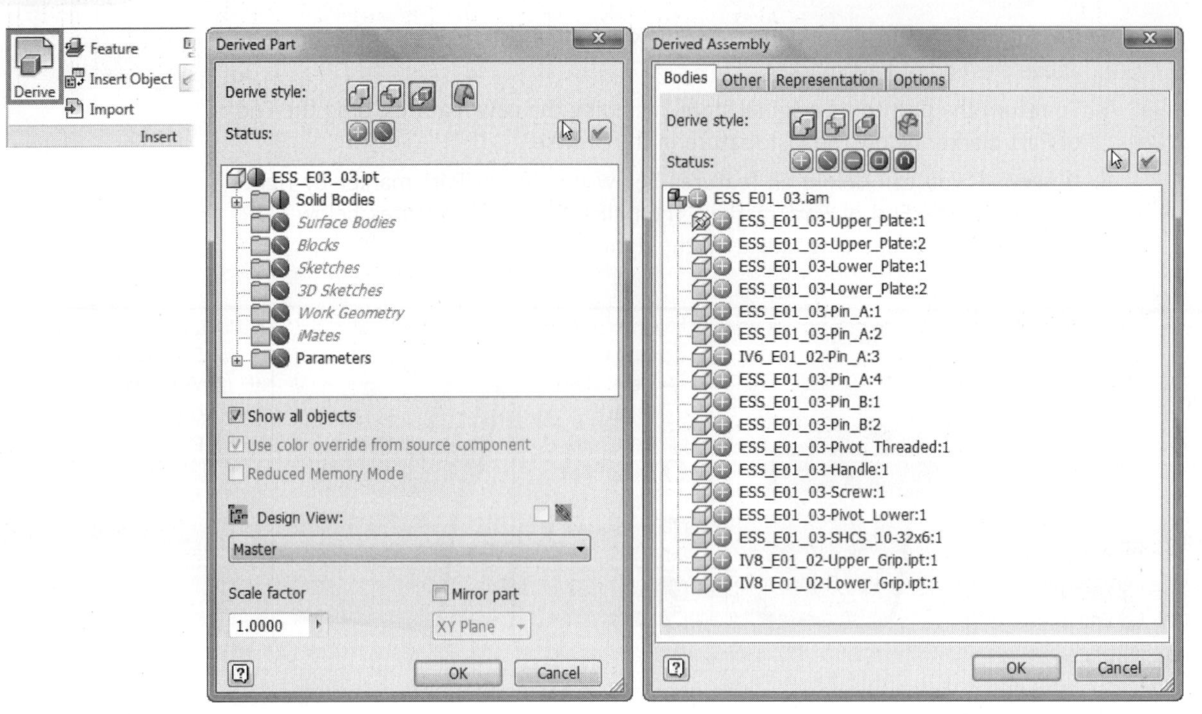

FIGURE 7.99

Derived Part Symbols

The symbols show you how the solid bodies will be handled, the status of geometry that will be contained in the derived part, and whether or not it will be included or excluded from the derived part. The Derived Part Symbols table shows the available symbols.

DERIVED PART SYMBOLS

Symbol	Function	Meaning
	Merge Seams	Single solid body merging out seams between planar faces.
	Retain Seams	Creates a single solid body part that retains the seams between planar faces.
	Solid Bodies	If the source contains a single body, it creates a single body part. If the source contains multiple visible solid bodies, select the required bodies to create a multi-body part. This is the default option.
	Create Surface	Creates a part with the selected bodies as base surfaces.
	Include	Indicates that the selected geometry will be included in the derived part.
	Exclude	Indicates that the selected geometry will not be included in the derived part. If a change is made to this type of geometry in the parent part, it will not be incorporated or updated in the derived part.
	Mixed	Indicates that the folder contains mixed included and excluded objects.

(Continued)

Symbol	Function	Meaning
	Select	Switches focus to the base part window so that you can take advantage of the selection commands and keep the Derived Part dialog open.
	Accept	Causes the Derived Part dialog to absorb the selections you made in the base part window, returns you to the Part environment, and highlights your selections in the dialog tree control.

Derived Part Dialog Box

The Derived Part dialog box contains the following options, as shown in the following chart.

Solid Body	Select this option to derive the part as a base solid.
Surface Bodies	Select this option to derive the part body as a work surface. The body is brought in, behaves, and appears as an Autodesk Inventor surface.
Sketches	Select this option to include any unconsumed 2D sketches from the original part.
3D Sketches	Select this option to include any unconsumed 3D sketches from the original part.
Work Geometry	Select this option to include any work features from the original file. They can then be used to create new geometry or to constrain a part in an assembly.
iMates	Select this option to include any iMates that exist in the original part.
Parameters	Select this option to include any parameters designated to be exported parameters in the original file.
Composite Features	Select this option to include any composite features that exist in the original part.

NOTE

If the original file consists only of surfaces, the surface option will be the only available option.

Scale Factor	Select a scale factor in percentage to scale the derived part. The default scale factor is 1.0 or the same size as the original file.
Mirror Part	Select this option to mirror the original part about the XY, XZ, or YZ origin work planes upon derived part creation. You can select the plane about which to mirror the derived part from the drop-down list.

Derived Assembly Symbols

The symbols available when deriving an assembly are different from the symbols available when deriving a part file. The following table shows the derived assembly symbols.

DERIVED ASSEMBLY SYMBOLS

Symbols	Function	Meaning
	Merge Seams	Single solid body merging out seams between planar faces.
	Retain Seams	Creates a single solid body part which retains the seams between planar faces.
	Solid Bodies	If the source contains a single body, it creates a single body part. If the source contains multiple visible solid bodies, select the required bodies to create a multi-body part. This is the default option.
	Composite Surface	Creates a part with the selected bodies as a single composite surface.
	Include	Indicates that the selected component will be included in the derived part.
	Exclude	Indicates that the selected component will not be included in the derived part. Changes to this type of component in the parent assembly will not be incorporated or updated in the derived part.
	Subtract	Indicates that the selected component will be subtracted from the derived part. If the subtracted component intersects another portion of an included part, the result will be a void or cavity in the derived part.
	Bounding Box	Indicates that the selected component will be represented as a bounding box. The bounding box size is determined by the extents of the component. The bounding box is used to represent a component as a placeholder and reduces memory. You can add features to a bounding box, and it will update when changes are made to the original part.
	Intersect	Intersects the selected component with the derived part. One component must have an Include status. If the component does not intersect the derived part, the result is not a solid.
	Select	Switches focus to the base assembly window so that you can take advantage of the selection commands and keep the Derived Assembly dialog open.
	Accept	Causes the Derived Assembly dialog to absorb the selections you made in the assembly environment, returns to the Part environment, and highlights your selections in the dialog tree control.

Scale Factor	Select a scale factor in percentage to scale the derived part. The default scale factor is 1.0 or the same size as the original file.
Mirror Part	Select this option to mirror the original part about the XY, XZ, or YZ origin work planes upon derived part creation. You can select the plane about which to mirror the derived part from the drop-down list.
Reduced Memory Mode	When checked on the Options Tab, Autodesk Inventor uses less memory by excluding bodies of the parts that are cached in memory. When the link is broken or suppressed the memory savings is removed.

The Other tab of the Derived Assembly dialog box contains options that allow you to select sketches, work geometry, surfaces, and exported parameters from the assembly file or any of the components that are in the assembly. Using the plus and minus icons, you can select which items you want included in the file.

The Representation tab of the Derived Assembly dialog box contains options that allow you to select available design view representation, positional representation, or level of detail representations present in the assembly.

The Options tab of the Derived Assembly dialog box are used to simplify, scale, or mirror an assembly. These options are also available in the Shrinkwrap command that will be covered in the next section.

Because you are using geometry or parts based on another file, any modifications to the original parts are incorporated into the derived component. If you modify a parent file, the update icon next to the derived component in the browser displays an update symbol, as shown in the following image on the left. The Local Update command on the Quick Access toolbar also becomes active, as shown in the following image in the middle. To update the file, click the Local Update command on the Quick Access toolbar and the update symbol for the derived part/assembly in the browser will change to what is shown in the following image on the right.

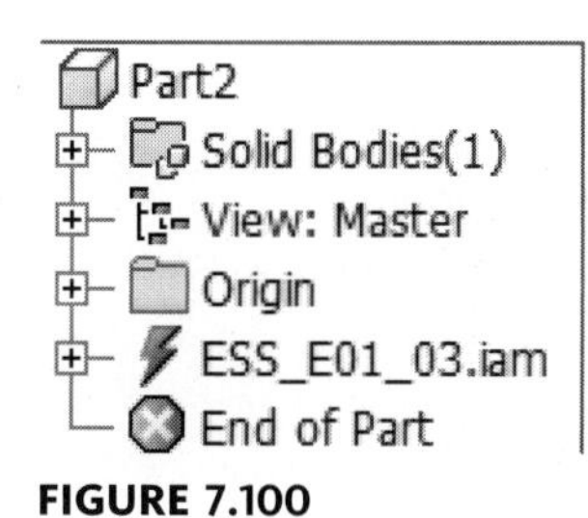

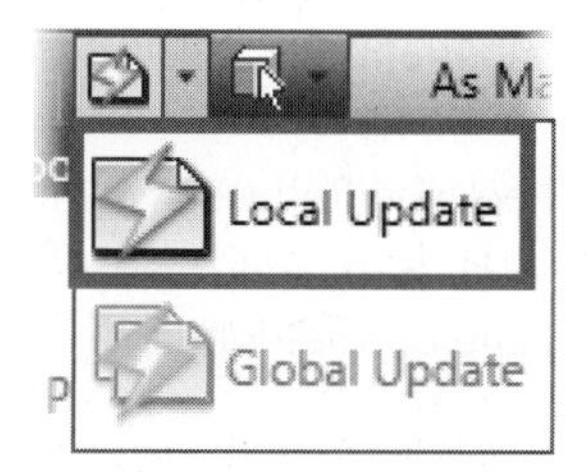

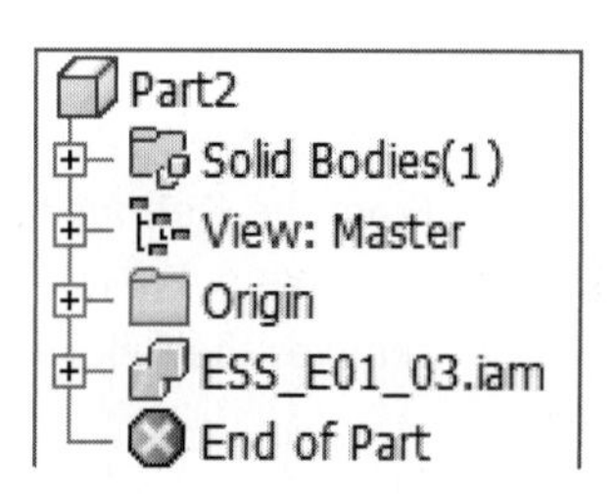

FIGURE 7.100

You can edit, break, or suppress the link to the parent file.

To edit the derived part, right-click on the derived parent component in the browser and select either Open Base Component, Edit Derived Assembly or Edit Derived Part from the menu and the same Derived Part dialog box will appear as the derived part was created with.

To break the link with the derived part and its original part, right-click on the derived component in the browser and from the menu click Break Link With Base Component, as shown in the following image on the left. Once the link is broken, the derived component icon will display a broken chain link to signify that the link to the parent file no longer exists. Once you have broken the link, you cannot reestablish it.

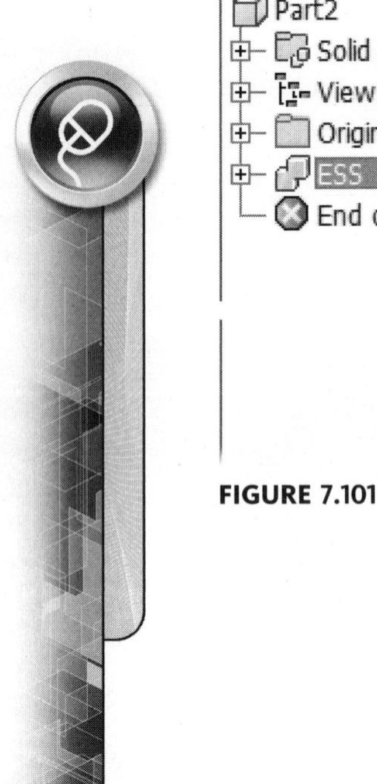

You can temporarily suppress the changes from the base assembly from affecting the derived part by clicking Suppress Link With Base Component from the menu. To reestablish the link, click Unsuppress Link With Base Component, as shown in the following image on the right.

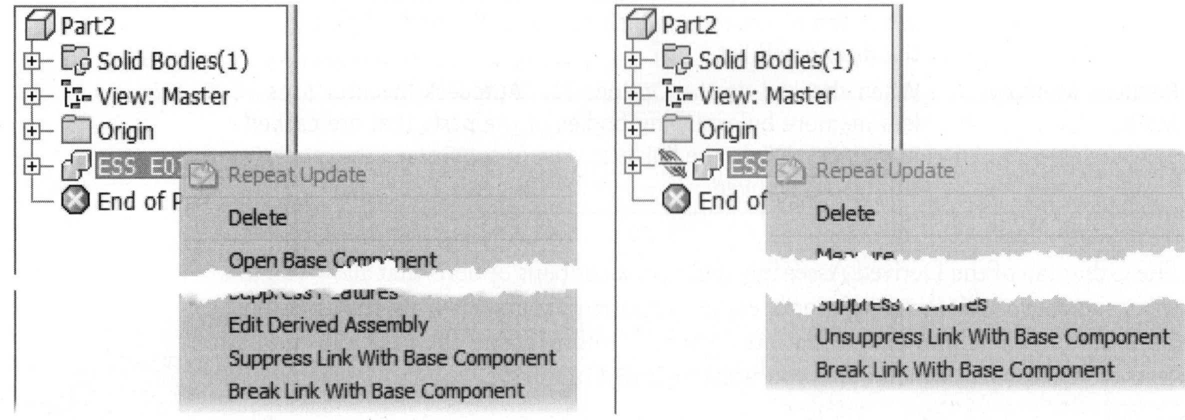

FIGURE 7.101

EXERCISE 7-7: CREATING A DERIVED PART

In this exercise, you create a mold cavity by deriving an assembly consisting of a part that you'll create a mold of and a mold base from which the part cavity will be removed. Assembly constraints have already been applied to the parts in the assembly to center the parts. You then edit the part and update the derived part.

1. Click the New command, click the English tab, and double-click *Standard (in).ipt*.
2. Exit the sketch environment by clicking Finish Sketch on the Exit panel.
3. Click the Derive command from the Manage tab > Insert panel.
4. In the Open dialog box, click the file *ESS_E07_07_Assembly.iam* from the Chapter 07 folder, and then click the Open button in the dialog box.
5. In the Derived Assembly dialog box, click the icon next to the entry *ESS_E07_07-Part:1* so the subtract symbol appears, as shown in the following image on the left.
6. From the Options tab, check the Reduced Memory Mode option.
7. Click OK to create the mold cavity and change the viewpoint so your screen resembles the following image on the right.

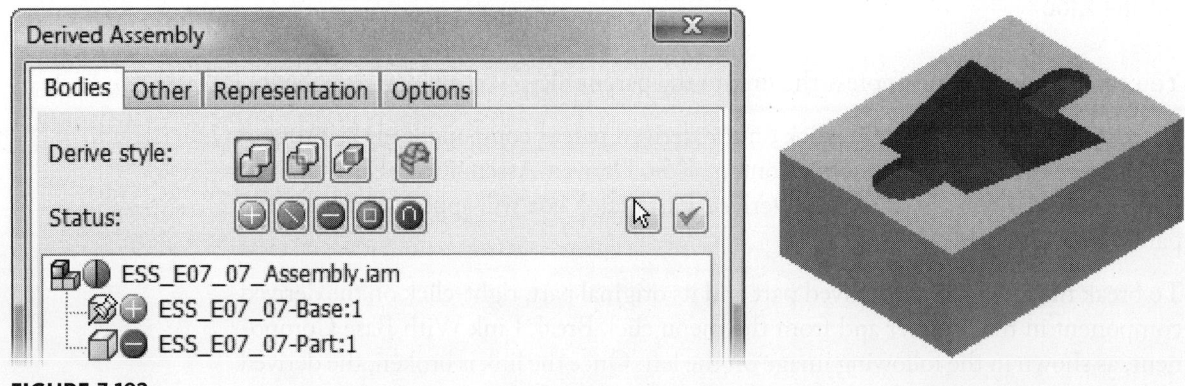

FIGURE 7.102

8. Next you create a hole for the material to flow into the cavity. Create a new sketch on the front-right face of the mold base.
9. Place a Point, Center Point on the midpoint of the top edge.
10. Start the Hole command, and follow these steps:
    - The point, center point should automatically be selected. Change the diameter to **.125 inches**.
    - Change the operation to To and check the Check to terminate feature on the extended face.
    - Select the inside circular face as shown in the following image.
    - Click OK to create the hole. The part should resemble the following image on the right.

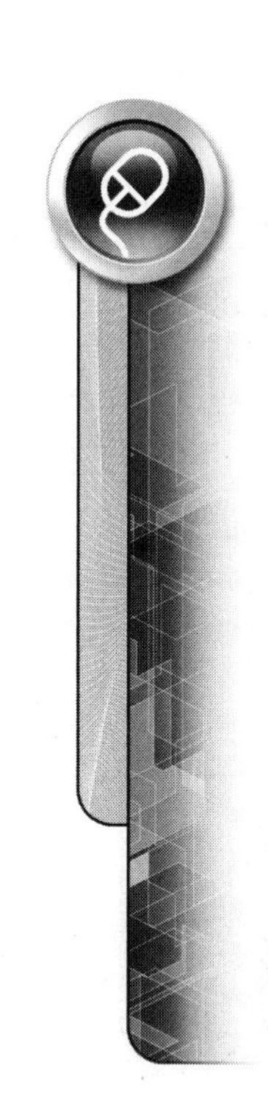

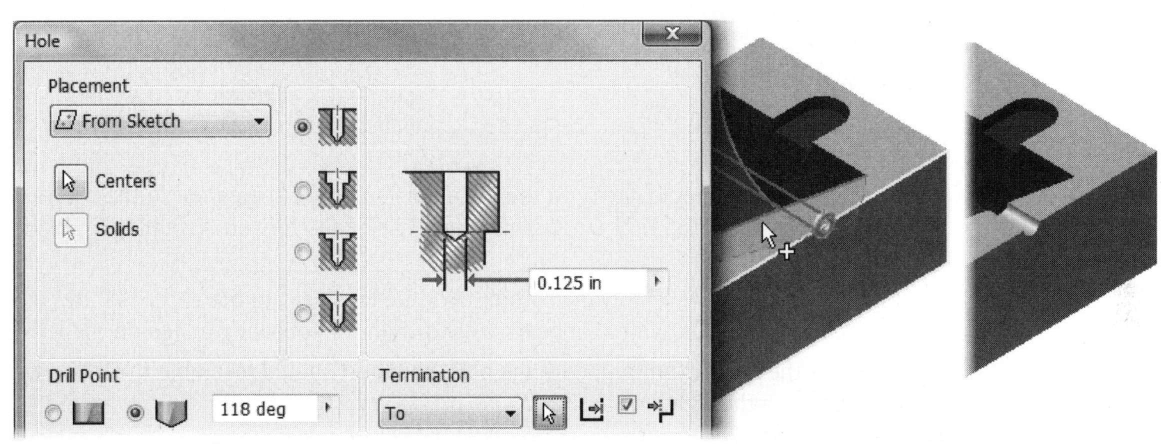

FIGURE 7.103

11. Save the file as *ESS_E07_07_MoldCavity.ipt* to the Chapter 07 folder.
12. Move the cursor in the browser over the entry *ESS_E07_07_Assembly.iam* and click Open Base Component from the menu. Notice that the hole in the derived part does not exist in the original part.
13. In the browser, double-click on *ESS_E07_07-Part:1* and edit Extrusion1 and Extrusion2 to the distance of **.375 inches**.
14. Click the Return command on the Return panel and save the assembly and changes to the part files.
15. Close the assembly file and verify that the file *ESS_E07_07_MoldCavity.ipt* is the current file.

16. In the browser, notice the red lightning bolt next to *ESS_E07_07_Assembly.iam.*
17. Click the Local Update command in the Quick Access toolbar, and the cavity should resemble the following image.

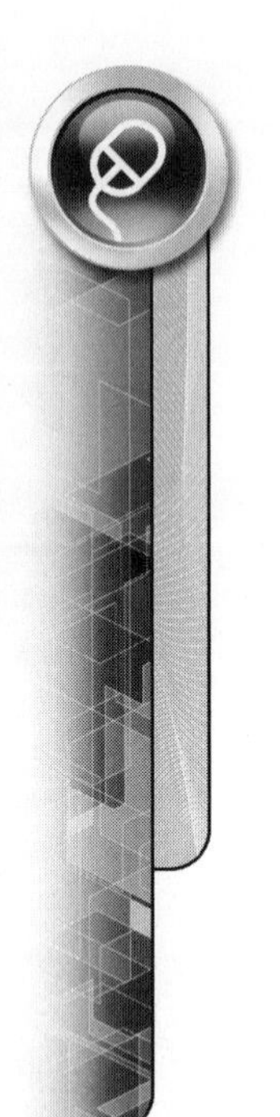

FIGURE 7.104

18. To change the mold cavity to the male portion, move the cursor in the browser over the entry *ESS_E07_07_Assembly.iam.* Click Edit Derived Assembly from the menu.
19. In the Derived Assembly dialog box, click the icon next to the entry *ESS_E07_07-Part:1* so the plus symbol appears, as shown in the following image on the left.
20. Click the OK button to update the part. Your part should resemble the following image on the right.

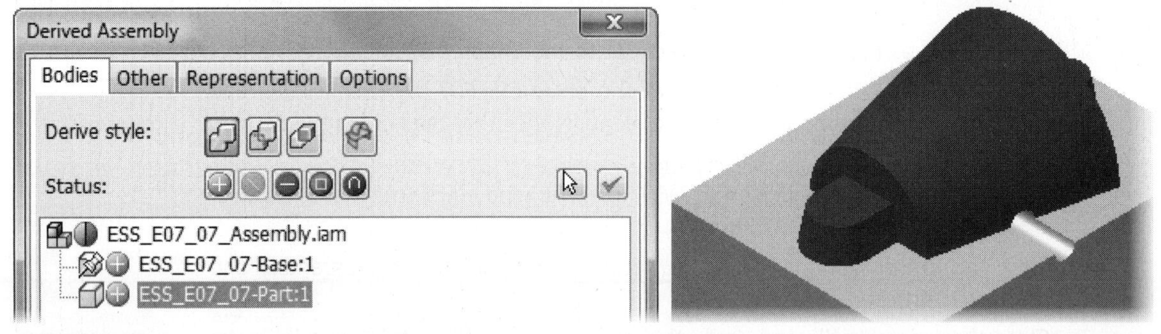

FIGURE 7.105

21. Close the file. Do not save changes. End of exercise.

SHRINKWRAP

Another method to push a derived part is from the Shrinkwrap command. The push technique creates a derived relation by exporting a part and maintains a relationship to this part. The Shrinkwrap command creates a stand-alone, single-part version of a model assembly and simplifies the assembly by removing geometry. Since the part is derived, any changes to the original assembly can be updated in the derived part. The Shrinkwrap command enables you to:

- Reduce detail of data to protect intellectual property and reduce file size.
- Create a substitute part for use with alternative representations.
- Create simplified data for complex purchased assemblies.

To create a shrinkwrap part, follow these steps:

1. Open an assembly that you want to create a simplified part from.
2. Click the Shrinkwrap command from the Assembly tab > Component panel, as shown in the following image on the left. You can also use the Shrinkwrap command to create a substitute level of detail, this is covered in Chapter 9.
3. Enter a name, select a template file, and specify a location for the derived part, as shown in the following image on the right.

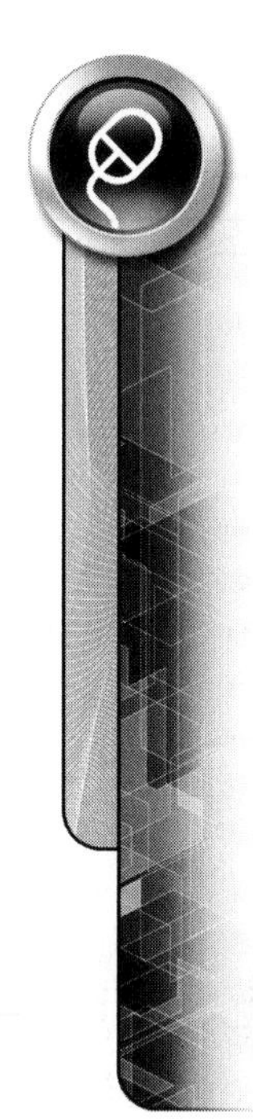

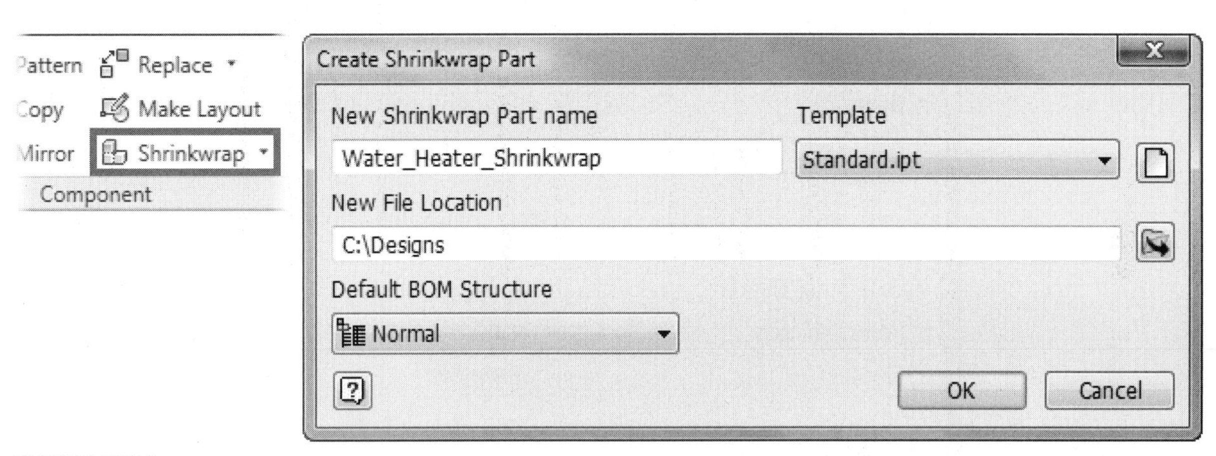

FIGURE 7.106

4. Click OK and the Assembly Shrinkwrap Options dialog box will appear, as shown in the following image.
5. Select the different options to remove the geometry. Click the Preview button to see the results of the existing options.
6. When done, click OK.

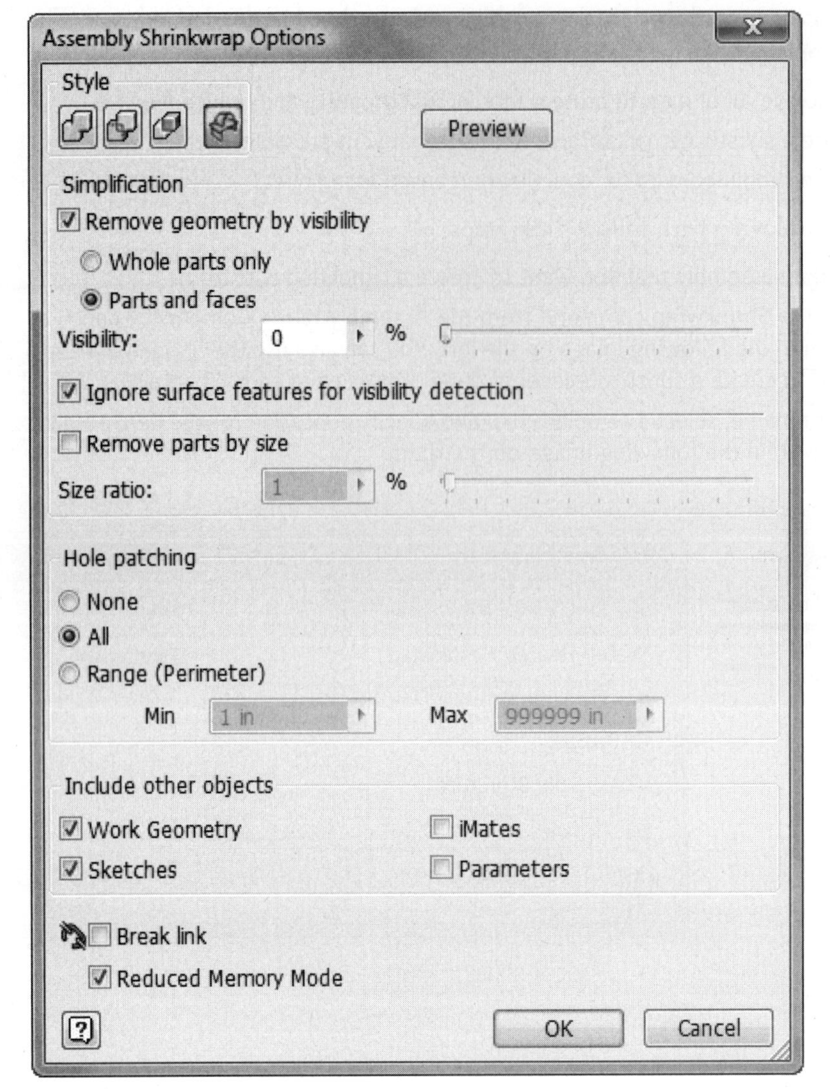

FIGURE 7.107

The options in the dialog box are explained below.

Style

Single solid body merging out seams between planar faces: Select to produce a single solid body without seams between planar faces. When you merge seams between faces, the face assumes a single color.

Solid body keep seams between planar faces: Select to produce a single solid body with seams between planar faces retained.

ASSEMBLY SHRINKWRAP OPTIONS

Symbols	Function	Meaning
	Single Solid	Creates a single solid body merging out seams between planar faces.
	Single Solid with Seams	Creates a single solid body part which retains the seams between planar faces.

(Continued)

Symbols	Function	Meaning
	Maintain each solid as a solid body	Creates a multi-body part, each part is created as a unique solid body.
	Single Composite Surface	Creates a part with the selected bodies as a single composite surface.

Simplification

Whole parts only: Whole parts which meet the visibility criteria are removed.

Parts and faces: Removes any face including entire parts which meet the visibility criteria.

Visibility percentage: A value of zero removes all parts or faces that are not visible in any view. Increasing the slider value removes more parts and faces.

Ignore surface features for visibility detection: Available if Remove geometry by visibility is enabled. If enabled, surface features do not impact visibility detection.

Remove parts by size: Check to enable the option to remove parts based on the size ratio. The ratio indicates the difference between the part bounding box and the assembly bounding box.

Hole Patching

None: No holes are removed.

All: Removes all holes that do not cross surface boundaries. Holes do not need to be round to be included; a void in material is seen as a hole.

Range: Specifies the circumference or perimeter of the holes to include or exclude. Holes do not need to be round to be included.

Include other objects

Work Geometry: When checked, any visible work features in the component are exported and can be derived.

Sketches: When checked, any visible and unconsumed 2D or 3D sketches in the component are exported and can be derived.

iMates: When checked, any iMates defined in the source assembly are exported and can be derived.

Parameters: When checked, any parameters in the source assembly are exported and can be derived.

Break link

Permanently disables any updates from the source component.

Reduced Memory Mode

When checked, a part is created using less memory by excluding source bodies from the cache.

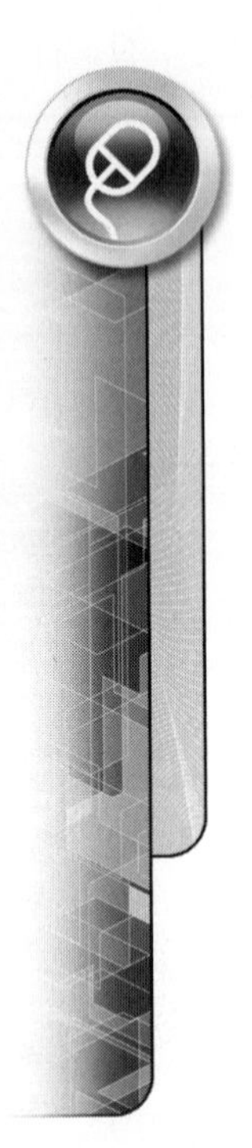

EXERCISE 7-8: SHRINKWRAP

In this exercise, you derive a simplified part from an assembly using the Shrinkwrap command.

1. Open *ESS_E07_08.iam* from the Chapter 07 folder.
2. Use the Free Orbit command to examine the assembly.
3. Return to the Home view.
4. Start the Shrinkwrap command from the Component panel.
5. Click OK to accept the default name and location.
6. Change the options, as highlighted in the following image.

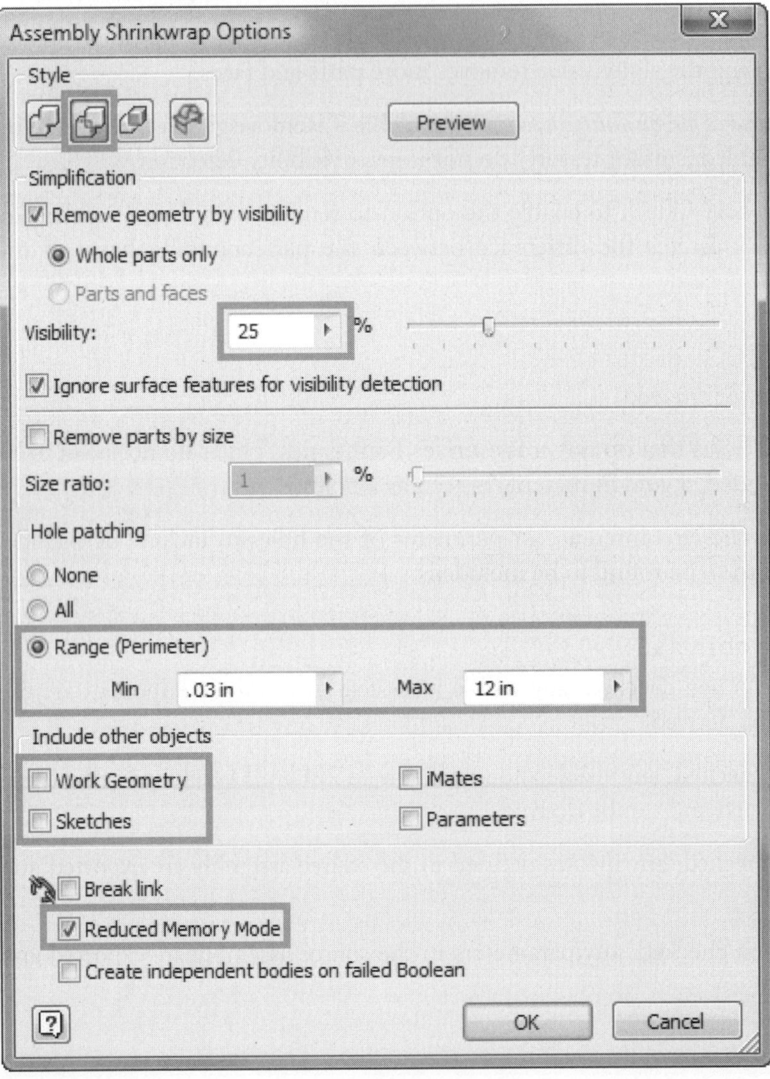

FIGURE 7.108

7. Click the Preview button to see the pending changes.
8. Try other options; click the Preview button to see the affects.
9. Put the options back to the settings in the previous dialog box.
10. Click OK to create the derived part.
11. Next you remove select components from the derived part. In the derived part, right-click in the browser on the entry *ESS_E07_08.iam* and click Edit Derived Assembly from the menu.

12. Exclude the highlighted parts, as shown in the following image on the left.
13. Click OK to complete the operation; rotate the viewpoint so your screen should resemble the following image on the right.

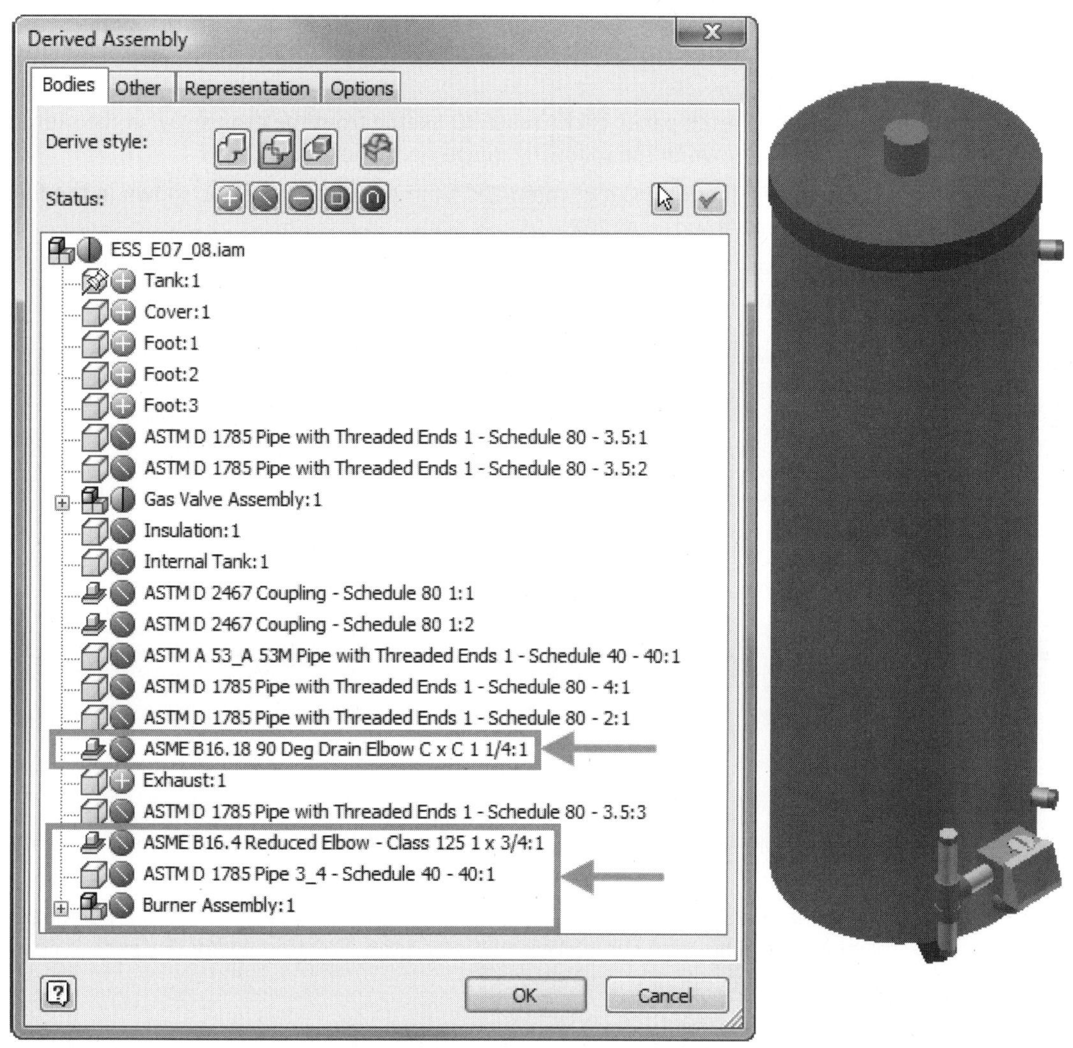

FIGURE 7.109

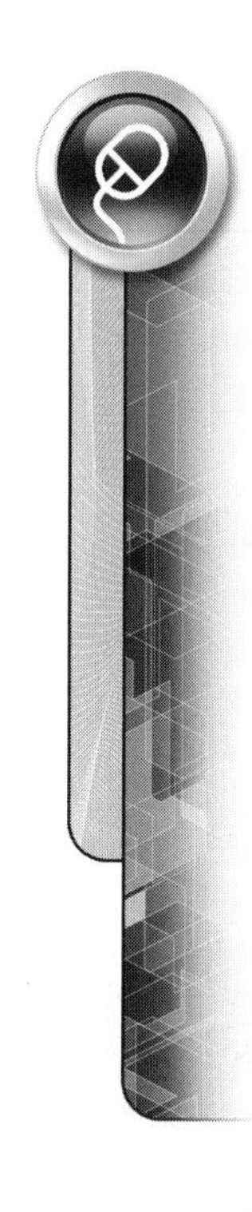

14. Change the display to wireframe and use the Free Orbit command to examine the part.
15. Practice changing the Edit Derived Assembly options.
16. Close the file. Do not save changes. End of exercise.

PLASTIC PART FEATURES

Autodesk Inventor has design specific tools that will reduce the efforts to create features that are found in plastic parts. This section will introduce how to create silhouette curves (parting line), grills, bosses, rests, hooks, and snaps and how to create fillets based on rules and lips features. If you have two mating parts where you need to place a boss or snap feature that must align to each other you can use a shared sketch and use the same point.

Silhouette Curves—Parting Line

While creating a part, it is common to split the part in its middle based on a parting line. The Silhouette Curve command creates a curve at the outermost boundary of the selected faces. This curve is used to create a surface that will be used as a split tool. To create a silhouette curve, follow these steps:

1. Open or create a part with the geometry that you want to create a silhouette curve on.
2. From the Sketch panel, click Create 3D Sketch from the drop arrow in the sketch panel, as shown in the following image on the left.
3. Click the Silhouette Curve command from the Draw panel, as shown in the following image in the middle.
4. The Create Silhouette Curve dialog box appears, as shown in the following image on the right.

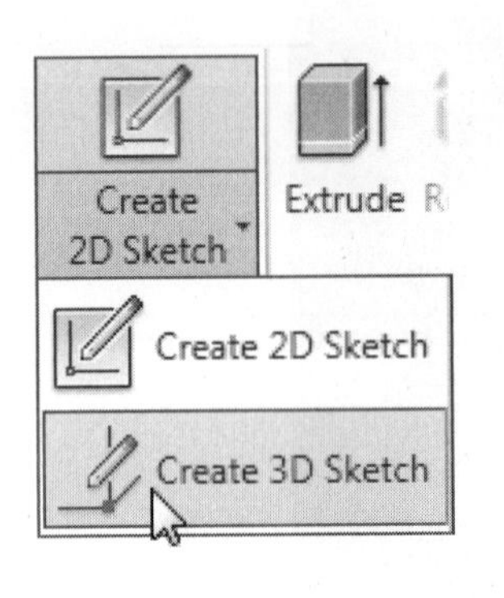

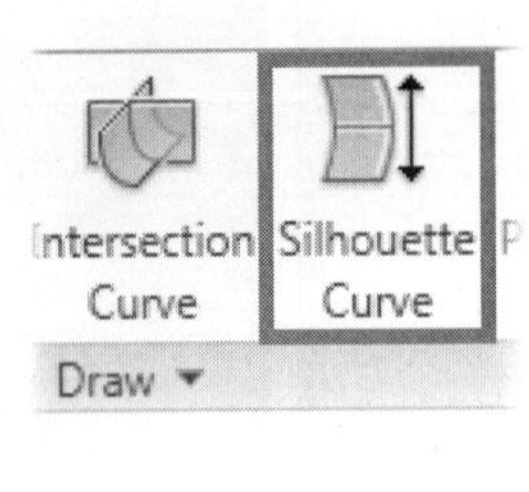

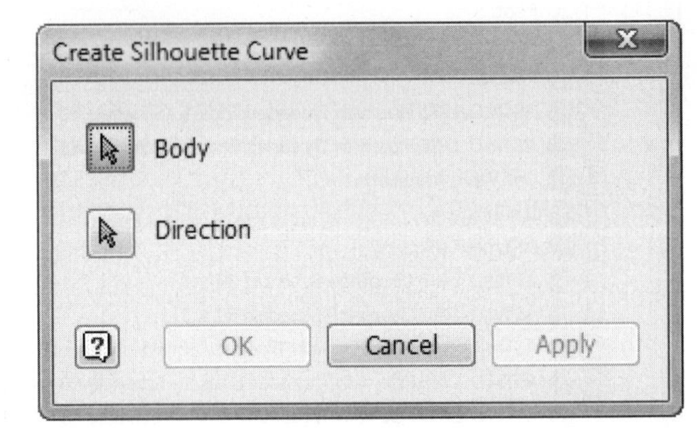

FIGURE 7.110

5. If multiple solid bodies exist in the file, select the solid body on which to create the curve.
6. Define the Direction by selecting a plane, edge, or axis to define the pull direction. The silhouette curve is created along the direction vector. Faces that lie in the pull direction are ignored.
7. Click OK to create the curve.

To split a part using the silhouette curve, follow these steps:

1. Create a silhouette curve.
2. Close open profiles with commands such as line and spline.
3. Use the Boundary Patch command to create a surface that is used to split the part.
4. If the surface does not touch all the bounding faces, extend the surface edges as needed.
5. Use the Split command to split the solid, as covered earlier in this chapter.

Grill

The Grill command creates a grill feature from a simplified sketch, and you define the details about the grill in the Grill dialog box. To create a grill feature follow these steps:

1. Create a 2D sketch and draw geometry that defines the boundary, an island, ribs, and spars as needed to define the grill (only sketch what is needed to define the grill, the geometry that defines the boundary is the only required

geometry). In the sketch, do not define thickness, as shown in the following image. See the online help for definitions of the grill features.

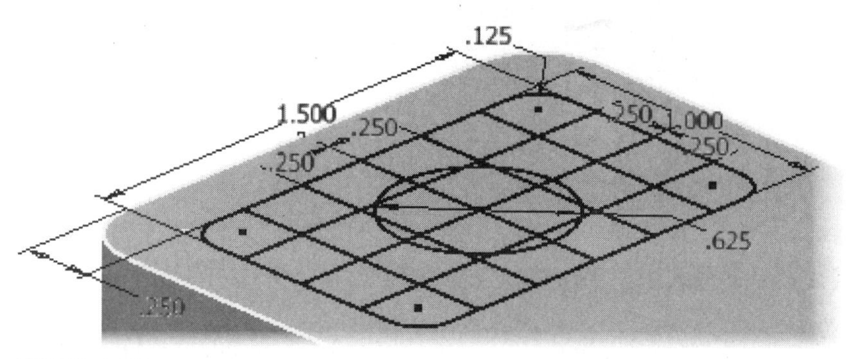

FIGURE 7.111

2. Click the Grill command from the Plastic Part panel, as shown in the following image on the left. The Grill dialog box will appear, as shown in the following image on the right.

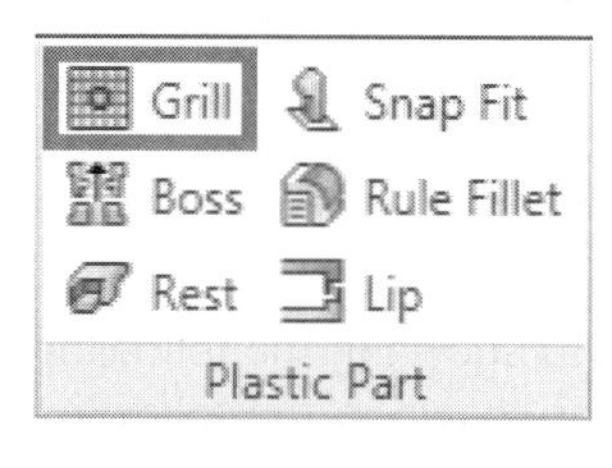

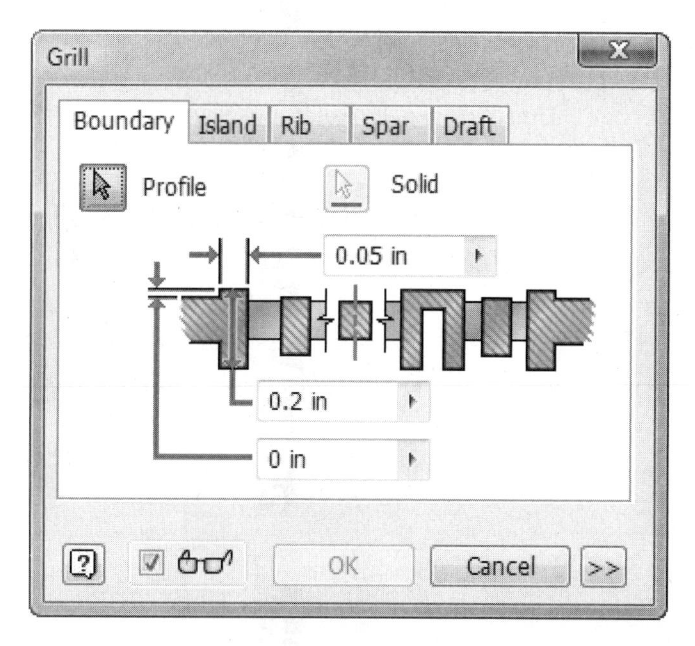

FIGURE 7.112

3. Click the closed profile that defines the boundary. Enter the values as needed.
4. Continue to select the required tabs, select the geometry, and enter the values as needed.
5. When done, click OK to create the grill feature. The following image shows a completed grill.

FIGURE 7.113

Boss

A boss is used to hold a fastener. You can create either a head or the thread portion of a boss.

NOTE

To align boss features on mating parts you can share a sketch with the point on it. The sketch needs to be shared before the solid bodies are moved apart with the Move Bodies command.

To create a boss feature, follow these steps:

1. Either create a 2D sketch and place Point, Center Point or place a work point to define the location of the boss.
2. Click the Boss command from the Plastic Part panel, as shown in the following image on the left. The Boss dialog box will appear, as shown in the following image on the right.

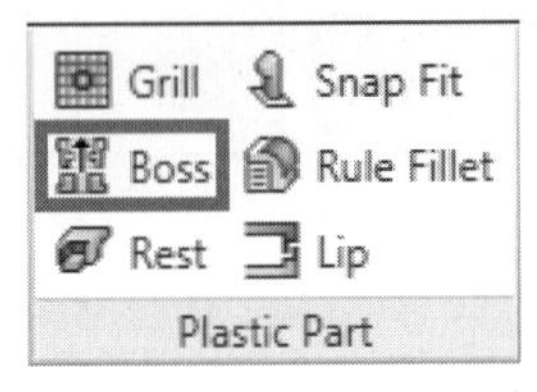

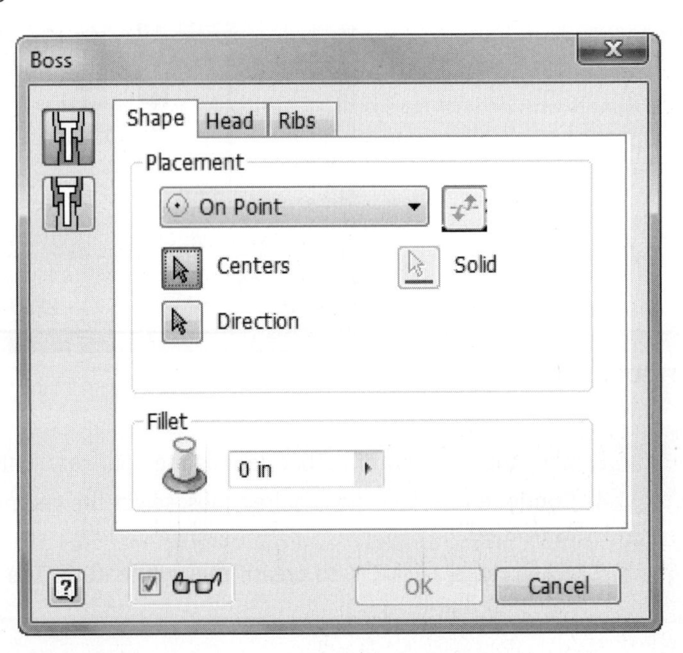

FIGURE 7.114

1. In the Boss dialog box, select or define the following:
   - Placement option
   - Direction
   - Boss type
   - Fillet
   - Head/Thread size
   - Ribs
2. When done, click OK to create the boss feature. The following image on the left shows a boss as a thread type (the shape and size of the boss is displayed but no thread is displayed) and the image on the right shows a boss as a head type.

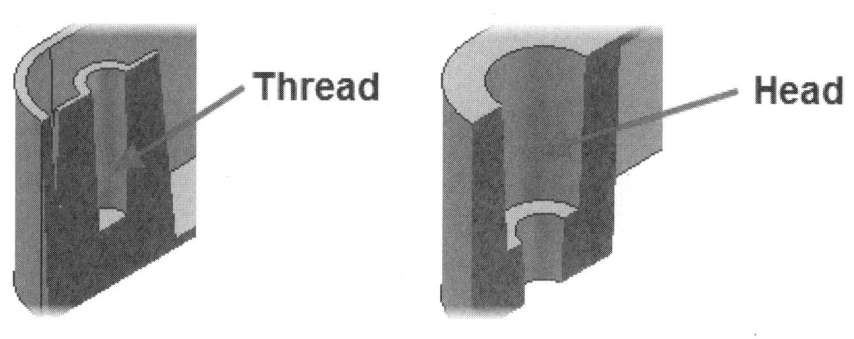

FIGURE 7.115

Rest

A rest is a feature of a plastic part that is applied to an angled or curved wall of a shelled part. It creates an area that goes partially inside and outside the solid. This area can be used to place another feature or part.

To create a rest feature follow these steps:

1. Create a 2D sketch at the location the rest will start.
2. Sketch, constrain, and dimension a closed profile that defines the boundary of the rest feature. The following image shows a sketch placed on a workplace that pierces the inside of the part.

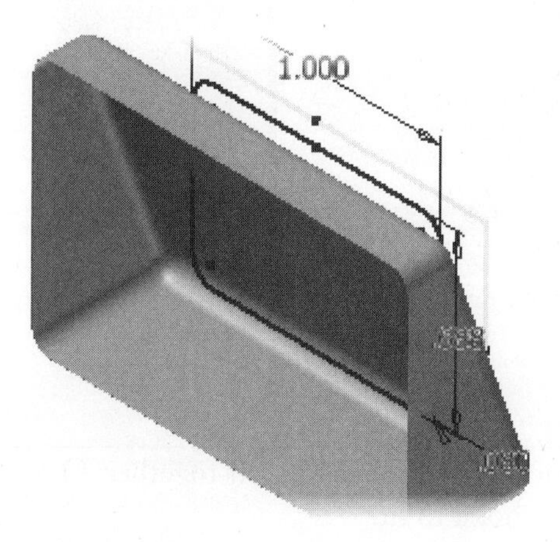

FIGURE 7.116

Click the Rest command from the Plastic Part panel, as shown in the following image on the left. The Rest dialog box will appear, as shown in the following image on the right.

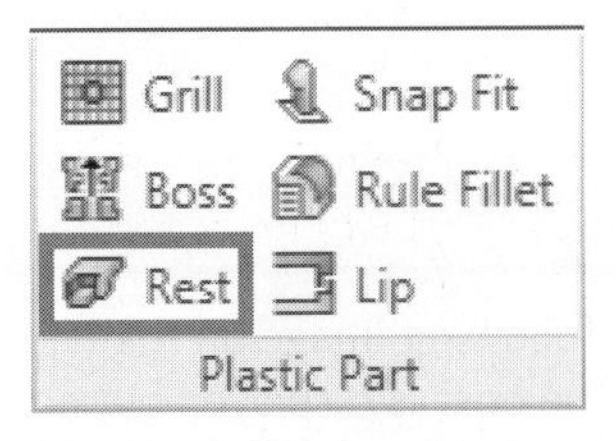

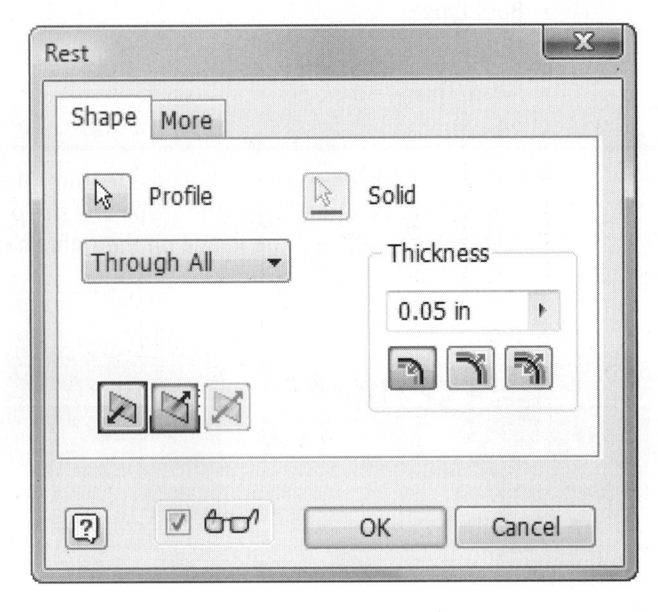

FIGURE 7.117

3. In the Rest dialog box, adjust the direction of the Rest and the thickness as needed.
4. From the More tab further adjust how the rest feature will be created.
5. When done, click OK to create the rest feature. The following images show different views of a cross-section of a rest feature.

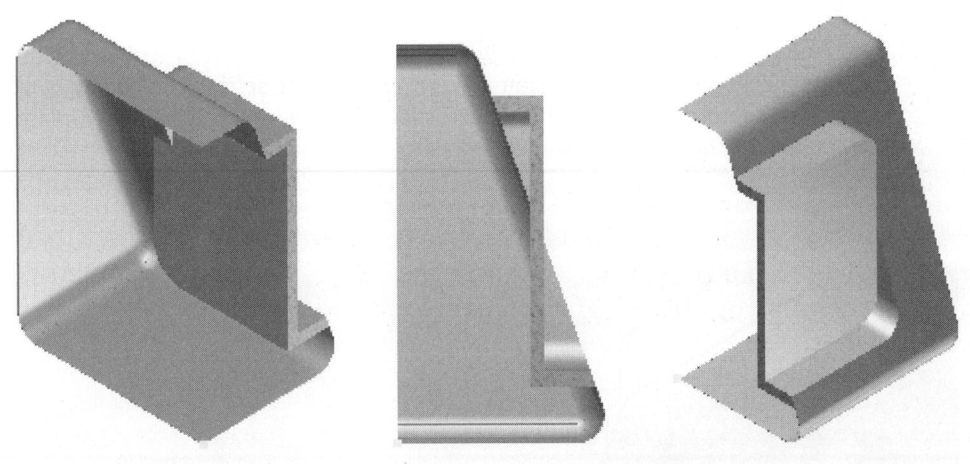

FIGURE 7.118

Snap Fit

A Snap Fit is a mechanism that is used in plastic parts to hold two parts together. One part has a hook and the other has a loop for the hook to fit.

NOTE

To align snap features on mating parts, you can share a sketch with the point on it. The sketch needs to be shared before the solid bodies are moved apart with the Move Bodies command.

To create a Snap Fit feature, follow these steps:

1. Either create a 2D sketch and place points or place a work point to define the location of the Snap Fit feature(s). The points are usually placed on the inside edges.
2. Click the Snap Fit command from the Plastic Part panel, as shown in the following image on the left. The Snap Fit dialog box will appear, as shown in the following image on the right.

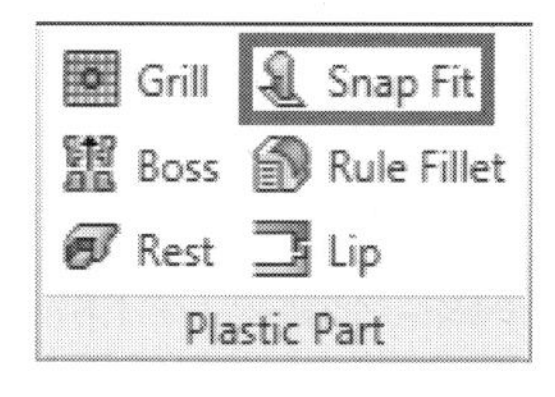

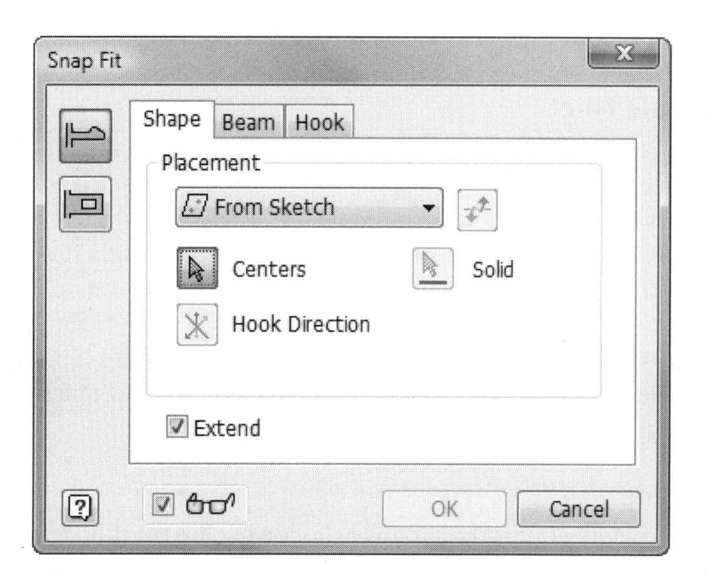

FIGURE 7.119

1. Select the point(s) on which to place the feature.
2. If you need to rotate the feature, click the Hook Direction button and then, in the graphics window, select the direction arrows. The feature will rotate in 90 degree increments, as shown in the following image on the left.
3. Select the type of feature to create, hook or loop.
4. Adjust the size of the feature by changing the values in the tabs.
5. Click OK to create the feature. The following image in the middle shows a hook feature. The image on the right shows a loop.

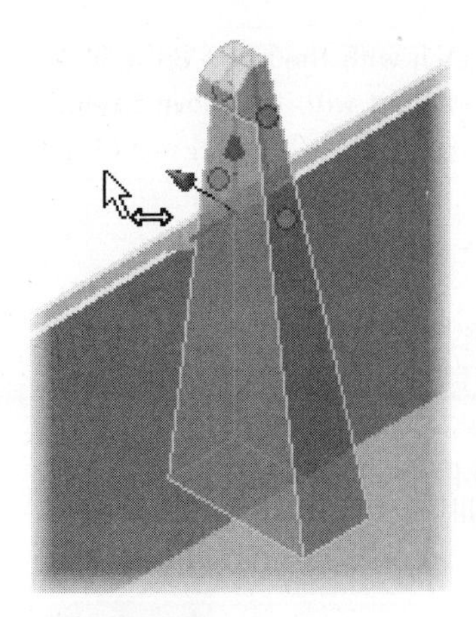

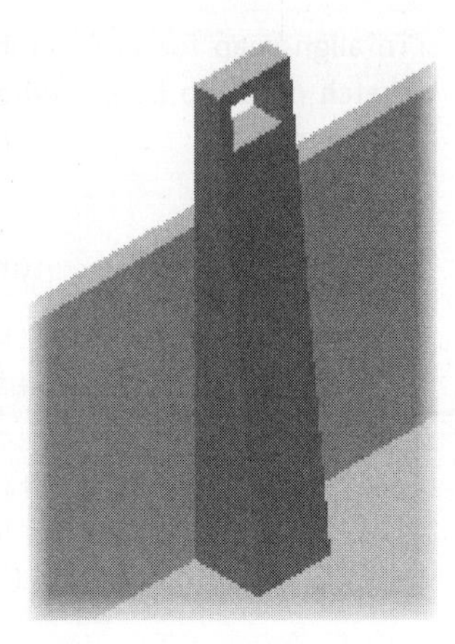

FIGURE 7.120

Rule Fillet

A rule-based fillet is different than the fillet command that was covered in Chapter 4. The fillets created in Chapter 4 were based on selected edges or face. Rule-based fillets are based on a set of rules that dictate when edges touch certain faces or features fillets will automatically be created. When creating a rule fillet, you specify a source and a rule.

Source

Select either a face(s) or feature(s) that when edges touch them fillets will be placed onto them.

Rule

A rule identifies edges to fillet. Following are the rule options:

Feature Rules

- *All Edges:* Fillets are created on all the edges that are created by the feature and on the edges that are intersected by the created features.
- *Against Part:* Edges created by the faces of the features and the faces of the part are filleted.
- *Against Features:* Edges created against a feature are filleted.
- *Free Edges:* Only the edges formed by the faces of the features in the source selection set are filleted.

Face Rules

- *All Edges:* Fillets are created on all the edges that touch the selected face(s).
- *Against Features:* Edges of a feature created against the selected face(s) are filleted.
- *Incident Edges:* Edges that touch the selected face(s) and are parallel to a selected axis are filleted.

NOTE

The Rule Fillet only applies fillets to geometry that exist above it in the browser.

1. Before creating a Rule Fillet ensure the geometry that you want to fillet exists.
2. Click the Rule Fillet command from the Plastic Part panel, as shown in the following image on the left. The Rule Fillet dialog box will appear, as shown in the following image on the right.

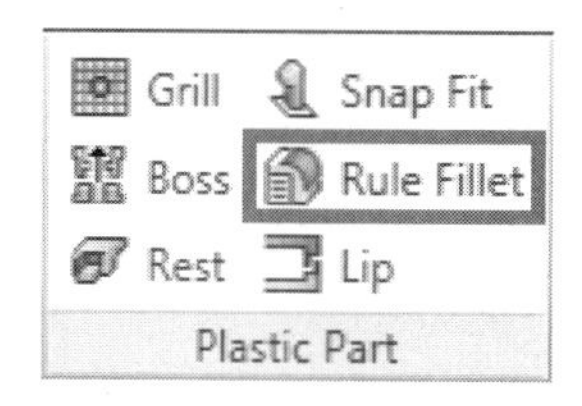

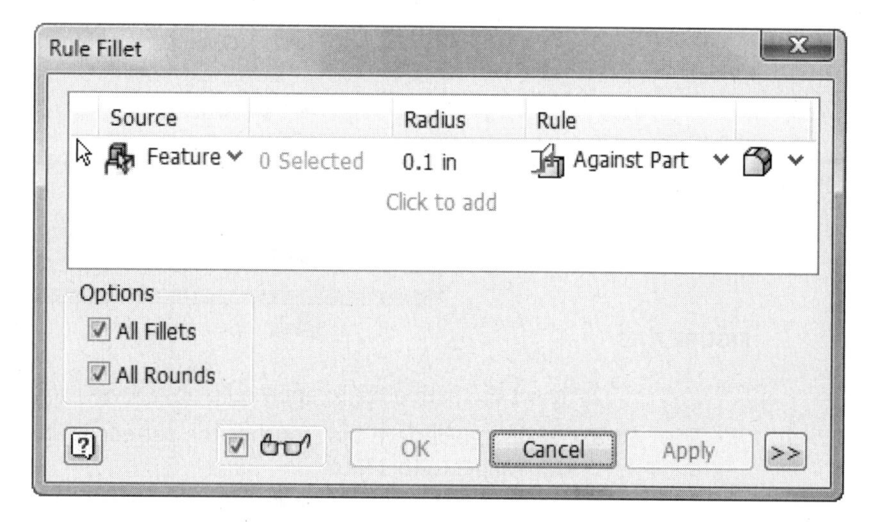

FIGURE 7.121

3. Select the source and apply a rule.
4. Create as many rules as needed.
5. Click OK to create the rule fillets. The following image shows an example of edges that were filleted that were incident to the top face.

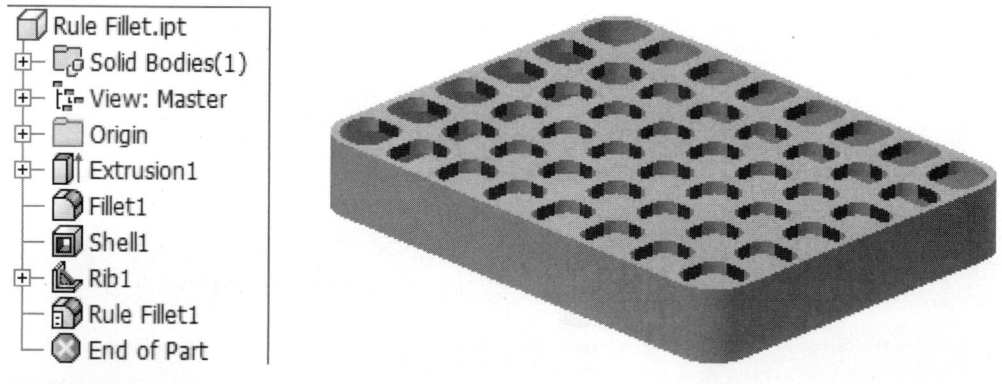

FIGURE 7.122

Lip

Lips and grooves are used to join two parts together at their parting lines. Lips add material and grooves remove material. To create a Lip feature (lip or groove) follow these steps:

1. Click the Lip command from the Plastic Part panel, as shown in the following image on the left. The Lip dialog box will appear, as shown in the following image on the right.

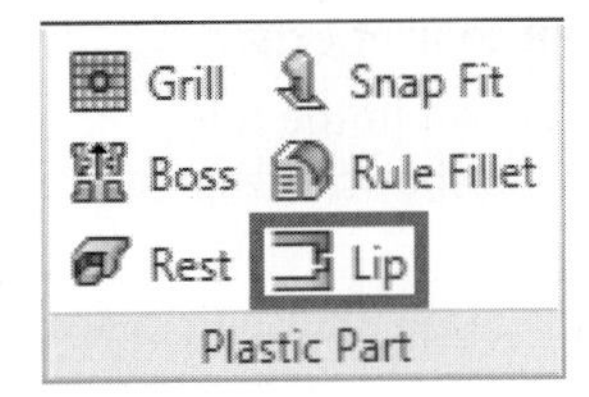

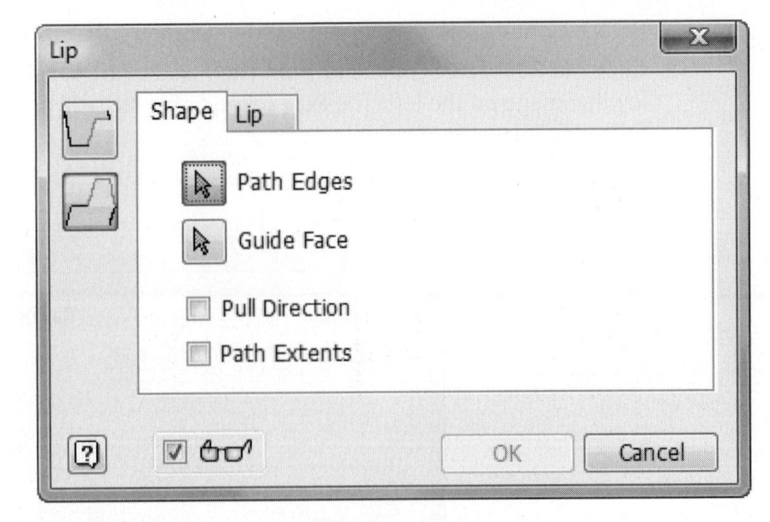

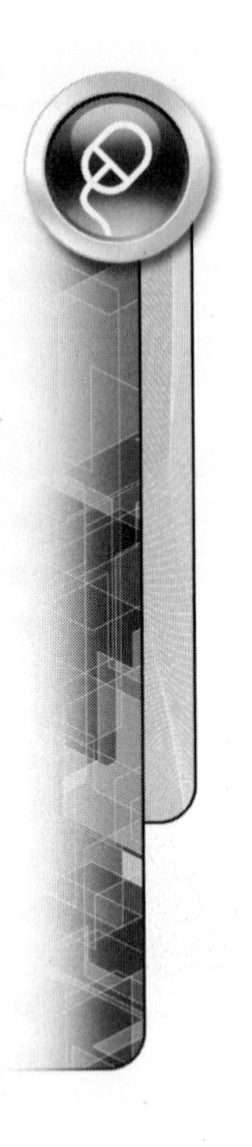

FIGURE 7.123

2. Select the path edge to apply a lip or groove.
3. Select a guide face that touches the pat edge. The feature will maintain a constant angle to this face.
4. A pull direction defines a direction that the feature parallels. When pull direction is selected, the Guide Face option is grayed out.
5. If needed define the oath extents that determine where the feature will stop.
6. Select the feature type, lip or groove.
7. Adjust the size of the feature by changing the values in the tabs.
8. Click OK to create the feature. The following image on the left shows a groove feature and the image on the right shows a lip feature.

FIGURE 7.124

EXERCISE 7-9: PLASTIC PART

In this exercise, you create features on a plastic part. To expedite the exercise, sketches have been created. After doing the exercise, feel free to redo the exercise with your own geometry. For clarity the images in this exercise are shown with the visual style Shaded with edges.

1. Open *ESS_E07_09.ipt* from the Chapter 07 folder.
2. Use the Free Orbit command to examine the part.
3. Return to the Home view.

4. Next you create a rest feature. In the browser, right-click on the entry Sketch_Rest and check Visibility from the menu.
5. Click the Rest command on the Plastic Part panel and make the following selections.
    - Change the Thickness to **.08 in**, as shown in the following image on the left.
    - Click OK to create the feature and rest feature should resemble the following image on the right.

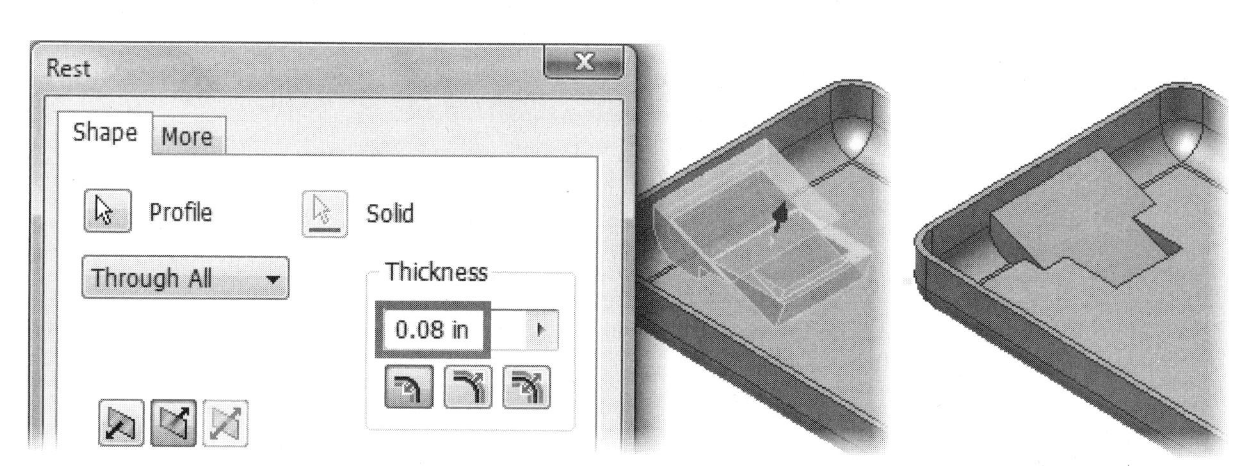

FIGURE 7.125

6. Use the Free Orbit command to examine the part and orient your view to see the opposite side of the part as shown in the following image on the left.
7. Next you create a grill feature. In the browser, right-click on the entry Sketch_Grill and check Visibility from the menu.
8. Click the Grill command from the Plastic Part panel and make the following selections:
    - On the Boundary tab, select the outside profile.
    - On the Island tab, select the inside circle.
    - On the Rib tab, select the seven inside horizontal and vertical lines.
    - Click OK to create the grill feature.
9. Return to the Home view. The grill feature should resemble the following image on the right.

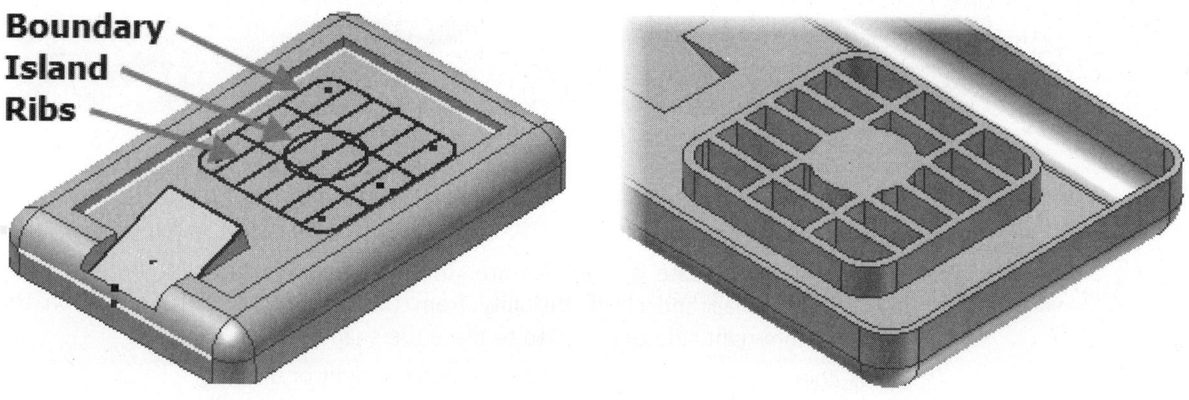

FIGURE 7.126

10. The grill's depth extends too far into the part. Edit the grill feature and from the Boundary and Rib tab change the depth to **.1** as shown in the following image on the left and in the middle. Click OK to complete the edit; when done, your screen should resemble the following image on the right.

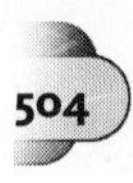

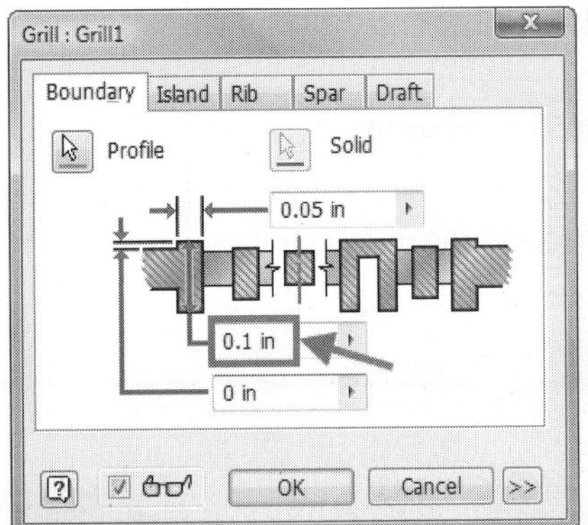

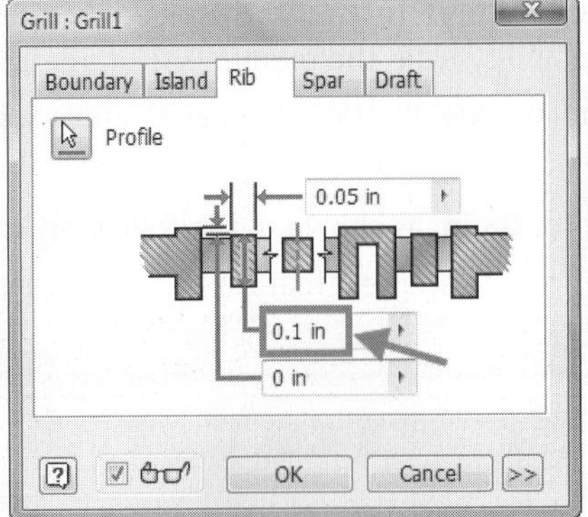

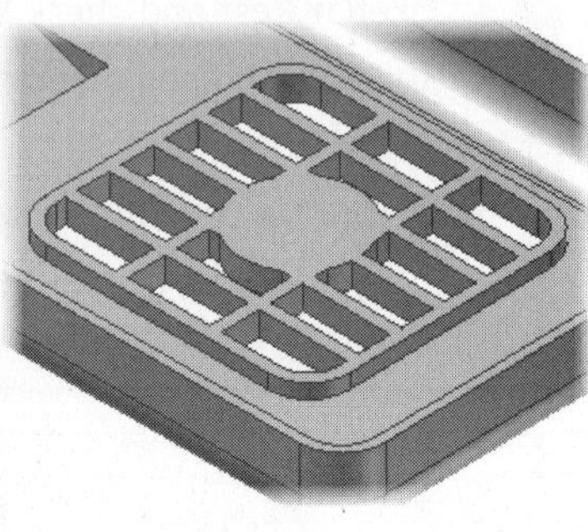

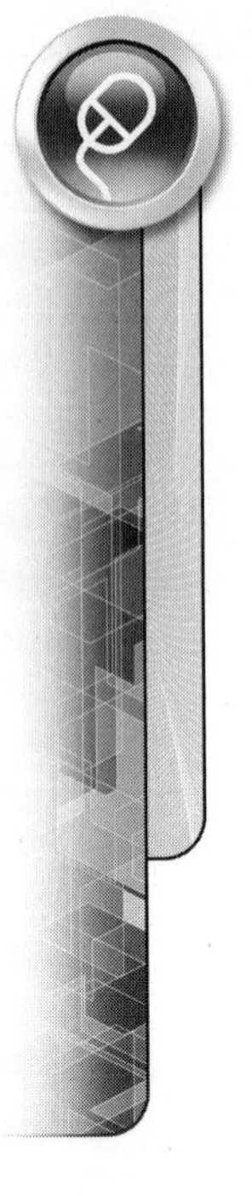

FIGURE 7.127

11. Next you create two bosses. In the browser, right-click on the entry Sketch_Boss and check Visibility from the menu. The two sketch points near the Rest feature will be used to place the bosses.
12. Click the Boss command from the Plastic Part panel and make the following selections.
    - Change the Boss type to Thread.
    - From the Thread tab change the top Thread diameter to **.3 in**, and uncheck Hole, as shown in the following image on the left.
    - Click OK to create the boss features and your model should resemble the following image on the right.
    - At this point you could add a hole feature such as a tapped hole to the bosses.

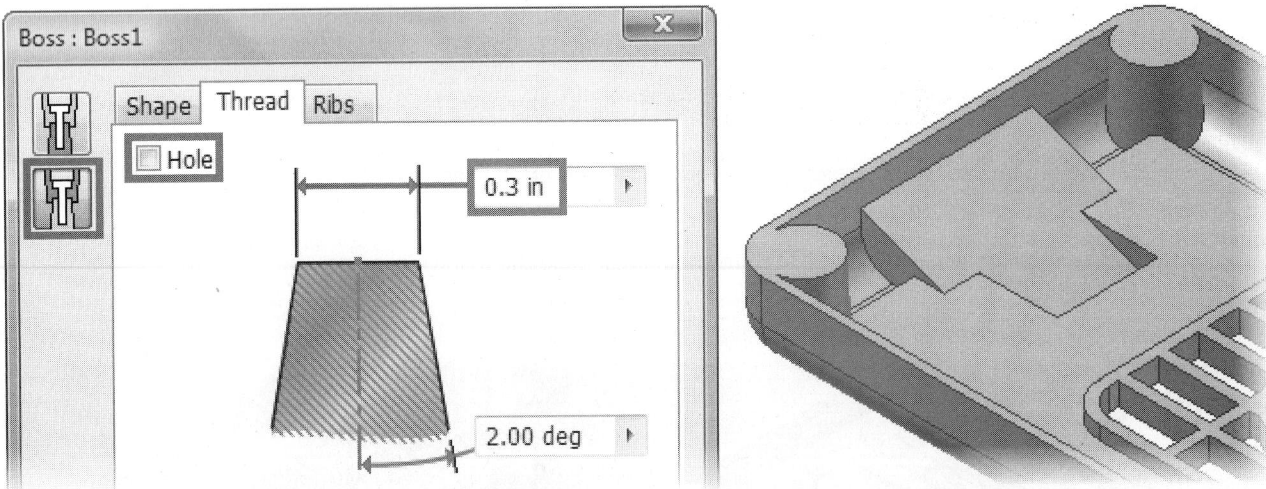

FIGURE 7.128

13. Next you create a snap feature. In the browser, right-click on the entry Sketch_Snap and check Visibility from the menu. The sketch point on the bottom-right side will be used to place the snap feature.
14. Zoom in on the center point on the bottom-right of the part.
15. Click the Snap Fit command from the Plastic Part panel and make the following selections:
    - In the graphics window rotate the snap by clicking the arrows in the middle of the preview of the snap feature until the hook points up and outward, as shown in the following image on the left.

- Click OK to create the feature.
- Rotate your view to see inside the part, as shown in the following image on the right.

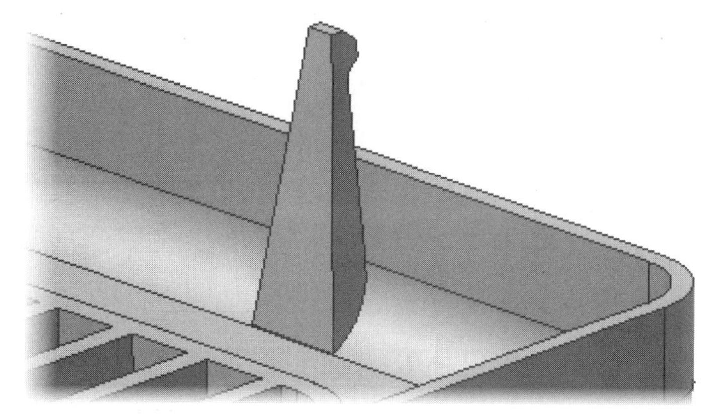

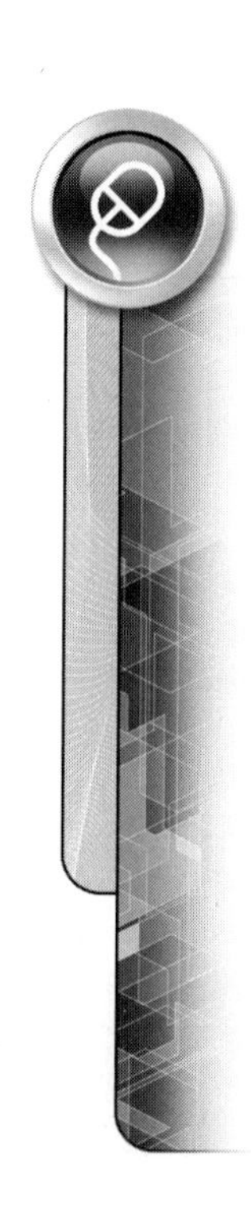

FIGURE 7.129

16. Change to the Home view.
17. Finally, create a Lip feature. Click the Lip command on the Plastic Part panel and make the following selections:
    - For the Path, select an outside edge and all the outside edges will automatically be selected.
    - For the Guide Face, select the top face adjacent to the selected edge.
    - Click the Groove type on the left column in the dialog box.
    - Click the Groove tab and enter the values in the following image to size the groove and give it an angle.
    - Click OK to create the lip feature.

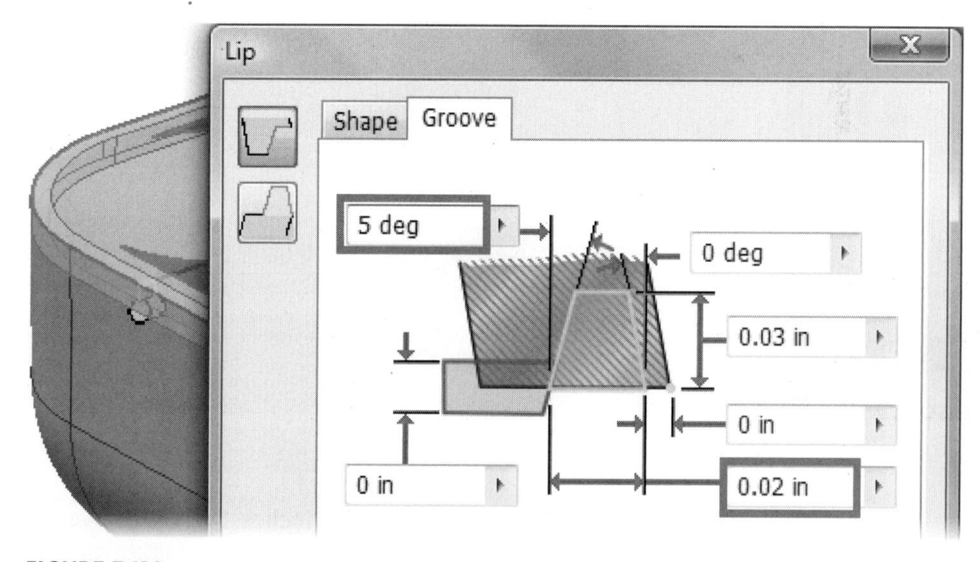

FIGURE 7.130

18. Use the Orbit command to examine the part and your part should resemble the following image.
19. Close the file. Do not save changes. End of exercise.

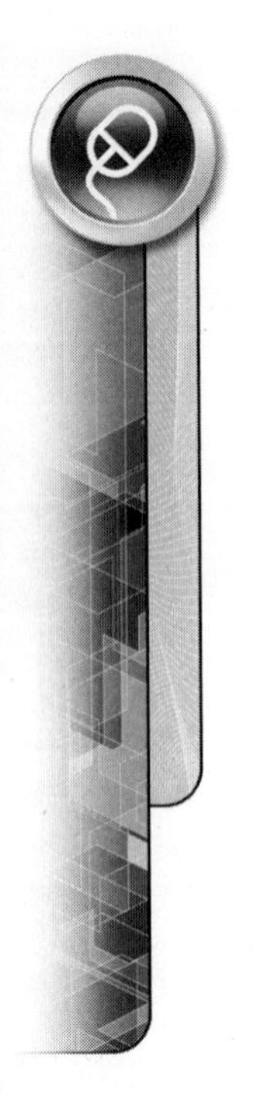

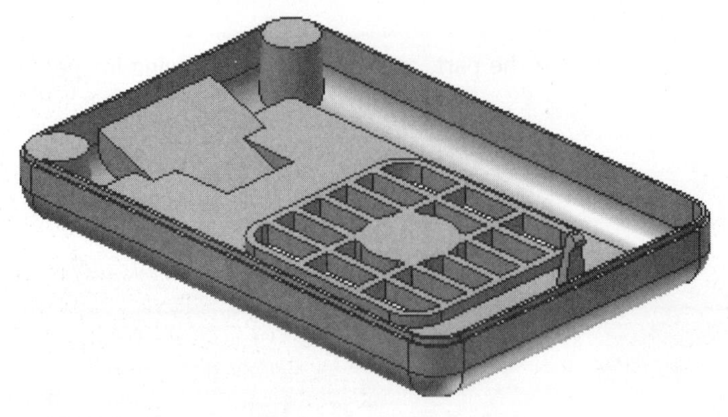

FIGURE 7.133

APPLYING YOUR SKILLS

Skill Exercise 7-1

In this exercise, you use the knowledge you gained through this course to create a joystick handle. You will use the loft, split, and a few plastic part commands.

1. Open *ESS_Skills_7-1.ipt* from the Chapter 07 folder.
2. Create a loft, for the sections use the three elliptical shapes in order from top to bottom and for the rails use the splines and your screen should resemble the following image on the left.
3. Create a **.1875 inch** fillet around the top edge, as shown in the following image on the right.

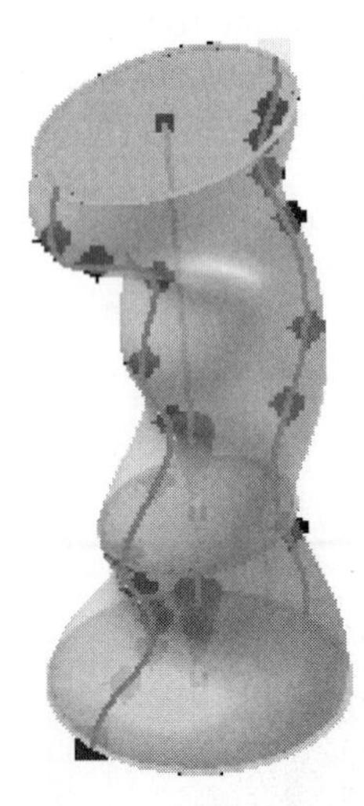

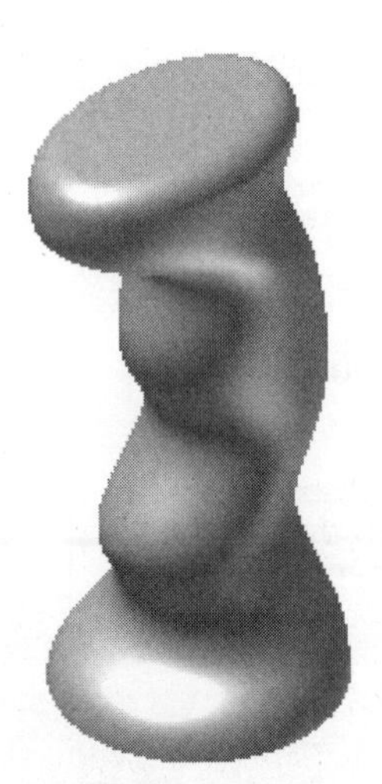

FIGURE 7.132

4. Shell the part to the inside with a thickness of **.0625 inches**. Do not remove any faces.
5. Next create a grill on the bottom of the part. Create and dimension a sketch, as shown in the following image on the left.
6. Create a grill with an island and ribs. Change the boundary and rib thickness and depth to **.1 inch**. The following image on the right shows the completed grill from an angled viewpoint.

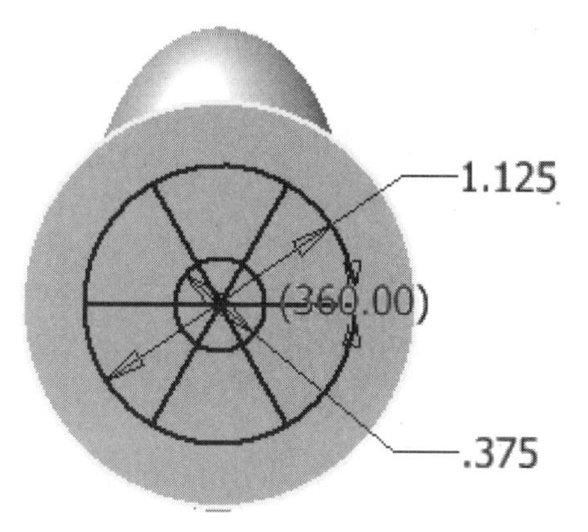

FIGURE 7.133

7. Next you create a silhouette curve. From a 3D sketch use the Silhouette Curve command to create a curve with the direction based on the XY origin plane.
8. Connect the splines on the top and bottom of the part with a 3D line.
9. Exit the 3D Sketch and use the Boundary Patch command on the Model tab > Surface panel and select all of the splines and lines that form the silhouette curve, as shown in the following image on the left.
10. Use the Split command and use the surface to split the solid into two solid bodies.
11. Turn off the visibility of the surface.
12. Use the Move Bodies command to move one of the solid bodies out **1.5 inches** in the Z direction, as shown in the following image in the middle.
13. Turn off the visibility of the front solid.
14. As desired add four **.2 inch** Thread Diameter bosses to accept the threads from a fastener.
15. Add features to the other solid body and, if desired, use the Make Components command to export the solids into an assembly.
16. Close the file. Do not save changes. End of exercise.

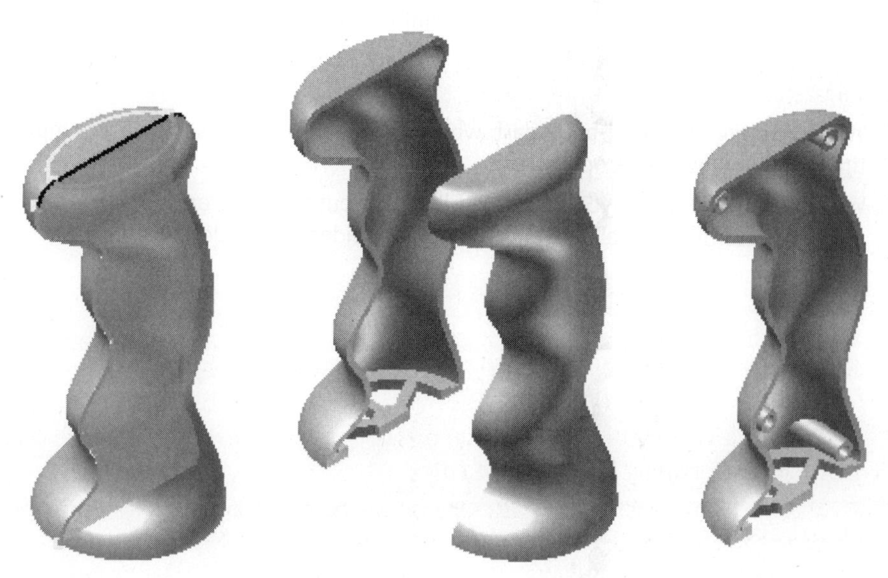

FIGURE 7.134

CHECKING YOUR SKILLS

Use these questions to test your knowledge of the material covered in this chapter.

1. True __ False __ When creating a rib or a web feature, you can only select a closed profile.
2. True __ False __ You can only emboss text on a planar face.
3. True __ False __ A sweep feature requires three unconsumed sketches.
4. True __ False __ In a 3D Sketch the Project Geometry command is used to project curves onto a circular face.
5. True __ False __ You can create a 3D curve with a combination of both 2D and 3D curves.
6. Explain how to create a 3D path using geometry that intersects with a part.
7. True __ False __ The easiest way to create a helical feature is to manually create a 3D path and then sweep a profile along this path.
8. True __ False __ You can control the twisting of profiles in a loft by defining point sets.
9. Explain how to split a part into two solid bodies and then save them to their own part file.
10. Explain the difference between suppressing and deleting a feature.
11. True __ False __ After mirroring a feature, the mirrored feature is always independent from the parent feature. If the parent feature changes, the mirrored feature will NOT reflect this change.
12. True __ False __ A derived part cannot be scaled.
13. True __ False __ The Shrinkwrap command can only patch voids created from a hole feature.
14. True __ False __ When creating a lip feature with the Lip command, you must first create a 3D path.
15. Explain how to reorder features.
16. True __ False __ When creating a rib network, each sketch object that will make up the rib network must exist on its own sketch.
17. True __ False __ When placing text on an arc with the Geometry Text command the text will start at the left quadrant of the arc.
18. When creating a sweep with the Path & Guide Surface option, what does the surface control?
19. While working in a multi-body part, which command do you use to export multiple solid bodies in one operation?
20. Which command allows you to create fillets without selecting each edge; for example, create fillets on all the edges that are incident to a selected face?

CHAPTER 8

iComponents and Parameters

INTRODUCTION

In this chapter, you will learn how to use advanced modeling techniques. Using advanced modeling techniques, you can automate the process of assembling files by predefining assembly constraints on your parts and subassemblies as they are created. You can combine like files in a single *.ipt* or *.iam* file that stores like parts or assemblies, and you can extract features from a part to insert versions of that feature into other files. You will also learn how to display dimensions in alternate formats, set up relationships between dimensions, and create parameters. The techniques discussed will help you to become more productive and efficient when working with Autodesk Inventor.

OBJECTIVES

After completing this chapter, you will be able to perform the following:

- Create iMates
- Change the display of dimensions
- Create relationships between dimensions
- Create parameters
- Create and place iParts
- Create and place iAssemblies
- Create and place iFeatures

iMATES

Another way to apply assembly constraints is to create iMates. An iMate holds information in the component or subassembly file on how it is to be constrained when placed in an assembly. Each component or subassembly holds half of the iMate information and, when combined with a component or subassembly that contains the other half of the iMate information, they form a pair or a complete iMate. If you want the two components to assemble automatically, an iMate needs to have the same name on both components or subassemblies that are being constrained together.

When an iMate with the same name exists in two components, you can automate the assembly constraint task for those two components. iMates are useful when similar components are switched in an assembly. You may, for example, have different pins that go into an assembly. The same iMate name can be assigned to each pin or corresponding hole. When the pin is placed into the assembly, it can be automatically constrained to the corresponding hole using one of two iMate options when placing the component. An iMate can only be used once in an assembly—once used, it is consumed.

To create an iMate, follow these steps:

1. Click the iMate command on the Manage tab > Author panel.
2. In the Create iMate dialog box, click the type of constraint to apply: mate, angle, tangent, or insert on the Assembly tab.
3. In the graphics window, select the geometry you want to use as the primary position geometry.

The following image shows the iMate tool on the left, the Create iMate dialog box in the middle, and a mate-axis constraint (the iMate) being applied to the part on the right.

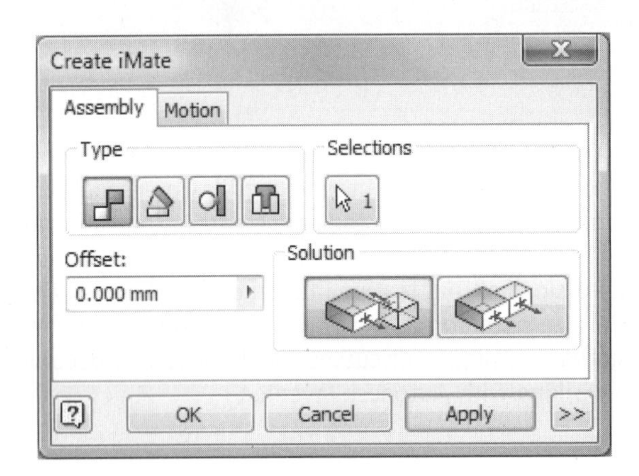

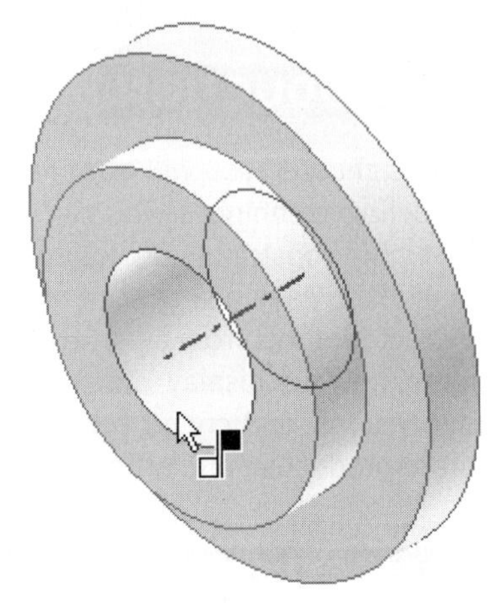

FIGURE 8.1

Click OK. An iMate glyph will appear on the component and in the browser, as shown in the following image.

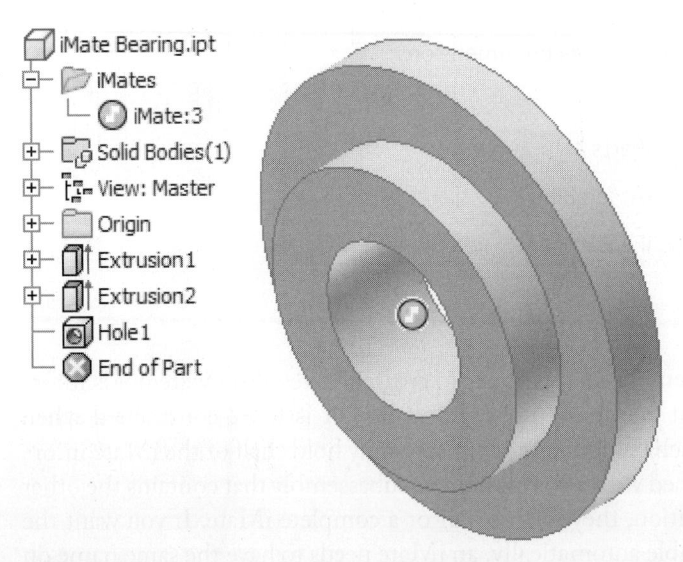

FIGURE 8.2

NOTE

Instead of clicking OK, you can click Apply and continue to create iMates.

The glyph shows the type and state of the iMate. When an iMate is created, it is given a default name according to the constraint type, such as *iInsert:1* or *iAngle:1*. The iMates can, however, be renamed to better reflect the conditions that they represent. Renaming iMates is a good idea. Providing a meaningful name that follows a common naming standard can aid when replacing components in an assembly. Matching iMate names can automatically place constraints between two components. When a component is replaced with another one that has the same iMate name, type, and offset or angle value, the constraint relationship will remain intact.

You can specify a name for an iMate using one of several methods.

- During iMate creation, expand the Create iMate dialog box, as shown in the following image, and enter a name in the field.

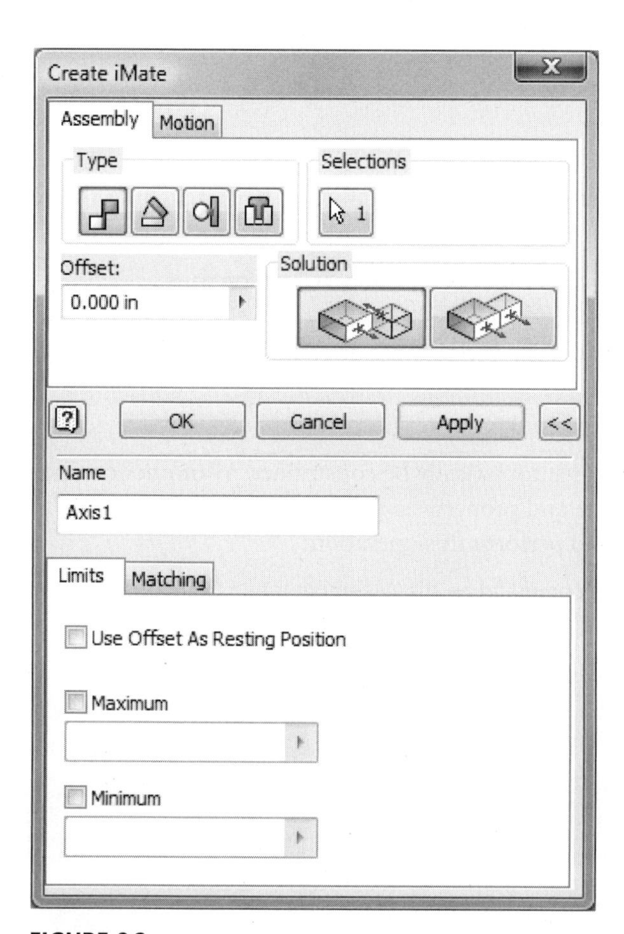

FIGURE 8.3

- Slowly double-click on the iMate name in the browser, and type in a new name.
- Right-click on the iMate in the browser, or on the iMate glyph in the graphics window, and select Properties from the menu. Type in a new name using the Name field.

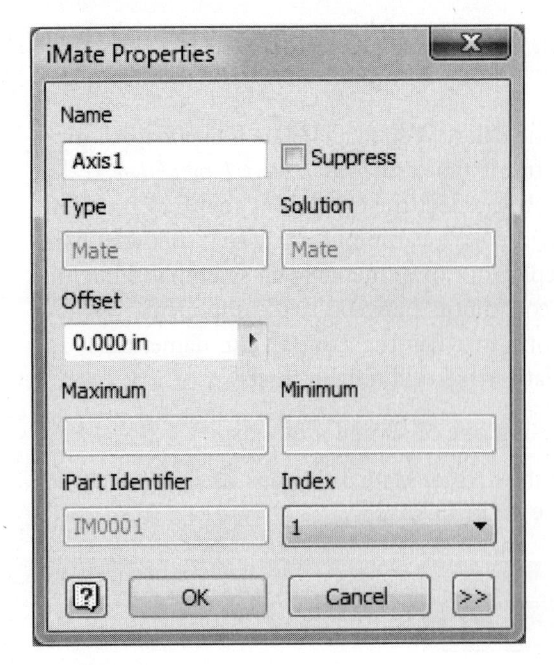

FIGURE 8.4

To assemble components that exist in the same assembly and have matching iMate types:

1. Use the Constrain command on the Assemble tab > Position panel.
2. Click an iMate on the component from the browser or graphics window, and then select a matching iMate from the browser or graphics window on another component.
3. Click the Apply button to create the constraint.

A component that has an iMate can automatically be constrained to another component that has the same iMate name and property as the component that exists in an assembly. Use the following steps to perform this operation:

1. Click the Place Component command on the Assemble tab > Component panel.
2. Select the component to place, and choose one of the iMates options in the lower portion of the Place Component dialog box, as shown in the following image.
3. Click the Open button.

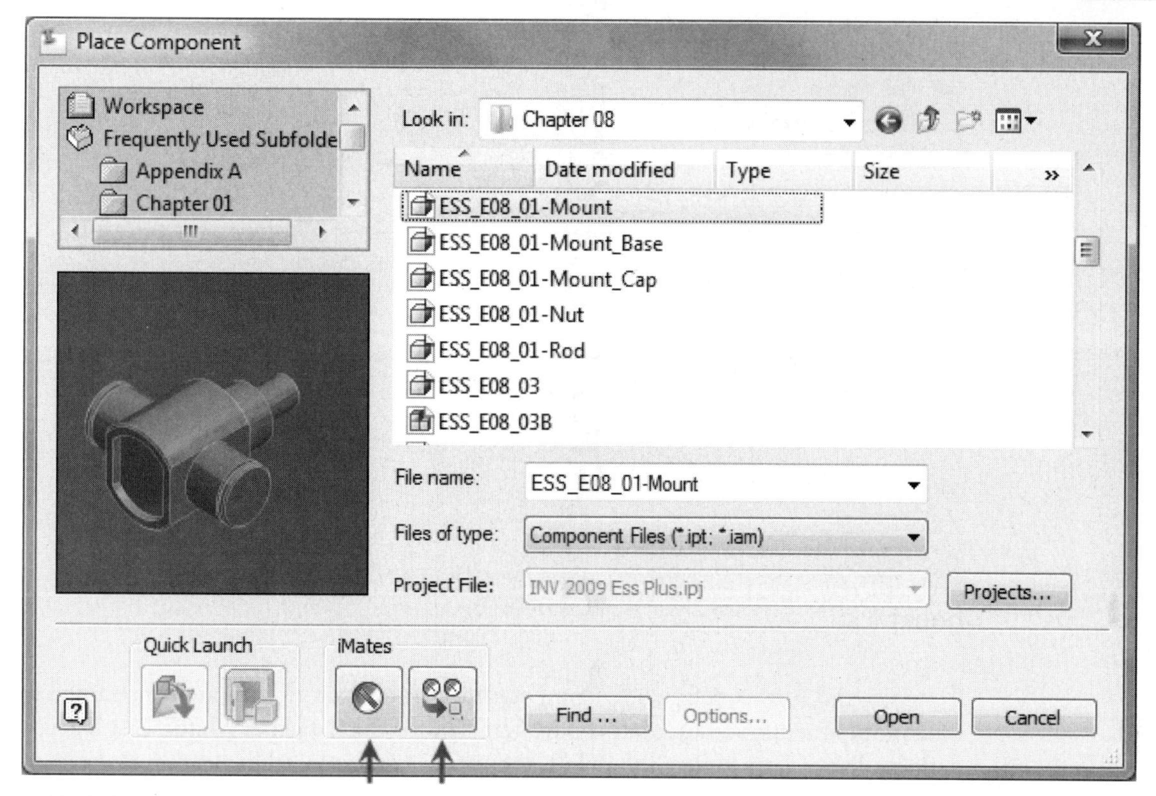

FIGURE 8.5

Interactively Place with iMates allows you to place one or more occurrences of a component that contains iMates. When using this mode, you can manually cycle through and apply unconsumed matching iMates or you can select to automatically place the component at all matching iMates.

The second option, Automatically Generate iMates on Place, can be used to place a single occurrence of a component that contains iMates, and have all of its iMates automatically apply to matching unconsumed iMates in other components of the assembly. If you use this option, the Place Component tool terminates once the single occurrence has been placed.

If a matching iMate on another component exists and has not yet been consumed, a preview showing how the component will be constrained will be displayed. You can left-click in the graphics window to accept the component placement. The new component will be constrained into position, the display of the assembly will be zoomed to the location of the component, and a consumed iMate icon will be shown in the browser. If multiple possible matches exist, you can continue to left-click in the graphics window to accept the additional placements, or you can right-click in the graphics window to access the iMate placement menu. The menu contains options to cycle through all valid iMate matches and component instances in the assembly. The iMate placement menu is shown in the following image.

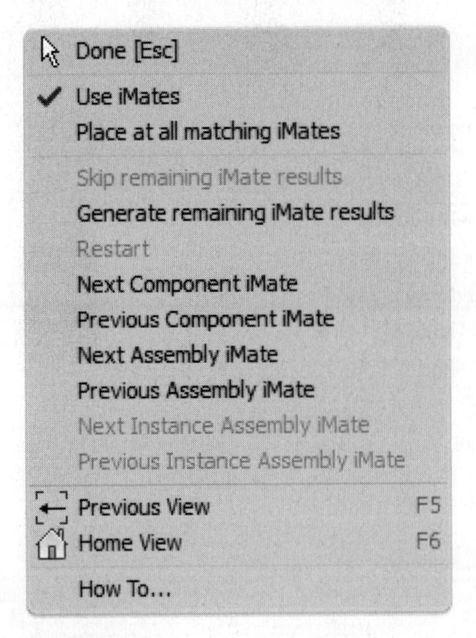

FIGURE 8.6

To assist and further refine how components that contain iMates are constrained to other components, you can define a Match List for the iMate. A Match List lets you define a list of names that the iMate should search for as valid matches to be constrained to. The Match List can be defined by selecting the Matching tab after expanding the Create iMate dialog box or by right-clicking an iMate in the browser and selecting Properties from the menu and expanding the iMate Properties dialog box. The following image shows the iMate Properties dialog box with the Match List panel available.

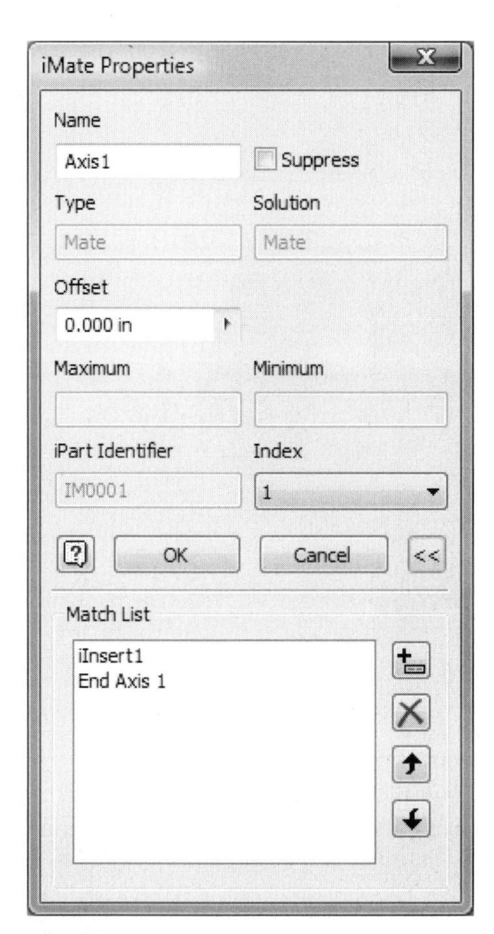

FIGURE 8.7

When you include names in the Match List, you can use the DELETE, move up, or move down keys to order the names in the list. The order that they appear is important when placing and automatically constraining the parts in an assembly. The match process begins at the top of the list, and if no matching name is found, proceeds to check the next name in the list, and continues this process until a match is found. If no match is found, the component is attached to the cursor and can be placed in the assembly in the same manner as a component that does not contain iMates.

Composite iMates

You can group multiple iMates into a single, composite iMate. In the following image of a standard drill transmission housing, three separate iMates have been combined into a single, composite iMate. This allows you to more easily constrain components using a single constraint where multiple constraints are needed. In the following image, the three iMates that were originally created on the part are combined into a single glyph. The browser also displays the newly created composite iMate. Expanding the composite iMate will display the original iMates.

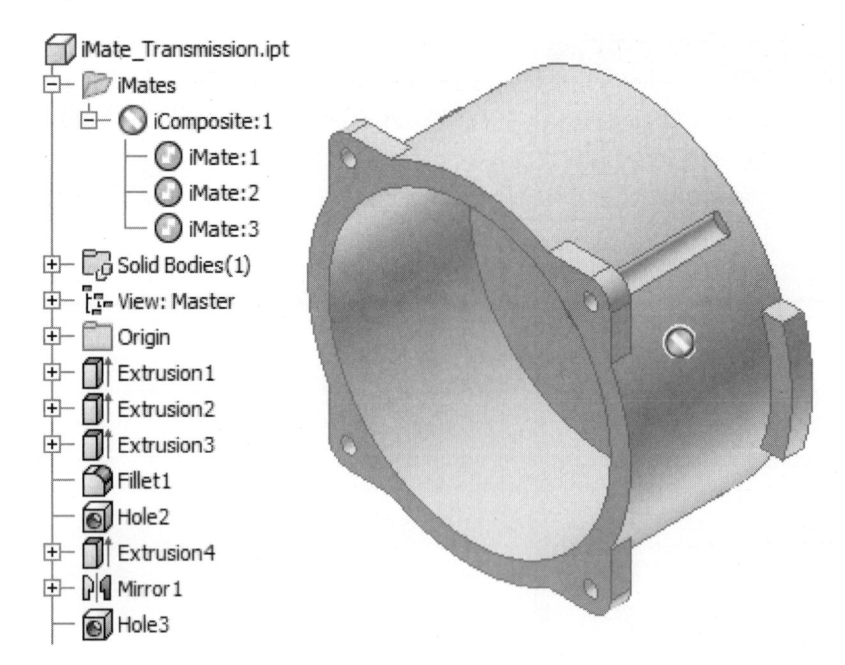

FIGURE 8.8

To create a composite iMate for any component, first create the individual iMates. Once created, press the CTRL key to select multiple iMates in the browser. Right-click on any iMate in the selection set, and select Create Composite from the menu.

Using Composite iMates

Similar to creating single iMates, there are two ways to orient components in your assemblies using composite iMates:

- Select one of the iMates options in the Place Component dialog box to automatically search for matching component iMates.
- Use the ALT + Drag shortcut to manually match composite iMates between two components.

To ensure a successful match between any two iMates in an assembly, check for the following criteria:

- The iMate type and values must match.
- The name at the top of the matching list is given priority when looking for a matching component in the assembly.
- The composites must have the same number and type of single iMates.
- The order of the single iMates that exist in the composite iMate must be identical to the other half of the composite iMate that is being matched.

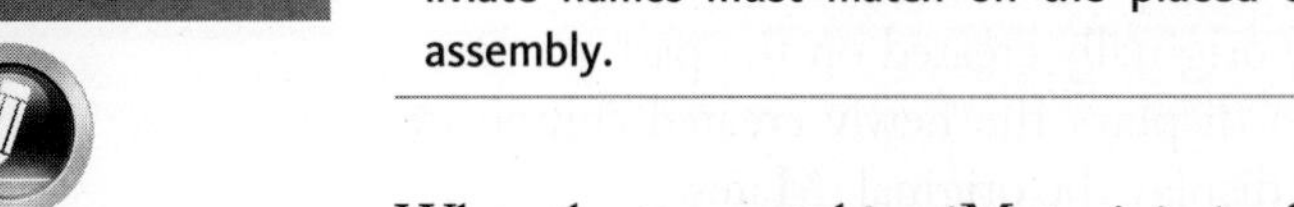

NOTE

iMate names must match on the placed component and the unconsumed iMate in the assembly.

When the two matching iMates join in the assembly, a single consumed iMate is created. Because the relationship is specific to two components with matching iMate halves, multiple occurrences cannot be placed.

The solution that is selected for iMates (mate and flush, inside and outside, or opposed and aligned) must be the same for the matching iMates.

ALT + Drag iMate Behavior

When using the ALT + Drag feature of constraining iMates on a component part, the other component part that contains the matching iMate solution will display the iMate glyph, as shown in the following image. To use this technique, press and hold the ALT key on your keyboard while selecting an iMate glyph. Then drag your cursor to another iMate glyph that is displayed on another component in the assembly.

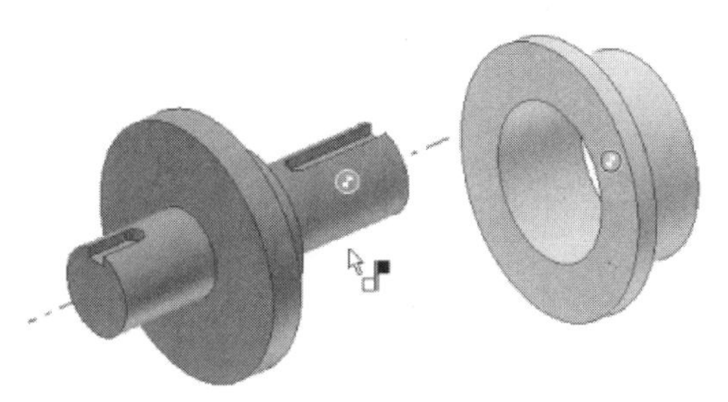

FIGURE 8.9

NOTE

iMate glyphs that are currently displayed will not display during an ALT + Drag feature if they do not match the selected iMate type.

EXERCISE 8-1: CREATING AND USING IMATES

In this exercise, you add iMates to an existing subassembly. You then use those iMates to position the subassembly. Next you open a part, combine existing iMates into a composite iMate, and automatically orient that part into the assembly using the composite iMate. You complete the exercise by replacing and automatically orienting a component in the assembly using iMates.

1. Open *ESS_E08_01.iam.*

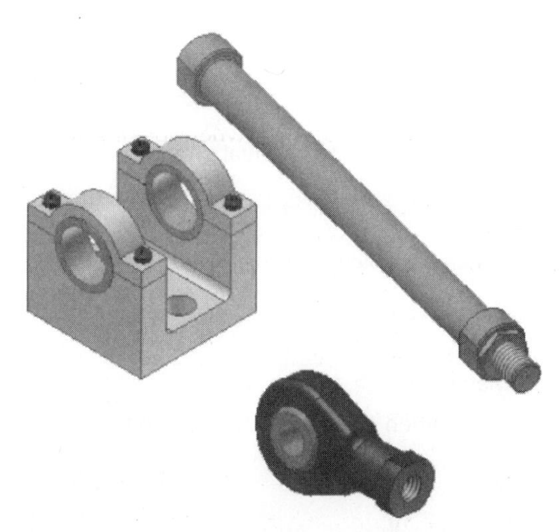

FIGURE 8.10

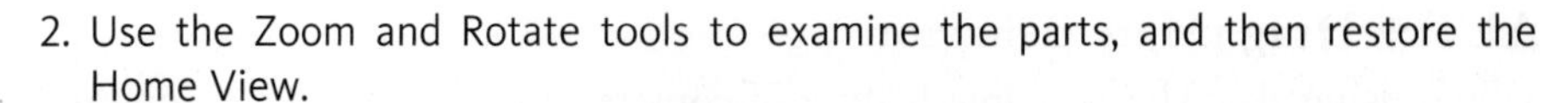

2. Use the Zoom and Rotate tools to examine the parts, and then restore the Home View.

   In the next few steps, you create iMates on the End subassembly (the red component) to automate the process of constraining this component to another component in the assembly. These iMates can also be used to automate the process of orienting the component in other potential assemblies.

3. In the browser, expand the Connector and End components.
4. In the browser, expand iMates under the Connector component. Notice that the End component has no iMates.

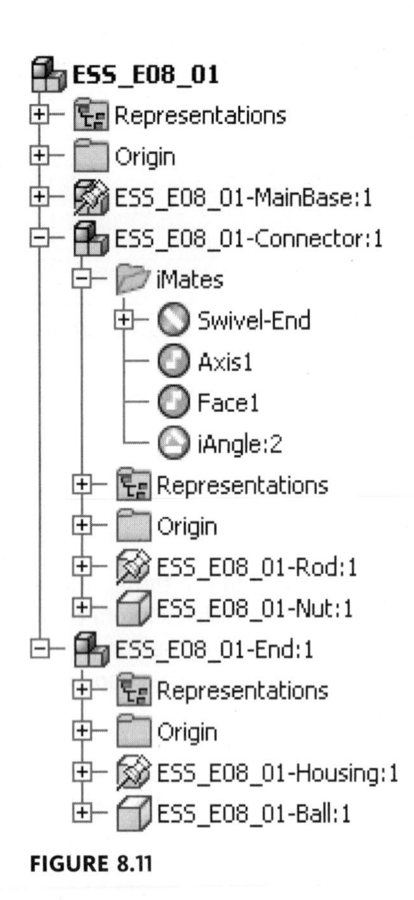

FIGURE 8.11

5. In the browser, double-click ESS_E08_01-End:1 to activate and edit the subassembly.

NOTE

iMates are created on individual components or subassemblies. You then use those iMates to position components in an assembly.

6. Click the iMate command from the Manage tab > Author panel.
7. Place your cursor over the tapped hole. When the axis is highlighted, click to select it.

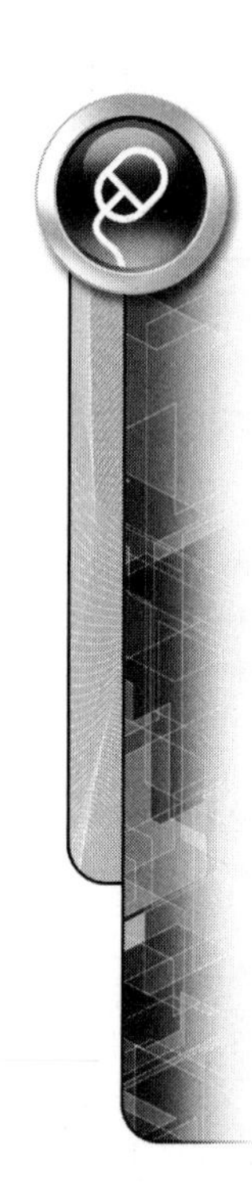

FIGURE 8.12

8. Expand the Create iMate dialog box.
9. Enter **Axis1** in the Name field, and click Apply to create the first iMate.

NOTE

You can rename the iMate after creation by slowly double-clicking the name in the browser or by accessing the Properties dialog box from the iMate once it is created.

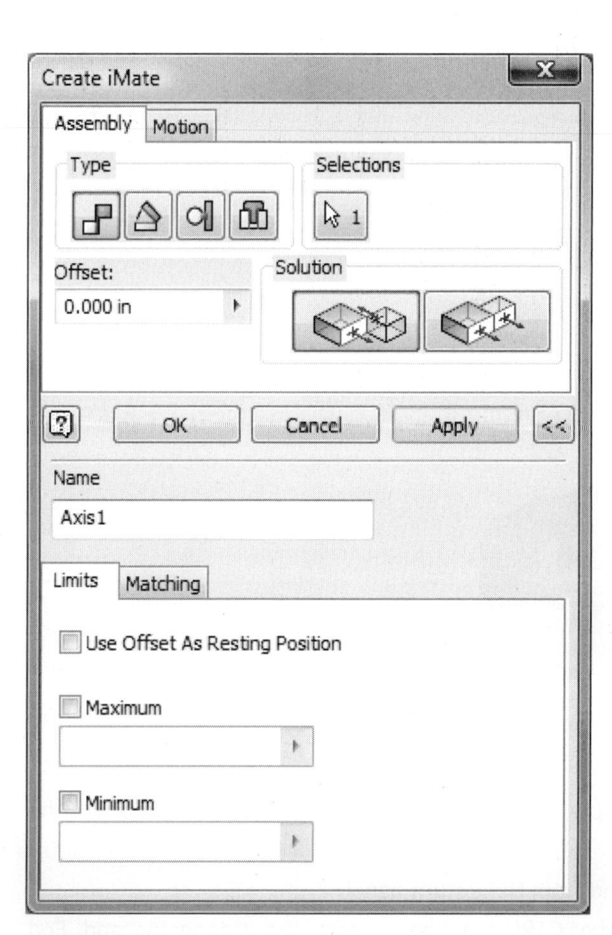

FIGURE 8.13

10. Place your cursor over the highlighted face, as shown in the following image. When the face is highlighted, click to select it.

FIGURE 8.14

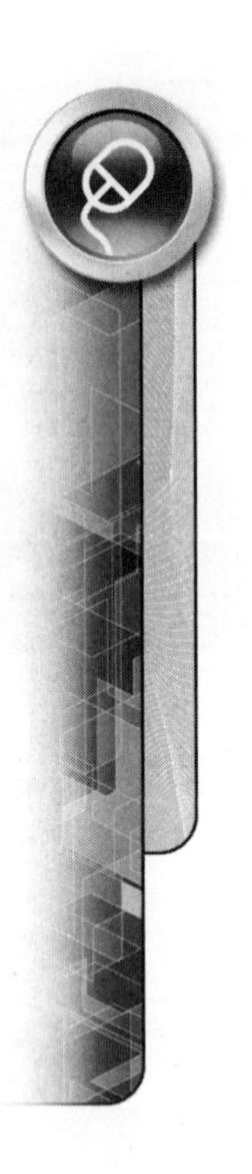

11. Enter **Face1** in the Name field of the Create iMate dialog box, and click Apply to create the second iMate.
12. In the Create iMate dialog box, select the Angle constraint button.
13. Place your cursor over the highlighted face, as shown in the following image. When the face is highlighted, click to select it, then click OK to create the third and final iMate.

NOTE You leave the default name of iAngle for this iMate.

FIGURE 8.15

Next return to the assembly, and use the iMates you just created to orient the End component.

14. Click the Return command from the Return panel.
15. In the browser, hold the CTRL key, and select the Connector and End components.
16. Right-click, and select iMate Glyph Visibility to turn on the iMate glyphs by placing a checkmark next to the option in the menu.

17. Click the Constrain command from the Assemble tab > Position panel.
18. Select the iMate on the End component—the one created by selecting the front face—as shown in the following image.

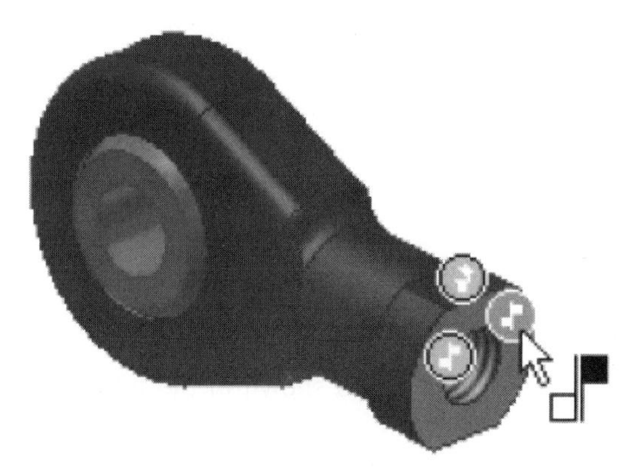

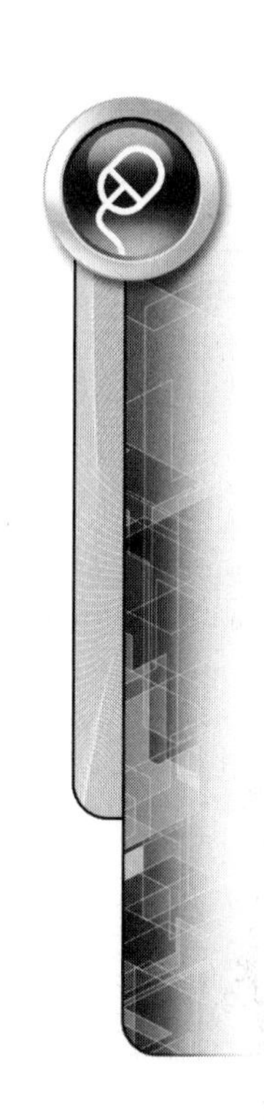

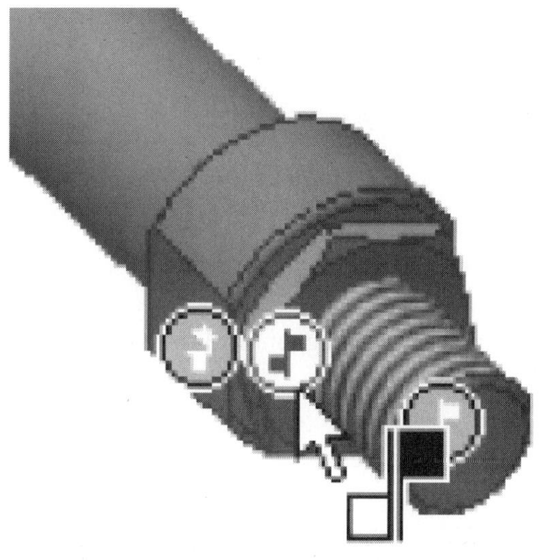

FIGURE 8.16

19. Select the iMate on the Connector component, as shown in the following image.

FIGURE 8.17

20. In the Place Constraint dialog box, click OK. You can also use the ALT + Drag shortcut rather than the Constrain command to create assembly constraints.
21. In the graphics window, click and drag the End component away from the Connector.

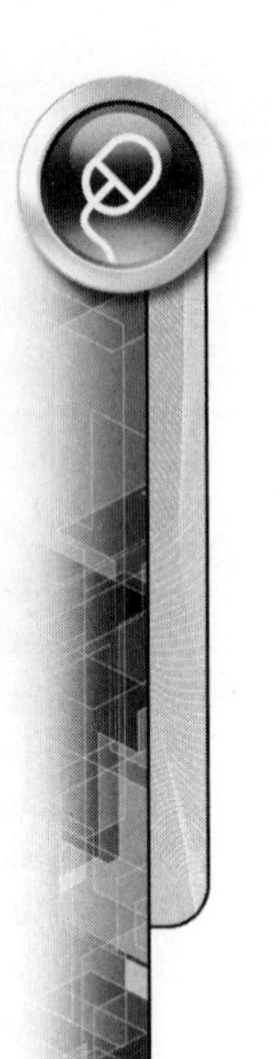

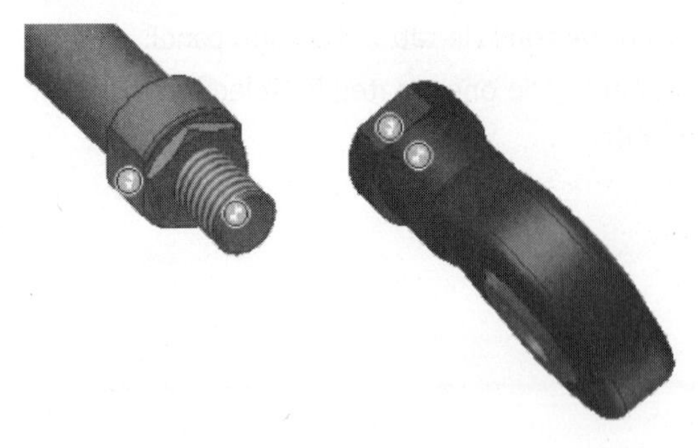

FIGURE 8.18

22. Press and hold the ALT key.
23. Select and drag the iMate on the End component, as shown in the following image.

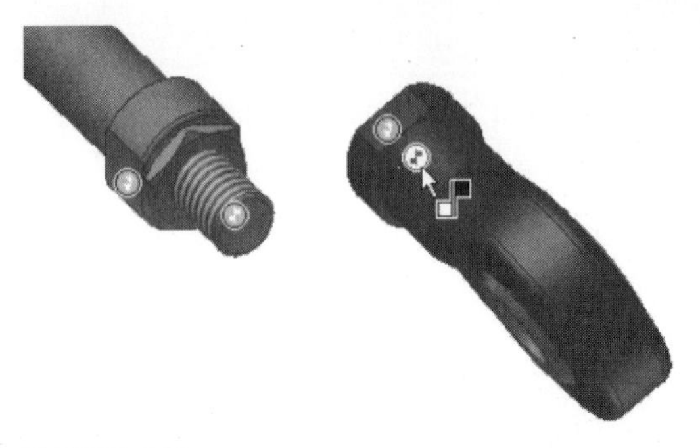

FIGURE 8.19

24. Drop the iMate on the Connector component iMate, as shown in the following image. Notice that after you select an iMate to drag, only iMate glyphs of the same type are visible.

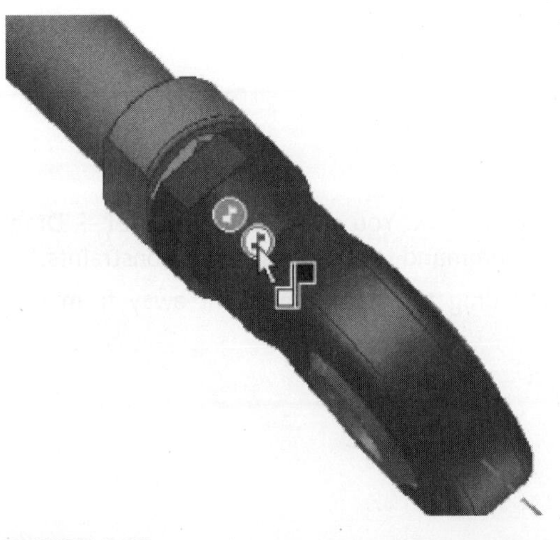

FIGURE 8.20

25. Use the Constrain command to place the final constraint using the third iMate. Click the Constrain command from the Assemble tab > Position panel.
26. Choose the Angle constraint type.
27. Select the iMate on the End component followed by the iMate on the Connector component, as show in the following image.
28. Click OK.

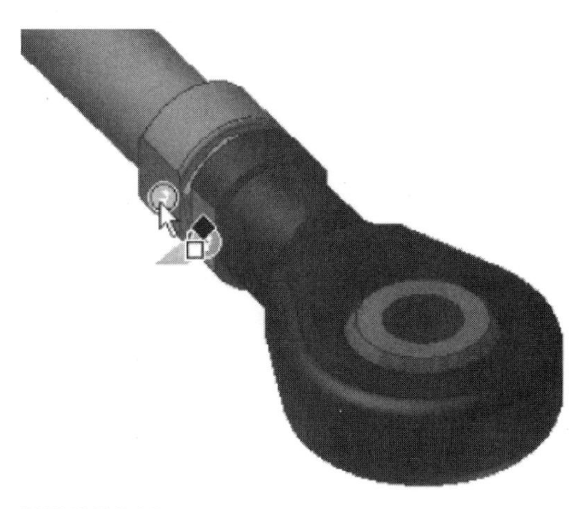

FIGURE 8.21

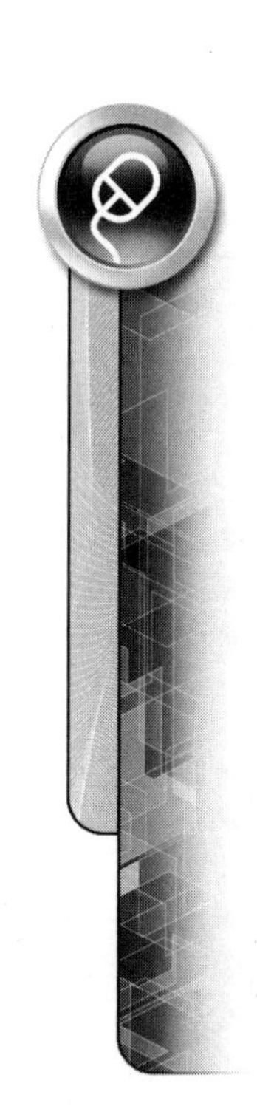

29. Select the Connector and End components.
30. Right-click, and select iMate Glyph Visibility to remove the checkmark.

    Next create a composite iMate on another part to match the composite iMate on the Connector component.

31. In the browser under ESS_E08_01-Connector:1, expand Swivel-End under the iMates folder.

    Notice that this composite iMate includes two iMates, as shown in the following image.

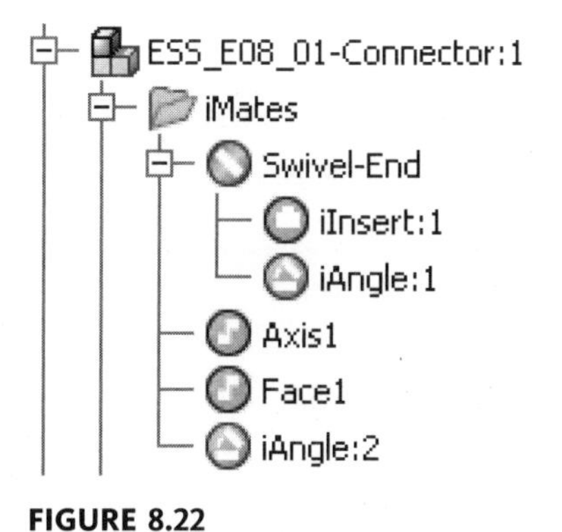

FIGURE 8.22

32. Open *ESS_E08_01-Mount.ipt* in a separate window.

FIGURE 8.23

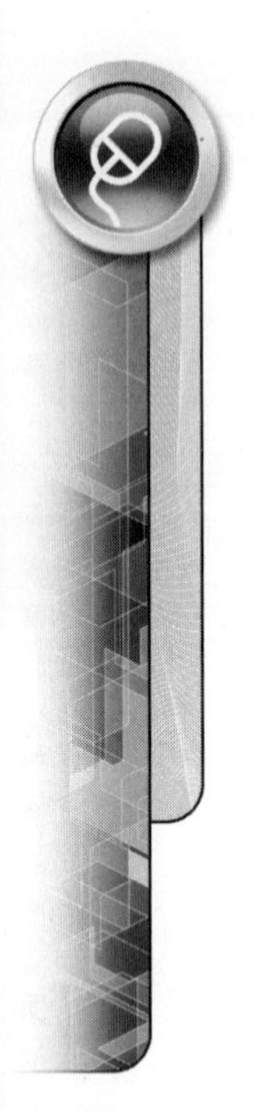

33. In the browser, expand iMates. Press and hold the CTRL key, and select the two iMates, as shown in the following image.

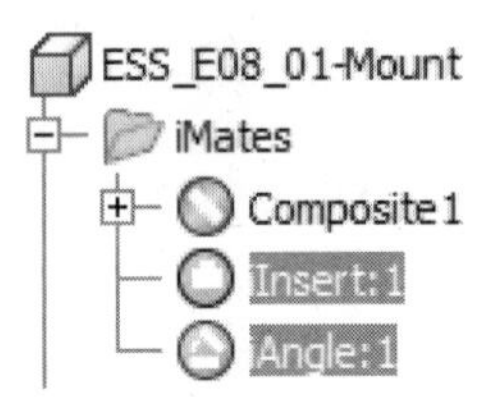

FIGURE 8.24

34. Right-click, and select Create Composite from the menu.

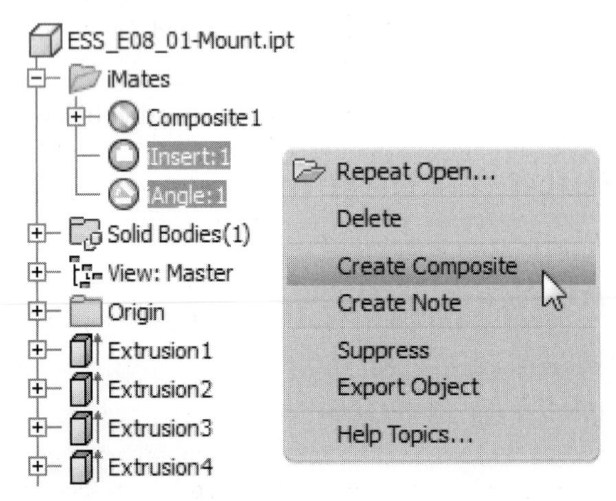

FIGURE 8.25

35. Right-click the new Composite iMate in the browser, and select Properties from the menu.
36. Change the Name field to **Swivel-End**. If the Suppress box is checked, unselect it.

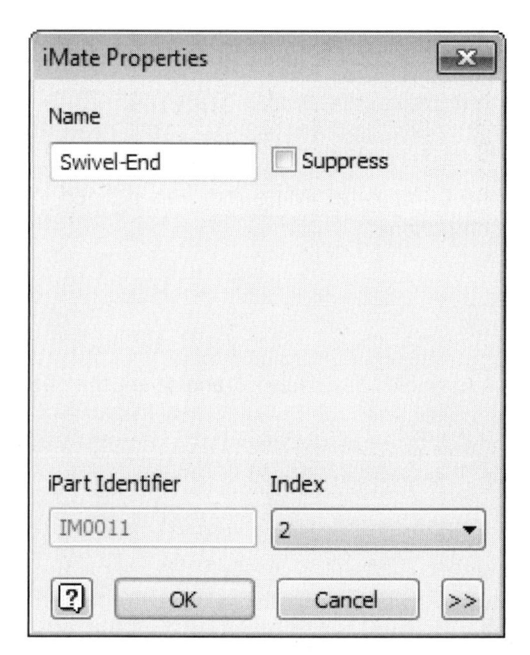

FIGURE 8.26

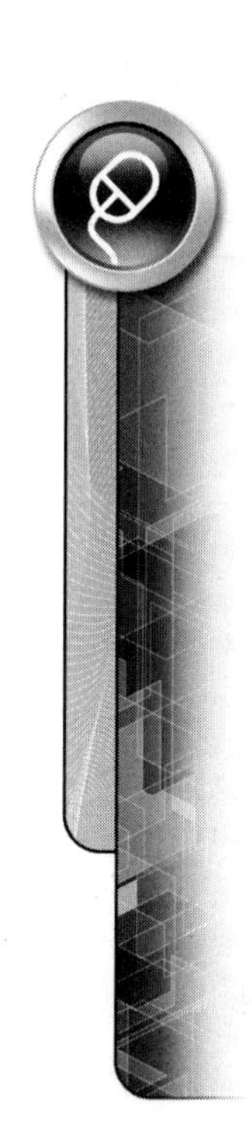

37. Click OK.

    Next save the modified Mount part with the new composite iMate under a different name.

38. On the Application menu, click Save As.
39. Enter a file name of **My_Mount**, and then click Save.
40. Close the window that contains the new *My_Mount.ipt* file without saving changes. Activate the ESS_E08_01.iam window if not already active.

    Next, you will insert the *My_Mount.ipt* file into the assembly, and automatically place the part using existing iMates.

41. Click the Place Component command on the Assemble tab > Component panel.
42. In the Place Component dialog box:
    a. Select *My_Mount.ipt*. Do not double-click.
    b. Select the Interactively Place with iMates option, which is located on the left portion of the iMates section in the Place Component dialog box, as shown in the following image.
    c. Click Open.

FIGURE 8.27

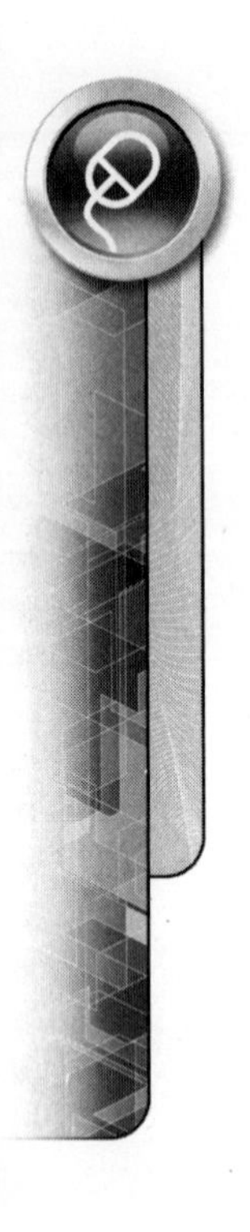

The part previews a constraint to the Composite1 iMate that was defined in both the MainBase and My_Mount components.

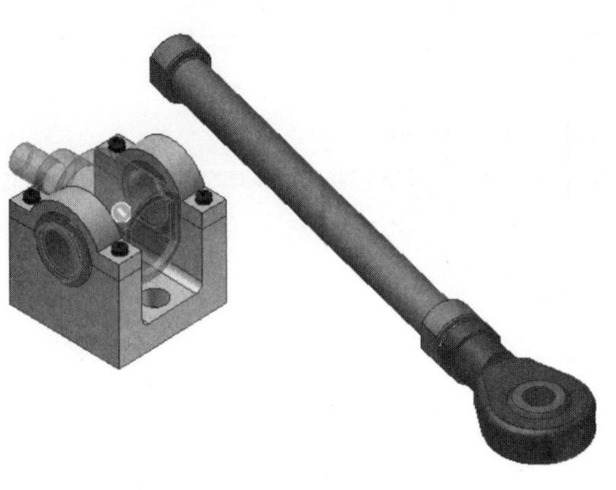

FIGURE 8.28

43. Right-click, and select Next Component iMate from the menu. The next match is previewed. Both previews look correct. Now you will complete the placement.

FIGURE 8.29

44. Click in the graphics window to accept the iMate result. The preview will go to the first match that it finds. In this case, go back to the first preview of the Composite1 iMate.

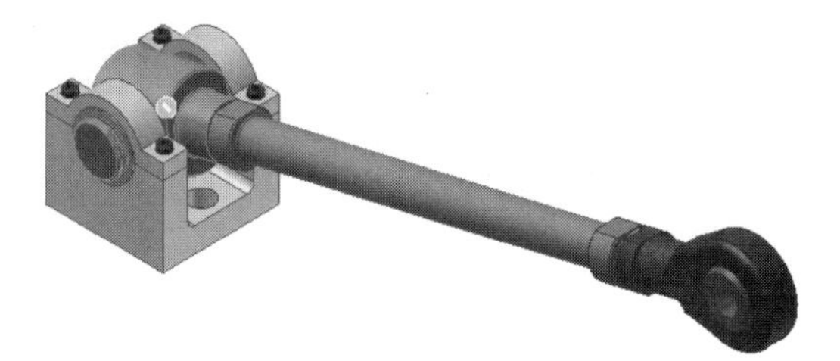

FIGURE 8.30

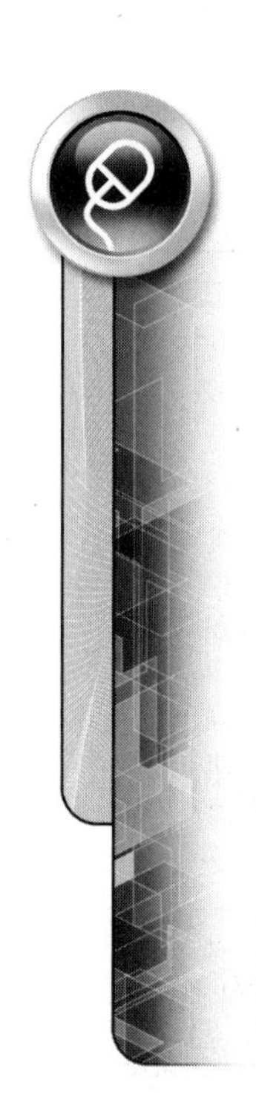

45. Click in the graphics window again to accept the iMate result.
46. Right-click in the graphics window, and select Done from the menu.

    Next, you replace a component with a component that has similar iMates.

47. In either the browser or the graphics window, right-click the End component, and select Component > Replace from the menu.
48. In the Place Component dialog box, select *ESS_E08_01-Clevis.ipt*. Click Open.
49. Click OK in the Possible Constraint Loss dialog box.
50. Click No when prompted to Save session edits to *ESS_E08_01-End.iam* prior to delete.

    The End component is replaced with the Clevis component, and the iMates are automatically applied.

FIGURE 8.31

51. Close all files. Do not save changes. End of exercise.

DIMENSION DISPLAY, RELATIONSHIPS, AND EQUATIONS

When creating part features, you may want to set up relationships between features and/or sketch dimensions. The length of a part may need to be twice that of its width, for example, or a hole may always need to be in the middle of the part. In Autodesk Inventor, you can use several different methods to set up relationships between dimensions. The following sections will cover these methods.

Dimension Display

When you create a dimension, it is automatically tagged with a label, or parameter name, that starts with the letter "d" and a number: for example, "d0" or "d27." The first dimension created for each part is given the label "d0." Each dimension that you place for subsequent part sketches and features is sequenced incrementally, one number at a time. If you erase a dimension, the next dimension does not go back and reuse the erased value. Instead, it keeps sequencing from the last value on the last dimension created. When creating dimensional relationships, you may want to view a dimension's display style to see the underlying parameter label of the dimension. Five options for displaying a dimension's display style are available.

Display as Value. Use to display the actual value of the dimensions on the screen.

Display as Name. Use to display the dimensions on the screen as the parameter name: for example, d12 or Length.

Display as Expression. Use to display the dimensions on the screen in the format of parameter# = value, showing each actual value: for example, d7 = 20 mm or Length = 50 mm.

Display as Tolerance. Use to display the dimensions on the screen that have a tolerance style for example, 40 ± .3.

Display as Precise Value. Use to display the dimensions on the screen and ignore any precision settings that are specified: for example, 40.3563123344.

To change the dimension display style, right-click in the graphics window. Click Dimension Display, as shown in the following image, or click the Tools tab > Document Settings from the Options panel and, on the Units tab, change the display style. After you select a dimension display style, all visible dimensions will change to that style. As you create dimensions, they will reflect the current dimension display style.

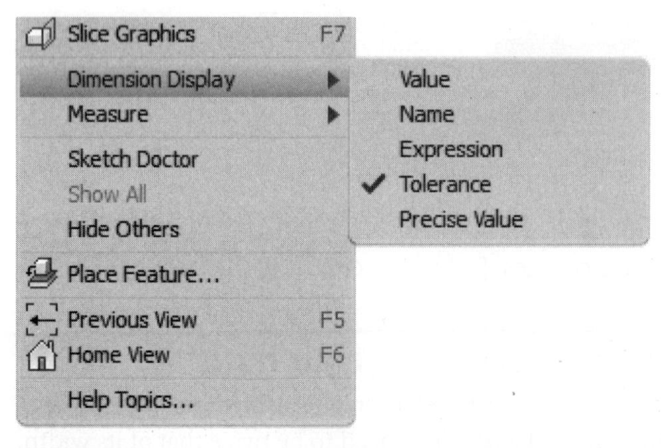

FIGURE 8.32

Dimension Relationships

Setting a dimensional relationship between two dimensions requires setting a relationship between the dimension you are creating and an existing dimension. When entering text in the Edit Dimension dialog box, enter the dimensions parameter name (d#) of the other dimension, or click the dimension with which you want to set the relationship in the graphics window. The following image on the left shows the Edit Dimension dialog box after selecting the 10 unit dimension to which the new dimension will be related. After establishing a relationship to another dimension, the dimension will have a prefix of fx:, as shown in the following image on the right.

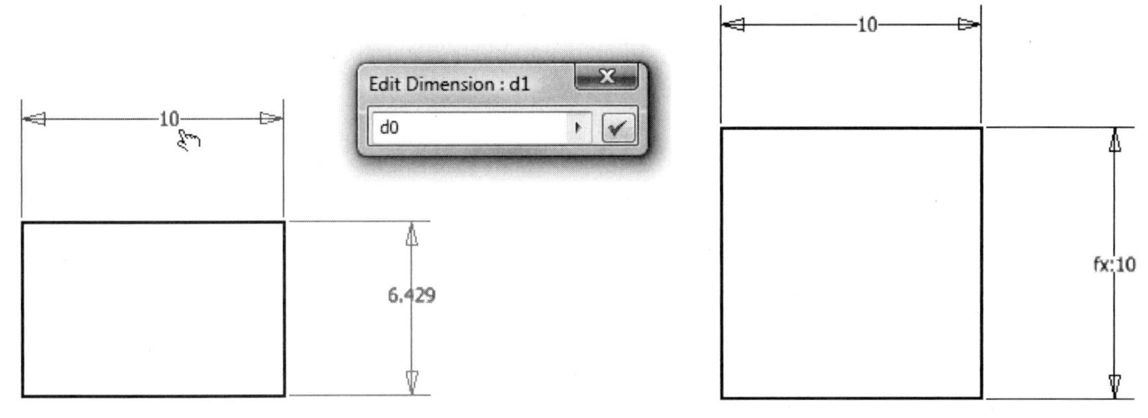

FIGURE 8.33

Equations

You can also use equations whenever a value is required: examples include (d9/4)*2 or 50 in + 19 in. When creating equations, Autodesk Inventor allows prefixes, precedence, operators, functions, syntax, and units. To see a complete listing of valid options, use the Help system, and search for Functions, Prefixes, and Algebraic Operators.

You can enter numbers with or without units; when no unit is entered, the default unit will be assumed. As you enter an equation, Autodesk Inventor evaluates it. An invalid expression will appear in red, and a valid expression will appear in black. For best results while using equations, include units for every term and factor in the equation.

To create an equation in any field, follow these steps:

1. Click in the field.
2. Enter any valid combination of numbers, parameters, operators, or built-in functions. The following image shows an example of an equation that uses both millimeters and inches for the units.
3. Press ENTER or click the green checkmark to accept the expression.

NOTE

Use ul (unitless) where a number does not have a unit. For example, use a unitless number when dividing, multiplying, or specifying values for a pattern count.

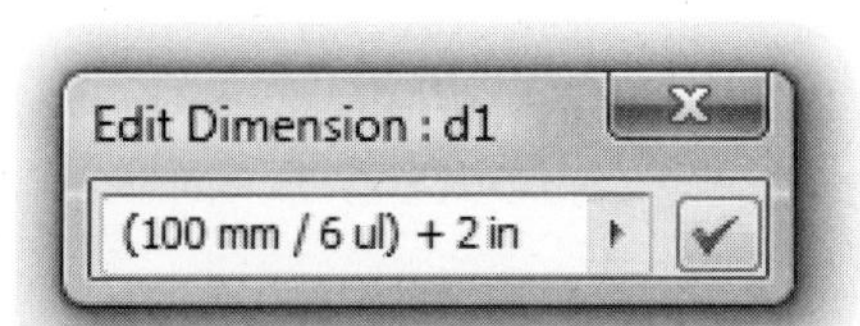

FIGURE 8.34

PARAMETERS

Another way to set up relationships between dimensions is to use parameters. A parameter is a user-defined name that is assigned a numeric value, either explicitly or through equations. You can use multiple parameters in an equation, and you can use parameters to define another parameter such as *depth* = *length* – *width*. You can use a parameter anywhere a value is required. There are four types of parameters: model, user, reference, and linked.

NOTE

Add-ins such as Simulation: Stress Analysis and Design Accelerator can automatically add parameter groups. Depending on the Add-in, you may or may not be able to remove those groups.

Model. This type is created automatically and assigned a name when you create a sketch dimension, feature parameter such as an extrusion distance, draft angle, or coil pitch, and the offset, depth, or angle value of an assembly constraint. Autodesk Inventor assigns a default name to each Model Parameter as you create it. The default name format is a "d" followed by an integer incremented for each new parameter. You can rename Model Parameters via the Parameters dialog box or specify a name upon initial creation in the Edit Dimension dialog box. This method is called "creating a parameter on the fly" and can be done in any field that you can specify a parametric dimensional value. The parameter that is being defined by the field will be renamed to the name specified at the beginning of the equation. For example, if you were defining a rectangular sketch and wanted to define a parameter named Length while dimensioning the sketch, you would enter "Length=10" (without the quotation marks) in the Edit Dimension dialog box. A parameter with the name of Length would be automatically created and available.

User. This type is created manually from an entry in the Parameters dialog box. You can add a numeric parameter, or if using rules-based design, you can add text or true-false user parameters.

Reference. This type is created automatically when you create a driven dimension. Autodesk Inventor assigns a default name to each reference parameter as you create it. The default name format is a "d" followed by an integer incremented for each new parameter. You can rename reference parameters via the Parameters dialog box.

Linked

This type is created via a Microsoft Excel spreadsheet and is linked into a part or assembly file. This is done using the Parameters command on the Manage tab > Parameters panel, as shown in the following image.

f_x
Parameters

FIGURE 8.35

After you click the Parameters command, the Parameters dialog box is displayed. The following image shows an example with some model, user, and reference parameters created. The Parameters dialog box is divided into two sections: Model Parameters and User Parameters. A third section called Reference Parameters is displayed if driven dimensions exist. The values from dimensions or assembly constraints used in the active document automatically fill the Model Parameters section. The User Parameters section is defined manually. You can change the names and equations of both types of parameters, and you can add comments by double-clicking in the cell and entering the new information. The column names for Model and User Parameters are the same, and the following sections define them.

Parameter Name. The name of the parameter will appear in this cell. To change the name of an existing parameter, click in the box, and enter a new name. When creating a new User Parameter, enter a new name after clicking the Add Numeric, Add Text, or Add True/False button. When you update the model, all dependent parameters update to reflect the new name.

Unit/Type. Enter a new unit of measurement for the parameter in this cell. With Autodesk Inventor, you can build equations that include parameters of any unit type. All length parameters are stored internally in centimeters; angular parameters are stored internally in radians. This becomes important when you combine parameters of different units in equations.

Equation. The equation will appear in this cell, and it will determine the value of the parameter. If the parameter is a discrete value, the value appears in rounded form to match the precision setting for the active document. To change the equation, click on the existing equation, and enter the new equation.

Nominal Value. The nominal tolerance result of the equation will appear in this cell, and it can only be modified by editing the equation.

Tol. (Tolerance). From the drop-down list, select a tolerance condition: upper, median, nominal, or lower.

Model Value. The actual calculated model value of the equation in full precision will appear in this cell. This value reflects the current tolerance condition of the parameter.

Key. Click to specify which defined parameters will be identified as a key parameter.

Export Parameters. Click to export the parameter to the Custom tab of the iProperties dialog box. The parameter will also be available when using the Derive command as well as in the bill of materials and Parts List Column Chooser dialog boxes.

Comment. You may choose to enter a comment for the parameter in this cell. Click in the cell, and enter the comment.

Parameters

Parameter Name	Unit/Type	Equation	Nominal Val	Tol.	Model Value	Key	Export	Comment
Model Parameters								
d0	mm	50 mm	50.000000		50.000000			
d1	mm	60 mm	60.000000		60.000000			
d2	mm	20 mm	20.000000		20.000000			
d3	mm	45 mm	45.000000		45.000000			
Center_Dist	mm	140 mm	140.000000		140.000000			
d5	mm	40 mm	40.000000		40.000000			
d7	deg	135 deg	135.000000		135.000000			
d8	deg	150 deg	150.000000		150.000000			
d9	mm	25 mm	25.000000		25.000000			
d10	mm	40 mm	40.000000		40.000000			
d11	deg	0 deg	0.000000		0.000000			
d12	mm	20 mm	20.000000		20.000000			
d13	mm	20 mm	20.000000		20.000000			
d14	mm	10 mm	10.000000		10.000000			
d15	mm	5 mm	5.000000		5.000000			
d16	mm	5 mm	5.000000		5.000000			
d17	mm	20 mm	20.000000		20.000000			
Shaft_Clearance...	mm	20 mm	20.000000		20.000000			
Brg_Seat_Dia	mm	28 mm	28.000000		28.000000			
d30	mm	40 mm	40.000000		40.000000			
d31	deg	0 deg	0.000000		0.000000			
d32	mm	-10 mm	-10.000000		-10.000000			
d33	mm	15 mm	15.000000		15.000000			
d34	mm	40 mm	40.000000		40.000000			
d35	mm	10 mm	10.000000		10.000000			
d36	deg	0 deg	0.000000		0.000000			
d37	mm	17 mm	17.000000		17.000000			
d43	mm	2 mm	2.000000		2.000000			
d44	mm	2 mm	2.000000		2.000000			
d45	mm	5.000 mm	5.000000		5.000000			
Reference Parameters								
d46	mm	120.000 mm	120.000000		120.000000			
User Parameters								
Shaft_Dia	mm	18 mm	18.000000		18.000000			
Bearing_Dia	mm	28 mm	28.000000		28.000000			
C:\Essentials Plus\Li...								
Link_Height	mm	60 mm	60.000000		60.000000			
Spindle_Dia	mm	22 mm	22.000000		22.000000			

Add Numeric | Update | Link | Immediate Update | Reset Tolerance | << Less | Done

FIGURE 8.36

User Parameters

User Parameters are ones that you define in a part or an assembly file. Parameters defined in one environment are not directly accessible in the other environment. If parameters are to be used in both environments, use a linked parameter via a common Microsoft Excel spreadsheet. You can use a parameter any time a numeric value is required. When creating parameters, follow these guidelines:

- Assign meaningful names to parameters, as other designers may edit the file and will need to understand your thought process. You may want to use the comments field for further clarification.
- The parameter name cannot include spaces, mathematical symbols, or special characters. You can use these to define the equation.

- Autodesk Inventor detects capital letters and uses them as unique characters. Length, length, and LENGTH are three different parameter names.
- When entering a parameter name where a value is requested, the upper- and lowercase letters must match the parameter name.
- When defining a parameter equation, you cannot use the parameter name to define itself; for example, Length = Length/2 would be invalid.
- Duplicate parameter names are not allowed. Model, User, and Spreadsheet-driven parameters must have unique names.

To create and use a User Parameter, follow these steps:

1. Click the Parameters command on the Manage tab > Parameters panel.
2. Click the appropriate Add button for the type of parameter you want to use (Numeric, Text, True/False) at the bottom of the Parameters dialog box.
3. Enter the information for each of the cells.
4. After creating the parameter(s), you can enter the parameter name(s) anywhere that a value is required. When editing a dimension, click the arrow on the right, and select List Parameters from the menu, as shown in the following image. All the available User Parameters and any named parameters will appear in a list similar to that in the image on the right. Click the desired parameter from the list.

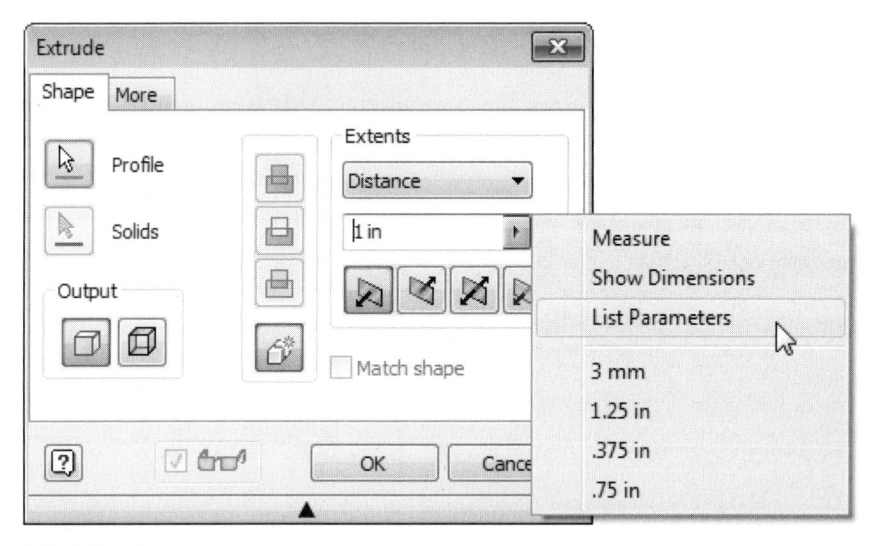

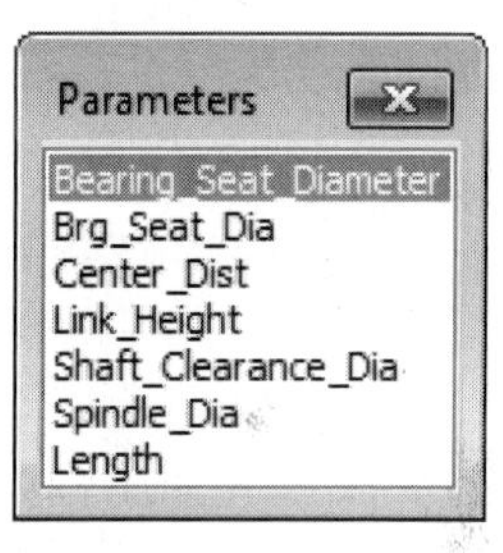

FIGURE 8.37

Linked Parameters

If you want to use the same parameter name and value for multiple parts, you can do so by creating a spreadsheet using Microsoft Excel. You then embed or link the spreadsheet into a part or assembly file through the Parameters dialog box. When you **embed** a Microsoft Excel spreadsheet, there is no link between the spreadsheet and the parameters in the Autodesk Inventor file, and any changes to the original spreadsheet will not be reflected in the Autodesk Inventor file. When you **link** a Microsoft Excel spreadsheet to an Autodesk Inventor part or assembly file, any changes in the original spreadsheet will update the parameters in the Autodesk Inventor file. You can link more than one spreadsheet to an Autodesk Inventor file, and you can link each spreadsheet to multiple part and assembly files. By linking a spreadsheet to an assembly file and to the part files that comprise the assembly, you can drive parameters from both environments from the same spreadsheet. There is no limit to the number of part or assembly files that can reference the same spreadsheet. Each linked spreadsheet appears in the browser under the 3rd Party folder.

When creating a Microsoft Excel spreadsheet with parameters, follow these guidelines:

- The data in the spreadsheet can start in any cell, but the cells must be specified when you link or embed the spreadsheet.
- The data can be in rows or columns, but they must be in this order: parameter name, value or equation, unit of measurement, and (if desired) a comment.
- The parameter name and value are required, but the other items are optional.
- The parameter name cannot include spaces, mathematical symbols, or special characters. You can use these to define the equation.
- Parameters in the spreadsheet must be in a continuous list. A blank row or column between parameter names eliminates all parameters after the break.
- If you do not specify a unit of measurement for a parameter, the default units for the document will be assigned when the parameter is used. To create a parameter without units, enter "ul" in the Units cell.
- Only those parameters defined on the first worksheet of the spreadsheet become linked to the Autodesk Inventor file.
- You can include column or row headings or other information in the spreadsheet, but they must be outside the block of cells that contains the parameter definitions.

The following image on the left shows three parameters that were created in rows with a name, equation, unit, and comment. The right side of the image shows the same parameters created in columns.

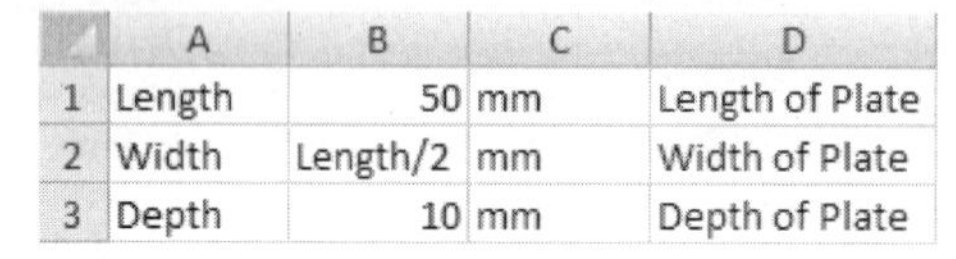

	A	B	C	D
1	Length	50	mm	Length of Plate
2	Width	Length/2	mm	Width of Plate
3	Depth	10	mm	Depth of Plate

	A	B	C
1	Length	Width	Depth
2	50	Length/2	10
3	mm	mm	mm
4	Length of Plate	Width of Plate	Depth of Plate

FIGURE 8.38

After you have created and saved the spreadsheet, you can create parameters from it by following these steps:

1. Click the Parameters command on the Manage tab > Parameters panel.
2. Click Link on the bottom of the Parameters dialog box.
3. The Open dialog box, similar to that shown in the following image, will appear.

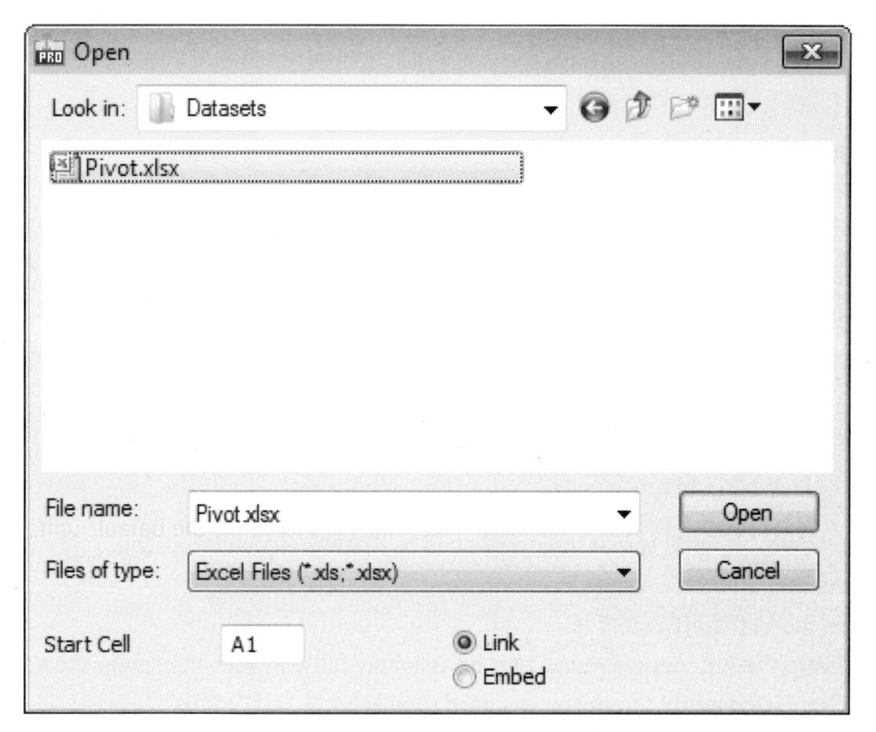

FIGURE 8.39

4. Navigate to and select the Microsoft Excel file to use.
5. In the lower-left corner of the Open dialog box, enter the start cell for the parameter data.
6. Select whether the spreadsheet will be linked or embedded.
7. Click the Open button: a section showing the parameters is added to the Parameters dialog box, as shown in the following image. If you embedded the spreadsheet, the new section will be titled Embedding #.
8. To complete the operation, click Done.

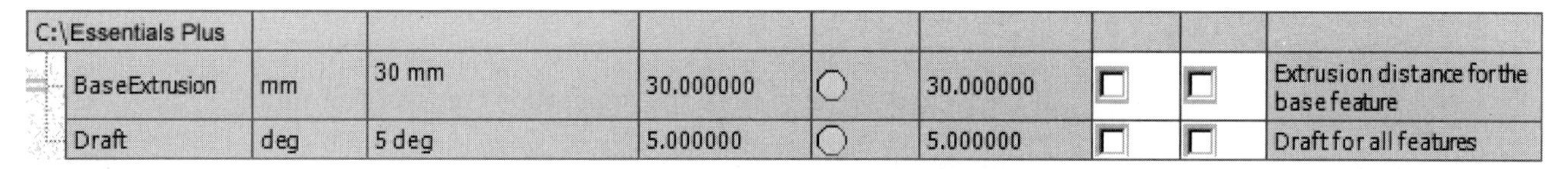

FIGURE 8.40

To edit the parameters that are linked, follow these steps:

1. Open the Microsoft Excel file.
2. Make the required changes.
3. Save the Microsoft Excel file.
4. Open the Autodesk Inventor part or assembly file that uses the spreadsheet.

You can also follow these steps when a spreadsheet has been linked. If you chose to embed the spreadsheet, the following steps must be used to edit the parameters:

1. Open the Autodesk Inventor part or assembly file that uses the spreadsheet.
2. Expand the 3rd Party folder in the browser.

3. Either double-click the spreadsheet or right-click and select Edit from the menu, as shown in the following image.

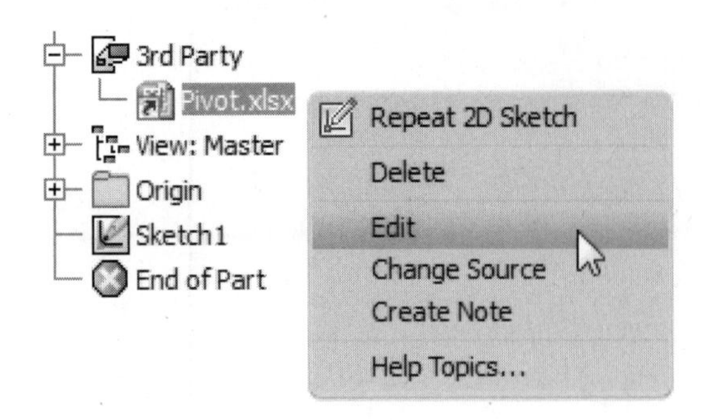

FIGURE 8.41

4. The Microsoft Excel spreadsheet will open in a new window for editing.
5. Make the required changes.
6. Save the Microsoft Excel file.
7. Activate the Autodesk Inventor part or assembly file that uses the spreadsheet.
8. Click the Update tool from the standard toolbar.

NOTE

If you embedded the spreadsheet, the changes will not be saved back to the original file but will only be saved internally to the Autodesk Inventor file.

EXERCISE 8-2: RELATIONSHIPS AND PARAMETERS

In this exercise, you create a sketch and dimension it. You then set up relationships between dimensions, and define User Parameters in both the model and an external spreadsheet.

1. Unless this step has already been performed, click the Application Options command on the Tools tab > Options panel, and then click on the Sketch tab. Check Autoproject Part Origin on Sketch Create to automatically have the origin point projected for the part.
2. Click OK or Close to close the Application Options dialog box.
3. Start a new file based on the default *Standard (in).ipt* file.
4. Draw the geometry, as shown in the following image. The lower-left corner should be located at the origin point, and the arc is tangent to both lines. The size should be roughly 25 in wide by 40 in high.

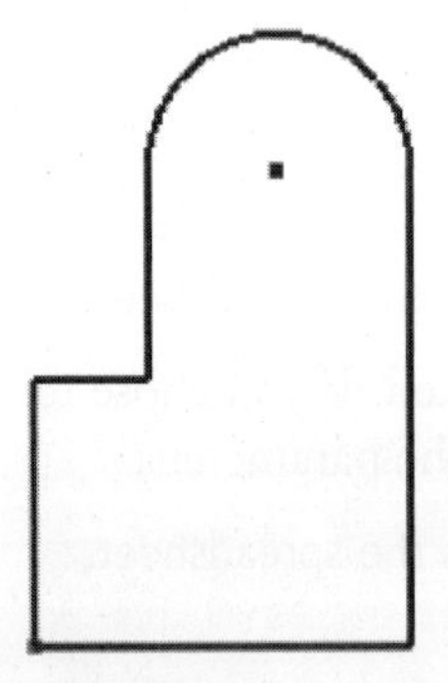

FIGURE 8.42

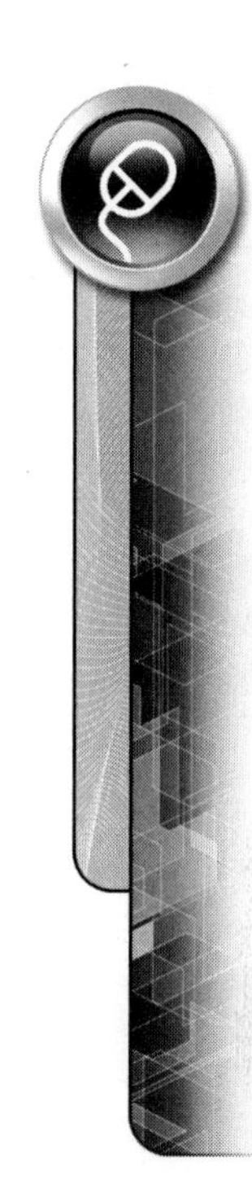

5. Add the four dimensions, as shown in the following image.

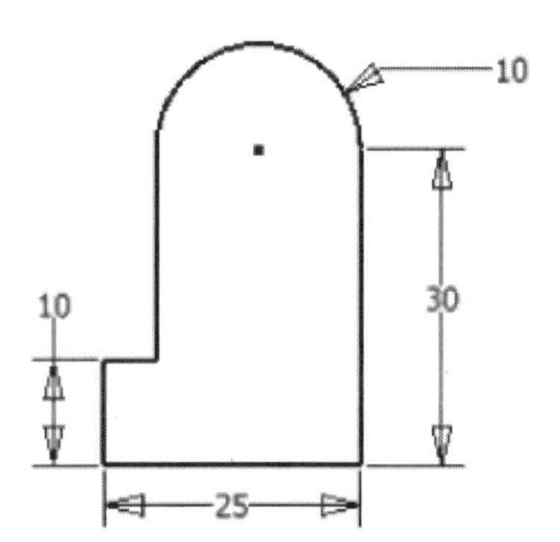

FIGURE 8.43

6. Double-click the 10 in radius dimension of the arc, and for its value, select the 25 in horizontal dimension, as shown in the following image.

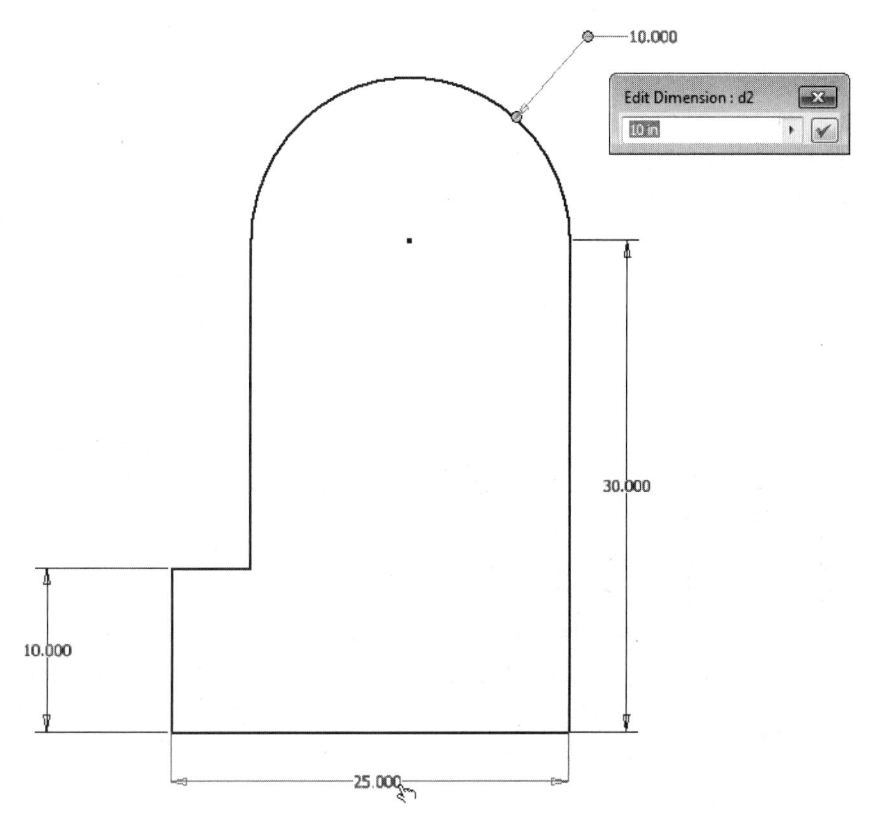

FIGURE 8.44

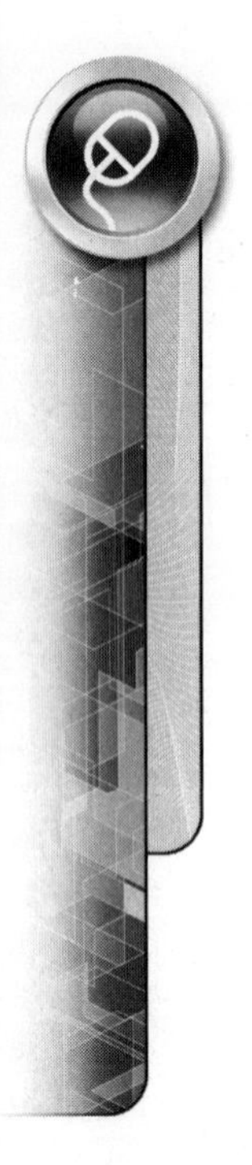

7. In the Edit Dimension dialog box, type **/4** after the d#, creating "d0/4." The 0 may be a different number depending upon the order in which your geometry was dimensioned. Click the green checkmark to accept the dimension value in the Edit Dimension dialog box.
8. Double-click the 10 in vertical dimension, and for its value, select the 30 in vertical dimension. Type **/2** after the d#. Click the green checkmark to accept the dimension value in the Edit Dimension dialog box.
9. Change the value of the 25 in horizontal dimension to **50 in** and the 30 in vertical dimension to **40 in**. When done, your sketch should resemble the following image. The fx: text that is displayed before the two dimensions denotes that they are equation driven or have a reference to a parameter.

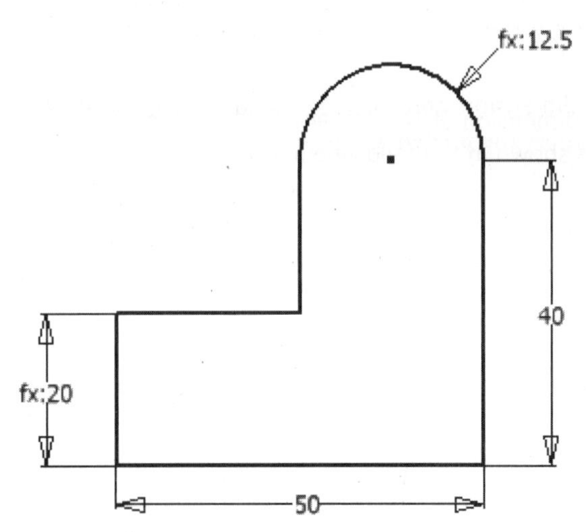

FIGURE 8.45

10. Next create parameters and drive the sketch from them. Click the Parameters command on the Manage tab > Parameters panel.
11. In the Parameter dialog box, create a User Parameter by clicking the Add Numeric button. Type in the following information:
    - Parameter Name = **Length**
    - Units = **in**
    - Equation = **75**
    - Comment = **Bottom length**
12. Create another User Parameter by selecting the Add Numeric button, and type in the following information:
    - Parameter Name = **Height**
    - Units = **in**
    - Equation = **Length/2**
    - Comment = **Height is half of the length**

    When done, the User Parameter area should resemble the following image.

User Parameters								
Length	in	75 in	75.000000	○	75.000000	☐	☐	Bottom length
Height	in	Length / 2 ul	37.500000	○	37.500000	☐	☐	Height is half of the length

FIGURE 8.46

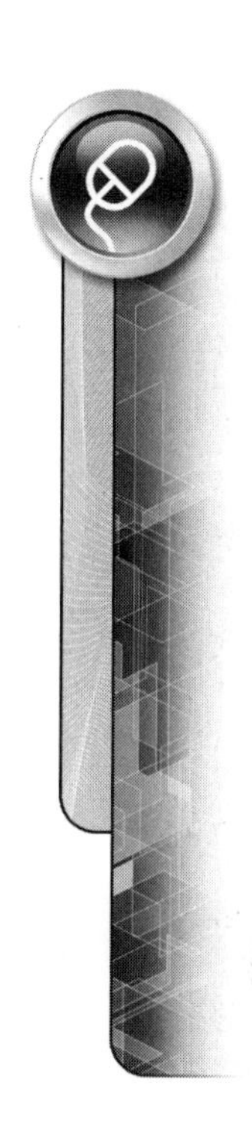

13. Close the Parameter dialog box by clicking Done.
14. Change the dimension display by right-clicking the graphics window, and then click Dimension Display > Expression.
15. Double-click the bottom horizontal dimension, and change its value to the parameter Length as follows: click the arrow; on the right side of the menu, click List Parameters, click Length from the list, and then click the checkmark in the dialog box.
16. Double-click the right vertical dimension, and change its value to Height. When done, your screen should resemble the following image.

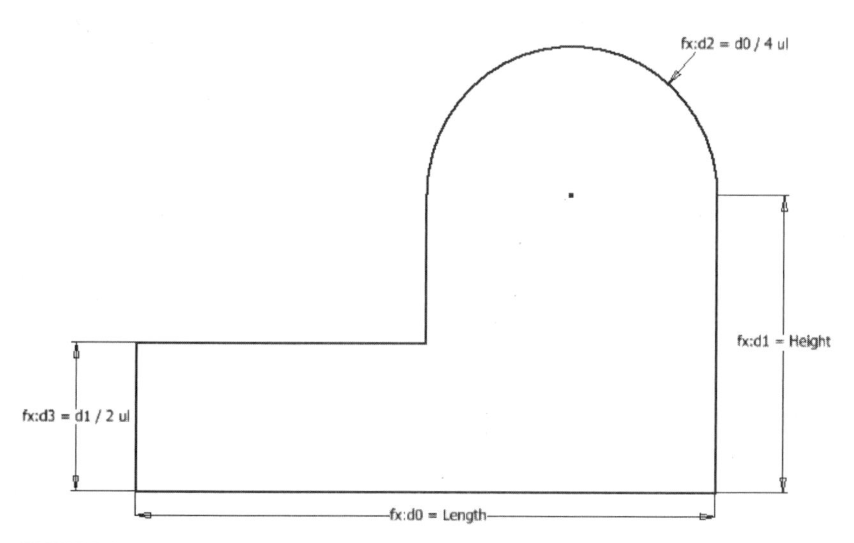

FIGURE 8.47

17. Click the Parameters command on the Manage tab > Parameters panel. From the User Parameters section, click on the equation cell of the parameter name Length, and change its value to **35**.
18. Click in the equation cell of the parameter name Height; change its value to **50** and the comment field to "**Height is half of the right side**." When done making the changes, the User Parameter area should resemble the following image.

User Parameters								
Length	in	35 in	35.000000	○	35.000000	☐	☐	Bottom length
Height	in	50 in	50.000000	○	50.000000	☐	☐	Height is half of the right side

FIGURE 8.48

19. To complete the changes, click the Done button.

20. Click the Finish Sketch command on Sketch tab > Exit panel, and the sketch should update to reflect the changes, as shown in the following image.

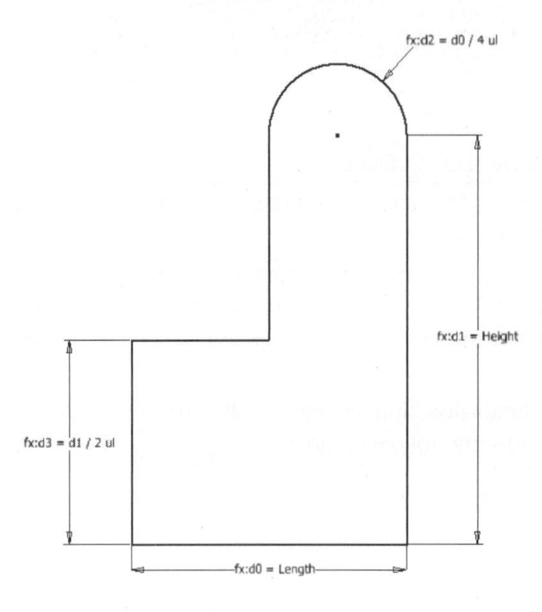

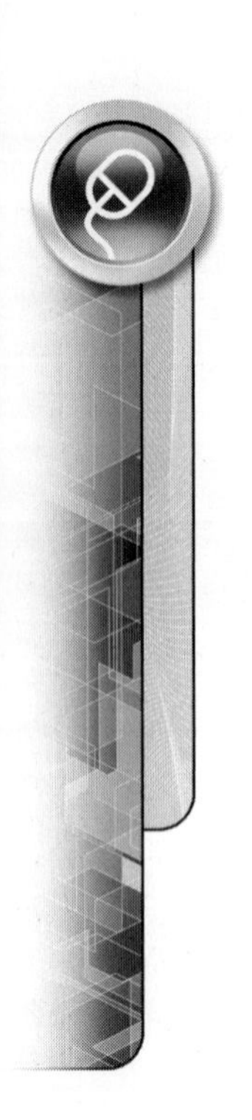

FIGURE 8.49

21. Save the file as *ESS_E08_02.ipt.*
22. Now create a spreadsheet that has two parameters. Create an Excel spreadsheet with the column names Parameter Name, Equation, Unit, and Comment, as shown in the following image. Enter the following data:
    - Parameter Name = BaseExtrusion
    - Equation = **30**
    - Units = **in**
    - Comment = **Extrusion distance for the base feature**
    - Parameter Name = **Draft**
    - Equation = **5**
    - Units = **deg**
    - Comment = **Draft for all features**

	A	B	C	D
1	Parameter Name	Equation	Unit	Comment
2	BaseExtrusion	30	in	Extrusion distance for the base feature
3	Draft	5	deg	Draft for all features

FIGURE 8.50

23. Save the spreadsheet as *ESS_08_Parameters.xlsx.*
24. Make Autodesk Inventor the current application, and then click the Parameters command.
25. In the Parameter dialog box, click the Link button and select the *ESS_08_Parameters.xlsx* file, but do not click Open yet.
26. For the Start Cell, enter **A2**. If you fail to do this, no parameters will be found.
27. Click the Open button, and a new spreadsheet area will appear in the Parameters dialog box, as shown in the following image. Click the Done button to complete the operation.

C:\Essentials Plus...								
BaseExtrusion	in	30 in	30.000000	○	30.000000	☐	☐	Extrusion distance for the base feature
Draft	deg	5 deg	5.000000	○	5.000000	☐	☐	Draft for all features

FIGURE 8.51

28. Change to the home view.
29. Click the Extrude command from the Model tab > Create panel, and click the More tab in the Extrude dialog box, as shown in the following image. For the Taper's value, enter **Draft** or click the arrow from the menu, click List Parameters, and then click the parameter Draft.

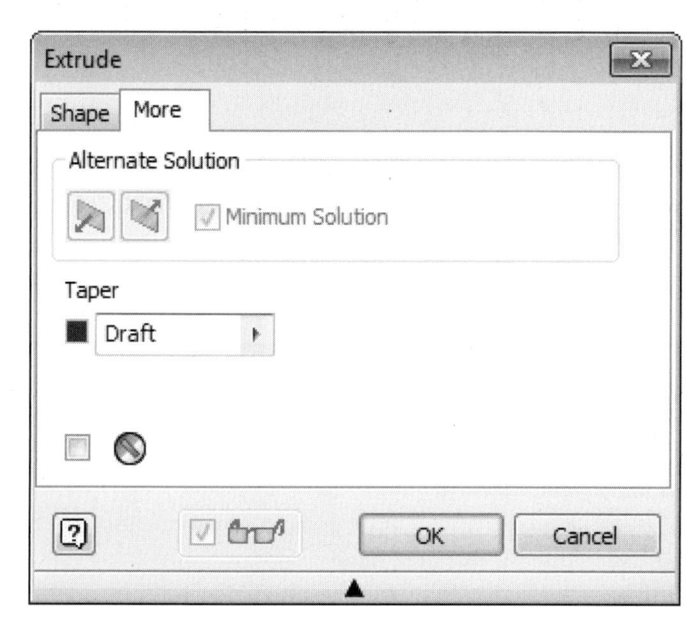

FIGURE 8.52

30. Click the Shape tab. For the Distance value, use the parameter BaseExtrusion, as shown in the following image.

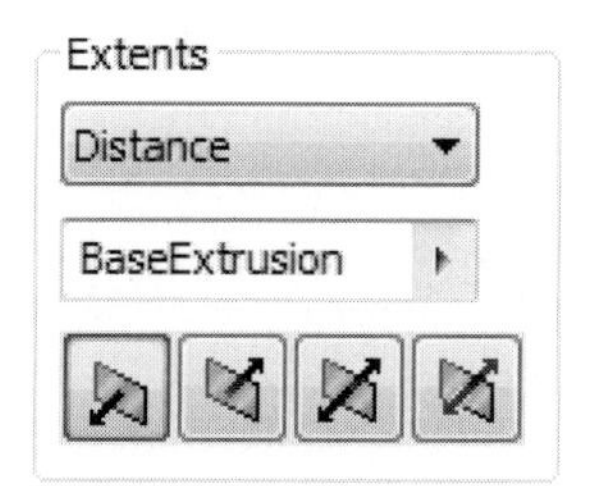

FIGURE 8.53

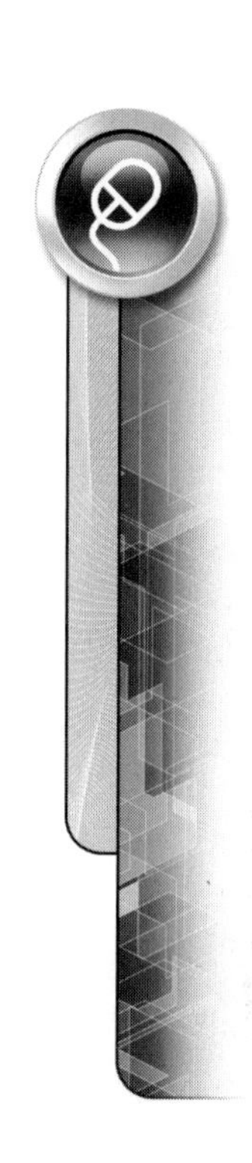

31. To complete the operation, click the OK button. Your model should resemble the following image.

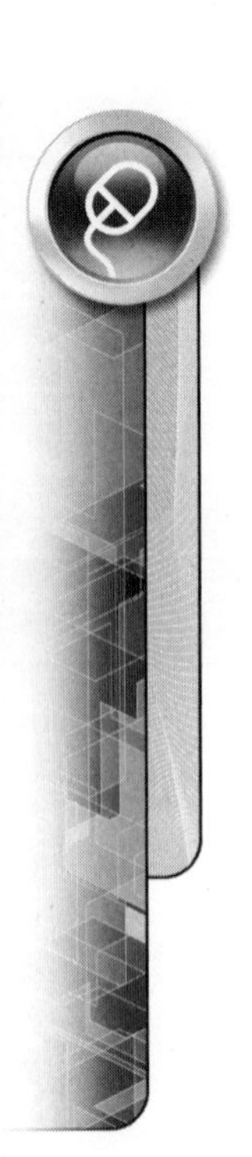

FIGURE 8.54

32. In the browser, expand the 3rd Party icon. Either double-click or right-click on the name *ESS_08_Parameters.xls*, and select Edit from the menu.
33. In the Excel spreadsheet, make the following changes: For the Parameter Name "BaseExtrusion," change the Equation to **50**. For the Parameter Name "Draft," change the Equation to **–3**.
34. Save the spreadsheet, and close Excel.
35. Make Autodesk Inventor the current application, and click the Local Update command from the Quick Access Bar. When done, your model should resemble the following image.

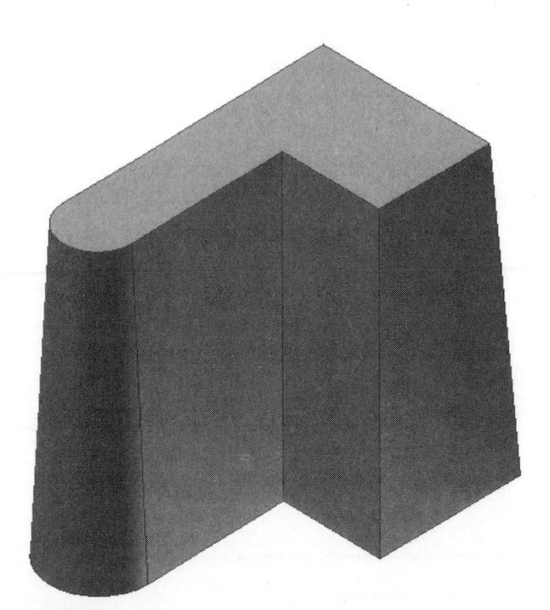

FIGURE 8.55

36. Close the file. Do not save changes. End of exercise.

iPARTS

iParts enable design intent and design knowledge to be shared and reused. They also enable you to store multiple part parameters and properties and then calculate unique part file versions based on certain configurations of the stored parameters and properties. Using the Parameters dialog box, you can store one value for each parameter. In the industry, configured parts have been referred to as tabulated parts, charted parts, or a family of parts. You can also use iParts to create part libraries that allow design data to be reused. An iPart is generated from a standard Autodesk Inventor part file (*.ipt*). When you activate the Create iPart command, the standard part file is converted into an iPart. You can add individual members, or configurations, to the iPart; it is then referred to as an iPart factory.

There are two stages to the use of iParts: authoring the iPart and placing an iPart version. In the authoring or creation stage, you design the part and establish all possible versions of the design in a table. The rows of the table describe the members of the iPart factory.

In the placement stage, you select a member from the iPart factory, and an iPart version is published and inserted into your assembly.

iParts allow you to suppress specific features, control iMates and thread properties, and add or modify file properties to the different members that are contained within the iPart factory. They also allow user-defined input for specific parameters. You can further enhance inputs to include an element of control. For instance, you can apply values within a predetermined range or from a list to a parameter. After creating an iPart, you can place it into an assembly, where you can select a specific member. The following sections explain how to create and then place an iPart into an assembly.

An example of an iPart is a simple bolt. The bolt has a number of different sizes that are associated with it. An iPart allows you to define these different sizes and configurations—the material, part properties, and so on—and have them reside within a single file, or factory. When placing the bolt in an assembly, you can select which size, or version/member, of the bolt, or iPart factory, you want to use. You can then constrain the placed member in the assembly like any other part file.

Creating iParts

You can create two types of iParts: standard and custom.

Standard iParts or Factories

You cannot modify iPart values; when placing a standard iPart into an assembly, you can only select the predefined members for placement. A standard iPart that you place in an assembly cannot have features added to it after placement.

Custom iParts or Factories

Custom iParts allows you to place a unique value for at least one variable. A custom iPart that you place in an assembly can have features added to it.

To create an iPart, follow these steps:

1. Create an Autodesk Inventor part or a sheet metal part.
2. Add the dimensions of the geometry to be changed to the iPart Author table.

3. For easier creation of the member table, use descriptive names for the values of the parameters.
4. If you do not use parameter names, you will need to determine what each parameter (d#) represents within the parts geometry.

TIP

Any parameters with a name other than d# are added automatically to the parameter table during the creation of the iPart factory.

5. Issue the Create iPart command on the Manage tab > Author panel, as shown in the following image, to add the members or configurations to the iPart Author table.

FIGURE 8.56

The iPart Author dialog box appears, as shown in the following image.

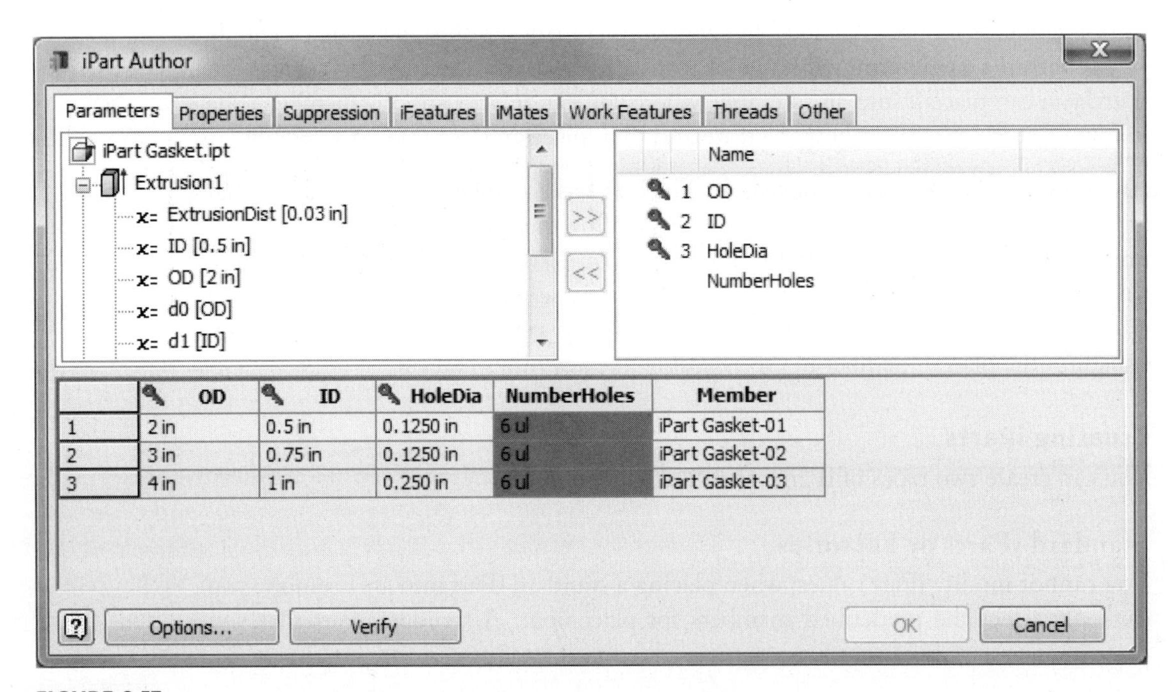

FIGURE 8.57

To create the iPart members, follow these steps:

6. Add to the right side of the dialog box, or the parameter list, all parameters and dimensions that will be configured. All named parameters are added automatically to the right side.

7. To add a d# dimension to the list on the right side, select it in the left column, and then click the Add parameters (>>) button. You can also double-click on the parameter to add it.
8. To remove a parameter from the right side of the dialog box, select its name, and click the Remove parameters (<<) button.
9. After you have added the parameters, you need to define the keys. Keys identify the column whose values are used to define the iPart member when the part is published or placed in an assembly. For example, setting a parameter as a primary key allows the designer placing the part to choose from all available values for that parameter in the selected items list.
10. To specify the key order, click an item in the key column of the selected parameters list to define it as a key, or right-click on the item and select the key sequence number. Selected keys are blue; items that are not selected as keys have dimmed key symbols. You can decide not to add primary keys or to add one or more additional keys as needed. You cannot specify custom table columns as keys.
11. Click on the other tabs in the dialog box to perform other specific operations. These items are not required to define the iPart factory. The following sections describe them in more detail.

Properties Tab. The Properties tab lists all of the summary, project, physical, and custom properties in the file. If you use the material property on this tab to control the material of your iParts, you must use the Material Column option in the table. This option is described in the Special Table Elements section later in this chapter. You must also set the current color of the iPart to As Material prior to saving the iPart. For part properties to be used in drawings, Bill of Materials (BOMs), and other downstream purposes, you have to include the properties in the table even if their values do not vary between members.

Suppression Tab. The Suppression tab allows you to suppress features of specific members of the iPart factory. If you enter Suppress in a cell for the feature, the feature will be suppressed when the version is calculated. When the cell contains Compute, the feature is not suppressed.

iFeatures Tab. On the iFeatures tab, select one or more table driven iFeatures included in the iPart. Control the iFeature member for each member of the iPart. They can also be suppressed as described above.

iMates Tab. On the iMates tab, select one or more iMates to include in the iPart. iMates are included in the iPart if their status is Compute. If their status is suppressed, they are not included. Control the iMate properties for each member using the iMates, Parameters, and Suppression tabs. On the iMates tab, you can perform the following actions:

- Define different offset values for different iPart members.
- Suppress iMates or Composite iMates for different iPart members.
- Change the matching name for different assembly configurations.
- Change the sequence of the iMates.

Since iMate names are not unique, each iMate is assigned an index tag in the iPart Author. Each iMate property, except the offset value, uses this tag to identify a specific iMate property in the table. Tags are not assigned to offset values because iMate offset values are parameters, which are unique by definition.

Work Features Tab. On the Work Features tab, select the user-defined work features to be included in the iPart. Origin work features are included automatically in an iPart factory and iPart members, but they are not listed. The work features are managed in the iPart if its table status is Include. If the status is Exclude, the work features are not managed in the iPart.

Threads Tab. You can control thread features in the iPart factory to create table-driven items for regular or tapered thread features. Use Family to identify the thread standard, Designation to control the size and pitch, and Class to define fit based upon the standard. You can assign the thread Direction as right or left in the table. You also have the option to modify the pipe diameter.

Other Tab. You can use the Other tab to create custom table items. For example, you can add a column that represents the name of the iPart member. You can create a column named Version to identify the appropriate member when it is placed in an assembly. You can also create custom values such as Color if you want to control the display style of the iPart.

Special Table Elements. You can right-click on a cell of a member of the table or a column label to access additional custom table capabilities. You can define the name of the iPart file to be used when it is published using the File Name Column option. Control the color of the iPart upon publishing using the Display Style Column option. You can also specify material of the iPart using the Material Column option. These options are available only for part properties and the table elements created using the Other tab. They are not available on parameter items.

Next add a member in the iPart table by right-clicking on a row number at the bottom of the dialog box and selecting Insert Row from the menu, as shown in the following image.

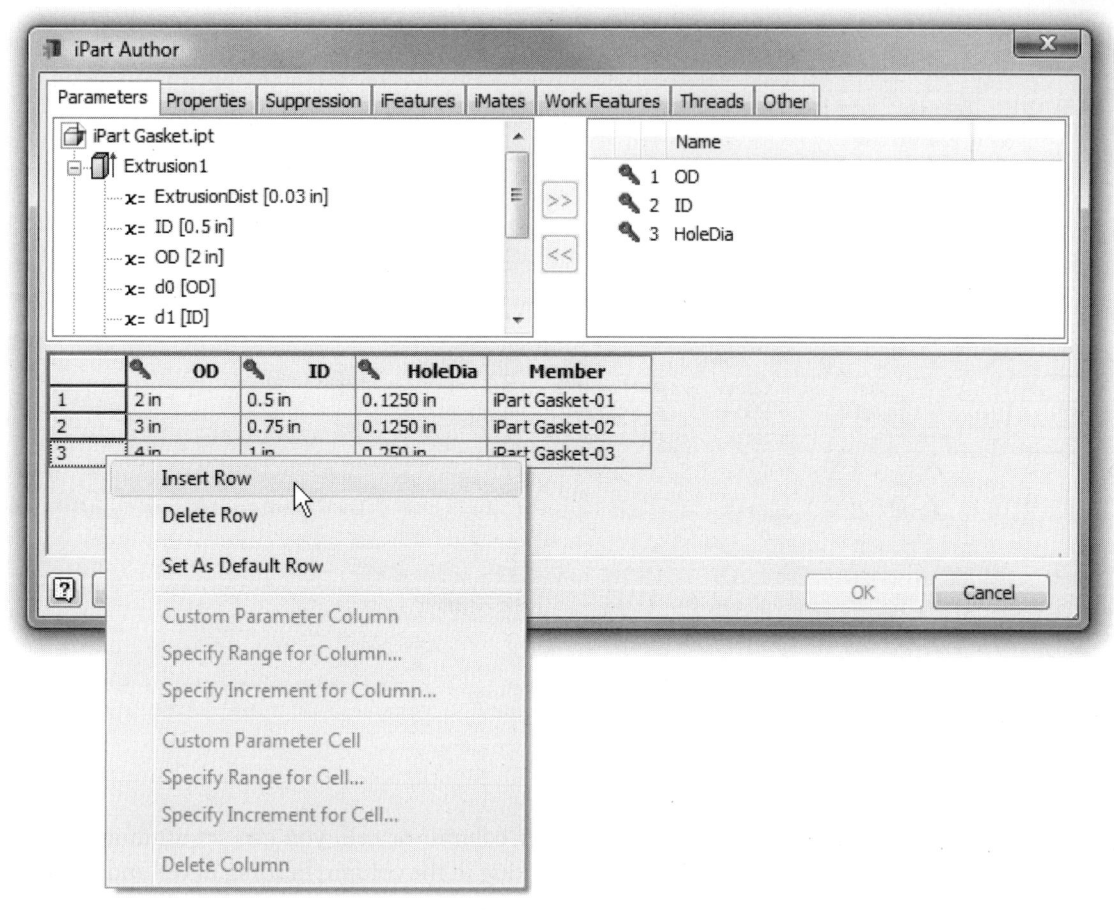

FIGURE 8.58

12. Edit the cell contents by clicking in the cell and typing new values or information as needed. To delete a row, right-click the row number, and select Delete Row from the menu. Each row that you add in the bottom pane of the dialog box represents an additional member within the iPart factory. The table functions similarly to a spreadsheet. In addition to adding or deleting members of the iPart factory, you can also use the table to change members by modifying cell values.

To allow the designer placing the iPart to specify a custom value for a given column, right-click on the column name, and select Custom Parameter Column from the menu, as shown in the following image. To make a specific cell custom, right-click in the individual cell, and select Custom Parameter Cell from the menu.

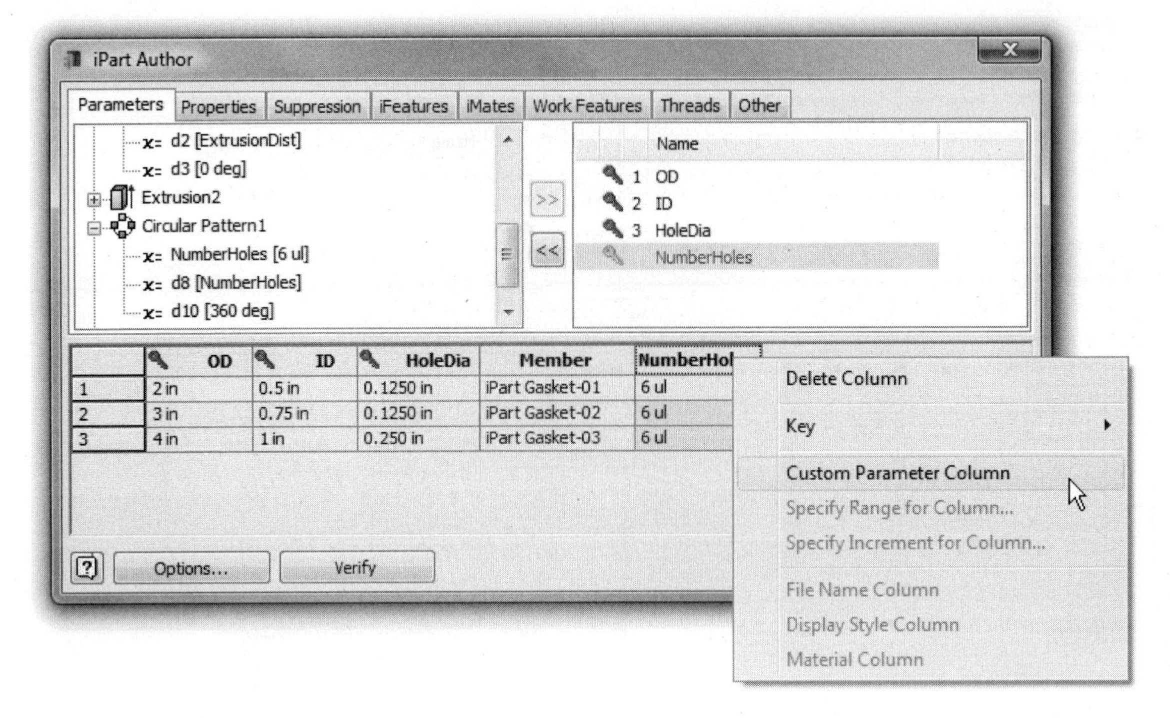

FIGURE 8.59

After designating a custom parameter column or cell, you can set a minimum and maximum range of values by right-clicking in the column heading or cell and selecting Specify Range for Column or Range for Cell from the menu. The Specify Range dialog box appears, as shown in the following image. Select the options as needed. Custom columns and cells are highlighted with a blue background in the table. You can also specify the increment that can be entered for a custom column or cell using the Specify Increment for Column option from the menu.

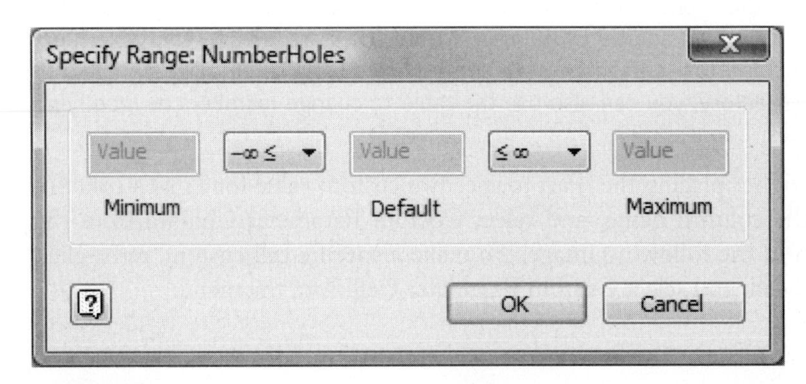

FIGURE 8.60

13. Set the member that will be the default by right-clicking on the row number and selecting Set As Default Row from the menu. You may select the other options as needed.

14. Click the Options button to edit the part number and member naming schemes for the iPart factory, as shown in the following image.

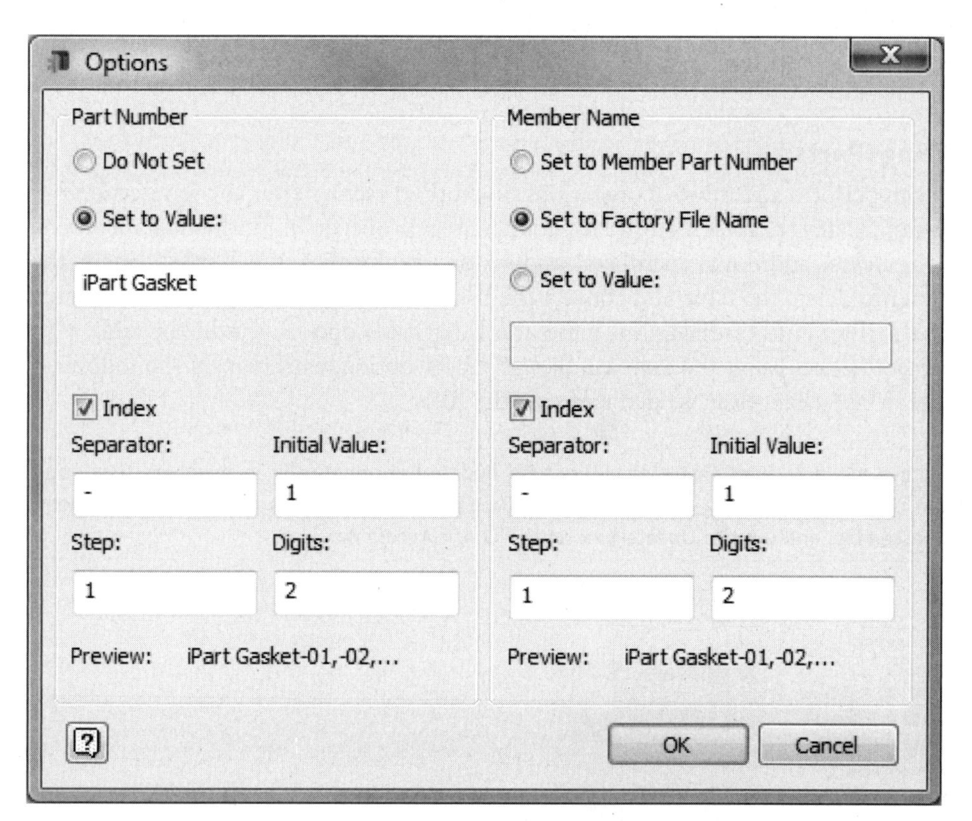

FIGURE 8.61

15. Click OK in the iPart Author dialog box when you have finished defining the contents of the table. The iPart Author dialog box closes, and the part is converted to an iPart factory. The table is saved, and a table icon appears in the browser, as shown in the following image.

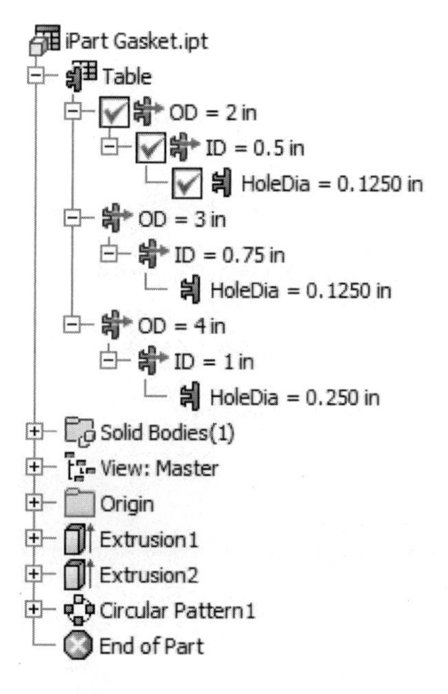

FIGURE 8.62

You can expand the Table icon in the browser to view the iPart members based on the member name or keys and values that you define. The active, or calculated, version appears with a checkmark, as shown in the previous image. You change the browser display to show the member name or keys by right-clicking and selecting either List by Member Name or List by Keys from the menu.

Editing iParts

You can perform a number of operations on an iPart factory after you have created it. You can delete the table, modify the parameters or properties for individual members, add or delete additional members, and so on. Right-click the Table icon in the browser to delete the table and convert the iPart factory back to a part, edit the table with the iPart Author dialog box using the Edit Table option, or edit the table with Microsoft Excel using the Edit via Spread Sheet option, as shown in the following image. Make changes as needed and save the file.

NOTE

Changes made to iPart factories will not be updated automatically in members that you have previously placed in assemblies. To update the iPart members in an assembly, open the assembly, and use the Update tool on the Quick Access Bar.

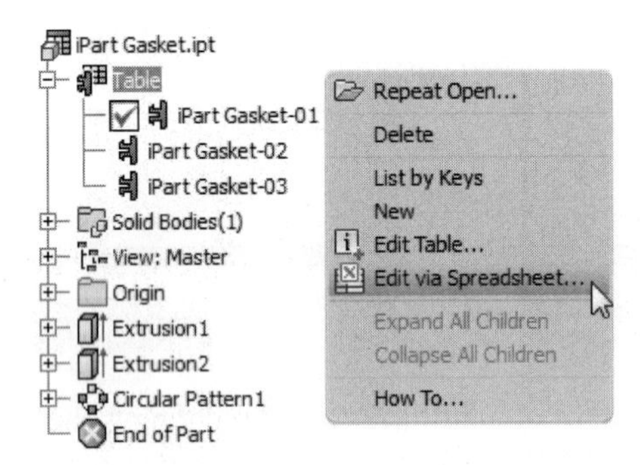

FIGURE 8.63

When editing the spreadsheet using Microsoft Excel, you can incorporate spreadsheet formulas, conditional statements, and multiple sheet data extractions, but you cannot modify spreadsheet formulas and conditional statements from the iPart Author dialog box. These types of cells are inactive and are highlighted in red when displayed in the iPart Author.

iPart Placement

You can place standard iParts in assemblies using the Place Component command. When you select a standard iPart factory, an additional Place Standard iPart dialog box appears. It allows you to use the Keys tab to identify an iPart member by selecting the key values, use the Tree tab to locate a member by expanding the key values, use the Table tab to identify an iPart member by selecting a row in the table, and place multiple instances of different iPart members.

To place a standard iPart into an assembly, follow these steps:

1. Start a new assembly or open an existing assembly in which to place the iPart.
2. Click the Place Component command, and navigate to and select the iPart to place in the assembly in the Place Component dialog box.

3. Click the Open button, and the Place Standard iPart dialog box will appear, as shown in the following image. If a custom cell(s) or column(s) exists, the Place Custom iPart dialog box appears.

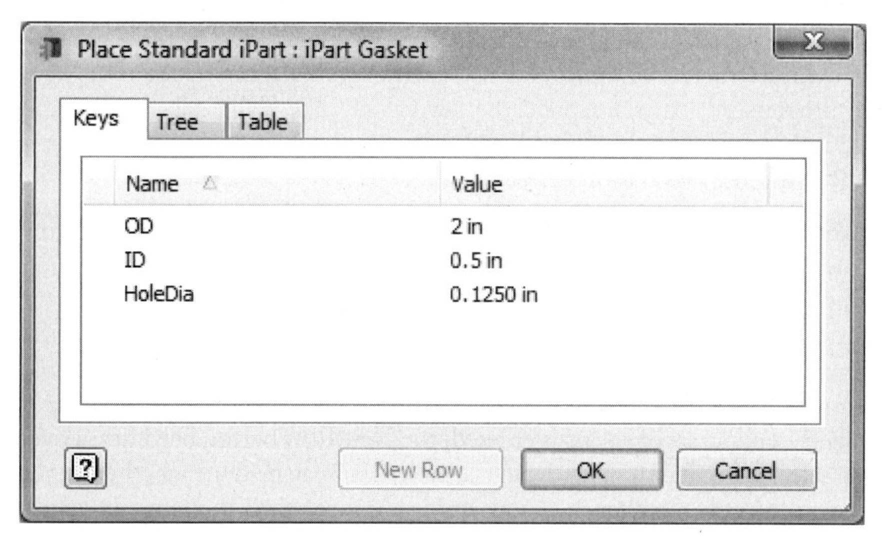

FIGURE 8.64

4. To select from the member list, select the Keys, Tree, or Table tab, and then select the member that defines the part you want to place.
5. If you are placing a custom part, select the Keys, Tree, or Table tab, and enter a value in the right side of the dialog box, as shown in the following image. The value must fall within the limits set in the iPart factory; otherwise, an alert appears.

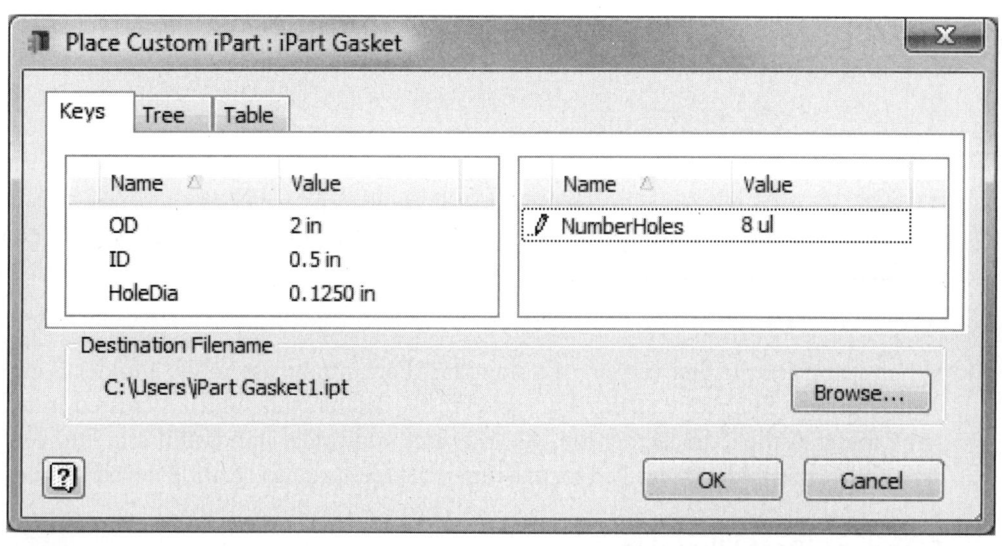

FIGURE 8.65

6. Place the part in the graphics window.
7. Continue placing instances of the iPart as needed.

8. To change an iPart in an assembly to a different configuration, expand the part in the browser, right-click on the table name, and select Change Component. Then select a new member from the Keys, Tree, or Table tab.

The Keys tab displays the primary and any secondary keys defined in the iPart factory. You can select the values of the keys to identify unique members. The Tree tab displays a hierarchical structure of the keys. If secondary keys exist, the values of the keys are filtered progressively as you expand the key values. The Table tab displays the entire table rather than just the keys. To identify a unique iPart member, select a row, and then click the OK button to place the iPart.

You can also dynamically create a new member for the factory when placing a standard iPart with the Table tab, as shown in the following image. In a row named "New" at the top of the table, you can enter values or select from a drop-down list if appropriate. As you enter values in the row, the table is reduced to display only the rows that match the values entered in the row. As you enter more values, the table continues to be filtered. If no member matches the entered values, a new member is created. When a unique set of values is entered, the New Row button becomes active, and you can select it to create a new row in the table. This can be done whether or not you are placing the new member in the assembly. Any columns that contain values that are not set are set to the same values as the default row.

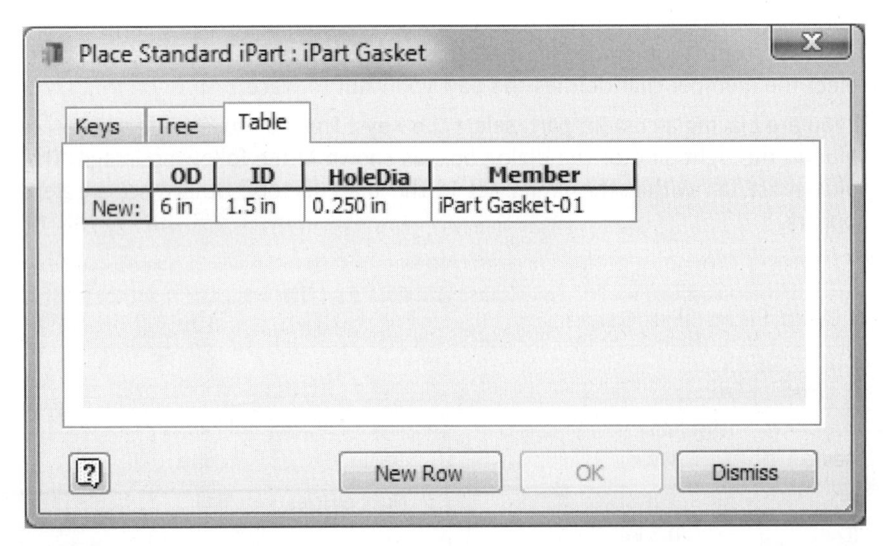

FIGURE 8.66

When you place the first version of a standard iPart into an assembly, a folder is created with the same name as the iPart factory. By default, this folder is created in the same folder as the iPart factory file. As you place additional standard iParts into your assemblies, this folder is checked for existing iPart files prior to creating new iPart files.

Standard iPart Libraries

In a collaborative design environment, the best way to manage iPart factories and the iParts published from those factories is by using libraries. Define a library within your project and place your iPart factories in this library's path. You can also specify that a standard iPart factory publish standard iParts to a different, or proxy, folder. To do this, create another library entry using the same name as the iPart factory library prefixed with an underscore, as shown in the following image, or right-click on the library's name, and click Add Proxy Path.

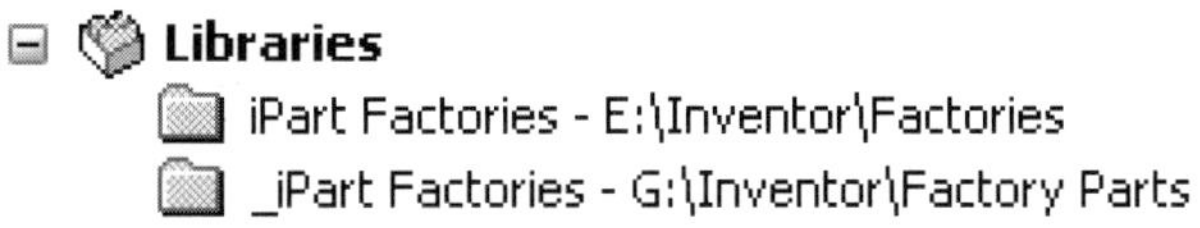

Libraries
iPart Factories - E:\Inventor\Factories
_iPart Factories - G:\Inventor\Factory Parts

FIGURE 8.67

If you have an iPart factory for bolts, for example, you could store the iPart factory in a library named Bolts, and then you could define an additional library where the individual members generated from the Bolt iPart factory will be stored. The library must be named _Bolts. Autodesk Inventor will automatically store all iParts published by the factory in the library directory specified by _Bolts. If you store an iPart factory in a workspace or a workgroup search path, the published iParts are stored in a subdirectory with the same name as the factory.

NOTE

It is recommended that you do not store iParts in the same folders as iPart factories.

Custom iParts

Custom iParts are placed into assemblies in the same way as standard iParts. If you select an existing custom iPart member file, that member of the part is placed into your assembly with no option to define the value of the custom parameters. When you select a custom iPart factory, the Place Custom iPart dialog box appears. As with the Standard iPart Placement dialog box, you can use the Keys tab to identify an iPart member by selecting the key value, the Tree tab to expand key values and select a member, or the Table tab to identify an iPart member by selecting a row in the table. The Place Custom iPart dialog box provides additional options so that you can enter values for custom parameters, define a destination and file name for the custom iPart by selecting Browse in the dialog box, and place multiple instances of different iPart members.

NOTE

Since each custom iPart is unique, you must provide different file names as you place different members.

After you have placed a custom iPart in an assembly, you can add more features to the iPart. The capability to add features to an iPart makes the custom iPart behavior similar to that of a derived part.

Auto-Capture

You can make changes automatically to an iPart or iAssembly, which are covered later in this chapter, using normal tools, without accessing the iPart or iAssembly Author table. You can automatically capture changes to the entire factory or to the active row. This is done using one of the two tools accessed from the Manage tab > Author panel, as shown in the following image. You can also use the Create iPart or iAssembly tools or edit the table via a spreadsheet.

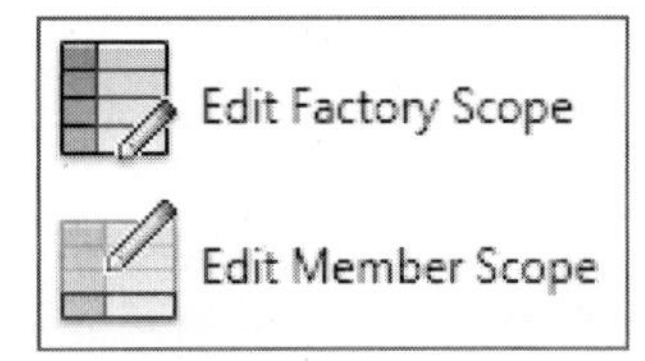

FIGURE 8.68

The available tools are as follows:

Edit Member Scope	Any edit that can be configured is automatically added as one or more new columns or will edit the member's cell value if a column for the change already exists in the table. If a new column is added, the new value of the object is added for the current member's row, and the original value is used for all other members.
Edit Factory Scope	This setting works similarly to Edit Member Scope, except that only columns are modified. New values are set for the entire column. Columns are not added or removed in this mode.
Create iPart/ iAssembly	Opens the iPart or iAssembly Author dialog box depending on which type of file is open.
Edit Using Spread Sheet	Opens the iPart or iAssembly table in a spreadsheet file for modification.

Edit Member Scope Example

For example, say that you have an iPart of a simple rectangle made up of a single extruded rectangle whose sketch has a "Length" and "Width" parameter that define the iPart table, as shown in the following image.

	Member	Part Number	Length
1	Rectangle-01	Rectangle-01	10 in
2	Rectangle-02	Rectangle-02	20 in
3	Rectangle-03	Rectangle-03	30 in

FIGURE 8.69

The factory is open in Autodesk Inventor, and Rectangle-02 is active. You have selected the Edit Member Scope tool, and you change the "Width" of the extrusion from 5 in to **8 in** by showing the dimensions and modifying them, making the change just as you would to any feature.

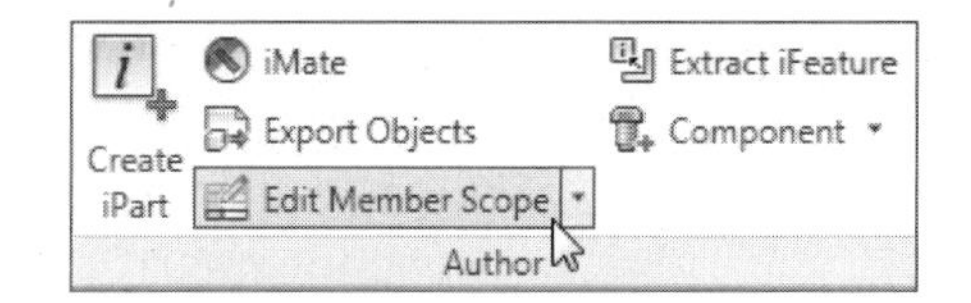

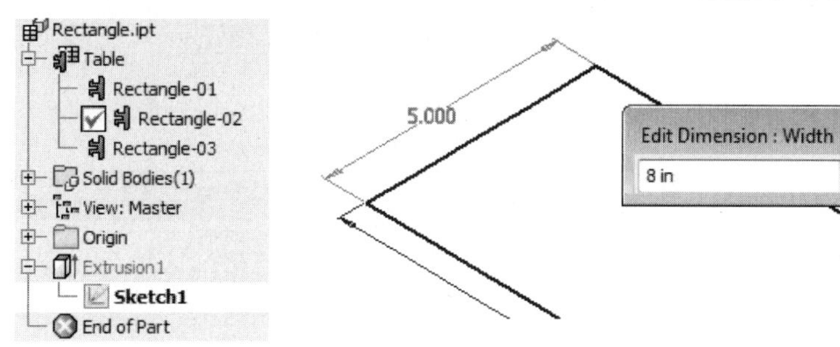

FIGURE 8.70

Because "Width" was not part of the original table, it is added as a new column to the table, and the original value of 5 in is set for all other members in the factory, as shown in the following image.

	Member	Part Number	Length	Width
1	Rectangle-01	Rectangle-01	10 in	5 in
2	Rectangle-02	Rectangle-02	20 in	8 in
3	Rectangle-03	Rectangle-03	30 in	5 in

FIGURE 8.71

Next you modify "Length" from 20 in to **25 in**. Because the table already has length configured, only the cell for Rectangle-02 is modified. The other members remain as defined.

	Member	Part Number	Length	Width
1	Rectangle-01	Rectangle-01	10 in	5 in
2	Rectangle-02	Rectangle-02	25 in	8 in
3	Rectangle-03	Rectangle-03	30 in	5 in

FIGURE 8.72

The images above show the table in intermediate states. The workflow of changing the width to 8 in and length to 25 in are done in a single edit. Both changes to the table happen simultaneously.

You can access the iPart/iAssembly tools on the Manage tab.

Drawings

When creating a drawing view from an iPart or an iAssembly, you can select the Model State tab of the Drawing View dialog box to access a particular member for the factory, as shown in the following image.

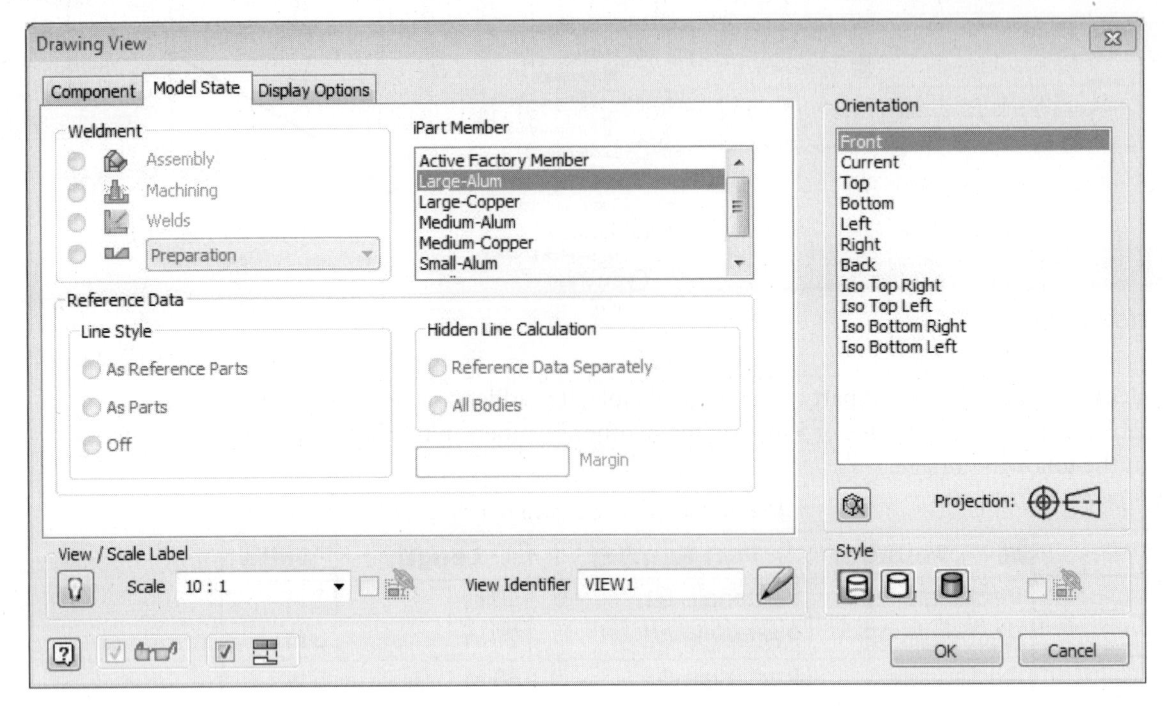

FIGURE 8.73

A General table command, available on the Annotate tab > Table panel, provides access to the Table dialog box, as shown in the following image. You can use this command to create a configuration table in a drawing. The style of the table is controlled using the Style and Standard Editor.

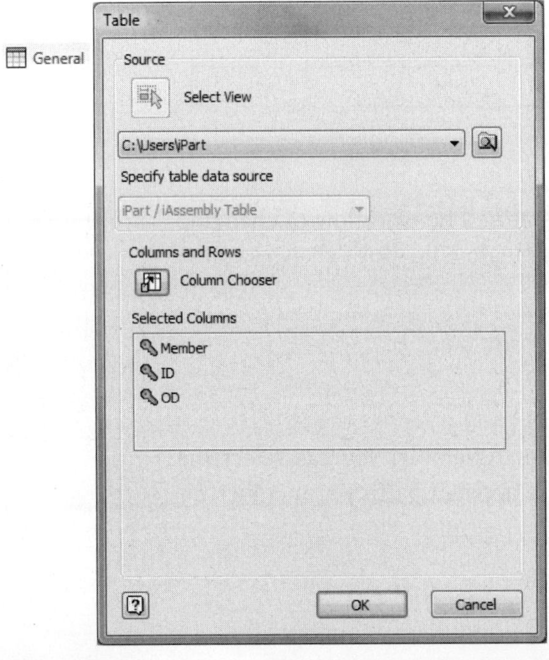

TABLE		
Member	Material	HoleDia
Large-Alum	Aluminum	8.1 mm
Medium-Alum	Aluminum	7.1 mm
Small-Alum	Aluminum	6.1 mm
Large-Copper	Copper	8.1 mm
Medium-Copper	Copper	7.1 mm
Small-Copper	Copper	6.1 mm

FIGURE 8.74

EXERCISE 8-3: CREATING AND PLACING iPARTS

In this exercise, you convert an existing part to a Standard iPart Factory. You then place Standard iParts from that factory into an assembly.

1. Open *ESS_E08_03.ipt.*
2. Click the Parameters command on the Manage tab > Parameters panel, and review the parameter names. In this exercise, you will use the parameter HoleDia to drive the iPart factory.
3. Click Done at the bottom of the Parameters dialog box.
4. From the Manage tab > Author panel, click Create iPart.
5. In the iPart Author dialog box, expand the Hole1 feature in the Parameters tab. Notice that the system parameter HoleDia has automatically been added to the list of table-driven items, and a column has been added to the table. User parameters and renamed system parameters are automatically added to the list.
6. HoleDia will be used as the primary key for this iPart Factory. Click the key next to HoleDia. It should turn blue, and the number "1" will be placed next to it.
7. You will also control the material from the iPart Factory. To add the Material property to the table, complete the following actions:
   - Click the Properties tab.
   - Collapse Summary.
   - Expand Physical.
   - Double-click Material. Notice that the material Aluminum, defined in the Properties dialog box in the Physical tab, is added to the table.

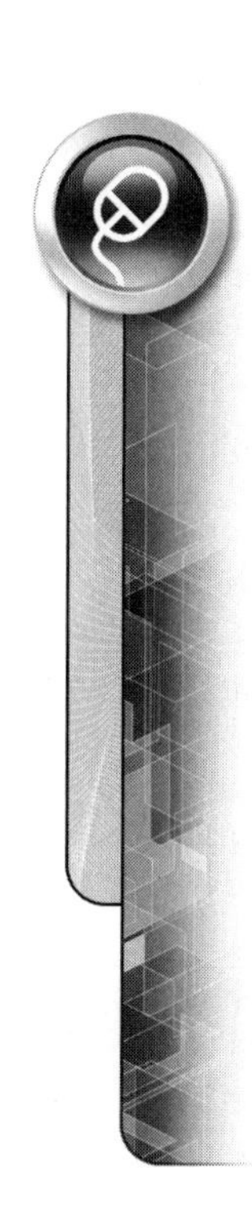

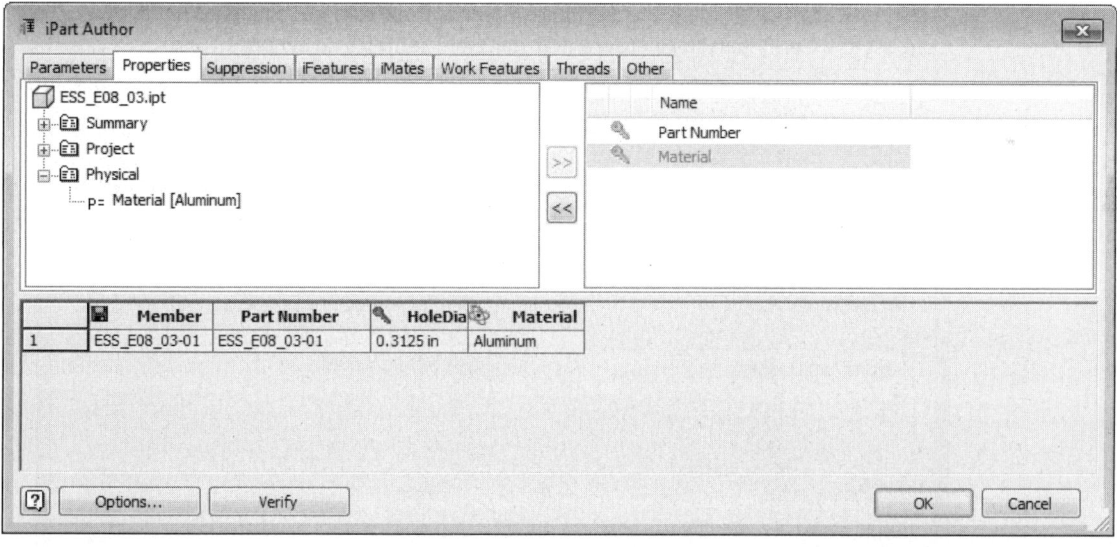

FIGURE 8.75

8. To identify a material when placing this Standard iPart, the Material property must also be defined as a key. To define Material as a secondary key, click the key next to Material. The number next to the key shows the key priority.

9. You can identify which column in the table is used to control the Material for each iPart version. In the table, right-click on the Material column label, and notice the checkmark next to Material Column. If no checkmark appears next to the Material Column option, select it. If one exists, do not deselect it. You'll also notice that a material icon is displayed on the column heading.

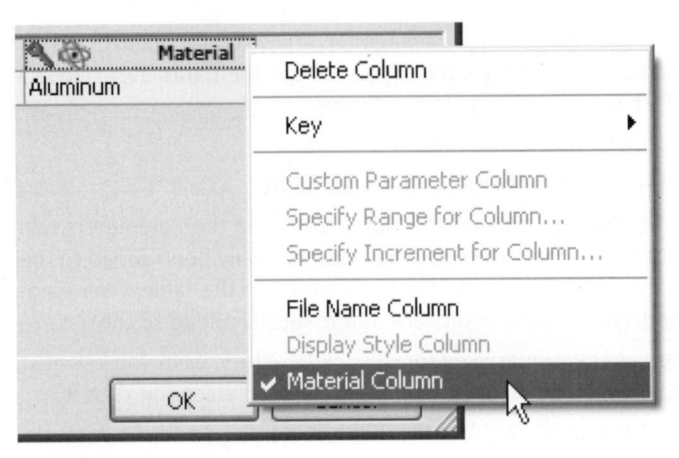

FIGURE 8.76

10. You will also control the Part Number of the Standard iPart when it is published. Click the Options button. Notice that the Set to Value: entry in the Part Number list is selected. Enter **Aluminum Bushing**, and then click OK.

NOTE

You can use the Options button to set rules and values for the Part Number and Member Name. You can further refine the part number by changing the character used for the separator, the initial value, the step increment, and the number of digits. Additionally, you can use the Verify button to ensure that the table contains no syntax errors.

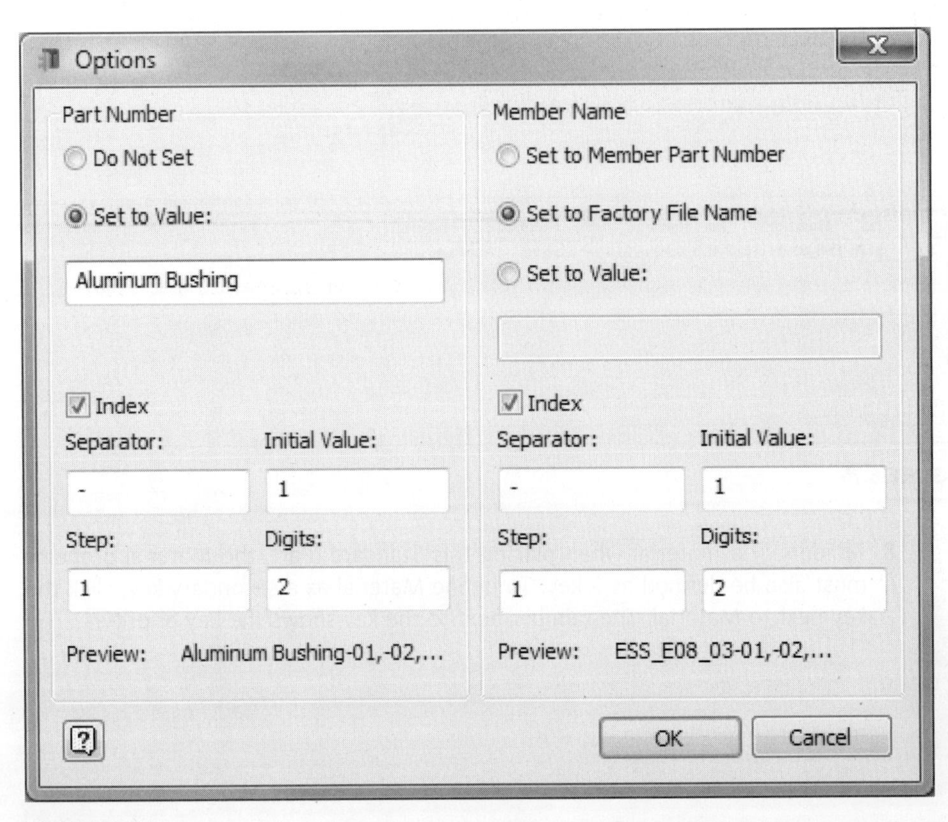

FIGURE 8.77

11. Click Yes when prompted to apply the part numbering scheme to all members.
12. Similar to the function of the Material Column, you can identify which column in the table is used to control the file name for each iPart version. In the table, right-click on the Member column label. There should be a checkmark next to the File Name Column option. If one exists, do not deselect it. An icon also appears in the column heading notifying you which column is used for the file name.
13. To define the file name for the first iPart version, double-click on the first cell in the Member column, enter **Large-Alum**, and then press ENTER.

	Member	Part Number	HoleDia	Material
1	Large-Alum	Aluminum Bushing-01	0.3125 in	Aluminum

FIGURE 8.78

14. Click OK when prompted that the file name has changed.
15. The iPart Factory will contain three different hole sizes and two different materials from which to choose. You will create the first three versions using the table at the bottom of the iPart Author dialog box. The last three versions are defined in a later step using Microsoft Excel to edit the embedded spreadsheet. To create a new row in the table, right-click on the first row label, and then click Insert Row.
16. Repeat the last step to insert a third row.
17. Double-click on individual cells, and enter the values, as shown in the following image. The row highlighted in green is the default row. This defines the default iPart version when you place an iPart from this factory into an assembly.

	Member	Part Number	HoleDia	Material
1	Large-Alum	Aluminum Bushing-01	0.3125 in	Aluminum
2	Medium-Alum	Aluminum Bushing-02	0.25 in	Aluminum
3	Small-Alum	Aluminum Bushing-03	0.125 in	Aluminum

FIGURE 8.79

18. To convert the part to a Standard iPart Factory using the data defined in the table, click OK at the bottom of the iPart Author dialog box.
19. In the browser, expand the Table. The browser represents your part as a Standard iPart Factory using a Table entry. You'll see each member listed by its Member Name.

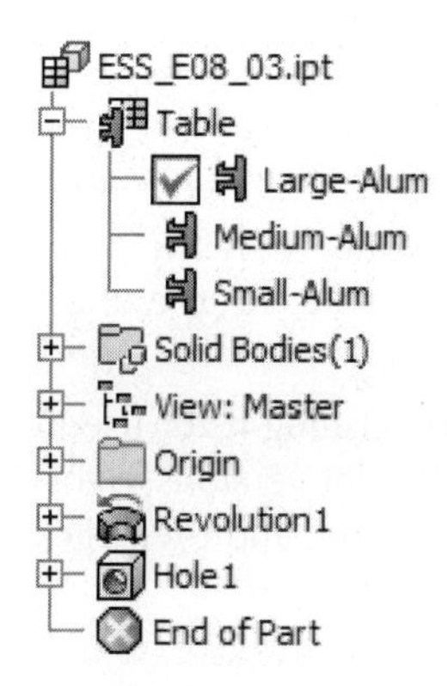

FIGURE 8.80

20. In the browser, right-click the table entry, and select List by Keys from the menu. Expand all entries in the browser. Each version is shown under the table sorted by the primary and secondary key names and their values, as shown in the following image. The active version has a checkmark next to it.

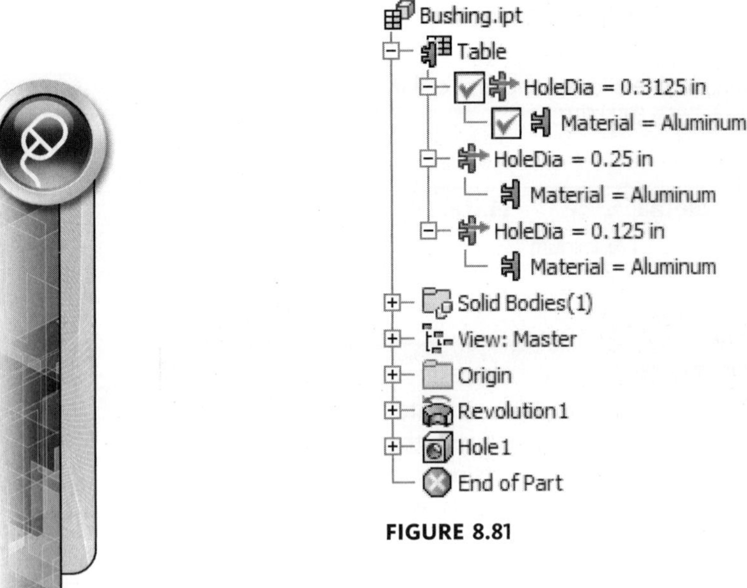

FIGURE 8.81

21. Notice that for each hole diameter in the table, only one material is defined, which is Aluminum. To add more versions to the table using Microsoft Excel, right-click the Table icon in the browser, and then click Edit via Spread Sheet.
    - Click OK when notified that the changes to the spreadsheet will take affect after the Excel process is closed.
    - Copy cells **A2** through **D4**, and paste them to cell **A5**.
    - Modify cells **A5** through **D7** to be consistent with the following image.

NOTE

You can also use the Edit using Spread Sheet tool from the iParts iAssemblies toolbar to access Excel.

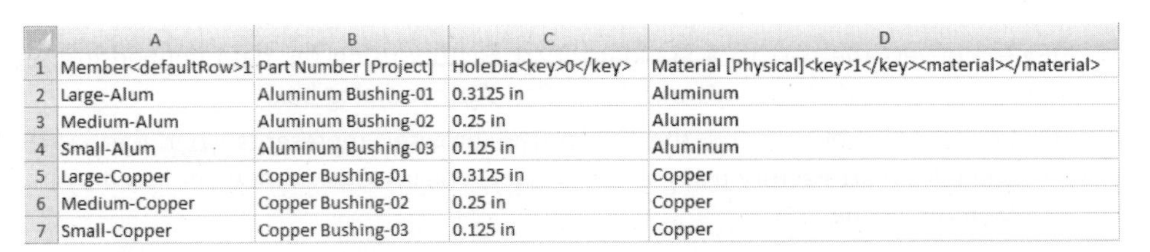

	A	B	C	D
1	Member<defaultRow>1	Part Number [Project]	HoleDia<key>0</key>	Material [Physical]<key>1</key><material></material>
2	Large-Alum	Aluminum Bushing-01	0.3125 in	Aluminum
3	Medium-Alum	Aluminum Bushing-02	0.25 in	Aluminum
4	Small-Alum	Aluminum Bushing-03	0.125 in	Aluminum
5	Large-Copper	Copper Bushing-01	0.3125 in	Copper
6	Medium-Copper	Copper Bushing-02	0.25 in	Copper
7	Small-Copper	Copper Bushing-03	0.125 in	Copper

FIGURE 8.82

22. When you finish modifying the contents of the spread sheet in Excel, click Save.
23. Close Microsoft Excel and return to Autodesk Inventor. Notice that the additional members with the Copper material have been added to the iPart Factory.
24. You can check each version in the iPart Factory prior to publishing it for use in your assemblies by changing the active version. To test different versions, right-click on different members in the browser under the Table icon, and then click Activate.
25. Make the Large-Aluminum member the active version.

26. To save the Standard iPart Factory with a different name, click Save As from the Application menu, and enter the file name Bushing. Click to save it in the same location as the exercise files.
27. Close the file. In the next portion of this exercise, you insert iParts from this iPart factory into an assembly.
28. Open *ESS_E08_03B.iam*.
29. On the Assemble tab > Component panel, click Place Component, select the *Bushing.ipt* file you just created, and then click Open.
30. When you select a Standard iPart Factory, the Place Standard iPart dialog box allows you to place different versions of the part. The default values are shown on the Keys tab. Place two default versions of the bushing by selecting the locations, as shown in the following image.

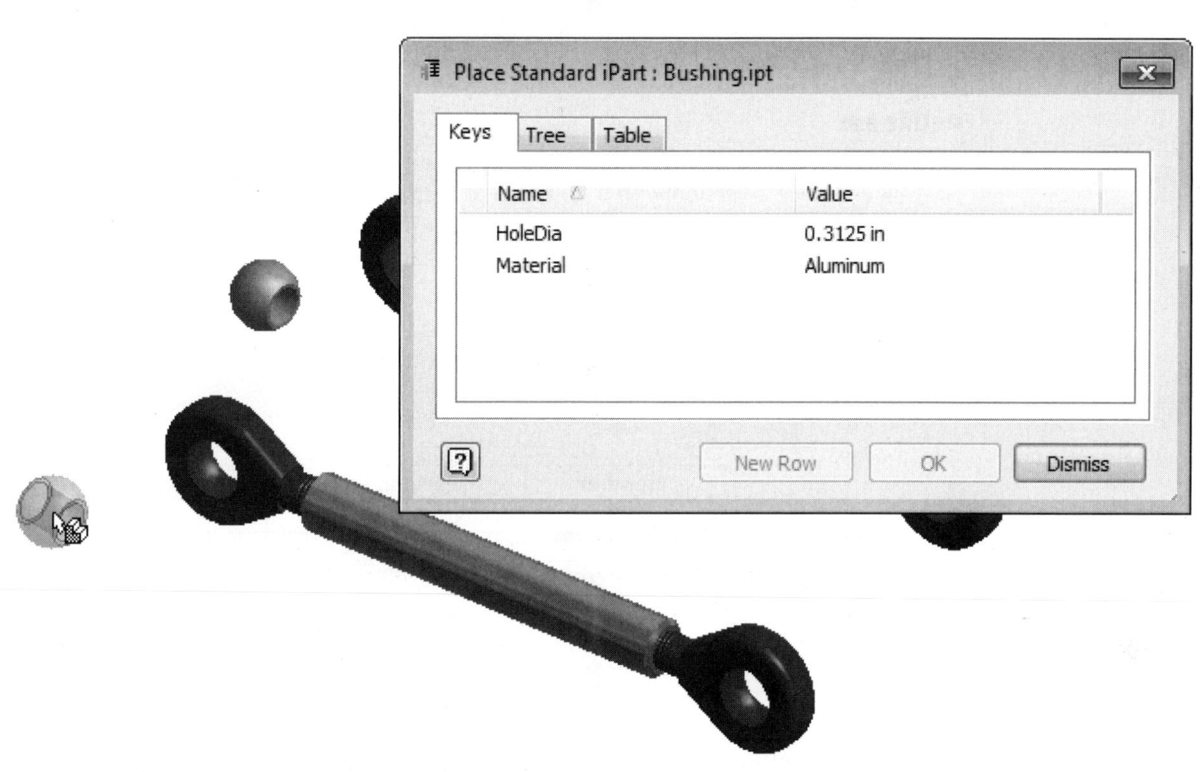

FIGURE 8.83

31. To place a version with a hole size of 0.125 in and a material of copper, click the value next to HoleDia, and then click 0.125 in from the list of available hole diameters, as shown the following image on the left.
32. Click the value next to Material, and then click Copper from the list of available Materials, as shown in the following image on the right.

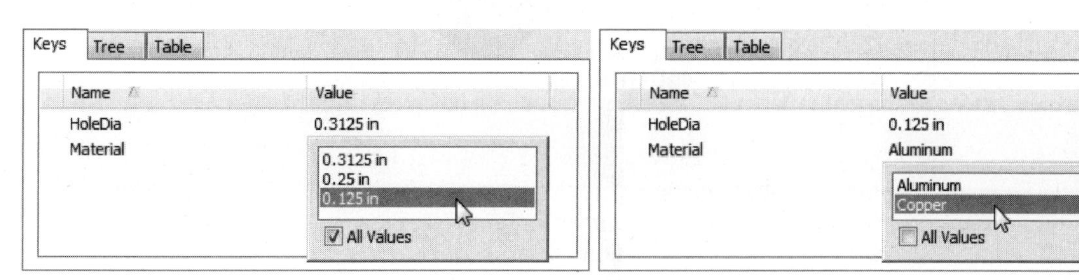

FIGURE 8.84

33. Click the placement location for the third iPart, as shown in the following image. Notice that the color of the iPart reflects the material you specified.

FIGURE 8.85

34. To place a version with a hole size of 0.25 in and a material of aluminum, click the Table tab, and click the second row in the table. Click a location for the fourth iPart, as shown in the following image.

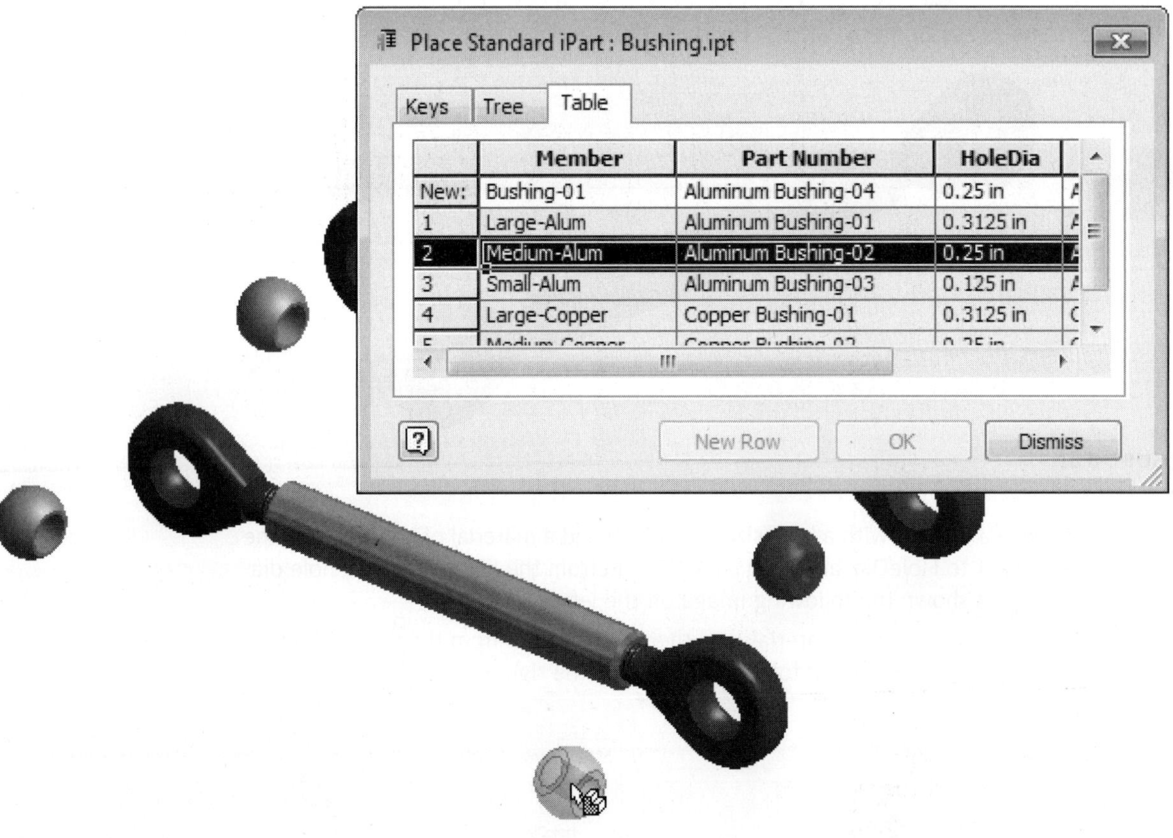

FIGURE 8.86

35. Click Dismiss in the Place Standard iPart dialog box.
36. To change the version of the iPart you just inserted, expand Medium-Alum:4 in the browser, right-click the Table, and click Change Component. Notice that the names of the iParts in the browser reflect the names you specified when defining the Member column in the iPart Factory.

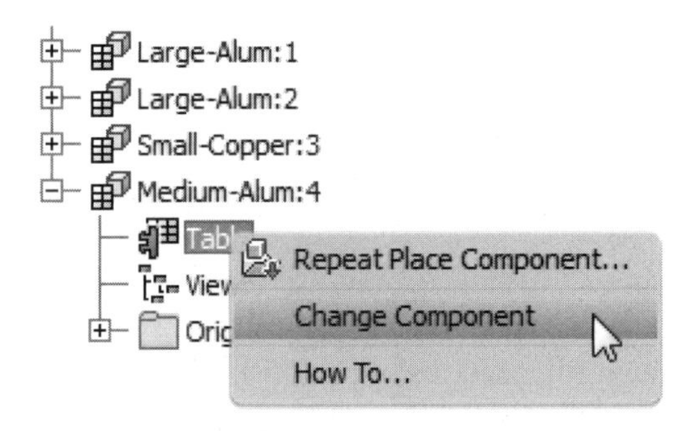

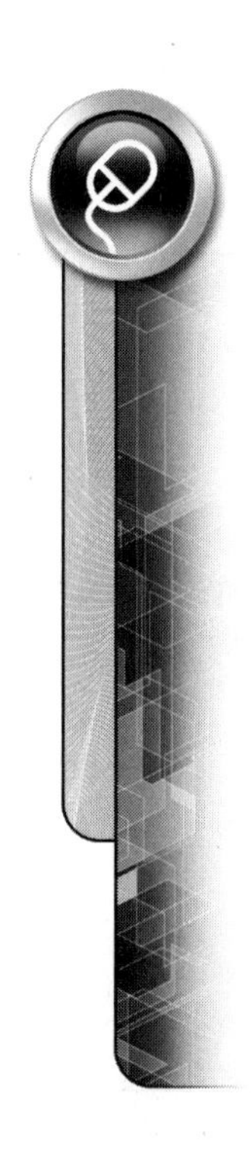

FIGURE 8.87

37. In the Place Standard iPart dialog box, change the HoleDia to **0.125 in** and Material to **Copper**.
38. Click OK in the Place Standard iPart dialog box, and verify that the bushing was replaced in the graphics window.
39. Close all open files. Do not save changes. End of exercise.

iASSEMBLIES

iAssemblies are used to group a set of similar designs in a table format. These are also commonly referred to as configurations, and they function in a similar method to iParts. An iAssembly factory is the primary definition of the family. The family is defined by a table and is a set of configurations of the same design. Each row of the table defines a unique product definition, which is called a configuration or member. When you use an iAssembly factory in another assembly, you can easily change from one configuration to another.

iAssembly members are stored as separate *.iam* files, similar to iParts. They are named in the table using the File name column designation, reference designation field, or key values. When member files are generated, they are stored in a subfolder with a name identical to that for the factory—similar to iParts, but with the *.iam* file type. You can also create a proxy search path for iAssemblies to designate a set of library locations to search for iAssembly factories and a set of corresponding locations to place populated members. Refer to the iPart section above for an example of a proxy path. You can leverage iMates to assist in assembling configurations and to make sure that the assembly conditions are maintained if you change between members of the iAssembly.

Creating iAssemblies

To create an iAssembly, use the Create iAssembly command on the Manage tab > Author panel, as shown in the following image.

FIGURE 8.88

After selecting the Create iAssembly command, the iAssembly Author dialog box is displayed, as shown in the following image. This dialog box displays the configurable items for an assembly and lists the different configurations of the assembly that have been defined—see the rows at the bottom of the dialog box. You can define what controls the iAssembly by using the seven tabs of the dialog box. Each tab has different properties that can be selected and used as configurable items.

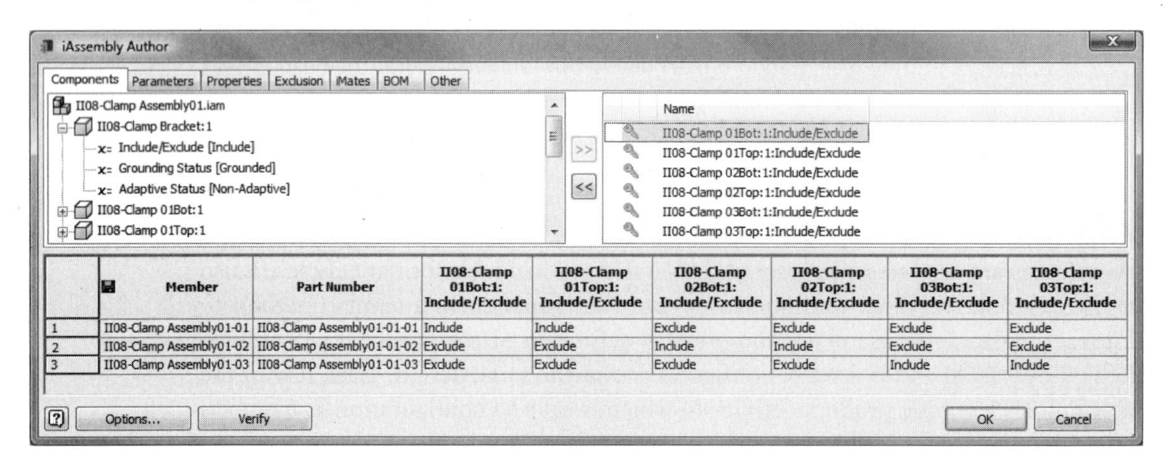

	Member	Part Number	II08-Clamp 01Bot:1: Include/Exclude	II08-Clamp 01Top:1: Include/Exclude	II08-Clamp 02Bot:1: Include/Exclude	II08-Clamp 02Top:1: Include/Exclude	II08-Clamp 03Bot:1: Include/Exclude	II08-Clamp 03Top:1: Include/Exclude
1	II08-Clamp Assembly01-01	II08-Clamp Assembly01-01-01	Include	Include	Exclude	Exclude	Exclude	Exclude
2	II08-Clamp Assembly01-02	II08-Clamp Assembly01-01-02	Exclude	Exclude	Include	Include	Exclude	Exclude
3	II08-Clamp Assembly01-03	II08-Clamp Assembly01-01-03	Exclude	Exclude	Exclude	Exclude	Include	Include

FIGURE 8.89

The iAssembly Author dialog box functions similar to the iPart Author dialog box. You can select a tab across the top to display objects that can be included for configuration in the iAssembly. You use the Add/Remove buttons to include objects from the top-left pane to the top-right pane of the dialog box. The list of configured objects resides in the top-right pane. The lower portion of the dialog box shows the configuration table that displays all of the configured items and defined rows, or members, of the iAssembly. You can use the Options button to set rules and values for the Part Number and Member Name. Additionally, you can use the Verify button to ensure that the table contains no syntax errors. When you have added all objects and defined all members, click OK to complete the authoring process. By right-clicking on a column in the table, you can specify which column you want to use as the file name column, designated by a disk icon in the column header. You can also specify that a column be used as a key column when establishing the iAssembly members.

In addition to creating the iAssembly factory with the iAssembly Author dialog box, you can also use Microsoft Excel or the Autocapture capabilities to create configurations. When using Autocapture, you use familiar tools for editing a design, and changes made to the assembly configuration are applied either to the current configuration or to the entire factory, depending on whether Edit Factory Scope or Edit Member Scope is selected.

When using Edit Factory Scope, the value or changes you make in the assembly are not automatically written to the configuration. If you have a value or settings change that you must click to open the iAssembly Author or to activate a different configuration member, a question dialog box is displayed. This dialog box lists the values in the active row that do not match the current values, and it prompts whether or not you want to update the current member's values with the modified values. Clicking Yes updates the configuration for the active member.

The Edit Member Scope setting will automatically apply changes and edits made to the assembly to the active configuration member. If you edit a property that is not listed as a configurable item in the table, Autocapture will add it to the table definition.

When you create a configuration of an assembly, the components in that assembly can be set to adaptive or flexible. However, a given component in a configuration can only be adaptive in one member of the configuration. All other members must be set to non-adaptive. If you want a component to adjust in size, you must control the size of the component using parameters and formulas or convert the part to an iPart and define the desired sizes. The actual iAssembly can be made flexible when placed in other assembly files, but it cannot be set to adaptive.

The Components tab lists the components of the assembly with the configurable items below. Items listed can be changed in the following ways: Include/Exclude, Grounding Status, Adaptive Status, and Table Replace.

The Parameters tab lists all assembly constraints, assembly features, assembly work features, iMates, component patterns, and other parameters, such as User Parameters, that can be included in the factory.

The Properties tab allows the inclusion and modification of summary, project, physical, and custom properties.

Similar to the Parameters list, the Exclusion tab shows all objects that can be excluded, including components, constraints, assembly features, assembly work features, iMates, Representations, and Component Patterns. Note that these elements can also be set on the Components tab, but the Exclusion list provides a more specific way to view and set the exclusion property.

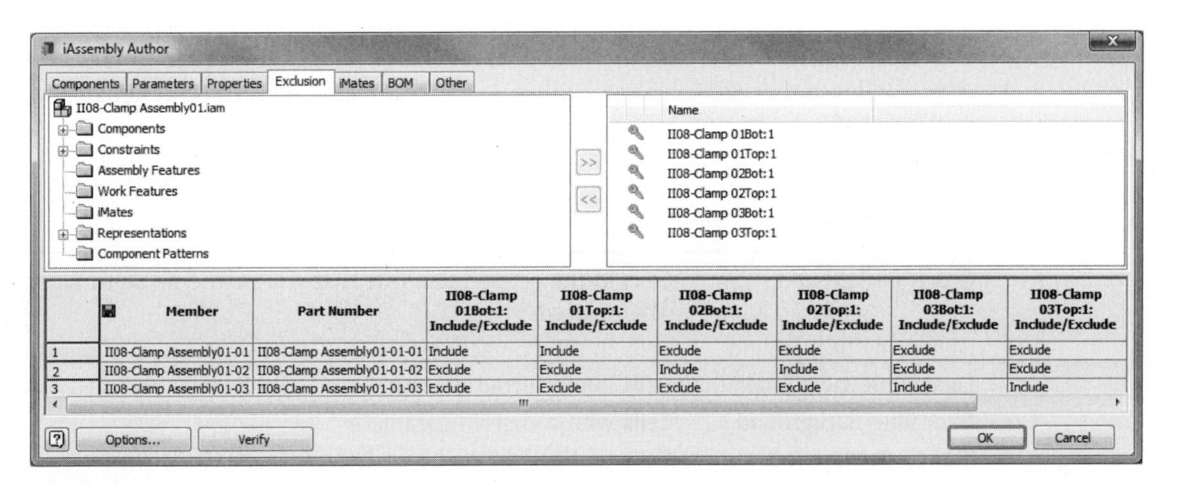

FIGURE 8.90

The iMates tab functions the same as it does for iParts. It lists each iMate with the offset value, include/exclude, matching name, and sequence number available for configuration.

The BOM tab can be used to work with bill of material specific properties, BOM Structure, and BOM Quantity. Two BOM viewing conditions exist for controlling what you see for an iAssembly Bill of Materials. You can view the BOM data from within an assembly that uses a specific iAssembly configuration as an occurrence of the design, or you can view BOM data from within the iAssembly file. In the first scenario, you can view all levels of its components and their details because the item number is a single configuration member.

If viewing the BOM from within the iAssembly, you can view only the top-level structure because the structure will vary based on each configuration member. As you vary the structure of the iAssembly, the item numbers will change from one configuration to another and cause issues for drawing documentation. To view the quantity information in the BOM, you have the option to display the Unit QTY values for a specific configuration member or all members, as shown in the following image.

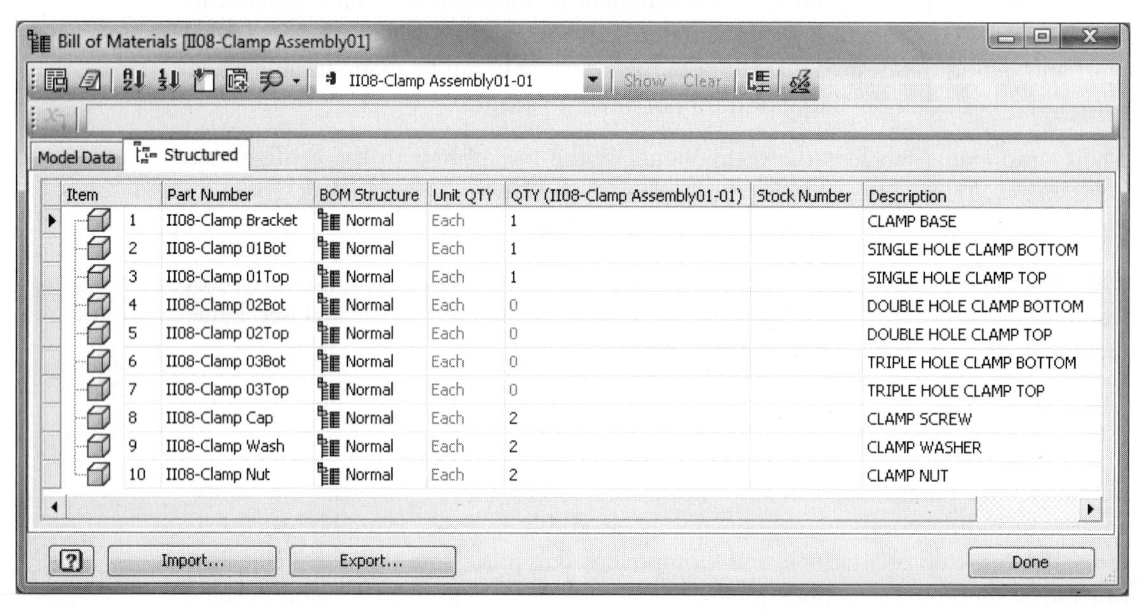

FIGURE 8.91

The Other tab functions the same as it does for iParts, and it can be used to specify a custom column that can contain a string value. Custom columns can be designated as keys or as a file name.

As the table is created, you will notice that different cell background colors provide information about the status of the cells as follows:

Green Background	Cells in the active table row that will be the default row when placed
Light Grey Background	Cells in the non-active row
Light Blue Background	Cells in a selected column
Dark Blue Background	Cells with a custom parameter
Mango Background	Cells that are driven by an Excel formula
Yellow Background	Cells that have an error

Placing iAssemblies

iAssemblies are placed into other assemblies using the Place Component command, similar to placing an iPart or any other regular component. When the selected component is an iAssembly file, the Place iAssembly dialog box is displayed, as shown in the following image.

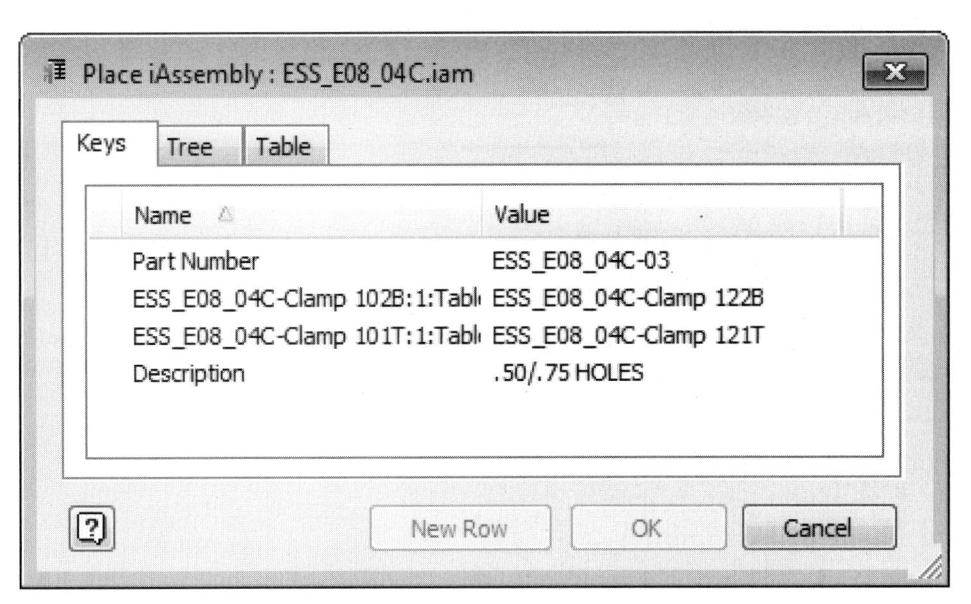

FIGURE 8.92

Before placing an occurrence of the iAssembly, select the configuration member from the Keys, Tree, or Table tabs. Each tab lists the same configurations, but they display the configuration members in different ways. These tabs function the same as when working with iParts. Refer to the iPart Placement section of this chapter for additional information. If iMates are used in the definition of the iAssembly and also in the iAssembly file itself, you can select one of the two iMates options in the Place Component dialog box to speed up the assembly process.

After an iAssembly configuration has been placed, you can change to another configuration by right-clicking the Table node in the browser and selecting Change Component, as shown in the following image.

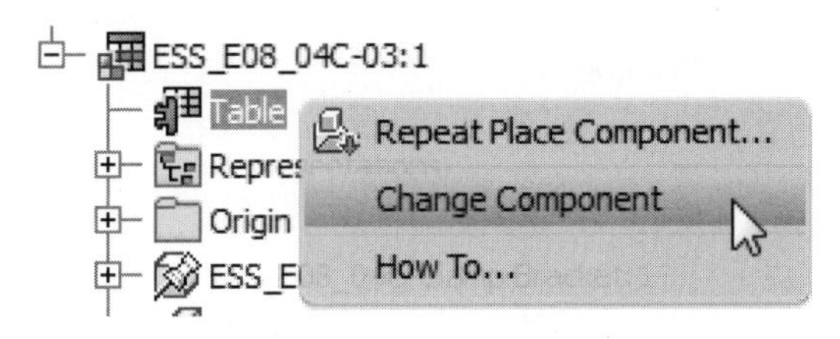

FIGURE 8.93

When a member of an iAssembly configuration is referenced, the member file with the specified and defined properties is generated. This file has either already been created—because it has been previously referenced—or a new member file is created and referenced. Instead of having the member files created when you select to use them in an assembly, you can pre-generate the member files of an iAssembly. This is done by opening the iAssembly and under the Table node, selecting one or more configurations, right-clicking, and selecting Generate Files from the menu, as shown in the following image. You can use this same method to update member files after changes have been made to an iAssembly, and the same method can be performed for iParts to pre-generate members of the iPart factory.

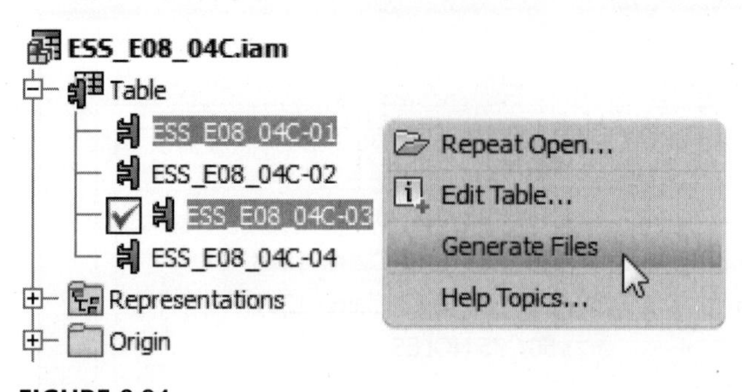

FIGURE 8.94

Documenting iAssemblies

Drawing views of iAssemblies are created using the Base View command. When an iAssembly file is selected, all of the iAssembly members are listed on the Model State tab of the Drawing View dialog box, as shown in the following image.

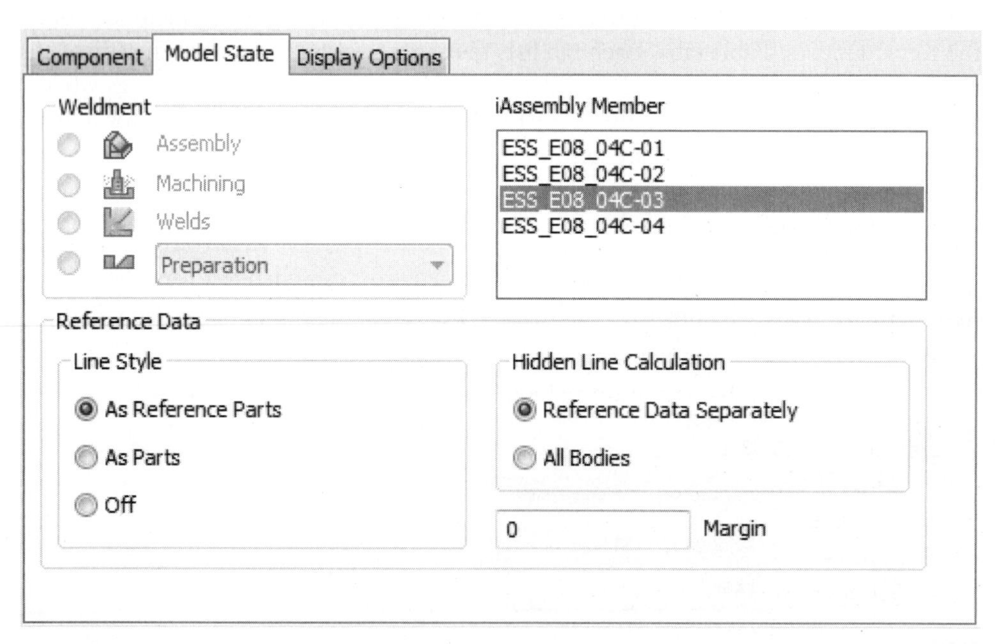

FIGURE 8.95

The created drawing view is based on the selected iAssembly member in that list. After the drawing view has been created, you can change it to a different iAssembly member by editing the view and selecting a different one from the list.

When adding a Parts List to a drawing that is based on an iAssembly, the quantity (QTY) column is based on the method used to create it. If you select an existing drawing view, the configuration member for that view is displayed. If you browse to an iAssembly file, the active configuration member in the file is displayed. Configuration members can be added to a Parts List by editing the existing Parts List and selecting the Member Selection button, as shown in the following image on the left. In the Select Member dialog box, shown in the following image on the right, you can select the configuration member's checkbox to include it in the Parts List.

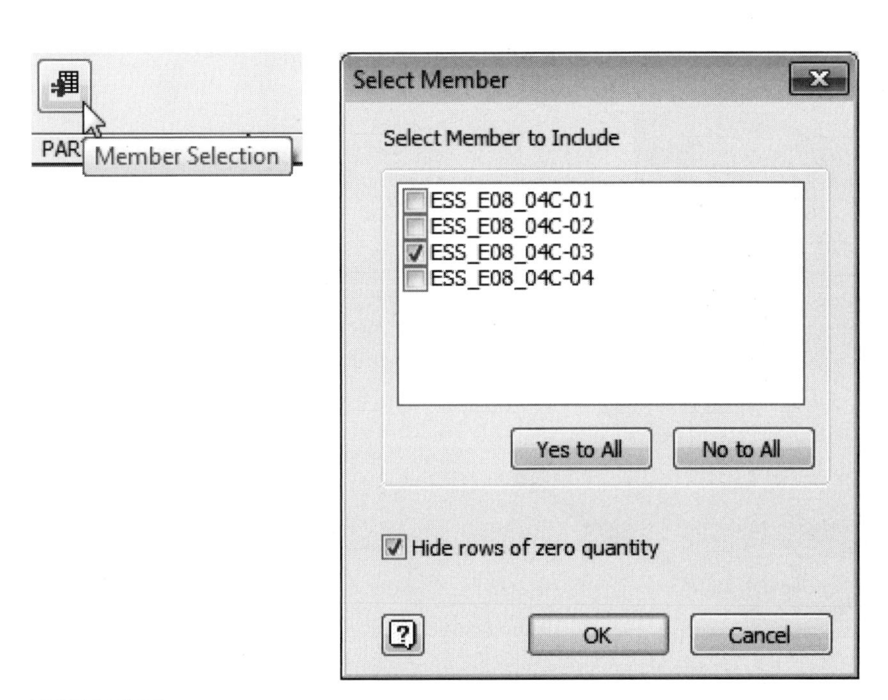

FIGURE 8.96

You can create tables that list iAssembly configuration values using the General table command. After selecting the General table command, you can choose what the table is based on by selecting an iAssembly drawing view or by choosing an iAssembly file. After selecting the source iAssembly, the Table dialog box lists the currently selected columns, as shown in the following image. By selecting Column Chooser, you can specify the attributes that you want to be included in the table.

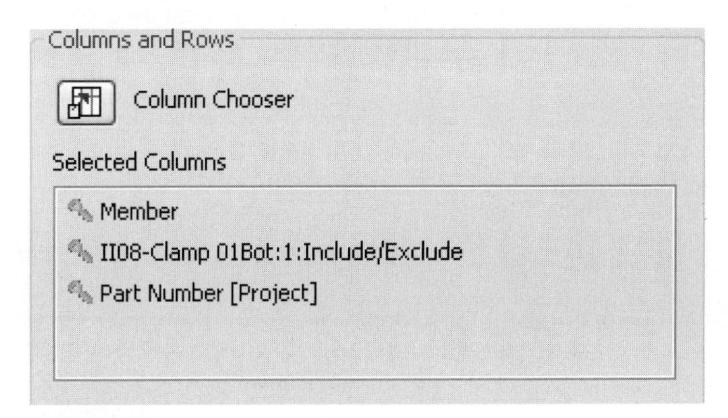

FIGURE 8.97

EXERCISE 8-4: WORKING WITH iASSEMBLIES

In this exercise, you will work with the functionality of iAssemblies. The exercise is meant to familiarize you with the functionality of iAssemblies and enable you to effectively create and work with them. First you will review the iAssembly Author tool.

1. Open *ESS_E08_04.iam.*

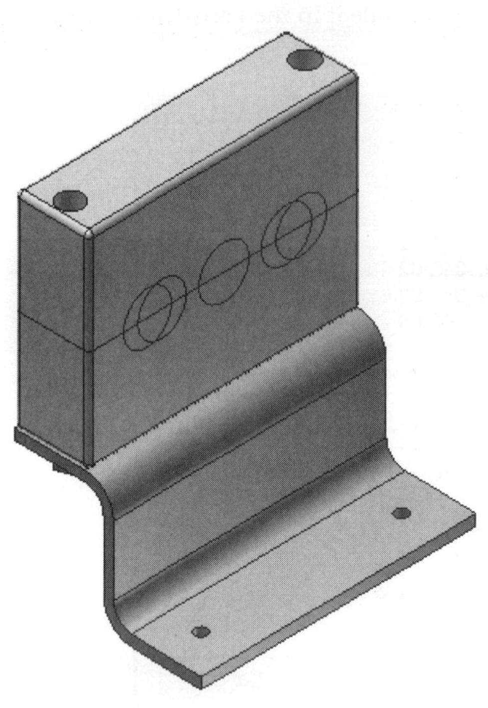

FIGURE 8.98

2. On the Manage tab > Author panel, click Create iAssembly.
3. Right-click row 1 in the table, and click Insert Row from the menu. Right-click row number 2, and click Insert Row to create a third row of information, as shown in the following image on the left. The table should appear similar to the following image on the right.

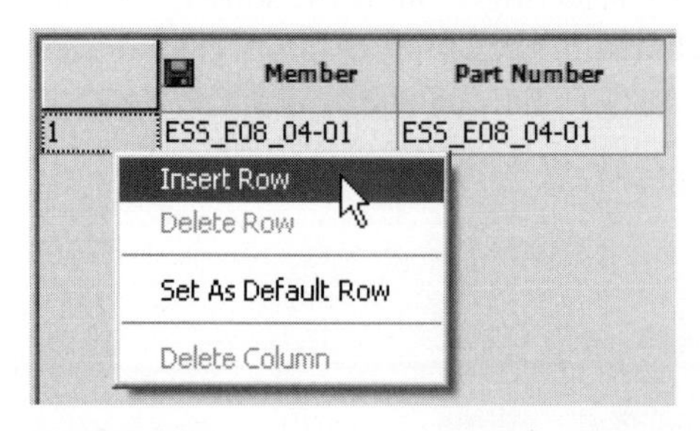

	Member	Part Number
1	ESS_E08_04-01	ESS_E08_04-01
2	ESS_E08_04-02	ESS_E08_04-02
3	ESS_E08_04-03	ESS_E08_04-03

FIGURE 8.99

4. Expand the ESS_E08_04-Clamp01Bot:1 component, and double-click the Include/Exclude [Include] property to add it to the right pane of the dialog box, as shown in the following image.

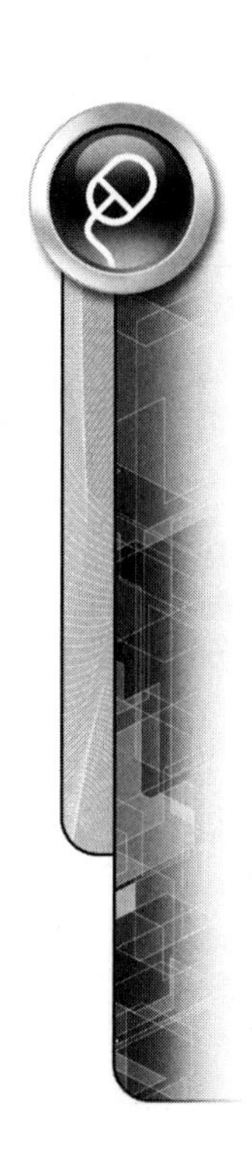

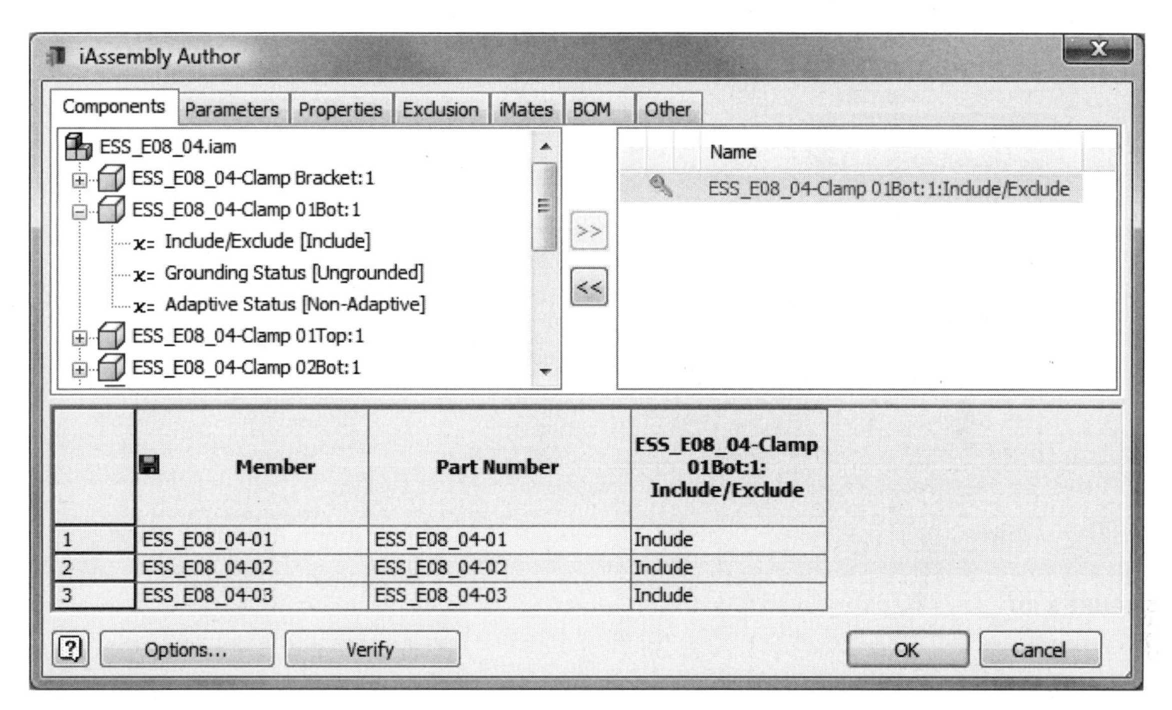

FIGURE 8.100

5. Continue adding the Include/Exclude property for the following parts:
   - ESS_E08_04-Clamp 01Top
   - ESS_E08_04-Clamp 02Bot
   - ESS_E08_04-Clamp 02Top
   - ESS_E08_04-Clamp 03Bot
   - ESS_E08_04-Clamp 03Top

As shown in the following image, notice that all components are available in the iAssembly table area.

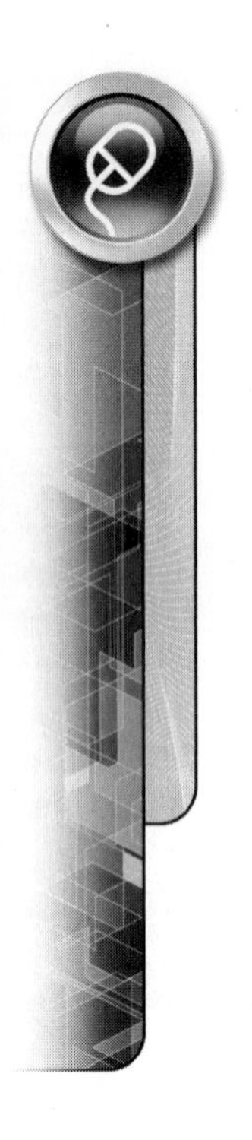

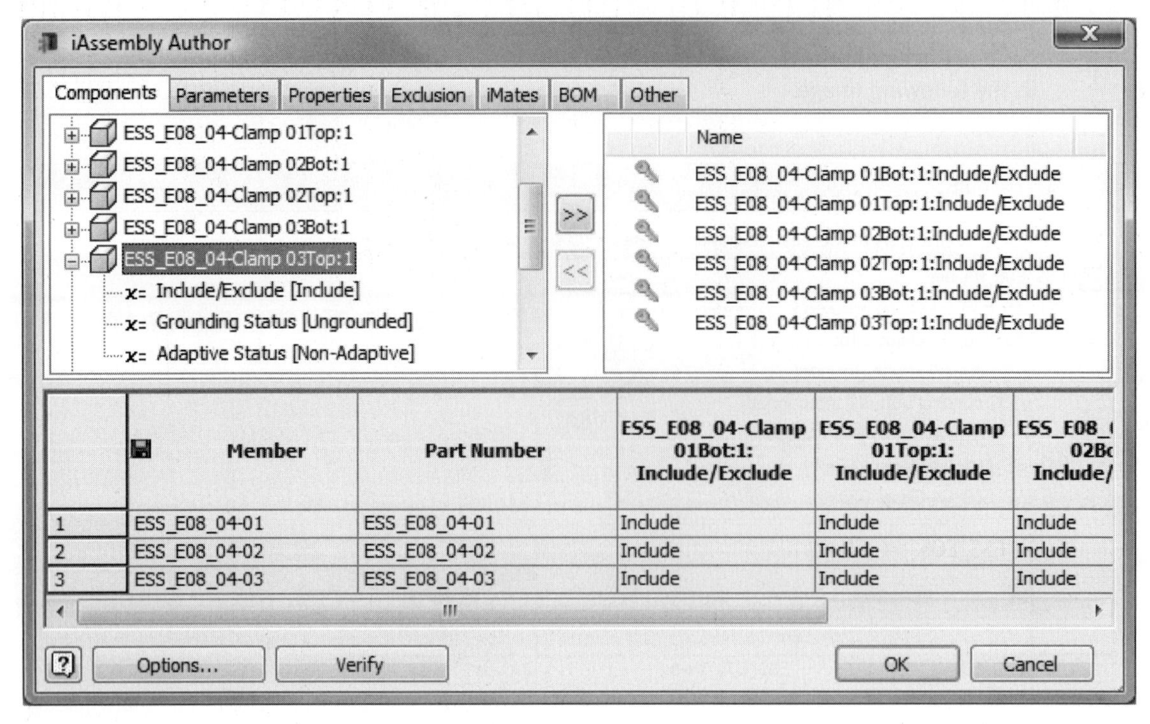

FIGURE 8.101

6. Now turn off or Exclude certain components, depending on the iAssembly member. For member ESS_08_04-01, set the Exclude property for the following parts, as shown in the following image:
   - ESS_E08_04-Clamp 02Bot
   - ESS_E08_04-Clamp 02Top
   - ESS_E08_04-Clamp 03Bot
   - ESS_E08_04-Clamp 03Top

	Member	Part Number	ESS_E08_04-Clamp 01Bot:1: Include/Exclude	ESS_E08_04-Clamp 01Top:1: Include/Exclude	ESS_E08_04-Clamp 02Bot:1: Include/Exclude	ESS_E08_04-Clamp 02Top:1: Include/Exclude	ESS I
1	ESS_E08_04-01	ESS_E08_04-01	Include	Include	Include	Exclude	Excl
2	ESS_E08_04-02	ESS_E08_04-02	Include	Include	Include	Include	Incl
3	ESS_E08_04-03	ESS_E08_04-03	Include	Include	Exclude	Include	Incl

FIGURE 8.102

7. For member ESS_08_04-02, set the Exclude property for the following parts, as shown in the following image:
   - ESS_E08_04-Clamp 01Bot
   - ESS_E08_04-Clamp 01Top
   - ESS_E08_04-Clamp 03Bot
   - ESS_E08_04-Clamp 03Top

	Member	Part Number	ESS_E08_04-Clamp 01Bot:1: Include/Exclude	ESS_E08_04-Clamp 01Top:1: Include/Exclude	ESS_E08_04-Clamp 02Bot:1: Include/Exclude	ESS_E08_04-Clamp 02Top:1: Include/Exclude	ESS Ir
1	ESS_E08_04-01	ESS_E08_04-01	Include	Include	Exclude	Exclude	Excl
2	ESS_E08_04-02	ESS_E08_04-02	Exclude	Include	Include	Include	Excl
3	ESS_E08_04-03	ESS_E08_04-03	Include	Include Exclude	Include	Include	Incl

FIGURE 8.103

8. For member ESS_08_04-03, set the Exclude property for the following parts:
   - ESS_E08_04-Clamp 01Bot
   - ESS_E08_04-Clamp 01Top
   - ESS_E08_04-Clamp 02Bot
   - ESS_E08_04-Clamp 02Top

The table should appear as shown in the following image.

	Part Number	ESS_E08_04-Clamp 01Bot:1: Include/Exclude	ESS_E08_04-Clamp 01Top:1: Include/Exclude	ESS_E08_04-Clamp 02Bot:1: Include/Exclude	ESS_E08_04-Clamp 02Top:1: Include/Exclude	ESS_E08_04-Clamp 03Bot:1: Include/Exclude	ESS_E08_04-Clamp 03Top:1: Include/Exclude
1	ESS_E08_04-01	Include	Include	Exclude	Exclude	Exclude	Exclude
2	ESS_E08_04-02	Exclude	Exclude	Include	Include	Exclude	Exclude
3	ESS_E08_04-03	Exclude	Exclude	Exclude	Exclude	Include	Include

FIGURE 8.104

9. Click OK to save the changes and exit the iAssembly Author dialog box. When you return to the model, notice the addition of the Table node in the Assembly browser.
10. Expand the table node, and double-click each node. The assembly should update to the single hole, double hole, and triple hole configurations.

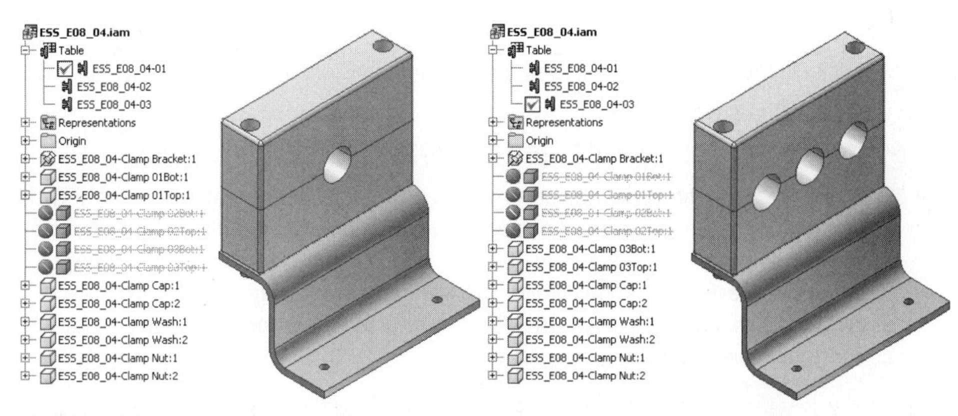

FIGURE 8.105

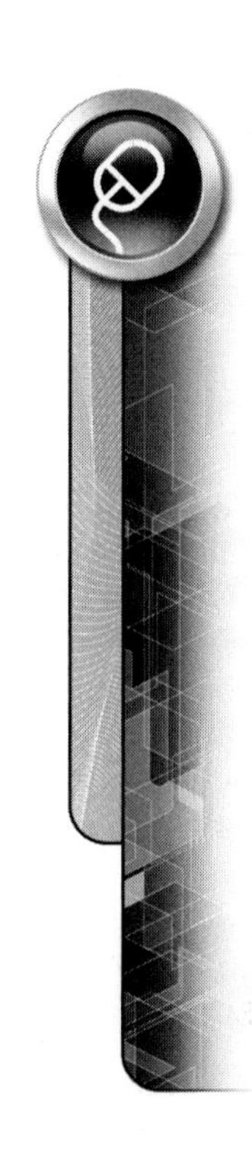

11. Save the file as *Clamp Assembly.iam.*
12. Open *ESS_E08_04-Frame Assembly.iam.*
13. Click the Place Component command, and select the *Clamp Assembly.iam* iAssembly file that you just created and click Open.
14. In the Place iAssembly dialog box, click the Table tab, and select one of the rows in the table.
15. Click a location in the graphics window to place the component.
16. Continue to place other members of the iAssembly that represent each type of clamp, as shown in the following image.

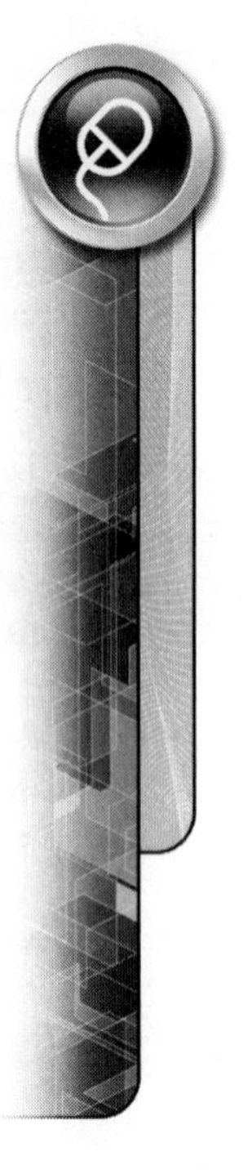

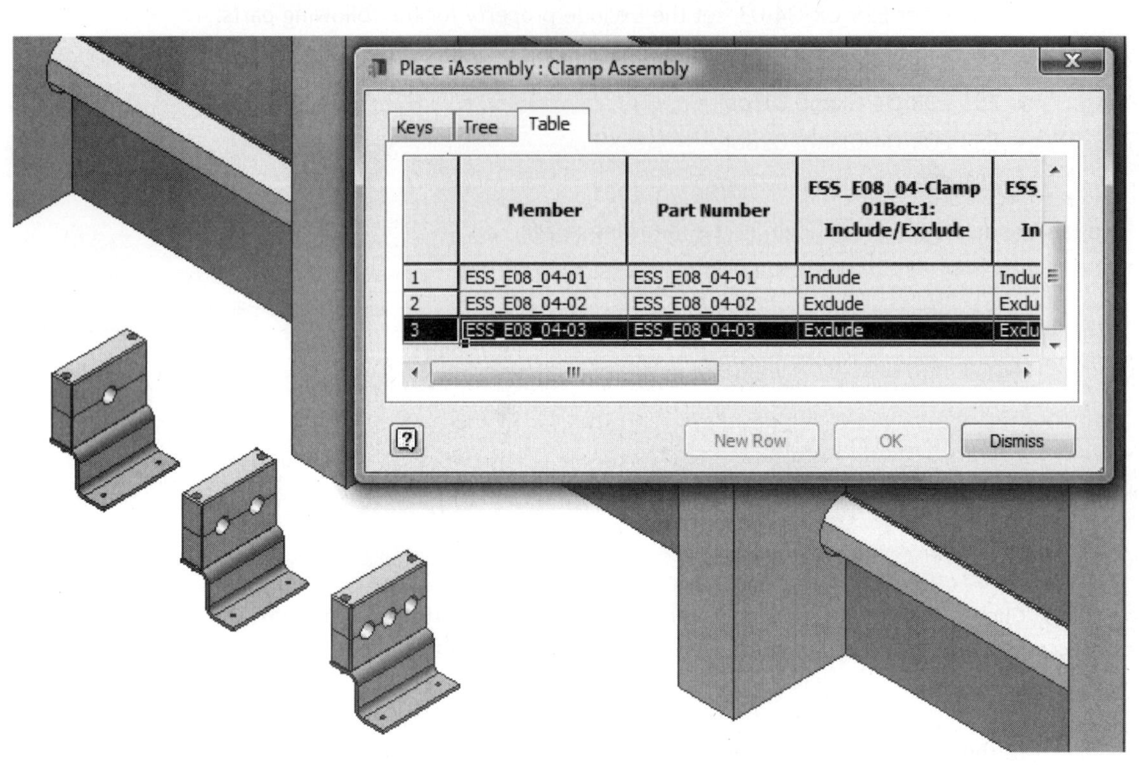

FIGURE 8.106

17. Click Dismiss in the Place iAssembly dialog box. Close all open files. Do not save changes.

Next, you work with the creation of an iAssembly using iParts.

18. Open *ESS_E08_04B.iam.* This assembly is similar to the previous assembly with one major difference: instead of assembling all components into a single assembly and turning off certain ones, the top and bottom clamp components of this assembly were created as iParts. In the following image, the four different clamp types are shown where iParts are used to control various diameter configurations.

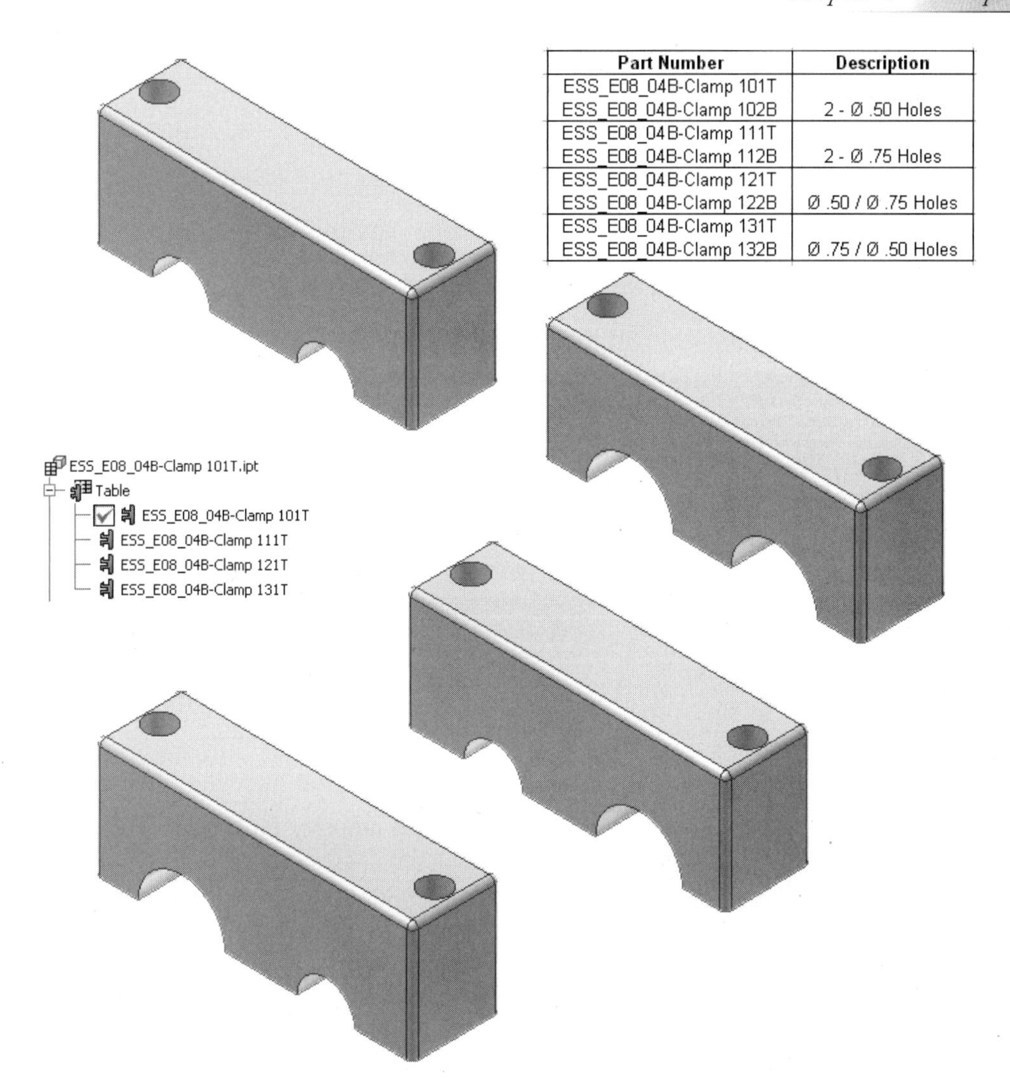

Part Number	Description
ESS_E08_04B-Clamp 101T ESS_E08_04B-Clamp 102B	2 - Ø .50 Holes
ESS_E08_04B-Clamp 111T ESS_E08_04B-Clamp 112B	2 - Ø .75 Holes
ESS_E08_04B-Clamp 121T ESS_E08_04B-Clamp 122B	Ø .50 / Ø .75 Holes
ESS_E08_04B-Clamp 131T ESS_E08_04B-Clamp 132B	Ø .75 / Ø .50 Holes

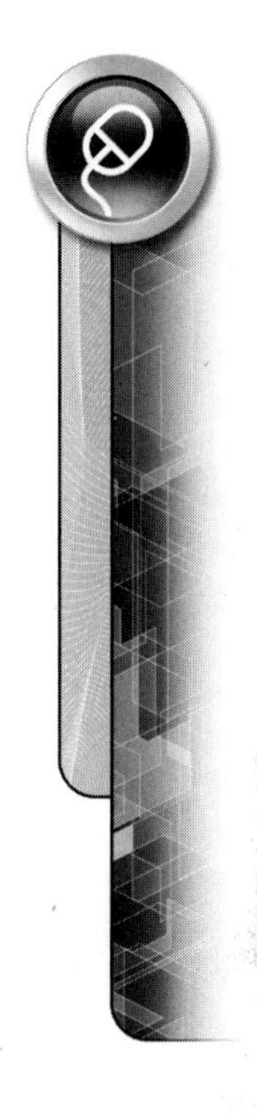

FIGURE 8.107

19. On the Manage tab > Author panel, click Create iAssembly.
20. In the iAssembly Author dialog box, right-click row 1 in the table, and click Insert Row from the menu.
21. Add two additional rows to the table so that you have four iAssembly members, as shown in the following image.

	Member	Part Number
1	ESS_E08_04B-01	ESS_E08_04B-01
2	ESS_E08_04B-02	ESS_E08_04B-02
3	ESS_E08_04B-03	ESS_E08_04B-03
4	ESS_E08_04B-04	ESS_E08_04B-04

FIGURE 8.108

22. On the Components tab, expand ESS_E08_04B-Clamp 102B (Bottom).
23. Double-click the Table Replace node to add the property to the right pane, as shown in the following image.

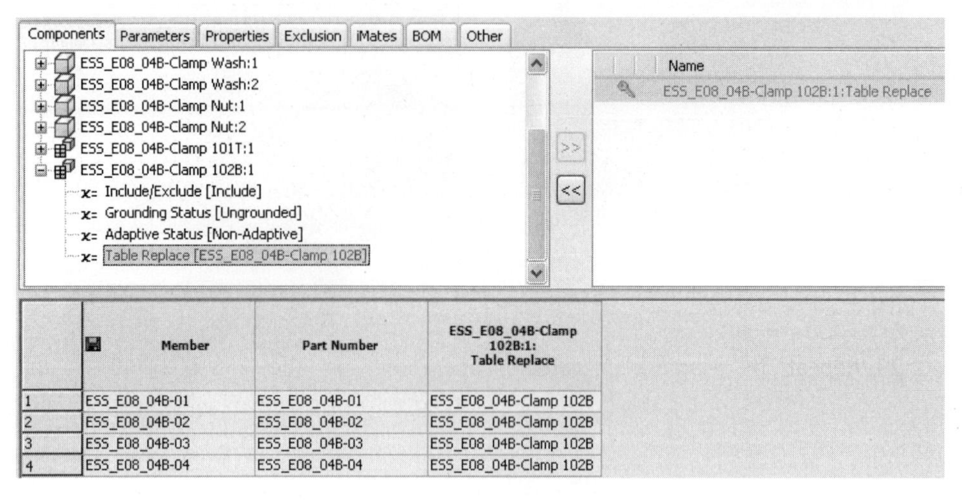

	Member	Part Number	ESS_E08_04B-Clamp 102B:1: Table Replace
1	ESS_E08_04B-01	ESS_E08_04B-01	ESS_E08_04B-Clamp 102B
2	ESS_E08_04B-02	ESS_E08_04B-02	ESS_E08_04B-Clamp 102B
3	ESS_E08_04B-03	ESS_E08_04B-03	ESS_E08_04B-Clamp 102B
4	ESS_E08_04B-04	ESS_E08_04B-04	ESS_E08_04B-Clamp 102B

FIGURE 8.109

24. Repeat the process to add the Table Replace property for the ESS_E08_04B-Clamp 101T (Top) component.
25. Click the Properties tab, and add the Description property to the iAssembly definition.
26. Add the description of .50/.50 HOLES to the description of the first member, as shown in the following image.

	Member	Part Number	ESS_E08_04B-Clamp 102B:1: Table Replace	ESS_E08_04B-Clamp 101T:1: Table Replace	Description
1	ESS_E08_04B-01	ESS_E08_04B-01	ESS_E08_04B-Clamp 102B	ESS_E08_04B-Clamp 101T	.50/.50 HOLES
2	ESS_E08_04B-02	ESS_E08_04B-02	ESS_E08_04B-Clamp 102B	ESS_E08_04B-Clamp 101T	
3	ESS_E08_04B-03	ESS_E08_04B-03	ESS_E08_04B-Clamp 102B	ESS_E08_04B-Clamp 101T	
4	ESS_E08_04B-04	ESS_E08_04B-04	ESS_E08_04B-Clamp 102B	ESS_E08_04B-Clamp 101T	

FIGURE 8.110

27. Add the remaining three descriptions to the table, as shown in the following image.

	Member	Part Number	ESS_E08_04B-Clamp 102B:1: Table Replace	ESS_E08_04B-Clamp 101T:1: Table Replace	Description
1	ESS_E08_04B-01	ESS_E08_04B-01	ESS_E08_04B-Clamp 102B	ESS_E08_04B-Clamp 101T	.50/.50 HOLES
2	ESS_E08_04B-02	ESS_E08_04B-02	ESS_E08_04B-Clamp 102B	ESS_E08_04B-Clamp 101T	.75/.75 HOLES
3	ESS_E08_04B-03	ESS_E08_04B-03	ESS_E08_04B-Clamp 102B	ESS_E08_04B-Clamp 101T	.50/.75 HOLES
4	ESS_E08_04B-04	ESS_E08_04B-04	ESS_E08_04B-Clamp 102B	ESS_E08_04B-Clamp 101T	.75/.50 HOLES

FIGURE 8.111

28. Select the ESS_E08_04B-02 member, and replace the following parts in the table, as shown in the following image:

    - ESS_E08_04B-Clamp 102B with ESS_E08_04B-Clamp 112B
    - ESS_E08_04B-Clamp 101T with ESS_E08_04B-Clamp 111T

	Member	Part Number	ESS_E08_04B-Clamp 102B:1: Table Replace	ESS_E08_04B-Clamp 101T:1: Table Replace	Description
1	ESS_E08_04B-01	ESS_E08_04B-01	ESS_E08_04B-Clamp 102B	ESS_E08_04B-Clamp 101T	.50/.50 HOLES
2	ESS_E08_04B-02	ESS_E08_04B-02	ESS_E08_04B-Clamp 112B	S_E08_04B-Clamp 101T	.75/.75 HOLES
3	ESS_E08_04B-03	ESS_E08_04B-03	ESS_E08_04B-Clamp 102B	ESS_E08_04B-Clamp 101T	.50/.75 HOLES
4	ESS_E08_04B-04	ESS_E08_04B-04	ESS_E08_04B-Clamp 102B	ESS_E08_04B-Clamp 111T	.75/.50 HOLES
				ESS_E08_04B-Clamp 121T	
				ESS_E08_04B-Clamp 131T	

FIGURE 8.112

29. Repeat the same replacement operations on the ESS_E08_04B-03 and ESS_E08_04B-04 members, as shown in the following image.

	Member	Part Number	ESS_E08_04B-Clamp 102B:1: Table Replace	ESS_E08_04B-Clamp 101T:1: Table Replace	Description
1	ESS_E08_04B-01	ESS_E08_04B-01	ESS_E08_04B-Clamp 102B	ESS_E08_04B-Clamp 101T	.50/.50 HOLES
2	ESS_E08_04B-02	ESS_E08_04B-02	ESS_E08_04B-Clamp 112B	ESS_E08_04B-Clamp 111T	.75/.75 HOLES
3	ESS_E08_04B-03	ESS_E08_04B-03	ESS_E08_04B-Clamp 122B	ESS_E08_04B-Clamp 121T	.50/.75 HOLES
4	ESS_E08_04B-04	ESS_E08_04B-04	ESS_E08_04B-Clamp 132B	ESS_E08_04B-Clamp 131T	.75/.50 HOLES

FIGURE 8.113

30. Click OK to save and exit the iAssembly Author dialog box.
31. A Table node is added to the Assembly browser. Expand the Table entry, and double-click on each assembly configuration to cycle through the different hole arrangements in the clamp, as shown in the following image.

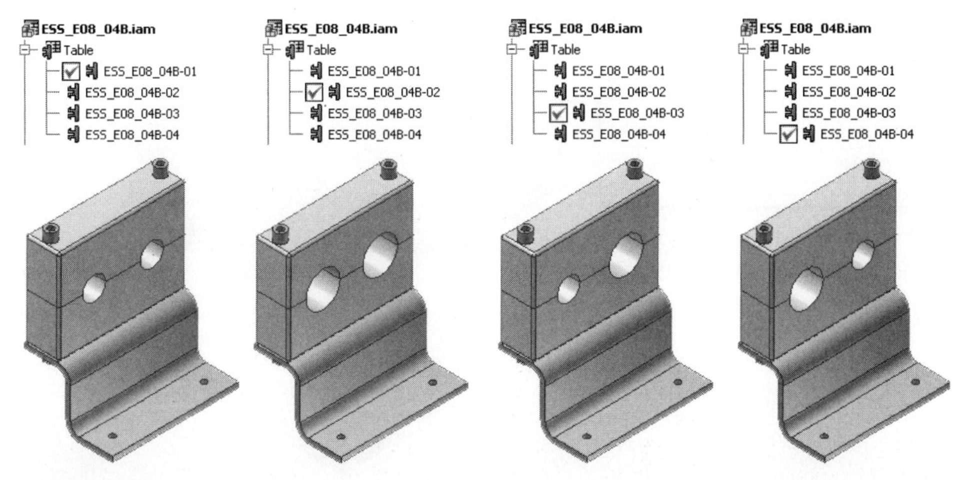

FIGURE 8.114

32. Close all open files. Do not save changes.
33. Open *ESS_E08_04C.iam*.

    This is the same clamp that is made up of iParts with which you just finished working. The *ESS_E08_04C-03* configuration is active.

    When an iAssembly is created, there is an iPart/iAssembly panel located on the Assemble tab. You will now work with the Edit Member Scope mode of the iParts/iAssemblies panel, as shown in the following image. When this mode is active and changes are made to the current iAssembly, the changes are present only in the iAssembly. The remaining iAssembly configurations are unaffected by changes.

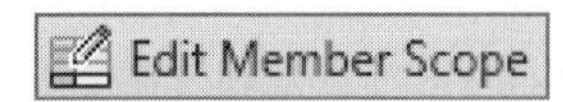

FIGURE 8.115

34. Set the current iAssembly to Edit Member Scope, as shown in the previous image.
35. On the Model tab > Modify Assembly panel, click Chamfer.
36. Chamfer both edges of both holes on both faces of the iAssembly clamp. Use a chamfer distance of **0.05** units, as shown in the following image.

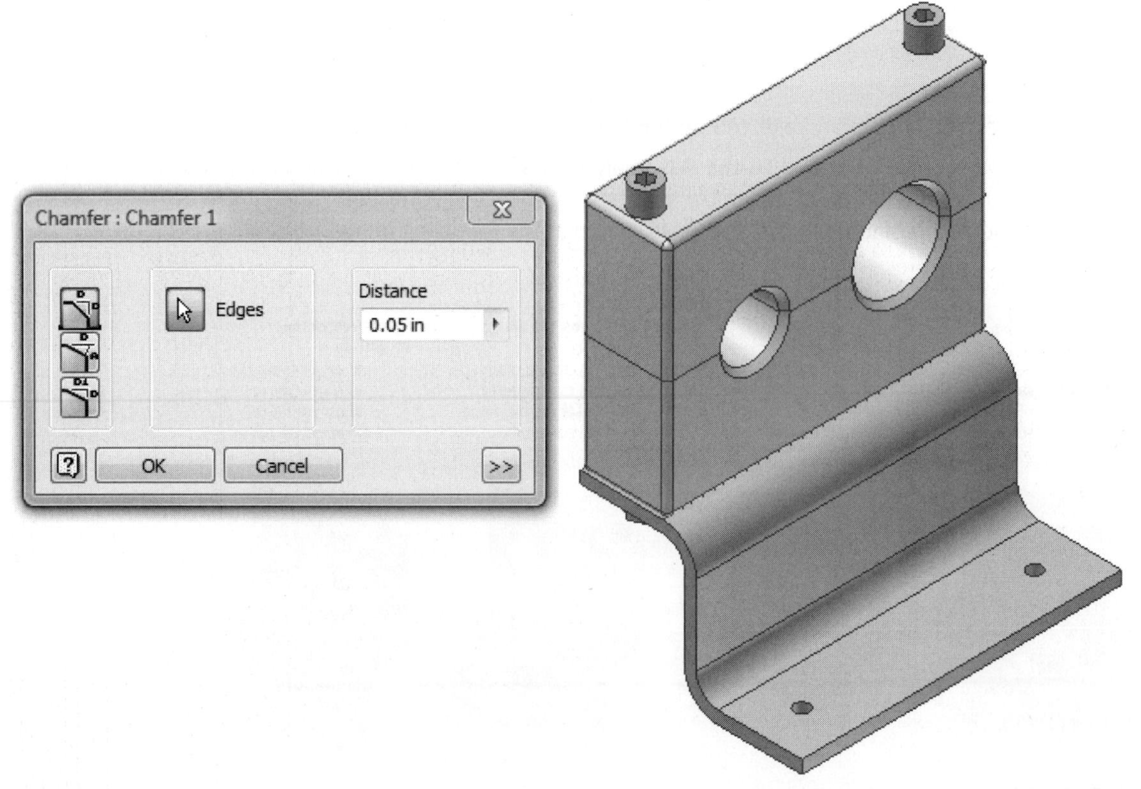

FIGURE 8.116

37. Double-click any of the other configurations of the iAssembly.

    Since Edit Member Scope was active, the chamfers are only available in the configuration that was active when they were created. Notice also that the Chamfer in the browser is excluded from the other configurations, as shown in the following image.

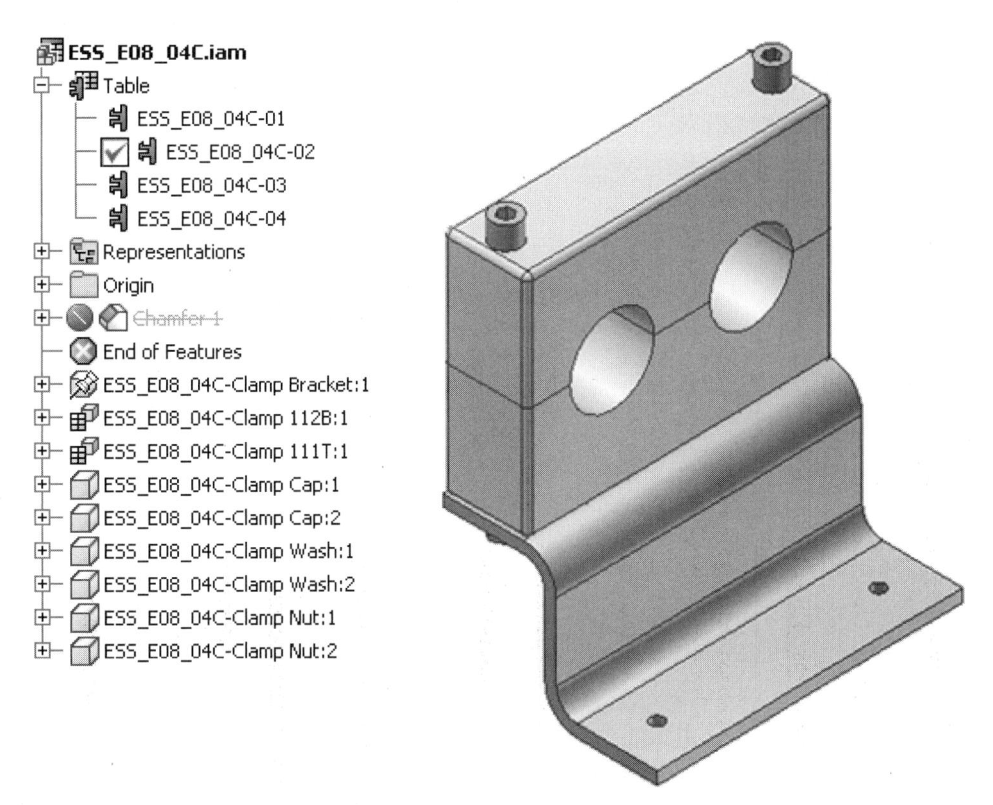

FIGURE 8.117

38. Close all open files. Do not save changes. End of exercise.

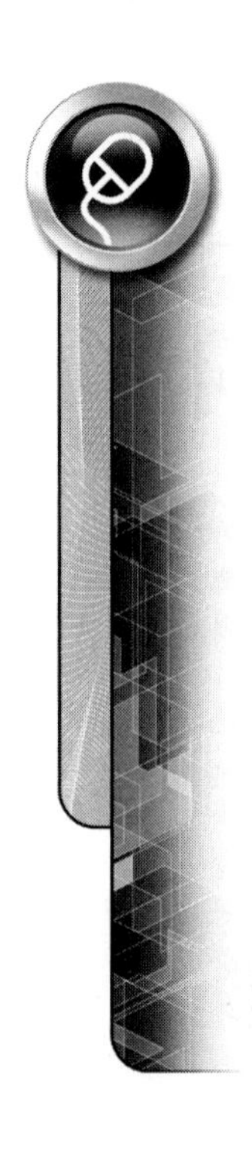

iFEATURES

iFeatures give you the ability to reuse single or multiple features from a part file in other Autodesk Inventor files. iFeatures capture the design intent built into a feature(s) that is going to be reused, such as the name or the size and position parameters. You can also embed or attach a file to the iFeature to be used as Placement Help when you place the iFeature into another design.

You can include reference edges as position geometry in your iFeatures. Reference edges allow you to capture additional design intent, but they require that the iFeature be positioned in the same way it was designed originally in every part in which it is placed. iFeatures are saved in their own type of file that has an *.ide* extension and a unique icon, as shown in the following image.

FIGURE 8.118

iFeatures are stored in a catalog. The catalog is a directory on your computer or on a server that is set to store all of the iFeatures that you create. You can browse the iFeature catalog at any time by clicking the View iFeature Catalog command on the Manage tab > Insert panel, at the bottom of the iFeature browser, as shown in the following image.

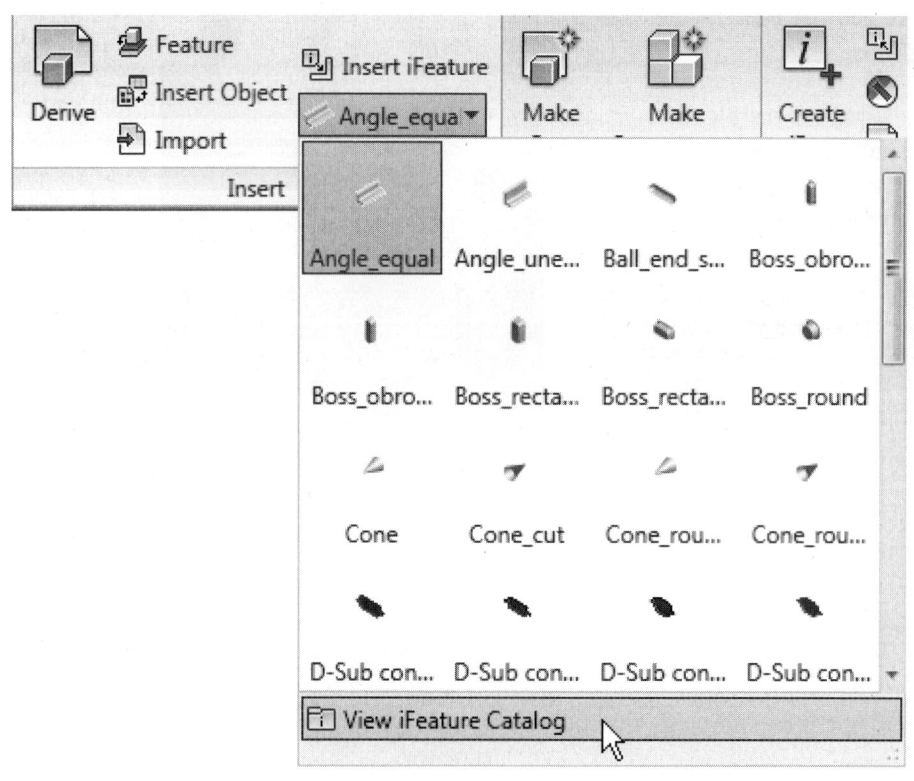

FIGURE 8.119

When the View iFeature Catalog command is selected, Windows Explorer opens to the directory specified in the iFeature root setting on the iFeature tab of the Application Options dialog box. In the iFeature root folder, you can view, copy, edit, or delete iFeatures from your catalog.

Create iFeatures

You create iFeatures using the Extract iFeature command on the Manage tab > Author panel, as shown in the following image.

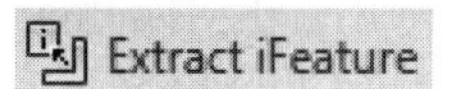

FIGURE 8.120

Once selected, the Extract iFeature dialog box appears, as shown in the following image.

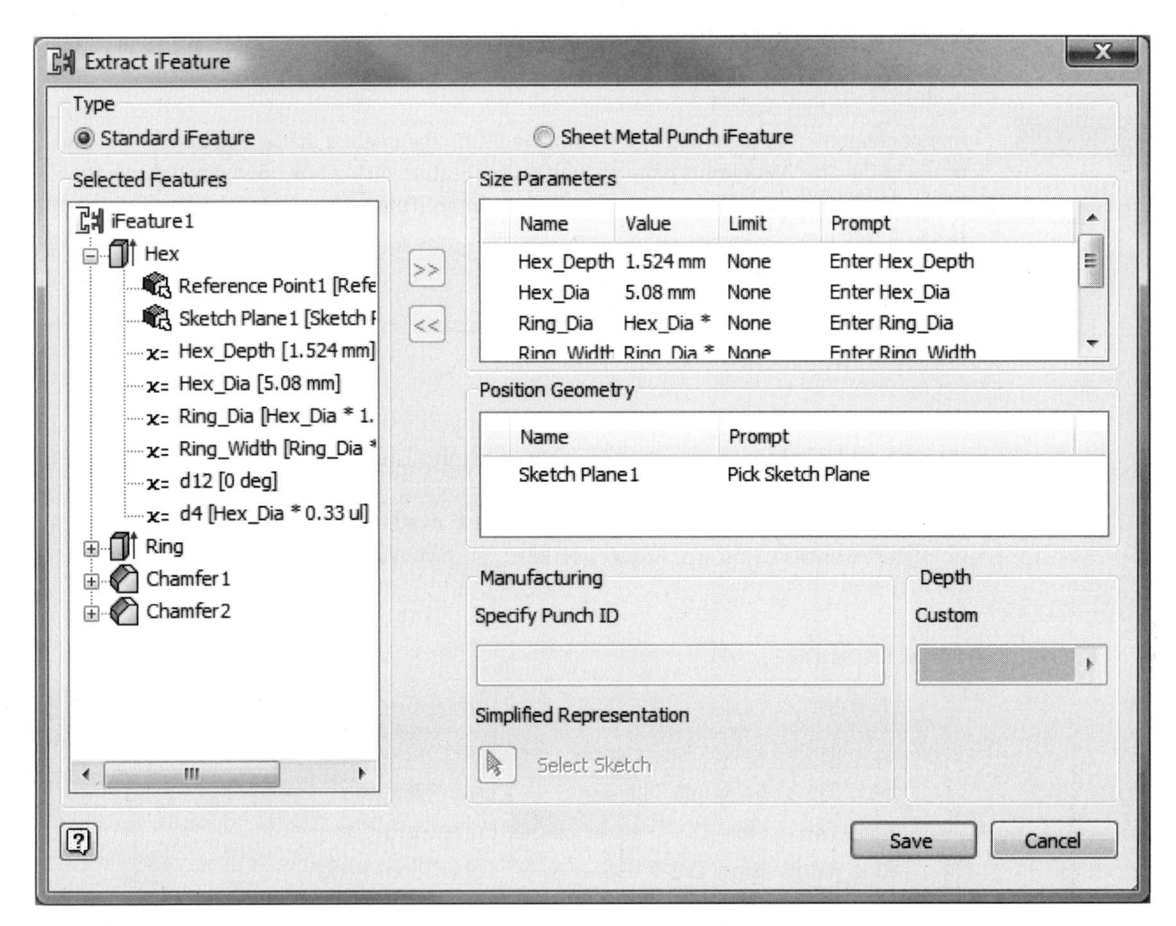

FIGURE 8.121

The Extract iFeature dialog box contains the following sections.

Type

In this area, you can select whether you want to create a Standard iFeature or a Sheet Metal Punch iFeature. In this chapter, we will focus on Standard iFeatures. Refer to Chapter 10 for information regarding the creation and use of the Sheet Metal Punch iFeature.

Selected Features

This area of the dialog box lists the features selected to be included in the iFeature. You can rename features in the list to more descriptive names to assist in working with the iFeature at a later time.

Use the Add parameters (≫) button and the Remove parameters (≪) button to move parameters from the highlighted features in the Selected Features list to the Size Parameters table.

Size Parameters

The Size Parameters table lists all of the parameters that will be used for interface when the iFeature is placed into another file. You can select the parameters by expanding the features listed in the Selected Features list or directly in the graphics window.

NOTE

Any parameters that have been given a name in the Parameters dialog box appear automatically in the Size Parameters pane of the Create iFeature dialog box upon iFeature creation. Renaming parameters that you want to include in an iFeature can speed up the process of creating them.

Name. Specify a descriptive name for the parameter. Use names that describe the purpose of the parameter.

Value. Place a value to be the default for the parameter when inserting your iFeature into a file. The value is restricted by settings in the Limit column.

Limit. Place restrictions on the values that are available for the parameter by using one of three options, None, Range, or List, as shown in the following image.

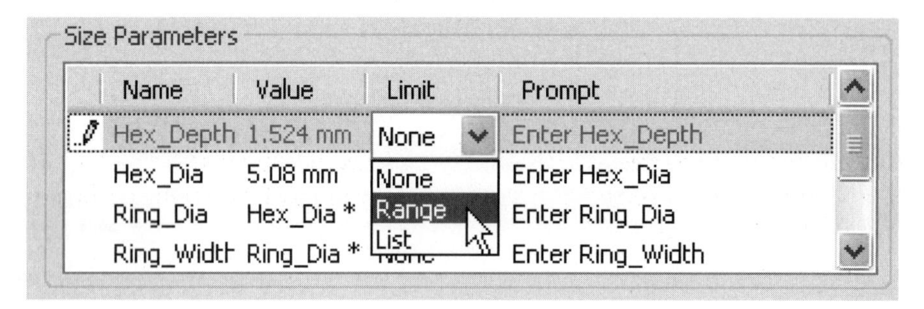

FIGURE 8.122

None specifies that no restrictions be placed on the Value field. Range gives you the ability to specify a minimum and maximum value, including less than, equal to, and infinity, as shown in the following image on the left. You can specify the default value to use. The List option, as shown in the following image on the right, allows you to predefine a list of values. Upon placement, you can choose from these values for the size parameters.

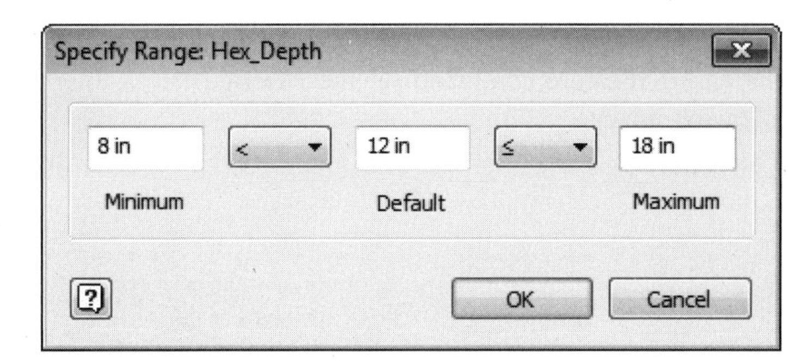

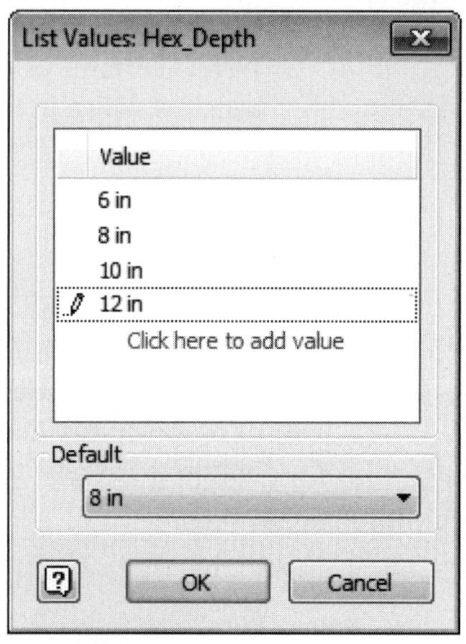

FIGURE 8.123

Prompt. Enter descriptive instructions to explain further why the parameter is used. The text entered in the prompt field appears in a dialog box during the iFeature placement.

Position Geometry. Specify the geometry of the iFeature that is necessary to position it on a part. You can add or remove geometry to or from the Position Geometry list in the Selected Features tree by right-clicking the geometry and selecting Expose Geometry. You remove geometry from the Position Geometry list by right-clicking it and selecting Remove Geometry.

Name. Describes the position geometry.

Prompt. Enter descriptive instructions to prompt a user for position geometry. The text entered in the prompt field appears in a dialog box during the iFeature positioning. You can customize the position geometry by right-clicking it and selecting one of the two additional options available on the menu.

Make Independent. Select this option to separate entries for geometry shared by more than one feature.

Combine Geometry. Select this option to combine listings of geometry shared by more than one feature in a single entry.

Manufacturing

This field is used when creating a Sheet Metal Punch iFeature. Refer to Chapter 10 for more information.

Depth

This field is used when creating a Sheet Metal Punch iFeature. Refer to Chapter 10 for more information.

Insert iFeatures

Insert iFeatures using the Insert iFeature command on the Manage tab > Insert panel, as shown in the following image.

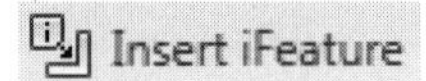

FIGURE 8.124

Once selected, the Insert iFeature dialog box appears, as shown in the following image. You can select tasks from the tree structure in the left pane of the dialog box or move forward and backward using the Next and Back buttons. Click the Browse button to navigate the iFeatures folder structure and select an iFeature to place.

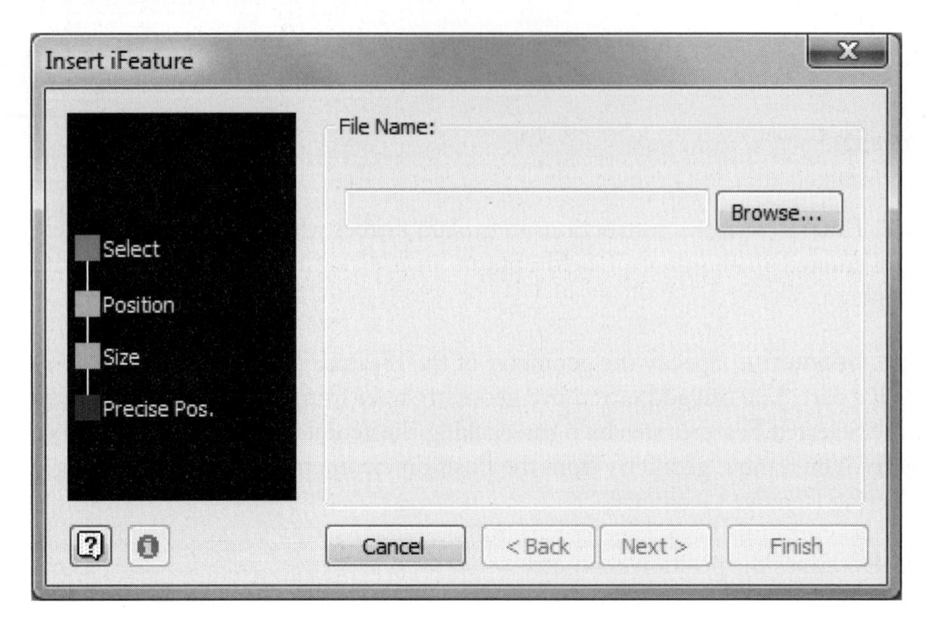

FIGURE 8.125

The Insert iFeature dialog contains the following sections.

Select

Choose the iFeature that you want to insert using the Browse button.

Position

This feature lists the names of the interface geometries specified during the creation process of the iFeature. After selecting the corresponding geometry in the part where you are placing the iFeature, click the arrowhead on the positioning symbol to either move or rotate the symbol, as shown in the following image.

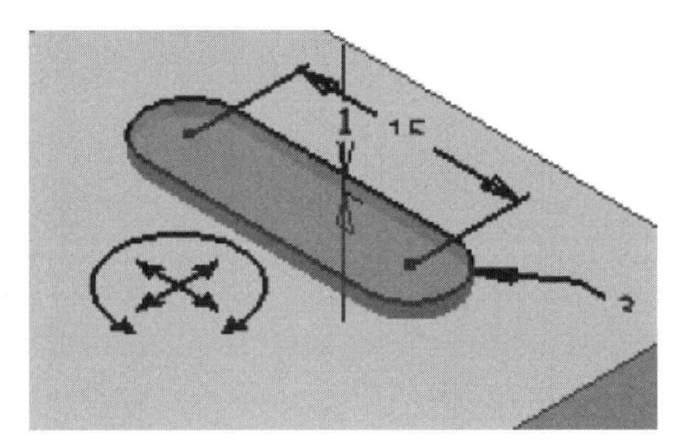

FIGURE 8.126

You can also specify a precise rotation value directly in the angle field of the dialog box. After the requirement has been satisfied, a checkmark will be placed in the left column, as shown in the following image.

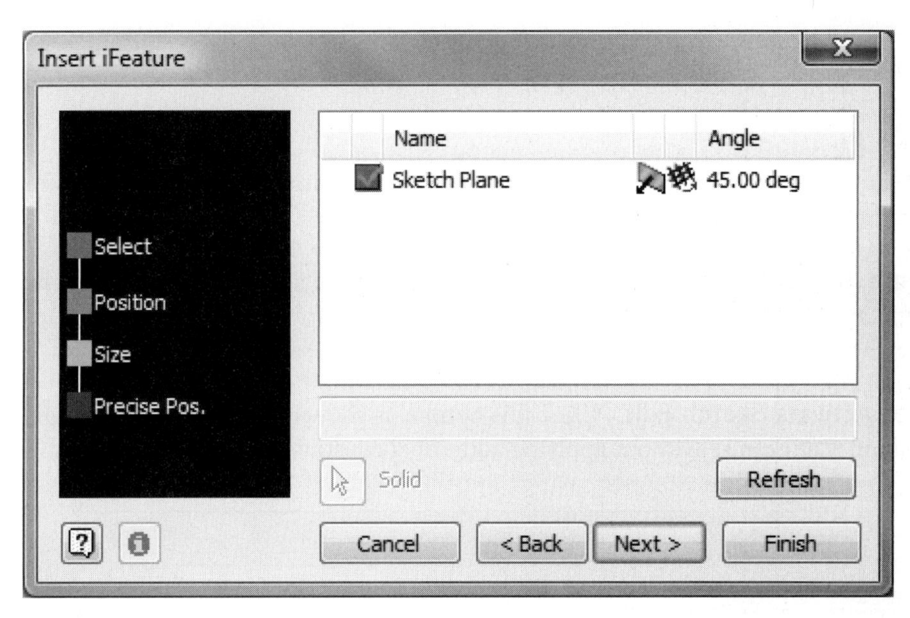

FIGURE 8.127

Name. This section lists the named interfacc geometry.

Angle. This section shows the default angle of the placement geometry on the iFeature.

Move Coordinate System. This section defines horizontal or vertical axes when the iFeature has horizontal or vertical dimensions or constraints included.

Size. This shows the names and default values specified for the iFeature, as shown in the following image. Click in the row to edit the values, and then click Refresh to preview the changes.

Name. This section lists the name of the parameter.

Value. This section lists the value of the parameter.

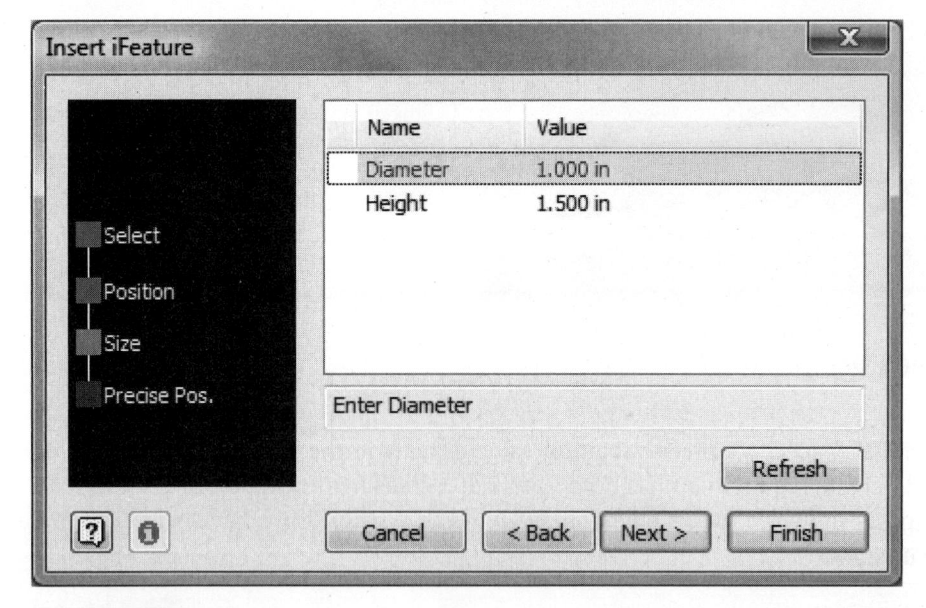

FIGURE 8.128

Precise Position

This feature further refines the position of the iFeature, using either dimensions or constraints after you have placed it.

Activate Sketch Edit Immediately. Click this option to activate the sketch of the iFeature and the 2D Sketch commands. You can then apply additional dimensions and/or constraints to position the iFeature on the part.

Do not Activate Sketch Edit. Click this option, as shown in the following image, to position the iFeature without applying additional constraints or dimensions.

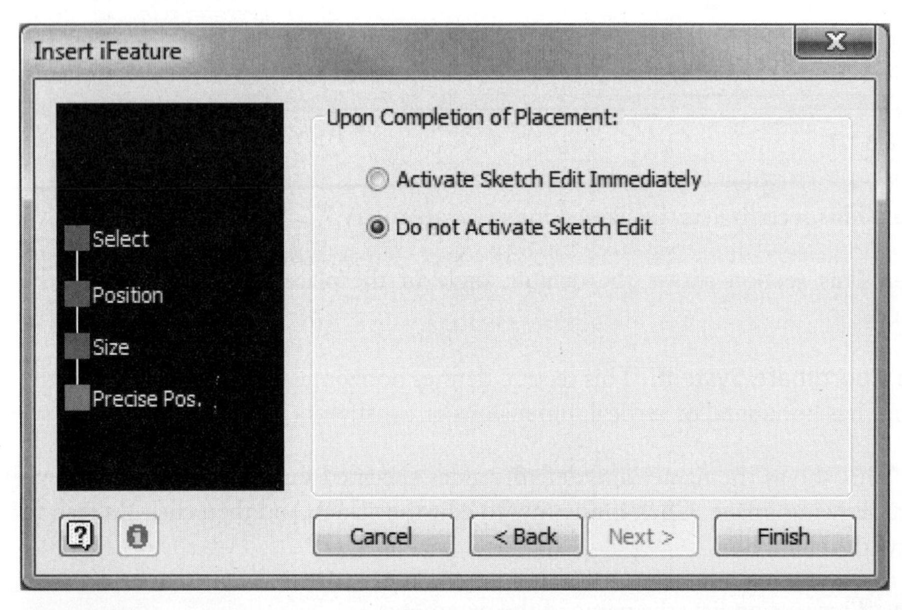

FIGURE 8.129

Editing iFeatures

You can edit an iFeature by opening the *.ide* file in Autodesk Inventor. Five commands are available in the iFeature environment when the file is opened: Edit iFeature, View Catalog, iFeature Author Table, Edit using Spread Sheet, and Change Icon.

The Edit iFeature command, as shown in the following image, opens the Edit iFeature dialog box. You cannot change which parameters are used to define the iFeature after it has been created, but you can modify the size parameters and position geometry by editing the properties for the name, value, limit, and prompt. The View Catalog command operates the same as noted previously. It opens Windows Explorer to the location specified in the iFeature root field on the iFeature tab of the Application Options dialog box.

FIGURE 8.130

The iFeature Author Table command, as shown in the following image, allows you to create iFeatures that are driven by a table.

FIGURE 8.131

Once selected, the iFeature Author dialog box appears, as shown in the following image. The table in the iFeature Author tool functions the same as it does when working with iParts. You can insert, delete, and specify the default row. You can also define custom parameter columns and include a range for the value. The main difference is that there is no option to set a column as a file name, display style, or material. If you are converting an iFeature to a table-driven iFeature that contains a parameter with a list of possible values, the table is populated automatically with the different sizes from the list as rows of the table. You can also edit the table via a spreadsheet. Refer to the tab descriptions in the iPart section of this chapter for additional information on how each tab of the iFeature Author dialog box functions.

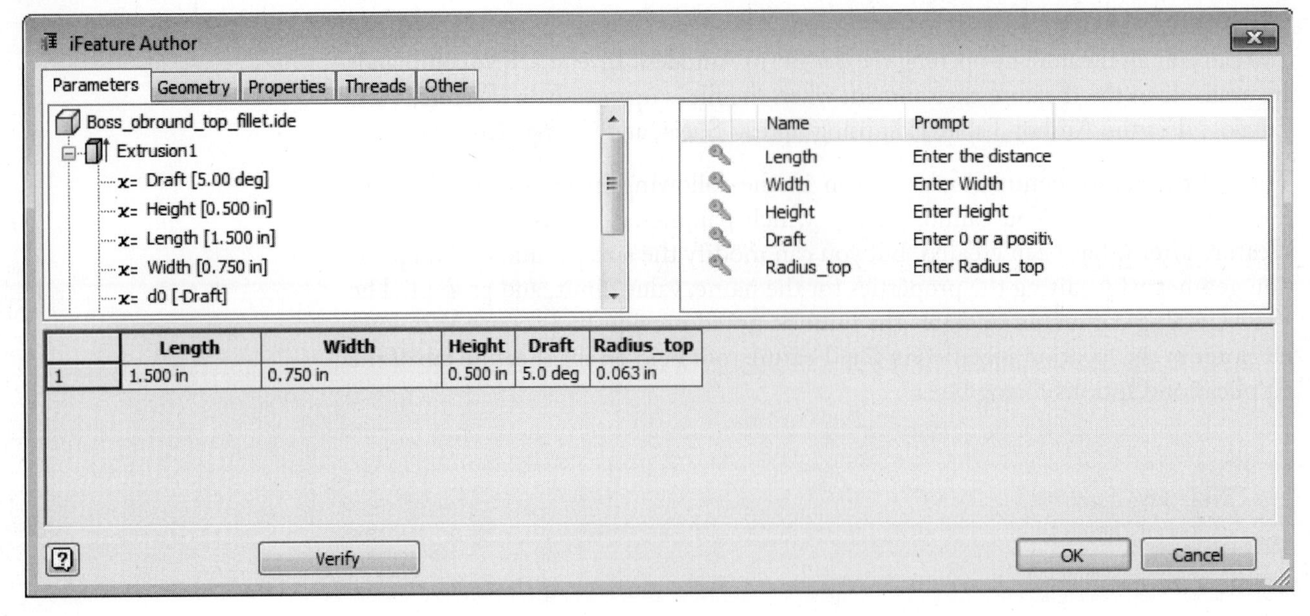

FIGURE 8.132

The Edit Using Spread Sheet command, shown in the following image, allows you to edit the iFeature table using Microsoft Excel instead of the iFeature Author table. The last command available is Change Icon. The Change Icon command allows you to create a custom icon for the iFeature that is displayed in the browser once it has been included in a part.

FIGURE 8.133

Inserting Table-Driven iFeatures

You insert table-driven iFeatures the same way that you would insert a typical iFeature. The wizard displays the key parameters as a drop-down list while defining the Size parameters, and any custom parameters will have the edit field available for entry.

The browser displays table-driven iFeatures in a similar manner to that used with iParts.

When you place a table-driven iFeature in a part file, a new *.ide* file is not created. The placement is similar to that of an iFeature that is not table driven. The file in which the iFeature is placed contains an independent copy of the iFeature that is not associative to the original *.ide* file. The spreadsheet that contains the table information is not visible in the browser, and you cannot edit it once you place it in the part file.

EXERCISE 8-5: CREATING AND PLACING iFEATURES

In this exercise, you will create a hex drive iFeature from an existing component. You will then insert the iFeature into another part file.

1. Open *ESS_E08_05.ipt*.
2. Click the Parameters command, and review the list of Model Parameters. The renamed diameter *(Hex_Dia)* and depth *(Hex_Depth)* parameters of the hex drive extrusion are the variables that will be extracted with the iFeature. The geometry of the raised ring around the hex drive is linked to the diameter of the hex drive.
3. In the Parameters dialog box, click Done.
4. From the Manage tab > Author panel, click Extract iFeature.
5. Click the features named *Hex* and *Ring* in the browser.

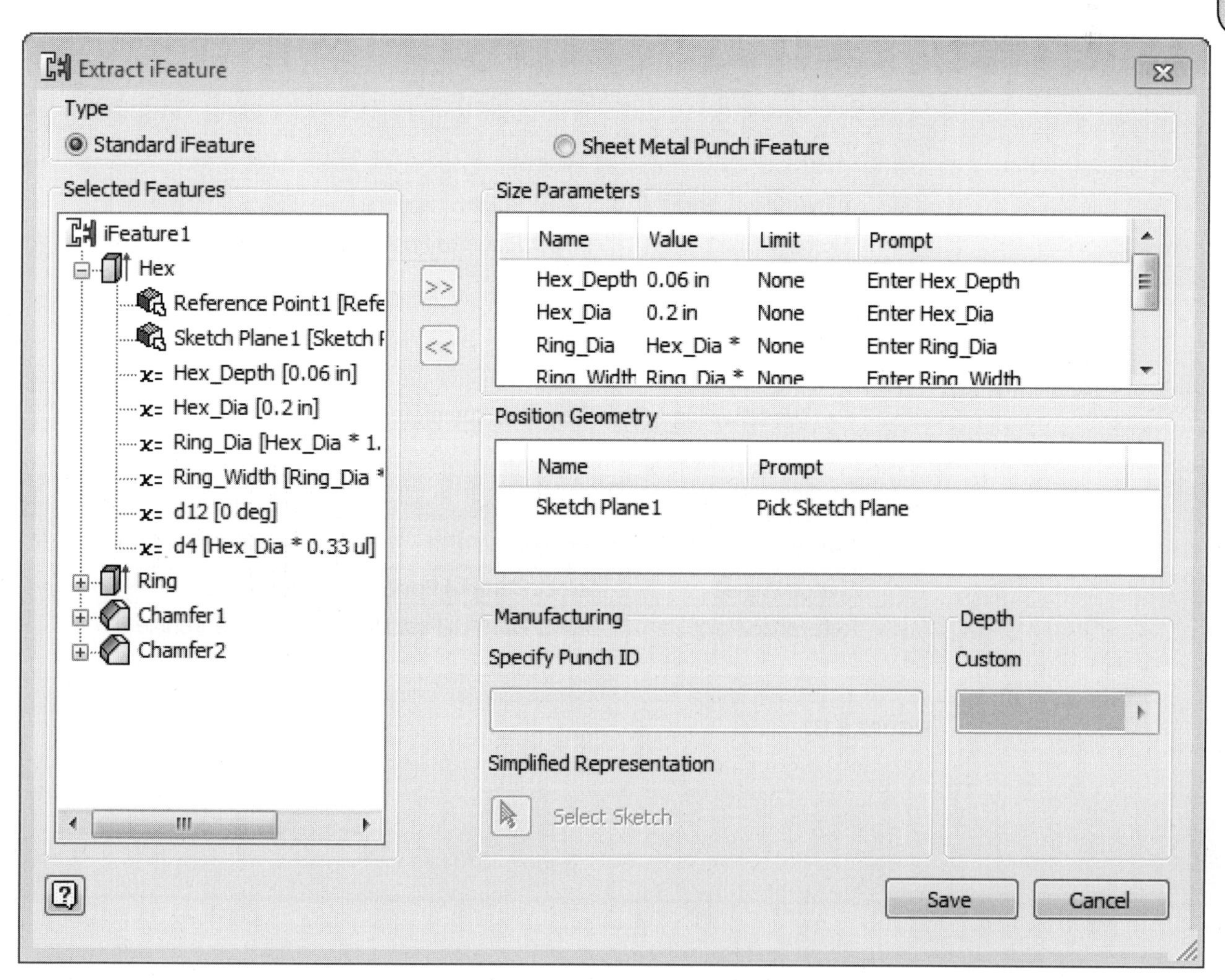

FIGURE 8.134

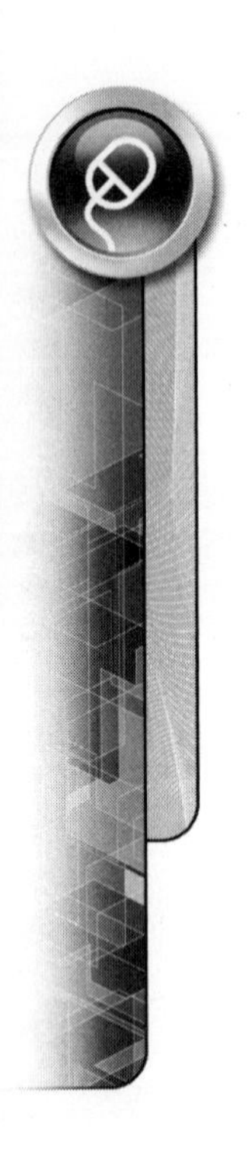

6. Click the *Pick Sketch Plane* prompt under Position Geometry, and enter **Select Plane to Position the Hex Drive** as the new prompt.
7. Expand *Hex* under Selected Features, right-click Reference Point1, and click Expose Geometry from the menu, as shown in the following image.

FIGURE 8.135

8. The point is added to the Position Geometry list. This point constrained the hex drive extrusion concentric to the circular face on the end of the cylinder. The point referenced will allow you to select model geometry to position the iFeature during placement in a new part. Click the *Pick Reference Point* prompt under Position Geometry.
   - Highlight the text, and replace it with **Select Point to Position Hex Drive Centerline** as the new prompt.
   - Click and drag the Reference Point1 position geometry below Sketch Plane1, as shown in the following image.

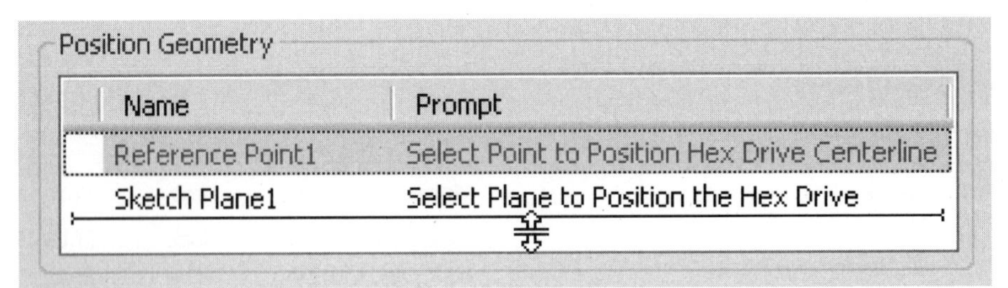

FIGURE 8.136

   - The following image shows the completed operation.

Position Geometry

Name	Prompt
Sketch Plane1	Select Plane to Position the Hex Drive
Reference Point1	Select Point to Position Hex Drive Centerline

FIGURE 8.137

9. Three parameters, *Ring_Height, Ring_Width,* and *Ring_Dia,* are linked to the hex drive diameter and are not required in the iFeature. To remove these parameters from the list, perform the following actions:
    - In the Size Parameters area, click *Ring_Height,* and click the Remove parameters button (≪).
    - In the Size Parameters area, click *Ring_Width,* and click the Remove parameters button (≪).
    - In the Size Parameters area, click *Ring_Dia,* and click the Remove parameters button (≪), as shown in the following image.

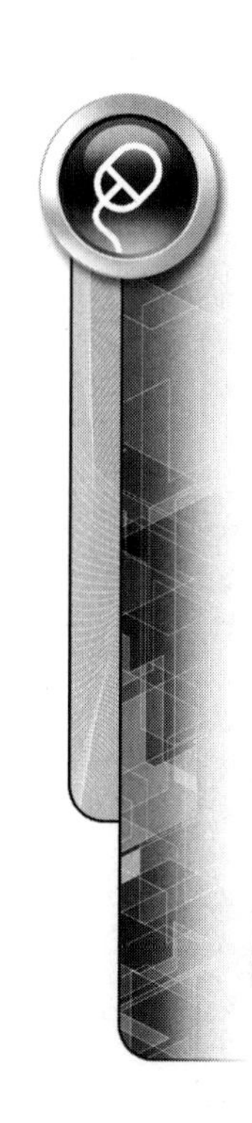

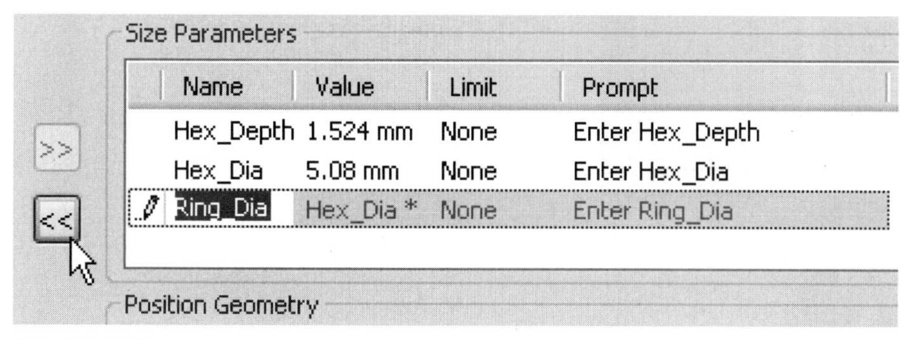

FIGURE 8.138

10. Only two parameters should now exist in the Size Parameters area: Hex_Depth and Hex_Dia.
11. Click Save, enter **Hex Drive** in the File name field, and place the file in the same location as the exercise files.
12. Close the file. Do not save changes.
13. Open *ESS_E08-05-Lock Post.ipt.*
14. Click the Insert iFeature command on the Manage tab > Insert panel.
    - Click the Browse button in the Insert iFeature dialog box.
    - Navigate to the directory where the exercises were installed.
    - Select *Hex Drive.ide.*
    - Click Open.

15. To locate Sketch Plane1, click the front face of the lock post, as shown in the following image.

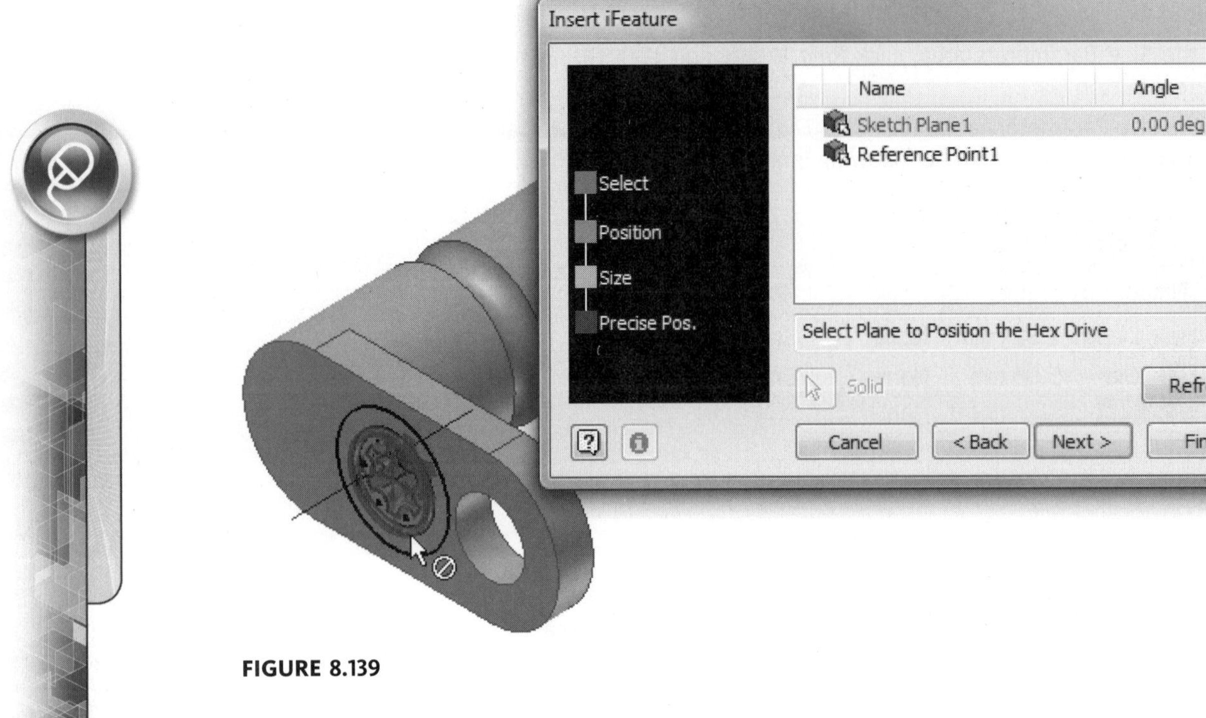

FIGURE 8.139

16. Click Reference Point1 in the Insert iFeature dialog box.
    - Click the left, outer circular edge of the lock post.
    - Click the point that is displayed on the part, as shown in the following image.

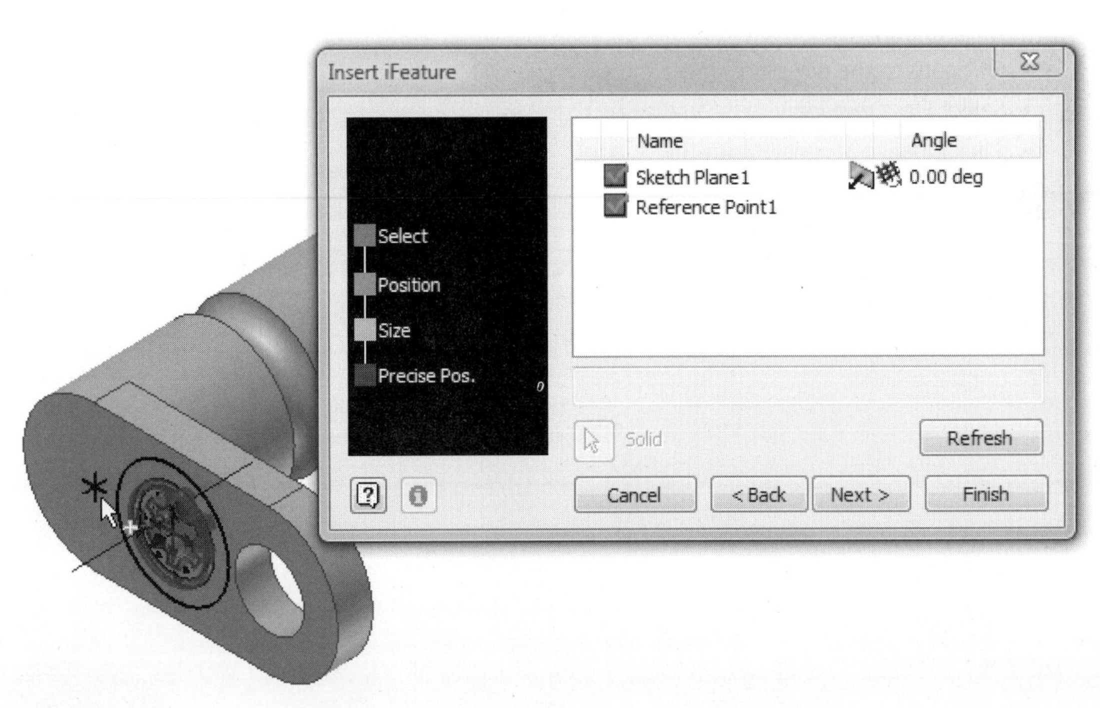

FIGURE 8.140

17. Click Next.
18. Leave the *Hex_Depth* and *Hex_Dia* variables as their default values.
19. Click Next.
20. Click Do not Activate Sketch Edit.
21. Click Finish.
22. The hex drive iFeature is placed with its center point coincident with the selected center point of the arc.

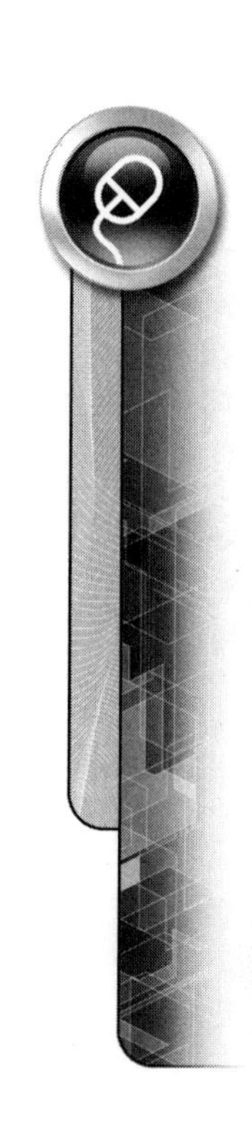

FIGURE 8.141

23. Close all open files. Do not save changes. End of exercise.

APPLYING YOUR SKILLS

Skill Exercise 8-1

In this exercise, you automate an iPart factory by controlling two iMates from the table. One version of the iPart has a single button, and the other has two buttons, as shown in the following image.

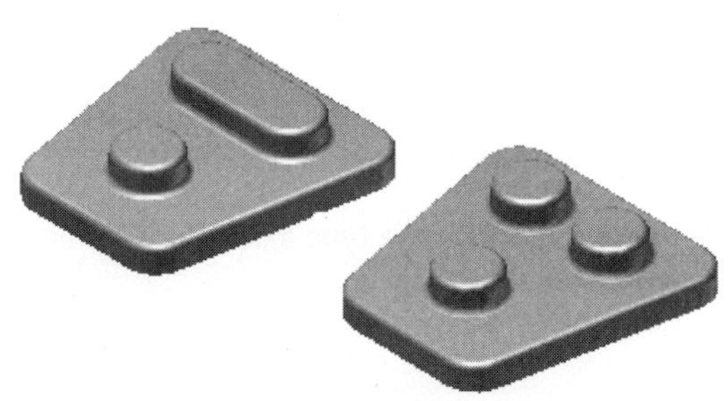

FIGURE 8.142

Place two different versions of the iPart in an assembly. Each version should be automatically positioned to the appropriate mating part using the automated iMates that you create, as shown in the following image.

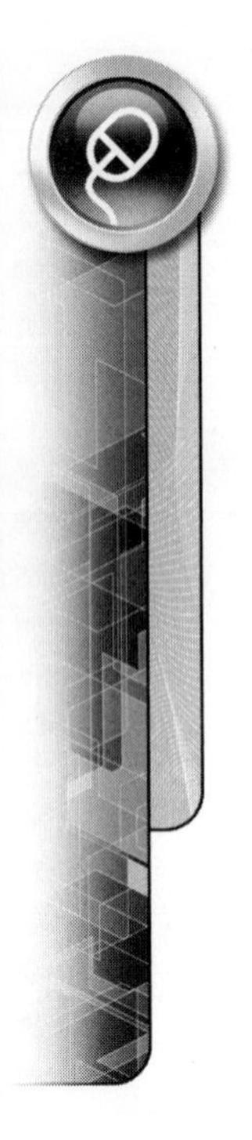

FIGURE 8.143

Start with an existing iPart factory of a rubber membrane for a remote keyless entry device.

1. Open ESS_E08_06.ipt and activate the single and double members of the iPart factory. Explore how this iPart factory uses feature suppression from the table to control the number of buttons at the top of the membrane.
2. In the following steps, you will use the button geometry to create a unique iMate for each iPart version.

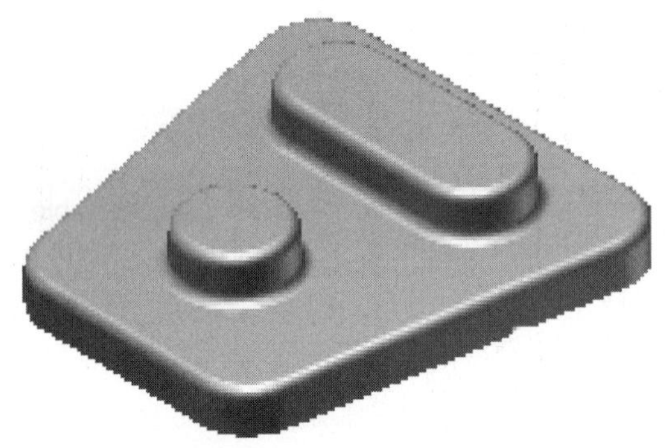

FIGURE 8.144

3. Use the Edit Member Scope functionality to add a mate type axial iMate named **Axis 1** to the cylindrical face as shown in the following image.

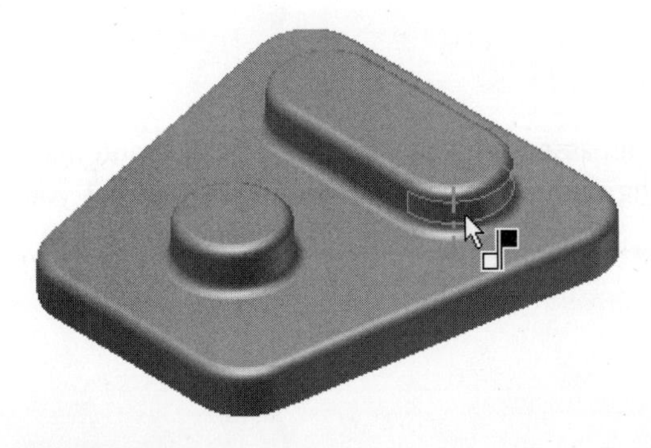

FIGURE 8.145

4. Activate the Double member and add a mate type axial iMate named **Axis2**, as shown in the following image.

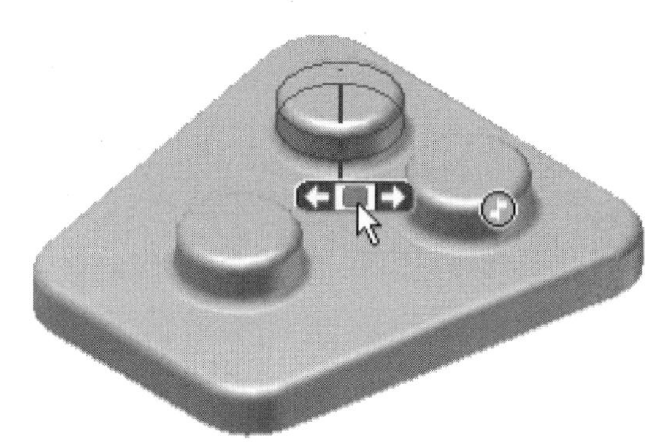

FIGURE 8.146

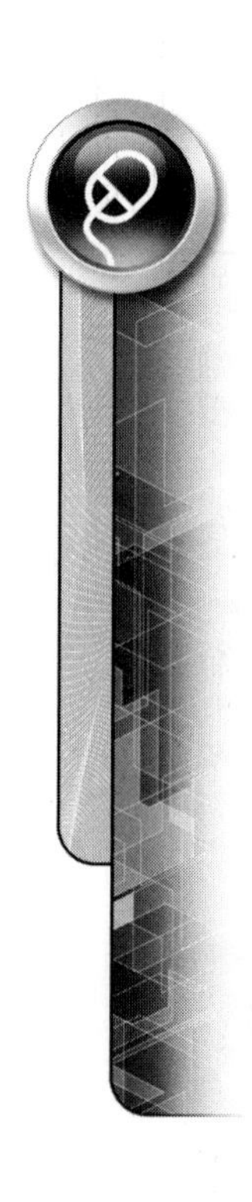

5. In the iPart Author dialog box, click the iMates tab, and note the Axis1 and Axis2 identifying labels IM0003 and IM0004, as shown in the following image.

FIGURE 8.147

6. In the table, verify the values for IM0003 and IM0004 with the following image. These values were automatically added and configured because Edit Member Scope was turned on.

	Member	Part Number	Extrusion3	Fillet3	Extrusion4	Fillet4	IM0003	IM0004
1	Single	Single	Compute	Compute	Suppress	Suppress	Compute	Suppress
2	Double	Double	Suppress	Suppress	Compute	Compute	Suppress	Compute

FIGURE 8.148

7. Verify your changes to the iPart table by activating the two iPart members from the browser.
8. Make the Single member active and save the iPart factory as *Membrane.ipt*.
9. Next open *ESS_E08_06B.iam*, an assembly that contains the mating cases for the Membrane iParts.
10. Use the Place Component command and the Interactively Place with iMates option to place the *Membrane.ipt* file. Use the Generate Remaining iMate Results to place and automatically constrain the previewed single member.
11. Change the member to the Double version and use the Place at All Matching iMates option from the menu to place the component.

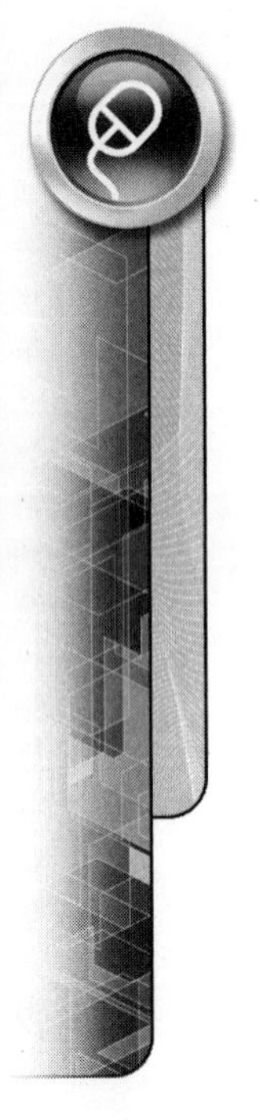

FIGURE 8.149

12. Close the file. Do not save changes. End of exercise.

CHECKING YOUR SKILLS

Use these questions to test your knowledge of the material covered in this chapter.

1. True ___ False ___ iMates are created while inside of a part file.
2. What needs to be selected to have a single component containing iMates be placed and automatically constrained in an assembly?
   a. Automatically Generate iMates on Place
   b. Place Component with iMate
   c. Interactively Place with iMates
   d. Use Composite iMate
3. True ___False ___ When creating parameters in a spreadsheet, the data items must be in the following order: parameter name, value or equations, unit of measurement, and, if needed, a comment.
4. What is the difference between a Model Parameter and a User Parameter?
5. True ___ False ___ Multiple versions of an iPart can be placed in an assembly.
6. True ___ False ___ When you make changes to an iPart factory, the changes are updated automatically in iParts that have been placed in assemblies.
7. Which tab of the Place iPart or iAssembly dialog box can be used to create a new member of the factory during placement in an assembly?
   a. Keys
   b. Table
   c. Tree
   d. None of the above
8. True ___ False ___ You can add features to standard iParts after they have been placed in an assembly.
9. True ___ False ___ Named parameters are added automatically as Size parameters during iFeature creation and cannot be removed.
10. What happens when you create a table-driven iFeature and one of the original parameters contains a list of values for the parameter?
11. True ___ False ___ iMates must be used as single iMates in order to facilitate easy assembly of components.

12. True ___ False ___ iMates are automatically assigned an index tag that is a unique value used to identify the iMate properties.
13. Explain what the following three (3) table cell background colors indicate when seen in the iPart author dialog box: Green, Dark Blue, and Red.
14. True ___ False ___ By defining a Proxy Path in the Libraries section of a project file, you can designate a specific location where you want generated iPart members to be stored on disc.
15. True ___ False ___ You cannot create an icon that will be displayed in the Browser when defining an iFeature.

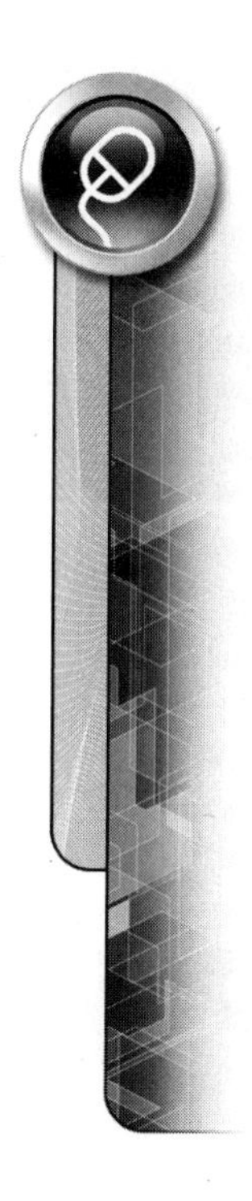

CHAPTER 9

Advanced Assembly Modeling Techniques

INTRODUCTION

In this chapter, you will learn how to use advanced assembly modeling techniques. Using these techniques, you can work more efficiently with assemblies. The techniques discussed will help you manage different views of your assemblies, improve the performance of your system, and analyze assemblies for interference between components. They will also aid in the creation of drawings from those assemblies.

OBJECTIVES

After completing this chapter, you will be able to perform the following:

- Create design view representations
- Create an assembly substitution
- Create flexible assemblies
- Create positional representations
- Create overlay drawing views
- Detect contact in assemblies
- Mirror an assembly
- Copy an assembly
- Create assembly work features
- Create assembly features
- Use the Frame Generator
- Locate commands related to the Content Center and Design Accelerator

DESIGN VIEW REPRESENTATIONS

While working in an assembly, you may want to save configurations that show the assembly in different states and from different viewing positions. Design view representations can store the following information:

- Component visibility: visible or not visible
- Sketch visibility: visible or not visible
- Component selection status: enabled or not enabled
- Color settings and style characteristics applied in the assembly
- Zoom magnification
- Viewing angle
- Display of origin work planes, origin work axes, origin work points, user work planes, user work axes, and user work points

You can use design view representations while working on the assembly and when creating presentation views or drawing views. You can also use them to reduce the number of items that display in an assembly, allowing large assembly files to be opened more quickly.

Access design view representations through the Representations folder > View node in the Assembly browser, as shown in the following image. Each assembly file contains a folder called Representations, as shown in the following image. Expanding this folder will display all of the representations that can be used. Notice the View, Position, and Level of Detail nodes. Expanding the View node will display all design view representations defined in the assembly model.

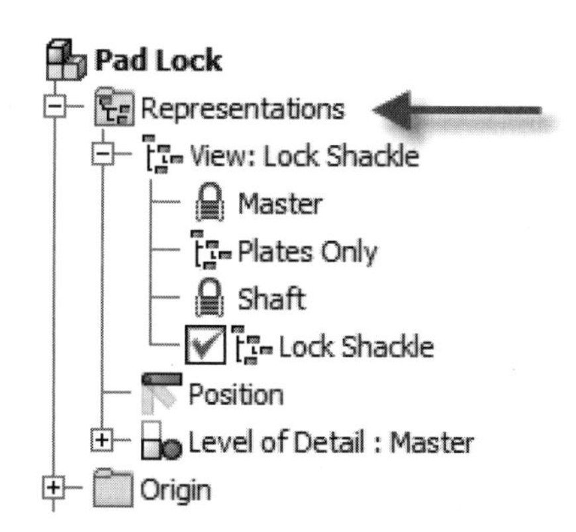

FIGURE 9.1

Creating a New Design View Representation

Begin the process of creating a new design view representation by expanding the Representations folder and the View: Default listing.

To create a design view representation, follow these steps:

1. In the browser, move your cursor over the View listing under the Representations folder and right-click.
2. Select New from the menu, as shown in the following image on the left.

Completing these steps will create a new design view representation called View1 and make it current. The current design view representation is identified by the presence of a checkmark next to its name in the browser, as shown in the following image on the right. The name of the current design view representation is also shown next to the View node under the Representations folder in the Assembly browser.

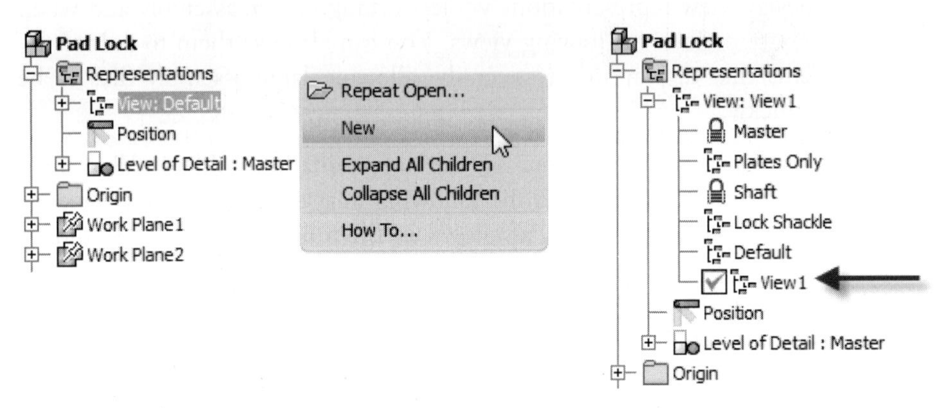

FIGURE 9.2

3. It is considered good practice to change the name of the design view representation to something more meaningful. As shown in the following image on the right, View1 has been renamed Alt - Color Blue.

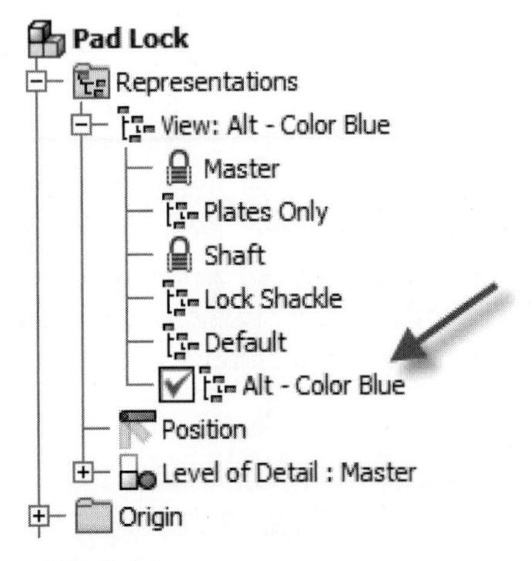

FIGURE 9.3

4. With a current design view representation set, various parts of the assembly can be modified. For this pad lock example, certain components of the assembly would be hidden such as the case, lock shackle, and dial. These changes are saved automatically to the current design view representation. In addition to

component visibility, you can also save the visibility state of sketches, work features, both user and origin, whether or not a component is enabled, color settings of components, the zoom magnification, and the viewing angle.

5. Double-clicking on the Master or any other design view representation name, will return the engine assembly back to its original representation. In the following image, the default representation of the pad lock is shown on the left, and an alternate representation displaying the combo subassembly is shown on the right.

FIGURE 9.4

6. Create additional design view representations as needed. You may want to create design view representations using alternate color schemes or to assist in the creation of presentation files or drawing views where particular components are disabled or hidden.

Once you are satisfied with all of the changes made to a design view representation, you can prevent any further changes to a representation by locking it. Right-clicking on a design view representation name will display the menu shown in the following image. Selecting Lock from this menu will add a padlock icon to the design view representation, as shown next to the Master design view representation in the following image. The Master design view is always locked.

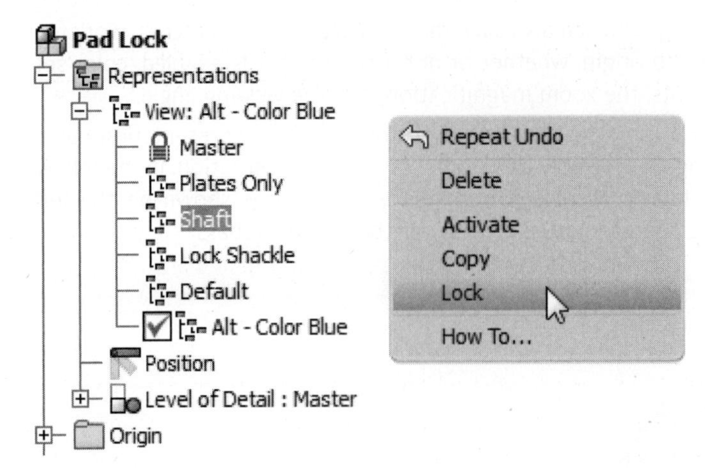

FIGURE 9.5

If you want to make additional changes to a locked design view representation, right-click on the name, and select Unlock from the menu.

Increasing Performance through Design View Representations

Through design view representations, additional control of the visibility of each subassembly is possible, allowing you to turn off the visibility of unimportant components to increase performance. The following sections describe additional design view representation controls, as shown in the following image.

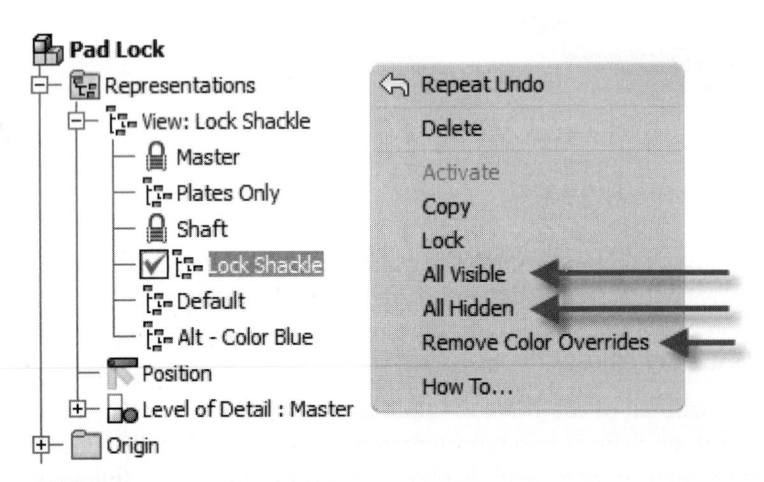

FIGURE 9.6

All Visible. Select this control to make all of the components in the assembly visible.

All Hidden. Select this control so that none of the components in the assembly will be visible. This can be beneficial for managing large assembly files. Opening a large assembly with all components hidden will allow you to manually turn on the visibility for only those components with which you need to work.

Remove Color Overrides. Select this control to restore the original colors of an assembly. This works well if you applied an assembly level color override to any of the components in an assembly.

NOTE

The All Hidden and All Visible design view representations will override any existing visibility settings in the subassembly. You cannot alter these representations or save new design view representations with the same name.

The design view representations applied to a subassembly from within an assembly are not associated with those created in the original subassembly. After a subassembly is placed in an assembly, changes or additional design view representations in the subassembly are not automatically reflected in the placed subassembly.

Creating Drawing Views from Design View Representations

To generate a drawing view based on a design view representation, activate the Drawing View dialog box, and click in the View section of the Representation area to select a valid design view, as shown in the following image.

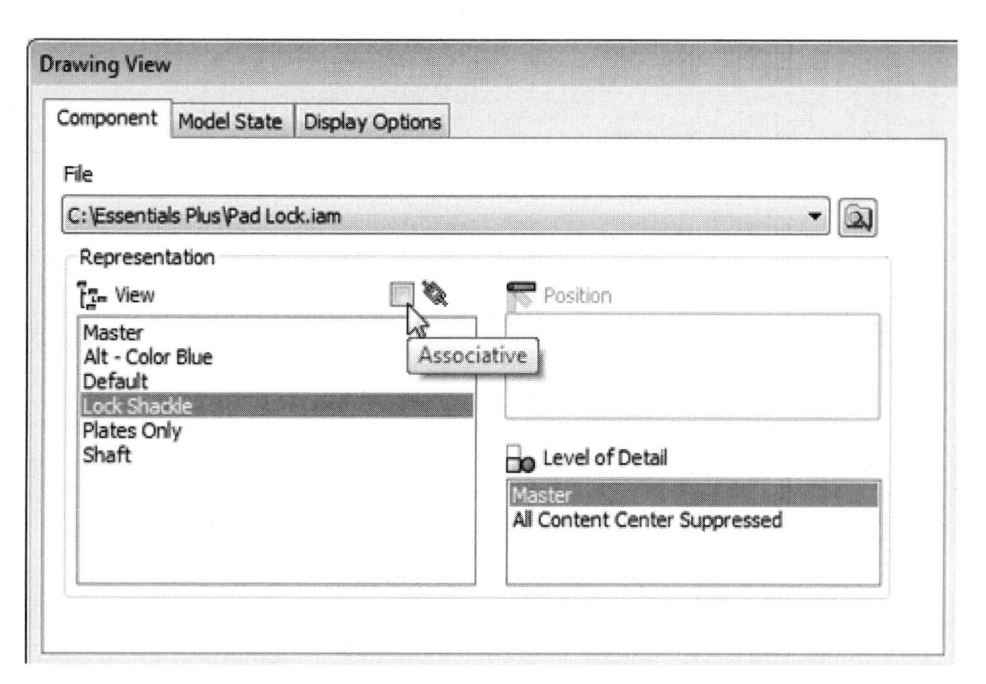

FIGURE 9.7

A projected view will be based on the design view representation from which its parent view was created. You can make drawing views generated from a design view representation associative by selecting the Associative checkbox, as shown in the previous image. This action will make the drawing view generated by the design view representation dependent upon the design view representation found in the assembly file. Any changes made to the design view representation in the assembly will update the drawing view.

Once a drawing view is generated based on a design view representation, you can change the drawing view or views if you select a different design view representation. To perform this operation, first select the drawing view, and then right-click. Select Apply Design View from the menu as shown in the following image on the left.

When the Apply Design View Representation dialog box appears, click in the list box, and select a new design view representation, as shown in the following image on the right. If multiple views of the assembly appear on the sheet, set the Apply status for each view. Set the Associative status by changing the value from No to Yes.

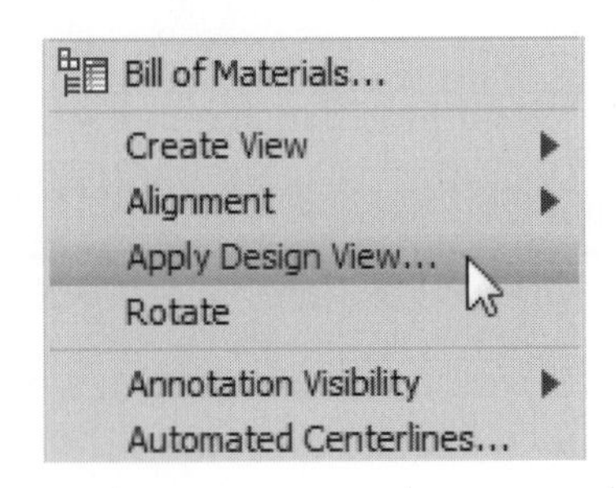

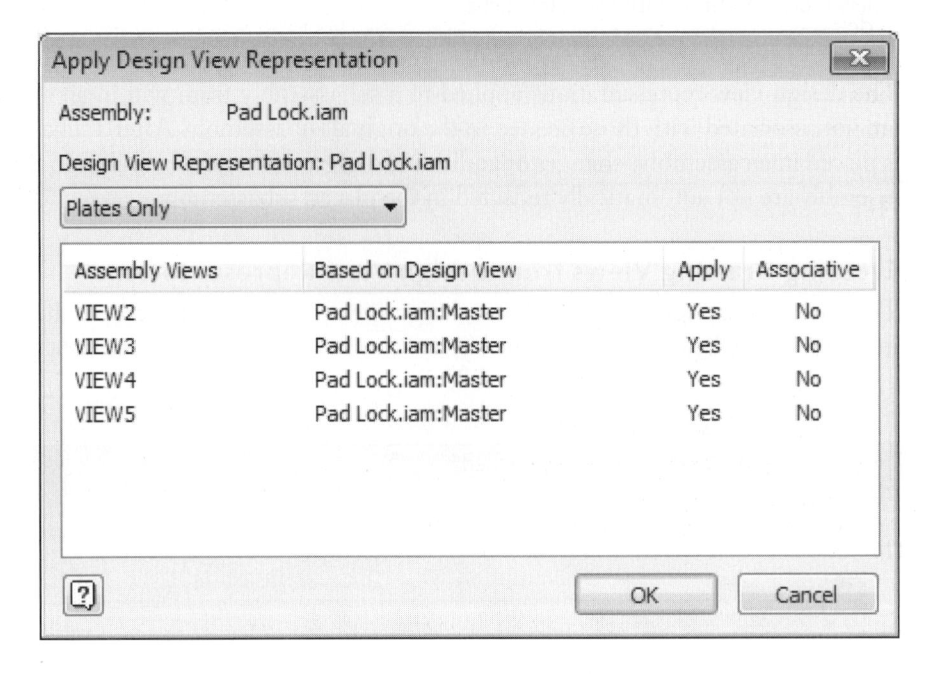

FIGURE 9.8

Levels of Detail Representations

The Levels of Detail Representations command can be used to improve capacity and performance in both the modeling and drawing environments. Levels of detail use component suppression to allow you to selectively choose which models are loaded into memory. The Suppress status of a component, accessed by right-clicking a component in either the browser or the graphics window, unloads the component(s) from memory, as shown in the following image on the left.

When a component is suppressed, the component's icon is modified in the browser, the top-level assembly name is modified to notify you which level of detail is active, and the component is not displayed in the canvas. If you place your cursor over or select a suppressed component in the browser, the component bounding box is previewed in the canvas, as shown in the following image on the right where two suppressed components have been selected.

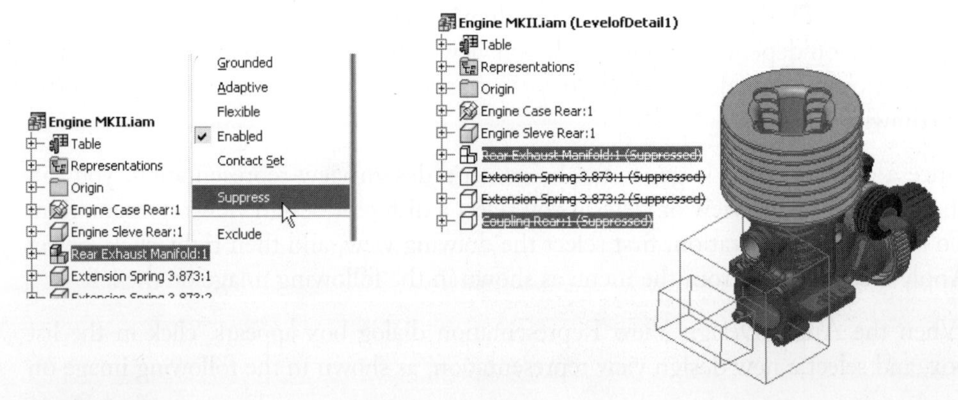

FIGURE 9.9

A Level of Detail node is available under the Representations folder in the browser, as shown in the following image. By default, four levels of detail are created in each assembly and can be activated to increase performance, if needed. The default levels of detail are Master, which is active by default, All Components Suppressed, All Parts Suppressed, and All Content Center Suppressed. You can activate a level of detail in the same way that you activate a design view representation or a positional representation: double-click on it in the browser.

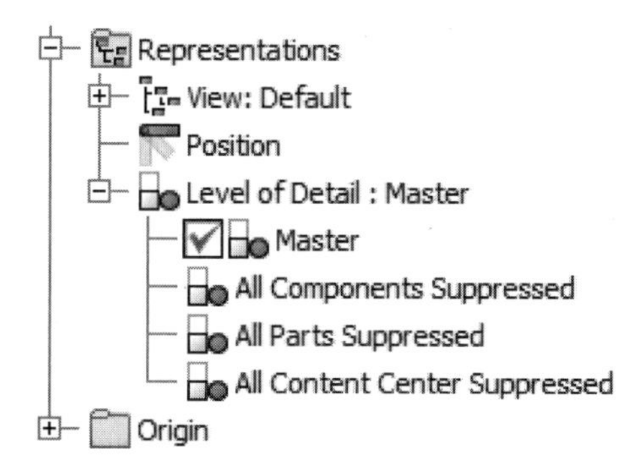

FIGURE 9.10

Similar to Design View Representations, the Master level of detail representation is active by default. You can make modifications to the status of children nodes of an assembly when this representation is active, but the changes are not maintained when the assembly file is saved. In order for the changes to be saved, a new or existing level of detail representation that you create must be active. Refer to the design view representation and positional representation sections of this chapter for more information about creating additional representations.

You can select a level of detail, design view, or positional representation to load when opening a file or placing a component in an assembly. This step is done by selecting the Options button in the File Open or Place Component dialog box and choosing the Level of Detail Representation that you want to load, as shown in the following image.

File Open Options
Design View Representation
Last Active
Positional Representation
Master
Level of Detail Representation
Master
Master
All Parts Suppressed
All Components Suppressed
All Content Center Suppressed
OK
Cancel

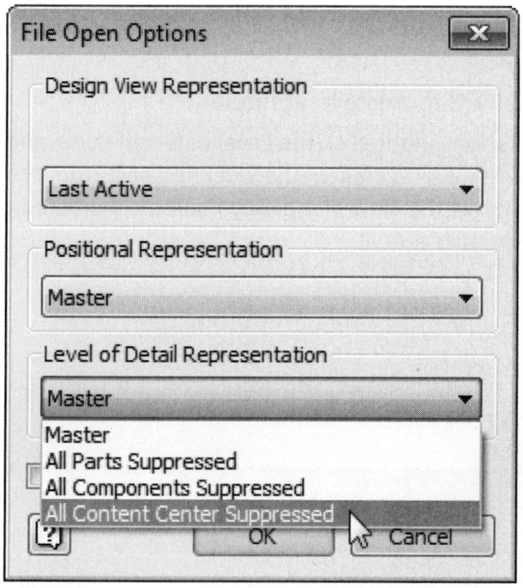

FIGURE 9.11

In a similar manner, when creating a drawing view of an assembly, you can select a level of detail representation from the Drawing View dialog box, as shown in the following image.

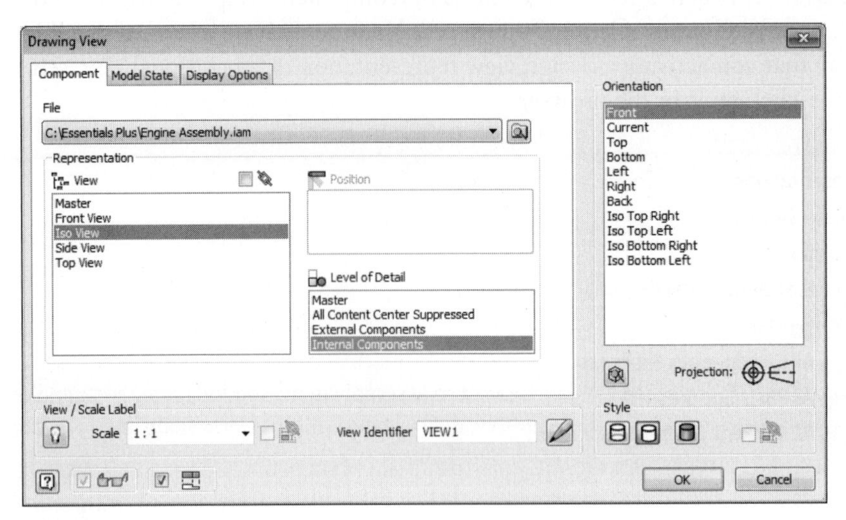

FIGURE 9.12

Assembly Substitution Level of Detail Representation

Another option for a Level of Detail is to create an assembly substitute. While working with an assembly you can use assembly substitution to reduce the memory that the assembly uses and reduce the amount of detail. The Level of Detail needs to be created at the top level of the open assembly. You need to open an assembly to create a substitute; you cannot create a substitute Level of Detail Representation of a subassembly in the active assembly. There is no limit to the number of substitutes that can be created. The iProperty data is maintained from the original assembly when a substitute is created. There are three methods for substituting assemblies; you can create a derived assembly, a shrinkwrap or substitute a subassembly for an existing part file.

Create a Substitute Level of Detail Representation Using Derive Assembly

To create an assembly substitute using the derived method follow these steps. The assembly constraints will be maintained.

1. Open an assembly for which you want to create a substitute.
2. In the browser expand Representations, right-click the Level of Detail node, and click New Substitute > Derive Assembly to create a new derived part from the assembly. The active Level of Detail at the time of creation contains the default body representation for the derive operation.

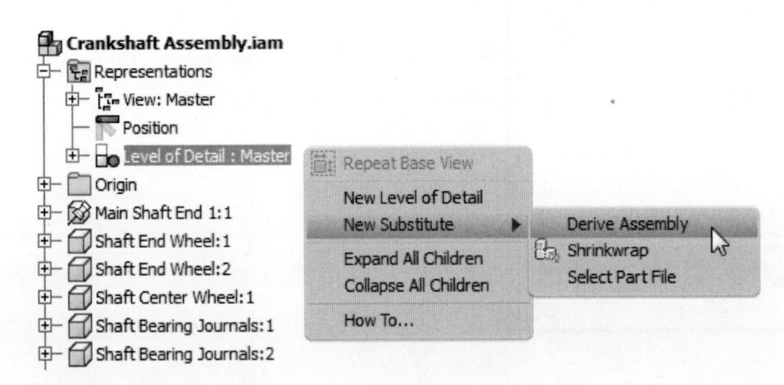

FIGURE 9.13

3. In the New Derived Substitute Part dialog box, enter a name for the component, select a template file to base the file on, select a location, and click OK.

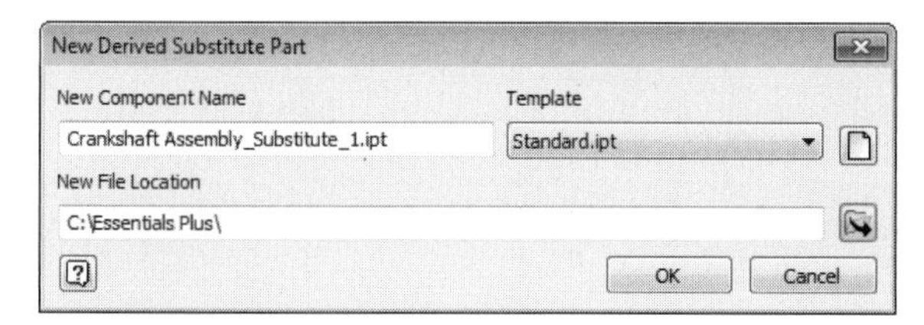

FIGURE 9.14

4. The Derived Assembly tool is opened.
5. On the Bodies tab, select the desired option for each component or the entire assembly. See Chapter 7 for more information about the derived options. The following image on the left shows the assembly in its original state. The middle image shows the Derived Assembly dialog box with the bounding box option selected for the entire assembly. The image on the right shows the resulting bounding box.
6. The Reduced Memory Mode option is on by default (available on the Options tab). This option uses less memory by not caching the source bodies. It is recommended to leave this option on.

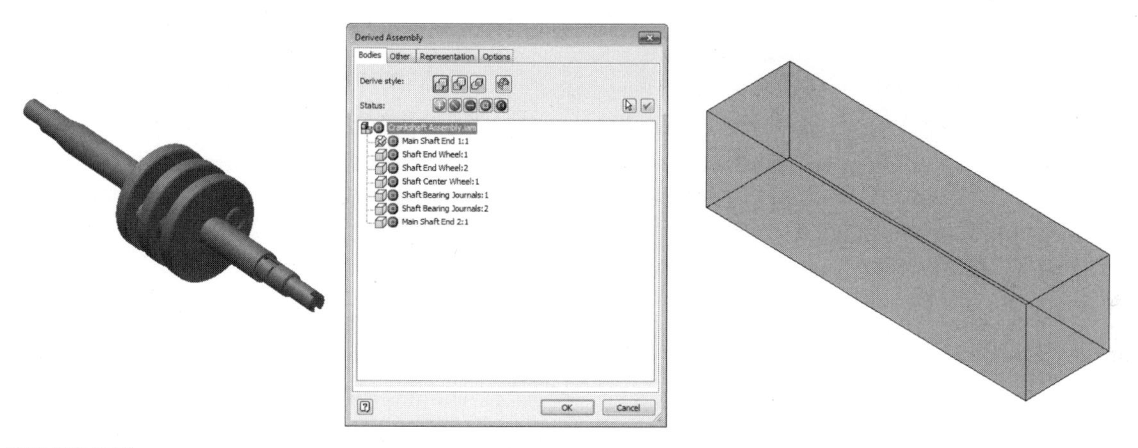

FIGURE 9.15

7. Click OK to create the derived part. A new entry under Level of Detail will appear in the browser as shown in the following image. If desired, you can rename the substitute. The part will appear in the graphics window as depicted in the last step.

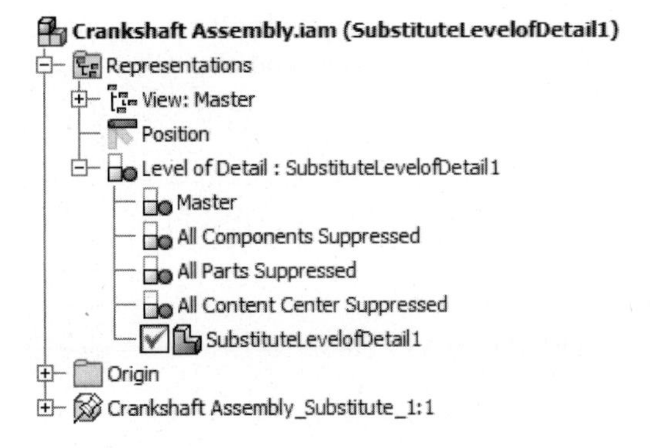

FIGURE 9.16

8. To edit the substitute make the derived part active by double-clicking the part in the browser, right-clicking the derived assembly, and clicking Edit Derived Assembly as shown in the following image. The Derived Assembly dialog will reappear.

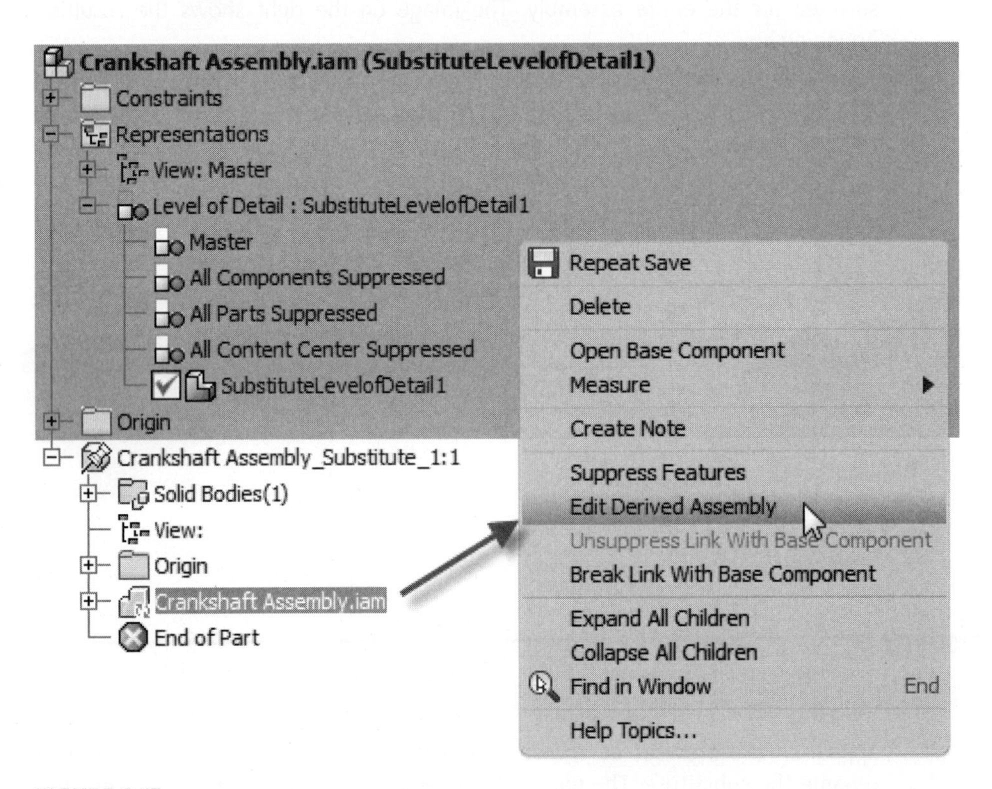

FIGURE 9.17

9. To use the substitute, make the assembly active for which you want to substitute. Expand the Representation folder. The following image shows the assembly with the substitute NOT active.

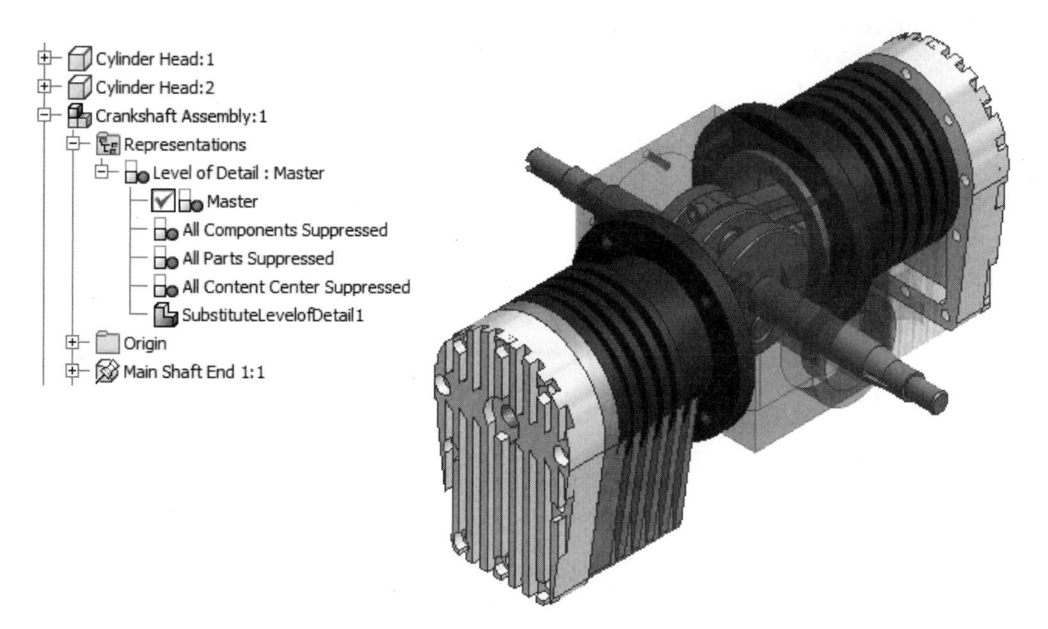

FIGURE 9.18

10. Make the substitution active by double-clicking the substitute's entry under the Level of Detail. The assembly constraints will be maintained.

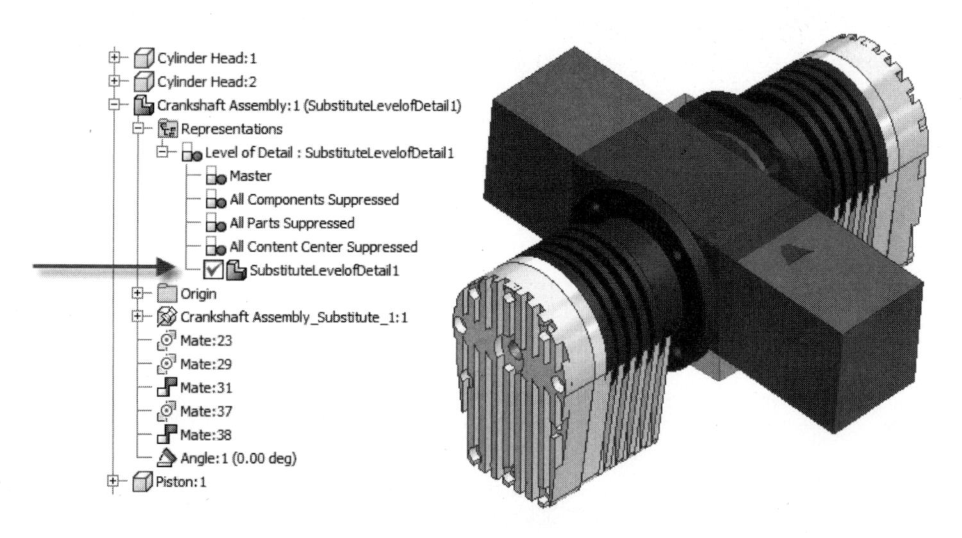

FIGURE 9.19

Create a Substitute Level of Detail Representation Using Shrinkwrap

A shrinkwrap part generated from an assembly is a simplified part representation that has enough detail to provide an accurate representation of the original models, but does not contain the components or features used to create them. Shrinkwrap parts are similar to those generated using the derived component tool. However, when you create a shrinkwrap file, you have additional options available to decrease the file size:

1. Open an assembly for which you want to create a shrinkwrap. You can also create a shrinkwrap when working in assembly by using the Shrinkwrap command.

2. In the browser, expand Representations, right-click the Level of Detail node, and click New Substitute > Shrinkwrap to create a new shrinkwrap part from the assembly.

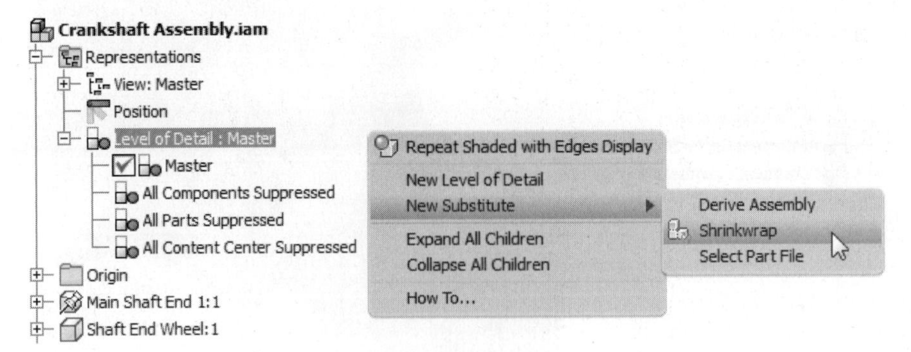

FIGURE 9.20

3. In the New Derived Substitute Part dialog box, enter a name for the component, select a template file to base the file on, select a location, and click OK.
4. The Assembly Shrinkwrap Options dialog box is opened.

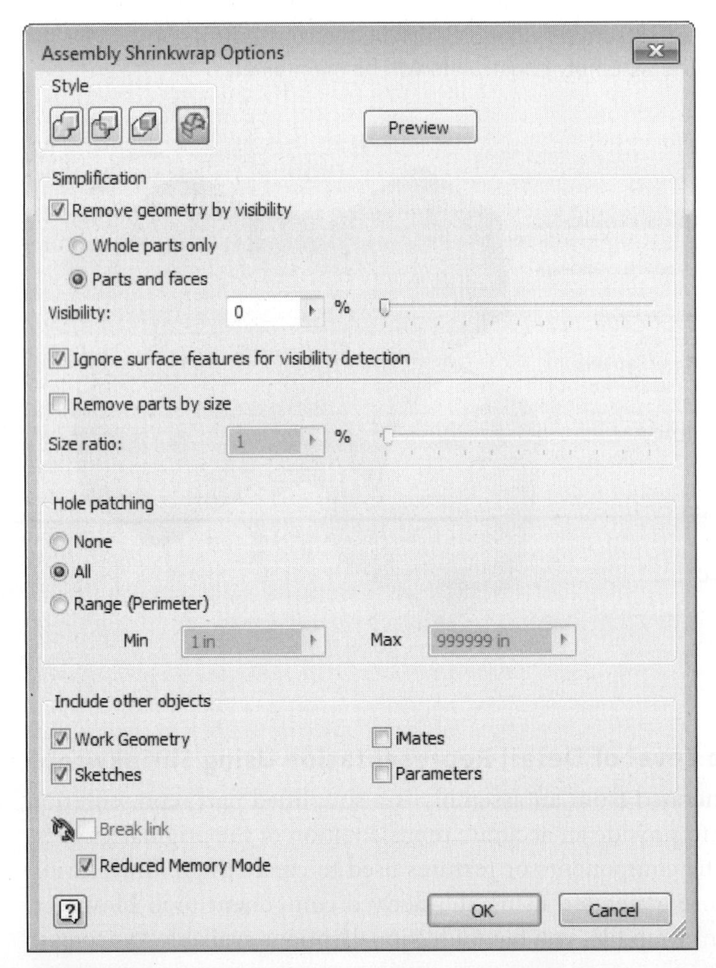

FIGURE 9.21

5. In the dialog box, select the desired options. You can choose a style for the shrinkwrap, which can be a solid, with or without seams, or a surface model. The Simplification area is used to remove geometry based on entire parts or faces. The visibility percentage slider can be used to remove parts based on the percentage of size that is visible in the graphics window. Hole patching is used to remove, or fill, openings in the new shrinkwrap part. The include other objects section provides the ability to select whether you want to include other types of objects in the shrinkwrap. Lastly, you can choose whether you want to break the link to the original model and use the reduced memory mode, previously discussed.
6. Click the Preview button to preview the shrinkwrap part before committing the changes to the model.
7. Click OK to create the shrinkwrap part. A new entry under Level of Detail will appear in the browser. If desired, you can rename the substitute. The new, simplified shrinkwrap part will appear in the graphics window.

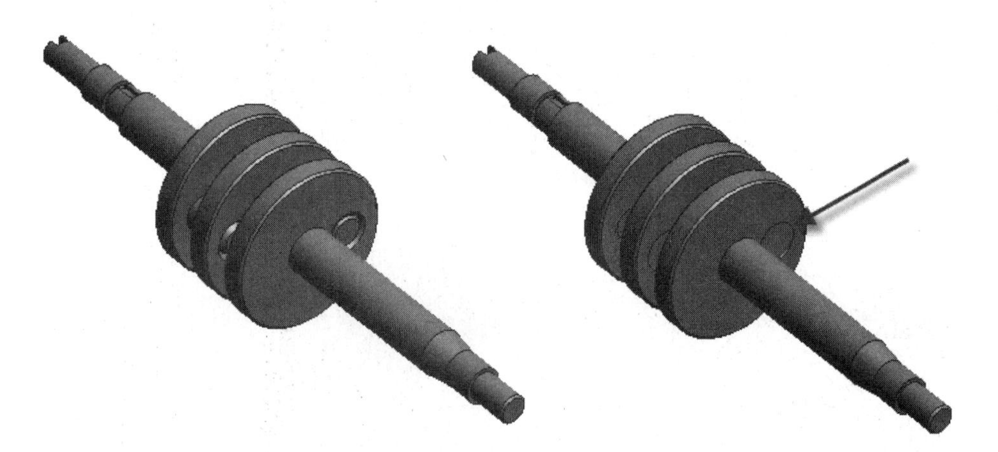

FIGURE 9.22

8. To use the shrinkwrap, make the assembly active for which you want to substitute the simplified part. Expand the Representation folder. Make the shrinkwrap active by double-clicking its name under the Level of Detail. The assembly constraints will be maintained.

Create a Substitute Level of Detail Representation Using a Part File

With this option you manually create or derive a part and use it to substitute an assembly. The assembly constraints may not be maintained.

1. Create the part using the same XYZ location and orientation that appears in the assembly. If you do not create the part this way, you will need to use Grip Snaps or apply assembly constraints to the origin work features to re-position the part. To avoid reorienting the part, project edges from the assembly or copy bodies from the assembly.
2. In the browser expand Representations, right-click the Level of Detail, and click New Substitute > Select Part File.

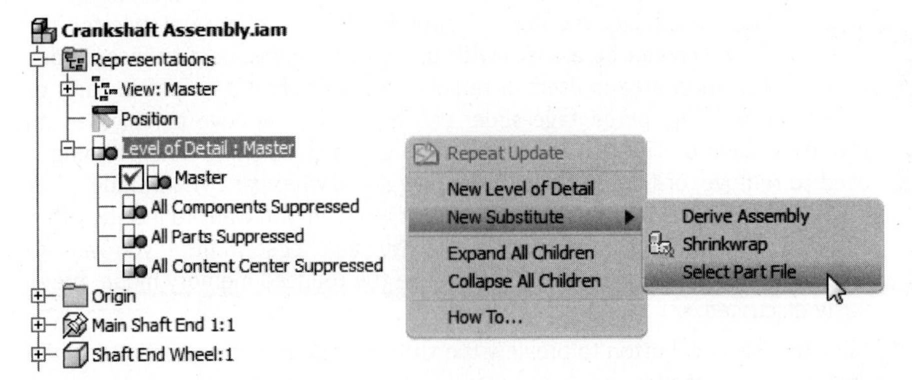

FIGURE 9.23

3. A dialog box will appear stating that the links to the original files will be disabled until the substitute is no longer active. Click Yes to close the dialog box. The following image shows a part being substituted. The entry in the browser was renamed.

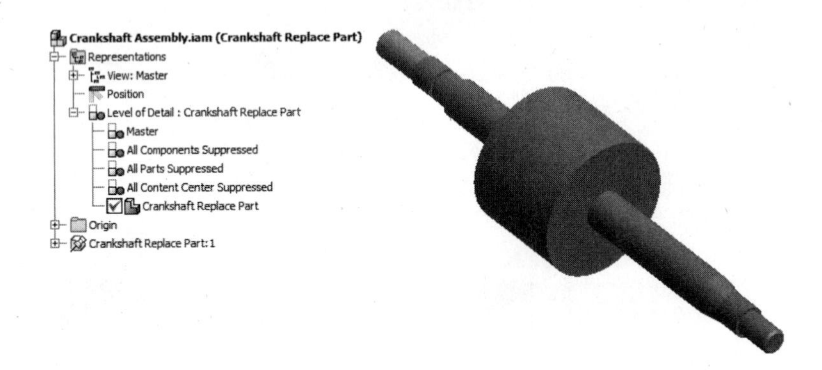

FIGURE 9.24

4. Save the file.

To switch between different levels of detail representations in different subassemblies, you can create a Level of Detail at the top-level assembly and make the Level of Details in the different subassemblies active.

FLEXIBLE ASSEMBLIES

As already discussed, you can set the adaptivity property of a subassembly. When you place numerous instances of the same subassembly in a main assembly model, all instances of the subassemblies will act together to display the motion. In the following image, a typical industrial shovel is being controlled by three hydraulic cylinder subassemblies. When one of the cylinders is set to adaptive, all cylinders display the same adaptive positions.

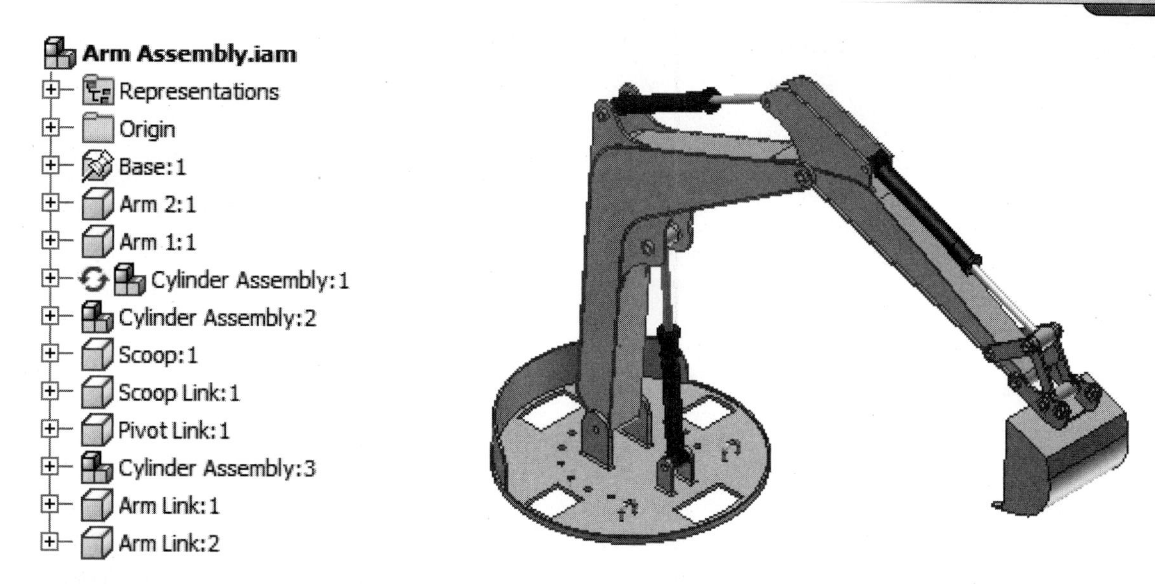

FIGURE 9.25

The following image illustrates the results of the adaptive cylinders. While these images may look acceptable, the problem with this solution is that when one cylinder opens or closes, the other two cylinders open or close to the same positional distance and direction. When working with assemblies of this nature, you want each cylinder to move independently.

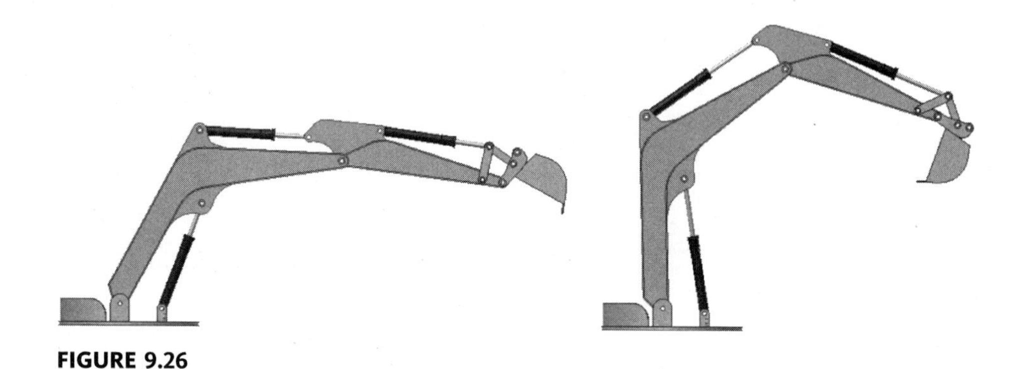

FIGURE 9.26

Instead of assigning the property of adaptivity that affects all cylinders, you can make each cylinder move independently by applying the Flexible property. In the following image, the Cylinder Assembly located in the Assembly browser has had the Adaptivity property removed. Right-clicking on this Cylinder Assembly will display the menu, as shown in the following image in the middle. Click on Flexible.

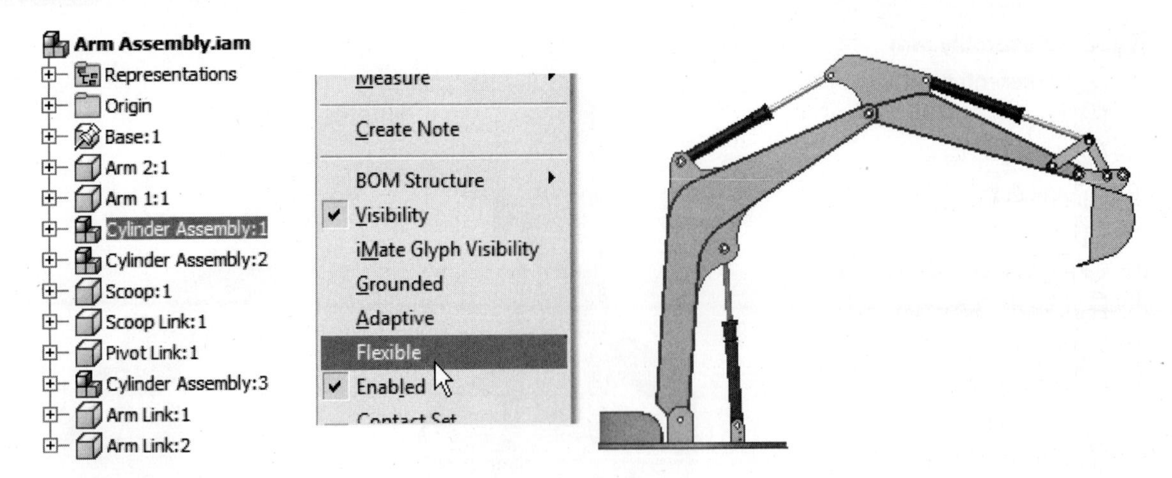

FIGURE 9.27

When you assign the Flexible property to the first occurrence of the cylinder assembly, an icon appears to represent flexibility for this subassembly, as shown in the following image. When the first cylinder moves, the others remain unaffected.

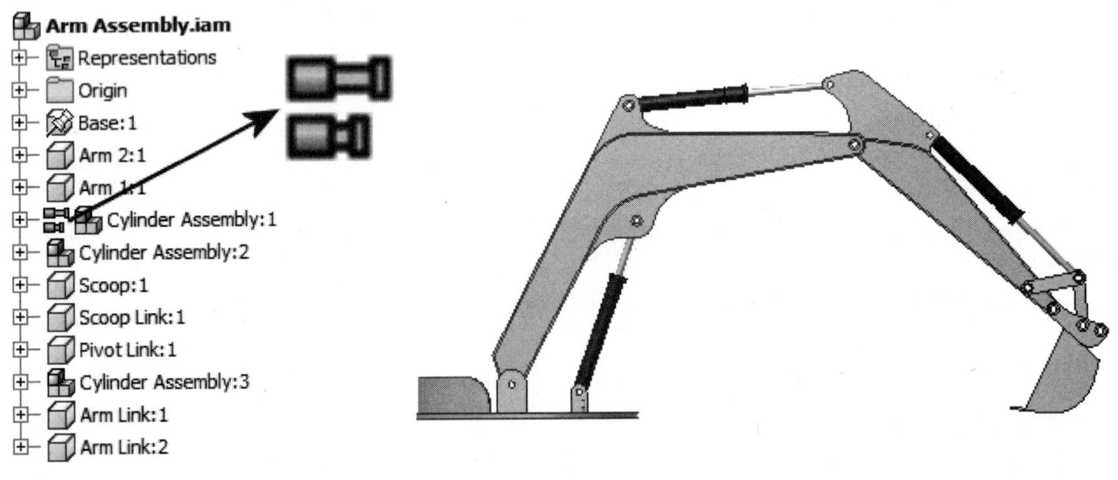

FIGURE 9.28

The following image shows all three cylinder assemblies set to Flexible. Now the hydraulic shovel assembly can move to any configuration because all three cylinders move independently of each other.

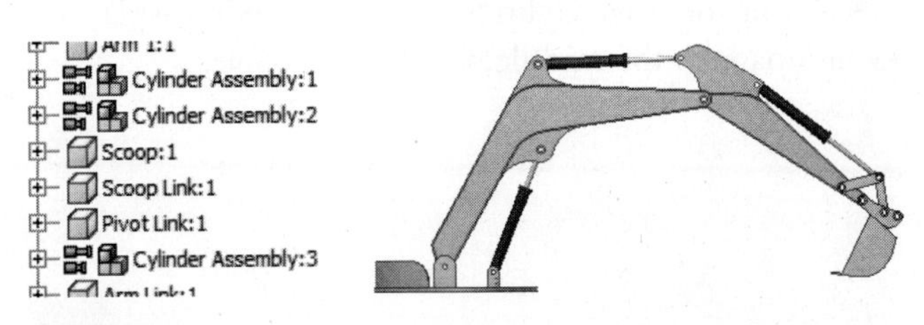

FIGURE 9.29

EXERCISE 9-1: FLEXIBLE ASSEMBLIES

In this exercise, you set the Flexible property for two subassemblies so that they can move independently in an assembly file.

1. Open *ESS_E09_01.iam*.

FIGURE 9.30

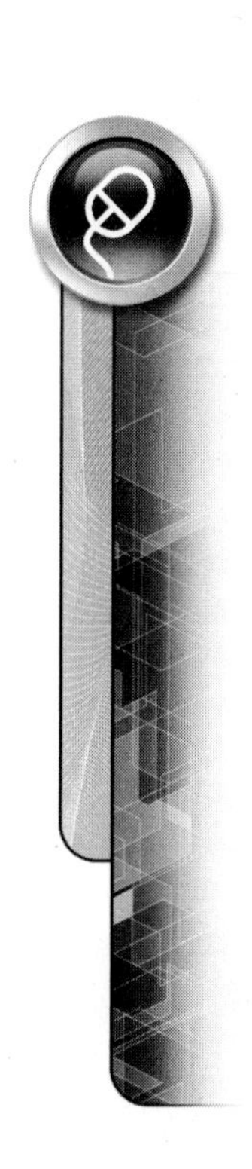

2. In the canvas, click and drag either of the Tube Assemblies. Notice that they are fully constrained and cannot move.
3. In the browser, double-click ESS_E09_01-TubeAssembly:1 to activate the subassembly for edit.
4. Expand ESS_E09_01-BaseTube:1. This component has an assembly constraint named Tube Reach with an offset distance set to 1.969 in that is preventing the base tube from moving.
5. In the browser, right-click the Tube Reach constraint below component ESS_E09_01-BaseTube:1 and click Suppress from the menu, as shown in the following image.

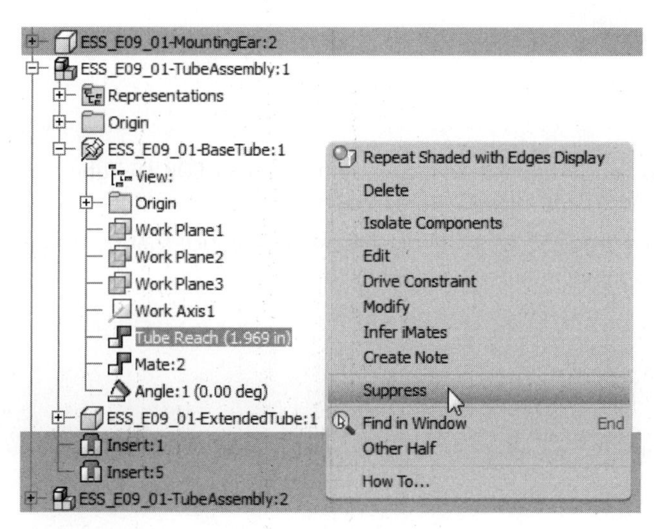

FIGURE 9.31

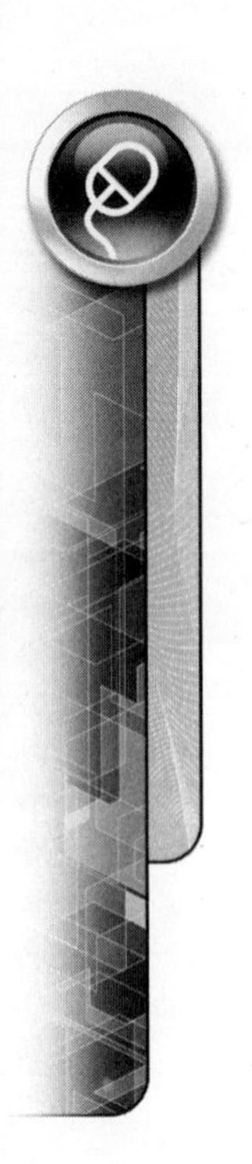

6. Click the Return command from the Assemble tab > Return panel.
7. In the browser, right-click ESS_E09_01-TubeAssembly:1 and click Flexible from the menu. Notice that a flexible icon is displayed next to the component, as shown in the following image.

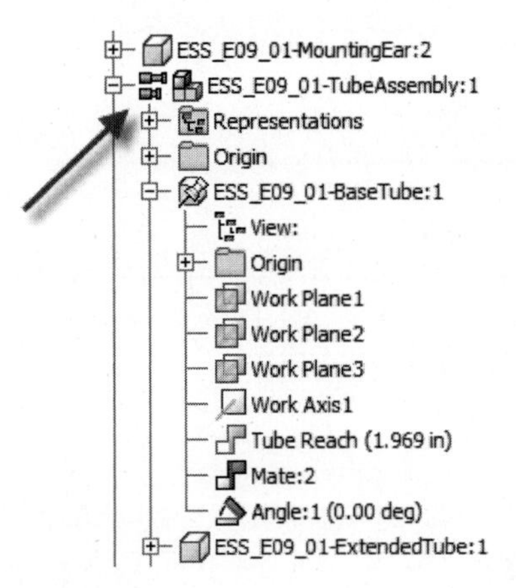

FIGURE 9.32

8. In the canvas, click and drag ESS_E09_01-TubeAssembly:1. Notice that with the subassembly set to flexible, it is free to move, but notice that the second occurrence of the subassembly remains in the same position.

FIGURE 9.33

9. To enable the second Tube Assembly to move, right-click ESS_E09_01-Tube Assembly:2 in the browser and click Flexible from the menu.
10. In the canvas, click and drag ESS_E09_01-TubeAssembly:2. Notice that both subassemblies are free to move independently of each other.

FIGURE 9.34

11. Close the file. Do not save changes. End of exercise.

POSITIONAL REPRESENTATIONS

Positional representations can be used to override aspects of an assembly and to create different positions of an assembly. All positional representations are saved with the assembly model. The gripper mechanism shown in the following image exemplifies how positional representations are created and applied. One of the gripper fingers is assembled using an angle constraint. The opposite gripper finger is assembled using another angle constraint controlled by a parameter. When the angle constraints are driven, both grippers move in opposition to each other. In this example, positional representations are used to show the gripper mechanism in its closed, middle, and open states.

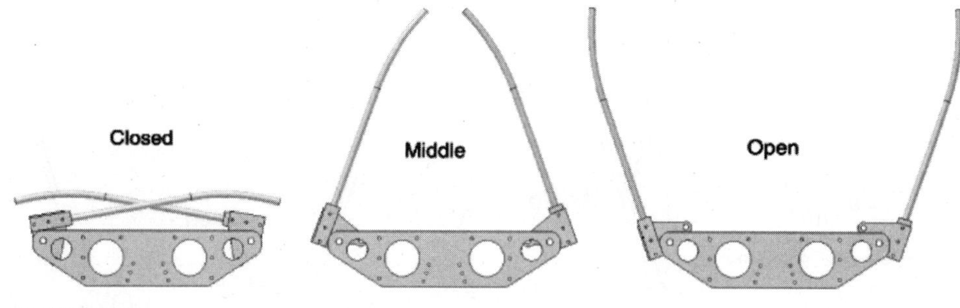

FIGURE 9.35

Creating Drawing Views from Positional Representations

When generating a base drawing view, you can specify the name of a positional representation for the view. To accomplish this, click on the Base View command on the Place Views tab > Create panel. This action will launch the Drawing View dialog box. Select the desired positional representation from the Position area in the dialog box. A drawing view will be created based upon the selected positional representation.

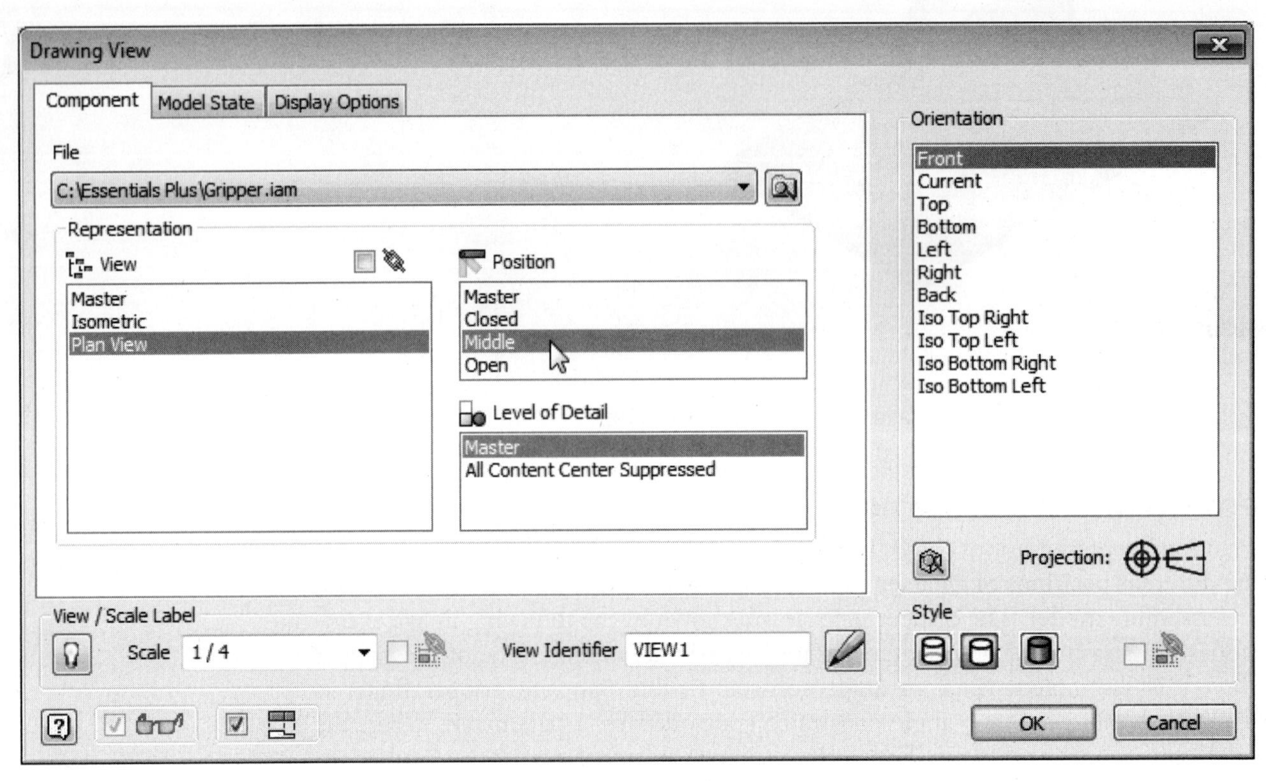

FIGURE 9.36

In the following image, three separate base views have been created from their corresponding positional representations: Closed, Middle, and Open.

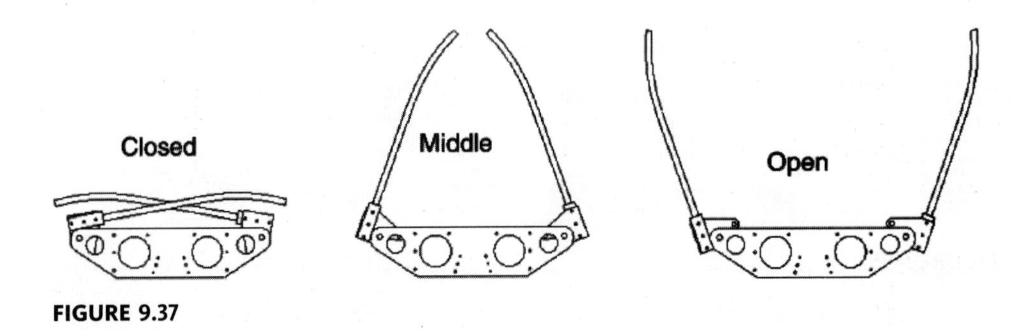

FIGURE 9.37

CREATING OVERLAY VIEWS

An alternate way to utilize positional representations in a drawing is to create an overlay view to show an assembly in multiple positions in a single view. Overlays are available for unbroken base, projected, and auxiliary views. Each overlay can reference a view representation independent of the parent view. Before creating an overlay view, create as many positional representations as you need to show your assembly in various positional states. It is considered good practice to create a number of view representations in order to focus on certain components and to reduce potential clutter in the drawing view.

Use the following steps to create overlay drawing views:

1. On the Place Views tab > Create panel, click the Base command, and position a view in your drawing. The following image illustrates a hydraulic cylinder set to its closed position. It would be advantageous to also show the cylinder in its fully opened position in this base view.

FIGURE 9.38

2. On the Place Views tab > Create panel, click the Overlay command, as shown in the following image on the left, and click the base view that was created in Step 1. If there is only one view on the sheet, you do not need to select it and the Overlay View dialog box appears immediately after clicking the Overlay command.
3. When the Overlay View dialog box is displayed, as shown in the following image on the right, click in the Positional Representation box, and select the desired representation. Click OK.

NOTE

A positional representation can be used only once per parent view.

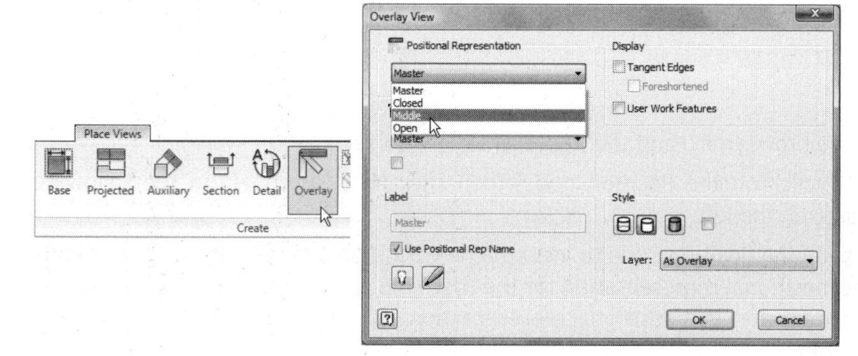

FIGURE 9.39

4. The new positional representation is added to the base view and is represented as a series of hidden or dashed lines, as shown in the following image. Dimensions can be added between the parent and the overlay.

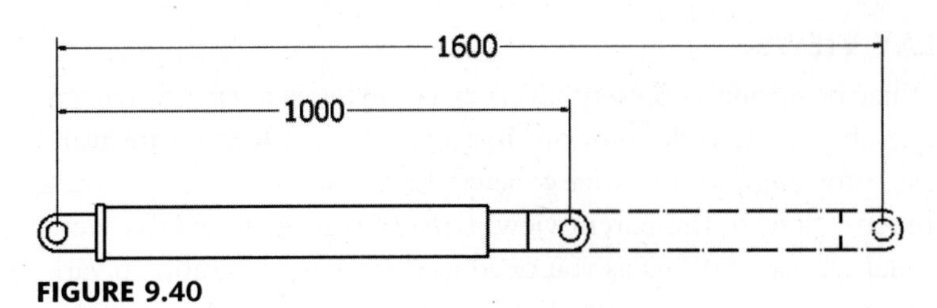

FIGURE 9.40

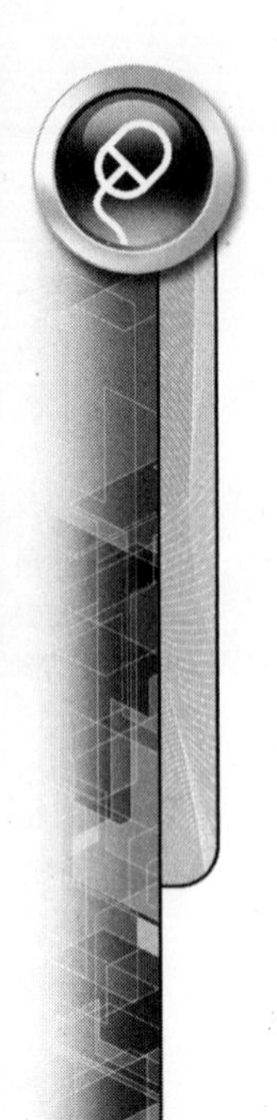

EXERCISE 9-2: POSITIONAL REPRESENTATIONS

In this exercise, you create positional representations of an assembly.

1. Open ESS_E09_02.iam.

   An angle constraint has been applied between one of the plates and the bracket that holds the gripper finger. The other side of the gripper finger is driven with a similar angle constraint that has a parameter assigned with a negative angle value, as shown in the following image on the left. In this way, the gripper fingers close and open in opposing directions, as shown in the following image on the right.

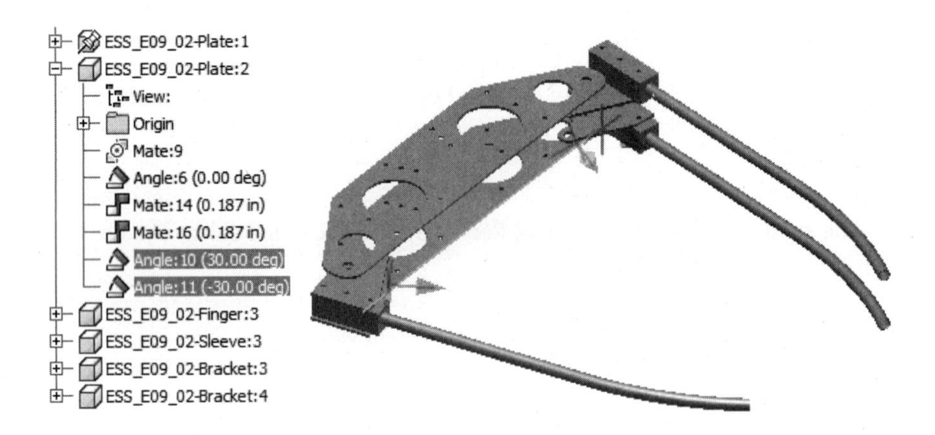

FIGURE 9.41

2. In the browser, expand the Representations folder.
3. Right-click on the Position node, then click New, as shown in the following image on the left.
4. Expand the Position node to view the new positional representation. Creating a new positional representation for the first time creates a default representation called Master. This positional representation displays the current position of the assembly.
5. The default name of the new positional representation is Position1. Rename it by slowly double-clicking on the name and entering **Closed**, as shown in the following image on the right.

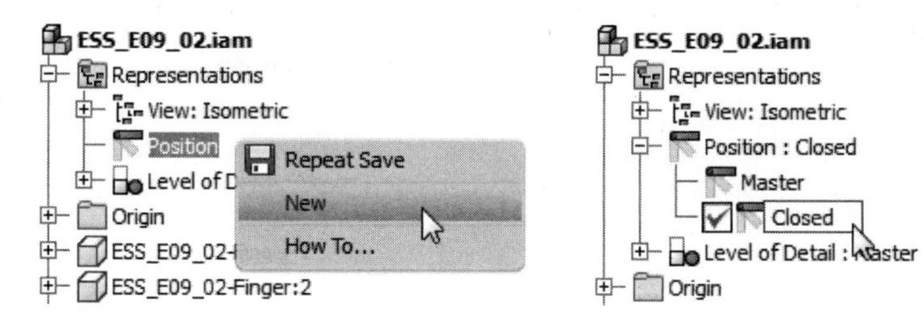

FIGURE 9.42

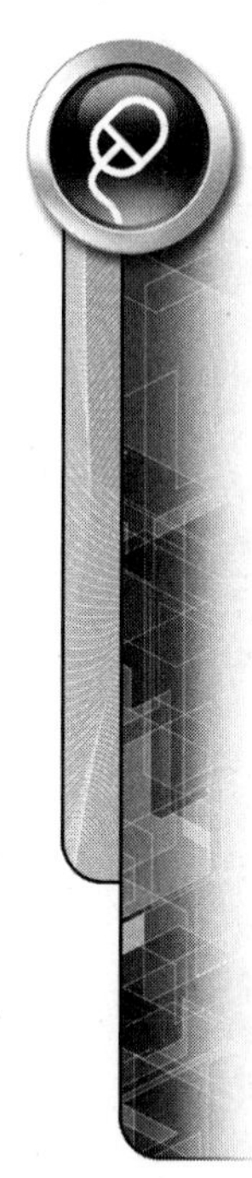

6. Next locate the constraint in the browser that will be affected by the positional representation. Expand *ESS_E09_02-Plate:2.*
7. Identify the constraint Angle:10 (30.00 deg). This is the constraint that will be affected by the positional representation.
8. In the browser, right-click the Angle:10 (30.00 deg) constraint, and select Override from the menu as shown in the following image on the left.

   The Override command is used to modify the solved state of the positional representation. You need to make two changes in order for the positional representation to function. First, the constraint needs to be enabled. Second, an offset value needs to be set.
9. On the Constraint tab of the Override Object dialog box, check the boxes next to Suppression and Value, as shown in the following image on the right.
10. Change the value to **100.00 deg**, as shown in the following image on the right.

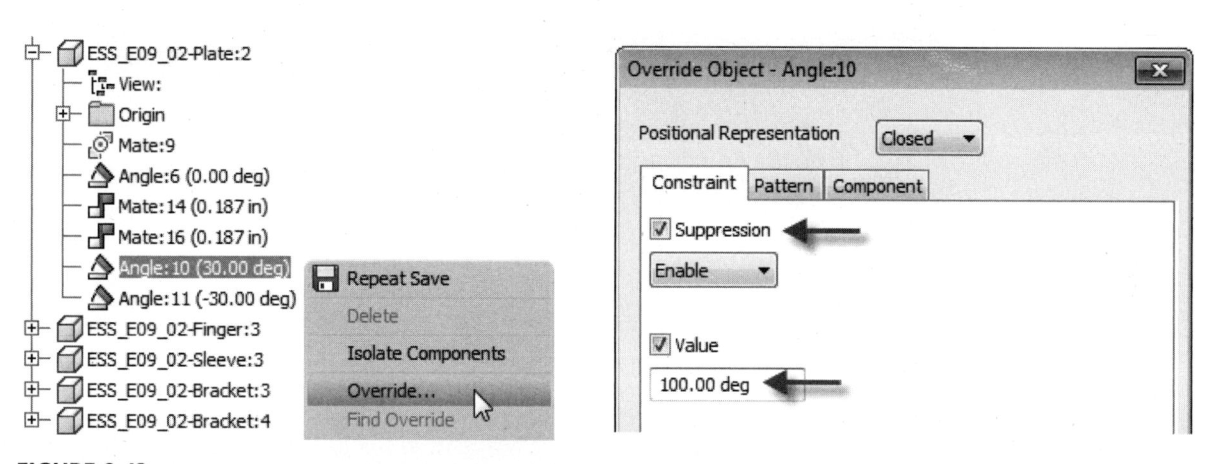

FIGURE 9.43

11. Click the OK button.

NOTE

The default master positional representation has the Angle:10 constraint set to 30 degrees. This master state displays the gripper somewhat open. Changing the angle to 100 degrees, as shown in the previous image, will close the gripper arm.

12. The Angle:10 constraint has been overridden, and this action closes the gripper, as shown in the following image. Notice in the browser that the text for the Angle:10 constraint is bold, emphasizing this change.

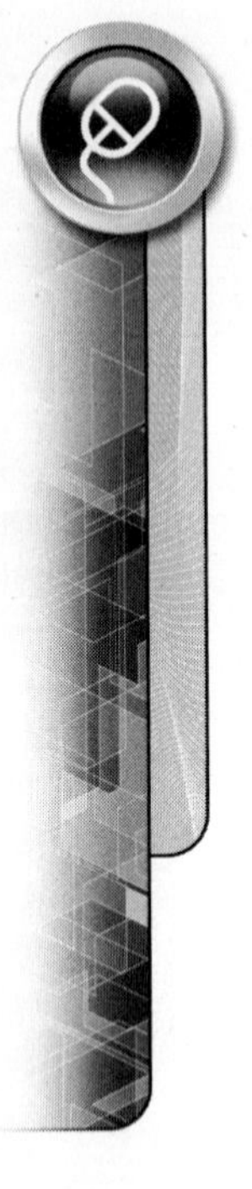

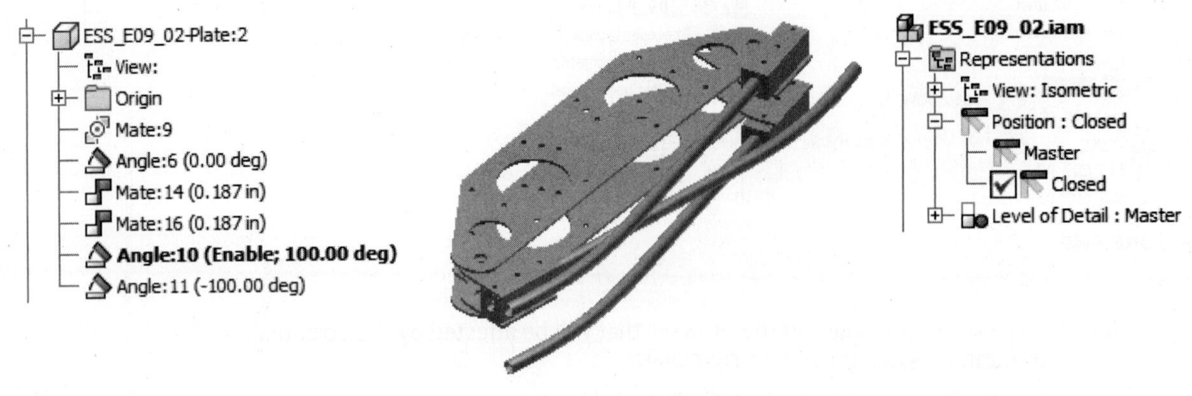

FIGURE 9.44

13. In the browser, expand the Representations folder and the Position node. Check the functionality of the positional representations by double-clicking on Master and Closed. The gripper assembly should update to reflect the active positional representation.
14. Switch from the Model browser to the Representations browser by clicking on the arrow next to Model and selecting Representations from the list, as shown in the following image on the left.
15. The Representations browser is displayed. Expand all items in this area of the browser, as shown in the following image on the right.

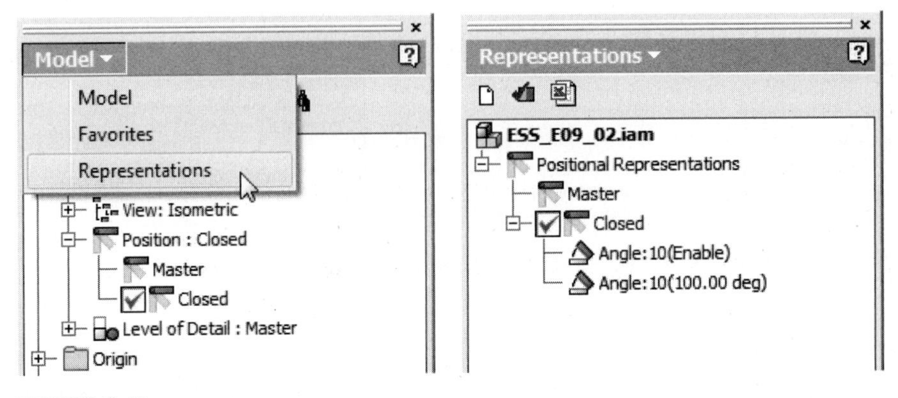

FIGURE 9.45

The Representations browser allows you to examine all positional representations associated with an assembly. It also shows the overrides for each positional representation.

16. Two additional positional representations will now be created to show the gripper in its middle and fully opened positions. Existing positional representations can be copied and then modified in order to display other versions of an assembly. In the Representations browser, check that the Positional Representations mode is expanded.
17. Right-click on the Closed positional representation, and select Copy from the menu as shown in the following image on the left.
18. A new positional representation called Closed1 has been created from the existing positional representation, as shown in the middle of the following image.

19. In the Representations browser, expand both positional representations. Notice that all overrides are copied into the new positional representation, Closed1, as shown in the following image on the right.

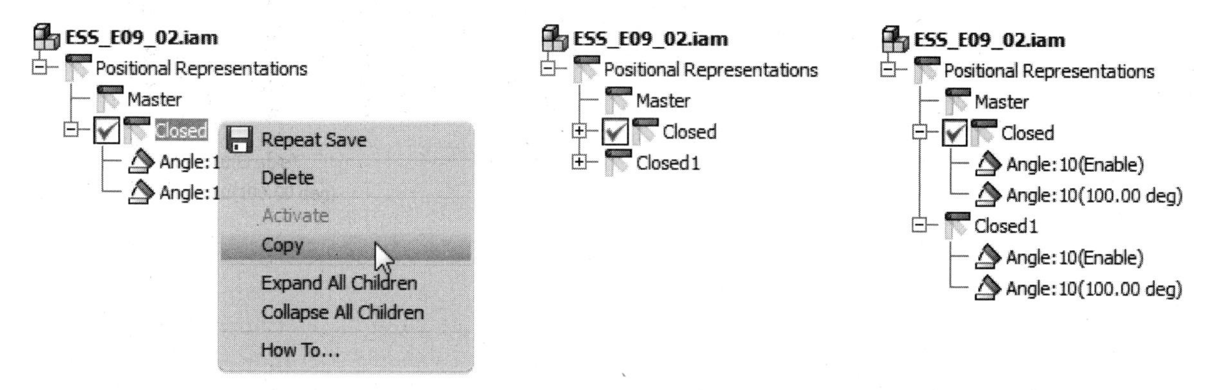

FIGURE 9.46

20. Create another positional reference copy. Base this copy on the existing Closed positional representation. When all copies have been made, three positional representations should be visible in the Representations browser, as shown in the following image on the left: Closed, Closed1, and Closed2.
21. In the Representations browser, rename Closed1 to **Middle** and Closed2 to **Open**. Your Representations browser should appear similar to the following image on the right.

To rename a positional representation, first single-click on its name, then single-click a second time. This will launch the rename mode.

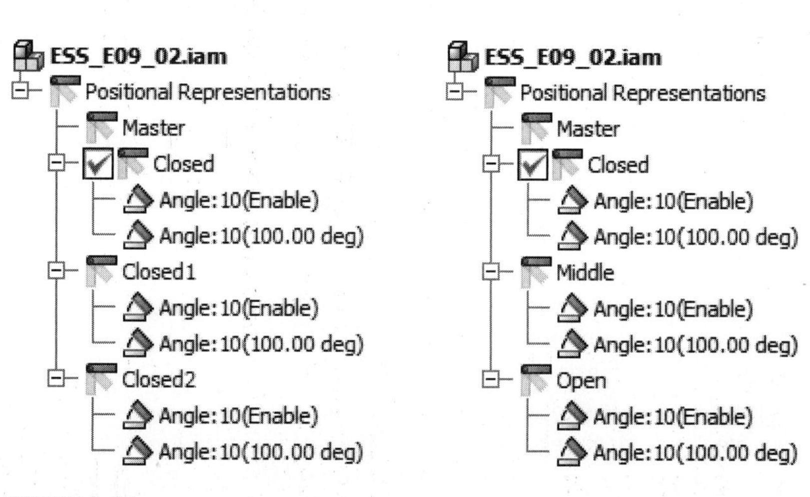

FIGURE 9.47

22. With the positional representations renamed, their values must now be modified in order to show different representations of the assembly. This step can be accomplished using a Microsoft Excel spreadsheet. To activate the Microsoft Excel spreadsheet, click on the Edit positional representation table button found at the top of the Representations browser, as shown in the following image on the left.
23. Clicking the Edit positional representation table button will open a Microsoft Excel table similar to the following image on the right. Notice that the master

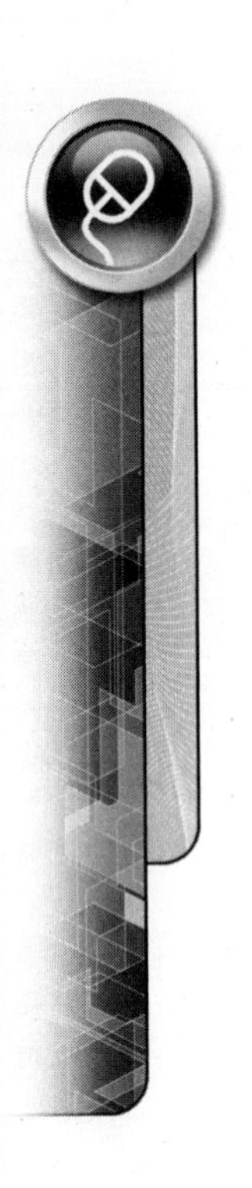

and all positional representations are displayed as a series of rows and columns. Notice also the degree values present in a separate column.

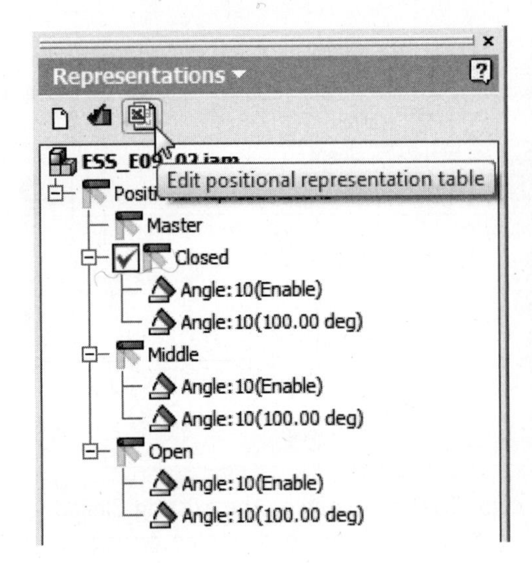

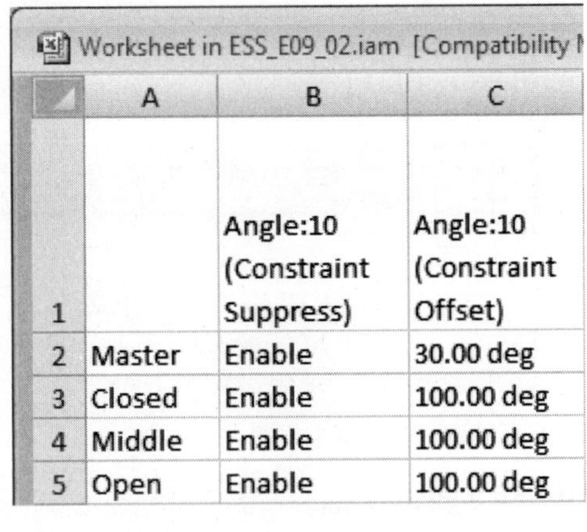
Worksheet in ESS_E09_02.iam [Compatibility M

	A	B	C
1		Angle:10 (Constraint Suppress)	Angle:10 (Constraint Offset)
2	Master	Enable	30.00 deg
3	Closed	Enable	100.00 deg
4	Middle	Enable	100.00 deg
5	Open	Enable	100.00 deg

FIGURE 9.48

24. Make the following modifications to this table:
    - For the Middle positional representation, change 100 degrees to **40** degrees.
    - For the Open positional representation, change 100 degrees to **0** degrees.

    Your table should appear similar to the following image on the left.
25. The changes you made in the Microsoft Excel spreadsheet need to be saved before you return back to the gripper assembly. From the Microsoft Excel File menu, click Close & Return to *ESS_E09_02.iam*, as shown in the following image on the right.

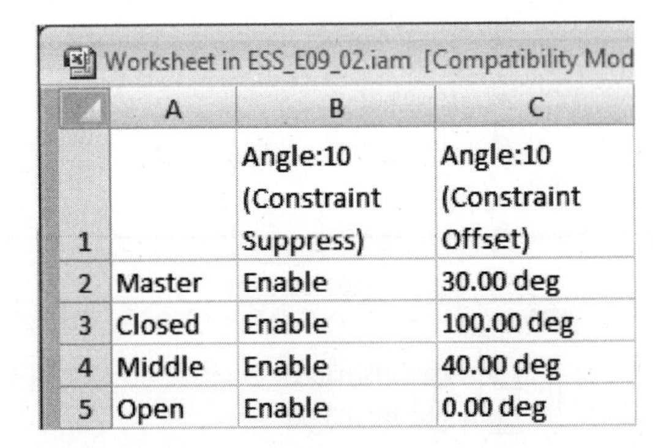
Worksheet in ESS_E09_02.iam [Compatibility Mod

	A	B	C
1		Angle:10 (Constraint Suppress)	Angle:10 (Constraint Offset)
2	Master	Enable	30.00 deg
3	Closed	Enable	100.00 deg
4	Middle	Enable	40.00 deg
5	Open	Enable	0.00 deg

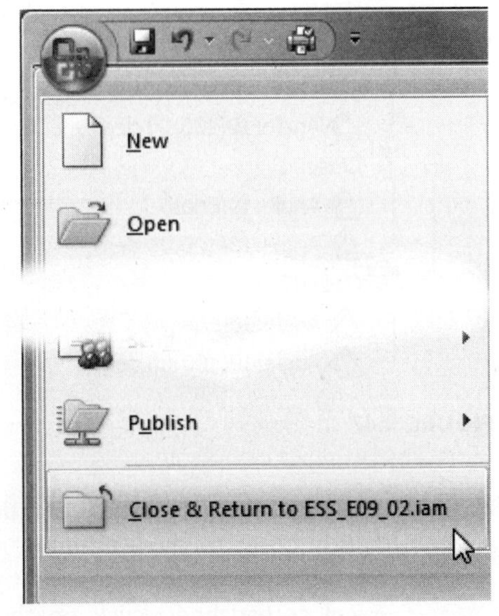

FIGURE 9.49

26. Saving the changes you made in the Microsoft Excel spreadsheet will update each angle value for the positional representations, as shown in the following image.

27. Activate each positional representation by double-clicking on its name in the Representations browser. Each should show a different state of the assembly, as shown in the following image.

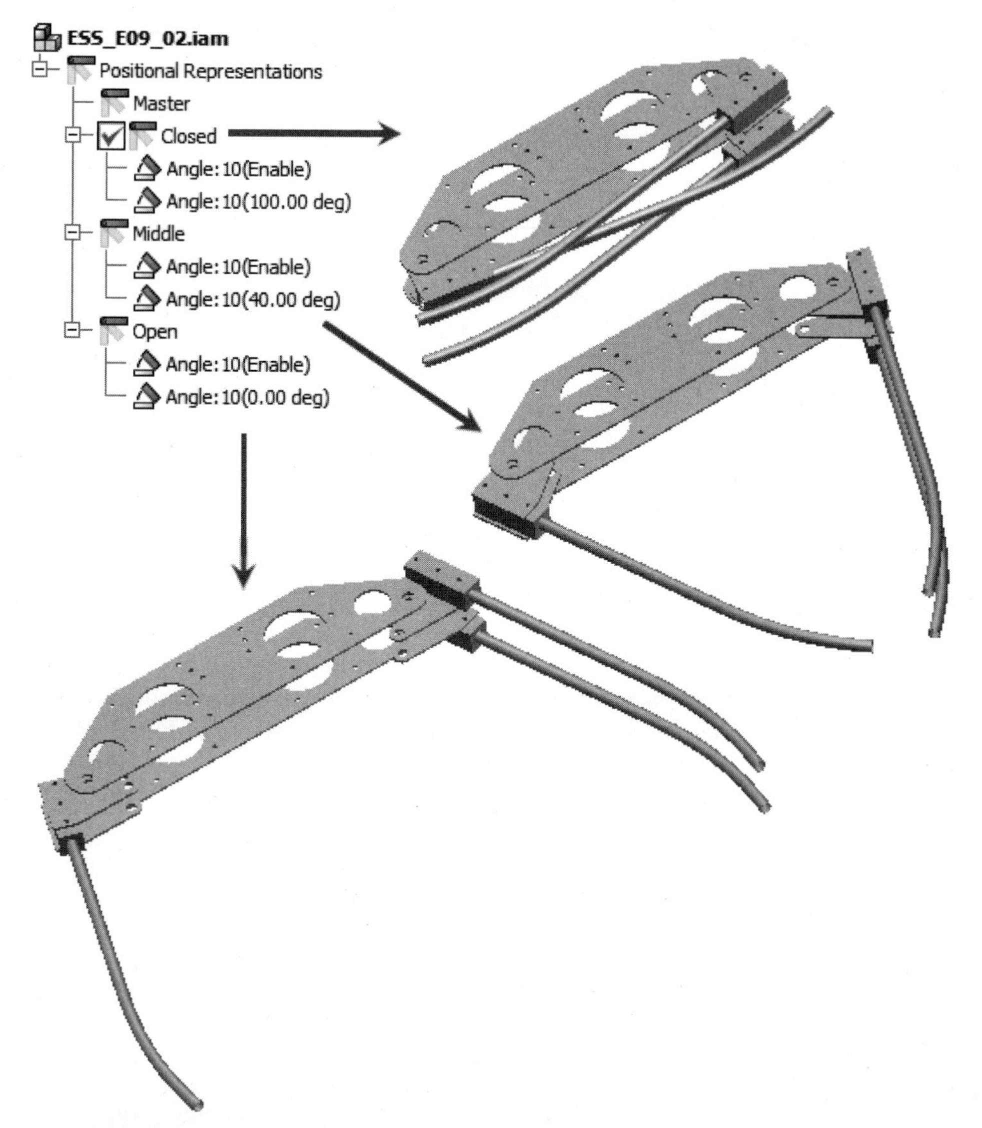

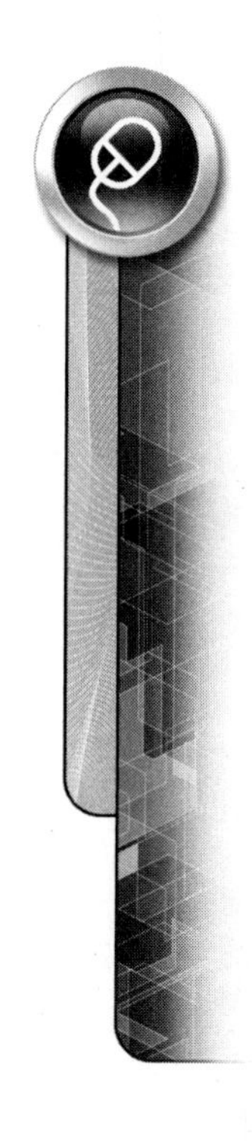

FIGURE 9.50

28. Double-click the Master positional representation to make it active.
29. From the File menu, click Save As.
30. Save the assembly file as *Gripper Assembly.iam*.
31. Create a new drawing file using the drawing template of your choice.
32. Click the Base command from the Place Views tab > Create panel, select the *Gripper Assembly.iam* file, choose Current for the orientation, and select the Open Positional representation. The view orientation can be set as desired.

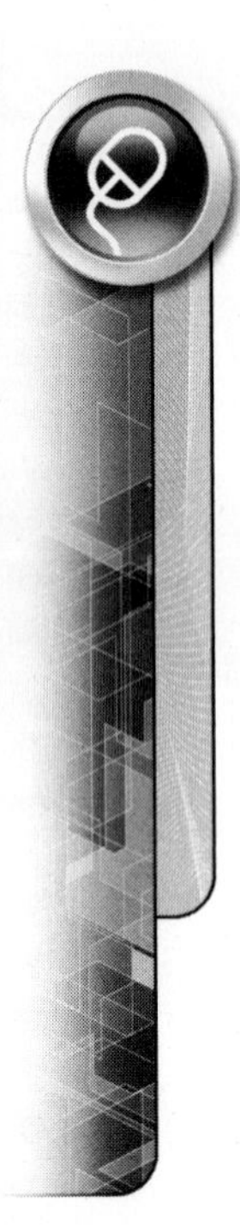

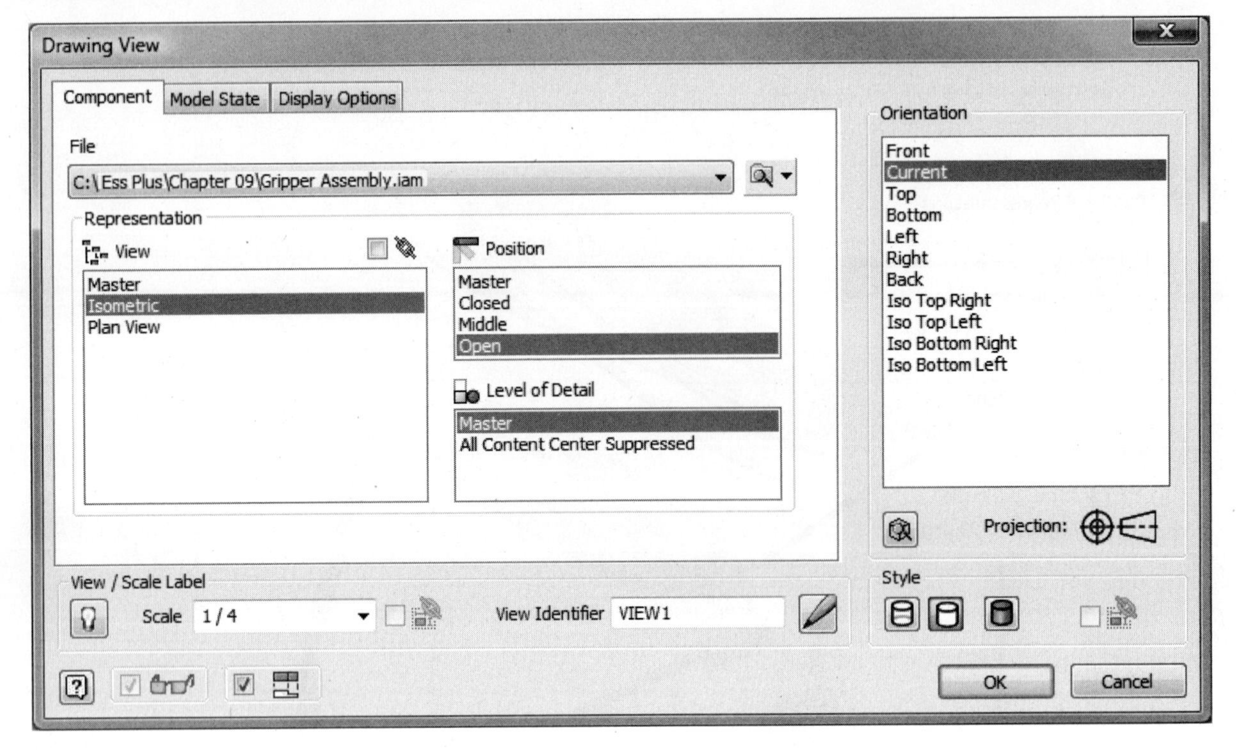

FIGURE 9.51

33. Click the drawing sheet to place the view on the sheet, then right-click and select Done from the menu. The base view should appear similar to the following image on the left.
34. Click the Overlay command from the Place Views tab > Create panel.
35. Select the Base view if necessary.
36. In the Overlay View dialog box, select the Middle positional representation, and click OK. The drawing view should look similar to the following image on the right.

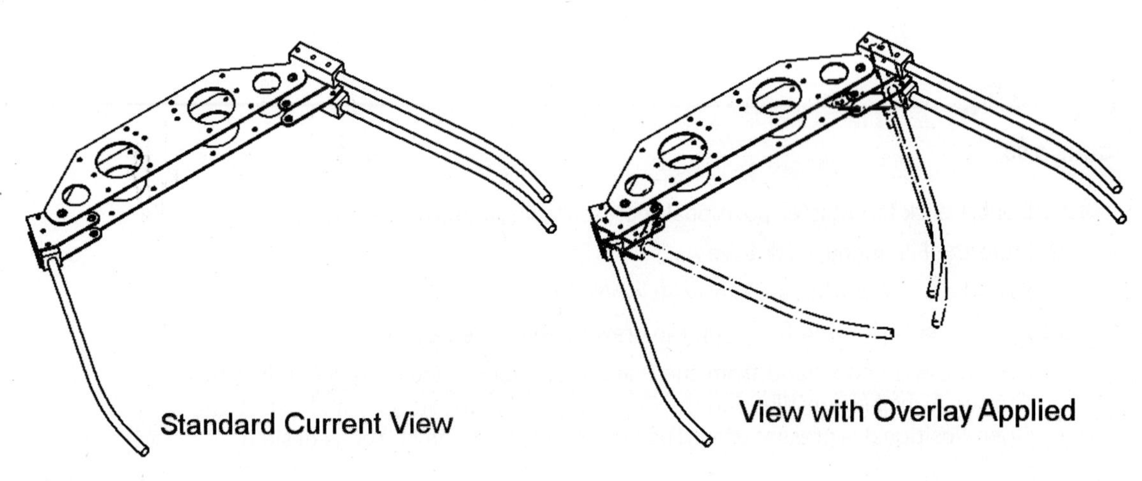

FIGURE 9.52

37. Add dimensions to the drawing view and experiment by placing the Closed positional representation with the Overlay command as desired.
38. Close all open files. Do not save changes. End of exercise.

CONTACT SOLVER

The Contact Solver determines how assembled components behave when a mechanical motion is applied. To get a better idea of what happens with the Contact Solver, study the three components shown in the following image. The base component, #1, has a single pin designed to run inside of the slot found on component #2. The second component, #2, has two pins in addition to a single slot. The two pins are designed to run inside of the slots found on component #3. The third component, #3, consists of two slots cut through its wall.

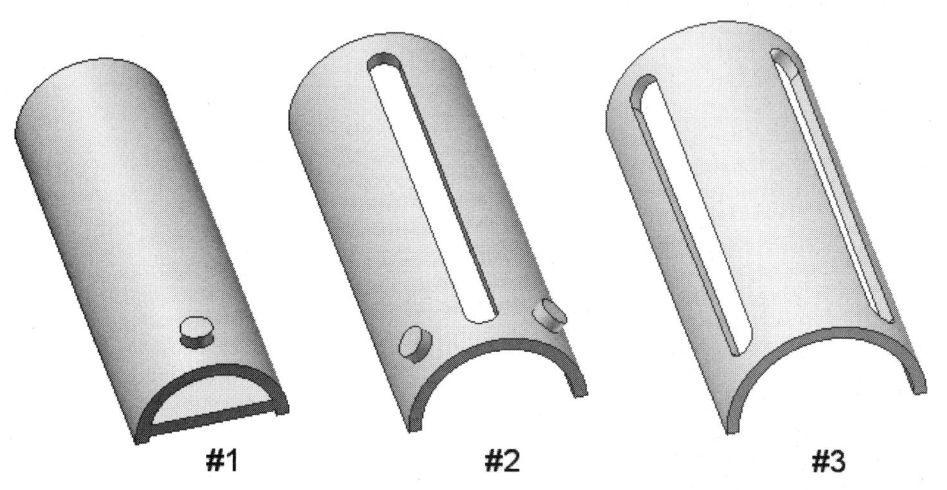

FIGURE 9.53

All three components are assembled with the freedom to translate along the common axis, as shown in the following image. Design intent describes that the pin features should stop at the end of a slot. However, the components continue to drag beyond the ends of the slots. The Contact Solver is based on your designation of certain components to be included in a contact set. This contact set will limit the motion of the objects, so they stop when the pin detects the end of a slot.

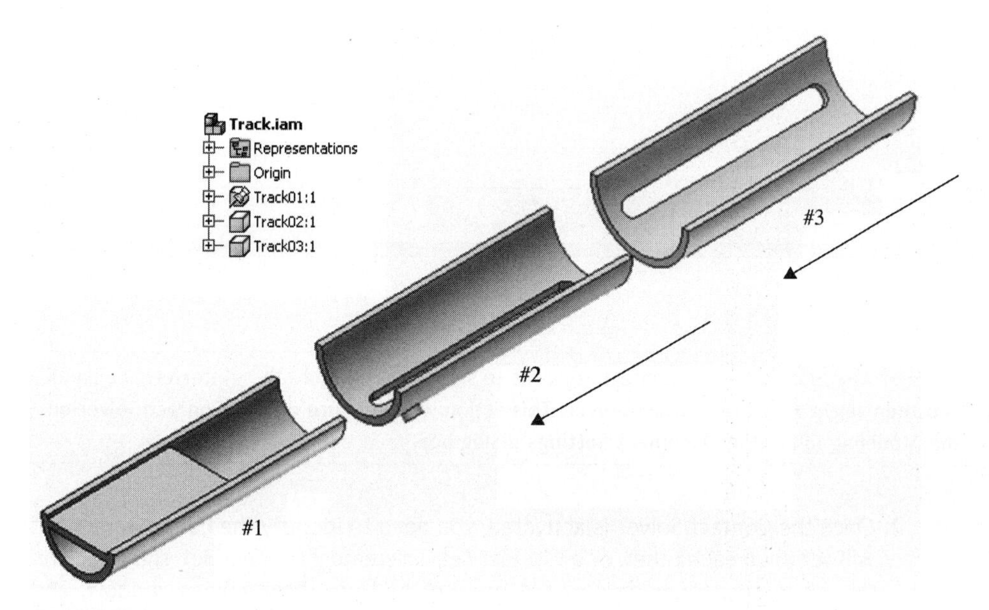

FIGURE 9.54

To designate components as part of a contact set and test a mechanism, follow these steps:

1. From the Tools tab > Options panel, click Document Settings. Click the Modeling tab, as shown in the following image. By default, the Contact Solver Off option is selected. You can change the option to All Components and set the contact solver to act on all components in the assembly. You can also change it to Contact Set Only and set it to act on specific, selected components.

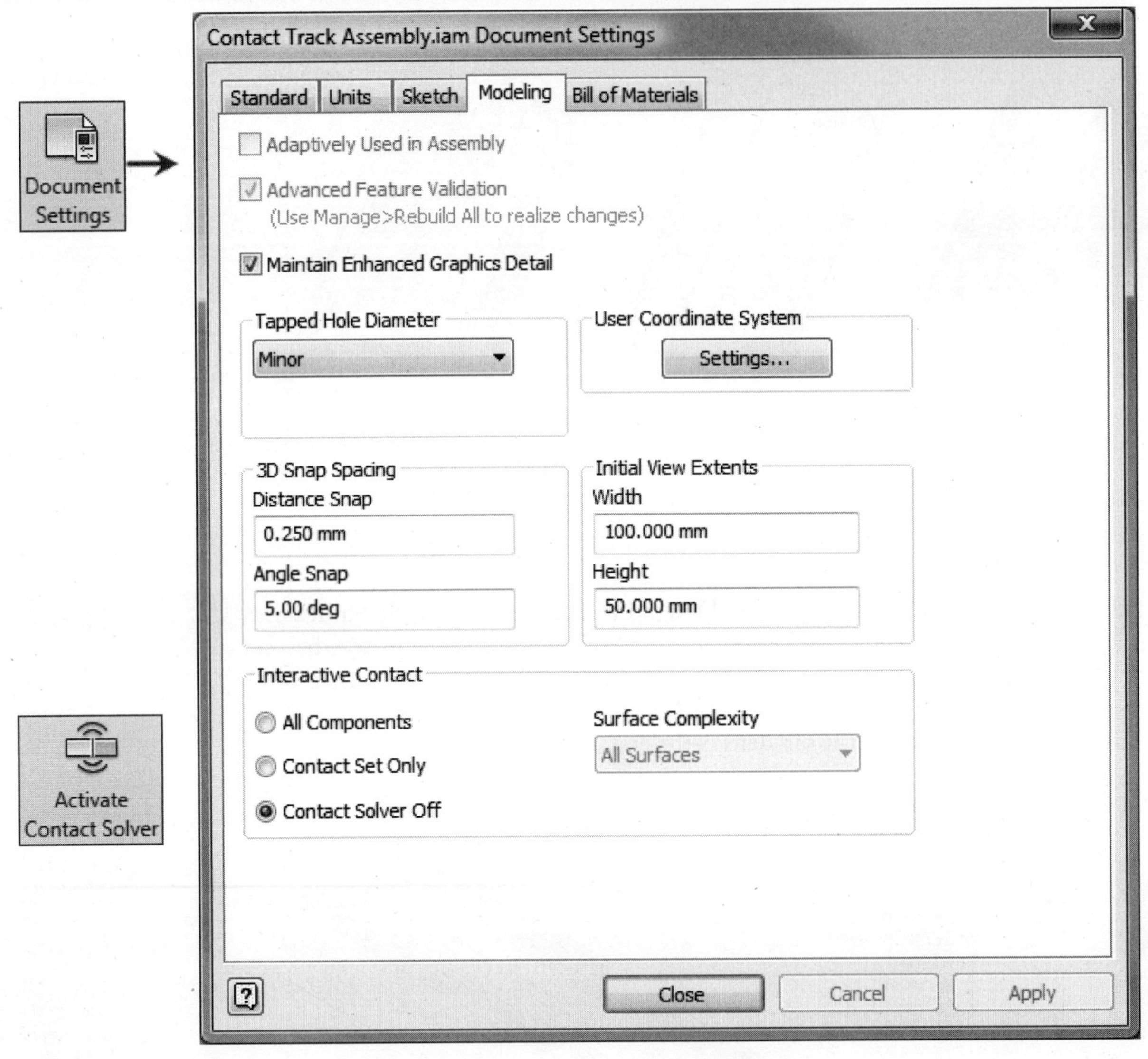

FIGURE 9.55

NOTE

Another way to activate the Contact Solver is to select the Inspect tab > Interference panel and then select Activate Contact Solver. This action will also turn on the Contact Solver on the Modeling tab of the Document Settings dialog box.

2. Once the Contact Solver is activated, you need to identify the components that will act upon each other, or a Contact Set. To create a Contact Set, right-click on one or more components in the browser or from the graphics window, and select Contact Set from the menu, as shown in the following image in the

middle. The assembly components defined as part of the Contact Set will be identified in the browser by a Contact Set Icon, as shown in the following image on the right.

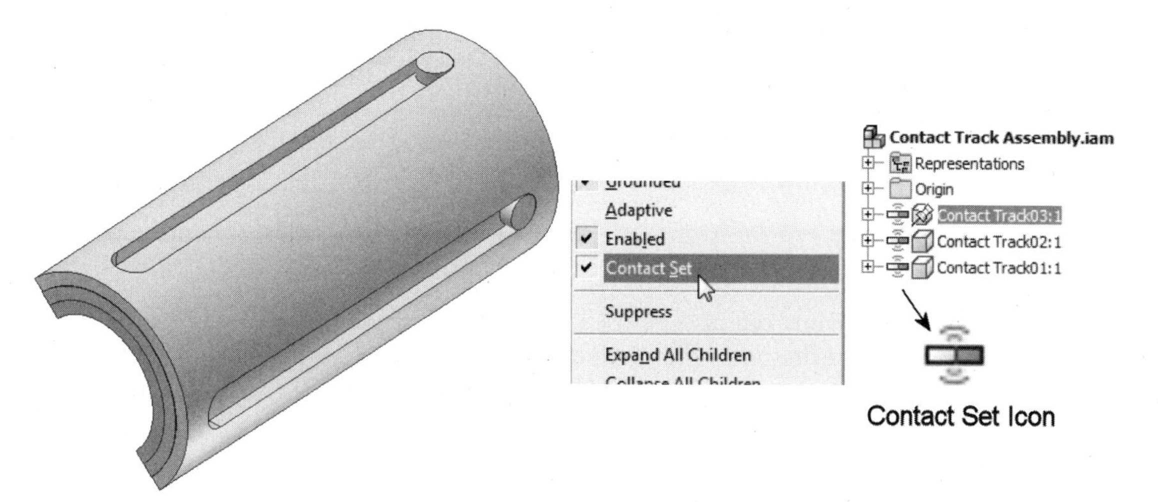

FIGURE 9.56

Another way of creating a Contact Set is to right-click on one or more components and select iProperties from the menu as shown in the following image on the left. When the iProperties dialog box appears, click on the Occurrence tab, and select Contact Set from the dialog box, as shown in the following image on the right.

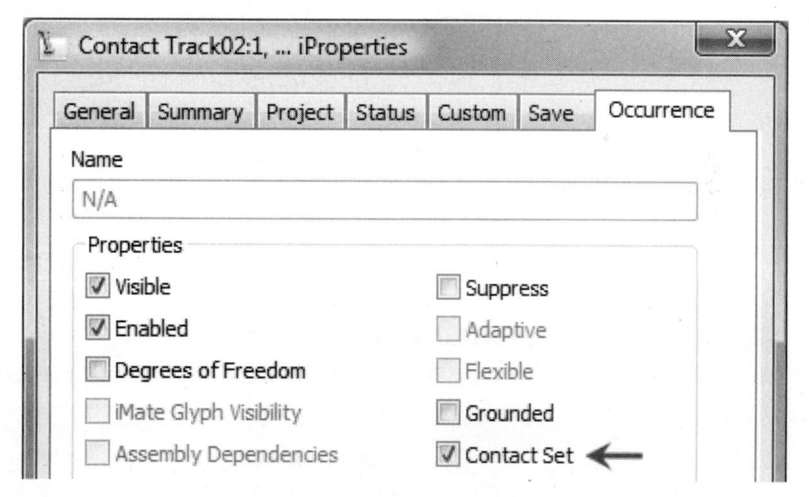

FIGURE 9.57

3. You can now test the mechanism based on the Contact Solver. As shown in the following image, begin dragging components through their intended range of motion. The pins should now stop when they contact the ends of the slots. If the pins do not stop, you may have to adjust the component positions and repeat the motion.

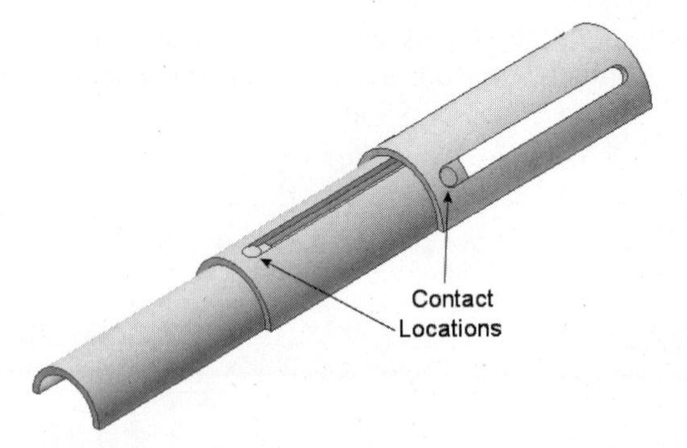

FIGURE 9.58

Drive constraints can also be used to produce intended motion. In the following image, a cam illustrates the motion of a Geneva drive mechanism as a pin comes into contact with a slot. Using the Contact Solver for these types of mechanisms usually produces dramatic results.

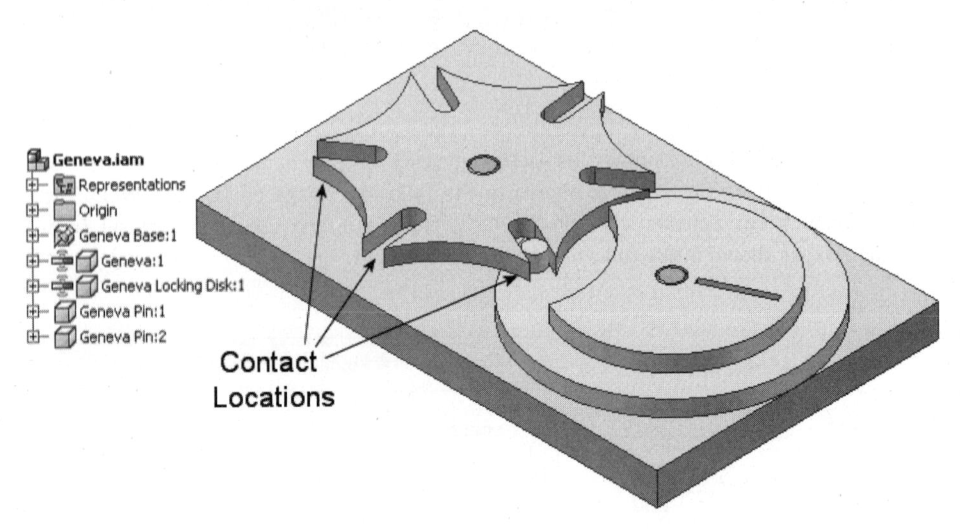

FIGURE 9.59

MIRRORING AN ASSEMBLY

To create right- and left-handed versions of an assembly and its components, use the Mirror command. Mirrored components can be created using one of two methods:

- **Create a new assembly:** A new assembly can be created by selecting the components of a source subassembly. A mirror copy is generated from the original subassembly. The new subassembly is created relative to a mirror plane.
- **Create instances of components:** Individual components can be selected in a current assembly to create mirrored instances or copies. Each new mirrored component generates a new file. The individual components are mirrored based on a mirror plane.

To begin, open the assembly that contains the components or subassemblies you want to mirror. The following image illustrates a portion of a hold-down clamping mechanism. The image represents half of the total assembly, and the Mirror command will be used to generate the other half of the clamp.

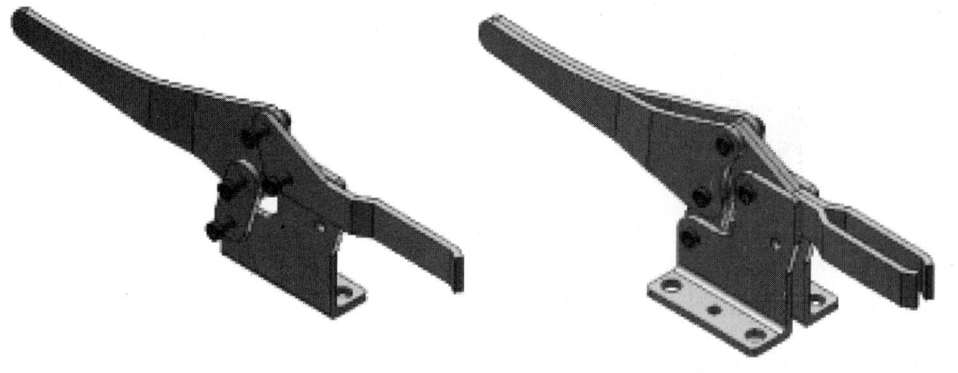

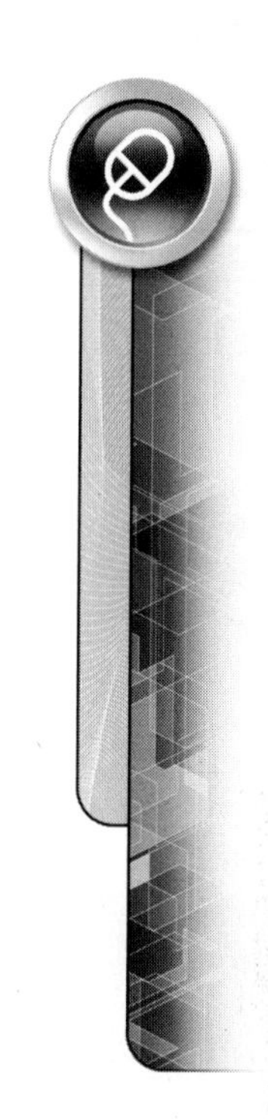

FIGURE 9.60

Mirroring Assembly Components

Follow this series of steps to create a mirrored group of components in an assembly:

1. From the Assemble tab > Component panel, click the Mirror command, as shown in the following image on the left. This will launch the Mirror Components Status dialog box as shown on the right.

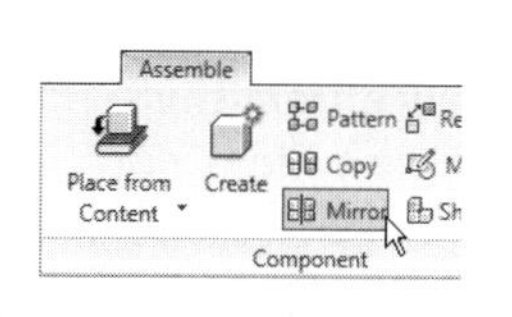

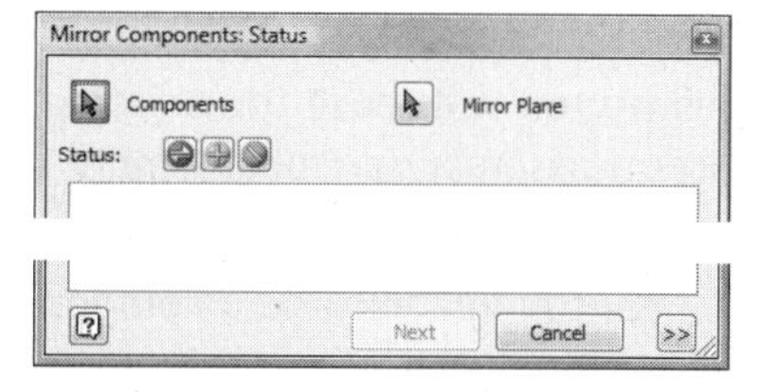

FIGURE 9.61

2. Select all components from the graphics screen or from the browser, as shown in the following image on the left. As each component is selected, it is added to the Mirror Components dialog box list as shown on the right.

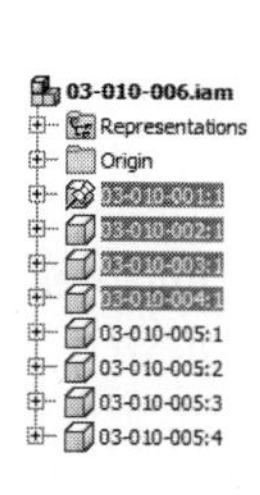

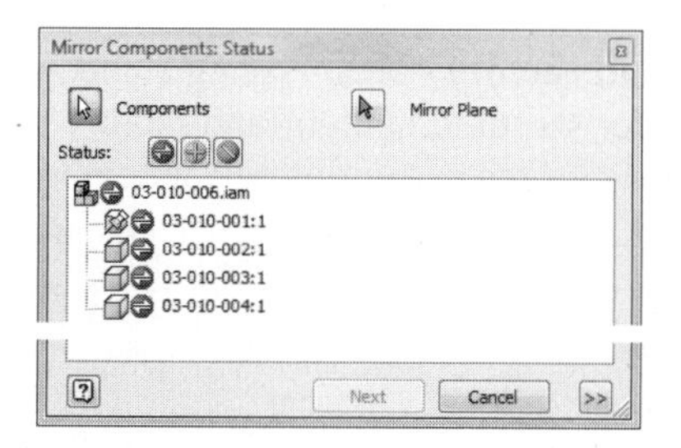

FIGURE 9.62

3. Next, click on the Mirror Plane button, if it is not already selected, and select a work plane or planar face from a part on the assembly or a plane in the Origin folder. Once the components and mirror plane are selected, you will see a preview of the components to be mirrored, as shown in the following image.

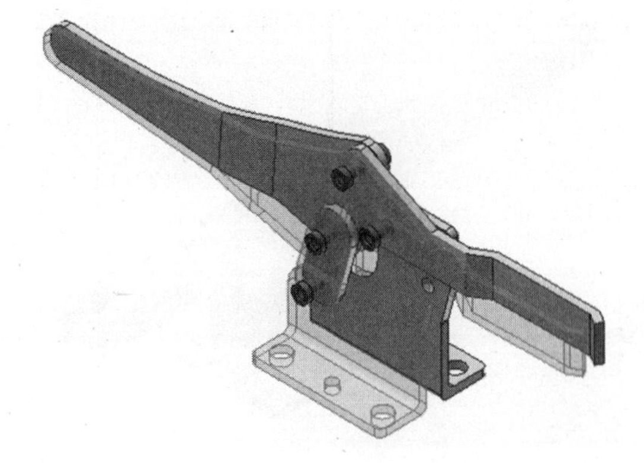

FIGURE 9.63

4. The following status buttons are available at the top in the Mirror Components dialog box. Click the status button next to a component to change its status based on the following information.

Symbol	Title	Function
	Mirrored	This button will create a mirrored instance of the component. A new file is created for the mirrored component.
	Reused	This button will reuse the existing instance from the current assembly or subassembly file. A new file is not created.
	Excluded	This button will exclude a component or subassembly in the mirror operation.
	Mixed Reused/ Excluded	This button indicates that a subassembly contains components with reused and excluded status. It could also mean that the reused subassembly is not complete.

5. Click on the More (>>) button to display previewing options and controls for handling content library and factory parts.
   - **Reuse Standard Content and Factory Parts:** Placing a check in this box will restrict the mirrored state for standard content and factory parts. Instances of the library parts are created in the current or new assembly file instead.
   - **Preview Components:** Place a check in each box to preview the mirrored, reused, or standard content on the canvas. Mirrored parts are previewed in green, reused parts in yellow.

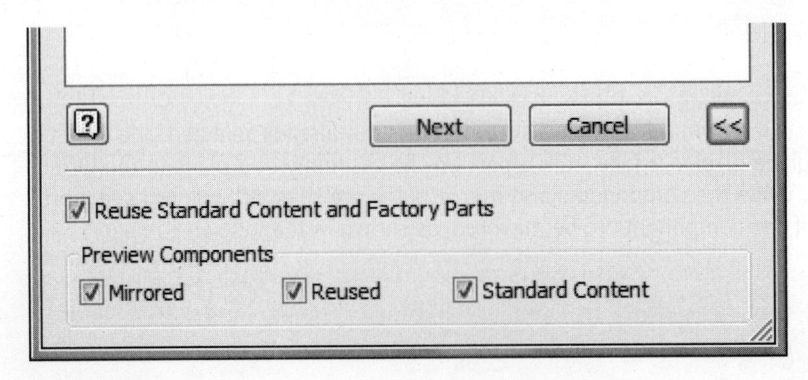

FIGURE 9.64

6. When you have finished making changes to settings, click the Next button to continue.
7. **The Mirror Components:** File Names dialog box appears. Use this dialog box to change the names of the mirrored components or to accept the default names, as shown in the following image. To change a part name, click on a current name under the New Name heading.

NOTE

You can also change the file location by right-clicking on Source Path under the File Location heading. However, it is considered good practice to keep the default file location in order to locate the mirrored components when reopening the assembly.

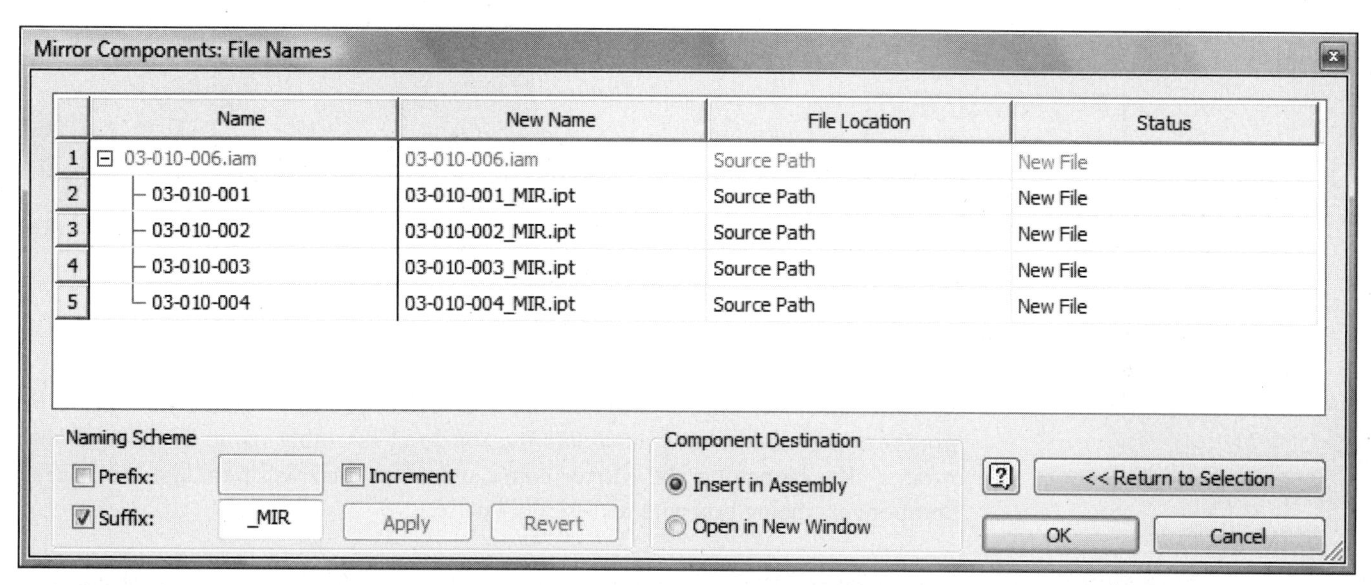

FIGURE 9.65

You can also control the listing of a prefix or suffix in the part name using the Naming Scheme area of the Mirror Components: File Names dialog box. Changing the _MIR listing to a different name activates the Apply button. Clicking on the Apply button will make the change to all component names under the New Name heading. If you want to return to the original values, you can activate the Revert button. Placing a check in the box next to Increment will increase all file name increments if a number is used as a Suffix instead of text.

You can also make changes in the Component Destination area. Clicking Insert in Assembly will place the mirrored components in the current assembly. Clicking Open in New Window will place the mirrored components in a new assembly file.

Clicking on the Return to Selection button will return you to the Mirror Components dialog box and allow you to change the status of the mirrored components or to select additional components. When you have finished making changes, click the OK button.

The results of the mirror operation are illustrated in the following image. All selected components were mirrored about a selected plane. The browser was also updated to reflect the new part additions and any additional instances of the existing components that were reused. In this image, the default _MIR suffix was added to the end of each part name.

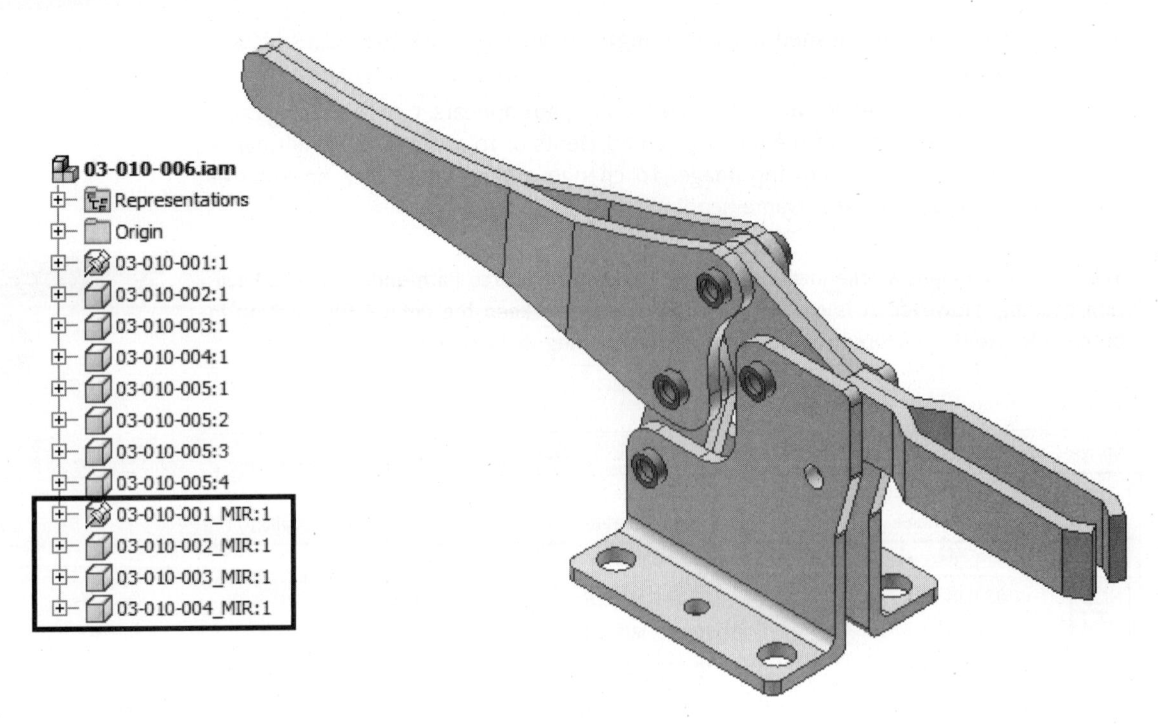

FIGURE 9.66

An entire assembly can be easily mirrored. To accomplish this task, follow the same process as described above but select the top-level assembly name in the assembly browser after launching the Mirror command. This step will populate the Mirror Components dialog box with all assembly parts.

NOTE

In addition to individual components and constraints, the following items will also be included in assembly mirror operations: Assembly Features, Patterns, iMates, and Work Features.

COPYING AN ASSEMBLY

You can copy assembly components or an entire assembly in the current design file. The process is the same as that outlined for mirroring components, but you click the Copy command, as shown in the following image on the left, from the Assemble tab > Component panel. The Copy Components dialog box is displayed as shown in the following image on the right.

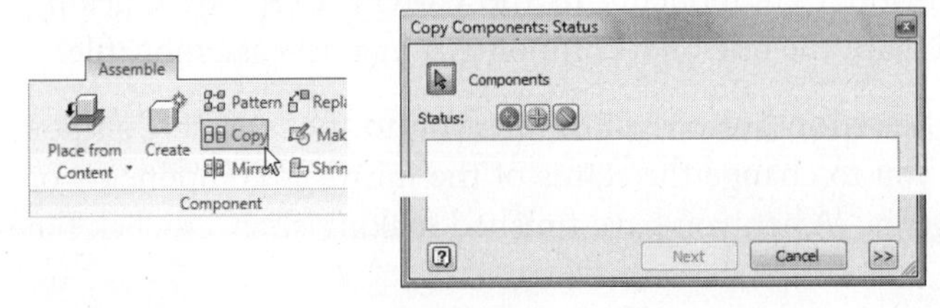

FIGURE 9.67

You then select the components or top-level assembly that you want to copy from the browser, as shown in the following image on the left. This will populate the Copy Components dialog box with the items to copy as shown in the following image on the right. This dialog box functions in a similar fashion to the Mirror Components dialog box previously discussed.

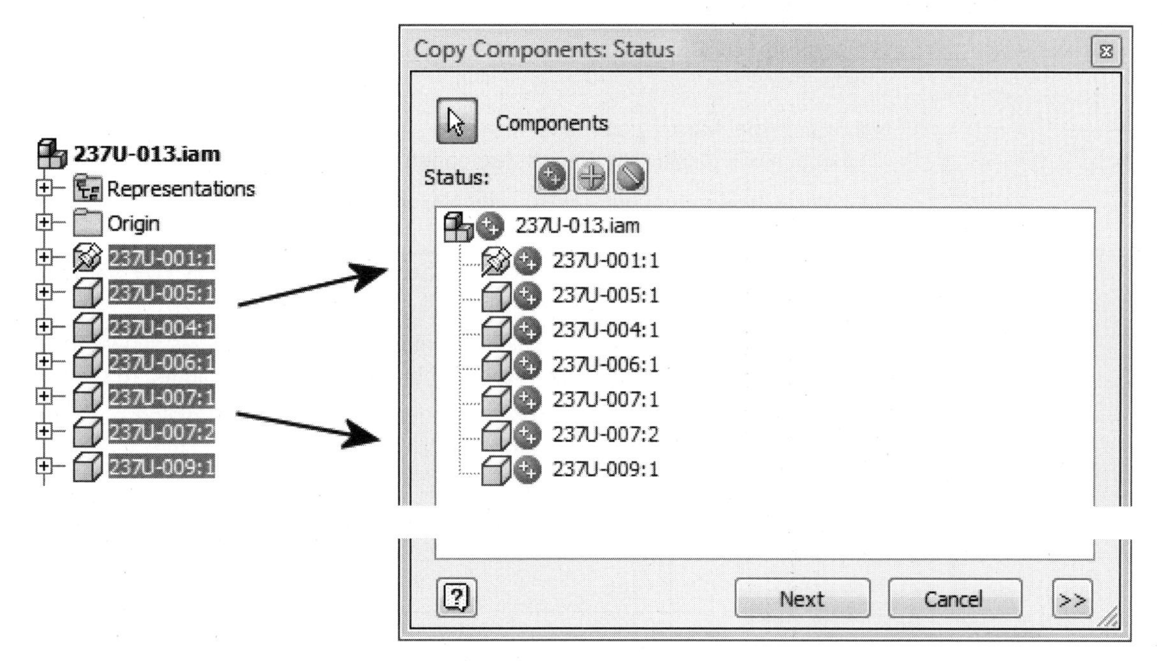

FIGURE 9.68

After you have finished selecting components and have made changes to the status buttons to copy, reuse, or exclude, the Copy Components: File Names dialog box can be used to change the names of the copied components or to keep the default names for the new files. The default names of the new files end with _CPY to stand out compared with already existing parts.

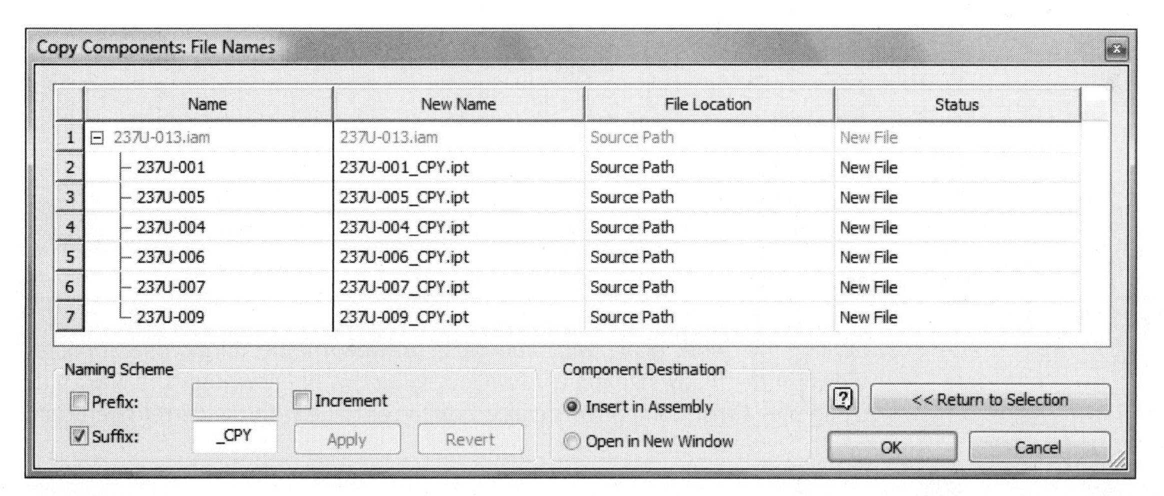

FIGURE 9.69

If you are using Autodesk Vault, there are some advanced tools that can be used to copy existing designs.

NOTE

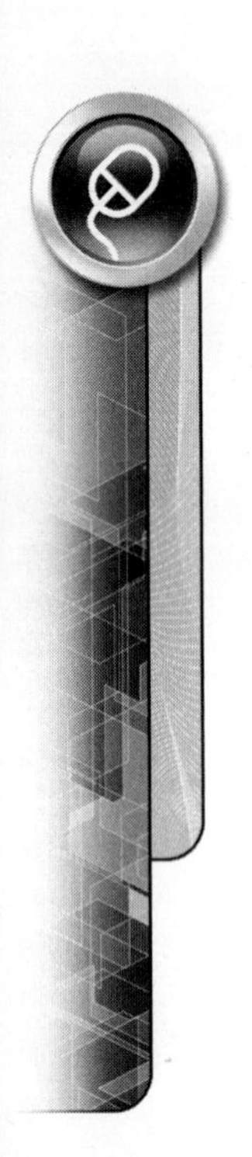

EXERCISE 9-3: MIRRORING ASSEMBLY COMPONENTS

In this exercise, you will create a mirrored copy of an assembly from an existing assembly.

1. Open *ESS_E09_03.iam*. This file consists of an assembly model and a predefined work plane.

   a. Click the Mirror command from the Assemble tab > Component panel, which will display the Mirror Components dialog box.

   b. With the Components button depressed, pick the top-level assembly from the browser (*ESS_E09_03.iam*), as shown in the following image on the left.

   c. Click the icon next to all the components except the ESS_E09_03-Body:1 component to the component status is set to reuse.

   d. Observe the results in the Mirror Components dialog box, as shown in the following image on the right. All assembly components are added to the list.

   e. Click the Mirror Plane button in the Mirror Components dialog box. Then select the visible work plane, on the canvas, as the mirror plane.

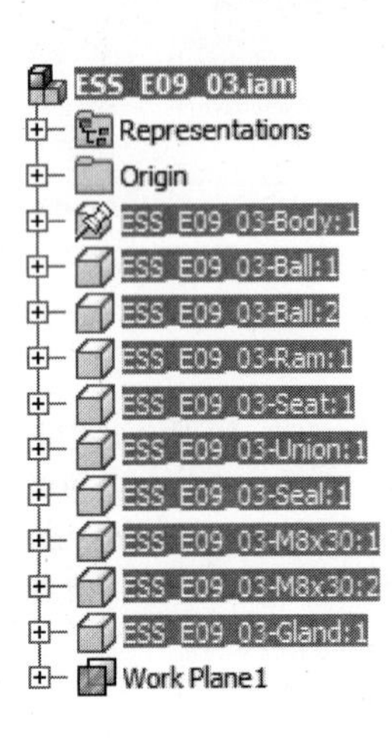

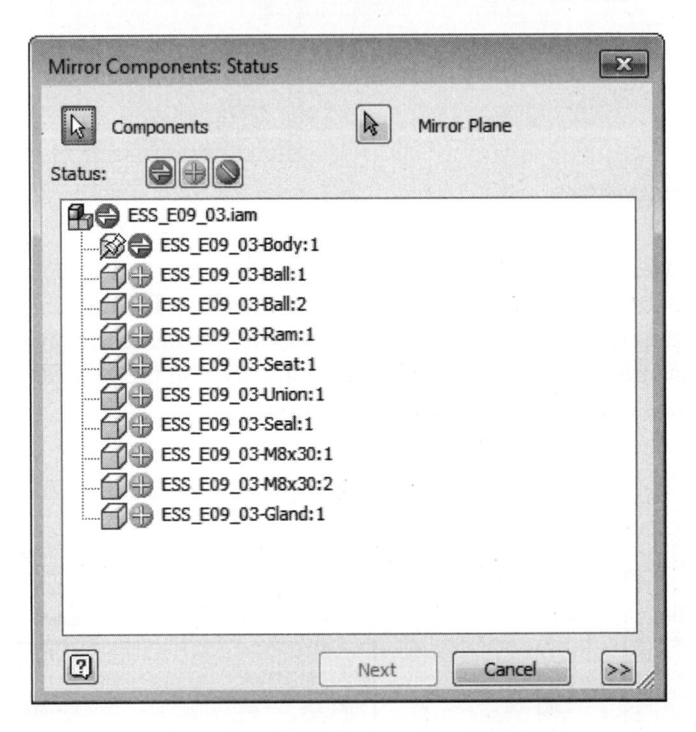

FIGURE 9.70

2. Clicking the work plane will produce a preview of the mirror operation, as shown in the following image. After reviewing this preview, click the Next button in the Mirror Components dialog box to continue with the mirror operation.

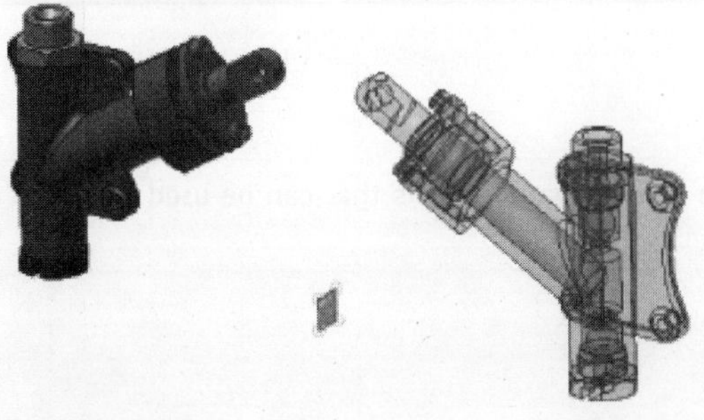

FIGURE 9.71

3. The Mirror Components: File Names dialog box is displayed, as shown in the following image. Notice that the mirrored part has its part number appended with an _MIR extension, which is designed to distinguish the mirrored parts from the original parts. Click the OK button in this dialog box to accept the default settings.

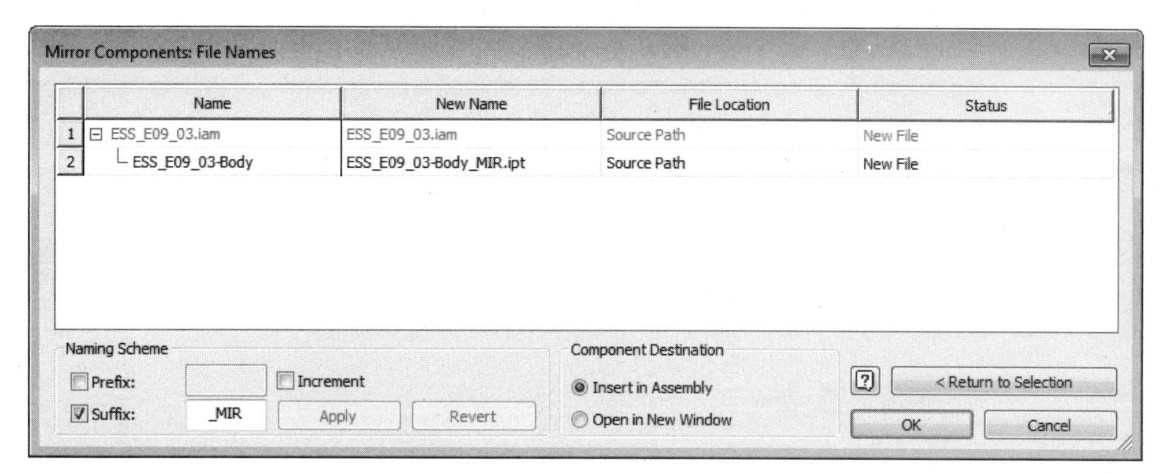

FIGURE 9.72

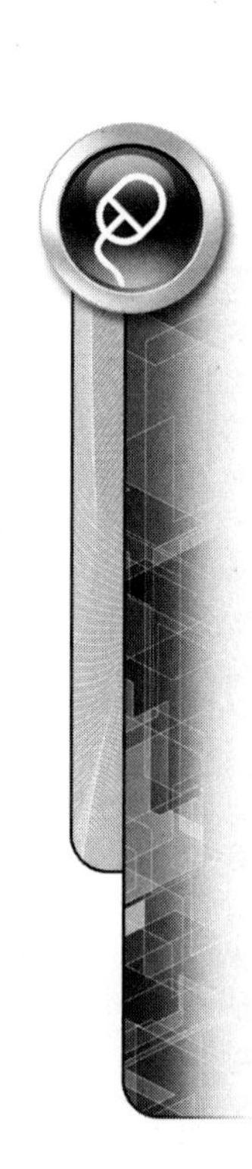

4. The results of the mirror assembly operation are shown in the following image.

FIGURE 9.73

5. Close all open files. Do not save changes. End of exercise.

ASSEMBLY WORK FEATURES

In the assembly environment, you can create work features to help you construct, position, and assemble components or subassemblies. You can create work planes and axes between parts in an assembly by selecting work features, faces, edges, or points on parts. These work features remain tied to each associated part and adjust accordingly as the assembly is modified. You can use assembly work features to parametrically position new components or subassemblies, to check for clearance in an assembly, and as construction aids. You can also use work planes to help you define section views of your assemblies.

Autodesk Inventor also allows you to globally turn off the visibility of work features. This is important in the assembly environment, where the display of work features from individual parts can quickly clutter the graphics window. Options for turning off work feature visibility are available by selecting the View tab > Visibility panel followed by Object Visibility, as shown in the following image.

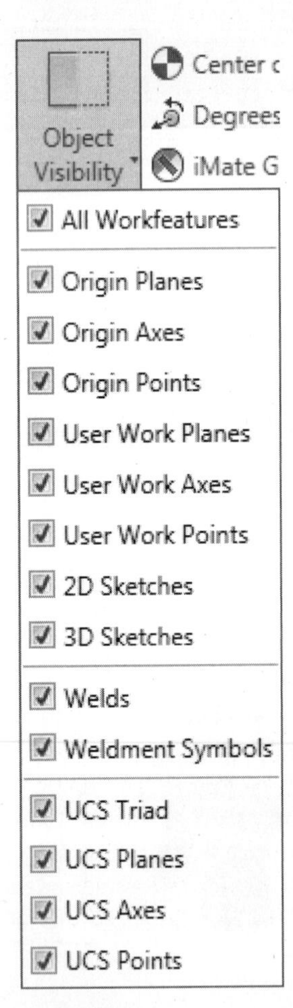

FIGURE 9.74

You can use these controls to turn off the visibility of work features by type. By default, all types of work geometry are initially selected for display. Thus, any work feature with its individual visibility turned on in the browser is visible in the assembly file.

To globally turn off the visibility of a particular type of work geometry, select it from the menu and clear the checkmark next to it. This action overrides the visibility setting for individual work features of that type in the assembly and in each part in the assembly. Although the components work features' visibility in the assembly is suppressed, their individual visibility control remains turned on in each of the component files. You can also control the visibility status of sketches, 3D sketches, welds, and weld symbols from the menu.

ASSEMBLY FEATURES

Assembly features are features that are defined in an assembly. They only affect a part when the part is viewed in the context of the assembly. Assembly features allow you to remove material from components after they have been assembled. Examples of material-removal processes include match-drilling operations and post-weld machining operations. Typical operations involving assembly features include cutting extrusions, drilling holes, and cutting chamfered edges.

In the following image of a post-machining operation, three blocks have been assembled through the use of mate-mate and mate-flush constraints. With all three items assembled, a feature in the form of two slots is then cut through the lower and upper faces of the assembly. A sketch is created on the front face of one of the parts, geometry is sketched and dimensioned, and the assembly feature is cut using an extrusion operation. Notice in the following image that in the completed part, a chamfer operation was applied at one end of the assembly.

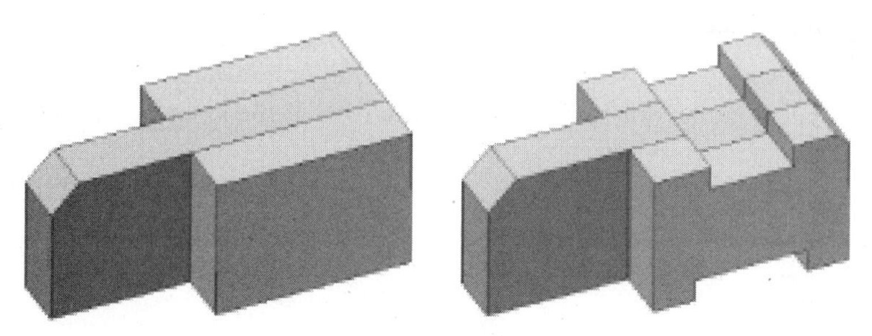

FIGURE 9.75

Assembly Sketches

Sketch geometry can be added in an assembly. You can sketch on a planar part face, a part's work plane, or an assembly work plane. Features can then be created from these sketches. Like creating features in part-modeling mode, you create a new sketch on a part or assembly plane or face—it does not matter which assembled part face or plane is selected for the sketch. This action will activate the Sketch tab.

When you create a sketch profile, geometry can be projected from various parts to the assembly, and constraints and dimensions are added, as shown in the following image.

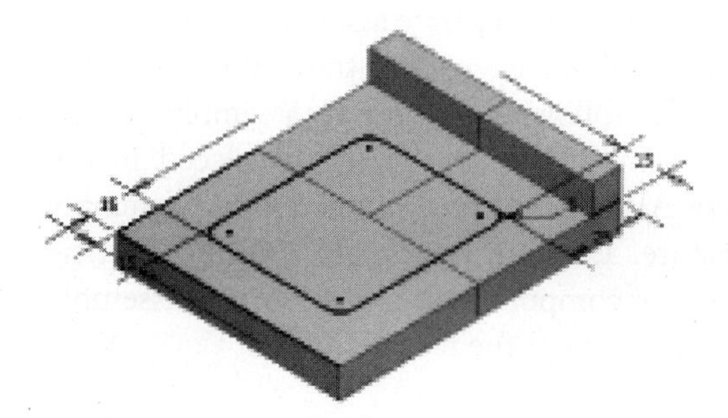

FIGURE 9.76

Creating Assembly Features

Once you have finished creating the desired sketch, return to the assembly mode. In the following image, clicking on the Model tab of the Ribbon displays the commands for creating assembly features. These commands include Extrude, Revolve, Hole, Sweep, Fillet, Chamfer, Move Face, Rectangular Pattern, Circular Pattern, and Mirror. The commands join the Work Plane, Work Axis, and Work Point commands that function inside of assembly models.

NOTE

Assembly features exist only at the assembly level. They do not affect the individual part files.

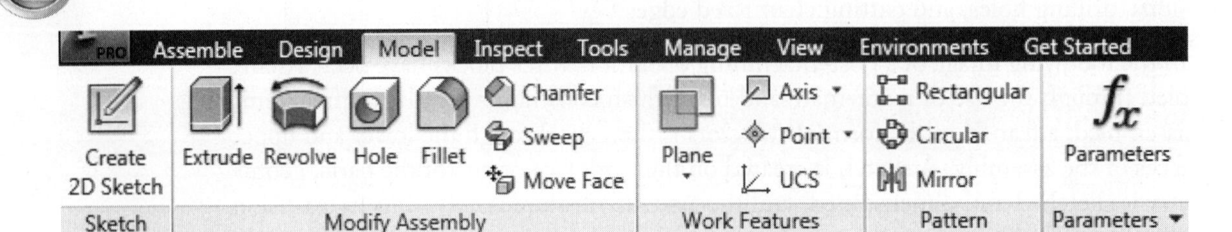

FIGURE 9.77

Use one of the assembly feature commands on the sketched geometry to create the desired assembly feature. In the following image, an extrusion operation is being performed by cutting the rectangular slot through the entire base of the assembly.

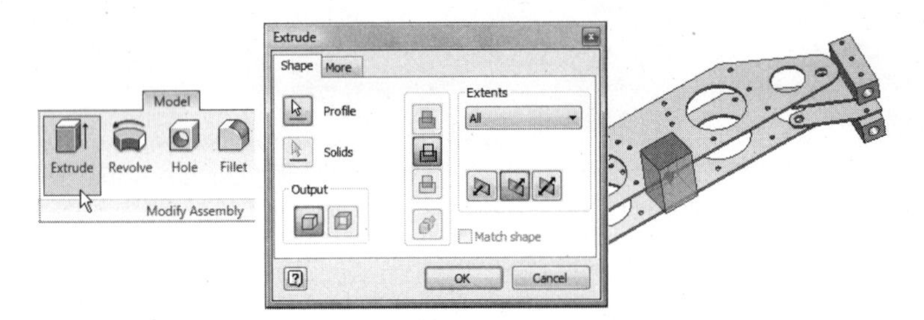

FIGURE 9.78

The result of the extrude cut operation on the assembly model is illustrated in the following image.

NOTE

When extruding an assembly feature, only the cut operation is available.

An End of Features node is present in the browser to separate assembly features from assembly components. When an assembly feature is created, the feature is added above all the assembly components. In the following image, an assembly feature named Extrusion 1 was added to the assembly, and the feature is placed in the browser above the End of Features node. All parts affected by the created assembly feature are listed below that assembly feature. These affected parts are called Participants, as they are all members of the group of components being cut by the assembly feature. They are shown in the browser as *children* of the feature.

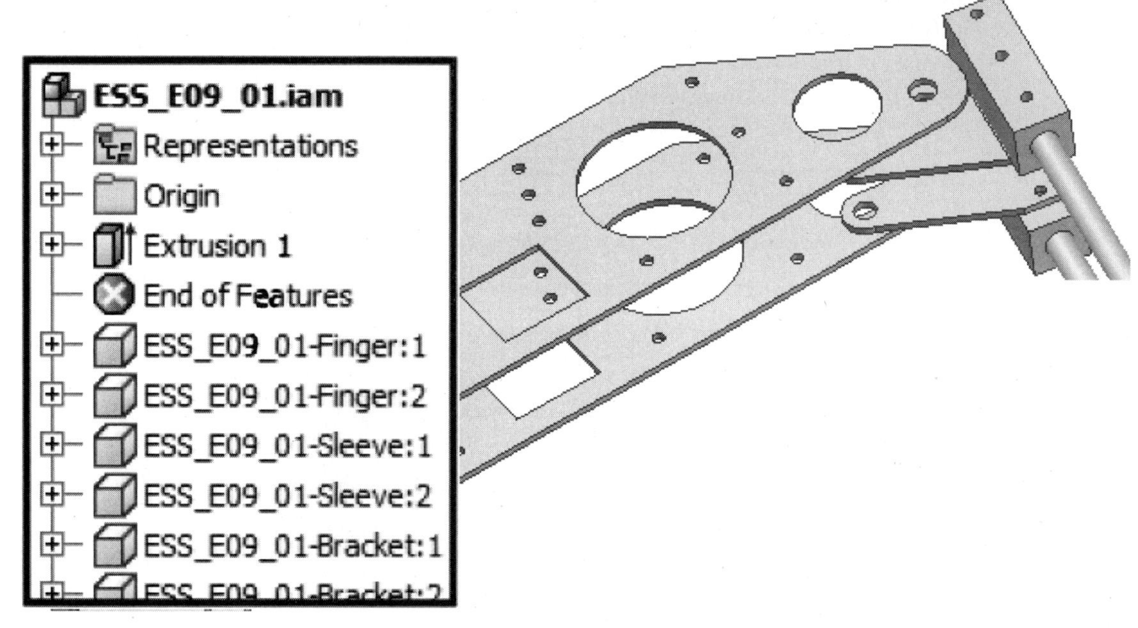

FIGURE 9.79

Removing and Adding Participants

When working with assembly features, it is possible to add and remove participants that have been affected by the assembly feature. To remove a Participant, right-click on a component listed as a Participant, and then select Remove Participant from the menu, as shown in the following image on the left. When a Participant has been removed from the assembly feature, the assembly feature no longer affects the component.

To add a Participant to an assembly feature, right-click the assembly feature in the browser, and select Add Participant from the menu, as shown in the following image on the right. You can then select the part to add from either the canvas or the browser.

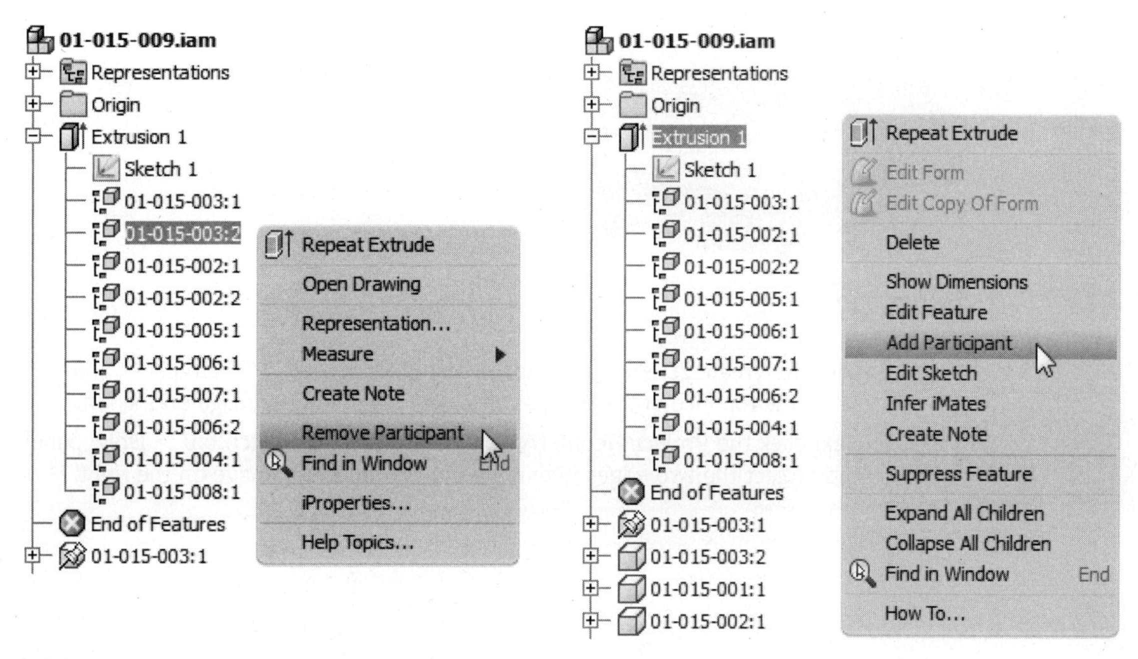

FIGURE 9.80

Assembly features can be suppressed by dragging the End of Features icon to a new position before the actual assembly feature, shown as Extrusion 1 in the following image on the left, or by right-clicking the feature and selecting Suppress Feature from the menu, as shown in the following image on the right.

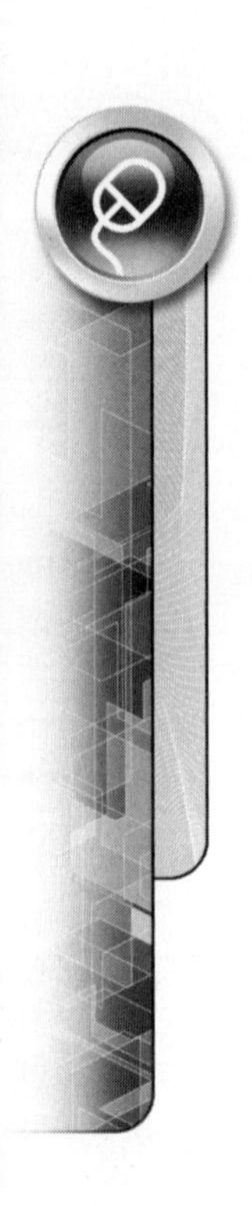

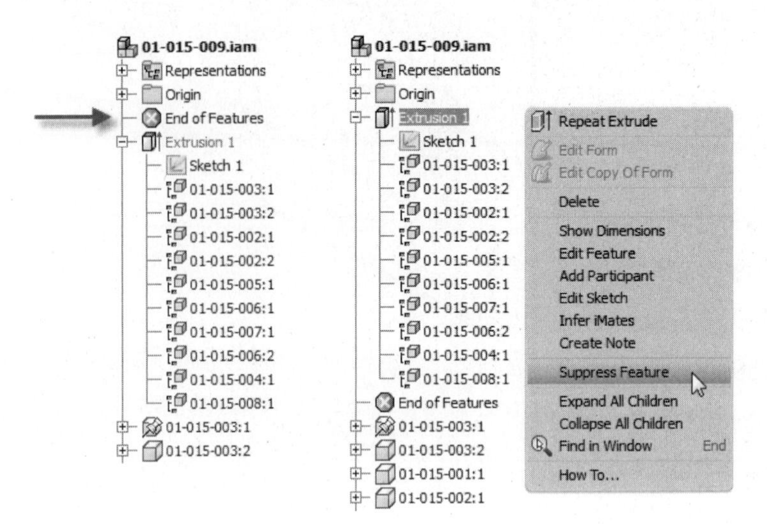

FIGURE 9.81

EXERCISE 9-4: CREATING ASSEMBLY FEATURES

In this exercise, you add a sketched assembly feature to an existing assembly, which is an array of index pockets spanning two assembly components. You then add an assembly hole feature and change the components that participate in the hole feature.

1. Open *ESS_E09_04.iam*. The mount assembly should appear similar to the one displayed in the following image. Use the Rotate and Zoom commands to examine the assembly. Right-click in the graphics window, and select Home View.

FIGURE 9.82

2. Click the Create 2D Sketch command on the Model tab > Sketch panel, and then select the top face of the base, as shown in the following image on the left.
3. Next click the Project Geometry command from the Sketch tab > Draw panel, and project the two edges highlighted in the following image on the right.

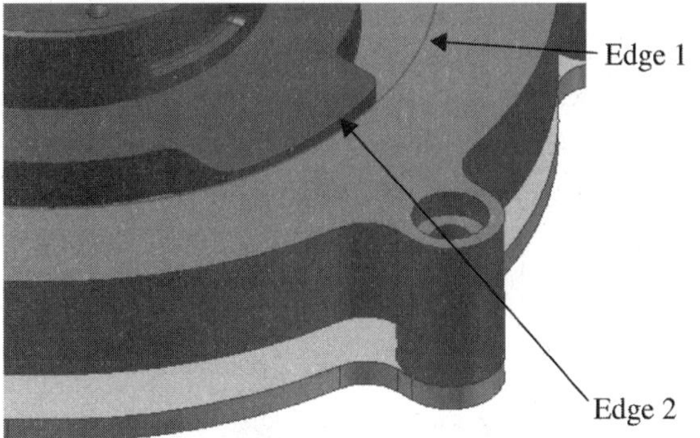

FIGURE 9.83

4. Click the Line command from the Sketch tab > Draw panel.
5. Select the Construction line style from the Sketch tab > Format panel, and then sketch the two lines, as shown in the following image on the left. The upper line is horizontal, and both lines are coincident to the center point and the inner projected edge.
6. Create a third construction line, as shown in the following image on the right. The line is parallel and coincident to the endpoint of the lower line, and it is coincident to the outer projected edge.

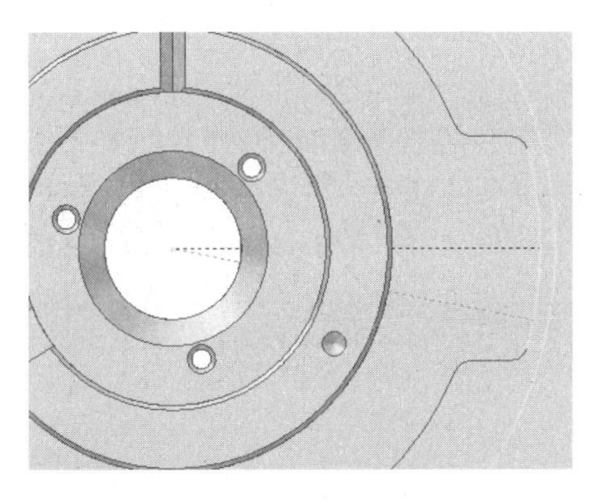

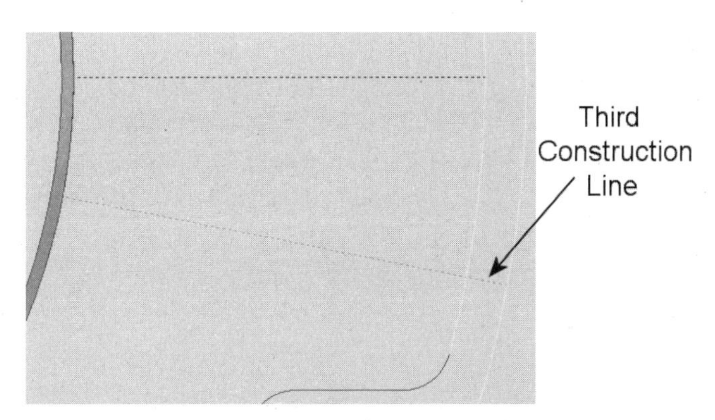

FIGURE 9.84

7. Click the Center Point Circle command from the Sketch tab > Draw panel. Select Normal for the line style, by deselecting the Construction style from the Sketch panel > Format panel. Sketch a circle centered on the midpoint of the short construction line, as shown in the following image on the left.
8. Click the Dimension command from the Sketch panel > Constrain panel, and add the two dimensions, 11° angle and .59 in diameter circle, as shown in the following image on the right.

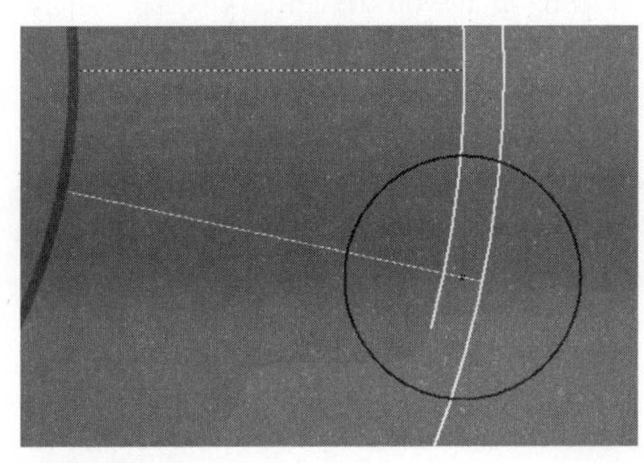

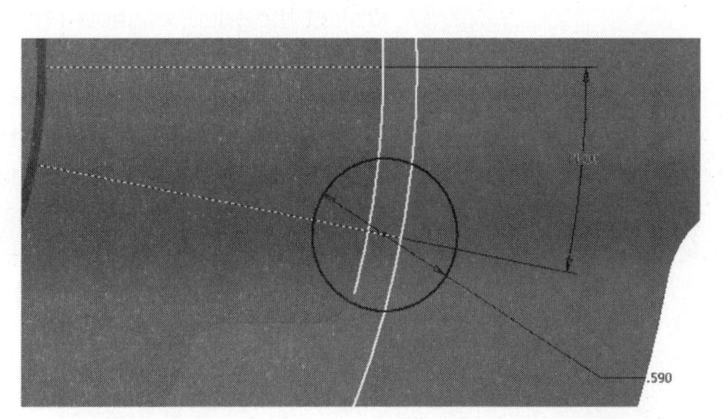

FIGURE 9.85

9. Click the Circular Pattern command from the Sketch tab > Pattern panel, and select the sketched circle. Click the Select Axis option in the Circular Pattern dialog box, and select the projected point at the center of the assembly.
10. Enter **3** in the Count edit box and **22 deg** in the Angle edit box as shown in the following image on the left.
11. Click OK to create the pattern. Your sketch should match the following image on the right.

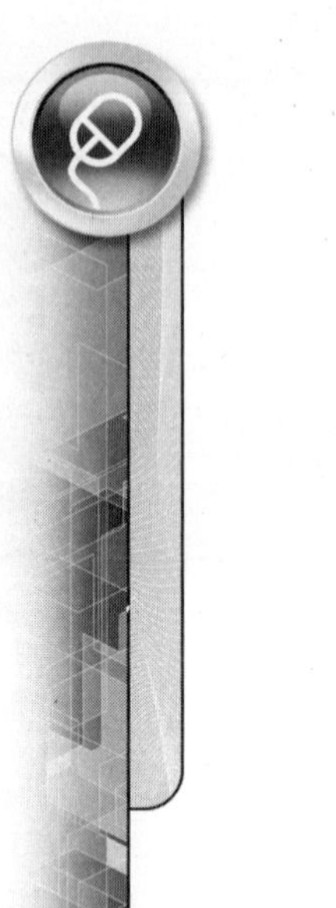

FIGURE 9.86

12. Right-click and select Home view from the menu; then right-click and select Finish 2D Sketch from the menu.
13. Click the Extrude command on the Model tab > Modify Assembly panel.
14. Click inside the three patterned circles as the areas to extrude. Enter a value of **.125 in** in the Depth edit box. By default, you will be performing a cut operation. The extrude feature previews, as shown in the following image on the left.
15. Click the OK button in the Extrude dialog box to create the feature. Your model should appear similar to the following image on the right.

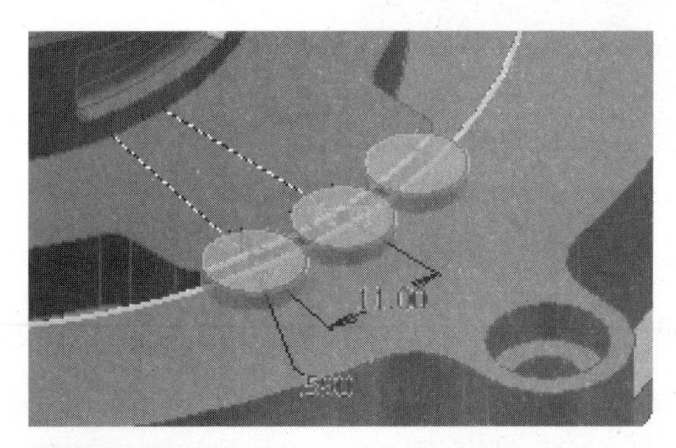

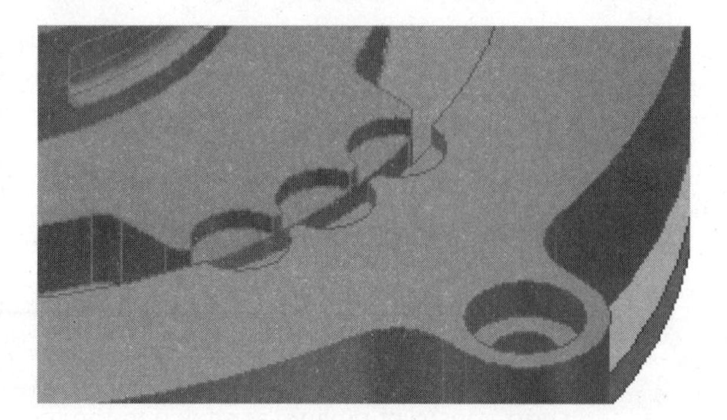

FIGURE 9.87

16. Create a new assembly sketch on the same face as the previous sketch.
17. Project the edge, as shown in the following image on the left.
18. Add a Point, Center Point on the left side of the part and place a horizontal constraint between the point, center point and the center of the projected edge. Dimension the distance from the center of the part to the point, center point, and enter **3.125 in** in the dimension edit box. Your sketch should match the one shown in the following image on the right.

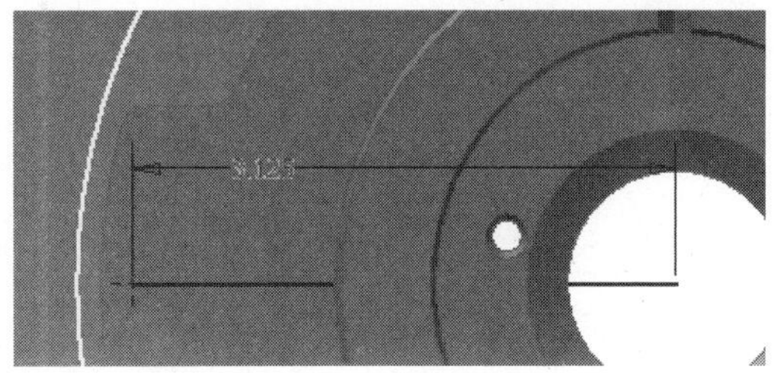

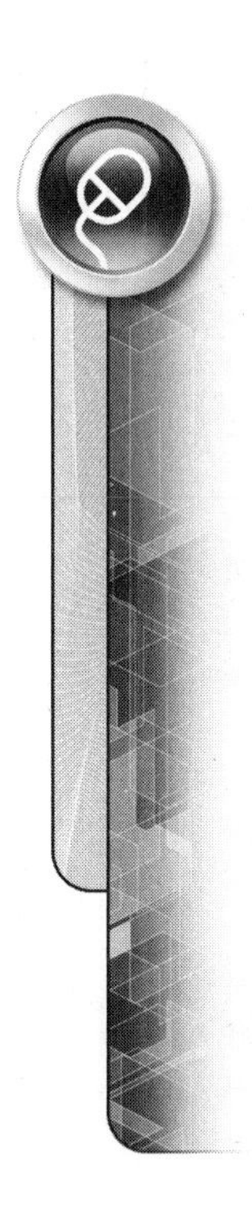

FIGURE 9.88

19. Right-click in the graphics window, and select Finish 2D Sketch from the menu.
20. Click the Hole command from the Model tab > Modify Assembly panel, and select Through All from the termination list.
21. Enter a value of **.20 in** as the hole diameter. Click the OK button to cut the hole. Notice that the hole feature cuts through all components in the assembly, as shown in the following image (the image is shown as a sectioned view for clarity).

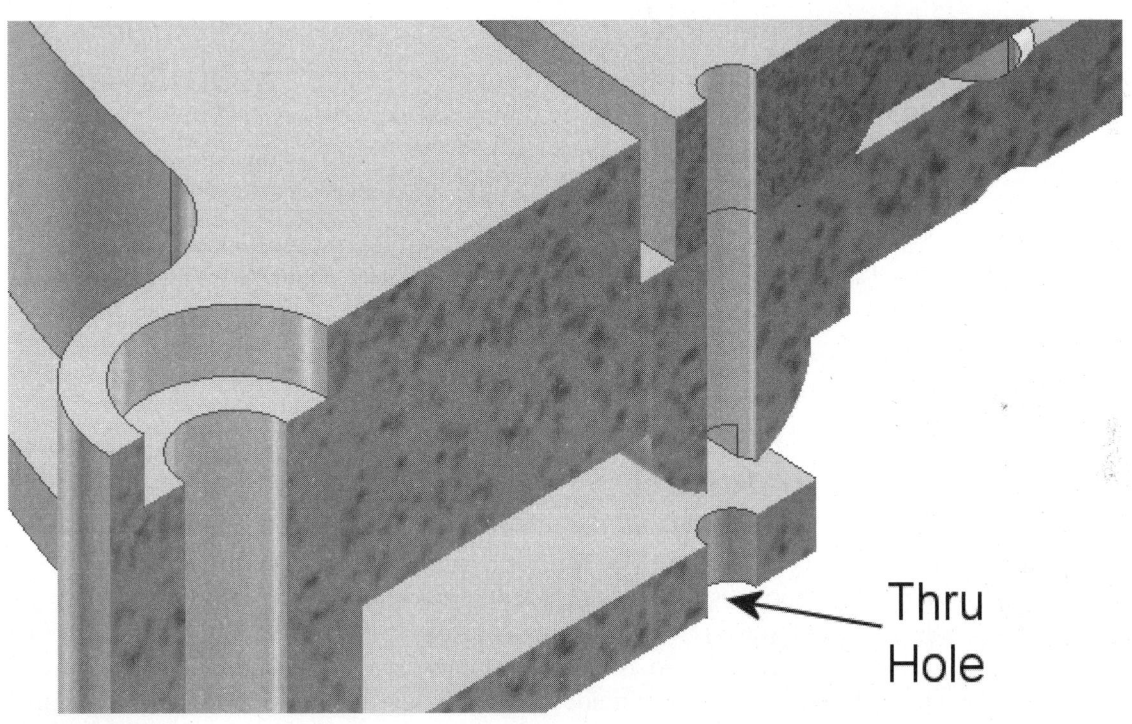

FIGURE 9.89

22. The hole is not required to cut through the bottom plate in the assembly. Expand Hole1 in the browser. All components affected by the feature are listed below the feature.
23. Right-click ESS_E09_04-BasePlate:1, and select Remove Participant from the menu. The assembly feature no longer cuts through the bottom plate, as shown in the following image.

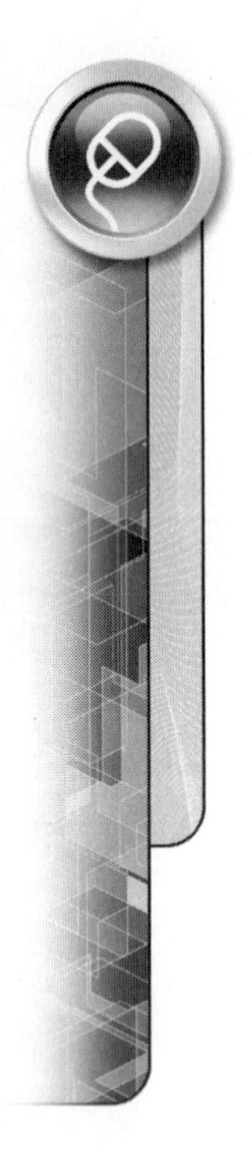

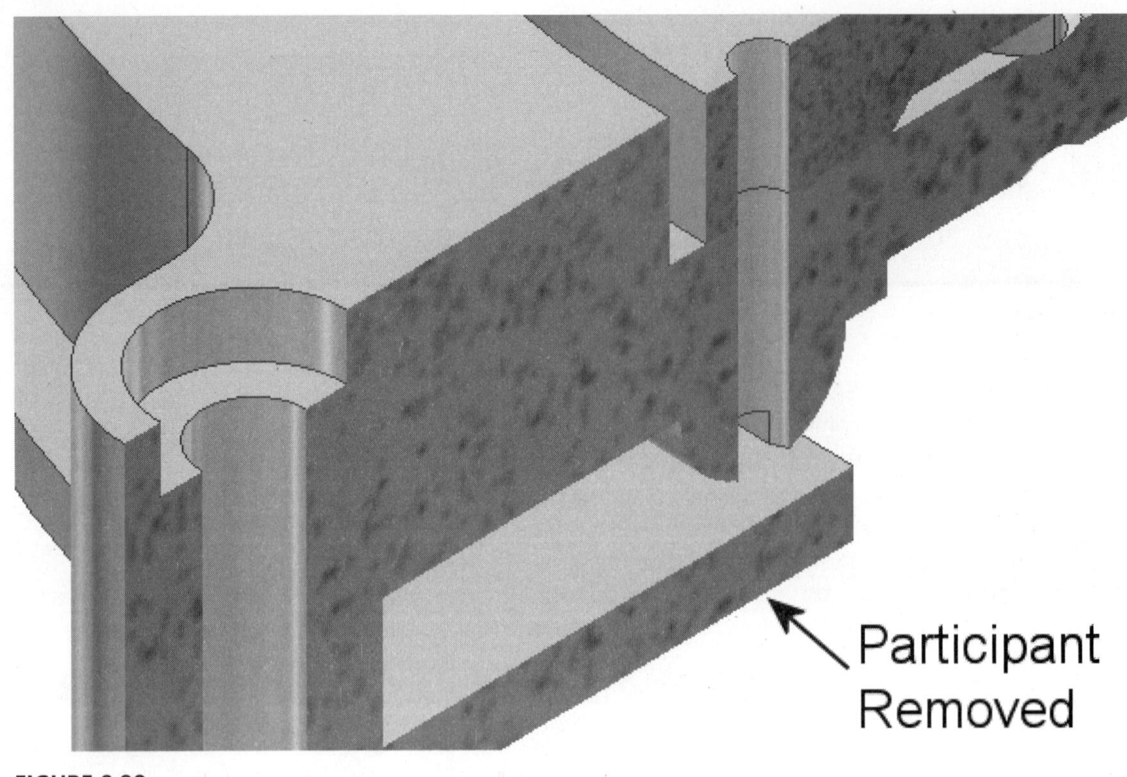

FIGURE 9.90

NOTE

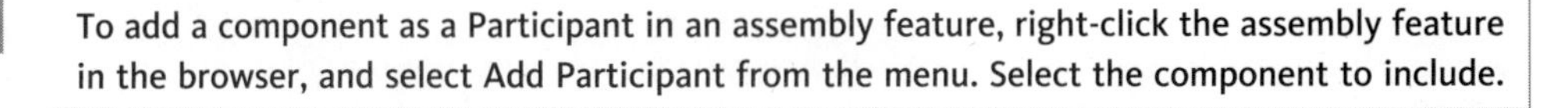

To add a component as a Participant in an assembly feature, right-click the assembly feature in the browser, and select Add Participant from the menu. Select the component to include.

24. Close all open files. Do not save changes. End of exercise.

THE FRAME GENERATOR

The Frame Generator allows you to design platforms, equipment racks, and other types of structural frames. You can easily create and edit structural members. The Frame Generator has its own set of tools, which can be accessed from the Design tab > Frame panel, as shown in the following image on the left. Before inserting structural members into an assembly, you must have a skeletal model to work from. The skeletal model acts as a guide in determining the direction in which the frame members will be placed in the assembly. An example of a structural frame assembly is shown in the following image on the right.

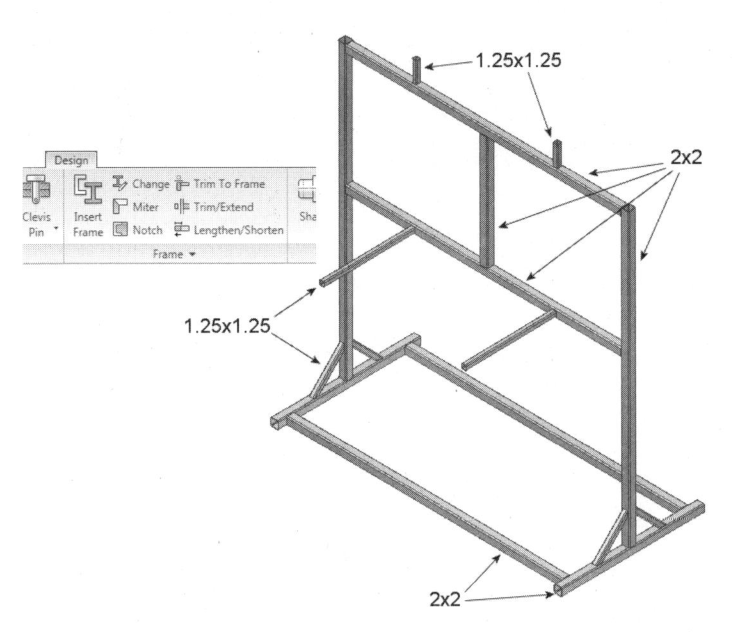

FIGURE 9.91

Since the Frame Generator operates in the assembly environment, any skeletal models used for generating the frames must first be created as a 2D or 3D sketch, or a solid body inserted into the assembly, and then inserted into the assembly model. A typical example of a skeletal model is shown in the following image.

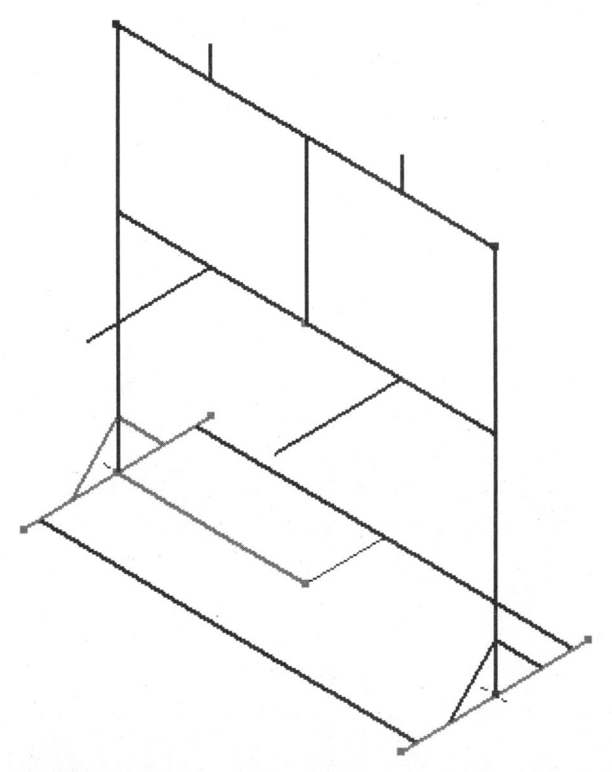

FIGURE 9.92

After the Frame Generator is loaded, a menu of commands is displayed, as shown in the following image.

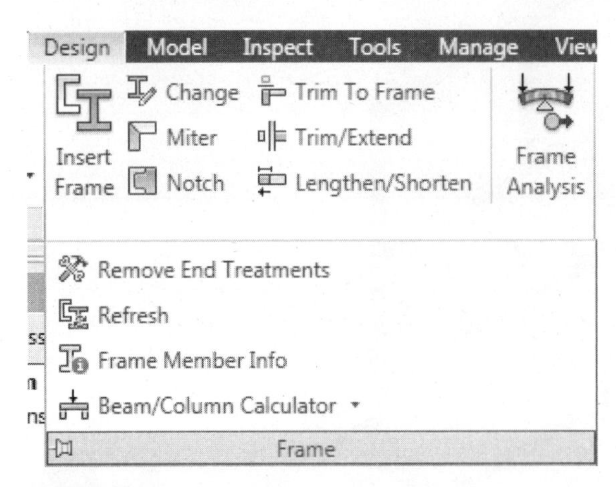

FIGURE 9.93

Clicking the Insert Frame command from the Design tab > Frame panel will display the Insert dialog box, as shown in the following image. The following changes were made in the Frame Member Selection area in this dialog box: Standard = ANSI; Family = ANSI AISC HSS (Rectangular) – Rectangular Tube; Size = 3 × 1 × 1/8. The remaining default settings were left as is. The edges of the skeletal frame, where you want the frame members to be placed, are then selected and you click either Apply or OK in the Insert dialog box.

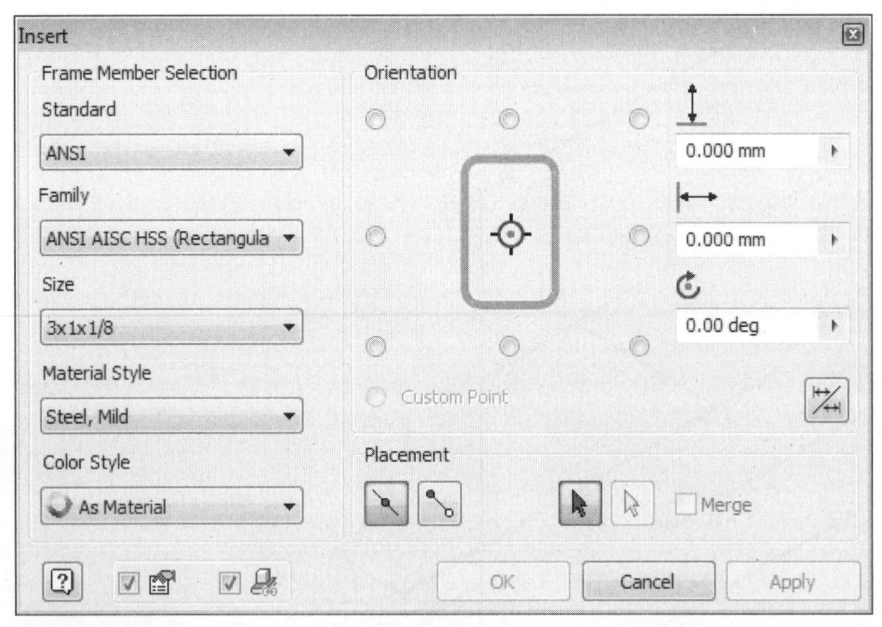

FIGURE 9.94

Clicking OK or Apply will launch the Create New Frame dialog box, as shown in the following image on the left. Here you define a new file location that will hold all of the frame members. Click OK in this dialog box to activate the Frame Member Naming dialog box, as shown in the following image on the right. You can keep the default display names and file names, or you can make changes to them that will be reflected in the browser.

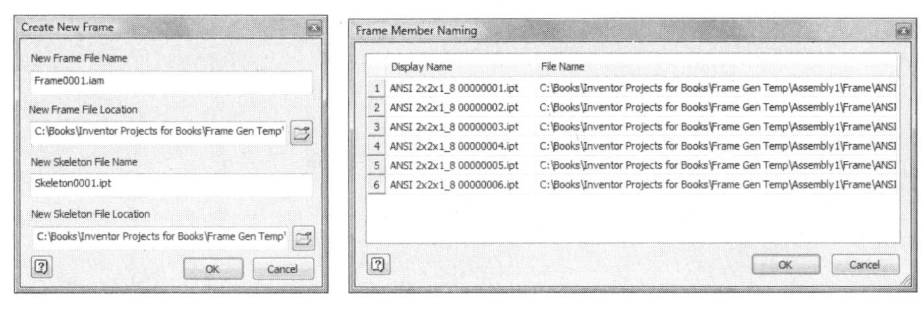

FIGURE 9.95

When you return to the assembly, all frame members should be properly placed. To aid in editing the frame members, turn off the skeletal frame, as shown in the following image.

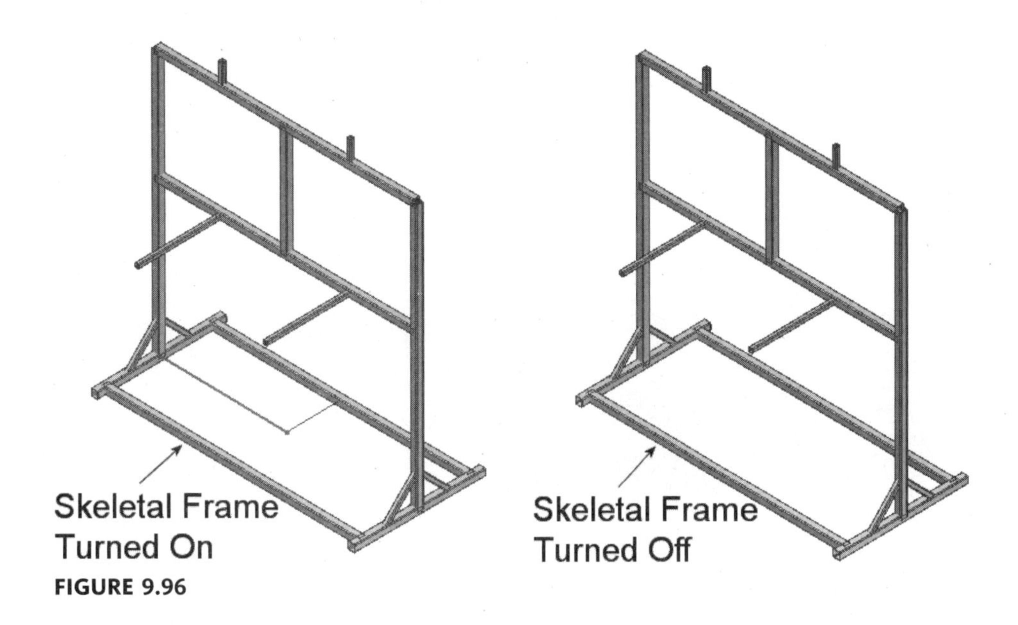

FIGURE 9.96

USING SOLIDS FOR FRAME GENERATION

The previous example of generating frame members relied on a wireframe skeletal model. You can also create skeletons from solids and surface bodies. In the following image of a truck bed extender, a solid model that outlines the basic shape of the extender was created, as shown in the following image on the left. Additional sketch lines were added for the vertical frame members. A new assembly file was created, and the solid skeleton was placed as the first component. As shown in the following image in the middle, frame members were generated by clicking the edges of the solid model in addition to the single line sketch objects. Once all frame members are generated, the solid skeleton is turned off, leaving the frame as shown in the following image on the right.

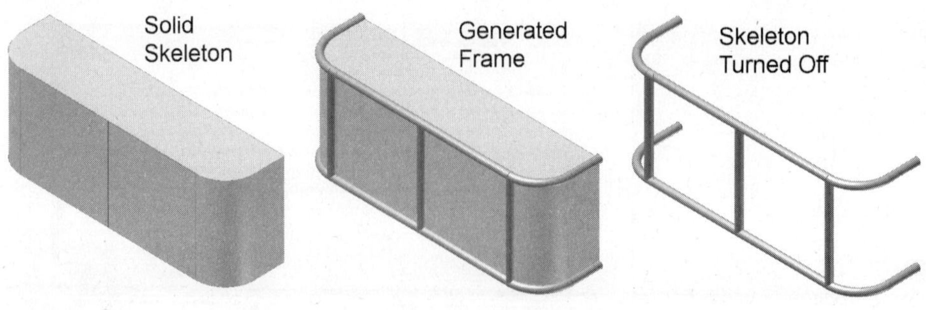

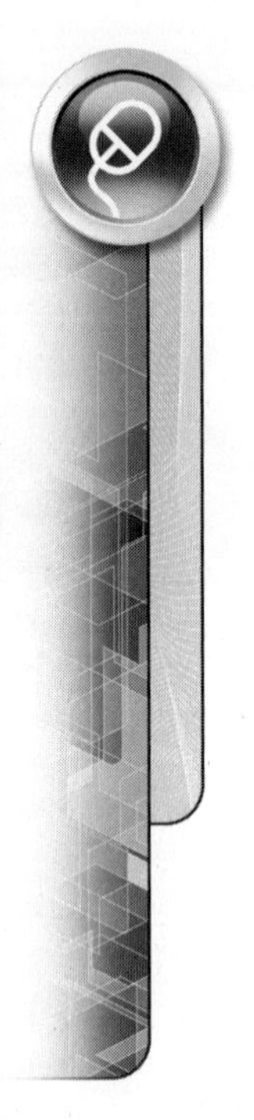

FIGURE 9.97

EXERCISE 9-5: USING THE FRAME GENERATOR

This exercise walks you through the creation of a frame. You will use an existing part file as a skeletal model when generating all frame members. You will then edit the frame members by trimming them to size. Finally, you will turn off the visibility of the skeleton to expose only the frame members.

1. Open *ESS_E09_05.iam*.
2. Click the Insert Frame command from the Design tab > Frame panel, as shown in the following image on the left.

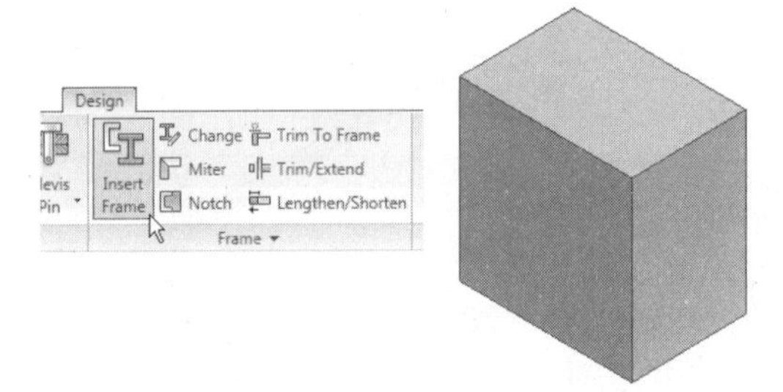

FIGURE 9.98

3. Make the following changes under the Frame Member Selection area of the Insert dialog box:
   - Standard = ISO
   - Family = ISO 657-1 hot rolled steel sections Part 1 - Equal-leg angles
   - Size = L20 × 20 × 3
   - Keep the remaining default settings
4. Under the Orientation area, select the insertion point of the structural member to be the lower-left corner of the shape. Keep all default offset values as shown in the following dialog box.
5. Click the top four edges of the skeletal frame, as shown in the following image on the right. If the frame members are not aligned, as shown in the following image on the right, click the Mirror Frame Member button to change the alignment. If mirroring does not fix the alignment, you may also have to experiment with different angle settings. When finished, click OK.

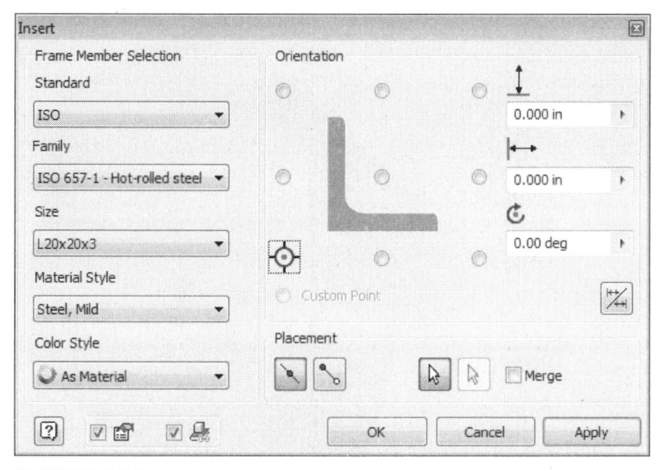

FIGURE 9.99

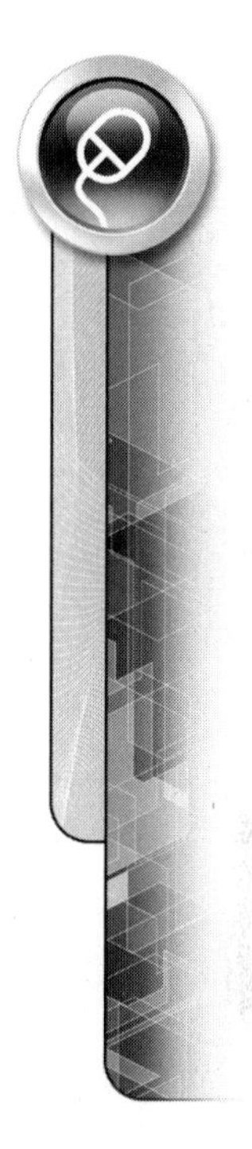

6. When the Create New Frame dialog box appears, as shown in the following image on the left, click the OK button to accept these defaults. These items dictate where the frame members will be located when created.
7. When the Frame Member Naming dialog box appears, click the OK button to accept the display names and file names.
8. The resulting frame members display, as shown in the middle of the following image, in their proper orientations. Notice also how the frame components are displayed in the Assembly browser, as shown in the following image on the right.

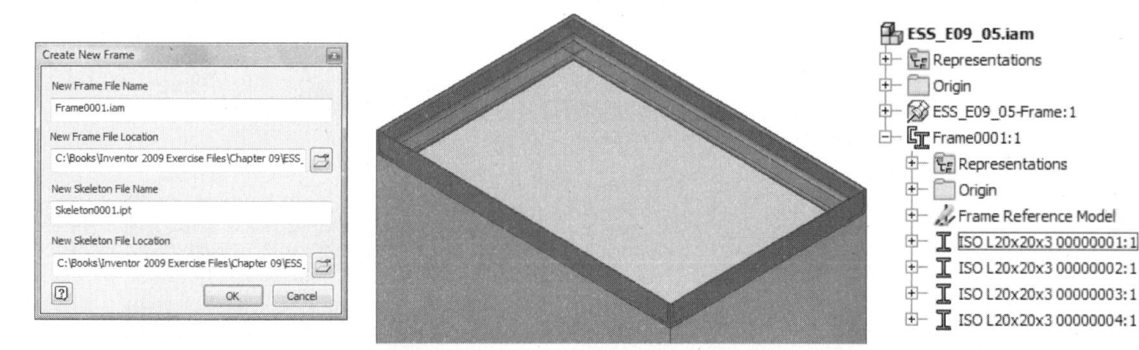

FIGURE 9.100

9. Use the Orbit command to adjust the viewpoint as shown in the following image.
10. Click the Insert Frame command from the Design tab > Frame panel.
11. In the Insert dialog box, notice that the previous settings are maintained. Keep all information the same under the Frame Member Selection area. Click the four bottom edges of the solid frame, as shown in the following image. Use the Mirror Frame Member button when needed to change any alignment issues.
12. When finished, click the Apply button and accept the default names. This will create the frame members and keep you in the Insert dialog box so that you can place other members.

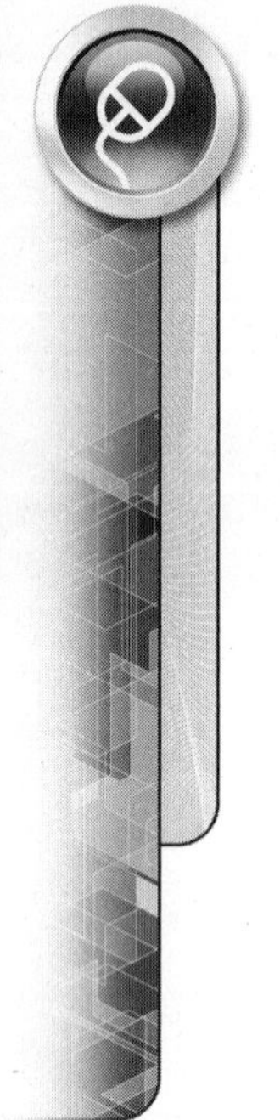

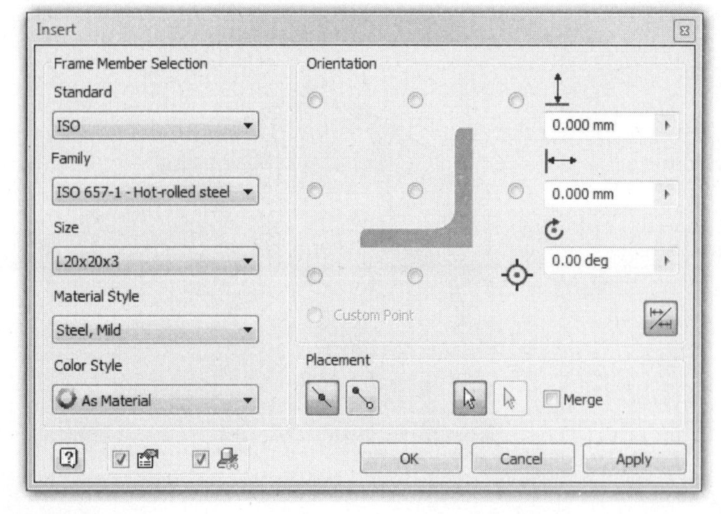

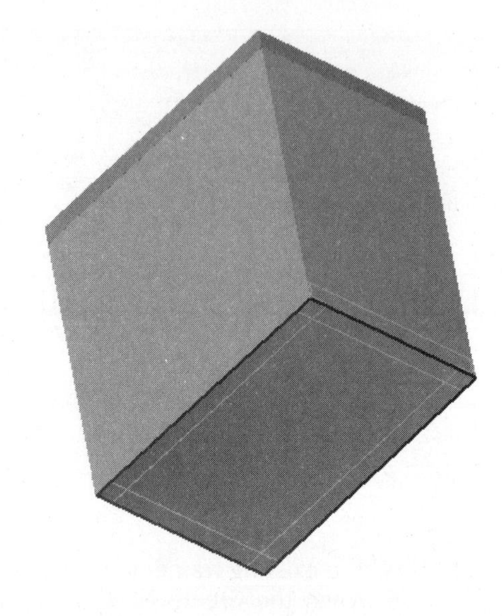

FIGURE 9.101

13. Press the F6 key to change the viewpoint to the Home View.
14. While still in the Insert dialog box, place the frame members using the vertical edges of the skeletal model. The frame members need to face inward. You will probably have to create each individual member separately. Use the Mirror Frame Member button and even a different angle setting to properly orient the components. When finished with each individual member, click the Apply button to remain in the Insert dialog box and accept the default names.
15. When you have finished placing all four vertical frame members, as shown in the following image, click the OK button to exit the Insert dialog box.

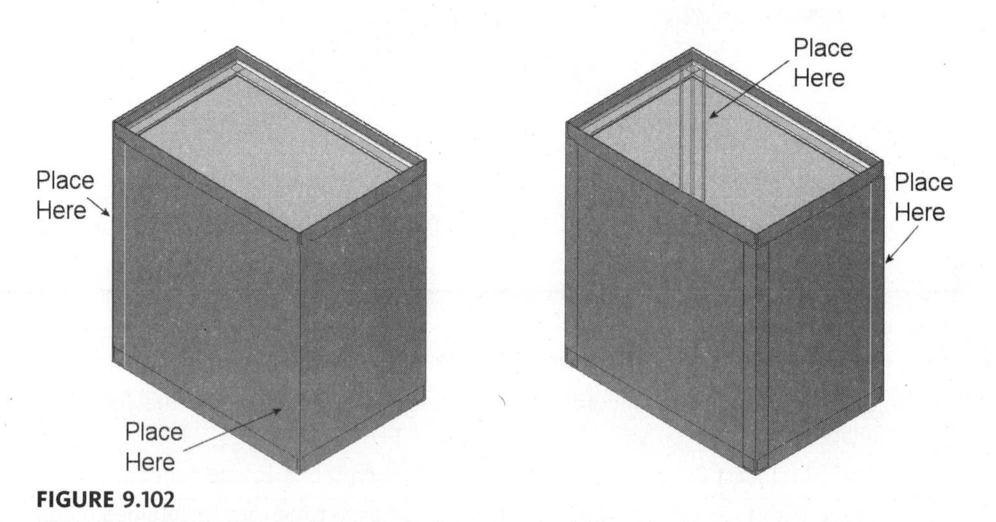

FIGURE 9.102

16. This completes the frame-generation process where an existing solid model was used to create the frame members, as shown in the following image on the left.
17. To turn off the display of the solid model, right-click *ESS_E09_05-Frame:1* in the browser, and pick Visibility from the menu to remove the checkmark, as shown in the following image in the middle. With the solid model turned off, the frame can be seen more clearly, as shown in the following image on the right.

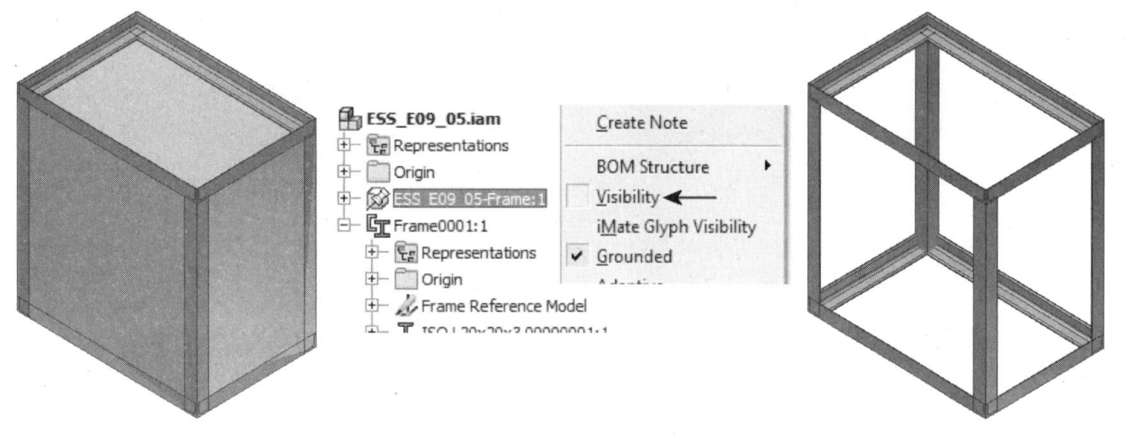

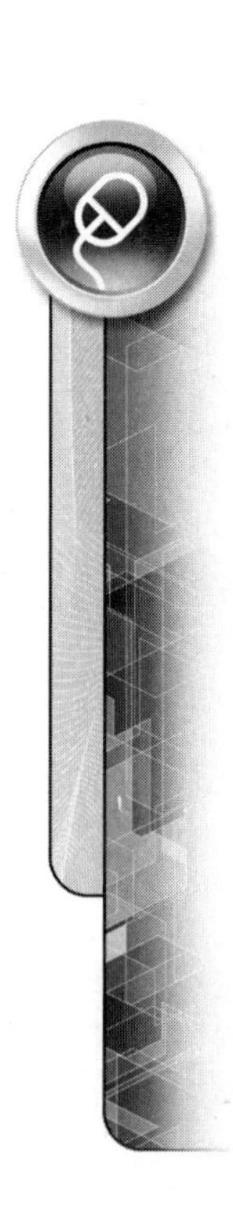

FIGURE 9.103

18. The existing frame has a number of members that overlap each other. Numerous frame members will need to be trimmed in order to produce the final assembly. Click the Trim to Frame command from the Design tab > Frame panel, as shown in the following image.
19. To trim a member to a frame, click the member labeled (A); this is the member to trim to. Next click the member labeled (B); this is the member that will be trimmed, as shown in the following image. Click Apply.

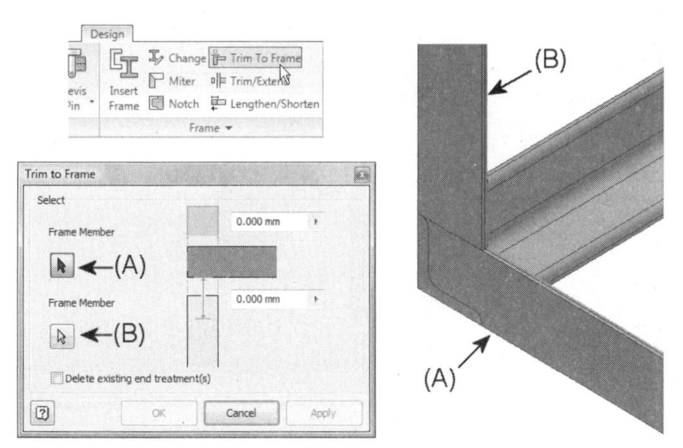

FIGURE 9.104

20. Continue picking individually, bottom frame members first, followed by vertical members and use the Apply button until you have trimmed all four vertical members. Your completed trim-to-frame operation should appear similar to the following image.

NOTE

As an alternative, you could use the Trim/Extend command to trim all members to a planar face.

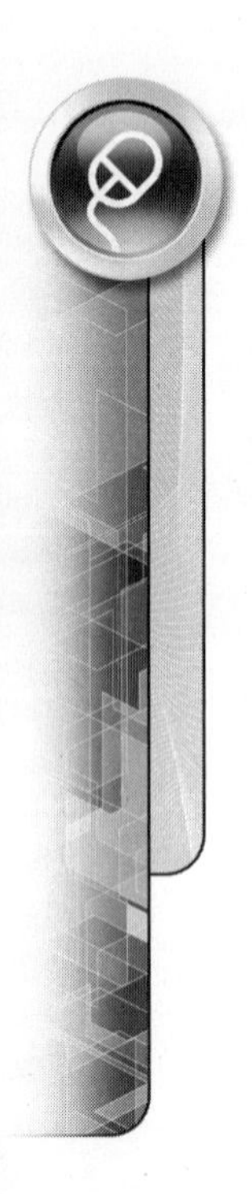

21. Close the Trim to Frame dialog box.

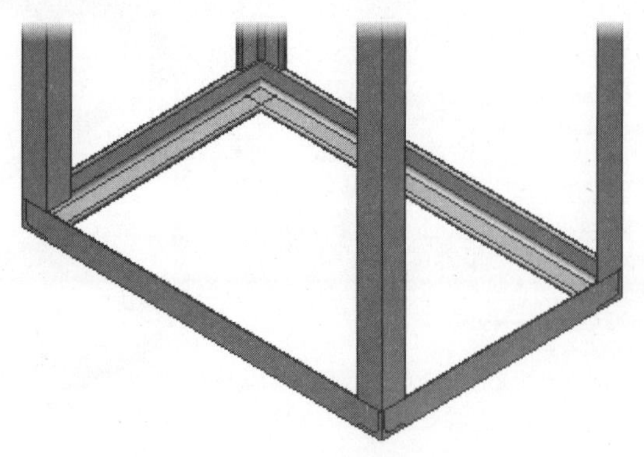

FIGURE 9.105

22. Next perform a mitering operation by clicking the Miter command from the Design tab > Frame panel, as shown in the following image on the left. Click member (A) and then member (B), as shown in the following image on the right, although the order for the miter operation is not critical. Click Apply.
23. Continue by mitering the remaining corners that represent the upper and lower frame and close the Miter dialog box.

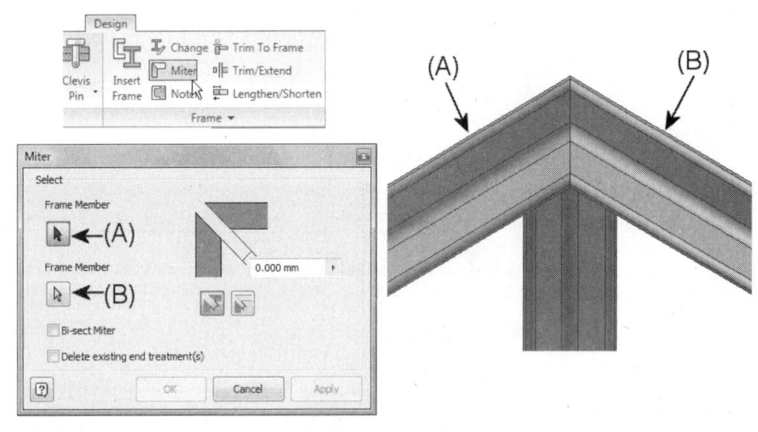

FIGURE 9.106

24. The finished cart frame is illustrated in the following image.

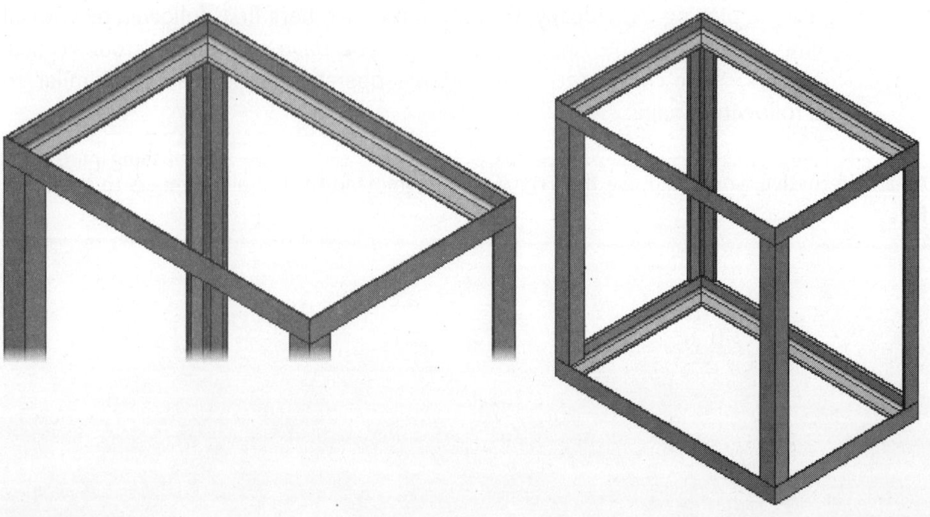

FIGURE 9.107

25. Close all open files. Do not save changes. End of exercise.

CONTENT CENTER

Autodesk Inventor's Content Center contains a number of standard components. It contains thousands of parts such as screws, nuts, bolts, washers, pins, and so on. You can place these standard components into an existing assembly using the Place from Content Center command that is found on the Assemble tab > Component panel, as shown in the following image on the left.

After the command is selected, the Place from Content Center dialog box opens, as shown in the following image on the right, and you can navigate between parts and features that are either included or published to the Content Center. Select the item you want to place, click OK, and place the part as you would any other assembly component.

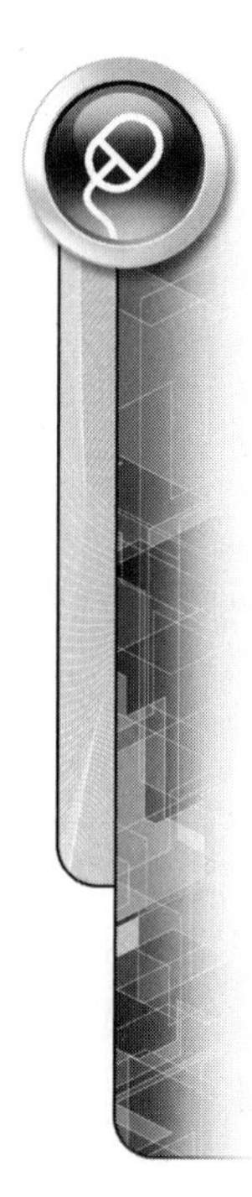

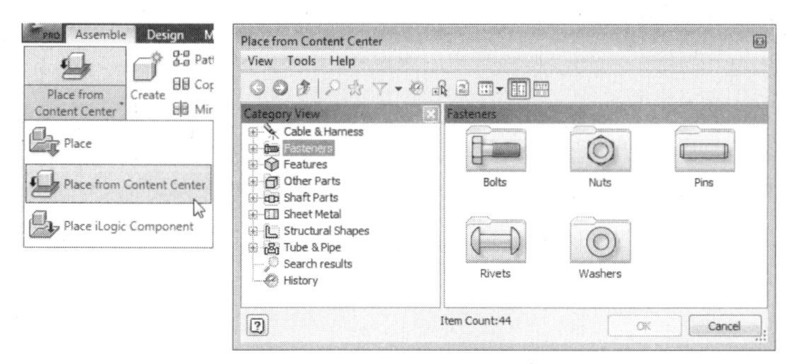

FIGURE 9.108

In addition to using the default items in the Content Center, you can publish your own features and parts to the Content Center. You publish parts and features using the Editor or Batch Publish commands, found on the Manage tab > Content Center panel as shown in the following image.

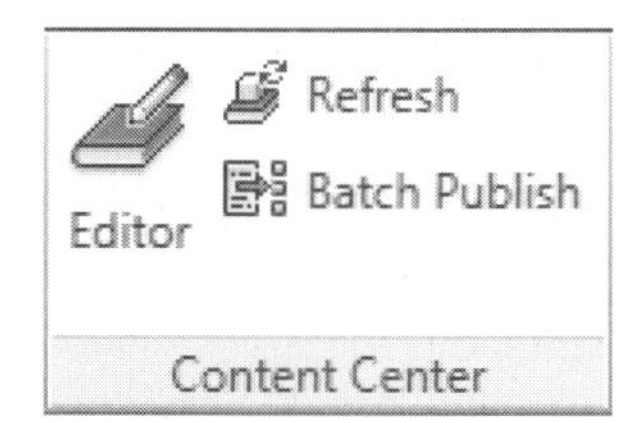

FIGURE 9.109

In order to publish content to the Content Center, a read/write library must exist and be added to the active project file.

NOTE

For more information about the Content Center, publishing features, publishing parts, or configuring libraries, refer to Autodesk Inventor's Help system.

EXERCISE 9-6: CONTENT CENTER

In this exercise, you learn how to search for content in the content center, add content to an assembly with and without the use of AutoDrop, and add content to your Favorites.

1. Open *ESS_E09_06.iam*. In this section of the exercise, you search for families of plain bearings that have specific sizes and are a specific industry standard.

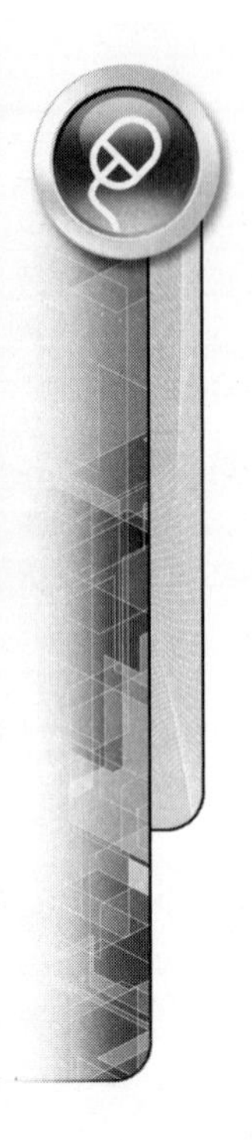

FIGURE 9.110

2. Click the Place from Content Center command on the Assemble tab > Component panel.

FIGURE 9.111

3. To begin searching for bearings of set sizes, in the Place from Content Center dialog box, click Tools menu > Search.
4. In the Quick Search pane, click Advanced Search.
5. In the Advanced Search dialog box, click Categories. In the Choose Categories list, expand the category for Shaft Parts and Bearings and select the Bearings-Plain check box to restrict the search to this family of parts then click OK.

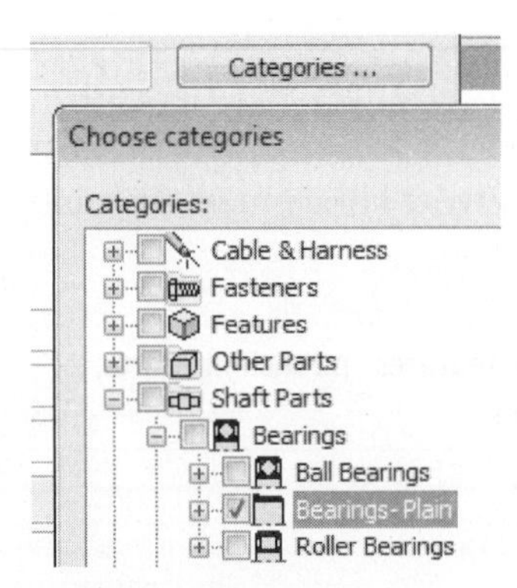

FIGURE 9.112

6. Click the Advanced tab.
7. To define a list of sizes to search for:
   - In the Parameter list, select Shaft Diameter.
   - In the Condition list, select In List.

- In the Value field, enter **24mm; 25mm**.
- Click Add To List, as shown in the following image.

Basic Advanced
Find items that match these criteria:
Define more criteria:
Remove
Parameter: Condition: Value:
Shaft Diameter In List 24mm; 25mm
Add to list

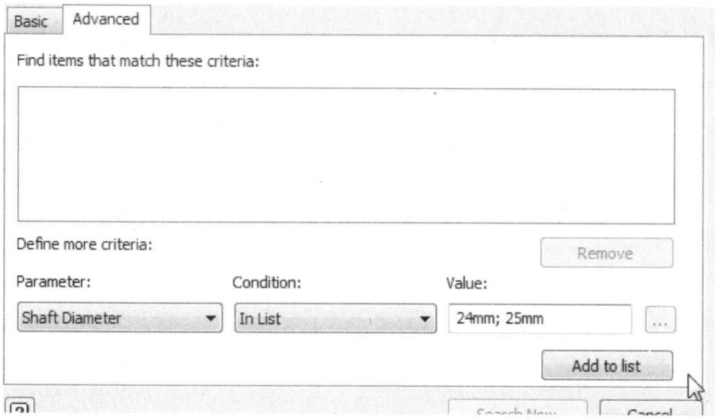

FIGURE 9.113

8. Click Search Now. The results contain a list of plain bearings with the specified sizes.

FIGURE 9.114

NOTE

Verify that you do not have any filters set for the content center. You can clear filters by selecting the Filter tool and verifying that there is no checkmark next to a standard, as shown in the following image.

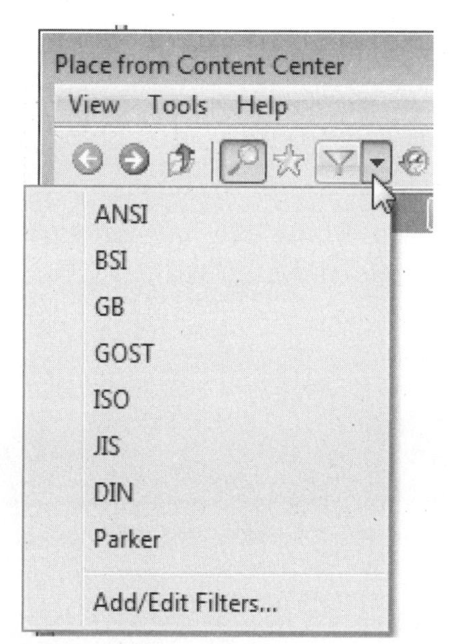

FIGURE 9.115

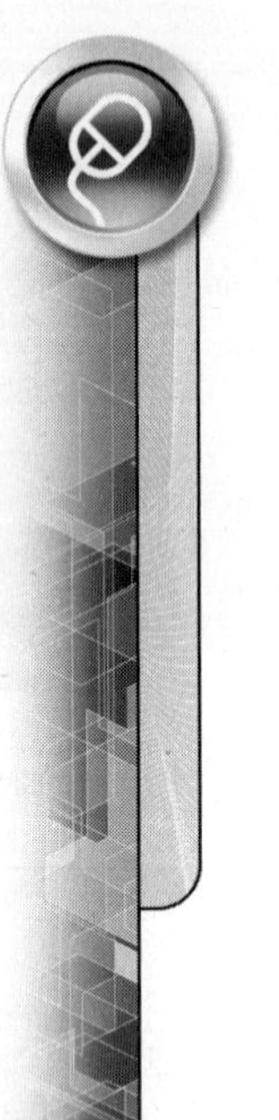

9. To add another condition to the search:
   - In the Parameter list, select Standard.
   - In the Condition list, select Contains.
   - In the Value field, enter DIN.
   - Click Add to List. The criteria for the standard is added as an *AND* search condition.

Parameter: Standard
Condition: Contains
Value: DIN
Add to list

FIGURE 9.116

10. Click Search Now. The results contain only plain bearing families with the listed sizes and for the DIN standard.

FIGURE 9.117

11. Click the Move to Content Center button in order to have these search results available for use. Next, you add content to an assembly with and without the use of AutoDrop.
12. Ensure that AutoDrop is turned on. If the AutoDrop button does not appear pushed in, click the button to turn it on.

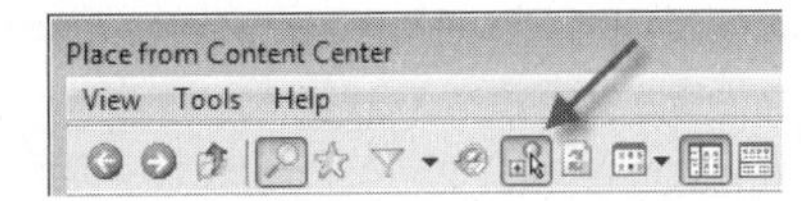

FIGURE 9.118

13. To begin adding a plain bearing to the assembly using AutoDrop:
    - In the Search Results pane, double-click DIN 1850-1 G.
    - In the graphics window, click the fillet edge of the shaft as shown.
    - On the AutoDrop toolbar, click Apply.
    - In the Table dialog box, click OK to accept the designated size.
    - Press ESC.

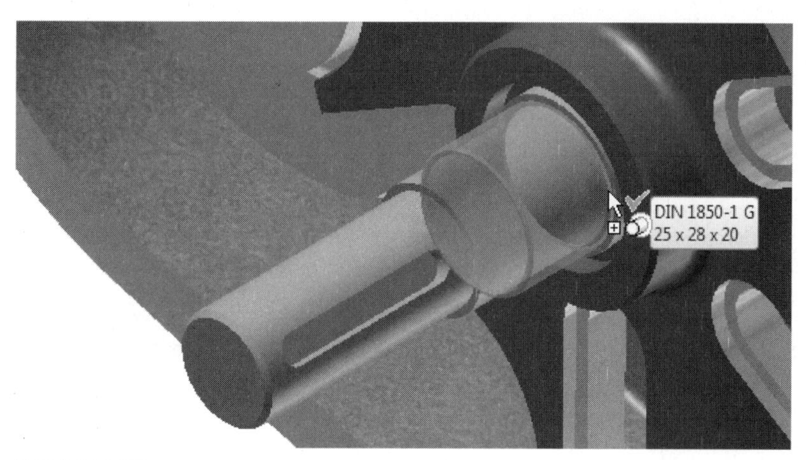

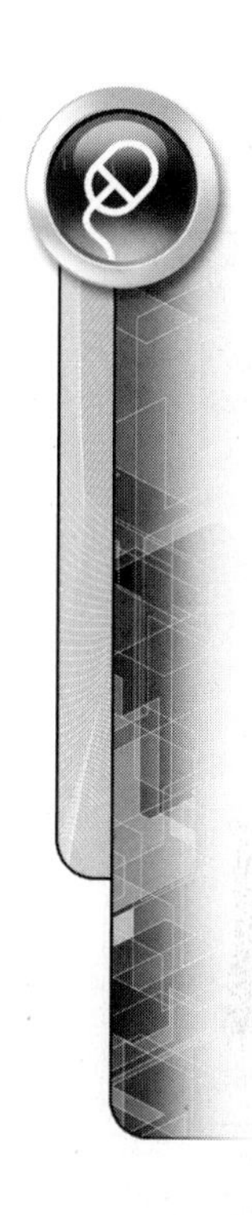

FIGURE 9.119

14. Click the Place from Content Center command on the Assemble tab > Component Panel.
15. To turn off AutoDrop, in the Place from Content Center dialog box, click Tools menu > AutoDrop to remove the checkmark.
16. To begin inserting a parallel key into the assembly without AutoDrop:
    - In the Category View tree list, expand the categories of Shaft Parts, Keys, Keys - Machine.
    - Click the Rounded category.
    - Double-click DIN 6885 A.

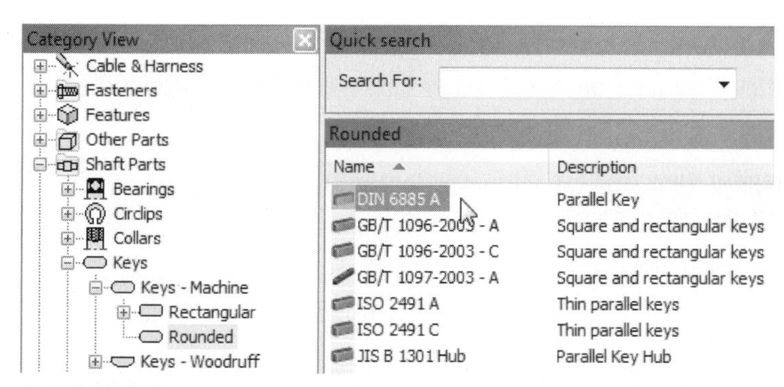

FIGURE 9.120

17. In the DIN 6885 A dialog box:
    - In the Shaft Diameter list, select 17 - 22.
    - In the Parallel Key Nominal Length list, select 40.
    - Click OK.

18. Click in the graphics window below the shaft to add a single instance of the parallel key. Press ESC.

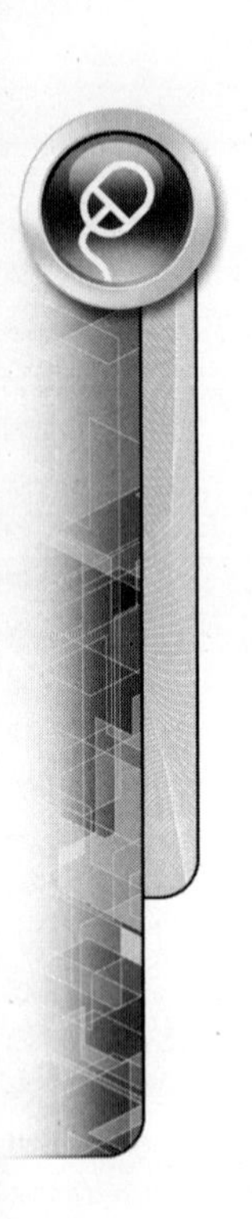

FIGURE 9.121

19. Rotate the model to view the shaft on the other side of the housing.

FIGURE 9.122

20. To insert another bearing from the search results and with AutoDrop turned off:
    - Click Place from Content Center command.
    - In the Category View tree list, click Search Results.
    - Double-click DIN 1850-6 T.
    - In the Size Designation list, select 24 x 30 x 30. Click OK.
21. Click in the graphics window below the shaft to add a single instance of the bearing. Press ESC.

FIGURE 9.123

22. Add an insert constraint between the identified edges of the bearing and the hole in the housing.

FIGURE 9.124

23. Press F6 to return to the Home view.
24. To begin inserting a circlip into the assembly with AutoDrop even though it is not turned on:
    - Click the Place from Content Center command.
    - In the Category View tree list, expand the categories of Shaft Parts and Circlips.
    - Click the External category.
    - Press and hold ALT. Double-click DIN 471 <= 9mm.
    - Select the circular edge at the bottom of the groove on the post as shown.

FIGURE 9.125

25. On the AutoDrop toolbar, click Apply. Press ESC.

FIGURE 9.126

26. Next, you will add a part to the Content Center favorites. You then change the browser display and add an instance using AutoDrop. Click the Place from Content Center command.
27. To add a part family to the favorites:
    - In the Category View tree list, expand the categories of Fasteners, Bolts, and select the Socket Head Category.
    - Right-click DIN EN ISO 4762. Click Add to Favorites.

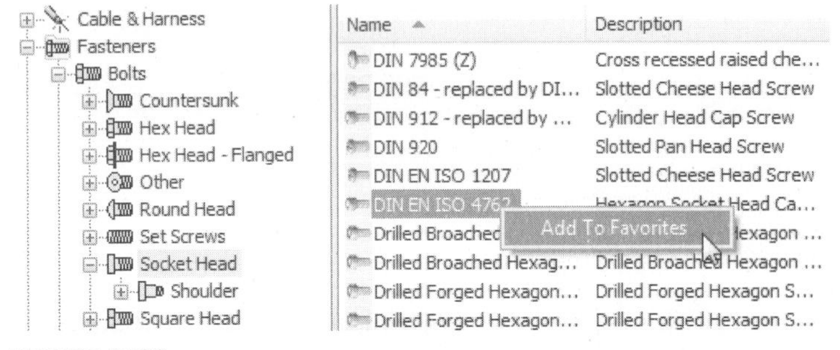

FIGURE 9.127

28. In the Place from Content Center dialog box, click Cancel.
29. In the browser, right-click ESS_E09_06-0208-06P:1. Click Visibility to turn on the visibility of the assembly cover.

FIGURE 9.128

30. In the browser under Model, click Favorites.

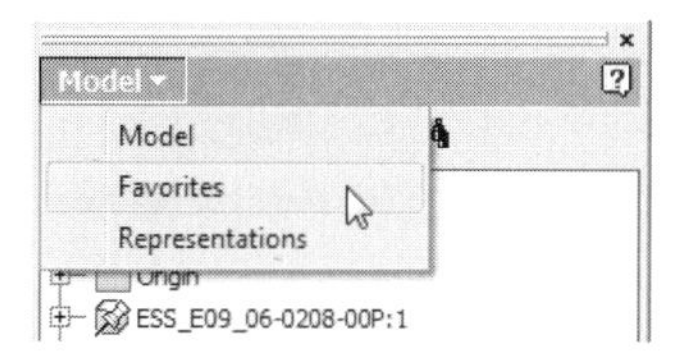

FIGURE 9.129

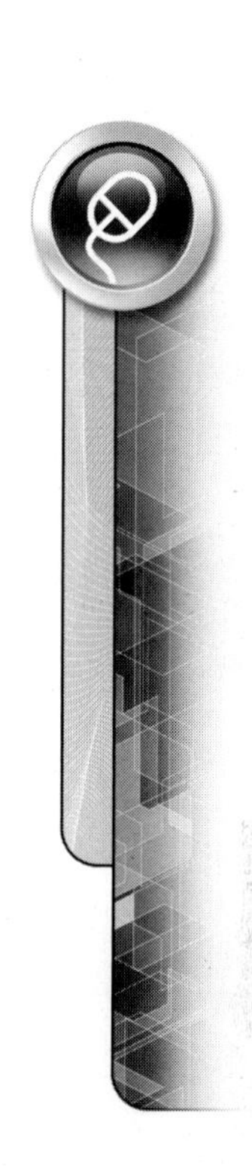

31. To begin adding an instance of the bolt:
    - In the browser, click and drag DIN EN ISO 4762 onto the canvas.
    - Click the inside edge of the hole that is located to the right of the shaft.

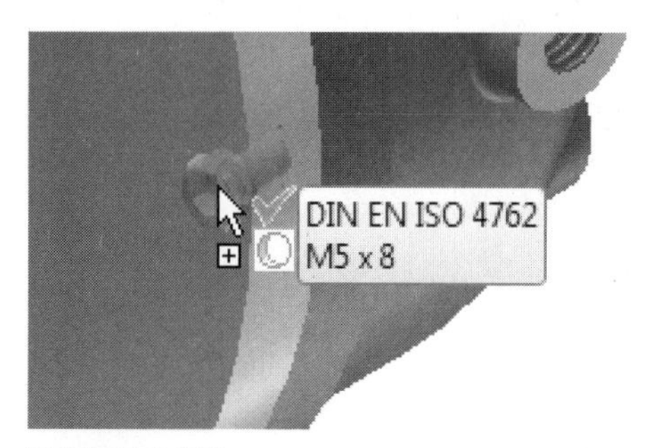

FIGURE 9.130

32. Click and drag the arrow grip to set its length to 16.

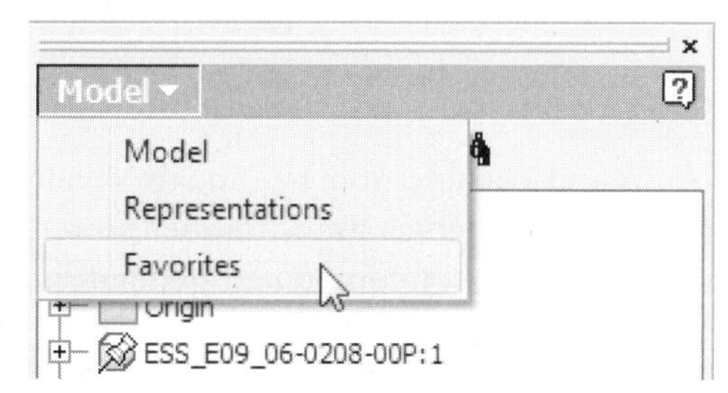

FIGURE 9.131

33. On the AutoDrop toolbar:
    - Verify that the Follow Pattern button is active.
    - Click Apply. Notice that the fastener was added to all instances of the hole contained in the component pattern.
    - Press ESC.

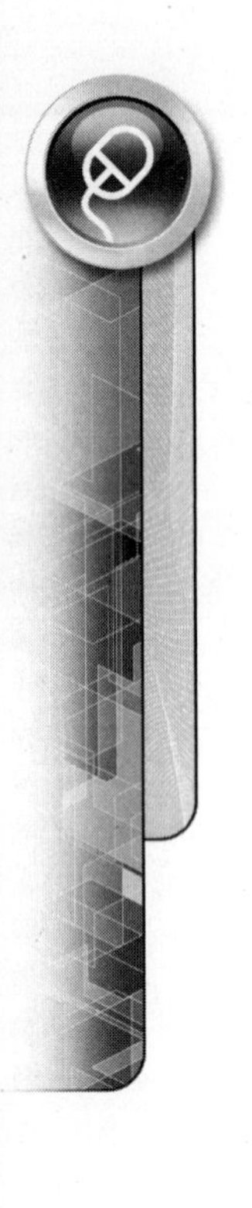

FIGURE 9.132

34. Close all open files. Do not save changes. End of exercise.

DESIGN ACCELERATOR

Autodesk Inventor's Design Accelerator commands enable you to quickly create complex parts and features based on engineering data such as ratio, torque, power, and material properties. The Design Accelerator consists of component generators, mechanical calculators, and the *Engineer's Handbook*.

While working in an assembly the Design Accelerator commands are accessed on the Design tab. The commands found under the tab are as shown in the following image.

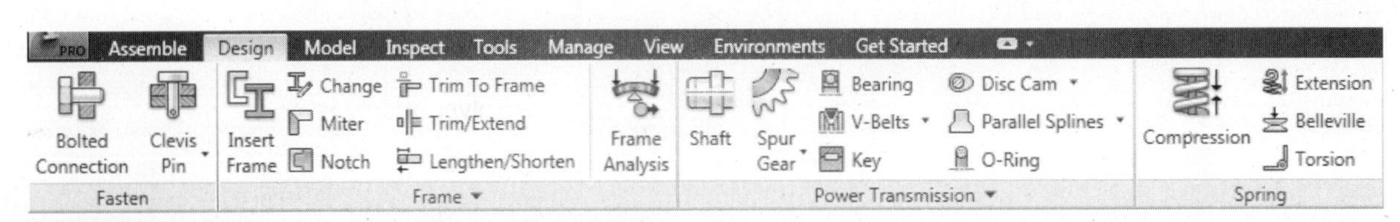

FIGURE 9.133

- Component and feature generators consist of bolted connections, shafts, gears, bearings, springs, belts, and pins.
- Mechanical calculators use standard mathematical formulas and theories to help you design components. The calculators available include weld, solder, hub joints, fit and tolerance, power screws, beams, and brakes.
- The *Handbook* contains engineering theory, formulas, and algorithms used in machine design.

To create parts and features with the Design Accelerator, follow these steps:

1. Open the assembly in which to place the part.
2. Activate the Design tab.
3. From the Design tab, select the operation you need to perform.
4. A wizard will walk you through the steps of the operation. An example of the Spur Gear Component Generator is shown in the following image.

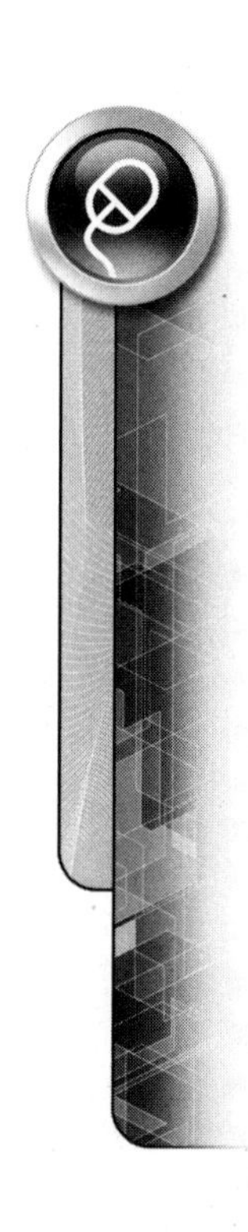

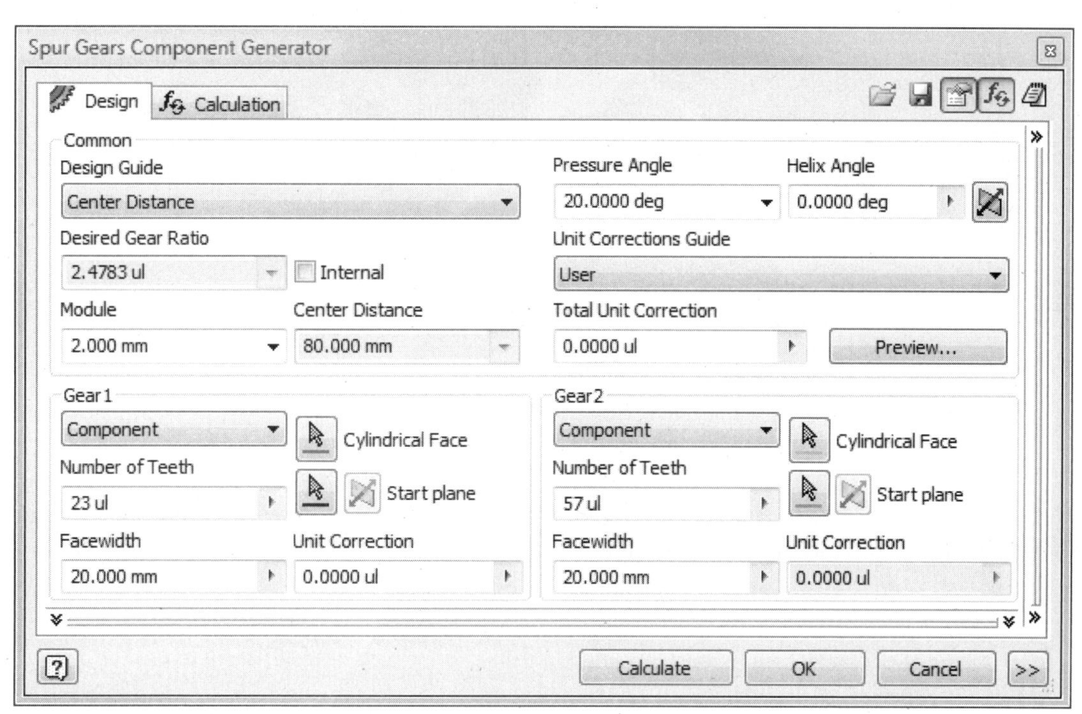

FIGURE 9.134

NOTE

For more information about the Design Accelerator, refer to Autodesk Inventor's Help system.

APPLYING YOUR SKILLS

Skill Exercise 9-1

In this exercise, you will create a contact set, and then use the Contact Solver to simulate intermittent motion.

1. Open *ESS_E09_07.iam*; see the following image on the left.
2. In the browser, right-click ESS_E09_07-0208-06P:1, and click Visibility to turn off the front plate, as shown in the following image on the right.

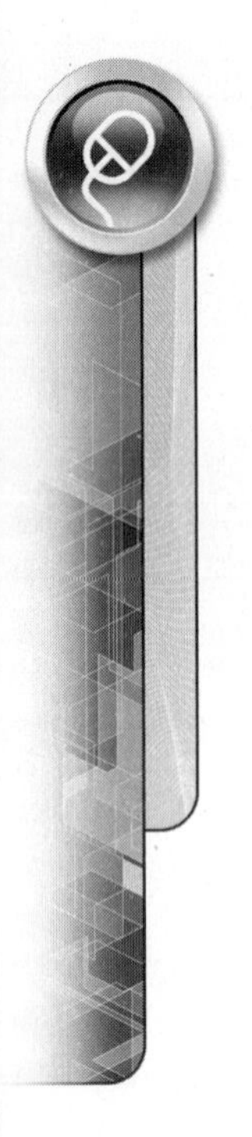

FIGURE 9.135

3. Expand the last entry in the browser, ESS_E09_07-0802-03P:1.
4. Right-click the Drive This angle constraint, and click Drive Constraint, as shown in the following image.
5. Click the Reverse arrow. Notice that the roller interferes with the other components in the assembly and has no effect on them. This is because the components are not included in a contact set.
6. Close the Drive Constraint dialog box.

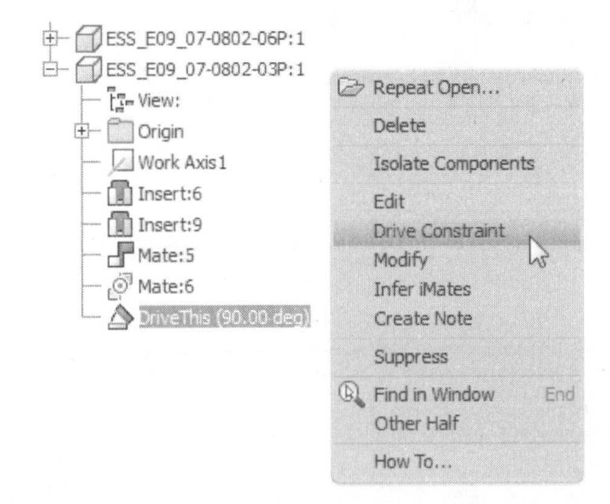

FIGURE 9.136

7. In the browser, select ESS_E09_07-0208-01P:1, DIN125-1 A A 7.4:1, ESS_E09_07-0802-03P:1, and ESS_E09_07-0802-06P:1.
8. Right-click one of the selected components, and select Contact Set from the menu, as shown in the following image on the left. Notice the contact set icon that is added next to the selected assembly components.

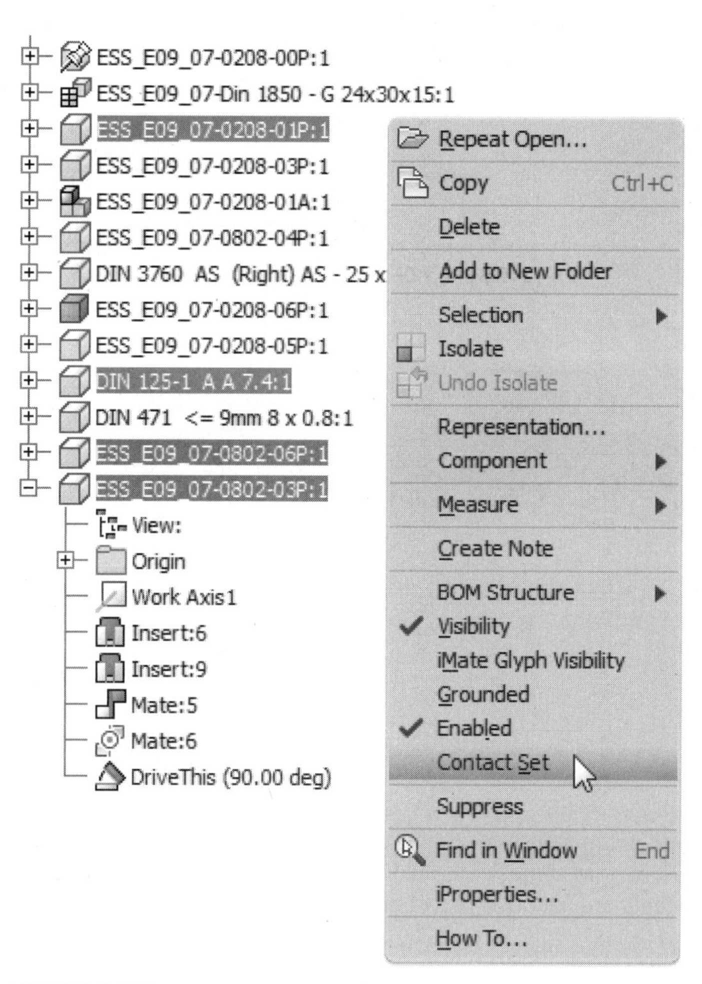

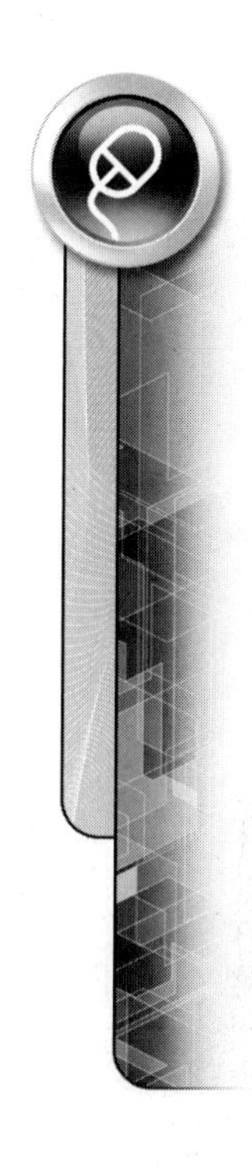

FIGURE 9.137

9. Right-click the Drive This angle constraint, and click Drive Constraint.
10. In the Drive Constraint dialog box, enter a Start value of **0**, and End value of **360**, an increment value of **3**, and Repetitions value of **2** as shown in the following image.

FIGURE 9.138

11. Click the Forward arrow. The roller moves to the start position and revolves until it contacts a slot in the follower component, as shown in the following image on the left. The motion continues while there is contact, as shown in the following image on the right.

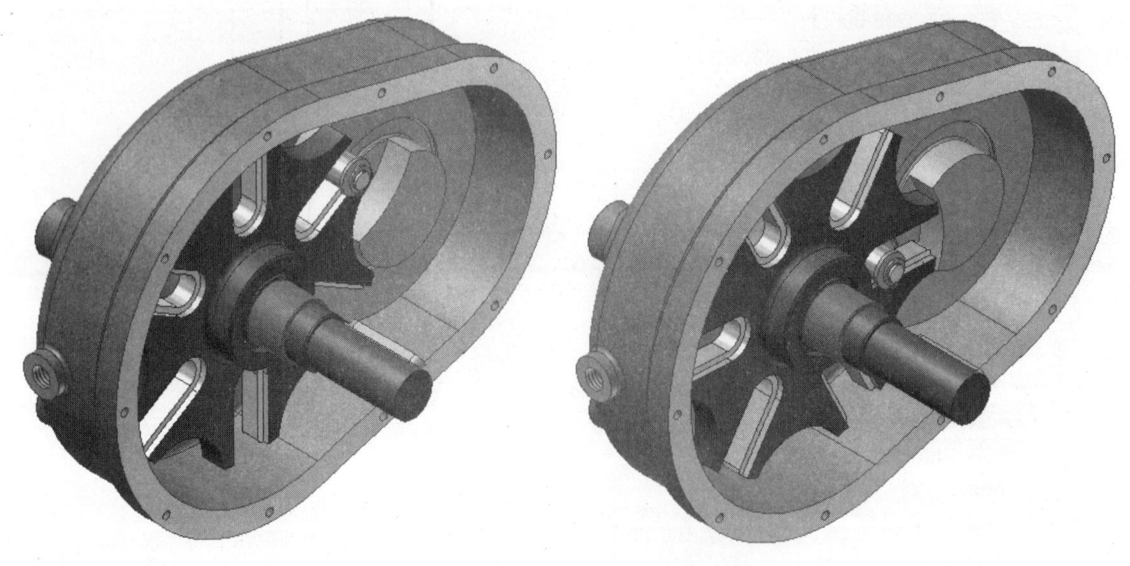

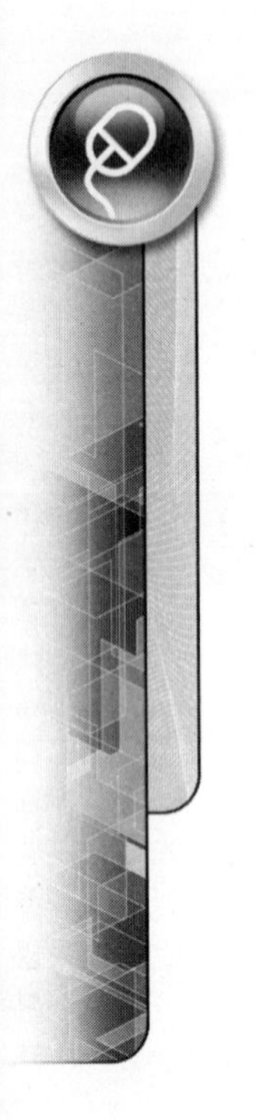

FIGURE 9.139

12. Close the Drive Constraint dialog box.
13. In the browser, right-click the Drive This angle constraint, and select Suppress from the menu.
14. Click and drag component ESS_E09_07-0802-03P:1. In addition to driving a constraint to work with components of a contact set, you can also "constrain-drag" components to view how components will behave when contact occurs and when the components are included in a contact set.
15. Close all open files. Do not save changes. End of exercise.

CHECKING YOUR SKILLS

Use these questions to test your knowledge of the material covered in this chapter.

1. True ___ False ___ A Design View Representation can control the display style, such as shaded or wireframe, of an assembly model.
2. What is the purpose of a positional representation?
3. When a new positional representation is added to the browser, how many are created by default?
   a. 1
   b. 2
   c. 3
   d. 4
4. When would you want to make an assembly have the Flexible property?
5. True ___ False ___ Assemblies can only be made flexible when used in another assembly as a subassembly.
6. When mirroring an assembly component, which icon would you use to determine that the component is being mirrored?
   a.
   b.
   c.
   d.

7. There are four levels of detail that are created/included with each assembly file. Explain why you may want to use the All Components Suppressed or the All Parts Suppressed levels of detail.
8. True __ False __ Locking a Design View Representation prevents further changes to a representation.
9. Explain what the Revert option of the Mirror Components: File Names dialog box does when mirroring components in an assembly.
10. True __ False __ When copying components in an assembly using the Copy command, the new components will have the suffix _CPY added to the their names and this cannot be changed.
11. True __ False __ When working with an assembly that has three occurrences of the same subassembly, with one occurrence set to Flexible that you have dragged to a new position in the canvas, the other two occurrences will be positioned the same as the occurrence that you have moved.
12. True __ False __ After you have created multiple drawing views based on a particular design view, you cannot change the design view that the drawing was based on.
13. True __ False __ If you have suppressed the work feature visibility of a component via the Object Visibility command, it is possible for the work feature to still be visible on the canvas.
14. True __ False __ Assembly features are defined in a part and affect multiple parts (or occurrences) in an assembly.
15. True __ False __ Once you have created Design View Representations in the model, they can be used to assist with the creation of presentation and/or drawing views.

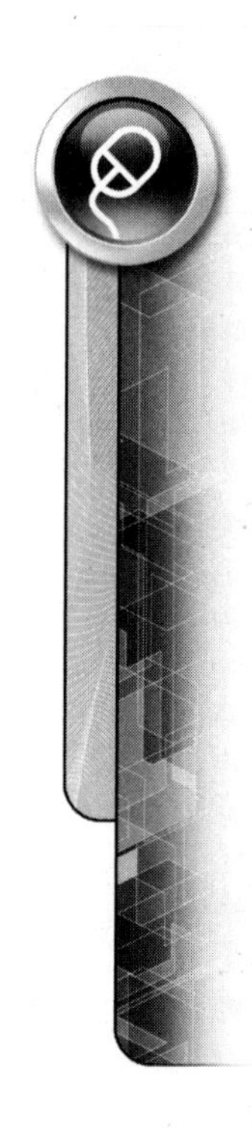

CHAPTER

10

Sheet Metal Design

INTRODUCTION

In this chapter, you will learn how to create sheet metal parts. Assemblies often require components that are manufactured by bending flat metal stock to form brackets or enclosures. Cutouts, holes, and notches are cut or punched from the flat sheet, and 3D deformations such as dimples or louvers are often formed into the flat sheet. The punched sheet is then bent at specific locations using a press brake or other forming tools to create a finished part. The sheet metal environment in Autodesk Inventor presents specialized tools for creating models in both the folded and unfolded states.

OBJECTIVES

After completing this chapter, you will be able to perform the following:

- Start the Autodesk Inventor sheet metal environment
- Modify settings for sheet metal design
- Create sheet metal parts
- Modify sheet metal parts to match design requirements
- Create sheet metal flat patterns
- Create drawing views of sheet metal parts

INTRODUCTION TO SHEET METAL DESIGN

A metal blank folded into a finished shape is a common component in a mechanical assembly. Examples include enclosures, guards, simple-to-complex bracketry, and structural members. Although the term sheet metal is often associated with these components, heavy plates can also be formed using similar methods.

Common sheet metal components include electronic and consumer product chassis and enclosures, lighting fixtures, support brackets, frame components, and drive guards.

Sheet metal fabrication can include a number of processes:

- Drawing
- Stamping
- Punching and cutting
- Braking
- Rolling
- Other, more complex operations

The tools in the Autodesk Inventor sheet metal environment enable you to create press brake or die-formed models. You can also create rolled shapes, including cones and cylinders. In addition, sheet metal parts can include formed features such as nail holes, lances, and louvers. Using lofts and surfacing tools, you can create parts that are fabricated through deformation processes such as drawing or stamping.

SHEET METAL PARTS

Sheet metal parts can be designed separately or in the context of an assembly. Since these parts are often used as supports or enclosures for other components, designing in the assembly environment can be advantageous.

Sheet Metal Design Methods

Sheet metal parts are most often created in the folded state, as shown in the following image on the left. The model is then unfolded into a flat sheet, as shown in the following image on the right, using the Create Flat Pattern command. To create the folded model, the first sketch is extruded the thickness of the sheet to create a face. Add additional faces or flanges to open edges of the part and add bends automatically between the features. Add cuts and other special sheet metal features to the part as required.

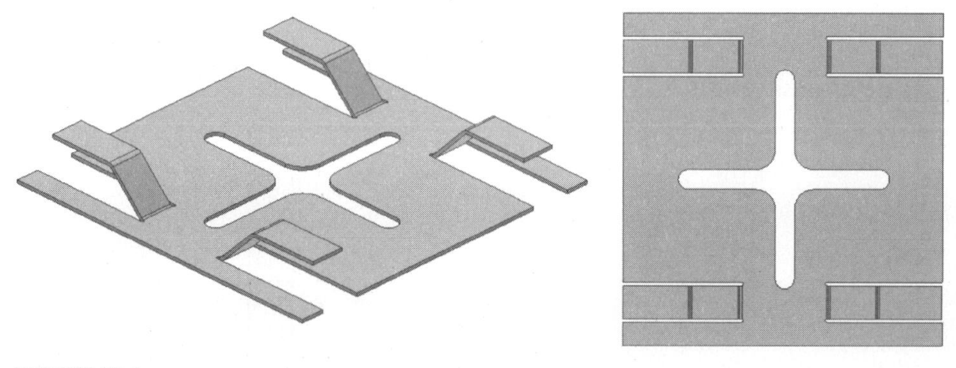

FIGURE 10.1

Autodesk Inventor's commands that create models with disjointed solids, or unconnected features, are very helpful when building sheet metal parts in an assembly, as shown in the following image. You can build separate faces of a sheet metal part quickly by referencing faces on other parts in the assembly. Additional sheet metal features can then join the distinct faces.

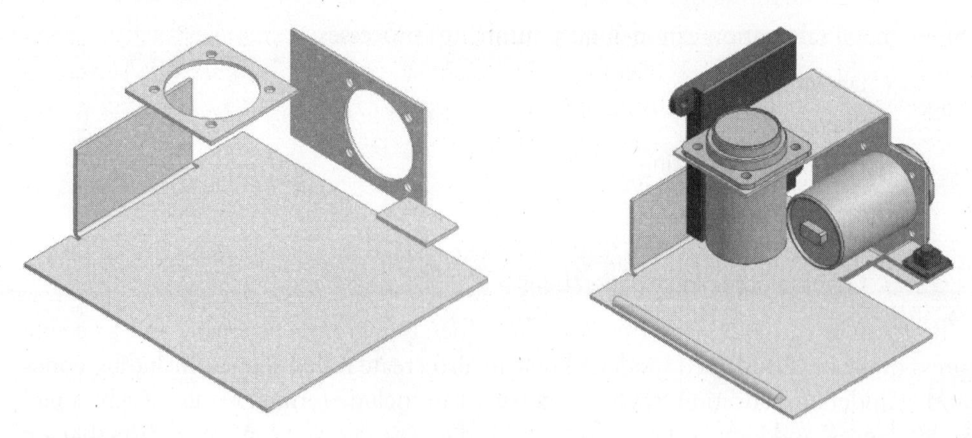

FIGURE 10.2

You can also create sheet metal parts from standard parts that have been shelled. All walls of the shelled part must be the same thickness and must match the thickness of the flat sheet as defined by the sheet metal rule. The solid corners of the shelled part can be ripped open to enable the model to unfold, and appropriate bends can be added along the edges between faces.

Creating a Sheet Metal Part

The first step in creating a sheet metal part is to select a sheet metal template with which to work. Autodesk Inventor comes with a sheet metal template named *Sheet Metal.ipt*, as shown in the following image.

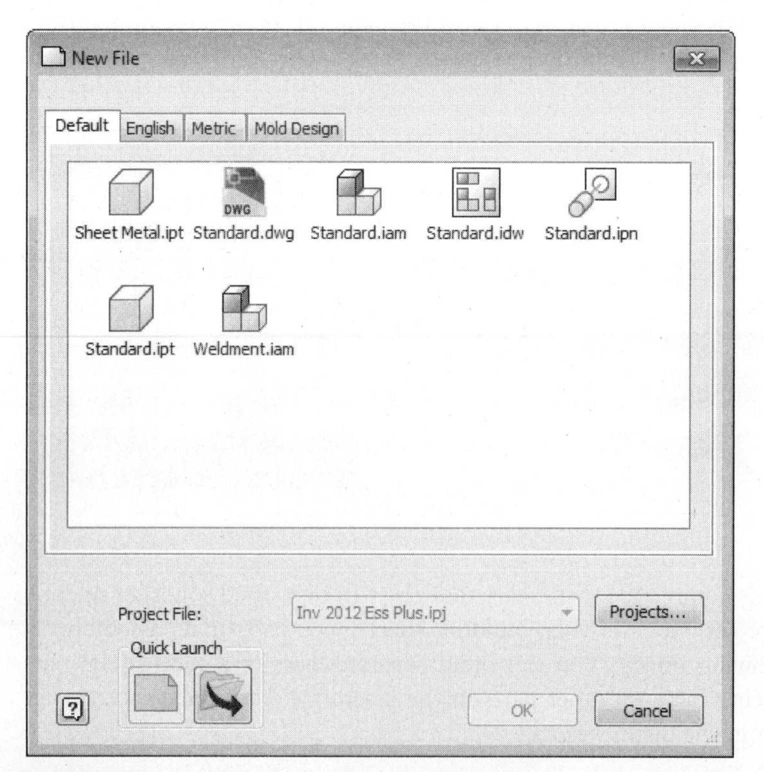

FIGURE 10.3

You can switch between the standard modeling environment and the sheet metal environment at any time by clicking either Convert to Standard Part or Convert to Sheet Metal from the Environments tab > Convert panel, as shown in the following image. You can use Autodesk Inventor's general modeling tools to add standard part features to a sheet metal part.

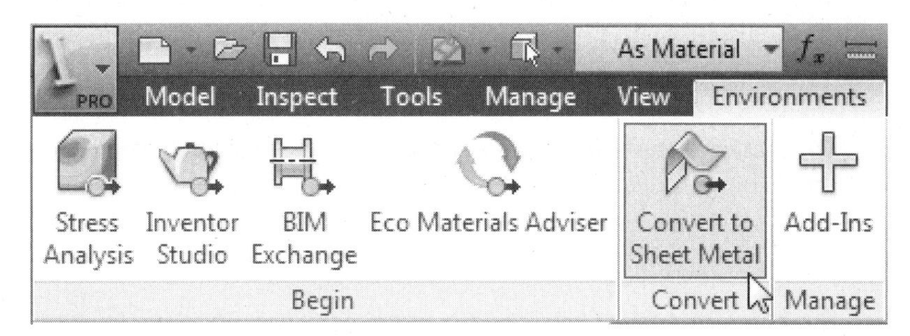

FIGURE 10.4

The creation of a sheet metal part usually begins by specifying the sheet metal defaults or rules such as sheet thickness, material, and bend radius. You then use sheet metal-specific commands to create parts by either adding faces at existing edges or connecting disjointed faces with bends. Use the optimized commands to add, cut, and clean up features of the part. Sheet metal faces and parts can be adaptive, and their size can be adjusted to meet design rules specified by assembly constraints.

SHEET METAL COMMANDS

After creating the new sheet metal document from a sheet metal template file, Autodesk Inventor's ribbon changes to reflect the sheet metal environment. As with standard parts, a sketch is created and becomes active. The sketch commands are common to both sheet metal and standard parts. The first sketch of a sheet metal part must be either a closed profile that is extruded the sheet thickness to create a sheet metal face or an open profile that is thickened and extruded as a contour flange. Use additional commands to add sheet metal features to the base feature. The following commands can be used to create sheet metal features. You can also add standard part features to a sheet metal part, but these features may not unfold when a flat pattern is generated from the folded model.

Use sheet metal commands to perform the following:

- Build a sheet metal part by adding faces along edges of existing faces.
- Create individual key faces of the sheet metal part and then connect these disjointed faces by adding bends between them.
- Extend faces automatically to create corner seams where face or flange edges meet.
- Cut shapes from faces with commands enhanced for sheet metal design.
- Add standard features such as chamfers and fillets, using commands that are optimized for working with thin sheets.
- Create a flat pattern model of the part with a single button click. This flat pattern model is updated automatically as features are added, removed, or edited.
- Create drawing views of the folded part and flat pattern to document your design for manufacturing.

Sheet Metal Rules

Sheet metal-specific parameters include the thickness and material of the sheet metal stock, a bend allowance factor to account for metal stretching during the creation of bends, and various parameters dealing with sheet metal bends and corners. The sheet metal-specific parameters of a part are stored in a sheet metal rule. Sheet metal rules are created and modified in the Style and Standard Editor found on the Manage tab > Styles and Standards panel. You can create additional sheet metal rules to account for various materials and manufacturing processes or material types. If you create sheet metal rules in a template file, they are available in all sheet metal parts based on that template. Sheet metal rules can also be created and managed using style libraries, which is the recommended workflow. Use of a style library makes your template files more lightweight and makes the management of sheet metal rules and other styles more robust.

The thickness of the sheet metal stock is the key parameter in a sheet metal part. Sheet metal commands such as Face, Flange, and Cut automatically use the Thickness parameter to ensure that all features are the same wall thickness, a requirement for unfolding a model. In the default sheet metal rule, all other parameters, such as Bend Radius, are initially based on the Thickness value.

To edit or create a sheet metal rule, follow these steps:

1. Click Styles Editor from the Manage tab > Styles and Standards panel. The Style and Standard Editor dialog box appears, as shown in the following image.

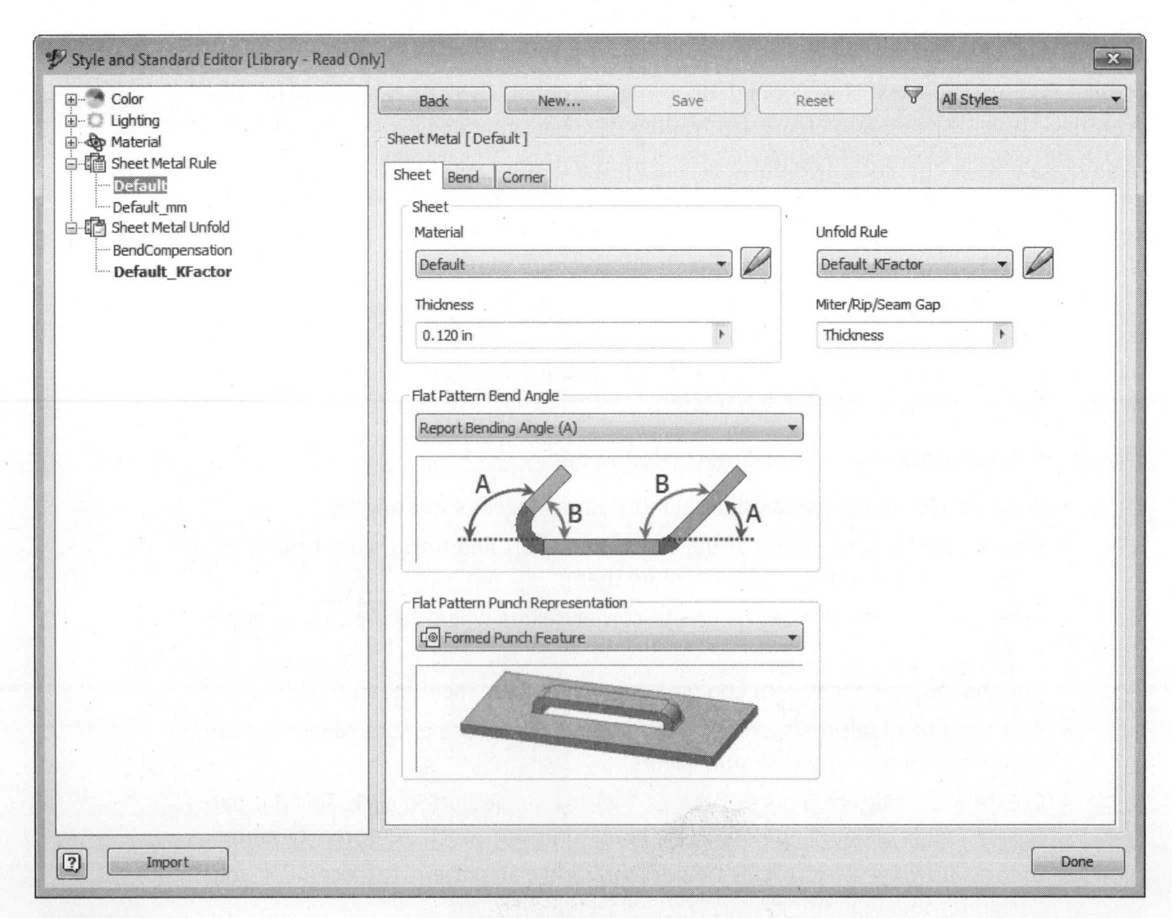

FIGURE 10.5

2. To create a new Sheet Metal Rule, select an existing rule and click the New button. This action creates a copy of the rule that is currently selected in the Sheet Metal Rule list.
3. Rename the rule and click the OK button.
4. Edit the values and settings on the Sheet, Bend, and Corner tabs to define the default feature properties of parts created with this sheet metal rule.
5. Click the Save button.
6. When multiple sheet metal styles exist in a part, set the active sheet metal rule in a part by selecting it in the Sheet Metal Defaults dialog box, shown below, which is accessed from the Sheet Metal Default command on the Sheet Metal tab > Setup panel, as shown in the following image. An alternate method of setting the active sheet metal rule is to double-click it from the list of Sheet Metal Rules in the Style and Standard Editor dialog box.

Changing to a different sheet metal rule or making changes to the active rule in the Style and Standard Editor updates the sheet metal part to match the new settings.

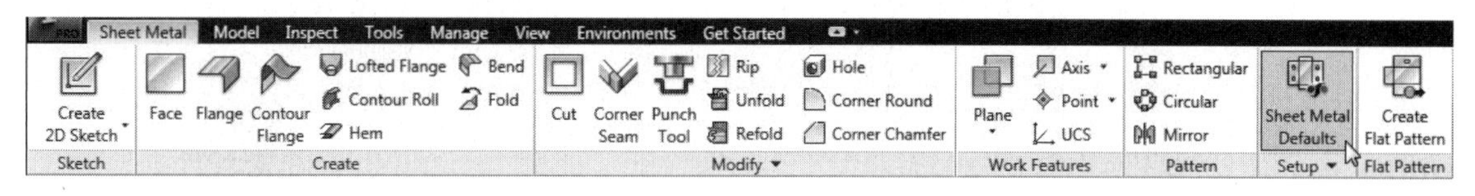

FIGURE 10.6

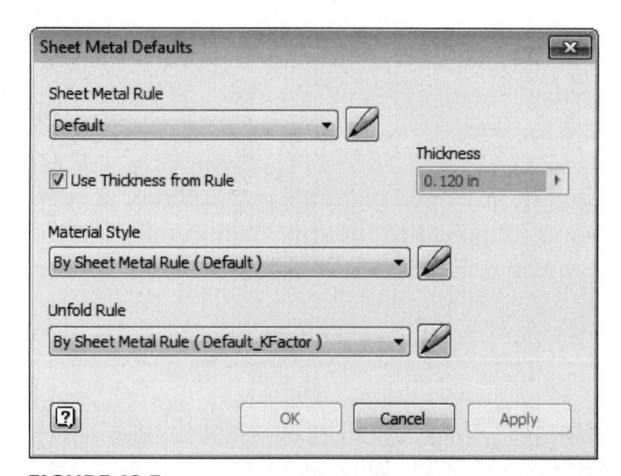

FIGURE 10.7

The following is a description of the settings for a sheet metal rule. The parameters with numeric values, such as Thickness and Bend Radius, are saved as Sheet Metal Parameters and are accessible in the Parameters dialog box, as shown in the following image. The parameter values are updated when you modify or activate a new sheet metal rule. Other model parameters and user-defined parameters can reference these parameters in equations.

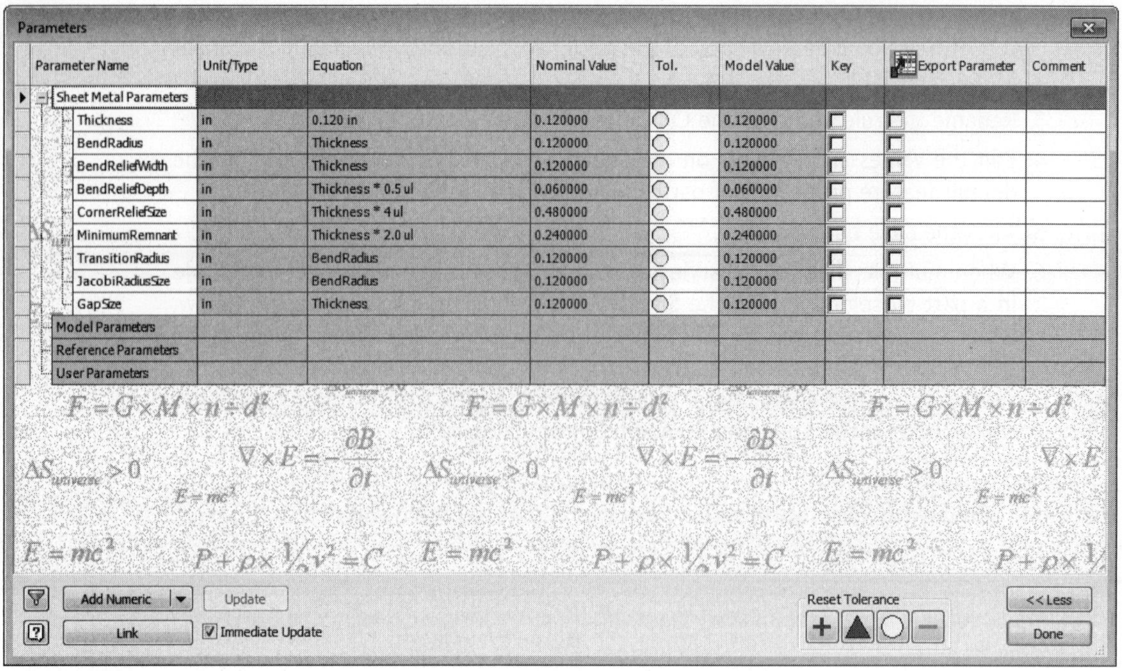
Parameters
Parameter Name
Unit/Type
Equation
Nominal Value
Tol.
Model Value
Key
Export Parameter
Comment
Sheet Metal Parameters
Thickness in 0.120 in 0.120000 0.120000
BendRadius in Thickness 0.120000 0.120000
BendReliefWidth in Thickness 0.120000 0.120000
BendReliefDepth in Thickness * 0.5 ul 0.060000 0.060000
CornerReliefSize in Thickness * 4 ul 0.480000 0.480000
MinimumRemnant in Thickness * 2.0 ul 0.240000 0.240000
TransitionRadius in BendRadius 0.120000 0.120000
JacobiRadiusSize in BendRadius 0.120000 0.120000
GapSize in Thickness 0.120000 0.120000
Model Parameters
Reference Parameters
User Parameters
Add Numeric
Update
Link
Immediate Update
Reset Tolerance
<< Less
Done

FIGURE 10.8

Sheet Tab

The Sheet tab contains settings for the sheet metal material, thickness, unfold rule, miter, rip, or seam gap, flat pattern bend angle, and the flat pattern punch representation. You can create multiple rules that contain different materials and thicknesses to be applied to a sheet metal model.

Material. Specify a material from the list of defined materials in the library or template file, and the material color setting is applied to the part. You can define new materials using the Material node of the Style and Standard Editor.

Thickness. Specify the thickness of the flat stock that will be used to create the sheet metal part.

Unfold Rule. Specify the rule or method used to calculate bend allowance, which accounts for material stretching during bending. The drop-down list provides access to the unfolding rules that are defined in the Sheet Metal Unfold node of the Style and Standard Editor.

Miter/Rip/Seam Gap. Specify a value to be used when creating features that require a miter, rip, or corner seam. The default gap is set to equal the Thickness parameter.

Flat Pattern Bend Angle. Specify the type of bending angle that you want to be reported. There are two options: bending angle or open angle. Based on the selection you choose, the appropriate angle will be used per the values designated in the dialog box.

Flat Pattern Punch Representation. Choose from one of four options to specify how you want a sheet metal punch to appear when the model is displayed as a flat pattern. These options allow you to display the punch as a formed punch feature, a 2D sketch representation, a 2D sketch representation with a center mark, or as a center mark only.

Bend Tab

The Bend tab contains settings for sheet metal bends, as shown in the following image. Most bend settings are typically defined as a function of the sheet thickness. Bend Radius refers to the inside radius of the completed bend. This setting is the default for all bends, but you can override it while creating any bend.

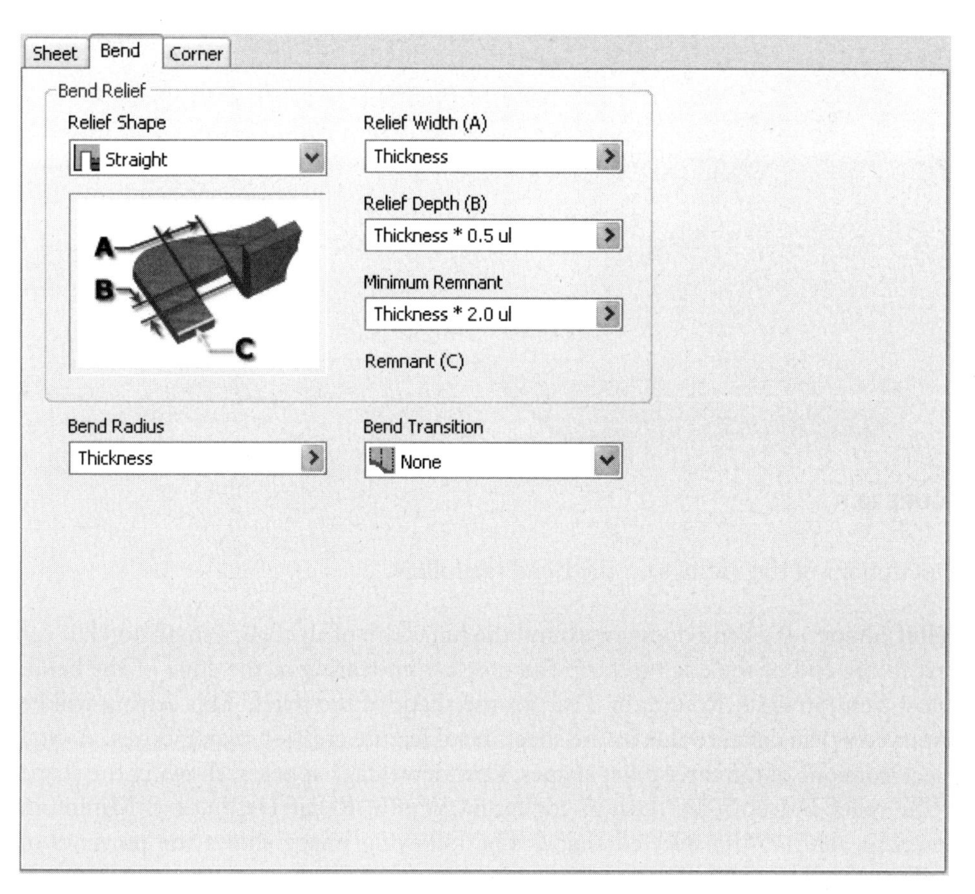

FIGURE 10.9

You can specify bend reliefs when a bend zone, the area deformed during a bend, does not extend completely beyond a face, as shown in the following image.

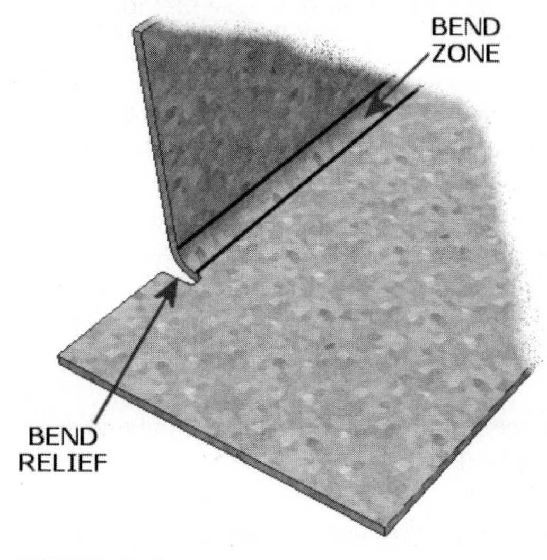

FIGURE 10.10

If bend reliefs are used, they are incorporated in the flat blank prior to folding. You can generate bend reliefs by punching or with laser, water-jet, or other cutting methods. Creating bend reliefs may increase production costs, and it is common practice with thin, deformable materials, such as mild steel, to add bends without bend reliefs. The material is allowed to tear or deform where the bend zone intersects the adjacent face, as shown in the following image.

FIGURE 10.11

Descriptions of the settings on the Bend tab follow.

Relief Shape. If a bend does not extend the full width of an edge, a small notch is cut next to the end of the edge to keep the metal from tearing at the edge of the bend. Select from Straight, Round, or Tear for the shape of the relief. This setting will be displayed as the default value in the sheet metal feature creation dialog boxes. As you select from one of the three relief shapes, a preview image appears, showing the shape of the relief and how the settings for Relief Width, Relief Depth, and Minimum Remnant apply to the selected shape. The following image shows the previews of the Straight, Round, and Tear shapes.

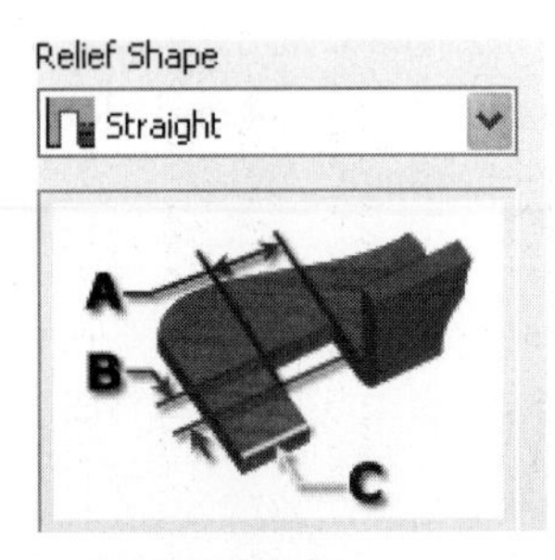

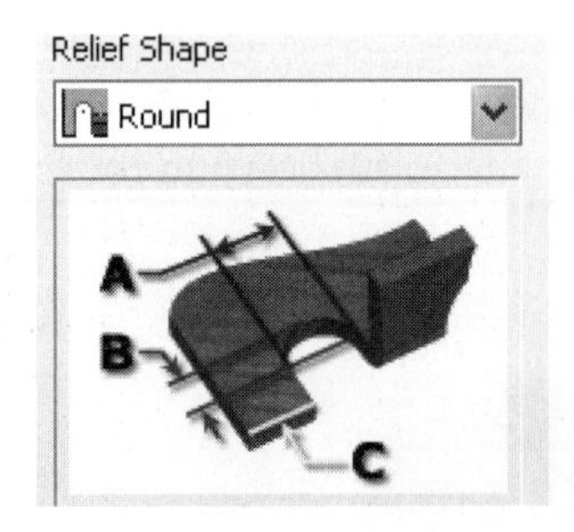

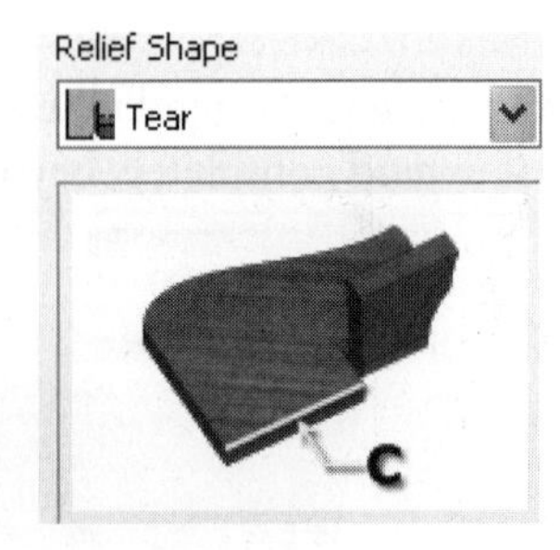

FIGURE 10.12

Relief Width. Specifies the width of the bend relief.

Relief Depth. Specifies the distance a relief is set back from an edge. Round relief shapes require this distance to be at least one-half of the Relief Width value.

Minimum Remnant. Specifies the distance from the edge at which, if a bend relief cut is made, the small tab of remaining material is also removed.

Bend Radius. Defines the inside bend radius value between adjacent, connected faces.

Bend Transition. Controls the intersection of edges across a bend in the flattened sheet. For bends without bend relief, the unfolded shape is a complex surface. Transition settings simplify the results, creating straight lines or arcs, which can be cut in the flat sheet before bending. The following image shows the five transition types.

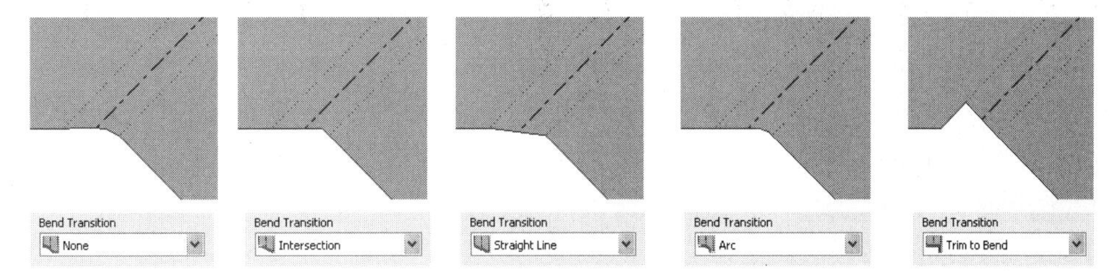

FIGURE 10.13

Corner Tab

A corner occurs where three faces meet. The corner seam feature controls the gap between the open faces and the relief shape at the intersection. As with bend reliefs, corner reliefs are added to the flat sheet prior to bending.

By using the Corner tab, as shown in the following image, you can set how corner reliefs will be applied to the model. You can designate the corner relief size, shape, and radius.

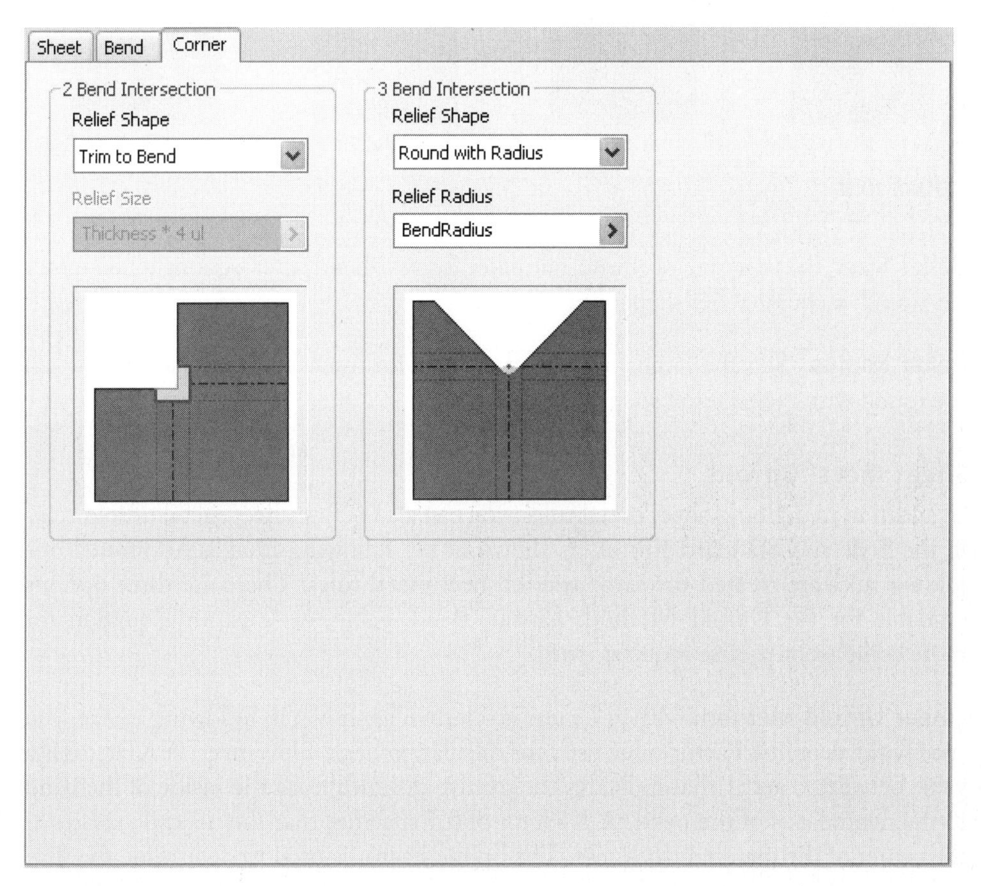

FIGURE 10.14

Relief Shape. Specify the shape of the corner relief for either two- or three-bend intersections. When a two-bend intersection is formed in your model, you can select from one of six corner relief options, as shown in the following image.

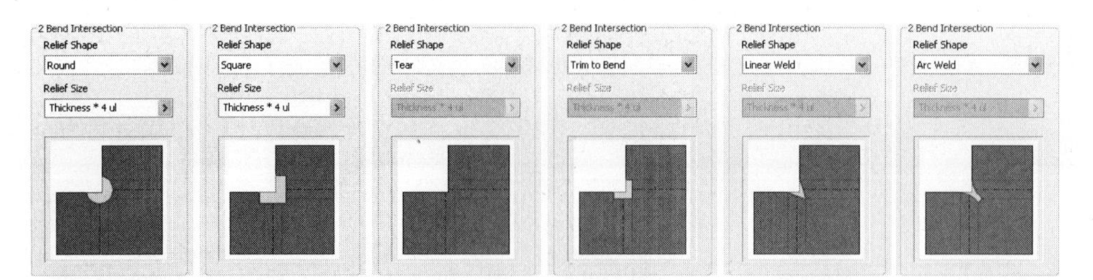

FIGURE 10.15

When a three-bend intersection, also referred to as a Jacobi corner, is formed in your model, four relief shapes can be used, as shown in the following image.

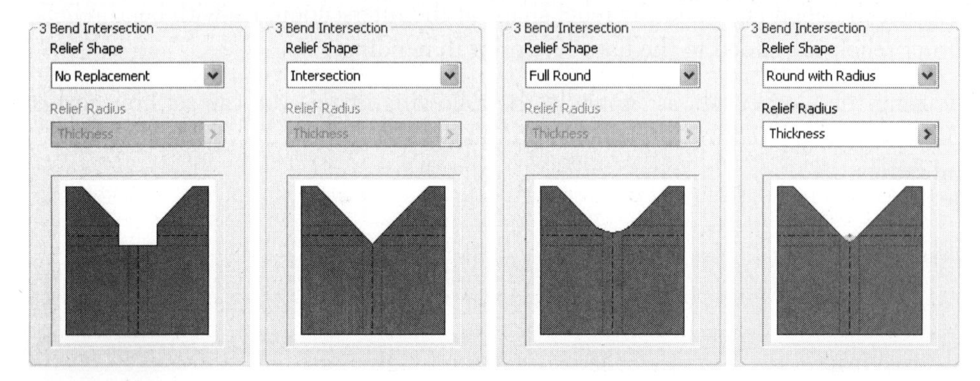

FIGURE 10.16

Relief Size. Sets the size of the corner relief for two-bend intersections when either the round or square relief shapes are selected.

Relief Radius. Sets the radius of the corner relief for three-bend intersections when the round with radius relief shape is selected.

Sheet Metal Unfold

In addition to defining sheet metal rules, you can also define sheet metal unfold rules in the Style and Standard Editor, as shown in the following image. Additional unfolding rules are created the same way as sheet metal rules. There are three options available for the Unfold Method: Linear, Bend Table, or Custom Equation for more complex or precise requirements.

Linear Unfold Method. When Linear is selected, as shown in following image, you specify the default KFactor value used for calculating bend allowances. A KFactor is a value between 0 and 1 that indicates the relative distance from the inside of the bend to the neutral axis of the bend. A KFactor of 0.5 specifies that the neutral axis lies at the center of the material thickness. The Spline Factor is used to determine the flattened size of specific sheet metal features and corresponds with specific sheet metal features when flattened—contour flanges, contour rolls, and lofted flanges—that result in a spline when converted to their nonformed state. The default value is

0.5, which can be adjusted up or down to represent your manufacturing requirements. The Unfold Method is combined with the Unfolding Rule specified on the Sheet tab of the sheet metal rule.

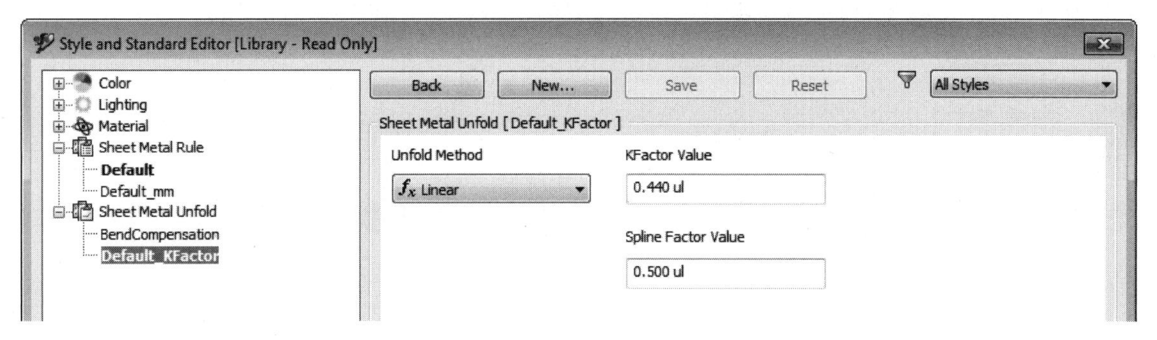

FIGURE 10.17

Bend Table Unfold Method. When Bend Table is selected from the Unfold Method list, you have the ability to create, import, or export a bend table. Autodesk Inventor can use a bend table that is a plain text file with a *.txt* extension. You can also create or edit a spreadsheet version of the bend table by using Microsoft Excel and then saving the table in text format.

Autodesk Inventor includes both metric (mm) and imperial unit (in) bend tables that you can modify, and each bend table is supplied in both *.txt* and *.xls* formats. The text file follows a rigid format; refer to the samples that ship with the product and are located in the *\Design Data\Bend Tables* directory (the exact location will depend on your operating system). The following image shows one of the sample bend tables that has been pasted into the Style and Standard Editor.

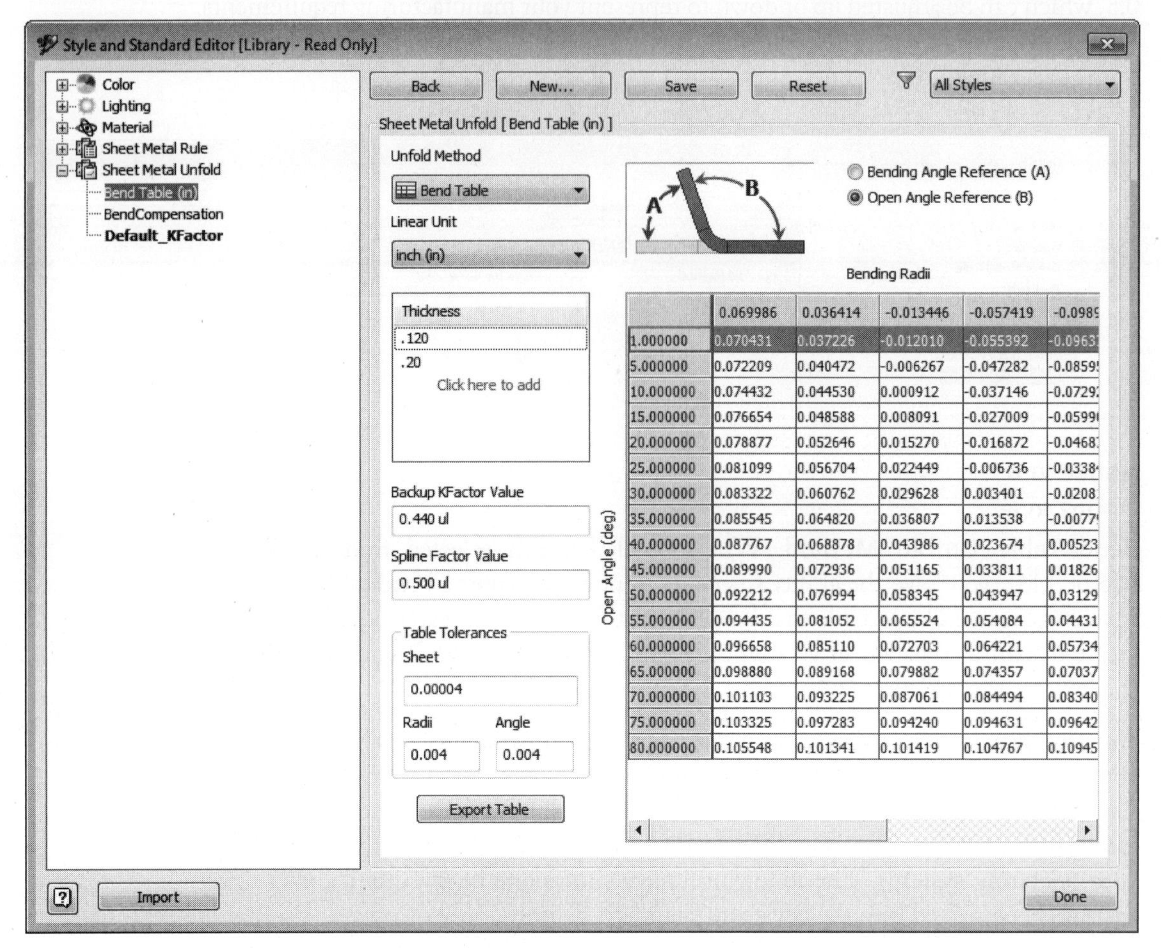

FIGURE 10.18

The angle information required in a bend table is shown in the following image. A flat sheet bent at an angle of 110° will have an open angle of 70°. The open angles are entered into the bend table. The sample bend tables use this method of entering the open angle, but by using the Unfold rules, you can declare how to read the angular values, whether they are open angles or bending angles, as shown in the previous image. Changing the control for the angle does not edit the bend table data, just how it is interpreted.

Unfolding rules are referenced by sheet metal rules that are created and saved in either a sheet metal template file or published to a style library for reuse in other sheet metal parts. Bend tables created in the Style and Standard Editor are stored as a Style Definition File in the *.styxml* format. It is possible to export the tables as an ascii.*txt* file. Exporting to *.txt* may be a requirement if you work in a mixed Autodesk Inventor version environment.

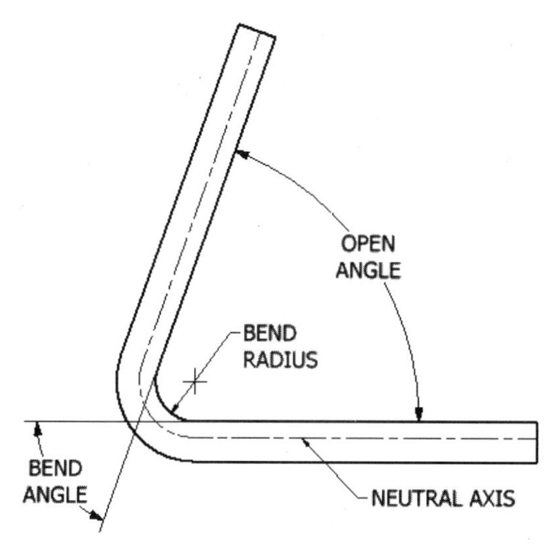

FIGURE 10.19

NOTE

When you open an older sheet metal part (prior to Inventor 2009), sheet metal styles are converted to sheet metal rules that have the same name as those defined in the older file. Material styles and unfold methods are also automatically converted to their Inventor 2009+ counterparts and are set up in the Style and Standard Editor.

Custom Equation Method. The Custom Equation should be used when you need additional control over the unfolding of your model. There are four equation types from which to select: Bend Compensation, Bend Allowance, Bend Deduction, and KFactor. Each type of equation enables you to enter different variables that can be provided as input for the unfolding technique. The image in the dialog box will change and display the definition of the variables that you enter into the table. In the equation table, you enter the custom equation and the bounding condition for the equation entered. Based on the equation type selected, you enter the variable that you want to solve for in the left column. The center column, custom equation, is where you enter the equation used to solve for the variable you entered. The bounding condition column designates the upper and lower limits for the equation in the middle column.

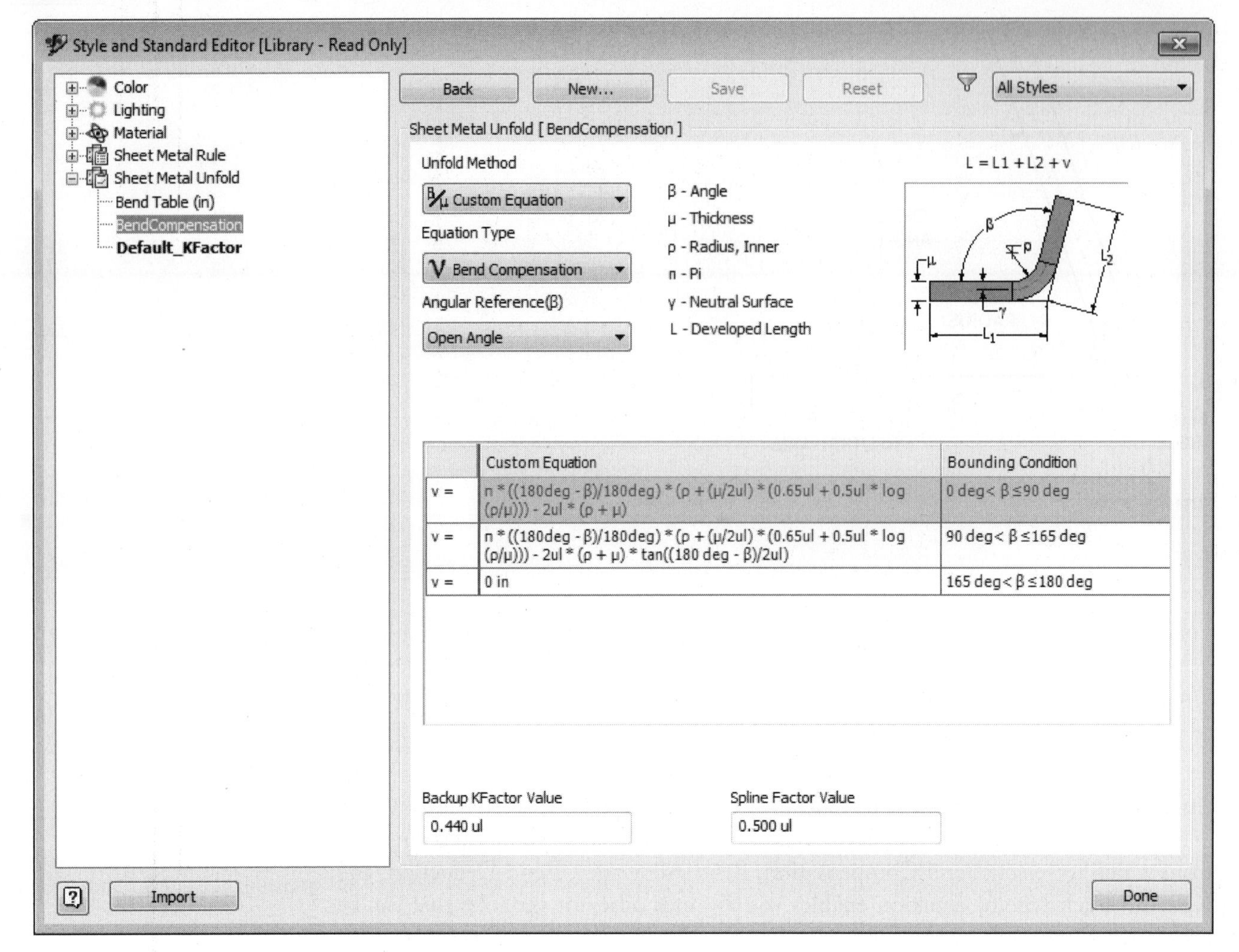

FIGURE 10.20

For additional information on creating or working with either bend tables or custom equations, refer to the Autodesk Inventor help system.

Sheet Metal Defaults

After you have defined sheet metal rules, you use the Sheet Metal Defaults command, as shown in the following image, to select the rule that is active for the sheet metal part.

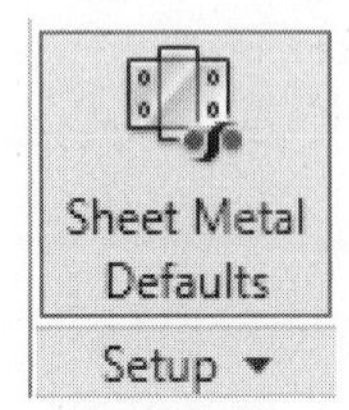

FIGURE 10.21

Once the command is selected, the Sheet Metal Defaults dialog box appears, as shown in the following image.

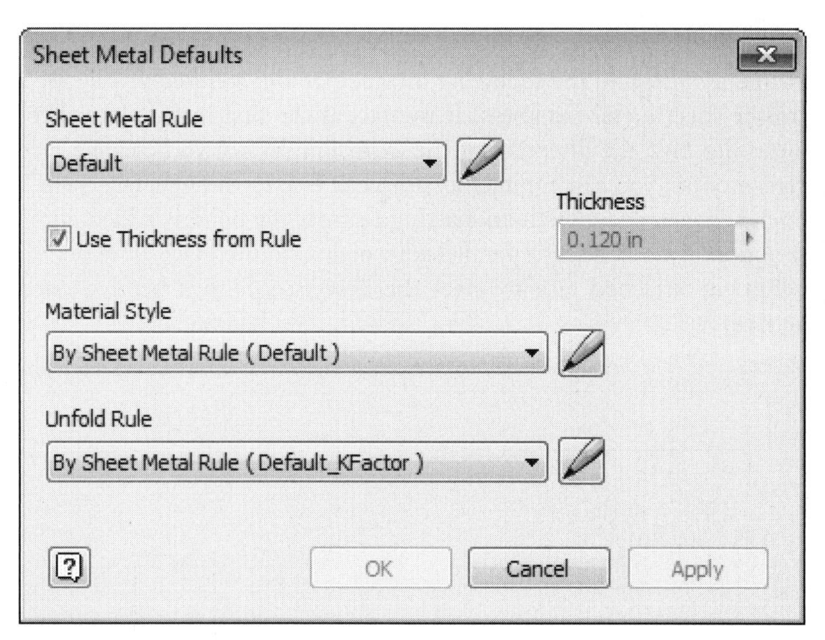

FIGURE 10.22

From this dialog box you can select a sheet metal rule that is defined in the Style and Standard Editor from the drop-down list. All settings defined in that rule are then applied to the sheet metal part that you are working on. The material style and unfolding rule can also be selected. By default, they are referenced from the sheet metal rule, and the text in the field tells you that the style and unfolding rule are defined By Sheet Metal Rule and then displays the specific name of the style or unfold rule in parentheses. You can change these settings so they are not the same as those defined in the sheet metal rule or you can click the Edit button (pencil icon) to be taken directly to the specific rule or style in the Style and Standard Editor to be modified. If you deselect the Use Thickness from Rule checkbox, the Thickness field becomes active. This allows you to specify a different thickness than that designated in the sheet metal rule. The ability to edit this field corresponds to the Thickness parameter in the Parameters dialog box. When the field is active in the sheet metal default, the field is also available for edit via the Parameters dialog box. When the checkbox is selected, as shown in the previous image, the Thickness parameter cannot be edited via the Parameters dialog box and is driven based on the value specified in the sheet metal rule.

The items displayed when selecting one of the drop-down lists in the Sheet Metal Defaults dialog box are controlled by the filter specified in the top right corner of the Style and Standard Editor, as shown in the following image. When Local Styles is selected, only those rules or styles that are stored inside of the active file are displayed. If All Styles is selected, you will see all styles that are available in both the active file and the styles library.

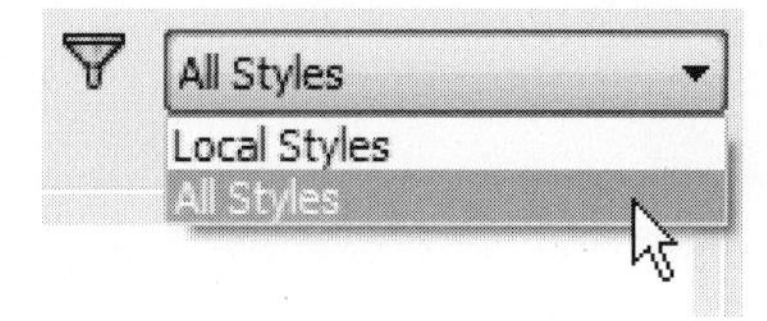

FIGURE 10.23

Face

The Face command, as shown in the following image, extrudes a closed profile for a distance equal to the sheet metal thickness. If the face is the first feature in a sheet metal part, you can flip only the direction of the extrusion. When you create a face later in the design process, you can connect an adjacent face to the new face with a bend. If the sketch shares an edge with an existing feature, the bend is added automatically. As an option, you can select a parallel edge on a disjointed face. This action will extend or trim the attached face to meet the new face, with a bend created between the two faces.

FIGURE 10.24

To create a sheet metal face, follow these steps:

1. Create a sketch with a single closed profile or a closed profile containing islands. The sketch is most often on a work plane created at either a specific orientation to other part features or by selecting a face on another part in an assembly.

NOTE **You can create a single face from multiple closed profiles.**

2. Click the Face command on the Sheet Metal tab > Create panel. The Face dialog box appears, as shown in the following image.

   If a single closed profile is available, it is selected automatically. If multiple closed profiles are available, you must select which one defines the desired face area.
3. If required, flip the thickness direction to extrude the profile.
4. If the face is not the first feature and the sketch does not share an edge with an existing feature, click the Edges button and select an edge to which to connect the face. The two faces are extended or trimmed as required, meeting at a bend. The bend is listed as a child of the new face feature in the browser. If the face attached to the selected edge is parallel to but not coplanar with the new face, the dialog box is expanded, and the double-bend options are available. An additional face is added to connect the two parallel faces. The orientation and shape of this face is determined by the selected double-bend options. The options for creating a double bend or full-radius bend can be accessed by clicking the More (<<) button. These options are discussed later in the section on bends.

   The Unfold Options and Bend tabs contain settings for overriding the default values in the sheet metal style. Overrides are applied to individual features.
5. Click OK.

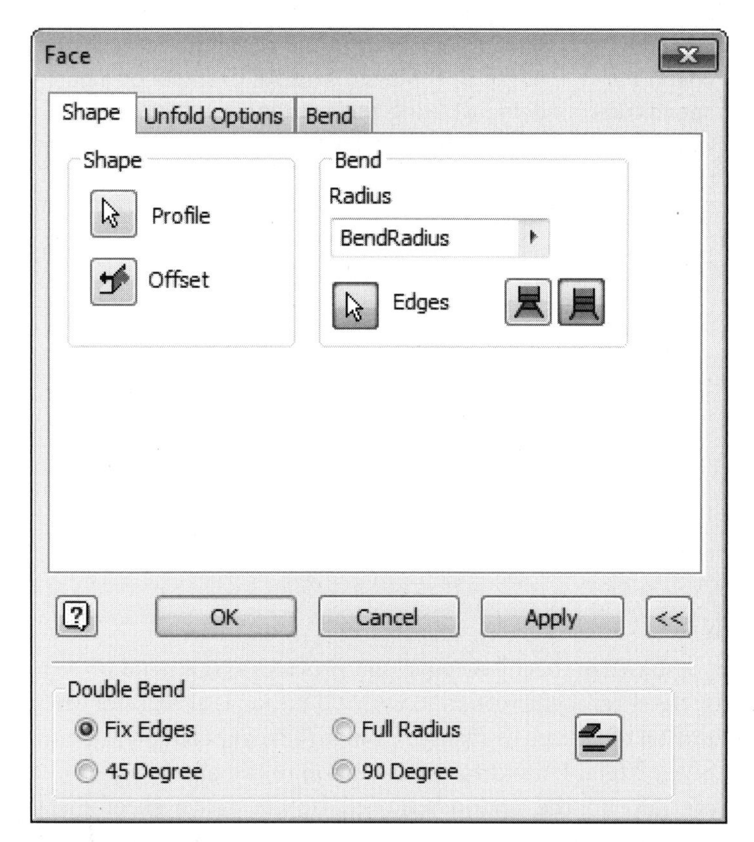

FIGURE 10.25

Shape Tab

By using the Shape tab, you can select the profile and direction of the face. You can also override the bend radius specified in the sheet metal rule, and you can select edges of other existing faces or edges that exist in the sheet metal part to apply a bend feature upon creation of the face. The following options are available on the Shape tab.

	Profile	Select a profile(s) to extrude a distance equal to the Thickness parameter.
	Offset	Toggle the direction for the creation of the feature.
Radius BendRadius	Radius	This field displays the default bend radius specified in the active sheet metal style. If the face will be attached to another edge of the part, you can specify the value for the bend radius to be used. You can modify the radius value on a per-feature basis.
	Edges	Select an edge to include in the bend. When selected, the edge is extended to match the edge of the face. A bend feature is created between the two faces.
	Extend Bend Aligned to Side Faces	Click to extend material along the faces of the edges connected by the bend instead of perpendicularly to the bend axis. An image of a part before this setting is applied, is shown in the following image on the left, and an example of the part when this option is active is shown on the right.

	Extend Bend Perpendicular to Side Faces	Click to extend material along the faces of the edges to the bend axis. An example of a part demonstrating this setting is shown in the middle of the following image.

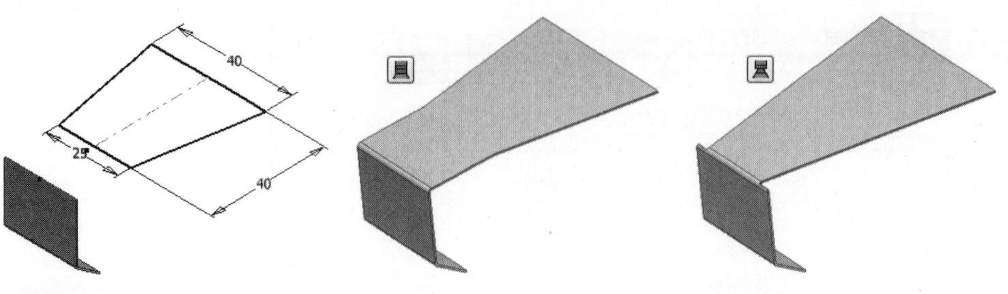

FIGURE 10.26

Unfold Options Tab

The Unfold Options tab, as shown in the following image, provides a command to override the unfold style set by the sheet metal rule, if necessary. This tab is also useful when working with other sheet metal tools such as Flange, Contour Flange, Hem, and Bend. Refer to the section on Sheet Metal Unfold for a description of the unfold style. The word Default will appear next to the option specified by the active sheet metal rule. This action is consistent in other dialog boxes, and it helps you to know how the active rule is defined without having to reopen the Sheet Metal Defaults dialog box.

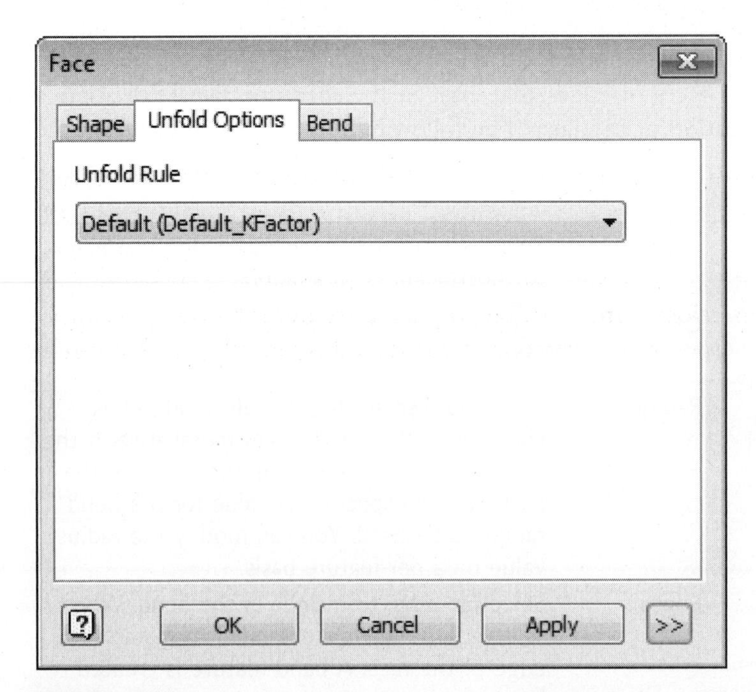

FIGURE 10.27

Bend Tab

The Bend tab, as shown in the following image, allows you to override the values set in the sheet metal rule, if necessary. As with the Unfold Options tab, the settings designated by the active sheet metal rule are displayed in parentheses with the word Default in front of them. This tab is also helpful when working with a number of the sheet metal tools such as Flange, Contour Flange, Hem, and Bend. These settings allow you to override how bends and bend reliefs are created on a per-feature basis. Refer to the section on sheet metal rules for a description of the settings.

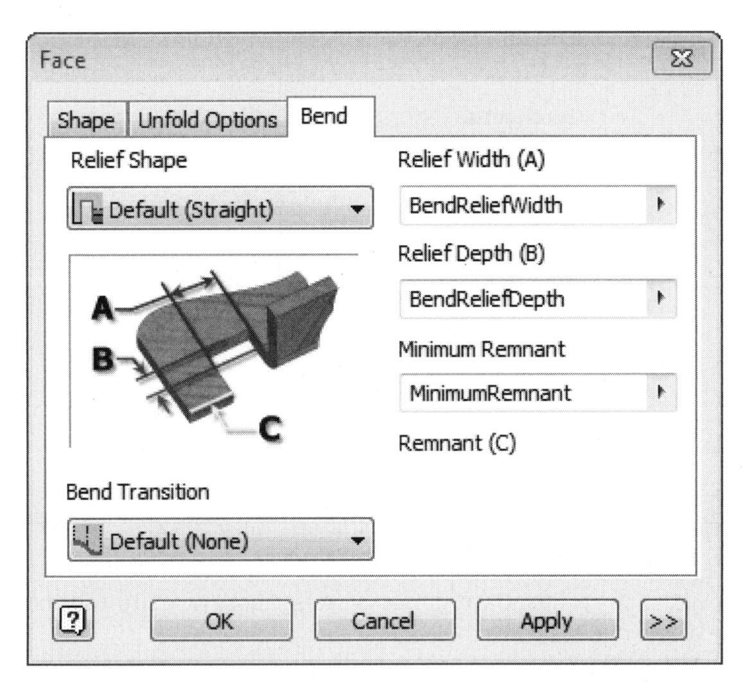

FIGURE 10.28

Contour Flange

Using the Contour Flange command shown in the following image, you create a contour flange from an open sketch profile. The sketch is extruded perpendicular to the sketch plane, and the open profile is thickened to match the sheet metal thickness. The profile does not require a sketched radius between line segments—bends are added at sharp intersections. Arc or spline segments are offset by the sheet metal thickness. A contour flange can be the first feature in a sheet metal part, or it can be added to existing features.

FIGURE 10.29

To create a contour flange, follow these steps:

1. Create a sketch with a single open profile. The sketch can contain line, arc, and spline segments, and it can be constrained to projected reference geometry to define a common edge between the contour flange and an existing face.

2. Click the Contour Flange command on the Sheet Metal tab > Create panel. The Contour Flange dialog box appears, as shown in the following image.

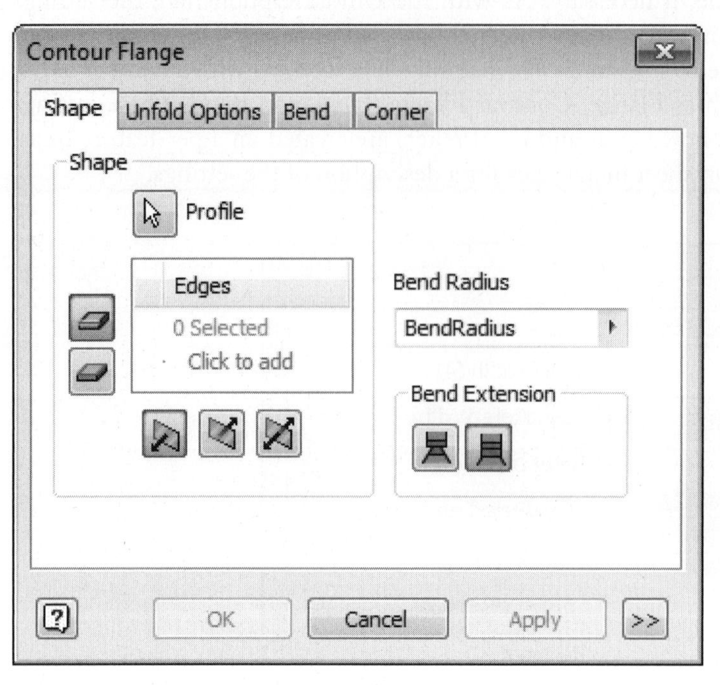

FIGURE 10.30

3. Click the open sketch profile to define the shape of the contour flange.
4. If required, flip the side to offset the profile or select both sides to use the selected profile as the mid-plane for the contour flange.
5. If the contour flange is the first feature in the part, enter an extrusion distance and direction. If the contour flange is not the first feature, you can select an edge perpendicular to the sketch plane to define the extrusion extents. If the sketch is attached to an existing edge, select that edge. If the sketch is not attached to any projected geometry, the contour flange is extended or trimmed to meet the selected edge. The selected edge is typically the closest edge to the end of the sketch. You can also use Loop Select Mode to have the flange created along all edges of a selected loop.
6. Apply any unfold, bend, or corner overrides on the Unfold Options, Bend, and Corner tabs.
7. If you select an edge to define the extrusion extents, you can select from five options, available by expanding the dialog box, to further refine the length of the contour flange. The options are described below.
8. Click OK.

Shape Tab

The Shape tab allows you to select the profile, edge, and offset direction of the contour flange. You can also override the bend radius that is specified in the sheet metal rule. The following options are available on the Shape tab.

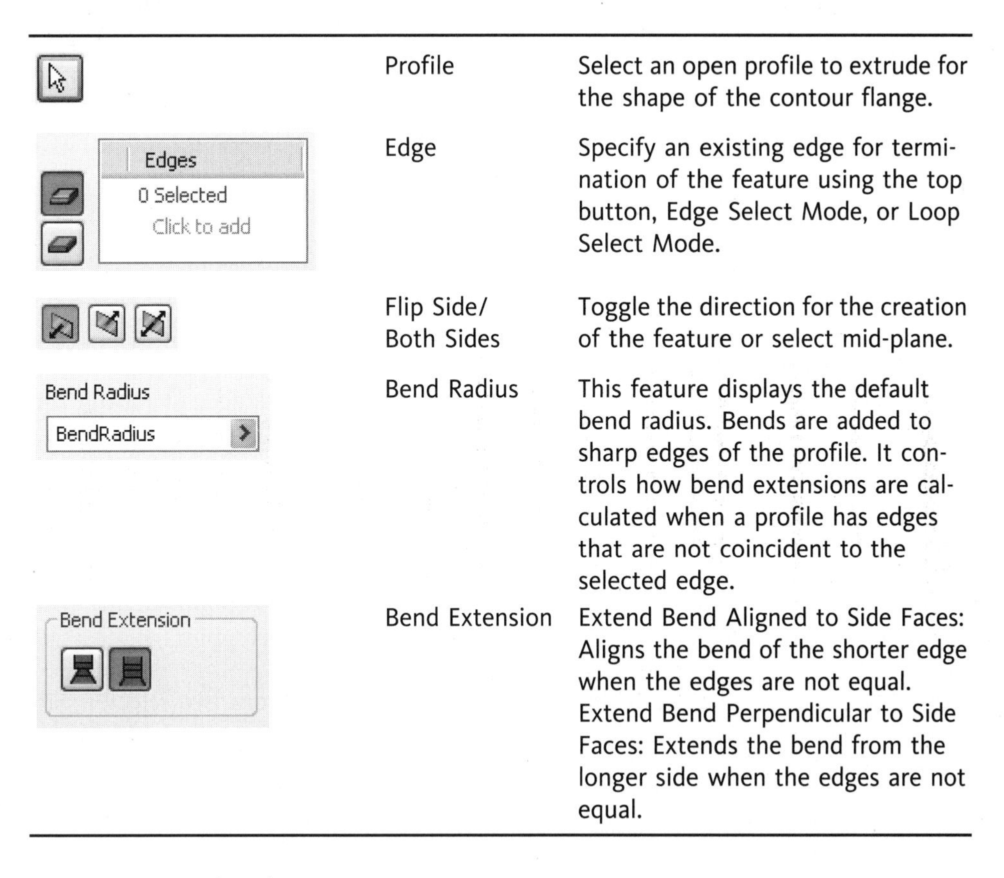

Option	Description
Profile	Select an open profile to extrude for the shape of the contour flange.
Edge	Specify an existing edge for termination of the feature using the top button, Edge Select Mode, or Loop Select Mode.
Flip Side/ Both Sides	Toggle the direction for the creation of the feature or select mid-plane.
Bend Radius	This feature displays the default bend radius. Bends are added to sharp edges of the profile. It controls how bend extensions are calculated when a profile has edges that are not coincident to the selected edge.
Bend Extension	Extend Bend Aligned to Side Faces: Aligns the bend of the shorter edge when the edges are not equal. Extend Bend Perpendicular to Side Faces: Extends the bend from the longer side when the edges are not equal.

More Button (<<)

The More button allows you to specify the type of extents for the feature, as shown in the following image.

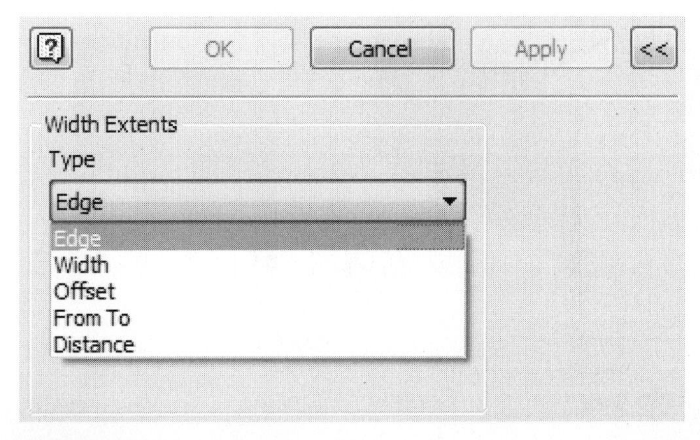

FIGURE 10.31

These extent types are also available for flange and hem features. However, the flange and hem features do not include the Distance option because the distance is specified in the main body of the dialog box. The five width extents options are shown in the following image, and their descriptions follow.

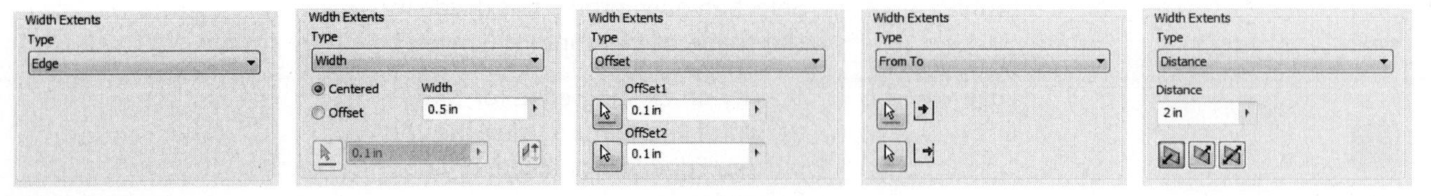

FIGURE 10.32

Edge. Select so that the contour flange extends the full length of the selected edge.

Width. Select so that a point defines one extent of the contour flange. The flange can be offset from this point and extended a fixed distance. You define the starting point by selecting an endpoint on the selected edge, a work point on a line defined by the edge, or a work plane perpendicular to the selected edge. You can also center the feature based on the selected edge and then specify the width of the created feature.

Offset. Similar to the Width option, select so that two selected points define both extents of the flange. You can specify an offset distance from each point.

From To. Select so that the width of the feature is defined from two selected part features. You can select work points, work planes, vertices, or planar faces that intersect the selected edge.

Distance. Select to enter a distance over which you want the contour flange to be extruded and specify a direction. The following image shows samples of the Edge, Width, and Offset extent types.

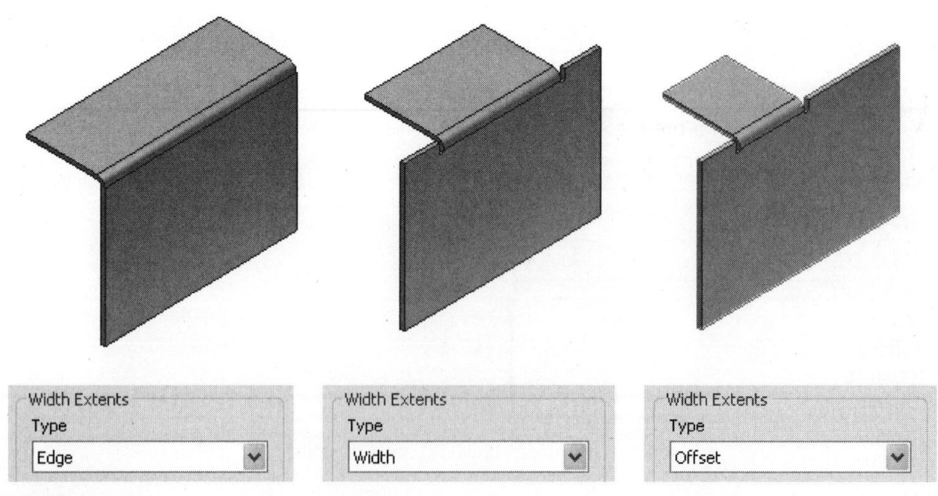

FIGURE 10.33

Flange

A sheet metal flange is a simple rectangular face created from existing face edges. A sketch is not required when creating a flange. The flange can extend the full length of the selected edge and can create automatic corner and miter features. The Flange command, as shown in the following image, adds a new sheet metal face and bend to an existing face. If multiple edges are selected for the flange feature, corner seams and miter features are also created. You set the flange length and the angle relative to the adjacent face in the Flange dialog box. The selected edge or edges define the bend location between the selected faces.

FIGURE 10.34

> **NOTE** A minimum of one sheet metal face must exist before creating a flange.

To create a flange, follow these steps:

1. Click the Flange command on the Sheet Metal tab > Create panel. The Flange dialog box appears, as shown in the following image.
2. Select an edge or edges of an existing face.
3. Enter the distance and angle for the flange in the Flange dialog box. The flange preview updates to match the current values.
4. If required, flip the direction for the flange, offset the thickness, and modify the datum used for the height measurement or the position of the bend for the feature.
5. Expand the dialog box, and select the appropriate extent type. See the Contour Flange section in this chapter for additional information on extents. When creating a flange feature, a Design option is available when the dialog box is expanded, as shown in the following image. When selected, the options that were introduced specifically in Autodesk Inventor 2008 are disabled. If you are working with a file created in a release prior to Autodesk Inventor 2009, the old method is selected, and the options that were used and available to create the feature in the previous release are available. Other options are disabled in the dialog box.

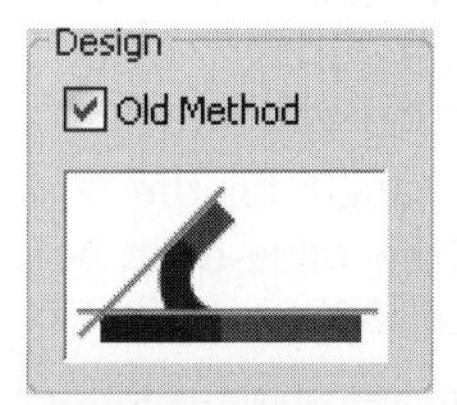

FIGURE 10.35

6. Click Apply to continue creating flanges, or click OK to apply the flange and exit the dialog box.

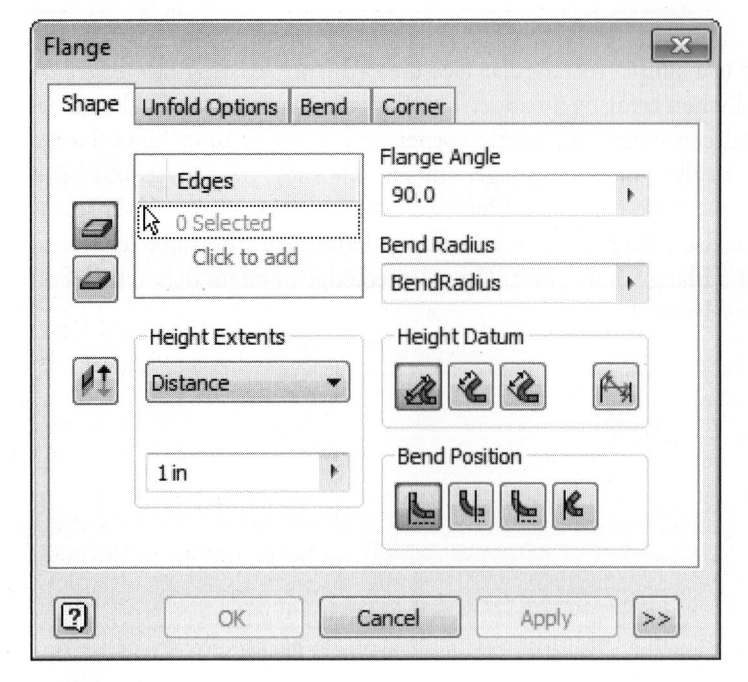

FIGURE 10.36

Shape Tab

The Shape tab allows you to select the edge, edge offset, flange distance, direction, and bend angle for the flange. You can also override the bend radius specified in the sheet metal style. The following options are available on the Shape tab.

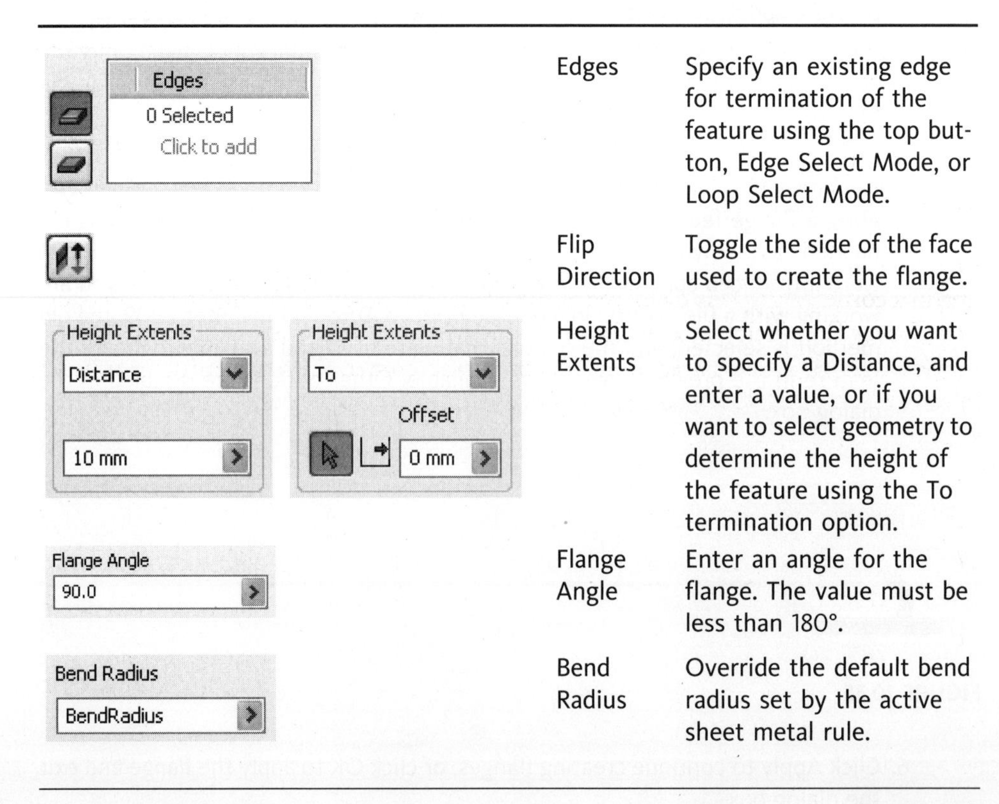

Option	Description
Edges	Specify an existing edge for termination of the feature using the top button, Edge Select Mode, or Loop Select Mode.
Flip Direction	Toggle the side of the face used to create the flange.
Height Extents	Select whether you want to specify a Distance, and enter a value, or if you want to select geometry to determine the height of the feature using the To termination option.
Flange Angle	Enter an angle for the flange. The value must be less than 180°.
Bend Radius	Override the default bend radius set by the active sheet metal rule.

(Continued)

	Height Datum	Select the desired method for calculating the datum for flange height. You can choose to have the flange length measured from the virtual intersection of the outer edges, the virtual intersection of the inner edges, or from the outer tangent of the bend feature. The fourth button is used to determine if the height value is measured to be aligned, or parallel to the bend angle, or orthogonal, measuring the flange normal to the adjacent face of the selected edge.
Bend Position	Bend Position	These buttons control the position of the bend in relationship to the selected model edge. You can choose one of four options: Inside of Bend Face Extents, Bend from the Adjacent Face, Outside of Base Face Extents, or Bend Tangent to Side Faces.

The options available on the Unfold Options and Bend tabs were discussed in the Face section. The Corner tab options were discussed in the Sheet Metal Rules section. The More button provides access to the Edge, Width, Offset, and From To extent types, and these are discussed in the Contour Flange section.

If you create or edit a flange or contour flange feature that has multiple edges converging in a corner, a glyph is displayed at the corner or corners. Clicking the glyph opens the Corner Edit dialog box, as shown in the following image. The Corner Edit command provides control over the geometry of the corner. These overrides can be set during feature creation or edit and control the type and size of relief, the overlap condition, and the miter gap.

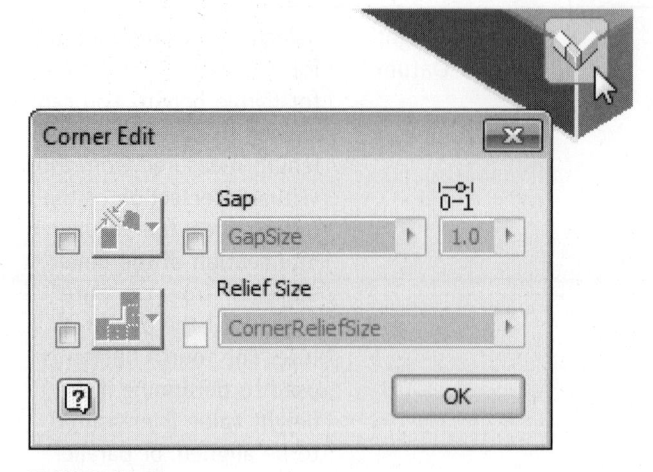

FIGURE 10.37

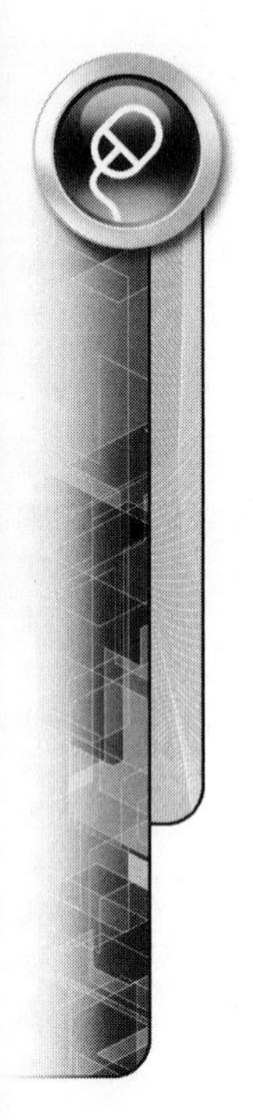

EXERCISE 10-1: CREATING SHEET METAL PARTS

In this exercise, you create sheet metal faces, flanges, and seams to build a small enclosure.

1. Open *ESS_E10_01.ipt*.
2. Click the Styles Editor from the Manage tab > Styles and Standards panel.
3. In the Style and Standard Editor:
   a. Expand Sheet Metal Rule.
   b. Right-click Default and click New Style from the menu.
   c. Enter Aluminum in the New Local Style dialog box and click OK.
   d. Double-click Aluminum in the Sheet Metal Rule list to make it active.
   e. Select Aluminum-6061 from the Material drop-down list.
   f. Enter a value of **0.0625** in the Thickness field.
   g. Select the Bend tab.
   h. Select Round from the Relief Shape drop-down list.
   i. Select the Corner tab.
   j. Select Round from the Relief Shape drop-down list under 2 Bend Intersection.
   k. Click Save.
   l. Click Done.

   The existing sheet metal face, as shown in the following image, is updated to reflect the revised sheet metal rule settings.

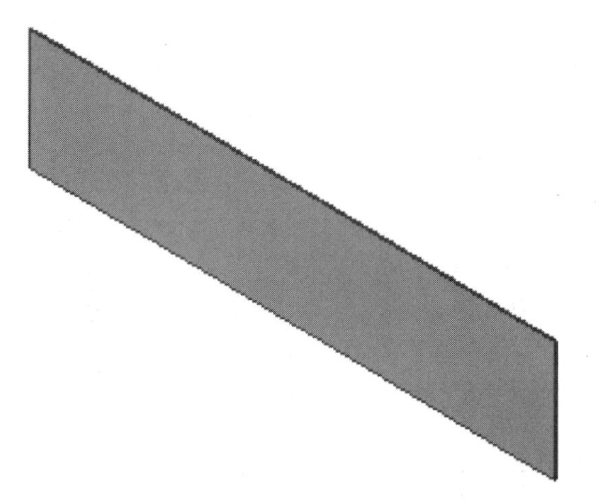

FIGURE 10.38

4. Create a new sketch on the top, long, thin face of the existing feature, as shown in the following image.

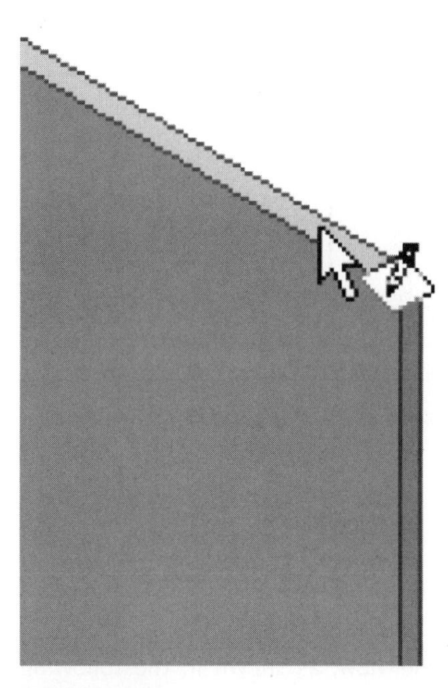

FIGURE 10.39

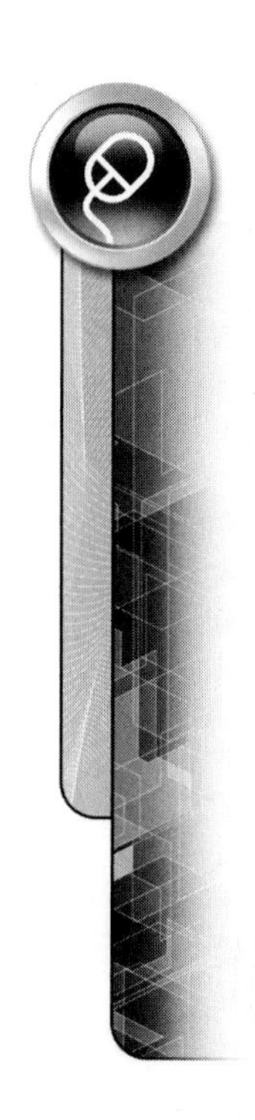

5. Click the Two-Point Rectangle command.
6. Click the upper horizontal edge when the coincident icon is displayed.
7. Create a rectangle coincident with the top edge of the profile, approximately as shown in the following image.

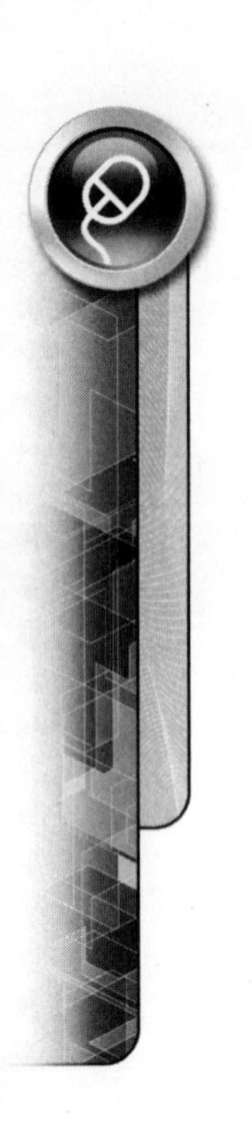

8. Add a **5 in** dimension to the vertical edge of the sketch rectangle, as shown in the following image.

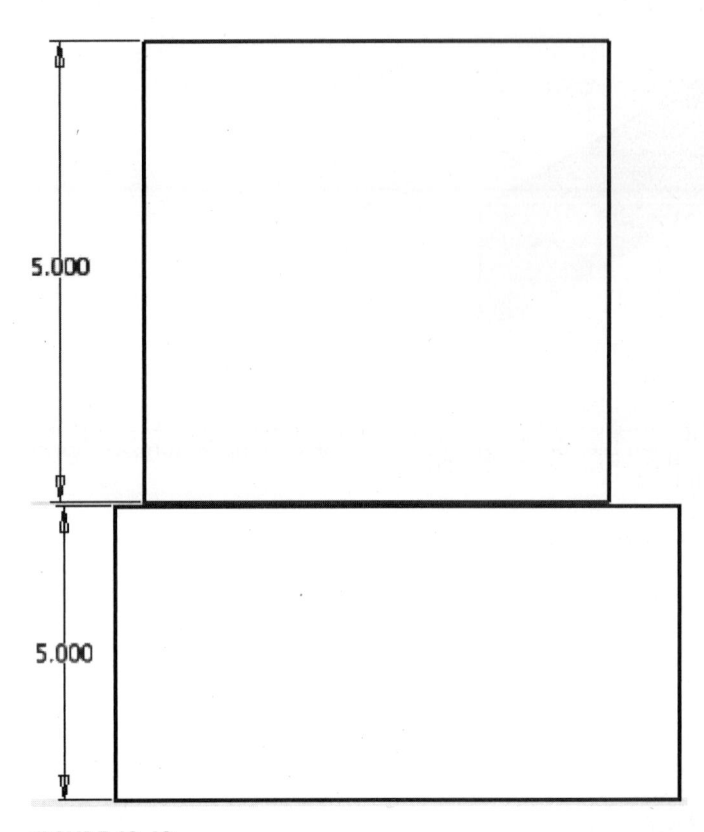

FIGURE 10.40

9. Click Finish Sketch to exit the sketch environment.
10. Click the Face command.
11. Select the rectangle profile.
12. Click OK to accept the default settings.
13. Zoom in on a corner of the bend, and examine the bend relief notch in the original part face.
14. Use Zoom, Rotate, and Pan to reorient the view, as shown in the following image.

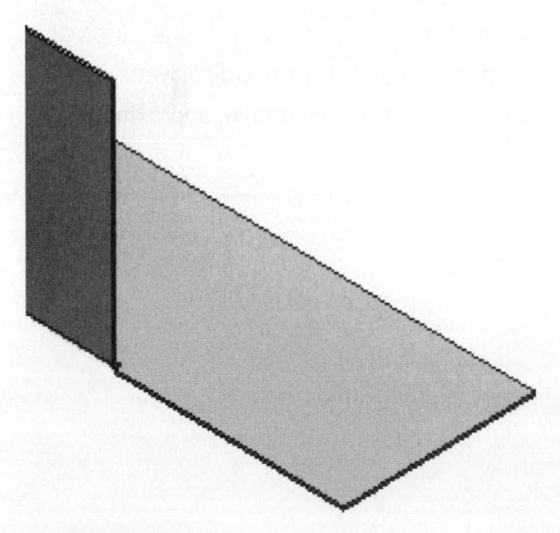

FIGURE 10.41

15. Click the Flange command.
16. In the graphics window, click the top edge on the horizontal surface, as shown in the following image.

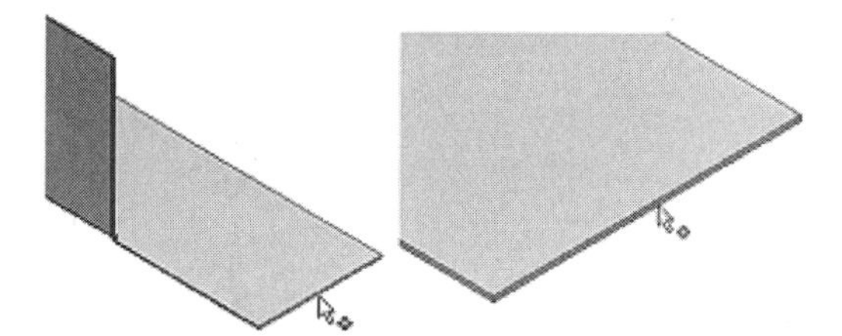

FIGURE 10.42

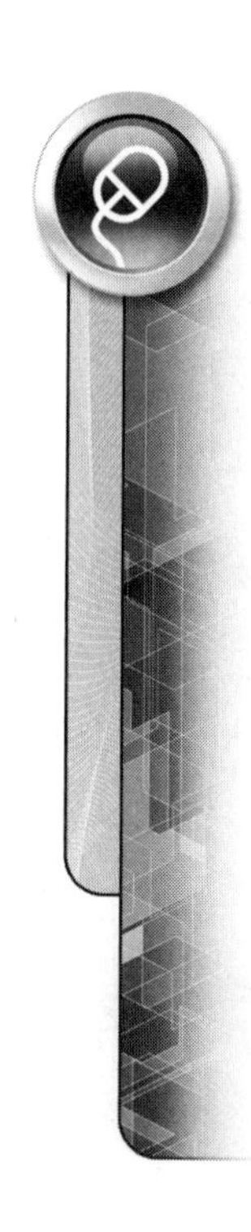

17. In the Flange dialog box:
    a. Under Height Extents, verify that Distance is selected, and then click the arrow next to the Distance value field.
    b. Select Show Dimensions.
    c. Click the first face that you created.
    d. Highlight all of the text in the Distance field.
    e. Click the displayed 5 in dimension, as shown in the following image.
    f. The dimension parameter name replaces the text in the Distance field.

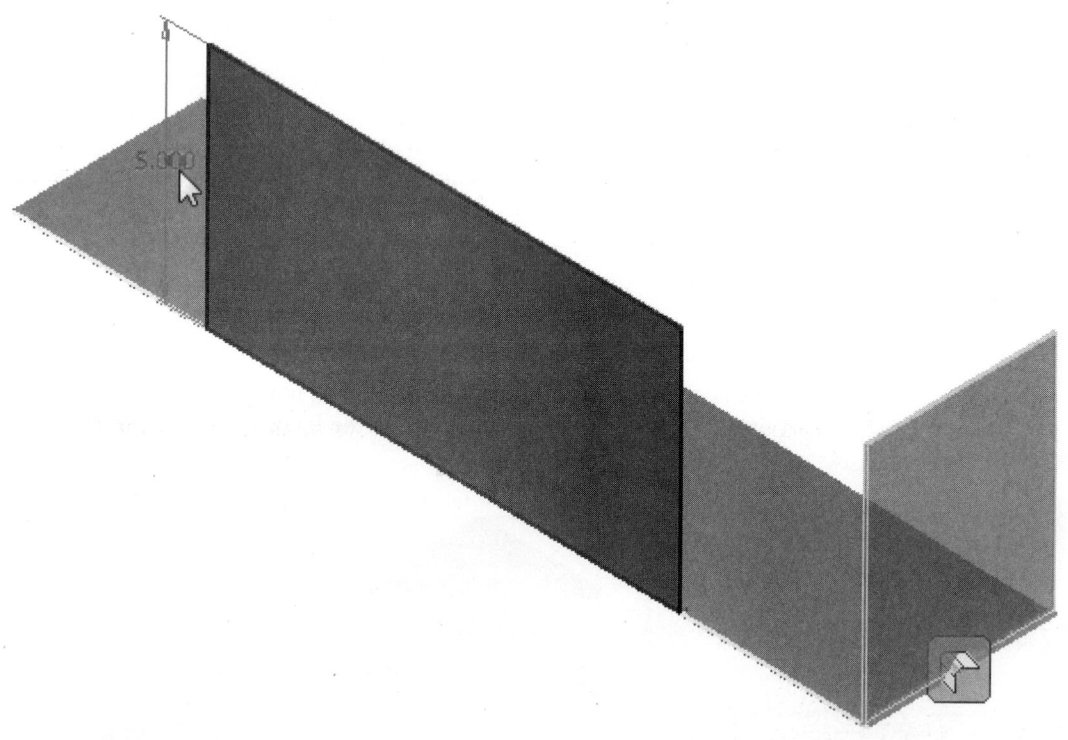

FIGURE 10.43

18. Click Apply to create the flange.
19. Rotate your view, as shown in the following image.

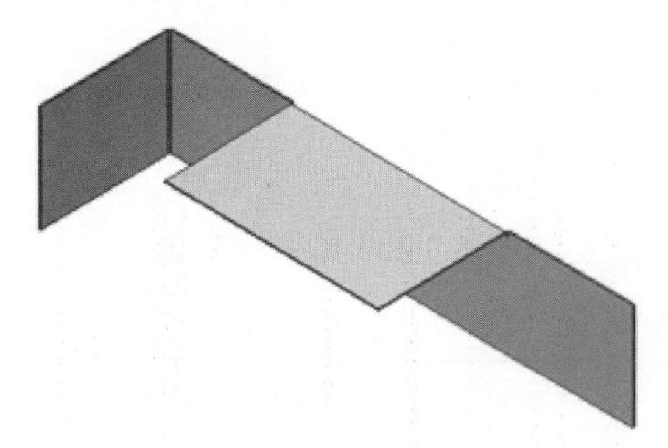

FIGURE 10.44

20. With the Flange dialog box still open, click the right-side (nearest) edge of the base feature, as shown in the following image.

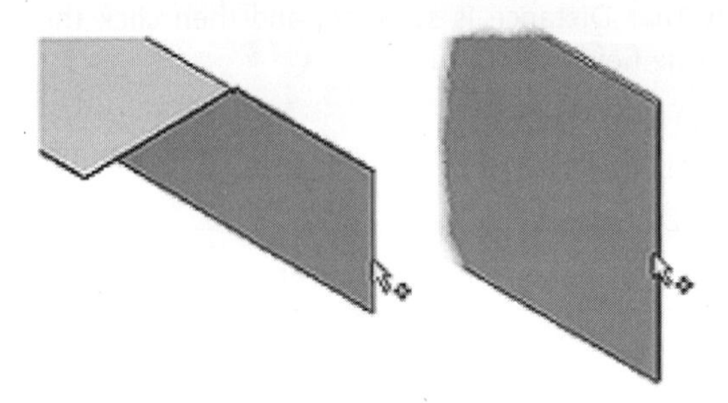

FIGURE 10.45

21. In the Flange dialog box:

    a. Verify that the parameter name is still shown under Height Extents. If not, click the arrow next to the Distance field, and select it from the list of recently used values.

    b. Complete the equation in the field by inserting /sin(60) after the dimension parameter name.

    c. Enter a value of **60** for the Flange Angle. Your dialog box should appear similar to the following image on the left.

    d. As the values are entered into the Flange dialog box, notice the sheet metal part updating to these values, as shown in the following image on the right.

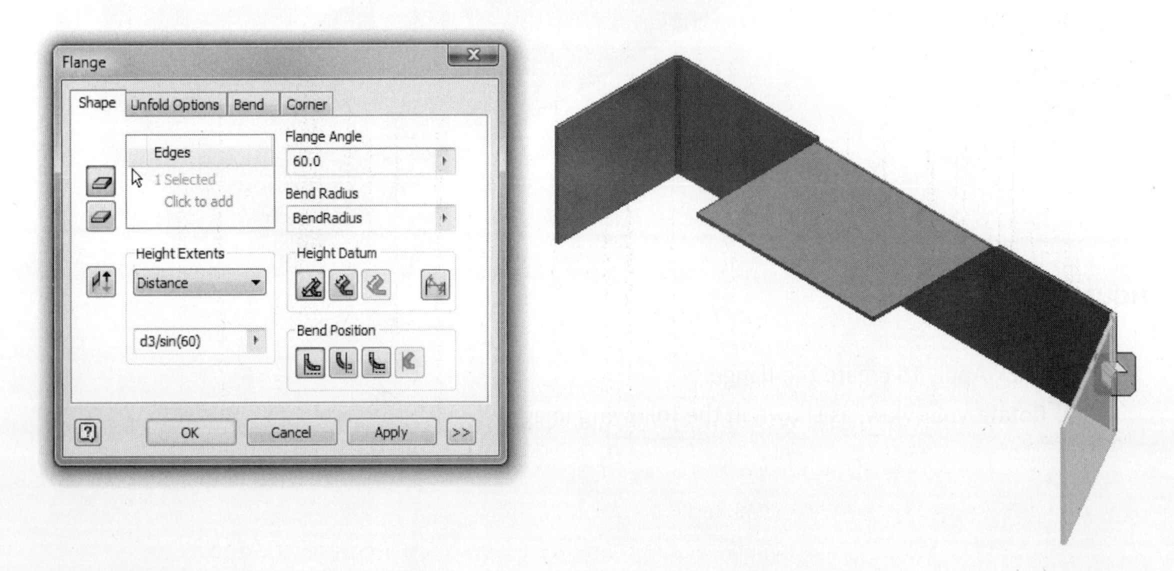

FIGURE 10.46

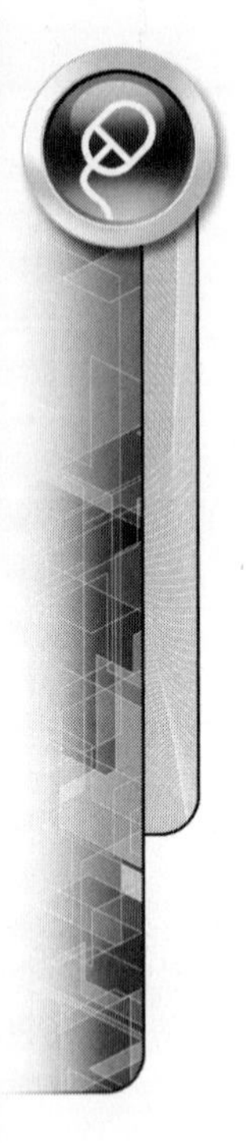

22. Click OK to complete the flange.

NOTE

The expressions entered for the two flange distances allow you to control the length of the face and the two flanges by editing the dimension of the first face you created.

23. Click the upper-left corner of the View Cube, as shown in the following image.

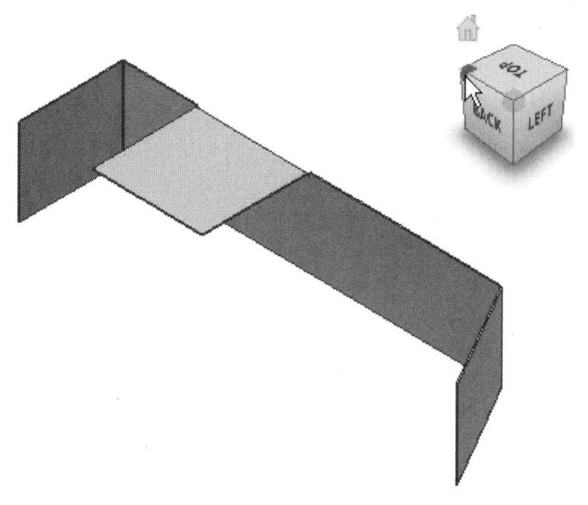

FIGURE 10.47

24. Click the Corner Seam command.
    a. Select the two edges, as shown in the following image.

NOTE

The order of selection and whether or not the top or bottom model edge is selected are not important. The specifics of this command will be covered later in the chapter.

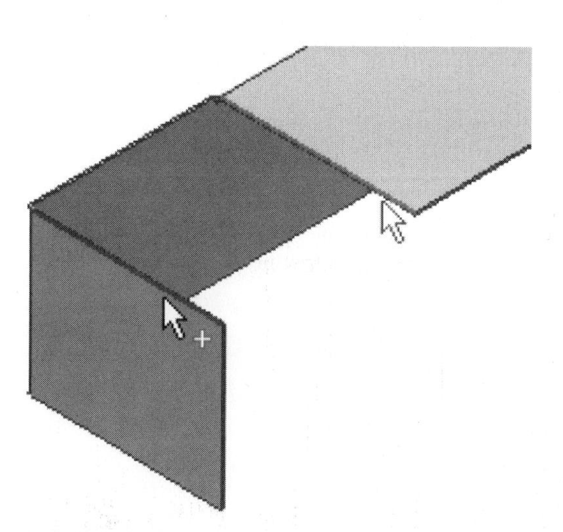

FIGURE 10.48

25. Verify that the GapSize parameter is entered in the Gap field. The GapSize is equal to the Thickness parameter as defined in the sheet metal rule.
26. Click the Face/Edge Distance option in the Seam area.
27. Click each of the three Seam buttons, and examine the resulting seam preview.
28. Click the No Overlap button.
29. Click Apply to create the corner seam.
30. Create a second corner seam between the opposite end of the face and the sloped flange using the same settings as the first corner seam.

31. Click the Flange command.
    a. Select the two bottom edges on the outer surfaces, as shown in the following image.

FIGURE 10.49

32. In the Flange dialog box:
    a. Enter a value of **.75** in the Distance field.
    b. Enter a value of **90** in the Flange Angle field.
    c. Zoom in on the lower left-hand corner of the flange where the preview is shown.
    d. Click the Height Datum and Bend Position buttons and observe the effect on the flange preview.
    e. Return the buttons to the default states: Bend from the intersection of the two outer faces and Inside of Bend Face Extents for the Bend Position.
    f. Click Apply.
33. Rotate the model so that you can see the angled flange.
34. Select the outside edge of the sloped flange, as shown in the following image.

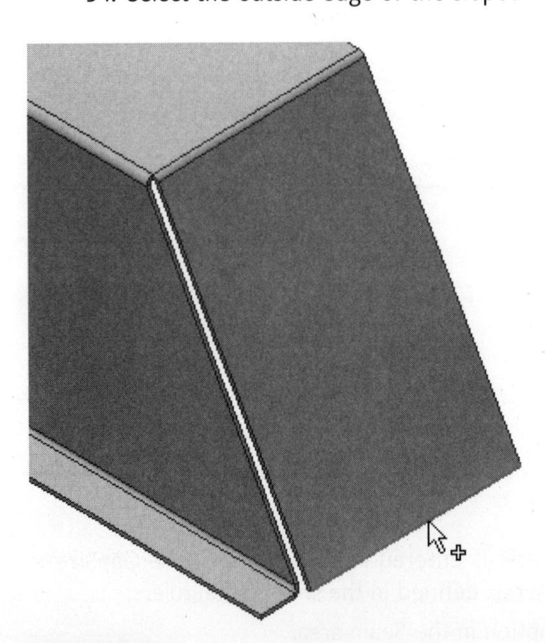

FIGURE 10.50

NOTE

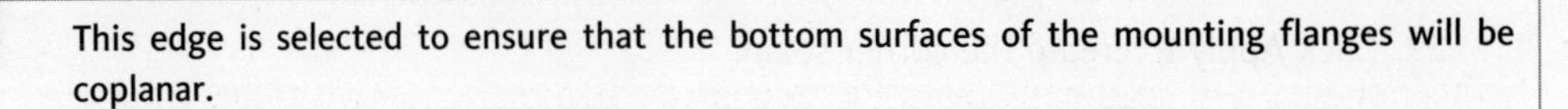

This edge is selected to ensure that the bottom surfaces of the mounting flanges will be coplanar.

35. Enter a value of **.75** in the Distance field.
36. Enter a value of **60** in the Flange Angle field.
37. Click OK to complete the flange.
38. Click the Sheet Metal Defaults command.
39. Select Default from the Sheet Metal Rule drop-down list. Notice that the material style is now set to Brass, Soft Yellow as set by the sheet metal rule in the Style and Standard Editor. The Thickness is set to 0.120 in, as shown in the following image.

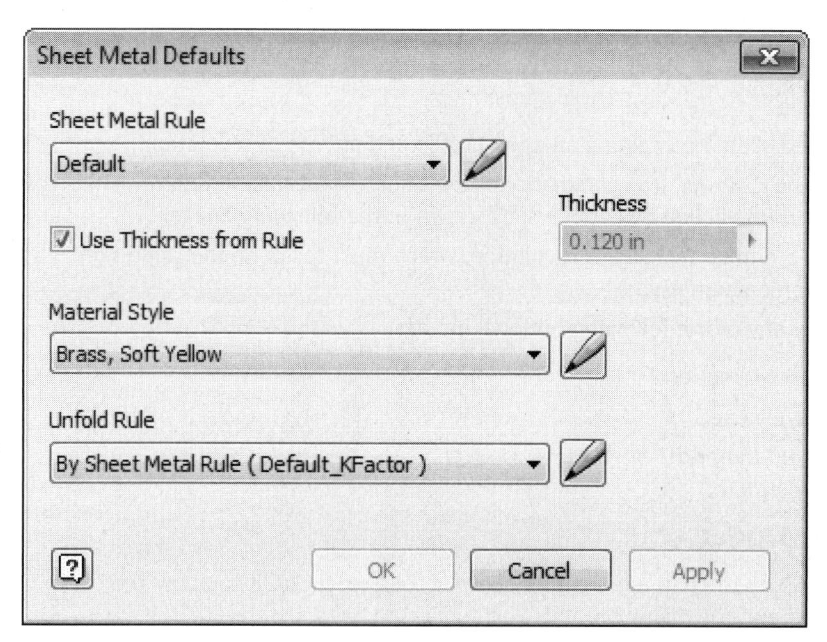

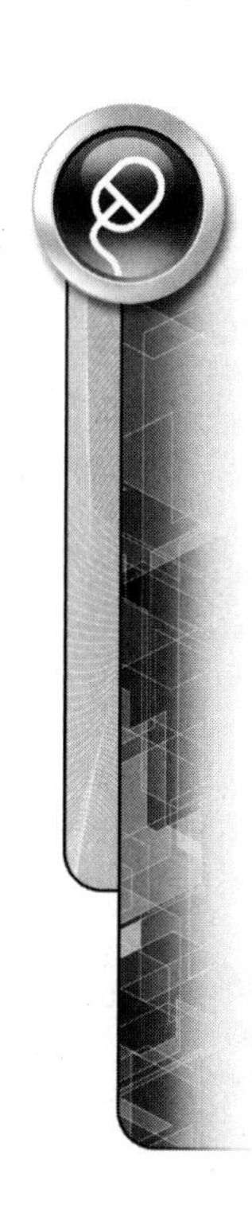

FIGURE 10.51

40. Click OK and the sheet metal part is updated to match the settings specified in the sheet metal rule, as shown in the following image.

FIGURE 10.52

41. Create additional sheet metal rules, faces, flanges, and corner seams as desired.
42. Close the file. Do not save changes. End of exercise.

Contour Roll

By using the Contour Roll command, as shown in the following image, you can create a contour roll from an open sketch profile. The profile does not require a sketched radius between line segments, and, like a contour flange, bends are added at sharp intersections. A contour roll can be the first feature in a sheet metal part, or it can be added to existing features to create rolled sheet metal components. The feature combines an open profile with an axis of revolution to revolve the profile at a specific, designated angle. After a contour roll feature has been created, it can be unrolled with the Unfold command so that you can add additional features to the part in a flat state. The feature can then be refolded using the Refold command.

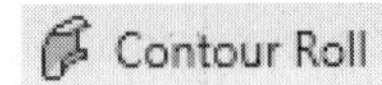

FIGURE 10.53

To create a contour roll, follow these steps:

1. Create a sketch containing an open profile and an axis of revolution.
2. Click the Contour Roll command on the Sheet Metal tab > Create panel. The Contour Roll dialog box appears, as shown in the following image.
3. Select a profile and axis of revolution, which must reside in the same sketch as the profile geometry.
4. Specify any of the following optional inputs:
   a. Offset direction
   b. Angle value
   c. Unroll Method
   d. Unfold Rule
   e. Bend Radius
5. Click Apply to continue creating contour rolls, or click OK to apply the contour roll and exit the dialog box.

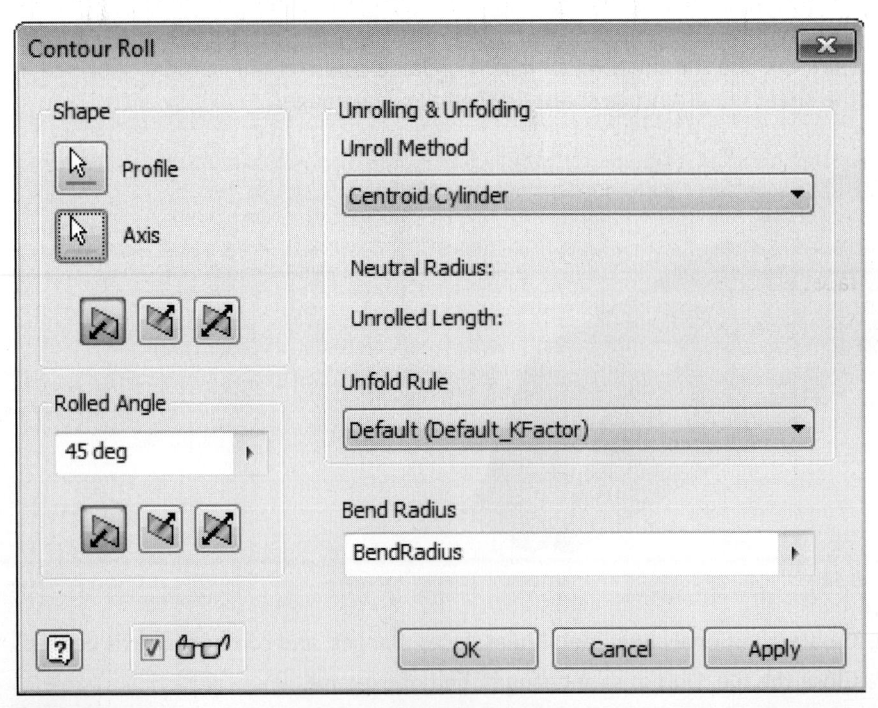

FIGURE 10.54

An example of two contour roll features, combined with two contour flange features, is shown in the following image. The contour roll features are shown in the two intermediate steps.

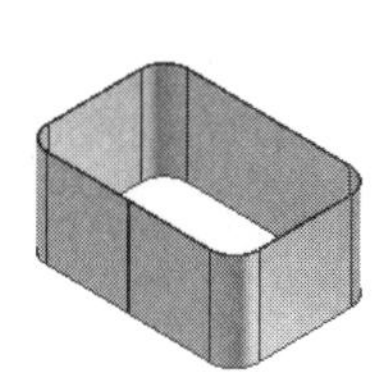 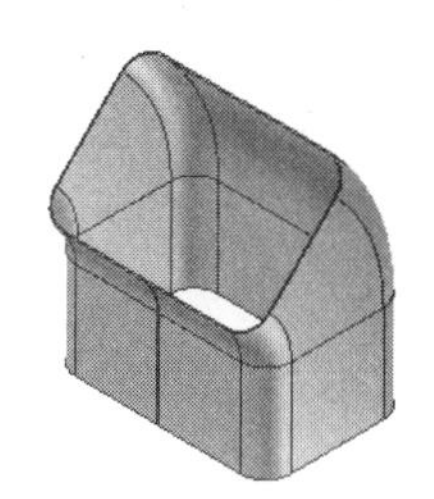 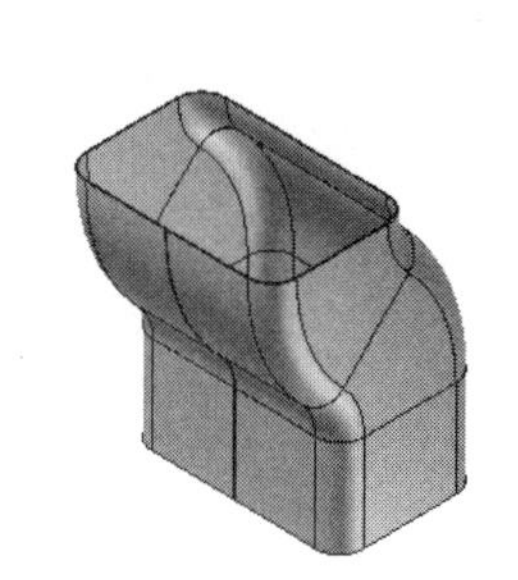 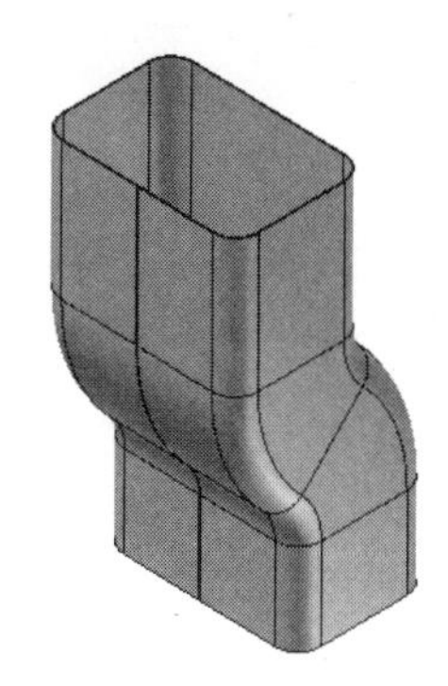

FIGURE 10.55

Lofted Flange

Using the Lofted Flange command shown in the following image, you create transitions from one shape to another. The bends that form the transition may be manufactured using either a die form process or a press brake process. The transition is comprised of two separate sketches that define the two transition shapes or profiles. Typically, the profiles are generated as closed profiles that define the end conditions for items such as duct work or ventilation hoods.

FIGURE 10.56

To create a lofted flange, follow these steps:

1. Create two sketches that contain closed profiles for the end conditions of the lofted flange.
2. Click the Lofted Flange command on the Sheet Metal tab > Create panel. The Lofted Flange dialog box appears, as shown in the following image.
3. Select the sketched profiles.
4. In the Lofted Flange dialog box, select either Die Formed or Press Brake from the Output section. If Press Brake is selected you can further refine the facets of the transition by entering a value for the chord tolerance, facet angle, or facet distance.
5. Enter a Bend Radius.
6. Click OK.

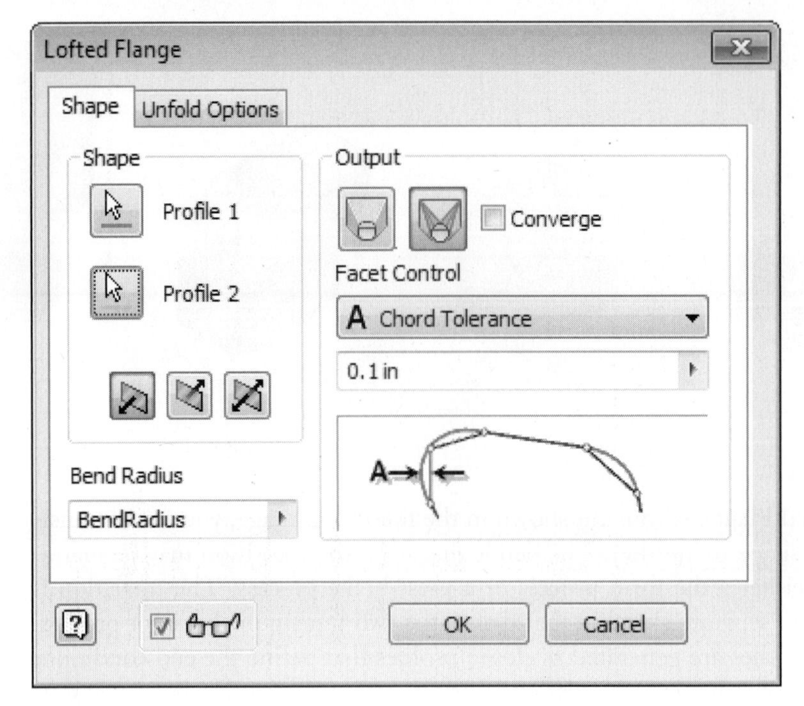

FIGURE 10.57

Rip

The Rip command, shown in the following image, can be used to rip open an entire bend plate. This command is different than the Rip option previously mentioned in the Corner Seam command. The Rip command can be used to rip a part in a straight line, based on a single sketch point, work point, midpoint of an edge, end points on face vertices, or you can use two sketch points to rip deformed rectangular profiles or cylindrical, elliptical, conic, or ruled bend plates.

FIGURE 10.58

To create rip features, follow these steps:

1. Click the Rip command on the Sheet Metal tab > Modify panel. The Rip dialog box appears, as shown in the following image.
2. Select one of the three available Rip Types: Single Point, Point to Point, or Face Extents. Depending on the option you select, you will need to satisfy the inputs accordingly.
3. If using Point to Point, you would select the face you want to rip, then the visible sketch point that defines the beginning of the rip location and a second point that defines the end location. If using Single Point, you select only the face and a single sketch point.
4. Optionally, you can modify the Gap Value to override the gap size as specified in the sheet metal rule.
5. Click Apply to continue ripping additional faces, or click OK to apply the rip feature and exit the dialog box.

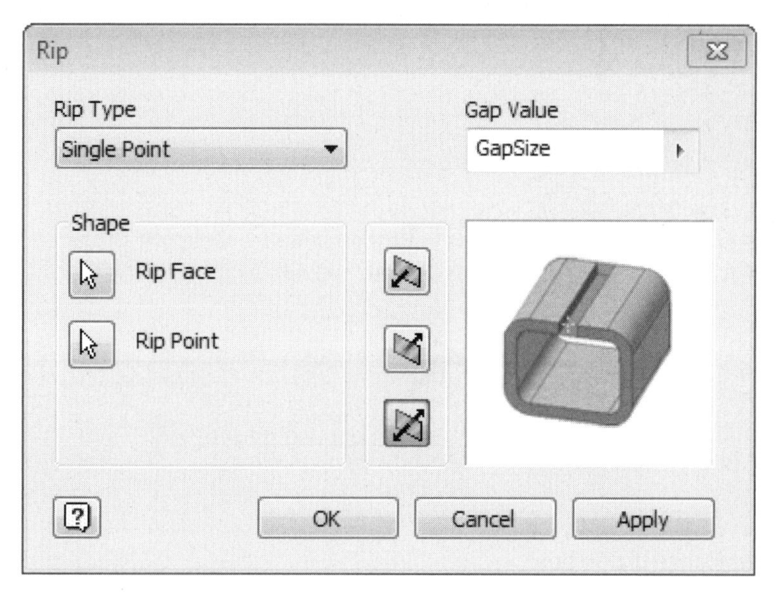

FIGURE 10.59

The following image shows an example of each Rip type: Single Point, Point to Point, and Face Extents, respectively from left to right.

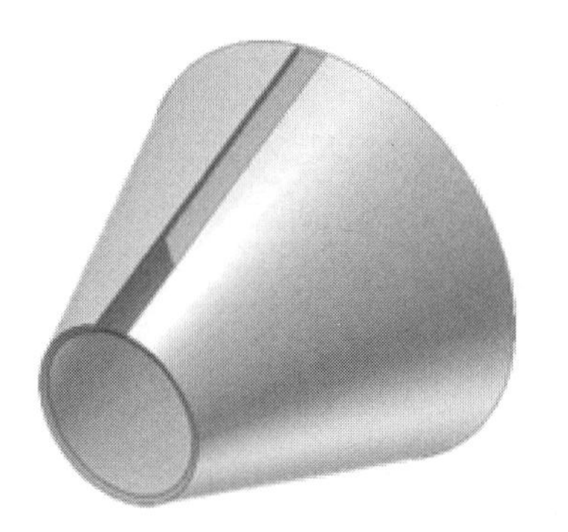
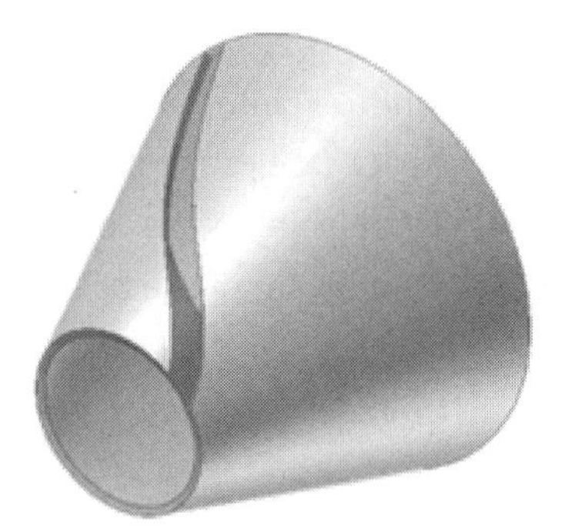

FIGURE 10.60

EXERCISE 10-2: LOFTED FLANGES, CONTOUR ROLL, AND RIPS

In this exercise, you create sheet metal parts using a combination of the lofted flange, contour roll, and rip features. The sheet metal file consists of several sketched profiles that will be used to create lofted flanges. You will use the Rip command to ensure that a flat pattern can be generated from the lofts. You will then create contour roll features to complete the construction of an oven hood.

1. Open *ESS_E10_02.iam*.
2. In the browser, double-click *ESS_E10_02-Hood:1*.
3. Click the Lofted Flange command.
4. Select the two profiles shown in the following image.

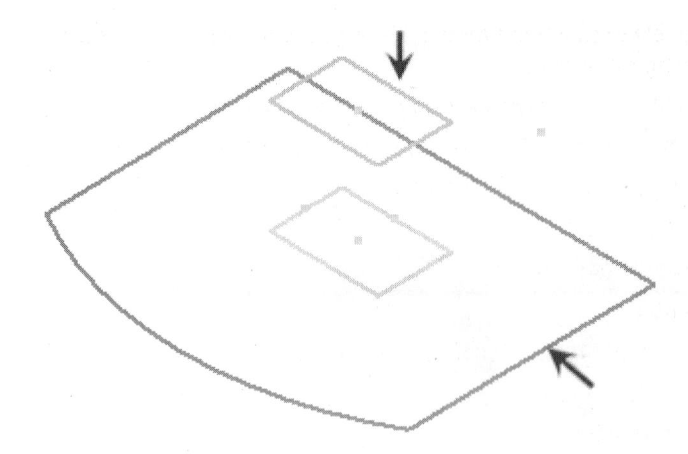

FIGURE 10.61

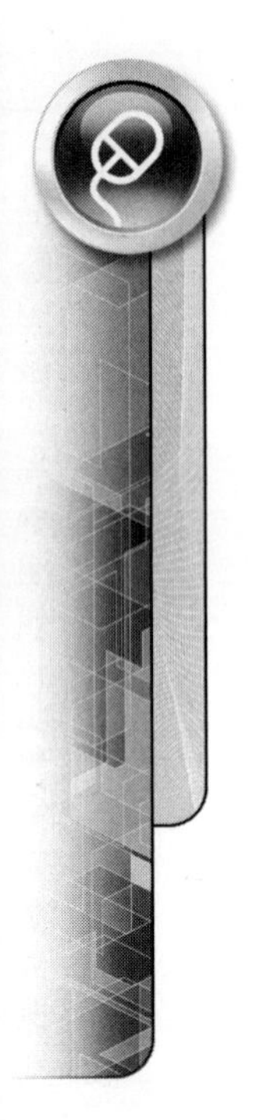

5. In the Lofted Flange dialog box:
   a. Select Die Formed from the Output area.
   b. Specify a bend radius of 2.0.
6. Click OK to create the lofted flange and close the dialog box.
7. Click the Rip command.
8. Rotate the model as shown in the following image.
9. In the Rip dialog box:
   a. Select Point to Point from the Rip Type list.
   b. Click the back face of the part.
   c. Select the midpoint of the top edge of the face.
   d. Select the midpoint of the bottom edge of the face.

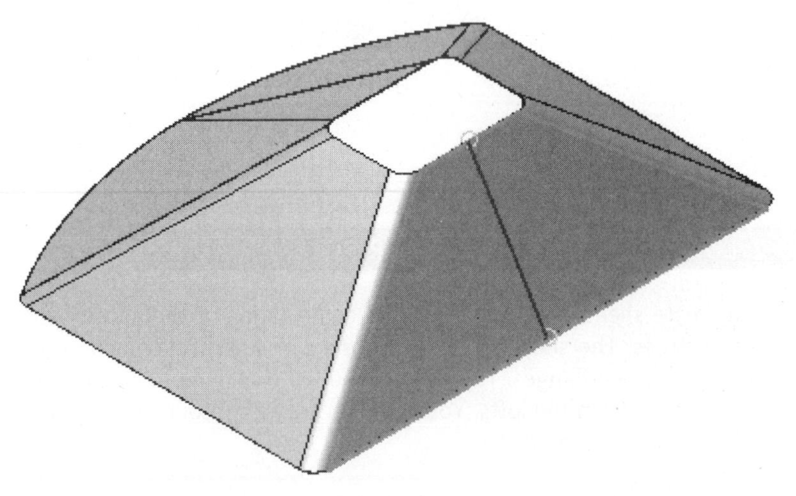

FIGURE 10.62

10. Click OK to complete the Rip feature.
11. On the Sheet Metal tab > Flat Pattern panel, click Create Flat Pattern to display the flat pattern in a separate window. This command will be covered in detail later in the chapter.

12. In the browser, double-click Folded Model. Next, you will create another lofted flange that will become the duct.
13. Click the *ESS_E10_02.iam* tab at the bottom of the canvas.
14. In the browser, right-click *ESS_E10_02-Duct:1* and select Visibility from the menu.
15. In the browser, double-click *ESS_10_02-Duct:1*.
16. Click the Lofted Flange command.
17. In the Lofted Flange dialog box:
    a. Select both profiles.
    b. Verify that Press Brake is selected in the Output area.
    c. Specify a bend radius of 2.0.
    d. Specify a chord tolerance of 1.0.
18. Click OK to create the lofted flange.
19. Click the Rip command.
20. In the Rip dialog box:
    a. Select Point to Point from the Rip Type list.
    b. Click the back face of the part.
    c. Select the midpoint of the top edge of the face.
    d. Select the midpoint of the bottom edge of the face.
21. Click OK to complete the Rip feature.

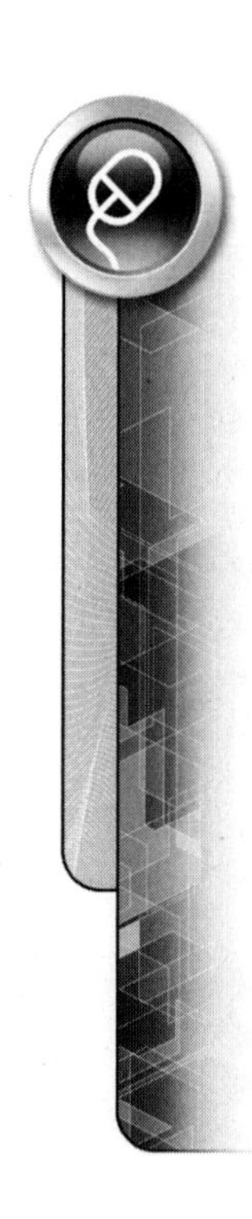

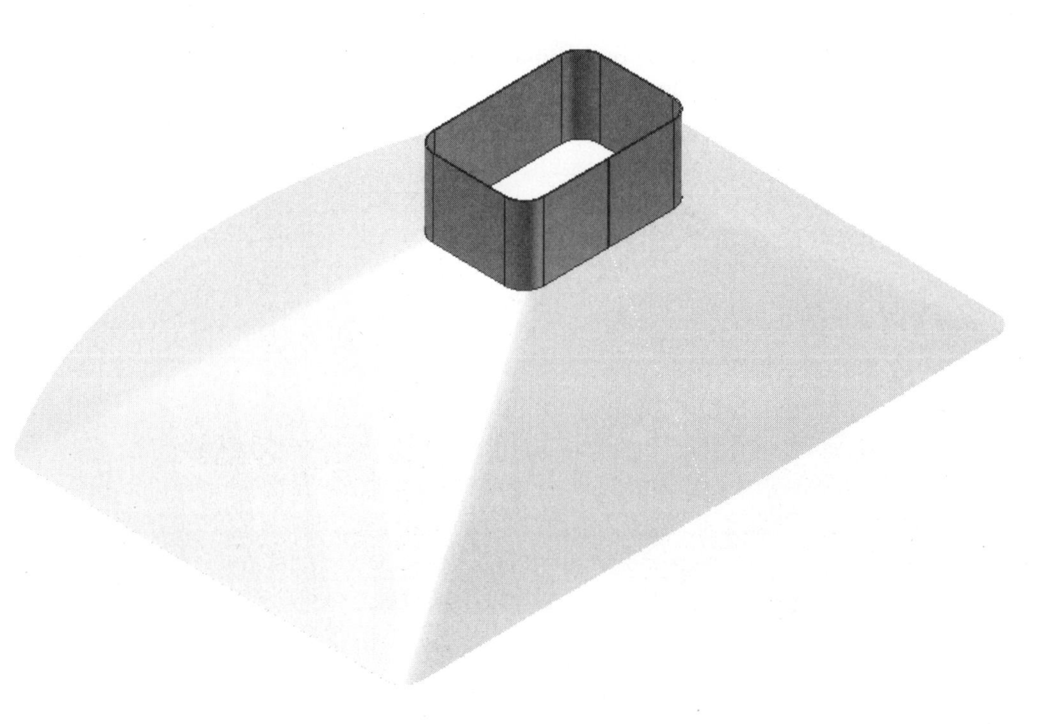

FIGURE 10.63

22. Create a sketch on the top face of the part.
23. Project only the outer edges of the face as indicated by arrows in the following image. When finished you should have a total of nine edges projected and a small gap where you created the rip feature in the previous steps.

NOTE

You may need to delete some edges that may be automatically projected based on your application option settings.

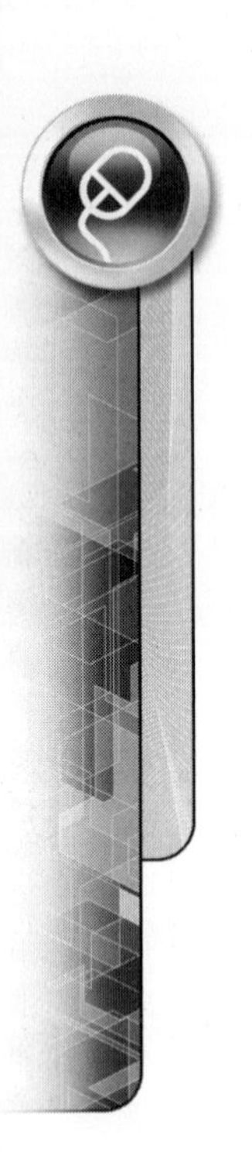

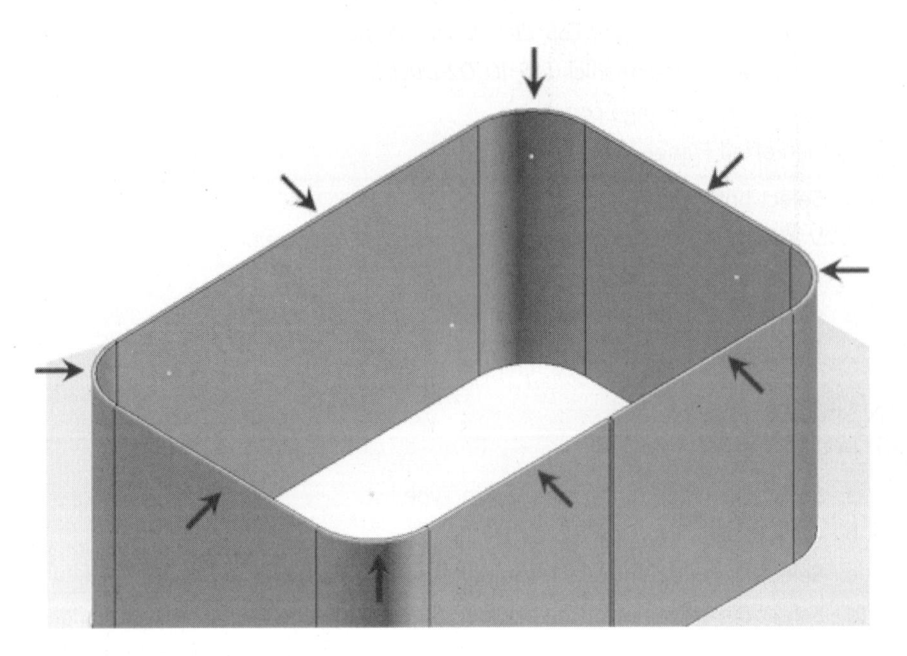

FIGURE 10.64

24. Sketch a centerline and add a symmetric dimension with a value of 8.00, as shown in the following image.

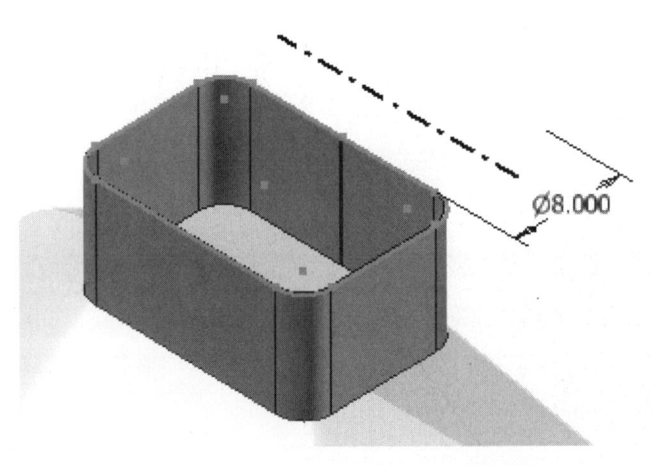

FIGURE 10.65

25. Click Finish Sketch.
26. Click the Contour Roll command.
27. In the Contour Roll dialog box, specify a Rolled Angle of 45° degrees.

28. Click OK to complete the contour roll feature.

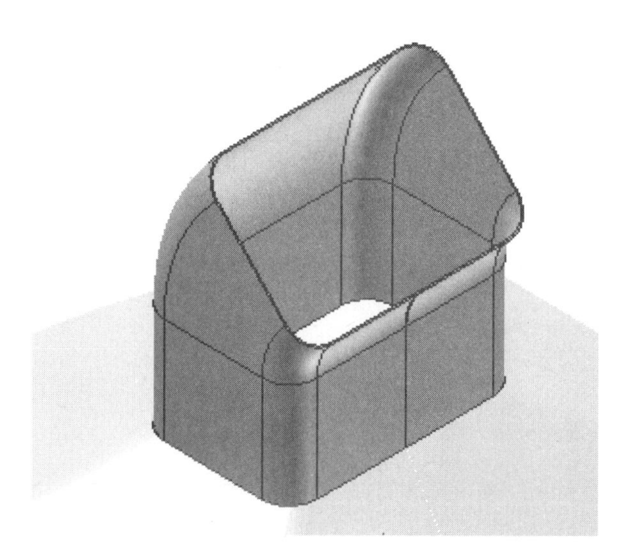

FIGURE 10.66

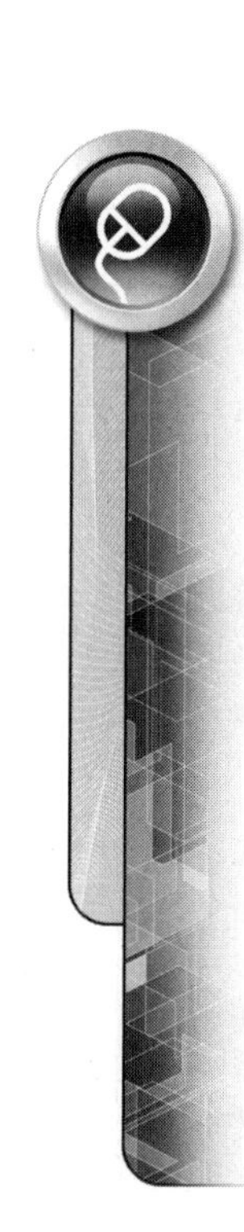

29. Create a sketch on the top face of the part.
30. Project the nine outer edges of the face as indicated by arrows in the following image.

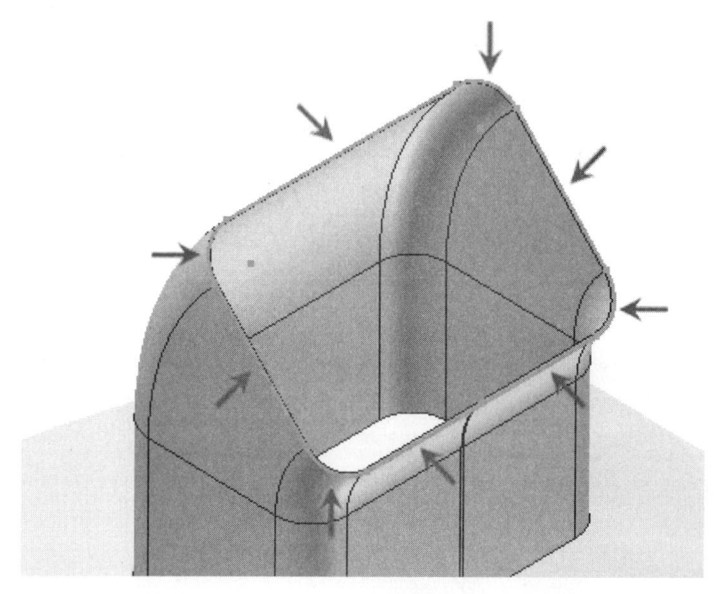

FIGURE 10.67

31. Sketch a centerline and add a symmetric dimension with a value of 8.00, as shown in the following image.

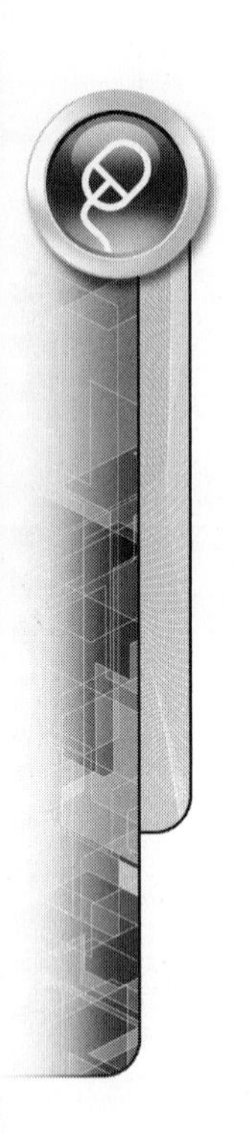

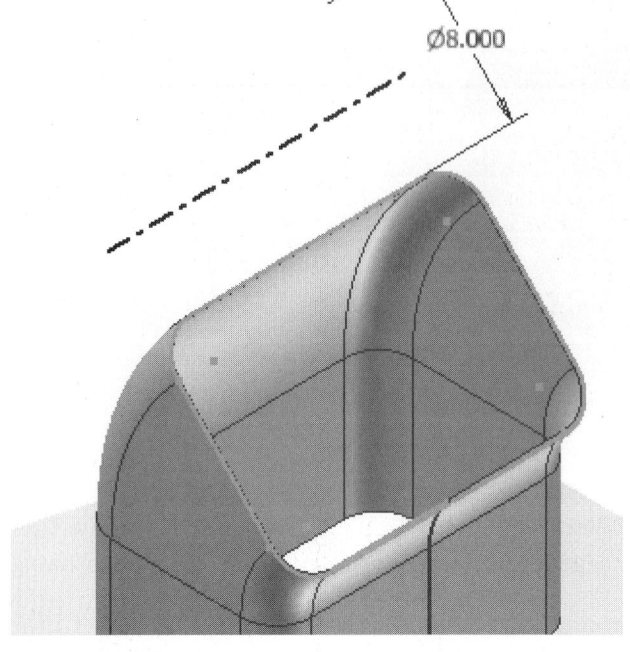

FIGURE 10.68

32. Click Finish Sketch.
33. Click the Contour Roll command.
34. In the Contour Roll dialog box, specify a Rolled Angle of 45°.
35. Click OK to complete the contour roll feature.

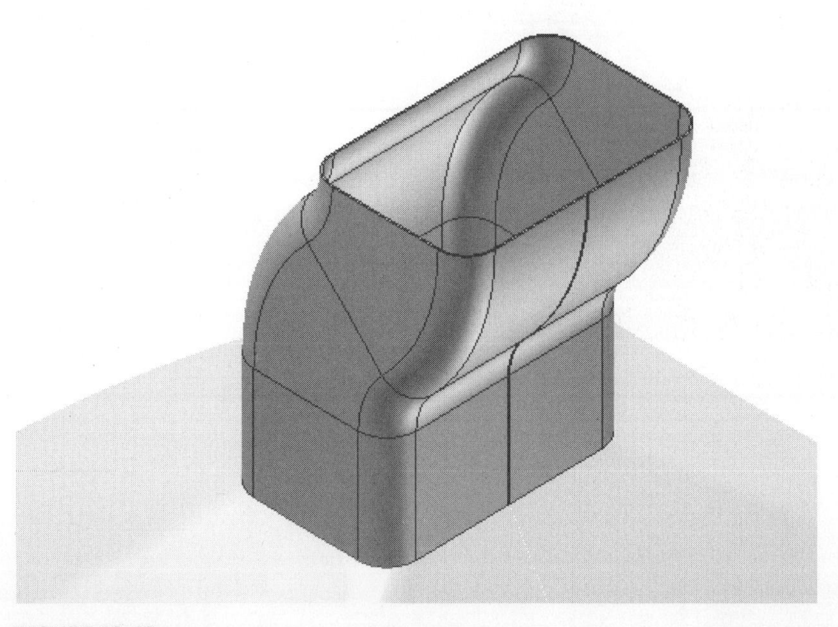

FIGURE 10.69

36. Click Create Flat Pattern to display the flat pattern in a separate window.

37. Click Front on the View Cube.

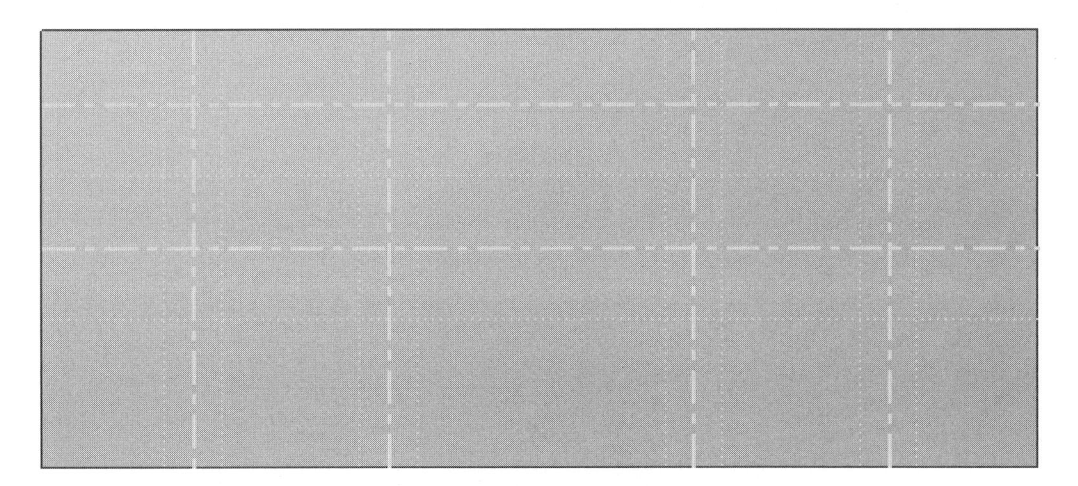

FIGURE 10.70

38. Click the Go to Folded Part command from the Flat Pattern tab > Folded Part panel.
39. Optionally, you can practice creating another lofted flange by performing the following steps:
    a. Create a sketch on the top face of the duct part.
    b. Project the nine outside edges onto the sketch.
    c. Finish the sketch.
    d. Create an offset work plane a distance of 24 from the top face of the duct.
    e. Create a new sketch on the offset work plane and project the nine outside edges from the step above onto the new sketch.
    f. Create a lofted flange between the two profiles using the die formed output type and a bend radius of 2.0.

NOTE

Use the Flip Side buttons, if necessary, to ensure that the direction of the loft is toward the inside of the duct.

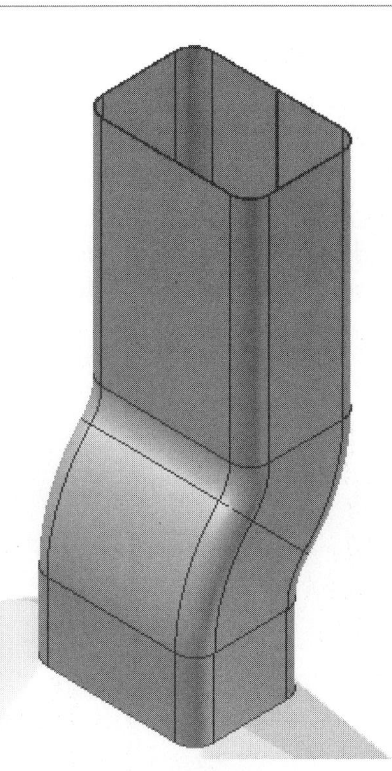

FIGURE 10.71

40. Click the *ESS_E10_02.iam* tab to return to the assembly.
41. In the browser, double-click *ESS_E10_02.iam* to activate the assembly, if it is not already active.

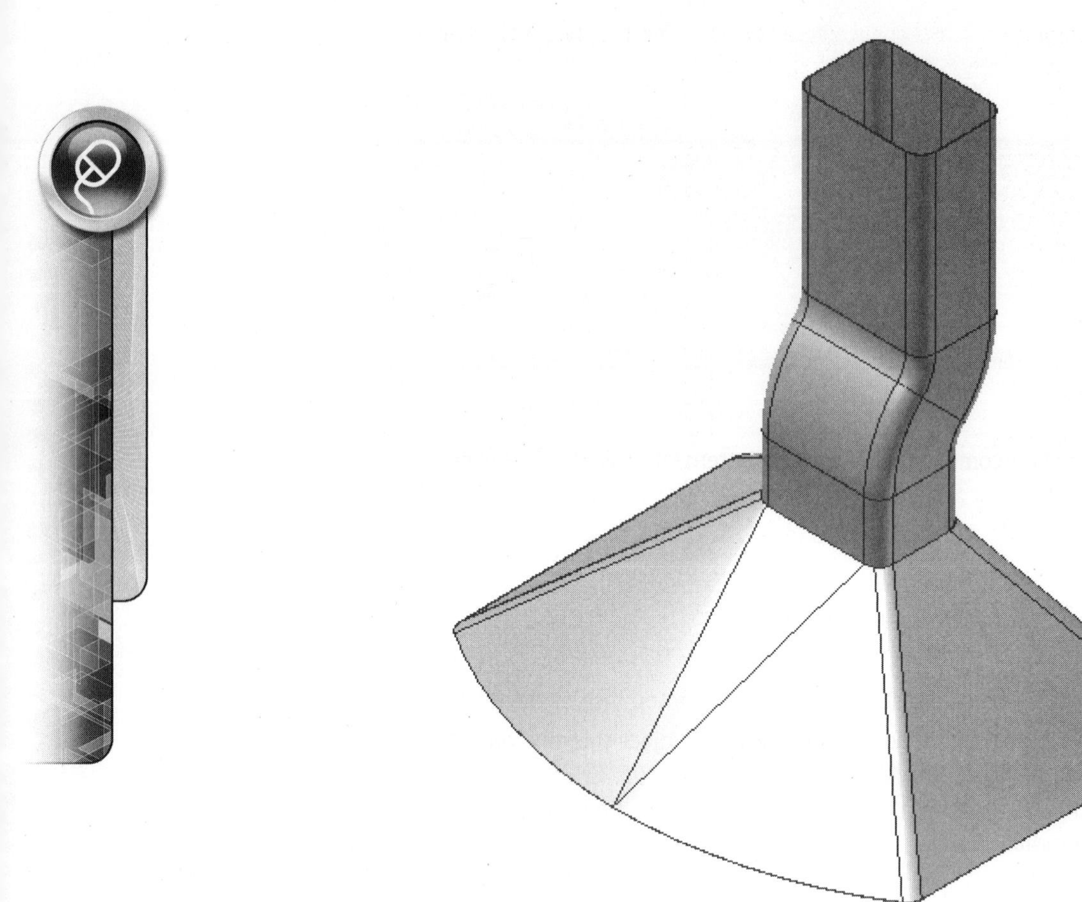

FIGURE 10.72

42. Close the file. Do not save changes. End of exercise.

Hem

Hems eliminate sharp edges or strengthen an open edge of a face. Material is folded back over the face with a small gap between the face and the hem. A hem does not change the length of the sheet metal part; the face is trimmed so that the hem is tangent to the original length of the face. Create hems using the Hem command, as shown in the following image.

FIGURE 10.73

A minimum of one sheet metal face must exist before creating a hem.

NOTE

To create a hem, follow these steps:

1. Click the Hem command on the Sheet Metal tab > Create panel. The Hem dialog box appears, as shown in the following image.

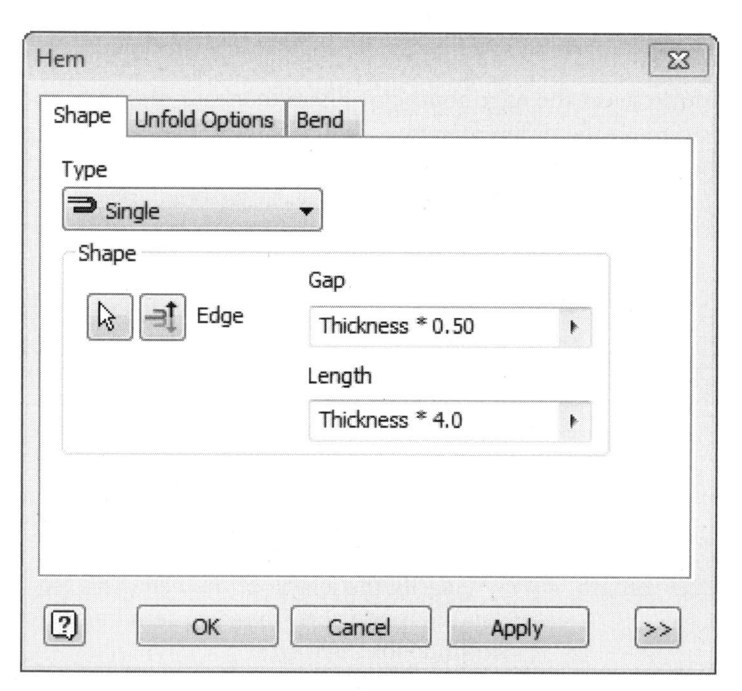

FIGURE 10.74

2. Select an open edge on a sheet metal face.
3. Select the hem type. Examples are shown in the following image:
   - Single: A 180° flange
   - Teardrop: A single hem in a teardrop shape
   - Rolled: A cylindrical hem
   - Double: Single hem folded 180° resulting in a double-thickness hem

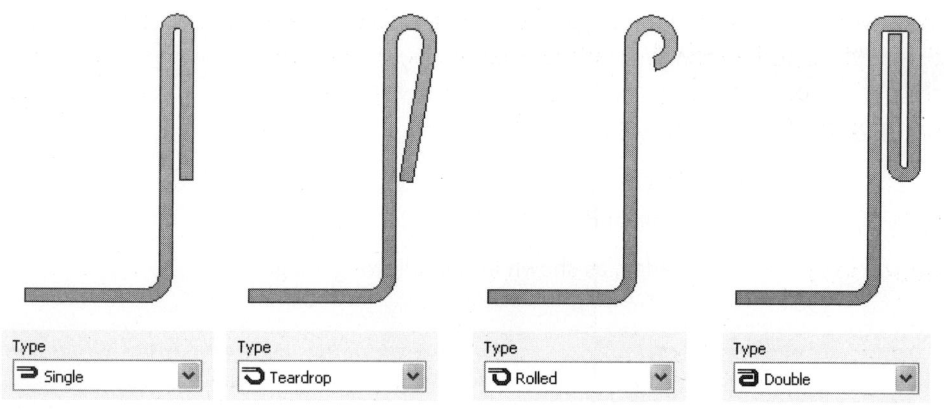

FIGURE 10.75

4. Enter values for the hem. Teardrop and rolled hems require radius and angle values, while single and double hems require gap and length values. The hem preview changes to match the current values.
5. Expand the dialog box by clicking the More (<<) button and selecting Edge, Width, Offset, or From To for the hem extents.
6. Click Apply to continue creating hems, or click OK to complete the hem and exit the dialog box.

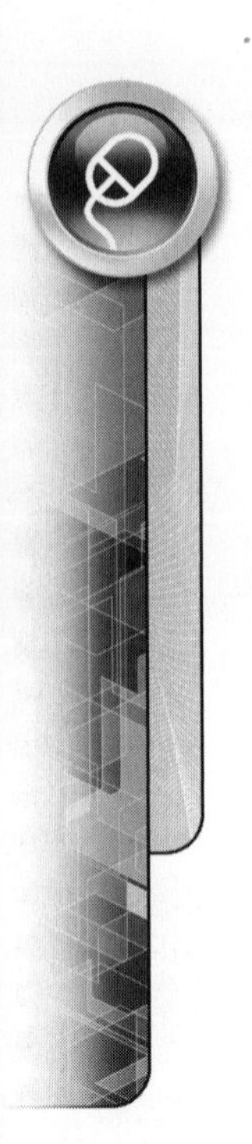

Shape Tab

The Shape tab allows you to select the edge and type of hem to create. Based on the type of hem that you want to create, different options will become active in the Hem dialog box. The following options are available on the Shape tab.

	Type	Select One of Four Hem Types
	Select Edge	Select the edge along which the hem will be created.
	Flip Direction	Toggle the direction in which the hem will be created.
Gap Thickness * 0.50	Gap	Specify the distance between the inside faces of the hem. This feature is available when you select a single or double hem type.
Length Thickness * 4.0	Length	Specify the length of the hem. This feature is available when you select a single or double hem type.
Radius BendRadius	Radius	Specify the bend radius to apply at the bend. This feature is available when you select a rolled or teardrop hem type.
Angle 190.0	Angle	Specify the angle applied to the hem. This feature is available when you select a rolled or teardrop hem type.

The options available on the Unfold Options and Bend tabs were covered in the Face section. The More (<<) button provides access to the Edge, Width, Offset, and From To extent types, which are discussed in the Contour Flange section.

EXERCISE 10-3: HEMS

In this exercise, you create two hem features.

1. Open *ESS_E10_03.ipt*.
2. Click the Hem command.
   a. Click the top edge, as shown in the following image.

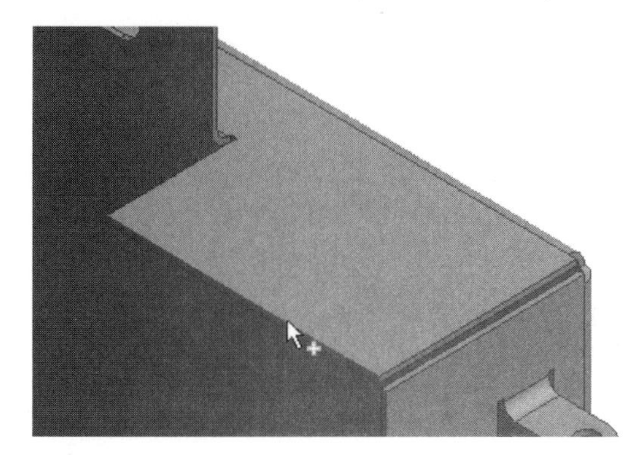

FIGURE 10.76

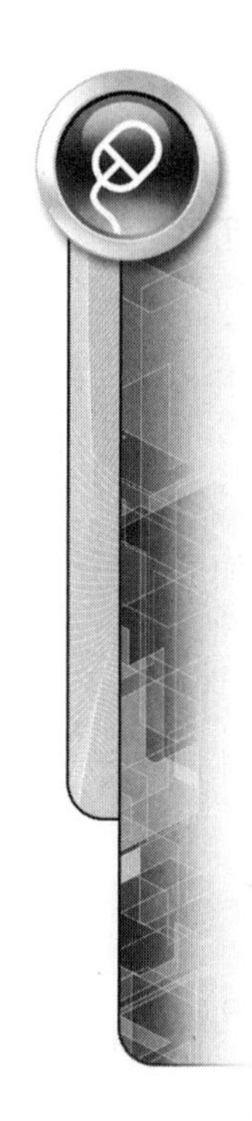

3. In the Hem dialog box:
   a. Enter **0.625 in** in the Length edit box.
   b. If the hem previews to the inside of the box, click the Flip Direction button.
   c. Enter **0.02 in** in the Gap edit box.
   d. Click OK to create the feature and close the Hem dialog box. Your display should appear similar to the following image.

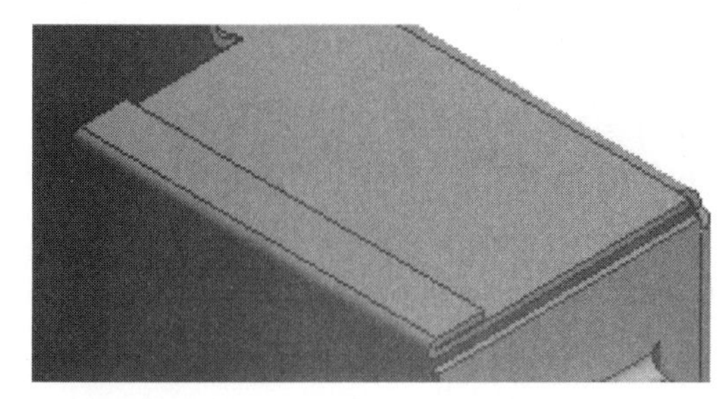

FIGURE 10.77

4. Use Zoom and Rotate to reorient your view and match the following image.

FIGURE 10.78

5. In the browser, right-click the Hem Extents Work plane.
6. Select Visibility from the menu.
7. Click the Hem command.

8. Click the edge, as shown in the following image.

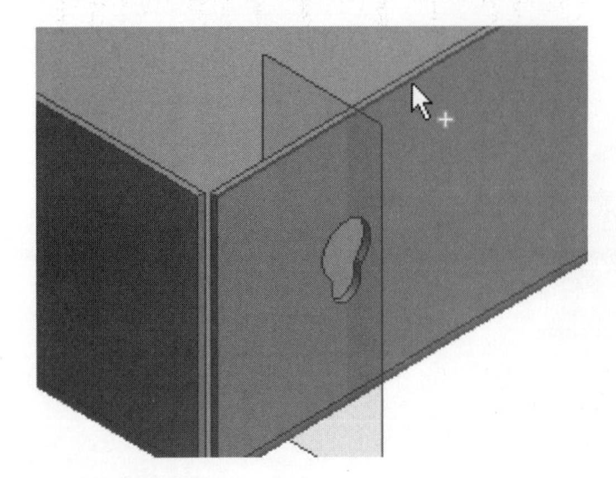

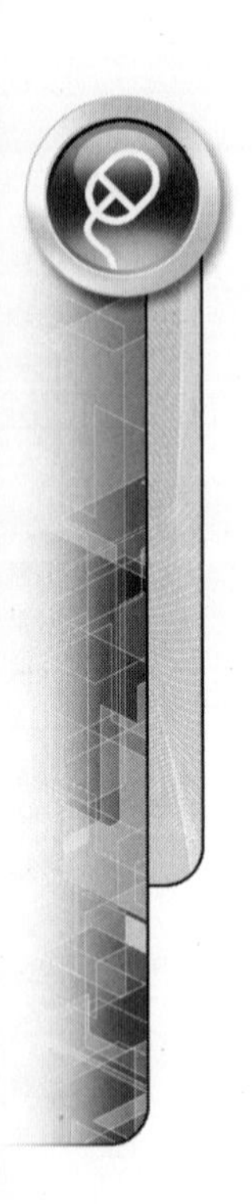

FIGURE 10.79

9. In the Hem dialog box:
   a. Select Rolled from the Type list.
   b. Enter **.08 in** in the Radius edit box.
   c. Enter **265°** in the Angle edit box.
   d. Click the Flip Direction button if the hem previews to the inside of the box.
   e. Click the More button to expand the dialog box.
   f. Select Offset from the Width Extents Type list.
   g. Click the selection arrow next to Offset1.
   h. In the graphics window, click the Hem Extents Work Plane.
   i. The second offset point is automatically set to the endpoint of the selected edge.
   j. Enter **0 in** in the Offset1 edit box.
   k. Enter **0.5 in** in the Offset2 edit box.
   l. Click OK to create the feature and close the Hem dialog box. Your display should appear similar to the following image.

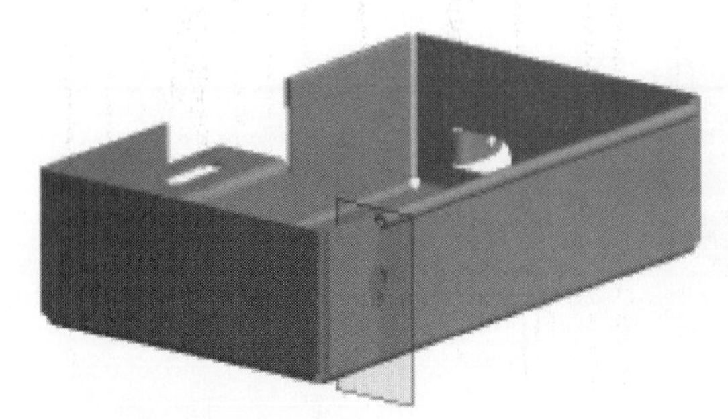

FIGURE 10.80

10. Experiment with the other Hem types and edges as desired.
11. Close the file. Do not save changes. End of exercise.

Fold

An alternate method for creating sheet metal features is to start with a known flat pattern shape and then add folds to the sketched lines on a face. The Fold command, as shown in the following image, can create sheet metal shapes that are difficult to create using the Face or Flange commands.

FIGURE 10.81

To create a fold, follow these steps:

1. Create a sketch on an existing face. Sketch a line between two open edges on the face.

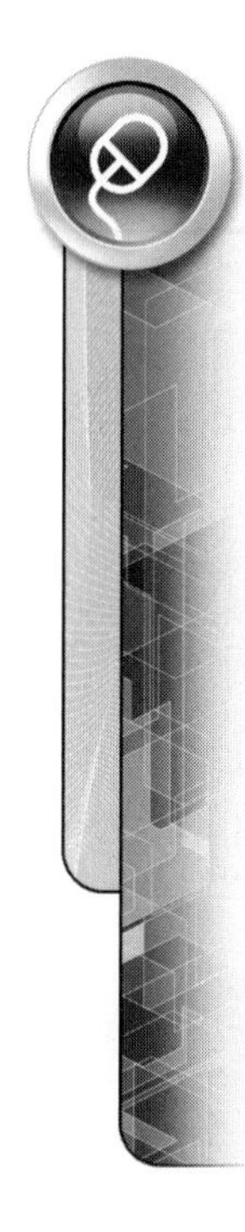

The sketched line endpoints must be coincident to the face edge.

2. Click the Fold command on the Sheet Metal tab > Create panel. The Fold dialog box appears, as shown in the following image.
3. Select the sketch or bend line. The fold direction and angle are previewed in the graphics window. The fold arrows extend from the face that will remain fixed. The face on the other side of the bend line will fold around the bend line.
4. If required, flip the fold direction and side.
5. Enter the angle of the fold.
6. Select the positioning of the fold with respect to the sketched line. The line can define the centerline, start, or end of the bend. The fold preview updates to match the current settings.
7. Make any needed changes on the Unfold Options or Bend tabs.

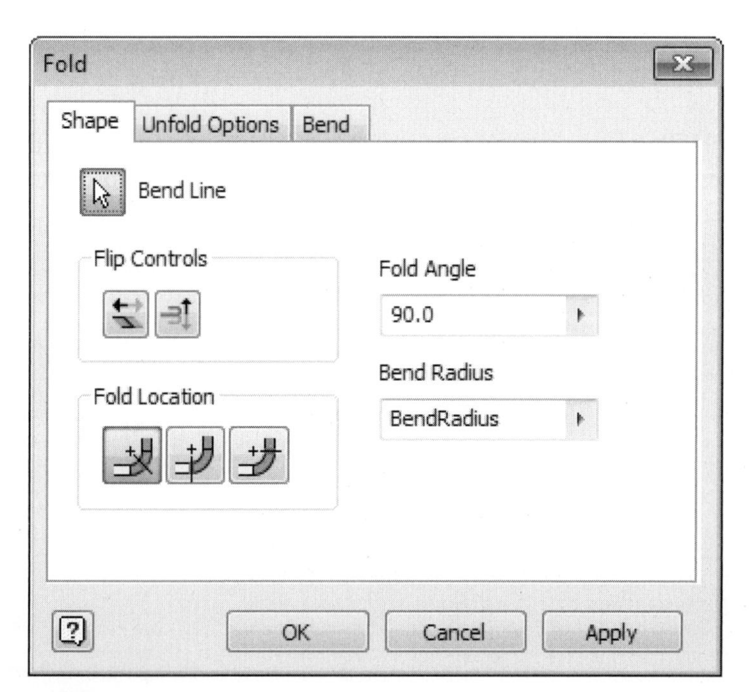

FIGURE 10.82

Shape Tab

The Shape tab allows you to select the bend line for the fold to be created. You can set the location of the selected bend line relative to the fold feature that determines the folded shape. The angle for the fold and direction are also specified on this tab, and you can also override the bend radius that is specified in the sheet metal rule. The following options are available on the Shape tab:

	Bend Line	Select a sketch line to use as the fold line.
	Flip Side	Flip the side used for the angle of the fold.
	Flip Direction	Toggle the direction that the fold will be created.
	Centerline of Bend	Determine the centerline of the fold from the selected sketch line.
	Start of Bend	Determine the start of the fold from the selected sketch line.
	End of Bend	Determine the end of the fold from the selected sketch line.
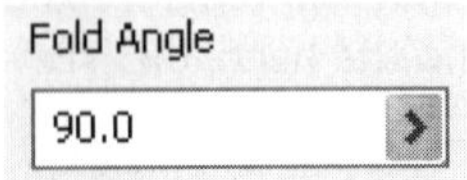 Fold Angle 90.0	Fold Angle	Specify the angle to apply to the fold.
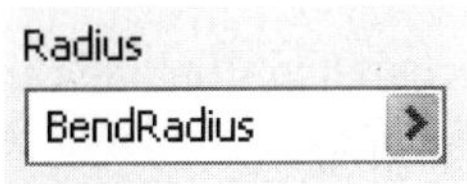 Radius BendRadius	Bend Radius	Override the default bend radius set by the active sheet metal rule.

The options available on the Unfold Options and Bend tabs were covered in the Face section.

Bend

You create bends with the Bend command, as shown in the following image, as child objects of other features when the feature connects two faces. You can also create bends as independent objects between disjointed faces.

FIGURE 10.83

A preview of a bend is shown in the following image, and the faces being joined are either trimmed or extended to connect the faces.

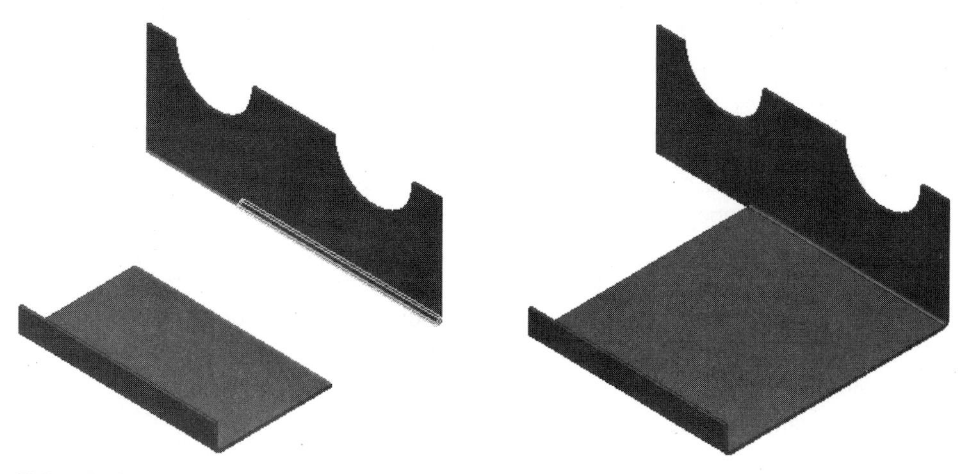

FIGURE 10.84

When you select parallel faces for the bend feature, a joggle or z-bend will be created, depending on the edge location. Joggles are often used to allow overlapping material. A sample of a joggle feature is shown in the following image. You can also create double bends when the two faces are parallel and the selected edges face the same direction.

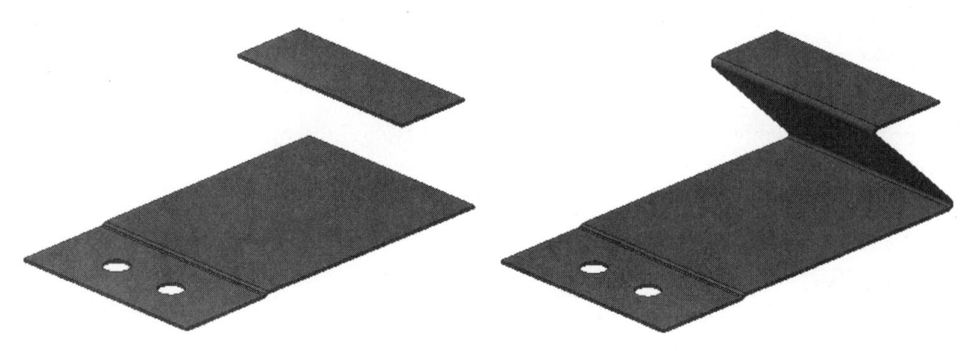

FIGURE 10.85

You can create the bend as two 90° bends with a tangent face between them, as shown in the following image on the left, or a single full-radius bend between the two faces, as shown in the following image on the right.

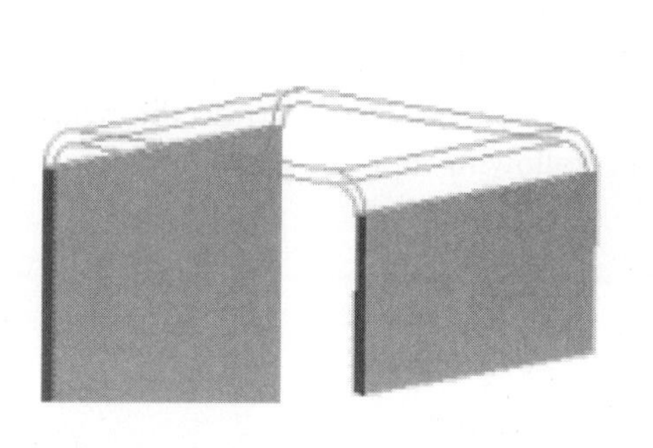

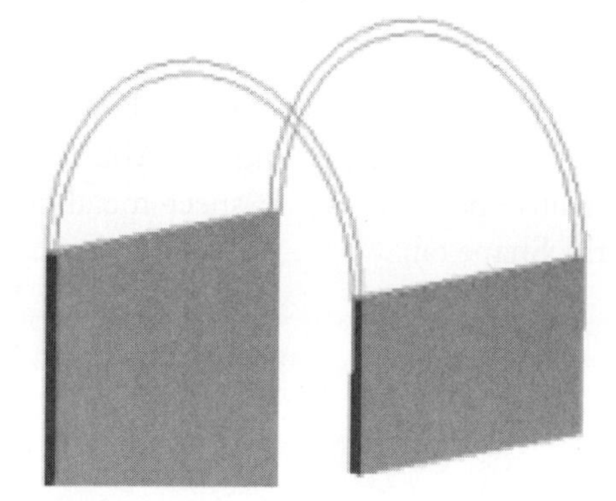

FIGURE 10.86

NOTE

To apply a bend, the sheet metal part must have two disjointed or sharp-cornered intersecting faces. An example of intersecting faces would be a shelled box that is being changed into a sheet metal part. The intersection of the box base and a wall is an edge that can be changed to a bend.

To create a bend, follow these steps:

1. Click the Bend command on the Sheet Metal tab > Create panel. The Bend dialog box appears, as shown in the following image.
2. Select the common edge of two intersecting faces, or select two parallel edges on disjointed, non-coplanar faces. If the two faces are parallel, the Double Bend options are available for selection. Depending on the position of the two faces, the allowed Double Bend options will be Fix Edges and 45 Degree or Full Radius and 90 Degree.
3. Make any changes on the Unfold Options or Bend tabs.
4. Click Apply to continue creating bends, or click OK to complete the bend and exit the dialog box.

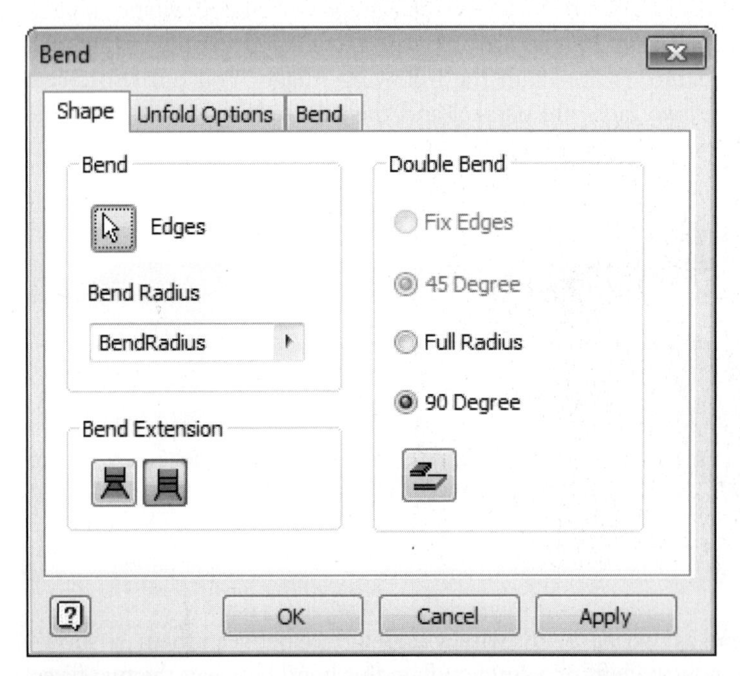

FIGURE 10.87

Shape Tab

The Shape tab allows you to select the edges between which to create the bend and specify the type of bend that you want to create. You can also override the bend radius specified in the sheet metal style. The following options are available on the Shape tab.

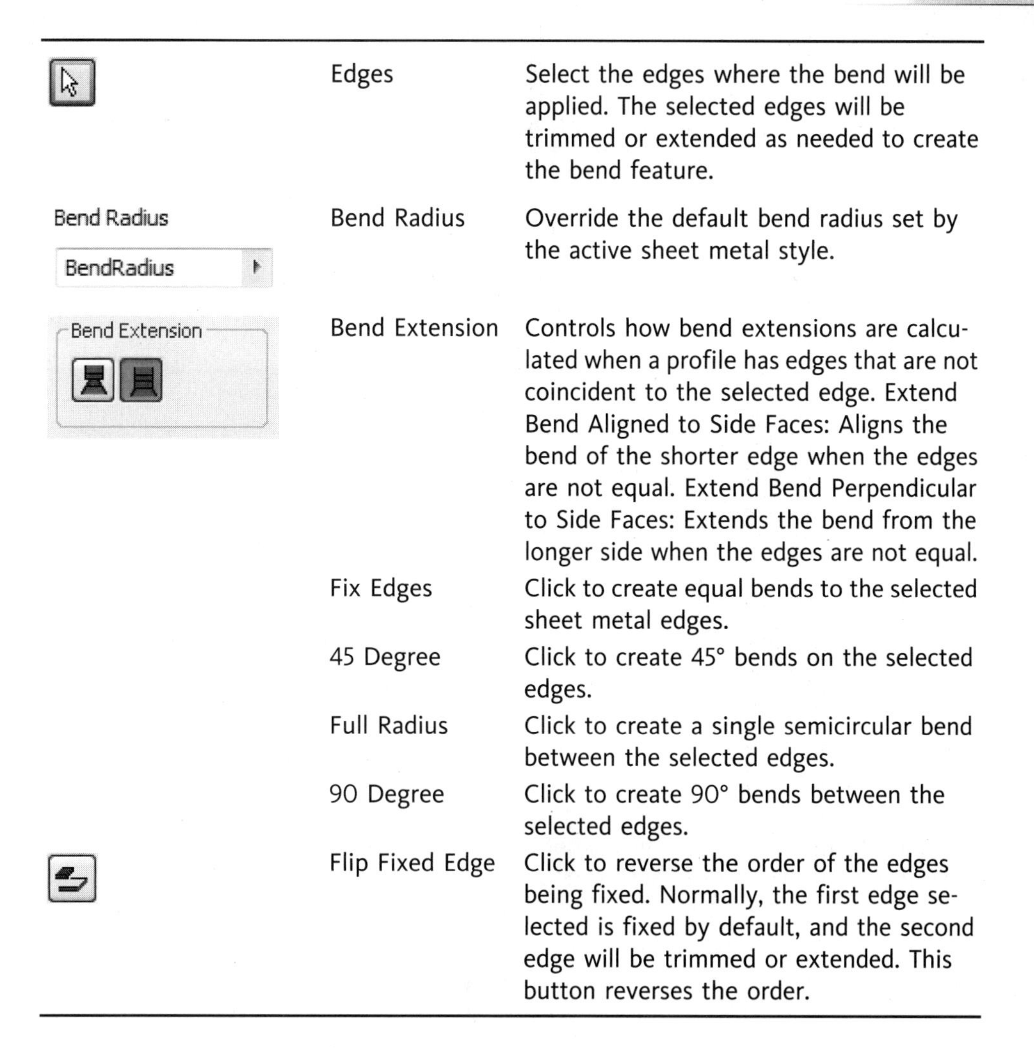

	Edges	Select the edges where the bend will be applied. The selected edges will be trimmed or extended as needed to create the bend feature.
Bend Radius BendRadius	Bend Radius	Override the default bend radius set by the active sheet metal style.
Bend Extension	Bend Extension	Controls how bend extensions are calculated when a profile has edges that are not coincident to the selected edge. Extend Bend Aligned to Side Faces: Aligns the bend of the shorter edge when the edges are not equal. Extend Bend Perpendicular to Side Faces: Extends the bend from the longer side when the edges are not equal.
	Fix Edges	Click to create equal bends to the selected sheet metal edges.
	45 Degree	Click to create 45° bends on the selected edges.
	Full Radius	Click to create a single semicircular bend between the selected edges.
	90 Degree	Click to create 90° bends between the selected edges.
	Flip Fixed Edge	Click to reverse the order of the edges being fixed. Normally, the first edge selected is fixed by default, and the second edge will be trimmed or extended. This button reverses the order.

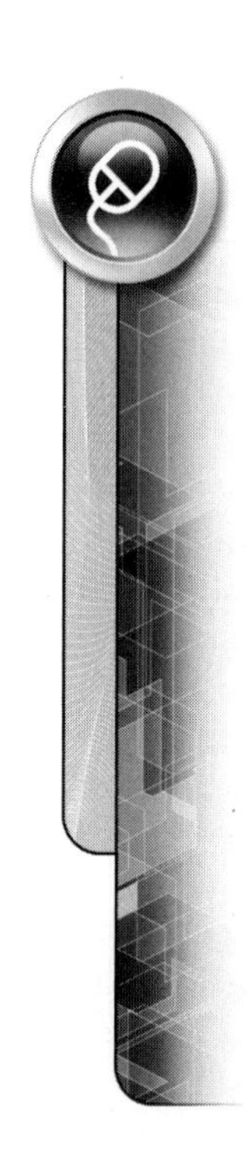

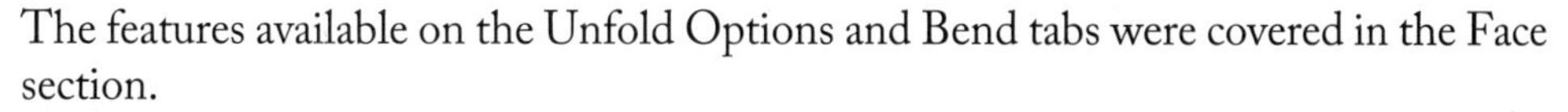

The features available on the Unfold Options and Bend tabs were covered in the Face section.

You can edit a bend that is listed under a feature in the browser at any time, allowing you to reselect the edges that define the bend. You can even edit the bend to connect two faces that were not initially joined when the bend feature was created.

EXERCISE 10-4: MODIFYING SHEET METAL PARTS

In this exercise, you complete the design of a sheet metal bracket within the context of an assembly and then modify the sheet metal part to eliminate component interference. You add bends to connect disjointed sheet metal faces and modify one bend to eliminate interference with a component in the assembly. You then add a hinge feature to the sheet metal part using general modeling tools.

1. Open *ESS_E10_04.iam*.
2. Apply the "Start" Design View Representation, as shown in the following image on the left. The assembly should appear, as shown in the following image on the right.

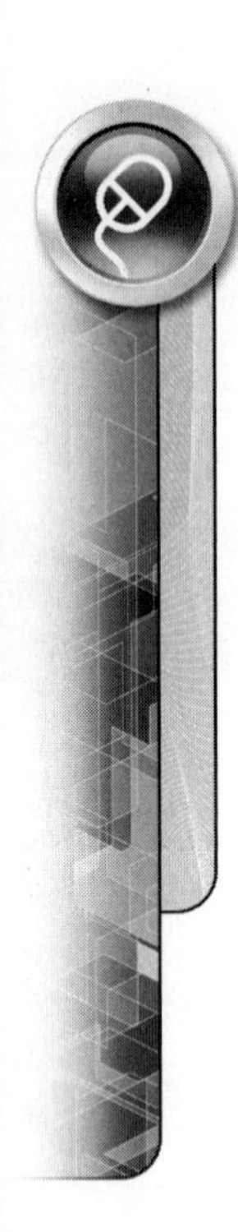

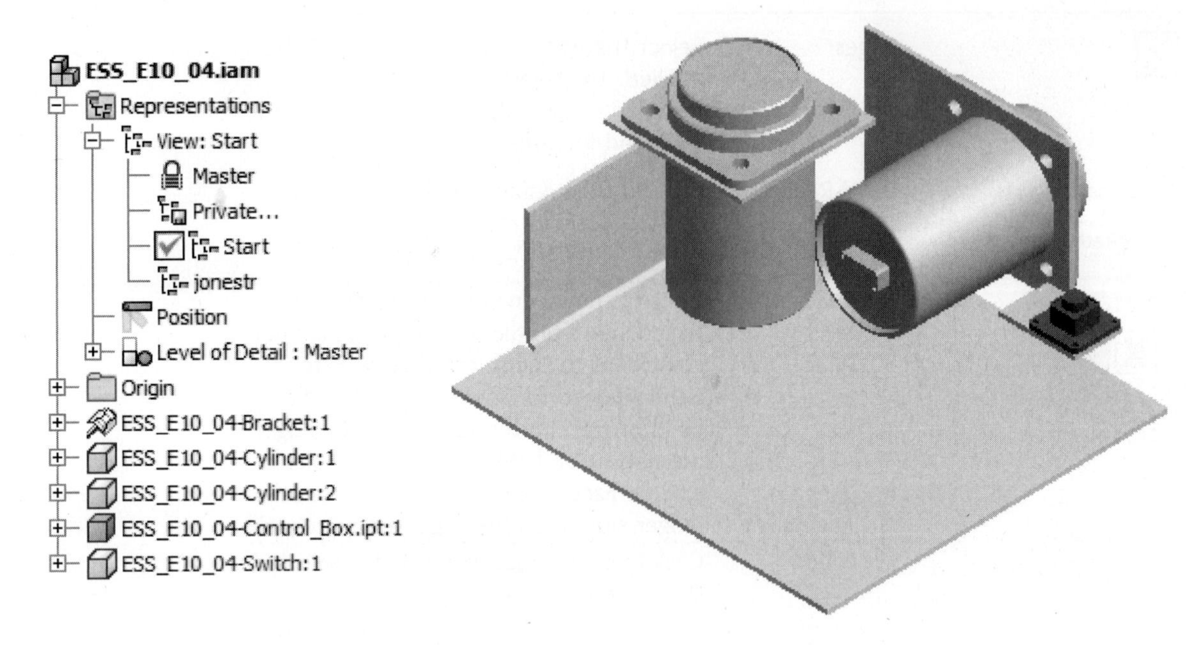

FIGURE 10.88

NOTE

The sheet metal bracket consists of several unconnected faces. Three faces act as mounting surfaces for assembly components; the two connected faces form the base of the bracket. The unique Autodesk Inventor method of creating sheet metal bends between disjointed faces is used to complete the sheet metal bracket.

3. In the browser, right-click *ESS_E10_04-Bracket:1.*
4. Select Edit from the menu.
5. Reorient the assembly, as shown in the following image.

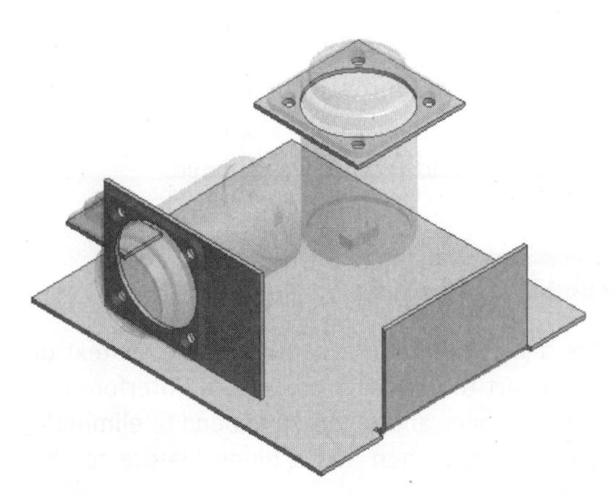

FIGURE 10.89

6. Click the Bend command.
   a. Click the two edges, as shown in the following image on the left.
7. Click Apply to place the bend. Your display should appear similar to the following image on the right.

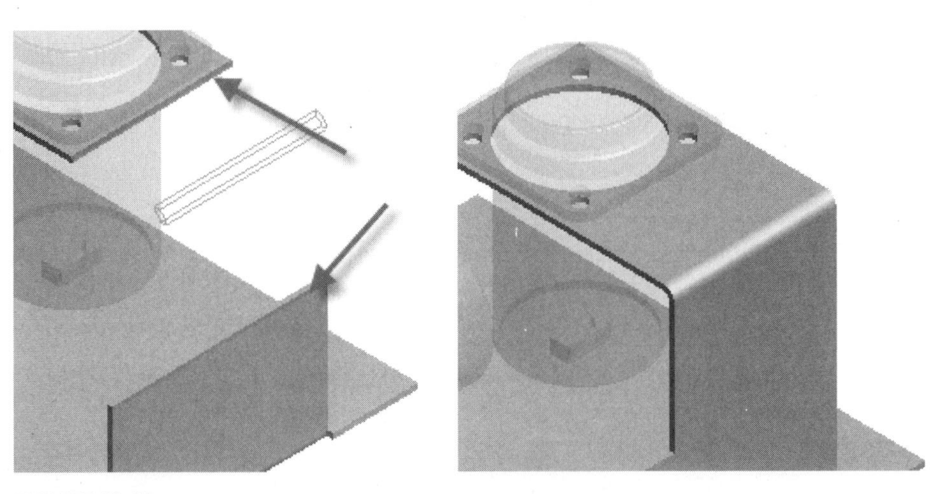

FIGURE 10.90

8. A second bend will now be placed. Click the two edges, as shown in the following image on the left.
9. Click Apply to place the bend. Your display should appear similar to the following image on the right.

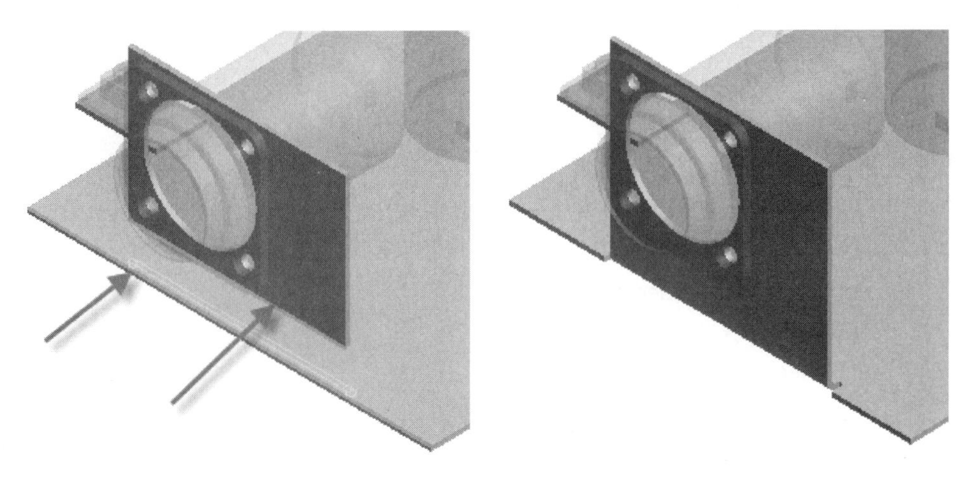

FIGURE 10.91

10. Reorient the view of the assembly, as shown in the following image.

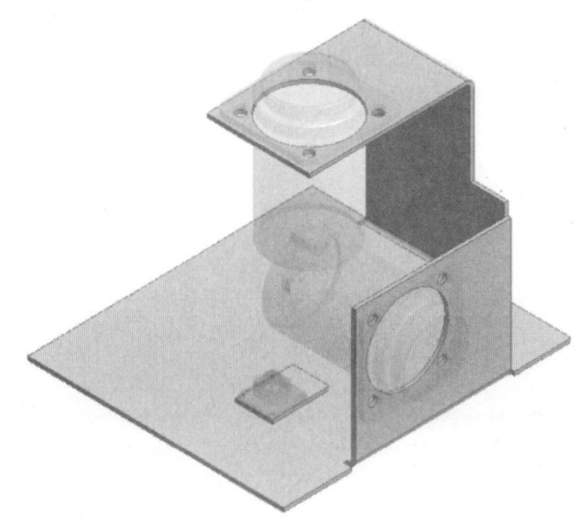

FIGURE 10.92

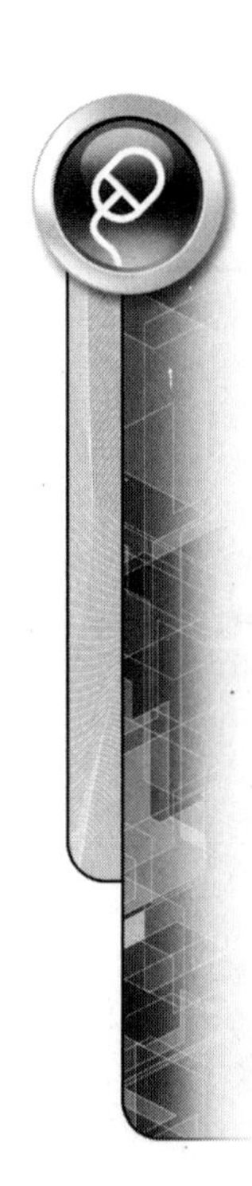

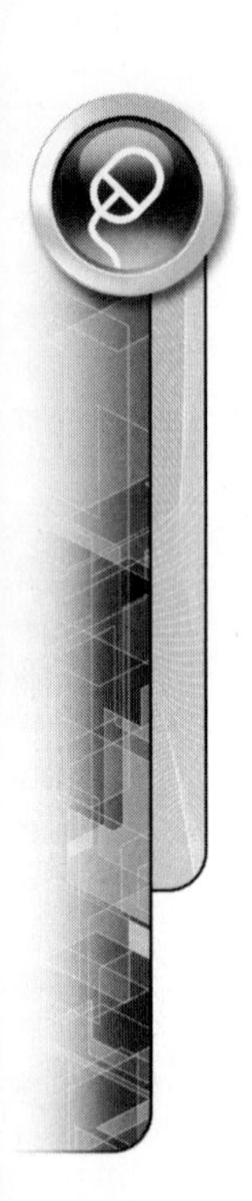

11. Click the back edge of the small face, shown as Edge 1 in the following image on the left.
    a. Click Edge 2, as shown in the following image on the left.
12. In the Bend dialog box:
    a. Click the Fix Edges option.
    b. Click OK to place the bend. Your display should appear similar to the following image on the right.
    c. The bend angles required to create the z-bend may be difficult to manufacture. You can easily edit the bend to return to the default 45° double bends.

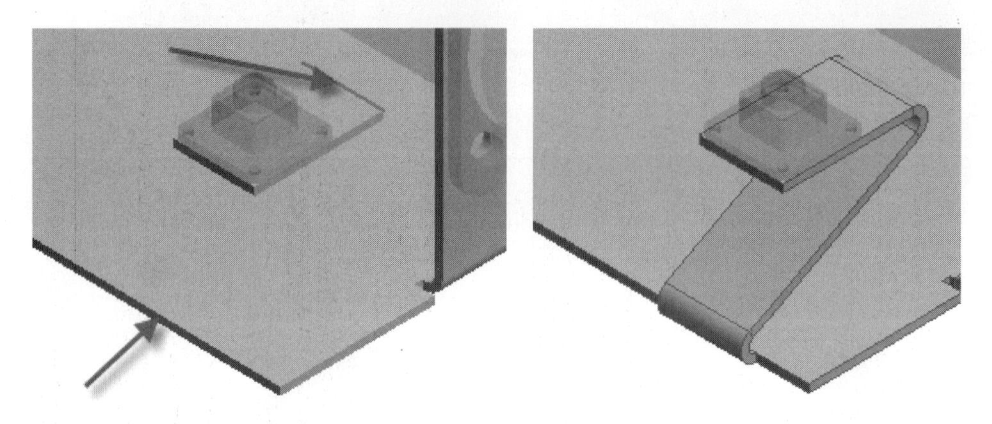

FIGURE 10.93

13. You will now edit the Z-shaped Bend in the browser:
    a. Expand the Folded Model node.
    b. Right-click the last bend listed.
    c. Select Edit Feature from the menu.
14. In the Bend dialog box:
    a. Click the 45 Degree option in the Double Bend area.
    b. Click OK to modify the bend. Your display should appear similar to the following image.

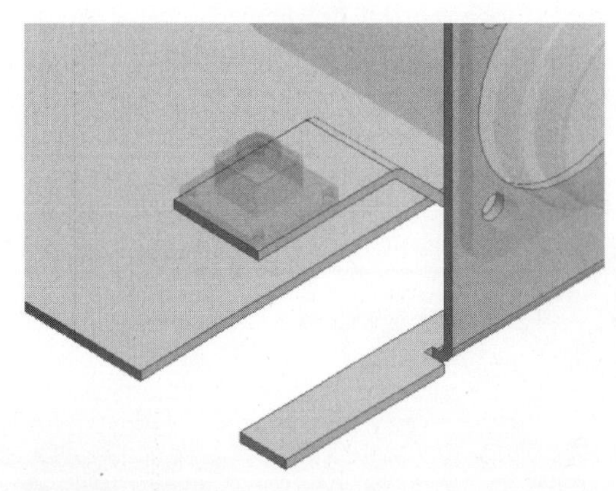

FIGURE 10.94

The small tab next to the vertical mounting face is not required.	NOTE

15. Click the large base face.
16. Click the Create Sketch tool, as shown in the following image.

FIGURE 10.95

17. Use Zoom and Pan to enlarge the view of the small tab.
18. Click the Line command.
19. Draw a line connecting the corner of the bend relief and the edge of the cutout for the 45° jog bend, as shown in the following image on the left.

Ensure that the line is constrained to the corner point of the bend relief and is perpendicular to the cutout edge.	NOTE

20. Click Finish Sketch, and then click the Cut command.

The details of the Cut command will be covered later in the chapter.	NOTE

21. Select the small tab as the profile.
22. Set the Extents to Distance and a value of Thickness.
23. Click OK to complete the cut.
24. Click Zoom All. Your display should appear similar to the following image on the right.

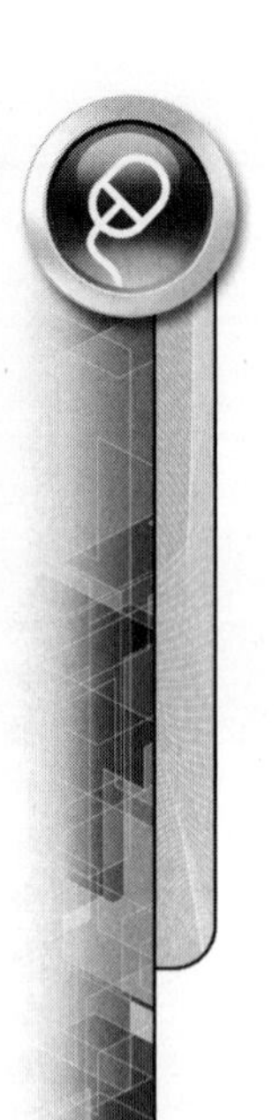

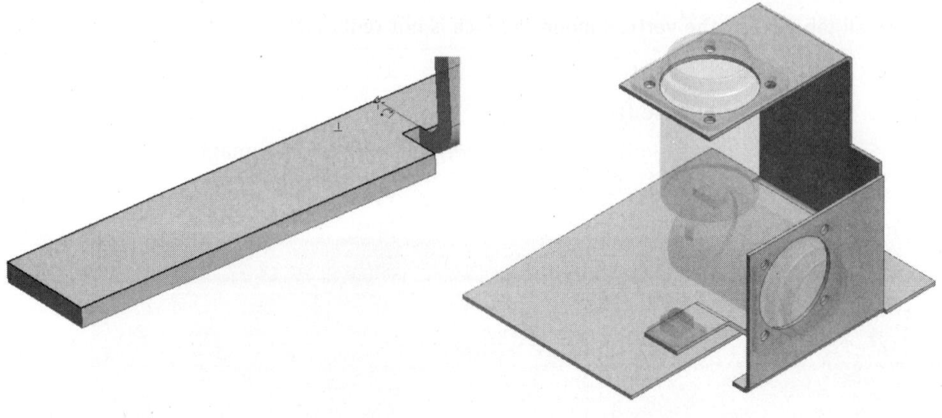

FIGURE 10.96

25. In the browser:
    a. Double-click *ESS_E10_04.iam* to return to the assembly environment.
    b. Right-click *ESS_E10_04-Control_Box.ipt:1.*
    c. Select Visibility from the menu.

NOTE

The Control Box interferes with the sheet metal part, as shown in the following image.

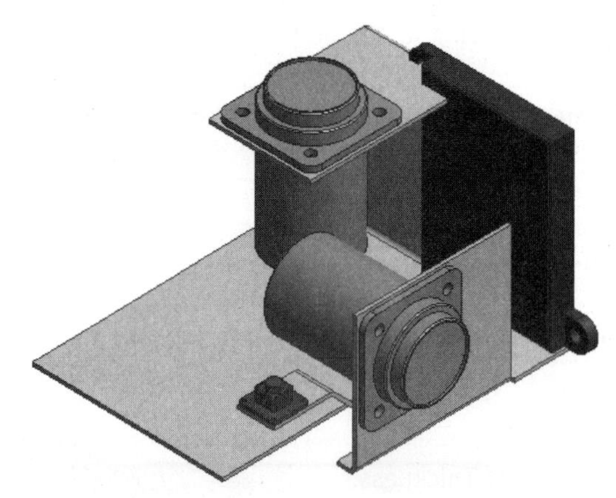

FIGURE 10.97

To eliminate the interference, edit the bend causing the conflict. In the graphics window:

26. Double-click *ESS_E10_04-Bracket:1* to activate it for editing.

27. Use Rotate and Zoom to orient the assembly, as shown in the following image.

FIGURE 10.98

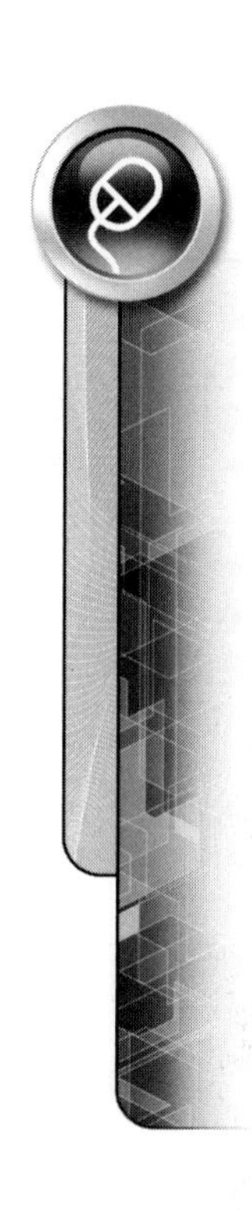

28. In the browser:
    a. Place the cursor over the first bend under the Folded Model node of *ESS_E10_04-Bracket:1*. The bend adjacent to the interference should be highlighted in the graphics window.
    b. Right-click this bend, and select Edit Feature from the menu.
29. In the Bend dialog box:
    a. Click the Edges button.
    b. Hold down the CTRL key, and deselect the two edges that define the bend, as shown in the following image on the left.
    c. Select the top edge of the face supporting the horizontal cylinder and the adjacent parallel edge on the face supporting the vertical cylinder, as shown in the following image on the right.

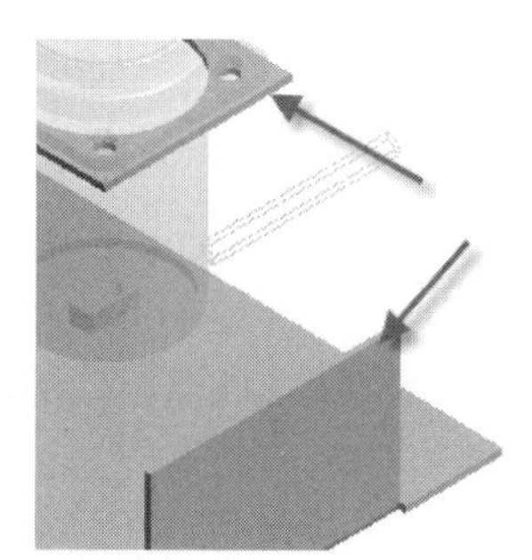

FIGURE 10.99

30. Click OK to create the modified bend.

    The control box no longer interferes with the sheet metal bracket. The vertical face adjacent to the control box can be modified to provide support for the control box. That modification is not covered in this exercise.

31. In the browser, double-click the top level assembly, *ESS_E10_04.iam*, to activate the assembly environment, as shown in the following image.

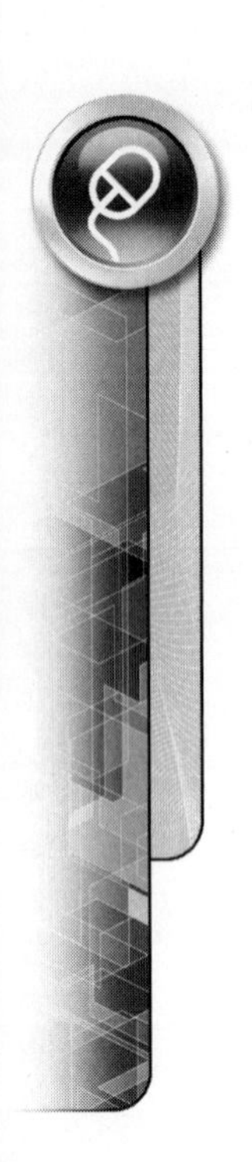

FIGURE 10.100

32. Close the file. Do not save changes. End of exercise.

Cut

The Cut command, as shown in the following image, is a sheet metal-specific implementation of the standard Extrude command. The Cut command always performs an extrude cut. The default extents are a blind cut at a distance equal to the sheet metal Thickness parameter. This action ensures that the cut extends only through the face containing the sketch and not through other faces that may be folded under the sketch face. You can enter a value that is less than the thickness of the part to create a cut that does not go all of the way through the material.

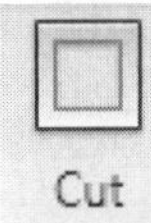

FIGURE 10.101

Most cuts are manufactured on the flat sheet stock before the sheet is bent to form the folded part, and cuts often cross bend lines. Because the part is modeled in the folded state, representing cuts that cross bend boundaries requires the bend be unfolded to represent the flat sheet. When the bend is refolded, the cut deforms around the bend to ensure that the extrusion remains perpendicular to the sheet metal faces on both sides of the bend and deforms throughout the bend, if required. When sketching a cut profile that will cross a bend, the Project Flat Pattern command, shown below, projects the unfolded flat pattern geometry onto the sketch face, as shown in the following images.

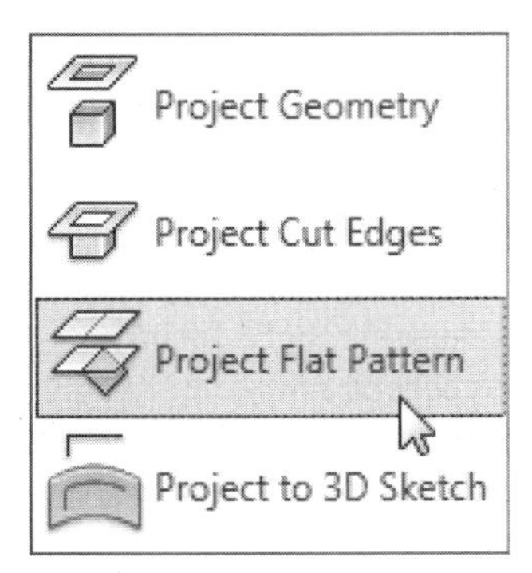

FIGURE 10.102

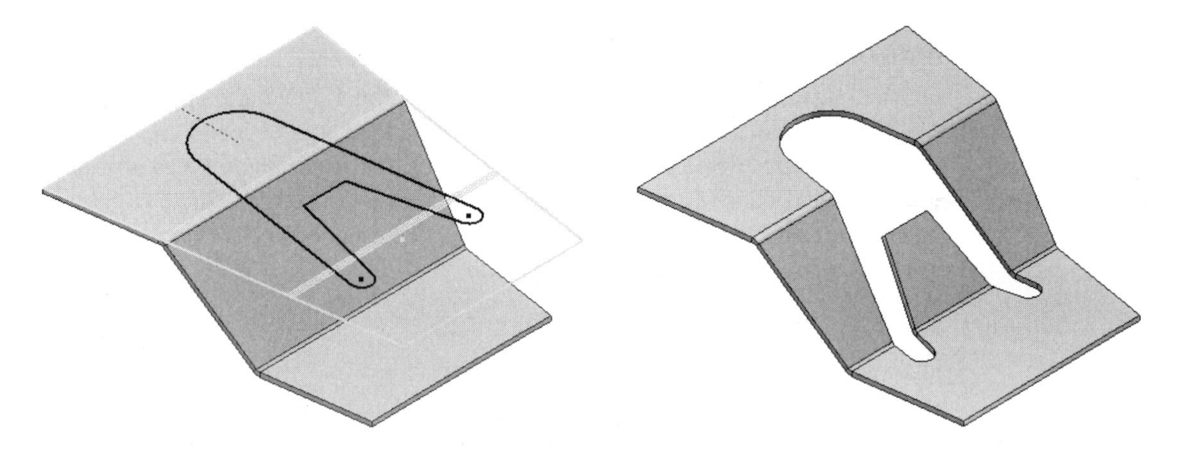

FIGURE 10.103

To create a cut feature, follow these steps:

1. Create a sketch on a sheet metal face that includes one or more closed profiles representing the area(s) to cut. If required, use the Project Flat Pattern command to project the unfolded geometry of connected faces onto the sketch.
2. Click the Cut command on the Sheet Metal tab > Modify panel. The Cut dialog box appears, as shown in the following image.
3. Select the profiles to cut.
4. If a profile crosses a bend, click Cut Across Bend.
5. Enter a distance for the extents of the cut.
6. Click OK to complete the cut and exit the dialog box.

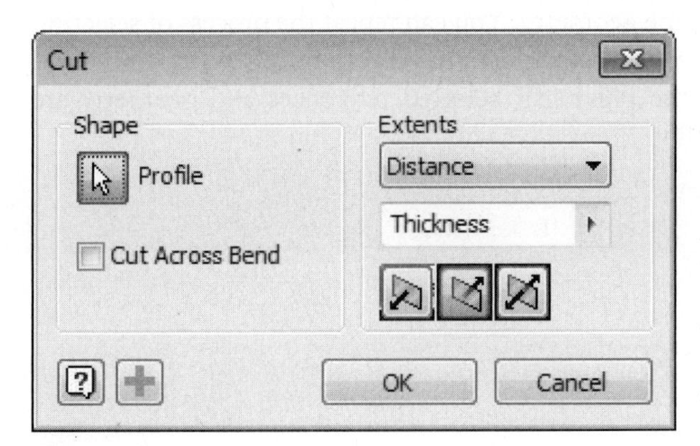

FIGURE 10.104

The following commands are available in the Cut dialog box:

	Profile	Select the profile(s) to be cut into the sheet metal part.
	Cut Across Bend	Click to project the profile across faces that are bent in the sheet metal part.
	Extents	Choose one of the typical extrusion options: Distance, To Next, To, From To, or All. If distance is selected, enter a distance for the extents of the cut or accept the default value of thickness to cut the feature using the thickness parameter defined in the sheet metal rule.
	Direction	Specify the direction for the cut to be created in the part.

Unfold/Refold

Using the Unfold and Refold commands, shown in the following image, you can select bends and rolls to flatten geometry. In many cases, it is easier to create certain sheet metal features while a model is in a flat state. Typically these types of features cross bend lines of a sheet metal part. You can use the Unfold command to flatten a formed state, create the necessary geometry and then use the Refold command to fold the geometry back to the original folded state.

FIGURE 10.105

To unfold sheet metal features, follow these steps:

1. Click the Unfold command on the Sheet Metal tab > Modify panel. The Unfold dialog box appears, as shown in the following image on the left.
2. Select one of the displayed temporary planes to be used as a stationary reference for the unfold operation.
3. Click highlighted bends or rolls to be unfolded. You can use the add all rolls or add all bends option to select all highlighted geometry.
4. Continue selecting sections of the model to unfold.
5. Click Apply to continue unfolding geometry, or click OK to apply the unfold feature and exit the dialog box.

The Refold command works in the same manner, but has the additional capability for you to not have to reselect the same geometry. You can repeat the process of selecting stationary planes and bends or rolls to be refolded, or you can simply select an Unfold feature in the browser. All of the previously selected references and geometry are automatically selected for refolding.

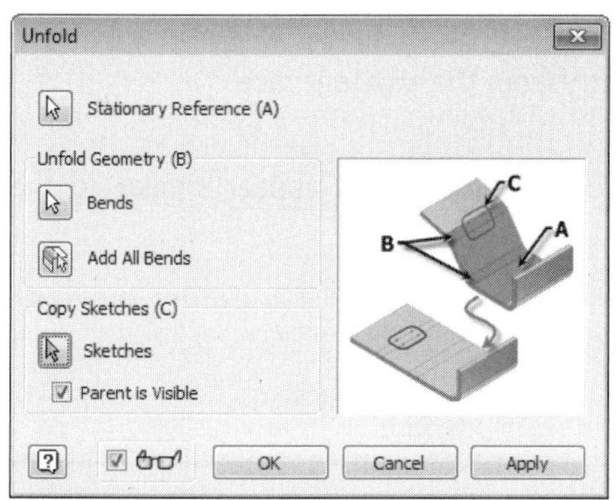

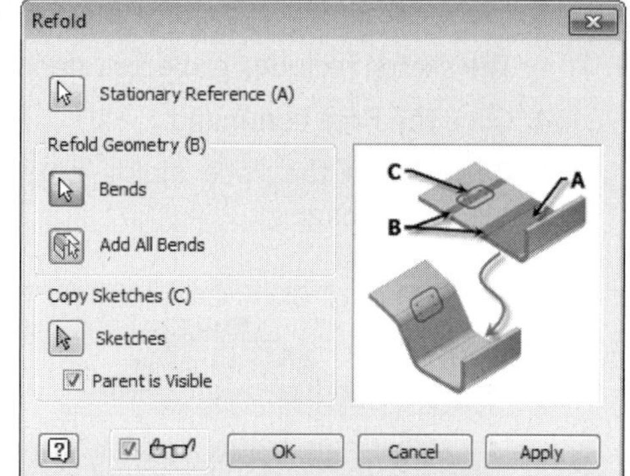

FIGURE 10.106

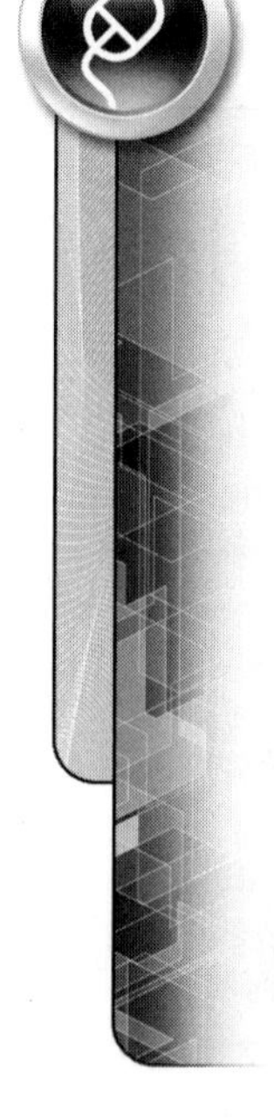

An example of a part being unfolded to add a stiffener across a transition and then having the model undergo the refold command is shown in the following image.

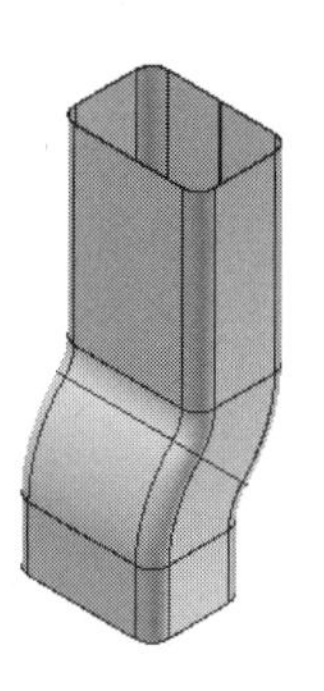
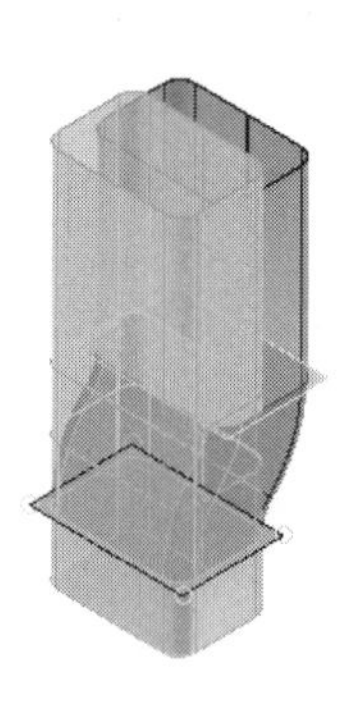
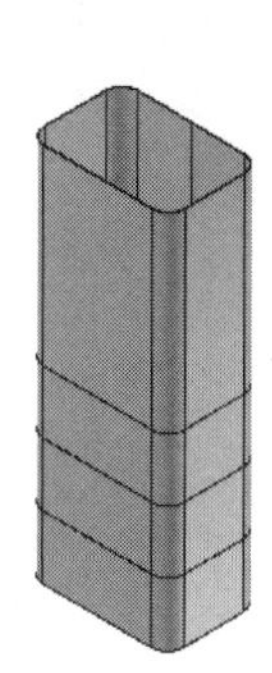
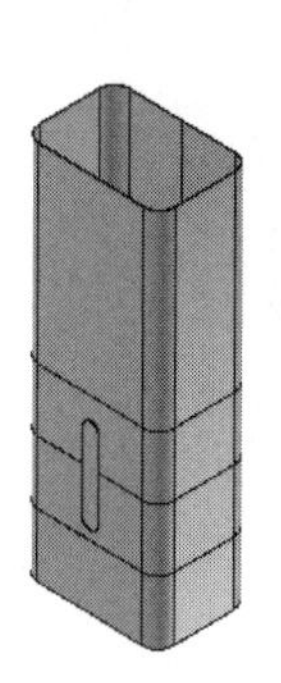
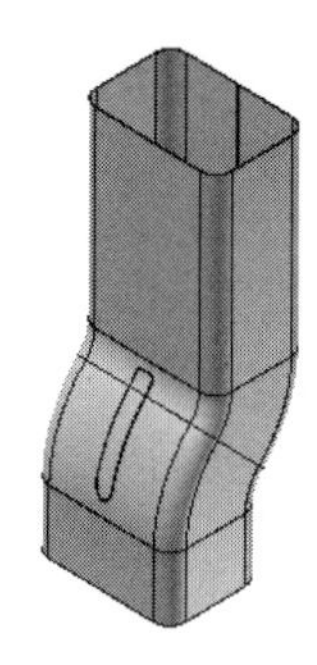

FIGURE 10.107

EXERCISE 10-5: CUT ACROSS BEND

In this exercise, you complete a sheet metal bracket using a fold and double bend. You then use the Project Flat Pattern command to create a cut across bends.

1. Open *ESS_E10_05.ipt*.
2. Use Zoom and Rotate to examine the part. Reorient your view to match the following image.

FIGURE 10.108

3. In the browser, right-click Face_Sketch, and select Visibility from the menu.

   The sketch includes projected geometry from the adjacent face.

4. Click the Face command.
5. Click OK in the Face dialog box. Your display should appear similar to the following image.

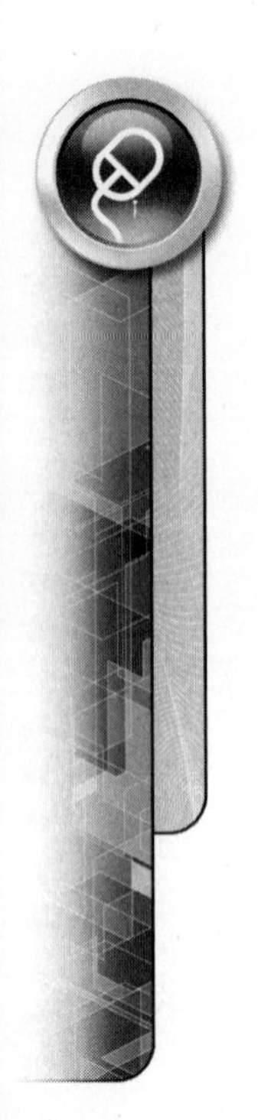

FIGURE 10.109

6. Reorient your view to match the following image.

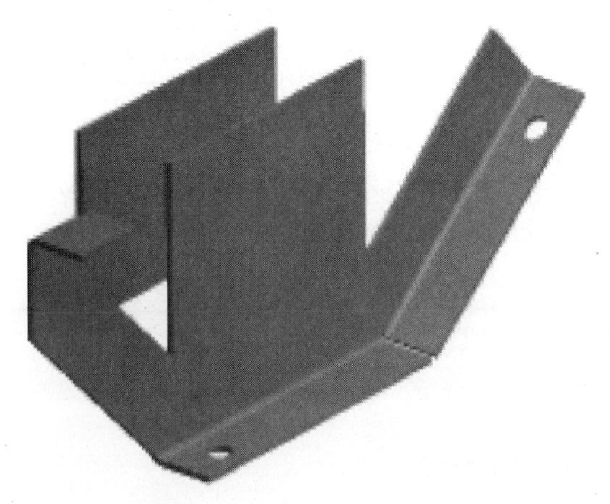

FIGURE 10.110

You will now create a double bend between the two parallel faces.

7. Click the Bend command.

8. Click the two edges, as shown in the following image.

FIGURE 10.111

9. In the Bend dialog box:
   a. Click Full Radius.
   b. Click 90 Degree.
   c. Click OK.
10. Reorient your view to resemble the following image. Create a new sketch on the face, shown highlighted in the following image.

FIGURE 10.112

11. Click the Line command.
12. Create a line, as shown in the following image.

FIGURE 10.113

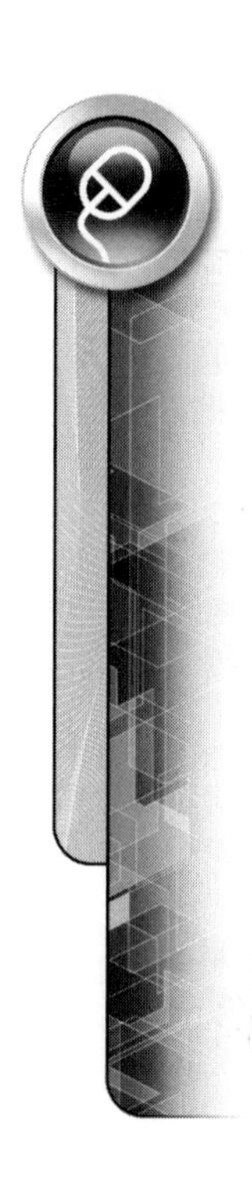

13. Right-click in the graphics window, and select Done.
14. Right-click in the graphics window, and select Finish 2D Sketch.
15. Click the Fold command, and select the sketched line.
16. In the Fold dialog box:
    a. If required, click the Flip direction button to match the following image.
    b. Click End of Bend for the Fold Location.

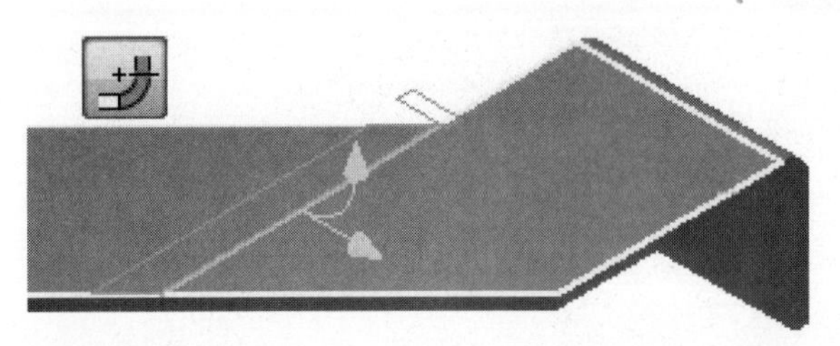

FIGURE 10.114

    c. Click OK.
    d. You will now use the Project Flat Pattern command to create a sketch, and you use this sketch to create a cut across bends.

17. Reorient your view to resemble the following image. Create a new sketch on the face, shown highlighted in the following image.

FIGURE 10.115

18. Click the Project Flat Pattern command.
19. Select the face highlighted in the following image.

FIGURE 10.116

20. Click the View Face command.
21. Click one of the projected flat pattern edges.
22. Create the sketch, as shown in the following image.

NOTE

The sketch is centered on the face by a horizontal constraint between the midpoint of the projected and vertical sketched line and a vertical constraint between the bottom horizontal projected edge and the center point of the arc.

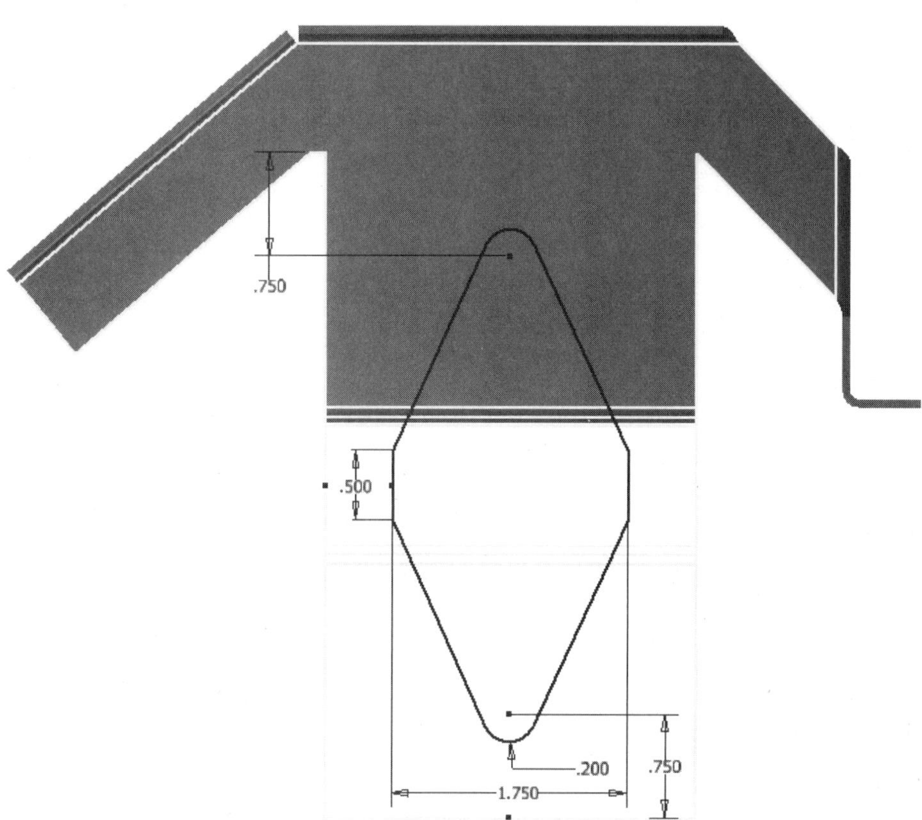

FIGURE 10.117

23. Right-click the graphics window, and select Finish 2D Sketch.
24. Click the Cut command.
25. Click inside the sketch if it is not already selected.
26. Ensure that Cut Across Bend is checked in the Cut dialog box.
27. Click OK. Your display should appear similar to the following image.

FIGURE 10.118

28. Close the file. Do not save changes. End of exercise.

Corner Seam

When three faces meet in a sheet metal part, a gap is required between two of the faces to enable unfolding. Using a box as an example, the walls of the box are connected to the floor, and gaps between the walls enable the box to be unfolded. The gap between adjacent faces is a corner seam, and you create it using the Corner Seam command, as shown in the following image. The Corner Seam command can work with two- and three-bend intersections. When two coplanar flanges meet, the flanges will result in a three-bend intersection, referred to as a Jacobi corner. If a three-bend intersection is detected, options will become active in the Corner Seam dialog box that enables you to define how the corner feature will be treated when a flat pattern is generated. You can also create a corner seam by ripping open a corner at an intersection between faces.

FIGURE 10.119

To create a corner seam, follow these steps:

1. Click the Corner Seam command on the Sheet Metal tab > Modify panel. The Corner Seam dialog box appears, as shown in the following image.
2. Select face edges that meet one of the following criteria:
   - Two open edges on faces that share a common connected edge: for example, two walls that share a connection to a floor.
   - Two nonparallel edges on coplanar faces: for example, the edges create a mitered joint with a gap between the extended faces.
   - A single edge at the intersection of two faces, such as the wall intersections of a shelled box.
3. Select the Seam option in the Corner Seam dialog box.
4. Enter a value for the corner seam Gap.
5. Make any changes to the Bend or Corner options.
6. Click Apply to continue creating corner seams, or click OK to complete the corner seam and exit the dialog box.

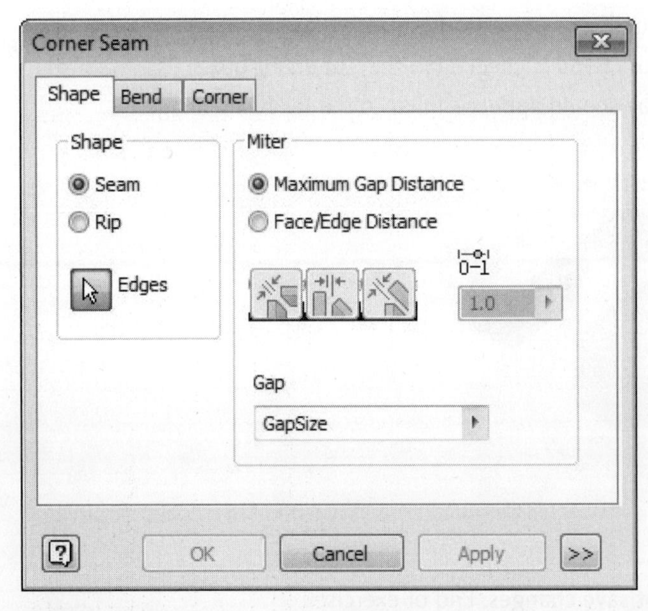

FIGURE 10.120

Shape Tab

The Shape tab allows you to select the edges where the corner seam will be created. You can also specify the orientation of the seam and whether or not to rip a corner. The following options are available on the Shape tab for miter corners. If you select coplanar faces, a miter corner is created and the options in the dialog box are displayed as shown in the previous image. If you select intersecting faces, a seam is created. The options available will change based on the selected geometry.

Icon	Option	Description
	Seam	Click to create a corner seam by extending or trimming existing coplanar or intersecting sheet metal faces.
	Rip	Click to rip open a square corner of a part.
	Edges	Select the edges of the model on which you will create a corner seam or rip.
	Maximum Gap Distance	Select to create a corner seam gap that is measured consistent with the use of a physical inspection gauge.
	Symmetric Gap	Creates a gap in between the selected edges where the distance is measured via a straight line between the inside edges of the seam.
	45 Degrees	Similar to the Symmetric Gap selection for the selected corners.
	Overlap	Creates a gap that is measured from the face of the first selected edge that will overlap the second selected edge.
	Reverse Overlap	Creates a gap that is measured from the face of the second selected edge that will overlap the first selected edge.
Gap GapSize	Gap	Enter a gap for the clearance distance between the edges. The default value is equal to the GapSize parameter.
0-1 1.0	Percentage Overlap	Enter a percentage of the flange thickness for the seam to overlap. This option can only be used with the overlap and reverse overlap types.

The options available on the Bend and Corner tabs were covered in the Face section and the Sheet Metal Rules section. On these tabs, you can override the default settings from the active sheet metal rule on a per-feature basis.

More Button

Additional options are available when you click the More (<<) button to expand the dialog box, as shown in the following image. The options are described as follows.

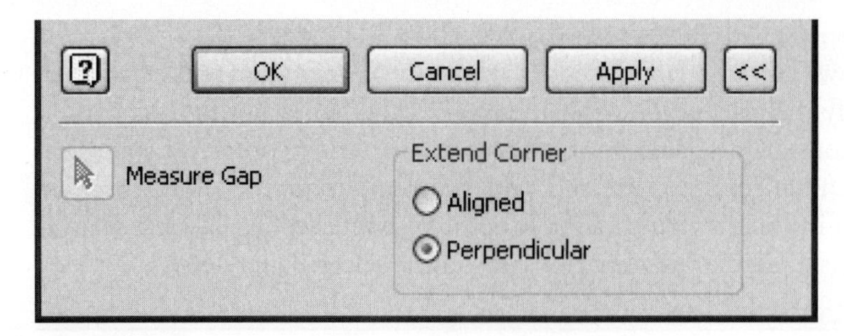

FIGURE 10.121

Measure Gap. Select two edges to measure the distance between them.

Aligned. Click to extend faces to match sloped edges adjacent to edges of the selected faces.

Perpendicular. Click to extend faces at the same height as the selected edge, ignoring any adjacent sloped edges.

Corner Round

The Corner Round command, as shown in the following image, is available when no sketch is active. It is a sheet metal-specific fillet tool. All edges other than those at open corners of faces—that is, all edges having a Length = Thickness—are filtered out. This action enables you to select these small edges easily without zooming in on the part.

FIGURE 10.122

To create a corner round, follow these steps:

1. Click the Corner Round command on the Sheet Metal tab > Modify panel. The Corner Round dialog box appears, as shown in the following image.
2. Enter a radius for the corner round.
3. Select the Corner or Feature Select Mode.
4. Select the corners or features to include.
5. Add additional corner rounds with different radii, and select corners or features for the additional corner rounds.
6. Click OK to complete the corner round and exit the dialog box.

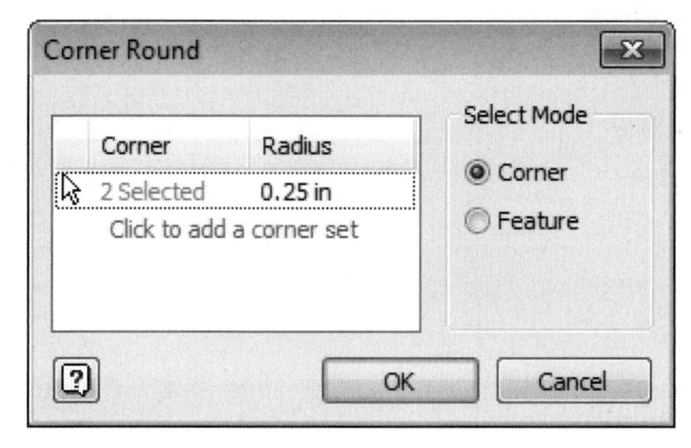

FIGURE 10.123

Select Mode

Selecting edges to which you apply a corner round is similar to selecting edges for the fillet feature. Two modes enhance the feature's use in the sheet metal environment: Corner and Feature.

Corner. Click to select individual thickness edges to be rounded or filleted.

Feature. Click to select all thickness edges of a feature automatically.

Corner Chamfer

The Corner Chamfer command, as shown in the following image, is a sheet metal-specific chamfer tool. As with the Corner Round command, all edges other than those at open corners of faces are filtered out. Because the edges are always discontinuous, the Edge Chain and Setback settings available with the standard Chamfer command are not available.

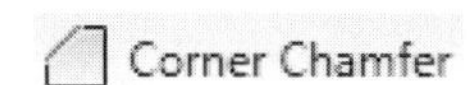

FIGURE 10.124

To create a corner chamfer, follow these steps:

1. Click the Corner Chamfer command on the Sheet Metal tab > Modify panel. The Corner Chamfer dialog box appears, as shown in the following image.
2. Select the chamfer style: One Distance, Distance and Angle, or Two Distances. See the following list for explanations of these styles.
3. Select the corners or edge and corner for Distance and Angle that you wish to include.
4. Enter the chamfer values.
5. Click OK to complete the corner chamfer and exit the dialog box.

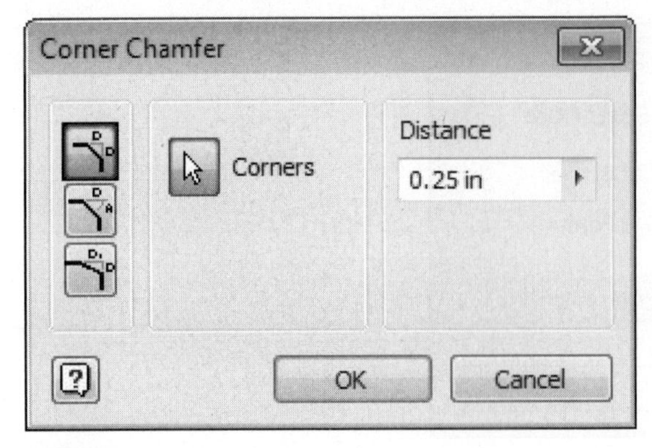

FIGURE 10.125

	One Distance	Creates a 45° chamfer on the selected edge. Determine the size of the chamfer by typing a distance in the dialog box. The value is offset from the two common faces.
	Distance and Angle	Creates a chamfer offset from a selected edge on a specified face at an angle from the number of degrees specified. Enter an angle and distance for the chamfer in the dialog box, click the face on which the angle is based, and specify the face edge to be chamfered.
	Two Distances	Creates a chamfer offset from two faces, each in the amount that you specify. Click a corner, and enter a value for Distance1 and Distance2. A preview image of the chamfer appears. To reverse the direction of the distances, click the Flip button.
	Corners	Click to select individual corners to chamfer.
	Edge	Click to select an edge for chamfers when using the Distance and Angle option.
	Flip	Click to flip the direction for the chamfer distance when using the Two Distances option.
	Distance	Enter a distance to be used for the chamfer feature.
	Angle	Enter an angle for the chamfer feature when using the Distance and Angle option.

PunchTool

Cuts and 3D deformation features, such as dimples and louvers, are usually added to the flat sheet metal stock in a turret punch before the sheet is bent into the folded part. Turret punch tools are positioned on the flat sheet by a center point corresponding to the tool center at an angle to a fixed coordinate system. The PunchTool command, as shown in the following image, places specially designed iFeatures that define the center point of the tool. A streamlined interface simplifies the selection and placement of the iFeatures.

FIGURE 10.126

A default Punch folder is installed under the top-level Catalog folder when you install Autodesk Inventor. A selection of punch tools is included in the folder. The PunchTool command lists all iFeatures in the designated punch folder. You can set the default Punch folder on the iFeature tab of the Application Options dialog box, as shown in the following image.

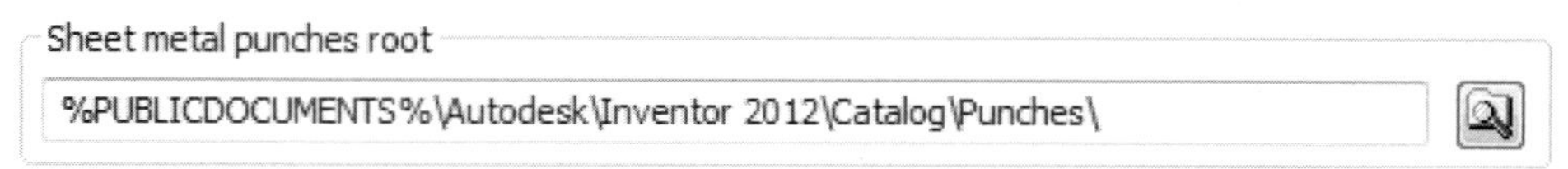

FIGURE 10.127

To qualify as a PunchTool, a saved iFeature must have a single point, center point in the sketch of the first feature included in the iFeature. The point corresponds to the center of the tool, and it will be used as the point of placement when you apply a PunchTool iFeature to a part face. Asymmetrical shapes must have the center point position controlled by geometric relationships or equations to ensure that it remains centered when the iFeature parameters are changed. Create the iFeature in a sheet metal part to ensure that parameters such as Thickness are saved with the iFeature.

You create a sheet metal punch iFeature in the same way that you create standard iFeatures. Select Extract iFeature from the Manage tab > Author panel, and the Extract iFeature dialog box is displayed, as shown in the following image. At the top of the dialog box, select Sheet Metal Punch iFeature from the Type area. This selection activates the Manufacturing, Depth, and Simplified Representation areas.

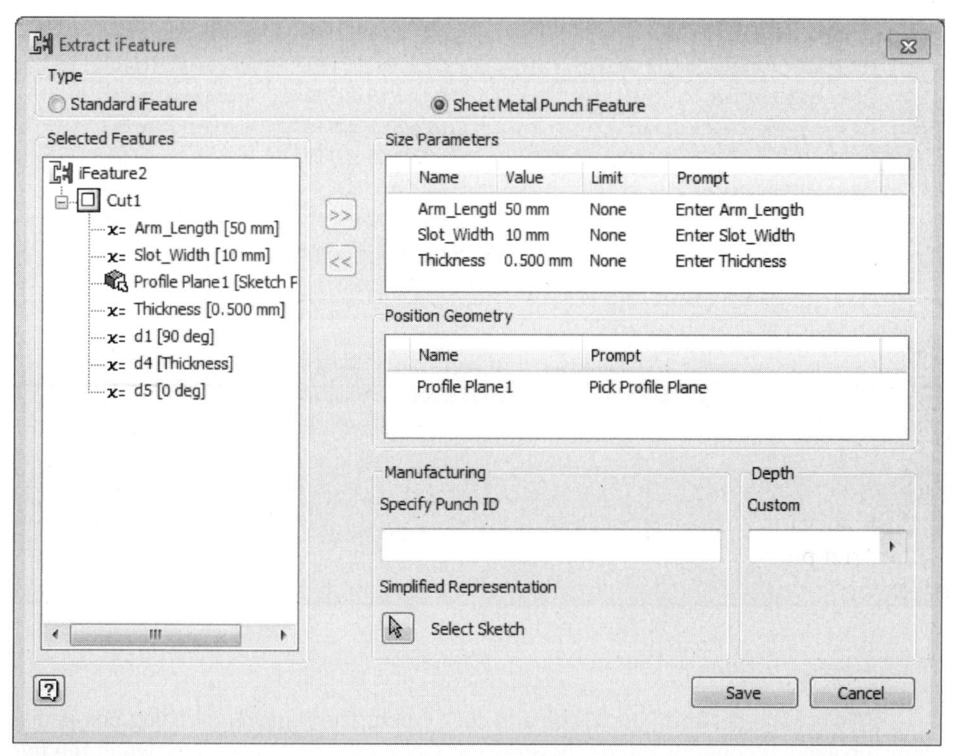

FIGURE 10.128

The Specify Punch ID field can be used to enter information, such as a vendor part number or designation, which will be stored with the data that defines the punch. The data can be extracted and used during drawing creation. The Custom field is used to define how far the punch penetrates past the selected surface. The Simplified Representation section allows you to select a sketch from the browser that will be used as an alternate representation for the punch in the flat pattern and when the flat pattern is placed on a drawing sheet. Similar to standard iFeatures, you can also create a table-driven iFeature to create a family of punches using the iFeature Author Table command; this command is located on the iFeature tab > iFeature panel when an extracted punch or standard iFeature is opened with Autodesk Inventor.

To place a PunchTool, follow these steps:

1. Create a sketch on a sheet metal face, and place at least one sketch point in the sketch. Point, Center Points are selected automatically as punch centers during PunchTool placement. You can manually add sketch points, line or curve endpoints, and arc centers as additional punch centers.
2. Click PunchTool on the Sheet Metal tab > Modify panel. The PunchTool Directory dialog box appears, as shown in the following image.

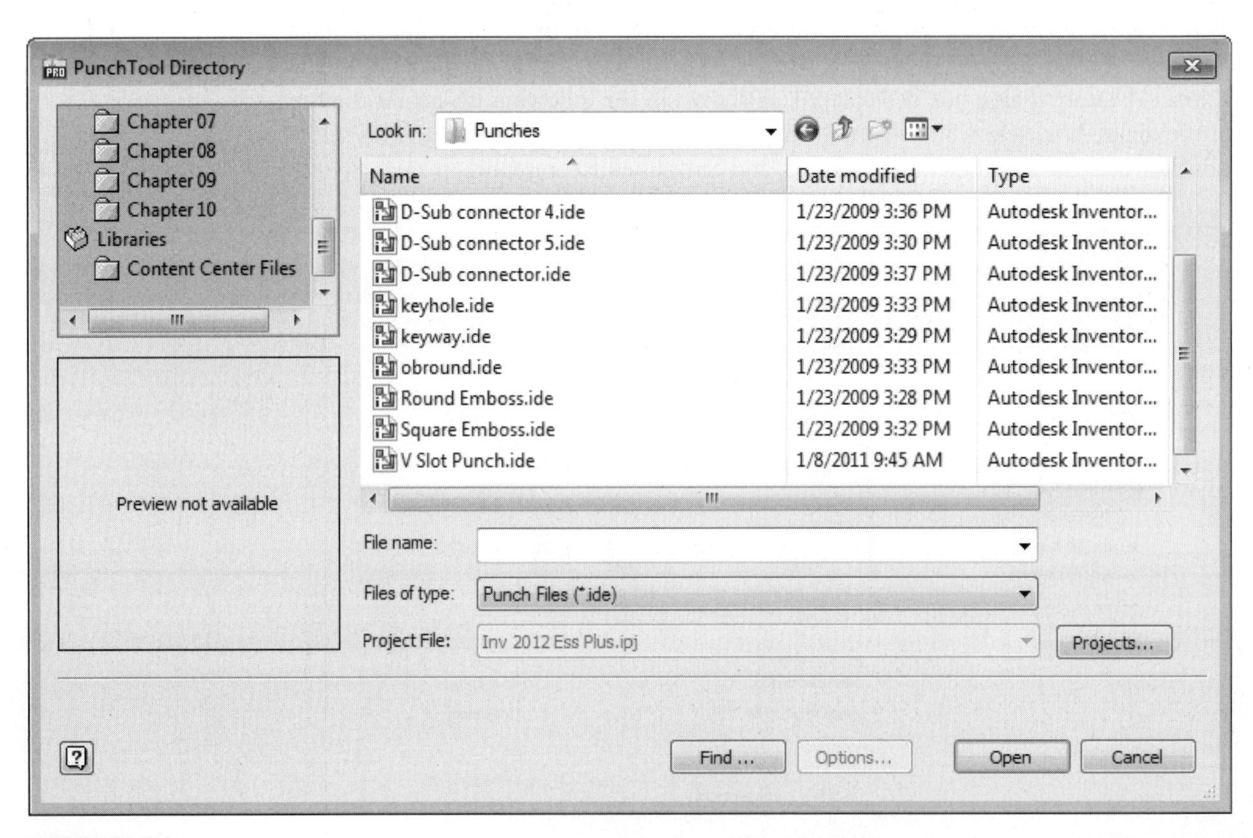

FIGURE 10.129

3. Select the desired PunchTool, and click Open. The PunchTool dialog box is displayed. It contains three tabs: Preview, Geometry, and Size, as shown in the following image.
4. All Point, Center Points are selected automatically. Hold down the SHIFT or CTRL key, select Point, Center Points to exclude them, and select other location geometry as required.
5. On the Geometry tab, select a rotation angle for all occurrences of the punch. At 0° rotation, the X-axis of the first feature in the saved iFeature is aligned with the X-axis of the current sketch.

6. On the Size tab, enter values for the PunchTool parameters.

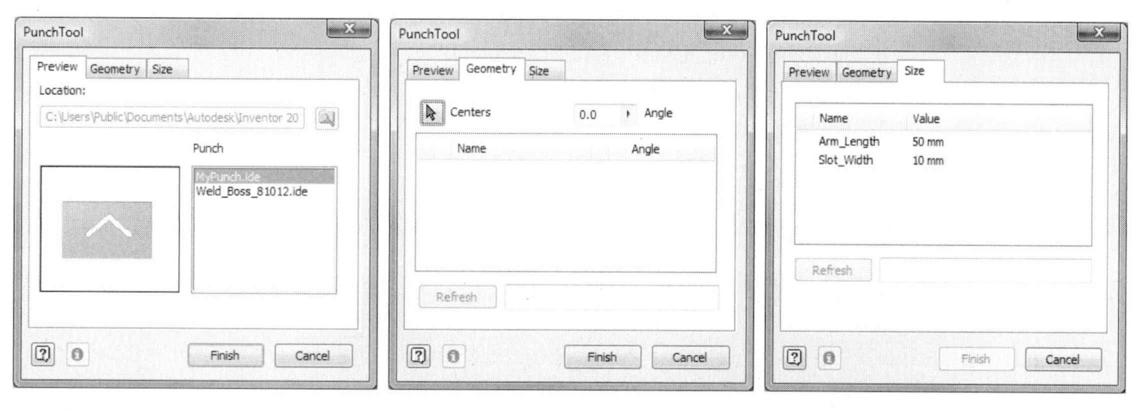

FIGURE 10.130

7. Click Finish to complete the PunchTool and exit the dialog box.

EXERCISE 10-6: PUNCH TOOL

In this exercise, you create an iFeature that can be used as a punch and then use the PunchTool command in another sheet metal part.

1. Open *ESS_E10_06.ipt*.

   You will complete the sketch and save the cutout as an iFeature. The visible sketch contains all geometry required for the cut. You will add construction geometry and a center point to define the center of the punch.
2. Click the View Face command.
3. Click the face containing the sketch.
4. Right-click Sketch2 in the browser, and then select Edit Sketch.
5. Click the Line command.
6. Select the Construction line style.
7. Sketch two lines, as shown in the following image. Endpoints for Line 1 are on the arc centers. Line 2 connects to the midpoint of Line 1.

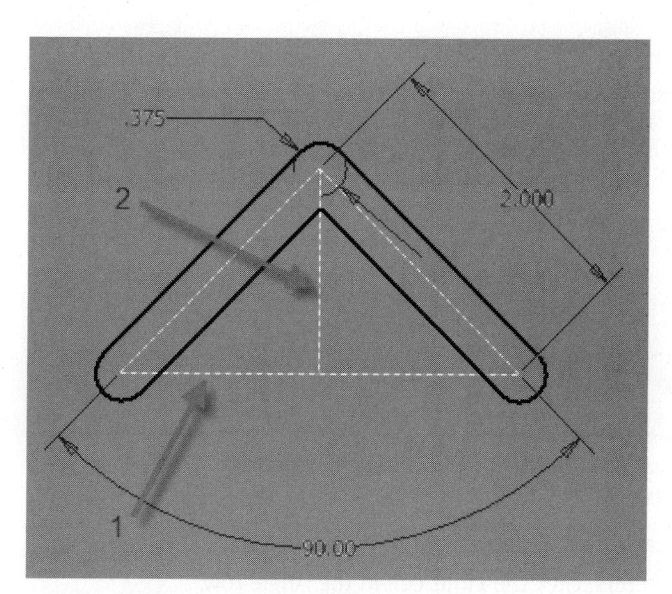

FIGURE 10.131

This iFeature is symmetrical in both X and Y. The midpoint of the vertical construction line is the center of the cut. For nonsymmetrical shapes, you must use equations to ensure that the center point remains at the center of the iFeature.

8. Click the Point, Center Point command.
9. Place a point at the midpoint of the vertical construction line, as shown in the following image.

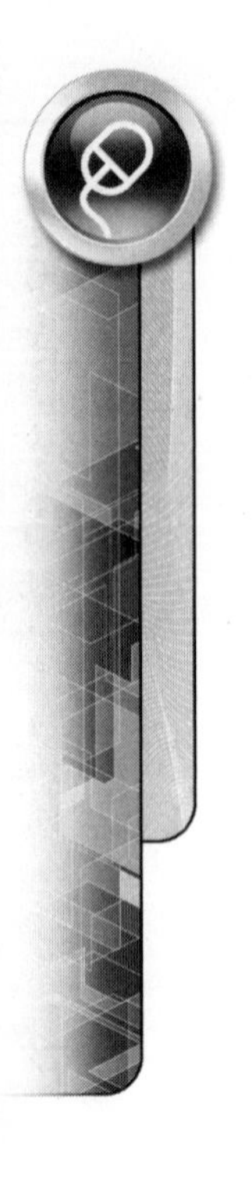

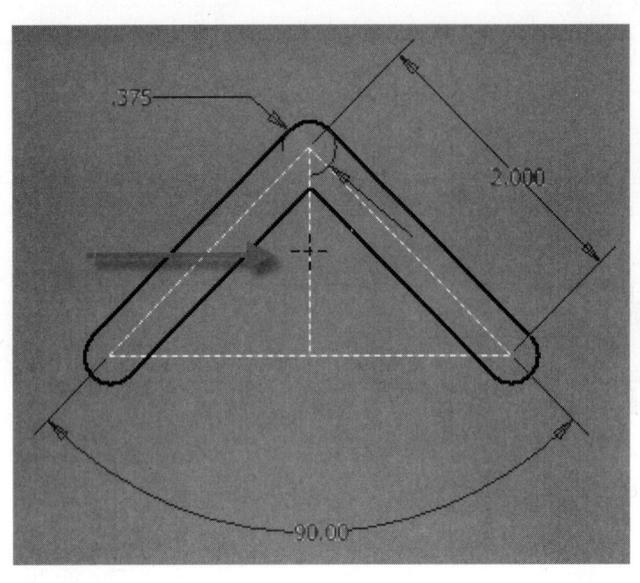

FIGURE 10.132

10. Press S to finish the sketch.
11. Click the Cut command from the Sheet Metal tab > Modify panel.
12. Click OK.

    You rename a parameter prior to saving the iFeature. Other required parameters have previously been renamed.

13. Click the Parameters command from the Manage tab > Parameters panel.
14. Click the model parameter name cell containing d1.
15. Enter **Angle** as the new parameter name.
16. Click Done.
17. Click the Extract iFeature command from the Manage tab > Author panel.
18. Click Cut1 in the browser. Your display should appear similar to the following image.

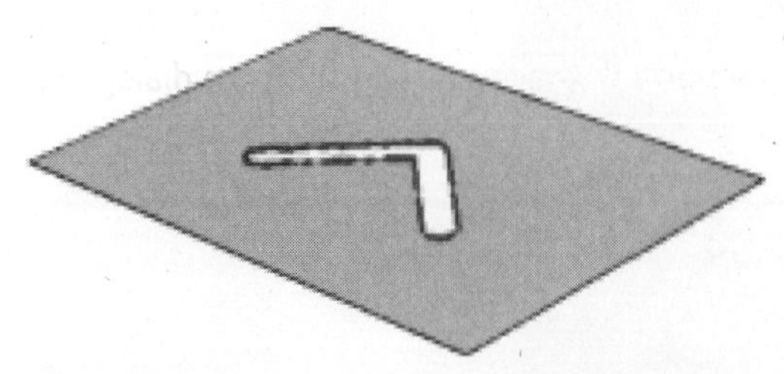

FIGURE 10.133

19. In the Extract iFeature dialog box:
    a. Select Sheet Metal Punch iFeature from the Type section.
    b. Under Size Parameters, click the Limit cell in the Angle row.
    c. Select Range from the list.

20. In the Specify Range:Angle dialog box:
    a. Select < (the less than operand) from the limit list between Minimum and Default.
    b. Select < (the less than operand) from the limit list between Default and Maximum.
    c. Enter **5** deg in the Minimum edit box.
    d. Leave Default set to 90.
    e. Enter **175** deg in the Maximum edit box.
    f. Click OK.
21. Click Save in the Extract iFeature dialog box.
22. Browse to the Catalog\Punches folder.
23. Enter **V Slot Punch** in the File name edit box.
24. Click Save. Click Yes if prompted that the location is not in the active project.

    Next place the punch feature into an existing part. You will create a sketch and use the V Slot punch tool.
25. Open *ESS_E10_06a.ipt*.
26. Create a new sketch on the top face of the part.
27. Click the View Face command, and select the sketch face.
28. Click the Point, Center Point command.
29. Place a point, as shown in the following image.

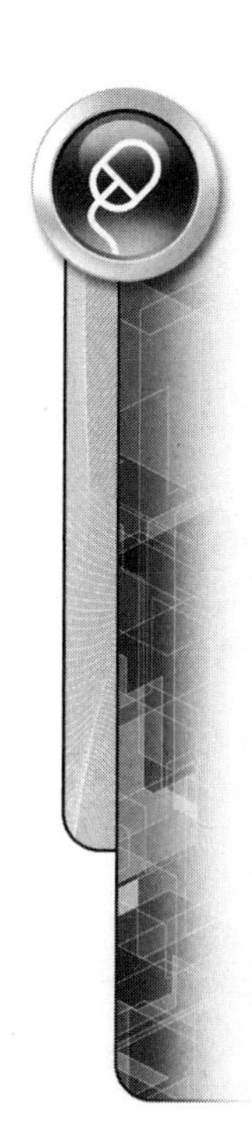

FIGURE 10.134

Next create a rectangular pattern of points.

30. Click the Rectangular Pattern command.
31. Click the Point.
32. Click the selection tool under Direction 1 in the Rectangular Pattern dialog box.

33. Click the edge, as shown in the following image.

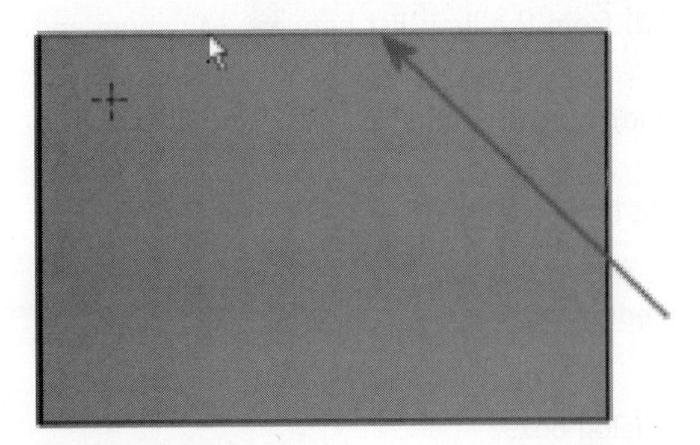

FIGURE 10.135

34. Enter **5** in the Direction 1 Count edit box.
35. Enter **2.75** in the Direction 1 Spacing edit box.
36. Click the selection tool under Direction 2.
37. Click the edge, as shown in the following image.

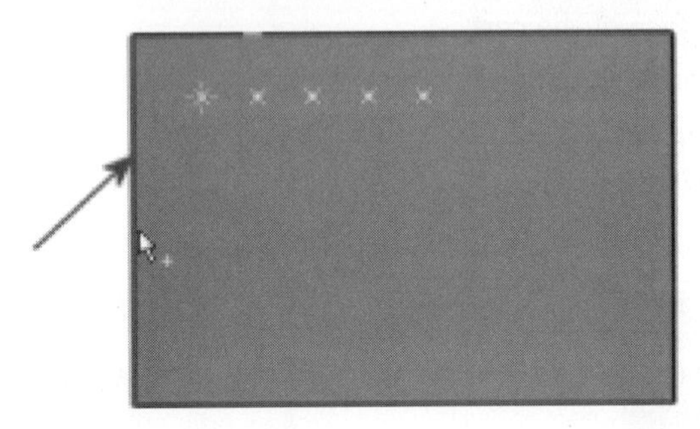

FIGURE 10.136

38. Enter **4** in the Direction 2 Count edit box.
39. Enter **2** in the Direction 2 Spacing edit box.
40. Click OK.

    Sketch a line to enable manual placement of PunchTool centers.
41. Click the Line command.
42. Sketch the line shown in the following image.

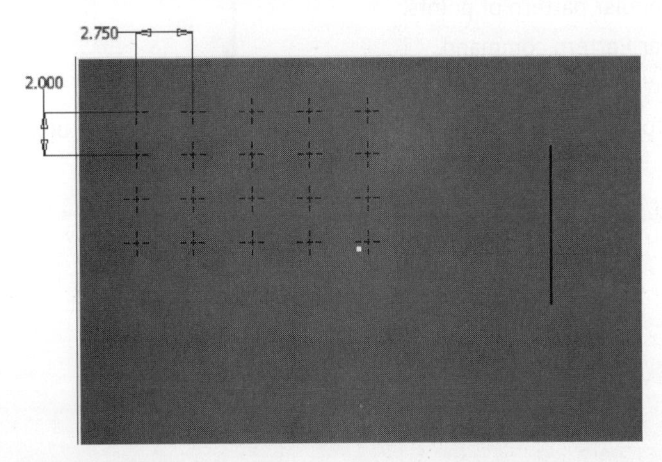

FIGURE 10.137

43. Press S to complete the sketch.
44. Click the PunchTool command from the Sheet Metal tab > Modify panel.
45. In the PunchTool Directory dialog box:
    a. Click V Slot *Punch.ide* in the File Name list.
    b. Click Open.
    c. The PunchTool shape is previewed at each point in the sketch.
46. Click the two line endpoints, as shown in the following image to add the line endpoints as tool centers.

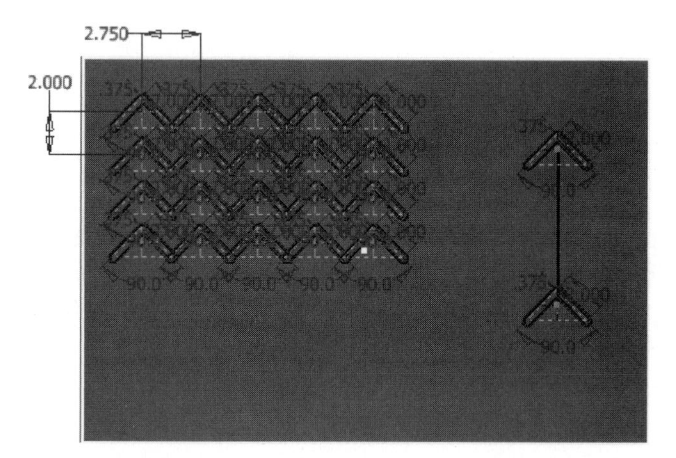

FIGURE 10.138

NOTE

The sketch plane is the only reference for this punch. Any additional geometry references must be satisfied before proceeding via the Geometry tab.

47. Click the Size tab of the PunchTool dialog box.
48. In the PunchTool dialog box:
    a. Enter **60** in the Angle Value cell.
    b. Enter **1** in the Arm_Length Value cell.
    c. Enter **0.25** in the Slot_Width Value cell.
    d. Click Refresh to update the preview with the new values.
    e. Click Finish. Your display should appear similar to the following image.

FIGURE 10.139

49. Close all open files. Do not save changes. End of exercise.

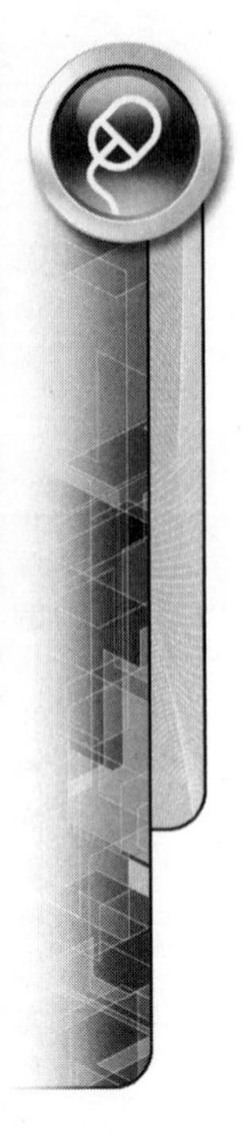

Flat Pattern

You can create a sheet metal flat pattern that unfolds all of the sheet metal features. The flat pattern represents the starting point for the manufacturing of the sheet metal part. The flat pattern can appear in a 2D drawing view, complete with lines indicating bend centerlines. The Create Flat Pattern command, as shown in the following image, creates a 3D model of the unfolded part.

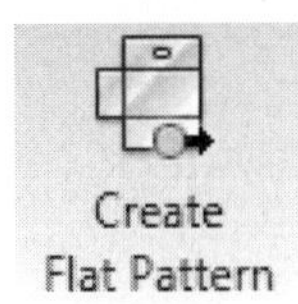

FIGURE 10.140

You can also export the flat pattern directly as a 3D (SAT file) or 2D (*.dxf* or *.dwg*) formats to be used to create machine tool programming for flat pattern punching or cutting. The following image on the left shows a sheet metal part in its formed state and the same part in its flat state on the right.

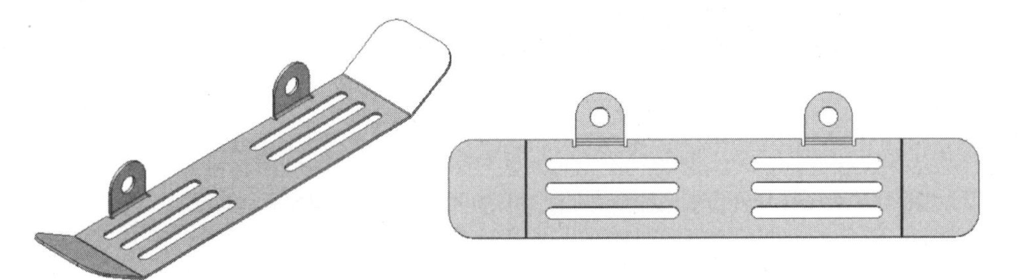

FIGURE 10.141

Manually modeling a flat pattern of a sheet metal part is straightforward, but with a complicated shape, the process can take considerable effort. Autodesk Inventor can create a 3D unfolded model of your sheet metal part quickly and update this model automatically as you modify the features that make up the part. You can create the flat pattern model at any time during the sheet metal modeling process.

The flat pattern model and icon are shown under the part name in the browser, as shown in the following image. The flat pattern model replaces the folded model in the graphics window, and you can toggle back and forth between the two model states by selecting either Folded Model or Flat Pattern in the browser. The flat pattern model is updated automatically as you modify sheet metal features. Double-click the flat pattern in the browser to view the current flat pattern shape after making changes to the folded model. You can also right-click on either node and select *Go To Folded Part* or *Go To Flat Pattern* from the menu to activate the selected model state.

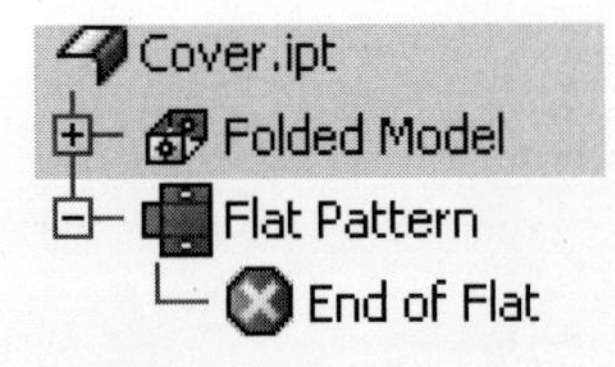

FIGURE 10.142

If you delete the flat pattern model from the browser, you will not be able to create a drawing view of the flat pattern. If you delete the flat pattern model after the creation of a flat pattern drawing view, the drawing view will be deleted.

Right-click the flat pattern in the browser, select Extents from the menu to display the dimensions of the smallest rectangle that encloses the flat pattern, as shown in the following image. Use this information for stock selection and for multiple flat pattern layouts on large sheets.

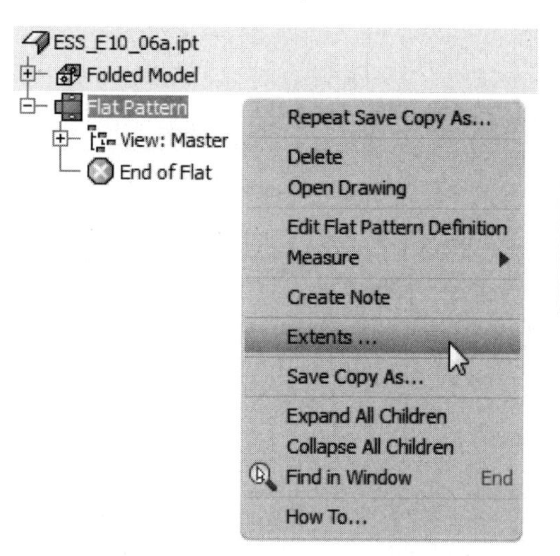

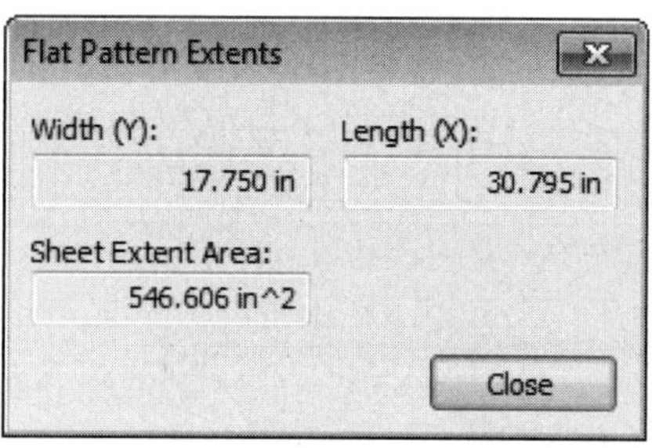

FIGURE 10.143

These values are also available when creating text on a drawing sheet. Select Sheet Metal Properties from the Type list and then choose flat pattern extents area, flat pattern extents length, or flat pattern extents width to include the parameters on your drawing sheet. In addition to including the flat pattern dimensions in text fields, you can also include them in a Parts List. To add them to a Parts List, you need to create custom iProperties that have the same name as the sheet metal properties: flat pattern area, flat pattern length, and flat pattern width. Examples of these custom iProperties are shown in the following image. Notice, from the tooltip, that the value is entered as a formula. If the extents of the flat pattern are changed, the values will update automatically.

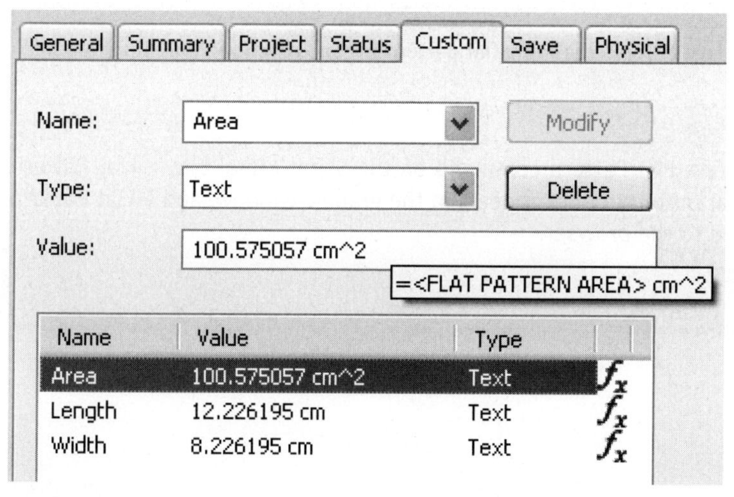

FIGURE 10.144

If you choose the Edit Flat Pattern Definition option from the right-click menu, a Flat Pattern dialog is displayed with three tabs: Orientation, Punch Representation, and Bend Angle as shown in the following image.

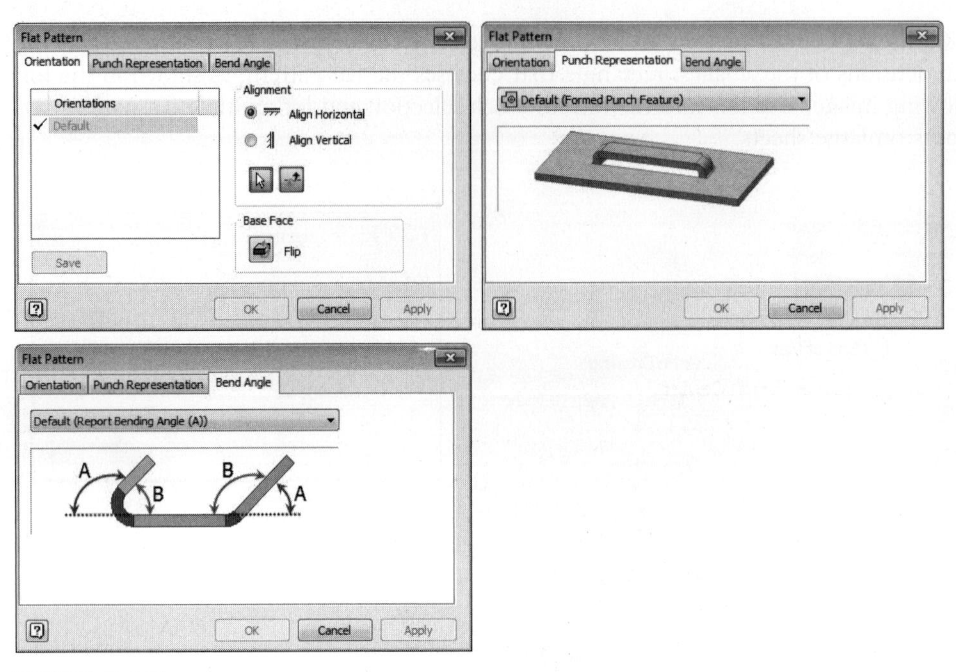

FIGURE 10.145

The Orientation tab lets you reorient the flat pattern. The Orientations list allows you to create, activate, delete, or rename flat pattern orientations using the right-click menu. The Alignment section allows you to select an edge and orient the flat pattern by making the edge align either horizontally or vertically. Typically, the flat pattern is created normal to the initial sketched face feature. You can flip the normal using the Flip command in the Base Face section. On the Punch Representation tab, you can override the display of punch features as defined in the active sheet metal style. The Bend Angle tab allows you to select whether the angle reported for bends is measured from the outside (option A) or inside (option B) face of the bend. To create a flat pattern, follow these steps:

1. Select a face that you expect will not be removed in future edits. The selected face will remain fixed, and all other faces will unfold from this face.

NOTE

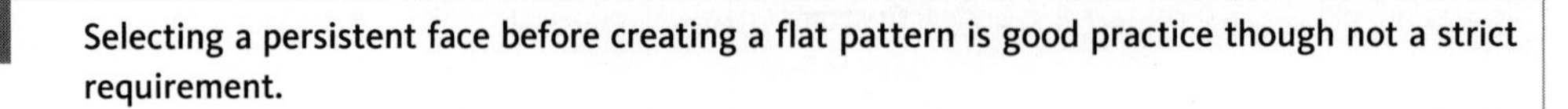

Selecting a persistent face before creating a flat pattern is good practice though not a strict requirement.

2. Click the Create Flat Pattern command on the Sheet Metal tab > Flat Pattern panel. The flat pattern model appears in the graphic window, and a Flat Pattern node is added to the browser.

3. Right-click Flat Pattern at the top of the browser. Menu items are available for saving the flat pattern, editing the flat pattern definition, and displaying the overall dimensions or extents of the flat pattern.
4. To document a flat pattern, start a new drawing. Create a base drawing view, and select Flat Pattern from the Sheet Metal View list in the Drawing View dialog box, as shown in the following image. When Flat Pattern is selected, the Recover Punch Center option can be selected to retrieve punch centers on the drawing view. The punch centers will be formatted based on settings in the active sheet metal rule or when overridden via the Edit Flat Pattern option.

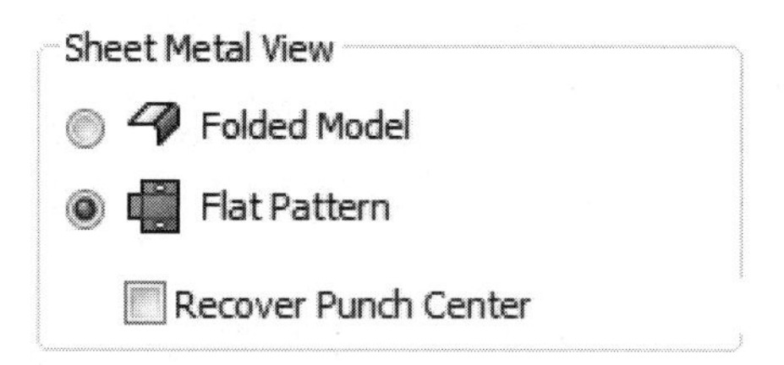

FIGURE 10.146

DETAILING SHEET METAL DESIGNS

You can create drawing views of the 3D model and the flat pattern when detailing a sheet metal part. The flat pattern drawing view enables all bend locations, bend extents, bend and corner reliefs, and cutouts to be located and sized with dimensions. The 3D model views describe the folded shape of the part and allow the forming tool operator to validate the folded part, as shown in the following image.

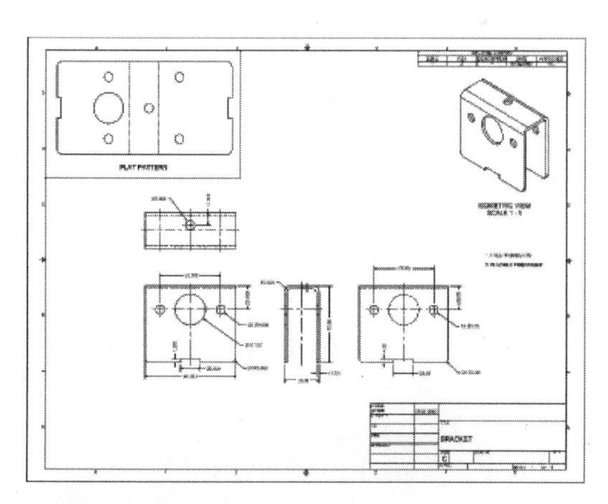

FIGURE 10.147

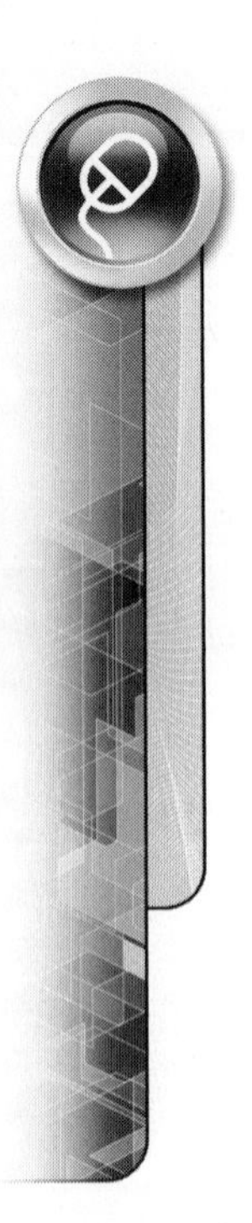

EXERCISE 10-7: DOCUMENTING SHEET METAL DESIGNS

In this exercise, you create a flat pattern model of a sheet metal part and observe the live nature of the flat pattern as you add or modify features. You obtain the flat pattern extents and create model and flat pattern drawing views of the sheet metal part.

1. Open *ESS_E10_07.ipt*.
2. Click the Create Flat Pattern command from the Sheet Metal tab > Flat Pattern panel. The flat pattern model of the sheet metal part is displayed in the graphics window, as shown in the following image.

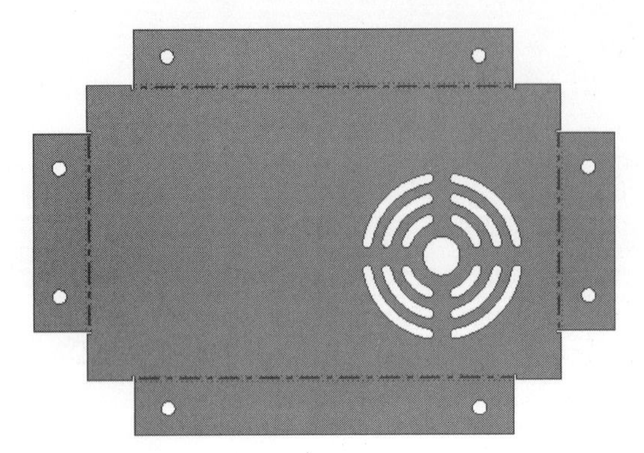

FIGURE 10.148

3. In the browser, double-click the Folded Model node to activate the folded model for edit.
4. Click the Corner Chamfer command from the Sheet Metal tab > Modify panel.
5. Place the cursor over the top-right corner of the large face.
6. When the edge of the top-right face is highlighted, click to select it, as shown in the following image on the left.
7. Select all four corners of the large face.
8. In the Corner Chamfer dialog box, enter a value of **0.25 in** in the Distance field, and verify that the One Distance creation method is selected.
9. Click OK to place the chamfers. Your display should appear similar to the following image on the right.

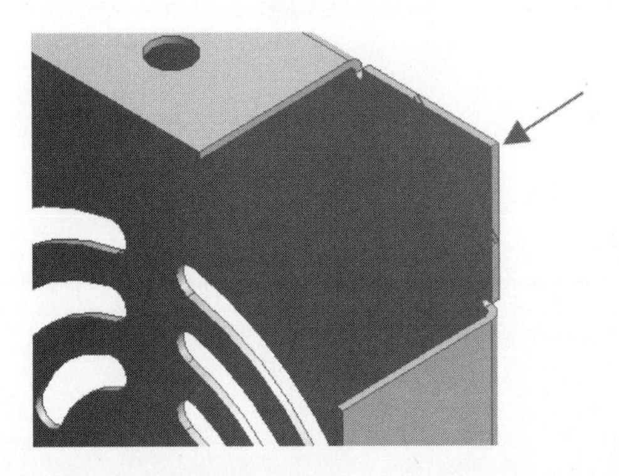

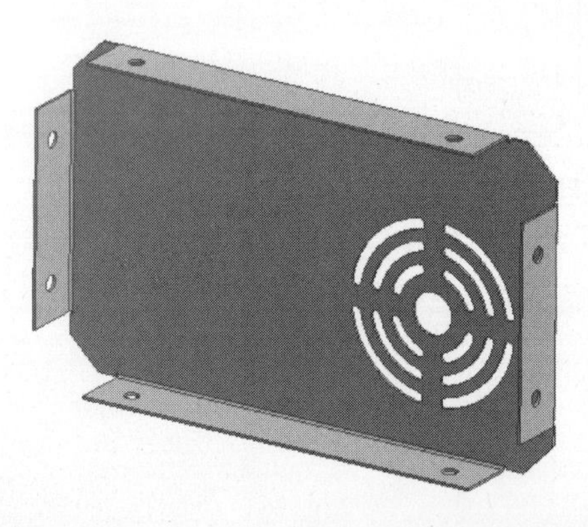

FIGURE 10.149

10. In the browser, right-click Flat Pattern, and select Go To Flat Pattern from the menu.
11. The flat pattern includes the chamfers, as shown in the following image.

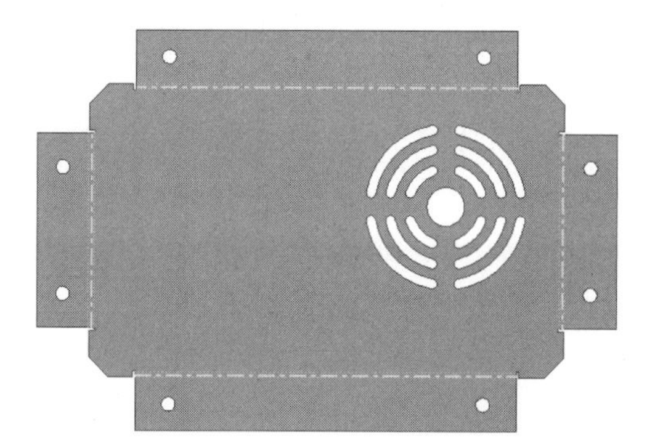

FIGURE 10.150

12. Activate the Folded Model by double-clicking the node in the browser.
13. Click the Corner Round command from the Sheet Metal tab > Modify panel.
14. In the Corner Round dialog box:
    a. Enter a value of **0.0625 in** for the radius.
    b. Click the Feature button under Select Mode.
15. Place the cursor over one of the four flanges.
16. Click when the flange is highlighted.

    The small corners at each end of the flange are highlighted as the flange is selected.
17. Select all four flanges.
18. Click OK to place the corner rounds.
19. Make the Home View current. Your display should appear similar to the following image.

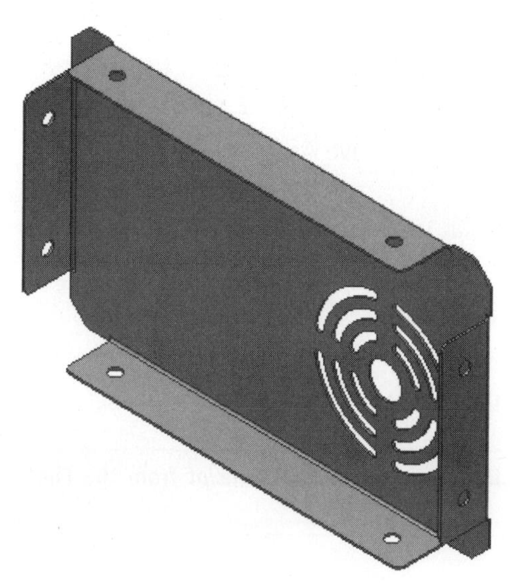

FIGURE 10.151

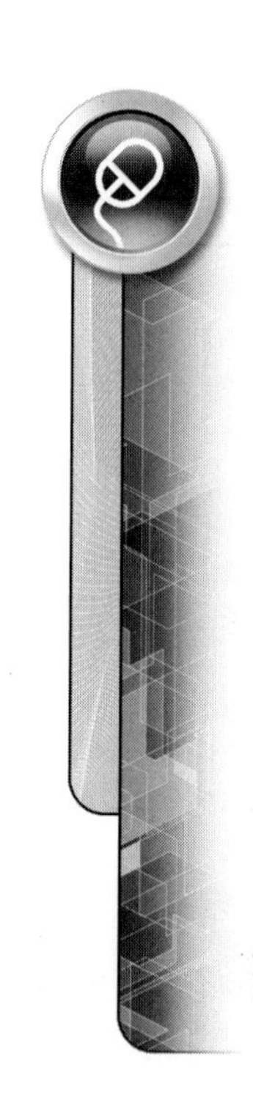

20. Click the Insert iFeature command from the Manage tab > Insert panel.
21. Browse to the Chapter 10 exercises folder, and select *Power_Plug.ide*.
22. Click Open to place the iFeature.
23. Click the large face on the part to place the iFeature.

    The placement plane is the only input for this iFeature; the plug dimensions are fixed.
24. Click the crossed arrows centered on the iFeature, and drag the shape to roughly place the iFeature, as shown in the following image on the left. Click to locate the feature.
25. Click Finish in the Insert iFeature dialog box. Your display should appear similar to the following image on the right.

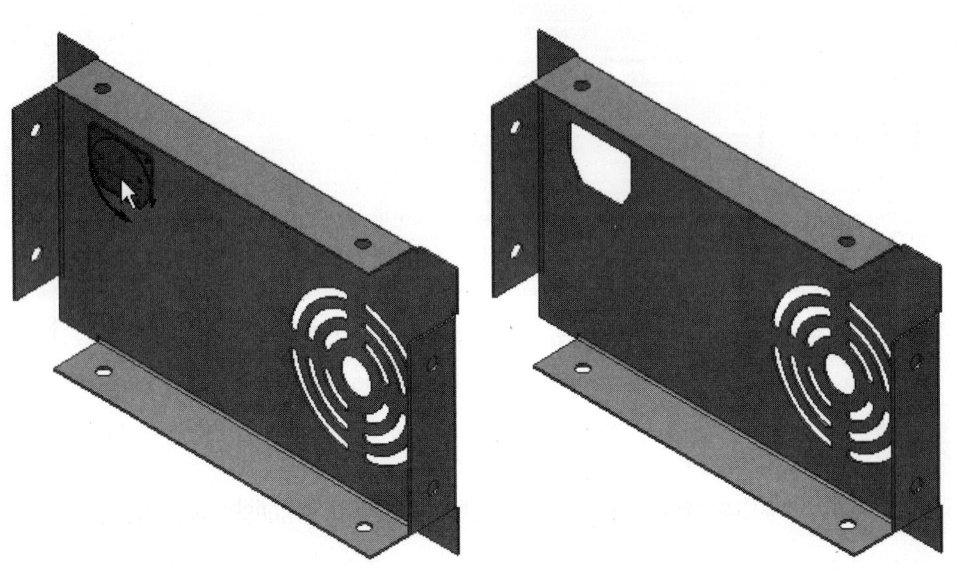

FIGURE 10.152

26. Double-click Flat Pattern in the browser.
27. In the browser, right-click Flat Pattern and select Extents.

    Dimensions for the flat pattern sheet size and area are shown in the Flat Pattern Extents dialog box.
28. Click Close to close the Flat Pattern Extents dialog box.
29. Double-click Folded Model in the browser.
30. Click the Inventor Application menu and select Save As.
31. Enter End_Plate.ipt as the file name.
32. Click Save.
33. Click the New command.
34. Click the English tab.
35. Double-click the *ANSI (in).idw* template.
36. Click the Base View command from the Place Views tab > Create panel.

NOTE

If you have more than one part or assembly file open, select *End_Plate.ipt* from the File drop-down list.

37. Select Folded Model from the Sheet Metal View list, select Bottom from the Orientation list, and click on the drawing sheet to place a view of the folded sheet metal part.

38. Create a top and right view, as shown in the following image.

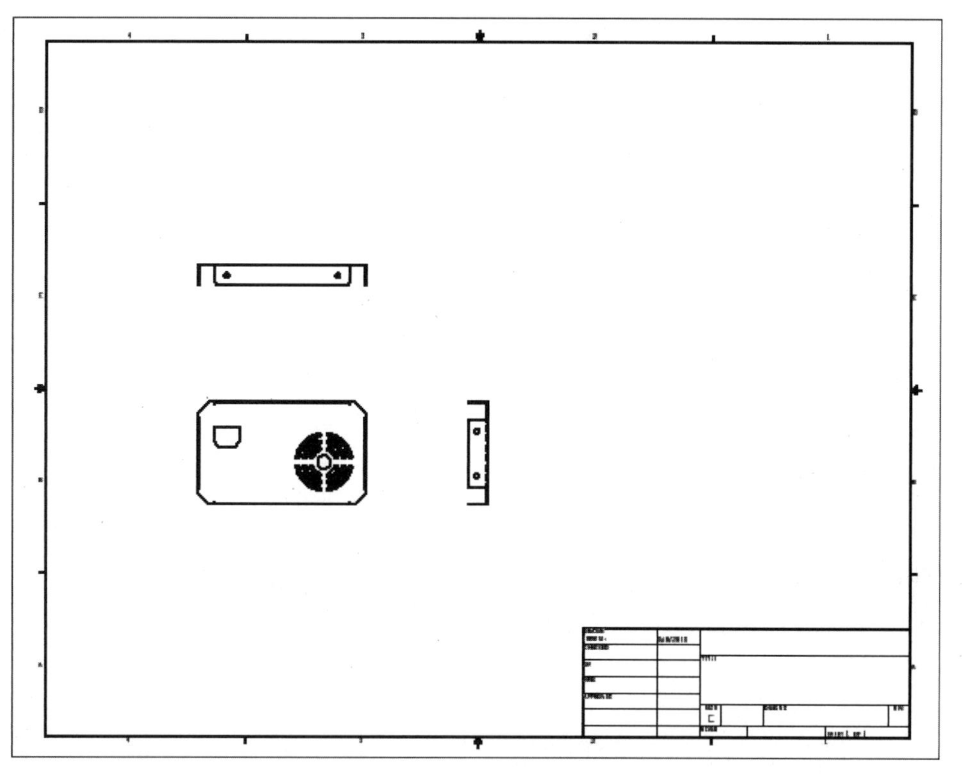

FIGURE 10.153

39. Click the Base View command from the Place Views tab > Create panel.

NOTE

If you have more than one part or assembly file open, select *End_Plate.ipt* from the File drop-down list.

40. In the Drawing View dialog box, select Flat Pattern from the Sheet Metal View list, and choose an option from the Orientation list to create a view similar to the following image.
41. Click on the drawing sheet to place the flat pattern view, as shown in the following image. Right-click and select Done from the menu.

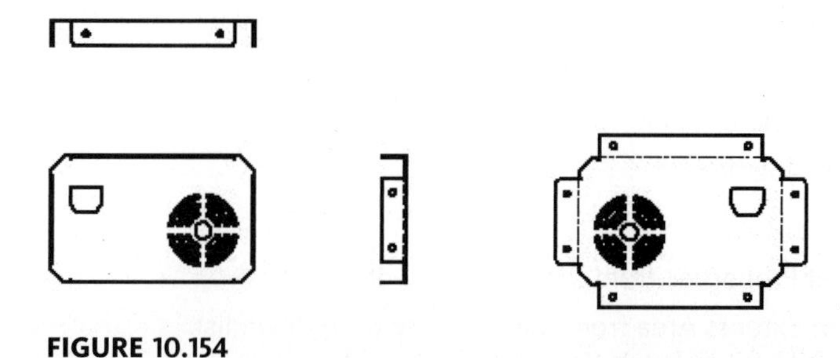

FIGURE 10.154

42. Click the Dimension command from the Annotate tab > Dimension panel.

43. Dimension both tab lengths in the flat pattern view, as shown in the following image.

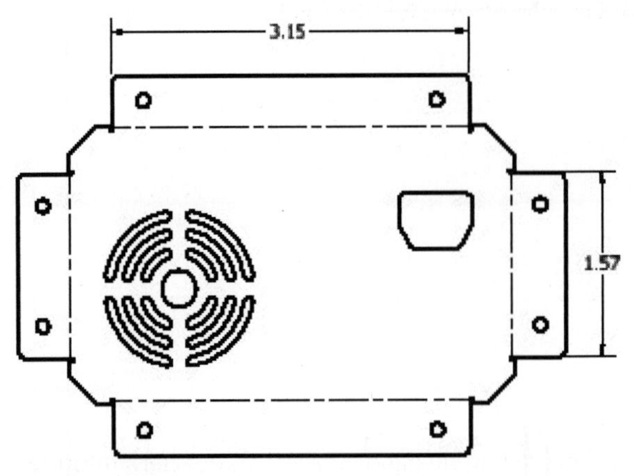

FIGURE 10.155

44. Click the Bend Notes command from the Annotate tab > Feature Notes panel.
45. Click each of the bend lines in the flat pattern view. The flat pattern should appear, as shown in the following image.

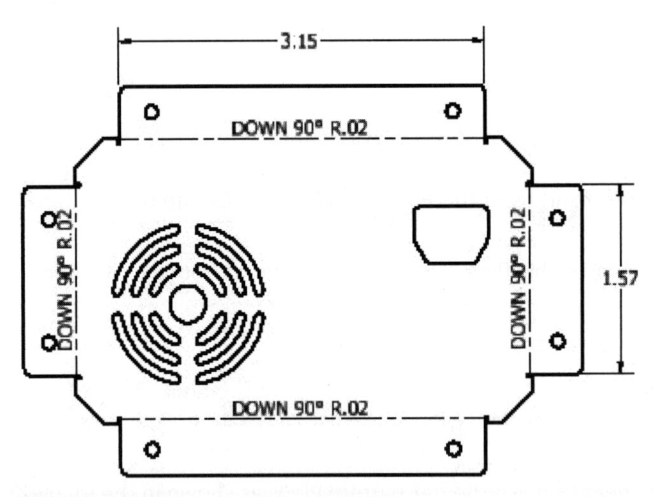

FIGURE 10.156

Next, you include the sheet metal length, width, and extents as a note on the drawing sheet.

46. Click the Text command from the Annotate tab > Text panel.
47. Click a location above the titleblock.
48. In the Format Text dialog box:
    - Type "Flat Pattern Extents:"
    - Select Sheet Metal Properties from the Type drop-down list.
    - Select Flat Pattern Extents Area from the Property drop-down list.
    - Click Add Text Parameter.
    - Press Enter to return to the next line of text.
49. Repeat the previous step to add the following line of text:
    - Type "Flat Pattern Length:"
    - Select Sheet Metal Properties from the Type drop-down list.
    - Select Flat Pattern Extents Length from the Property drop-down list.

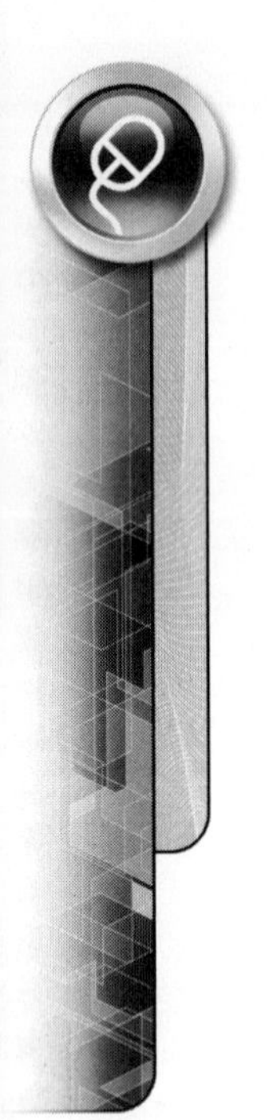

- Click Add Text Parameter.
- Press Enter to return to the next line of text.

50. Repeat the previous step to add the following line of text:
    - Type "Flat Pattern Width:"
    - Select Sheet Metal Properties from the Type drop-down list.
    - Select Flat Pattern Extents Width from the Property drop-down list.
    - Click Add Text Parameter.

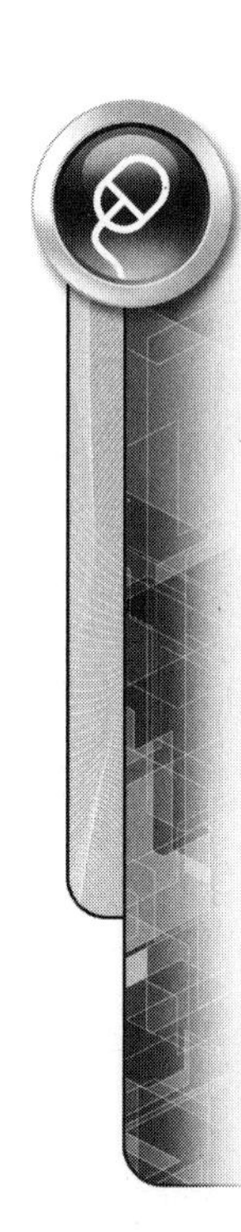

Flat Pattern Extents 15.589 in^2
Flat Pattern Length 4.813 in
Flat Pattern Width 3.239 in

FIGURE 10.157

51. Click OK to close the Format Text dialog box and view the flat pattern information in the appropriate units on the drawing sheet, as shown in the following image.

PARTS LIST					
ITEM	QTY	PART NUMBER	LENGTH	WIDTH	AREA
1	1	End_Plate	122.262 mm	82.262 mm	100.575057 cm^2

FIGURE 10.158

52. Save the drawing as *End_Plate.idw*.
53. Close all open files. Do not save changes. End of exercise.

APPLYING YOUR SKILLS

Skill Exercise 10-1

Using the knowledge you gained in this chapter, create the sheet metal part shown in the following image. As part of the exercise, create a drawing that details both the flat and folded states of the part.

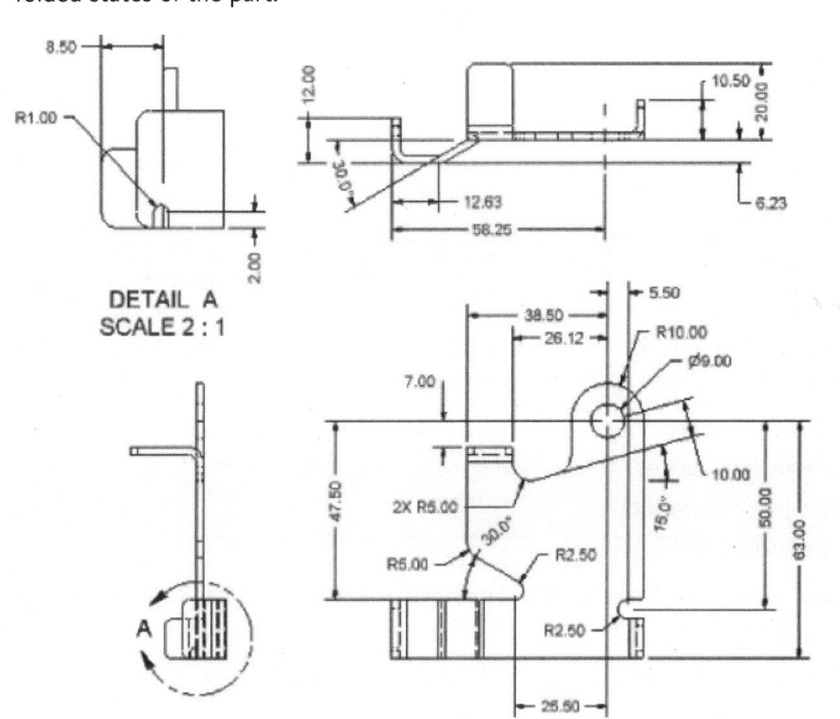

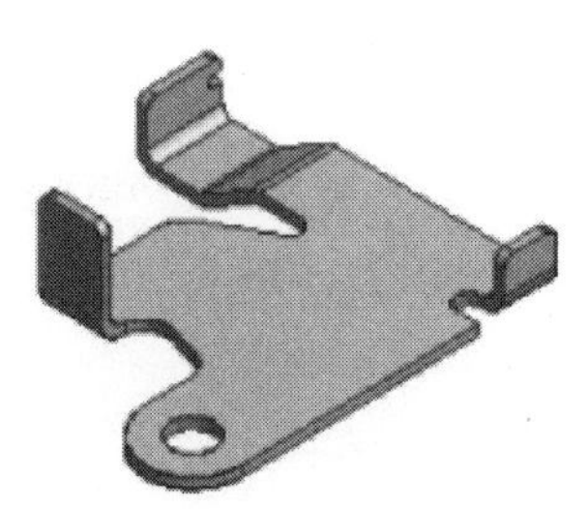

NOTES:
1. PART IS 0.75 in THICK.
2. ALL ROUNDS ARE 0.75 in. UNLESS OTHERWISE SPECIFIED.
3. INSIDE BEND RADII 0.75 in.

FIGURE 10.159

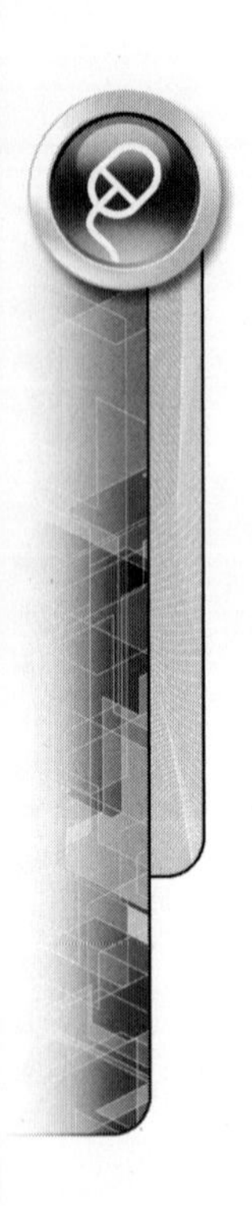

CHECKING YOUR SKILLS

Use these questions to test your knowledge of the material covered in this chapter.

1. The base feature of a sheet metal part is most often a:
   a. Revolve
   b. Face
   c. Extrude
   d. Flange
2. What is the correct procedure to change the edges connected by a bend feature?
   a. Suppress the existing bend, and add a new bend.
   b. Delete the existing bend, and add a new bend.
   c. Edit the bend, and select the new edges.
   d. Create a corner seam between the desired faces.
3. Which command would you use to create a full-length rectangular face off of an existing face edge?
   a. Flange
   b. Extrude
   c. Face
   d. Bend
4. What action is required to update a flat pattern model?
   a. Right-click on the flat pattern in the browser, and select Update.
   b. Erase the existing flat pattern, and re-create it.
   c. Create a new flat pattern drawing view.
   d. None, the flat pattern is updated automatically.
5. True __ False __ Sheet metal parts can contain features created with Autodesk Inventor modeling tools.
6. True __ False __ Sheet Metal Style settings cannot be overridden; a new style must be created for different settings.
7. True __ False __ During the creation of a sheet metal face, the face can extend to meet another face and connect to it with a bend.
8. True __ False __ A sheet metal cut that used the Distance extents option must be created equal to the Thickness parameter.
9. What is a common term for a three-bend intersection?
10. True __ False __ When creating a sheet metal punch, you can select an alternate 2D representation to display in flat patterns and drawing views.
11. Which Project Geometry option is most helpful when creating a cut across a bend in a sheet metal part?
    a. Project Geometry
    b. Project Cut Edges
    c. Project Flat Pattern
    d. Project to 3D Sketch
12. True __ False __ You can apply multiple Sheet Metal Rules to features of a sheet metal part.

13. True ___ False ___ After creating a sheet metal and using the Unfold command to unfold it, you want to refold it into its formed state. In order to do so, you must select a stationary plane and the bends or folds in the same order as when they were unfolded.
14. Which command would be the most efficient to use if you wanted to open the corners of a shelled box that you need to create a flat pattern from?
    a. Cut
    b. Extrude
    c. Rip
    d. Corner Chamfer
    e. Corner Round
15. Which command is a sheet metal specific fillet tool that automatically filters out all edges that are not located at open corners of faces?
    a. Contour Roll
    b. Corner Round
    c. Corner Chamfer
    d. Corner Seam
    e. Hem

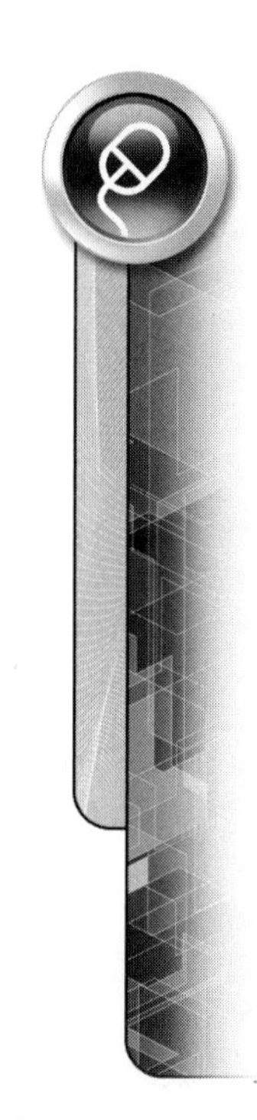

INDEX

D

E

F

G

T

U

V

W

Z